T0265337

ACTIVE OB STARS: STRUCTURE, EVOLUTION, MASS-LOSS, AND CRITICAL LIMITS

IAU SYMPOSIUM No. 272

COVER ILLUSTRATION: Logo of the IAUS 272, designed by Sylvain Cnudde

The logo is an abstract representation of an active OB star, with an equatorial disk and a polar wind. It is reproduced with the permission of its author, who is gratefully thanked.

Other pictures in this book have been taken by François Cochard and Olivier Thizy. They are reproduced with the permission of their authors, who we thank very much.

INTERNATIONAL ASTRONOMICAL UNION

UNION ASTRONOMIQUE INTERNATIONALE

ACTIVE OB STARS: STRUCTURE, EVOLUTION, MASS-LOSS, AND CRITICAL LIMITS

PROCEEDINGS OF THE 272th SYMPOSIUM OF THE INTERNATIONAL ASTRONOMICAL UNION HELD IN PARIS, FRANCE JULY 19–23, 2010

Edited by

CORALIE NEINER
Observatoire de Paris-Meudon, France

GREGG WADE
Royal Military College, Canada

GEORGES MEYNET
Observatoire de Genève, Switzerland

and

GERALDINE PETERS
University of Southern California, USA

CAMBRIDGE
UNIVERSITY PRESS

University Printing House, Cambridge CB2 8BS, United Kingdom

One Liberty Plaza, 20th Floor, New York, NY 10006, USA

477 Williamstown Road, Port Melbourne, VIC 3207, Australia

314-321, 3rd Floor, Plot 3, Splendor Forum, Jasola District Centre, New Delhi - 110025, India

79 Anson Road, #06-04/06, Singapore 079906

Cambridge University Press is part of the University of Cambridge.

It furthers the University's mission by disseminating knowledge in the pursuit of education, learning and research at the highest international levels of excellence.

www.cambridge.org
Information on this title: www.cambridge.org/9780521198400

First published 2011

A catalogue record for this publication is available from the British Library

ISBN 978-0-521-19840-0 Hardback

Table of Contents

Part 1. OPENING THE SYMPOSIUM

Part 2. RAPID ROTATION AND MIXING IN ACTIVE OB STARS

Section A. Review and contributed talks

Chairs: Georges Meynet and Geraldine Peters

Section B. Posters

Part 3. WINDS AND MAGNETIC FIELDS OF ACTIVE OB STARS

Section A. Review and contributed talks

Chairs: Jean-Claude Bouret and Lydia Cidale

Section B. Posters

Part 4. POPULATIONS OF OB STARS IN GALAXIES

Section A. Review and contributed talks

Chairs: Hideyuki Saio and Gregg Wade

Section B. Posters

Part 5. CIRCUMSTELLAR ENVIRONMENT OF ACTIVE OB STARS

Section A. Review and contributed talks

Chairs: Douglas Gies and Richard Townsend

Section B. Posters

Part 6. PERIODIC VARIATIONS AND ASTEROSEISMOLOGY OF OB STARS

Section A. Review and contributed talks

Chairs: Juan Fabregat and Thomas Rivinius

Section B. Posters

Part 7. 'NORMAL' AND ACTIVE OB STARS AS EXTREME CONDITION TEST BEDS

Section A. Review and contributed talks

Chairs: Marc Gagné and Eduardo Janot Pacheco

Section B. Posters

Part 8. LAST MINUTE CONTRIBUTION

Part 9. CLOSING THE SYMPOSIUM

Preface

Early-type (OB) stars dominate the ecology of the universe as cosmic engines via their extreme output of radiation and matter, not only as supernovae but also during their entire lifetime with far-reaching consequences. Active OB stars are massive and intermediate-mass stars that display strong variability on various time scales due to such phenomena as mass outflows, rapid rotation, pulsations, magnetism, binarity, radiative instabilities, and the influence of their circumstellar environment. This concerns in particular classical and Herbig Be, Bp, β Cep, Slowly Pulsating B Stars (SPB), B[e] and O stars, as well as massive binaries such as the Be X-ray binaries and those that harbor O-type subdwarf companions.

Research in the domain of active OB stars has been progressing very rapidly in the last decade and is entering a new era thanks to the advent of new space and ground-based instrumentation. Space asteroseismology (MOST, CoRoT, Kepler) allows us to study the internal structure of massive stars and their rotation; efficient high-resolution spectropolarimetry (Narval, Espadons) provides clues about magnetic fields and the confinement of the circumstellar environment; interferometry (e.g. VLTI, CHARA) allows us to probe the shape of these environments and investigate differential rotation; multi-object spectroscopy on very large telescopes (e.g. Giraffe@VLT) allows us to study the effects of stellar environment and evolution on the active OB stars and provides information on the distribution of surface velocities and abundances; large galactic surveys (e.g. IPHAS, INTEGRAL) allow the detection of large numbers of emission line objects and massive X-ray binaries with great potential for studies of galactic structure. Active OB stars studies have also taken a leap forward thanks to state-of-the-art modeling (Monte-Carlo radiative transfer, asteroseismic models, models of magnetospheres, and disk models, including rapid rotation effects, multi-dimensional calculations). Moreover, massive stars are inherently extreme objects, in terms of rotation, mass loss, radiation fields and in some cases magnetic fields, and thus can serve as testbeds for extreme conditions. From a theoretical viewpoint, this is likely to be important for understanding the first massive star generations in the universe and will provide strong clues on the physics of fast rotation, which is a key ingredient in many current models such as the collapsars model to explain the long soft Gamma Ray Bursts. Alternatively, B stars have historically provided the astronomical community with the best calibrations of fundamental parameters for upper main sequence stars (e.g. $T_{\rm eff}$, $\log g$, masses, radii, chemical composition). Since activity of various types is commonplace in the B stars it is important to assess the uncertainty in the parameters of upper main sequence stars that results from the presence of stellar activity.

The major progress currently obtained in the field of active OB stars will thus help to answer long-standing as well as new questions such as:

- What is the role of magnetic field, rotation, metallicity and mass loss in the evolution of OB stars? In particular, how does it influence their late stages (neutron stars, black holes, GRBs)? How do surface abundances evolve?
- What is the role of magnetic fields, rotation, and pulsations in the activity of OB stars, in particular on their circumstellar environment? What is responsible for wind clumping and the formation of a disk (Be phenomenon) or clouds?
- What is the internal structure of active and near-critically rotating OB stars? How is the angular momentum transported? Is there a magnetic field, and is it of fossil or dynamo origin?

• Under what conditions do active OB stars become Be stars? What causes LBV outbursts? What happens when a star reaches critical rotational velocity?

• What are the statistical properties of the various populations of OB stars? What is the incidence of magnetic fields? What is the distribution of intrinsic velocities? What are the properties of mass loss?

• How do recent advances in theory and computation of model atmospheres and synthetic spectra, and in the observation of such spectra, make it possible to discover and study new and puzzling phenomena in active OB stars and test theoretical models of these phenomena?

In 1999, IAU Colloquium 175 held in Alicante (Spain) concentrated on "The Be Phenomenon in Early-Type Stars". In 2005, an international meeting held in Sapporo (Japan) was dedicated to "Active OB-stars: Laboratories for Stellar and Circumstellar Physics". In 2010 in Paris the IAU Symposium 272 was dedicated to discussion of the structure, evolution, mass loss, and critical limits of early-type stars, four axes of research that are currently providing important clues about the physics of these objects. The meeting allowed fruitful exchange by bringing together scientists working in the fields of O stars, B stars, Bp stars, Be stars and Herbig Be stars, at wavelengths that span the electromagnetic spectrum (especially X-ray, UV, optical, and IR) with emphasis on forefront observational techniques (e.g. spectropolarimetry, interferometry, asteroseismology). The meeting thus resulted in further important progress in our understanding of active OB stars and gave rise to new projects that will be undertaken in the second decade of the 21st century.

The key topics of the IAU Symposium 272 were:

• the internal structure of active OB stars: pulsations, rotation, magnetism, transport processes

• their evolution: stellar environment, formation, binaries, late stages (including magnetars and GRBs)

• their circumstellar environment: disks, magnetospheres, the Be phenomenon, wind, clumping

• active OB stars as extreme condition test beds: critical rotation, mass loss, radiation fields

• 'normal' OB stars as calibrators: fundamental parameters, astronomical quantities

• populations of OB stars: population studies, tracers of galactic structure, cosmic history

The IAU Symposium 272 was held in Paris from July 19 to 23, 2010. We acknowledge the financial support of our sponsors listed on page *xxi* of these Proceedings, as well as the very active support of the members of the LOC in preparing the numerous details associated with such a symposium. It is a great pleasure to acknowledge in particular Michèle Floquet, the co-chair of the LOC, for whom this symposium was a "last hurrah" before her retirement.

Coralie Neiner and the SOC members
Meudon, September 12, 2010

THE ORGANIZING COMMITTEE

Scientific

Jean-Claude Bouret (France)
Lydia Cidale (Argentina)
Juan Fabregat (Spain)
Marc Gagné (USA)
Douglas Gies (USA)
Eduardo Janot-Pacheco (Brazil)
Georges Meynet (Switzerland)
Coralie Neiner (chair, France)
Geraldine Peters (USA)
Thomas Rivinius (ESO)
Hideyuki Saio (Japan)
Richard Townsend (USA)
Gregg Wade (Canada)

Local

Evelyne Alecian (co-chair)
Bertrand de Batz
Francisco Espinosa Lara
Michle Floquet (co-chair)
Anne-Laure Huat
Anne-Marie Hubert
Bernard Leroy
Olga Martins
Coralie Neiner
Annick Oger

Acknowledgements

The symposium is sponsored and supported by the IAU Divisions IV (Stars) and V (Variable Stars), by the IAU Commissions No. 27 (Variable Stars), No. 36 (Theory of Stellar Atmospheres) and No. 42 (Close Binary Stars), and by the IAU Working Groups on "Active B Stars", "Ap and Related Stars" and "Massive Stars".

The Local Organizing Committee operated under the auspices of the Paris Observatory.

Funding by the
International Astronomical Union,
Région Ile de France,
Institut National des Sciences de l'Univers (INSU) du CNRS,
Ministère de l'enseignement supérieur et de la recherche,
Programme National de Physique Stellaire (PNPS),
GEPI laboratory,
and
Scientific Council of the Paris Observatory,
as well as contributions by
A&A and EDP Sciences
is gratefully acknowledged.

CONFERENCE PHOTOGRAPH

1. Norbert Przybilla
2. Daniela Korcakova
3. Dietrich Baade
4. Hideyuki Saio
5. Stanislav Štefl
6. Michèle Floquet
7. Stéphane Mathis
8. Juan Gutiérrez-soto
9. Omar Delaa
10. Dominik Bomans
11. Jean-Paul Zahn
12. Jean Zorec
13.
14. Paula Marchiano
15. Erika Grundstrom
16. Vladimir Strelnitski
17.
18.
19. Rodolfo Vallverdu
20.
21. Joachim Puls
22. Jean-Claude Bouret
23. Artemio Herrero
24. Philippe Stee
25. Christopher Tycner
26. Chris Engelbrecht
27. Maria-Fernanda Nieva
28. Sashin Moonsamy
29. Bernard Leroy
30. Lydia Cidale
31. Stan Owocki
32. Paul Dunstall
33. Jorick Vink
34. Cyril Georgy
35. Vitaly Goranskij
36. Virginia McSwain
37. Miriam Garcia
38. Brankica Surlan
39. Fabio Frescura
40. Anahí Granada
41. André Maeder
42. Nancy Remage Evans
43. David Cohen
44. Sylvia Ekstrom
45. Olivier Thizy
46. Viktor Votruba
47. Philip Peters
48. Jose Groh
49. Geraldine Peters
50. Elena Barsukova
51. Selma de Mink
52. Yuuki Moritani
53. Gail Schaefer
54.
55. Olivier Chesneau
56. Robbie Halonen
57. Huib Henrichs
58. Brian Van Soelen
59. Nazhatulshima Ahmad
60. Ashley Ames
61. Véronique Petit
62. John Wisniewski
63. Vincent Duez
64. Mary Oksala
65. Jan Polster
66. Bertrand de Batz
67.
68. Georges Meynet
69. Pavel Koubsky
70. Ming Zhao
71. Wenjin Huang
72. Nicholas Hill

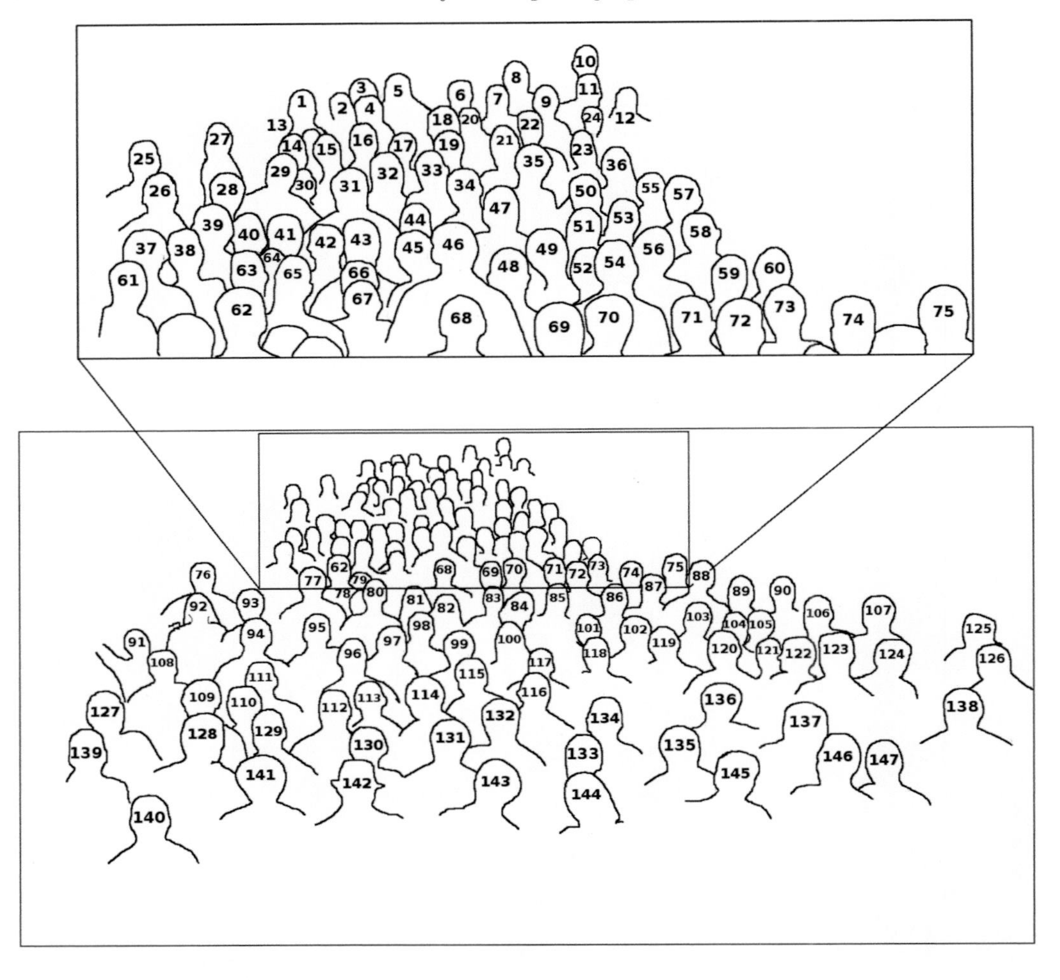

73. Catherine Lovekin
74. Paul KT
75. Francois Cochard
76. Gregg Wade
77. Zachary Draper
78. Rachael Roettenbacher
79. Finny Oktariani
80. Douglas Gies
81. Amber Marsh
82. Carol Jones
83. Chloé Fourtune-Ravard
84. Kostas Markakis
85. Anne-Laure Huat
86. Myron Smith
87. Matteo Cantiello
88. Christopher Russell
89. Kerstin Weis
90. Thomas Rivinius
91. Evelyne Alecian
92. Coralie Neiner
93. Oleg Kochukhov
94. Vitalii Makaganiuk
95. Thibaut Decressin
96. Nathan Smith
97. Juan Fabregat
98. Brenda Perez-Rendon
99. Anatoly Miroshnichenko
100. Evgenia Koumpia
101. Alceste Bonanos
102. Richard Townsend
103. Alejandro Garcia Varela
104. Marc Gagné
105. Mélanie Godart
106. Rémi Fahed
107. Laurent Mahy
108. Gabriela Michalska
109. Jiri Krticka
110. Gérard Muratorio
111. Jason Grunhut
112. Matthew Austin
113. Gloria Koenigsberger
114. Derck Massa
115. Atsuo Okazaki
116. René Oudmaijer
117. Delia Volpi
118. Ines Brott
119. Thierry Morel
120. Adrian Potter
121. Beatriz Sabogal
122. Laura Penny
123. Alex Lobel
124. Daniel Lennon
125. Sergio Simon-Diaz
126. Michel Cure
127. Alex Fullerton
128. Thierry Semaan
129. Marc-Antoine Dupret
130. Anne-Marie Hubert
131. Eduardo Janot-Pacheco
132. Alexandre David-Uraz
133. Romain Selier
134. Raman Prinja
135. Alexander Kholtygin
136. André-Nicolas Chené
137. Michelle Edwards
138. Kalju Annuk
139. Aaron Sigut
140. Jonas Debosscher
141. Mohammad Heydari-Malayeri
142. Matthew Shultz
143. Chien-De Lee
144. Yamina Touhami
145. Daniel Reese
146. Francisco Espinosa Lara
147. Astrid Lamberts

Participants

Nazhatulshima **Ahmad**	University of Malaya, Malaysia	n_ahmad@um.edu.my
Yael **Aidelman**	Universidad Nacional de La Plata, Argentina	yaelaidelman@hotmail.com
Evelyne **Alecian**	LAOG, France	evelyne.alecian@obspm.fr
Ashley **Ames**	Central Michigan University, USA	ames1ap@cmich.edu
Kalju **Annuk**	Tartu Observatory, Estonia	annuk@aai.ee
Matthew **Austin**	UCL, UK	mattyjim2@gmail.com
Dietrich **Baade**	ESO-Garching, Germany	dbaade@eso.org
Daniela **Barria**	Universidad de Concepcion, Chile	dbarria@astro-udec.cl
Elena **Barsukova**	Special Astrophysical Observatory, Russia	bars@sao.ru
Bertrand **de Batz**	Observatoire de Paris-Meudon – GEPI, France	B.deBatz@obspm.fr
Philippe **Bendjoya**	Lab. H. Fizeau -UNS-OCA, France	bendjoya@unice.fr
David **Bohlender**	National Research Council of Canada, Canada	david.bohlender@nrc-cnrc.gc.ca
Dominik **Bomans**	Astro. Inst. Ruhr-University Bochum, Germany	bomans@astro.rub.de
Alceste **Bonanos**	National Observatory of Athens, Greece	bonanos@astro.noa.gr
Jean-Claude **Bouret**	LAM, France & Goddard SFC, USA	jean-claude.bouret@oamp.fr
Ines **Brott**	SIU, Netherlands	ines.brott@gmx.net
Matteo **Cantiello**	Argelander Institute for Astronomy, Germany	cantiello@astro.uni-bonn.de
Alex **Carciofi**	Universidade de Sao Paulo, Brazil	carciofi@usp.br
Peter **de Cat**	Royal Observatory of Belgium, Belgium	Peter.DeCat@oma.be
André-Nicolas **Chene**	National Research Council - HIA, Canada	andre-nicolas.chene@nrc-cnrc.gc.ca
Olivier **Chesneau**	Observatoire de la Cote Azur, France	Olivier.Chesneau@obs-azur.fr
Lydia **Cidale**	Instituto de Astrofisica La Plata, Argentina	lydia@fcaglp.unlp.edu.ar
François **Cochard**	Shelyak Instruments, France	francois.cochard@shelyak.com
David **Cohen**	Swarthmore College, USA	dcohen1@swarthmore.edu
Alicia **Cruzado**	FCAGLP, UNLP, Argentina	acruzado@fcaglp.unlp.edu.ar
Michel **Cure**	Departamento de Fisica y Astronomia, Chile	michel.cure@uv.cl
Augusto **Damineli**	IAG-USP, Brazil	damineli@astro.iag.usp.br
Alexandre **David-Uraz**	Universite de Montreal, Canada	alexandre@astro.umontreal.ca
Jonas **Debosscher**	Instituut voor Sterrenkunde-KULeuven, Belgium	jonas@ster.kuleuven.be
Thibaut **Decressin**	Argelander-Institut für Astronomie, Germany	decress1@astro.uni-bonn.de
Omar **Delaa**	Observatoire de la Cote dAzur, France	Omar.Delaa@obs-azur.fr
Armando **Domiciano de Souza**	Observatoire de la Cote d Azur, France	Armando.Domiciano@oca.eu
Zachary **Draper**	University of Washington, USA	zhd@u.washington.edu
Guillaume **Dubus**	LAOG, France	Guillaume.Dubus@obs.ujf-grenoble.fr
Vincent **Duez**	Argelander-Institut für Astronomie, Germany	vduez@astro.uni-bonn.de
Paul **Dunstall**	Queens University Belfast, UK	pdunstall01@qub.ac.uk
Marc-Antoine **Dupret**	Université de Liège, Belgium	MA.Dupret@ulg.ac.be
Michelle **Edwards**	Gemini Observatory, Chile	medwards@gemini.edu
Sylvia **Ekström**	Geneva Observatory, Switzerland	sylvia.ekstrom@unige.ch
Chris **Engelbrecht**	University of Johannesburg, South Africa	chrise@uj.ac.za
Francisco **Espinosa Lara**	Observatoire de Paris-Meudon – GEPI, France	francisco.espinosa@obspm.fr
Christopher **Evans**	UK ATC, UK	cje@roe.ac.uk
Nancy Remage **Evans**	SAO, USA	nevans@cfa.harvard.edu
Juan **Fabregat**	Universidad de Valencia, Spain	juan.fabregat@uv.es
Rémi **Fahed**	Université de Montréal, Canada	fahed@astro.umontreal.ca
Michèle **Floquet**	Observatoire de Paris-Meudon – GEPI, France	michele.floquet@obspm.fr
Chloé **Fourtune-Ravard**	Univ Paris 7, France & RMC Canada, Canada	c.fourtune.ravard@gmail.com
Fabio **Frescura**	U Witwatersrand, South Africa	fabio.frescura@wits.ac.za
Alex **Fullerton**	Space Telescope Science Institute, USA	fullerton@stsci.edu
Marc **Gagné**	West Chester University, USA	mgagne@wcupa.edu
Alejandro **García Varela**	Universidad de los Andes, Colombia	josegarc@uniandes.edu.co
Miriam **Garcia**	Instituto de Astrofisica de Canarias, Spain	mgg@iac.es
Cyril **Georgy**	Geneva Observatory, Geneva University, Switzerland	Cyril.Georgy@unige.ch
Douglas **Gies**	Georgia State University, USA	gies@chara.gsu.edu
Mélanie **Godart**	ASTA, University of Lige, Belgium	melanie.godart@ulg.ac.be
Vitaly **Goranskij**	Sternberg Astron. Inst., Moscow, Russia	goray@sai.msu.ru
Anahí **Granada**	Instituto de Astrofísica La Plata, Argentina	granada@fcaglp.unlp.edu.ar
Jose **Groh**	Max-Planck-Institute for Radioastronmy, Germany	jgroh@mpifr-bonn.mpg.de
Erika **Grundstrom**	Vanderbilt University, USA	erika.grundstrom@vanderbilt.edu
Jason **Grunhut**	Royal Military College of Canada, Canada	Jason.Grunhut@rmc.ca
JUan **Gutiérrez-Soto**	IAA, Spain	juan.gutierrez@obspm.fr
Robbie **Halonen**	The University of Western Ontario, Canada	rhalonen@uwo.ca
Xavier **Haubois**	Universidade de Sao Paulo-IAG, Brazil	xhaubois@astro.iag.usp.br
Huib **Henrichs**	University of Amsterdam, Netherlands	h.f.henrichs@uva.nl
Artemio **Herrero**	Instituto de Astrofisica de Canarias, Spain	ahd@iac.es
Anthony **Herve**	Université de Liège, Belgium	herve@astro.ulg.ac.be
Mohammad **Heydari-Malayeri**	Paris Observatory, France	m.heydari@obspm.fr
Nicholas **Hill**	University of Wisconsin-Madison, USA	nhill@astro.wisc.edu
Wenjin **Huang**	University of Washington, USA	hwenjin@astro.washington.edu
Anne-Marie **Hubert**	Observatoire de Paris-Meudon – GEPI, France	anne-marie.hubert@obspm.fr
Robert **Izzard**	Argelander-Institut für Astronomie, Germany	izzard@astro.uni-bonn.de
Eduardo **Janot-Pacheco**	IAG - USP, Brazil	janot@astro.iag.usp.br
Andressa **Jendreieck**	IAG - USP, Brazil	ajendreieck@usp.br
Carol **Jones**	The University of Western Ontario, Canada	cejones@uwo.ca
Alexander **Kholtygin**	Saint-Petersburg University, Russia	afkholtygin@gmail.com
Oleg **Kochukhov**	Uppsala University, Sweden	oleg.kochukhov@fysast.uu.se
Gloria **Koenigsberger**	Instituto de Ciencias Fisicas, UNAM, Mexico	gloria@astro.unam.mx
Daniela **Korcakova**	Astronomical Institute AVCR, Czech Republic	kor@sunstel.asu.cas.cz
Pavel **Koubsky**	Astronomical Institute, Ondrejov, Czech Republic	koubsky@sunstel.asu.cas.cz
Evgenia **Koumpia**	University of Athens, Greece	koumpia@astro.noa.gr
Jiri **Krticka**	Masaryk University, Czech Republic	krticka@physics.muni.cz
Paul **KT**	Christ University, India	paul.kt@christuniversity.in
Astrid **Lamberts**	LAOG, France	astrid.lamberts@obs.ujf-grenoble.fr
Norbert **Langer**	Argelander-Institut fuer Astronomie, Germany	nlanger@astro.uni-bonn.de
Chien-De **Lee**	IANCU, Taiwan	m959009@astro.ncu.edu.tw
Daniel **Lennon**	ESA - STScI, USA	lennon@stsci.edu

Bernard **Leroy**	Observatoire de Paris-Meudon – LESIA, France	Bernard.Leroy@obspm.fr
Alex **Lobel**	Royal Observatory of Belgium, Belgium	alobel@sdf.lonestar.org
Catherine **Lovekin**	LESIA, Observatoire de Paris, France	catherine.lovekin@obspm.fr
André **Maeder**	Geneva Observatory, Switzerland	andre.maeder@unige.ch
Laurent **Mahy**	University of Lige, Belgium	mahy@astro.ulg.ac.be
Vitalii **Makaganiuk**	Uppsala University, Sweden	vitaly.makaganiuk@gmail.com
Paula **Marchiano**	Fac. Cs. Astronómicas y Geofísicas, Argentina	pmarchiano@fcaglp.unlp.edu.ar
Kostas **Markakis**	University of Athens, Greece	markakis@astro.noa.gr
Amber **Marsh**	Lehigh University, USA	anm506@lehigh.edu
Christophe **Martayan**	ESO, Chile	cmartaya@eso.org
Olga **Martins**	Observatoire de Paris-Meudon – GEPI, France	olga.martins@obspm.fr
Derck **Massa**	STScI, USA	massa@stsci.edu
Stéphane **Mathis**	CEA DSM IRFU SAp, France	stephane.mathis@cea.fr
Meghan **McGill**	The University of Western Ontario, Canada	mmcgill8@uwo.ca
Ginny **McSwain**	Lehigh University, USA	mcswain@lehigh.edu
Anthony **Meilland**	Max-Planck Institut für RadioAstronomie, Germany	meilland@mpifr-bonn.mpg.de
Georges **Meynet**	Geneva Observatory, Switzerland	georges.meynet@unige.ch
Gabriela **Michalska**	Wroclaw University, Poland	michalska@astro.uni.wroc.pl
Florentin **Millour**	MPIFR, Germany	fmillour@mpifr.de
Selma **de Mink**	Astronomical Institute Utrecht, Netherlands	sedemink@gmail.com
Anatoly **Miroshnichenko**	University of North Carolina Greensboro, USA	a_mirosh@uncg.edu
Sashin **Moonsamy**	University of the Witwatersrand, South Africa	sashin.moonsamy@gmail.com
Thierry **Morel**	University of Liège, Belgium	morel@astro.ulg.ac.be
Yuuki **Moritani**	Kyoto University, Japan	moritani@kusastro.kyoto-u.ac.jp
Maria **Muratore**	Univ. Nacional de La Plata, Argentina	fmuratore@carina.fcaglp.unlp.edu.ar
Gérard **Muratorio**	OAMP LAM, France	gerard.muratorio@oamp.fr
Yael **Naze**	Dept AGO, Université de Lège, Belgium	naze@astro.ulg.ac.be
Coralie **Neiner**	Observatoire de Paris-Meudon – GEPI, France	coralie.neiner@obspm.fr
Maria-Fernanda **Nieva**	Max Planck Institute for Astrophysics, Germany	fnieva@mpa-garching.mpg.de
Annick **Oger**	Observatoire de Paris-Meudon – GEPI, France	olga.martins@obspm.fr
Atsuo **Okazaki**	Hokkai-Gakuen University, Japan	okazaki@elsa.hokkai-s-u.ac.jp
Mary **Oksala**	University of Delaware, USA	meo@udel.edu
Finny **Oktariani**	Hokkaido University, Japan	finny@astro1.sci.hokudai.ac.jp
Rene **Oudmaijer**	University of Leeds, UK	r.d.oudmaijer@leeds.ac.uk
Stan **Owocki**	University of Delaware, USA	owocki@udel.edu
Laura **Penny**	College of Charleston, USA	pennyl@cofc.edu
Antonio **Pereyra**	Observatonio Nacional, Brazil	pereyra@on.br
Brenda **Perez-Rendon**	DIFUS, Universidad de Sonora, Mexico	brenda@cajeme.cifus.uson.mx
Geraldine **Peters**	University of Southern California, USA	gjpeters@mucen.usc.edu
Véronique **Petit**	Royal Military College of Canada, Canada	veronique.petit.1@gmail.com
Jan **Polster**	Astronomical institute AV CR, Czech Republic	jpolster@email.cz
Adrian **Potter**	University of Cambridge, UK	atp27@cam.ac.uk
Raman **Prinja**	University College London, UK	rkp@star.ucl.ac.uk
Norbert **Przybilla**	Dr. Remeis Obs. Bamberg, Germany	przybilla@sternwarte.uni-erlangen.de
Joachim **Puls**	Observatory, University Munich, Germany	uh101aw@usm.uni-muenchen.de
Gregor **Rauw**	Liége University, Belgium	rauw@astro.ulg.ac.be
Daniel **Reese**	LESIA, Observatoire de Paris, France	daniel.reese@obspm.fr
Thomas **Rivinius**	ESO Chile, Chile	triviniu@eso.org
Jose **Ricardo Rizzo**	Centro de Astrobiologia, Spain	ricardo@cab.inta-csic.es
Rachael **Roettenbacher**	Lehigh University, USA	rmr207@lehigh.edu
Christopher **Russell**	University of Delaware, USA	crussell@udel.edu
Beatriz **Sabogal**	Universidad de los Andes, Colombia	bsabogal@uniandes.edu.co
Hideyuki **Saio**	Tohoku University, Japan	saio@astr.tohoku.ac.jp
Hugues **Sana**	Astronomical Institute Anton Pannekoek, Netherlands	hsana@uva.nl
Gail **Schaefer**	CHARA Array of Georgia State University, USA	schaefer@chara-array.org
Romain **Selier**	LERMA - Observatoire of Paris, France	romain.selier@obspm.fr
Thierry **Semaan**	Observatoire de Paris Meudon, France	thierry.semaan@obspm.fr
Matthew **Shultz**	Queens University, Canada	matt.shultz@gmail.com
Aaron **Sigut**	The University of Western Ontario, Canada	asigut@uwo.ca
Sergio **Simon-Diaz**	Instituto de Astrofisica de Canarias, Spain	ssimon@iac.es
Myron **Smith**	Catholic University of America, USA	msmith@stsci.edu
Nathan **Smith**	UC Berkeley, USA	nathans@astro.berkeley.edu
Philippe **Stee**	Observatoire de la Cote d Azur - CNRS, France	philippe.stee@obs-azur.fr
Stanislav **Stefl**	European Southern Observatory, Chile	sstefl@eso.org
Vladimir **Strelnitski**	Maria Mitchell Observatory, USA	vladimir@mmo.org
Brankica **Surlan**	Astronomical Institute Ondrejov, Czech Republic	brankica74@sunstel.asu.cas.cz
Mairan **Teodoro**	IAG-USP, Brazil	mairan.teodoro@gmail.com
Olivier **Thizy**	Shelyak Instruments, France	olivier.thizy@shelyak.com
Yamina **Touhami**	Georgia State University, USA	yamina@chara.gsu.edu
Richard **Townsend**	University of Wisconsin-Madison, USA	townsend@astro.wisc.edu
Christopher **Tycner**	Central Michigan University, USA	c.tycner@cmich.edu
Rodolfo **Vallverdú**	Observatorio Astronómico de La Plata, Argentina	rodolfo@fcaglp.unlp.edu.ar
Walter **Van Rensbergen**	Vrije Universiteit Brussel, Belgium	wvanrens@vub.ac.be
Brian **Van Soelen**	University of the Free State, South Africa	vansoelenb@ufs.ac.za
Jorick **Vink**	Armagh Observatory, UK	jsv@arm.ac.uk
Delia **Volpi**	Royal Observatory of Belgium, Belgium	Delia.Volpi@oma.be
Viktor **Votruba**	Astronomical Institute AV CR, Czech Republic	votruba@physics.muni.cz
Gregg **Wade**	Royal Military College of Canada, Canada	Gregg.Wade@rmc.ca
Kerstin **Weis**	Astr. Inst. Ruhr-University-Bochum, Germany	kweis@astro.rub.de
John **Wisniewski**	University of Washington, USA	jwisnie@u.washington.edu
Ruslan **Yudin**	Pulkovo Observatory, Russia	ruslan.yudin@hotmail.com
Jean-Paul **Zahn**	Observatoire de Paris, France	jean-paul.zahn@obspm.fr
Ming **Zhao**	Jet Propulsion Lab, USA	ming.zhao@jpl.nasa.gov
Oleksandra **Zhukova**	Kiev Observatory, Ukraine	a-zhukova@ukr.net
Jean **Zorec**	Institut d'Astrophysique de Paris, France	zorec@iap.fr
Nataliya **Zubreva**	Kiev Observatory, Ukraine	n-zubreva@ukr.net

Address by the Scientific and Local Organizing Committees

Dear colleagues,

It was a pleasure to welcome all of you to the 272dn Symposium of the International Astronomical Union in Paris and to write on behalf of the Scientific and Local Organizing Committees. It has been a long time since our last meetings on active OB stars in Alicante and Sapporo. We hope that, in this new occasion, the symposium fulfilled your expectations, both scientifically and socially. We have counted on very active members of the SOC, who prepared an ambitious scientific programme as well as interactive discussions. Motivated people in the LOC also presented us with a pleasant social programme, and we hope this will also contribute to nice memories of this symposium. I would like to thank in particular the main members of the LOC Evelyne Alecian, Bertrand de Batz, Michèle Floquet and Bernard Leroy, as well as our secretaries Olga Martins and Annick Oger. However, the success of this symposium mostly comes from all of you who contributed in various ways by bringing your knowledge, results, questions, and answers to a fruitful discussion. We thus thank all of you for your participation!

We are deeply grateful to several institutions for their financial and organizing support: the International Astronomical Union, Région Ile de France, Institut National des Sciences de l'Univers (INSU) du CNRS, Ministère de l'enseignement supérieur et de la recherche, Programme National de Physique Stellaire (PNPS), GEPI laboratory, Scientific Council of the Paris Observatory, A&A, and EDP Sciences.

Thank you very much!

Coralie Neiner, SOC chair and LOC member
Paris, 12 September 2010

Active OB stars: structure, evolution, mass loss, and critical limits
Proceedings IAU Symposium No. 272, 2010
C. Neiner, G. Wade, G. Meynet & G. Peters, eds.

doi:10.1017/S1743921311009914

Active OB Stars - an introduction

Dietrich Baade[1], Thomas Rivinius[2], Stanislas Štefl[2], and Christophe Martayan[2]

[1]European Organisation for Astronomical Research in the Southern Hemisphere, Karl-Schwarzschild-Str. 2, 85748 Garching. b. München, Germany
email: dbaade@eso.org

[2]European Organisation for Astronomical Research in the Southern Hemisphere, Casilla 19001, Santiago 19, Chile
email: triviniu@eso.org, sstefl@eso.org, and cmartaya@eso.org

Abstract. Identifying seven activities and activity-carrying properties and nine classes of Active OB Stars, the OB Star Activity Matrix is constructed to map the parameter space. On its basis, the occurrence and appearance of the main activities are described as a function of stellar class. Attention is also paid to selected combinations of activities with classes of Active OB Stars. Current issues are identified and suggestions are developed for future work and strategies.

Keywords. stars: activity, stars: binaries, stars: circumstellar matter, stars: early-type, stars: emission-line, Be, stars: magnetic fields, stars: mass loss, stars: oscillations, stars: rotation

---

1. Active OB Stars: the concept

1.1. *The activities*

The term *Active B Stars* was introduced in 1994 when the IAU *Working Group on Be Stars* was renamed *Working Group on Active B Stars*. The name was to capture all physical processes that might be active in Be stars and so be required to understand Be and other active B stars, similar to Richard Thomas' standing characterization of Be stars as the crossroads of OB stars. This paper considers every intrinsically variable OB star an Active OB Star.

The potential (and actual) variabilities are the same as everywhere else in the Hertzsprung-Russel Diagram (HRD): Stars may pulsate and so vary in temperature and shape. If they possess a magnetic field, there may be associated activity, and rotation will periodically modulate observables, especially if diffusion has led to locally varying surface abundances and/or the magnetic field is strong enough to trap an otherwise present wind. Very young stars may still be working on the dispersal of their natal disk, and somewhat more evolved stars may return some of the originally accreted mass to the ambient space through radiatively driven winds and/or discrete mass-loss events. Both fossil and newly lost circumstellar matter may cause variable extinction. The associated large-scale dynamics as well as local variations in temperature, density or excitation would manifest themselves also in variable profiles of spectral lines with circumstellar origin.

1.2. *The stellar classes*

With the objective of merely providing a first crude chart of the Active OB Star territory, this section tries to use the smallest amount of information necessary. All in all, this territory is presently divided into nine major, multiply overlapping areas (see Fig. 1).

At both very young and advanced ages, OB stars are variable so that they are Active OB Stars on account of their evolutionary phase. They are then respectively known as **Herbig Be stars** and **supergiants** or even **Luminous Blue Variables (LBVs)**.

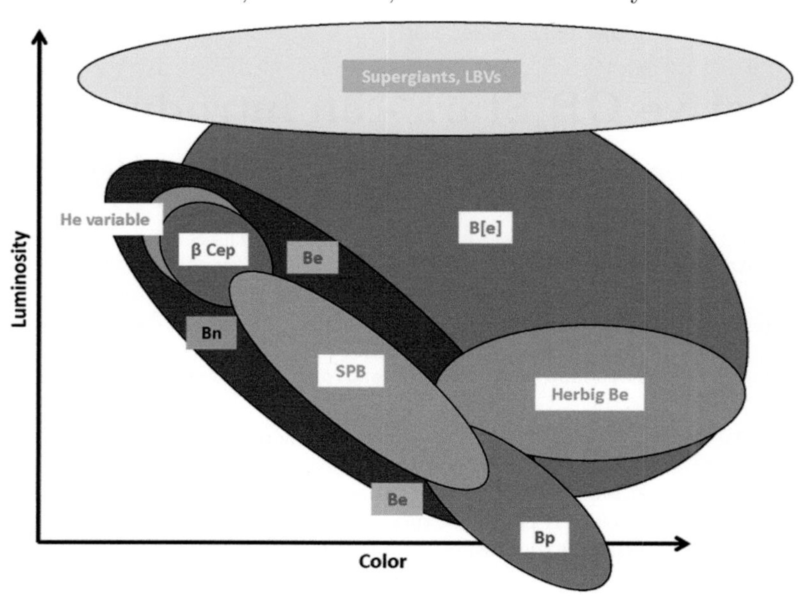

Figure 1. Schematic HRD of the various classes of Active OB Stars. The B[e] blob would be grossly oversized if B[e] stars did not include pre-main sequence *and* post-main sequence stars but just the latter. Within a couple of years, major revisions can be expected from much expanded observing efforts.

Two classes of Active B Stars are defined primarily by the presence and appearance of circumstellar emission lines: **Be stars** (sometimes also dubbed "classical Be stars") and **B[e] stars**. Be stars are characterized by rapid rotation. The appearance of their emission lines depends strongly on aspect angle, implying that the circumstellar matter, from which they arise, forms rotationally flattened equatorial structures. Be stars are further distinguished from B[e] stars by not having any appreciable amounts of circumstellar dust while many B[e] stars exhibit infrared excesses that can only be explained by warm dust. The notation [e] indicates that forbidden low-excitation emission lines are common in B[e] stars. But such lines are not a mandatory attribute of the class.

More work is needed to find out whether the classes of Be and B[e] stars require further subdivision. This is especially true for the B[e] stars, which have been identified among pre-main sequence stars as well as supergiants whereas more recent work has questioned very young ages of B[e] stars. Among the Be stars, the later spectral sub-classes are often said to be less active on all time scales.

Be stars have somewhat inconspicuous cousins, namely rapidly rotating B stars not known to have shown optical line emission. For the lack of the latter, and not considering interferometric techniques, such stars cannot be recognized unless they are viewed roughly equator on and reveal their extremely broad spectral lines. In the MK notation they are known as **Bn** or **Bnn stars**. Unlike Be stars, they do not seem to pulsate in low-order ($\ell = |m| = 2$) nonradial pulsation (NRP) modes while higher-order modes with $m \sim 10$ appear common. However, this distinction is still pending confirmation from larger samples, which should also examine whether there is any variation with spectral type.

There are two classes of Active OB Stars that have no defining properties other than their pulsations: The classical β **Cephei stars** between B0 and B2 and the **Slowly Pulsating B (SPB) stars** between B2 and B7. Both types occur along the main sequence and actually overlap around B2, with the two domains being properly delineated only now by space observatories such as BRITE, CoRoT, Kepler or MOST. More disjunct seem to be the ranges in periods: 0.1-0.25 d for β Cep stars but 0.5-3 d for SPB stars.

Magnetic OB stars, too, come in a hot and a cool variety: the **helium-variable stars** $\leqslant$ B2 and the **Bp stars** $\geqslant$ B7. Typical longitudinal field strengths are around several kG. They are most easily recognized through rotationally propagating surface abundance anomalies, which at the hot end mostly involve helium and rare earths at the cool one and may artificially widen, or even create, the gap in spectral type. Of?p stars and τ Sco-like stars present further magnetic sub-groups.

Various kinds of activities can easily co-exist. In the HRD, more or less the entire area apportioned to OB stars is covered by one or more Active OB Star tiles. However, these tiles are porous because Active OB Stars are only an estimated 5-25% minority. With ever more sensitive observations, this fraction keeps increasing.

The expected wealth of new observations is likely to demand major revisions, especially of the definitions of B[e], pulsating, and magnetic stars. Some activities extend over larger regions than the quoted spectral-type based names suggest: In addition to Be stars, there are Oe and Ae stars although the latter mostly manifest themselves through narrow absorption lines when the line of sight passes through the disk. Bp stars continue into the A-types domain, and there are yellow supergiants. Herbig Ae stars are more numerous than Herbig Be stars. Concerning pulsations, the boundary region between radiative and convective envelopes, which on the main sequence is not far from the A/B subdivision in spectral type, should let stars on either side have their own pulsation properties.

2. The OB-Star Activity Matrix

From the above compilations of variables and properties, the OB-Star Activity Matrix can be constructed. Fig. 2 shows it filled with what is believed to be today's knowledge. Most of the rest of this paper consists of a discussion (and sometimes a justification) of the contents of the Activity Matrix. This is partly organized in an 'across'- and 'down'-like manner, with more emphasis on the vertical dimension because it is the one less often explored as most (especially observational) studies are object oriented. Finally, various combinations of links between selected matrix elements are examined.

3. Analytical tools and models

The past two decades have seen progress in the amount and quality of observations that would have been difficult to predict. But the understanding of Active OB Stars has probably benefited still more strongly from the vast improvements in the flexibility of models, which now affords their tailoring to specific observations. Also, while just 10 years ago most model calculations would assume slow rotation at most, the effects of extreme rotation are now widely considered.

Within the suite of models, stellar evolution assumes a central position because it is the one that puts all else into a common perspective. It does not stop at the level of single objects but also quantifies the chemical, radiative, and mechanical output of OB-star populations, which often dominate the corresponding global balances of their environments up to entire galaxies. The interface to explosive models terminating the evolution of many OB-stars is similarly important.

Evolutionary and mass-loss models continue to intensify their links. Mass loss from massive stars can exceed 50% in the first few million years with drastic effects on evolution. Radiative-wind models incorporate more and more of the observations that picture

	Pulsation	Magnetic field	Wind	Mass loss events	CSM	Very fast rotation	Dependencies on metallicity
Supergiants, LBVs	(X)	(X)	X	(X)	X	(X)	X
B[e] stars			(X)		X		
Be stars	X	(X)	((X))	X	X	X	X
Bn stars	X					X	
β Cephei stars	X	(X)	(X)		(X)		
SPB stars	X	(X)					
He-variable stars		X	X		X	(X)	
Bp stars		X					
Herbig Be stars		(X)	X		X		

Figure 2. The OB-Star Activity Matrix. Crosses: combinations of property and type of star that are typical. Single brackets: the combination is not generally established. Double brackets: doubtful cases. Not marked: combinations do not exist or are not known to exist.

mass loss from OB stars as a very dynamic, sometimes violent process. Not any less important are smaller-scale clumps. The increased escape probabilities for photons between such clumps has led to downward revisions of mass loss rates by factors of $\sim$5.

Evolution models have especially advanced in the handling of mixing processes so that relatively subtle surface abundance patterns can be predicted, which may encapsulate much of a star's history. This strongly broadens the practical scope of stellar atmosphere and radiation-transport models that have been developed to ever higher perfection.

Asteroseismology has been around as a term for a long time. With the coming into operation of wide-angle space photometers many of the promises will now come true for many types of Active OB Stars - as the first results suggest, after much additional work. The eventual combination of stellar evolution models with surface abundances and asteroseismologically measured internal structures stands high chances to lay the most important foundation to the understanding of Active OB Stars.

Another application of complex data inversion techniques is the incipient construction of images from interferometric observations with multiple telescopes and baselines.

OB stars shape their environments through winds and radiation and partly build their own circumstellar structures, e.g., through discrete mass-loss events (LBVs, Be stars) or in the form of magnetospheres. Much of this is becoming accessible to interferometers and so demands radiative and dynamical models to be combined and extended to higher spatial dimensionality. The current rate of progress is impressively high.

Much of this work has been enabled not just by ever cheaper hardware but especially by the parallelization of codes that now often run on hundreds of CPUs. The partial replacement of CPUs by Graphics Processing Units is opening another new phase.

4. Binaries and binarity

The Activity Matrix lists neither binaries as a distinct class of Active OB Stars nor binarity as a distinct type of intrinsic variability. However, it is important to recall what situations can arise in double stars that are not likely in single stars.

Most prominent to name are binaries, in which the primary has entered one of the classes of Active OB Stars as the result of binary evolution. For instance, some Be stars have hot sub-dwarf companions and probably owe their fast rotation to mass transfer from the companion.

The orbital motion and associated modification of the gravitational potential may trigger single events as well as trigger or suppress long-term behaviors. For instance, the ejection of a circumstellar disk around the primary in δ Sco has been tentatively attributed to the periastron passage. It is hoped that the next periastron passage in 2011 will clarify this issue. If the disk ejection is indeed caused by the varying gravitational pull, one may wonder whether the transmitter process is forced nonradial pulsation. On the other hand, β Cephei stars seem to be rare in, or even absent from, binaries with orbital periods of less than a couple of days.

If an OB star with circumstellar structures is orbited by a compact object, periodic dives of the companion through that matter ideally probe its properties. Conversely, the response of the latter can reveal much detail about the nature especially of exotic objects; but there are also effects on the large-scale dynamics, e.g., truncation or precession of disks. In other OB binaries, both stars may have a wind, the collision of which establishes a similarly interesting laboratory.

Last but not least, there are also special features like the jet in β Lyrae, which deserve attention because of their seemingly rare occurrence. But it may also happen that a star appears very exceptional only as long as the role of a companion has not been appreciated. For instance, β Cep was claimed to break the rule that all Be stars rotate rapidly, until it was realized that the emission lines actually belong to a nearby second star.

The most important property of binaries is, of course, their enabling role in the quantitative measurement of fundamental stellar parameters. Because activities will increase the noise at least of the measurements, many Active OB Stars cannot be primary sources of such information. But comparisons with not-active OB stars will still be essential.

Since the fraction of binaries is high among early-type stars, the possibility of some observed properties being due to binarity must always be considered. But there does not seem to be any intrinsic variability limited to binaries, and there is no obvious class of Active OB Stars with much above-average binarity rate.

5. Activities as a function of stellar class

5.1. *Pulsation*

Of all activities observed in Active OB Stars, pulsation is the most frequently and widely detected one. From ground-based observations, which were limited in number and especially sensitivity, it seemed that β Cep stars pulsate in nonradial p modes and often also radially, Be stars in g- as well as in p-modes, SPB stars in g-modes, and Bn stars in p-modes. Supergiants were mostly suspected to pulsate in g-modes with p-modes occasionally being a possibility.

With the advent of space-borne wide-field precision photometry, this chapter in the history of Active OB Stars is presently undergoing so strong and rapid revisions that it is advisable to neither repeat outdated past wisdom nor speculate about unconfirmed new insights. The still relatively few pulsation spectra published so far are partly very rich and do not in all cases find straight-forward matches in theoretical models, especially in the presence of very fast rotation. Also the mapping of the occurrence of various pulsational properties has seriously commenced only with these observations.

An interesting question that has not been addressed for a fair while is: High-m nonradial modes are only seen in broad-lined stars. Is this so because such modes are difficult to detect in narrow lines or because rapid rotation favors their excitation?

Another theoretical expectation that is still awaiting attempts of an observational verification is that nonradial g-modes may transport angular momentum between different radial zones of a star, with possible evolutionary ramifications.

5.2. *Magnetism*

On average, magnetic fields are more difficult to detect than pulsations are, especially on small spatial scales, so that magnetism might actually be the premiere activity of Active OB Stars. Detections have been reported for almost any kind of OB stars; increases in the sensitivity and diversity of searches may wash out the current classification scheme and propose new criteria. But some reports have provoked strong responses questioning claimed significances. Unfortunately, the current plans for ELTs do not have polarimetry among their main science drivers so that studies of magnetic fields (and other applications of spectropolarimetry) may not benefit from the large gain in light-collecting power.

Because magnetic fields below the current detection threshold of 50-100 G for the mean field strength may still have significant dynamical effects (and thereby make convenient attributes of various toy models), methods other than polarimetry may have to be resorted to. For instance, it is conceivable that space photometry may place constraints on the strength and structure of magnetic fields: If some variability can be attributed to rotation, star spots are a reasonable guess for the explanation, and the associated light curve would give hints at their size, contrast, and latitude. Non-periodic variations would require still more interpretative caution but could also be related to magnetic activity.

Although the rich spectrum of observations demonstrates that actual OB stars do not generally find it difficult to exhibit strong, large-scale magnetic fields, theoretical 'confirmation' is missing. In fact, there may be no other topic in Active OB Star research, where observations and models are farther apart from one another although only two basic mechanisms seem to require consideration: dynamo processes and the freezing in of magnetic fields present in the parental cloud. Although sufficiently effective dynamos are difficult to develop in mostly radiative envelopes, the clustering of magnetic stars along the main sequence may give this hypothesis a slight edge over the fossil alternative. Interesting may also be the recent report that, in Bp stars, magnetic fields only become detectable after 30% of the main-sequence evolution has been completed.

5.3. *Structured winds: Photospheric heritage?*

Observations find more and more structure in the mass loss from OB stars. Extended *IUE* campaigns have established a long time ago that Discrete Absorption Components (DACs) in wind lines often repeat on rotational time scales. The concept of Corotating Interaction Regions has enabled models to reproduce the observations very well. However, the origin and nature of the perturbations at the base of the wind have remained unidentified. The velocity fields and temperature modulations of the wide-spread nonradial pulsations are obvious candidates. With very few exceptions, observations have been insufficient to demonstrate a satisfactory match of the time scales, let alone periods. Perhaps, long-term photometry from space can contribute further cases.

Large-scale magnetic field structures would a priori lead to rotational modulations (if any). In OB stars with unambiguously detected magnetic fields, periodic effects on the wind can be clearly seen. If the field strength reaches kG levels, the wind is trapped in a magnetosphere. However, magnetic fields are not so far generally detected; indirect support may come from space-based precision photometry.

Groundbased time-resolved spectropolarimetry of emission lines has identified irregularly outward-moving knots, which have led to the concept of porous winds with embedded clumps. Unlike DACs, these small-scale structures significantly affect the escape probability of wind-driving photons, reducing mass loss rates by a factor of ~5.

Inspired by the case of Be stars, equatorial density enhancements have been searched for also in rapidly rotating O stars. The results were mostly negative. Therefore, disk formation in the more extremely rotating Be stars may not just be a property of winds in the presence of very fast rotation, as various models have tried to picture. In particular, the angular momentum necessary for rotationally supported disks is not supplied by a wind alone. So far, this is only achieved by viscous decretion. This model has the additional important strength that it also explains the subsequent angular momentum transport from the inner to the outer disk.

Once formed, disks of Be stars only last for ~10 years. Observations show that disk dispersal progresses inside-out so that stellar radiation pressure may be involved in this process. In fact, insufficiently sampled and coordinated ground-based and *IUE* observations suggest that the strength or even presence of winds from Be stars may depend on the presence of a disk: Winds only set in after a significant disk has been built up, and they cease before the optical line emission from the disk has fully vanished. Alternatively, if during active phases disk replenishments were frequent, the wind might be fed with debris left over from this process. Either ansatz might also explain the puzzling fact that, in Be stars, stellar winds are observed down to cooler spectral types than in non-Be stars. More importantly, it may be useful to identify those wind variabilities that Be and other OB stars have in common. Although (early-type) Be stars exhibit strong nonradial pulsations, the latter may not be an important source of wind variability.

IUE observations also suggest that the winds of Be stars are strongest at intermediate inclinations and, therefore, stellar latitudes. This is consistent with the wind material being supplied by the disk-building process. By contrast, recent interferometric observations of Be stars report polar structures that are being identified as winds. In the absence of velocity information, this latter step seems to be a very large on.

6. Metallicity

In Table 2, column "Dependencies on metallicity" is special in that it actually presents a third dimension, which reports whether the frequency or properties of a class of Active OB Stars have been observed to depend on metallicity. Metallicity is, therefore, perhaps the most valuable diagnostic because a model is vastly more convincing if it reproduces observations over a range in metallicity. This makes it important to take advantage of the present 8-10 m class telescopes and the planned generation of ELTs, which bring the Milky Way's satellite galaxies and their metallicity levels within the reach of fairly high spectral resolution. Since at the same time the number of stars in such galaxies is larger per unit area in the sky, multi-object spectroscopy can be used to good advantage.

Low metallicity reduces the efficiency of the linear-momentum transfer from radiation to matter. The resulting basic dependency of mass loss rates on metallicity is well established in luminous stars, which are observable even beyond the local group.

The accompanying reduced loss of angular momentum is one reason to expect low-metallicity stars to rotate more rapidly, with the other reason being that at lower opacity stars have smaller radii if all else is equal. Both circumstances imply a higher fractional critical rotation rate. On this ground, it has been argued that, because of their rapid rotation, Be stars should be more frequent in low-metallicity environments, which is, in fact, observed. However, in a strict sense, this does not permit the sometimes heard conclusion to be drawn that low metallicity makes more Be stars. Rapid rotation only is a necessary condition for the Be phenomenon; according to all what is known today, rotation alone is not sufficient for a B star to become a Be star: In the Galaxy, the number of Bn stars is about as large as the one of Be stars. Therefore, one can only claim that, down to SMC metallicity levels, whatever other properties are needed for the Be phenomenon, their net dependency on metallicity is not strongly negative. This also leads to the following: It has been demonstrated that the beating of nonradial pulsation modes can drive mass loss events in Galactic Be stars. But pulsation responds negatively to reduced metallicity so that one would expect such a mechanism to be less effective in the SMC. Since the SMC is more abundant in Be stars than the Galaxy, this might be another hint that ω/ω_c increases at low metallicity.

The prediction by theory of a metallicity dependence of pulsations is still without observational counterpart for comparison. And the sensitivity or not to metallicity of dynamo processes or fossil magnetic fields is hardly explored at all.

7. Classical disputes about classical Be stars

Previous meetings with a large fraction of Be star contents sometimes became quite lively on the grounds of widely differing opinions. Today, most of these issues can be considered closed. Others have arisen but are discussed less sanguinely:

- *Can low-velocity, low-excitation and high-velocity, high-excitation spectral lines be put into one common radial scheme?* No, they cannot. There is a cool, geometrically thin disk with a fast and hot wind at higher latitudes.
- *Can transitions (observed in a few stars) from a Be to a Be-shell phase (with narrow absorption lines sumperimposed to very broad emission lines) and back to a Be phase (without such additional absorption lines) be explained by a radial restructuring of the atmosphere?* No, they cannot. The most plausible model involves disk precession in a binary star. In single stars, warping or flaring of the disk are alternatives.
- *Is the photospheric line-profile variability due to rotation or to nonradial pulsation?* It could be both with pulsation being the dominant effect by far. Not surprisingly, space photometry finds evidence also of rotation time scales. But the cross identification in line profile variations is still pending.
- *Are the variations in the violet-to-red ratio of the strength of emission line components due to orbital aspect variations?* No, they are not. They are the signature of global, one-armed oscillations of a Keplerian disk around a rotationally distorted star.
- *Are all Be stars products of binary evolution?* Yes, some are. But there is no evidence of an overabundance of binaries among Be stars. (If they were even underabundant one might be driven to wonder whether the rapid rotation comes about because the initial angular momentum does not have to be shared with the orbital motion of a companion.)
- *Is rotationally induced polar enhancement the reason that Be stars exhibit winds down to later spectral types than do normal B-type stars?* It depends. At the same polar $T_{\rm eff}$ of a rotating and a non-rotating star, the difference is small for pole-on stars. At other viewing angles, the comparison is made difficult by the range in $T_{\rm eff}$ of the rapid rotator. The *IUE* database appears more consistent with the wind being fed by radiative ablation of the circumstellar disk (cf. Sect. 5.3).

8. Selected combinations of specific properties

8.1. *Rotation + pulsation + mass loss (events) in Bn/Be stars*

For every broad-lined Be star, the *Bright Star Catalogue* contains between 0.5 and 2 Bn stars with about the same spectral type and magnitude. In view of the 'saturation' in rapidly rotating stars of conventionally used optical lines to measure rotation velocities, it is not certain that $v \sin i$ values measure the same thing in Be and Bn stars. But it is reasonable to suspect that the differences are not dramatic (even though potentially systematic, also because of circumstellar contaminations in Be stars).

While (early-type) Be stars pulsate in low-order g-modes *and* higher-order p-modes, only the latter seem to be present in Bn stars. At the same time, only Be stars develop circumstellar disks whereas Bn stars do not exhibit emission lines. This invites the suspicion that the low-order modes play a role in the disk-building process. It is supported by two empirical facts: (i) The variability of optical emission lines suggests that individual events may dominate the mass loss from Be stars. (ii) In a very small number of Be stars evidence has been found that such events are driven by the beating of nonradial pulsation modes with the same ℓ, m values.

However, reality is more complex: More common are Be stars, in which intensive long-term spectroscopic monitoring could not find a second low-order mode. And yet, these Be stars do undergo outbursts. If Be stars rotate close to the critical rate, outbursts not assisted by beating NRP modes pose no fundamental problem because any process with associated velocities of the order of the turbulence could let the total velocity exceed the critical threshold. An alternative, which existing observations may not be able to address fully, is that in such stars the beat periods are of the order of some years, i.e., the time between consecutive outbursts, which can be relatively equally spaced. In order to study this matter more broadly, photometric databases such as *OGLE* or *MACHO* could be searched for Be stars with regularly spaced outbursts. Since the primary pulsation frequency normally is easily found, the difference between this frequency and the one of the outbursts would correspond to a second pulsation frequency to be searched for - if the outburst frequency equals a beat frequency.

The outbursts of Be stars are reminiscent of classical limit-cycle systems: Some mechanism drives the star quasi-peridocially to the limit of stability. It is the beating of NRP modes in some cases, but may also be angular-momentum transfer from the core to the envelope. The angular-momentum redistribution, too, may be caused by NRP modes (which, however, should be prograde), or it may be accomplished by magnetic fields, meridional circulation, other processes or any combination thereof. There is no reason to believe that only one mechanism converts Bn stars to Be stars. The mix of such processes may depend on mass; ditto for the threshold in ω/ω_c. In fact, it has been reported that the fractional critical velocity required for the Be phenomenon increases towards lower masses. The Bn-to-Be number ratio increases in the same direction, which is a consistent observation but requires cautious interpretion becauses cooler stars are less effective in ionizing disks, which may, then, remain undetected because of weak line emission.

8.2. *Magnetic fields + rotation*

Magnetic fields brake the rotation of young convective stars. Consistent with that, Bp stars are throughout slow rotators. Therefore, sometimes the reasoning is inverted, and it is argued that rapidly rotating stars, e.g., Be stars, do not possess significant magnetic fields. The logic of this conclusion is impeccable: If A implies B, not-B implies not-A. However, recently a new population of extremely rapidly (nearly critically) rotating helium-variable stars was discovered. Perhaps, their evolutionary history is at least as exciting as their present properties. The contradiction between rapid rotation and strong magnetic field could be resolved if either or both properties had been acquired only during the course of time. The reported delayed occurrence of magnetic fields in Bp stars could be interesting also in this context.

8.3. *Pulsations + magnetic fields*

Pulsations and magnetic fields co-exist in various Active OB Stars. Interestingly, there seem to be no known combinations of kGauss fields with pulsation. This may be an issue of observational sensitivity. But it could also result from magnetic fields inhibiting the equalization on spherical surfaces of opacity, mean molecular weight, etc. It would be very attractive to investigate strongly magnetic stars for pulsations from space.

Magnetic fields at a level of up to a few 100 Gauss have been reported for a number of β Cep as well as SPB stars. But there are no deep enough seismological analyzes that would permit the detection of any magnetic effects on the pulsations. There has been the suggestion that, similar to the rapidly rotating Ap stars, β Cep might be an oblique magnetic rotator, in which the pulsational symmetry axis is aligned with the magnetic rather than the rotation axis. Because this report has remained without independent confirmation, also in other stars, it may be uncertain. (But magnetic axes seem to be mostly tilted with respect to the rotation axes.)

9. The fate of Active OB Stars

Can any terminal evolutionary stages of stars be traced back to specific types of Active OB Stars? It seems fairly certain that Bp stars evolve to magnetic white dwarfs. There is also ample evidence that stars with initial masses above $\sim 8\ M_\odot$ end up as core-collapse supernovae, but already the nature and temporal sequence of the intermediate stages (LBVs and the various flavors of Wolf-Rayet stars) still give rise to discussion.

The collapsar model is widely supported as one explanation of long Gamma-Ray Bursters (GRBs). The required rapid rotation renders Oe/Be stars with initial masses above 20 $M_\odot$ interesting candidates. But rapid rotation may also result from mass transfer in binaries. In either case, low metallicity would help to maximally preserve the accumulated angular momentum. In fact, evidence is emerging that low-metallicity environments do favor GRB events.

Still more uncertain is it whether (rapidly rotating) magnetic helium-variable stars play a role as progenitors of magnetars and Soft Gamma Repeaters. The small number of known helium-variable stars is a serious obstacle.

10. A wish list for future work

The progress in the understanding of Active OB Stars is driven through many channels, and the highest priority must be to preserve and enlarge this diversity. Still, there are some developments that may have a particularly broad impact:

- Hot stars emit most of their flux in the UV. The variability of Active OB Stars requires monitoring on many time scales, which is not easily done with present large space-borne facilities. But an *IUE*-like satellite that is small enough to be of no use outside the Local Group could close one of the largest gaps in the arsenal of observing facilities. Since many of the *IUE* observations were interpreted on the basis of the understanding of 20 years ago, a first valuable step could consist of a re-analysis of pertinent archival data.
- Non-active members of star clusters present the most reliable basis for the study of Active OB Stars because distances, ages, metallicities, etc. can be derived with good accuracy. Only statistics obtained from star clusters can empirically guide stellar evolution models. Star clusters may also help identify the switches that do or do not let OB stars branch off main-track evolution towards the various types of Active OB stars. Much of the existing work has been done using imaging instruments. But multi-object spectroscopy, or at least slit-less spectroscopy, obviously has a much higher potential. Pure

photometry reaches hard limits if observations of extreme rotators are not corrected for gravity darkening. However, photometry alone cannot determine v_{eq} and i.

• Recent years have seen spectacular progress from interferometric observations of Active OB Stars. The potential of such work is very far from having been exhausted. Spectrointerferometry and first simple images from closure phase techniques show the way.

• The 'poor man's substitute for interferometry' is (spectro-)polarimetry, which sometimes can yield nearly equivalent insights. Regrettably, it is broadly underutilized.

• Satellites suitable for asteroseismology (e.g., BRITE, CoRoT, Kepler, MOST) have already produced amazing results for selected stars. Broader statistics is very much wanted. (Like searches for extra-solar planets, with which such facilities are often shared, statistics benefits from relatively large telescope apertures. However, this limits the maximal brightness of observable targets and so excludes objects, for which the also needed long series of spectral observations are available and/or feasible.)

With numerical models moving ever closer to reproducing individual Active OB Stars, also the reverse becomes increasingly more true: Observations without advanced modeling support are losing in value.

11. Outlook

Maybe, in 25+ years, there will be an IAU Symposium *Active OB Stars Explained!*. The following attempts to give a selective preview of the agenda and work it may be based on; the admitted Be-centricity may prove to be extendable to other classes of Active OB stars:

• *The difference(s) between Be and Bn stars*: Their dominating pulsation modes seem to be different. Therefore, seismology should be the method of choice.

• *The evolutionary stage of Be stars*: Since the dominating NRP modes excited in (early-type) Be stars seem to be deep-reaching g-modes, they may hold one key to the answer, depending on the number of modes found by space photometry. Since at the length of Be-star periods frequency spectra are dense, sufficiently precise identifications may be challenging so that cross-calibration with open-cluster data could be important.

• *The threshold of v_{eq}/v_{crit} for the Be phenomenon*: This will reveal how important the contributions by other processes to the release of energy and angular momentum need to be. A pre-requisite is the determination of the inclination angle, i, from precision photometry (direct rotational modulation? rotational splitting of NRP frequencies?), photospheric line-profile modeling (taking into account rotational gravity darkening and rotationally modified angular NRP eigenfunctions), fitting of circumstellar emission lines, and possibly also interferometry. For comparison, v_{eq}/v_{crit} needs to be determined for a number of Bn stars as well (which have large values of i by definition). Asteroseismology will also show whether extreme rotators spin differentially.

• *The role of $\ell, m = 2, +2$ g-modes in the mass loss from Be stars*: Results for the previous point will be valuable also in this context. But (the already beginning) observations of the behavior of pulsational frequencies and amplitudes during outbursts will provide the key. The main technical challenge will be the proper timing of such observations. Stars with mass loss events triggered by the beating of pulsations modes and stars with frequent mass ejections should be the first targets. However, the analysis must remember the selection biases associated with such a strategy.

• *The relevance of magnetic fields for the Be phenomenon*: Since conventional spectropolarimetry seems to have reached its limits without delivering a significant number of detections, the technique might have to team up with interferometry (at baselines of several 100 m) to search for smaller-scale structures. Candidates might result from high-precision photometry. High-energy phenomena associated with magnetic processes might provide an additional avenue.

• *The fraction of the Be star population due to binary evolution*: Cool giants and white dwarfs have been searched for with limited success. The main remaining possibility are hot sub-dwarfs. Interstellar HI would prevent their detection from the combined spectral energy distribution unless the Be star is so cool that the flux excess becomes visible at longer wavelengths. Other techniques are high-S/N observations of high-excitation lines forming in the region of the disk facing such a companion, radial-velocity measurements, and interferometry. If low-order NRP is indeed important for the Be phenomenon, searches for them in Be stars spun up in binary systems may make an important contribution to the unresolved issue of the homogeneity of the Be phenomenon. How do B stars spun up in binaries become Be stars if rapid rotation alone is not a sufficient condition?

• *The disk or stellar origin of the winds of Be stars*: Spectroscopic monitoring of optical disk emission lines and UV resonance lines over one or two disk life cycles in a small number of Be stars should clarify this issue.

• *The disk-truncation mechanism in Be stars*: The combination of spectro-interferometry at IR to mm wavelengths and models for the dynamics and radiative transfer in disks will give the answer.

• *The nature(s) of B[e] stars*: Parallaxes from *GAIA* will be an important step towards understanding the reported diversity. With some luck, objects observable with interferometers may be identified.

• *The wind structure of OB supergiants*: Simultaneous UV and optical spectroscopy, spectroplarimetry, and IR and mm observations designed such as to permit discrimination between processes scaling with ρ and ρ^2 will give important guidance to model calculations. If winds from Be stars do not originate from the photosphere but nevertheless share properties with the winds of more luminous stars, interpretative efforts may perhaps abstract more strongly from putative photospheric print-throughs.

• *The origin of magnetic fields in radiative atmospheres*: Seismological comparisons of magnetic and non-magnetic stars and observations in star clusters will constrain the structure and evolutionary phase of magnetic stars. But most of the load will be carried by detailed model calculations.

• *The progenitors and descendants of Active OB Stars*: This is the area, in which the Active OB Star community has the least autonomy. But it also offers a large opportunity for external impacts.

To almost all topics, observations of Active OB Stars in galaxies with different metallicities will make fundamental contributions, and accurate abundance analyzes will often yield important clues on the time-integrated evolutionary effects in individual stars.

Perhaps the most important steps forward can be expected from numerical model calculations. Simulations of many activities will be possible in three dimensions, and time dependencies will be included. It will be possible to weave extensive model grids even around specific sets of observations and thereby place tight limits on the derived quantitative parameters. Real interferometric images may be within reach.

Un(?)fortunately, there is no way to predict genuinely new discoveries. Massive Population III stars may be one of them. In any event, new developments will probably force the title of that future IAU Symposium to be: *Active OB Stars: Ever new challenges.*

References

On purpose, this paper does not provide any references: What is not expanded on in the following papers probably does not matter for the scope of this Symposium.

Participants in the auditorium.

Active OB stars: structure, evolution, mass loss, and critical limits
Proceedings IAU Symposium No. 272, 2010
C. Neiner, G. Wade, G. Meynet & G. Peters, eds.

doi:10.1017/S1743921311009926

Rapid rotation and mixing in active OB stars – Physical processes

Jean-Paul Zahn

LUTH, Observatoire de Paris, CNRS UMR 8102, Université Paris Diderot
5 place Jules Janssen, 92195 Meudon, France
email: Jean-Paul.Zahn@obspm.fr

Abstract. In the standard description of stellar interiors, O and B stars possess a thoroughly mixed convective core surrounded by a stable radiative envelope in which no mixing occurs. But as is well known, this model disagrees strongly with the spectroscopic diagnostic of these stars, which reveals the presence at their surface of chemical elements that have been synthesized in the core. Hence the radiation zone must be the seat of some mild mixing mechanisms. The most likely to operate there are linked with the rotation: these are the shear instabilites triggered by the differential rotation, and the meridional circulation caused by the changes in the rotation profile accompanying the non-homologous evolution of the star. In addition to these hydrodynamical processes, magnetic stresses may play an important role in active stars, which host a magnetic field. These physical processes will be critically examined, together with some others that have been suggested.

Keywords. stars: early-type, stars: interiors, stars: magnetic fields, stars: rotation

---

1. The need for non-standard stellar models

Until recently, stellar models ignored the possibility that some mixing could occur in the radiation zones of stars. But the situation is evolving because there is increasing evidence for such mixing, of which we shall give here only two examples.

It is well known that some A-type stars display anomalies in their surface composition, when they are compared to other, called 'normal' stars. These peculiarities were ascribed successfully to radiative acceleration and gravitational settling, by Michaud (1970) and his collaborators. But it turns out that these microscopic processes are so efficient that they would produce surface anomalies that are much more pronounced than those that are observed. For instance, helium would disappear from the surface of A-type stars in about one million years, as was pointed out by Vauclair *et al.* (1974). Since this is not observed, Vauclair *et al.* (1978) suggested that some mild turbulence - other than thermal convection - must operate near the surface to smooth the composition gradients. This was again emphasized and illustrated by Richer & Michaud (2000).

Another proof of such mixing is the overabundance, observed at the surface of massive stars, of chemical elements that are synthesized in the nuclear core. The overabundance of ^{4}He and ^{14}N, can only be explained if the radiative envelope has undergone some mixing (cf. Meynet & Maeder 2000). Interestingly, these overabundances seem to be correlated with the rotation velocity of the stars (Herrero *et al.* 1992); this property was again stressed by Maeder *et al.* (2009), who re-discussed the data gathered by Hunter *et al.* (2008).

There are thus strong indications that radiation zones undergo some kind of mixing, and several teams are now taking that mixing into account in their stellar evolution code. There are two ways of dealing with the problem. One is to model the mixing by an *ad hoc* diffusion, which is allowed to depend on one or more parameters, and to adjust these parameters to fit the observations. Our preference goes instead to the physical approach, which strives to implement as well as possible the physical processes that are likely to cause the mixing. We believe that they have been identified: these are the large scale circulation required by the transport of angular momentum, and the turbulence generated by the shear of differential rotation.

2. Rotational mixing

2.1. *Meridional circulation*

In its original treatment (Eddington 1925; Vogt 1925), the meridional circulation was ascribed to the fact that the radiative flux is no longer divergence-free in a rotating star, due to the centrifugal force. The characteristic time of the circulation was derived by Sweet (1950), and has since been named the Eddington-Sweet time: $t_{\rm ES} = t_{\rm KH}(GM/\Omega^2 R^3)$, with $t_{\rm KH} = GM^2/RL$ being the Kelvin-Helmholtz time. R, M, L designate respectively the radius, mass and luminosity, Ω the angular velocity and G the gravitational constant. Sweet's result suggested that rapidly rotating stars should be well mixed by this circulation and thus would remain homogeneous; therefore they would not evolve to the giant branch, contrary to what is observed.

However, these early studies overlooked the fact that the circulation carries angular momentum and thus modifies the rotation profile: starting from arbitrary initial conditions, the star undergoes a transient phase which lasts indeed about an Eddington-Sweet time, after which it settles into a quasi-stationary regime where the circulation is governed solely by the torques applied to the star. For instance, when the star loses angular momentum through a strong wind, the circulation adjusts precisely such as to transport that momentum to the surface (Zahn 1992). The resulting rotation is then non-uniform, and a baroclinic state sets in, with the density varying with latitude along isobars. When the angular velocity depends on the radial coordinate only, as we shall assume for reasons that will be explained later on, the density perturbation is given by $\delta\rho(r,\theta) = \tilde{\rho}(r)P_2(\cos\theta)$, and the baroclinic equation takes the simple form

$$\frac{\tilde{\rho}}{\bar{\rho}} = \frac{1}{3}\frac{r^2}{\bar{g}}\frac{d\Omega^2}{dr}, \tag{2.1}$$

with $\bar{\rho}$ and $\bar{g}$ being the mean density and gravity on the level surface, and $P_2(\cos\theta)$ the Lengendre polynomial of degree 2. On the other hand, when the star does not exchange angular momentum, the circulation would die altogether, as predicted by Busse (1982), if it had not to compensate the effects of structural adjustments (contraction, expansion) as the star evolves, and also the weak turbulent transport down the gradient of angular velocity that will be discussed next.

2.2. *Shear turbulence caused by differential rotation*

Since the rotation regime that results from the applied torques is not uniform, the shear of that differential rotation is prone to various instabilities, which generate turbulence and therefore mixing. Here we shall consider only those instabilities that apparently play a major role, namely the shear instabilities, also called Kelvin-Helmholtz instabilities.

2.2.1. *Turbulence produced by the vertical shear*

Let us first examine the instability produced by the vertical shear, $\Omega(r)$. This instability is very likely to occur, because the Reynolds number characterizing such flows in stellar interiors is extremely high, due to the large sizes involved. However the stable entropy stratification acts to hinder the shear instability: in the absence of thermal dissipation, it occurs only if locally

$$\frac{N_T^2}{(\mathrm{d}V_\mathrm{h}/\mathrm{d}z)^2} \leqslant Ri_\mathrm{c}, \tag{2.2}$$

where V_h is the horizontal velocity, z the vertical coordinate, and N_T the buoyancy frequency defined by $N_T^2 = (g\delta/H_P)(\nabla_\mathrm{ad} - \nabla)$, with the classical notations and $\delta = -(\partial \ln \rho/\partial \ln T)_P$. This condition is known as the *Richardson criterion*; Ri_c, the critical Richardson number, is of the order of unity and it depends somewhat on the rotation profile.

In a stellar radiation zone, this criterion is modified because the perturbations are no longer adiabatic, due to thermal diffusion. When the radiative diffusivity K exceeds the turbulent diffusivity $D_\mathrm{v} = w\ell$ (ℓ and w represent the size and the vertical rms. velocity of the largest eddies), the instability criterion takes the form (Dudis 1974; Zahn 1974)

$$\frac{N^2}{(\mathrm{d}V_\mathrm{h}/\mathrm{d}z)^2}\left(\frac{w\ell}{K}\right) \leqslant Ri_\mathrm{c}. \tag{2.3}$$

From the largest eddies that fulfill this condition, one can derive the turbulent diffusivity D_v acting in the vertical direction in the radiation zone of a star. However this instability criterion (2.3) holds only in regions of uniform composition, where the stability is enforced solely by the temperature gradient; when the molecular weight μ increases with depth, it seems at first sight that one should replace this criterion by the original one, expression (2.2), where now the buoyancy frequency is controlled by the gradient of molecular weight:

$$N^2 \approx N_\mu^2 = \frac{g\varphi}{H_P}\frac{\mathrm{d}\ln\mu}{\mathrm{d}\ln P},$$

with $\varphi = (\partial \ln \rho/\partial \ln \mu)_{P,T}$. However, as Meynet and Maeder (1997) pointed out, this condition is so severe that it would prevent any mixing in massive main-sequence stars, contrary to what is observed. We shall see below how that stabilizing action of μ-gradients can be overcome.

2.2.2. *Turbulence produced by the horizontal shear*

Likewise, the horizontal shear $\Omega(\theta)$ will also generate turbulence, and this turbulence will probably be highly anisotropic, due to the vertical stratification, with much stronger transport in the horizontal than in the vertical direction, i.e. $D_\mathrm{h} \gg D_\mathrm{v}$. This shear instability tends to suppress its cause, namely the differential rotation in latitude, and it will thus lead to a 'shellular' rotation state, where the angular velocity depends on the radial coordinate only: $\Omega \sim \Omega(r)$.

Such anisotropic turbulence interferes with the meridional circulation, turning the advective transport into a vertical diffusion (Chaboyer & Zahn 1992). To lowest order, the vertical velocity of the circulation is given by $u_r(r,\theta) = U(r)P_2(\cos\theta)$, where P_2 is the Legendre polynomial of degree 2; then the resulting diffusivity is

$$D_\mathrm{eff} = \frac{1}{30}\frac{(rU)^2}{D_\mathrm{h}}, \tag{2.4}$$

when $D_\mathrm{h} \geqslant rU$. Unfortunately, a reliable prescription for that horizontal diffusivity D_h is still lacking, in spite of recent attempts to improve it (see Maeder 2003; Mathis *et al.* 2004).

Another property of such anisotropic turbulence is that, by smoothing out chemical inhomogeneities on level surfaces, it reduces the stabilizing effect of the vertical μ-gradient. The Richardson criterion for the vertical shear instability then involves the horizontal diffusivity D_h, and the vertical component of the turbulent viscosity can be derived as before (Talon & Zahn 1997):

$$D_{\rm v} = Ri_{\rm c} \left[\frac{N_T^2}{K + D_{\rm h}} + \frac{N_\mu^2}{D_{\rm h}} \right]^{-1} \sin^2\theta \left(\frac{{\rm d}\Omega}{{\rm d}\ln r} \right)^2 . \tag{2.5}$$

2.3. *Rotational mixing of type I*

The two transport processes that have just been discussed (meridional circulation and shear-induced turbulence) are both linked with the differential rotation. Therefore, when modeling the evolution of a star including these mixing processes, it is necessary to calculate also the evolution of its rotation profile $\Omega(r)$ (since Ω is a function of r only, due to the anisotropic turbulence mentioned above). Then all perturbations separate in r and colatitude θ, as illustrated already above for the vertical component of the meridional velocity: $u_r(r,\theta) = U(r)P_2(\cos\theta)$. For a detailed account of how this mixing may be implemented in stellar evolution codes, we refer to Zahn (1992), Maeder & Zahn (1998) and Mathis & Zahn (2004).

We first examine the simplest case, that we call 'rotational mixing of type I', where the angular momentum is transported by the same processes that are responsible for the mixing, namely meridional circulation and turbulent diffusion. The angular velocity then obeys the following advection/diffusion equation, obtained by averaging over latitude:

$$\frac{\partial}{\partial t}\left[\rho r^2 \Omega\right] = \frac{1}{5r^2}\frac{\partial}{\partial r}\left[\rho U r^2 \Omega\right] + \frac{1}{r^2}\frac{\partial}{\partial r}\left[\rho \nu_{\rm v} r^4 \frac{\partial \Omega}{\partial r}\right] + \text{applied torques}, \tag{2.6}$$

with the turbulent viscosity $\nu_{\rm v} \approx D_{\rm v}$ given by (2.5). In spite of the fact that this equation is one-dimensional only, it captures the advective character of the angular momentum transport by the meridional circulation: depending on the sense of the circulation, angular momentum may be transported up the gradient of Ω, which is never the case when the effect of meridional circulation is modeled just as a diffusive process, as it is done most often.

At first sight, and neglecting evolutionary effects, the circulation is governed mainly by the applied torques. When the star loses little angular momentum, or none, it settles into a regime of differential rotation where a weak inward advection of angular momentum compensates the turbulent diffusion directed outwards. Integrating (2.6) over r, we have then

$$U(r) \approx -5\nu_{\rm v} \frac{d\ln\Omega}{dr}, \tag{2.7}$$

which shows that the strength of the meridional circulation scales with the vertical component of the turbulent diffusivity. In the limit $\nu_{\rm v} \to 0$, the circulation would vanish altogether, as was first pointed out by Busse (1982).

On the other hand, when the star loses angular momentum, the circulation adjusts itself such as to transport precisely that amount towards the surface (Zahn 1992). The balance reads then

$$\frac{3}{8\pi}\frac{\partial J(r,t)}{\partial t} = \frac{1}{5}\rho r^4 \Omega U + \rho \nu_{\rm v} r^4 \frac{\partial \Omega}{\partial r}, \tag{2.8}$$

where J is the flux of angular momentum. Near the surface, angular momentum is transported mainly by the meridional circulation, since $\partial\Omega/\partial r \to 0$, but this holds also in the bulk of the radiation zone.

However, as the star evolves, the inner regions contract and the envelope expands; this results in readjustments of the angular velocity, associated with transport of angular momentum, which are accounted for by the l.h.s. of (2.6).

Massive main-sequence stars belong to the first category examined above, that with relatively modest angular momentum loss, and their models have been seriously improved by the implementation of rotational mixing (Maeder & Meynet 2000). The theoretical isochrones agree with the observed ones, and such rotational mixing accounts well for the observed enhancement of He and N at the surface of early-type stars (Talon *et al.* 1997; Meynet & Maeder 2000 and subsequent papers). Finally, combined with a suitable description of the mass loss, this type of mixing also predicts the observed proportion of blue and red giants in open clusters.

Note that this good agreement between observations and models of massive stars is obtained without having to invoke other processes than meridional circulation and shear turbulence to transport angular momentum and thus shape the rotation profile. This is not the case for solar-type stars, where gravity waves (or rather gravito-inertial waves) probably participate in what may be called 'rotational mixing of type II'. As it has been shown by Talon & Charbonnel (2003 - see also Charbonnel & Talon 2005), these waves, which are emitted at the base of the convection zone of such stars, are able to extract angular momentum from their radiative interior, and to enforce there the nearly uniform rotation detected through helioseismology. To what extend this wave transport also occurs in massive stars remains for the moment an open question, for lack of observational constraints on their rotation profile; the situation should improve soon thanks to asteroseismology.

3. Magnetism of massive stars: most likely of fossil origin

When we were describing the effect of rotation on the structure and evolution of massive stars, we made implicitly the assumption that magnetism plays a negligible role in them. This approach is no longer valid for the active stars to which this symposium is dedicated, because these are the seat of strong magnetic fields.

What we know about stellar magnetism will be addressed in several contributions during this symposium; we shall just recall here the salient facts. Magnetic fields are observed at the surface of all solar-like low-mass stars, namely stars that possess a thick outer convection zone. These fields are variable on the scale of months or years, and they are highly structured; most probably they are generated through dynamo action driven by the convective motions.

In contrast, among the more massive stars, that do not possess such convection zones, only a small fraction, less than 5% according to Power *et al.* (2008), are hosting magnetic fields: these are the so-called Ap and Bp stars. These fields display much simpler topologies, and they seem unchanged over long timescales (at least when compared to human life span); that is why they are believed to be of fossil origin (Wade *et al.* 2009), resulting from the contraction, during star formation, of the primeval, Galactic field. Moreover, all magnetic stars manifest anomalies of their surface composition, and most of them are the seat of slow uniform rotation.

Since they are rooted in the radiative interior, fossil fields evolve on a considerably longer timescale than dynamo fields; the Ohmic decay time for a dipolar field is given by $R^2/2\pi\eta$, R being the radius and η the magnetic diffusivity. This time amounts to 10 Gyr for the Sun (Cowling 1957), and it scales with mass M and radius roughly as $M^{3/2}/R^{1/2}$. Therefore these fields are extremely long lived, even if they are of high spherical degree (the decay rate varies as $\ell(\ell+1)$, ℓ being the order of the multipole). Unless the fields are destroyed by some instability, a possibility that we shall examine next.

3.1. *MHD instabilities in stellar radiation zones*

Magnetic fields are liable to various instabilities, which could perhaps lead to mixing and may therefore have an impact on stellar evolution. These instabilities have been discussed by Spruit (1999) in a comprehensive review; he concluded that the most likely to play a role in stellar radiation zones are those studied by Tayler and his collaborators in the 1970's. These affect axisymmetric poloidal and toroidal fields and they are linked to the classical pinch and kink instabilities.

Describing the fully non-linear regime of the instability represents a difficult task. Spruit (2002) suggested that the instability should saturate when the turbulent magnetic diffusivity η_t, that supposedly is accompanying it, yields a zero growth rate. We had an opportunity to verify his prediction by examining the results of actual non-linear simulations performed with the 3-dimensional ASH code (for Anelastic Spherical Harmonics). The purpose of this calculation was to study the interaction of a fossil magnetic field with the solar tachocline (Brun & Zahn 2006), in order to check whether such a field could prevent this transition layer from expanding into the deep interior, as it had been suggested by Gough and McIntyre (1998).

We started the simulations with a deeply buried poloidal field of dipole type and a uniformly rotating radiation zone, on the top of which we imposed the differential rotation of the convection zone. As time proceeds, the poloidal field diffuses upward and the differential rotation expands downward; their interaction produces a toroidal field which is antisymmetric with respect to the equatorial plane. Once the poloidal field reaches the convection zone, it imprints the differential rotation of that region on the radiative interior below, according to Ferraro's law (Ferraro 1937). Since this is not observed, we concluded that such a fossil poloidal field, as it was postulated by Gough and McIntyre, does not exist in the Sun.

The benefit of carrying out these simulations with a 3-dimensional code was to capture the non-axisymmetric instabilities that affect the large scale magnetic field. We observed indeed the instabilities that had been described by Tayler and his collaborators. The first instability to appear was that of the initial poloidal field, with an azimuthal number $m \approx 40$. It was followed by that of the toroidal field, once that field had been generated through shearing the poloidal field by the differential rotation. The second instability, of azimuthal order $m = 1$, is clearly that studied by Pitts and Tayler (1985), although here it occurred in the presence of both the toroidal field and of the poloidal field, a configuration which is deemed to be more stable than that of just a toroidal field.

This instability saturates when its energy reaches that of the fossil field, but it is not clear whether this is just a coincidence. An unexpected result was to find that the Ohmic decline of the poloidal field is not accelerated by the instability. The 3-dimensional perturbations associated with this instability behave as Alfvén waves, rather than as turbulence, and they do not produce the turbulent diffusion invoked by Spruit. We conclude therefore that they cannot achieve either any mixing of the stellar material.

But other instabilities may arise in magnetic stars, and this was illustrated recently by a series of numerical simulations performed by Braithwaite (Braithwaite & Spruit 2004; Braithwaite & Nordlund 2006; Braithwaite 2009). He showed that an arbitrary, small-scale initial field relaxes into a twisted torus configuration which combines a poloidal and a toroidal field. The quest for such stable configurations is pursued actively, both through numerical simulation and through theory (cf. Duez & Mathis 2010); we shall hear more of it during this symposium.

3.2. *Other causes for the magnetic field*

The properties of the magnetic field in Ap-Bp stars strongly suggest that the field is of fossil origin: their simple topology, their constancy in time, the fact that the surface rotates uniformly. But other possible causes are worth considering.

3.2.1. *A dynamo in the radiation zone?*

When Spruit examined the MHD instabilities in stellar radiation zones, he made an interesting conjecture, namely that such instability could regenerate the toroidal field which initially triggers it, and would thus operate a dynamo, much as that driven by the convective motions in a convection zone. However the dynamo loop cannot work as he describes it (cf. Spruit 2002): it requires some type of α-effect to generate a mean electromotive force, as is well-known in dynamo theory (Parker 1955). As in the case of the Sun, the only way to check whether this α-effect is actually present is through numerical simulations.

But so far these have yielded conflicting results: Braithwaite (2006) claims to observe dynamo action, with field reversals, contrary to us, who don't see any regeneration of either the poloidal or the toroidal field (Zahn *et al.* 2007). The main difference between our simulations lies in the way the equations are solved. Our code is of pseudo-spectral type, which allows us to reach, with enhanced diffusivities, a magnetic Reynolds number of 10^5 for a resolution of 128x256x192. That should be more than sufficient to detect a dynamo, if it exists at all. Braithwaite uses instead a 6th order finite difference scheme, with a resolution of 64x64x33; his numerical diffusivity is tuned to ensure stability and it is not easy to infer from it the effective Reynolds number. Moreover, it appears that his calculations have not been run long enough beyond the transient phase to establish without any doubt that a dynamo is at work. Clearly further simulations are needed to settle this issue.

3.2.2. *A dynamo operated through thermal convection?*

Recent 3-D simulations have taught us that convective regions very easily generate magnetic fields. A good example is the simulation of core convection in a rotating A-type star of two solar masses which has been performed by Brun *et al.* (2005). They introduce a seed field in a core with fully developed convection, and observe its growth to an amplitude where its energy is comparable to that of the flows. The result does not seem to depend much on the parameters used. So there is little doubt that massive stars can produce a dynamo field in their convective core.

The question which remains is whether such a field would be visible at the surface. If the field has reversals as observed in the Sun, say of period P, it would penetrate only to a distance $d = (\eta P)^{1/2}$, which amounts to 1 km for $P = 10$ years. Only the DC component would reach the surface, and this after a global diffusion time exceeding a few Gyr. Admittedly, flux tubes could rise from the core through magnetic buoyancy, but this was not observed in the simulations by Brun *et al.*.

Another possibility is that a local dynamo could operate in the subsurface convection zones due to the ionization of He and Fe, as it was suggested by Cantiello *et al.* (2009). We will certainly hear more about that during this symposium.

Thus one cannot rule out that the magnetism of Ap-Bp stars is due to a dynamo generated by thermal convection, although it would then be difficult to explain why the observed fields are so constant in time - at least since they have been observed, which is at most a few decades.

3.3. *Impact of magnetic fields on stellar structure*

In order to have an impact on the overall structure of a star, the magnetic force (often called Lorentz force although it was Laplace who first gave its expression) ought to be able to compete with the two main forces that govern its hydrostatic equilibrium, namely gravity and the pressure force. This would occur if the Alfvén velocity $V_A = B/\sqrt{4\pi\rho}$ equaled the sound speed; in the Sun that would require the field to be of order $2\,10^8$ G, which is considerably stronger than any field observed on the surface of main sequence stars.

However, it is quite possible that such extremely strong fields are hiding in the deep interior of some stars. Presumably the only way to detect them would be through asteroseismology. The frequencies of the oscillation eigenmodes - either p or g-modes, depending on the restoring force (pressure or buoyancy) - are split and displaced by a magnetic field, much like what occurs under the effect of rotation, and this may be used to estimate the magnetic field. In massive stars, where only few eigenmodes are observed, moreover of low spherical degree, the task is rather difficult, except perhaps for the slowly pulsating B stars, where high order g-modes could bear the signature of a 100 kG field located near the boundary of the convective core (Hasan *et al.* 2005). The effect of a dipole field on the p-modes in a rotating star was considered in full detail by Bigot *et al.* (2000), with application to the rapidly rotating Ap stars.

3.3.1. *Strong magnetic fields can suppress thermal convection*

It is well known that thermal convection can generate magnetic fields, but magnetic fields can also suppress the convective instability. This can be seen by examining the stability of a stratified fluid in presence of magnetic field. A perturbation described by its displacement $\vec{\xi} \propto \exp\left[st+i\vec{k}\cdot\vec{x}\right]$ has a growth-rate s that obeys the following dispersion relation, where we neglect all kind of dissipation (thermal, Ohmic, viscous):

$$s^2 = \left(\frac{k_h}{k}\right)^2 N_t^2 - (\vec{k}\cdot\vec{V}_A)^2, \qquad N_t^2 = \left[\frac{g}{H_P}(\nabla - \nabla_{\rm ad})\right], \tag{3.1}$$

N_t being the buoyancy frequency. Here g is the local gravity, H_P the pressure scale height, $\vec{k}$ the wavenumber, k_h its horizontal component, and $\nabla = \partial \ln T/\partial \ln P$ designates as usual the logarithmic temperature gradient.

At first sight it would seem that, to suppress convection, it requires a field strength of

$$\frac{B^2}{4\pi\rho} = V_A^2 \gtrsim N_t^2 R^2 \tag{3.2}$$

according to (3.1), which would translate into around 10^7 G (10^3 Tesla) in the radiation zone of a massive star, or 10^3 G near its surface. However that threshold is lowered by two orders of magnitude when diffusion is taken in account; the stability condition then becomes (Chandrasekhar 1961)

$$\frac{B^2}{4\pi\rho} = V_A^2 \gtrsim \frac{\eta}{K} N_t^2 R^2, \tag{3.3}$$

with η and K being the Ohmic and thermal diffusivities (typically $\eta/K \approx 10^{-4}$ in stellar interiors). The instability is of double-diffusive type, with heat diffusing much faster than the magnetic field.

In the sunspots we see a striking proof of that inhibition of thermal convection by a field of kilogauss strength. That is also why we observe surface inhomogeneities in the Ap-Bp stars: due to the presence of magnetic field, there is no convection to smooth out such inhomogeneities, as in normal stars where convection exists in the thin superadiabatic layers due to the ionization of hydrogen and helium.

3.3.2. *Magnetic fields amplify the loss of angular momentum by stellar winds*

A well established fact is that solar-like stars spin down as they evolve. At the ZAMS, some of such stars have equatorial velocities that reach 100 km/s, while that of the Sun is only 2 km/s. The only way stars can achieve this is by losing mass, through a wind whose mechanism has been elucidated by Parker (1958). In the absence of magnetic field this is not a very efficient process, but the picture changes when one includes the magnetic field, as was pointed out by Schatzman (1962). The wind is then forced to rotate with the star up to a distance where the wind speed exceeds the Alfvén velocity: the lever arm is then no longer the radius R, but the so-called Alfvén radius, which in the present Sun is about $R_A \approx 15\,R$.

This mechanism works also for massive stars, where the magnetic field interacts likewise with their radiation-driven wind; it has been studied in detail in a series of papers by ud-Doula and Owocki (ud-Doula & Owocki 2002; Owocki & ud-Doula 2004; ud-Doula *et al.* 2006, 2008). They introduce a 'wind magnetic confinement parameter' $\eta* = B_{\rm eq}^2 R^2/\dot{M} v_\infty$, which measures the ratio between the magnetic field energy density (at the equator) and the kinetic energy density of the wind (v_∞ is its terminal speed and $\dot{M}$ the mass loss rate). They show that for $\eta^* \gg 1$ the latitude dependent Alfvén radius $R_A(\theta)$ scales with the confinement parameter according to a rather simple law:

$$\left[\frac{R_A(\theta)}{R}\right]^{2q-2} - \left[\frac{R_A(\theta)}{R}\right]^{2q-3} = \eta^*[4 - 3\sin^2(\theta)], \tag{3.4}$$

where θ is the colatitude, and q the magnetic exponent. For a dipole field $q = 3$ and therefore $R_A \propto {\eta^*}^{1/4}$.

3.3.3. *Magnetic fields inhibit the rotational mixing*

We recalled above that in the absence of magnetic field, the radiation zone of a rotating star undergoes mild mixing through internal motions - a combination of turbulence and large-scale flows - that are due to several causes. These are mainly the loss of angular momentum by a wind, eventually the coupling with an accretion disk, and the structural adjustments the star undergoes as it evolves (such as the moderate core contraction and envelope expansion on the main sequence).

Let us examine now what occurs in the presence of magnetic field. One effect of such a field is to enhance the angular momentum loss, and at first sight this should speed up the rotational mixing. But the field acts also in the deep interior: when it is axisymmetric with respect to the rotation axis, it tends to render the rotation uniform along the field lines of the meridional field (Ferraro 1937). To estimate the strength of the axisymmetric field above which that rotational mixing is inhibited, we refer to the horizontally averaged angular momentum transport equation; we neglect the turbulent transport, since there is no shear anymore, but include the torque exerted by the Lorentz force $\vec{j} \times \vec{B}/c$:

$$\rho \frac{d}{dt}\left(r^2\overline{\Omega}\right) = \frac{1}{5r^2}\partial_r\left(\rho r^4 \overline{\Omega} U\right) + \overline{\vec{B}_p \cdot \vec{\nabla}(rB_\varphi/4\pi)}, \tag{3.5}$$

with $\vec{B}_p$ and B_φ being the meridional and azimuthal components of the magnetic field.

Introducing the characteristic time $t_{\rm AM}$ for the angular momentum evolution (due to a wind, or to structural adjustments), we see that the Lorentz torque balances the l.h.s.

when

$$B_p B_\varphi \gtrsim B_{\rm crit}^2 = 4\pi\rho\,\frac{R^2\Omega}{t_{\rm AM}}. \tag{3.6}$$

If we apply this result to a 15 $M_\odot$ star, taking $\rho = 0.07$ g/cm^3, $R = 4.5\,10^{11}$ cm, $R\,\Omega = 3\,10^7$ cm/s and setting somewhat arbitrarily $t_{\rm AM} = 10^6$ yr, we find that the critical field strength for suppressing rotational mixing is of the order of $B_{\rm crit} \approx 600$ Gauss, a level which is probably exceeded in the interior of Ap-Bp stars. This crude estimate needs to be confirmed by more thorough calculations, for which the formalism has been derived by Mathis and Zahn (2005).

When the field is non-axisymmetric, as is often observed, uniform rotation tends to be enforced everywhere in the radiation zone. Little more is known about this oblique rotator model, as can be seen in Mestel (1999), and this is clearly a promising field for future research.

4. Conclusion and perspectives

Despite all the work that has been accomplished since the 1970's, notably by Leon Mestel, Roger Tayler and their collaborators, we still do not fully understand the origin of magnetism in massive stars and the role it plays in their evolution. But let me summarize the few results that are now firmly established.

Only a minority of massive stars host magnetic fields - less than 5% according to the latest survey. In those stars, the magnetic field is probably of fossil origin, and it is deeply rooted in the radiative interior. We don't know why magnetic stars are so few: presumably all stars had initially a fossil field, which they gathered from the interstellar medium during the process of formation, but it must have been later destroyed in most of them, presumably through MHD instabilities.

It has been known since the 1960's that magnetic fields enhance the loss of angular momentum through stellar winds (Schatzman 1962), which explains why most magnetic stars are slow rotators. One would thus be tempted to conclude that rotational mixing is enhanced by this angular momentum loss; but it turns out on the contrary that the internal motions - either large scale circulations or turbulence - are frozen by the interior field. The field itself can become unstable, but the instability saturates in a regime that is wave-like and does not lead to turbulent diffusion. Thus one suspects that magnetic stars will not show signs of internal mixing - we are eagerly waiting observational data that will tell us whether this prediction is correct.

The study of stellar magnetism is still in its infancy, but the situation is improving rapidly. The new generation of extremely powerful spectro-polarimeters is delivering much wanted observational constraints of high quality, while the steadily progressing computer resources allow numerical simulations with a increasing degree of realism. There will be much to discover and to explain, for the young astrophysicists entering this exciting field!

References

Bigot, L., Provost, J., Berthomieu, G., Dziembowski, W. A. *et al.* 2000, *A&A*, 356, 218
Braithwaite, J. 2006, *A&A*, 449, 451
Braithwaite, J. 2009, *MNRAS*, 397, 763
Braithwaite, J. & Nordlund, Å. 2006, *A&A*, 450, 1077
Braithwaite, J. & Spruit, H. C. 2004, *Nature*, 431, 819
Brun, A. S., Browning, M. K., & Toomre, J. 2005, *ApJ*, 629, 461

Brun, A. S. & Zahn, J.-P. 2006, *A&A*, 457, 665

Busse, F. H. 1982, *ApJ*, 259, 759

Cantiello, M., Langer, N., Brott, I., de Koter, A. *et al.* 2009, *A&A*, 499, 279

Chaboyer, B. & Zahn, J.-P. 1992, *A&A*, 253, 173

Chandrasekhar, S. 1961, *Hydrodynamic and hydromagnetic stability*, International Series of Monographs on Physics (Oxford, Clarendon)

Charbonnel, C. & Talon, S. 2005, *Science*, 309, 2189

Cowling, T. G. 1957, *Magnetohydrodynamics* (Interscience Publishers, Inc., New York)

Duez, V. & Mathis, S. 2010, *A&A*, 517A, 58

Dudis, J. J. 1974, *Journal of Fluid Mechanics*, 64, 65

Eddington, A. S. 1925, *Observatory*, 48, 73

Ferraro, V. C. A. 1937, *MNRAS*, 97, 458

Gough, D. O. & McIntyre, M. E. 1998, *Nature*, 394, 755

Hasan, S. S., Zahn, J.-P., & Christensen-Dalsgaard, J. 2005, *A&A*, 444, L29

Herrero, A., Kudritzki, R. P., Vilchez, J. M., Kunze, D. *et al.* 1992, *A&A*, 261, 209

Hunter, I., Brott, I., Lennon, D. J., Langer, N. *et al.* 2008, *ApJ* (Letters), 676, 29

Maeder, A. 2003, *A&A*, 399, 263

Maeder, A., Meynet, G., Ekström, S., & Georgy, C. 2009, *Communications in Asteroseismology*, 158, 72

Maeder, A. & Meynet, G. 2000, *ARAA*, 38, 143

Maeder, A. & Zahn, J.-P. 1998, *A&A*, 334, 1000

Mathis, S., Palacios, A., & Zahn, J.-P. 2004, *A&A*, 425, 243

Mathis, S. & Zahn, J.-P. 2004, *A&A*, 425, 229

Mathis, S. & Zahn, J.-P. 2005, *A&A*, 440, 653

Mestel, L. 1999, *Stellar Magnetism*, International series of monographs on physics (Oxford, Clarendon)

Meynet, G. & Maeder, A. 1997, *A&A*, 321, 465

Meynet, G. & Maeder, A. 2000, *A&A*, 361, 101

Michaud, G. 1970, *ApJ*, 160, 641

Richer, J., Michaud, G., & Turcotte, S. 2000, *ApJ*, 529, 338

Owocki, S. P. & ud-Doula, A. 2004, *ApJ*, 600, 1004

Parker, E. N. 1955, *ApJ*, 122, 293

Parker, E. N. 1958, *ApJ*, 128, 664

Pinsonneault, M. H., Kawaler, S. D., Sofia, S., & Demarque, P. 1989, *ApJ*, 338, 424

Pitts, E. & Tayler, R. J. 1985, *MNRAS*, 216, 139

Power, J., Wade, G. A., Aurière, M., Silvester, J. *et al.* 2008, *Contributions of the Astronomical Observatory Skalnate Pleso*, 38, 443

Schatzman, E. 1962, *Annales d'Astrophysique*, 25, 18

Spruit, H. C. 1999, *A&A*, 349, 189

Spruit, H. C. 2002, *A&A*, 381, 923

Sweet, P. A. 1950, *MNRAS*, 110, 548

Talon, S. & Charbonnel, C. 2003, *A&A*, 405, 1025

Talon, S. & Zahn, J.-P. 1997, *A&A*, 317, 749

Talon, S., Zahn, J.-P., Maeder, A., & Meynet, G. 1997, *A&A*, 322, 209

ud-Doula, A. & Owocki, S. P. 2002, *ApJ*, 576, 413

ud-Doula, A., Owocki, S. P., & Townsend, R. H. D. 2008, *MNRAS*, 385, 97

ud-Doula, A., Townsend, R. H. D., & Owocki, S. P. 2006, *ApJ* (Letters), 640, L191

Vauclair, G., Vauclair, S., & Michaud, G. 1978, *ApJ*, 223, 920

Vauclair, G., Vauclair, S., & Pamjatnikh, A. 1974, *A&A*, 31, 63

Vogt, H. 1925, *AN*, 223, 229

Wade, G. A., Silvester, J., Bale, K., Johnson, N. *et al.* 2009, in: S. V. Berdyugina, K. N. Nagendra, & R. Ramelli (eds.), *Stellar pulsation: challenges for theory and observation*, ASP-CS 405, p. 499

Zahn, J.-P. 1974, in: P. Ledoux, A. Noels, & A. W. Rodgers (eds.), *Stellar Instability and Evolution*, IAU Symposium 59, p. 185
Zahn, J.-P. 1992, *A&A* 265, 115
Zahn, J.-P., Brun, A. S., & Mathis, S. 2007, *A&A*, 474, 145

Discussion

OWOCKI: In your estimate of the field needed to suppress differential rotation and mixing, you assumed a fixed angular momentum loss time. But that time should itself depend on field strength (and mass loss rate). If one includes this, it would give a different scaling and estimate for the required field

ZAHN: You are right: one should indeed include that dependence in deriving the scaling.

BOMANS: You stressed the importance of fossil magnetic fields. Observations imply low strength ordered magnetic fields in lower mass galaxies, which means low metallicity galaxies. Would you therefore expect to see low magnetic field OB stars in these galaxies?

ZAHN: From the crude scaling of the magnetic field as density to the power 2/3, one should expect a lower field in low metallicity galaxies. But will that scaling hold over so many orders of magnitude?

PULS: Did I understand correctly that your recent calculations show that the Spruit-Taylor dynamo does not work?.

ZAHN: Yes, as we explain in Zahn *et al.* (2007), we observe the Taylor instability, but there is no dynamo action, i.e. the mean axisymmetric field is not regenerated.

PULS: However, present evolutionary codes include this mechanism. What would happen if the fields were more correctly described?

ZAHN: This is precisely what we want to find out.

Active OB stars: structure, evolution, mass loss, and critical limits
Proceedings IAU Symposium No. 272, 2010
C. Neiner, G. Wade, G. Meynet & G. Peters, eds.

doi:10.1017/S1743921311009938

Mixing of CNO-cycled matter in pulsationally and magnetically active massive stars

Norbert Przybilla[1] and Maria-Fernanda Nieva[2]

[1]Dr. Remeis-Observatory & ECAP, Sternwartstr. 7, D-96049 Bamberg, Germany
email: przybilla@sternwarte.uni-erlangen.de

[2]Max-Planck-Institut für Astrophysik, Postfach 1317, D-85741 Garching, Germany
email: fnieva@mpa-garching.mpg.de

Abstract. We report on the abundances of helium, carbon, nitrogen and oxygen in a larger sample of Galactic massive stars of $\sim$7-20 $M_\odot$ near the main sequence, composed of apparently normal objects, pulsators of β-Cephei- and SPB-type, and magnetic stars. High-quality spectra are homogeneously analysed using sophisticated non-LTE line-formation and comprehensive analysis strategies. All the stars follow a previously established tight trend in the N/C-N/O ratio and show normal helium abundances, tracing the nuclear path of the CNO-cycles quantitatively. A correlation of the strength of the mixing signature with the presence of magnetic fields is found. In conjunction with low rotation velocities this implies that magnetic breaking is highly efficient for the spin-down of some massive stars. We suggest several objects for follow-up spectropolarimetry, as the mixing signature indicates a possible magnetic nature of these stars.

Keywords. stars: abundances, stars: activity, stars: atmospheres, stars: early-type, stars: evolution, stars: magnetic fields

---

1. Introduction

Active and normal OB stars occupy the same region in the Hertzsprung-Russell diagram (HRD). While the observational characteristics of the different forms of activity are well-defined, a comprehensive understanding of the physical drivers of activity has yet to be established. Quantitative analyses of samples of active and normal stars at high precision are without doubt the starting point for progress to be made.

Two types of active OB stars are of interest in the following, *pulsators* of β Cephei-type and slowly-pulsating B stars (SPBs), and *magnetic stars*. Their surface characteristics will be compared to those of apparently normal objects. Crucial improvements over previous work are achieved by the introduction of sophisticated modelling and a novel analysis technique, which allows in particular abundance uncertainties to be reduced to about 10-20% (1-σ statistical), in contrast to typical values of a factor $\sim$2 in literature.

We focus on surface abundances of C, N and O, which may be altered already on the main sequence due to mixing with CNO-cycled matter in the course of the evolution of rotating massive stars (e.g. Maeder & Meynet 2000). The aim is twofold: I) to test whether mixing signatures correlate with the presence of magnetic fields (Morel *et al.* 2008), and II) to test whether the occurrence of pulsations is related to CNO mixing.

2. Observations and Analysis

Our observational sample consists of 29 bright and apparently slowly rotating early B-type stars in nearby OB associations and in the field. High-resolution ($R = 40$-$48\,000$)

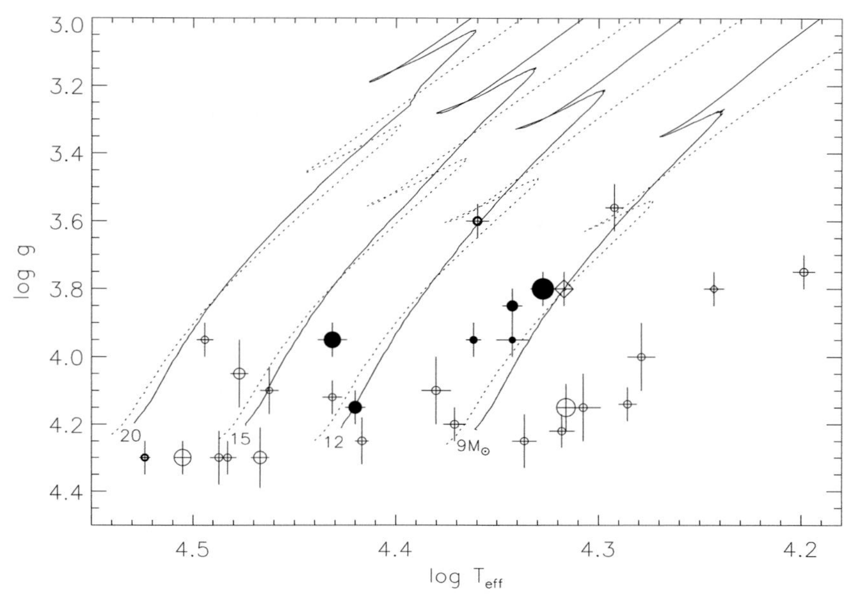

Figure 1. The sample stars in the $\log T_{\rm eff} - \log g$-plane. Normal stars are marked by circles, β Cep-type pulsators by dots (candidates by thick circles) and SPBs by diamonds. The symbol size encodes the N/C mass ratio, varying between ~0.3 (cosmic value, smallest symbols) and ~1.2. Evolution tracks for rotating (full lines, $v_{\rm ini} = 300\,{\rm km\,s^{-1}}$) and non-rotating stars (dotted lines) are overlaid (Meynet & Maeder 2003). The tracks were computed with $Z = 0.02$, while the sample stars have $Z = 0.014$, implying a downward shift of the theoretical ZAMS.

and high-S/N spectra ($S/N = 250$-800) with wide wavelength coverage were obtained with the Echelle spectrographs FOCES, FEROS and FIES, and from the ELODIE archive. Details of the observations and the data reduction are discussed by Nieva & Przybilla (in prep.) and Nieva, Simón-Díaz & Przybilla (in prep.).

The spectra were analysed using a hybrid non-LTE approach (Nieva & Przybilla 2007, 2008; Przybilla *et al.* 2008b) and a sophisticated analysis methodology. State-of-the-art atomic input data were used in the modelling. In contrast to previous work, multiple hydrogen lines, the helium lines, multiple metal ionization equilibria and the stellar energy distributions were reproduced *simultaneously* in an iterative approach to determine the stellar atmospheric parameters. Chemical abundances for a wide range of elements were derived from analysis of practically the entire observable spectrum per element. The rewards of such a comprehensive, but time-consuming procedure were unprecedentedly small statistical error margins and largely reduced systematics (Nieva & Przybilla 2010).

The positions of the sample stars in the $\log T_{\rm eff} - \log g$-plane are displayed in Fig. 1. All stars are relatively unevolved objects on the main sequence in the mass range of ~7-20 $M_\odot$. While the bulk of the objects are normal stars, a small number are known pulsators, either of β Cephei-type or SPBs. Three stars are magnetic, β Cep (Henrichs *et al.* 2000), ζ Cas (Neiner *et al.* 2003) and τ Sco (Donati *et al.* 2006), one is a candidate magnetic star, δ Cet (Schnerr *et al.* 2008; Hubrig *et al.* 2009). The majority of the stars have CNO abundances compatible with cosmic values (Przybilla *et al.* 2008b; Nieva & Przybilla, in prep.), about a third of the objects shows signs of mixing with CNO-cycled matter. We will concentrate on this aspect and its relation to the presence of magnetic fields in the following. Fundamental parameters of the sample stars and also

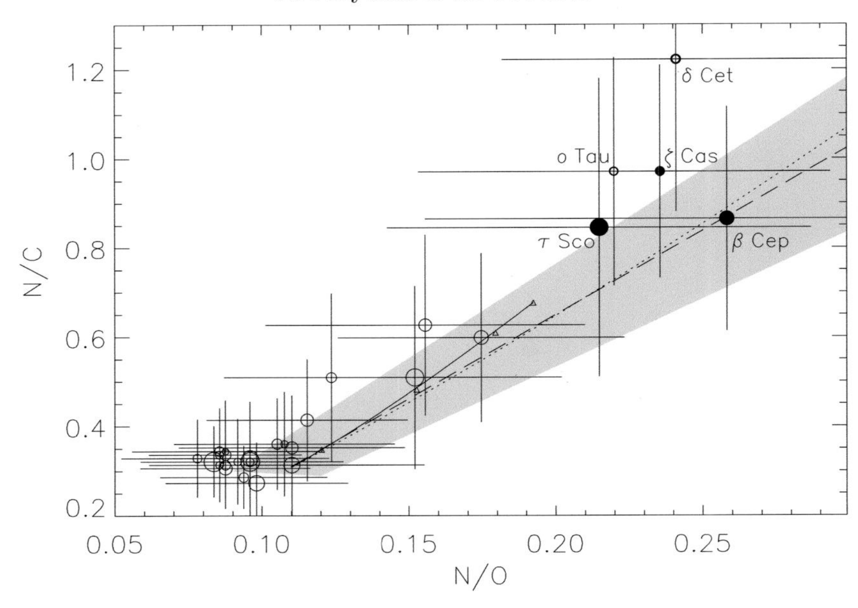

Figure 2. N/C vs. N/O abundance ratios (by mass) for our sample stars. The symbol size encodes the stellar mass, magnetic stars are marked by filled symbols (candidates by thick circles). Predictions for 9 $M_\odot$ models are indicated by the full line, the triangles marking ratios reached at the end of core-H burning for $v_{\rm ini}$ equivalent to 30, 50, 70 and 90% of the star's breakup velocity (bottom to top, Ekström *et al.* 2008). The dotted line describes a magnetic 15 $M_\odot$ model of Maeder & Meynet (2005) with $v_{\rm ini} = 300\,{\rm km\,s^{-1}}$. The long-dashed line corresponds to a slope 3.77 calculated analytically for the nuclear path of the CN-cycle and initial CNO abundances as in the evolution models, while the grey area spans the full range of theoretical slopes using different references for solar abundances and the cosmic abundance standard (upper envelope).

further details on the modelling and analysis are discussed by Nieva & Przybilla (these proceedings).

3. Signatures of chemical mixing in early B-type stars

Observational indications of superficial abundance anomalies for carbon, nitrogen, and oxygen (and the burning product helium) in OB-type stars on the main sequence are known for a long time. They have found a theoretical explanation by evolution models for rotating stars (e.g. Maeder & Meynet 2000; Heger & Langer 2000), where meridional circulation and turbulent diffusion provide the means to transport CNO-cycled matter from the stellar core to the surface. The interaction of rotation with a magnetic dynamo may increase the transport efficiency even further (Maeder & Meynet 2005).

The changes of CNO abundances relative to each other follow a simple pattern. At the beginning of hydrogen burning, the ^{14}N enhancement comes from the ^{12}C destruction via the CN cycle, and the oxygen content remains about constant. A fraction of this processed material is mixed with matter of pristine composition in the course of the subsequent evolution. A simple calculation shows that the mixing signature follows the nuclear path with slope $d(N/C)/d(N/O) \approx 4$ when abundances are expressed in mass fraction, see Przybilla *et al.* (2010) for details. Stellar evolution models follow this trend and make quantitative predictions concerning the enrichment with nuclear-processed matter. The mixing signature gets more pronounced with increasing rotational velocity and increasing mass (not shown here), and magnetic models predict the strongest mixing, see Fig. 2. The

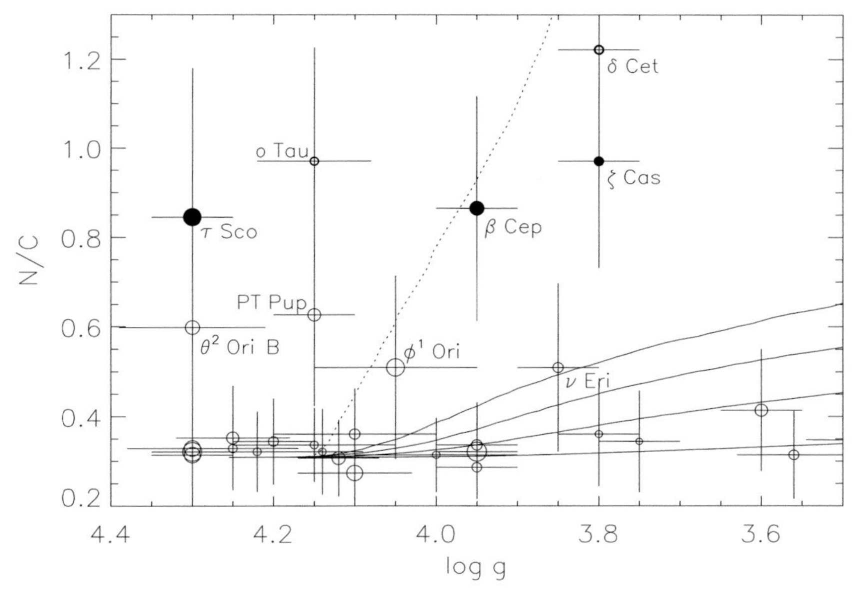

Figure 3. N/C abundance ratios on the main sequence as function of $\log g$. The same symbols as in Fig. 2 are used, the tracks for $v_{\rm ini}$ of 30, 50, 70 and 90% breakup velocity (bottom to top) are separated in this diagram. Several additional stars with their mixing signature indicating a possible magnetic nature are also identified. The difference in metallicity between the models ($Z = 0.02$) and the sample stars ($Z = 0.014$) implies a shift of the tracks to the left in this case.

high-quality analysis of our sample stars facilitates the predicted trend to be recovered for the first time (overplotting data from previous studies would fill all areas in the N/O-N/C plane, see Fig. 1 of Przybilla *et al.* 2010, with the error bars exceeding the range of Fig. 2 in many cases). About half of the sample clusters around [N/O $\approx$ 0.09, N/C $\approx$ 0.32], which corresponds to unaltered cosmic values, several more stars show signatures apparently compatible with the rotating models, and five stars show a very high degree of mixing. Of these all but one are known (candidate) magnetic stars, suggesting a magnetic nature for the remaining one, o Tau, as well. Note that the agreement between observations and theory would be even better if the models would account for cosmic ($Z = 0.014$, Przybilla *et al.* 2008b) instead of solar abundances ($Z = 0.02$, Grevesse & Noels 1993). Helium abundances for all the sample stars are compatible with cosmic/solar values.

However, while Fig. 2 provides a crucial test for the quality of the analyses it does not facilitate to decide which models are more realistic than others, or to identify the mixing mechanism(s). Tighter constraints can be put via the $\log g$-N/C diagram (see Fig. 3), which relates the mixing signature to the evolutionary state of the sample stars. Indeed, the majority of objects follows the predicted trends for rotating stars, showing only mild enrichment (N/C $\lesssim$ 0.4) during their evolution from the zero-age main sequence (ZAMS, at the highest gravities) to the giant stage (lowest gravities). Truely fast rotators (seen pole-on) seem to be absent, which is likely a relic of one of the selection criteria for the sample assembly (stars with low $v \sin i$).

The magnetic stars stand out again with pronounced mixing signatures (N/C > 0.8), complemented by a group of stars at intermediate values of N/C $\approx$ 0.5-0.7. Some of the signatures found in the more evolved magnetic stars like β Cep and ζ Cas may be explained by evolution models accounting for a magnetic dynamo. However, these models do not predict the magnetic field to reach the stellar surface. The magnetic fields are probably of fossil origin instead. It is likely that the enhanced transport efficiency (when compared to models accounting for rotation only) is retained in that case, as the basic physics behind is not sensitive to the origin of the magnetic field. However, this has to be confirmed by detailed modelling.

Other objects, in particular τ Sco, challenge present stellar models seriously. The observed mixing signature of τ Sco may be explained by homogeneous evolution, but this has also to be confirmed by further computations. Homogeneous evolution would require a highly-efficient spin-down mechanism, as the star is a truely slow rotator at present (Donati *et al.* 2006), like several other magnetic stars with pronounced mixing signature (Morel *et al.* 2008). One may speculate on magnetic breaking due to angular-momentum losses by a magnetically confined line-driven stellar wind or magnetic coupling to the accretion disc during the star-forming process in the case of a fossil field. Even though the topic is not understood theoretically in a comprehensive way, spin-down times of the order of 1 Myr (Ud-Doula *et al.* 2009; Townsend *et al.* 2010) or even less (Mikulášek *et al.* 2008) are reported for some magnetic massive stars, possibly leading to the required slow rotation already close to the ZAMS.

After having established the connection of pronounced mixing signatures with the presence of magnetic fields, one may turn the argumentation and use the chemical fingerprint as a selection criterion for the search of magnetic fields in massive stars. Based on their mixing signatures we therefore suggest the following objects for follow-up spectropolarimetric observations: o Tau as a high-potential target and θ^2 Ori B, ϕ^1 Ori, PT Pup and ν Eri (weak constraints for the latter two were presented by Hubrig *et al.* 2009).

Alternative explanations of the mixing signatures may be required in the case of non-detections. Case A mass-transfer in close binary systems – i.e. near the end of the core H-burning phase of the primary – is one of the most promising mechanisms. This can produce stars highly enriched in CNO-cycled products close to the ZAMS. Such stars would be fast rotators because of angular momentum transfer accompanying the mass overflow. The observational identification of such systems may be challenging, though, as the companion of the visible OB star is likely a low-mass helium star in a wide orbit, see Wellstein *et al.* (2001) for details.

Finally, we want to comment briefly on the question whether the occurrence of pulsations is in any way related to CNO mixing. The pulsations of β Cep stars and SPBs are excited via the κ-mechanism because of a subsurface opacity bump due to the ionization of iron-group elements (Pamyatnykh 1999). Yet, an additional factor is required to explain the occurrence of pulsators and non-pulsators in the same regions of the HRD. Additional changes in the opacities due to modified CNO abundances as a consequence of mixing with nuclear-processed matter may be a possibility. However, our results do not support this hypothesis, pulsations occur independently of mixing signatures, see Fig. 1.

4. Summary and Outlook

Theory predicts surface abundances of massive stars to follow a tight relation when mixing with CNO-cycled material occurs. The nuclear path in the N/O-N/C plane is solely determined by the initial CNO abundances. This provides a powerful criterion for judging the quality of quantitative analyses. Few early-type massive stars can be expected to deviate from this nuclear path, either due to atomic diffusion (which is normally not operational in OB-type stars) or because of pollution with material from advanced nuclear burning phases, e.g. accreted in the course of a binary supernova (Przybilla *et al.* 2008a; Irrgang *et al.* 2010). Only observational data passing this test is well-suited for being used in the verification of different stellar evolution models.

The high-quality analysis of our sample stars indicates that most objects can be explained well by existing models for rotating stars. It also facilitates the identification of a connection of pronounced mixing signatures with the presence of magnetic fields – initially suggested by Morel *et al.* (2008) – at high confidence. In reverse, pronounced mixing signatures may be used to define high-potential targets for spectropolarimetric surveys that aim at the detection of magnetic stars. Several sample stars are suggested for follow-up observations.

Further studies, observational as well as theoretical, are required to develop a comprehensive understanding of chemical mixing in massive stars. High-quality analyses have to be extended to larger samples of stars, covering a wide range of rotational velocities. Further contraints to stellar evolution models can be obtained by inclusion of stars at different metallicity, like e.g. objects in the Magellanic Clouds. Once the ongoing surveys as the MiMeS project (Wade *et al.*, these proceedings) will detect more magnetic stars it will become feasible to develop a better overview on the rôle that magnetic dynamos and/or fossil fields play in the course of the evolution of massive stars.

Acknowledgements

We would like to thank A. Maeder, G. Meynet, V. Petit, S. Simón-Díaz and G. Wade for fruitful discussion in preparation of this work. NP acknowledges travel support by the *Deutsche Forschungsgemeinschaft*, project number PR 685/3-1.

References

Donati, J.-F., Howarth, I. D., Jardine, M. M., Petit, P. *et al.* 2006, *MNRAS*, 370, 629

Ekström, S., Meynet, G., Maeder, A., & Barblan, F. 2008, *A&A*, 478, 467

Grevesse, N. & Noels, A. 1993, in: N. Prantzos, E. Vangioni-Flam, & M. Casse (eds.), *Origin and Evolution of the Elements*, p. 15

Heger, A. & Langer, N. 2000, *ApJ*, 544, 1016

Henrichs, H. F., de Jong, J. A., Donati, J.-F., Catala, C. *et al.* 2000, in: M. A. Smith, H. F. Henrichs, & J. Fabregat (eds.), *IAU Colloq. 175: The Be Phenomenon in Early-Type Stars*, ASP-CS 214, p. 324

Hubrig, S., Briquet, M., De Cat, P., Schöller, M. *et al.* 2009, *AN*, 330, 317

Irrgang, A., Przybilla, N., Heber, U., Nieva, M. F. *et al.* 2010, *ApJ*, 711, 138

Maeder, A. & Meynet, G. 2000, *ARAA*, 38, 143

Maeder, A. & Meynet, G. 2005, *A&A*, 440, 1041

Meynet, G. & Maeder, A. 2003, *A&A*, 404, 975

Mikulášek, Z., Krtička, J., Henry, G. W., Zverko, J. *et al.* 2008, *A&A*, 485, 585

Morel, T., Hubrig, S., & Briquet, M. 2008, *A&A*, 481, 453

Neiner, C., Geers, V. C., Henrichs, H. F., Floquet, M. *et al.* 2003, *A&A*, 406, 1019

Nieva, M. F. & Przybilla, N. 2007, *A&A*, 467, 295

Nieva, M. F. & Przybilla, N. 2008, *A&A*, 481, 199

Nieva, M.-F. & Przybilla, N. 2010, in: C. Leitherer, P. Bennett, P. Morris, & J. van Loon (eds.), *Astronomical Society of the Pacific Conference Series*, ASP-CS 425, p. 146

Pamyatnykh, A. A. 1999, *AcA*, 49, 119

Przybilla, N., Nieva, M. F., Heber, U., & Butler, K. 2008a, *ApJ* (Letters), 684, L103

Przybilla, N., Nieva, M.-F., & Butler, K. 2008b, *ApJ* (Letters), 688, L103

Przybilla, N., Firnstein, M., Nieva, M. F., Meynet, G. *et al.* 2010, *A&A*, 517, A38

Schnerr, R. S., Henrichs, H. F., Neiner, C., Verdugo, E. *et al.* 2008, *A&A*, 483, 857

Townsend, R. H. D., Oksala, M. E., Cohen, D. H., Owocki, S. P. *et al.* 2010, *ApJ* (Letters), 714, L318

Ud-Doula, A., Owocki, S. P., & Townsend, R. H. D. 2009, *MNRAS*, 392, 1022

Wellstein, S., Langer, N., & Braun, H. 2001, *A&A*, 369, 939

Active OB stars: structure, evolution, mass loss, and critical limits
Proceedings IAU Symposium No. 272, 2010
C. Neiner, G. Wade, G. Meynet & G. Peters, eds.

doi:10.1017/S174392131100994X

3D MHD simulations of subsurface convection in OB stars

Matteo Cantiello[1], Jonathan Braithwaite[1], Axel Brandenburg[2,3], Fabio Del Sordo[2,3], Petri Käpylä[2,4] and Norbert Langer[1]

[1]Argelander-Institut für Astronomie der Universität Bonn, Auf dem Hügel 71, D–53121 Bonn, Germany email: cantiello@astro.uni-bonn.de [2]NORDITA, AlbaNova University Center, Roslagstullsbacken 23, SE-10691 Stockholm, Sweden [3]Department of Astronomy, AlbaNova University Center, Stockholm University, SE–10691 Stockholm, Sweden [4]Department of Physics, Gustaf Hällströmin katu 2a (PO Box 64), FI-00014, University of Helsinki, Finland

Abstract. During their main sequence evolution, massive stars can develop convective regions very close to their surface. These regions are caused by an opacity peak associated with iron ionization. Cantiello *et al.* (2009) found a possible connection between the presence of sub-photospheric convective motions and small scale stochastic velocities in the photosphere of early-type stars. This supports a physical mechanism where microturbulence is caused by waves that are triggered by subsurface convection zones. They further suggest that clumping in the inner parts of the winds of OB stars could be related to subsurface convection, and that the convective layers may also be responsible for stochastic excitation of non-radial pulsations. Furthermore, magnetic fields produced in the iron convection zone could appear at the surface of such massive stars. Therefore subsurface convection could be responsible for the occurrence of observable phenomena such as line profile variability and discrete absorption components. These phenomena have been observed for decades, but still evade a clear theoretical explanation. Here we present preliminary results from 3D MHD simulations of such subsurface convection.

Keywords. convection, hydrodynamics, waves, stars: activity, stars: atmospheres, stars: evolution, stars: magnetic fields, stars: spots, stars: winds, outflows

1. Introduction

Hot luminous stars show a variety of phenomena in their photospheres and in their winds which still lack a clear physical interpretations at this time. Among these phenomena are photospheric turbulence, line profile variability (LPVs), discrete absorption components (DACs), wind clumping, variable or constant non-thermal X-ray and radio emission, chemical composition anomalies, and intrinsic slow rotation. Cantiello *et al.* (2009) argued that a convection zone close to the surface of hot, massive stars, could be responsible for various of these phenomena. This convective zone is caused by a peak in the opacity due to iron recombination and for this reason is referred as the "iron convection zone" (FeCZ). A physical connection may exist between microturbulence in hot star atmospheres and a subsurface FeCZ. The strength of the FeCZ is predicted to increase with increasing metallicity Z, decreasing effective temperature T and increasing luminosity L, and all three predicted trends are reflected in observational data of microturbulence obtained in the context of the VLT-FLAMES survey of massive stars (Evans *et al.* 2005). Moreover recent measurements of microturbulence (Fraser *et al.* 2010) are in agreement with the results of Cantiello *et al.* (2009). This suggests that microturbulence corresponds to a physical motion of the gas in hot star atmospheres. This motion may then be connected to wind clumping, since the empirical microturbulent velocities are comparable to the local sound speed at the stellar surface.

The FeCZ in hot stars might also produce localized surface magnetic fields (Cantiello *et al.* 2009). Such magnetic fields may become buoyant and reach the surface, creating magnetic spots. This could explain the occurrence of DACs (discrete absorption components in UV absorption lines), also in very hot main sequence stars for which pulsational instabilities are not predicted. Moreover there may be regions of the upper HR diagram for which the presence of the FeCZ influences, or even excites, non-radial stellar pulsations. Interestingly stochastic excitation of non-radial pulsations has been recently found in massive stars (Belkacem *et al.* 2009; Degroote *et al.* 2010).

The FeCZ could also turn out to directly affect the evolution of hot massive stars. If it induces wind clumping, it may alter the stellar wind mass-loss rate. Such a change would also influence the angular momentum loss. In addition, magnetic fields produced by the iron convection zone could lead to an enhanced rate of angular momentum loss. These effects become weaker for lower metallicity, where the FeCZ is less prominent or absent.

Finally, the consequences of the FeCZ might be strongest in Wolf-Rayet stars. These stars are so hot that the iron opacity peak, and therefore the FeCZ, can be directly at the stellar surface, or — to be more precise — at the sonic point of the wind flow (Heger & Langer 1996). This may relate to the very strong clumping found observationally in Wolf-Rayet winds (Lepine & Moffat 1999; Marchenko *et al.* 2006), and may be required for an understanding of the very high mass-loss rates of Wolf-Rayet stars (Eichler *et al.* 1995; Kato & Iben 1992; Heger & Langer 1996).

2. Simulations of subsurface convection

Convection in the FeCZ is relatively inefficient: the transport of energy is dominated by radiation, which usually accounts for more than 95% of the total flux. This region of the star is very close to the photosphere, above which strong winds are accelerated. The continuous loss of mass from the stellar surface moves the convection region to deeper layers, revealing to surface material that has been processed in the FeCZ. In rotating stars, the associated angular momentum loss might also drive strong differential rotation in the region of interest. It is clear that, under these circumstances, the mixing length theory can only give a qualitative picture of the convective properties in these layers. In order to study the effects induced by the presence of subsurface convection at the stellar surface in a more quantitative way, we perform local 3D MHD calculations of convection. In these simulations we can vary the relative importance of the background radiative flux and include the effects of rotation and shear, in order to model conditions as similar as computationally permitted to subsurface convection in OB stars.

2.1. *Computational model*

The setup is similar to that used by Käpylä *et al.* (2008). A rectangular portion of a star is modeled by a box situated at colatitude θ. The dimensions of the computational domain are $(L_x,\, L_y,\, L_z) = (5,\, 5,\, 5)d$ where d is the depth of the convectively unstable layer. Our (x, y, z) correspond to (θ, ϕ, r) in spherical polar coordinates. The box is divided into three layers, an upper cooling layer, a convectively unstable layer, and a stable overshoot layer (see below). The following set of equations for compressible magnetohydrodynamics is being solved:

$$\frac{\partial \boldsymbol{A}}{\partial t} + Sx\frac{\partial \boldsymbol{A}}{\partial y} = \boldsymbol{U} \times \boldsymbol{B} - \eta\mu_0 \boldsymbol{J} - SA_y\hat{\boldsymbol{x}}, \tag{2.1}$$

$$\frac{\mathrm{D}\ln\rho}{\mathrm{D}t} = -\boldsymbol{\nabla}\cdot\boldsymbol{U}, \tag{2.2}$$

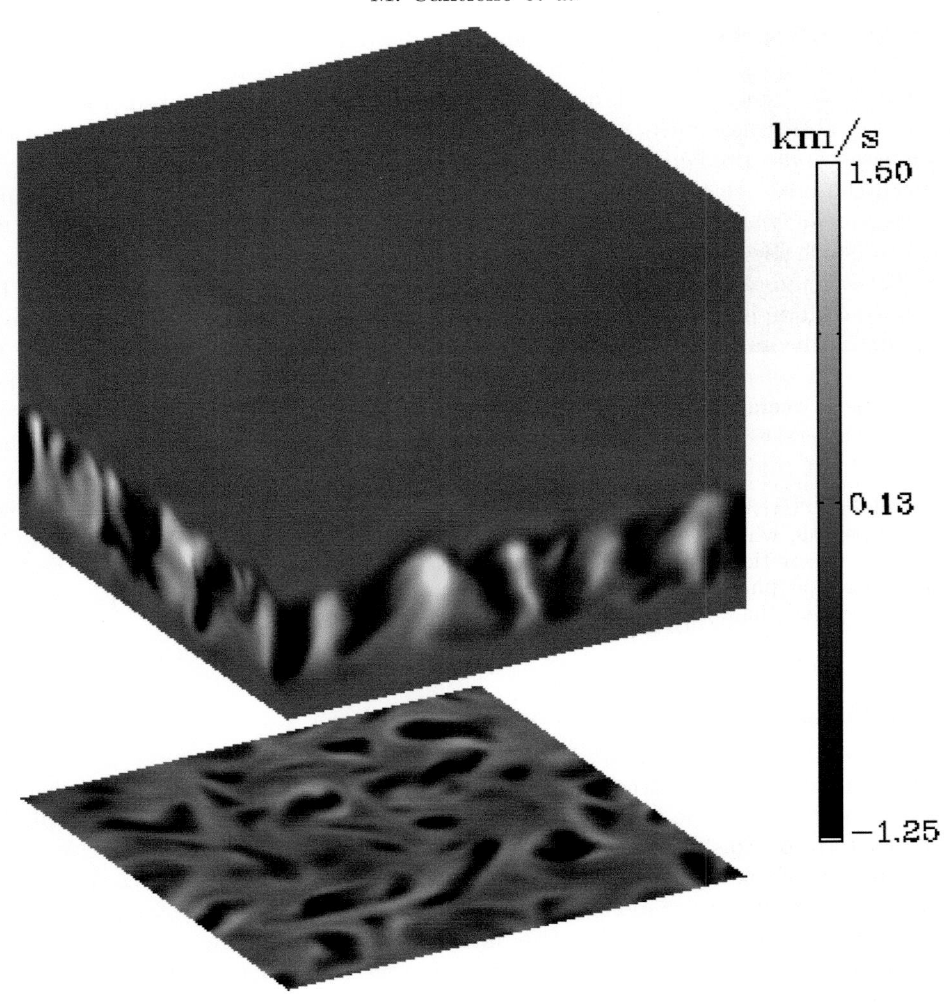

Figure 1. Simulation of subsurface convection. Starting from the top, the box is divided into three layers: a radiative layer with an upper cooling boundary, a convectively unstable layer and another stable layer at the bottom. The snapshot shows values of vertical velocity. The plane below the box shows the vertical velocity field at the lower boundary of the convective layer.

$$\frac{\mathrm{D}\boldsymbol{U}}{\mathrm{D}t} = -\frac{1}{\rho}\boldsymbol{\nabla} p + \boldsymbol{g} - 2\boldsymbol{\Omega}\times\boldsymbol{U} + \frac{1}{\rho}\boldsymbol{J}\times\boldsymbol{B} + \frac{1}{\rho}\boldsymbol{\nabla}\cdot 2\nu\rho\mathbf{S} - SU_x\hat{\boldsymbol{y}}, \tag{2.3}$$

$$\frac{\mathrm{D}e}{\mathrm{D}t} = -\frac{p}{\rho}\boldsymbol{\nabla}\cdot\boldsymbol{U} + \frac{1}{\rho}\boldsymbol{\nabla}\cdot K\boldsymbol{\nabla} T + 2\nu\mathbf{S}^2 + \frac{\eta}{\rho}\mu_0\boldsymbol{J}^2 - \frac{e-e_0}{\tau(z)}, \tag{2.4}$$

where $\mathrm{D}/\mathrm{D}t = \partial/\partial t + (\boldsymbol{U}+\overline{\boldsymbol{U}}_0)\cdot\boldsymbol{\nabla}$, and $\overline{\boldsymbol{U}}_0 = (0, S\,x, 0)$ is an imposed large-scale shear flow in the y-direction. The magnetic field is written in terms of the magnetic vector potential, $\boldsymbol{A}$, with $\boldsymbol{B} = \boldsymbol{\nabla}\times\boldsymbol{A}$, $\boldsymbol{J} = \mu_0^{-1}\boldsymbol{\nabla}\times\boldsymbol{B}$ is the current density, μ_0 is the vacuum permeability, η and ν are the magnetic diffusivity and kinematic viscosity, respectively, K is the heat conductivity, ρ is the density, $\boldsymbol{U}$ is the velocity, $\boldsymbol{g} = -g\hat{\boldsymbol{z}}$ is the gravitational acceleration, and $\boldsymbol{\Omega} = \Omega_0(-\sin\theta, 0, \cos\theta)$ is the rotation vector. The fluid obeys an ideal gas law $p = (\gamma - 1)\rho e$, where p and e are pressure and internal energy, respectively, and $\gamma = c_\mathrm{P}/c_\mathrm{V} = 5/3$ is the ratio of specific heats at constant pressure and volume, respectively. The specific internal energy per unit mass is related to the temperature via

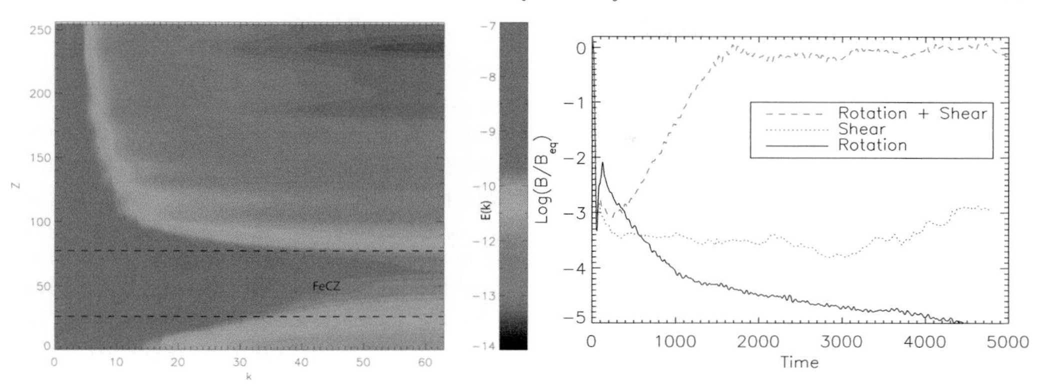

Figure 2. Left: energy spectrum of the velocity field in our run of subsurface convection. On the x-axis is the spatial wavenumber k, on the y-axis the z coordinate of the computational domain in gridpoints. The color bar on the right indicates the value of the energy per wavenumber $E(k)$. The location of the convection zone in the z-coordinate is shown by the dashed lines, the rest of the domain is radiative. Just above and below the convection zone kinetic energy is associated with overshooting, while at further distances the transport of energy is facilitated by gravity waves. **Right:** evolution of the magnetic field as function of time (in code units) in three different runs with different physics. The value on the y-axis is the logarithm of the ratio between the magnetic field and the equipartition value (the value corresponding to equipartition between kinetic and magnetic energy). Dynamo action reaching equipartition is found in simulations including both rotation and shear.

$e = c_{\rm V}T$. The rate of strain tensor S is given by

$$\mathsf{S}_{ij} = \tfrac{1}{2}(U_{i,j} + U_{j,i}) - \tfrac{1}{3}\delta_{ij}\boldsymbol{\nabla}\cdot\boldsymbol{U}. \tag{2.5}$$

The last term of Eq. (2.4) describes cooling at the top of the domain. Here $\tau(z)$ is a cooling time which has a profile smoothly connecting the upper cooling layer and the convectively unstable layer below, where $\tau \to \infty$.

The positions of the bottom of the box, bottom and top of the convectively unstable layer, and the top of the box, respectively, are given by $(z_1, z_2, z_3, z_4) = (-0.5, 0, 1, 4.5)d$, where d is the depth of the convectively unstable layer. Initially the stratification is piecewise polytropic with polytropic indices $(m_1, m_2, m_3) = (3, 0.9, 3)$, which leads to a convectively unstable layer between two stable layers. The cooling term leads to a stably stratified isothermal layer at the top. All simulations with rotation use $\theta = 0°$ corresponding to the north pole.

Stress-free boundary conditions are used in the vertical (z) direction for the velocity,

$$U_{x,z} = U_{y,z} = U_z = 0, \tag{2.6}$$

where commas denote partial derivatives, and perfect conductor conditions are used for the magnetic field, i.e.

$$B_{x,z} = B_{y,z} = B_z = 0. \tag{2.7}$$

In the x and y directions periodic boundary conditions are used. The simulations were performed with the Pencil Code†, which uses sixth order explicit finite differences in space and third order accurate time stepping method. Resolutions of up to $128^2 \times 256$ mesh points were used.

† `http://pencil-code.googlecode.com/`

2.2. *Preliminary results*

As a preliminary study we performed low resolution simulations (128x128x256), where the density contrast between the bottom of the convective layer and the top of the domain is only $\sim$20. This is about ten times smaller than in the case of the FeCZ. Moreover the ratio of the convective to radiative flux is about 0.3, higher than in the FeCZ case. This is because smaller values of convective flux result in steady convection at the low resolution of these preliminary runs. Therefore, at this stage, the velocities of convective motions cannot directly be compared to the velocities obtained by mixing length theory. However already in these preliminary runs we could follow the excitation and propagation of gravity waves above the convective region. In the right panel of Fig. 2 we show the kinetic energy spectrum (as function of the spatial wavenumber k) in the horizontal plane, as function of depth. The maximum of energy is found in the convective region for those wavenumbers k corresponding to the number of resolved convective cells (about 5 along one of the horizontal directions). Energy is also transported up to the top layer by gravity waves, where the maximum of the energy is deposited in those wavelengths that are resonant with the scale of convective motions, as predicted, for example, by Goldreich & Kumar (1990).

Käpylä *et al.* (2008) found excitation of a large scale dynamo in simulations of turbulent convection including rotation and shear. Our computational setup is very similar, so it's not surprising that we can confirm this result. Dynamo action reaching equipartition is found in our simulations that include shear and rotation (see left panel of Fig. 2), with magnetic fields on scales larger than the scale of convection.

2.3. *Discussion*

The connection found by Cantiello *et al.* (2009) between the presence of sub-photospheric convective motions and microturbulence in early-type stars is intriguing. The recent identification of solar-like oscillations in hot massive stars (Belkacem *et al.* 2009; Degroote *et al.* 2010) and further measurements of microturbulence (Fraser *et al.* 2010) also point toward a picture in which the FeCZ influences surface properties of OB stars.

We performed 3D MHD simulations of convection to investigate the excitation and propagation of gravity waves above a subsurface convection zone. Analytical predictions of Goldreich & Kumar (1990) on the spatial scale at which the maximum of energy is injected in gravity waves seem to be confirmed in our preliminary calculations. Further investigation is required to understand if the subsurface convection expected in OB stars excites gravity waves of the required amplitude to explain the observed microturbulence in massive stars. In particular we need higher resolution to increase the Reynolds number of our simulations and be able to decrease the ratio of convective to radiative flux, which appears to be an important parameter in determining the convective velocities (Brandenburg *et al.* 2005).

Magnetic fields reaching equipartition values are found in simulations of turbulent convection if rotation and shear are present (Käpylä *et al.* 2008). Since massive stars are usually fast rotators, it could be that the interplay between convection, rotation and shear is able to drive a dynamo in OB stars. Indeed our simulations of subsurface convection including rotation and shear show the excitation of dynamo action, with magnetic fields reaching equipartition. This means that fields up to a kG could be present in the FeCZ. These magnetic fields might experience buoyant rise and reach the surface of OB stars, where they could have important observational consequences. In particular it has already been suggested that the discrete absorption components observed in UV lines of massive stars could be produced by low amplitude, small scale magnetic fields at the stellar surface (Kaper & Henrichs 1994). Further study is needed to investigate the amplitude and geometry of magnetic fields reaching the stellar surface, as well as their effects on the photosphere.

References

Belkacem, K., Samadi, R., Goupil, M.-J., Lefèvre, L. *et al.* 2009, *Science*, 324, 1540
Brandenburg, A., Chan, K. L., Nordlund, Å., & Stein, R. F. 2005, *AN*, 326, 681
Cantiello, M., Langer, N., Brott, I., de Koter, A. *et al.* 2009, *A&A*, 499, 279
Degroote, P., Briquet, M., Auvergne, M., Simón-Díaz, S. *et al.* 2010, *A&A*, 519, A38
Eichler, D., Bar Shalom, A., & Oreg, J. 1995, *ApJ*, 448, 858
Evans, C., Smartt, S., Lennon, D., Dufton, P. *et al.* 2005, *The Messenger*, 122, 36
Fraser, M., Dufton, P. L., Hunter, I., & Ryans, R. S. I. 2010, *MNRAS*, 404, 1306
Goldreich, P. & Kumar, P. 1990, *ApJ*, 363, 694
Heger, A. & Langer, N. 1996, *A&A*, 315, 421
Kaper, L. & Henrichs, H. F. 1994, *Ap&SS*, 221, 115
Käpylä, P. J., Korpi, M. J., & Brandenburg, A. 2008, *A&A*, 491, 353
Kato, M. & Iben, Jr., I. 1992, *ApJ*, 394, 305
Lépine, S. & Moffat, A. F. J. 1999, *ApJ*, 514, 909
Marchenko, S. V., Moffat, A. F. J., St-Louis, N., & Fullerton, A. W. 2006, *ApJ* (Letters), 639, L75

Active OB stars: structure, evolution, mass loss, and critical limits
Proceedings IAU Symposium No. 272, 2010
C. Neiner, G. Wade, G. Meynet & G. Peters, eds.

doi:10.1017/S1743921311009951

Rotation rates of massive stars in the Magellanic Clouds

Laura R. Penny[1] **and Douglas R. Gies**[2]

[1]Dept. of Physics & Astronomy, College of Charleston,
Charleston, SC, USA
email: pennyl@cofc.edu

[2]Dept. of Physics & Astronomy, Georgia State University,
Atlanta, GA, USA
email: gies@phy-astr.gsu.edu

Abstract. We present the results of our survey of the projected rotational velocities of 161 O-type stars in the Magellanic Clouds from archival FUSE observations. The evolved and unevolved samples from each environment are compared through the Kolmogorov-Smirnov test to determine if the distribution of equatorial rotational velocities is metallicity dependent for these massive objects. Stellar interior models predict that massive stars with SMC metallicity will have significantly reduced angular momentum loss on the main sequence compared to their Galactic counterparts. Our statistical results find some support for this prediction but also show that even at Galactic metallicity, evolved and unevolved massive stars have fairly similar fractions of stars with large V sin i. What is more compelling are the few evolved objects in the Magellanic Clouds with rotational velocities that approach or even exceed those predicted from the evolutionary models.

Keywords. stars: rotation, stars: early-type, stars: evolution, ultraviolet: stars

---

1. Introduction and Methodology

Here we present the results of our project to test the treatment of angular momentum in the new stellar interior models through a large scale survey of projected rotational velocities of O-type stars in three metallicity environments: the Milky Way ($Z_{MW} = 0.020$), LMC ($Z_{LMC} = 0.007$), and SMC ($Z_{SMC} = 0.004$). The Far Ultraviolet Spectrographic Explorer (FUSE) archive at the Multimission Archive at Space Telescope (MAST) contains spectra of 161 LMC and SMC stars with spectral classes between B2 - O2. The targets are 120 evolved (luminosity classes I, II, & III) and 41 unevolved (IV & V luminosity classes) stars in these low Z environments. These observations, along with 97 archival spectra of Galactic O-type stars, were obtained from MAST. Target star spectra are cross-correlated with that of a relatively narrow-lined template star. The CCF is the sum of the square of the differences between the test spectrum and the reference spectrum shifted in velocity from -1000 km s^{-1} to $+1000$ km s^{-1} at 10 km s^{-1} intervals. The functions are then rectified, inverted, and Gaussian fitted to obtain an estimate of the full-width at half-maximum (FWHM). Then using the calibrations developed from stars with known projected rotational velocities, a $V \sin i$ value is determined from the Gaussian width of each fitted CCF (for more details see Penny & Gies 2009).

2. Analysis

We divide the data into the following samples: SMC unevolved (32 stars), SMC evolved (19 stars), LMC unevolved (36 stars), and LMC evolved (42 stars). We also create samples

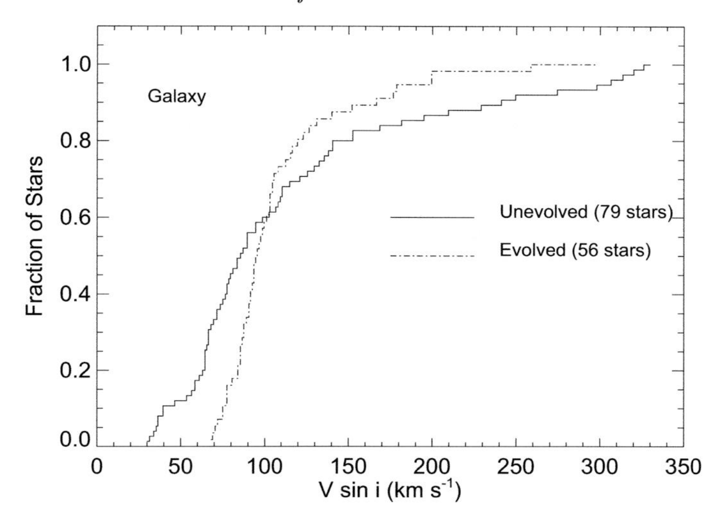

Figure 1. Cumulative distribution functions of $V \sin i$ values for unevolved and evolved stars in the Milky Way.

for Galactic unevolved (79 stars) and evolved (56 stars) from Howarth *et al.* (1997). Following the convention of earlier studies (Penny 1996, Howarth *et al.* 1997, Mokiem *et al.* 2006, Mokiem *et al.* 2007), unevolved refers to stars with luminosity classes V & IV, while luminosity classes I & II are termed evolved. Class III stars are omitted from both groups. For each sample we create a CDF and plot these in comparison with other samples to determine the similarity. For each comparison a K-S statistic, D, and its corresponding probability, p, are calculated and these are presented in Table 1. Plots of the comparisons made are presented in Figures 1 – 6.

First we examine the evolved and unevolved Galactic samples to determine the level of difference we would expect in an environment where the stars do slow down as they evolve (Fig. 1). We see that the maximum variance in their CDF is 0.30 and this occurs at $V \sin i$ of 83 km s^{-1}. This large a D value results in a p of 0.5%, but we should not be misled by this low value. At these smaller velocities the source of the difference between the evolved and unevolved samples is not a result of angular momentum loss, but from the larger amount of macroturbulence that is present in the evolved stars' photospheres, broadening their line profiles. At the higher velocities where we expect to see evidence of spin down, the largest divergence is ≈ 0.12, which would give us a significantly higher p of 71%. This is not say that the evolved stars in the Galactic sample have not slowed down, but that this effect may be more subtle than we originally expected.

In the LMC, the same comparison of unevolved to evolved stars has a very different appearance (Fig. 2). The CDF of both samples appear very similar, and the derived $D = 0.15$ and corresponding $p = 75\%$ supports the null hypothesis that both are drawn from the same parent distribution. This result is in agreement with the suggestion of Wolff *et al.* (2008), using data from Hunter *et al.* (2008), that massive stars in the LMC have similar $V \sin i$ distributions regardless of their $\log g$ values. Examining Fig. 2, we notice that there is almost no divergence at low $V \sin i$ values, and a smaller variance at larger $V \sin i$ than we see in the Milky Way samples. The very high p value here is primarily due to the presence of slow rotators among both the unevolved and evolved samples, unlike the Galactic evolved group. At high $V \sin i$ values, the CDFs of evolved to unevolved differ by ≈ 0.09 which is only slightly smaller than the value for the Galactic samples. It is at the low velocity end of the CDF where the LMC samples differ from

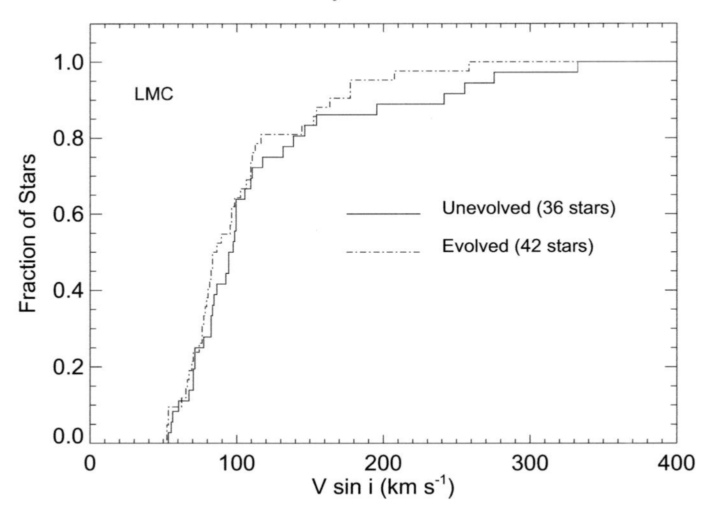

Figure 2. Cumulative distribution functions of $V \sin i$ values for unevolved and evolved stars in the LMC.

those of the Galactic samples. The good agreement between the low ends of the two CDFs for the LMC stars suggests that development of macroturbulence with evolution is not as large a factor in the photospheres of the massive stars in that environment and that the processes that lead to macroturbulent broadening may have a metallicity dependence.

The CDFs of the same populations in the SMC are presented in Figure 3. Although the p value from the KS test is 23%, which would indicate that the samples are drawn from the same parent population, the maximum difference D is 0.29, which is very close to that from the Galactic samples. The larger p value from the similar D reflects the much smaller sample sizes in the SMC. At the high velocity end of the CDF, the fraction of evolved stars with $V \sin i$ above 200 km s^{-1} is 11% compared to 13% for the unevolved. This does support the hypothesis that stars at SMC metallicity will not slow down as they evolve on the MS. At the other end of the CDF, there is a slightly larger fraction of unevolved stars with $V \sin i$ below ≈ 70 km s^{-1}, but this difference is much smaller than we see in the Galactic comparison. The real divergence between the evolved and unevolved CDFs comes at the intervening velocities. The maximum divergence, 0.29, occurs at $V \sin i = 107$ km s^{-1}. The fractions of stars with $V \sin i$ at or below this value for the evolved and unevolved samples are 0.84 and 0.55, respectively. This trend was also seen by Mokiem *et al.* (2006), who surmised that the initial rotational velocity distribution in the SMC might vary significantly from that in the Galaxy. Why we see a disparity in the CDFs of the evolved and unevolved samples in between $80 - 190$ km s^{-1} only in the SMC samples is not clear. We emphasize that the K-S test results accept the null hypothesis that the unevolved and evolved samples in the SMC are drawn from the same parent distribution.

A primary purpose of this project is to examine observationally the amount of angular momentum loss during the MS lifetimes of massive stars and the effects of metallicity on this loss. Comparing the evolved to unevolved samples in Figures 1–3, we see that the D statistic at high $V \sin i$ values range from 0.12, 0.09, 0.02 in the Galaxy, LMC, and SMC, respectively. Taken by itself this is suggestive of a trend with decreasing Z. However, for the LMC and SMC comparisons, D statistics of 0.09 and 0.02 result in p values of 99.6% and 100%, indicating there is no statistical difference between the evolved and

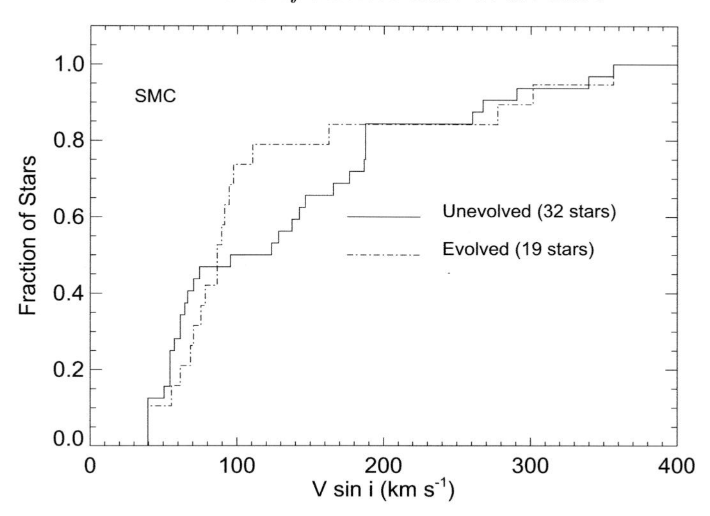

Figure 3. Cumulative distribution functions of $V \sin i$ values for unevolved and evolved stars in the SMC.

unevolved CDFs at high $V \sin i$. But this is also true in the Galaxy where we do expect to see the significant spin down between the evolved and unevolved samples. Here the corresponding p for a D of 0.12 is 71%, far above the 5% cutoff. Statistically the loss of angular momentum for these H-burning stars is not so different between these three metallicity environments.

We also are interested in whether the initial $V \sin i$ distribution is the same in the three environments. In Figure 4, we plot the CDFs of all three dwarf samples. Statistically all three have p values that are above the cut-off of 5%, and certainly at the high $V \sin i$ end all three look extremely similar, indicating that the maximum rotational velocities are very similar. Below 200 km s^{-1}, the behavior of the three distributions varies. Again we see that the shape of the CDF for the SMC dwarfs is dissimilar from that of the counterparts in the Galaxy and the LMC between 80 – 190 km s^{-1}. Below 80 km s^{-1}, the fraction of slowly rotating stars varies between the three environments, with the Galactic sample in between the SMC and LMC. In fact the largest divergence between the LMC and SMC CDFs comes at $V \sin i \approx 65$ km s^{-1}, resulting in a p value just above the statistically significant level. Hunter *et al.* (2008) showed an analogous plot, but for objects with $M < 25 M_{\odot}$. Similarly they find good agreement between the LMC and Galactic samples, with the SMC sample lying slightly beneath the other two, especially in the region below 200 km s^{-1}. It is interesting that there are no very slow rotators in our LMC dwarf sample. At these low velocities, the dominating effect must not depend upon metallicity since the Galaxy's metallicity, with significantly higher Z, is situated between that of the two low Z samples. We conclude that the initial velocity distribution in our three unevolved samples are statistically indistinguishable.

Finally we compare the evolved stars in each environment to examine the relative fractions of stars with large $V \sin i$ (Fig. 5). Looking at the distributions at the high velocity end, we see that a slightly larger fraction of evolved stars in the SMC have $V \sin i$ values larger than 200 km s^{-1} than in the Galaxy or LMC, which might support theoretical predictions. But the large D values that are found between the LMC and SMC samples are again at the low velocity range, ≈ 80 km s^{-1}. The maximum differences between the Galaxy and both the LMC and SMC CDFs occur near this velocity, with both of the low metallicity environments having a larger fraction of stars with smaller

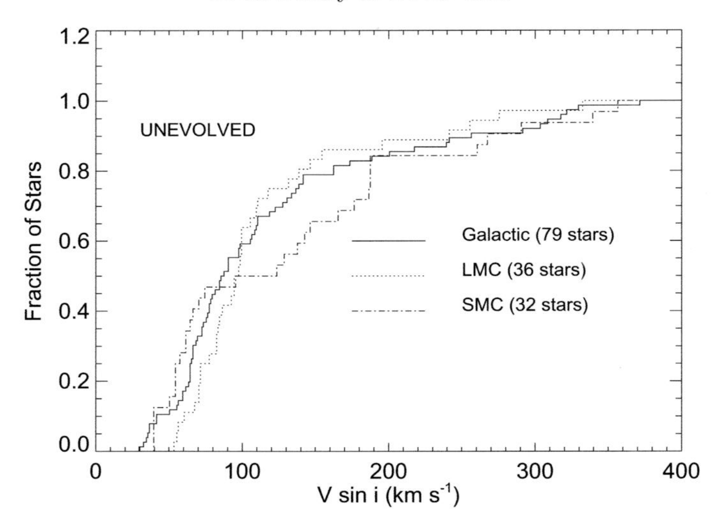

Figure 4. Cumulative distribution functions of $V \sin i$ values for unevolved (luminosity classes IV - V) stars in the SMC, LMC and Galaxy.

values compared to the Milky Way sample. In fact the trend is very supportive of our earlier suggestion that metallicity plays an important role in macroturbulent broadening in evolved O-type stars. A recent paper by Cantiello *et al.*(2009) discusses the origin of atmospheric turbulence in massive stars by sub-surface convection zones that are driven by Fe-peak element ionizations. In their simulations, the threshold luminosity for the occurrence of an iron convective zone is ten times lower at Z_{MW} than that for Z_{SMC}. Our results support their prediction that turbulence will increase with metallicity. The SMC evolved sample has the largest fraction of slow rotators, followed by the LMC and then the Galaxy. We note that the D values for the SMC vs. Galaxy and LMC vs. Galaxy are extremely similar, but result in differing p values owing to the smaller SMC sample size. The large divergence at small $V \sin i$ values between the Galaxy and LMC, and possibly SMC, leads us to reject the hypothesis that they are both from the same parent distribution. Again we stress that the differences that cause this are not the fractions of stars with large $V \sin i$, but those with small values.

In conclusion, we find some support for the new stellar interior model predictions that massive stars in lower metallicity environments will remain at almost constant rotation rates throughout their MS lifetimes. But we also see that this effect, loss of angular momentum during the MS, is more subtle than previously reported even at the higher Galactic metallicity. We have suggested that metallicity may play an important role in the development of macroturbulence in the photospheres of the evolved, massive stars. Finally we note that there are several evolved stars in the SMC and LMC with very large $V \sin i$ values. Notable amongst these are: AV 321 (O9 Ib, 357 km s^{-1}) and SK 190 (O8 Iaf, 302 km s^{-1}).

References

Cantiello, M., Langer, N., Brott, I., de Koter, A. *et al.* 2009, *Communications in Asteroseismology* 158, 61

Howarth, I. D., Siebert, K. W., Hussain, G. A. J., & Prinja, R. K. 1997, *MNRAS*, 284, 265

Hunter, I., Lennon, D. J., Dufton, P. L., Trundle, C. *et al.* 2008, *A&A*, 479, 541

Mokiem, M. R., de Koter, A., Evans, C. J., Puls, J. *et al.* 2006, *A&A*, 456, 1131

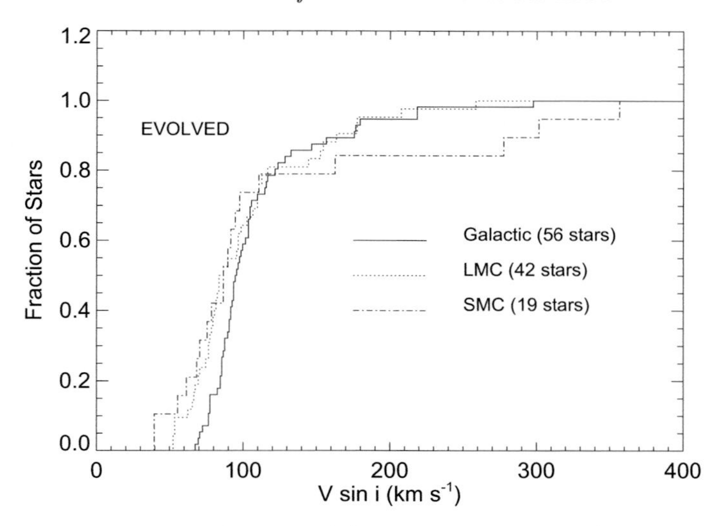

Figure 5. Same as Figure 4 but for evolved (luminosity classes I -II) stars in the same three environments.

Table 1. Kolmogorov-Smirnov Statistics

Sample 1	n_1 stars	$< V \sin i >$ (km s^{-1})	Sample 2	n_2 stars	$< V \sin i >$ (km s^{-1})	D	p (%)
			Unevolved Vs. Evolved Within Each Environment				
Galactic unevolved	79	129.8	Galactic evolved	56	109.5	0.296	0.5
LMC unevolved	36	132.0	LMC evolved	42	118.6	0.151	74
SMC unevolved	32	116.2	SMC evolved	19	101.8	0.289	23
			Unevolved Samples				
LMC unevolved	36	132.0	Galactic unevolved	79	129.8	0.180	37
SMC unevolved	32	116.2	Galactic unevolved	79	129.8	0.172	47
SMC unevolved	32	116.2	LMC unevolved	36	132.0	0.295	8
			Evolved Samples				
LMC evolved	42	118.6	Galactic evolved	56	109.5	0.321	1
SMC evolved	19	101.8	Galactic evolved	56	109.5	0.297	13
SMC evolved	19	101.8	LMC evolved	42	118.6	0.137	95

Mokiem, M. R., de Koter, A., Evans, C. J., Puls, J. *et al.* 2007, *A&A*, 465, 1003
Penny, L. R. 1996, *ApJ*, 463, 737
Penny, L. R. & Gies, D. R. 2009, *ApJ*, 700, 844
Penny, L. R., Sprague, A. J., Seago, G., & Gies, D. R. 2004, *ApJ*, 617, 1316
Wolff, S. C., Strom, S. E., Cunha, K., Daflon, S. *et al.* 2008, *AJ*, 136, 1049

Active OB stars: structure, evolution, mass loss, and critical limits
Proceedings IAU Symposium No. 272, 2010
C. Neiner, G. Wade, G. Meynet & G. Peters, eds.

doi:10.1017/S1743921311009963

Interferometric studies of rapid rotators

Ming Zhao[1], John D. Monnier[2] and Xiao Che[2]

[1]Jet Propulsion Laboratory, 4800 Oak Grove Dr, Pasadena, CA 91101
email: `ming.zhao@jpl.nasa.gov`

[2]Dept. of Astronomy, University of Michigan, 500 Church St., Ann Arbor, MI 48109

Abstract. Stellar rotation, like stellar mass and metallicity, is a fundamental property of stars. Rapid rotation distorts the stellar photosphere and affects a star's luminosity, abundances and evolution. It is also linked to stellar wind and mass loss. The distortion of the stellar photosphere due to rapid rotation causes the stellar surface brightness and effective temperature to vary with latitude, leading to a bright pole and a dark equator - a phenomenon known as 'Gravity Darkening'. Thanks to the development of long baseline optical interferometry in recent years, optical interferometers have resolved the elongation of rapidly rotating stars, and have even imaged a few systems for the first time, directly confirming the gravity darkening effect. In this paper, we review the recent interferometric studies of rapid rotators, particularly the imaging results from CHARA-MIRC. These sub-milliarcsecond resolution observations permit the determination of the inclination, the polar and equatorial radius and temperature, as well as the fractional rotation speed of several rapid rotators with unprecedented precision. The modeling also allows the determination of the true effective temperatures and luminosities of these stars, permitting the investigation of their true locations on the HR diagram. Discrepancies from standard models were also found in some measurements, suggesting the requirement of more sophisticated mechanisms such as non-uniform rotation in the model. These observations have demonstrated that optical interferometry is now sufficiently mature to provide valuable constraints and even model-independent images to shed light on the basic physics of stars.

Keywords. stars: fundamental parameters, stars: rotation, stars: imaging, techniques: interferometric

---

1. Introduction

Stellar rotation, like stellar mass and metallicity, is a fundamental property of stars. For decades, stellar rotation was generally overlooked in stellar models and was regarded to have a trivial influence on stellar evolution because most stars are slow rotators, such as the Sun (Maeder & Meynet 2000). Although the effects of rotation on solar type stars are indeed relatively mild, they are more prominent on hot stars. Studies have shown that a large fraction of hot stars are rapid rotators with rotational velocities more than 120 km s^{-1} (Abt & Morrell 1995; Abt *et al.* 2002). Virtually all the emission-line B (Be) stars are rapid rotators with rotational velocities of $\sim$ 90% of breakup (Frémat *et al.* 2005). Rapid rotation affects stellar brightness distribution, internal flows, stellar evolution, and circumstellar environments (Maeder & Meynet 2000) - almost all aspects of stellar astrophysics.

Traditionally, stellar rotation is studied using spectra line profiles, doppler imaging, photometry and polarimetry. Techniques such as asteroseismology also become popular in recent years (e.g., Gizon & Solanki 2004). These methods are all very powerful in characterizing stellar structure and/or rotation. On the other hand, they also have constraints and sometimes even require good knowledge of the targets. For instance, Vega shows narrow lines typical of slow rotation (Vsini$\sim$21 km s^{-1}). However, some weak metal spectra lines are "flat-bottomed" (Gulliver *et al.* 1994). The "flat-bottomed" line profiles have led to inaccurate determination of its inclination angle (Hill *et al.* 2004) until Vega was confirmed to be a pole-on rapid rotator by interferometry (Peterson *et al.* 2006; Aufdenberg *et al.* 2006), and rotation-induced macro turbulence of 7-10 km s^{-1} was included in the model (Yoon *et al.* 2008; Hill *et al.* 2010). This suggests that detailed studies of stars require measurements from multiple techniques to provide multi-angle views and disentangle degeneracies of parameters.

In the past few years, optical interferometers have resolved the elongated photospheres of rapidly rotating stars for the first time. More excitingly, interferometric observations have lead to the first images of main sequence stars other than the Sun, showing the elongated photospheres and uneven surface brightness distributions due to rapid rotation. The emergence of these high angular resolution observations of hot stars has shined a spotlight on critical areas of stellar evolution and basic astrophysics that demand our attention. Measurements from interferometers can determine the oblateness, geometry, rotational speed, and surface brightness distribution of rapid rotators, and even provide model-independent images of stars. These measurements can directly and independently provide constraints to other measurements such as spectroscopy and asteroseismology, providing better understanding of the structure and physics of the targets. In this paper, we review the recent interferometric studies of rapid rotators, particularly the imaging results from CHARA-MIRC.

The paper is organized as follows. We briefly discuss the effects of rapid rotation in §2, and give a minimal overview of the basics of interferometry in §3. We highlight the recent interferometric studies in §5 and present some CHARA imaging results in §6. Finally, we give some future prospects in §7.

2. Effects of rapid rotation

Stars that are rapidly rotating have many unique characteristics. The centrifugal force from rapid rotation distorts their photospheres and causes them to be oblate. In the standard Roche approximation (i.e., point mass distribution and uniform rotation), the maximum equatorial to polar radius ratio due to the rotational distortion has a maximum value of 1.5. Recent studies also suggest that this ratio can be larger than 1.5 with quadrupolar moment and/or differential rotation (Zahn *et al.* 2010; MacGregor *et al.* 2007).

The distortion of stellar surface also causes the surface brightness and the effective temperature (T_{eff}) to vary with latitude, and the equatorial temperatures are predicted to be much cooler than their polar temperatures, a phenomenon known as "Gravity Darkening" (von Zeipel 1924a,b). The standard gravity darkening law suggests $T_{eff} \propto g^{\beta}$, where g is the local gravity and β is the gravity darkening coefficient (von Zeipel 1924a). β equals to 0.25 for stars with radiative envelopes (von Zeipel 1924a), while it equals to 0.08 for stars with convective envelopes (Lucy 1967). For stars in the transition stage between radiative and convective envelopes, the value of β can be in between (Claret 2003).

Large rotation can modify the interior angular momentum of stars, causing internal flows and even differential rotation (e.g., Jackson *et al.* 2004). Recent stellar models show that rapid rotation can also cause abundance anomalies by rotation-induced mixing (Pinsonneault 1997). Rapid rotation also affects stellar luminosity and evolution, and increase stellar lifetime, which can cause scatters to the H-R diagram and may even alter the Mass-Luminosity relation (Maeder & Meynet 2000). In addition, stellar rotation can affect circumstellar environments through enhanced stellar winds and mass loss (e.g.,

Maeder *et al.* 2007), and it is even linked to Gamma-Ray bursts (MacFadyen *et al.* 2001; Burrows *et al.* 2007). A more detailed review can be found in Maeder & Meynet (2000).

3. Basics of interferometric studies

3.1. *Interferometric observables*

Interferometers combine light from distant objects using multiple telescopes and measure the interference of light. The light collected by telescopes is combined to form interference fringes similar to those of the Young's Double Slit experiment. The fringe contrast (also known as the fringe visibility, V) along with the associated phase are then measured by detectors. These two quantities together, also known as the complex visibility, are the basic observables in interferometry.

The visibility amplitude (i.e., the fringe contrast mentioned above) measures the angular scale of a target along the projected direction of a telescope-pair on the sky. The angular resolution is determined by the projected baseline between the two telescopes by $\lambda/2B_{proj}$, where λ denotes the wavelength and B_{proj} denotes the projected baseline. A point source gives unit visibility, while a resolved source gives much lower visibility. Therefore, by measuring a target along different directions on the sky, we can determine its shape and know if it is larger in one direction than the other.

Phase is another important quantity in the complex visibility as it carries information about the brightness distribution of the source. However, the phase of the light wave coming from a distant source is always corrupted as atmospheric turbulence always induces random and extra phase shifts. This makes the phases of complex visibilities useless. Nevertheless, we can combine 3 telescopes in a closed triangle, and the extra phase seen by each telescope can be canceled in a closed form (see Monnier 2003, for details). This phase closure, or closure phase, carries some intrinsic phase information of the target. It is immune to any phase shifts induced by the atmosphere and many other systematic errors as well. Because of this, it was widely applied to very long baseline interferometry in radio to compensate poor phase stability (Monnier 2003). Closure phase is also a good quantity for stable and precise measurements, and is widely used in aperture synthesis imaging for phase calibration (e.g., Zhao *et al.* 2008).

In addition, differential phase is another widely used quantity in interferometry. It measure the phase difference at two wavelengths, and therefore can also eliminate the phase error caused by atmospheric turbulence. Differential phase is sensitive to chromatic signatures and shift of photo-centers of the targets due to e.g., star spots, non-radial pulsations, and rotation, etc. (see e.g., Jankov *et al.* 2001)

3.2. *Interferometric Imaging*

Using visibilities and closure phases, one can construct models for a stellar object. A model-independent image on the other hand, is more intuitive and can provide more straightforward information.

According to the van Cittert-Zernike theorem (see e.g., Thompson *et al.* 1986), the complex visibility of an object is directly linked to the Fourier transform of its intensity distribution. Thus, the intensity distribution can be inverted by the inverse Fourier transform of its complex visibilities. However, in reality this inversion process is not straightforward as the actual sampling of the complex visibility space is always sparse and discrete. One needs a reasonable number of samplings to get enough information of the original intensity distribution, and a large number of sample points can greatly help the reconstruction of the image. In ground-based interferometry, this problem can be slightly improved by the Earth's rotation, as the baselines' projected directions change while the Earth rotates, providing more samplings of the visibility space.

Because of the limited visibility and phase information, special imaging algorithms are thus required for image reconstruction. The reconstruction generally has many possible solutions. Thus some reasonable constraints are applied to regulate the solution, for instance, limiting the field-of-view of the image, making the imaging as smooth as possible, and providing a *priori* information to constrain the image.

Popular image reconstruction algorithms include CLEAN (Högbom 1974), the Maximum Entropy Method (MEM) (Narayan & Nityananda 1986), and the Markov Chain Monte-Carlo imaging (MCMC) (e.g., Ireland *et al.* 2006), etc. CLEAN is widely used in radio interferometry, while MEM and MCMC is more commonly adopted in optical/IR interferometry. Other imaging algorithms such as MIRA (Thiébaut 2008) are also emerging and can produce robust reconstructions. More details of optical interferometric imaging can be found in Malbet *et al.* (2010); Cotton *et al.* (2008) and references therein. All the images we present in §6 are reconstructed using the MACIM package, and details can be found in Ireland *et al.* (2006).

4. Modeling of rapid rotators

Here we briefly introduce the standard Roche - von Zeipel model commonly used in interferometry for rapid rotators. The basic assumptions of the Roche model (Cranmer & Owocki 1995) are point mass concentrated in the center of star and uniform rotation. These assumptions, together with the von Zeipel's theorem (see §2), allow us to construct a 2-dimensional map of a rapid rotator with six parameters: the stellar radius and temperature at the pole (R_{pol}, T_{pol}), the angular rotation rate as a fraction of breakup (ω), the gravity darkening coefficient β, the inclination angle i, and the position angle (east of north) ϕ of the star. The detailed prescription of the model can be found in Cranmer & Owocki (1995) and Aufdenberg *et al.* (2006).

Similar prescriptions can also be found in other literatures such as Domiciano de Souza (2009). The constructed 2D sphere is then projected onto the sky according to the position angle ϕ, and the inclination i. The intensity and limb darkening at each point of the stellar surface is calculated using stellar atmosphere models such as the Kurucz models (Kurucz 1993) and the PHOENIX models (Aufdenberg *et al.* 2006) as a function of local effective temperature, gravity, viewing angle, and wavelength.

5. Highlights of early observations

The development of long baseline optical interferometry in recent years has provoked many observations on nearby rapid rotators since the first observation in 2001 (van Belle *et al.* 2001). We highlight some of the results below.

<u>*Altair*</u>. Altair (α Aquilae, A7IV-V, HR 7557, $V = 0.77$, $H = 0.102$) was the first rapid rotator observed by interferometer (van Belle *et al.* 2001) . It is a member of the famous "Summer Triangle" (the other two stars are Vega and Deneb). Altair shows very large projected velocity, varying from 200 km s^{-1} to 242 km s^{-1} (e.g., Abt & Morrell 1995). Using the Palomar Testbed Interferometer (PTI), van Belle *et al.* (2001) found that Altair is elongated and one dimension is $\sim 14\%$ larger than the other (see the left panel of Fig. 1). Using NPOI, Ohishi *et al.* (2004) found that Altair has asymmetric surface brightness distribution and its pole is brighter than other parts of the star. Using the same data, Peterson *et al.* (2006) and Domiciano de Souza *et al.* (2005) independently

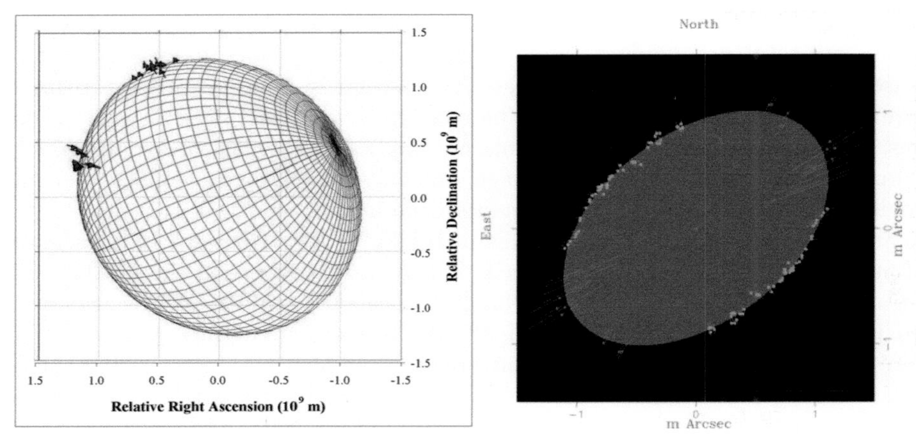

Figure 1. *Left*: PTI measurements of Altair (van Belle *et al.* 2001). The orientation of the star was flipped by 180° in the original paper and is corrected in this plot (van Belle 2009, private communication). The red triangles show the measured sizes of Altair at different directions. *Right*: VLTI/VINCI measurements of Achernar (Domiciano de Souza *et al.* 2003). The colored dots show the angular diameter measurements along the stellar surface, indicating the star is strongly elongated.

modeled Altair using the gravity darkening model and confirmed the oblateness and asymmetric structure of Altair due to rapid rotation.

Regulus. Regulus (α Leo, HR3982, V = 1.35, H = 1.658) is a B7V (Johnson & Morgan 1953) or B8 IVn (Gray *et al.* 2003) star in a triple system. The rapid rotation of the primary was first measured by McAlister *et al.* (2005) with the CHARA array in the K' band. Combining the visibilities with spectroscopic measurements, McAlister *et al.* (2005) found Regulus is seen nearly equator-on, and its photosphere is very elongated due to rapid rotation, with an R_{eq}/R_{pol} ratio of 1.32. Recently, Gies *et al.* (2008) discovered that the primary is also a spectroscopic binary with a white dwarf company.

Achernar. The most extreme case of rapid rotation known to date comes from the brightest Be star Achernar (α Eri, HR 472, V=0.50, H =0.865). Using the VINCI instrument at the Very Large Telescope Interferometer (VLTI), Domiciano de Souza *et al.* (2003) measured the rotational flattening of Achernar to have a R_{eq}/R_{pol} ratio of 1.56 (left panel of Fig. 1), significantly higher than the limit given by the Roche approximation (1.5). This deviation suggests the basic assumption of Roche approximation may not be valid in this type of stars and differential rotation may exist (e.g., Jackson *et al.* 2004). Recent studies also suggest that an equatorial circumstellar disk/envelope could exist and contribute partially to the observed large flattening of Achernar (Vinicius *et al.* 2006; Kanaan *et al.* 2008). In addition, Kervella & Domiciano de Souza (2006) also found strong polar emission that can account for $\sim 5\%$ of the near-IR flux of Achernar. These discoveries have made Achernar an extreme and ideal test case for theoretical models, and have evoked extensive discussions in the field.

Vega. Perhaps the most compelling results of rapid rotators come from the photometric standard Vega (α Lyr, HR 7001, A0V). For decades, Vega was found to be anomalously brighter than the average A0V stars by 0.5 mag in V (Millward & Walker 1985). Early interferometric studies ruled out the binary hypothesis and also found that Vega's radius is significantly larger (60% larger) than that of Sirius, a A1V star. These phenomenon implied that Vega may be a rapid rotator seen pole-on. The flat-bottomed appearance of some weak metal lines of Vega, in addition, also suggests the same hypothesis.

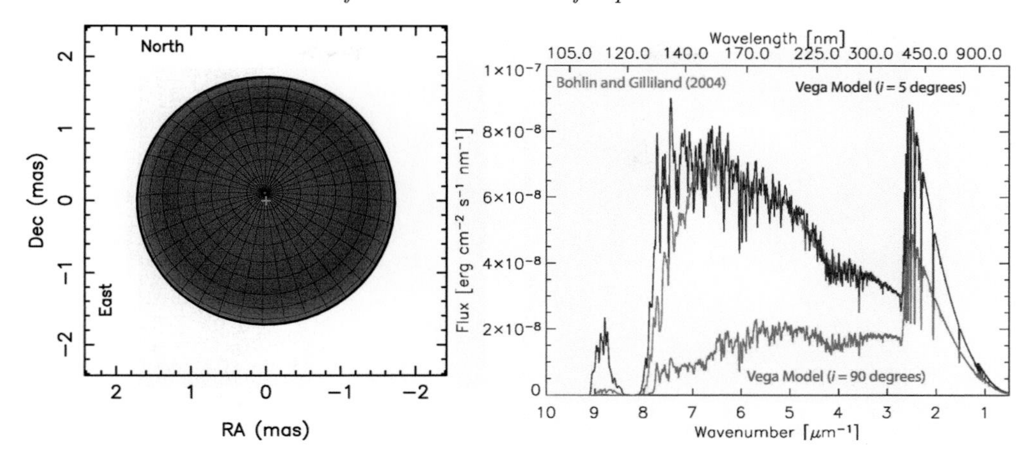

Figure 2. *Left*: Gravity darkening model of Vega from Peterson *et al.* (2006) using NPOI, showing that Vega is a nearly pole-on star. *Right*: The SED model of Vega, seen at the pole ($i = 5^o$, blue) and the equator ($i = 90^o$, red) respectively (Aufdenberg *et al.* 2006). The SEDs at the polar and the equatorial regions are significantly different due to the large temperature differences.

The mystery was eventually uncovered by interferometric measurements from both the CHARA array (Aufdenberg *et al.* 2006) and the Navy Prototype Optical Interferometer (NPOI) (Peterson *et al.* 2006). By measuring the extra darkening at the stellar limbs caused by gravity darkening (Aufdenberg *et al.* 2006) and the closure phase (Peterson *et al.* 2006) respectively, these two studies both confirmed that Vega is indeed a nearly pole-on rapid rotator ($i \sim 5^o$) rotating at $\sim 91\% - 93\%$ of its angular break-up rate (Fig. 2). They also show that the polar radii of Vega is over 23% shorter than that of the equator, while the temperature at the poles is over 2000K hotter than that of the equator.

6. Imaging of rapid rotators

The interferometric studies mentioned above have confirmed the oblateness of rapid rotators and the general picture of von Zeipel's gravity darkening law for the first time, greatly improved our understandings of rapid rotators. On the other hand, discrepancies with the basic assumptions such as solid rotation and point mass distribution of stars are also raised. Indeed, more sophisticated hydrodynamical models have shown that non-solid body rotation such as differential rotation and meridional flows can exist in these stars (e.g., Jackson *et al.* 2004; MacGregor *et al.* 2007). To shed light on these issues, and further advance our knowledge of the physics of rapid rotation, "model-independent" images are necessary to test wide variety of models and to provide more intuitive knowledge of those stars.

Most of the previous interferometric studies were entirely based on model-fitting of interferometric visibilities with a few baselines, and therefore are not able to generate model-independent images. However, thanks to the recent development of imaging combiners such as CHARA-MIRC (Monnier *et al.* 2004) and VLTI-AMBER (Petrov *et al.* 2007), imaging stellar surfaces with only a few milli-arcseconds across has become true.

Up to date, imaging studies of rapid rotators have been carried out for five stars (Altair, Alderamin, Rasalhague, Regulus, Caph) and four of them have been successfully imaged. All of the images were obtained with CHARA-MIRC at the near-IR H band. Below we highlight the results from these studies.

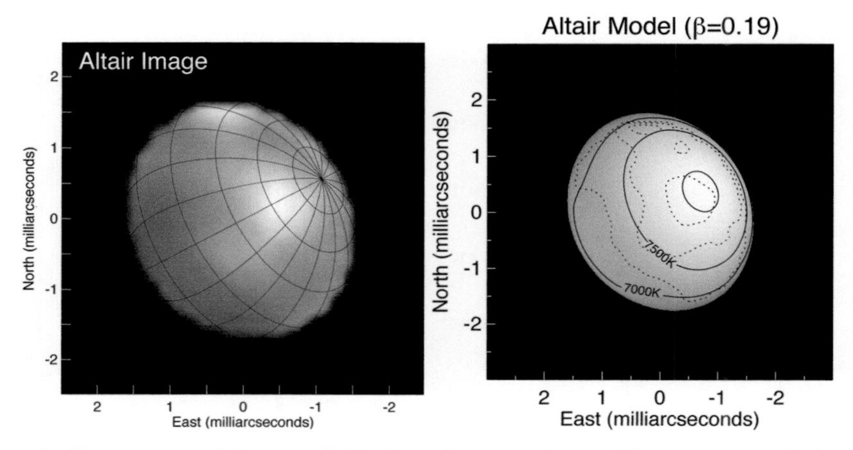

Figure 3. Reconstructed image of Altair and comparison with its gravity darkening model (Monnier *et al.* 2007). The angular resolution is $\sim$ 0.64 milliarcseconds. The latitudes and longitudes are overplotted to show the geometry of the star (left). The bright feature in the upper-most limb of the star is artifact that is beyond the resolution limit. The right panel shows the gravity darkening model for Altair, with the temperature contours from both the model (solid lines) and the image (dotted lines). The brightest spot of the stellar surface is shifted slightly to the center because of limb darkening and the fact the pole of Altair is seen at the limb.

Altair. The first image was obtained for Altair by Monnier *et al.* (2007). The left panel of Fig. 3 shows the reconstructed image. The image shows that the stellar photosphere of Altair is well-resolved and appears oblate, with a bright polar area and dark equatorial area - directly confirming the gravity darkening phenomenon. The right panel of Fig. 3 shows the gravity darkening model of Altair fitted to compare with the image. The model shows Altair is medium inclined, with an inclination of 57° and is rotating at 92.3% of its angular break-up rate. The temperature at the poles is 8450K, while that of the equator is only 6860K, suggesting the existence of convective equator and radiative poles. The model also prefers a non-standard β of 0.19. The contours of the model match with the image very well in general except at the equator where the image appears to be darker and colder.

Alderamin. The star Alderamin (α Cephei, HR 8162, V=2.46, H=2.13) was classified as an A7 IV-V star in early studies, and was recently classified as an A8V main sequence star by Gray *et al.* (2003). It is one of the few A stars (including Altair) that are found to have chromosphere activities (Walter *et al.* 1995; Simon *et al.* 2002). van Belle *et al.* (2006) studied Alderamin using the CHARA array and found it is rotating close to break-up and is elongated. Observations from Zhao *et al.* (2009) yielded an image of Alderamin and an improved model. As shown in Fig. 4, Alderamin is also elongated, with a bright pole at the bottom and a dark equator in the middle. The top of the image appears bright again due to the gravity darkening effect. The model of Alderamin suggests it is inclined by 56° and is rotating at $\sim$92.6% of its angular break-up rate. The temperature at the poles is $\sim$2400 K higher than at the equator, while its radius at the equator is 26% larger than at the poles. The non-standard β value of 0.22, and the low temperature at the equator, plus the evidence of strong chromosphere activities directly linked to convection implies the equator of Alderamin is probably also convective. This similar feature of both Alderamin and Altair indicates that rapid rotators can have both radiative and convective envelopes and a clear boundary between convective stars and radiative stars may not exist.

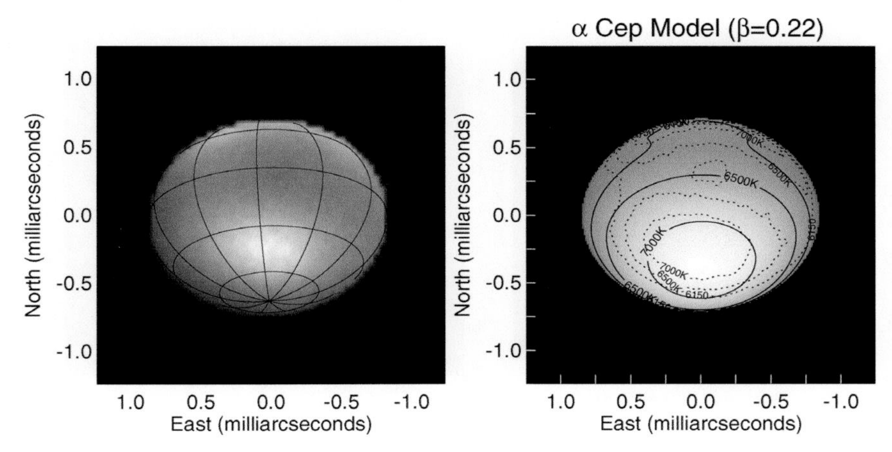

Figure 4. Reconstructed image of Alderamin and comparison with its gravity darkening model. The angular resolution is $\sim$ 0.68 milliarcseconds. The latitudes and longitudes are overplotted to show the geometry of the star. The right panel shows the corresponding gravity darkening model for Alderamin, with the temperature contours from both the model (solid lines) and the image (dotted lines). (Zhao *et al.* 2009)

Rasalhague. Rasalhague (α Ophiuchi, HR 6556, $V = 2.09$, H=1.66) is a nearby subgiant binary system (Wagman 1946). The primary is a A5IV sub-giant (Gray *et al.* 2001). Several groups have tried to study the orbit of the system (e.g., Mason *et al.* 1999, etc.), and it was lately determined to have a period of $\sim$ 8.6 yrs and a semi-major axis between 0.4" - 0.5".

Zhao *et al.* (2009) have tried extensively to reconstruct an image for Rasalhague but failed to find a reliable solution due to the fact it is seen nearly pole-on ($i = 88^o$), which makes the star nearly symmetric and thus hard to constrain by closure phases (closure phases are only sensitive to asymmetries). Figure 5 shows the model of Rasalhague. The right panel shows the degeneracy between β and the inclination angle which is also caused by the symmetric brightness distribution of Rasalhague. Fortunately, this degeneracy can be partially lifted with the aid of Vsini measurements, although the scatter of various measurements from literatures is large. The model shows that the photosphere of Rasalhague is also very elongated and has bright poles and a dark equator. Its radius at the equator is $\sim$20% larger than at the poles. It is rotating at 88.5% of its angular break-up speed and the poles are $\sim$1840 K hotter than the equator.

Regulus. Regulus was first studied by McAlister *et al.* (2005) with the CHARA array (see §5). We revisited Regulus with CHARA-MIRC and created an image along with a refined model for it (Che *et al.*, in preparation). The model is consistent with that of McAlister *et al.* (2005), showing that Regulus is also nearly equator-on ($i = 85.7^o$) and is rotating at a high angular rate of 95.3% of its breakup, causing its radius at the equator 29% larger than that of the poles. The temperature at pole is about 14300K, about 3000K hotter than that of the equator ($\sim$ 11260 K). As a late B type star with temperatures well above 11000K, Regulus should be radiative and β should be 0.25. However, our best-fit model still prefers a non-standard β of 0.18, implying that the standard von Zeipel's law may not be full-filled in this case and non-uniform rotation may exist in Regulus as well.

Caph. Caph (β Cassiopeiae, HR21, V = 2.27, H = 1.585) is a late type F2III-IV star that has evolved off the main sequence (Rhee *et al.* 2007). Its Vsini has been measured to be between 70 km s^{-1} to 83 km s^{-1}, relatively low compared to other rapid rotators. Our observations in 2007 and 2009 have also allowed us to reconstruct an image and

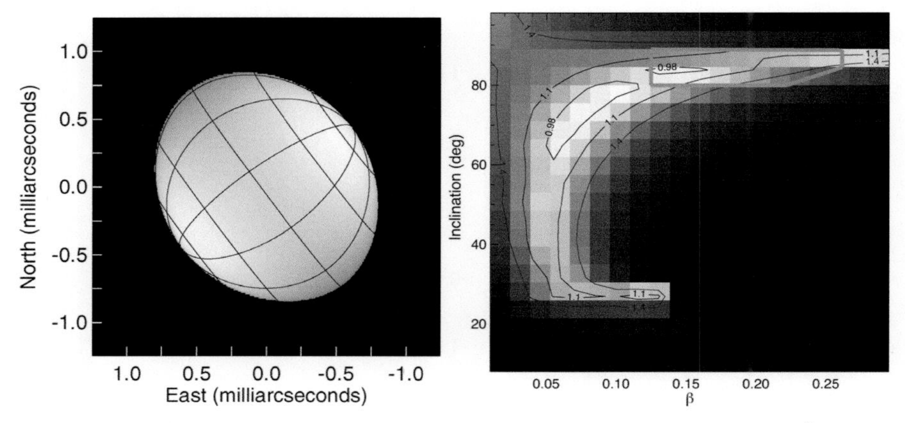

Figure 5. *Left*: The gravity darkening model of Rasalhague. *Right*: Reduced χ^2 space of inclination and β. The green box indicates the parameter space constrained with Vsini measurements.

model for it (Che *et al.*, in preparation). The results show that Caph is rotating at 93.6% of its angular breakup speed, and is seen nearly pole-on ($i = 18^o$) like Vega. Therefore, although its oblateness ratio is about 26%, its appearance on the sky is not elongated. Its polar temperature is only about 7200K and it has a β of about 0.15.

To summarize all the interferometric observations to date, we list the targets and their physical parameters in Table 1. The table shows that all the best-fit β values (Regulus, Altair, Alderamin, Caph) are lower than the standard of 0.25, suggesting the basic assumptions of the standard model may be hard to full-fill in most cases (even for some B type stars like Regulus), and non-uniform rotation is necessary to be included at this level to address the issue.

With the parameters from the gravity darkening models, we are able to calculate the true T_{eff}s and luminosities of these stars and compare with their apparent values. This also allows us to locate their true positions on the HR diagram. Figure 6 shows the HR diagram for four stars: Altair, Vega, Alderamin, and Rasalhague, and marks their apparent positions and true positions on the diagram. Non-rotation equivalents of these stars are also marked for comparison with non-rotating stellar evolution models. These figures suggest that the apparent position of a rapid rotator is strongly dependent on its inclination, i.e., the viewing angle. For a $\sim 90^o$ inclination star such as Rasalhague

Table 1. Summary of interferometric observations of rapid rotators (ranked by spectral type)

Star	inclination deg	T_{pol} K	R_{pol} $R_\odot$	T_{eq} K	R_{eq} $R_\odot$	ω %	β
Achernar[a]	50	20000	8.3	9500	12	-	0.25 (fixed)
Regulus[e]	85.7 ± 0.8	14315 ± 370	3.17 ± 0.03	11261 ± 170	4.08 ± 0.39	95.3 ± 1.2	0.178 ± 0.010
Vega[1,b]	4.7 ± 0.3	10150 ± 100	2.26 ± 0.07	7900 ± 400	2.78 ± 0.02	91 ± 3	0.25 (fixed)
Rasalhague[d]	87.7 ± 0.4	9300 ± 150	2.39 ± 0.01	7460 ± 100	2.87 ± 0.02	88.5 ± 1.1	0.25 (fixed)
Altair[c]	57.2 ± 1.9	8450 ± 140	1.63 ± 0.01	6860 ± 150	1.48 ± 0.01	92.3 ± 0.6	0.190 ± 0.012
Alderamin[d]	55.7 ± 6.2	8588 ± 300	2.16 ± 0.04	6574 ± 300	2.74 ± 0.04	94.1 ± 2.0	0.216 ± 0.021
Caph[e]	18.2 ± 3.8	7236 ± 111	2.99 ± 0.14	6058 ± 93	3.77 ± 0.17	93.6 ± 5.4	0.149 ± 0.017

References: *a*. Domiciano de Souza *et al.* (2003); *b*. Aufdenberg *et al.* (2006); *c*. Monnier *et al.* (2007); *d*. Zhao *et al.* (2009); *e*. Che *et al.*, in preparation

[1]**Note**: Peterson *et al.* (2006) have also determined consistent physical parameters for Vega, and thus are not listed here.

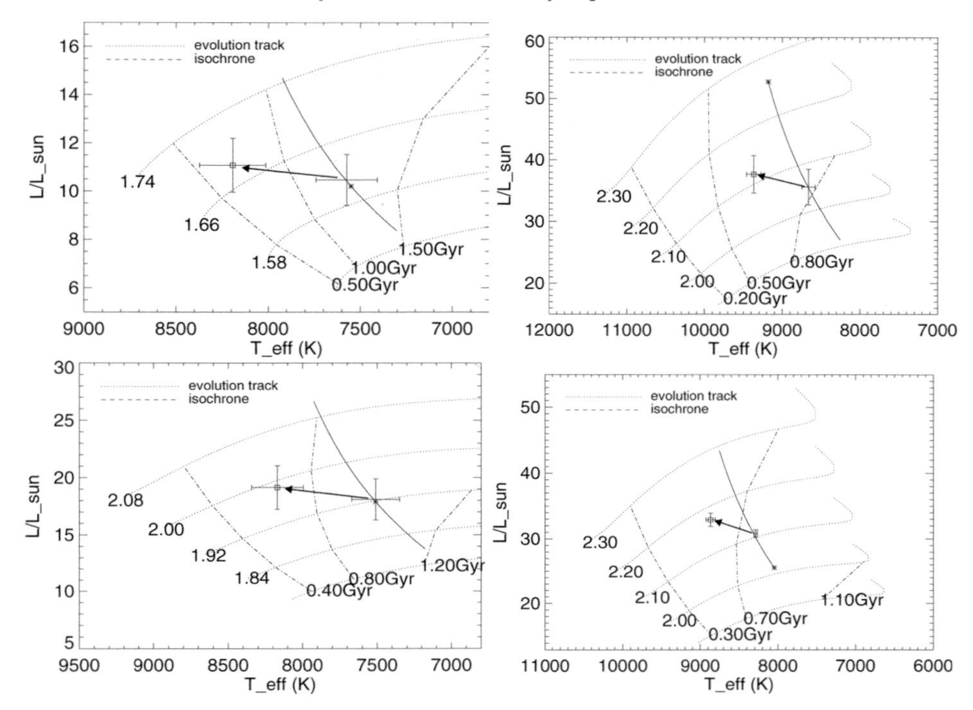

Figure 6. Locations of Altair (top left), Vega (top right), Alderamin (bottom left) and Rasalhague (bottom right) on the HR diagram. The Y^2 models (Demarque *et al.* 2004) are used to calculate the evolution tracks and the isochrones. The solid lines show the positions of each star on the H-R diagram as a function of inclination. The error bars across the inclination curves mark the true T_{eff}s and true luminosities of the stars, while the asterisks on the curves indicate their apparent T_{eff}s and luminosities which are dependent on their inclinations. Lastly, the squares with error bars in the left mark the non-rotating equivalent position for each star to allow comparison to the isochrones.

(bottom right panel), its apparent position is at the lower end of the curve; and for a 0° inclination star such as Vega (top right panel), its apparent position is at the higher end of the curves. The inclination-dependent temperature and luminosity of rapid rotators can also bring scatters and errors to the HR diagram.

7. Conclusions and Future Prospects

We have presented and discussed the interferometric observations of seven rapid rotators studied to date, ranging from spectra type B3 to F2. These measurements and images have directly confirmed the basic picture of the standard model, but have also discovered discrepancies, suggesting more sophisticated mechanisms such as non-uniform rotation are necessary to be taken into account. Overall, these observations have greatly improved our knowledge of rapid rotators, demonstrating that optical interferometry is now mature to provide valuable constraints and even model-independent images to shed light on the basic physics of stars.

For the near future, multi-wavelength interferometric observations in both near-IR and visible are already on the horizon. Since gravity darkening is much more prominent at shorter wavelengths, observations with optical combiners such as CHARA-VEGA, CHARA-PAVO, and VLTI-MATISSE will be able to study stars in unprecedented details together with valuable spectral information. The current sample size will also be increased toward fainter and smaller stars with higher angular resolution in shorter wavelengths. In addition, differential interferometry (Domiciano de Souza 2009) will also allow us to measure the differential rotation of rapid rotators and shed light on the issues of non-uniform rotation.

Interferometric observations can be even more valuable when combined with other techniques to provide multi-dimensional views of stars. The well determined geometry and rotational speed of a rapid rotator by interferometry can help to constrain models from other techniques. Such joint studies have already been applied in some spectroscopic studies (e.g., Yoon *et al.* 2008), and most recently, in asteroseimology studies as well (Monnier *et al.* 2010). More joint studies with interferometry and other techniques is a promising direction in the near future, and is expected to further advance our knowledges of the underlying physics of stellar rotation.

References

Abt, H. A., Levato, H., & Grosso, M. 2002, *ApJ*, 573, 359
Abt, H. A. & Morrell, N. I. 1995, *ApJS*, 99, 135
Aufdenberg, J. P., Mérand, A., Coudé du Foresto, V., Absil, O. *et al.* 2006, *ApJ*, 645, 664
Bohlin, R. C. & Gililand, R. L. 2004, *AJ*, 127, 3508
Burrows, A., Dessart, L., Livne, E., Ott, C. D. *et al.* 2007, *ApJ*, 664, 416
Claret, A. 2003, *A&A*, 406, 623
Cotton, W., Monnier, J., Baron, F., Hofmann, K.-H. *et al.* 2008, in: (eds.), *Society of Photo-Optical Instrumentation Engineers (SPIE) Conference Series* 7013
Cranmer, S. R. & Owocki, S. P. 1995, *ApJ*, 440, 308
Demarque, P., Woo, J.-H., Kim, Y.-C., & Yi, S. K. 2004, *ApJS*, 155, 667
Domiciano de Souza, A., Kervella, P., Jankov, S., Abe, L. *et al.* 2003, *A&A*, 407, L47
Domiciano de Souza, A., Kervella, P., Jankov, S., Vakili, F. *et al.* 2005, *A&A*, 442, 567
Domiciano de Souza, A. 2009, in: J.-P. Rozelot & C. Neiner (eds.), *The Rotation of Sun and Stars*, Lecture Notes in Physics 765 (Berlin Springer Verlag), p. 171
Frémat, Y., Zorec, J., Hubert, A.-M., & Floquet, M. 2005, *A&A*, 440, 305
Gies, D. R., Dieterich, S., Richardson, N. D., Riedel, A. R. *et al.* 2008, *ApJ* (Letters), 682, L117
Gizon, L. & Solanki, S. K. 2004, *Solar Phys.*, 220, 169
Gray, R. O., Corbally, C. J., Garrison, R. F., McFadden, M. T. *et al.* 2003, *AJ*, 126, 2048
Gray, R. O., Napier, M. G., & Winkler, L. I. 2001, *AJ*, 121, 2148
Gulliver, A. F., Hill, G. & Adelman, S. J. 1994, *ApJ* (Letters), 429, L81
Hill, G., Gulliver, A. F. & Adelman, S. J. 2004, in: J. Zverko, J. Ziznovsky, S. J. Adelman, & W. W. Weiss (eds.), *The A-Star Puzzle*, IAU Symposium 224, p. 35
Hill, G., Gulliver, A. F., & Adelman, S. J. 2010, *ApJ*, 712, 250
Högbom, J. A. 1974, *A&AS*, 15, 417
Ireland, M. J., Monnier, J. D., & Thureau, N. 2006, in: (eds.), *Society of Photo-Optical Instrumentation Engineers (SPIE) Conference Series* 6268
Jackson, S., MacGregor, K. B., & Skumanich, A. 2004, *ApJ*, 606, 1196
Jankov, S., Vakili, F., Domiciano de Souza, Jr., A., & Janot-Pacheco, E. 2001, *A&A*, 377, 721
Johnson, H. L. & Morgan, W. W. 1953, *ApJ*, 117, 313
Kanaan, S., Meilland, A., Stee, P., Zorec, J. *et al.* 2008, *A&A*, 486, 785
Kervella, P. & Domiciano de Souza, A. 2006, *A&A*, 453, 1059
Kurucz, R. L. 1993, VizieR Online Data Catalog, 6039
Lucy, L. B. 1967, *ZfA*, 65, 89
MacFadyen, A. I., Woosley, S. E., & Heger, A. 2001, *ApJ*, 550, 410
Maeder, A. & Meynet, G. 2000, *ARAA*, 38, 143
Maeder, A., Meynet, G., & Ekström, S. 2007, in: A. Vallenari, R. Tantalo, L. Portinari, & A. Moretti (eds.), *From Stars to Galaxies: Building the Pieces to Build Up the Universe*, ASP-CS 374, p. 13

MacGregor, K. B., Jackson, S., Skumanich, A., & Metcalfe, T. S. 2007, *ApJ*, 663, 560
Malbet, F., *et al.* 2010, in: W. C. Danchi, F. Delplancke & J. K. Rajagopal (eds.), *Optical and Infrared Interferometry*, SPIE Conference Series 7734, 138
Mason, B. D., Martin, C., Hartkopf, W. I., Barry, D. J. *et al.* 1999, *AJ*, 117, 1890
McAlister, H. A., ten Brummelaar, T. A., Gies, D. R., Huang, W. *et al.* 2005, *ApJ*, 628, 439
Millward, C. G. & Walker, G. A. H. 1985, *ApJS*, 57, 63
Monnier, J. D. 2003, *Reports on Progress in Physics*, 66, 789
Monnier, J. D., Berger, J.-P., Millan-Gabet, R., & ten Brummelaar, T. A. 2004, in: W. A. Traub (eds.), *New Frontiers in Stellar Interferometry*, SPIE Conference Series 5491, p. 1370
Monnier, J. D., Zhao, M., Pedretti, E., Thureau, N. *et al.* 2007, *Science*, 317, 342
Monnier, J. D., Townsend, R. H. D., Che, X., Zhao, M. *et al.* 2010, *ApJ*, 725, 1192
Narayan, R. & Nityananda, R. 1986, *ARAA*, 24, 127
Ohishi, N., Nordgren, T. E., & Hutter, D. J. 2004, *ApJ*, 612, 463
Petrov, R. G., Malbet, F., Weigelt, G., Antonelli, P. *et al.* 2007, *A&A*, 464, 1
Peterson, D. M., Hummel, C. A., Pauls, T. A., Armstrong, J. T. *et al.* 2006, *Nature*, 440, 896
Pinsonneault, M. 1997, *ARAA*, 35, 557
Rhee, J. H., Song, I., Zuckerman, B., & McElwain, M. 2007, *ApJ*, 660, 1556
Simon, T., Ayres, T. R., Redfield, S., & Linsky, J. L. 2002, *ApJ*, 579, 800
Thiébaut, E. 2008, in: M. Schöller, W. C. Danchi & F. Delplancke (eds.), *Optical and Infrared Interferometry*, SPIE Conference Series 7013, p. 43
Thompson, A. R., Moran, J. M., & Swenson, G. W. 1986, *Interferometry and synthesis in radio astronomy*, New York, Wiley-Interscience
van Belle, G. T., Ciardi, D. R., ten Brummelaar, T., McAlister, H. A. *et al.* 2006, *ApJ*, 637, 494
van Belle, G. T., Ciardi, D. R., Thompson, R. R., Akeson, R. L. *et al.* 2001, *ApJ*, 559, 1155
Vinicius, M. M. F., Zorec, J., Leister, N. V., & Levenhagen, R. S. 2006, *A&A*, 446, 643
von Zeipel, H. 1924a, *MNRAS*, 84, 665
von Zeipel, H. 1924b, *MNRAS*, 84, 684
Wagman, N. E. 1946, *AJ*, 52Q, 50
Walter, F. M., Matthews, L. D., & Linsky, J. L. 1995, *ApJ*, 447, 353
Yoon, J., Peterson, D. M., Zagarello, R. J., Armstrong, J. T. *et al.* 2008, *ApJ*, 681, 570
Zahn, J.-P., Ranc, C., & Morel, P. 2010, *A&A*, 517, A7+
Zhao, M., Gies, D., Monnier, J. D., Thureau, N. *et al.* 2008, *ApJ* (Letters), 684, L95
Zhao, M., Monnier, J. D., Pedretti, E., Thureau, N. *et al.* 2009, *ApJ*, 701, 209

Discussion

HUANG: I have a question on Regulus 2010 results. Is this imaging technique accurate enough to constrain the β value? There are several parameters entangled here, $v \sin i$, mass, β.

ZHAO: The β values are constrained by the gravity darkening models. Under the assumption of the Roche - von Zeipel law, they can be constrained pretty well as long as the stars are well resolved and not symmetric (i.e., not pole-on or equator on).

GIES: Observers use both critical velocity of Ω/Ω_{crit} and V_{eq}/V_{crit}. These will be different (reason Huang was concerned about the result for Regulus). And the one used here is the angular velocity.

TYCNER: Are the models you show fitted/constrainted by the reconstructed images or the actual data? If fitted/constrained by the data why not fitted/constrain by the image?

ZHAO: The models I showed are fitted by the actual data. It is always better to fit directly to the data than to the images. The images are not exact representations of the stars and, strictly speaking, they are also parameter-free models.

Active OB stars: structure, evolution, mass loss, and critical limits
Proceedings IAU Symposium No. 272, 2010
C. Neiner, G. Wade, G. Meynet & G. Peters, eds.

doi:10.1017/S1743921311009975

Effects of fast rotation on the wind of Luminous Blue Variables

Jose H. Groh

Max-Planck-Institut für Radioastronomie, Auf dem Hügel 69, D-53121 Bonn, Germany
email: jgroh@mpifr.de

Abstract. While theoretical studies have long suggested a fast-rotating nature of Luminous Blue Variables (LBVs), observational confirmation of fast rotation was not detected until recently. Here I discuss the diagnostics that have allowed us to constrain the rotational velocity of LBVs: broadening of spectral lines and latitude-dependent variations of the wind density structure. While rotational broadening can be directly detected using high-resolution spectroscopy, long-baseline near-infrared interferometry is needed to directly measure the shape of the latitude-dependent photosphere that forms in a fast-rotating star. In addition, complex 2-D radiative transfer models need to be employed if one's goal is to constrain rotational velocities of LBVs. Here I illustrate how the above methods were able to constrain the rotational velocities of the LBVs AG Carinae, HR Carinae, and Eta Carinae.

Keywords. stars: rotation, stars: atmospheres, stars: emission-line, techniques: interferometric

1. Introduction

Tremendous advancement on understanding the evolution of massive stars has been achieved in the last decades (e. g., Maeder & Meynet 1994), in particular after the inclusion of the effects of rotation in evolutionary models (Maeder & Meynet 2000b). These models predict the existence of a short-lived, transitional stage, usually referred to as the Luminous Blue Variable (LBV) phase (Conti 1984), when the star has a high mass-loss rate ($\dot{M} \sim 10^{-5} - 10^{-3}\ M_{\odot}\,\mathrm{yr}^{-1}$). In the current picture of stellar evolution, LBVs are rapidly evolving massive stars in the transitory phase from being an O-type star to becoming a Wolf-Rayet star (Humphreys & Davidson 1994; Meynet & Maeder 2003).

Although the effects of rotation on classical O-type stars have been documented for almost a century, their more active descendants, the Luminous Blue Variables, have eluded direct detection of fast rotation until recently, when the prototype LBV AG Carinae was found to be a fast rotator (Groh *et al.* 2006). Obtaining the rotational velocity (v_{rot}) of LBVs is challenging because of the presence of (a) an optically-thick wind caused by the large mass-loss rate of the star and (b) time variability of the stellar radius due to the S-Dor type variability.

The presence of a dense wind causes the formation of an extended photosphere, making the use of a radiative transfer code mandatory to obtain information about the hydrostatic core. Time variability caused by the S-Dor cycles, characterized by changes of up to a factor of 10 in the hydrostatic radius of LBVs, imply variations of v_{rot} as a function of time. The issue is that LBVs are apparently closer to the critical rotational velocity for break-up (v_{crit}) only during visual minimum epochs of the S-Dor cycle (Groh *et al.* 2006; Groh *et al.* 2010, in prep.). Unfortunately, LBVs spend most of their time at visual maximum, requiring continuous spectroscopic and photometric monitoring (and a great deal of patience!) to catch the star at the right time and obtain their *maximum* value of v_{rot}.

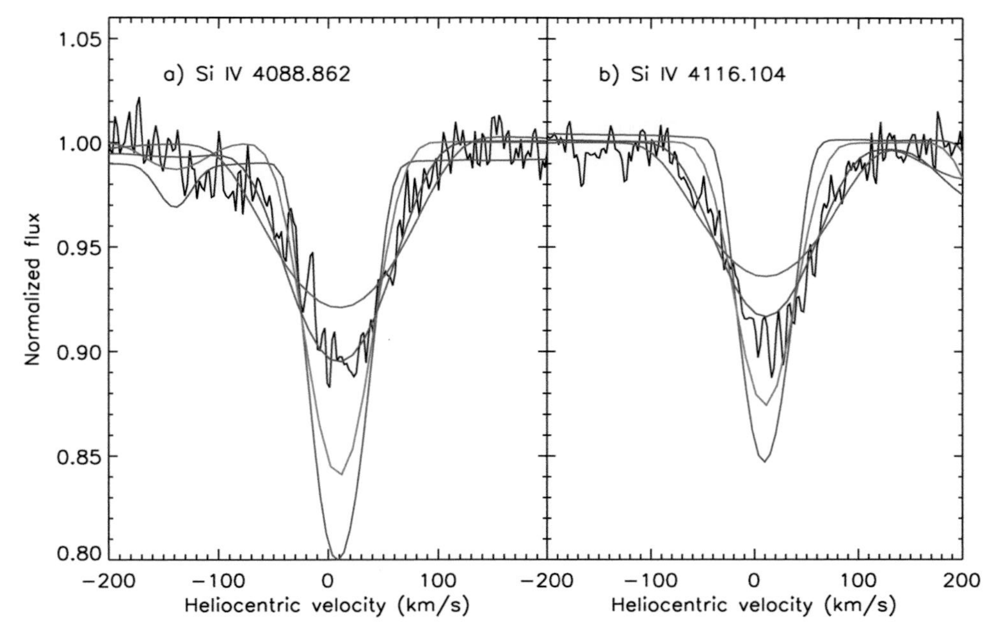

Figure 1. Comparison between the Si IV λλ 4088–4116 line profiles of HR Car observed on 2009 May 3rd with synthetic model spectra computed with the 2-D code from Busche & Hillier (2005) for different rotational velocities, assuming an inclination angle of $i = 30°$: 0 km s^{-1} (green), 75 km s^{-1} (red), 150 km s^{-1} (blue), and 200 km s^{-1} (purple). From Groh *et al.* (2009a).

Here I discuss the diagnostics that have allowed us to constrain the rotational velocity of LBVs: broadening of spectral lines and latitude-dependent variations of the wind density structure.

2. Rotational broadening in LBVs

As in WR stars, the high mass-loss rate of LBVs also cause a ionization stratification of the wind, and spectral lines from ions of different ionization potentials will form at different distances r from the star. The most common lines present in the ultraviolet and optical spectra of LBVs, such as Hα and other Balmer lines, He I, N II, Fe II, and Si III lines, are formed at very large r. Since the azimuthal velocity component is proportional to r^{-1}, these lines do not show any evidence of rotational broadening, explaining why this phenomenon went undetected for so long.

Lines from high-ionization species, which are formed closer to the stellar photosphere, need to be observed to detect rotational broadening. Thus, in addition to observing LBVs at the right time, one has to look at the right lines. It turns out that Si IV $\lambda\lambda 4088-4116$ lines are the highest ionization lines present in the optical spectrum of most LBVs and, therefore, should be formed closest to the stellar photosphere and are ideal for deriving $v_{\rm rot}$ in LBVs.

To obtain realistic values of $v_{\rm rot}$, a 2-D radiative transfer model which takes into account the dynamical effects of rotation on the wind velocity field are required. We used an updated version of the 2-D code from Busche & Hillier (2005) to constrain the rotational velocities of the LBVs AG Carinae (Groh *et al.* 2006) and HR Carinae (Groh *et al.* 2009a). Figure 1 shows the Si IV 4088 line profiles of HR Car observed during visual minimum (2009 May), compared to synthetic model spectra. Assuming an inclination angle of $i = 30°$, we derived $v_{\rm crit} \simeq 170 \pm 20$ km s^{-1}.

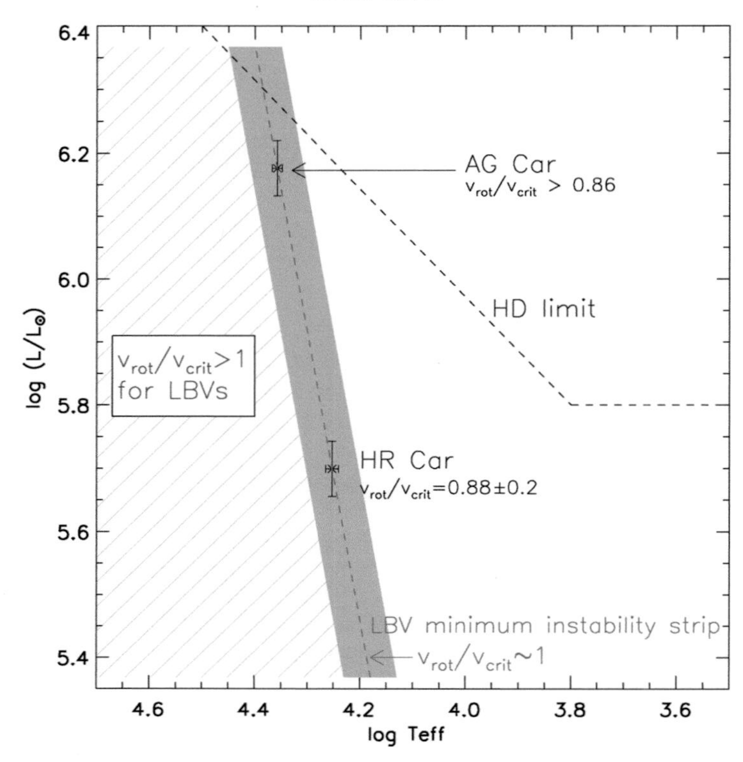

Figure 2. HR diagram showing the position of the LBVs AG Car and HR Car during visual minimum (Groh *et al.* 2009a,b).The location of the forbidden region for strong-variable LBVs (green hatched region) and the Humphreys-Davidson limit (black dashed line, Humphreys & Davidson 1994) are shown. From Groh *et al.* (2009a).

How large is $v_{\rm rot}$ compared to $v_{\rm crit}$ for HR Car? Assuming an admittedly uncertain mass of $M = 25\ M_{\odot}$, our CMFGEN model of HR Car predicts an Eddington parameter of $\Gamma \simeq 0.80$, and we can derive $v_{\rm rot}/v_{\rm crit} \simeq 0.88 \pm 0.2$. Therefore, similar to AG Car , for which we found $v_{\rm rot}/v_{\rm crit} > 0.86$ (Groh *et al.* 2006), HR Car is also likely rotating very close to, if not *at*, its critical velocity for break-up at visual minimum.

We have proposed that rotation plays a key role in determining the location of LBVs in the HR diagram during visual minimum. Figure 2 presents the position of AG Car and HR Car in the HR diagram based on their updated stellar parameters (Groh *et al.* 2009a,b). Previous works have recognized that, during visual minimum, LBVs are located in a well-defined region of the HR diagram, the so-called "LBV minimum instability strip" (Wolf 1989; Clark *et al.* 2005). Our results suggest that the LBV minimum instability strip is considerably steeper than previously determined, being characterized by $\log{(L/L_{\odot})} =$ 4.54 $\log{(T_{\rm eff}/{\rm K})}$ - 13.61 (red dashed line in Fig. 2). More importantly, we suggest that the LBV minimum instability strip corresponds to the region where critical rotation is reached for LBVs with strong S-Dor type variability. When LBVs are evolving towards maximum, the star moves away from the LBV minimum instability strip (to the right in the HR diagram), and $v_{\rm rot}/v_{\rm crit}$ decreases considerably (Groh *et al.* 2006). The region in the HR diagram on the left side of the LBV minimum instability strip would be populated by unstable LBV stars with $v_{\rm rot}/v_{\rm crit} > 1$, which would make it a "forbidden region" for LBVs. Indeed, no confirmed strong-variable LBV is seen in this region (van Genderen 2001; Clark *et al.* 2005).

3. Latitude-dependent variations of the wind density structure

A second way of constraining $v_{\rm rot}$ in LBVs is by probing the latitudinal dependence of the wind density structure, which is thought to occur when rotation is significant (e. g., Owocki *et al.* 1996). If a star is surrounded by a reflection nebula with a well-determined 3-D geometry, the latitudinal variation of the wind density can in principle be probed through the reflected stellar spectrum on the nebula. This interesting technique was applied to reflected spectra in the Homunculus nebula around Eta Car, suggesting that the star possesses a faster and denser polar wind (Smith *et al.* 2003). However, such a configuration is unfortunately not seen in many LBVs, and different techniques need to be employed.

Another neat way of probing latitudinal variations of the wind is by *directly* resolving the geometry of the wind photosphere, which is most easily achieved in the K-band using near-infrared long-baseline interferometry. This technique was applied to Eta Car using observations gathered with the VINCI (van Boekel *et al.* 2003) and AMBER (Weigelt *et al.* 2007) instruments on the Very Large Telescope Interferometer from the European Southern Observatory. The interferometric measurements are consistent with an ellipsoidal shape projected on the sky. Both papers analyzed the measurements using simple geometrical models and interpreted their results as evidence for a dense prolate wind generated by fast rotation, as theoretically predicted by Owocki *et al.* (1996, 1998).

In Groh *et al.* (2010) we analyzed the published K-band continuum interferometric data of Eta Car, with the goal of constraining $v_{\rm rot}$ and the spatial orientation of the rotation axis of Eta Car based on the effects of rotation on the wind density structure. This ultimately determines the geometry of the K-band emitting region. We computed 2-D latitude-dependent wind models for the primary star (hereafter $\eta_{\rm A}$) in Eta Car using an updated version of the 2-D radiative transfer code of Busche & Hillier (2005) and via a methodology similar to that described in Groh *et al.* (2008). The reader is referred to Busche & Hillier (2005), Groh *et al.* (2006, 2008, 2009a, 2010), and Driebe *et al.* (2009) for further details on the code and its applications.

The 2-D code allows for the specification of any arbitrary latitude-dependent variation of the wind density ρ and v_∞. For the latitudinal variation of ρ, we adopted the predictions from Owocki *et al.* (1998) for gravity-darkened (GD) line-driven prolate winds,

$$\frac{\rho(\theta)}{\rho_0} \propto \sqrt{1 - W^2 \sin^2 \theta} \ , \qquad (3.1)$$

where θ is the colatitude angle (0°= pole, 90°= equator), ρ_0 is the density at the pole, and W is the ratio of $v_{\rm rot}$ to its critical velocity ($v_{\rm crit}$). For consistency, we also assumed the Owocki *et al.* (1998) predictions for $v_\infty(\theta)$. We assumed the same parameters of $\eta_{\rm A}$ derived by Hillier *et al.* (2001): stellar temperature $T_\star$ = 35,310 K (at Rosseland optical depth $\tau_{\rm Ross}$ = 150), effective temperature $T_{\rm eff}$= 9,210 K (at $\tau_{\rm Ross}$ = 2/3), luminosity $L_\star = 5 \times 10^6\ L_\odot$, $\dot{M} = 10^{-3}\ M_\odot\,{\rm yr}^{-1}$, wind terminal velocity v_∞ = 500 km s^{-1}, a clumping volume-filling factor $f = 0.1$, and distance $d = 2.3$ kpc. The atomic model and abundances are described in Hillier *et al.* (2001).

We calculated a grid of prolate models with the inclination angle i ranging from 0° to 90° in steps of 1°, and $W = v_{\rm rot}/v_{\rm crit}$ ranging from 0 to 0.99 in steps of 0.01. For a desired position angle (PA) orientation on the sky, the K-band image, visibilities, and closure phases (CP) were computed. Figure 3 shows the reduced χ^2 values of the fit of the model predictions to the VLTI/VINCI visibilities as a function of W and i, for prolate models with PA = 130°. Being conservative, models with CP $\geqslant$ 5° are ruled out, since Weigelt *et al.* (2007) measured CP = $0 \pm 2°$. Thus, to fit simultaneously the

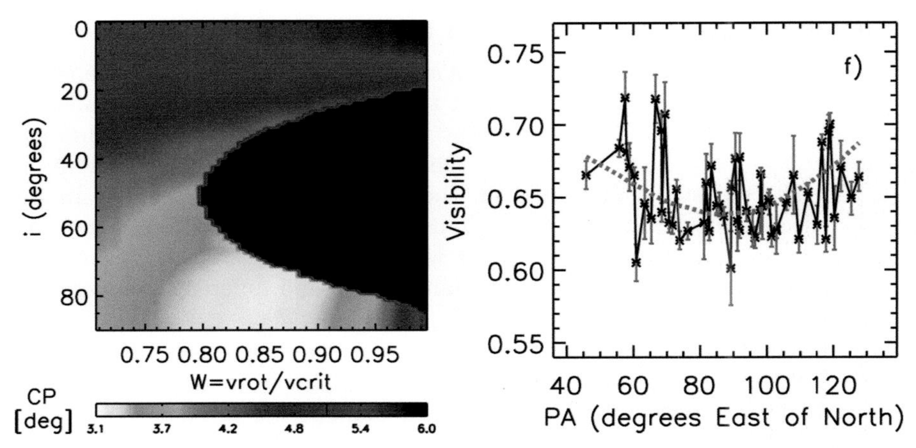

Figure 3. *Left panel:* χ^2 values of the fit of the VINCI visibilities rejecting models with CP $\geqslant 5°$, as a function of W and i, for prolate models with PA = 130° *Right:* VINCI visibilities for the 24 m baseline as a function of telescope PA (connected black asterisks) compared to one of the best fit models, with $W = 0.92$, $i = 80°$, and PA = 108° (red dotted line). Note that the projected baseline length of the VINCI measurements changes as a function of PA.

VLTI/VINCI visibilities and the VLTI/AMBER CPs, $W = 0.77$–0.92 and $i = 60° - 90°$ are needed (Figure 3). Strikingly, this inclination is significantly higher than that of the Homunculus ($i_{\rm Hom} = 41°$; Smith 2006), which has been assumed so far to align with the *current* rotation axis of $\eta_{\rm A}$. We find that, based on the available interferometric data, if $\eta_{\rm A}$ has a prolate wind, its rotation axis is not necessarily aligned with the Homunculus polar axis.

4. Short Summary and Future Prospects

Constraining the rotational velocity of LBVs is crucial for understanding how mass and angular momentum are lost during this short, unstable evolutionary phase, and how that can possibly affect the fate of a massive star. Here we have discussed the two diagnostics that have allowed us to obtain, for the first time, rotational velocities for the most famous and well-observed LBVs such as AG Car, HR Car, and Eta Car. The two diagnostics are complementary, given that rotational broadening requires observations of high-ionization lines which are usually present in the blue region of the optical spectra. Effects of rotation on the wind structure, on the other hand, can be currently probed in the near-infrared, allowing observations of obscured stars.

The comparison between the value of $v_{\rm rot}$ obtained using these diagnostics might allow us to verify whether the LBVs indeed have denser, faster polar winds as theoretically predicted, and to obtain inclination angles. Indeed, we are analyzing a large dataset of interferometric observations of LBVs in the near-infrared for which the determination of the rotational velocity seems promising.

Acknowledgments

This paper would not be possible if I had not benefited from a very fruitful collaboration and continuous support from John Hillier. This work is an excellent example of a joint effort by observers and theoreticians that has characterized the massive star field in the last years, and I am more than happy to thank Augusto Damineli, Tom Madura, Stan Owocki, Gerd Weigelt, Rodolfo Barba, Eduardo Fernandez-Lajus, Roberto Gamen,

Alessandro Moises, Gladys Solivella, and Mairan Teodoro, for sharing their knowledge with me. I also acknowledge financial support from the Max-Planck-Gesellschaft.

References

Busche, J. R. & Hillier, D. J. 2005, *AJ*, 129, 454
Clark, J. S., Larionov, V. M., & Arkharov, A. 2005, *A&A*, 435, 239
Conti, P. S. 1984, in: A. Maeder & A. Renzini (eds.), *Observational Tests of the Stellar Evolution Theory*, IAU Symposium 105, p. 233
Driebe, T., Groh, J. H., Hofmann, K.-H., Ohnaka, K. *et al.* 2009, *A&A*, 507, 301
Groh, J. H., Damineli, A., Hillier, D. J., Barbá, R. *et al.* 2009a, *ApJ* (Letters), 705, L25
Groh, J. H., Hillier, D. J., & Damineli, A. 2006, *ApJ* (Letters), 638, L33
Groh, J. H., Hillier, D. J., Damineli, A., Whitelock, P. A. *et al.* 2009b, *ApJ*, 698, 1698
Groh, J. H., Madura, T. I., Owocki, S. P., Hillier, D. J. *et al.* 2010, *ApJ* (Letters), 716, L223
Groh, J. H., Oliveira, A. S., & Steiner, J. E. 2008, *A&A*, 485, 245
Hillier, D. J., Davidson, K., Ishibashi, K., & Gull, T. 2001, *ApJ*, 553, 837
Humphreys, R. M. & Davidson, K. 1994, *PASP*, 106, 1025
Maeder, A. & Meynet, G. 1994, *A&A*, 287, 803
Maeder, A. & Meynet, G. 2000a, *A&A*, 361, 159
Maeder, A. & Meynet, G. 2000b, *ARAA*, 38, 143
Meynet, G. & Maeder, A. 2003, *A&A*, 404, 975
Owocki, S. P., Cranmer, S. R., & Gayley, K. G. 1996, *ApJ* (Letters), 472, L115
Owocki, S. P., Cranmer, S. R., & Gayley, K. G. 1998, *Ap&SS*, 260, 149
Smith, N. 2006, *ApJ*, 644, 1151
Smith, N., Davidson, K., Gull, T. R., Ishibashi, K. *et al.* 2003, *ApJ*, 586, 432
van Boekel, R., Kervella, P., Schöller, M., Herbst, T. *et al.* 2003, *A&A*, 410, L37
van Genderen, A. M. 2001, *A&A*, 366, 508
Weigelt, G., Kraus, S., Driebe, T., Petrov, R. G. *et al.* 2007, *A&A*, 464, 87
Wolf, B. 1989, *A&A*, 217, 87

Discussion

MEYNET: I have a question regarding the HR diagram where the forbidden region is in the left of the LBV minimum instability strip. I would imagine that the star evolves from the blue to red, encounters the instability, suffers strong mass loss which brings the stars to the right because of the optical thickness of the wind.

GROH: The optical thickness of the wind is included in our calculations for the positions of the LBVs in the HR diagram. Apparently, the stars are very close to the Eddington limit modified by rotation at the LBV minimum instability strip, but in the case of AG Car and HR Car we do not see a runaway increase in mass loss, which is actually comparable to other LBVs. That is related to the very important issue of how much rotation increases the mass-loss rate and certainly deserves future studies.

PULS: 1. You referred to the critical velocity for break-up. Since you are dealing with LBVs (Γ close to unity), did you include the effects of the Eddington limit modified by rotation ($\Omega\Gamma$ limit)? 2. Which kind of models are underlying your analysis?

GROH: First, yes, we include the effects of rotation on the Eddington limit following Maeder & Meynet (2000a). In these analyses we have used an updated version of the 2-D code from Busche & Hillier (2005) using a prescribed latitudinal variation of the wind density as a function of $v_{\rm rot}$ as in Owocki *et al.* (1996).

Active OB stars: structure, evolution, mass loss, and critical limits
Proceedings IAU Symposium No. 272, 2010
C. Neiner, G. Wade, G. Meynet & G. Peters, eds.

doi:10.1017/S1743921311009987

Massive stellar models: rotational evolution, metallicity effects

Sylvia Ekström[1], Cyril Georgy[1], Georges Meynet[1], André Maeder[1], and Anahí Granada[1,2]

[1]Geneva Observatory, University of Geneva
Maillettes 51 - Sauverny, CH-1290 Versoix, Switzerland

[2]Instituto de Astrofísica de La Plata, Universidad Nacional de La Plata,
Paseo del Bosque S/N, La Plata, Buenos Aires, Argentina

Abstract. The Be star phenomenon is related to fast rotation, although the cause of this fast rotation is not yet clearly established. The basic effects of fast rotation on the stellar structure are reviewed: oblateness, mixing, anisotropic winds. The processes governing the evolution of the equatorial velocity of a single star (transport mechanisms and mass loss) are presented, as well as their metallicity dependence. The theoretical results are compared to observations of B and Be stars in the Galaxy and the Magellanic Clouds.

Keywords. stars: evolution, stars: rotation, stars: winds, outflows, stars: mass loss, stars: emission-line, Be

1. Introduction

Why care about rotation? Just because stars do rotate! A look at the velocity distribution established by Huang & Gies (2006) reveals that the peak in the probability density occurs at $\upsilon_{\rm eq} \simeq 200$ km/s, which represents a ratio $\upsilon/\upsilon_{\rm crit} \simeq 0.5 - 0.6$, *i.e.* a substantial fraction of the keplerian velocity.

The effects of rotation in stars were studied since the works of von Zeipel and Eddington in the years 1924-1925. In the end of the 60s, they were included in polytropic or simplified stellar models (Roxburgh *et al.* 1965; Roxburgh & Strittmatter 1966; Faulkner *et al.* 1968; Kippenhahn & Thomas 1970; Endal & Sofia 1976). About 30 years later, stellar models became more sophisticated and also benefited from the inclusion of rotational effects (Pinsonneault *et al.* 1989; Deupree 1990; Fliegner & Langer 1994; Chaboyer *et al.* 1995; Meynet 1996).

Since the end of the 90s, more or less extended grids of rotating stellar models were computed (Langer *et al.* 1997; Meynet & Maeder 1997; Siess & Livio 1997; Heger *et al.* 2000). Those grids showed that the inclusion of the effects of rotation improved the adequation between models and massive stars observations in many aspects:

- the surface abundances of light elements (Heger & Langer 2000; Meynet & Maeder 2000);
- the predicted surface velocities in clusters (Martayan *et al.* 2006a,b; Meynet & Maeder 2000);
- the blue- to red-supergiants number ratio in the SMC (Maeder & Meynet 2001);
- the WR populations number with metallicity (Meynet & Maeder 2003, 2005; Vink & de Koter 2005);
- the rotation rates of pulsars (when strong coupling is assumed, Heger *et al.* 2005);
- the SN types and GRB progenitors (Meynet & Maeder 2005; Yoon *et al.* 2006; Georgy *et al.* 2009).

Roughly summarised, rotating stars are expected to present

(*a*) a modified gravity:

- the surface characteristics become dependent on the colatitude considered;
- there is a mass loss enhancement and anisotropy;

(*b*) chemical species and angular momentum transport mechanisms:

- the behaviour on the HR diagram is modified;
- the nucleosynthesis is altered;
- there is a surface enrichment;
- the mass loss is modified;
- the rotation profile evolves during the star life, becoming steeper (when no magnetic fields are considered).

We will go through the different points in this review.

2. Surface characteristics

2.1. *Gravity and shape*

Because of rotation, the effective gravity is modified, becoming a function of the rotation velocity Ω and of the colatitude θ of the star:

$$\vec{g}_{\rm eff} = \vec{g}_{\rm eff}(\Omega, \theta) = \left(-\frac{GM}{r^2} + \Omega^2 r \sin^2\theta \right) \vec{e}_r + \Omega^2 r \sin\theta \cos\theta \, \vec{e}_\theta .$$

We immediately see that at the pole ($\theta = 0^o$) the effective gravity is just the gravitation acceleration $-GM/r^2$. At the equator ($\theta = 90^o$) the centrifugal force adds a sustaining term $\Omega^2 r \sin^2\theta$. In these two cases, the effective gravity is still radial, while at intermediate θ, the term $\Omega^2 r \sin\theta \cos\theta$ does not vanish and implies that the effective gravity is no more radial.

In the frame of the Roche model, the maximal oblateness allowed when the star rotates at the critical velocity† is $R_{\rm eq,crit} = 1.5\, R_{\rm pol,crit}$.

Recently, interferometry has allowed to determine the deformation of fast rotating stars. A first evaluation of the oblateness of Achernar by Domiciano de Souza *et al.* (2003) showed a larger ratio $R_{\rm eq}/R_{\rm pol}$ than the one allowed in the Roche model. However, more recent observations of the same star (Vinicius *et al.* 2006; Carciofi *et al.* 2008) have revised this ratio and found a lower value, more compatible with these theoretical expectations.

2.2. *Flux and effective temperature*

Since the flux is related to the effective gravity (von Zeipel 1924; Owocki *et al.* 1996, 1998; Maeder 1999), it becomes also dependent on colatitude:

$$\vec{F} = \vec{F}(\Omega, \theta) \simeq -\frac{L}{4\pi\, G\, M^\star}\, \vec{g}_{\rm eff}(\Omega, \theta)$$

with $M^\star = M \left(1 - \frac{\Omega^2}{2\pi\, G\, \rho_m}\right)$ the so-called *reduced mass* which takes into account the reduction of the gravitational potential by rotation ($\rho_{\rm m}$ is the mean density inside the considered isobar).

† The critical velocity is reached when $\vec{g}_{\rm eff}$ vanishes because the centrifugal force at the equator counterbalances the gravity exactly (see Section 3.2).

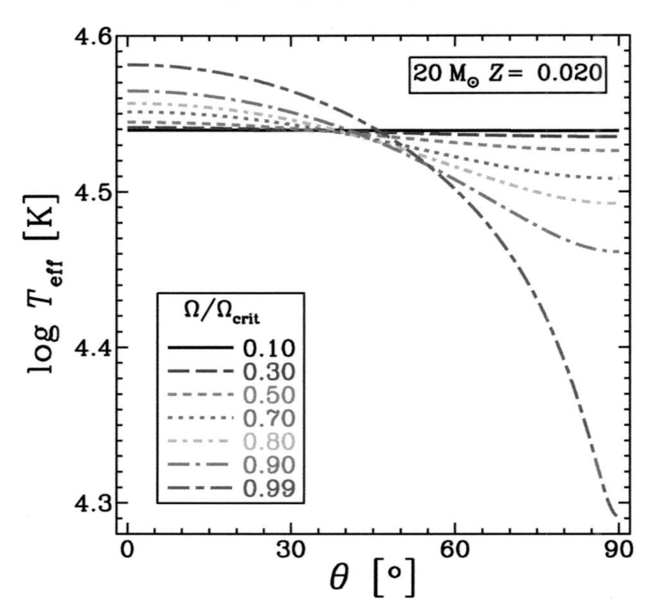

Figure 1. Variation of the effective temperature with the colatitude for various initial rotation rates (figure from Ekström *et al.* 2008).

The Stefan-Boltzmann law $F = \sigma T^4$ implies a dependence on the colatitude for the effective temperature as well:

$$T_{\text{eff}} = T_{\text{eff}}(\Omega, \theta) = \left[\frac{L}{4\pi\sigma \, G \, M^{\star}} \, \vec{g}_{\text{eff}}(\Omega, \theta) \right]^{1/4} .$$

Recent interferometric observations, like for example Monnier *et al.* (2007) on Altair or Zhao *et al.* (2009) on α Cephei and α Ophiuchi, have provided a possibility to test this relation. Let's have a close look at Altair: the rotation rate $\Omega/\Omega_{\text{crit}}$ of this star is evaluated between 0.90 and 0.92, and the temperature difference between the pole and the equator is between 1.19 and 1.32. For such a rotation rate, the theoretical models (cf. Fig. 1) predict a difference of 1.26-1.33, in good agreement with the observational value.

2.3. *Mass loss*

According to the works of Owocki & Gayley (1997), Maeder & Meynet (2000) and Petrenz & Puls (2000), rotation enhances the mass loss by a factor:

$$\frac{\dot{M}(\Omega)}{\dot{M}(0)} = \left[\frac{(1 - \Gamma_{\text{Edd}})}{\left(1 - \frac{\Omega^2}{2\pi \, G \, \rho_m} - \Gamma_{\text{Edd}}\right)} \right]^{\frac{1}{\alpha}-1}$$

where Γ_{Edd} is the Eddington factor, *i.e.* the ratio of the luminosity of the star to the Eddington luminosity $L_{\text{Edd}} = \frac{\kappa_{\text{s}}}{4\pi \, c \, G \, M}$, with κ_{s} the electron-scattering opacity.

It also changes the geometry of the mass flux (Maeder 2002, 2009), which is no longer constant on the whole surface of the star. The mass loss by surface unit at a given latitude θ follows the relation:

$$\frac{\text{d}\dot{M}(\theta)}{\text{d}\sigma} \sim A(\alpha, k) \left(\frac{L}{4\pi \, G \, M^{\star}} \right)^{\frac{1}{\alpha}-\frac{1}{8}} \frac{g_{\text{eff}}(\theta)^{1-\frac{1}{8}}}{(1 - \Gamma_{\Omega}(\theta))^{\frac{1}{\alpha}-1}}$$

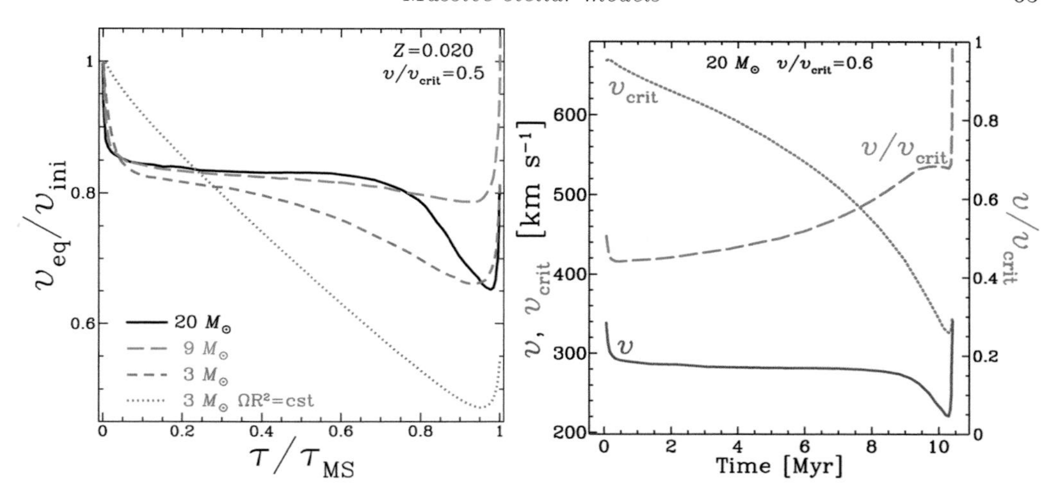

Figure 2. *Left:* evolution of the equatorial velocity (normalised to the initial velocity) during the main sequence for different mass domains. *Right:* evolution of the equatorial velocity, the critical velocity and the ratio $v/v_{\rm crit}$ during the main sequence.

where $A(\alpha, k) = (k\alpha)^{1/\alpha}\left(\frac{1-\alpha}{\alpha}\right)^{(1-\alpha)/\alpha}$ is a function of the force multiplier parameters which characterise the stellar opacity.

If we consider that A = cst, a star rotating for example at $\Omega/\Omega_{\rm crit} = 0.95$ will have $\dot{M}(\rm pol) = 3.2\,\dot{M}(\rm eq)$. However, if one accounts for the change of the force multiplier parameters, *i.e.* a variation in the opacity regime caused by the drop of $T_{\rm eff}$ at the equator, the equatorial mass loss can be enhanced at a given point, driving the formation of a decretion disc (Owocki 2004).

On the observations side, interferometry again sheds a new light in this topic, showing features that can be interpreted as polar enhanced winds (Kervella & Domiciano de Souza 2006; Meilland *et al.* 2007), as well as discs around active stars (Meilland *et al.* 2007; Schaefer *et al.* 2010).

The geometry of the wind may leave an imprint on the circumstellar medium. There are some indications of asymmetry detected in spectropolarimetry observations of some supernovae (see for example SN 2007rt by Trundle *et al.* 2009). This aspect is actually under study with 2- and 3D simulations and will be the subject of a future paper (Walder *et al.* in prep).

3. Rotational evolution

3.1. *Two competing processes*

The evolution of the surface velocity of a star is the result of the competition between two processes:

(*a*) the **mass loss**, which removes angular momentum at the surface and thus decelerates the rotational velocity;

(*b*) the **transport** of angular momentum inside the star, which brings some internal angular momentum to the surface (mainly through the *meridional circulation*) and may counterbalance the loss by the winds.

Both processes are dependent on the mass of the star: the more massive the star, the stronger winds and larger meridional currents it experiences (see Fig. 2, *left*). At solar metallicity typically, around 20 $M_\odot$ and above, the winds contribution wins so the star

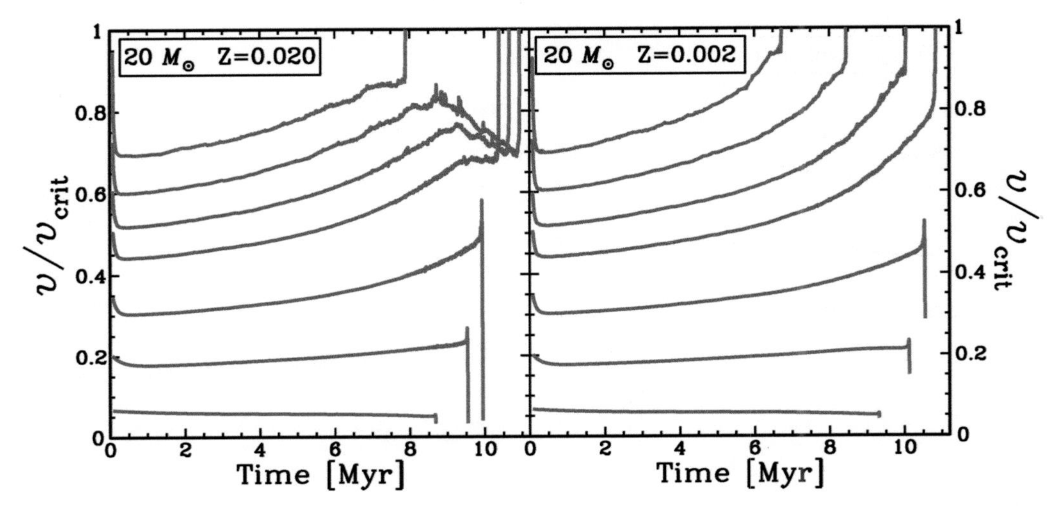

Figure 3. Evolution of the $v/v_{\rm crit}$ ratio during the main sequence for 20 $M_\odot$ models at $Z = 0.020$ and $Z = 0.002$.

decelerates during the main sequence (MS). Around 9 $M_\odot$, both processes are in equilibrium so the surface velocity remains almost constant during the largest part of the MS. Around 3 $M_\odot$, there are almost no mass loss through winds, but the core-envelope coupling through the meridional circulation is so weak that the surface velocity decreases during the MS. Note that recent observations by Huang *et al.* (2010, see also Poster S1-05 in these proceedings) show a quicker deceleration of the less massive of the B stars than predicted by current theoretical models, more compatible with a regime of local angular momentum conservation $\Omega\, R^2 = \rm cst$.

3.2. *Critical limit*

One thing is the evolution of the surface velocity, another thing is the evolution of the ratio to the critical limit $v_{\rm crit} = \sqrt{\frac{2}{3}\frac{G\,M}{R_{\rm pol}}}$. During the MS, the mass may be reduced by mass loss mechanisms, and the radius steadily inflates, so the critical limit drops, as shown in Fig. 2, *right*. The result is that although the surface may decelerates, the star may encounter the critical limit at a moment in its main sequence lifetime. Comparing the left panels of Fig. 3 and 4, we see that the conditions to reach the critical limit during the MS are met if the star has not a too high mass, and also if it is not a too slow rotator at birth. For each mass domain, there is a minimal initial $v/v_{\rm crit}$ ratio allowing for the reaching of the critical limit.

Once at the critical limit, the star remains close to it. It experiences phases of mass ejection (slowing the surface below the critical value) followed by quiescent phases, during which it slowly re-accelerates toward the critical limit.

It is possible now for theoretical models to evaluate the mass ejected in the form of a disc (see Poster S1-03 by Georgy *et al.* and S1-06 by Krtička *et al.* in these proceedings). This 'mechanical' mass loss seems to occur at a lower rate than the one that can be measured around Be stars (Rinehart *et al.* 1999; Stee 2003). However, in the models, the mechanical mass loss is averaged on a much longer timestep than the period of mass ejection observed in Be stars. The instantaneous mass-loss rate is expected to be higher than the average one given by the models.

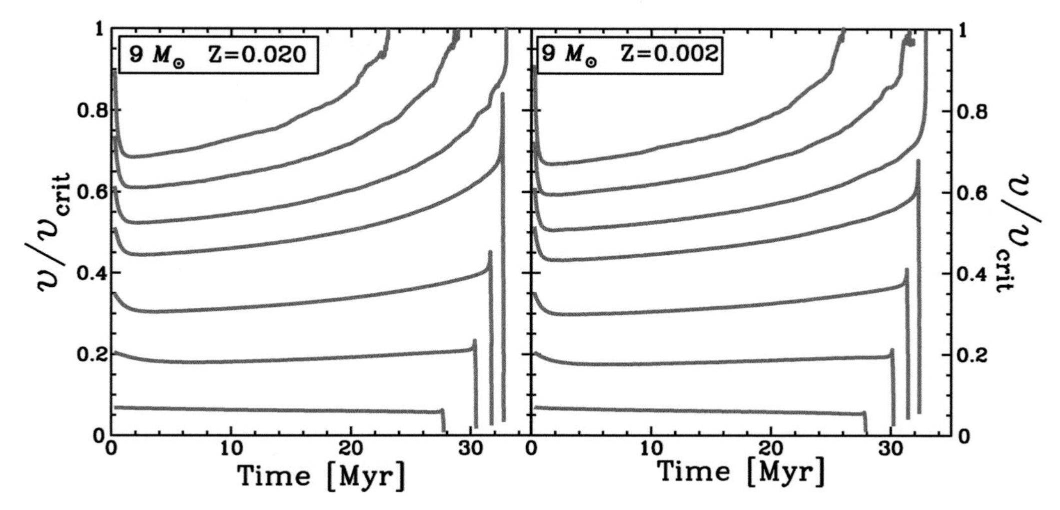

Figure 4. Evolution of the $v/v_{\rm crit}$ ratio during the main sequence for 9 $M_\odot$ models at $Z = 0.020$ and $Z = 0.002$.

3.3. *Metallicity effects*

Both the radiative mass loss rate and the strength of the meridional circulation depend on the metallicity. A low metallicity reduces both mechanisms: there are less metal lines to interact with the photons and form a radiatively driven wind and the greater compactness of the star reduces the amplitude of the meridional circulation.

The net effect on the evolution of the equatorial velocity depends strongly on the mass domain considered: for the most massive stars, the reduction of the winds is the dominant effect, so the stars reach more easily the critical limit (see Fig. 3). On the contrary, for less massive stars for which the winds are not too strong anyway, the reduction of the meridional circulation is dominant, so the stars have more difficulties to reach the critical limit (see Fig. 4).

3.4. *Be phenomenon*

Fast rotation is supposed to be linked with the Be phenomenon. It is however not yet clear whether it may explain by itself the origin of the equatorial disc observed around Be stars. Observations show a ratio $v/v_{\rm crit} \approx 0.70 - 0.80$ (Porter 1996; Chauville *et al.* 2001; Tycner *et al.* 2005). Maybe this ratio is true, but there may also be alternative explanations: first, as shown by Townsend *et al.* (2004), there is a saturation effect in the widening of the lines by rotation, so the true velocity may be underestimated. Second, the Be phenomenon seems to consist of ejection phases followed by quiescent phases during which the stars may rotate with a lower rate.

Observations show a metallicity trend in the appearance of the Be phenomenon (see for example Maeder *et al.* 1999 or Wisniewski & Bjorkman 2006). Two scenarios are evoked to explain the fast rotation of the Be stars: the binary channel (see for example McSwain & Gies 2005) (where the fast rotation arises from the accretion of angular momentum from the mass-donor companion), or the single star channel (Ekström *et al.* 2008) (where the surface acceleration is a natural evolution due to the core-envelope coupling). Any such scenario should present this metallicity trend to be valid. To study the single star evolution scenario, Ekström *et al.* (2008) computed 112 stellar models (4 masses, 4 metallicities and 7 rotation rates) with the Geneva code. To get populations numbers, we convolved the models with a Salpeter IMF, using the velocity distribution of Huang & Gies (2006).

A good agreement between the theoretical population ratio Be/(B+Be) and the observed one is obtained modulo two adjustments:

(*a*) the Be phenomenon should appear already at 70% of the critical velocity;

(*b*) the velocity distribution should count more fast rotators at birth in low-metallicity environments.

Point (*a*) can be sustained by several mechanisms. For example, sub-surface convective motions have been shown to be able to give the needed impulse to eject matter from the surface (Maeder *et al.* 2008; Cantiello *et al.* 2009), as also would non-radial pulsations do (McSwain *et al.* 2008; Owocki 2004). The binarity of the star could play a role, here not by spinning-up the star but by adding a gravitational pull allowing to launch the matter in the disc (Kervella *et al.* 2008). Point (*b*) presents contradictory observational supports. While Martayan *et al.* (2007) do find a higher mean $\Omega/\Omega_{\rm crit}$ ratio in the SMC compared to the LMC and the Galaxy, Penny & Gies (2009) draw opposite conclusions, finding no clear evidence for any difference in the velocity distribution at different metallicities.

Note that the stellar models have been computed without the effects of magnetic fields, the inclusion of which could change the picture. It may be that taking into account the strong core-envelope coupling brought by the magnetic fields relieves the theoretical prediction from the two aforementioned adjustments.

3.5. *LBVs*

Until now, the critical velocity we have considered was determined only by the centrifugal contribution $\vec{g}_{\rm rot}$ against $\vec{g}_{\rm grav}$ (Ω-limit). However, a critical point appears whenever $\vec{g}_{\rm tot} = \vec{g}_{\rm grav} + \vec{g}_{\rm rot} + \vec{g}_{\rm rad} = 0$. If $\vec{g}_{\rm rot} = 0$ or is negligible, the star may meet the classical Eddington-limit when $\vec{g}_{\rm rad} = \vec{g}_{\rm grav}$. If $\vec{g}_{\rm rad}$ is negligible, the star may encounter the Ω-limit when $\vec{g}_{\rm rot} = \vec{g}_{\rm grav}$: this describes the first critical velocity considered until now:

$$v_{\rm crit,1} = \sqrt{\frac{2}{3}\frac{GM}{R_{\rm pol,crit}}}$$

with $R_{\rm pol,crit}$ the polar radius when the star is at the critical limit.

If all three terms are significant, the star may meet the $\Omega\Gamma$-limit when $\vec{g}_{\rm tot} = 0$. It means that for a given Ω, there is a maximum luminosity given by:

$$L_{\Gamma\Omega} = \frac{4\pi cG\,M}{\kappa}\left(1 - \frac{\Omega^2}{2\pi G\,\rho_{\rm m}}\right)$$

with κ the total opacity. Inversely, depending on the Eddington factor of the star, a second critical velocity can be defined:

$$v_{\rm crit,2} = \sqrt{\frac{9}{4}v^2_{\rm crit,1}\frac{1-\Gamma_{\rm max}}{V'(\omega)}\frac{R_{\rm e}^2(\omega)}{R^2_{\rm pol,crit}}}$$

where the quantity $V'(\omega) = \frac{V(\omega)}{\frac{4\pi}{3}R^3_{\rm pol,crit}}$ is the ratio of the actual volume of a star with rotation $\omega = \Omega/\Omega_{\rm crit}$ to the volume of a sphere of radius $R_{\rm pol,crit}$.

Note that for $\Gamma_{\rm Edd} < 0.639$, $v_{\rm crit,2}$ is not defined. Above this value, $v_{\rm crit,2}$ becomes smaller than $v_{\rm crit,1}$, so the star encounters the $\Omega\Gamma$-limit before the Ω-limit. This could be the case of some known LBVs (Groh *et al.* 2006, 2009), that present both a high Eddington factor and a high rotation rate. With a $\Gamma_{\rm Edd} \simeq 0.8$, they are probably meeting their second critical velocity already with $\Omega/\Omega_{\rm crit} \simeq 0.85 - 0.9$.

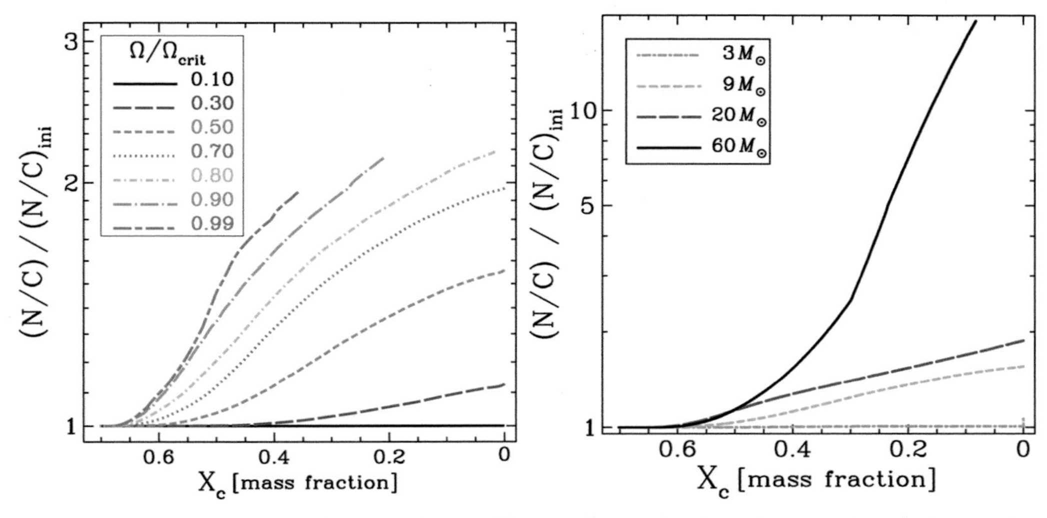

Figure 5. *Left:* evolution of the surface N/C ratio (normalised to the initial one) during the main sequence (expressed as the central mass fraction of hydrogen X_c) for 9 $M_\odot$ models at $Z = 0.020$ with varying the initial ratio $\Omega/\Omega_{\rm crit}$. *Right:* same as *left*, but for different mass domains.

4. Internal mixing

4.1. *Diffusion of chemical species*

The radiative zones inside a star are supposed to rotate differentially. This causes shear turbulence which in turn drives some turbulent mixing of chemical species. The higher the rotation rate, the stronger is the mixing (see Fig. 5, *left*).

We can evaluate the diffusion time: $\tau_{\rm mix} \simeq \frac{R^2}{D}$ with D the diffusion coefficient. With a relation between the radius and the mass, we get $\tau_{\rm mix} \propto M^{-1.8}$ while the main sequence lifetime is $\tau_{\rm MS} \propto M^{-0.7}$. We see that the more massive the star, the stronger is the mixing, leading to a surface enrichment already on the main sequence (see Fig. 5, *right*).

4.2. *Metallicity effects*

Low-metallicity stars are more compact than higher-metallicity ones. The meridional currents are less efficient, so the Ω-profile inside the star is steeper. The mixing is thus stronger, while the mixing time is shorter, so we expect a strong surface enrichment on the main sequence.

4.3. *Surface abundances*

The surface abundances are expected to be modified by mixing. A good signature of mixing is the C, N, and O abundances: their respective ratios are expected to change, but their sum C+N+O is supposed to remain constant. Only at very low Z, primary nitrogen production may occur, leading to a net metallicity increase. A careful abundance analysis made by Przybilla *et al.* (2010) in the solar neighbourhood shows that the observed mixing follows very well the trend expected from CNO nuclear reactions (see their contribution in the same proceedings). The adequation with stellar models is also good, but the observations are not very constraining yet.

It is very important to keep in mind that the mixing is a function of the rotation rate, of course, but also (as shown above) of the mass, the metallicity and the age, as well as other characteristics as the binarity for example. The study of the rotational mixing in stars should imply a separate analysis, in order to discriminate between several parameters. For example, while Hunter *et al.* (2008), studying field and clusters stars in the SMC, find many outliers from the expected path in the N/C vs $v_{\rm rot}$, a re-analysis by Maeder *et al.* (2009) of a sub-group of the same stars, restrained to a narrow mass range and belonging to the same cluster (*i.e.* having the same age), shows that most of the outliers shifted back to the expected path. Most of the outliers left cannot be used to test the rotation-induced mixing, because they are evolved stars or known binaries. For the very few remaining, we indeed need another explanation than pure rotational mixing. There are several possibilities that shall be the subject of further studies, as the influence of magnetic fields inside the star or magnetic braking at the surface. However, the need for a yet unknown process is evoked (see the contribution of Ines Brott in the same proceedings).

5. Conclusion

Rotation is linked to many very interesting types of objects as Be stars or LBVs. The inclusion of its effects in theoretical stellar models is essential to understand those objects and their direct environment. While rotating models are in a better adequation with observations, there is still a lot of work to do in order to improve our understanding of the physical processes at work.

Theoretical and numerical developments are continuously on-going, but theory alone is like a car without a driver. Observations are highly needed:

• *larger surveys* of well identified objects would allow to make a separate analysis of sub-groups in masses, ages, ..., with a significant statistics;

• a larger number of objects studied by *interferometric measurements* would allow to put constraints on the surface characteristics of the models;

• the progress in *asteroseismology* should provide constraints on the internal structure of the stars;

• precise observations of the *circumstellar environment* could help to constrain the mass loss mechanisms, which are a key ingredient in massive stars.

References

Cantiello, M., Langer, N., Brott, I., de Koter, A. *et al.* 2009, *A&A* 499, 279

Carciofi, A. C., Domiciano de Souza, A., Magalhães, A. M., Bjorkman, J. E. *et al.* 2008, *ApJ* (Letters) 676, L41

Chaboyer, B., Demarque, P., & Pinsonneault, M. H. 1995, *ApJ* 441, 865

Chauville, J., Zorec, J., Ballereau, D., Morrell, N. *et al.* 2001, *A&A* 378, 861

Deupree, R. G. 1990, *ApJ* 357, 175

Domiciano de Souza, A., Kervella, P., Jankov, S., Abe, L. *et al.* 2003, *A&A* 407, L47

Ekström, S., Meynet, G., Maeder, A., & Barblan, F. 2008, *A&A* 478, 467

Endal, A. S. & Sofia, S. 1976, *ApJ* 210, 184

Faulkner, J., Roxburgh, I. W., & Strittmatter, P. A. 1968, *ApJ* 151, 203

Fliegner, J. & Langer, N. 1994, in: L. A. Balona, H. F. Henrichs, & J. M. Le Contel (eds.), *Pulsation; Rotation; and Mass Loss in Early-Type Stars*, IAU Symposium 162, p. 147

Georgy, C., Meynet, G., Walder, R., Folini, D. *et al.* 2009, *A&A* 502, 611

Groh, J. H., Damineli, A., Hillier, D. J., Barbá, R. *et al.* 2009, *ApJ* (Letters) 705, L25

Groh, J. H., Hillier, D. J., & Damineli, A. 2006, *ApJ* (Letters) 638, L33

Heger, A. & Langer, N. 2000, *ApJ* 544, 1016

Heger, A., Langer, N., & Woosley, S. E. 2000, *ApJ* 528, 368

Heger, A., Woosley, S. E., & Spruit, H. C. 2005, *ApJ* 626, 350

Huang, W. & Gies, D. R. 2006, *ApJ* 648, 580

Huang, W., Gies, D. R., & McSwain, M. V. 2010, *ApJ* 722, 605
Hunter, I., Brott, I., Lennon, D. J., Langer, N. *et al.* 2008, *ApJ* (Letters) 676, L29
Kervella, P. & Domiciano de Souza, A. 2006, *A&A* 453, 1059
Kervella, P., Domiciano de Souza, A., & Bendjoya, P. 2008, *A&A* 484, L13
Kippenhahn, R. & Thomas, H.-C. 1970, in: A. Slettebak (eds.), *Stellar Rotation*, IAU Colloquium 4 (Gordon and Breach Science Publishers), p. 20
Langer, N., Fliegner, J., Heger, A., & Woosley, S. E. 1997, *Nuclear Physics A* 621, 457
Maeder, A. 1999, *A&A* 347, 185
Maeder, A. 2002, *A&A* 392, 575
Maeder, A. 2009, *Physics, Formation and Evolution of Rotating Stars*, Astronomy and Astrophysics Library (Springer Berlin Heidelberg)
Maeder, A., Georgy, C., & Meynet, G. 2008, *A&A* 479, L37
Maeder, A., Grebel, E. K., & Mermilliod, J.-C. 1999, *A&A* 346, 459
Maeder, A. & Meynet, G. 2000, *A&A* 361, 159
Maeder, A. & Meynet, G. 2001, *A&A* 373, 555
Maeder, A., Meynet, G., Ekström, S., & Georgy, C. 2009, *Communications in Asteroseismology* 158, 72
Martayan, C., Frémat, Y., Hubert, A.-M., Floquet, M. *et al.* 2006a, *A&A* 452, 273
Martayan, C., Frémat, Y., Hubert, A.-M., Floquet, M. *et al.* 2007, *A&A* 462, 683
Martayan, C., Hubert, A. M., Floquet, M., Fabregat, J. *et al.* 2006b, *A&A* 445, 931
McSwain, M. V. & Gies, D. R. 2005, *ApJS* 161, 118
McSwain, M. V., Huang, W., Gies, D. R., Grundstrom, E. D. *et al.* 2008, *ApJ* 672, 590
Meilland, A., Stee, P., Vannier, M., Millour, F. *et al.* 2007, *A&A* 464, 59
Meynet, G. 1996, in: C. Leitherer, U. Fritze-von-Alvensleben, & J. Huchra (eds.), *From Stars to Galaxies: the Impact of Stellar Physics on Galaxy Evolution*, ASP-CS 98, p. 160
Meynet, G. & Maeder, A. 1997, *A&A* 321, 465
Meynet, G. & Maeder, A. 2000, *A&A* 361, 101
Meynet, G. & Maeder, A. 2003, *A&A* 404, 975
Meynet, G. & Maeder, A. 2005, *A&A* 429, 581
Monnier, J. D., Zhao, M., Pedretti, E., Thureau, N. *et al.* 2007, *Science* 317, 342
Owocki, S. P. 2004, in: A. Maeder & P. Eenens (eds.), *Stellar Rotation*, IAU Symposium 215, p. 515
Owocki, S. P., Cranmer, S. R., & Gayley, K. G. 1996, *ApJ* (Letters) 472, L115
Owocki, S. P., Cranmer, S. R., & Gayley, K. G. 1998, in: A. M. Hubert & C. Jaschek (eds.), *B[e] stars*, Astrophysics and Space Science Library 233, p. 205
Owocki, S. P. & Gayley, K. G. 1997, in: A. Nota & H. Lamers (eds.), *Luminous Blue Variables: Massive Stars in Transition*, ASP-CS 120, p. 121
Penny, L. R. & Gies, D. R. 2009, *ApJ* 700, 844
Petrenz, P. & Puls, J. 2000, *A&A* 358, 956
Pinsonneault, M. H., Kawaler, S. D., Sofia, S., & Demarque, P. 1989, *ApJ* 338, 424
Porter, J. M. 1996, *MNRAS* 280, L31
Przybilla, N., Firnstein, M., Nieva, M. F., Meynet, G. *et al.* 2010, *A&A* 517A, 38
Rinehart, S. A., Houck, J. R., & Smith, J. D. 1999, *AJ* 118, 2974
Roxburgh, I. W., Griffith, J. S., & Sweet, P. A. 1965, *ZfA* 61, 203
Roxburgh, I. W. & Strittmatter, P. A. 1966, *MNRAS* 133, 345
Schaefer, G. H., Gies, D. R., Monnier, J. D., Richardson, N. *et al.* 2010, in: T. Rivinius & M. Curé (eds.), *The Interferometric View on Hot Stars*, Rev. Mexicana AyA Conference Series 38, p. 107
Siess, L. & Livio, M. 1997, *ApJ* 490, 785
Stee, P. 2003, *A&A* 403, 1023
Townsend, R. H. D., Owocki, S. P., & Howarth, I. D. 2004, *MNRAS* 350, 189
Trundle, C., Pastorello, A., Benetti, S., Kotak, R. *et al.* 2009, *A&A* 504, 945
Tycner, C., Lester, J. B., Hajian, A. R., Armstrong, J. T. *et al.* 2005, *ApJ* 624, 359
Vinicius, M. M. F., Zorec, J., Leister, N. V., & Levenhagen, R. S. 2006, *A&A* 446, 643
Vink, J. S. & de Koter, A. 2005, *A&A* 442, 587

von Zeipel, H. 1924, *MNRAS* 84, 665
Wisniewski, J. P. & Bjorkman, K. S. 2006, *ApJ* 652, 458
Yoon, S.-C., Langer, N., & Norman, C. 2006, *A&A* 460, 199
Zhao, M., Monnier, J. D., Pedretti, E., Thureau, N. *et al.* 2009, *ApJ* 701, 209

Discussion

S. Owocki: I would like to emphasize an important difference between reaching the two critical rotation speeds identified in the Maeder & Meynet analysis. The first critical speed applies to low luminosity stars like Be stars, and should lead to a circumstellar decretion disk that is ejected mechanically at the equator. But for the second critical speed, which is modified by the radiative forces associated with $\Gamma > 0.65$, the mass loss should be mainly over the poles, not equator. In effect, the rapid rotation and associated equatorial gravity darkening forces the stellar luminosity to emerge over a smaller surface area over the poles, so that even if $\Gamma \lesssim 1$ the local flux over the poles can exceed the Eddington value, leading to a radiatively driven, bipolar, prolate mass loss. In short, in low luminosity, $\Gamma << 1$ Be stars, critical rotation leads to the observed equatorial disk, while in $\Gamma \lesssim 1$ LBVs it leads to bipolar nebulae.

O. Chesneau: A comment to Stan: If the mass loss of LBVs is essentially prolate, and directed toward the pole, how can you explain why the environment of LBVs is dominated by ring-like structures (see comment of K. Weis)?

S. Owocki: Well, in slowly rotating LBVs you should get spherical mass loss that will likely appear ring-like. Perhaps Kerstin can comment on how relatively common these rings are vs. bipolar LBVs.

K. Weis: As Stan already mentioned, LBVs create polar winds which lead to larger number of bipolar nebulae and not only spherical ring nebulae. See my talk on Thursday.

Active OB stars: structure, evolution, mass loss, and critical limits
Proceedings IAU Symposium No. 272, 2010
C. Neiner, G. Wade, G. Meynet & G. Peters, eds.

doi:10.1017/S1743921311009999

Testing models of rotating stars

Adrian T. Potter and Christopher A. Tout

Institute of Astronomy, The Observatories, Madingley Road, Cambridge CB3 0HA, England
email: apotter@ast.cam.ac.uk

Abstract. The effects of rapid rotation on stellar evolution can be profound but we are only now starting to gather the data necessary to adequately determine the validity of the many proposed models of rotating stars. Some aspects of stellar rotation, particularly the treatment of angular momentum transport within convective zones, still remain very poorly explored. Distinguishing between different models is made difficult by the typically large number of free parameters in models compared with the amount of available data. This also makes it difficult to determine whether increasing the complexity of a model actually results in a better reflection of reality. We present a new code to straightforwardly compare different rotating stellar models using otherwise identical input physics. We use it to compare several models with different treatments for the transport of angular momentum within convective zones.

Keywords. stars: evolution, stars: general, stars: rotation

1. Introduction

It has been known for many years that rapid rotation can cause significant changes in the evolution of stars. Not only does it cause a broadening of the main sequence in the HR diagram but it also produces enrichment of a number of different elements at the stellar surface (Hunter *et al.* 2009). With new large scale surveys such as VLT-FLAMES now reaching maturity, the amount of data available for rotating stars is growing rapidly (e.g. Evans *et al.* 2005, 2006). Any rotating stellar model must be able to match the observed changes in chemical enrichment and structure. The treatment of rotation and its induced chemical mixing in stars has changed dramatically over the past two decades. The model of Zahn (1992) has formed the basis for much of the work and many variations from the original model have been used during this time to generate different sets of stellar evolution tracks (e.g. Talon *et al.* 1997; Meynet & Maeder 2000; Maeder 2003). There are alternate formalisms (e.g. Heger *et al.* 2000) that treat the physical processes very differently though no direct comparison of the predictions of each model using the same stellar evolution code has been made. Particular emphasis is often placed on the treatment of meridional circulation. Whereas those models based on Zahn (1992) treat meridional circulation as an advective process, Heger *et al.* (2000) treat it as a diffusive process. This may lead to significantly different predictions for each model even though both treatments are frequently quoted in the literature for their predictions of the effect of rotation on stellar evolution.

One of the most poorly explored features of stellar rotation models is the treatment of angular momentum transport within convective zones. All current 1D models appear to treat convective zones as a rotating solid body. This isn't necessarily correct and there are strong reasons to explore alternatives such as uniform specific angular momentum. This potentially has dramatic consequences for the evolution because a star with uniform specific angular momentum through its convective core has much more total angular momentum for a given surface velocity. Uniform specific angular momentum also results in a strong shear layer at a convective boundary that can drive additional transport of chemical elements.

Now that the amount of data is increasing rapidly because of large scale surveys such as VLT-FLAMES it is important to make a direct comparison of the different models for stellar rotation on a common platform with otherwise identical input physics. Using the Cambridge stellar evolution code (Eggleton 1971; Pols *et al.* 1995; Stancliffe & Eldridge 2009) we have produced a code that is able to model rotating stars in 1D under the shellular rotation hypothesis of Zahn (1992). The code, *RoSE* (Rotating Stellar Evolution), can be easily programmed to run with different models for stellar rotation and can model convective zones under a variety of different assumptions. This allows us to compare a variety of models for stellar rotation and determine what, if any, observable traits could be used to distinguish between them. We foresee two possibilities; either there will be clear observational tests to eliminate certain models or the models will show no testable difference in which case a simplified model could be formulated to produce identical evolutionary predictions.

2. Physics of rotating stars

There are a number of physical effects due to rotation that must be taken into account to form a complete model. The first is the distortion to the geometry and change in effective gravity because of the centrifugal force. By suitable averaging over isobars it is possible to evaluate the effect of the distortion of the star on the equations of hydrostatic and radiative equilibrium. The modified equations are

$$\frac{dP}{dm} = -\frac{Gm}{4\pi r^4} f_p \tag{2.1}$$

and

$$\frac{d\ln T}{d\ln P} = \frac{3\kappa PL}{16\pi acmGT^4} \frac{f_T}{f_p} \tag{2.2}$$

where P is the pressure, T is the temperature, L is the luminosity, κ is the opacity, a is the radiation constant, c is the speed of light and G is the gravitational constant. f_p and f_T are typically of order unity and vary with angular velocity. For non-rotating stars r and $m(r)$ are the radius from the centre and mass inside r. In the rotating case their exact definitions are modified sligtly due to distortion of the star because of the centrifugal force. For the full derivation see Endal & Sofia (1976). Under this framework Endal & Sofia (1976) also derive an expression for the effective gravity modified by the action of the centrifugal force. For moderate rotation rates the distortion to the star is small. It becomes large only when the rotation rate approaches critical (Tassoul 1978).

Von Zeipel (1924) showed that the thermal flux through a point in a star is proportional to the effective gravity. Because the centrifugal force is stronger at the equator than at the poles, the effective gravity depends heavily on the co-latitude and hence so does the thermal flux. This leads to a violation of thermal equilibrium along isobars and gives rise to meridional circulation. The presence of such a circulation has been considered for a long time and is commonly approximated by Eddington-Sweet circulation (Sweet 1950). Zahn (1992) proposed an alternative but similar treatment of the meridional circulation based on energy conservation along isobars.

Differential rotation is expected to arise in stars because of hydrostatic structural evolution and meridional circulation. Stars are therefore subject to a number of hydrodynamic shear instabilities. The proposed hypothesis by Zahn (1992) is that because of the strong stratification present in massive stars, the turbulent mixing caused by these instabilities is much stronger horizontally than vertically. This leads to a situation where the angular velocity variations along isobars are negligible compared to vertical variations. This is referred to as shellular rotation where we describe the angular velocity profile by $\Omega = \Omega(r)$.

In 1D evolution models convective zones are currently assumed to be in solid body rotation. This may be caused by strong magnetic fields induced by dynamo action (Spruit 1999) but there is no conclusive evidence why this must be the case. Certainly in the Sun we see latitudinal variations in the angular velocity throughout the outer convective layer (Schou *et al.* 1998). Standard mixing length theory suggests that a rising fluid parcel should conserve its angular momentum before mixing it with the surrounding material after rising a certain distance. This would lead to uniform specific angular momentum rather than solid body rotation. This is supported by a recent MLT-based closure model for rotating stars (Lesaffre *et al.* 2010) and by 3D hydrodynamic simulations (Arnett & Maekin 2009). In reality magnetic fields are likely to play some part in the transport of angular momentum but it is uncertain whether these are strong enough to enforce solid body rotation. The asymptotic behaviour of the rotation profile in convective zones could have a profound effect on the evolution of the star, first because the total angular momentum content of a star for a given surface rotation increases dramatically for uniform specific angular momentum and secondly because uniform angular momentum in the convective zone produces a layer of strong shear at the boundary with the radiative zone which drives additional chemical mixing. To explore the different possible behaviours we have introduced, in *RoSE*, the capacity to vary the distribution of angular momentum in convective zones.

With all of these effects and a suitable spherical average, as described by Zahn (1992), the angular velocity evolves according to

$$\frac{\partial (r^2\Omega)}{\partial t} = \frac{1}{5r^2}\frac{\partial (r^4\Omega U)}{\partial r} + \frac{1}{r^2}\frac{\partial}{\partial r}\left(D_{\rm shear} r^4 \frac{\partial \Omega}{\partial r}\right) + \frac{1}{r^2}\frac{\partial}{\partial r}\left(D_{\rm conv} r^{(2+n)} \frac{\partial r^{(2-n)}\Omega}{\partial r}\right) \quad (2.3)$$

and the chemical abundances evolve according to

$$\frac{\partial c_i}{\partial t} = \frac{1}{r^2}\frac{\partial}{\partial r}\left((D_{\rm shear} + D_{\rm eff} + D_{\Omega=0})\, r^2 \frac{\partial c_i}{\partial r}\right), \quad (2.4)$$

where Ω is the angular velocity, $U(r)(3\cos^2\theta - 1)/2$ is the radial component of the meridional circulation, c_i is the mass fraction of element i and n determines the distribution of specific angular momentum in convective zones. $D_{\rm conv}$, $D_{\rm shear}$ and $D_{\rm eff}$ are the various diffusion coefficients describing the various effects of rotation. $D_{\Omega=0}$ is the diffusion coefficient for chemical mixing derived for the non-rotating case. $D_{\rm conv}$ is non-zero only in convective zones and $D_{\rm shear}$ and $D_{\rm eff}$ are non-zero only in radiative zones. The major difference between different models (e.g. Talon *et al.* 1997; Heger *et al.* 2000; Meynet & Maeder 2000; Maeder 2003) is the treatment of the meridional circulation, U, the diffusion coefficients, $D_{\rm shear}$, $D_{\rm eff}$ and $D_{\rm conv}$, and the steady power law distribution of angular momentum in convective zones determined by n.

With all of these effects, *RoSE* is capable of producing stellar evolution tracks for various different models very quickly and it can be readily adapted to include additional effects. On a single 2.83 GHz processor we have produced a grid of stellar evolution tracks between the ZAMS and base of the giant branch for 15 masses and 10 rotation rates in 22 hr. More rapid computation may be acheived by sacrificing some numerical accuracy.

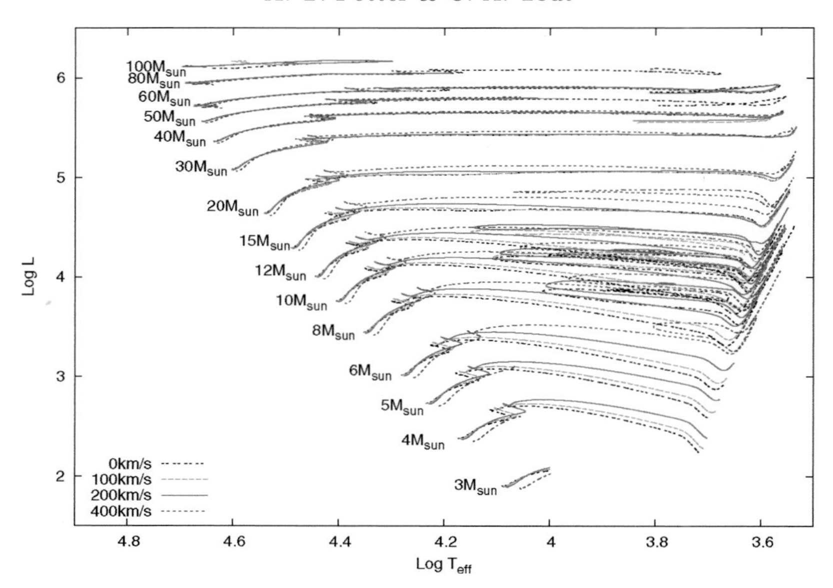

Figure 1. Stellar evolution tracks for stars between $4M_{\odot}$ and $100M_{\odot}$ calculated with *RoSE*. Tracks for the 4 different models are almost indistinguishable in the HR diagram.

3. Results

For our preliminary results we have run grids for four different rotating models. The grid contained the following masses and initial equatorial rotation velocities,

$$M/\mathrm{M}_{\odot} = \{4, 5, 6, 8, 10, 12, 15, 20, 30, 40, 50, 60, 80, 100\} \tag{3.1}$$

and

$$v_{\mathrm{ini}}/\mathrm{km\,s^{-1}} = \{0, 20, 50, 100, 150, 200, 250, 300, 400, 500\}. \tag{3.2}$$

For the four different models we used the format of equations (2.3) and (2.4) and the formulation for D_{shear} from Talon *et al.* (1997), D_{h} from Maeder (2003) and U from Meynet & Maeder (2000). We then use the additional parameters listed below.

Model 1: $n = 2$ and $D_{\mathrm{conv}} = D_{\mathrm{mlt}}$
Model 2: $n = 0$ and $D_{\mathrm{conv}} = 10^{16}\,\mathrm{cm^2\,s^{-1}}$
Model 3: $n = 0$ and $D_{\mathrm{conv}} = 10^{12}\,\mathrm{cm^2\,s^{-1}}$
Model 4: $n = 0$ and $D_{\mathrm{conv}} = D_{\mathrm{lctp}}$

We derive D_{mlt} from the typical turbulent velocity and time scales of mixing length theory and D_{lctp} is from Lesaffre *et al.* (2010). In the non-rotating limit $D_{\mathrm{lctp}} \to D_{\mathrm{mlt}} \approx 10^{16} cm^2 s^{-1}$. Traditionally, the HR diagram has been the testing ground for variations in stellar models. However, in this case, the four different models predict almost indistinguishible tracks in the HR diagram with variation only at the most massive, most rapidly rotating stars. Even then, the change is small. This suggests that the evolution tracks for rotating stars in the HR diagram are affected almost entirely by the evolution of the angular momentum distribution in radiative zones and not convective zones. We have plotted some of the tracks predicted by the models in figure 1.

There are a number of additional observational tests to distinguish between different stellar models. One of the most favoured is to look at the nitrogen enrichment produced by rotation by means of a Hunter diagram (Hunter *et al.* 2009) as shown in figure 2.

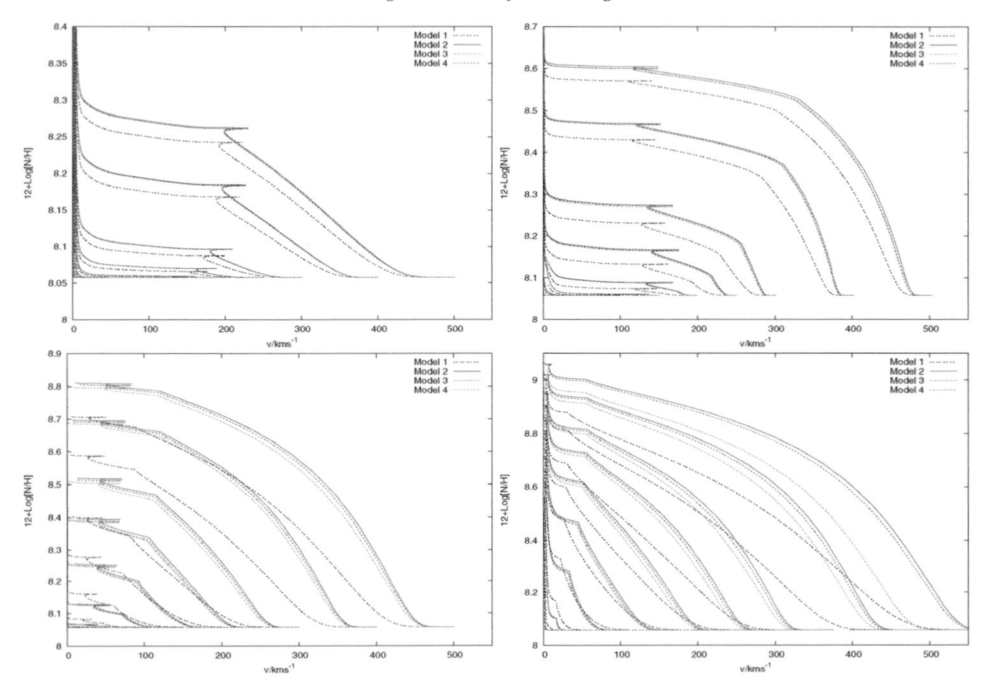

Figure 2. Nitrogen enrichment with a range of initial rotation rates for the 4 different models. Top left is for $10M_\odot$, top right for $20M_\odot$, bottom left for $40M_\odot$ and bottom right for $60M_\odot$. As expected the enrichment is much greater for more massive stars. There is much more enrichment for models in which the core has uniform specific angular momentum, an effect that becomes progressively more pronounced for higher masses.

When we do this we see there are very clear differences between the predictions of each model. Models 2, 3 and 4 which have uniform specific angular momentum throughout their convective zones have significantly more enrichment for all masses and rotation rates but with more pronounced differences for higher mass rapid rotators. However, we note that it is far more difficult to distinguish between these three models with $n = 0$. They are almost indistinguishable for masses below $20M_\odot$ and the difference is small even for the $40\,M_\odot$ model. For the highest mass stars we do see more enrichment in the stars with higher $D_{\rm conv}$ but $D_{\rm conv}$ is four orders of magnitude lower in model 3 than model 2. Therefore slighter changes in the diffusion coefficient are unlikely to be dectable by this analysis as indicated by the lack of any signigifcant difference between the results of models 2 and 4.

4. Conclusion

There are many different models currently used to make predicitions regarding the change to stellar evolution as a result of rotation. These models often use very different physics and there are certain aspects of the models, such as angular momentum transport by convection, that remain almost totally unexplored. Now that the available data on rotating stars is starting to increase rapidly, thanks to large scale surveys such as VLT-FLAMES, it is important to compare different models and the data to determine whether we can narrow down the number of possible models. Furthermore, by comparing the predictions of the different models, we can provide observational tests that will allow us to eliminate more of the potential models.

We have shown that, if convection results in uniform specific angular momentum, there is significantly more enrichment in the surface nitrogen abundance of the star compared to similar models where the core is kept in solid body rotation. This does not result in any distinct change in the HR diagram for any rotation rate and masses between $3\,M_{\odot}$ and $100\,M_{\odot}$ suggesting that evolution tracks in the HR diagram are influenced by the evolution of angular velocity in radiative zones but not convective zones. The amount of enrichment for models with uniform angular momentum in convective zones is only weakly dependent on the diffusion coefficient and changes between the results with $D_{\rm conv} = 10^{12}\,{\rm cm^2\,s^{-1}}$ and $D_{\rm conv} = 10^{16}\,{\rm cm^2\,s^{-1}}$ become significant only for masses above about $60 M_{\odot}$. By calculating the outcome of these models using our new stellar rotation code, *RosE*, we can be sure that the variations in the models come exclusively from the differences in the model for rotation and not from the numerical method or differences in the rest of the stellar evolution package.

In the future we plan to compute additional models to provide other observational tests to distinguish between them. It will also be important to use the predicitions of each model in detailed population syntheses such as the one by Izzard *et al.* (2009) in order to suitably compare the predicitions with the data.

Acknowledgements

ATP thanks the STFC for his studentship and CAT thanks Churchill college for his fellowship.

References

Arnett, W. D. & Meakin, C. 2009, in: K. Cunha, M. Spite, & B. Barbuy (eds.), *Chemical Abundances in the Universe: Connecting First Stars to Planets*, IAU Symposium 265, p. 106
Eggleton, P. P. 1971, *MNRAS*, 151, 351
Endal, A. S. & Sofia, S. 1976, *ApJ*, 210, 184
Evans, C. J., Smartt, S. J., Lee, J.-K., Lennon, D. J. *et al.* 2005, *A&A*, 437, 467
Evans, C. J., Lennon, D. J., Smartt, S. J., & Trundle, C. 2006, *A&A*, 456, 623
Heger, A., Langer, N., & Woosley, S. E. 2000, *ApJ*, 528, 368
Hunter, I., Brott, I., Langer, N., Lennon, D. J. *et al.* 2009, *A&A*, 496, 841
Izzard, R. G., Glebbeek, E., Stancliffe, R. J., & Pols, O. R. 2009, *A&A*, 508, 1359
Lesaffre P., Chitre S. M., Tout C. A., & Potter A. T., 2010, private communication
Maeder, A. 2003, *A&A*, 399, 263
Meynet, G. & Maeder, A. 2000, *A&A*, 361, 101
Pols, O. R., Tout, C. A., Eggleton, P. P., & Han, Z. 1995, *MNRAS*, 274, 964
Schou, J., Antia, H. M., Basu, S., Bogart, R. S. *et al.* 1998, *ApJ*, 505, 390
Spruit, H. C. 1999, *A&A*, 349, 189
Stancliffe, R. J., & Eldridge, J. J. 2009, *MNRAS*, 396, 1699
Sweet, P. A. 1950, *MNRAS*, 110, 548
Talon, S., Zahn, J.-P., Maeder, A., & Meynet, G. 1997, *A&A* 322, 209
Tassoul J.-L. 1978, *Theory of rotating stars*, Princeton Series in Astrophysics, Princeton: University Press
von Zeipel, H. 1924, *MNRAS*, 84, 665
Zahn, J.-P. 1992, *A&A*, 265, 115

Active OB stars: structure, evolution, mass Loss, and critical limits
Proceedings IAU Symposium No. 272, 2010
C. Neiner, G. Wade, G. Meynet & G. Peters, eds.

doi:10.1017/S1743921311010003

Discussion – Rapid rotation and mixing in active OB stars

Geraldine J. Peters[1] **& Georges Meynet**[2]

[1]Space Sciences Center/Department of Physics & Astronomy, University of Southern California, Los Angeles, CA 90089-1341, USA
email: gjpeters@mucen.usc.edu

[2]Geneva Observatory, University of Geneva, Maillettes 51, 1290 Sauverny, Switzerland
email: Georges.Meynet@unige.ch

Abstract. The general discussion following Session 1: Rapid Rotation and Mixing in Active OB Stars is summarized. Topics that focus on observational and theoretical issues are included.

Keywords. stars: abundances, stars: rotation, stars: evolution, stars: emission-line, Be

1. Introduction

This is the first of six general discussion sessions that followed each of the sections of IAUS-272. We decided to divide the allotted time of 30 minutes equally between observational and theoretical topics. Geraldine Peters presided over the former and Georges Meynet chaired the latter.

2. Observational Topics

The nitrogen abundance as a marker of photospheric mixing of CNO-processed material from the interior of an OB star was mentioned throughout this session. Since the determination of the N abundance is a straightforward way to confront theory with observation, it is important to confirm the errors that exist in the abundance analyses for both the slowly-rotating and fast-rotating stars, especially the Be stars. Sources of error might include the choice of model atmospheres, input parameters, and the observations themselves. The opportunity today to obtain very high quality spectra of OB stars in our galaxy and the Magellanic Clouds is unprecedented. As I see it, the greatest challenge in interpreting the optical data is correcting for the flux from the disk. In the UV one is faced with coverage and blending issues, continuum placement, and in the case of the Be stars possible line emission from circumstellar material and sometimes pervasive shell spectra. Of course the abundances that are determined are only as good as the model atmospheres. How well do the models represent real stars? This is quite an important issue for rapidly-rotating OB stars. Perhaps the least certain input parameter is the microturbulence. It has been known for some time now that the value for V_{turb} determined from optical lines (e.g. O II) is larger than that found from the UV lines. For a typical B star with modest microturbulence the difference is $\sim$3 km s^{-1} (cf. Fig. 1). An error of the latter size produces an uncertainty of $\sim$0.1–0.4 dex in the abundance (Hunter *et al.* 2007) depending on which line in which star is being analyzed and the upper limit is approaching the theoretical predictions for the nitrogen enrichment in the photosphere.

Daniel Lennon, who spoke on behalf of the VLT-FLAMES Survey, reaffirmed that it is very difficult to determine nitrogen abundances in the Be stars, or any fast rotator, due to their broad lines. Very high S/N is needed and this presents problems. The error

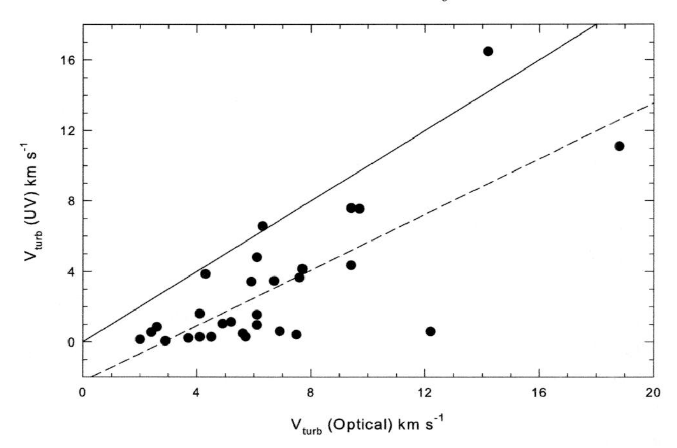

Figure 1. The microturbulence parameter determined for the same B star from optical (Gies & Lambert 1992) and UV(Proffitt & Quigley 2001) lines. The lower *dashed* line is a linear regression fit to the data, while the upper *solid* line represents a one-to-one correlation. The implied V_{turb} from the UV study is systematically smaller and suggests that V_{turb} may be depth-dependent.

bars for an analysis of an individual star will be large. But it helps to observe a large sample of stars, and this is where the VLT-FLAMES project is contributing. With a large sample one can simulate a population based upon what you know about the error bars and feed this into the simulation. This is the focus of Inès Brott's work. However if one looks at the parameter space and number of stars in each mass bin it is clear that some regions are sparsely populated. VLT-FLAMES II (The Tarantula Survey) will address this problem by gathering a larger sample of stars in the LMC. According to Lennon, the objective of the VLT-FLAMES Surveys is not to understand each individual star but to "understand all of the stars all of the time, not some of the stars some of the time – and this is a challenge". Peters commented that it is unfortunate that a new UV spacecraft is not forthcoming as many strong N II and N III lines are found in this spectral region. Lennon pointed out that HST/COS and HST/STIS are currently available and urged the audience to submit proposals.

The VLT-FLAMES group was then asked by Georges Meynet why there are systematic differences between the nitrogen abundances that they found and those determined by Norbert Przybilla and Kresimir Pavlovski. Lennon replied that their sample included a large number of intrinsically-slow rotators that tended to show large nitrogen abundances. There were too many stars with low values of $v$sin i for it to be an inclination effect. Many slow rotators showed a high N abundance, contrary to the predictions for simple rotational mixing. Improved NLTE calculations and further study are warranted. It was then pointed out that the calculations must not be that bad because the lowest N abundance is consistent with that of the baseline abundance for the LMC.

Norbert Przybilla recommended that the VLT-FLAMES group reinvestigate the N abundance from their FLAMES I database using N III. Peters pointed out that many good N III reside in the UV. Lennon then mentioned that there are good N III lines in the optical and that the VLT-FLAMES group is looking at both species in FLAMES II.

3. Theoretical Topics

The discussion was initiated by a question of Georges Meynet on mass loss: mass loss, in addition to be often clumpy in space also frequently presents some "clumpiness" in time. Indeed, mass loss, in many circumstances does not appear as a continuous flow but as shell ejections or outbursts. This seems to be the case for Be stars which eject matter in their equatorial disks in a discontinuous way. This is also the case for Luminous Blue Variables (LBV) which suffer very strong outbursts. Thus the question is what can be the physical cause of such behaviors? Can it be related to change of surface abundances?

Gloria Koenigsberger made the following addendum to that question: a handful of LBVs eruptions are apparently not conserving their bolometric luminosity. The bolometric luminosity increases. What can be the physical cause for these changes? This is a key question to answer for understanding these objects. Olivier Chesneau mentioned different works supporting the idea that the collective effects of non radial pulsations can play a role in stochastic mass ejections.

Stan Owocki emphasized the difference between baseline-type theories for wind mass loss and theories for shell ejections or eruptions. He noted that the CAK theory for line driven winds is very successful in explaining inferred mass loss of hot stars, and constitutes a good theoretical framework to study the effects of rotation, magnetic fields, pulsations, etc. However when stars approach critical limits, either the Eddington limit or a critical rotation limit, it becomes difficult to make firm predictions. Stan makes the analogy of water flowing over the edge of a fountain. When regular and not perturbed, this overflow is just at the steady rate set by the water source. But on a windy day the ripples and waves on the water surface lead to splashes and sprays, making the overflow quite variable and difficult to predict in detail. Analogously, when stars approach or reach a critical limit, small fluctuations can have dramatic effects, making the associated flows dynamical and variable, and hard or even impossible to predict. Nonetheless, when averaged over such fluctuations, the net outflow depends on whatever interior processes drive the star towards the critical limit, whether Eddington limit, critical rotation, or even Roche lobe overflow. If these interior driving processes are known and understood, then the averaged mass loss can be robustly predicted, since it is governed by the need to remove mass in order to evolve back from the critical limit and recover some more stable regime. For instance at critical rotation, mass loss will be governed by the necessity for the star to remove an excess of angular momentum. Near the Eddington limit, similar processes likely occur, though it is still not clear what causes the giant eruptions like the one of η Car. But again internal evolution probably plays a major role, and further study is needed to understand the nature of that.

Nathan Smith noted that in order to explain the giant LBV eruptions which may eject 15 $M_{\odot}$ with energies of the order of 10^{50} ergs, surface effects such as the reaching of some critical rotation or opacity limits may not be sufficient. Some processes occurring in deeper interior layers might need to be invoked for the larger eruptions.

Coming now to another point, Nathan Smith noted that the circumstellar nebula around LBVs offer a kind of archival record of what happened in the past, allowing one to trace back the history of mass ejection and changes in surface abundances. Determinations of the abundances in different parts of the nebula around η Car indicates that these shell eruptions can produce large and rather sudden changes in the surface abundances of a star, and may at least temporarily change its surface rotation rate after an eruption. Concerning η Car, he recalls that the N/O ratio has passed from solar value to values around 20 after the shell eruptions, producing a change of this ratio in an incredibly short interval of time of a few thousand years (Smith & Morse 2004). Thus he was wondering whether similar processes, although perhaps less extreme in amplitude compared to η Car, could also possibly be invoked to explain the puzzling positions of some stars in the Hunter diagram (N/H versus $v \sin i$ plane) as mentioned in the talk by Inès Brott, for instance those stars presenting strong surface enrichments and very low surface velocities.

References

Gies, D. R. & Lambert, D. L. 1992, *ApJ*, 387, 673
Hunter, I., Dufton, P. L., Smartt, S. J., Ryans, R. S. I. *et al.* 2007, *A&A*, 466, 277
Proffitt, C. R. & Quigley, M. F. 2001, *ApJ*, 548, 429
Smith, N. & Morse, J. A. 2004, *ApJ*, 605, 854

Active OB stars: structure, evolution, mass loss, and critical limits
Proceedings IAU Symposium No. 272, 2010
C. Neiner, G. Wade, G. Meynet & G. Peters, eds.

doi:10.1017/S1743921311010015

Radiation driven winds with rotation: the oblate finite disc correction factor

Ignacio Araya,[1] Michel Curé,[1] Anahí Granada[2] & Lydia S. Cidale[3]

[1]Departamento de Física y Astronomía, Facultad de Ciencias, Universidad de Valparaíso, Chile
[2]Facultad de Ciencias Astronómicas y Geofísicas, Universidad Nacional de La Plata, Argentina
[3]Instituto de Astrofísica La Plata, CCT La Plata-CONICET-UNLP, Argentina
[4]Observatoire de Geneve. Université de Geneve, Suisse

Abstract. We have incorporated the oblate distortion of the shape of the star due to the stellar rotation, which modifies the finite disk correction factor (f_D) in the m-CAK hydrodynamical model. We implement a simplified version for the f_D allowing us to solve numerically the non–linear m-CAK momentum equation. We solve this model for a classical Be star in the polar and equatorial directions. The star's oblateness modifies the polar wind, which is now much faster than the spherical one, mainly because the wind receives radiation from a larger (than the spherical) stellar surface. In the equatorial direction we obtain slow solutions, which are even slower and denser than the spherical ones. For the case when the stellar rotational velocity is about the critical velocity, the most remarkable result of our calculations is that the density contrast between the equatorial density and the polar one, is about 100. This result could explain a long-standing problem on Be stars.

Keywords. stars: rotation, stars: winds, outflows, stars: early-type, stars: emission-line, Be

1. Introduction

Pelupessy *et al.* (2000) formulated the wind momentum equation for sectorial line driven winds including the finite disk correction factor for an oblate rotating star with gravity darkening for both the continuum and the lines. They calculated models with line–force parameters around the bi–stability jump at 25 000 K. In this case, from the pole to the equator, the mass flux increases and the terminal velocity decreases. Their results showed a wind density contrast $\rho(equator)/\rho(pole)$ (hereafter ρ_e/ρ_p) of about a factor 10 independent of the rotation rate of the star.

In this work, we implement an approximative version for the oblate finite–disk correction factor, f_O, allowing us to solve numerically the non linear m–CAK momentum. We solve then this equation for a *classical Be* star, for polar and equatorial directions. In this study we do not take into account the bi–stability jump.

2. Oblate Factor

In order to incorporate the oblate distortion of the shape of the star due to the stellar rotation to the m–CAK hydrodynamic model, we implement an approximative function. In view of the behaviour of f_O and f_D we approximate its ratio f_O/f_D via a sixth order polynomial interpolation in the inverse radial variable $u = -R_\star/r$, i.e., $f_O = Q(u)\, f_D$. In this form, we assure that the topology found by Curé (2004) is maintained by the f_D term, but it is modified by the incorporation of the $Q(u)$ polynomial. With this approximation we can solve numerically the non–linear m–CAK differential equation.

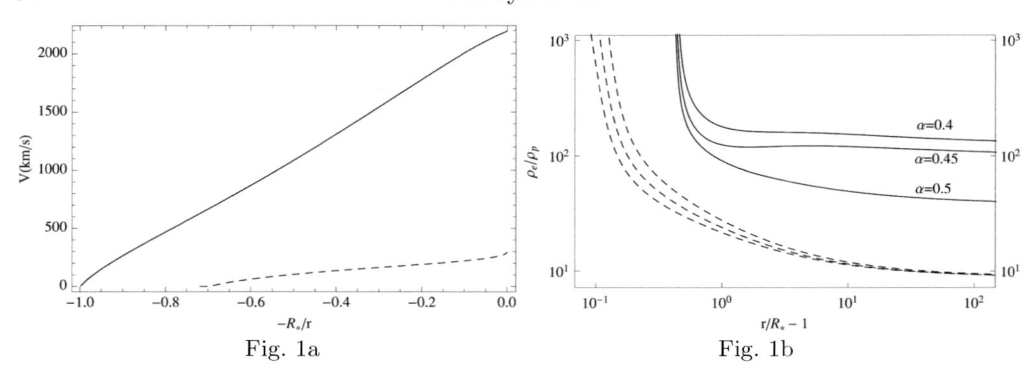

Fig. 1a Fig. 1b

Figure 1. a) Velocity versus u for equator (*dashed line*) and pole (*solid line*) with $\alpha = 0.4$ and $\delta = 0.07$. b) Solid lines: Density constrasts for $\omega = 0.99$ and $\delta = 0.07$. Dashed lines: Spherical cases.

3. Results

We solve the oblate m–CAK equations for a classical Be star with the following stellar parameters: $T_{eff} = 25\,000$ K, $log\, g = 4.03$, and $R/R_{\odot} = 5.3$ (Slettebak *et al.* 1980) and line–force parameter $k = 0.3$. For the other parameters, we have used two different values for δ, namely: $\delta = 0.07$ and $\delta = 0.15$, and three values for α, i.e., $\alpha = 0.4, 0.45, 0.55$. In this study, we have considered *the same value of* α*,* k *and* δ for the pole and equator, i.e. without taking into account the bi–stability jump. We show only solutions for $\omega = 0.99$ at the equator and pole. For the cases where $\alpha = 0.4$ and $\alpha = 0.45$, the density contrasts exceed a factor 100, values which are in agreement with observations (Lamers & Waters 1987).

4. Summary and Conclusions

The oblate correction factor has been implemented in an approximative form. The factor Q_r in the oblate correction factor certainly modifies the topology of the hydrodynamical differential equation and we suspect from first calculations that other critical points may exists and, therefore, more solutions might be present. We recover the observed density contrast only when the rotational velocity of the star is near the break-up velocity, confirming other theoretical works (see e.g., Townsend *et al.*, 2004). The use of a set of self–consistent line force parameters is necessary to understand the wind dynamics of these rapid rotators. The full version of the oblate correction factor will be implemented to study the topology of the wind.

References

Curé, M. 2004, *ApJ*, 614, 929
Lamers, H. J. G. L. M. & Waters, L. B. F. M. 1987, *A&A*, 182, 80
Pelupessy, I., Lamers, H. J. G. L. M., & Vink, J. S. 2000, *A&A*, 359, 695
Slettebak, A., Kuzma, T. J., & Collins, II, G. W. 1980, *ApJ*, 242, 171
Townsend, R. H. D., Owocki, S. P., & Howarth, I. D. 2004, *MNRAS*, 350, 189

Active OB stars: structure, evolution, mass loss, and critical limits
Proceedings IAU Symposium No. 272, 2010
C. Neiner, G. Wade, G. Meynet, G. Peters, eds.

doi:10.1017/S1743921311010027

Where do Be stars stand in the picture of rotational mixing?

Paul R. Dunstall[1], **Ines Brott**[2], **Philip L. Dufton**[1] **and Chris J. Evans**[3]

[1]Department of Physics and Astronomy, The Queens University of Belfast, BT7 1NN, Northern Ireland, UK
email: pdunstall01@qub.ac.uk

[2]Astronomical Institute, Utrecht University, Princetonplein 5, NL-3584CC,Utrecht, Netherlands

[3]UK Astronomy Technology Centre, Royal Observatory Edinburgh, Blackford Hill, Edinburgh, EH9 3HJ, UK

Abstract. Atmospheric parameters and photospheric abundances have been estimated for 60 Be-type stars located in 4 fields over the Magellanic Clouds. Particular attention has been given to the absolute nitrogen abundances to test theories of rotational mixing, an important factor in the evolutionary status of B-type stars, Hunter *et al.* (2008). The analysis used the non-LTE atmospheric code TLUSTY and required the implementation of a procedure to compensate for possible contamination due to the presence of a circumstellar disc. Through comparison with evolutionary models of fast rotating B-type stars and projected rotational velocity distributions our results support the theory that Be-type stars are typically faster rotators than B stars, but the measured nitrogen enhancements appear to be significantly less than expected for Be stars rotating with velocities greater than 70% of their critical velocity

Keywords. stars: emission-line, Be, stars: atmospheres, stars: rotation, stars: evolution, stars: abundances, (galaxies:) Magellanic Clouds

1. Introduction

Be-type stars are defined as B-type stars that shows or has previously shown prominent Balmer emission features, indicating the presence of a circumstellar disc. Although the mechanism of the phenomenon is still unclear, such stars are believed to rotate with velocities close to breakup velocity. Nitrogen can be used to determine the degree of mixing due to the rate of rotation.

2. Analysis and Discussion

Rotationally broaden profiles were used to obtain vsini estimates. The non-LTE code TLUSTY was used to estimate the effective temperature, surface gravity and micro-turbulence of 30 Be stars which presented observable nitrogen absorption lines. Effective temperature, surface gravities shown in Fig. 1. A procedure was developed to compensate for possible disc contamination, requiring the following assumptions:

- **Contamination is a constant featureless continuum**
- **Uniform over spectral range observed**

A comparison was made between B and Be star vsini distributions. That supported previous results yielding a 17% fraction of Be stars. Vsini distributions were modeled with Gaussian functions in order to obtain the intrinsic rotational velocity distribution (Figure. 2).

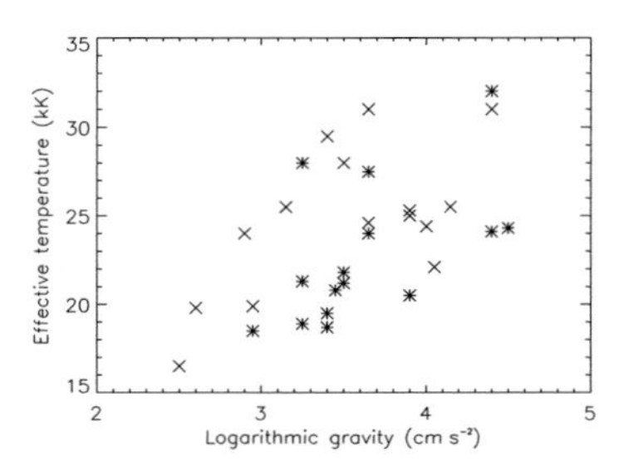

Figure 1. Effective temperature and surface gravity estimates plotted for the SMC (crosses) and LMC (stars) targets.

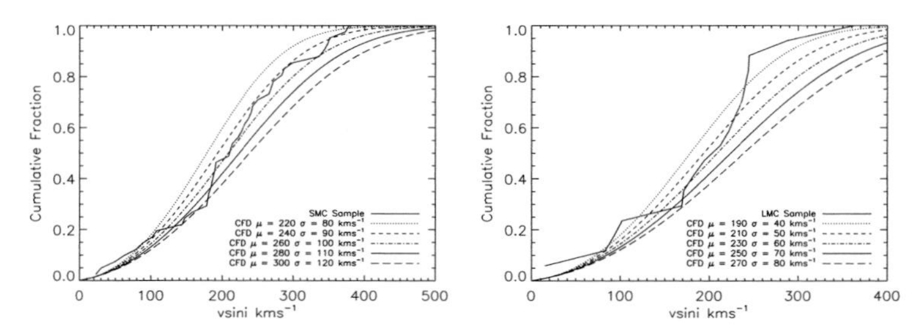

Figure 2. Cumulative fraction distributions for a range of Gaussian distributions based on the central velocity and width of the sample. SMC (**left**) LMC (**right**)

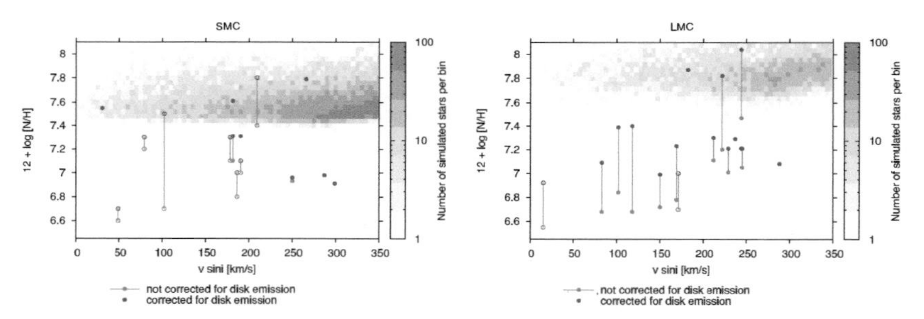

Figure 3. Nitrogen abundances and projected rotational velocities plotted for the SMC (**left**) and LMC (**right**) Be star samples before and after contamination correction. Also shown are the results of a simulation for evolutionary models of B stars with>70% breakup velocity (Brott *et al.*, in prep.).

Projected rotational velocities and absolute nitrogen abundances are plotted against a simulation of the expected nitrogen abundances and current vsini for B stars initially rotating with >70% of the breakup velocity (Fig. 3).

Our principle conclusions can be summarized as follows:

- Be stars appear to rotate faster than B stars, with more in the SMC.
- Nitrogen abundances appear inconsistent with initial rotational velocities close to breakup velocities.

Reference

Hunter, I., Lennon, D. J., Dufton, P. L., Trundle, C. *et al.* 2008a, *A&A*, 479, 541

Active OB stars: structure, evolution, mass loss, and critical limits
Proceedings IAU Symposium No. 272, 2010
C. Neiner, G. Wade, G. Meynet & G. Peters, eds.

doi:10.1017/S1743921311010039

About the Luminous Blue Variable He3-519

Anthony Hervé[1] and Jean-Claude Bouret[2]

[1]GAPHE, AGO, Université de Liège, Allée du 6 Août 17, Bât. B5c, 4000 Liège, Belgium
email: herve@astro.ulg.ac.be

[2]LAM, 38 Frédéric Joliot-Curie, 13388 Marseille, cedex 13, France
email: jean-claude.bouret@oamp.fr

Abstract. Luminous Blue Variables (LBVs) are massive stars, in a transition phase, from being O-type stars and rapidly becoming Wolf-Rayet objects. LBVs possess powerful stellar winds, high luminosities and show photometric and spectroscopic variability. We present the stellar and wind parameters of He3-519 obtained by the modeling of UVES observations with the model atmosphere code CMFGEN. We compare our results to previous studies in order to find mid-time scale variability of the stellar parameters and finally, we use stellar evolution models to determine the evolutionary status of this star.

Keywords. stars: early-type, stars: winds, outflows, stars: variables: other

1. Introduction

In 1994, He3-519 was reclassified as a LBV by Smith *et al.* (1994) with a WN11 spectral type. This LBV, surrounded by a nebulae of size 1', is located at 8 kpc from the sun and at 20' from AG Car. In 2008, we obtained and reduced seven new spectra obtained over two nights with the UVES spectrograph. Then we derived new stellar and wind parameters from the fit of the observational spectrum with the atmosphere model code CMFGEN (Hillier & Miller 1998). We compared them to the work of Smith *et al.* (1994), and we report a mid-time scale variability. Finally, we study its stellar evolution status with Geneva models in order to determine its initial mass of and its position in the HR diagram.

2. Tools and results

For our study, we use the code CMFGEN which computes line-blanketed, NLTE models, through a super-level approach, thus allowing the inclusion of many energy levels from ions of several species. The code solves the radiative transfer and statistical equilibrium equations in the comoving frame of the fluid, for a spherically symmetric outflow. After convergence of the model, a formal solution of the radiative transfer equation is computed in the observer's frame, thus providing the synthetic spectrum for comparison to observations. The code does not solve the full hydrodynamics, but rather assumes the density structure computed with TLUSTY (Lanz & Hubeny, 2003) in the deep layers, and the wind is described with a standard-velocity law. The velocity law is connected to the hydrostatic density structure from TLUSTY above (approximately) the sonic point. A simple, parametric treatment of wind clumping is implemented in CMFGEN. It assumes a void interclump medium and the clumps to be small compared to the photon mean free path. Others physical processes observed in stellar winds are included in the code, e.g. X-rays.

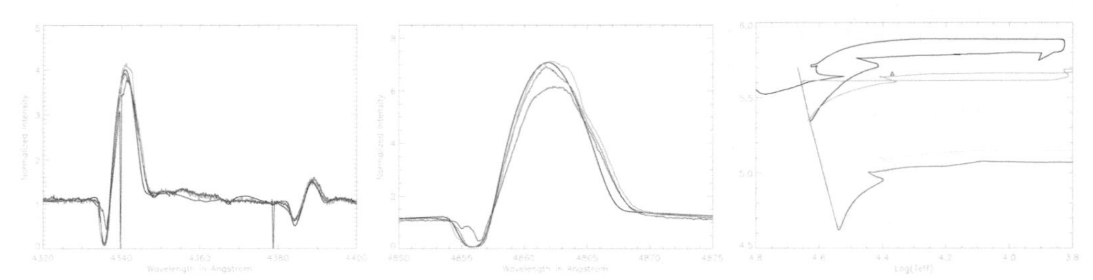

Figure 1. Left: Our best fit (in black) of H_γ λ 4340 and HeI λ 4387 of the UVES data (in red) and the spectra of Smith obtained in march 1991 and december 1993 (in green and blue, respectively). Center: The same as on the left, but for H_β λ 4861. Right: The upper part of the HR diagram. He3-519 (in red) is close to the evolution of a massive star of $40M_\odot$ (in green) without initial rotation. But, the comparison of chemical abundance lead us to choose the model with rotation (in black). The evolution tracks of a $25M_\odot$ with and without rotation (in blue and yellow, respectively) are added for comparison.

Table 1. Stellar and wind parameters of He3-519 compared to the most famous LBV, AG Car

	T_{eff} (K)	V_∞ ($km.s^{-1}$)	$\dot{M}$ (10^{-5} $M_\odot$ yr^{-1})	f	$\frac{He}{H}$ (in number)	$\frac{N}{H}$ (in number)	
He3-519 (2008)	23,500	350	5.5	0.3	0.33	$6.32.10^{-4}$	our work
He3-519 (1993)	27,200	365	12.0	no	0.55	*	Smith *et al.* (1994)
AG Car (1987-1990)	26,200	300	1.5	0.1	*	*	Groth *et al.* (2009)
AG Car (2003)	14,300	150	6.0	0.25	0.43	$1.40.10^{-4}$	Groth *et al.* (2009)

We have determined new stellar parameters for He3-519 (Fig.1, Tab.1). This LBV possesses a powerful stellar wind with a mass loss rate estimated about 5.5 10^{-5} M.yr^{-1} and a low terminal velocity of 350 kms^{-1}. Futhermore P Cygni wings of Hβ and Hγ show evidence of clumping. The analysis of the abundances of the principal elements confirmed the evolved status of this star. (2.0 and 3.44 for N). Nevertheless, the lack of O and C lines in our data, doesn't allow us to measure the abundances for these elements. As for others LBVs like AG Car, we suspect a mid-time scale variability lower than 15 years (Fig. 1). The difference between our results and those from previous studies (Smith *et al.* 1994) can't be blamed only on an evolution of the stellar atmosphere code but instead are real. With the evolution code of Geneva and our stellar parameters, we determined a possible evolution pattern. Our results are most compatible with the evolution of a $40M_\odot$ initial mass (nevertheless, for LBVs, it is difficult to determine the mass of the star by spectroscopy). In the near future, it will be important to re-observe He3-519 in order to constrain the knowledge of its variability cycle. Observations in the UV band can also help us to determine the abundance of oxygen and carbon and compare them to the evolution model.

References

Groh, J. H., Hillier, D. J., Damineli, A., Whitelock, P. A. *et al.* 2009, *ApJ*, 698, 1698
Hillier, D. J. & Miller, D. L. 1998, *ApJ*, 496, 407
Smith, L. J., Crowther, P. A., & Prinja, R. K. 1994, *A&A*, 281, 833
Lanz, T. & Hubeny, I. 2003, *ApJS*, 146, 417

Active OB stars: structure, evolution, mass loss, and critical limits
Proceedings IAU Symposium No. 272, 2010
C. Neiner, G. Wade, G. Meynet & G. Peters, eds.

doi:10.1017/S1743921311010040

Results from a recent stellar rotation census of B stars

Wenjin Huang[1], Douglas R. Gies[2] and M. Virginia McSwain[3]

[1]Department of Astronomy, University of Washington, Box 351580, Seattle, WA 98195-1580
email: hwenjin@astro.washington.edu

[2]Center for High Angular Resolution Astronomy, Department of Physics and Astronomy
Georgia State University, P.O. Box 4106, Atlanta, GA 30302-4106;
email: gies@chara.gsu.edu

[3]Department of Physics, Lehigh University, 16 Memorial Drive East, Bethlehem, PA 18015;
email: mcswain@lehigh.edu

Abstract. In an analysis of the rotational properties of more than 1100 B stars (~660 cluster and ~500 field B stars), we determine the projected rotational velocity (V sin i), effective temperature, gravity, mass, and critical rotation speed for each star. The new data provide us a solid observational base to explore many hot topics in this area: Why do field B stars rotate slower than cluster B stars? How fast do B stars rotate when they are just born? How fast can B stars rotate before they become Be stars? How does the rotation rate of B stars change with time? Does the evolutionary change in rotation velocity lead to the Be phenomenon? Here we report the results of our efforts in searching for answers to these questions based on the latest B star census.

Keywords. line: profiles, stars: early-type, stars: emission-line, Be, stars: fundamental parameters, stars: rotation

1. Observation and Analysis

Modern theoretical studies (Heger & Langer 2000; Meynet & Maeder 2000) predict that stellar rotation strongly influences the evolution of massive OB stars. Rapid rotation can trigger strong interior mixing, extend the core hydrogen-burning lifetime, significantly alter the luminosity, and change the chemical composition of the stellar surface over time. Spectroscopic investigation on a large and homogeneous star sample can provide the key data for us to measure rotational properties of massive stars and their evolutionary state and, therefore, is an ideal way to test our knowledge of the interior structure of a massive rotating star. Our first spectroscopic survey was carried out for about 500 cluster B stars in 2000 and 2001 (Huang & Gies 2006a). More recently, we accomplished two similar spectroscopic observing campaigns in 2006 (obtained ~230 cluster B stars) and in 2008 (obtained ~ 370 field B stars). In addition to these observed stars, about 100 field B stars from the Indo-U. S. stellar spectra Library are also included in our analysis.

For each star in our sample, we derived $V \sin i$ from a line profile fit using the realistic models with the gravitational darkening effect in mind. Then we determined $T_{\rm eff}$ and $\log g$ values by fitting the Hγ profile, and estimated $\log g_{\rm polar}$ which is not affected by stellar rotation and, therefore, can be used as a good indicator of evolutionary status of a rotating star. We also estimated stellar mass according to $T_{\rm eff}$ and $\log g_{\rm polar}$, and we then calculated the critical velocity, $V_{\rm crit}$, for each star. The details of these steps can be found in Huang & Gies (2006a,b) and Huang, Gies & McSwain (2010). Among the 1100 B stars, we found about 160 radial velocity variables which were not used to derive the statistical results summarized below.

2. Results

We summarize our major findings below:

1) The mean $V \sin i$ for the field sample (441 stars) is slower than that for the cluster sample (557 stars), confirming results from previous studies (Abt, Levato, & Grosso 2002; Huang & Gies 2006a; Wolff *et al.* 2007). By comparing stars with similar evolutionary status, we find that the stars in these two samples have similar rotational properties when plotted as a function of $\log g_{\rm polar}$. Thus we conclude that the overall slowness of rotation of field B stars is mainly due to the presence of a larger fraction of more evolved stars than found among cluster B stars.

2) The rotation distribution curves based on young stars with $\log g_{\rm polar} > 4.15$ suggest that massive stars are born at various rotation rates, including some very slow rotators ($V_{\rm eq}/V_{\rm crit} < 0.1$). The mass dependence suggests that higher mass B stars may preferentially experience angular momentum loss processes during and after formation. The low mass stars are born with more rapid rotators than high mass stars if the stellar rotation rate is evaluated by $V_{\rm eq}/V_{\rm crit}$.

3) The statistics based on the normal B (non-Be) stars in our sample indicates that low mass B stars ($< 4M_\odot$) may require a high threshold of $V_{\rm eq}/V_{\rm crit} > 0.96$ to become Be stars. As stellar mass increases, this threshold decreases, dropping to 0.64 for B stars with $M > 8.6M_\odot$. This implies that the mass loss processes leading to disk formation may be very different for low and high mass Be stars.

4) Comparing with modern evolutionary models of rotating stars (for 3 and $9M_\odot$ from Ekström *et al.* 2008), the apparent evolutionary trends of $< V \sin i/V_{\rm crit} >$ are in good agreement for the high mass B stars, but the data for low mass B stars shows a more pronounced spin-down trend than predicted.

5) Predictions for the fractions of rapid rotators and Be stars produced by the redistribution of angular momentum with evolution agree with observations for the higher mass B stars but vastly overestimate the Be population for the lower mass stars. The greater than expected spin-down of the lower mass stars explains this discrepancy and suggests that most of the low mass Be stars were probably spun up recently.

Acknowledgments

This material is based upon work supported by the National Science Foundation grant AST-0606861 (DRG). WJH thanks G. Wallerstein and the Kenilworth Fund of the New York Community Trust for partial financial support of this study. WJH is very grateful for partial finance support from NSF grant AST-0507219 to J. G. Cohen. MVM is grateful for support from NSF grant AST-0401460 as well as Lehigh University. This research has made use of the Simbad database, operated at CDS, Strasbourg, France, and of the Webda database, operated at the Institute for Astronomy of the University of Vienna.

References

Abt, H. A., Levato, H., & Grosso, M. 2002, *ApJ*, 573, 359
Ekström, S., Meynet, G., Maeder, A., & Barblan, F. 2008, *A&A*, 478, 467
Heger, A. & Langer, N. 2000, *ApJ*, 544, 1016
Huang, W. & Gies, D. R. 2006a, *ApJ*, 648, 580
Huang, W. & Gies, D. R. 2006b, *ApJ*, 648, 591
Huang, W., Gies, D. R., & McSwain, M. V. 2010, *ApJ* 722, 605
Meynet, G. & Maeder, A. 2000, *A&A*, 361, 101
Wolff, S. C., Strom, S. E., Dror, D., & Venn, K. 2007, *AJ*, 133, 1092

Active OB stars: structure, evolution, mass loss, and critical limits
Proceedings IAU Symposium No. 272, 2010
C. Neiner, G. Wade, G. Meynet & G. Peters, eds.

doi:10.1017/S1743921311010052

Mass and angular momentum loss of fast rotating stars via decretion disks

Jiří Krtička[1], **Stan P. Owocki**[2] **and Georges Meynet**[3]

[1]Masaryk University, Kotlářská 2, CZ-611 37 Brno, Czech Republic
email: krticka@physics.muni.cz

[2]Bartol Research Institute, University of Delaware, Newark, DE 19716, USA
email: owocki@bartol.udel.edu

[3]Geneva Observatory, CH-1290 Sauverny, Switzerland
email: Georges.Meynet@unige.ch

Abstract. The spinup of massive stars induced by evolution of the stellar interior can bring the star to near-critical rotation. In critically rotating stars the decrease of the stellar moment of inertia must be balanced by a net loss of angular momentum through an equatorial decretion disk. We examine the nature and role of mass loss via such disks. In contrast to the usual stellar wind mass loss set by exterior driving from the stellar luminosity, such decretion-disk mass loss stems from the angular momentum loss needed to keep the star near and below critical rotation, given the interior evolution and decline in the star's moment of inertia. Because the specific angular momentum in a Keplerian disk increases with the square root of the radius, the decretion mass loss associated with a required level of angular momentum loss critically depends on the outer radius for viscous coupling of the disk, and can be significantly less than the spherical, wind-like mass loss commonly assumed in evolutionary calculations.

Keywords. stars: mass loss, stars: rotation, stars: evolution

1. Angular momentum loss at the critical limit

Evolutionary models of massive stars show that the stellar rotation can increase with age on the main sequence (Meynet *et al.* 2006). In stars with moderately rapid initial rotation, and with only moderate angular momentum loss from a stellar wind, this spinup from internal evolution can even bring the star to critical rotation. Since any further increase in rotation rate is not dynamically allowed, the further contraction of the interior must then be balanced by a net loss of angular momentum through an induced mass loss.

We examine the scenario that such mass loss occurs through an equatorial, viscous decretion disk.

2. Analytic estimate

Let us assume rigid body rotation of the star. In this case the norm of the total stellar angular momentum J is given by the product of the stellar moment of inertia I and the rotation angular frequency Ω, $J = I\Omega$. The change of the moment of inertia during stellar evolution is $\dot{J} = \dot{I}\Omega + I\dot{\Omega}$. Once the star reaches the critical rotation frequency, $\Omega = \Omega_{\rm crit}$, the spin-up ends ($\dot{\Omega} = 0$), requiring a shedding of angular momentum given by $\dot{J} = \dot{I}\Omega_{\rm crit}$. Assuming this occurs via mass loss at a rate $\dot{M}$ through a Keplerian decretion disk, the angular momentum loss is set by the outer disk radius $R_{\rm out}$, given by $\dot{J}_{\rm K}(R_{\rm out}) = \dot{M} v_{\rm K}(R_{\rm out}) R_{\rm out}$, where the Keplerian velocity is $v_{\rm K}(r) = \sqrt{GM/r}$, and M is the stellar mass. Setting $\dot{J}_{\rm K}(R_{\rm out})$ equal to the above $\dot{J}$ required by a momentum of

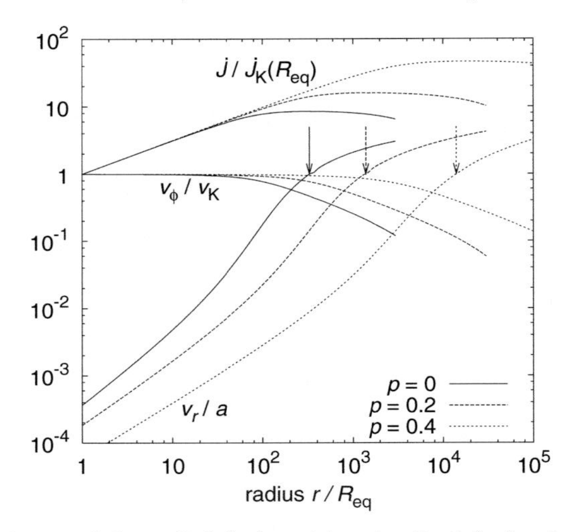

Figure 1. The dependence of the radial (v_r) and longitudinal (v_ϕ) velocities, and the angular momentum loss rate ($\dot{J}$) in units of equator release angular momentum loss rate $\dot{J}_K(R_{eq})$ on the radius in a viscous disk (a is the sound speed). Calculated for $T_{eff} = 30\,000\,K$, $M = 50\,M_\odot$, $R = 30\,R_\odot$ and different disk temperature laws $T = T_0(R_{eq}/r)^p$. Close to the star the anomalous viscosity keeps the matter in the disk in orbits which are nearly Keplerian ($v_\phi/v_K \approx 1$).

inertia change $\dot{I}$, we find the required mass loss rate is (R_{eq} is the equatorial radius)

$$\dot{M} = \frac{\dot{I}}{R_{eq}^2}\sqrt{\frac{R_{eq}}{R_{out}}}, \qquad (2.1)$$

As R_{out} gets larger the required mass loss rate is smaller.

3. Numerical disk models

To obtain a disc structure, we solve stationary hydrodynamic equations in cylindrical coordinates assuming axial symmetry. In analogy with the Shakura & Sunyaev (1973) accretion disks, we model an outward angular momentum transfer via anomalous viscosity. A maximum disk angular momentum loss is obtained in the case when the disk has its outer edge at the radius where $\dot{J}$ is maximum (see Fig. 1). An analytic approximation of the maximum angular momentum loss can be obtained (Krtička *et al.*, in preparation).

4. Radiative ablation

As the radiative force may drive large amount of mass out of the hot stars via line-driven wind it may also effectively set the outer disk radius. As the outer disk radius gets lower, the radiative ablation may decrease the efficiency of the disk angular momentum loss (Krtička *et al.*, in preparation).

Acknowledgements

This work was supported by grant GA ČR 205/08/0003.

References

Meynet, G., Ekström, S., & Maeder, A. 2006, *A&A*, 447, 623
Shakura, N. I. & Sunyaev, R. A. 1973, *A&A*, 24, 337

Active OB stars: structure, evolution, mass loss, and critical limits
Proceedings IAU Symposium No. 272, 2010
C. Neiner, G. Wade, G. Meynet & G. Peters, eds.

doi:10.1017/S1743921311010064

Mass loss in 2D rotating stellar models

Catherine Lovekin[1] **and Robert G. Deupree**[2]

[1]LESIA, Observatoire de Paris
5 Place Jules Janssen, 92195 Meudon, France
email: clovekin@lanl.gov

[2]Institute for Computational Astrophysics, Saint Mary's University
923 Robie St, Halifax, NS B3H 3C3, Canada

Abstract. Radiatively driven mass loss is an important factor in the evolution of massive stars. The mass loss rates depend on a number of stellar parameters, including the effective temperature and luminosity. Massive stars are also often rapidly rotating, which affects their structure and evolution. In sufficiently rapidly rotating stars, both the effective temperature and surface flux vary significantly as a function of latitude, and hence mass loss rates can vary appreciably between the poles and the equator. In this work, we discuss the addition of mass loss to a 2D stellar evolution code (ROTORC) and compare evolution sequences with and without mass loss.

Keywords. stars: evolution, stars: mass loss, stars: rotation

---

1. Introduction

Many massive stars are known to be rotating, and this causes a variation in temperature from pole to equator. Since radiatively driven mass loss depends sensitively on the temperature and local flux, these differences can affect the distribution of mass lost. These differences can cause small changes in the structure of a star, which can accumulate over the course of the evolution. In this paper, we investigate the 2D mass loss of rotating stars.

2. Models

We calculate mass loss rates for 2D stellar structure models calculated using ROTORC (Deupree 1990, Deupree 1995). To calculate the mass loss, we use the local effective temperature and luminosity at each surface zone. The mass loss rates are calculated using the prescription of Vink *et al.* (2001). The mass loss rate in each zone is weighted by the area of the zone to calculate the mass lost. Mass loss can also be calculated using the overall effective temperature and luminosity of the model, giving a pseudo-1D mass loss calculation. In this case, the mass lost is divided evenly among the angular zones.

3. Mass Loss

The 2D mass loss rates were compared to a pseudo-1D calculation based on the overall effective temperature and luminosity of the star. In both cases, the total amount of mass lost is the same. The 2D calculation, using the local temperature and luminosity has a much higher mass loss rate at the pole than the 1D case, while the mass loss rate at the equator is lower.

We have also compared the calculated variation in mass loss rates as a function of angle to that predicted using Castor, Abbott & Klein (1975, hereafter CAK) mass loss

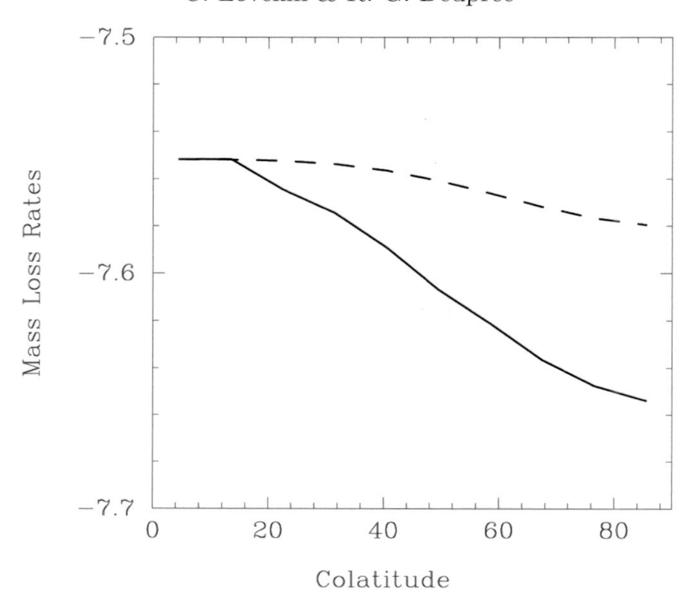

Figure 1. Calculated mass loss rates for a 20 $M_\odot$ rotating at 275 km/s for the 2D case (solid) and the CAK predictions (dashed). The mass loss rates in the 2D calculation decrease towards the equator much more rapidly than the CAK predictions.

rates and assuming a von Zeipel (1924) gravity darkening law. The CAK mass loss rates are normalized to be the same as the Vink mass loss rates at the pole. As shown in Figure 1, our calculated mass loss rates are considerably lower at the equator than the CAK predictions. Work is in progress to determine how much of this difference arises from the models, and how much arises from the different mass loss rates.

We have compared the interior structure of models that have had mass removed by both the 2D calculation and the pseudo 1D calculation. As discussed above, the type of calculation does not affect the total amount of mass removed, merely the distribution of this mass. A comparison of the 1D and 2D mass loss distributions results in differences in the core density of approximately 3.5%.

4. Conclusions

Preliminary results indicate that a full 2D calculation of mass loss using the local effective temperature and luminosity can significantly affect the distribution of mass loss in rotating main sequence stars. More mass is lost from the pole than predicted by 1D models, while less mass is lost at the equator. This change in the distribution of mass loss will affect the angular momentum loss, the surface temperature and luminosity, and even the interior structure of the star. After a single mass loss event, these effects are small, but can be expected to accumulate over the course of the main sequence evolution.

References

Castor, J. I., Abbott, D. C., & Klein, R. I. 1975, *ApJ*, 195, 157
Deupree, R. G. 1990, *ApJ*, 357, 175
Deupree, R. G. 1995, *ApJ*, 439, 357
Vink, J. S., de Koter, A., & Lamers, H. J. G. L. M. 2001, *A&A*, 369, 574
von Zeipel, H. 1924, *MNRAS*, 84, 665

Active OB stars: structure, evolution, mass loss, and critical limits
Proceedings IAU Symposium No. 272, 2010
C. Neiner, G. Wade, G. Meynet & G. Peters, eds.

doi:10.1017/S1743921311010076

Models of classical Be stars with gravity darkening

Meghan A. McGill, T. A. Aaron. Sigut and Carol E. Jones
Department of Physics and Astronomy, The University of Western Ontario,
London, Ontario, Canada, N6A 3K7
emails: mmcgill8@uwo.ca, asigut@uwo.ca, cejones@uwo.ca

Abstract. Classical Be stars are rapidly rotating, hot stars that possess an equatorial disk formed from gas released by the central star. The mechanism driving the stellar mass loss has yet to be fully explained, but the rapid rotation of the central B star is believed to be crucial. Rapid rotation also produces gravity darkening, and we have now extended our disk models to include these effects. In this contribution, we focus on the effect of gravity darkening on the thermal structure of a circumstellar disk.

Keywords. stars: emission-line, Be, circumstellar matter, stars: rotation

1. Introduction

A Be star disk re-processes the stellar radiation from the central star producing an emission line spectrum, particularly in the Hα line of hydrogen. A continuum excess is observed from the infrared to the radio, and Be systems can also exhibit linear polarization caused by electron scattering in the flattened disk (Porter & Rivinius 2003). The major focus of Be star research is to explain how these disks are formed.

In a rapidly rotating star, the centrifugal acceleration causes the magnitude of the local gravity to decrease towards the equator. This distorts the shape of the star, making the radius larger at the equator than at the pole, and it also produces a decrease in the local temperature near the equator and an increase near the pole (Collins, 1966). This is illustrated in Fig. 1. The critical rotation velocity, $v_{\rm crit}$, is defined as the equatorial rotation speed at which the equator is unbound, and at critical rotation, the equatorial radius is 1.5 times the polar radius. While the changes in the central star due to rapid rotation are well known, there has been little systematic work on how rapid rotation changes the temperature structure of the circumstellar disk.

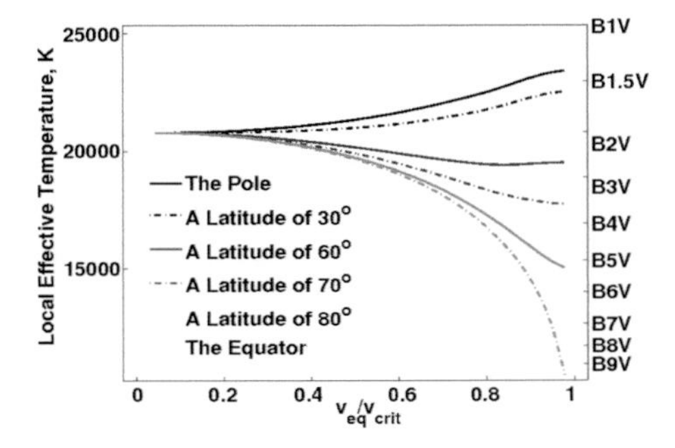

Figure 1. The divergence in the stellar surface temperatures at different latitudes with increasing rotation.

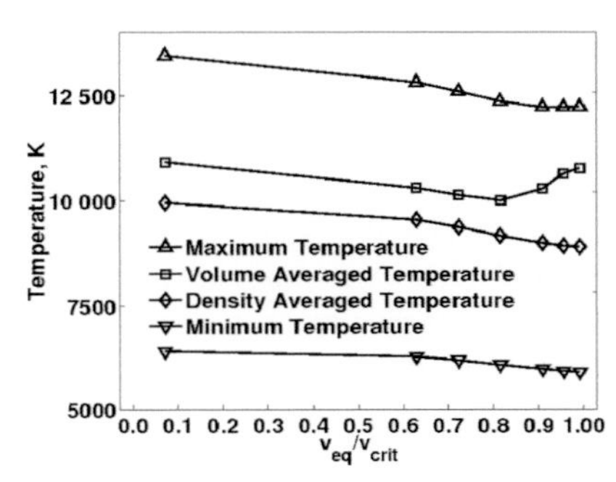

Figure 2. The change in several disk temperature diagnostics with increasing rotation.

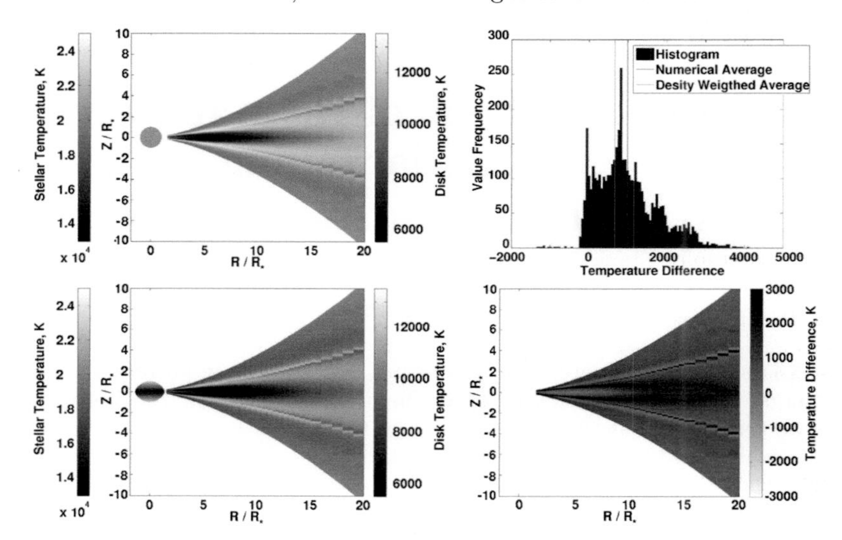

Figure 3. Top left: The temperature structure of a star and disk in the non-rotating case. Bottom left: The temperature structure of a star and disk for a star rotating at $v_{\rm frac} = 0.90$. Top right: A histogram of the temperature differences between the two models. Positive numbers indicate that the non-rotating model is hotter. Bottom right: The temperature differences in the disk between the non-rotating model and the rotating model.

2. Calculations

The BEDISK code of Sigut & Jones (2007) has been modified to include gravity darkening. To illustrate these effects, we have calculated the thermal structure of a disk with an equatorial density of $\rho(R) = 5 \cdot 10^{-11}(R_*/R)^{3.5}$ g cm^{-3} and of solar composition. The central star is assumed to be a B2V star with $R_p = 5.33\,R_\odot$ and $M = 9.11\ M_\odot$. The temperature structure of this disk was computed for rotation rates ranging from 0% to 99% of the critical rotation rate.

3. Temperature Results

The bulk of the disk becomes cooler as stellar rotation increases. This is shown in Fig. 2 and Fig. 3. The decrease in the average disk temperature is ≈ 800 K as $v_{\rm frac}$ increases to 0.8. With faster rotation, the increasing temperature of the stellar pole begins to provide additional heating for the disk. At high rotation rates the density-averaged temperature continues to decrease, but the volume-averaged temperature increases as shown in Fig. 2. A detailed comparison of the disk temperature structure between a non-rotating star and one rotating at $v_{\rm frac} = 0.9$ is shown in Fig. 3. The inner cool core experiences no significant temperature change, but it does grow larger. The outer portions of the disk and regions far from the mid-plane decrease in temperature. The most interesting behaviour is in the hot sheet of material above the cool core, this sheet moves higher up in the disk, becomes narrower and increases slightly in temperature. More results, including those from additional spectral types, will be presented in McGill *et al.* 2010 (in preparation).

References

Collins, II, G. W. 1966, *ApJ* 146, 914
Porter, J. M. & Rivinius, T. 2003, *PASP*, 115, 1153
Sigut, T. A. A. & Jones, C. E. 2007, *ApJ*, 668, 481

Active OB stars: structure, evolution, mass loss, and critical limits
Proceedings IAU Symposium No. 272, 2010
C. Neiner, G. Wade, G. Meynet & G. Peters, eds.

doi:10.1017/S1743921311010088

Mixing in two magnetic OB stars discovered by the MiMeS collaboration

Thierry Morel

Institut d'Astrophysique et de Géophysique, Université de Liège, 4000 Liège, Belgium

Abstract. Recent observational and theoretical arguments suggest that magnetic OB stars may suffer more mixing than their non magnetic analogs. We present the results of an NLTE abundance study revealing a lack of CN-cycled material at the surface of two magnetic stars discovered by the MiMeS project (NGC2244 #201 and HD 57682). The existence of a strong magnetic field is therefore not a sufficient condition for deep mixing in main-sequence OB stars.

Keywords. stars: magnetic fields, stars: early-type, stars: abundances, stars: fundamental parameters, stars: individual (NGC2244 #201, HD 57682)

---

1. Introduction

The existence of a sizeable population of slowly-rotating and unevolved, yet N-rich, B stars in the Magellanic clouds (Hunter *et al.* 2008) and in the Galaxy (e.g., Gies & Lambert 1992) urges the need to (re)address the impact that magnetic fields may have on mixing of the internal layers in OB stars. This is mainly motivated by two considerations. First, the Geneva evolutionary models incorporating magnetic fields generated through dynamo action predict a greater amount of mixing and hence higher CNO abundance anomalies (Maeder & Meynet 2005). Second, indication for a high incidence of a nitrogen excess in magnetic B stars was found by Morel *et al.* (2008) who found 8 out of the 10 magnetic stars in their sample to be N rich by a factor ~3. To further characterize the mixing properties of magnetic OB stars, we present the first results of an NLTE abundance study of a number of stars recently detected by the MiMeS collaboration.

2. Observations and results

High-resolution ($R \sim 46{,}000$) FIES spectra of the four O9–B2 IV–V targets (NGC 2244 #201, Par 1772, NU Ori and HD 57682) were obtained in late 2009 at the Nordic Optical Telescope. Spectropolarimetric observations of these stars revealed a field with a strength in the range 500–2000 G (Alecian *et al.* 2008; Grunhut *et al.* 2009; Petit *et al.* 2008). Spectral synthesis of HD 57682 using the NLTE, unified code CMFGEN provides $T_{\rm eff} = 34500 \pm 1000$ K and $\log g = 4.0 \pm 0.2$ dex (Grunhut *et al.* 2009).

The atmospheric parameters are derived purely on spectroscopic grounds: $\log g$ from fitting the collisionally-broadened wings of the Balmer lines, $T_{\rm eff}$ from ionisation balance of various species (He, N, Ne and/or Si) and the microturbulence, ξ, from requiring the abundances yielded by the O II lines to be independent of their strength. The abundances are computed using Kurucz atmospheric models, the NLTE line-formation codes DETAIL/SURFACE and classical curve-of-growth techniques. Here we present the results for the two narrow-lined stars NGC 2244 #201 and HD 57682. The two fast rotators remain to be analysed using spectral synthesis techniques. The atmospheric parameters and abundances are provided in Table 1 where they can be compared with the values obtained following exactly the same methodology for the magnetic, N-rich star τ Sco.

Table 1. Atmospheric parameters and elemental abundances of NGC 2244 #201 and HD 57682 (on a scale in which $\log \epsilon$[H]=12). Results obtained for τ Sco using exactly the same tools and techniques are shown for comparison (Hubrig *et al.* 2008). The number of used lines is given in brackets. A blank indicates that no value could be determined. The solar [N/C] and [N/O] ratios are −0.60±0.08 and −0.86±0.08 dex, respectively (Asplund *et al.* 2009).

	NGC 2244 #201	HD 57682	τ Sco
$T_{\rm eff}$ (K)	27000±1000	33000±1000	31500±1000
$\log g$	4.20±0.15	4.00±0.15	4.05±0.15
ξ (km s^{-1})	3±3	5±5[a]	2±2
$v \sin i$ (km s^{-1})	22±2	25±4	8±2
He/H	0.072±0.023 (9)	0.106±0.030 (10)	0.085±0.027 (9)
$\log \epsilon$(C)	8.22±0.13 (6)	8.20±0.19 (6)	8.19±0.14 (15)
$\log \epsilon$(N)	7.68±0.13 (20)	7.52±0.25 (8)	8.15±0.20 (35)
$\log \epsilon$(O)	8.63±0.18 (31)	8.31±0.21 (14)	8.62±0.20 (42)
$\log \epsilon$(Ne)	8.02±0.12 (7)	7.95±0.17 (1)	7.97±0.10 (5)[b]
$\log \epsilon$(Mg)	7.29±0.20 (1)	7.37±0.18 (1)	7.45±0.09 (2)
$\log \epsilon$(Al)	6.20±0.13 (3)	6.07±0.21 (1)	6.31±0.29 (3)
$\log \epsilon$(Si)	7.41±0.25 (5)	7.47±0.32 (5)	7.24±0.14 (9)
$\log \epsilon$(S)	7.30±0.19 (1)		7.18±0.28 (3)
$\log \epsilon$(Fe)	7.33±0.13 (20)		7.33±0.31 (13)
[N/C]	−0.54±0.14	−0.68±0.30	−0.04±0.25
[N/O]	−0.95±0.21	−0.79±0.19	−0.47±0.29

Notes:
[a] Assumed values.
[b] From Morel & Butler (2008).

3. Discussion

These two main sequence stars do not show evidence for contamination of their surface layers by core-processed material. In the case of NGC 2244 #201, the CNO logarithmic abundance ratios are consistent with the solar values, fully confirming the results of Vrancken *et al.* (1997). The results for HD 57682 are more uncertain owing to the weakness of the spectral lines and their strong $T_{\rm eff}$ sensitivity, but there is no indication for significant departures from the solar ratios either. It thus appears that these two stars do not display the N excess observed in other magnetic B stars (Morel *et al.* 2008). We conclude that the relationship between magnetic fields and an N excess may only be statistical and that other (still elusive) parameters may control the amount of mixing experienced by B-type stars. This is particularly well illustrated by considering the dichotomy between the CNO abundance ratios of HD 57682 and τ Sco (Table 1) despite having similar characteristics in terms of location in the HR diagram and rotation rate.

References

Alecian, E., Wade, G. A., Catala, C., Bagnulo, S. *et al.* 2008, *A&A*, 481, L99
Asplund, M., Grevesse, N., Sauval, A. J., & Scott, P. 2009, *ARAA*, 47, 481
Gies, D. R. & Lambert, D. L. 1992, *ApJ*, 387, 673
Grunhut, J. H., Wade, G. A., Marcolino, W. L. F., Petit, V. *et al.* 2009, *MNRAS*, 400, L94
Hubrig, S., Briquet, M., Morel, T., Schöller, M. *et al.* 2008, *A&A*, 488, 287
Hunter, I., Brott, I., Lennon, D. J., Langer, N. *et al.* 2008, *ApJ* (Letters), 676, L29
Maeder, A. & Meynet, G. 2005, *A&A*, 440, 1041
Morel, T., Hubrig, S., & Briquet, M. 2008, *A&A*, 481, 453
Morel, T. & Butler, K. 2008, *A&A*, 487, 307
Petit, V., Wade, G. A., Drissen, L., Montmerle, T. *et al.* 2008, *MNRAS*, 387, L23
Vrancken, M., Hensberge, H., David, M., & Verschueren, W. 1997, *A&A*, 320, 878

Active OB stars: structure, evolution, mass loss, and critical limits
Proceedings IAU Symposium No. 272, 2010
C. Neiner, G. Wade, G. Meynet & G. Peters, eds.

doi:10.1017/S174392131101009X

Evolutionary effects of stellar rotation of massive stars in their pre-supernova environments

Brenda Pérez-Rendón[1], Horacio Pineda-León[2], Alfredo Santillán[3] and Liliana Hernández-Cervantes[4]

[1]Departamento de Investigación en Física. Universidad de Sonora. Hermosillo, Sonora, México
email: brenda@cajeme.cifus.uson.mx

[2]Posgrado en Ciencias Físicas, DIFUS. Universidad de Sonora. Hermosillo, Sonora, México
[3]DGSCA, UNAM. Cd. Universitaria, México, D.F.
[4]Instituto de Astronomía, UNAM. Cd. Universitaria, México, D.F.

Abstract. Massive main sequence stars are fast rotators. Stellar rotation affects massive stellar rotation due to rotationally induced mixing processes, the increase of mass loss rates, etc. and also affects the circumstellar medium due to their interaction with the stellar wind. The parameters of stellar winds depends on stellar parameters so the wind parameters change as the star evolves, coupling the evolution of circumstellar medium to the star itself. In this work we used a stellar code to build models of two massive stars (30 and 40 $M_{\odot}$) and we used their wind parameters to simulate the hydrodynamics of their surrounding gas with the ZEUS-3D code in order to explore the effects of stellar rotation in the pre-supernova environments.

Keywords. stars: rotation, stars: evolution, stars: circumstellar matter

---

1. Introduction

The strong mass loss in massive stars affects significantly their stellar evolution. When the stars pass through evolutionary phases they develop different stellar wind parameters which may produce a wide variety of structures in the surrounding gas. At the end of its life, the star will explode in a SN explosion and the resulting shock wave will interact with this modified medium. In this work we built up several stellar evolution models (using the STERN code, Langer & El Eid, 1986) and their corresponding circumstellar gas evolution. We calculate four stellar models with masses of 30 and 40$M_{\odot}$ in two sequences of stellar models that are calculated with the same input physics, differing only in the value of their initial rotational velocity in order to compare the effect of rotation in stellar evolution and their circumstellar pre-SN environment. We calculated a non-rotating model with an initial rotational velocity of $V_{rot} = 0$ km/s (**model A**) and a rotating model with $V_{rot} = 200$ km/s (**model B**) as the models used by Pérez-Rendón *et al.* (2008). Then we have used the values of mass loss rate and stellar wind velocity as inner boundary conditions in an explicit hydrodynamical code ZEUS-3D (Stone & Norman, 1992) to simulate the hydrodynamical evolution of the circumstellar medium. These numerical simulations are done in the same way as García-Segura *et al.* (1996), Pérez-Rendón *et al.* (2009).

2. Rotational effects

The stellar evolution of **model A** (30 $M_{\odot}$) was computed from the ZAMS up to oxygen core exhaustion prior to Si burning (pre-SN stage). After the RSG stage, this star

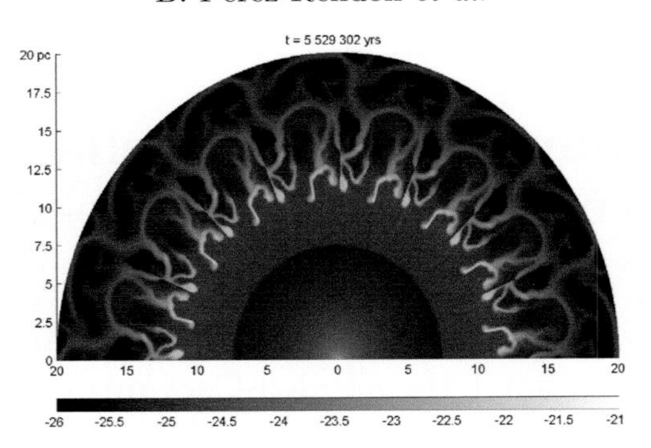

Figure 1. Circumstellar medium pre-SN in 30 $M_{\odot}$, model B. The figure shows the logarithm of gas density (g cm^{-3}).

evolves to the blue side of the HRD as a Luminous Blue Supergiant (LBSG), performing a blue loop. **Model B** (30 $M_{\odot}$) stellar evolution was computed from the ZAMS until Ne exhaustion in the core. The rotating star has a shorter RSG phase than the non-rotating star, but the mass loss rate is enhanced and the rotating star evolves to a Wolf-Rayet H-rich (WNL) star. In our 30 $M_{\odot}$ models the stellar evolution was different and each model built up different pre-supernova media. The distributions of density, temperature, chemical composition and velocity field of the pre-supernova environment are different in both cases, for the same initial mass.

In both models, the mass loss during the MS built up a circumstellar wind-blown cavity surrounding the star, bordered by a thin, dense, cold shell (Weaver *et al.*, 1977) with a radius greater than 30 pc. In the RSG stage, a shell of shocked RSG wind starts to build up closer to both star models (R $<$ 5 pc). During the post-RSG evolution the stellar wind velocity increases to 500 km/s in the LBSG case (**model A**), and to 1500 km/s in the WR case (**model B**). We observe the formation of a "blue" shell around the non-rotating star and the WR shell built by the rotating star. These shells are unstable due to Vishniac and/or Rayleigh-Taylor instabilities while they propagates outwards. The post-RSG shell eventually hits the RSG shell with different velocities, forming a swept "blue+RSG" unstable shell around the $30 M_{\odot}$ (**model A**) star at a radius of 6pc, and the **model B** star pre-SN environment is a fragmented "WR+RSG" shell around the WR progenitors at a distance of approximately 12 pc from the star, as is shown in Fig. 1. In the 40 $M_{\odot}$ models, the rotation does not affect the evolutionary track of the star (both models become a WR star at the pre-SN stage) but it affects the CSM morphology and chemical environment. This work will be discussed in a forthcoming paper.

Acknowledgements

This work has been partially supported from CONACyT proyect 104651, Intercambio UNAM-UNISON U38P170 and DGAPA–UNAM grant IN121609.

References

García-Segura, G., Langer, N., & Mac Low, M.-M. 1996, *A&A*, 316, 133
Langer, N. & El Eid, M. F. 1986, *A&A*, 167, 265
Pérez-Rendón, B., García-Segura, G., & Langer, N. 2008, *Revista Mexicana de Fisica Supplement* 54, 101
Pérez-Rendón, B., García-Segura, G., & Langer, N. 2009, *A&A*, 506, 1249
Stone, J. M. & Norman, M. L. 1992, *ApJS*, 80, 753
Weaver, R., McCray, R., Castor, J., Shapiro, P. *et al.* 1977, *ApJ*, 218, 377

Active OB stars: structure, evolution, mass loss, and critical limits
Proceedings IAU Symposium No. 272, 2010
C. Neiner, G. Wade, G. Meynet & G. Peters, eds.

doi:10.1017/S1743921311010106

The Nitrogen Abundance in Be Stars Determined from UV Spectra

Geraldine J. Peters

Space Sciences Center/Department of Physics & Astronomy, University of Southern California, Los Angeles, CA 90089-1341, USA
email: gjpeters@mucen.usc.edu

Abstract. A NLTE abundance analysis of the *IUE* spectra of 8 Be stars with $v \sin i < 150$ $km\ s^{-1}$ reveals no evidence for nitrogen enrichment from mixing of core and photospheric material.

Keywords. stars: abundances, stars: emission-line, Be, stars: rotation, stars: evolution

1. Overview of Project

The abundance of nitrogen in Be stars is of interest because contemporary models for the structure and evolution of rapidly rotating OB stars predict a photospheric enrichment due to the mixing of the processed material from the stars core with the original surface material (Meynet & Maeder 2000, Maeder & Meynet 2000). But the analyses of Be star spectra using standard spectrum synthesis techniques are confronted with some challenges that are not encountered in abundance studies of sharp-lined, non-emission B stars, including the treatment of blended, rotationally broadened lines, assessment of the microturbulence, correction for disk emission and possible shell absorption, and latitudinal variation of $T_{\rm eff}$ and log g. Results are presented from a spectrum synthesis study of high dispersion FUV spectra of Be stars from the *IUE* spacecraft. The Hubeny/Lanz NLTE codes TLUSTY/SYNSPEC (Hubeny 1988, Hubeny & Lanz 1995) and the Lanz & Hubeny (2007) model atmospheres for B stars were employed. The FUV offers an advantage over the optical region as there is far less influence from disk emission and some FUV lines of N II-III are intrinsically stronger. However shell lines can sometimes significantly blend with the lines that are the focal point of the study and make it difficult to place the continuum. This study focused on Be stars with values of $v \sin i <$ 150 km s^{-1} in which the effect of latitudinal parameter variation is minimized (Frémat *et al.* 2005).

2. Results

The Be stars considered in this study are in the B1-B3 range and include 11 Cam (B3 Ve), FW CMa (B3 Ve), 16 Peg (B3 Ve), ω CMa (B2.5 Ve), 31 Peg (B2 IVe), μ Cen (B2 IV-Ve), MX Pup (B1.5 IVe), and χ Oph (B1.5 Ve). The starting values for $T_{\rm eff}$, log g, and $v \sin i$ were from Frémat *et al.* (2005). $T_{\rm eff}$ and log g were then adjusted based upon the observed strengths of the Si II and Si IV features. For the cooler Be stars log g was assessed from the outer red wing of Lyα. $V \sin i$ was determined from several moderately-strong, isolated lines. The most uncertain parameter is the microturbulence. Based upon several published abundance studies of B star spectra in the UV, I adopted a $V_{\rm turb}$ of 4 km s^{-1} for stars earlier than B2 and $1-2$ km s^{-1} for the cooler objects. Some useful spectral features for determining the carbon and nitrogen abundance include C III 1176 Å and N III 1183.0, 1184.5 Å for Be stars with spectral types of B2 or earlier and C II 1323.9 Å, C III 1247.4 Å, and N I 1243.2 Å in the B2–B3 stars.

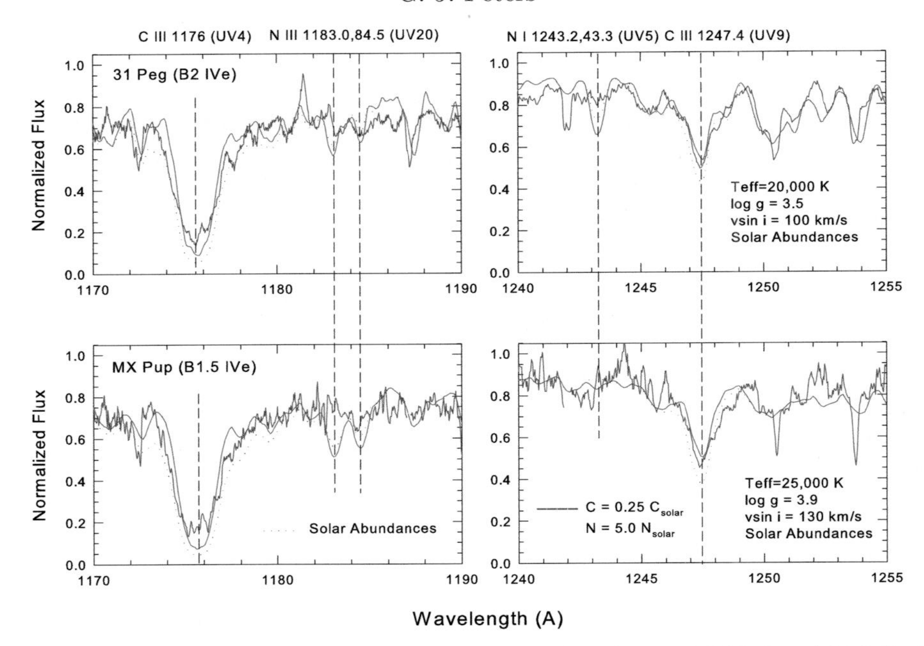

Figure 1. Comparison between the observed spectra of 31 Peg and MX Pup (*thick line*) and model spectra computed with the stellar parameters indicated on the plot. Model spectra are for solar abundances (*dotted line*) and elevated nitrogen but reduced carbon (*thin line*).

Representative comparisons between the model calculations and observations are shown in Fig. 1. The observed spectra were produced from coadding good SWP HIRES images that were taken with the large aperture. Nitrogen enrichment in the B2-B3 stars would be revealed by a stronger N I at 1243 Å , while in the hotter stars the N III doublet at 1183.0, 1184.5 Å would be prominent. From the spectra shown in Fig. 1 it is clear that the presence of an elevated nitrogen abundance is not indicated. In fact often the observed N lines are weaker than one would predict even with solar abundances. C III 1176 Å tends to be weak in most stars and its profile suggests emission filling. Overall the Si III lines are also weak and indicate the possible presence of a 20000 K circumstellar plasma that primarily radiates in the doubly-ionized species. Given the errors it appears that the N abundance in the Be stars is normal. None of the program Be stars show clear evidence for an elevated nitrogen abundance. Expected mixing is apparently suppressed, and this study lends no support for Be star models based upon critical rotation, a condition where an especially high N abundance is predicted.

Acknowledgements

This project received support from NASA Grants NNX07AH56G, NNX06AD34G, NNX07AF89G, and the USC WiSE program.

References

Frémat, Y., Zorec, J., Hubert, A.-M., & Floquet, M. 2005, *A&A*, 440, 305
Hubeny, I. 1988, *Computer Physics Communications* 52, 103
Hubeny, I. & Lanz, T. 1995, *ApJ*, 439, 875
Lanz, T. & Hubeny, I. 2007, *ApJS*, 169, 83
Maeder, A. & Meynet, G. 2000, *ARAA*, 38, 143
Meynet, G. & Maeder, A. 2000, *A&A*, 361, 101

Active OB stars: structure, evolution, mass loss, and critical limits
Proceedings IAU Symposium No. 272, 2010
C. Neiner, G. Wade, G. Meynet & G. Peters, eds.

doi:10.1017/S1743921311010118

Spectroscopic and interferometric approach for differential rotation in massive fast rotators

Juan Zorec[1], Yves Frémat[2], Omar Delaa[3], Armando Domiciano de Souza[3], Philippe Stee[3], Denis Mourard[3], Lydia S. Cidale[4], and Christophe Martayan[5]

[1]Institut d'Astrophysique de Paris, UMR 7095 du CNRS, Univ.P&MC, France (zorec@iap.fr)
[2]Royal Observatory of Belgium, 3 av. Circulaire, 1180 Brussels, Belgium
[3]Laboratoire Fizeau, UNS-OCA-CNRS UMR6203, Parc Valrose, 06108 Nice Cedex 02, France
[4]Facultad de Ciencias Astronómicas y Geofísicas, Universidad Nacional de La Plata, Argentina
[5]European Organization for Astronomical Research in the Southern Hemisphere, Chile

Abstract. The coupling between the convective region in the envelope and rotation can produce a surface latitudinal differential rotation that may induce changes of the stellar geometry and on the spectral line profiles that it may be scrutinized spectroscopically and by interferometry.

Keywords. stars: rotation, techniques: spectroscopic, techniques: interferometric

1. Motivations

Clement (1979) using a conservative distribution of the angular velocity, and Maeder *et al.* (2008), with a "shellular" distribution, have shown that in fast rotating early-type stars the envelope layers laying from 1/3 to 1/4 of the stellar radius beneath the surface are unstable to convection. The baroclinic balance relation can then be solved by sketching the coupling between convection and rotation, assuming that the specific entropy is a function of the specific angular momentum j, Ω^2 or the specific rotation energy $\epsilon_\Omega = \varpi^2\Omega^2$. Fig. 1 shows the curves of constant Ω for $S = S(\Omega^2)$ which can closely reproduce the differential rotation in the convective zone of the Sun, and $S = S(\epsilon_\Omega)$ that reproduces the differential rotation profiles obtained with models for radiative stars. According to Maeder (2009) the shape of the surface of a star with non-conservative rotation law can be described with:

$$\Phi_{\rm G}(\theta) - \frac{1}{2}\Omega^2(\varpi, z)\varpi^2 + \frac{1}{2}\int_0^\theta \varpi^2(\nabla\Omega^2.d\mathbf{s}) = \Phi_{\rm G}(0)\ . \qquad (1.1)$$

where $\Phi_{\rm G}(\theta)$ is the gravitational potential; $d\mathbf{s}$ in an arbitrary displacement over the stellar surface; ϖ is the distance to rotation axis; $\Omega(\theta)$ is the angular velocity as a function on the co-latitude θ. For a simplified Maunder velocity law $\Omega_\odot(\theta) = \Omega_o(1+\alpha\cos^2\theta)$ the contours of stars shown in Fig. 1 were obtained for $\eta_o = \Omega_o^2/R_{\rm e}^3/GM = 0.8$ and $\alpha = 0.3$. We note that although $\eta_o \neq 1$ the stars approach the critical flattening $R_{\rm e}/R_{\rm p} = 1.5$ obtained for the critical rotation in the Roche approximation of the surface equipotential.

2. Spectroscopic and Interferometric signatures

With a modified FASTROT (Frémat *et al.* 2005) code we can obtain synthetic line profiles produced by differentially rotating model atmospheres, where the gravity

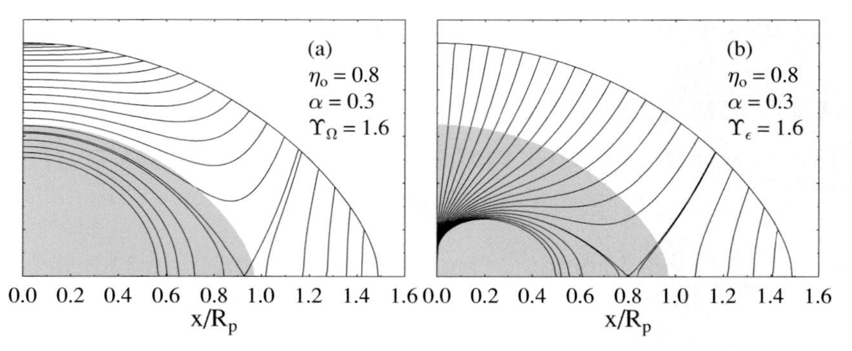

Figure 1. $\Omega(r, \theta) =$ constant curves. (**a**) $S = S(\Omega^2)$; (**b**) $S = S(\varpi^2\Omega^2)$. Solutions are not valid in the shaded zones. $\Upsilon_{\Omega,\epsilon} = [2\alpha_{\Omega,\epsilon} GM]/[1 - \alpha_{\Omega,\epsilon}(\Omega^2, \epsilon)]$; $\alpha_{\Omega,\epsilon} = (1/C_P)[dS/d(\Omega^2, \epsilon)]$

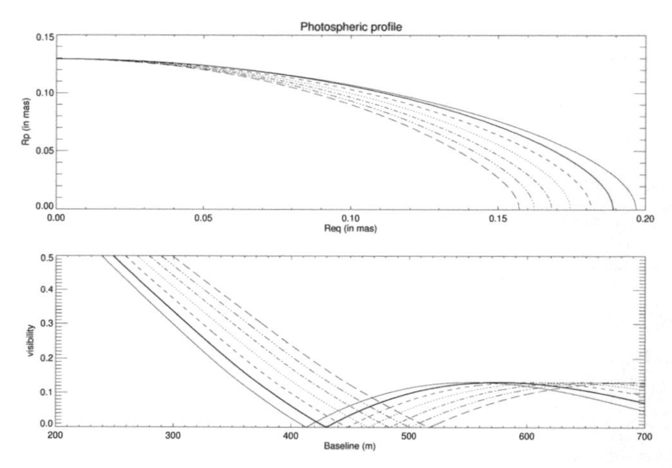

Figure 2. Apparent angular dimension of a B2V-type star seen at d = 300 pc having a surface rotational parameter $\eta_o = 0.7$ at the equator and deformed according to several values of the parameter α in the latitudinal differential rotation given by $\Omega_\odot(\theta) = \Omega_o(1 + \alpha \cos^2 \theta)$. Visibility curves for the respective stellar deformations induced by the surface differential rotation.

darkening effect is calculated consistently with the shape changes of the star according to the adopted Maunder rotation law. The Fourier transform of the line profiles show that the ratio of the first two zeroes is a fairly nicely resolved quantity to estimate the differential rotation parameter α. We have also calculated the visibility curves of differential-rotation deformed stars [see Fig. 2 as function of the baseline seen by the VEGA/CHARA for a B stars at 300 pc (Mourard *et al.* 2009)]. This calculation show that differences carried on the visibility curves in the visual spectral range by the geometrical deformations shown in the figure can be resolved with the VEGA/CHARA interferometric array. The combined spectroscopic and interferometric analysis could then lead to valuable constraints to the modeling of the surface differential rotation and the internal differential rotation beneath the surface.

References

Clement, M. J. 1979, *ApJ*, 230, 230
Frémat, Y., Zorec, J., Hubert, A.-M., & Floquet, M. 2005, *A&A*, 440, 305
Maeder, A., Georgy, C., & Meynet, G. 2008, *A&A*, 479, L37
Maeder, A. 2009, *Physics, Formation and Evolution of Rotating Stars*, (Springer)
Mourard, D., Clausse, J. M., Marcotto, A., Perraut, K. *et al.* 2009, *A&A*, 508, 1073

Questions in the audience. From left to right and top to bottom: Olivier Chesneau, Artemio Herrero (and Jean-Paul Zahn), Myron Smith, Véronique Petit, Marc Gagné, Norbert Langer, Christopher Tycner (and Erika Grundstrom), Matteo Cantiello (and Stéphane Mathis).

Active OB stars: structure, evolution, mass loss, and critical limits
Proceedings IAU Symposium No. 272, 2010
C. Neiner, G. Wade, G. Meynet & G. Peters, eds.

doi:10.1017/S174392131101012X

Observations of magnetic fields in hot stars

Véronique Petit

Dept. of Geology & Astronomy, West Chester University, West Chester, PA 19383, USA
email: VPetit@wcupa.edu

Abstract. The presence of magnetic fields at the surfaces of many massive stars has been suspected for decades, to explain the observed properties and activity of OB stars. However, very few genuine high-mass stars had been identified as magnetic before the advent of a new generation of powerful spectropolarimeters that has resulted in a rapid burst of precise information about the magnetic properties of massive stars. During this talk, I will briefly review modern methods used to diagnose magnetic fields of higher-mass stars, and summarize our current understanding of the magnetic properties of OB stars.

Keywords. stars: early-type, stars: magnetic fields, techniques: polarimetric

1. Introduction

Hot, massive stars have an enormous impact on their galactic environment. Energy and momentum are injected into the surrounding interstellar medium by their powerful stellar winds and radiation fields during their short, but nonetheless colourful, lives and by the dramatic core-collapse supernovae that mark their deaths. In this way, they seed the interstellar medium with the products of their nucleosynthesis, to be recycled into the next generations of stars and planets, hence playing a key role in the chemical enrichment of the Universe. These rapidly-evolving stars thereby drive the chemistry, structure and evolution of galaxies, dominating the ecology of the Universe - not only as supernovae, but also during their entire lifetimes - with far-reaching consequences.

The evolution of a massive star is strongly determined by its rotation, as well as the mass lost through its stellar wind, both of which can be influenced by the presence of a magnetic field. A field can couple different layers of a star's interior, hence modifying internal differential rotation (Maeder and Meynet 2005). If a field has a large-scale component that extends outside the stellar surface, it can also channel a stellar wind, creating a structured wind - a magnetosphere - which will modify the rate and geometry of mass loss. Furthermore, if the field couples the rotating surface of the star with its outflowing stellar wind, both effects will result in a different angular momentum loss (via the outflowing stellar wind) than that of a non-magnetic star (ud-Doula *et al.* 2009). As angular momentum and mass loss are determining factors in stellar evolution calculations, it is crucial that the effect of magnetic field be understood properly in order to correctly use evolutionary tracks and isochrones when interpreting, for example, large datasets of OB associations (see Evans, these proceedings).

In the last decade, our knowledge of the basic statistical properties of massive star magnetic fields has significantly improved, in part due to a new generation of powerful spectropolarimetric instrumentation. In this paper I will review modern methods used to diagnose magnetic fields of higher-mass stars, and briefly summarize our current understanding of the magnetic properties of OB stars.

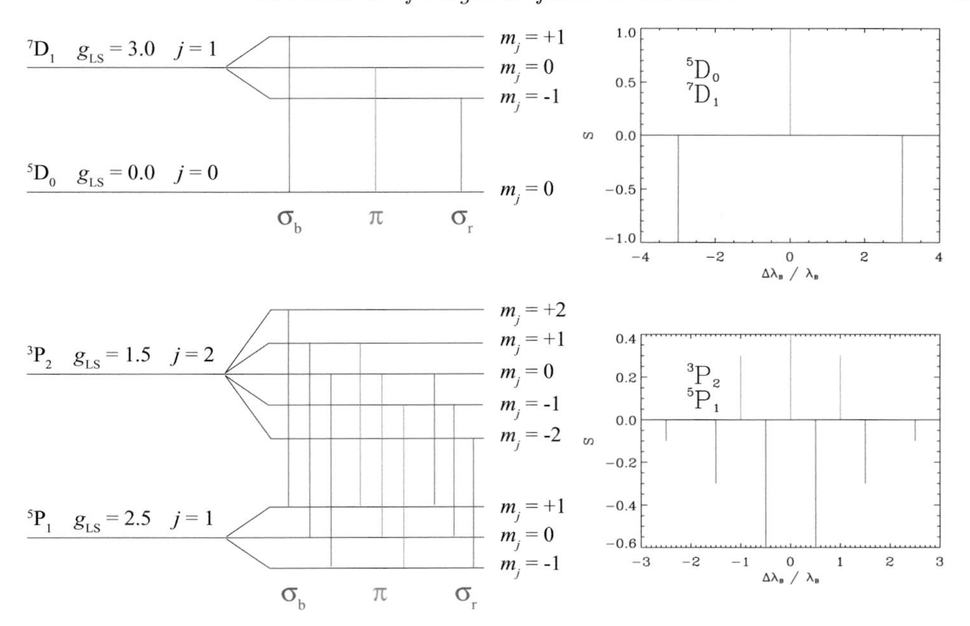

Figure 1. *Left:* Zeeman splitting level separation and admissible sub-transitions for the transitions $^7D_1$ - $^5D_0$ and $^3P_2$ - $^5P_1$. The sub-transitions that correspond to $\Delta m_j = -1, 0, +1$ are named σ_r, π and σ_b, respectively. The Landé factors of each level are computed under *LS* coupling. *Right:* Zeeman patterns corresponding to the illustrated transitions. The component separations are expressed in terms of Lorentz units (λ_B). By convention, the π components are illustrated upward, and the σ components downward.

2. Zeeman effect

The best way to directly detect a stellar magnetic field is by the Zeeman effect. When light passes through a medium and forms a spectral line by a atomic transition, the radiative transfer will be modified by the presence of the field, which split the energy levels in multiple components.

The first effect will be a splitting of the spectral line in multiple component. The width of the splitting $\delta\lambda_z$ is proportional to the modulus of the field:

$$\Delta\lambda_z \propto \lambda_0 \bar{g} |\vec{B}|, \tag{2.1}$$

where λ_0 is the wavelength of the unperturbed transition and $\bar{g}$ is the effective Landé factor of the transition. The splitting is therefore a scalar quantity that is not affected by the orientation of the field. However, the typical spectral separation will only be about 1-2 $\mathrm{km\,s^{-1}}$ per kilogauss in the optical domain. So, unless the field is quite strong, the splitting will be less than the typical Doppler broadening of the line profiles of a hot star.

In addition to the line splitting, the multiple Zeeman components of the transition, illustrated in Fig. 1 will have different polarisation states. The effect of the field can be decomposed in two contributions. The component of the field that is along the line of sight, the longitudinal field $B_{||}$, will partially circularly polarise each of the σ components. The polarisation of the σ_b component will be orthogonal to that of the σ_r component. As the Stokes V parameter is the subtraction of the two orthogonal states, we will observe a net change in Stokes V across the line profile. It is important to keep in mind that the circular polarisation produced is not only dependent on the field strength, but also on the orientation of the field with respect to the observer.

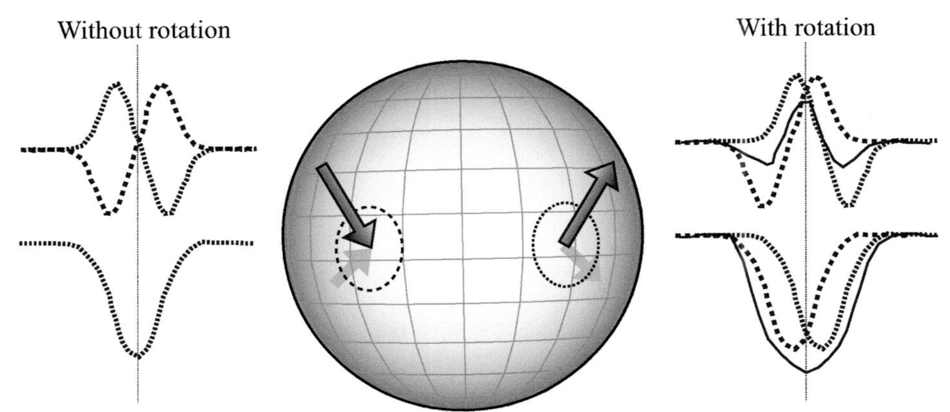

Figure 2. Illustration of the effect of rotation on local line profiles. In the case without rotation (left), two local field vectors with opposite orientations (indicated by the arrows) will produce overlapping Stokes I profiles. The local Stokes V profiles will cancel out (dotted and dashed lines). When the star is rotating the local profiles will not overlap because of the induced Doppler shifts The Stokes V profiles will not cancel out perfectly, leaving some net circular polarisation (solid line).

The component of the field that is perpendicular to the line of sight, the transverse field $B_{\perp}$, will produce linear polarisation along the transverse field axis for the σ components and perpendicular to the field axis for the π components. There will therefore be a change in the Stokes parameters Q and U across the line profile. The amount of polarisation in each linear Stokes parameters depends on the orientation of the transverse field on the plane of the sky. It is important to note than for the same field strength in the longitudinal and transverse directions, the linear polarisation produced will be substantially weaker than the circular polarisation.

Circular polarisation observations are therefore a really good choice in order to detect magnetic fields. Furthermore, given the accessibility of the four Stokes parameters (I, V, Q and U), it is in principle possible to characterize completely a magnetic field's strength and orientation. However, it is not as simple as it looks. It is important at this point to remember that the light coming from a spatially unresolved star is in fact the combination of the light produced at each point on the visible stellar surface - which can posses different emission properties, as well as different local field strengths and orientations.

As the polarisation produced by a certain point on the stellar surface is sensitive to the orientation of the local field vector at that point, some cancellation can occur in the disc-averaged profile with another point with opposite field orientation, hence making the global observed polarisation more difficult to detect and more difficult to interpret.

The rotation of the star is also important to consider. If the star is not seen pole-on, the line-of-sight components of the rotation velocity will introduce a Doppler shift distribution across the stellar disk. Hence, the contribution to the polarisation coming from two regions of opposite local field orientation might not cancel out perfectly, as shown in Fig. 2. Therefore, a polarisation signal across the line profile may still be seen even if the net magnetic field over the stellar disk in the line-of-sight direction is null. The required line-of-sight projected rotation is only of a few $\mathrm{km\,s^{-1}}$.

For rotating stars, the morphology of the Stokes V profile therefore contains information about the distribution of magnetic field strength and orientation over the visible stellar disc. Our ability to access this information depends not only on the structure of

Table 1. Non-exhaustive list of spectropolarimetric instruments.

Instrument	Telescope	Resolving power	Resolution element ($km\,s^{-1}$)
FORS1 & FORS2 [1]	VLT (8.2 m)	< 2 000	150
ISIS	WHT (4.2 m)	< 10 000	30
Spectropolarimeter	DAO/Plaskett (1.8 m)	< 10 000	30
ESPaDOnS	CFHT (3.6 m)	65 000	5
Narval	TBL (2m)	65 000	5
SEMPOL/UCLES	AAT (3.9 m)	70 000	4
HARSPol	ESO-3.6 (3.6 m)	70 000	4
NES	SAO/BTA (6 m)	60 000	5

Notes:
[1] FORS1 has been decommissioned, and replaced by FORS2, which has similar characteristics.

the field and the rotation of the star, but also the resolving power of the instrument used to observe it. In practice, we need to divide the spectropolarimetric instruments in two categories, depending on their capacity to resolve the line profile in the spectral domain. If the resolution elements of the instrument are wider than the Doppler shifts introduced by rotation, all the spatial information is lost. Therefore, lower resolution instrument are only sensitive to the net longitudinal field component across the stellar disk. Table 1 lists the spectropolarimetric instruments currently in use, as well as the resolution and the width of their resolution elements.

3. Low-resolution instruments

Fig. 3 (left) shows an example of a FORS1 spectrum (low-resolution) of a strongly magnetic Bp star (kG level), where changes of the circular polarisation across the line profiles can be clearly seen. In order to interpret this polarisation in term of a interesting magnetic quantity, the method described by Bagnulo *et al.* (2002) is generally used. The idea is to determine the longitudinal component value that would provide such a Stokes V profile if it was *local*, in the weak field approximation (then the splitting is negligible). It has been shown (see Landstreet 1982) that this approximation will result in a line-strength weighted mean of the longitudinal field integrated over the stellar disc. This value is usually simply referred to as the global longitudinal field B_l in the literature. In order to interpret this global longitudinal field in terms of the magnetic field at the stellar surface, one must then make hypothesis about the structure of the field and also about the emission properties of the stellar disc. For example, a 50 G measurement of the global longitudinal field could reflect a surface magnetic field of relatively simple topology (i.e. relatively little global cancellation) with a strength of order 100 G, or alternatively a rather highly structured field (i.e. lots of global cancellation) with a much greater strength.

As it does not make assumptions about the surface topology of the field itself, the global longitudinal field value is a really useful value to find rotational periods, as the visible field changes as the star rotates (e.g. Bychkov *et al.* 2005) . It is also possible to perform some modelling, by assuming some geometry (e.g. Landstreet & Mathys 2000). This value is also useful as a basis for statistical studies (e.g. Landstreet *et al.* 2007, Kholtygin *et al.*, these proceedings).

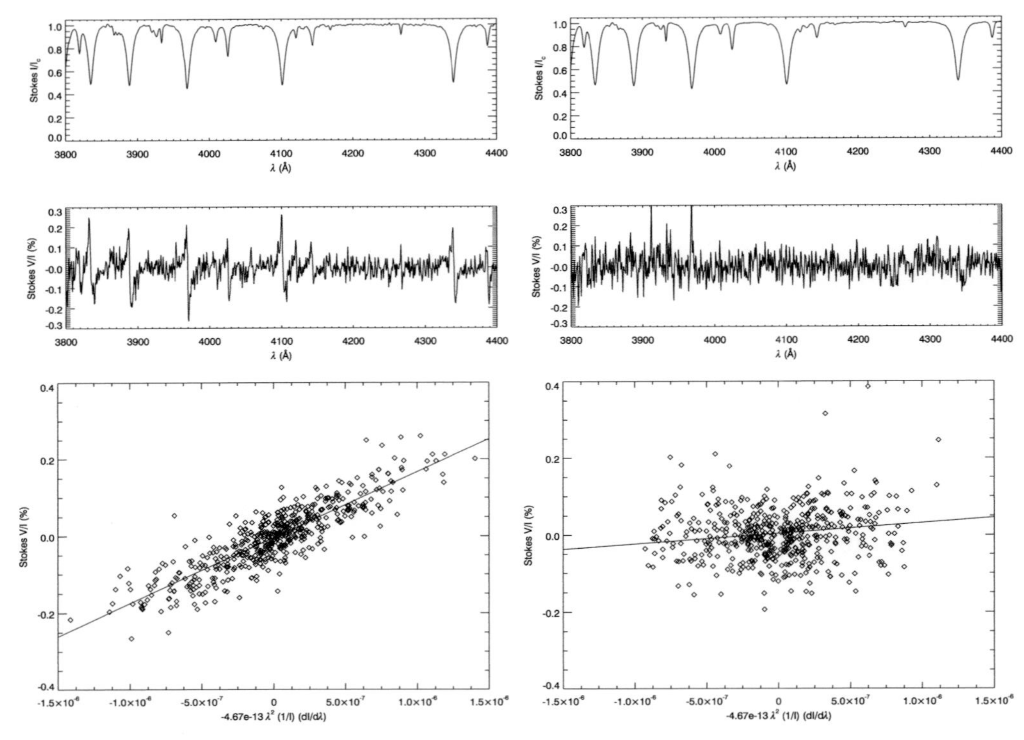

Figure 3. FORS1 observations of the B-type stars NGC 3766 170 (left) and NGC 3766 94 (right). *Top:* Stokes I/Ic spectrum. *Middle:* Stokes V/Ic spectrum. *Bottom:* The global longitudinal field B_l is proportional to the slope of the least-squares linear fit to the observed data. The results are $B_l = 1\,710 \pm 32$ G (53σ) and $B_l = 276 \pm 55$ G (5σ), respectively. From McSwain (2008).

The observed circular polarisation V/I is related to the global longitudinal field by:

$$\frac{V}{I} = -\bar{g}\, C_z\, \lambda^2\, \frac{1}{I}\, \frac{\mathrm{d}I}{\mathrm{d}\lambda}\, B_l, \qquad C_z = \frac{e}{4\pi m_e c^2}\ (\simeq 4.67 \times 10^{-13}\,\text{Å}^{-1}), \tag{3.1}$$

where e is the electron charge, m_e the electron mass and c the speed of light (Bagnulo *et al.* 2002). The strategy is to take each point in the spectrum and plot the value of V/I versus $(\mathrm{d}I/\mathrm{d}\lambda)/I$ and solve for the slope, which yields a value for B_l (see Fig. 3 bottom left).

In some cases, the magnetic field present is quite evident, just by a quick glance at the polarisation spectrum. However, as the we are trying to detect magnetic fields at the limit of the instrument's capacity, we need to rely on statistics (for example, see Fig. 3, right panels). It is customary to use the derived global longitudinal field value as a detection diagnostic, by looking at the statistical significance of the field detection. Therefore, a realistic and complete treatment of the uncertainties is required. Novel methods that have been put forward recently (e.g. Rivinius *et al.* 2010) have shown that B_l error bar of FORS1 data may have been significantly underestimated by some authors (see Rivinius *et al.*, these proceedings). Re-analysis of archived FORS1 observations has identified spurious detections in the literature (Bagnulo *et al.* in prep, Fullerton *et al.*, these proceedings). This could explain reported magnetic field detections that have not been confirmed by high-resolution observations (see Silvester *et al.* 2009).

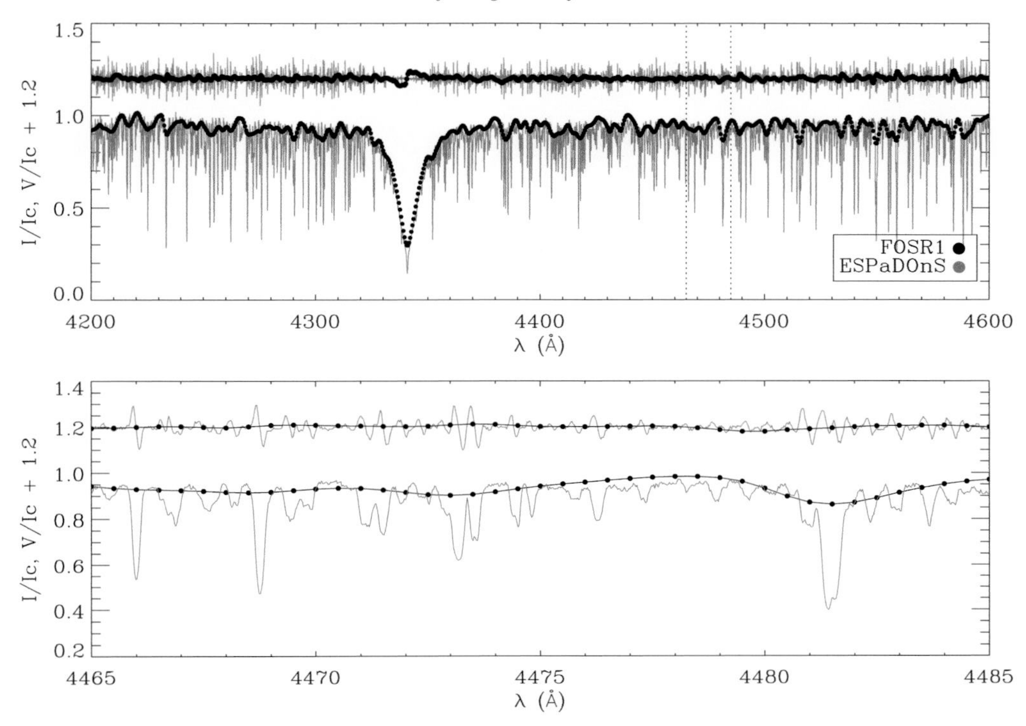

Figure 4. FORS1 and ESPaDOnS observations of HD 94660. *Top:* portion of the Stokes I and Stokes V (shifted vertically by 1.2) spectra, observed by FORS1 at VLT (in black) and by ESPaDOnS (in grey). *Bottom:* enlargement of a part of the spectrum (as shown by dotted lines on the top panel) to illustrate the higher resolution of ESPaDOnS. Note the multitude of spectral lines, and their associated Stokes V signatures, that are clearly evident in the high-resolution spectrum, but that blend together almost of invisibility in the low-resolution spectrum.

4. High-resolution instruments

Fig. 4 shows a comparison of two spectra of the magnetic Bp star HD 94660, obtained with the low resolution instrument FORS1 at VLT (in black) and the high-resolution intrument ESPaDOnS at CFHT (in grey). With the higher resolution, the multitude of metallic lines can be resolved. Information about the velocity of features in the line profile can therefore be translated into a spatial location on the stellar disk. One also has the ability to clearly identify which lines are present in the spectrum, and to diagnose the field using specific, clearly resolved features.

With a rotation resolved time-series of high-resolution spectra in circular or circular and linear polarisation (i.e. Stokes IV or Stokes $IVQU$), it is possible to perform a magnetic modelling that solves the radiative transfer problem simultaneously for detailed distribution of both magnetic field and other relevant quantities (e.g. chemical abundance) over a star's surface (Piskunov & Kochukhov 2002). The result of such an analysis is a map of abundance in diverse chemical elements as well as a map of the surface magnetic field (e.g. Kochukhov *et al.*, these proceedings). Fig 5 show such a map of the surface field of the Ap star α^2 CVn (Kochukhov & Wade 2010).

Because the polarisation induced in spectral lines by the Zeeman effect is relatively weak (less than 0.1% for a 1 kG field for a strong metallic spectral of a star rotating with a $v \sin i$ of 50 km s^{-1}), it is not always possible to measure the polarisation induced in individual spectral lines. As most of the high-resolution spectropolarimeters currently

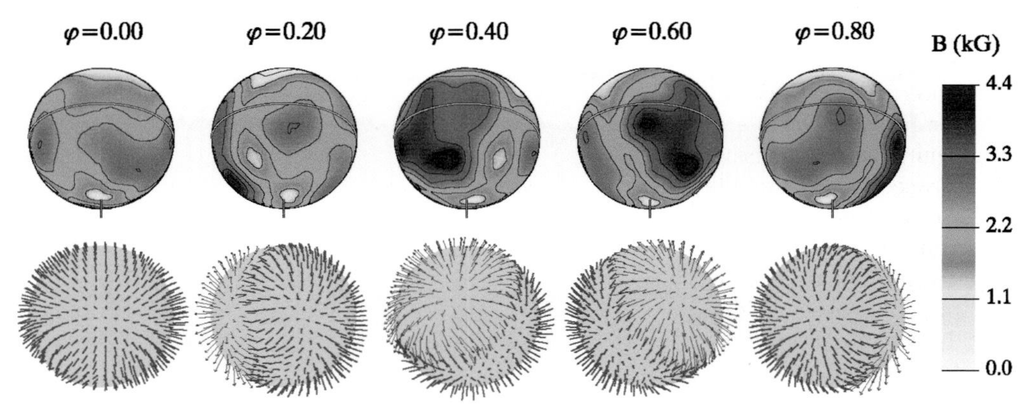

Figure 5. Surface magnetic field distribution of the Ap star α^2 CVn derived from an high--resolucion MuSiCoS time-serie of all four Stokes parameters. The star is shown at the five rotational phases indicated at the top of the figure. The upper row represent the distribution of field strength, with contour of equal magnetic field strength plotted every 0.5 kG. The lower panel shows the orientation of the magnetic vectors. From Kochukhov & Wade (2010).

available are echelle spectrometers, they cover a large spectral domain. It is then possible to increase the magnetic sensitivity by the simultaneous use of many lines present in a spectrum. Currently, the most widely-used and powerful technique is the Least-Squares Deconvolution (LSD) procedure, as described in Donati *et al.* (1997). The main assumption of the LSD method is that the shape of the line profile in intensity and circular polarisation is roughly the same for all the lines, and that this shape is scaled by a weight in order to give the observed profiles. These weights are proportional to the line depth for Stokes I and proportional to the product of the line depth, the Landé factor and the wavelength for Stokes V. From a list of spectral lines, a small window is cut out of the observed Stokes I and V spectrum and is converted in velocity space for each spectral line. One could imagine that all these windows are then averaged together, in order to produce a mean, high s/n profile. In practice, the shape that best fits the ensemble of single line windows is extracted analytically by the least-squares method.

The presence of a signal in the circular polarisation LSD profile is usually diagnosed by looking at the deviation of the signal from zero. This is quantified by the probability that a deviation as large as the one encountered can be produced by random noise. Generally, a signal is considered definitively detected when this probability gets lower than 1×10^{-5} (i.e. 0.001%).

It is possible to obtain a value equivalent to the global longitudinal field measured by low-resolution instruments by calculating the first moment of the Stokes V profile, as described by Donati *et al.* (1997) and modified by Wade *et al.* (2000):

$$B_l = -2.14 \times 10^{11} \frac{\int vV(v)\mathrm{d}v}{\bar{g}\,\lambda\, c \int(1 - I(v))\mathrm{d}v} \;\; [\mathrm{G}]. \tag{4.1}$$

As high-resolution instruments can detect net circular polarisation across the line profile even when the global longitudinal field is null, it is important to note that the detection of the Stokes V signature provides a more robust field detection diagnostic - a qualitatively new diagnostic that is only available from high-resolution data.

The question is now, what is the meaning of the derived LSD shape? The general assumption as been that the LSD profile is equivalent to a real spectral line. It has been shown however that for hot stars this assumption is crude and gives satisfactory results only in a few circumstances, for example at field strengths lower than $\sim$ 2 kG when considering Stokes V (Kochukhov *et al.* 2010). Some extra care must therefore be taken when interpreting LSD profiles to derive magnetic properties.

Nevertheless, LSD is a powerful technique to detect weak magnetic fields in hot stars (e.g. Henrichs *et al.*, these proceedings; Grunhut *et al.*, these proceedings; Petit *et al.*, these proceedings), and also to some extent characterize the surface field with the means of phase-resolved Stokes V time series (for example, see the complex surface field of τ Sco by Donati *et al.* 2006b) We can also add some information to the formal χ^2 statistics used for detection, by using knowledge of the shape of an expected deviation in Stokes V. With multiple noisy observations, it is possible to pick up an underlying signal by computing the odd ratios of the no magnetic field model (M_0) to the inclined dipole model (M_1), in a Bayesian framework (Fig. 6). As the exact rotation phases of stars are not generally known in advance, observation needs to be compared with the observed Stokes V profiles to a rotation independent geometry (see Petit *et al.* 2008 for the dipolar oblique rotator case). Furthermore, by performing a Bayesian parameter estimation for the dipole model, it is possible to obtain an estimate of the dipole field strengths admissible by observations. This is useful for preliminary analysis of sparse dataset (e.g. Grunhut *et al.* 2009), or to derive upper limits for a non-detection (e.g. Fullerton *et al.*, these proceedings; Petit *et al.*, these proceedings; Shultz *et al.*. these proceedings).

5. A family portrait of stellar magnetism

In the sun and essentially all other cool, low-mass stars, vigorous magnetic activity results from the conversion of convective and rotational mechanical energy into magnetic energy, generating highly structured and variable magnetic fields whose properties correlate strongly with stellar mass, age and rotation rate (e.g. Hartmann & Noyes 1987). Although the dynamo mechanism that drives this process is not understood in detail, its basic principles are well established.

The magnetic fields of higher-mass stars (above about 2 solar masses, in which the energy flux from the convective envelope begins to vanish, and in which no fully-convective pre-main sequence Hayashi phase is experienced) are qualitatively different from those of cool, low-mass stars. They are detected in only a small fraction of stars, they are structurally much simpler, and frequently much stronger, than the fields of cool stars. Most remarkably, their characteristics show no clear correlation with basic stellar properties,in particular mass and rotation rate. (e.g. Kochukhov & Bagnulo 2006; Landstreet *et al.* 2007).

The weight of opinion currently holds that these puzzling characteristics reflect a fundamentally different field origin than that of cool stars: that the observed fields are not generated by dynamos, but rather that they are fossil fields; i.e. remnants of field accumulated or generated during star formation (e.g. Braithwaite 2009; Duez & Mathis 2010). Although the fossil paradigm provides a useful framework for interpreting the large-scale magnetic fields of higher-mass stars, other dymano related process could also be at work on smaller scales, or within the convective core of the star (see Mathis *et al.*, these proceedings; Cantiello, these proceedings).

Historically, it has been assumed that magnetic fields in OB stars are very rare, and perhaps altogether absent in stars with masses above 8 solar masses. However, the increasing discoveries of fields in early B-type stars on the main sequence and pre-main sequence and in both young and evolved O-type stars show convincingly that fossil fields can and do exist in stars with masses as large as 45 solar masses. Given that the detected fields are sufficiently weak (0.3-1.5 kG) to have remained undetected by previous

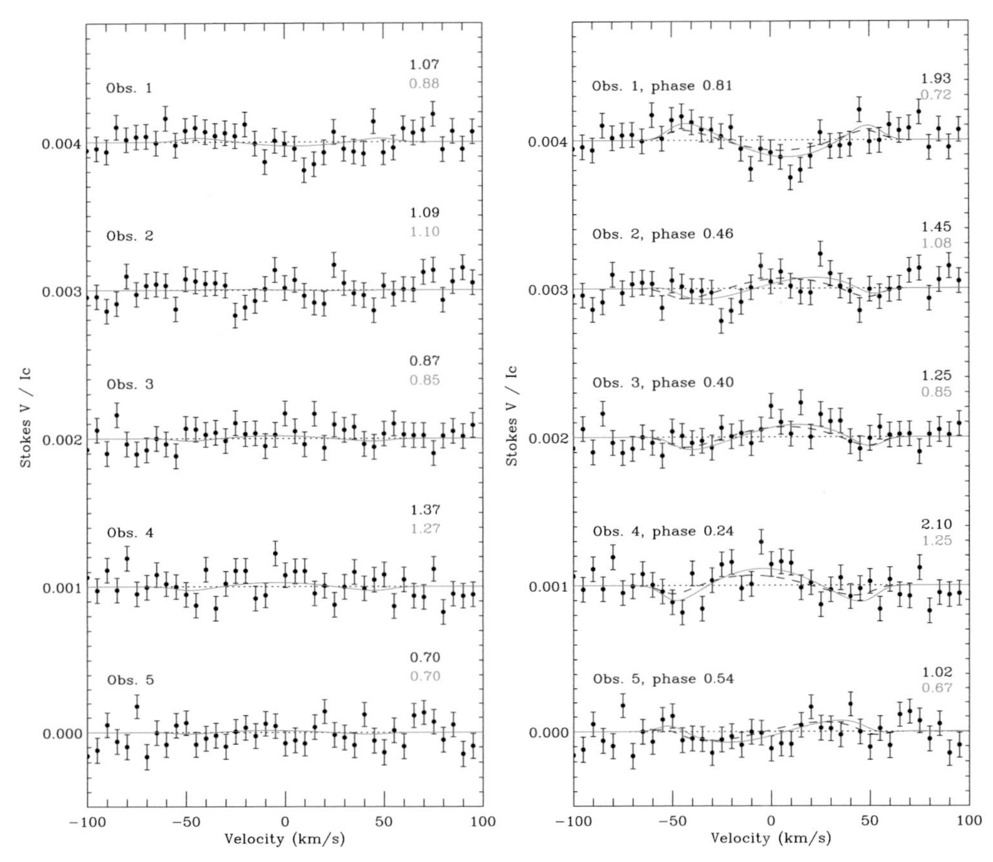

Figure 6. Simulated Stokes V data for a non-magnetic star (left) and for a dipolar field of 125 G for 5 randomly chosen rotational phases (right). The underlying magnetic profile is shown in dashed lines. The dotted line shows the no magnetic field model (M_0) and its associated reduced χ^2 is indicated at the top. The full line show the best dipole configuration (model M_1) for the combined observations and its χ^2 is indicated at the bottom. The odds ratio for the non-magnetic case is $\log(M_0/M_1) = 0.295$. The odds ratio for the magnetic case is $\log(M_0/M_1) = -9.05$ (i.e. the magnetic model is 9 orders of magnitude more preferred in the second case). The underlying magnetic signal can therefore be detected, even if none of the individual observation leads to a formal detection. From Petit *et al.* in prep.

generations of instrumentation, and that recent observational results suggest that the fraction of magnetic stars increases toward higher masses (see Fig. 7; Power *et al.* 2007), it may well be that magnetic fields are far more common in OB stars than has been supposed. Some preliminary studies of incidence in OB clusters have been performed, although with limited samples. The incidence of magnetic fields in OB stars seems indeed widespread, from a tenth to a third of the massive star population (Petit *et al.* 2008, and these proceedings; McSwain 2008). Large scale spectropolarimetric surveys - like the Magnetism in Massive Stars Large Program (Wade *et al.*, these proceedings) - will provide a sample large enough to derive more precise incidences.

The magnetic fields of ApBp stars are closely tied with abundance anomalies at their surface. In fact, it has been shown that all firmly classified Ap/Bp stars show detectable surface magnetic fields (Aurière *et al.* 2007). The first magnetic hot OB stars to be discovered were main sequence helium-weak and helium-strong stars (Borra & Landstreet 1979, Borra, Landstreet & Thompson 1983), objects which also display strong

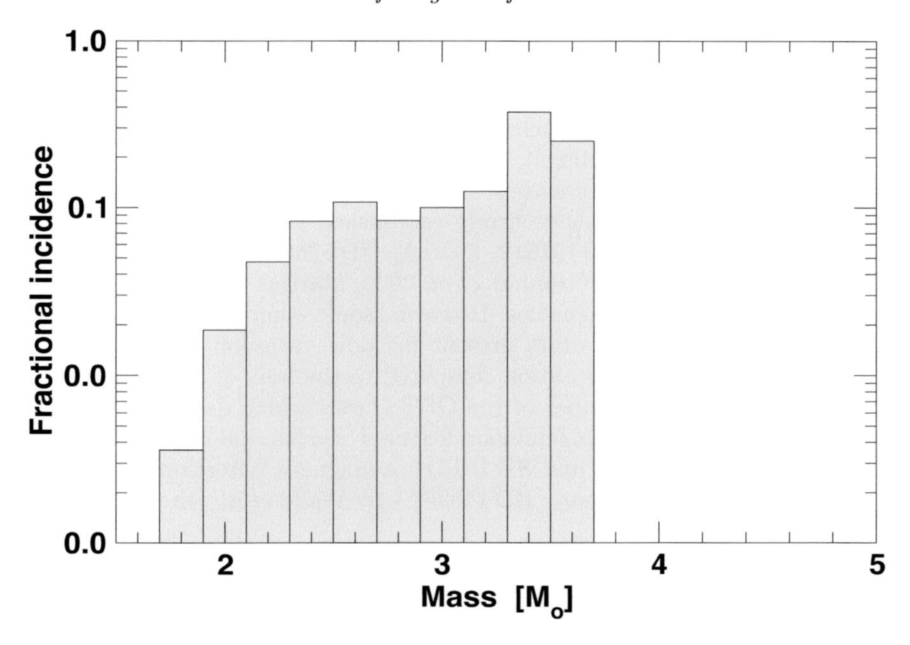

Figure 7. Inferred incidence of magnetic fields versus mass for all intermediate-mass stars within 100 pc of the sun. A trend of increasing incidence with stellar mass is seen. The number of stars per bin decreases with mass, due to the IMF. For example, the 2 M$_\odot$ bin contains a total of 438 stars (8 magnetic stars) whereas the 3.6 M$_\odot$ bin contains 11 stars (3 magnetic stars). From Power (2007).

photospheric chemical abundance anomalies, thought to be the extension of the ApBp phenomenon to higher temperatures. However, magnetic fields are also found in chemically normal B-type stars of surface temperature similar to He-strong stars, a coexistence that does not occur for ApBp stars. Furthermore, magnetic fields are found in a broad range of high-mass objects; emision-line stars, pulsating stars, X-ray bright stars, etc.

Although some magnetic detections have been reported for classical Be stars, none have been currently confirmed. Therefore, large-scale magnetic fields do not seem related to the Be phenomenon, nor the Oe phenomenon (e.g. Fullerton *et al.*, these proceedings). This is not surprising considering the theoretical difficulties in a Keplerian rotation profile from a magnetically-supported disc (Owocki 2004). However, a rigid rotation disk can be produced by a large scale magnetic field. The prototype for stars with such discs is the magnetic B-type star σ Ori E (see Oksala *et al.*, these proceedings), for which the peculiar photometric variation, the *periodic* variation of its Hα line profile, and its UV and X-ray variability have been coherently explained by a strongly magnetically confined wind, co-rotating with the star (Townsend *et al.* 2005). This scenario is strengthened by the recent discovery of similar stars, for example HR 7355 (Oksala *et al.* 2010, and these proceedings; Rivinius *et al.* 2010, and these proceedings) and HD 142184 (Grunhut *et al.*, these proceedings).

A large incidence of magnetic stars in slowly pulsating B-type stars and β Cephei-like stars was reported by Hubrig *et al.* (2006, 2009). However, according to Silvester *et al.* (2009), only 1 SPB star (ξ^1 CMa, see Fourtune-Ravard *et al.*, these proceedings) and 1 β Cep star (16 Peg) were confirmed in a sub-sample (containing 12 SPB stars and 7 β Cep-like stars) of Hubrig *et al.* sample - where 8 stars were claimed detected. Outside of the study of Silvester et al, there are only a few cases of confirmed magnetic fields in

β Cephei and SPB stars, namely β Cep itself (Henrichs *et al.* 2000), V2052 Oph (Neiner *et al.* 2003b) and ζ Cas (Neiner *et al.* 2003a).

Magnetic fields are also found in hot stars presenting periodic variability in their UV wind lines, for example σ Lupi (Henrichs *et al.*, these proceedings) and τ Sco (Donati *et al.* 2006b; Petit *et al.*, these proceedings).

Confirmed magnetic O-type stars are still rare. To date, repeated detections of circular polarization within line profiles have firmly established the presence of magnetic fields in 5 O-type stars (θ^1 Ori C, HD 191612, ζ Ori A, HD 57682 and HD 108; Donati *et al.* 2002, 2006a; Bouret *et al.* 2008; Grunhut *et al.* 2009; Martins *et al.* 2010). These O-type stars show a wide variety of properties. However, some common characteristics can be tentatively identified. All these stars present periodic variation of Hα. Most of these stars present anomalously slow rotation compared to the bulk of the O star population. HD 191612 and HD 108 are member of the Of?p class - which displays strong C III 4650 emission lines and narrow P Cygni/emission features (see Nazé *et al.* 2008, and these proceedings). In addition to HD 108 and HD 191612, a magnetic detection has been reported in the third prototypical Of?p star - HD 148937 - by Wade *et al.* (these proceedings).

References

Aurière, M., Wade, G. A., Silvester, J., Lignières, F. *et al.* 2007, *A&A* 475, 1053
Bagnulo, S., Szeifert, T., Wade, G. A., Landstreet, J. D. *et al.* 2002, *A&A* 389, 191
Bagnulo, S., Hensberge, H., Landstreet, J. D., Szeifert, T. *et al.* 2004, *A&A* 416, 1149
Borra, E. F. & Landstreet, J. D. 1979, *ApJ* 228, 809
Borra, E. F., Landstreet, J. D., & Thompson, I. 1983, *ApJS* 53, 151
Bouret, J.-C., Donati, J.-F., Martins, F., Escolano, C. *et al.* 2008, *MNRAS* 389, 75
Braithwaite, J. 2009, *MNRAS* 397, 763
Bychkov, V. D., Bychkova, L. V., & Madej, J. 2005, *A&A*, 430, 1143
Donati, J.-F., Babel, J., Harries, T. J., Howarth, I. D. *et al.* 2002, *MNRAS* 333, 55
Donati, J.-F., Howarth, I. D., Bouret, J.-C., Petit, P. *et al.* 2006a, *MNRAS* 365, L6
Donati, J.-F., Howarth, I. D., Jardine, M. M., Petit, P. *et al.* 2006b, *MNRAS* 370, 629
Donati, J.-F., Semel, M., Carter, B. D., Rees, D. E. *et al.* 1997, *MNRAS*, 291, 658
Duez, V. & Mathis, S. 2010, *A&A* 517, A58
Grunhut, J. H., Wade, G. A., Marcolino, W. L. F., Petit, V. *et al.* 2009, *MNRAS* 400, L94
Hartmann, L. W. & Noyes, R. W. 1987, *ARAA* 25, 271
Henrichs, H. F., de Jong, J. A., Donati, J.-F., Catala, C. *et al.* 2000, in: M. A. Smith, H. F. Henrichs, & J. Fabregat (eds.), *IAU Colloq. 175: The Be Phenomenon in Early-Type Stars*, ASP-CS 214, p. 324
Hillier, D. J. & Miller, D. L. 1998, *ApJ* 496, 407
Hubrig, S., Briquet, M., Schöller, M., De Cat, P. *et al.* 2006, *MNRAS* 369, L61
Hubrig, S., Briquet, M., De Cat, P., Schöller, M. *et al.* 2009, *AN* 330, 317
Kochukhov, O. & Bagnulo, S. 2006, *A&A* 450, 763
Kochukhov, O., Makaganiuk, V., & Piskunov, N. 2010, *A&A* 524, 5
Kochukhov, O. & Wade, G. A. 2010, *A&A* 513, A13
Landstreet, J. D. 1982, *ApJ* 258, 639
Landstreet, J. D., Bagnulo, S., Andretta, V., Fossati, L. *et al.* 2007, *A&A* 470, 685
Landstreet, J. D. & Mathys, G. 2000, *A&A* 359, 213
Maeder, A. & Meynet, G. 2005, *A&A* 440, 1041
Martins, F., Donati, J.-F., Marcolino, W. L. F., Bouret, J.-C. *et al.* 2010, *MNRAS*, 407, 1423
McSwain, M. V. 2008, *ApJ* 686, 1269
Nazé, Y., Walborn, N. R., & Martins, F. 2008, *Rev. Mexicana AyA* 44, 331
Neiner, C., Geers, V. C., Henrichs, H. F., Floquet, M. *et al.* 2003a, *A&A* 406, 1019
Neiner, C., Henrichs, H. F., Floquet, M., Frémat, Y. *et al.* 2003b, *A&A* 411, 565
Oksala, M. E., Wade, G. A., Marcolino, W. L. F., Grunhut, J. *et al.* 2010, *MNRAS* 405, L51

Owocki, S. P. 2004, in: A. Maeder & P. Eenens (eds.), *Stellar Rotation*, IAU Symposium 215, p. 515

Petit, V., Wade, G. A., Drissen, L., Montmerle, T. *et al.* 2008, *MNRAS* 387, L23

Piskunov, N. & Kochukhov, O. 2002, *A&A* 381, 736

Power, J. 2007, MSc. thesis, Queen's University (Canada)

Rivinius, T., Szeifert, T., Barrera, L., Townsend, R. H. D. *et al.* 2010, *MNRAS* 405, L46

Silvester, J., Neiner, C., Henrichs, H. F., Wade, G. A. *et al.* 2009, *MNRAS* 398, 1505

Townsend, R. H. D., Owocki, S. P., & Groote, D. 2005, *ApJ* (Letters) 630, L81

ud-Doula, A., Owocki, S. P. & Townsend, R. H. D. 2009, *MNRAS* 392, 1022

Wade, G. A., Donati, J.-F., Landstreet, J. D., & Shorlin, S. L. S. 2000, *MNRAS* 313, 823

Active OB stars: structure, evolution, mass loss, and critical limits
Proceedings IAU Symposium No. 272, 2010
C. Neiner, G. Wade, G. Meynet & G. Peters, eds.

doi:10.1017/S1743921311010131

The MiMeS project: overview and current status

Gregg A. Wade[1], Evelyne Alecian[2], David A. Bohlender[3], Jean-Claude Bouret[4], David H. Cohen[5], Vincent Duez[6], Marc Gagné[7], Jason H. Grunhut[1], Huib F. Henrichs[8], Nick R. Hill[9], Oleg Kochukhov[10], Stéphane Mathis[11], Coralie Neiner[12], Mary E. Oksala[13], Stan Owocki[13], Véronique Petit[7], Matthew Shultz[1], Thomas Rivinius[14], Richard H. D. Townsend[9], Jorick S. Vink[15] and the MiMeS collaboration†

[1]Kingston, Canada, [2]LOAG, France, [3]HIA, Canada, [4]LAM, France, [5]Swarthmore, USA, [6]Argelander, Germany, [7]West Chester, USA, [8]Amsterdam, Netherlands, [9]Madison, USA, [10]Uppsala, Sweden,[11]CEA, France, [12]Paris Observatory, France, [13]Delaware, USA, [14]ESO, Chile, [15]Armagh, UK

Abstract. The Magnetism in Massive Stars (MiMeS) Project is a consensus collaboration among many of the foremost international researchers of the physics of hot, massive stars, with the basic aim of understanding the origin, evolution and impact of magnetic fields in these objects. At the time of writing, MiMeS Large Programs have acquired over 950 high-resolution polarised spectra of about 150 individual stars with spectral types from B5-O4, discovering new magnetic fields in a dozen hot, massive stars. The quality of this spectral and magnetic matériel is very high, and the Collaboration is keen to connect with colleagues capable of exploiting the data in new or unforeseen ways. In this paper we review the structure of the MiMeS observing programs and report the status of observations, data modeling and development of related theory.

Keywords. stars: magnetic fields, stars: early-type, stars: formation, stars: evolution, stars: winds, outflows, techniques: polarimetric

---

1. Introduction

Massive stars are those stars with initial masses above about 8 times that of the sun, eventually ending their lives in catastrophic supernovae. These represent the most massive and luminous stellar component of the Universe, and are the crucibles in which the lion's share of the chemical elements are forged. These rapidly-evolving stars drive the chemistry, structure and evolution of galaxies, dominating the ecology of the Universe - not only as supernovae, but also during their entire lifetimes - with far-reaching consequences.

The Magnetism in Massive Stars (MiMeS) Project represents a comprehensive, multidisciplinary strategy by an international team of top researchers to address the big questions related to the complex and puzzling magnetism of massive stars. MiMeS has been awarded "Large Program" status by the Canada-France-Hawaii Telescope (CFHT) and the Télescope Bernard Lyot (TBL), resulting in a total of 1230 hours of time allocated to the Project with the high-resolution spectropolarimeters ESPaDOnS and Narval from late 2008 through 2012. This commitment of the observatories, their staff, their resources and expertise is being used to acquire an immense database of sensitive measurements of the optical spectra and magnetic fields of massive stars, which will be combined with

† www.physics.queensu.ca/∼wade/mimes

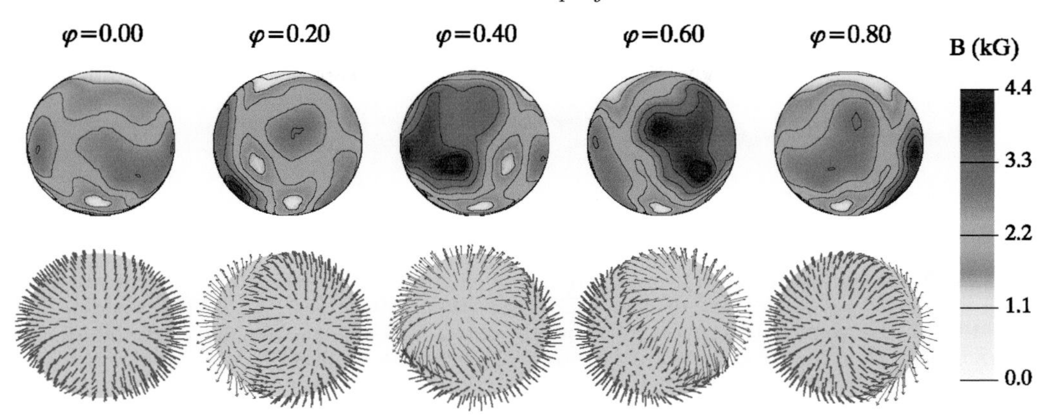

Figure 1. Magnetic Doppler Imaging (MDI) of the B9Vp star α^2 CVn (HD 112413; Kochukhov & Wade 2010), illustrating the reconstructed magnetic field orientation (lower images) and intensity (upper images) of this star at 5 rotation phases. The maps were obtained from a time-series of 21 Stokes *IVQU* spectral sequences. Maps similar to these will be constructed for other stars in the MiMeS Targeted Component.

a wealth of new and archival complementary data (e.g. optical photometry, UV and X-ray spectroscopy), and applied to address the 4 main scientific objectives of the MiMeS Project:

(*a*) To identify and model the physical processes responsible for the generation of magnetic fields in massive stars;

(*b*) To observe and model the interaction between magnetic fields and hot stellar winds;

(*c*) To investigate the role of the magnetic field in modifying the bulk rotation and differential rotational profiles of massive stars;

(*d*) To understand the overall impact of magnetic fields on massive star evolution, and the connection between magnetic fields of non-degenerate massive stars and those of neutron stars and magnetars, with consequential constraints on stellar evolution, supernova astrophysics and gamma-ray bursts.

2. Structure of the Large Programs

To address these general problems, we have devised a two-component observing program that will allow us to obtain basic statistical information about the magnetic properties of the overall population of hot, massive stars (the Survey Component), while simultaneously providing detailed information about the magnetic fields and related physics of individual objects (the Targeted Component).

Targeted Component: The MiMeS Targeted Component (TC) will provide data to map the magnetic fields and investigate the physical characteristics of a sample of known magnetic stars of great interest, at the highest level of sophistication possible. The roughly 25 TC targets have been selected to allow us to investigate a variety of physical phenomena, and to allow us to directly and quantitatively confront the predictions of stellar evolution theory, MHD magnetised wind simulations, magnetic braking models, etc.

Each TC target is to be observed many times with ESPaDOnS and Narval, in order to obtain a high-precision and high-resolution sampling of the rotationally-modulated circular (and sometimes linear) polarisation line profiles. Using state-of-the-art tomographic reconstruction techniques such as Magnetic Doppler Imaging (Piskunov & Kochukhov

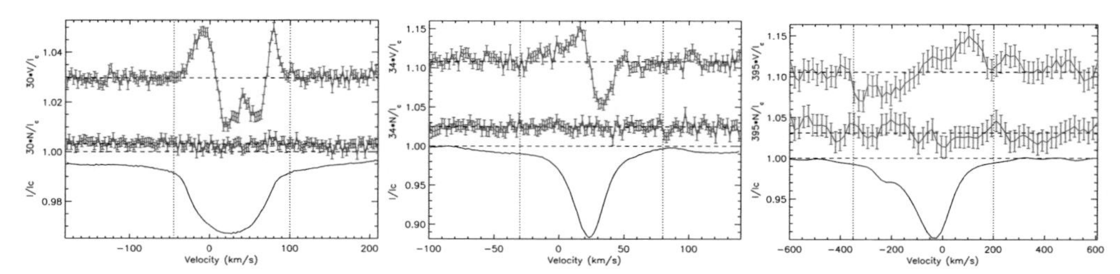

Figure 2. Least-Squares Deconvolved profiles of 3 hot stars in which magnetic fields have been discovered by the MiMeS Project: HD 61556 (B5V, left), HD 57682 (O9IV, middle) and HD 148937 (Of?p, right). The curves show the mean Stokes I profiles (bottom curve), the mean Stokes V profiles (top curve) and the N diagnostic null profiles (middle curve). Each star exhibits a clear magnetic signature in Stokes V. To date, a dozen new magnetic stars have been discovered through the MiMeS Survey Component.

2002), detailed maps of the vector magnetic field on and above the surface of the star will be constructed (e.g. see Fig. 1). In combination with new and archival complementary data, detailed analyses will be undertaken to model their evolutionary states, rotational evolution and wind structure and dynamics.

Survey Component: The MiMeS Survey Component (SC) provides critical missing information about field incidence and statistical field properties for a much larger sample of massive stars. It will also serve to provide a broader physical context for interpretation of the results of the Targeted Component. From an extensive list of potential OB stars compiled from published catalogues, we have generated an SC target sample of about 200 targets, covering the full range of spectral types from B4-O4, which are selected to sample broadly the parameter space of interest, while being well-suited to field detection. Our target list includes pre-main sequence Herbig Be stars, field and cluster OB stars, Be stars, and Wolf-Rayet stars.

Each SC target has been, or will be, observed once or twice during the Project, at very high precision in circular polarisation (e.g. see Figs. 2 & 3). From the SC data we will measure the bulk incidence of magnetic massive stars, estimate the variation of incidence versus mass, derive the statistical properties (intensity and geometry) of the magnetic fields of massive stars, estimate the dependence of incidence on age and environment, and derive the general statistical relationships between magnetic field characteristics and X-ray emission, wind properties, rotation, variability, binarity and surface chemistry diagnostics.

Of the 1230 hours allocated to the MiMeS LPs, about one-half is assigned to the TC and one-half to the SC.

3. Precision magnetometry of massive stars

For all targets we exploit the longitudinal Zeeman effect in metal and helium lines to detect and measure magnetic fields in the line-forming region. Splitting of a spectral line due to a longitudinal magnetic field into oppositely polarised σ components produces a variation of circular polarisation across the line (commonly referred to as a (Stokes V) Zeeman signature or magnetic signature; see Fig. 2.). The amplitude and morphology of the Zeeman signature encode information about the strength and structure of the global magnetic field. For some TC targets, we will also exploit the transverse Zeeman effect to constrain the detailed local structure of the field. Splitting of a spectral line by a transverse magnetic field into oppositely polarised π and σ components produces a

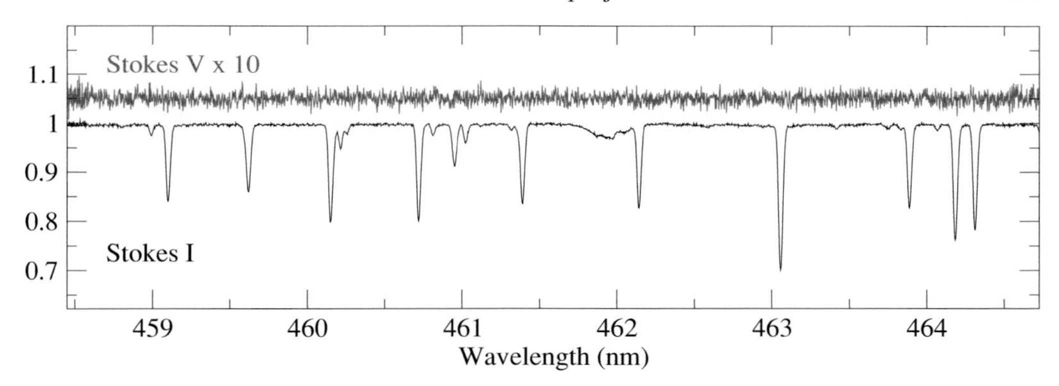

Figure 3. Detail of the MiMeS SC spectrum of the sharp-lined β Cep star δ Ceti (= HD 16582). The peak S/N per 1.8 km/s pixel is 1000 (for an exposure time of 280 s), typical for an SC observation. LSD analysis yields no evidence of a magnetic field, with a 1σ longitudinal field error bar of just 10 G.

variation of linear polarisation (characterized by the Stokes Q and U parameters) across the line (e.g. Kochukhov *et al.* 2004, review by Petit in these proceedings).

3.1. *Survey Component*

For the SC targets, the detection of magnetic field is diagnosed using the Stokes V detection criterion described by Donati *et al.* (1997), and the "odds ratio" computed using the powerful Bayesian estimation technique of Petit *et al.* (2008). After reduction of the polarised spectra using the Libre-Esprit optimal extraction code (see Fig. 3), we employ the Least-Squares Deconvolution (LSD; Donati *et al.* 1997) multi-line analysis procedure to combine the Stokes V Zeeman signatures from many spectral lines into a single high-S/N mean profile (see Fig. 2), enhancing our ability to detect subtle magnetic signatures. Least-Squares Deconvolution of a spectrum requires a line mask to describe the positions, relative strengths and magnetic sensitivities of the lines predicted to occur in the stellar spectrum. In our analysis we employ custom line masks that we tailor interactively to best reproduce the observed stellar spectrum, in order to maximise our sensitivity to weak magnetic fields.

The exposure duration required to detect a Zeeman signature of a given strength varies as a function of stellar apparent magnitude, spectral type and projected rotational velocity. This results in a large range of detection sensitivities for our targets. The SC exposure times are based on an empirical exposure time relation derived from real ESPaDOnS observations of OB stars, and takes into account detection sensitivity gains resulting from LSD and velocity binning, and sensitivity losses from line broadening due to rapid rotation. Exposure times for our SC targets correspond to the time required to definitely detect (with a false alarm probability below 10^{-5}) the Stokes V Zeeman signature produced by a surface dipole magnetic field with a specified polar intensity. Although our calculated exposure times correspond to definite detections of a dipole magnetic field, our observations are also sensitive to substantially more complex field topologies.

Examples of SC targets in which magnetic fields have been discovered by MiMeS are shown in Fig. 2.

3.2. *Targeted Component*

Zeeman signatures are detected repeatedly in all spectra of TC targets. The spectropolarimetric timeseries are interpreted using several magnetic field modeling codes at our disposal. For those stars for which Stokes V LSD profiles will be the primary model

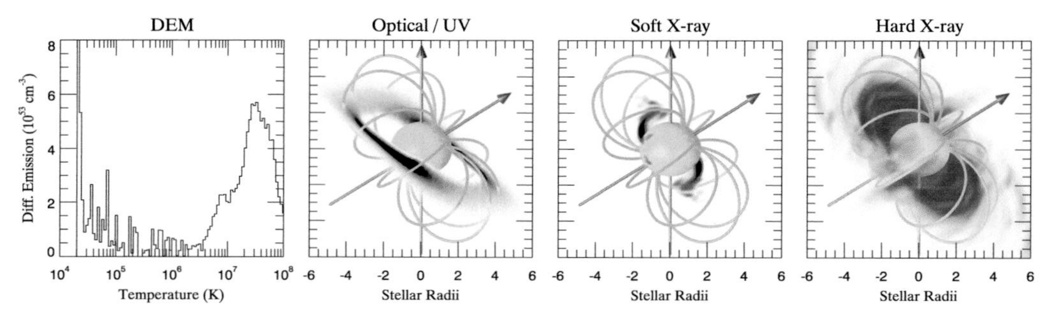

Figure 4. Example of the spectral and spatial emission properties of a rotating massive star magnetosphere modeled using Rigid Field Hydrodynamics (Townsend *et al.* 2007). The stellar rotation axis (vertical arrow) is oblique to the magnetic axis (inclined arrow), leading to complex plasma flow produced by radiative acceleration, Lorentz forces and centripetal acceleration. The consequent heated plasma distribution in the stellar magnetosphere (illustrated in colour/grey scale) shows broadband emission, and is highly structured both spatially and spectrally. Magnetically confined winds such as this are responsible for the X-ray emission and variability properties of some OB stars, and models such as this are being constructed for the MiMeS Targeted Component.

basis, modeling codes such as those of Donati *et al.* (2006) or Alecian *et al.* (2008) will be employed. For those stars for which the signal-to-noise ratio in individual spectral lines is sufficient to model the polarisation spectrum directly, we will employ the Invers10 Magnetic Doppler Imaging code to simultaneously model the magnetic field, surface abundance structures and pulsation velocity field (Piskunov & Kochukhov 2002, Kochukhov *et al.* 2004). The resultant magnetic field models will be compared directly with the predictions of fossil and dynamo models (e.g. Braithwaite 2006, 2007, Duez & Mathis 2010, Arlt 2008).

Diagnostics of the wind and magnetosphere (e.g. optical emission lines and their linear polarisation, UV line profiles, X-ray photometry and spectroscopy, radio flux variations, etc.) are being modeled using both the semi-analytic Rigidly-Rotating Magnetosphere approach, the Rigid-Field Hydrodynamics (Townsend *et al.* 2007) approach and full MHD simulations using the ZEUS code (e.g. Stone & Norman 1992; ud Doula *et al.* 2008; Townsend, Hill, Rivinius, Grunhut in these proceedings; see also Fig. 4).

4. Project status

At the time of writing, MiMeS Large Programs have acquired over 950 spectra of over 150 individual stars. About 45% of the spectra correspond to the SC, while 65% correspond to TC targets. Following their acquisition in Queued Service Observing mode at the CFHT, ESPaDOnS polarised spectra are immediately reduced by CFHT staff using the Upena pipeline feeding the Libre-Esprit reduction package and downloaded to the dedicated MiMeS Data Archive at the Canadian Astronomy Data Centre (CADC) in Victoria, Canada. Narval data are similarly reduced by TBL staff then downloaded to the MiMeS Collaboration's Wiki site at Observatoire de Paris, France. All reduced spectra are carefully normalized to the continuum using custom software tailored to hot stellar spectra. Each reduced spectrum is then subject to an immediate quick-look analysis to verify nominal resolving power, polarimetric performance and S/N. The quality of the ~ 420 SC spectra acquired to date is very high: the median peak SNR of the Stokes I spectra is 1150 (per 1.8 km/s spectral pixel) and the median longitudinal field error bar

(1σ) derived from the Stokes V spectra is just 37 G. A typical SC spectrum is illustrated in Fig. 3.

Preliminary LSD profiles are extracted using our database of generic hot star line masks to perform an initial magnetic field diagnosis and further quality assurance. Ultimately, each spectrum will be processed by the MiMeS Massive Stars Pipeline (MSP; currently in production) to determine a variety of critical physical data for each observed target, in addition to the precision magnetic field diagnosis: effective temperature, surface gravity, mass, radius, age, variability characteristics, projected rotational velocity, radial velocity and binarity, and mass loss rate. These meta-data, in addition to the reduced high-quality spectra, will be uploaded for publication to the MiMeS Legacy Database†.

A large variety of MiMeS results are presented in these proceedings, both observational (Fourtune-Ravard *et al.*, Grunhut *et al.*, Henrichs *et al.*, Oksala *et al.*, Petit *et al.*, Rivinius *et al.*, Shultz *et al.* and Wade *et al.*) and theoretical (Duez *et al.*, Mathis *et al.*, Townsend *et al.*, Hill *et al.*).

Acknowledgements

The MiMeS CFHT Large Program (2008B-2012B) is supported by both Canadian and French Agencies, and was one of 4 such programs selected in early 2008 as a result of an extensive international expert peer review of many competing proposals. The MiMeS TBL Large Program (2010B-2012B) was allocated in the context of the competitive French National Program for Stellar Physics (PNPS).

Based on observations obtained at the Canada-France-Hawaii Telescope (CFHT) which is operated by the National Research Council of Canada, the Institut National des Sciences de l'Univers of the Centre National de la Recherche Scientifique of France, and the University of Hawaii. Also based on observations obtained at the Bernard Lyot Telescope (TBL, Pic du Midi, France) of the Midi-Pyrénées Observatory, which is operated by the Institut National des Sciences de l'Univers of the Centre National de la Recherche Scientifique of France.

The MiMeS Data Access Pages are powered by software developed by the CADC, and contains data and meta-data provided by the CFH Telescope.

References

Alecian, E., Catala, C., Wade, G. A., Donati, J.-F. *et al.* 2008, *MNRAS*, 385, 391
Arlt, R. 2008, *Contributions of the Astronomical Observatory Skalnate Pleso* 38, 163
Braithwaite, J. 2006, *A&A* 453, 687
Braithwaite, J. 2007, *A&A* 469, 275
Donati, J.-F., Semel, M., Carter, B. D., Rees, D. E. *et al.* 1997, *MNRAS* 291, 658
Donati, J.-F., Howarth, I. D., Jardine, M. M., Petit, P. *et al.* 2006, *MNRAS* 370, 629
Duez, V. & Mathis, S. 2010, *A&A* 517A, 58
Kochukhov, O., Bagnulo, S., Wade, G. A., Sangalli, L. *et al.* 2004, *A&A* 414, 613
Kochukhov, O. & Wade, G. A. 2010, *A&A* 513A, 13
Petit, V., Wade, G. A., Drissen, L., Montmerle, T. *et al.* 2008, *MNRAS* 387, L23
Piskunov, N. & Kochukhov, O. 2002, *A&A* 381, 736
Stone, J. M. & Norman, M. L. 1992, *ApJS* 80, 753
Townsend, R. H. D., Owocki, S. P., & Ud-Doula, A. 2007, *MNRAS* 382, 139
ud-Doula, A., Owocki, S. P., & Townsend, R. H. D. 2008, *MNRAS* 385, 97

† The MiMeS Project is undertaken within the context of the broader MagIcS (Magnetic Investigations of various Classes of Stars) collaboration, www.ast.obs-mip.fr/users/donati/magics).

Active OB stars: structure, evolution, mass loss, and critical limits
Proceedings IAU Symposium No. 272, 2010
C. Neiner, G. Wade, G. Meynet & G. Peters, eds.

doi:10.1017/S1743921311010143

Modeling the magnetosphere of the B2Vp star σ Ori E

Mary E. Oksala[1,2], Gregg A. Wade[2], Rich H. D. Townsend[3], Oleg Kochukhov[4] and Stan P. Owocki[5]

[1]Department of Physics and Astronomy, University of Delaware, Newark, DE, USA
[2]Department of Physics, Royal Military College of Canada, Kingston, Ontario, Canada
[3]Department of Astronomy, University of Wisconsin-Madison, Madison, WI, USA
[4]Department of Physics and Astronomy, Uppsala University, Uppsala, Sweden
[5]Bartol Research Institute, University of Delaware, Newark, DE, USA

Abstract. This paper presents results obtained from Stokes I and V spectra of the B2Vp star sigma Ori E, observed by both the Narval and ESPaDOnS spectropolarimeters. Using Least-Squares Deconvolution, we investigate the longitudinal magnetic field at the current epoch, including period analysis exploiting current and historical data. σ Ori E is the prototypical helium-strong star that has been shown to harbor a strong magnetic field, as well as a magnetosphere, consisting of two clouds of plasma forced by magnetic and centrifugal forces to co-rotate with the star on its 1.19 day period. The Rigidly Rotating Magnetosphere (RRM) model of Townsend & Owocki (2005) approximately reproduces the observed variations in longitudinal field strength, photometric brightness, Hα emission, and various other observables. There are, however, small discrepancies between the observations and model in the photometric light curve, which we propose arise from inhomogeneous chemical abundances on the star's surface. Using Magnetic Doppler Imaging (MDI), future work will attempt to identify the contributions to the photometric variation due to abundance spots and due to circumstellar material.

Keywords. stars: magnetic fields, stars: rotation, stars: early-type, stars: circumstellar matter, stars: individual (HD 37479), techniques: polarimetric

1. Introduction

The term "magnetosphere" was coined to describe "the region in the vicinity of the earth in which the Earth's magnetic field dominates all dynamical processes" (Gold 1959). The Earth's magnetic field is distorted by the solar wind, compressed on the side closest to the sun and pulled out on the opposite side into the magnetotail. As observations improved, it was shown that each planet with a magnetic field also possessed a similar magnetosphere to the Earth's, all created by the solar wind's effect on the magnetic field. Eventually, "magnetosphere" was expanded to include stellar environments where winds and magnetic fields interact, e.g. the solar wind and magnetic field.

Unlike planetary magnetospheres, stellar magnetospheres expand outward, as the wind is coming directly from the star itself. We can understand solar system magnetospheres in great detail, describing the complex plasma physics, as we are able to obtain direct observations by satellites. In the stellar case, we must use remote observations to infer physical properties (i.e., temperature, density), augmented by MagnetoHydroDynmaical (MHD) simulations.

There is a distinctive difference between solar and massive-star magnetospheres. In lower mass stars like the sun, an α-Ω dynamo combines a convective envelope with differential rotation to produce complex, small-scale, time variable field structures. Globally, the field is quite weak. The solar wind is pressure-driven with a relatively weak mass loss rate ($\sim 10^{-14}$ $M_\odot$/yr). More massive stars have mostly radiative envelopes. This major difference makes magnetic field generation by a dynamo difficult to explain, so the question of the origin of magnetic fields in massive stars remains open. Massive-star fields are also quite different from their solar counterparts. Their fields are generally steady, large scale, and sometimes very strong. Massive-star winds are radiatively driven with strong mass loss rates (on average $\sim 10^{-7}$ $M_\odot$/yr). Hereafter, this paper will focus solely on massive-star magnetospheres, specifically in the strong magnetic field case.

2. Massive star magnetospheres

As the wind interacts with the magnetic field, material from the stellar wind can be perturbed, channeled, torqued, and even confined. Massive star magnetospheres are generally structured and dynamic. Material is released in reconnection events, falls back onto the star and creates collisional shocks. Rotation complicates the physical picture, but it also allows for regular modulation of observables, through which the physical properties of these magnetospheres can be identified. The magnetospheric properties depend on the relative strengths of the wind and magnetic energy densities of the star. ud-Doula & Owocki (2002) define a wind confinement parameter to quantify this struggle between the wind strength and magnetic strength,

$$\eta_\star = \frac{B_\star^2 R_\star^2}{\dot{M} v_\infty}. \tag{2.1}$$

For $\eta_\star < 1$, the wind overwhelms the field, but for $\eta_\star > 1$, the magnetic field is dominant near the star, channeling wind material into a magnetosphere. The Alfvén radius, R_A, is where the magnetic and wind energy densities become equal. For a strong magnetic confinement ($\eta_\star \gg 1$) and a dipole field configuration, $R_A \sim \eta_\star^{-1/4} R_\star$ (ud-Doula & Owocki 2002). ud-Doula *et al.* (2008) also define a Kepler co-rotation radius ($R_K = (v_{rot}/v_{crit})^{-2/3}\ R_\star$). If the Alfvén radius is farther out than the Kepler radius a region of confinement is created. Wind material is spun up by the magnetic field and supported by centrifugal forces. Above R_A, the field is too weak to confine material; below R_K, it is no longer supported by centrifugal forces and so falls back on the star.

3. The RRM model and σ Ori E

In many magnetic Bp stars, the magnetic field is strong, rotation is fast, and $\eta_\star \gg 1$. The field spins up and confines the channeled wind keeping it rigidly rotating well beyond the Kepler co-rotation radius, but also held down against the net outward centrifugal force relative to gravity. For this strong field limit, Townsend & Owocki (2005) developed an analytical model, the Rigidly Rotating Magnetosphere (RRM) model, wherein wind plasma settles at the local minima of the effective (gravitational + centrifugal) potential. The plasma then accumulates in the magnetosphere and is forced to co-rotate with the star.

σ Ori E (HD 37479) is a helium-strong B2Vp star with a 10 kG global dipole magnetic field. It is an oblique rotator with two plasma clouds rigidly rotating with the star (Landstreet & Borra 1978) with rotational speeds greater than the surface speed of $v \sin i = 150$ km s^{-1}. Observations show modulation according to the 1.19 d rotation period in longitudinal magnetic field (Landstreet & Borra 1978), Hα emission (Walborn 1974), He line strength (Pedersen & Thomsen 1974), photometry (Hesser *et al.* 1976), UV line strength (Smith & Groote 2001), 6 cm radio emission (Leone & Umana 1993), and linear

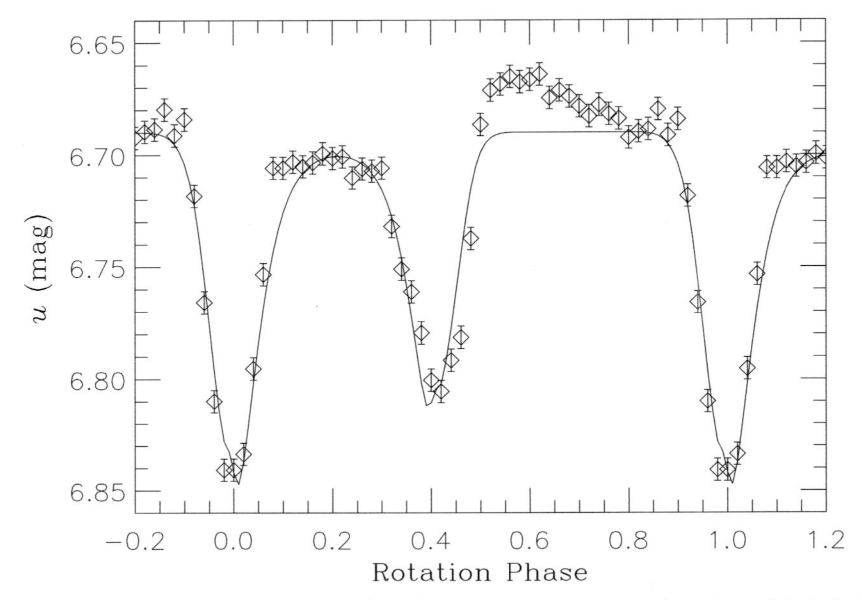

Figure 1. Figure from Townsend *et al.* (2005). Observed (diamonds) and modeled (solid line) Strömgren u-band light curves of σ Ori E, phased on the star's 1.19 day rotation period.

polarization (Kemp & Herman 1977). σ Ori E also has anomalous surface abundances of He and Si (Reiners *et al.* 2000). The RRM model reproduces the photometric, magnetic field, and Hα observations relatively well. It only accounts for the circumstellar effects that create the observed double eclipse pattern, but the stellar photosphere also contributes to the brightness variations. This leads to discrepancies (Figure 1) between the observed and simulated photometric light curve, which we are currently investigating.

4. Spectropolarimetric observations

We obtained a total of 18 high resolution (R=65000) broadband (370-1040 nm) circular polarization spectra of σ Ori E. Sixteen of these spectra were obtained in November 2007 from the Narval spectropolarimeter attached to the 2.2m Bernard Lyot telescope at the Pic du Midi Observatory in France. The remaining 2 spectra were obtained in February 2009 from the spectropolarimeter ESPaDOnS attached to the 3.6-m Canada-France-Hawaii Telescope, as part of the Magnetism in Massive Stars (MiMeS) Large Program (Wade *et al.*, these proceedings).

We used the method of Least-Squares Deconvolution (LSD) to derive values of longitudinal field for each spectrum. LSD describes the stellar spectrum as the convolution of a mean Stokes I or V profile, representative of the average shape of the line profile, and a line mask, describing the position, strength and magnetic sensitivity of all lines in the spectrum. A mask was used for a star with $T_{\rm eff}$=23000 K, log g = 4.0, and solar abundances except for enhanced helium. From the LSD mean Stokes I and V profiles, we calculate the longitudinal magnetic field, B_ℓ:

$$B_\ell = -2.14 \times 10^{11} \frac{\int vV(v)dv}{\lambda gc \int [1 - I(v)]dv} \qquad (4.1)$$

(Wade *et al.* 2000), where λ is the weighted average wavelength and g is the weighted

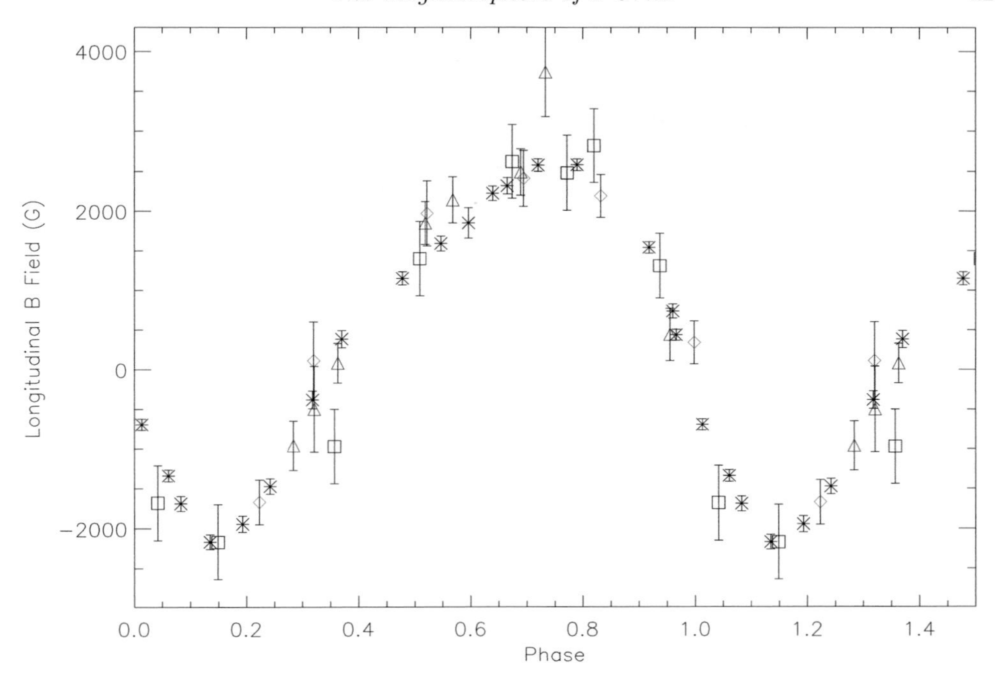

Figure 2. New longitudinal field measurements for each individual spectrum of σ Ori E (asterisks) along with 1σ error bars. In addition, we have plotted measurements from Bohlender *et al.* (1987), as triangles (hydrogen) and diamonds (helium), and Landstreet & Borra (1978) (squares) each with 1σ error bars.

average Landé factor in the mask. I_c is the continuum value of the intensity profile. The integral is evaluated over the full velocity range of the mean profile.

When compared with historical data from Landstreet & Borra (1978) and Bohlender *et al.* (1987), our new measurements agree surprisingly well, indicating stability of the global magnetic structure over three decades. The new measurements are also much more precise, decreasing the error bars by a factor of four versus the historical data. The spectropolarimetric data presented in Figure 2 allow a long baseline to study the periodicity of the field as compared with the period derived from photometric studies. Using the Scargle periodogram, the entire set of magnetic data gives a period of 1.190842 $\pm$ 0.000004 days. The baseline of the data is long enough that the period can be determined down to a third of a second. This is strictly an average period over the time frame of the observations. As Townsend *et al.* (2010) have shown, σ Ori E is spinning down due to magnetic braking at a rate of 77 milli-seconds per year. The associated spindown time calculated from observations, 1.34 Myr, agrees remarkably well with the theoretical prediction of 1.4 Myr by ud-Doula *et al.* (2009).

5. Magnetic Doppler Imaging

From rotationally modulated line profiles, the surface abundance and magnetic field vectors of a star can be mapped using a technique called Magnetic Doppler Imaging (MDI), originally developed by Piskunov & Kochukhov (2002) for the Ap star α^2 CVn. MDI was adapted from Doppler Imaging (DI), which reconstructs features on the surfaces of stars by inverting a time series of high-resolution spectral line profiles. The code

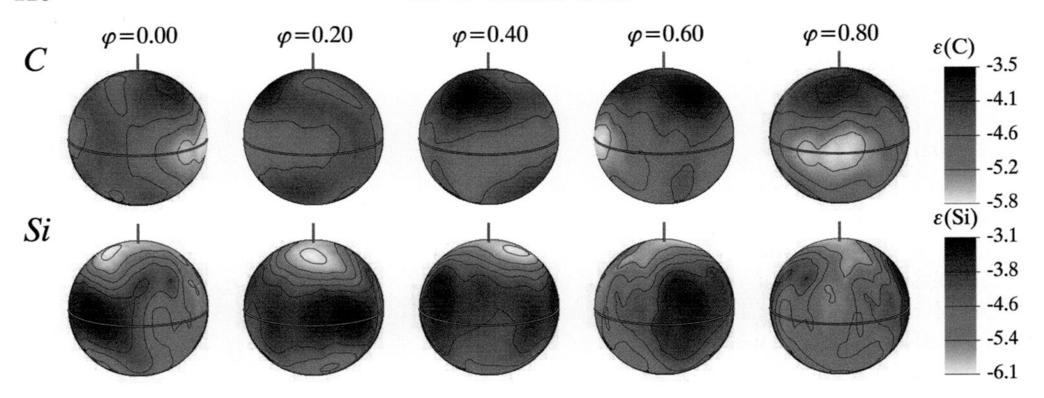

Figure 3. The chemical abundance distribution of σ Ori E derived from Stokes I and V profiles of the C and SI lines. The star is shown at 5 equidistant rotational phases viewed at the inclination angle, $i = 75°$ and $v \sin i = 150$ km s^{-1}. The scale gives abundance as ε(Elem) corresponding to $\log(N_{Elem}/N_{tot})$. The rotation axis is vertical.

compares observational Stokes parameters with synthetic parameters, using least squares minimization. MDI solves the radiative transfer equation to obtain the synthetic line profiles, assuming an unpolarized continuum and LTE. The code can be used for arbitrarily complex fields and abundance distributions.

This paper applies MDI to the Bp star σ Ori E. The maps created from circularly polarized line profiles show a dipole magnetic field with a polar strength of 9.6 kG, although higher-pole magnetic field geometries and small scale fields cannot be identified without Stokes Q and U information. The surface of the star (as shown in Figure 3) shows spots in carbon and silicon. The helium lines of σ Ori E show NLTE effects, especially at longer wavelengths. Currently, MDI is fitting helium lines with a partial NLTE implementation to include departure coefficients calculated from the NLTE atmosphere code TLUSTY (Lanz & Hubeny 2007). It may be possible to fit other elements from σ Ori E's spectrum, however rapid rotation and lack of available lines affect MDI's ability to calculate accurate maps.

6. Synthetic photometry

Krtička *et al.* (2007) used DI maps of HD 37776 to study the effects of spectroscopic spots on the surface of a star on the redistribution of flux. The maps of silicon and helium used show that both elements contribute to changes in the continuum of the star's emergent flux. In their paper, LTE atmospheres are used to calculate radiative fluxes and simulate light curves for each element. The authors note that NLTE effects in lines may change the shape of the predicted light curve. Matching the observed photometric light curve requires a combination of each elemental light curve. This method has been shown to be quite successful at reproducing the light curve of HD 37776, as the star does not show an eclipse pattern expected from a magnetosphere.

For other stars like σ Ori E with magnetospheres, Magnetic Doppler Imaging (MDI) can be used in the same way to create synthetic light curves for specific elements, taking into account the effects of chemical spots on the total flux emitted due to bound-free transitions. In §3, we mentioned that the RRM model can provide a synthetic light curve for the circumstellar effects on brightness. To fully account for the contributions to the emergent flux from σ Ori E, we should merge both the light curves synthesized from the RRM model and from MDI. Ideally, when we combine both the photospheric and circumstellar brightness variations, we should be able reproduce the details of the observed photometric light curve.

7. Summary

Massive star magnetospheres are structured, co-rotate with the star, and produce rotationally modulated observables. The RRM model of the circumstellar environment of σ Ori E can reproduce observations relatively well, except for unexplained differences in the photometric light curve. From new spectropolarimetric observations we show that σ Ori E has a strong (9.6 kG), dipole magnetic field that new observations show has remained steady in structure over 3 decades. Its longitudinal magnetic field varies from -2.3 kG to 2.5 kG. The period derived from these longitudinal field strength variations is 1.190842 ± 0.000004 days. This precise period is the average over the time baseline, as σ Ori E's rotation period gains 77 milli-seconds each year due to magnetic braking. We can use MDI to map the surface abundance of σ Ori E and synthesize a photospheric brightness light curve. By merging the light curves from both the MDI and RRM models, we expect to properly account for both the circumstellar and photospheric effects on the star's brightness. Using this combined approach, we hope to reproduce the details of the observed photometric light curve of σ Ori E.

Acknowledgements

MEO and RHDT acknowledge support from NASA grant grant *LTSA*/NNG05GC36G. O.K. is a Royal Swedish Academy of Sciences Research Fellow supported by grantsfrom the Knut and Alice Wallenberg Foundation and the Swedish Research Council.

References

Bohlender, D. A., Landstreet, J. D., Brown, D. N., & Thompson, I. B. 1987, *ApJ*, 323, 325
Gold, T. 1959, *Journal of Geophysical Research*, 64, 1665
Hesser, J. E., Walborn, N. R., & Ugarte, P. P. 1976, *Nature*, 262, 116
Kemp, J. C., & Herman, L. C. 1977, *ApJ*, 218, 770
Krtička, J., Mikuláček, Z., Zverko, J., & Žižňovský, J. 2007, *A&A*, 470, 1089
Landstreet, J. D. & Borra, E. F. 1978, *ApJ* (Letters), 224, 5
Lanz, T. & Hubeny, I. 2007, *ApJS*, 169, 83
Leone, F. & Umana, G. 1993, *A&A*, 268, 667
Pedersen, H. & Thomsen, B. 1977, *A&AS*, 30, 11
Piskunov, N. & Kochukhov, O. 2002, *A&A*, 381, 736
Reiners, A., Stahl, O., Wolf, B., Kaufer, A. *et al.* 2000, *A&A*, 363, 585
Smith, M. A. & Groote, D. 2001, *A&A*, 372, 208
Townsend, R. H. D. & Owocki, S. P. 2005, *MNRAS*, 357, 251
Townsend, R. H. D., Owocki, S. P., & Groote, D. 2005, *ApJ* (Letters), 630, 81
Townsend, R. H. D., Oksala, M. E., Cohen, D. H., Owocki, S. P. *et al.* 2010, *ApJ* (Letters), 714, 318
ud-Doula, A. & Owocki, S. P. 2002, *ApJ*, 576, 413
Ud-Doula, A., Owocki, S. P., & Townsend, R. H. D. 2008, *MNRAS*, 385, 97
Ud-Doula, A., Owocki, S. P., & Townsend, R. H. D. 2009, *MNRAS*, 392, 1022
Wade, G. A., Donati, J.-F., Landstreet, J. D., & Shorlin, S. L. S. 2000, *MNRAS*, 313, 851
Walborn, N. R. 1974, *ApJ* (Letters), 191, 95

Active OB stars: structure, evolution, mass loss, and critical limits
Proceedings IAU Symposium No. 272, 2010
C. Neiner, G. Wade, G. Meynet & G. Peters, eds.

doi:10.1017/S1743921311010155

The rapid magnetic rotator HR 7355†

Thomas Rivinius[1], Rich H. D. Townsend[2], Stanislas Štefl[1] and Dietrich Baade[3]

[1]ESO, Chile; [2]UW Madison, USA; [3]ESO, Germany

Abstract. For early type magnetic stars slow, at most moderate rotational velocities have been considered an observational fact. The detection of a multi-kilogauss magnetic field in the B2Vpn star with $P \approx 0.52\,d$ and $v \sin i \approx 300\,\mathrm{km/s}$ has brought down this narrative. We have obtained more than 100 high-resolution, high-S/N echelle spectra in 2009. These spectra provide the most detailed description of the variability of any He-strong star to date. The circumstellar environment is dominated by a rotationally locked magnetosphere out to several stellar radii, causing hydrogen emission. The photosphere is characterized by surface chemical abundance inhomogeneities, with much stronger amplitudes, at least for helium, than slower rotating stars like σ Ori E. The highly complex rotational line profile modulations of metal lines are probably a consequence the equatorial gravity darkening of HR 7355, and thus may offer an independent measurement of the von Zeipel parameter β.

Keywords. stars: early-type, stars: magnetic fields

1. Introduction

The Helium-strong star HR 7355 has recently been found to host a strong magnetic field. The short period of $P = 0.52$ d puts it among the shortest period non-degenerate magnetic stars known, while its $v \sin i \approx 310\mathrm{km\,s}^{-1}$ is the highest for any known non-degenerate magnetic star.

HR 7355 is a star for which effects like gravity darkening and oblate deformation cannot be ignored anymore. This means traditional analysis methods to derive stellar parameters will, at best, give uncertain results, and at worst misleading ones.

The presence of photospheric abundance pattern is intriguing, as in a hot star rotationally induced meridional flows should dilute such pattern quickly. While it has been suggested that the magnetic field would inhibit this circulation, HR 7355 is the first case where this can be tested observationally: not only are there abundance variations across the stellar surface, but the amplitude of the equivalent width of the HeI lines is much larger that in the similar, though less rapidly rotating, Bp star σ Ori E.

2. Observations

Apart from archival and literature data this work is based on high-resolution echelle spectra obtained in 2009 with UVES at the 8.2m Kueyen telescope on Cerro Paranal. The instrument was used in its DIC2 437/760 setting, which gives a blue spectrum from about 375 to 498 nm and a red spectrum from about 570 to 950 nm, with a small gap at 760 nm. The slit-width was 0.8 arcsec, giving a resolving power of about $R = 50\,000$ over the entire spectrum. The UVES observations were done on three chunks, in April, July, and September 2009.

Exposure times were between 120 seconds in April and 30 seconds July to September, where the shorter exposures were repeated four times. Since the period is short we did

† Based on observations under ESO programs 081.D-2005, 383.D-0095.

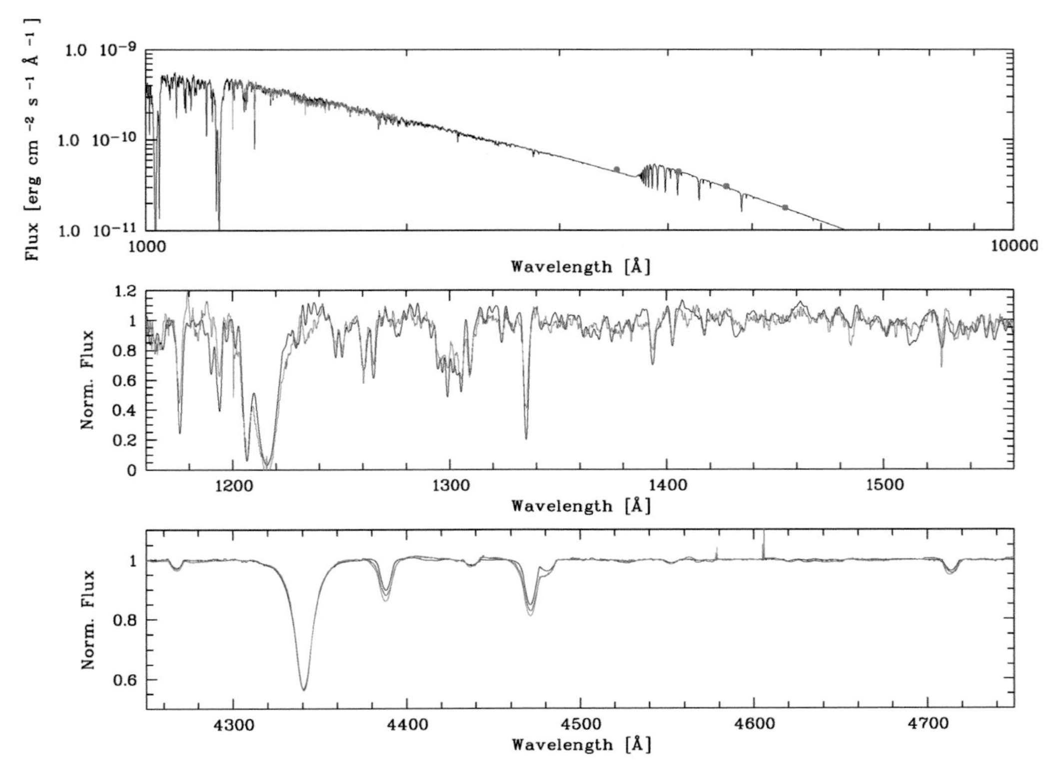

Figure 1. The B3-model vs. observed data: Upper panel: UV IUE short-wavelength spectrophotometry and Strömgren visual photometry vs. modeled fluxes. Middle panel: IUE spectrum vs. modeled UV-line profiles. Lower panel: All UVES (red) and FORS2 (blue) spectra averaged vs. modeled visual line profiles.

not average shorter exposures, but treat them as taken at distinct phases. In total, we have 104 blue and 112 red spectra. The typical S/N of a single spectrum is about 275 in the blue region and 290 in the red.

3. Stellar parameters

In order to determine the stellar parameters, we made use of spectral synthesis, both for line strength and profiles, as well as for absolute fluxes. The code we used is the third, re-programmed version of Townsend's (1997) BRUCE and KYLIE suite, which in the further we will call "B3".

In a first step, the profile of C II 4267, the strongest of all metal lines, was used to obtain the **projected rotational velocity** of $v \sin i = 310 \pm 5\,\mathrm{km\,s^{-1}}$, in good agreement with Oksala *et al.* (2010). Together with the rotational period of $P = 0.5214404\,\mathrm{d}$ the **equatorial radius** then becomes $R_{\star,\mathrm{equ.}} \sin i = 3.19\mathrm{R}_{\odot}$. Assuming that a rapidly rotating star is constrained by five independent parameters, usually $v_{\mathrm{eq}}, i, T_{\mathrm{eff}}, M_{\star}, R_{\star}$, this leaves only three of them to be determined, namely the mass, the effective temperature, and the inclination.

The meaning of the **effective temperature** T_{eff} in a rapidly rotating star is not straightforward, so we note that T_{eff} here is the uniform black-body temperature, that a star of the same surface area would need to have the same full solid angle luminosity as the actual gravity darkened star. The T_{eff} of HR 7355 is constrained using de-reddened

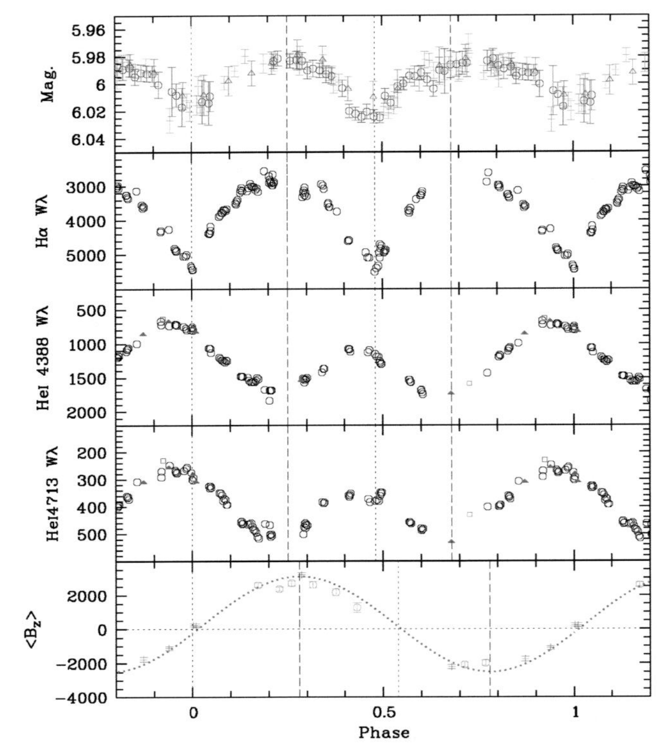

Figure 2. Phased observational data: Photometric data from Oksala *et al.* (2010), and ASAS data from Mikulášek *et al.* (2010) (uppermost), EWs from UVES (middle), with FORS and FEROS W_λ measurements of He I 4388, 4713 added, as well as FORS (Rivinius *et al.* (2010)) and ESPaDOns magnetic data (lowermost panel).

flux-calibrated UV spectra (from IUE) and visual photometry. These data are fitted best with $T_{\mathrm{eff}} = 17\,000\,\mathrm{K}$ (see Fig. 1).

This temperature seems incompatible with the spectral classification of B2. However, due to chemical peculiarity the helium lines are much stronger than they would be for solar abundances, biasing classifications towards the spectral type with the strongest He I lines, which is B2.

Since P_{rot} and $v \sin i$ are well constrained, the **inclination** effectively determines the stellar equatorial radius and the geometric projection onto the line of sight, i.e. the area of the star seen. Together with the above derived T_{eff} and the Hipparcos distance, the flux levels of the SED for 17 000 K constrain the inclination, therefore. For $T_{\mathrm{eff}} = 17\,000$ models with $i = 60^\circ$ are just so compatible with the flux curve for the closest possible distance; better fitting inclinations result in less plausible parameters elsewhere.

Although this still does not constrain the **stellar mass** in a straightforward observational way, all the other parameters are sufficiently known to leave only a narrow range of acceptable evolutionary track masses. This is because we can, at least in first order, expect that the luminosity of a star of a given mass does not change with its rotational velocity. The model has a luminosity of about $L_\star = 1000\,\mathrm{L}_\odot$, and such a luminosity is only compatible with a mass of about $6 \pm 0.5\,\mathrm{M}_\odot$, not with a higher one.

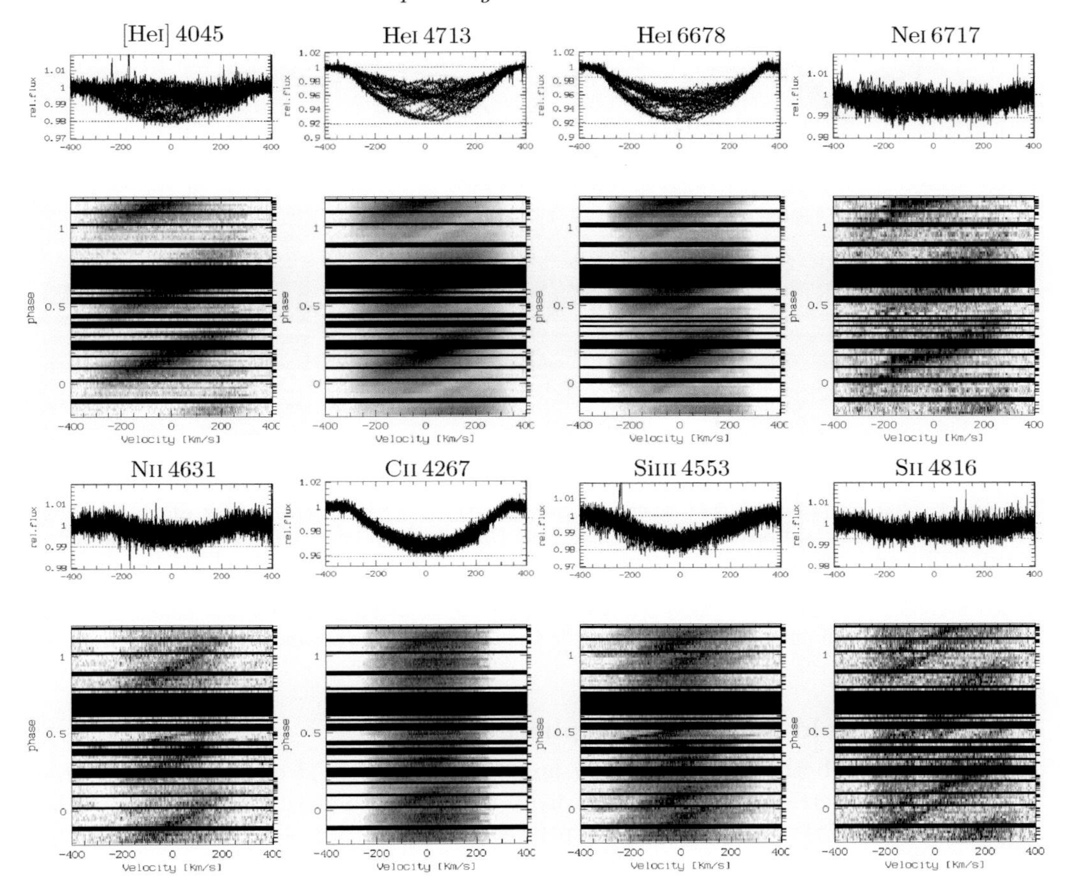

Figure 3. Phased variations of photospheric Helium and metal lines constructed with 64 bins.

4. Variability

For the **ephemeris** we adopt the epoch given by Rivinius *et al.* (2010), $T_0(\mathrm{MJD}) = 54\,940.33$, which is the mid-date of an occultation of the star by the magnetospheric material, seen in the Balmer line spectroscopy, and as such very well defined. The best period proved to be the value published by Oksala *et al.* (2010), $P = 0.5214404$ d, who could rely on a twice as long time-base. Figure 2 shows previously published data together with equivalent width measured in the UVES data, phased with the ephemeris given here.

The Hα equivalent widths are fully dominated by variations of the circumstellar environment, there is no indication for a photospherically intrinsic variation of this or any other hydrogen line's strength. The equivalent width of helium lines like HeI 4713 and 4388 are clearly of a double wave character. Although the strongest EW points in both half-cycles are on a similar level, the weakest EW values differ by a factor of two.

In terms of line profiles, the most obvious photospheric variation is that of the HeI lines. The variations are very strong, in percentage of the line strength much stronger than in other He-strong stars like σ Ori E. Figure 2 shows a factor of three to four between the weak and strong states of HeI lines. Other than the figure suggests, however, a detailed look at the spectra shows that the variation is not as plain as consisting of two enhanced polar regions and a depleted belt only, in particular not in the metal lines, which hardly show EW variations, but clear line profile variability (see Fig. 3).

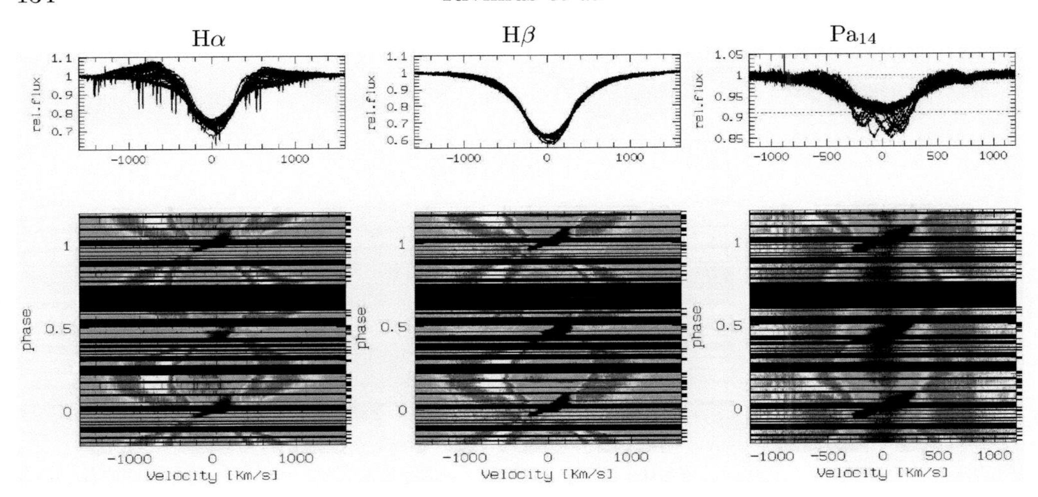

Figure 4. Colour-coded, phased observational variations of the circumstellar magnetosphere. Hα and Hβ are the residuals after subtracting the B3 model; the residual level in the line center is unity to better than 1 %. The respective original spectra are overplotted for Hα and Hβ. Emission clearly seen in residuals, in fact up to Hδ. Paschen lines show emission clearly as well, shown here is Pa_{14}; note that these are not residuals and the different abscissa scale. 128 phase bins were used for all panels in order to sample the fast variations around the occultation phases in sufficient detail.

5. Magnetic field and magnetosphere

Constraints on both the inclination i and the obliqueness β can be obtained from the observed properties of the magnetic field. Applying standard diagnostics of stellar magnetic fields gives $r = B_{\min}/B_{\max} = \cos(\beta + i)/\cos(\beta - i) = -0.78$. This means that $\beta + i \lesssim 140°$, for which $i = \beta = 70°$. Since we have determined $i = 60°$ above, we conclude that both i and β are between 60 and 80°, in such a way that $\beta + i$ is about 130 to 140°. The field strength at the magnetic poles would then be 11 kG. As will be shown in the next paragraph things are probably more complicated than in a pure dipole field, but the above values should at be at least a good approximation.

The measured magnetic curve, when fitted with a sine, has an offset of +0.3 kG, magnetic nulls expected to occur at $\phi = 0$ and $\phi = 0.54$ (dotted lines in lowermost panel of Fig. 2). Yet, the cloud transits are observed to occur at $\phi = 0.0$ and $\phi = 0.46$ (dotted lines in upper panels of Fig. 2). Similarly, the magnetic poles pointing towards the observer would be expected for $\phi = 0.28$ and 0.78 in case of a sinusoidal variation (dashed lines in lowermost panel of Fig. 2), but spectroscopically, i.e. by maximal He-enhancement, are rather seen at $\phi = 0.25$ and 0.68 (dashed lines in upper panels of Fig. 2). Both these offsets could be explained with a misaligned magnetic dipole wrt. to the stellar center, which would display in a non-sinusoidal variability curve of the magnetic field. Unfortunately, there are not enough magnetic measurements to address this point in detail.

The two occultations occur at phases $\phi = 0.0$ and $\phi = 0.46$ (Fig. 4). The passing of the lobe is fast, about $\Delta\phi = 0.13$ from $-v \sin i$ to $+v \sin i$. If the magnetospheric lobes reached down to the star, this would rather be $\Delta\phi = 0.5$. As the crossing time decreases quadratically with distance, there is no absorbing material directly above the star until about $2\,R_\star$. A quarter of a cycle later, the emitting material is seen next to the star, and since it is in magnetically bound corotation there is a linear relation between velocity and distance from the stellar surface. The emission has an inner edge at about $600\,\mathrm{km\,s^{-1}} \approx 2v \sin i$, which again points to an empty region inside $2\,R_\star$. The outer edge of the emission is at a velocity of about $4 \times v \sin i$ in Hα and $3 \times v \sin i$ in Pa_{14}, which gives outer geometrical limits for the lobe emission of about $4\,R_\star$ and $3\,R_\star$, respectively.

The theoretical line profiles from the B3 modeling are good approximations of the photospheric profile, as seen in Fig. 1. The residuals (Fig. 4) represent the clean circumstellar emission and it is possible to measure Balmer decrements when the emitting material is next to the star. While the values for D_{54} would be in agreement with logarithmic particle densities between 11.7 and 12.5 per cm^3, the densities from D_{34} are somewhat higher, at 12.2 to 12.8 per cm^3. In any case, these values are close to the optically thick limit, above which the decrements become independent of density.

6. Conclusions

Apart from its high $v \sin i$, HR 7355 is a rather typical member of the class of He-strong stars. Due to its relative proximity and brightness it is an ideal target to study the interplay of rapid rotation effects, like gravity darkening and meridional circulation, with magnetic effects, like chemical peculiarities and inhibition of the meridional circulation and other non-rigid motions. Indeed it seems very unlikely that a relatively strong and ordered field as seen in HR 7355 could give rise to so many distinct zones of Helium and metal abundances.

The presented data, more than 100 high-quality spectra filling the phase diagram, provide an excellent base to apply techniques like Doppler imaging, rigidly rotating magnetosphere models, and Monte-Carlo radiative transfer models.

References

Mikulášek, Z., Krtička, J., Henry, G. W., de Villiers, S. N. *et al.* 2010, *A&A*, 511, L7
Oksala, M. E., Wade, G. A., Marcolino, W. L. F., Grunhut, J. *et al.* 2010, *MNRAS*, 405, L51
Rivinius, T., Szeifert, T., Barrera, L., Townsend, R. H. D. *et al.* 2010, *MNRAS*, 405, L46
Townsend, R. H. D. 1997, *MNRAS*, 284, 839

Discussion

GROH: Could you comment on the expected angular extension of the Hα and Brγ line forming regions and whether those could be resolved by interferometric observations?

RIVINIUS: Due to the rigidly rotating magnetosphere, $R_{\max} = v_{\max}/(v \sin i)$, so for both regions we expect about 3 to 4 stellar radii. For HR 7355 this is well sub-milliarcsecond, but for HR 5907 (see Grunhut *et al.*, this volume) at least the spectrally dispersed phase signature is probably in reach of AMBER/VLTI and the CHARA combiners.

WADE: Due to the peculiarities hydrogen may be non-uniformly distributed in the atmosphere. Could the apparent departures from dipolar field topology be due to an effect like this?

RIVINIUS: I would be surprised, since a) already the timing of the magnetosphere occultations, i.e. without looking at the magnetic field at all (except polarity) requires a non-sinusoidal (i.e. non-dipole) field signature and b) also with the ESPaDOnS data, derived from metal and He-lines, a similar argument can be made.

Active OB stars: structure, evolution, mass loss, and critical limits
Proceedings IAU Symposium No. 272, 2010
C. Neiner, G. Wade, G. Meynet & G. Peters, eds.

doi:10.1017/S1743921311010167

Structure in the winds of O-Type stars: observations and inferences

Alex W. Fullerton

Space Telescope Science Institute, 3700 San Martin Drive, Baltimore, MD 21218, USA
Herzberg Institute of Astrophysics, 5071 West Saanich Road, Victoria, BC V9E 2E7, Canada
email: fullerton@stsci.edu

Abstract. This review describes the observational evidence for structure in the winds of O-type stars due to large-scale perturbations and small-scale inhomogeneities. Despite considerable progress, a comprehensive theoretical framework that explains the origin. properties, and coexistence of wind structure on different spatial scales has yet to be constructed and incorporated into model atmosphere analyses. Consequently, it is not yet possible to assess the effect of non-stationary structures on different wind diagnostics in a rigorous way, with the result that accurate empirical determinations of mass-loss rates remain elusive.

Keywords. stars: early-type, stars: winds, outflows, stars: mass loss, ultraviolet: stars

1. Introduction

Modern understanding of continuous mass loss from early-type stars is based on the elegant theory of line-driven stellar winds that was introduced by Lucy & Solomon (1970), developed by Castor, Abbott, & Klein (1975), and refined by, e.g., Pauldrach, Puls, & Kudritzki (1986; see also subsequent papers II – XV in this series). In this theory, the momentum required to accelerate material outward is transferred from the intense radiation field via scattering in many spectral lines. The standard model envisages the outflow as being (a) spherically symmetric; (b) smooth, with a monotonically increasing velocity; and (c) stationary. Our confidence in the standard model stems from its ability to explain the observational diagnostics of these outflows: P Cygni profiles; optical emission lines; and excess continuum emission at infrared and radio wavelengths. By any measure, the standard model of line-driven winds provides a rich harvest of physical insight and predictive power.

However, there are many indications that the winds of O-type stars contain velocity and density structure beyond that envisaged by the standard model. Some key observational manifestations of additional structure are illustrated in Fig. 1, which shows *Copernicus* spectra of the N V resonance doublet for ζ Ophiuchi (top) and ζ Puppis (bottom). Inset panels provide schematic comparisons of the expected wind-profile morphology for weak (ζ Oph) and saturated lines (ζ Pup). Neither of the predictions reproduce the shape of the observed line profile. These failures point to the existence of wind structure on large and small spatial scales that is not incorporated in the standard model.

2. Diagnostics of Large-Scale Wind Structure

2.1. *Discrete Absorption Components*

Snapshot observations of unsaturated P Cygni profiles of UV resonance lines obtained with *Copernicus* and *IUE* showed that narrow, localized optical depth enhancements near the blue edge of the absorption trough are common among the O stars; see the

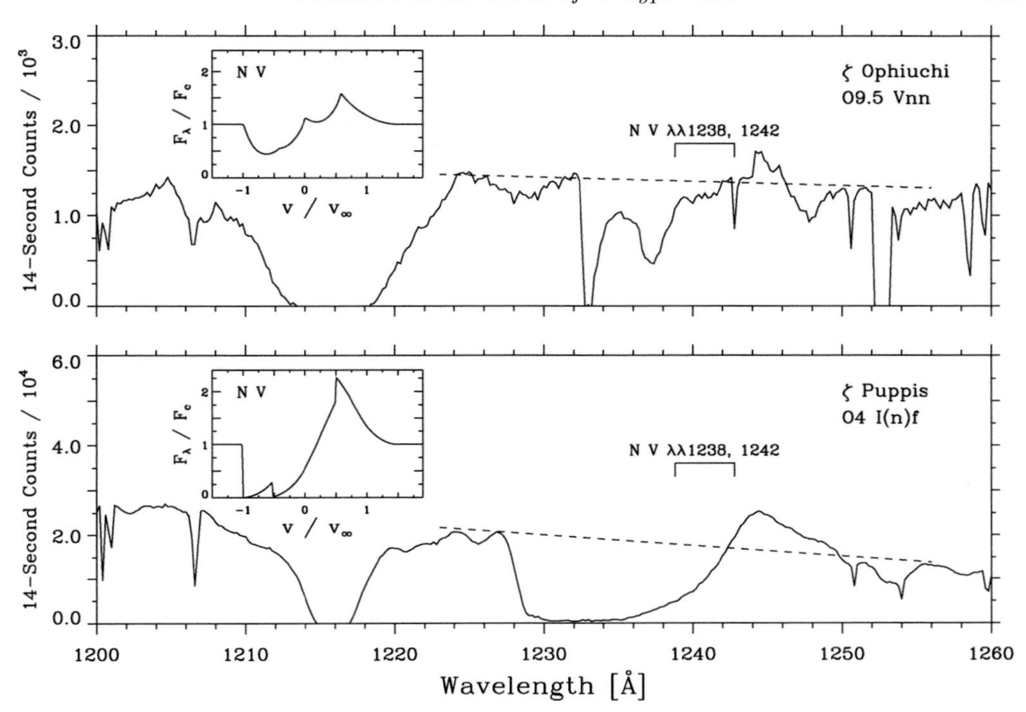

Figure 1. *Copernicus* spectra of the N V λλ1238, 1242 resonance doublet of ζ Oph (top) and ζ Pup (bottom). Inset panels show predictions from the standard model. Inspired by Fig. 1 of Morton (1976).

N V profiles of ζ Oph in Fig. 1. From a large survey of 203 Galactic O stars observed with *IUE*, Howarth & Prinja (1989) estimated that ∼97% of all O stars exhibit "narrow absorption components" (NACs) at any one time. Such a high incidence implies that the phenomenon is essentially universal.

The presence of NACs implies the existence of additional large-scale wind structure beyond that predicted by the standard model. Consider that the Sobolev optical depth of the wind in the radial direction is

$$\tau_{rad}(r) \;\propto\; q_i(r)\,\rho(r)\left(\frac{\mathrm{d}v}{\mathrm{d}r}\right)^{-1} \tag{2.1}$$

where q_i is the local ion fraction of the species responsible a particular wind line; ρ is the local density; and $\mathrm{d}v/\mathrm{d}r$ is the local velocity gradient. Since the NACs represent enhanced optical depth, at least one of these parameters must be different from its standard value: either the ion fraction is increased, the density is increased, or the velocity gradient is shallower at the location of the NAC; or all three changes are combined, since the variables are not independent. Furthermore, since the NACs are often very deep (see Fig. 1), the absorbing gas responsible for them must be optically thick *and* cover a substantial fraction of the stellar disk. Thus, we infer that the NACs are due to changes in the velocity field, density, or ionization structure of the smooth wind that are localized in velocity, but occur on spatial scales that are comparable to the stellar radius.

Since NACs are always present in unsaturated wind profiles of a given star, they were initially interpreted in terms of persistent structures in the wind; see, e.g., Lamers *et al.* (1982). However, further analysis of sporadic time series (see, e.g., Prinja & Howarth 1986) showed that NACs vary on time scales of a day or less. Intensive monitoring with

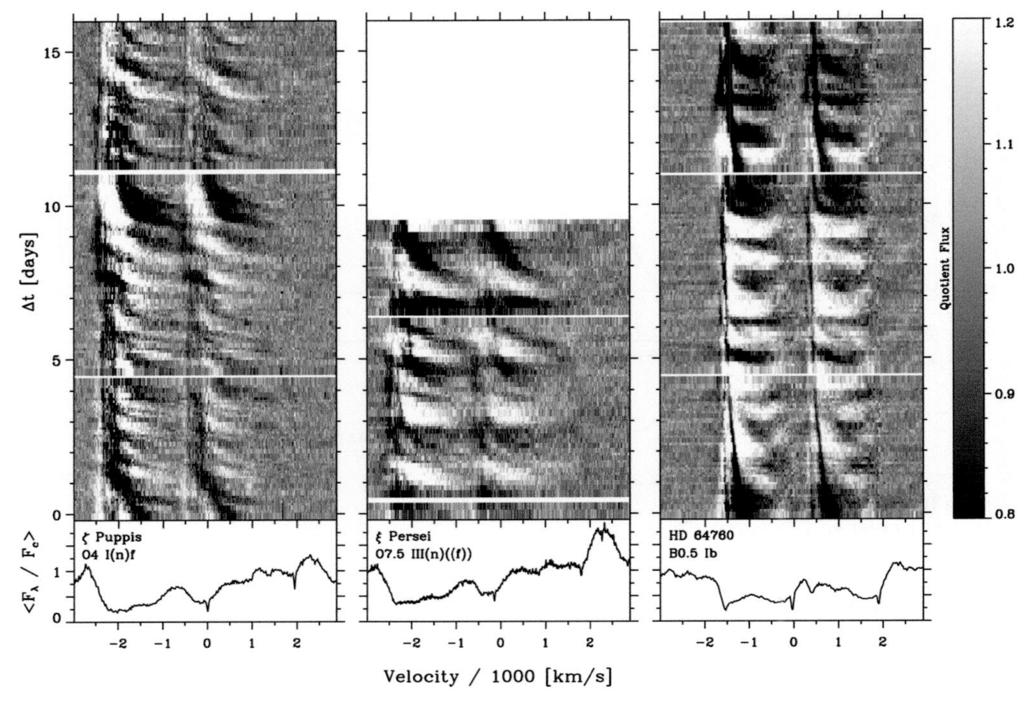

Figure 2. Spectroscopic time series observations of the Si IV λλ 1394, 1403 resonance doublet of ζ Pup (left). ξ Per (middle), and HD 64760 (right). The time series of ζ Pup and HD 64760 were obtained during the *IUE* "MEGA Campaign", while the observations of ξ Per were obtained with *IUE* in 1994 October. In all cases the dynamic spectra have been normalized by the mean spectrum from the time series.

IUE of ξ Persei [O7.5 III(n)((f)); Prinja *et al.* 1987] and 68 Cygni [O7.5 III:n((f)); Prinja & Howarth 1988] demonstrated the fundamentally dynamic and repetitive nature of the phenomenon. As illustrated in Fig. 2 for three key objects, broad, weak absorptions are initially detected at quite small velocities. These absorptions narrow and strengthen as they accelerate blueward through the absorption trough, until they eventually merge with the persistent, high-velocity NAC. Consequently, the NACs are now recognized as the end products of the more general phenomenon of "discrete absorption components" (DACs), which move through a large swath of the absorption trough. With this understanding, we *infer* from the ubiquity of NACs that DACs must also be universally present in unsaturated wind lines of O stars.

Although comparatively few stars have been observed intensively, the following general properties have been identified by tenacious observational effort (e.g., Kaper *et al.* 1996; Prinja *et al.* 2002): (a) DACs accelerate slowly through the absorption trough compared to the expectations based on the mean flow of the wind; (b) DACs recur on intervals related to the estimated stellar rotation period ($P_{\rm rot}$); and (c) although DACs are always present in multi-epoch observations of the same star and the pattern of variability is always similar, the variations do not seem to be phase locked over intervals of several months.

As a direct test of the connection between DAC activity and the rotation period of the underlying star, ζ Pup and HD 64760 were monitored for ∼16 consecutive days in 1995 January as part of the *IUE* "MEGA Campaign" (Massa *et al.* 1995). These exquisite time series are shown in Fig. 2 along with an equally heroic effort on ξ Per that

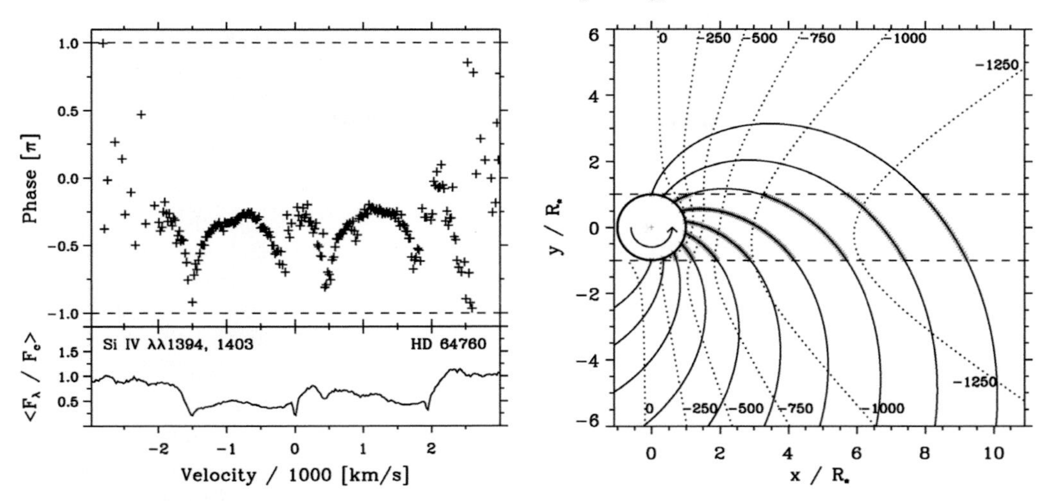

Figure 3. Left: Phase of the 1.2-day PAM as a function of position in the absorption trough of the Si IV doublet of HD 64760. Right: Schematic illustration of the origin of phase bowing from Fullerton *et al.* (1997).

included co-ordinated ground-based observations (de Jong *et al.* 2001). The duration of the observations covered 3–4 consecutive repetitions of the estimated $P_{\rm rot}$ for each star; and in all cases, variation related to this period were detected. However, each star is a little different. For ζ Pup, the DACs recur every 0.80 days, and the entire profile is modulated every $\sim$5.2 days $\sim P_{\rm rot}$ (Howarth *et al.* 1995) while for ξ Per the DACs recur every 2.086 days $\simeq P_{\rm rot}/2$ (de Jong *et al.* 2001). Curiously, DACs of HD 64760 recur only once, which indicates that their recurrence time is longer than $P_{\rm rot}$. However, a strikingly different type of variation is immediately apparent in Fig. 2 in the form of periodic absorption modulations.

2.2. *Periodic Absorption Modulations*

The periodic component of wind variability exhibited by HD 64760 is visible in Fig. 2 as light and dark bands that recur every 1.2 days (Fullerton et al. 1997). These variations are due to *modulations* of the optical depth in the absorption trough, not enhancements. These periodic absorption modulations (PAMs) do not move sequentially through the absorption trough, but instead affect a broad range of velocities nearly simultaneously. Consequently, PAMs are a distinctly different phenomenon from DACs, though obviously they can coexist.

Although the PAMs of HD 64760 are thus far unique, they provide many insights into the geometry of repetitive wind structures through the phenomenon of "phase bowing". Phase bowing can be seen especially well in Fig. 2 near ΔT=2 days, where curved, upward-facing bright features corresponding to a single phase of the modulation evolve simultaneously toward higher and lower velocities in the absorption trough. It is even more evident in the distribution of the sinusoidal phase constant associated with the modulations as a function of position in the absorption trough (Fig. 3), which shows a maximum at an intermediate velocity. Owocki *et al.* (1995) explained this behavior in terms of the way that an azimuthally extended structure exits the line of sight to a distant observer. As shown schematically in Fig. 3, such structure first exits the column of material projected against the disk of the star at an intermediate velocity, and thereafter exits at both large and small line-of-sight velocities simultaneously, as observed.

Thus, phase bowing is now believed to be a robust diagnostic of the presence of longitudinally extended – probably spiral-shaped – structures in the wind. This geometry is not uniquely associated with PAMs. For example, by carefully comparing the behaviour of the phase constant in different wind lines, de Jong et al. (2001) detected phase bowing associated the DACs of ξ Per.

2.3. *Co-Rotating Interaction Regions*

Following the suggestion of Mullan (1984), cyclical variability in hot-star winds is now attributed to the behaviour of azimuthally extended perturbations known as co-rotating interaction regions (CIRs). CIRs are a well known phenomenon in the solar wind that are caused by differences in the initiation of the wind at different locations. These changes in boundary conditions cause "fast" and "slow" streams to emerge from adjacent regions of the stellar surface along nearly radial trajectories. However, as the star rotates, the "fast" wind from one sector overtakes the "slow" wind that previously emerged from an adjacent sector. The resultant interaction produces a shock and a shallow, plateau-like velocity gradient. Although material continues to flow through the interaction zone, its slower speed also causes a localized density enhancement (compared with an unstructured wind). The net result is a long-lived, spiral-shaped structure in the wind that rotates with a period determined by the surface phenomenon responsible for creating the "fast" and "slow" wind streams in the first place.

CIRs provide a natural explanation for essentially all the properties of cyclical wind variability; see, e.g., the hydrodynamic simulations of Cranmer & Owocki (1996), Dessart (2004), and Lobel & Blomme (2008). Enhanced optical depth is readily produced by the velocity plateau and, secondarily, the density enhancement caused by the interaction region. The systematic narrowing of the DAC as it propagates blueward through the absorption trough is accounted for by projection effects along the spiral-shaped CIR as it rotates into the line of sight, while "phase bowing" can be attributed to the behaviour of the spiral as it exits the near hemisphere of the wind. Most importantly, the slow apparent motion of the DAC does not imply that material is also moving much more slowly than expected, because the DAC is produced by different material flowing through the interaction region at different times. The apparent acceleration is determined by the shape of the CIR, which in turn is determined by the processes at the stellar surface that are responsible for the emergence of "fast" and "slow" wind streams. This coupling between the CIR and the stellar surface ultimately determines the recurrence period of cyclical wind variations in a particular star.

Thus, in the current paradigm, photospheric processes are ultimately responsible for large-scale wind variability. The ubiquity of DACs (or at least NACs) requires a ubiquitous "trigger" mechanism: i.e., a process or processes that organize the stellar surface into azimuthal sectors that alternately produce "fast" and "slow" outflows of material. Motivated by the expected temperature variations associated with nonradial pulsations (NRP) of modest degree (see, e.g., Kaufer *et al.* 2006), hydrodynamic simulations use the artifice of one or more bright spots distributed around the star to produce these streams. Of the "usual suspects" – pulsations or magnetic fields – it does seem more likely that multi-mode pulsation can provide the necessary coverage of the parameter space occupied by O- and B-type stars, if indeed a single"trigger" is responsible for creating CIRs.

Lobel & Blomme (2008) also show that retrograde NRP can lead to recurrence times that are substantially longer than the rotation period of the star. For the particular case of HD 64760 (Fig. 2), they show that the long interval between the recurrence of DACs can be accommodated. Thus, it may be that the interplay between the pattern speed of multimode NRP and the rotation of the stellar surface prevent DACs from being phase-locked over long intervals, even though the patterns are similar in multi-epoch observations (see, e.g., Kaper *et al.* 1999). In the specific case of HD 64760, the relationship between the DACs and the PAMs needs clarification, since it is not obvious how multiple instances of CIRs with extremely different time scales can coexist in the wind.

3. Diagnostics of Small-Scale Wind Structure

In addition to the large-scale spiral structures implied by the cyclical variability of wind lines, there is compelling evidence that the winds of O-type stars are inhomogeneous over much smaller distances.

3.1. *Transient Emission-Line Substructures*

The most direct indication of small-scale density enhancements in stellar winds comes from observations of narrow, transient substructures in strong emission lines (e.g, Hα, He II λ4686). These lines are predominantly formed by recombination in the wind, which is a two-body or "density squared" (hereafter ρ^2) process. Consequently, the narrow bumps of excess emission trace regions of enhanced density that are localized in velocity (which probably implies being localized in space, too). The bumps move systematically away from line center over the course of a few hours, which is broadly consistent with the flow time and suggests that the inhomogeneities are carried along by the wind. Similar transients have been observed extensively in the much broader emission lines formed in the winds of Wolf-Rayet stars; see, e.g., Moffat *et al.* (1988). However, it is quite difficult to detect transient features in the narrower and weaker emission profiles of O-stars and to date they have only been observed in the wind profiles of ζ Pup (O4 I(n)f; Eversberg *et al.* 1998, Lépine & Moffat 2008) and HD 93129A (O2 If*; Lépine & Moffat 2008).

3.2. *Black Troughs of Saturated UV Resonance Lines*

Strongly saturated resonance lines of O supergiants typically exhibit extended intervals of zero residual flux; see, e.g., the N V resonance doublet of ζ Pup in Fig. 1. These "black troughs" are impossible to understand in the context of the standard model, which predicts that a saturated P Cygni profile will have no residual flux only at its blue edge, because all other wavelengths have a component of emission directed toward a distant observer. Lucy (1982) recognized that "black troughs" are caused by scattering in a wind with a multiply non-monotonic velocity field, which leads to enhanced back-scattering and a systematic reduction in the blue-shifted (forward-scattered) emission; see Fig. 4. Thus, "black troughs" are a signature of non-monotonic velocity structure in a stellar wind, which is presumably accompanied by localized density inhomogeneities. As discussed by Puls *et al.* (1993), a substantial amount of small-scale velocity structure is required to get extremely black profiles; see also the discussion by Sundqvist *et al.* (2010).

3.3. *Soft Blue Edges of UV Resonance Lines*

A second morphological peculiarity of the absorption troughs of P Cygni profiles is that their blue edges do not rise to the continuum as steeply as expected. This "softness" of the blue edge is particularly evident in saturated lines, but is commonly seen in weaker lines too; see Fig. 1. In the standard model, the most blue-shifted absorption in a strong P Cygni profile denotes the terminal velocity of the wind (v_∞), which is the largest velocity along the line of sight to the star (provided that the ion responsible for the resonance transition exists throughout the wind). Since, by definition, no wind material exists at larger velocities to scatter light from the star, the transition to the local stellar

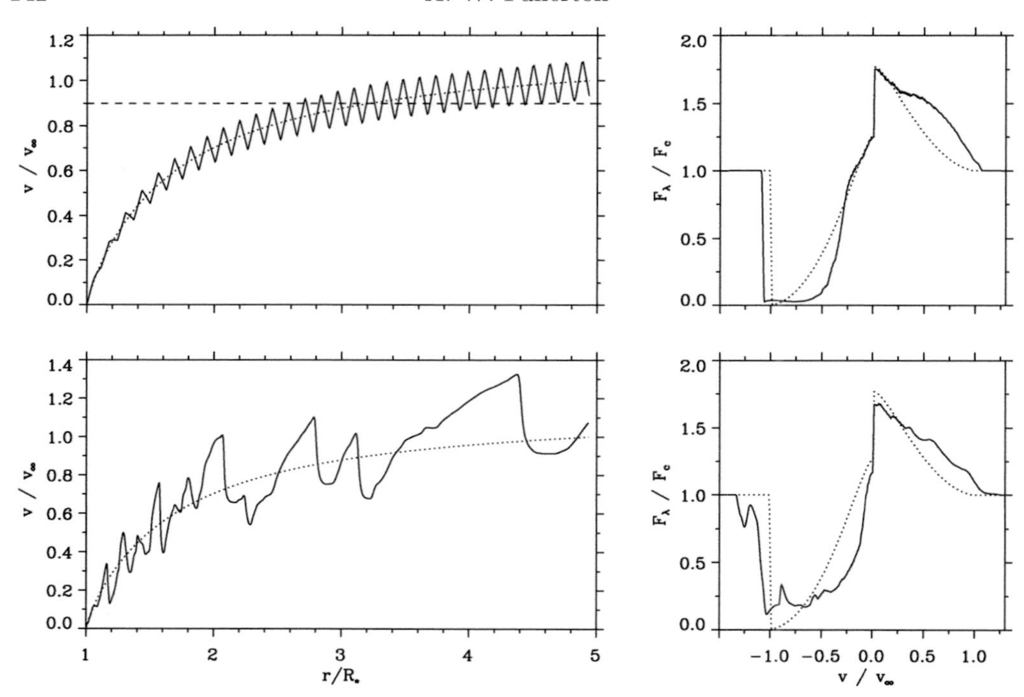

Figure 4. Upper: Schematic illustration of the creation of a black trough from multiple scattering interactions in a non-monotonic velocity field. In the left-hand panel, a sawtooth variation is imposed on the smooth, monotonic velocity profile (dotted). The dashed horizontal line indicates that a photon interacts with the sawtooth profile many times. The right-hand panel shows that preferential back-scattering in the structured wind creates extended black troughs, which are not completely black due to "vorosity". Since structure occurs in spherical shells for this 1D simulation, excess red-shifted emission is seen from back-scattering in the far hemisphere of the wind. Lower: Velocity structure from a 1-D radiation hydrodynamics simulation of the line-deshadowing instability. In this case, there are not enough interactions between a continuum photon and the wind to backscatter all the flux and the profile is not saturated. Adapted from Puls *et al.* (1993).

continuum must necessarily be abrupt. Instead, the observed "softness" of the blue edge indicates that there is a distribution of material at velocities in excess of the terminal velocity; i.e., there is a dispersion of velocities about the mean velocity field. This velocity dispersion also implies localized variations in density within the wind.

3.4. *Discrepant Mass-Loss Estimates*

Large, systematic differences in estimates of the mass-loss rate ($\dot{M}$) from diagnostics with different density dependencies provide the least direct evidence for small-scale structure, but constitute the strongest indication that such structure is a universal property of O-star winds. The key distinction is between (a) scattering of continuum photons by UV resonance lines, which is linearly proportional to density (ρ); and (b) emission processes (recombination lines, free-free continuum emission), which are proportional to ρ^2. However, only $\dot{M}q_i$ can be determined from wind profiles of unsaturated UV resonance lines. Reliable values of q_i are notoriously difficult to estimate theoretically or measure empirically for the most commonly observed species, which means that $\dot{M}$ is poorly constrained. As a result, recent estimates of $\dot{M}$ for O-type stars have all relied on ρ^2 diagnostics, especially Hα.

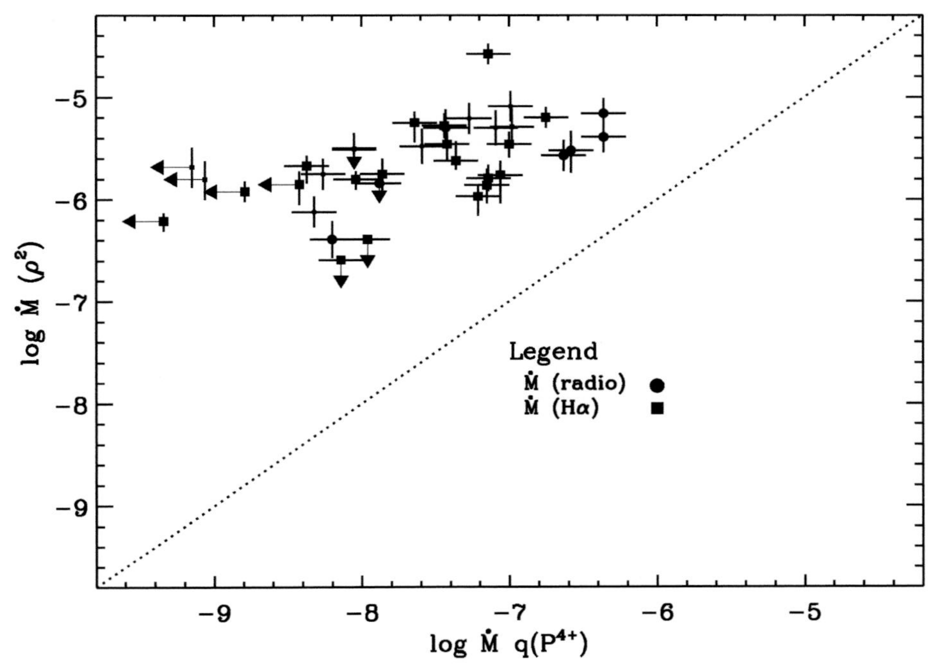

Figure 5. Comparison of $\dot{M}$ determined from ρ^2 diagnostics and the P V resonance lines of 40 Galactic O-type stars. Stars with temperature class between O4 and O7.5 are closest to the 1–1 correlation line, but have discrepant estimates of $\dot{M}$ by a factor of 20. From Fullerton, Massa, & Prinja (2006).

The launch of *FUSE* in 1999 changed the situation dramatically by providing access to the rich suite of resonance-line diagnostic between ~900 and 1200 Å. For O-type stars, the P V λλ1122, 1128 doublet is crucially important, for two reasons. First, since P is not a cosmically abundant element, the absorption troughs of the P V P Cygni profiles are never saturated and $\dot{M}q_i$ can be measured accurately. Second, state-of-the-art model atmosphere calculations based on the standard model predict that P^{4+} is the dominant ion in the winds of mid- to late- O-type stars. For these stars, q_i is known *a priori* to lie between ~0.5 and 1. Consequently, simple analysis of P V wind profiles should yield measurements of $\dot{M}$ that agree with determinations from ρ^2 diagnostics to within a factor of ~2 for at least some O stars.

However, Fig. 5 shows that measurements of $\dot{M}$ from ρ^2 diagnostics for a "blue-ribbon" sample of 40 Galactic O-type stars are *always* greater than measurements from the P V resonance lines *by a factor of at least 20* (Fullerton, Massa, & Prinja 2006). Detailed model atmosphere calculations indicate similar levels of disagreement between Hα and the P V doublet for O-type supergiants in the Galaxy (Bouret *et al.* 2005), Large Magellanic Cloud (Crowther *et al.* 2002; Massa *et al.* 2003) and Small Magellanic Cloud (Hillier *et al.* 2003).

The simplest resolution of this "P V problem" is to assume that hot-star winds are significantly "clumped" on small scales, in which case ρ^2 diagnostics necessarily over estimate the amount of material in the wind when they are interpreted within the context of the standard model. Thus, we *infer* from the discrepancy illustrated in Fig. 5 that O-star winds are inhomogeneous on small spatial scales; and, as a corollary, that previous estimates of $\dot{M}$ are too large. Strictly speaking, the discrepancy only exists for the subset of O stars for which P^{4+} is the dominant ion. However, there is no reason to suppose that the mechanism responsible for clumping the wind is limited to this subset, so the inference is more generally valid.

3.5. *The Origin and Nature of Clumping*

The strong line-deshadowing instability (LDI) is an intrinsic property of the line-driving mechanism (Owocki & Rybicki 1984). As such, it provides a unifying framework to interpret the observational signatures of small-scale structure. In particular, numerical simulations of the nonlinear growth of the LDI by Owocki, Castor, & Rybicki (1988) and Feldmeier (1995) show that small amounts of material are accelerated to velocities in excess of v_∞, exactly as required to explain the "soft" blue edges of P Cygni profiles. In 1D simulations, subsequent interactions between this fast moving gas and slower material at larger radii compress the wind into a series of dense shells, which necessarily make the velocity profile non-monotonic (black troughs; Fig. 4). Exploratory simulations in 2D by Dessart & Owocki (2005) suggest that these shells fragment laterally into many small clumps, which manifest themselves as transient emission-line substructures and significantly bias determinations of $\dot{M}$. Shocks and collisions between dense shells also produce X-rays distributed throughout the wind; see, e.g., the article by Cohen & Leutenegger in these proceedings.

Although the LDI provides a firm theoretical basis to understand the origin of small-scale structure, there are still many uncertainties concerning the size and shape of the clumps. Fortunately, work to reconcile estimates of $\dot{M}$ from different diagnostics provides some general constraints on these properties. Initial attempts to incorporate clumping in model atmosphere calculations relied on a two-component model, in which the wind was assumed to consist of (a) optically thin clumps characterized by enhanced density compared to predictions of the standard model; and (b) vacuum. Consistent estimates of $\dot{M}$ can be obtained with this approach, essentially by changing the prevailing ionization equilibrium of an element (e.g., P) in favor of lower ionization stages in denser environments. Very small volume filling factors are derived for the clumps in some cases (e.g., Bouret *et al.* 2005), which implies that the wind is mostly vacuum and that the clumps are very dense.

However, this approach does not adequately treat radiation transfer in spectral lines because it does not contain any length scales. Consequently, there is no way of testing whether the clumps are optically thick, which they must be for some of the implied overdensities. The character of the radiation transfer in the wind changes once the clumps become optically thick, because (a) radiation is trapped within the clumps; and (b) channels open up between the clumps that permit radiation from the star to leak out; i.e., the wind becomes "porous". The leakage implied by porosity *reduces* the effective optical depth of the wind (Owocki & Cohen 2006; Oskinova *et al.* 2007) in a particular transition. Oskinova et al. (2007) recognized that resonance transitions become optically thick before other transitions, and consequently a clumped wind could be a porous medium for the P V doublet but not for Hα. By allowing continuum light to "leak" around clumps that are optically thick in resonance lines, Oskinova et al. (2007) obtained reasonable fits to both the P V doublet and the Hα profile of ζ Pup with only a modest reduction in the "smooth wind" value of $\dot{M}$.

Owocki (2008) realized that the non-monotonic velocity field produced by the LDI has gaps in the velocity field that continuum photons can also leak through. This "porosity in velocity space" (dubbed "vorosity" by Owocki) is distinct from porosity in physical space, but also reduces the effective optical depth of the wind in different lines. In an impressive exploratory study, Sundqvist *et al.* (2010) confirmed that both porosity (due to optically thick clumps) and vorosity (due to velocity gaps between clumps) play a similar role in reducing the effective optical depth of the wind in intrinsically strong transitions. By accounting for "leakage" in velocity and physical space, they showed that it is in principle possible to obtain concordance between different mass-loss diagnostics.

However, Sundqvist *et al.* (2010) could not obtain satisfactory agreement between different diagnostics using the wind structure predicted by the current generation of radiation-hydrodynamics simulations of the LDI. Compared with these models, it appears that (a) structure must be initiated deeper in the wind; and (b) clumps need a greater internal velocity gradient. Further work to simulate the time-dependent structure of line-driven winds is required to determine whether these shortcomings are fundamental or simply artifacts of the current generation of models. An interesting twist might be to incorporate deep-seated perturbations from shallow convection zones to trigger the onset of clumping nearer to the photosphere (Cantiello *et al.* 2009).

4. What Does a O-Star Wind Look Like?

How should we picture the wind of an O-type star? The idea of a smooth, spherically expanding outflow is well entrenched in our minds, and certainly provides a useful point of departure for thinking about O-star winds. However, both theory and observations have encouraged us to consider the wind as an ensemble of distinct clumps, each with its own size, velocity, and propensity to interact with each other and the low-density medium that surrounds them. At the same time, variability studies have convinced us that large-scale spiral waves are embedded in the outflow and that these structures are continually renewed by photospheric processes that control the emergence of the wind.

Since observations suggest that each of these pictures contains an element of truth, the current challenge is to synthesize them into a coherent physical theory. There are a variety of issues that must be resolved, but foremost is to demonstrate that large-scale CIRs can coexist with the LDI. Initial attempts to include CIRs in time-dependent radiation hydrodynamics simulations by Owocki (1999) met with mixed success, with the survival of a CIR depending on the precise treatment of the nonlocal line force. At present, further progress on this fundamental issue appears to be limited by the numerical techniques required to treat this difficult, two-dimensional problem.

An observational challenge is to identify definitively the "trigger mechanism" responsible for CIRs. If, e.g., DACs are the observational manifestation of CIRs and CIRs are produced by temperature variations on the stellar surface due to NRP, then the ubiquity of DACs implies that *all* O stars should be non-radial pulsators. This is a testable prediction, which receives some support from the widespread occurrence of photospheric line profile variations (Fullerton *et al.* 1996), though the much sought connection between photospheric activity and DACs has proven difficult to demonstrate explicitly (see, e.g., de Jong *et al.* 2001). This difficulty is now compounded by the lack of long-term spectroscopic monitoring capability in space. *Our community desperately needs access to a modern version of IUE to make continued progress!*

Understanding the time-dependent structure of hot-star winds is a fascinating problem for observers and theoreticians alike. However, this research is also important in a broader astrophysical context, because we cannot properly quantify the biases associated with different $\dot{M}$ diagnostics until we have a clearer understanding of the origin and nature of wind structure on all spatial scales. Initial indications are that the changes in mass flux associated with large-scale structure in the form of CIRs are rather small, (at least for HD 64760; Lobel & Blomme 2008), whereas changes in the *interpretation* of mass-loss diagnostics due to the presence of small-scale structure are large (factor of 5?) and very uncertain. Until a unified treatment of all the different forms of wind structure becomes available, the mass-loss rates of O-type stars will necessarily remain uncertain by unacceptably large factors.

References

Bouret, J.-C., Lanz, T., & Hillier, D. J. 2005, *A&A*, 438, 301
Cantiello, M., Langer, N., Brott, I., de Koter, A. *et al.* 2009, *A&A*, 499, 279
Castor, J. I., Abbott, D. C., & Klein, R. I. 1975, *ApJ*, 195, 157
Cranmer, S. R. & Owocki, S. P. 1996, *ApJ*, 462, 469
Crowther, P. A., Hillier, D. J., Evans, C. J., Fullerton, A. W. *et al.* 2002, *ApJ*, 579, 774
de Jong, J. A., Henrichs, H. F., Kaper, L., Nichols, J. S. *et al.* 2001, *A&A*, 368, 601
Dessart, L. 2004, *A&A*, 423, 693
Dessart, L. & Owocki, S. P. 2005, *A&A*, 437, 657
Eversberg, T., Lepine, S., & Moffat, A. F. J. 1998, *ApJ*, 494, 799
Feldmeier, A. 1995, *A&A*, 299, 523
Fullerton, A. W., Gies, D. R., & Bolton, C. T. 1996, *ApJS*, 103, 475
Fullerton, A. W., Massa, D. L., & Prinja, R. K. 2006, *ApJ*, 637, 1025
Fullerton, A. W., Massa, D. L., Prinja, R. K., Owocki, S. P. *et al.* 1997, *A&A*, 327, 699
Hillier, D. J., Lanz, T., Heap, S. R., Hubeny, I. *et al.* 2003, *ApJ*, 588, 1039
Howarth, I. D. & Prinja, R. K. 1989, *ApJS*, 69, 527
Howarth, I. D., Prinja, R. K., & Massa, D. 1995, *ApJ* (Letters), 452, L65
Kaper, L., Henrichs, H. F., Nichols, J. S., Snoek, L. C. *et al.* 1996, *A&AS*, 116, 257
Kaper, L., Henrichs, H. F., Nichols, J. S., & Telting, J. H. 1999, *A&A*, 344, 231
Kaufer, A., Stahl, O., Prinja, R. K., & Witherick, D. 2006, *A&A*, 447, 325
Lamers, H. J. G. L. M., Gathier, R., & Snow, Jr., T. P. 1982, *ApJ*, 258, 186
Lépine, S. & Moffat, A. F. J. 2008, *AJ*, 136, 548
Lobel, A. & Blomme, R. 2008, *ApJ*, 678, 408
Lucy, L. B. 1982, *ApJ*, 255, 278
Lucy, L. B. & Solomon, P. M. 1970, *ApJ*, 159, 879
Massa, D., Fullerton, A. W., Nichols, J. S., Owocki, S. P. *et al.* 1995, *ApJ* (Letters), 452, L53
Massa, D., Fullerton, A. W., Sonneborn, G., & Hutchings, J. B. 2003, *ApJ*, 586, 996
Morton, D. C. 1976, *ApJ*, 203, 386
Moffat, A. F. J., Drissen, L., Lamontagne, R., & Robert, C. 1988, *ApJ*, 334, 1038
Mullan, D. J. 1984, *ApJ*, 283, 303
Oskinova, L. M., Hamann, W.-R., & Feldmeier, A. 2007, *A&A*, 476, 1331
Owocki, S. P. 1999, in: B. Wolf, O. Stahl, & A. W. Fullerton (eds.), *IAU Colloq. 169: Variable and Non-spherical Stellar Winds in Luminous Hot Stars*, Lecture Notes in Physics, Berlin Springer Verlag 523, p. 294
Owocki, S. P. 2008, in: W.-R. Hamann, A. Feldmeier, & L. M. Oskinova (eds.), *Clumping in Hot-Star Winds*, p. 121
Owocki, S. P. & Cohen, D. H. 2006, *ApJ*, 648, 565
Owocki, S. P. & Rybicki, G. B. 1984, *ApJ*, 284, 337
Owocki, S. P., Castor, J. I., & Rybicki, G. B. 1988, *ApJ*, 335, 914
Owocki, S. P., Cranmer, S. R., & Fullerton, A. W. 1995, *ApJ* (Letters), 453, L37
Pauldrach, A., Puls, J., & Kudritzki, R. P. 1986, *A&A*, 164, 86
Prinja, R. K. & Howarth, I. D. 1986, *ApJS*, 61, 357
Prinja, R. K. & Howarth, I. D. 1988, *MNRAS*, 233, 123
Prinja, R. K., Howarth, I. D., & Henrichs, H. F. 1987, *ApJ*, 317, 389
Prinja, R. K., Massa, D., & Fullerton, A. W. 2002, *A&A*, 388, 587
Puls, J., Owocki, S. P. & Fullerton, A. W. 1993, *A&A*, 279, 457
Sundqvist, J. O., Puls, J., & Feldmeier, A. 2010, *A&A*, 510, A11

Discussion

MEYNET: You showed a very impressive difference between the ρ^2 diagnostics and the P^{4+} UV diagnostics for $\dot{M}$. Does this difference decrease (for instance) when the metallicity decreases?

FULLERTON: Very similar discrepancies are seen in samples of O-type stars observed with *FUSE* in both the LMC (Crowther *et al.* 2002, Massa *et al.* 2003) and SMC (Hillier *et al.* 2003). Although these samples are limited, they don't provide any evidence to suggest that the degree of clumping depends strongly on the overall metallicity of the star.

LOBEL: Thank you for this clear review. You made the point that it is difficult to reconcile large-scale structures in the wind with the small-scale structures. It appears to me that we should separate the large-scale wind structures that cause DACs from those that produce the rotational modulations. Could you comment on the "phase bowing" in the modulations, for example, observed in *IUE* spectra of HD 64760? Some of its modulations do not reveal the phase bowing and are almost completely flat. Instead of the large-scale wind spirals we need for DACs, the modulations therefore require almost 'spoke-like' density enhancements in wind regions of $r \leqslant 10\ R_\star$. Hydrodynamic modelling has recently shown that the modulations can be caused by pressure waves at the surface, instead of the spots or brighter equatorial regions that cause the large-scale CIRs.

FULLERTON: I'll look forward to learning more about your simulations that include perturbations from pressure waves! As for "phase bowing": I agree that some occurrences of the PAMs exhibit more curvature than others in dynamic spectra (e.g., Fig. 2); but I think that most of the PAMs are curved. My analysis of "phase bowing" is based on the distribution of the sinusoidal fitting constant as a function of position in the absorption trough shown in Fig. 3, which is admittedly less sensitive to any peculiarities associated with an individual occurrence of the modulation.

KHOLTYGIN: What can you say about the small-scale structure of large-scale structures? Is it possible that large-scale streams is the stellar wind of OB stars consist of small clumps?

FULLERTON: Very little is known about the structure of the CIRs from observations, but I think it is plausible that the "fast" and "slow" streams responsible for them could be composed of small blobs of material. One implication would be that the onset of clumping occurs very deep, which is not what simulations of the LDI currently predict.

MASSA: I would like to point out that the Cosmic Origins Spectrograph on *HST* can obtain spectra in the *FUSE* range with roughly the same sensitivity as *FUSE* and resolving power of 8,000 – 10,000. Calibration observations are being obtained at this time.

Active OB stars: structure, evolution, mass loss, and critical limits
Proceedings IAU Symposium No. 272, 2010
C. Neiner, G. Wade, G. Meynet & G. Peters, eds.

doi:10.1017/S1743921311010179

Modeling the winds and magnetospheres of active OB stars

Richard H. D. Townsend

Department of Astronomy, University of Wisconsin-Madison, Madison, WI 53706, USA

Abstract. After briefly reviewing the theory behind the radiative line-driven winds of OB stars, I examine the processes that can generate structure in them; these include both intrinsic instabilities, and surface perturbations such as pulsation and rotation. I then delve into wind channeling and confinement by magnetic fields as a mechanism for forming longer-lived circumstellar structures. With a narrative that largely follows the historical progression of the field, I introduce the key insights and results that link the first detection of a magnetosphere, over three decades ago, to the recent direct measurement of magnetic braking in a number of active OB stars.

Keywords. magnetohydrodynamics: MHD, stars: early-type, stars: emission-line, stars: magnetic fields, stars: mass loss, stars: rotation, stars: winds, outflows

---

1. Line-Driven Winds

Stars lose mass in a radiation-driven wind whenever the effective Eddington parameter — representing the ratio of radiative to gravitational acceleration — exceeds unity in their outer layers. In OB stars, however, the pertinent parameter is not the usual

$$\Gamma_{\rm e} = \frac{\kappa_{\rm e} L_*}{4\pi G M_* c} \tag{1.1}$$

associated with continuum electron-scattering opacity (here denoted by $\kappa_{\rm e}$), but instead that associated with *line* opacity arising in resonance-scattering transitions. Within the Castor, Abbott & Klein (1975, hereafter CAK) formalism for these line-driven winds, the effective Eddington parameter is given by

$$\Gamma_\ell = \frac{1}{1-\alpha} \Gamma_{\rm e} \bar{Q} \left(\frac{{\rm d}v/{\rm d}r}{\rho c \bar{Q} \kappa_{\rm e}} \right)^\alpha \tag{1.2}$$

where the notation follows Owocki (2004), and in particular $\bar{Q}$ is a measure of the total line opacity in the wind, in units of the electron scattering opacity $\kappa_{\rm e}$ (see Gayley 1995). The term in parentheses is the reciprocal of the Sobolev optical depth; through the appearance of the spatial velocity gradient ${\rm d}v/{\rm d}r$, it represents the degree to which the wind is able to Doppler-shift the lines out of their own shadow.

The inverse dependence of the radiative acceleration on the density ρ naturally introduces a negative feedback loop that helps to self-regulate the wind. If too much mass is launched from the surface, the acceleration decreases to a point where Γ_ℓ drops below unity; the wind then stalls and falls back to the stellar surface. Accordingly, the mass-loss rate $\dot{M}$ of a radiatively driven wind is not a free parameter, but instead established by this self-regulation process. That is, $\dot{M}$ is an *eigenvalue* of the system, established by the requirement that a smooth velocity profile links the subsonic outflow at the stellar surface to the supersonic outflow at the nominal outer (far-star) boundary. In the idealized

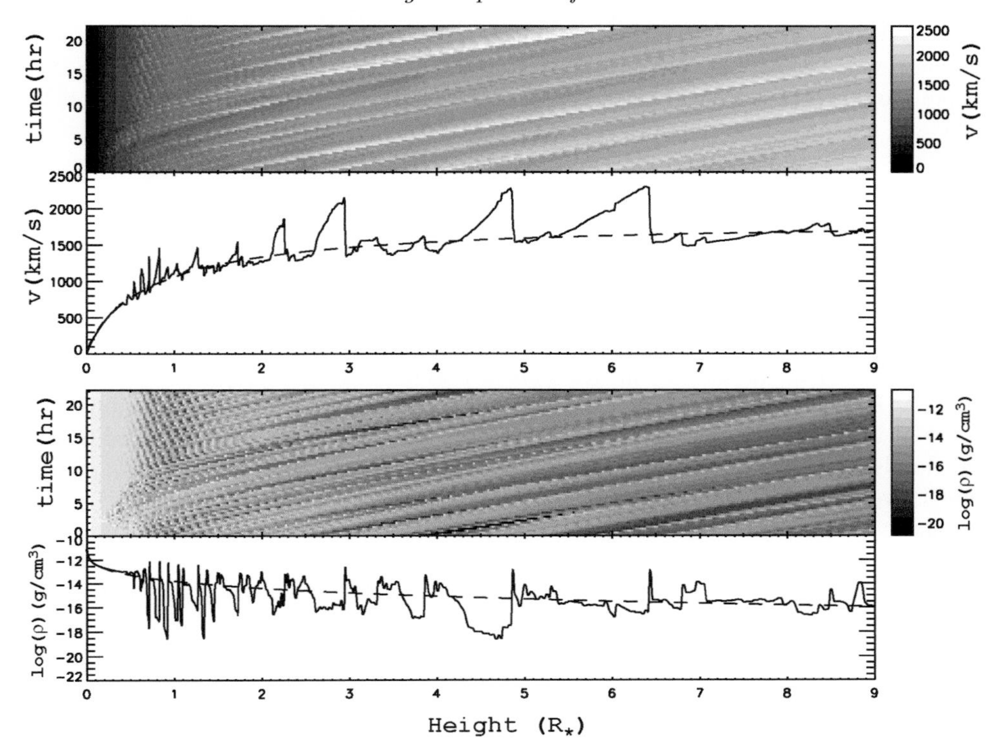

Figure 1. Results of 1-D hydrodynamical simulations of the line-deshadowing instability, calculated using the Smooth-Source-Function formulation of Owocki (1991). The line plots show the spatial variation of velocity (upper) and density (lower) at a fixed, arbitrary time snapshot. The corresponding grey scales show both the time (vertical axis) and height (horizontal axis) evolution. The dashed curve shows the corresponding smooth, steady CAK model (figure courtesy of Stan Owocki).

point-star case, this eigenvalue is given by CAK theory as

$$\dot{M}_{\rm CAK} = \frac{L_*}{c^2}\frac{\alpha}{\alpha - 1}\left(\frac{\bar{Q}\Gamma_{\rm e}}{1-\Gamma_{\rm e}}\right)^{(1-\alpha)/\alpha}, \tag{1.3}$$

and solution of the equation-of-motion for $\dot{M} = \dot{M}_{\rm CAK}$ gives a wind terminal velocity of

$$v_\infty = \sqrt{\frac{\alpha}{1-\alpha}\frac{2GM_*}{R_*}}. \tag{1.4}$$

The run of wind velocity with radius can be well-approximated by a so-called β-law,

$$v(r) = v_\infty\left(1 - \frac{R_*}{r}\right)^\beta \tag{1.5}$$

where β is typically around 0.5. Owocki (2004) provides an excellent review of the formalism leading to these various results.

2. Wind Structure

The CAK theory outlined above presumes a smooth wind outflow from a spherically symmetric star. In reality, we know from observations of OB stars that this cannot be the case; there is ample evidence for wind structure at both small and large scales (see, e.g., the review by Alex Fullerton). Where does this structure come from?

A simple linear perturbation analysis of radiative driving (e.g., Lucy & Solomon 1970) predicts a perturbed line force that is directly proportional to the perturbed velocity; this simply reflects the amount of additional opacity that is Doppler shifted out of its own shadow at the blue edge of line profiles. As a result, line-driven winds are linearly unstable, with small disturbances amplified exponentially. The typical growth timescale is $\tau_{\rm grow} \sim 10^{-2}\,\tau_{\rm wind}$, where $\tau_{\rm wind} = R_*/v_\infty$ is the wind flow timescale; therefore, this so-called *line-deshadowing instability* (LDI) is potentially a very powerful source of wind structure.

A more-detailed analysis indicates that the LDI operates only for velocity perturbations whose physical length scale is shorter than the Sobolev length $\ell_{\rm sob} = v_{\rm th}/({\rm d}v/{\rm d}r)$ (see Owocki & Rybicki 1984). At larger scales, the perturbed line force is proportional to the perturbed velocity *gradient* (as one might expect from eqn. 1.2), and the wind is stable. Accordingly, the LDI is primarily responsible for the generation of *small-scale* structure in line-driven winds.

Numerical simulation of the LDI can be computationally expensive, as the Sobolev approximation — which assumes a smooth wind at length scales $< \ell_{\rm sob}$ — cannot be used. 1-D hydro simulations by Feldmeier & Owocki (1998) indicate that the instability breaks up a smooth CAK wind solution into a sequence of reverse shocks, where fast, low-density wind material runs into the back of slower-moving, high-density material (see Fig. 1). These wind shocks are considered a likely source for the soft, broad-lined X-ray emission observed in many single OB stars (e.g., Owocki & Cohen 2006, and references therein). Extending the simulations to 2-D, Dessart & Owocki (2005) find that the shell-like shocks produced by the LDI are disrupted by Rayleigh-Taylor instabilities, and the wind structure rapidly becomes incoherent down to angular scales approaching the grid scale. Thus, it appears difficult for the LDI to play any significant role in generating *large-scale* wind structure.

A more plausible origin for this structure is by imprinting from the star itself. OB stars can exhibit inhomogeneities in their surface properties (e.g., temperature, velocity, abundances, magnetic fields), which seed large-scale, non-axisymmetric disturbances in their wind outflows. Rotational modulation of these disturbances seems a promising explanation for cyclical wind variability seen in UV resonance lines of active OB stars, such as episodic blueward-migrating discrete absorption components (DACs; e.g., Kaper & Henrichs 1994). Prinja *et al.* (1995) discovered a novel form of variability in the wind of the B0.5Ib supergiant HD 64760, consisting of 1.2 d-periodic absorption modulations superimposed over a longer-timescale DAC pattern. The phase-bowed structure of these modulations led Owocki *et al.* (1995) to propose that they arise in co-rotating, mutually interacting wind streams rooted in stellar-surface variations. Subsequent work by Cranmer & Owocki (1996) and Fullerton *et al.* (1997) fleshed out this idea of co-rotating interaction regions (CIRs; see also Mullan 1984); however, these authors were not able to determine the nature of the surface variability responsible for forming the CIRs in HD 64760. A significant step forward came with the discovery by Kaufer *et al.* (2006) that the star's wind-sensitive Hα line is variable on a 6.8 d period, corresponding to the beat period between photospheric non-radial pulsation modes. This represents the first real evidence of the long-sought 'photospheric connection' between surface and wind variability. Nevertheless, the link between this 6.8 d period and the 1.2 d CIR period still remains unclear.

Rotation also plays a more-direct role in generating large-scale wind structure, through the action of the centrifugal force. A simple 1-D extension of eqn. (1.3) gives a

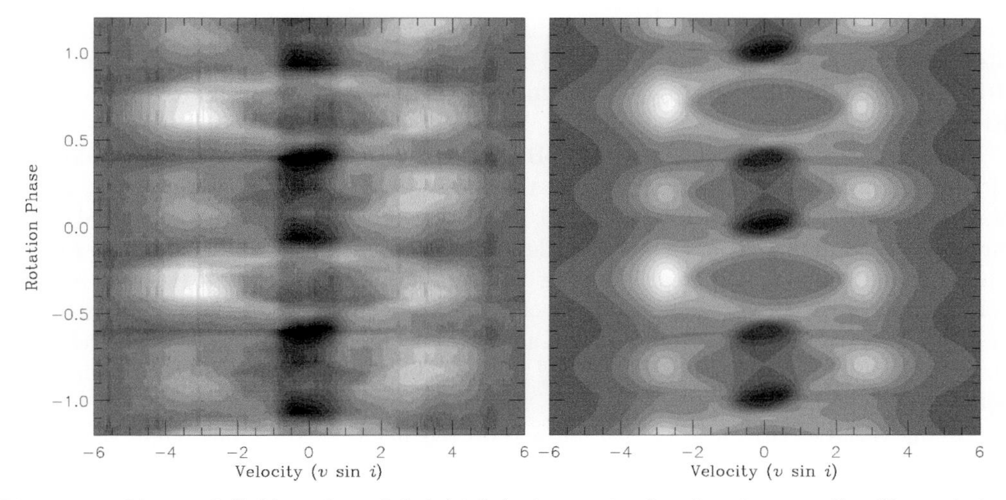

Figure 2. Observed (left) and modeled (right) time-series for the circumstellar Hα emission of σ Ori E, phased on the star's 1.2 d rotation period. White indicates emission relative to the background photospheric profile, and black indicates absorption. The velocity axis is expressed in units of the star's projected rotation velocity, $v \sin i = 160\,\mathrm{km\,s^{-1}}$.

latitude-dependent effective mass-loss rate

$$\dot{M}_{\mathrm{CAK}}(\theta) = \frac{FR_*^2}{c^2}\frac{\alpha}{\alpha-1}\left(\frac{\bar{Q}\Gamma_{\mathrm{e}}}{1-\Gamma_{\mathrm{e}}}\right)^{(1-\alpha)/\alpha}, \tag{2.1}$$

where now

$$\Gamma_{\mathrm{e}}(\theta) = \frac{\kappa_{\mathrm{e}} F}{g_{\mathrm{eff}}\, c}, \tag{2.2}$$

and F and g_{eff} are the local radiative flux and effective gravity, respectively.

A naive interpretation of eqn. (2.1) suggests an equatorially enhanced mass loss (i.e., a disk-like outflow), due to the decrease in g_{eff} toward the equator, and the consequent increase in Γ_{e}. However, as first observed by Cranmer & Owocki (1995), Γ_{e} is *invariant* across the surface of a radiative-envelope star, owing to the von Zeipel (1924) gravity-darkening law $F \propto g_{\mathrm{eff}}$. Thus, the bracketed term in eqn. (2.1) does not depend on latitude (although it *does* depend on rotation rate; see, e.g., Maeder & Meynet 2000). The latitude dependence of $\dot{M}$ in fact arises via the FR_*^2 term, which is strongest at the *poles*. Hence, line-driven mass loss from rapidly rotating stars should be bi-polar rather than disk-like. Observational evidence that this is indeed the case continues to mount (e.g., Smith *et al.* 2003).

Of course, this narrative excludes the possible variation of the opacity parameter $\bar{Q}$ with latitude. The interplay between gravity darkening and the bistability jump can enhance $\bar{Q}$ sufficiently that an equatorially enhanced wind ensues. This has been proposed as a mechanism for forming the dense outflow disks of B[e] stars (e.g., Lamers & Pauldrach 1991).

3. Circumstellar Structure

In addition to transient wind structures that continually advect outward, active OB stars can harbor plasma in circumstellar structures that persist over many wind flow times and/or rotation cycles. Historically, the focus has been on the equatorial Keplerian disks of classical Be stars, as discussed in detail in the review by Alex Carciofi. These disks seem unlikely to be formed by wind outflows, because the predominantly radial line driving supplies none of the angular momentum necessary to place material into bound Keplerian orbits.

However, there are other classes of OB star that exhibit the signatures of persistent circumstellar material. One particularly remarkable object is the helium-strong B2Vpe star σ Orionis E, which was discovered by Walborn (1974) to show twin-peaked Hα emission, modulated on a 1.2 d cycle identified with the rotation period (see Fig. 2). Subsequent investigations (see Groote & Hunger 1982, and references therein) revealed corresponding rotational modulation in a panoply of observables, and their detection of a $\sim$ 10 kG dipole magnetic field led Landstreet & Borra (1978) to conclude that the star's variability arises from '*hot gas [...] trapped in a magnetosphere above the magnetic equator*'.

Similar variable Hα emission has been reported in other stars belonging to the He-strong class; notable recent examples include δ Ori C (Leone *et al.* 2010) and HR 7355 (Rivinius, these proceedings; see also the contribution by Bohlender). Among the He-strong stars that appear not to show emission, many nevertheless exhibit variable UV C IV resonance lines consistent with trapped circumstellar material (see Shore & Brown 1990). Common to *all* of the He-strong stars is the presence of a strong (multi-kG), ordered (typically, dipole) magnetic field.

Outside of the He-strong class, magnetospheres have been discovered around a number of other OB stars. These include θ^1 Ori C, the first O-type star to be discovered as magnetic (Wade *et al.* 2006); HD 191612, another magnetic O star with a surprisingly long rotation period of 538 d (Donati *et al.* 2006a); the archetypal early-type pulsator β Cephei (Henrichs *et al.* 2000); and τ Sco, an unusual B0 star known for some time to be a strong, hard X-ray source (Donati *et al.* 2006b). In fact, all of these stars emit X-rays from their magnetospheres, for reasons discussed in §5 below.

4. Rigid-Field Models

Given the richness of the observational data, it is straightforward to determine empirically that the magnetospheric plasma around σ Ori E is predominantly confined into a pair of co-rotating clouds situated above the intersections between magnetic and rotational equators (e.g., Landstreet & Borra 1978; Groote & Hunger 1982). But what is the physical reason for such a plasma distribution? Nakajima (1981, 1985) was the first to move successfully beyond phenomenological explanations, by applying a rigid-field formalism originally developed by Michel & Sturrock (1974) for modeling the Jovian magnetosphere.

As their name implies, rigid-field magnetosphere models rest on the assumption that field lines are completely rigid in the frame of reference that co-rotates with the star. (I demonstrate in §6 that this is a reasonable assumption for He-strong stars such as σ Ori E). The frozen-flux condition of ideal magnetohydrodynamics (MHD) then constrains plasma to flow along fixed trajectories determined by the field topology; in essence, the field lines behave like rigid pipes. Ignoring for the moment any radiative line driving, the only forces capable of accelerating the plasma are pressure gradients and the tangential (field-parallel) components of the gravity and the centrifugal force arising from enforced co-rotation. Acting in tandem, these forces will bring the plasma on any given field line into magnetohydrostatic equilibrium. In the isothermal case at temperature T,

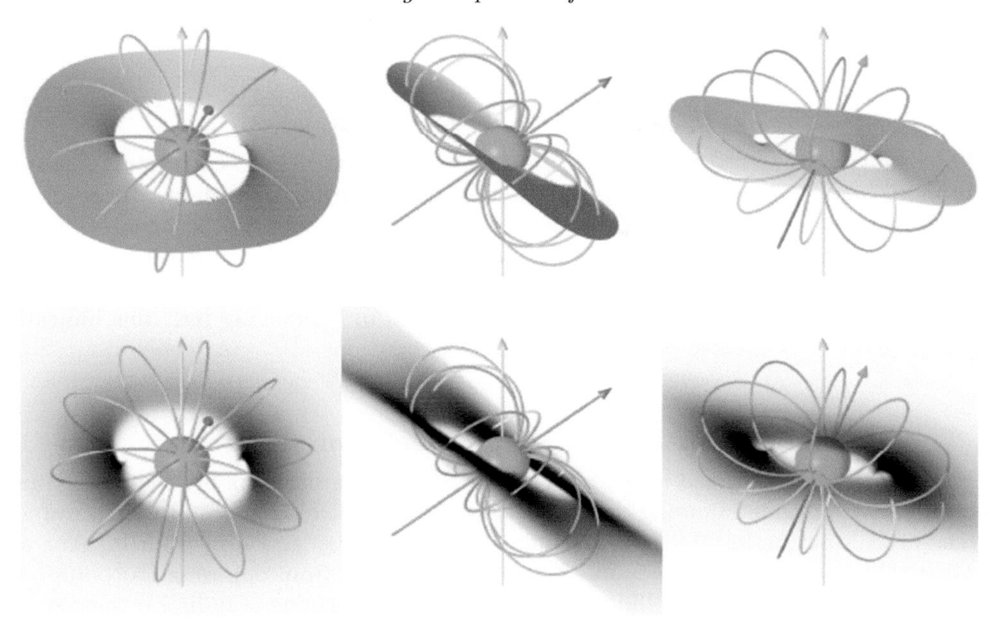

Figure 3. Upper panels: the accumulation surface for a dipole-field star with magnetic obliquity $\beta = 55^\circ$, viewed from an inclination $i = 75^\circ$ and at three different azimuths. Although the surface formally extends to infinity, for clarity it has been truncated at a radius $6\,R_*$. Lower panels: the plasma column density predicted by the RRM model, for the same configurations in the upper panel (although without any truncation). The lower-left panel shows especially well the concentration of plasma into a pair of clouds, situated above the intersections between magnetic and rotational equators.

the equilibrium density distribution is given by the simple expression

$$\rho(s) = \rho_0 \exp\{-[\Phi(s) - \Phi_0]\mu m_{\rm H}/kT\}. \tag{4.1}$$

Here, s is the arc distance measured along the field line from some (arbitrary) reference point $s = 0$; Φ is the effective (centrifugal plus gravitational) potential; and ρ_0 and Φ_0 are the density and effective potential at the reference point. In the frame of reference aligned with the rotation axis, the effective potential in spherical polar coordinates is

$$\Phi = -\frac{GM_*}{r} - \frac{1}{2}\Omega^2 r^2 \sin^2\theta, \tag{4.2}$$

where Ω is the rotation angular frequency.

Equation (4.1) indicates that the plasma density will be maximal in the vicinity of local minima of Φ, as sampled along a given field line. This is consistent with our physical intuition, which tells us that the plasma will tend to migrate toward the lowest points (i.e., the most negative effective potential). Given that the density drops off exponentially away from the potential minima, the plasma in the magnetosphere will largely be confined to surfaces formed by the loci of the minima. For a dipole field, these so-called 'accumulation surfaces' take the form of warped, tilted disks, whose mean normal lies somewhere between the magnetic and rotational axes (see upper panels of Fig. 3). On the innermost field lines no potential minima exist, because the centrifugal force remains too weak to support plasma against the inward pull of gravity. Therefore, the accumulation surfaces do not extend all the way down to the stellar surface, but are instead truncated at a radius commensurate with the Kepler co-rotation radius

$$R_{\rm Kep} = \sqrt[3]{\frac{GM_*}{\Omega^2}}. \tag{4.3}$$

The beauty of the rigid-field formalism is that it is applicable to completely arbitrary field topologies, and requires extremely modest computational resources. However, because it provides no prescription for how much plasma populates each field line, the normalizing density ρ_0 remains undetermined. Thus, the rigid-field models of Nakajima and others (e.g., Preuss *et al.* 2004) are unable to predict the relative distribution of plasma across the accumulation surfaces. As I discuss in §7, moving past this limitation required insights gleaned from numerical MHD simulations.

5. The Magnetically Confined Wind Shock Paradigm

The *magnetically confined wind shock* (MCWS) paradigm is arguably the most important narrative for understanding the magnetospheres of active OB stars. It was introduced by Babel & Montmerle (1997a,b) to understand the X-ray emission from IQ Aur and θ^1 Ori C. The concept is simple: line-driven wind streams from opposing footpoints of a closed dipole magnetic loop collide at the loop summit, forming standing reverse shocks situated above and below the magnetic equator. Sandwiched between these shocks are regions of hot ($\sim 10^6 - 10^7$ K) plasma, which emit X-rays as they cool by radiative recombination; and, in the magnetic equatorial plane, a dense disk-like accumulation of plasma that has had sufficient time to cool back down to near-photospheric temperatures.

Underpinning this narrative is the same rigid-field assumption of earlier models (§4). However, a novel element is the recognition that the process filling the magnetosphere is the radiative line driving reviewed in §1. This is key to being able to model the distribution of plasma throughout the magnetosphere.

As a brief historical aside, the MCWS paradigm was applied to θ^1 Ori C to explain the star's hard ($\gtrsim 1$, keV) X-ray emission, *before* the ~ 1.1 kG magnetic field was detected. This underscores the notion that the presence of a magnetic field is often signalled through a proxy diagnostic — something to bear in mind, given the high observational cost of direct field detection via spectropolarimetry.

6. Magnetohydrodynamical Simulations

The MCWS paradigm was put to the test by ud-Doula & Owocki (2002) and Owocki & ud-Doula (2004), who undertook extensive MHD simulations of radiative line-driven winds in the presence of dipole magnetic fields. Beyond confirming the basic wind-shock narrative, these simulations led to a pivotal characterization of the wind/field interaction. Locally, the relative importance of the wind and field in determining flow dynamics can be characterized by the ratio

$$\eta = \frac{\mathcal{E}_{\rm field}}{\mathcal{E}_{\rm wind}} = \frac{B^2/8\pi}{\rho v^2/2} \tag{6.1}$$

between the energy densities $\mathcal{E}$ of the field and the wind. The wind dominates the field when $\eta \ll 1$, and vice versa when $\eta \gg 1$. To determine the overall outcome of this competition, it is useful to write the above expression in the form

$$\eta(r) \sim \frac{B_*^2 R_*^2}{\dot{M} v_\infty} \left(\frac{r}{R_*}\right)^{-2n}. \tag{6.2}$$

This is obtained under the assumption of an n-pole field (dipole: $n = 2$; quadrupole: $n = 3$; etc.) with strength B_* at the stellar surface, together with a wind that is everywhere at the terminal velocity v_∞. Based on this expression, ud-Doula & Owocki (2002) introduced the *magnetic confinement parameter*

$$\eta_* \equiv \frac{B_*^2 R_*^2}{\dot{M} v_\infty} \tag{6.3}$$

as a key determinant of global magnetosphere structure. Substituting this into eqn. (6.2) and setting $\eta(r) \sim 1$ gives an Alfvén radius — at which the flow transitions from field-dominated to wind-dominated — of

$$R_{\rm Alf} \sim R_* \eta_*^{1/2n}. \tag{6.4}$$

MHD simulations over a range of η_* values confirm that, qualitatively, field lines remain closed and relatively unperturbed from their force-free configuration in the inner parts of the magnetosphere ($r \lesssim R_{\rm Alf}$), but are ripped open by the wind in the outer parts ($r \gtrsim R_{\rm Alf}$). In the specific case of θ^1 Ori C, the simulations reveal a moderately confined wind with field lines closed out to $\approx 2\,R_*$, consistent with the Alfvén radius for a dipole field ($n = 2$) and a confinement parameter $\eta_* \approx 16$.

Although the initial MHD simulations by ud-Doula and Owocki focused on isothermal flow, later calculations incorporate an explicit energy equation to follow the shock heating of material and its subsequent radiative cooling. This allows quantitative predictions of the X-ray emission, suitable for comparison against observations. Focusing again on θ^1 Ori C, Gagné *et al.* (2005a,b) found that the distribution of X-ray-emitting plasma predicted by the non-isothermal MHD simulations is in good agreement with that determined from *Chandra* HETGS spectroscopy of the star — namely, the bulk of this plasma is situated at $\sim 1.8\,R_*$, just inside the closed-field regions of the magnetosphere.

7. The Rigidly Rotating Magnetosphere Model

In spite of their successes, MHD simulations are not a panacea. They are computationally expensive in 2-D, and even more so in the 3-D required to model systems such as oblique-dipole rotators. Most problematically, the computational cost becomes prohibitive in the limit of large η_*, because the timestep necessary to ensure numerical stability is tiny.

However, in this limit the field lines are almost completely rigid, and the formalism reviewed in §4 can be applied — *if* the normalizing density ρ_0 can be determined. In the *rigidly rotating magnetosphere* (RRM) model introduced by Townsend & Owocki (2005), this is done by adopting a result from the MHD simulations by Owocki & ud-Doula (2004), namely that the stellar-surface mass flux onto a field line tilted by an angle Θ to the local surface normal is well approximated by

$$\dot{m} = \frac{\dot{M}_{\rm CAK}}{4\pi R_*^2} \cos\Theta. \tag{7.1}$$

The total mass of plasma populating a field line, after a filling time of Δt†, is $m = (\dot{m}_{\rm N} + \dot{m}_{\rm S})\Delta t$ (where the subscripts on $\dot{m}$ refer to the Northern- and Southern-hemisphere footpoints). Thus, given the above expression for $\dot{m}$, the normalizing density ρ_0 can be fixed by requiring that the volume integral of eqn. (4.1) along a field line equal m.

† See the Appendix of Townsend & Owocki (2005), for a discussion of how this filling time might be established.

For each of accumulation surfaces shown in the upper panels of Fig. 3, the corresponding plasma distributions predicted by the RRM model are shown in the lower panels. As expected, the plasma is confined close to these surfaces, supported against the inward pull of gravity by the centrifugal force. Its distribution across the surfaces is quite non-uniform, however, being concentrated into two clouds situated above the intersections between magnetic and rotational equators — exactly the configuration inferred empirically from the observations.

The RRM model is generally applicable to stars having strong magnetic fields and low mass-loss rates, since these both contribute toward a large η_* (see eqn. 6.3). Practically speaking, the He-strong stars fall into this category; for instance, the archetype σ Ori E has $\eta_* \approx 10^5$. Townsend *et al.* (2005) demonstrate that the RRM model can successfully reproduce the Hα and photometric variability of this particular star (see Fig. 2; also, Oksala, these proceedings). Other stars exhibiting similar Hα emission (e.g., δ Ori C, HR 7355) seem promising candidates for application of the model, although it is likely that the model will have to be modified to account for certain features such as the time-skewed emission in HR 7355 (see Fig. 4 of Rivinius, these proceedings).

8. The Rigid-Field Hydrodynamics Approach

The RRM model focuses on the accumulation of cooled post-shock plasma. Although it incorporates the mass-flux scaling of Owocki & ud-Doula (2004), it does not provide any explicit description of the wind upflow into the magnetosphere, nor the shocks responsible for heating plasma to X-ray-emitting temperatures. These limitations provided the motivation for Townsend *et al.* (2007) to develop the *rigid-field hydrodynamics* (RFHD) approach for dynamical modeling of stellar magnetospheres.

This approach adopts the same rigid-field assumption as the RRM model, but replaces the condition of magnetohydrostatic equilibrium (eqn. 4.1) with hydrodynamical simulations of the wind flow feeding into the magnetosphere from field footpoints. Each field line is treated as a separate 1-D flow system with varying cross-sectional area. Forces acting on the plasma include gravity, pressure gradients, the centrifugal force and the radiative line force. Energy losses due to optically thin radiative cooling and inverse Compton scattering are included, and in more-recent calculations (see Hill & Townsend, these proceedings) energy transport due to thermal conduction is also modeled. The inclusion of these energetic processes allows predictions of magnetospheric X-ray emission to be made. Therefore, the RFHD approach nicely complements MHD simulations; the latter is valid at small and intermediate values of the confinement parameter η_*, and the former at large values.

Formally, the Alfvén radius is at infinity in RRM and RFHD models. However, in order to avoid overestimating the extent of the magnetosphere (and the amount of emission produced by it), it is necessary to truncate the circumstellar density distribution at the radius where the magnetic and plasma energy densities are equal. The latter is typically dominated not by the kinetic energy of the wind, but by the centrifugal potential energy of the material on the accumulation surfaces. Hence, it is always the case that the truncation radius $R_{\rm trunc}$ is somewhat less than the (finite) $R_{\rm Alf}$ defined by eqn. (6.4).

9. Magnetic Braking

Recent MHD simulations by ud-Doula *et al.* (2008, 2009) have investigated the interaction between field and wind in rotating, aligned-dipole stars. Toward larger values of

η_*, the simulations reveal the accumulation of a dense equatorial disk, exactly as predicted by RRM and RFHD models. At the outer, truncation edge of the disk, the field is stretched out into radial configurations by the centrifugally assisted wind. As they advect away from the star, the wind and field carry angular momentum with them, contributing to the gradual braking of the star's rotation.

A key outcome of the ud-Doula *et al.* (2009) study is a parametrization of the characteristic spin-down timescale $\tau_{\rm spin}$ due to magnetic braking. For a dipole field,

$$\tau_{\rm spin} \sim \tau_{\rm mass} \frac{3}{2} k \eta_*^{-1/2}, \tag{9.1}$$

where $k \approx 0.1$ is a dimensionless measure of the star's moment of inertia, and $\tau_{\rm mass} \equiv M_*/\dot{M}$ is the mass-loss timescale. Applying this expression to σ Ori E gives an estimated spin-down timescale of 1.4 Myr — in remarkably good agreement with the direct measurement $\tau_{\rm spin} = 1.34^{+0.10}_{-0.09}$ Myr by Townsend *et al.* (2010).

The ability for direct comparisons between the theory and observation of magnetic braking is a very exciting development. Amongst non-degenerate objects, active OB stars are unique in this respect; for other types of star, magnetic braking can only be characterized indirectly, by studying populations across a range of ages. However, many uncertainties still remain. For instance, it is not clear how to extend the parametrization above to handle oblique-dipole fields or higher-order fields. With the recent measurement of braking in other active OB stars — in particular, the quadrupole-field HD 37776 (Mikulášek *et al.* 2008) — the ball is clearly in the theoreticians' court.

10. The Magnetosphere Zoo

The magnetospheres of active OB stars can appear like a zoo — the creatures look very different from each other, and it can be difficult to understand how they all derive from a common ancestor. To help dispel any possible confusion, I offer the following 'field guide' for those wanting to know what to expect when searching for or studying magnetospheres:

(i) Does the star have a measured magnetic field?
Yes — go to (ii).
No — either keep looking for a field (example: ζ Pup), or give up and choose another star.

(ii) Is the confinement parameter η_* (eqn. 6.3) greater than unity?
Yes — go to (iii).
No — the star's wind is not magnetically confined; the star won't have a magnetosphere (example: ζ Ori; see Bouret *et al.* 2008).

(iii) Is the Kepler radius $R_{\rm Kep}$ (eqn. 4.3) less than the Alfvén radius $R_{\rm Alf}$ (eqn. 6.4)?
Yes — there will be an accumulation of plasma between $R_{\rm Kep}$ and $R_{\rm Alf}$; Hα emission is a possibility (example: σ Ori E).
No — go to (iv).

(iv) Does the star have a large mass-loss rate ($\gtrsim 10^{-7}\,{\rm M}_\odot\,{\rm yr}^{-1}$)?
Yes — even though the centrifugal force is insufficient to support plasma, a transient accumulation of plasma out to $R_{\rm Alf}$ can occur; Hα emission is a possibility (example: θ^1 Ori C).
No — the mass-loss rate is too low for any accumulation to occur; no Hα emission is expected (example: β Cep; see Favata *et al.* 2009).

Having established what to expect, the next step is to choose a suitable modeling approach (e.g., MHD; RRM; RFHD), and attempt to reproduce the observations.

Acknowledgements

I acknowledge support from NASA grant *LTSA*/NNG05GC36G.

References

Babel, J. & Montmerle, T. 1997a, *A&A*, 323, 121
Babel, J. & Montmerle, T. 1997b, *ApJ* (Letters), 485, L29
Bouret, J.-C., Donati, J.-F., Martins, F., Escolano, C. *et al.* 2008, *MNRAS*, 389, 75
Castor, J. I., Abbott, D. C., & Klein, R. I. 1975, *ApJ*, 195, 157
Cranmer, S. R. & Owocki, S. P. 1995, *ApJ*, 440, 308
Cranmer, S. R. & Owocki, S. P. 1996, *ApJ*, 462, 469
Dessart, L. & Owocki, S. P. 2005, *A&A*, 437, 657
Donati, J.-F., Howarth, I. D., Bouret, J.-C., Petit, P. *et al.* 2006a, *MNRAS*, 365, L6
Donati, J.-F., Howarth, I. D., Jardine, M. M., Petit, P. *et al.* 2006b, *MNRAS*, 370, 629
Favata, F., Neiner, C., Testa, P., Hussain, G. *et al.* 2009, *A&A*, 495, 217
Feldmeier, A. & Owocki, S. 1998, *Ap&SS*, 260, 113
Fullerton, A. W., Massa, D. L., Prinja, R. K., Owocki, S. P. *et al.* 1997, *A&A*, 327, 699
Gagné, M., Oksala, M. E., Cohen, D. H., Tonnesen, S. K. *et al.* 2005a, *ApJ*, 628, 986
Gagné, M., Oksala, M. E., Cohen, D. H., Tonnesen, S. K. *et al.* 2005b, *ApJ*, 634, 712
Gayley, K. G. 1995, *ApJ*, 454, 410
Groote, D. & Hunger, K. 1982, *A&A*, 116, 64
Henrichs, H. F., de Jong, J. A., Donati, J.-F., Catala, C. *et al.* 2000, in: M. A. Smith, H. F. Henrichs, & J. Fabregat (eds.), *IAU Colloq. 175: The Be Phenomenon in Early-Type Stars*, ASP-CS 214, p. 324
Kaper, L. & Henrichs, H. F. 1994, *Ap&SS*, 221, 115
Kaufer, A., Stahl, O., Prinja, R. K., & Witherick, D. 2006, *A&A*, 447, 325
Lamers, H. J. G. & Pauldrach, A. W. A. 1991, *A&A*, 244, L5
Landstreet, J. D. & Borra, E. F. 1978, *ApJ* (Letters), 224, L5
Leone, F., Bohlender, D. A., Bolton, C. T., Buemi, C. *et al.* 2010, *MNRAS*, 401, 2739
Lucy, L. B. & Solomon, P. M. 1970, *ApJ*, 159, 879
Maeder, A. & Meynet, G. 2000, *A&A*, 361, 159
Michel, F. C. & Sturrock, P. A. 1974, *Planetary & Space Science*, 22, 1501
Mikulášek, Z., Krtička, J., Henry, G. W., Zverko, J. *et al.* 2008, *A&A*, 485, 585
Mullan, D. J. 1984, *ApJ*, 283, 303
Nakajima, R. 1981, *Sci. Rep. Tohoku Univ. Eighth Ser.*, 2, 130
Nakajima, R. 1985, *Ap&SS*, 116, 285
Owocki, S. P. 1991, in: *NATO ASIC Proc. 341: Stellar Atmospheres – Beyond Classical Models* (Dordrecht, D. Reidel Publishing Co.), p. 235
Owocki, S. 2004, in: M. Heydari-Malayeri, P. Stee, & J.-P. Zahn (eds.), *EAS Publications Series* 13, p. 163
Owocki, S. P. & Rybicki, G. B. 1984, *ApJ*, 284, 337
Owocki, S. P., Cranmer, S. R., & Fullerton, A. W. 1995, *ApJ* (Letters), 453, L37
Owocki, S. P. & ud-Doula, A. 2004, *ApJ*, 600, 1004
Owocki, S. P. & Cohen, D. H. 2006, *ApJ*, 648, 565
Preuss, O., Schüssler, M., Holzwarth, V., & Solanki, S. K. 2004, *A&A*, 417, 987
Prinja, R. K., Massa, D., & Fullerton, A. W. 1995, *ApJ* (Letters), 452, L61
Shore, S. N. & Brown, D. N. 1990, *ApJ*, 365, 665
Smith, N., Davidson, K., Gull, T. R., Ishibashi, K. *et al.* 2003, *ApJ*, 586, 432
Townsend, R. H. D. & Owocki, S. P. 2005, *MNRAS*, 357, 251
Townsend, R. H. D., Owocki, S. P., & Groote, D. 2005, *ApJ* (Letters), 630, L81
Townsend, R. H. D., Owocki, S. P., & Ud-Doula, A. 2007, *MNRAS*, 382, 139

Townsend, R. H. D., Oksala, M. E., Cohen, D. H., Owocki, S. P. *et al.* 2010, *ApJ* (Letters), 714, L318
ud-Doula, A. & Owocki, S. P. 2002, *ApJ*, 576, 413
Ud-Doula, A., Owocki, S. P., & Townsend, R. H. D. 2008, *MNRAS*, 385, 97
Ud-Doula, A., Owocki, S. P., & Townsend, R. H. D. 2009, *MNRAS*, 392, 1022
Wade, G. A., Fullerton, A. W., Donati, J.-F., Landstreet, J. D. *et al.* 2006, *A&A*, 451, 195
Walborn, N. R. 1974, *ApJ* (Letters), 191, L95
von Zeipel, H. 1924, *MNRAS*, 84, 665

Active OB stars: structure, evolution, mass loss, and critical limits
Proceedings IAU Symposium No. 272, 2010
C. Neiner, G. Wade, G. Meynet & G. Peters, eds.

doi:10.1017/S1743921311010180

Dynamics of fossil magnetic fields in massive star interiors

Stéphane Mathis

Laboratoire AIM, CEA/DSM-CNRS-Université Paris Diderot, IRFU/SAp Centre de Saclay, F-91191 Gif-sur-Yvette, France; email: stephane.mathis@cea.fr

LESIA, Observatoire de Paris, CNRS, Université Paris Diderot, UPMC, 5 place Jules Janssen, 92190 Meudon, France

Abstract. In this talk, I review the different MHD processes, which take place in massive star interiors. First, I describe MHD instabilities, which act on magnetic fields in stellar radiation zones, and the dynamo action in massive stars that give strong indications in favor of a fossil origin of the fields observed at the surface of these stars. Then, I discuss the study of MHD turbulent relaxation processes, which are now examined in stellar interiors, to describe initial conditions for fossil magnetic fields. Finally, I focus on the state of the art of the modeling of the interaction between differential rotation, fossil magnetic field, meridional circulation, and turbulence.

Keywords. magnetohydrodynamics: MHD, plasmas, stars: magnetic fields, stars: rotation, stars: evolution

1. Introduction

Magnetic fields are now detected more and more often at the surface of main-sequence (and Pre Main-Sequence) intermediate mass and active massive stars, which have an external radiative envelope. Indeed, strong fields (300 G to 30 kG) are observed in some fraction of Herbig stars (Alecian *et al.* 2008), A stars (the Ap stars, see Aurière *et al.* 2007), as well as in B stars and in a handful of O stars (see the MiMeS program results discussed by G. Wade in this volume and Grunhut *et al.* 2009). Furthermore, non convective neutron stars display fields strength of $10^8 - 10^{15}$ G. Magnetic fields in stably stratified non convective stellar regions thus deeply modify our vision of massive stars evolution since their formation (Commerçon *et al.* 2010) to their late stages, for example for gravitational supernovae. Thus, these drive stellar internal dynamics for the transport of angular momentum and the resulting rotation history, and chemicals mixing (Maeder & Meynet 2000; Mathis & Zahn 2005).

The large-scale, ordered nature (often approximately dipolar) of such magnetic fields and the scaling of their strengths as a function of their host properties (according to the flux conservation scenario) favour a fossil hypothesis (even if a dynamo is present in the convective core, *c.f.* Brun *et al.* 2005), whose origin has to be investigated. One of the fundamental question is then the understanding of the topology of these large-scale magnetic fields. To have survived since the star's formation or the PMS stage, a field must be stable on a dynamic (Alfvén) timescale. It was suggested by Prendergast (1956) that a stellar magnetic field in stable axisymmetric equilibrium must contain both poloidal (meridional) and toroidal (azimuthal) components, since both are unstable on their own (Tayler 1973; Wright 1973). This was confirmed recently by numerical simulations by Braithwaite & Spruit (2004); Braithwaite & Nordlund (2006); Braithwaite (2008) who showed that initial stochastic helical fields evolve on an Alfvén timescale into stable configurations: axisymmetric and non-axisymmetric mixed poloidal-toroidal fields were found. This phenomenon well known in plasma physics is a MHD turbulent relaxation (*i.e.* a self-organization process involving magnetic reconnections in resistive MHD). In this short paper, we present our present physical understanding of such mechanism in stellar interiors focusing on the axisymmetric case. Then, the field interaction with differential rotation and meridional circulation is discussed.

2. Relaxed non force-free configurations

Here, we focus on the minimum energy non force-free MagnetoHydroStatic (MHS) equilibrium (the balance between gravity, the pressure gradient and the Lorentz force) that a stably stratified radiation zone can reach. Several reasons inclined us to focus on such equilibria instead of force-free ones, which are often studied in plasma laboratory experiments. First, Reisenegger (2009) shows us that no configuration can be force-free everywhere. Although there do exist "force-free" configurations, these induce discontinuities such as current sheets, which are unlikely to appear in nature except in a transient manner. Second, non force-free equilibria have been identified in plasma physics as the result of MHD relaxation (Montgomery & Phillips 1988). Third, as shown by Duez & Mathis (2010), this family of equilibria is a generalization of Taylor states (force-free relaxed equilibria in plasma laboratory experiments; see Taylor 1974) in a stellar context, where the stable stratification of the medium plays a crucial role.

The axisymmetric magnetic field $\boldsymbol{B}(r,\theta)$ is expressed as a function of a poloidal flux $\Psi(r,\theta)$, a toroidal potential $F(r,\theta)$, and the potential vector $\boldsymbol{A}(r,\theta)$ so that it is divergence-free by construction:

$$\boldsymbol{B} = \frac{1}{r\sin\theta}\left(\boldsymbol{\nabla}\Psi \times \hat{\mathbf{e}}_\varphi + F\,\hat{\mathbf{e}}_\varphi\right) = \boldsymbol{\nabla}\times\boldsymbol{A}, \tag{2.1}$$

where in spherical coordinates the poloidal component ($\boldsymbol{B}_{\rm P}$) is in the meridional plane $(\hat{\mathbf{e}}_r, \hat{\mathbf{e}}_\theta)$ and the toroidal component ($\boldsymbol{B}_{\rm T}$) is along the azimuthal direction ($\hat{\mathbf{e}}_\varphi$). Given the field strengths in real stars, the ratio of the Lorentz force to gravity is very low: stellar interiors are thus in a regime where $\beta = P/P_{\rm Mag} \gg 1$, $P_{\rm Mag} = B^2/(2\mu_0)$ being the magnetic pressure. Then, we identify the invariants governing the evolution of the reconnection phase, that leads to relaxed states in the non force-free case. The first one is the magnetic helicity $\mathcal{H} = \int_{\mathcal{V}} \boldsymbol{A}\cdot\boldsymbol{B}\,{\rm d}\mathcal{V}$, which is an ideal MHD invariant known to be roughly conserved at large scales during relaxation. The second one is the mass encompassed in poloidal magnetic surfaces $M_\Psi = \int_{\mathcal{V}} \Psi\,\rho\,{\rm d}\mathcal{V}$ (ρ is the density), conserved because of the stable stratification, which inhibits the radial movements and thus the transport of mass and flux in this direction. Next, we assume a selective decay during relaxation (*c.f.* Biskamp 1997), in which the magnetic energy $E_{\rm mag} = \int_{\mathcal{V}} \frac{\mathbf{B}^2}{2\mu_0}{\rm d}\mathcal{V}$ (μ_0 being the vaccum magnetic permeability), and thus the total energy (internal+gravific+magnetic), decays much faster than $\mathcal{H}$ and M_Ψ, so that they can be considered constant on an energetic decay e-folding time. This is due to the stable stratification and to the different orders of spatial derivatives involved in the variation of $E_{\rm mag}$ and $\mathcal{H}$. The reached equilibrium is thus the one of minimum energy for given magnetic helicity and mass encompassed in magnetic flux tubes. This can be determined applying a variational method where we minimize E with respect to $\mathcal{H}$ and M_Ψ as described by Woltjer (1959) and Duez & Mathis (2010). This leads to the purely dipolar MHS barotropic state (in the *hydrodynamic* meaning of the term, *i.e.* isobar and iso-density surfaces coincide and the field is explicitly coupled with stellar structure through $\boldsymbol{\nabla}\times(\boldsymbol{F}_{\mathcal{L}}/\rho) = \boldsymbol{0}$, where $\boldsymbol{F}_{\mathcal{L}}$ is the

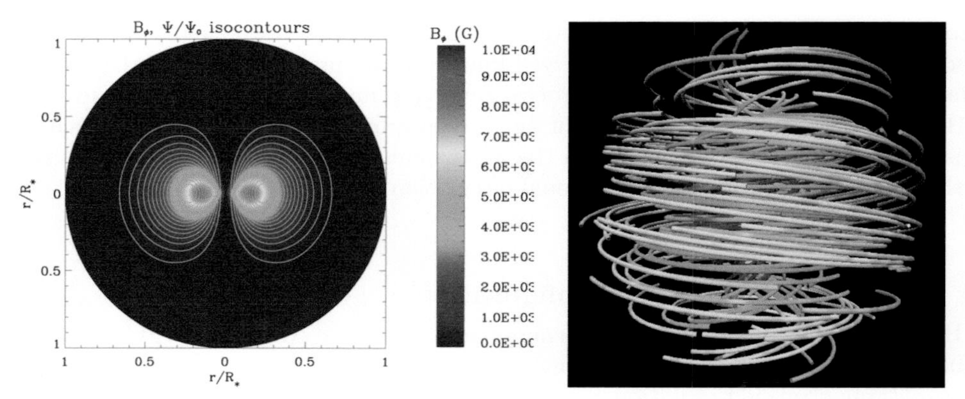

Figure 1. Left: toroidal magnetic field strength in colorscale (arbitrary field's strength) and normalized isocontours of the poloidal flux function (Ψ) in meridional cut for the lowest energy equilibrium configuration ($\lambda_1^1 \simeq 33$); the neutral line is located at $r \simeq 0.23\, R_*$. Right: magnetic field lines representing this mixed field configuration in 3-D looking from the side (the colorscale is a function of the density). Taken from Duez *et al.* (2010a, courtesy The Astrophysical Journal).

Lorentz force) for Ψ:

$$\Psi(r,\theta) = -\mu_0\beta_0\lambda_1^1\frac{r}{R}\left\{j_1\left(\lambda_1^1\frac{r}{R}\right)\int_r^R\left[y_1\left(\lambda_1^1\frac{\xi}{R}\right)\overline{\rho}\xi^3\right]\mathrm{d}\xi + y_1\left(\lambda_1^1\frac{r}{R}\right)\int_0^r\left[j_1\left(\lambda_1^1\frac{\xi}{R}\right)\overline{\rho}\xi^3\right]\mathrm{d}\xi\right\}\sin^2\theta, \tag{2.2}$$

where $\overline{\rho}$ is the density in the non-magnetic case, R the upper boundary confining the magnetic field, and β_0 is related to the surface field intensity. λ_1^1 is the first eigenvalue allowing to verify the boundary conditions at R where we cancel both radial and latitudinal components of the field to avoid any current sheets. The functions j_l and y_l are respectively the spherical Bessel functions of the first and the second kind. The toroidal magnetic field is then given by $F(\Psi) = \lambda_1^1\Psi_1/R$. Furthermore, this state is ruled by the following helicity-energy relation $\mathcal{H} = \frac{2\mu_0 R}{\lambda_1^1}\left(E_{\rm mag} - \frac{1}{2}\beta_0 M_\Psi\right)$, which generalizes the one known in plasma physics for Taylor states (which are recovered if we do not take into account M_Ψ) to the stellar non force-free case. In the case of a stably stratified $n = 3$ polytrope (a good approximation to an upper main-sequence star radiative envelope) where we set $R = 0.85\, R_*$ (R_* is the radius of the star), we have $\lambda_1^1 \simeq 32.95$ (represented in Fig. 1). This is a generalization of Prendergast's equilibrium taking into account compressibility.

Let us now compare this analytical configuration to those obtained using numerical simulations (see Braithwaite & Spruit 2004; Braithwaite & Nordlund 2006; Braithwaite 2008). Braithwaite and collaborators performed numerical magnetohydrodynamical simulations of the relaxation of an initially random magnetic field in a stably stratified star. Then, this initial magnetic field is found to relax on the Alfvén time scale into a stable MHS equilibrium mixed configuration consisting of twisted flux tube(s). Two families are then identified: in the first, the equilibria configurations are roughly axisymmetric with one flux tube forming a circle around the equator, such as the present analytical configuration; in the second family, the relaxed fields are non-axisymmetric consisting of one or more flux tubes forming a complex structure. Whether an axisymmetric or non-axisymmetric equilibrium forms depends on the initial condition chosen for the

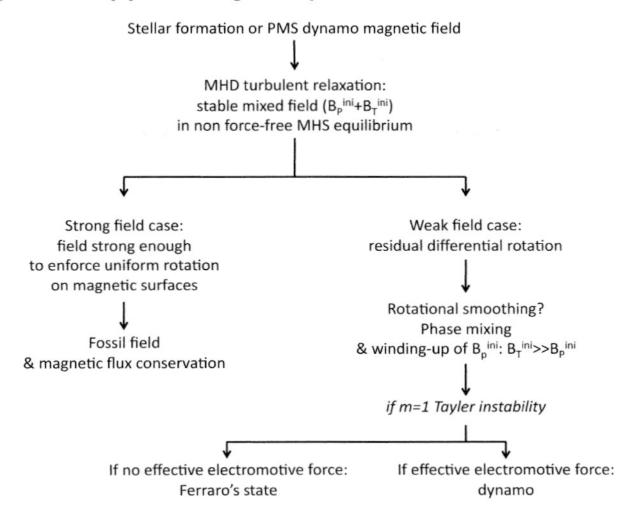

Figure 2. Interaction of fossil magnetic field and differential rotation.

radial profile of the initial stochastic field strength $||\mathbf{B}|| \propto \overline{\rho}^p$: a centrally concentrated one evolves into an axisymmetric equilibrium as in our configuration while a more spread-out field with a stronger connection to the atmosphere relaxes into a non-axisymmetric one; using an ideal-gas star modeled initially with a polytrope of index $n = 3$, the threshold is $p \approx 1/2$. Moreover, as shown in Fig. 7 in Braithwaite (2008), the selective decay of the magnetic helicity ($\mathcal{H}$) and of the magnetic energy ($E_{\rm mag}$) assumed here occurs and the transport of flux and mass in the radial direction is inhibited because of the stable stratification and the mass encompassed in poloidal magnetic surfaces is conserved (*i.e.* M_Ψ). The obtained configuration is of course non force-free.

Finally, note that this analytical configuration for which $E_{\rm mag;P}/E_{\rm mag} \approx 5.23 \times 10^{-2}$ (where $E_{\rm mag;P} = \int_{\mathcal{V}} \boldsymbol{B}_{\rm P}^2/(2\mu_0)\,{\rm d}\mathcal{V}$) verifies the stability criterion derived by Braithwaite (2009) for axisymmetric configurations: $\mathcal{A}\,E_{\rm mag}/E_{\rm grav} < E_{\rm mag;P}/E_{\rm mag} \leqslant 0.8$, where $E_{\rm grav}$ is the gravitational energy in the star, and $\mathcal{A}$ a dimensionless factor whose value is ~ 10 in a main-sequence star and $\sim 10^3$ in a neutron star, while we expect $E_{\rm mag}/E_{\rm grav} < 10^{-6}$ in a realistic star (see for example Duez *et al.* 2010b). This analytical solution is thus similar to the axisymmetric non force-free relaxed solution family obtained by Braithwaite & Spruit (2004) and Braithwaite & Nordlund (2006). Its stability has now been demonstrated by Duez *et al.* (2010a, see also Duez *et al.* in this volume).

These configurations can thus be relevant to model initial equilibrium conditions for evolutionary calculations involving large-scale fossil fields in stellar radiation zones (see for example Mathis & Zahn 2005). We here restrict ourselves to the non-rotating case, but results also apply to radiative regions in a state where rotation is uniform (Woltjer 1959); the case of MHD relaxation with differential rotation has now to be studied.

3. Interaction with differential rotation and meridional circulation

Once the initial non force-free magnetic configuration (axi or non-axisymmetric) has been established by the initial MHD turbulent relaxation processes, this interacts with

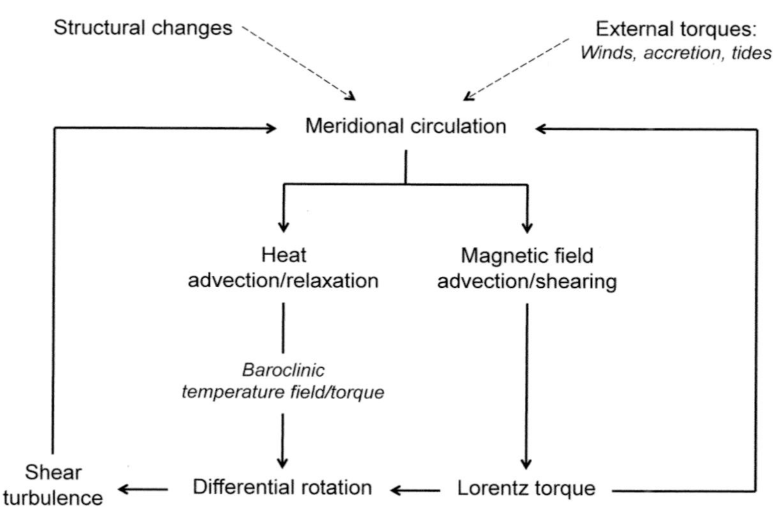

Figure 3. The transport loop in a differentially rotating magnetized stellar radiation zone

differential rotation. Then, two cases are possible as described by Spruit (1999). In the first case, if the field is strong, the rotation becomes uniform on magnetic surfaces due to Alfvén waves phase mixing, which damps the differential rotation; in the axisymmetric case this leads to the Ferraro's state where $\Omega = f(\Psi)$ and to a uniform rotation in the non-axisymmetric case (the oblique rotators case for example). Then, the field can just be modified by structural adjustments. In the second case, if the field is weak, it could first become axisymmetric if it is non-axisymmetric because of rotational smoothing and then, because of phase mixing, this leads to the Ferraro's state. This picture could be modified by magnetic instabilities, if during the first step of the phase mixing, the residual differential rotation on each magnetic surface is able to generate a strong toroidal component of the field that becomes unstable and if this instability becomes able to trigger a dynamo action through an α-effect; this question remains open (Spruit 2002; Braithwaite 2006; Zahn *et al.* 2007). The critical value of the field that gives the limit between the weak and the strong field regimes has been given in Moss (1992); Aurière *et al.* (2007). A summary is given in Fig. 2.

Let us now take into account the meridional circulation. To understand its interaction with the other dynamical processes (the differential rotation and the shear induced turbulence) in presence of a fossil magnetic field, we shall adopt the picture of rotational transport as described by Busse (1981); Zahn (1992); Rieutord (2006) and Decressin *et al.* (2009) and generalise it to the magnetic case. As described in those works, meridional circulation in radiation zones are driven by applied torques (internal like the Lorentz torque or external like those induced by stellar winds), structural adjustments during stellar evolution, and turbulent transport. In the case where all these sources vanish, the meridional circulation dies after an Eddington-Sweet time and the star settles in a baroclinic state described by the thermal wind equation. If we apply this picture to the case of radiation zones with a fossil magnetic field, we thus understand that the meridional circulation (if we consider a star without structural adjustments and external torques) will be mainly driven by the residual magnetic torque until the phase mixing leads the star to a torque-free state. Then, the meridional circulation advection of angular momentum balances the residual Lorentz torque (see Mestel *et al.* 1988).

Acknowledgements

S. Mathis thanks the organizers for this very nice conference and A. Maeder, G. Meynet, J.-P. Zahn, N. Langer, V. Duez, M. Cantiello, and the MiMeS collaboration for fruitful discussions. This work was supported in part by PNPS (CNRS/INSU).

References

Alecian, E., Wade, G. A., Catala, C., Bagnulo, S. *et al.* 2008, *A&A*, 481, L99
Aurière, M., Wade, G. A., Silvester, J., Lignières, F. *et al.* 2007, *A&A*, 475, 1053
Biskamp, D. 1997, *Nonlinear Magnetohydrodynamics*, (Cambridge, UK: Cambridge University Press)
Braithwaite, J. 2006, *A&A*, 449, 451
Braithwaite, J. 2008, *MNRAS*, 386, 1947
Braithwaite, J. 2009, *MNRAS*, 397, 763
Braithwaite, J. & Spruit, H. C. 2004, *Nature*, 431, 819
Braithwaite, J. & Nordlund, Å. 2006, *A&A*, 450, 1077
Brun, A. S., Browning, M. K., & Toomre, J. 2005, *ApJ*, 629, 461
Busse, F. H. 1981, *Geophysical and Astrophysical Fluid Dynamics*, 17, 215
Commerçon, B., Hennebelle, P., Audit, E., Chabrier, G. *et al.* 2010, *A&A*, 510, L3
Decressin, T., Mathis, S., Palacios, A., Siess, L. *et al.* 2009, *A&A*, 495, 271
Duez, V. & Mathis, S. 2010, *A&A*, 517, A58
Duez, V., Braithwaite, J., & Mathis, S. 2010a, *ApJ* (Letters) 724, L34
Duez, V., Mathis, S., & Turck-Chièze, S. 2010b, *MNRAS*, 402, 271
Grunhut, J. H., Wade, G. A., Marcolino, W. L. F., Petit, V. *et al.* 2009, *MNRAS*, 400, L94
Maeder, A. & Meynet, G. 2000, *ARAA*, 38, 143
Mathis, S. & Zahn, J.-P. 2005, *A&A*, 440, 653
Mestel, L., Tayler, R. J., & Moss, D. L. 1988, *MNRAS*, 231, 873
Montgomery, D. & Phillips, L. 1988, *Phys. Rev. A*, 38, 2953
Moss, D. 1992, *MNRAS*, 257, 593
Prendergast, K. H. 1956, *ApJ*, 123, 498
Reisenegger, A. 2009, *A&A*, 499, 557
Rieutord, M. 2006, *A&A*, 451, 1025
Spruit, H. C. 1999, *A&A*, 349, 189
Spruit, H. C. 2002, *A&A*, 381, 923
Tayler, R. J. 1973, *MNRAS*, 161, 365
Taylor, J. B. 1974, *Phys. Rev. Lett.*, 33, 1139
Woltjer, L. 1959, *ApJ*, 130, 405
Wright, G. A. E. 1973, *MNRAS*, 162, 339
Zahn, J.-P. 1992, *A&A*, 265, 115
Zahn, J.-P., Brun, A. S., & Mathis, S. 2007, *A&A*, 474, 145

Discussion

M. Gagné: We know that $5-8\,M_{\odot}$ pre-main sequence stars go through a fully convective phase before they establish a radiative outer envelope on the main sequence. I wonder how long these stars might be in such convective state and how that might change the initial fossil field.

S. Mathis: This is one point that must be strongly investigated in a near future. Moreover, we also know that pre-neutron stars go through a convective phase and the phenomenon is thus the same for such objects. Note that the variational method presented here does not depend on the initial condition that are chosen.

Active OB stars: structure, evolution, mass loss, and critical limits
Proceedings IAU Symposium No. 272, 2010
C. Neiner, G. Wade, G. Meynet & G. Peters, eds.

doi:10.1017/S1743921311010192

Magnetic Doppler imaging of early-type stars

Oleg Kochukhov[1], Thomas Rivinius[2], Mary E. Oksala[3] and Iosif Romanyuk[4]

[1]Dept. Physics and Astronomy, Uppsala University, Box 516, 75120 Uppsala, Sweden
[2]European Southern Observatory, Casilla 19001, Santiago 19, Chile
[3]Dept. of Physics and Astronomy, University of Delaware, Newark, DE 19716, USA
[4]Special Astrophysical Observatory of RAS, Nizhnij Arkhyz 369167, Russia

Abstract. Doppler imaging of early-type magnetic stars is the most advanced method to interpret their line profile variations. DI allows us to study directly a complex interplay between chemical spots, magnetic fields, and the mass loss. Here we outline the general principles of the surface mapping of stars, discuss adaption of this technique to early-type stars and present several recent examples of the abundance and magnetic mapping performed for rapidly rotating early-B stars. In particular, we present the first Doppler images for the very fast rotating He-rich star HR 7355 and a reconstruction of magnetic field for the well-known Bp star σ Ori E. We also present new magnetic maps for the He-strong star HD 37776, which possesses one of the most complex magnetic field topologies among the upper main sequence stars.

Keywords. polarization, stars: chemically peculiar, stars: magnetic fields, stars: atmospheres, stars: individual (HR 7355, σ Ori E, HD 37776)

---

1. Introduction

The presence of strong magnetic fields on the stellar surfaces leads to the formation of horizontal inhomogeneities, which manifest themselves in the periodic variation of brightness, flux distribution, and spectral line profiles. The late-type stars with convective envelopes possess small-scale, rapidly evolving magnetic fields, which are associated with dark cool spots. The geometry of these horizontal structures changes noticeably on the time-scales from days to months due to the emergence of new magnetic flux and stellar activity cycles. In these stars the dynamo mechanism responsible for the generation of magnetic fields is closely connected to the stellar rotation rate and differential rotation in the interior (Donati & Landstreet 2009).

The hot stars with radiative envelopes exhibit spots of a different nature. A small fraction of A and B-type stars are slow rotators and show kG-strength magnetic fields on their surfaces. These fields appear to be stable and typically (but not always!) are dominated by a dipolar component (Landstreet 1992). The stabilizing effect of magnetic fields, the lack of the surface convection zone, and a slow rotation provide an ideal environment for the vertical segregation of chemical elements under the influence of atomic diffusion (Leblanc *et al.* 2009). Consequently, the magnetic A and B stars, also known as Ap/Bp stars, have a strikingly non-solar chemical composition of their atmospheres. Anisotropy of the diffusion processes introduced by the magnetic field also yields a strongly inhomogeneous distribution of chemical elements, both with height in the atmosphere (Kochukhov *et al.* 2006) and horizontally across the surface (Kochukhov *et al.* 2004).

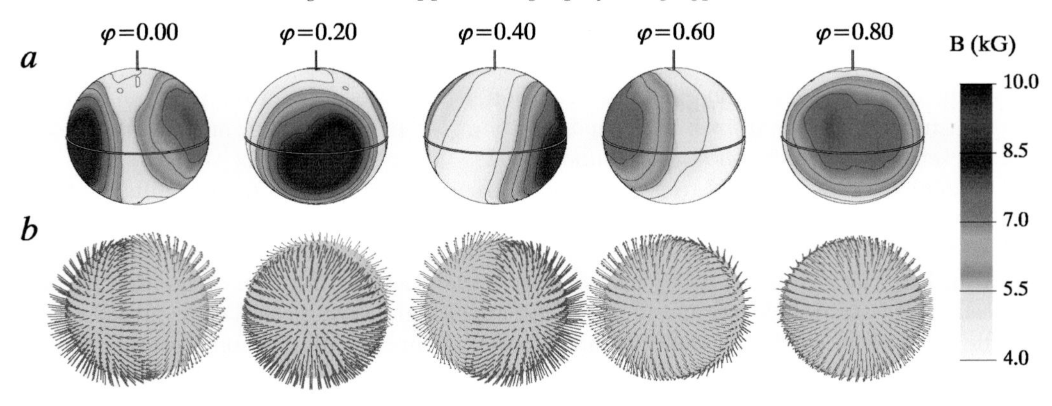

Figure 1. Surface magnetic field distribution reconstructed for σ Ori E with magnetic Doppler imaging. The star is shown at five rotational phases as indicated above the spherical plots. The aspect corresponds to the inclination angle $i = 75^{\circ}$ and vertically oriented rotational axis. **a)** The distribution of the magnetic field strength. The thick line shows the stellar rotational equator. Rotational axis is indicated by the short vertical bar. **b)** The orientation of the magnetic field vectors. In these vector maps the light arrows show the field vectors pointing outside the stellar surface and the dark arrows correspond to the vectors pointing inwards. The arrow length is proportional to the field strength.

Among the hottest and most massive stars, the Bp phenomenon is represented by the He-rich objects. These stars exhibit all of the phenomenology typical of the cooler magnetic Ap stars but, in addition to that, have strong radiately driven winds which interact with the global magnetic field, feeding an active and dynamic magnetosphere (Townsend *et al.* 2007). The properties of the magnetospheric material and its distribution around the star are directly related to the configuration and strength of the stellar magnetic field. Moreover, one has to understand and model in detail the stellar flux redistribution due to chemical spots in order to properly use the stellar brightness variation for the diagnosis of circumstellar environments. Thus, understanding the physical processes taking place at the surfaces and in the extended magnetospheres of early B-type stars requires one to develop a sophisticated surface models of the chemical star spots and magnetic fields.

In this contribution we review the progress in obtaining detailed chemical and magnetic maps of early-type stars with the help of the Doppler imaging (DI) technique and its recent extension to magnetic fields – magnetic DI.

2. The principles of magnetic Doppler imaging

Star spots leave a characteristic signature in the stellar line profiles. The spectrum of a spot usually differs substantially from the radiation from the rest of the stellar surface. Hence, the disk-integrated flux profile will contain a distortion – a dip (a deviation towards less flux) for chemical overabundance or a bump (a deviation towards more flux) for a cool spot – at the velocity shift corresponding to the spot's longitude on the stellar surface. As the star rotates, this distortion moves within the line profile, allowing to track individual spots. A numerical technique to reconstruct a two-dimensional map of the stellar surface from the line profile time-series is called Doppler imaging (DI).

DI has been successfully applied to mapping abundance and temperature spots on the stellar surfaces (Khokhlova *et al.* 1986; Vogt *et al.* 1987) and was extended to the problem of reconstructing vector magnetic field topologies (Brown *et al.* 1991; Piskunov & Kochukhov 2002). Magnetic DI is considerably more challenging compared to its non-magnetic counterpart due an increased number of degrees of freedom for vector maps and a typically poor signal-to-noise ratio of the stellar spectropolarimetric observations. Nevertheless, when all four Stokes parameter spectra with dense enough rotational phase coverage are available for the magnetic inversion, it is possible to obtain a unique solution without introducing an *a priori* information about the field geometry, such as multipolar parameterization (Kochukhov & Piskunov 2002).

Different DI codes applied to the magnetic mapping of hot stars mainly differ in their treatment of the line formation and the degree of self-consistency in modeling magnetic fields and associated inhomogeneities. The mapping technique known as Zeeman Doppler imaging (e.g., Donati *et al.* 2006) uses a simple analytical description of the line shapes, limited to the weak-field regime and applied to a single average Stokes V profile (LSD profile) without considering either chemical spots or linear polarization. Another family of codes (e.g., Kochukhov & Wade 2010), used in the studies discussed below, models the polarized radiative transfer in detail for all four Stokes parameters and carries out a self-consistent mapping of spots and magnetic fields. The latter methodology is, however, much more demanding computationally and currently can be applied only to a few narrow spectral intervals at a time.

3. Applications to early-type stars

3.1. σ *Ori E*

σ Ori E (HD 37479) is a prototype He-strong star with magnetospheric structures and is by far the most studied object in its class. It exhibits synchronous modulation of brightness, UV resonance lines, and Hα emission with a period of 1.19 d, all pointing to an accumulation of the circumstellar material in two corotating clouds situated at the intersection between the equator of a roughly dipolar magnetic field and rotational equator (Groote & Hunger 1982). Considerable theoretical and multi-wavelength observational effort has been put forth in the past few years to develop a physically realistic model of the interaction between the magnetic field and the wind in this star (e.g., Townsend *et al.* 2005, this meeting). But despite the importance of σ Ori E for our understanding of the hot-star magnetospheres, relatively little emphasis was put on collecting precise information about its magnetic field and the stellar surface structures. σ Ori E is known to have ~ 10 kG dipolar-like magnetic field (Landstreet & Borra 1978) and high-contrast surface abundance inhomogeneities (Reiners *et al.* 2000), yet the previous analyses of these surface structures did not proceed beyond simple parametric models of abundance spots (Veto *et al.* 1991) and dipolar fits to the longitudinal magnetic field measurements (Townsend *et al.* 2005). We have collected high-resolution, wide wavelength coverage circular polarization spectra of σ Ori E with the ESPaDOnS and NARVAL polarimeters. These observations are interpreted with the DI code of Piskunov & Kochukhov (2002) to obtain a new, more realistic and detailed model of the magnetic field geometry and study distribution of chemical spots. Fig. 1 shows preliminary magnetic field structure determined from the Stokes I and V profiles of the He I 6678 Å line. Evidently, the magnetic field of σ Ori E is approximately dipolar, in agreement with previous studies. However, magnetic DI map also reveals a considerable field strength difference between the two poles, suggesting the presence of a non-negligible quadrupolar component.

3.2. *HR 7355*

The unusual properties of HR 7355 (HD 182180) were pointed out by Rivinius *et al.* (2008), who found this star to be a very fast rotating emission-line object with an enhanced He absorption and variable spectral lines. With its rotation period of 0.521 d

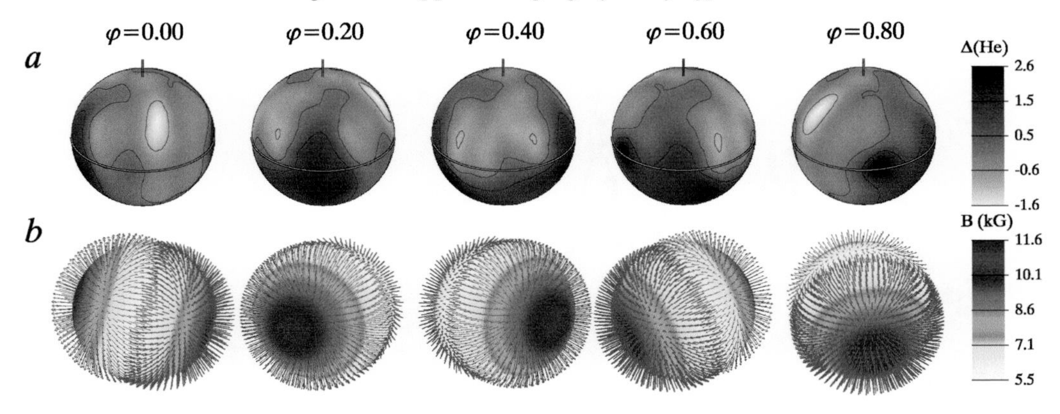

Figure 2. a) Doppler imaging map showing the He abundance distribution derived for HR 7355. The He abundance is given on a logarithmic scale relative to the Sun. **b)** The dipolar magnetic field geometry adopted for HR 7355. This plot shows distribution of the magnetic field strength and orientation of the field vectors.

and $v_{\rm e} \sin i > 300$ km s^{-1}, HR 7355 stands out among Bp stars, which are generally expected to have slow to moderate rotation only. Recently, Oksala *et al.* (2010) and Rivinius *et al.* (2010) discovered a variable longitudinal magnetic field with the strength between -2.2 and $+3.2$ kG. The roughly sinusoidal longitudinal field curve indicates a dipole-dominated magnetic topology with a polar strength exceeding 10 kG. The presence of such an exceptionally strong magnetic field on a rapid rotator confirms the outstanding nature of HR 7355, pushing the boundaries of the Bp star phenomenon into a previously unexplored domain. Strong spectroscopic variability and rapid rotation make HR 7355 an ideal target for Doppler imaging. We are carrying out DI modeling of several absorption features in the UVES time-series spectra of this star with the aim to investigate the surface distributions of He, Si, and C. The preliminary map of the He abundance obtained from the He I 4713 Å line is presented in Fig. 2. This appears to be the first published Doppler image for such a rapidly rotating spotted star. The spectropolarimetric observations of HR 7355 do not yet have a phase coverage sufficient for magnetic DI. Therefore, our abundance mapping employed a dipolar field model with the parameters $B_{\rm p} = 11.6$ kG and $\beta = 78^{\rm o}$ inferred from the longitudinal magnetic field observations. In the future we plan to perform simultaneous mapping of the magnetic field and chemical spots in this remarkable star.

3.3. *HD 37776*

HD 37776 (V901 Ori) is a B2IV He-strong star rotating with a period of ≈ 1.5387 d. It shows variation of He and metal lines (Bohlender 1994), as well as a prominent rotational modulation of the brightness in different photometric bands (Adelman 1997). It is also one of the few spotted stars for which the rotational spin-down has been directly measured (Mikulášek *et al.* 2008) and the light variations were explained with a physically realistic star-spot model (Krtička *et al.* 2007). Uniquely for an early-type chemically peculiar star, HD 37776 shows a complex double-wave longitudinal magnetic field variation (Thompson & Landstreet 1985), suggesting that its magnetic field topology significantly departs from an oblique dipolar field common for other Bp stars. Previous studies of the magnetic field structure of HD 37776 gave inconclusive results. Bohlender (1994) fitted the longitudinal field curve with an axisymmetric multipolar magnetic geometry, dominated by a quadrupolar component. Khokhlova *et al.* (2000) used a simplified version of the magnetic DI to derive a range of possible magnetic field geometries, consisting of a

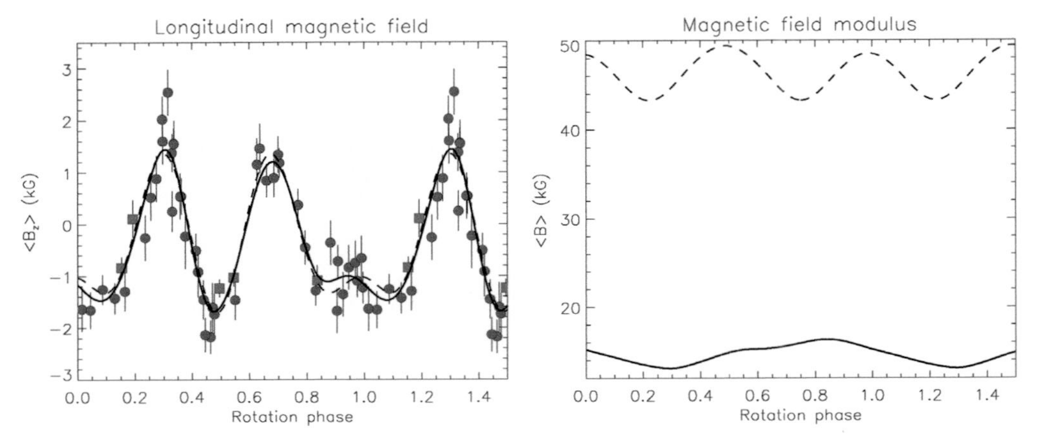

Figure 3. *Left panel:* comparison of the mean longitudinal magnetic field of HD 37776 measured by Thompson & Landstreet (1985) (circles) and Donati *et al.* (1997) (squares) with the fit achieved by the axisymmetric multipolar model (dashed curve) and by the direct magnetic inversion (solid curve). *Right panel:* the mean field modulus variation predicted by the multipolar model (dashed curve) and by the magnetic DI field map (solid curve).

non-aligned dipole and quadrupole. None of the previous studies could find a magnetic field topology successfully reproducing both the longitudinal magnetic field measurements and the polarization in line profiles. At the same time, all studies agree that an exceptionally strong, 50–100 kG, non-dipolar field is needed to describe the magnetic observations of HD 37776 if the field topology is parameterized with low-order multipoles. If true, this would make HD 37776 the most magnetized non-degenerate star known. We carried out a new analysis of the magnetic field structure of HD 37776, combining the traditional fitting of the longitudinal magnetic field curve with a more sophisticated magnetic DI study of the circular polarization observations in the He I 5876 Å line collected with the Russian 6-m telescope at the Special Astrophysical Observatory. Adopting an intermediate inclination angle ($i = 50^{\circ}$), we find that $\approx$ 100 kG octupolar magnetic topology is required to describe the double-wave longitudinal field curve (Fig. 3). However, this multipolar model fails to match the circular polarization spectra, predicting Stokes V profile of both a wrong amplitude and an incorrect shape for the significant fraction of the rotation cycle. This clearly illustrates the fundamental limitation of the magnetic modeling based on $\langle B_z \rangle$ measurements alone.

To develop a new comprehensive magnetic field model of HD 37776, we combined interpretation of the longitudinal field measurements and radiative transfer modeling of the He I 5876 Å line profiles in a single, self-consistent magnetic DI inversion, as was previously done for Ap stars (Kochukhov *et al.* 2002). Abandoning the assumption of a low-order multipolar field, we obtained a model of the magnetic field topology which for the first time successfully reproduced all magnetic observations available for HD 37776. These DI results, illustrated in Fig. 4, suggest that the field geometry of this star is highly complex and non-axisymmetric, lacking any resemblance to a low-order multipolar structure. This field is, however, much weaker than suggested by the previous studies. The field strength does not exceed $\approx$ 30 kG at the surface, yielding the phase-averaged magnetic field modulus of only $\approx$ 14.5 kG. The new magnetic field topology of HD 37776 has important implications for understanding different aspects of the variability of this unique star and for our understanding of the early B-type star magnetism in general.

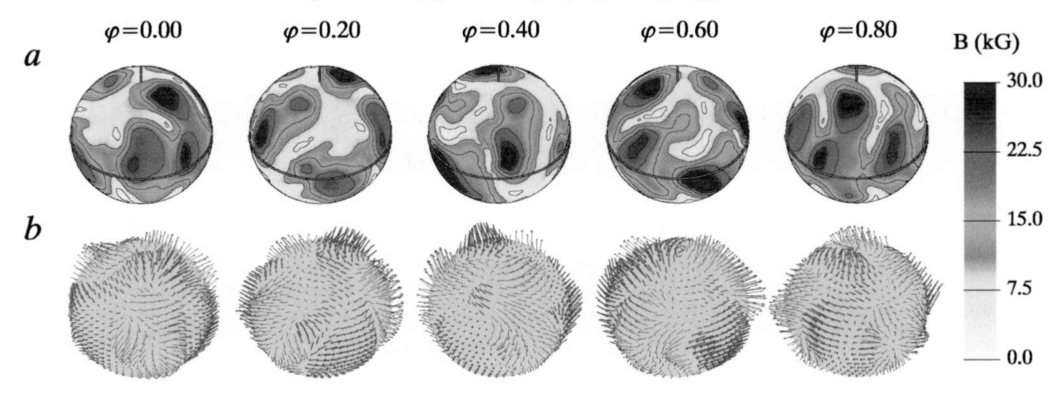

Figure 4. Surface magnetic field structure of HD 37776 derived with magnetic Doppler imaging. The format of this figure is similar to Fig. 1.

Acknowledgements

O.K. is a Royal Swedish Academy of Sciences Research Fellow supported by grants from the Knut and Alice Wallenberg Foundation and the Swedish Research Council.

References

Adelman, S. J. 1997, *A&AS*, 125, 65

Bohlender, D. A. 1994, in: L. A. Balona, H. F. Henrichs, & J. M. Le Contel (eds.), *Pulsation; Rotation; and Mass Loss in Early-Type Stars*, IAU Symposium 162, p. 155

Brown, S. F., Donati, J.-F., Rees, D. E., & Semel, M. 1991, *A&A*, 250, 463

Donati, J.-F. & Landstreet, J. D. 2009, *ARAA*, 47, 333

Donati, J.-F., Howarth, I. D., Jardine, M. M., Petit, P. *et al.* 2006, *MNRAS*, 370, 629

Donati, J.-F., Semel, M., Carter, B. D., Rees, D. E. *et al.* 1997, *MNRAS*, 291, 658

Groote, D. & Hunger, K. 1982, *A&A*, 116, 64

Khokhlova, V. L., Rice, J. B., & Wehlau, W. H. 1986, *ApJ*, 307, 768

Khokhlova, V. L., Vasilchenko, D. V., Stepanov, V. V., & Romanyuk, I. I. 2000, *Astron. Lett.*, 26, 177

Kochukhov, O., Drake, N. A., Piskunov, N., & de la Reza, R. 2004, *A&A*, 424, 935

Kochukhov, O. & Piskunov, N. 2002, *A&A*, 388, 868

Kochukhov, O., Piskunov, N., Ilyin, I., Ilyina, S. *et al.* 2002, *A&A*, 389, 420

Kochukhov, O., Tsymbal, V., Ryabchikova, T., Makaganyk, V. *et al.* 2006, *A&A*, 460, 831

Kochukhov, O. & Wade, G. A. 2010, *A&A*, 513, A13

Krtička, J., Mikuláček, Z., Zverko, J., & Žižňovský, J. 2007, *A&A*, 470, 1089

Landstreet, J. D. 1992, *A&AR*, 4, 35

Landstreet, J. D. & Borra, E. F. 1978, *ApJ* (Letters), 224, L5

Leblanc, F., Monin, D., Hui-Bon-Hoa, A., & Hauschildt, P. H. 2009, *A&A*, 495, 937

Mikulášek, Z., Krtička, J., Henry, G. W., Zverko, J. *et al.* 2008, *A&A*, 485, 585

Oksala, M. E., Wade, G. A., Marcolino, W. L. F., Grunhut, J. *et al.* 2010, *MNRAS*, 405, L51

Piskunov, N. & Kochukhov, O. 2002, *A&A*, 381, 736

Reiners, A., Stahl, O., Wolf, B., Kaufer, A. *et al.* 2000, *A&A*, 363, 585

Rivinius, T., Szeifert, T., Barrera, L., Townsend, R. H. D. *et al.* 2010, *MNRAS*, 405, L46

Rivinius, T., Štefl, S., Townsend, R. H. D., & Baade, D. 2008, *A&A*, 482, 255

Thompson, I. B. & Landstreet, J. D. 1985, *ApJ* (Letters), 289, L9

Townsend, R. H. D., Owocki, S. P., & Groote, D. 2005, *ApJ* (Letters), 630, L81

Townsend, R. H. D., Owocki, S. P., & Ud-Doula, A. 2007, *MNRAS*, 382, 139

Veto, B., Hempelmann, A., Schoeneich, W., & Stahlberg, J. 1991, *AN*, 312, 133

Vogt, S. S., Penrod, G. D., & Hatzes, A. P. 1987, *ApJ*, 321, 496

Active OB stars: structure, evolution, mass loss, and critical limits
Proceedings IAU Symposium No. 272, 2010
C. Neiner, G. Wade, G. Meynet & G. Peters, eds.

doi:10.1017/S1743921311010209

Discussion – Winds and magnetic fields of active OB stars

Jean-Claude Bouret[1,2] **and Lydia Cidale**[3,4]

[1]Laboratoire d'Astrophysique de Marseille,
38 rue Joliot-Curie, 13388 Marseille Cedex 13, France

[2]NASA Goddard Space Flight Center, Greenbelt, 20771 MD, USA
email: jean-claude.bouret@oamp.fr

[3]Facultad de Ciencias Astronómicas y Geofísicas, Universidad Nacional de La Plata

[4]Instituto de Astrofísica La Plata, CCT La Plata-CONICET-UNLP
Paseo del Bosque S/N, La Plata, Argentina
email: lydia@fcaglp.unlp.edu.ar

Abstract. The discussion on winds and magnetic fields of active OB stars was carried out by S. Owoki, G. Wade, M. Cantiello, O. Kochukhov, M. Smith, C. Neiner, T. Rivinius, H. Henrichs and R. Townsend. The topics were the ability to detect small and large scale magnetic fields in massive stars and the need to consider limits on photometric variability of the star surface brightness.

Keywords. stars: early-type, stars: magnetic fields, stars: mass loss

---

Discussion

The discussion of session 2 was first driven by the invited talk about observations of magnetic fields by V. Petit, who reviewed modern spectropolarimetric methods used to diagnose magnetic fields of massive stars. The discussion was mostly focused on large and small scale magnetic fields, their strengths, and on the possibility of detecting great complex fields.

Stan Owocki compared the large complex magnetic fields that are actually detected on massive stars with spectropolarimetry with those strong fields seen at small scale in the Sun. He explained that in the case of very strong fields well spread over the star, the rotational modulation allowed us to see a modulation-signal. However, he wondered what would happen if some stars present small complex magnetic fields, as solar-like spots, or have large and complex fields with different scales and different polarization, and the two fields canceled on the disc. How could we see those? The answer to that issue is really important because people, for a long time, have been wondering whether magnetic fields might play an important role in the Be phenomena. Brown & Cassinelli proposed for instance to use a magnetic field moment arm to have a sub-critical rotation and that moment arm then swept up the matter to reach Keplerian orbits. On the other hand, numerical simulations (Owocki and collaborators) rather suggest that large scale magnetic fields are really incompatible with Keplerian discs because these discs have shear. This is in stark contrast with observational results (presented in some posters at this symposium) bringing strong evidence that the disc of Be stars are generally Keplerian. We note however that, at the same time, very strong complex field might play an important role in ejecting material from the star into the orbit.

Another issue that was mentioned was the long-term V/R variations observed in many Be discs. If they really are interpreted as one-arm modes, they imply the precession of features as the material is going around the star every day while the one-arm actually goes around the star every few years. The precession of orbits thus indicates the existence of very small perturbation of 1/r potential. Since the perturbation amplitude is on the scale of the rotation period over the precession period, it means that there could have very large magnetic forces and large scale forces, other than the 1/r gravity as a potential acting on the material. Therefore, it is clear that it is of great importance to make the distinction between what could be observed at large scale versus small scale or in other words to state why fields cannot be detected if/because they are too complex.

From the observational point of view, G. Wade answered that in upper main sequence stars, and perhaps to some extent in evolved stars, it is possible to establish the existence of strong stable large scale magnetic fields, which are oblique to stellar rotation axes and do not appear to correlate strongly with the physical properties of the stars. This phenomenon was well-established in the '40 by Babcock, and strengthened by people like J. Landstreet and the Russian team during the '70 and '80. G. Wade also stated that nowadays this phenomenon was globally understood and details were being investigated. However, it is clear that although the measured quantity is a vector, for sufficiently complex fields or sufficiently weak fields, it is possible that magnetic fields could remain below the detection threshold. Therefore, the field could be certainly present but still remain undetectable. Those kind of fields, that had been illustrated in the talks by Kochukhov and Petit (this volume), are examples of the kind of complex dynamo generated magnetic fields observed routinely in active type stars. It is worth mentioning that the activity expressed by these stars is significantly greater than the one seen in the Sun.

G. Wade made the point that, at the present time, the state-of-art is that "we are barely able to detect a magnetic field similar to that of the Sun in a Sun-like star and, probably in most cases, barely might be highly stretch. However, we are certainly not able to detect a similar field in a B-type star". While thousands of lines can be worked with when multi-line-techniques are used in late-type stars, like G-type stars, the reality is that only tens of lines can be used in the case of a B star.

G. Wade's conclusion is that it is absolutely possible that fields of that sorts could be highly below the detection threshold. He also commented that M. Cantiello and himself were discussing about the degree of complexity in fields that could be predicted with Cantiello's sub-surface convection models. It is very likely that fields like these are probably not detectable in B-type stars either, using today techniques. Furthermore, Cantiello gave orders of magnitude for the amplitude of complex magnetic fields. He explained that when looking at the equipartition fields in the convective regions the magnetic field strength got up to Kilogauss, but if the field got to the surface by buoancy instability, the density had basically to be rescaled and the magnetic fields got up to hundreds, maybe two hundreds gauss (unless there was something special, like erupting like-tubes where the fields are stronger). He remarked that the average field would have been up to hundreds of gauss, but not bigger than that and could present very complex scales. In addition, G. Wade commented, for example, that in the case of our Sun the field was complex, and the strongest regions had fields of the order of Kilogauss. He explained that since those magnetic fields could not be possibly detected, there was no reason for believing that such fields did exist. He stated that it was certainly true that it wouldn't have been detectable using today techniques. Nevertheless, this point can be regarded as over-pessimistic as magnetic fields in solar-type stars have already been detected by P. Petit.

In order to give an answer to the magnetic field scale, O. Kochukhov invited theoreticians to provide observers with a power spectrum because it would help to calculate the Stoke parameters. He explained that if the scale of the magnetic fields was small, smaller than the mean photon path-length, magnetic fields would not be observable. However, if the spectrum contained features at that scale, of a few degrees, it would be possible to do something. In relation to this last comment, T. Rivinius brought in mind the case of γ Cas which is a rapid rotator. He speculated that if it just was a matter of a few degrees, or how quick the star rotates, γ Cas should be easily resolved in a few degrees. To this particular problem, C. Neiner explained that if the magnetic fields were of the order of a Kilogauss, like those fields found in M. Smith's model, they would had been detected by now, even if the scale of the field was small because of the very big $v \sin i$ of that star. She added that they had observed γ Cas with stellar high-resolution spectropolarimeters, as NARVAL and ESPaDOnS, during many nights and they were not able to detect any field in the photospheric lines. Still, she stated that there might be very marginal magnetic signatures in the Fe II lines. So, if there is a magnetic field in that star it is in the disc and not in the star. Moreover, G. Wade pointed out that there were families of field topologies, that were going to be undetectable in any data set and so that came down to absence of evidence being not evidence of absence. Even though, theoreticians would be able to provide with a model, the field could be hidden in the noise and certainly remain undetectable.

On the other hand, M. Smith mentioned that, particularly for γ Cas, they had four lines of independent evidence of magnetic field. One of them obtained under the bolometric light-curve, which was a very robust reproducible result. The real issue with this star would be then to explain the production of X-ray of 10^8 degrees, 12 Kev at least, without a magnetic field! There are now eight stars sharing the same global properties, one of which gives plasma temperature twice that of γ Cas. Therefore, M. Smith argued that we are looking in a fairly, rare, a very small subgroup of stars and might not be easy to extrapolate σ Ori's and even the HD 37776 successes other than by the photometric signature that went out to 2 or 3 Kilogauss.

Rich Townsend remarked that rotation modulation and surface spot leading to photometric variability were not necessary signs of magnetic field. It is worth mentioning that some works presented in this volume, showing also spectroscopic variability due to surface inhomogeneities, are not associated with magnetic fields. In contrast to the previous examples, H. Henrichs mentioned the particular case of σLup, a B1 V star with a magnetic field that phases exactly with the 3 days photometric period, and the maximum field coincides with the maximum bright.

The need to consider the limits on photometric variability of the surface brightness of massive stars was highlighted by Owocki. He argued, in particular, that the photometric variation is directly related to the presence of CIRs in the wind of these stars and the best way to make a CIR is to have a bright spot on the stellar surface. Therefore, he suggested that high-precision photometric data like those delivered by MOST (or CoRoT) could put tight constraints on Cantiello's model and wondered if that model could predict a bright magnetic spot, versus the case of the Sun where the magnetic spots are actually dark.

Cantiello explained that the envelope calculation gave spots in excess of 200 K respect to the rest of the photosphere, but of course, it would also depend on how many spots were there. It was difficult to define the filling factor of the magnetic field to determine the factor of fluctuation in the luminosity.

Another important reason to look at photometric variations is related to the topology of the wind. In order for a magnetic field to channel the wind, a large scale magnetic field is needed and it must have a large η_* (as defined in a series of papers by Owocki and Ud-doula) but in order to have an effect in perturbing the base of the wind, it should be on smaller scales and more complex. In any case, the signature would be in brightness variations, thus the importance of having very small limits for the detection of photometric variations.

R. Townsend indicated that, probably, the limit for periodical photometric variations that we could get from ground based observations is around 10 millimagnitudes. He also remarked that with space satellite (MOST, CoRoT) we could see all sort of interesting variations that showed off nowadays.

The session on winds and magnetic fields ended on these comments.

Active OB stars: structure, evolution, mass loss, and critical limits
Proceedings IAU Symposium No. 272, 2010
C. Neiner, G. Wade, G. Meynet & G. Peters, eds.

doi:10.1017/S1743921311010210

Searching for massive star magnetospheres†

David Bohlender and Dmitry Monin

Herzberg Institute of Astrophysics, National Research Council of Canada,
5071 West Saanich Road, Victoria BC, Canada V9E 2E7

Abstract. We review the status of a long-term program to search for stellar magnetospheres in Bp stars. A few new σ Ori E analogues discovered during the course of this investigation are briefly discussed and other stars that may be worthy of further study are noted.

Keywords. stars: chemically peculiar, stars: circumstellar matter, stars: early-type, stars: emission-line, stars: magnetic fields, stars: individual (σ Ori E, HD 176582)

1. Introduction

Most magnetic Bp stars that show evidence for magnetospheric material in the form of variable Hα emission or shell absorption and satellite UV lines are also non-thermal radio sources. The primary example is the prototypical helium-strong star σ Ori E. Because of this, we have been conducting a long-term program on the Dominion Astrophysical Observatory's (DAO) 1.2-m and 1.8-m telescopes to search for variable Hα profiles and magnetic fields in other northern Bp stars known to be radio sources. Table 1 lists our current program objects, selected from compilations of radio emission surveys of Ap and Bp stars by Linsky *et al.* (1992), Leone *et al.* (1994) and Drake *et al.* (2006).

Table 1. Bright Bp stars with non-thermal radio emission.

HD	Name	Spectral Type	$v \sin i$ (km s^{-1})	6 cm Flux (mJy)	B_e Field Extrema (kG)	Hα Variability?[1]	Period (d)
12447	α Psc	A2 SiSrCr	75	0.36 ± 0.09	-0.5 to 0.4	?	1.491
19832	56 Ari	B8 Si	80	0.45 ± 0.12	-0.4 to 0.4	No	0.72790
35298	V1156 Ori	B6 He-wk	60?	0.29 ± 0.10	-4.5 to 4.5	?	1.8111
35502		B6 SrCrSi	80?	2.97 ± 0.10	-3.0 to 0.0	Emission	0.8538?
36313	V1093 Ori	B8 He-wk Si	35	0.49 ± 0.06	-1.5 to -1.1	SB2?	0.592?
36485	δ Ori C	B3 He-str	32	0.95 ± 0.07	-3.0 to -1.7	Emission	1.47775
37150	HR 1906	B3Vv	190	0.51 ± 0.07	$\leqslant$ 0.5	?	?
37642	V1148 Ori	B9 He-wk Si	100	0.60 ± 0.07	-3 to 2.7	?	1.0787?
79158	36 Lyn	B9 He-wk	48	0.45 ± 0.05	-1.5 to 1.0	Shell episodes	3.83475
124224	CU Vir	B9 Si	130	4.07 ± 0.14	-1.0 to 1.0	No	0.5207
164429	HR 6718	B9 SiCrSr	95?	0.30 ± 0.05	-0.8 to 0.8	?	1.0820
170000	ϕ Dra	A0 Si	75	0.45 ± 0.05	-0.2 to 0.6	?	1.7165
171247	HR 6967	B8 Si	60	3.04 ±0.09	0.0 to 1.2?	?	3.91227
176582	HR 7185	B5 He-wk	100	0.46 ± 0.05	-2.0 to 2.0	Emission	1.58199
196178	HR 7870	B8 Si	50	3.00 ± 0.07	-1.4 to -0.4	?	$\approx$ 10?

[1] A question mark indicates that phase coverage is insufficient to draw a conclusion about Hα variability.

2. Results

We first discovered variable Hα emission with a 1.47775 day period in the helium-strong star δ Ori C (Leone *et al.* 2010). While its emission variability is quite similar to that of σ Ori E, in contrast to the prototypical helium-strong star, δ Ori C's low $v \sin i$

† Based on observations acquired at the Dominion Astrophysical Observatory, Herzberg Institute of Astrophysics, National Research Council of Canada.

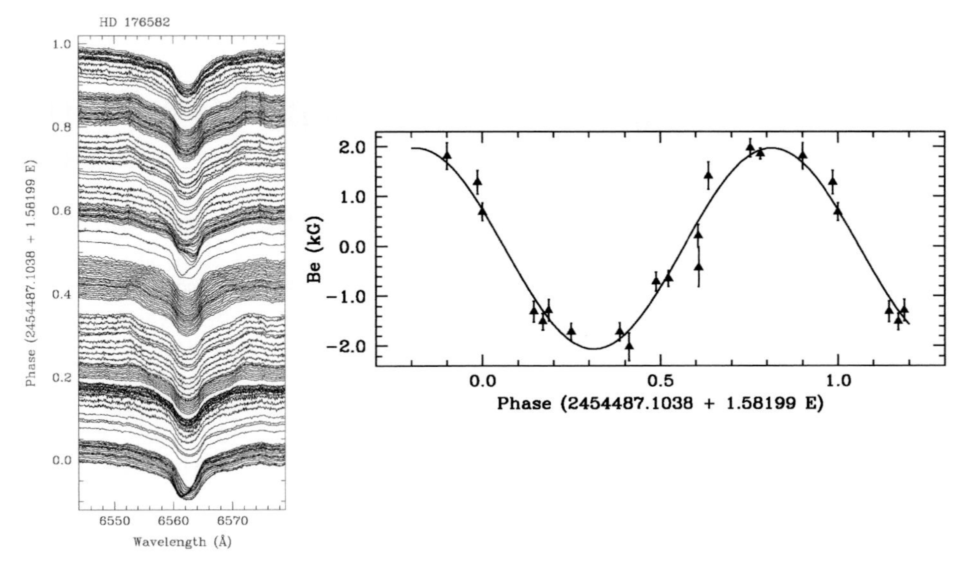

Figure 1. Hα emission and magnetic field variability of HD 176582. Note that shell absorption episodes occur when the magnetic equator crosses the line of sight (near phases 0.0 and 0.5).

and approximately constant -2.5 kG magnetic field show that it has a small inclination and a relatively small magnetic obliquity.

The helium-weak star 36 Lyn shows less obvious evidence of magnetospheric material. Two brief "shell" episodes occur for about 10% of its rotation period just as the magnetic equator crosses the line of sight to the observer (Wade *et al.* 2006; Smith *et al.* 2006).

The magnetic helium-weak star HD 35502 also has variable Hα emission. DAO magnetic field measurements suggest that it has a strong, non-reversing magnetic field with a -3 kG extremum. The star is a rapid rotator and we are continuing to acquire data to try to establish a unique rotation period among several possibilities near 0.85 days.

We have recently discovered that the helium-weak star HD 176582 is also a magnetic Hα emission variable. DAO polarimeter data show a strong, approximately dipolar magnetic field with field extrema of ± 2 kG. Figure 1 shows our phase-resolved Hα and magnetic field data; the Hα variability is weaker but very similar in appearance to that of σ Ori E.

3. Conclusion

A few other stars in Table 1 warrant additional observations. Our DAO magnetic field observations of HD 35298 show that it has a very strong ≈ 5 kG field but our small set of Hα spectra for the star do not sample its rotation period very well. We have also detected magnetic fields in the non-thermal radio sources HD 164429 and HD 171247.

References

Drake, S. A., Wade, G. A., & Linsky, J. L. 2006, in: A. Wilson (eds.), *The X-ray Universe 2005*, ESA-SP 604, p. 73

Leone, F., Trigilio, C., & Umana, G. 1994, *A&A*, 283, 908

Leone, F., Bohlender, D. A., Bolton, C. T., Buemi, C. *et al.* 2010, *MNRAS*, 401, 2739

Linsky, J. L., Drake, S. A., & Bastian, T. S. 1992, *ApJ*, 393, 341

Smith, M. A., Wade, G. A., Bohlender, D. A., & Bolton, C. T. 2006, *A&A*, 458, 581

Wade, G. A., Smith, M. A., Bohlender, D. A., Ryabchikova, T. A. *et al.* 2006, *A&A*, 458, 569

Active OB stars: structure, evolution, mass loss, and critical limits
Proceedings IAU Symposium No. 272, 2010
C. Neiner, G. Wade, G. Meynet & G. Peters, eds.

doi:10.1017/S1743921311010222

A family of stable configurations to model magnetic fields in stellar radiation zones

Vincent Duez[1], Jonathan Braithwaite[1] and Stéphane Mathis[2]

[1]Argelander Institut für Astronomie, Universität Bonn, Auf dem Hügel 71, D-53111 Bonn, Germany, email: vduez@astro.uni-bonn.de, jonathan@astro.uni-bonn.de

[2]Laboratoire AIM, CEA/DSM-CNRS-Université Paris Diderot, IRFU/SAp Centre de Saclay, F-91191 Gif-sur-Yvette, France, email: stephane.mathis@cea.fr

Abstract. We conduct 3D magneto-hydrodynamic (MHD) simulations in order to test the stability of the magnetic equilibrium configuration described by Duez & Mathis (2010). This analytically-derived configuration describes the lowest energy state for a given helicity in a stellar radiation zone. The necessity of taking into account the non force-free property of the large-scale, global field is here emphasized. We then show that this configuration is stable. It therefore provides a useful model to initialize the magnetic topology in upcoming MHD simulations and stellar evolution codes taking into account magneto-rotational transport processes.

Keywords. stars: magnetic fields, magnetohydrodynamics: MHD

---

1. Introduction

The large-scale, ordered nature of magnetic fields detected at the surface of some Ap, O and B type stars and the scaling of their strengths (according to the flux conservation scenario) favour a fossil hypothesis, whose origin is not yet elucidated. To have survived since the stellar formation, a field must be stable on a dynamic (Alfvén) timescale. It was suggested by Prendergast (1956) that a stellar magnetic field in stable axisymmetric equilibrium must contain both poloidal (meridional) and toroidal (azimuthal) components, since both are unstable on their own (Tayler 1973; Wright 1973). This was confirmed recently by numerical simulations (Braithwaite & Spruit 2004; Braithwaite & Nordlund 2006) showing that an arbitrary initial field evolves on an Alfvén timescale into a stable configuration; axisymmetric mixed poloidal-toroidal fields were found. On the other hand, magnetic equilibria models displaying similar properties have been re-examined analytically by Duez & Mathis (2010). We here address the question of their stability.

2. The model

We deal with initially confined non force-free magnetic configurations (*i.e.* with a non-zero Lorentz force) in equilibrium inside a conductive fluid in absence of convection. Several reasons inclined us to focus on non force-free equilibria. First, Reisenegger (2009) reminds us that no configuration can be force-free everywhere. Although there do exist "force-free" configurations, they must be confined by some region or boundary layer with non-zero or singular Lorentz force. Second, non force-free equilibria have been identified in plasma physics as the result of relaxation (self-organization process involving magnetic reconnections in resistive MHD), *e.g.* by Montgomery & Phillips (1988). Third, as shown by Duez & Mathis (2010), this family of equilibria is a generalization of Taylor states (force-free relaxed equilibria; see Taylor 1974) in a stellar context, where the stratification of the medium plays a crucial role. The equilibrium obtained is described in detail in

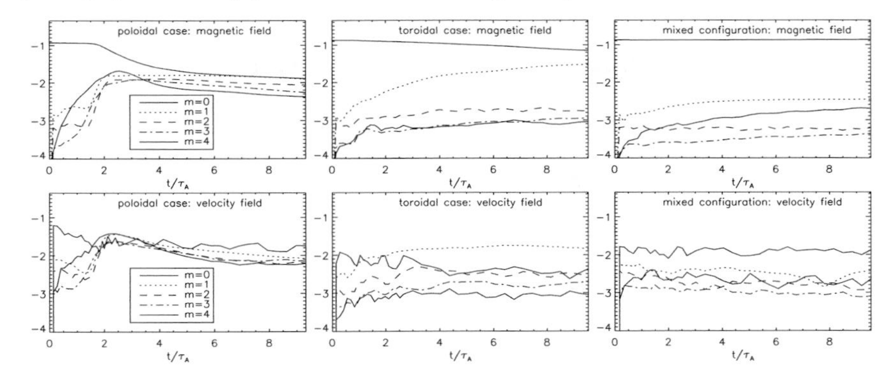

Figure 1. Time evolution of the (log) amplitudes in azimuthal modes $m = 0$ to 4 averaged over the stellar volume of the magnetic field (top row) and the velocity field (bottom row) in the simulations with the purely poloidal field (left), purely toroidal field (middle) and the mixed field (right). Initially, all the magnetic energy is in the $m = 0$ mode since the initial conditions are axisymmetric.

Duez & Mathis (2010) as the lowest energy configuration conserving the invariants of the problem (during the relaxation phase) which are the magnetic helicity (preventing the rapid energy decay) and the mass enclosed in magnetic poloidal flux surfaces (to account for the non force-free property and which is due to the stable stratification).

3. Stability numerical analysis

We use the Stagger code (Nordlund & Galsgaard 1995). We model the star as a self-gravitating ball of ideal gas ($\gamma = 5/3$) with radial density and pressure profiles initially obeying the polytropic relation $P \propto \rho^{1+(1/n)}$, with n = 3, therefore stably stratified, as in Braithwaite & Nordlund 2006. The configuration is then submitted to white perturbations (1% in density). The dynamical evolution of the mixed configuration is compared to its purely poloidal and its purely toroidal components, whose behaviour are well known to be unstable due to kink-type instabilities. The magnetic and velocity amplitudes are plotted on Fig. 1. As we can see, in contrast to these unstable configurations, the mixed poloidal-toroidal one does not exhibit any sign of instability, even for high azimuthal wavenumbers (up to about 40). This is the first time the stability of an analytically-derived stellar magnetic equilibrium has been confirmed numerically. This result has strong astrophysical implications: the configuration provides a good initial condition to stellar magnetohydrodynamic modelling and simulations; furthermore it could help to appreciate the internal magnetic structure of neutron stars and magnetic white dwarfs.

References

Braithwaite, J. & Nordlund, Å. 2006, *A&A*, 450, 1077
Braithwaite, J. & Spruit, H. C. 2004, *Nature*, 431, 819
Duez, V. & Mathis, S. 2010, *A&A*, 517, A58
Montgomery, D. & Phillips, L. 1988, *Phys. Rev. A*, 38, 2953
Nordlund, Å. & Galsgaard, K. 1995, *A 3D MHD code for Parallel Computers*, (Astronomical Observatory, Copenhagen University), http://www.astro.ku.dk/~aake/papers/95.ps.gz
Prendergast, K. H. 1956, *ApJ*, 123, 498
Tayler, R. J. 1973, *MNRAS*, 161, 365
Taylor, J. B. 1974, *Phys. Rev. Lett.*, 33, 1139
Wright, G. A. E. 1973, *MNRAS*, 162, 339

Active OB stars: structure, evolution, mass loss, and critical limits
Proceedings IAU Symposium No. 272, 2010
C. Neiner, G. Wade, G. Meynet & G. Peters, eds.

doi:10.1017/S1743921311010234

The constant magnetic field of ξ^1 CMa: geometry or slow rotation?

Chloé Fourtune-Ravard[1,2], Gregg A. Wade [2], Wagner L. F. Marcolino[3,4], Matthew Shultz[5], Jason H. Grunhut[2,5], Huib F. Henrichs[6] and the MiMeS Collaboration

[1]Université Paris Diderot-Paris 7, UFR de Physique, France
email: c.fourtune.ravard@gmail.com

[2]Department of Physics, Royal Military College of Canada, Ontario, Canada

[3]LAM-UMR 6110, CNRS & Univ. de Provence, France

[4]Observatòrio Nacional, Rio de Janeiro, Brazil

[5]Department of Physics, Engineering Physics & Astronomy, Queen's University, Canada

[6]Astronomical Institute Anton Pannekoek, University of Amsterdam, the Netherlands

Abstract. We report recent observations of the sharp-lined magnetic β Cep pulsator ξ^1 CMa (= HD 46328). The longitudinal magnetic field of this star is detected consistently, but it is not observed to vary significantly, during nearly 5 years of observation. In this poster we evaluate whether the constant longitudinal field is due to intrinsically slow rotation, or rather if the stellar or magnetic geometry is responsible.

Keywords. stars: magnetic fields, stars: rotation, stars: early-type, stars: individual (ξ^1 CMa)

1. Introduction, observations and stellar parameter determination

ξ^1 CMa is known to be a B0.5 pulsator with sharp lines. The longitudinal magnetic field of this star, first detected by Hubrig *et al.* (2006) with FORS1, is detected consistently in over 5 years of observations, but has remained approximately constant at Bz $\sim$375 G. Within the rigid rotator paradigm, two explanations can explain this behaviour: either the star rotates very slowly, or the stellar or magnetic geometry is responsible for the constant value of the longitudinal magnetic field. We acquired 18 Stokes V ESPaDOnS spectra ($370 \leqslant \lambda \leqslant 1000$ nm, $R = 65,000$, S/N$\sim$ 1000 per 1.8 km/s pixel) with the aim of precisely studying the magnetic field, investigating the rotational period and geometry.

We employed CMFGEN to determine physical and wind parameters of ξ^1 CMa. The luminosity was computed using the parallax of van Leeuwen (2007; 2.36 mas), which provides a good fit to the IUE (SWP+LWR) low resolution and large aperture data. An E(B-V) of 0.04 was also used. We find $T_{\rm eff} = 27500 \pm 2000$ K, $\log g = 3.50 \pm 0.20$, $L/L_\odot = 38370$, $R/R_\odot = 8.6$ and $M/M_\odot \sim 9.0$. The projected rotational velocity is constrained to be $v \sin i \leqslant 15$ km/s.

2. Magnetic field and rotational period

Longitudinal magnetic field measurements were inferred using Least-Squares Deconvolution (LSD) with a line mask carefully customised to the spectrum of ξ^1 CMa. All spectra yield definite detections of Stokes V profiles and flat diagnostic N profiles with longitudinal field uncertainties of $\sim$7 G. A straight-line fit to the longitudinal field measurements extracted from Stokes V gives a reduced χ^2 of 4.8, while analogous measurements extracted from diagnostic N give reduced χ^2 of just 1.1. This points to weak variability of Stokes V that is not present in N.

ξ^1 CMa is a well-known β Cep pulsator that displays monoperiodic radial mode photometric and line profile variability with a period of 0.209 days (Heynderickx *et al.* 1994; Saesen *et al.* 2006). Our spectra sample the full pulsational cycle and reveal a peak-to-peak radial velocity variation of 38 km.s^{-1}. We perfomed a period search of the Stokes V longitudinal field measurements using a Lomb-Scargle algorithm, detecting significant power at 4.2680 days. When the longitudinal field measurements are phased with this period they describe a sinusoidal variation with amplitude ~30 G and reduced χ^2 of 1.2.

3. Hα emission and UV wind line morphology

We observed emission in the Hα line profile. We extracted the emission profile from each spectrum by first using Hβ to construct a photospheric template, then subtracting a model photospheric profile from the Hα profile. The derived emission profile is approximately constant and characterised by a FWHM of ~120 km.s^{-1}. The UV C IV and Si IV wind lines of ξ^1 CMa show no variability in IUE spectra acquired in 1978 and 1979. They are remarkably similar to those of the magnetic star β Cep at phases of maximum emission (i.e. when the star is viewed closest to the magnetic pole). This could imply that we currently view ξ^1 CMa near its magnetic pole as well. Such a configuration is consistent with either a long rotation period or a pole-on geometry. Nevertheless, the lack of variability observed in the wind lines leads us to prefer the pole-on geometry model.

4. Conclusions: magnetic field, stellar geometry and rotation

The lack of any secular change in the field during the period of observation, in combination with the very high precision of the magnetic measurements, suggests either that the rotational period of the star is remarkably long, or that the stellar geometry is such that the disc-integrated line-of-sight component of the field remains approximately constant. The latter model is more consistent with the observed stability of the Hα and UV line emission, the UV line morphology, and the period detected in the longitudinal field measurements. If β Cep has $i = 60°$ and $\beta = 85°$, the magnetic pole comes within 35° of the line of sight. If we accept the arguments above, the magnetic pole of ξ^1 CMa must come similarly close to the line of sight, and furthermore it must remain there. This would all hold together if the 4.26 day period is in fact the stellar rotation period. This would require very low inclination (5 to 10°) to be consistent with the low vsini, and in that case the weak modulation of the longitudinal field can be matched by a 1450 G dipole with obliquity $\beta = 25°$. This would imply, in fact, that the magnetic pole is never more than 30° from the line of sight. An accurate determination of the $v \sin i$ of this star is critical to confirming this view, as effectively any non-zero value of $v \sin i$ would imply that the rotational period is shorter than the total span of the observations.

References

Heynderickx, D., Waelkens, C., & Smeyers, P. 1994, *A&AS*, 105, 447

Hubrig, S., Briquet, M., Schöller, M., De Cat, P. *et al.* 2006, *MNRAS*, 369, L61

Van Leeuwen, F. 2007, *Hipparcos, the New Reduction of the Raw Data*, Astrophysics and Space Science Library 350

Saesen, S., Briquet, M., & Aerts, C. 2006, *Communications in Asteroseismology*, 147, 109

Active OB stars: structure, evolution, mass loss, and critical limits
Proceedings IAU Symposium No. 272, 2010
C. Neiner, G. Wade, G. Meynet & G. Peters, eds.

doi:10.1017/S1743921311010246

Is the wind of the Oe star HD 155806 magnetically confined?

Alex W. Fullerton[1], **Véronique Petit**[2], **Stefano Bagnulo**[3], **Gregg A. Wade**[4], **and the MiMeS Collaboration**

[1]Space Telescope Science Institute, 3700 San Martin Drive, Baltimore, MD 21218, USA
email: fullerton@stsci.edu

[2]Dept. of Geology & Astronomy, West Chester University, West Chester, PA 19383, USA
email: VPetit@wcupa.edu

[3]Armagh Observatory, College Hill, Armagh, BT61 9DG, Northern Ireland
email: sba@arm.ac.uk

[4]Dept. of Physics, Royal Military College of Canada, Kingston, ON, K7K 4B4, Canada
email: Gregg.Wade@rmc.ca

Abstract. Spectropolarimetric observations of HD 155806 – the hottest Galactic Oe star – were obtained with CFHT/ESPaDOnS to test the hypothesis that disk signatures in its spectrum are due to magnetic channeling and confinement of its stellar wind. We did not detect a dipole field of sufficient strength to confine the wind, and could not confirm previous reports of a magnetic detection. It appears that stellar magnetism is not responsible for producing the disk of HD 155806.

Keywords. stars: early-type, stars: emission-line, Be, stars: winds, outflows, stars: magnetic fields, stars: individual (HD 155806)

---

1. Why Are There Oe Stars?

Oe stars are a rare subset of the O-type stars that exhibit double-peaked or central emission in their Balmer lines. This emission-line morphology is distinct from signatures of stellar winds, and is conventionally attributed to a circumstellar disk. Although Oe stars are usually considered to be a continuation of the Be phenomenon toward hotter spectral types, it is difficult to understand how stable disks can coexist with the increasingly strong stellar winds typical of O-type stars.

A plausible explanation is that the disk is maintained by a large-scale (e.g., dipolar) magnetic field that channels outflowing wind material toward the magnetic equator, in which case the disk and the wind are really a single entity. A straightforward test of this hypothesis is to search for dipolar magnetic fields of sufficient strength to confine the wind. Since HD 155806 is the hottest Galactic Oe star currently known (O7.5 V[n]e according to Walborn 1973), it should have the strongest wind; and would therefore require the largest magnetic field to confine it. Moreover, its comparatively narrow photospheric lines and brightness enhance the detectability of large-scale magnetic fields. The hypothesis of magnetic confinement was bolstered when Hubrig et al. (2007) reported the detection of a magnetic field with longitudinal strength of $\langle B_z \rangle = -155 \pm 37$ G, though Hubrig et al. (2008) did not detect a significant field in subsequent observations.

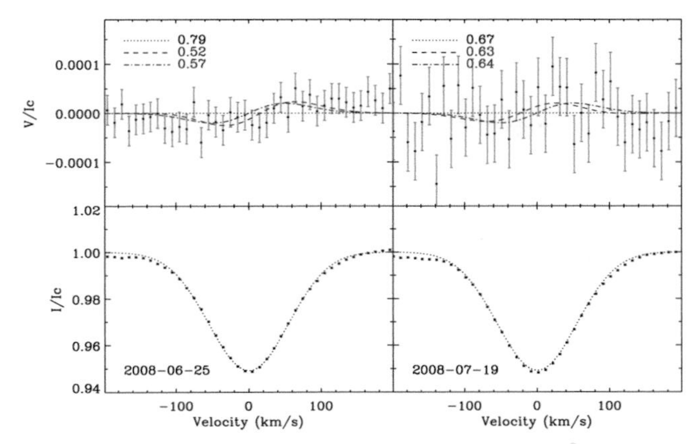

Figure 1. *Lower:* Mean Stokes I absorption line profiles of HD 155806 obtained with ESPaDOnS. *Upper:* Stokes V profiles. The best fit of a dipole model to an individual observation is shown in dashed lines and its corresponding likelihood is labelled, while the best fit to both observations is shown in dot-dashes lines. Formally, the individual fits give $\langle B_z \rangle = (-55 \pm 19, 8 \pm 37)$ G, but neither is more significant than the fit for the null hypothesis of "no field" (dotted line).

2. Observations and Analysis

We used CFHT/ESPaDOnS to obtain high S/N spectropolarimetric observations of HD 155806 on 2008-06-25 and 2008-07-19. The lower panels of Fig. 1 show the mean absorption line profiles obtained with the least-squares deconvolution technique (Donati et al. 1997), while the upper panels show the distribution of circularly polarized light across the mean profiles. Since deviations as large as the ones observed would occur (73%, 17%) of the time for the observations in (2008 June, 2008 July), these data provide no evidence for the presence of a magnetic field. In addition, our reanalysis of the 5 VLT/FORS1 grism observations discussed by Hubrig *et al.* (2008) resulted in non-detections. **Thus, we cannot confirm the detection of a significant longitudinal field in HD 155806.** A Bayesian analysis of the ESPaDoNs data shows that a dipolar magnetic field cannot be stronger than (199, 425) G with a confidence of (63.3, 95.4)%.

3. Implications

For HD 155806, a wind magnetic confinement parameter of $\eta_\star \equiv B_{\rm eq}^2 R_\star^2 / \dot{M} v_\infty \geqslant 1$ (ud-Doula & Owocki 2002) implies that a dipolar field of *at least* 235 G is required to confine the wind. Since upper limits on the *non-detection* of a magnetic field in the ESPaDOnS spectra are comparable, we conclude that the disk of HD 155806 is not caused by magnetic confinement of its wind. Similarly, Nazé, *et al.* (2010) did not detect the hard X-ray component in HD 155806 that should be produced in such a hot star by the collision of channelled streams of wind emerging from opposite magnetic hemispheres. Evidently the disk signatures in the spectrum of HD 155806 require a different explanation.

References

Donati, J.-F., Semel, M., Carter, B. D., Rees, D. E. *et al.* 1997, *MNRAS*, 291, 658
Hubrig, S., Yudin, R. V., Pogodin, M., Schöller, M. *et al.* 2007, *AN*, 328, 1133
Hubrig, S., Schöller, M., Schnerr, R. S., González, J. F. *et al.* 2008, *A&A*, 490, 793
Nazé, Y., Rauw, G., & Ud-Doula, A. 2010, *A&A*, 510A, 59
ud-Doula, A. & Owocki, S. P. 2002, *ApJ*, 576, 413
Walborn, N. R. 1973, *AJ*, 78, 1067

Active OB stars: structure, evolution, mass loss, and critical limits
Proceedings IAU Symposium No. 272, 2010
C. Neiner, G. Wade, G. Meynet & G. Peters, eds.

doi:10.1017/S1743921311010258

The slow winds of A-type supergiants

Anahí Granada[1,2,4], Michel Curé[3] and Lydia S. Cidale[1,2]

[1]Facultad de Ciencias Astronómicas y Geofísicas, Universidad Nacional de La Plata, Argentina
[2]Instituto de Astrofísica La Plata, CCT La Plata-CONICET-UNLP, Argentina
[3]Universidad de Valparaíso, Chile
[4]Observatoire de Genève. Université de Genève, Suisse

email: granada@fcaglp.unlp.edu.ar

Abstract. The line driven- and rotation modulated-wind theory predicts an alternative *slow* solution, besides from the standard m-CAK solution, when the rotational velocity is close to the critical velocity. We study the behaviour of the winds of A-type supergiants (Asg) and show that under particular conditions, e.g., when the δ line-force parameter is about 0.25, the slow solution could exist over the whole star, even for the cases when the rotational speed is slow or zero. We discuss density and velocity profiles as well as possible observational conterparts.

Keywords. stars: supergiants, stars: winds, outflows, stars: mass loss

1. Introduction

The theory of radiation driven winds (CAK, Castor *et al.* 1975) and later m-CAK (Pauldrach *et al.* 1986) succeeded in describing terminal velocities (V_∞) and mass losses ($\dot{M}$) of hot stars apart from predicting the wind momentum luminosity (WML) relationship. This relationship was empirically found for the most of luminous O-type stars (Puls *et al.* 1996) and extended to lower luminosity objects by Kudritzki *et al.* (1999), who found that it depends on the spectral type. Particularly, Asg show V_∞ values a factor 3 lower than theoretical ones (Achmad *et al.* 1997) and V_∞ decreases when increasing the escape velocity (V_{esc}) (Verdugo *et al.* 1998a), in clear contradiction with the CAK theory (fast solution). Moreover, Kudritzki *et al.* (1999) found that Hα profile of these stars can be modeled with β velocity laws, for $\beta > 1$. These observational discrepancies with CAK and m-CAK theories could be related to a change in the parameter α along the wind due either to a change in the ionization of the wind or to a decoupling of the line-driven ions in the wind from the ambient gas (Achmad *et al.* 1997). In 2004, Curé (2004) revisited the theory of steady fast-rotating line-driven winds and found that for $\omega = V/V_{crit} > 70\%$ there exists another hydrodynamical or *slow* solution, which is denser and slower than the standard m-CAK solution. For B-type stars he obtained velocity distributions, as well as the critical point and $\dot{M}$, that can be matched by a velocity law with $\beta > 1$. Therefore, we propose here that the winds of Asg could be related to the *slow*

Table 1. Stellar and Wind Parameters

Mod	Teff [K]	log g	R [$R_\odot$]	ω	α	k	δ	α_{eff}	$V\infty_{pol}$ [km s^{-1}]	$F_{m,pol}$ [$M_\odot$/(yr sr)]	$V\infty_{eq}$ [km s^{-1}]	$F_{m,eq}$ [$M_\odot$/(yr sr)]	$\dot{M}$ [$M_\odot$/yr]
1 (s)	13000	1.73	68	0.4	0.51	0.03	0.23	0.28	160	4.78×10^{-5}	142	5.78×10^{-5}	6.41×10^{-4}
2 (f)	10000	2.0	60	0.4	0.49	0.07	0.15	0.34	350	1.14×10^{-5}	291	1.54×10^{-5}	1.59×10^{-4}
3 (s)	10000	2.0	60	0.4	0.49	0.07	0.26	0.23	207	9.79×10^{-8}	181	1.21×10^{-7}	1.32×10^{-6}
4 (s)	9500	1.7	80	0.4	0.49	0.07	0.26	0.23	168	5.94×10^{-7}	149	7.40×10^{-7}	8.04×10^{-6}
5 (s)	9000	1.7	100	0.4	0.49	0.07	0.26	0.23	188	3.08×10^{-7}	165	3.84×10^{-7}	4.18×10^{-6}
6 (s)	9000	1.7	120	0.4	0.49	0.07	0.26	0.23	206	3.57×10^{-7}	178	4.49×10^{-7}	4.87×10^{-6}

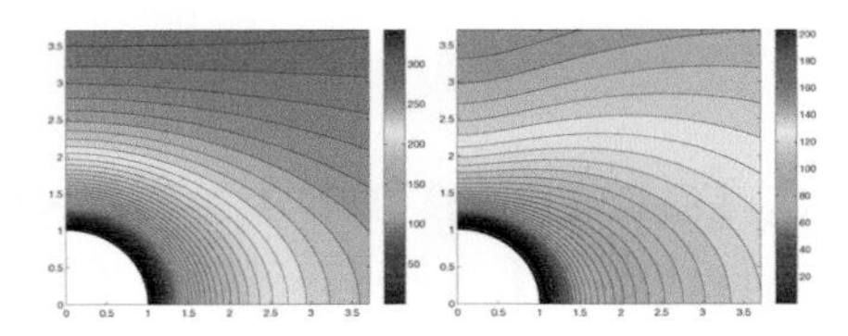

Figure 1. Wind velocity distributions. Left: Model 2 (Fast). Right: Model 3 (Slow).

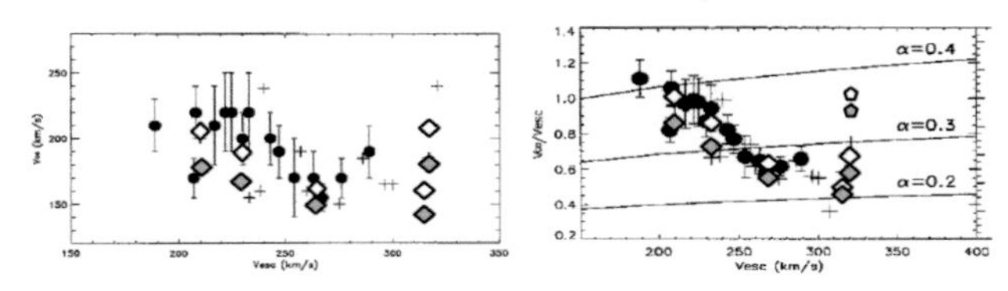

Figure 2. The white/grey romboids (penthagons) correspond to equator/pole slow (fast) solutions, black symbols correspond to Figs. 2 and 3 of Verdugo et al. (1998b).

hydrodynamical solutions described by Curé (2004). We demonstrate numerically that for Asg, given a particular set of wind parameters, k, α and δ, there exists a *slow solution* at all latitudes for low rotational velocities ($\omega < 0.4$), and even for the non-rotating case.

2. Results and Conclusions

We calculated numerically wind solutions for different rotational velocities and different stellar and wind parameters. Table 1 lists some of the models we computed for low values of ω: those marked with an "s" in the first column present a slow solution at all latitudes, while the remaining corresponds to the fast solution at all latitudes ("f"). Figure 1 displays the behaviour of the fast and slow solutions as a function of latitude for models with the same stellar and wind parameters, except δ. $\dot{M}$ and V_{∞} obtained with models 1, 3, 4, 5 and 6 are in good agreement with those observed in Asg (Verdugo *et al.* 1998a; Kudritzki *et al.* 1999). Instead, the fast solution leads to higher $\dot{M}$ and V_{∞}. We find that the solutions for slow winds (see Figure 2, in white/grey romboids) match observational data of some Asg stars (black symbols, Verdugo *et al.* 1998b).

By solving the 1-D hydrodynamic equation of rotating line-driven winds of Asg for different sets of line force parameters, we found a slow wind regime over all latitudes when increasing the parameter δ, which describes changes in the ionization stage of the wind. These slow solutions trace the observational trends found by Verdugo *et al.* (1998a). Our result supports these authors' hypothesis, stating that the negative slope of V_{∞}/V_{esc} vs. V_{esc} could be related to the degree of ionization and wind density.

References

Achmad, L., Lamers, H. J. G. L. M., & Pasquini, L. 1997, *A&A*, 320, 196

Castor, J. I., Abbott, D. C., & Klein, R. I. 1975, *ApJ*, 195, 157

Curé, M. 2004, *ApJ*, 614, 929

Kudritzki, R. P., Puls, J., Lennon, D. J., Venn, K. A. *et al.* 1999, *A&A*, 350, 970

Pauldrach, A., Puls, J., & Kudritzki, R. P. 1986, *A&A*, 164, 86

Puls, J., Kudritzki, R.-P., Herrero, A., Pauldrach, A. W. A. *et al.* 1996, *A&A*, 305, 171

Verdugo, E., Talavera, A., & Gmez de Castro, A. I. 1998a, *Ap&SS*, 263, 263

Verdugo, E., Talavera, A., & Gmez de Castro, A. I. 1998b, in: W. Wamsteker, R. Gonzalez Riestra & B. Harris (eds.), *Ultraviolet Astrophysics Beyond the IUE Final Archive*, ESA-SP 413, p. 293

Active OB stars: structure, evolution, mass loss, and critical limits
Proceedings IAU Symposium No. 272, 2010
C. Neiner, G. Wade, G. Meynet & G. Peters, eds.

doi:10.1017/S174392131101026X

Searching for magnetic fields in the descendants of massive OB stars

Jason H. Grunhut[1], Gregg A. Wade[1], David A. Hanes[1], Evelyne Alecian[2]

[1]Kingston, Canada; [2]Grenoble, France

Abstract. We present the results of a recent survey of cool, late-type supergiants - the descendants of massive O- and B-type stars - that has systematically detected magnetic fields in these stars using spectropolarimetric observations obtained with ESPaDOnS at the Canada-France-Hawaii Telescope. Our observations reveal detectable, often complex, Stokes V Zeeman signatures in Least-Squares Deconvolved mean line profiles in a significant fraction of the observed sample of $\sim$30 stars.

Keywords. instrumentation: polarimeters, techniques: spectroscopic, stars: magnetic fields, stars: supergiants

1. Introduction

Supergiants are the descendants of massive O and B-type main sequence stars. Unlike their main sequence progenitors, cool supergiants are characterized by a helium-burning core and a deep convective envelope.

Due to their extended radii, low-atmospheric densities, slow rotation and long convective turnover times, supergiants provide an opportunity to study stellar magnetism at the extremes of parameter space.

In fact, observations of late-type supergiants show characteristics consistent with magnetic activity, such as luminous X-ray emission and flaring, and emission in chromospheric UV lines - phenomena suggesting the presence of dynamo-driven magnetic fields.

Motivated by the activity-related puzzles of late-type supergiants, the near complete lack of direct constraints on their magnetic fields, and recent success of measuring fields of red and yellow giants (e.g. Aurière *et al.* 2008), we have initiated a program to search for direct evidence of magnetic fields in these massive, evolved stars. Here we summarize the recent results of Grunhut *et al.* (2010).

2. Observations

Circular polarization (Stokes V) spectra were obtained with the high-resolution (R$\sim$68000) ESPaDOnS and NARVAL spectropolarimeters at the Canada-France-Hawaii Telescope and Bernard Lyot Telescope, as part of a large survey investigating the magnetic properties of late-type supergiants.

To date, we have observed more than 30 stars: 4 A-type stars, 8 F-type stars, 11 G-type stars, 7 K-type stars, and 3 M-type stars.

3. Magnetic Field Diagnosis & Results

We applied the Least-Squares Deconvolution (LSD; Donati *et al.* 1997) technique to all our data in order to increase the S/N and detect weak Zeeman signatures.

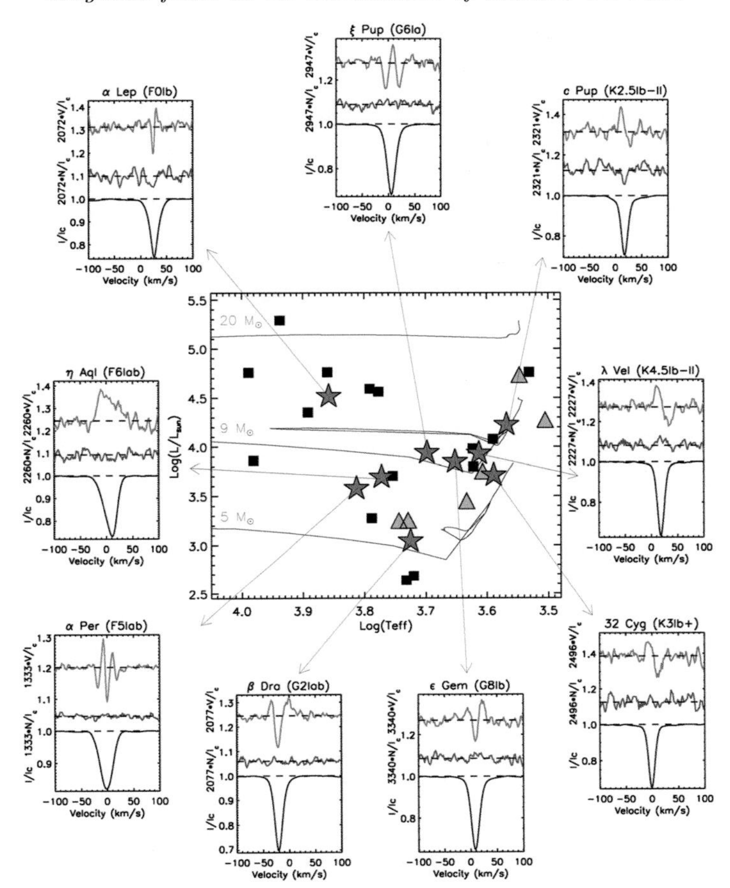

Figure 1. HR diagram showing all observed supergiants. Black squares indicate stars for which no Zeeman signatures were detected, orange triangles indicate stars with suggestive Zeeman signatures, while red stars represent the supergiants with clear Zeeman signatures. Surrounding the HR diagram are illustrative mean Stokes V (top), diagnostic null (middle), and unpolarized Stokes I (bottom) LSD profiles of the 9 stars with clear detections.

In Fig. 1 we present all stars with clear Zeeman signatures detected in Stokes V. Also shown in Fig. 1 is the placement of all observed stars on an HR diagram.

Our investigation shows that many late-type supergiants host detectable Stokes V Zeeman signatures, which are frequently complex. Overall, we find that approximately 1/3 of our sample reveal detectable Zeeman signatures in Stokes V. However, we find no clear differences between classical activity indicators (such as Ca II H&K or Hα emission) of those stars with or without detections. However, we do find a weak correlation between the Ca II core equivalent width and the magnetic field strength for those stars with multiple observations. In addition, we also see clear temporal variability of the Stokes V profiles for those targets with multiple observations.

References

Aurière, M., Konstantinova-Antova, R., Petit, P., Charbonnel, C. *et al.* 2008, *A&A*, 491, 499
Donati, J.-F., Semel, M., Carter, B. D., Rees, D. E. *et al.* 1997, *MNRAS*, 291, 658
Grunhut, J. H., Wade, G. A., Hanes, D. A., & Alecian, E. 2010, *MNRAS* 408, 2290

Active OB stars: structure, evolution, mass loss, and critical limits
Proceedings IAU Symposium No. 272, 2010
C. Neiner, G. Wade, G. Meynet & G. Peters, eds.

doi:10.1017/S1743921311010271

A MiMeS analysis of the magnetic field and circumstellar environment of the weak-wind O9 sub-giant star HD 57682

Jason H. Grunhut[1], **Gregg A. Wade**[1], **Wagner L. F. Marcolino**[2], **Véronique Petit**[3], **and the MiMeS Collaboration**

[1]Kingston, Canada; [2]Marseille, France; [3]West Chester, USA

Abstract. I will review our recent analysis of the magnetic properties of the O9IV star HD 57682, using spectropolarimetric observations obtained with ESPaDOnS at the Canada-France-Hawaii telescope within the context of the Magnetism in Massive Stars (MiMeS) Large Program. I discuss our most recent determination of the rotational period from longitudinal magnetic field measurements and Hα variability - the latter obtained from over a decade's worth of professional and amateur spectroscopic observations. Lastly, I will report on our investigation of the magnetic field geometry and the effects of the field on the circumstellar environment.

Keywords. instrumentation: polarimeters, techniques: spectroscopic, stars: magnetic fields, stars: rotation, stars: individual (HD 57682)

---

1. Introduction

The presence of strong, globally-organized magnetic fields in hot, massive stars is rare. To date, only a handful of massive O-type stars are known to host magnetic fields. In 2009, Grunhut *et al.* reported the discovery of a strong magnetic field in the weak-wind O9IV star HD 57682 from the presence of Zeeman signatures in mean Least-Squares Deconvolved (LSD) Stokes V profiles. Their analysis of the IUE and optical spectra determined the following atmospheric and wind properties: $T_{eff} = 34.5\,\mathrm{kK}$, $\log(g) = 4.0 \pm 0.2$, $R = 7.0^{+2.4}_{-1.8}\,R_{\odot}$, $M = 17^{+19}_{-9}\,M_{\odot}$, and $\log(\dot{M}) = -8.85 \pm 0.5\,M_{\odot}\,\mathrm{yr}^{-1}$.

2. Temporal Variability

Both the longitudinal magnetic field and Hα equivalent width of HD 57682 are strongly variable. In addtion to our 17 ESPaDOnS observations, we've also utilized Hα observations from amateur spectroscopy from the BeSS database, as well archival ESO UVES and FEROS observations dating back over a decade. A period search of these data resulted in a period of ∼31 d, consistent with the rotational period estimated by Grunhut *et al.* (2009). However, the magnetic data could not be reasonably phased with this period. Ultimately, adopting a period of 63.58 d (twice the period obtained from the Hα data) resulted in a coherent phasing of all the data at our disposal, as shown in Fig. 1.

The longitudinal magnetic field appears to vary sinusoidally, consistent with a magnetic field dominated by a strong dipolar component. The Hα equivalent width shows a double-wave pattern with peak emission occurring at the magnetic crossover phases (i.e. when the longitudinal field is null).

The photometric light curve from Hipparcos shows no apparent variability. This likely indicates that the column density of the magnetically confined plasma is relatively low at eclipse phases.

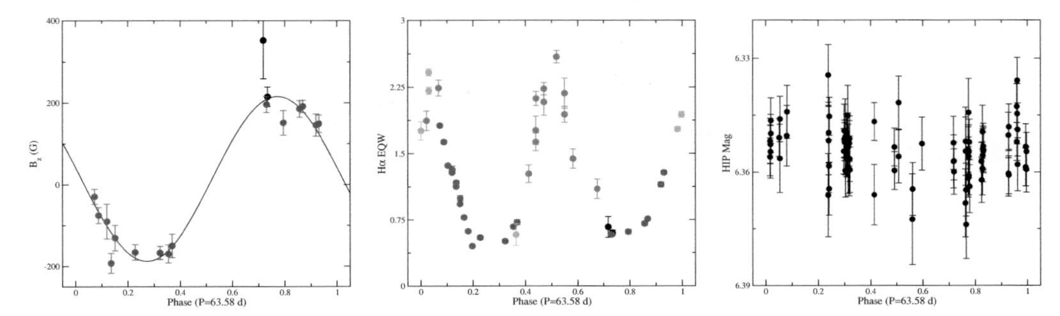

Figure 1. Phased longitudinal magnetic field measurements (left), Hα equivalent width variation (middle), and Hipparcos photometry (right), for HD 57682. Different colours indicate different epochs of observations.

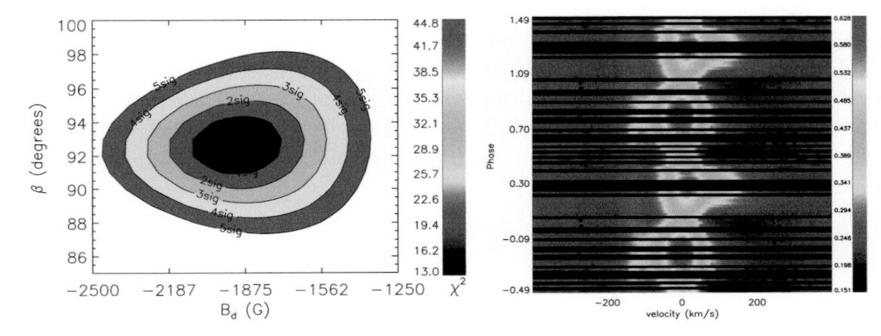

Figure 2. Left: χ^2 landscape as a function of dipole field strength (B_d) and obliquity angle (β). Shown are the intervals corresponding to 1, 2, 3, 4, and 5σ using $i = 22°$. **Right:** Phased Hα residual variations relative to an LTE model. Note the variability is likely due to a magnetically confined wind.

3. Magnetic Geometry and Circumstellar Environment

Using the longitudinal component of the magnetic field, measured from the mean LSD Stokes V and I profiles, we are able to fit a dipole model, characterized by the magnetic field strength at the poles (B_d) and the angle of obliquity of the magnetic axis relative to the rotation axis (β). Using the $v \sin i$ and radius as determined by Grunhut *et al.* (2009) and assuming rigid rotation, we can infer an inclination angle $i = 22^{+17}_{-9}{}^{\circ}$. Using this value we obtain the χ^2 landscape as a function of B_d and β shown in Fig. 2. Taking into account the range of possible inclination angles, we find that $B_d = 1 - 3\,\mathrm{kG}$ and $\beta = 94° \pm 5°$.

In Fig. 2 we also show the residual variations of Hα phased with the adopted rotational period. We conclude based on the characteristics of this dynamic spectrum that the magnetic field is exerting strong confinement (confinement parameter $\eta* \sim 10^3 - 10^5$; ud-Doula & Owocki 2002) on the weak wind of HD 57682, resulting in the Hα variability. However, the slow rotation is likely unable to centrifugally support a stable magnetosphere. Therefore, the plasma that is present likely has a relatively short residence time in the magnetosphere, and must therefore be continually replenished (see Townsend *et al.* these proceedings).

References

Grunhut, J. H., Wade, G. A., Marcolino, W. L. F., Petit, V. *et al.* 2009, *MNRAS*, 400, L94

ud-Doula, A. & Owocki, S. P. 2002, *ApJ*, 576, 413

Active OB stars: structure, evolution, mass loss, and critical limits
Proceedings IAU Symposium No. 272, 2010
C. Neiner, G. Wade, G. Meynet & G. Peters, eds.

doi:10.1017/S1743921311010283

Discovery of the most rapidly-rotating, non-degenerate, magnetic massive star by the MiMeS collaboration

Jason H. Grunhut[1], Gregg A. Wade[1], Thomas Rivinius[2], Wagner L. F. Marcolino[3], Richard H.D. Townsend[4], and the MiMeS Collaboration

[1]Kingston, Canada; [2]ESO Chile; [3]Marseille, France; [4]Madison, USA

Abstract. We discuss the recent detection of a strong, organized magnetic field in the bright, broad-line B2V star, HD 142184, using the ESPaDOnS spectropolarimeter on the CFHT as part of the Magnetism in Massive Stars (MiMeS) survey. We find a rotational period of 0.50833 days, making it the fastest-rotating, non-degenerate magnetic star ever detected. Like the previous rapid-rotation record holder HR 7355 (also discovered by MiMeS: Oksala *et al.* 2010, Rivinius *et al.* 2010), this star shows emission line variability that is diagnostic of a structured magnetosphere.

Keywords. techniques: spectroscopic, stars: magnetic fields, stars: individual (HD 142184)

---

1. Introduction & observations

Magnetic fields are unexpected in hot, massive stars due to the lack of convection in their outer envelopes. However, a small number of massive B stars host strong, organized magnetic fields, such as the chemically peculiar He strong stars (see Bohlender & Monin, these proceedings) like the archetypical star σ Ori E (see Oksala *et al.*, these proceedings) and the recently discovered B2V star HR 7355 (Oksala *et al.* 2010; Rivinius *et al.* 2010). These stars are rapidly rotating and host strong magnetic fields that are coupled to a co-rotating magnetosphere (Townsend *et al.* 2005).

Twenty-one high-resolution ($R \sim 68000$) observations of the variable B2V star HD 142184 were obtained with the ESPaDOnS spectropolarimeter at the Canada-France-Hawaii telescope between February and March 2010. These initial observations clearly show the presence of Zeeman signatures in the circular polarization, Stokes V Least-Squares Deconvolved (LSD), mean line profiles, indicative of a magnetic field.

We also obtained six low-resolution ESO-VLT FORS observations in addition to 27 UVES observations in April 2010 to follow up the ESPaDOnS observations.

2. Rotational period, variability, and magnetic field geometry

Using Hipparcos photometry, we find a single-wave period of 0.50831 ± 0.00002 d (see Fig. 1), which differs from the double-wave photometric light curve of HR 7355 or σ Ori E, likely indicating a different geometry of the magnetic field. From our longitudinal magnetic fields measurements, we confirm this period, which we take to be the rotational period of this star (see Fig. 1), making it the fastest rotating, non-degenerate, magnetic massive star! However, the period is sufficiently imprecise that the relative phasing between our current data and the Hipparcos data can be offset by so much as 0.5 cycles. Therefore, we have adopted a period of 0.50833 d so that the relative phasing between the

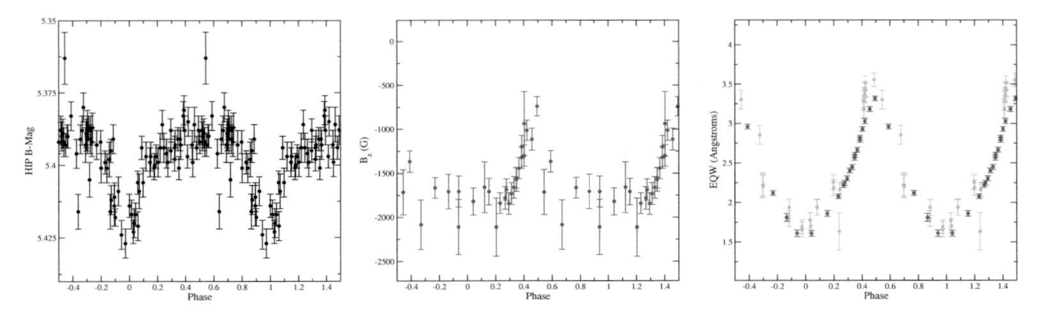

Figure 1. Phased Hipparcos (left), longitudinal magnetic field (middle), and Hα equivalent width measurements for HD 142184. Different colours correspond to the different instruments (red=ESPaDOnS, blue=FORS, green=UVES).

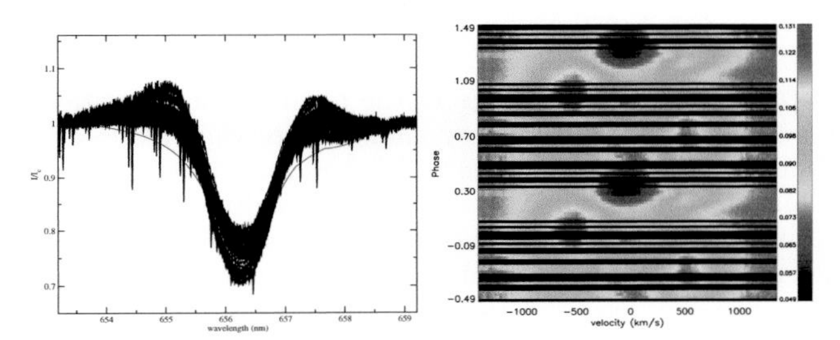

Figure 2. Left: Hα (black) profiles for different nights compared to a NLTE model profile (red). **Right:** Phased residual Hα variations relative to the NLTE model.

photometric minimum and the peak of the longitudinal field curve differs by 0.5 cycles - consistent with the predictions of semi-analytical models for a rotationally supported magnetosphere (Townsend 2008; Townsend & Owocki 2005).

From NLTE model fits and a Fourier analysis to the ESPaDOnS spectra, we find that $T_{\rm eff} = 19 \pm 2\,{\rm kK}$, $\log(g) = 2.95 \pm 0.04$, and $v \sin i = 270 \pm 10\,{\rm km\,s^{-1}}$.

In addition to the photometric and magnetic periodicity, we also find that Hα varies with the same period, as shown in Fig. 1. Hα shows line profile variations of emission extending to high velocities, as shown in Fig. 2. The double-lobed pattern and equivalent width variations strongly suggests that HD 142184 hosts a structured magnetosphere similar to σ Ori E and HR 7355 consisting of co-rotating, magnetically confined clouds of stellar wind plasma.

Assuming rigid rotation, we infer that the inclination $i \sim 75°$. From fits to the longitudinal field curve shown in Fig. 1 we estimate that HD 142184 hosts a mainly dipole magnetic field, with a strength at its pole of $\sim 20\,{\rm kG}$, and a magnetic axis nearly aligned with the rotation axis. Currently, the inclination is poorly constrained, which results in a large possible range for the dipole field strength. However, using the predictions of the Rigidly Rotating Magnetosphere model (Townsend & Owocki 2005), we expect to better constrain the geometry of HD 142184 based on the variations shown in Fig. 1.

References

Oksala, M. E., Wade, G. A., Marcolino, W. L. F., Grunhut, J. *et al.* 2010, *MNRAS*, 405, L51
Rivinius, T., Szeifert, T., Barrera, L., Townsend, R. H. D. *et al.* 2010, *MNRAS*, 405, L46
Townsend, R. H. D. & Owocki, S. P. 2005, *MNRAS*, 357, 251
Townsend, R. H. D., Owocki, S. P., & Groote, D. 2005, *ApJ* (Letters), 630, L81
Townsend, R. H. D. 2008, *MNRAS*, 389, 559

Active OB stars: structure, evolution, mass loss, and critical limits
Proceedings IAU Symposium No. 272, 2010
C. Neiner, G. Wade, G. Meynet & G. Peters, eds.

doi:10.1017/S1743921311010295

The magnetic field of the B1/B2V star σ Lup

Huib F. Henrichs[1], Katrien Kolenberg[2], Benjamin Plaggenborg[1], Stephen C. Marsden[3], Ian A. Waite[4], Gregg A. Wade[5] and the MiMeS collaboration

[1]Astronomical Institute Anton Pannekoek, University of Amsterdam, Science Park 904, 1098XH Amsterdam, Netherlands, email: h.f.henrichs@uva.nl

[2]Institut für Astronomie, Universität Wien, Türkenschanzstrasse 17, A-1180 Vienna, Austria

[3]Anglo-Australian Observatory, PO Box 296, Epping, NSW 1710, Australia

[4]Faculty of Sciences, University of Southern Queensland, Toowoomba, Qld 4350 Australia

[5]Dept. of Physics, Royal Military College of Canada, Kingston, Canada

Abstract. The ultraviolet stellar wind lines of the photometrically periodic variable early B-type star σ Lupi were found to behave very similarly to what has been observed in known magnetic B stars, although no periodicity could be determined. AAT spectropolarimetric measurements with SEMPOL were obtained. We detected a longitudinal magnetic field with varying strength and amplitude of about 100 G with error bars of typically 20 G. This type of variability supports an oblique magnetic rotator model. We fold the equivalent width of the 4 usable UV spectra in phase with the well-known photometric period of 3.019 days, which we identify with the rotation period of the star. The magnetic field variations are consistent with this period. Additional observations with ESPaDOnS attached to the CFHT strongly confirmed this discovery, and allowed to determine a precise magnetic period. Like in the other magnetic B stars the wind emission likely originates in the magnetic equatorial plane, with maximum emission occurring when a magnetic pole points towards the Earth. The 3.0182 d magnetic rotation period is consistent with the photometric period, with maximum light corresponding to maximum magnetic field. No helium or other chemical peculiarity is known for this object.

Keywords. stars: magnetic fields, techniques: polarimetric, stars: atmospheres, stars: individual (σ Lup), stars: early-type, stars: winds, outflows, stars: rotation

---

Introduction and Analysis

In nearly all magnetic OB stars the dipole component is dominant. As the rotation and magnetic axis do not coincide in general, these objects act as oblique rotators. The outflowing stellar wind is perturbed by the surface magnetic field, and is periodically modified. In fact, the discovery of a number of magnetic early-type stars was preceded by the discovery of strictly periodic wind variability as observed in the UV, which appeared to be the strongest indirect indicator for the presence of a magnetic field. By this method three magnetic B stars have been found: β Cep (Henrichs *et al.* 2000), ζ Cas, and V2052 Oph (Neiner *et al.* 2003a, b), with rotation periods of 12 d, 5.4 d and 3.6 d, respectively.

In our search for stellar wind variability in the IUE archives, we found that the B1/B2V star σ Lup had variable UV wind lines (Fig. 1a), similar to other magnetic B stars. This prompted us to observe this star with SEMPOL at the AAT, with follow-up observations in the frame of the MiMeS collaboration (http://www.physics.queensu.ca/~wade/mimes). We describe here the discovery of the magnetic field and its further analysis.

The main stellar parameters of σ Lup are (Levenhagen and Leister 2006): $V = 4.4$, $\log(L/L_{\odot}) = 3.76 \pm 0.06$, $M/M_{\odot} = 9.0 \pm 0.5$, $R/R_{\odot} = 4.8 \pm 0.5$, $T_{\rm eff} = 23000 \pm 500$ K, $v\sin i = 80 \pm 14$ km s^{-1}, and a photometric period $P_{\rm phot} = 3.0186 \pm 0.0004$ d.

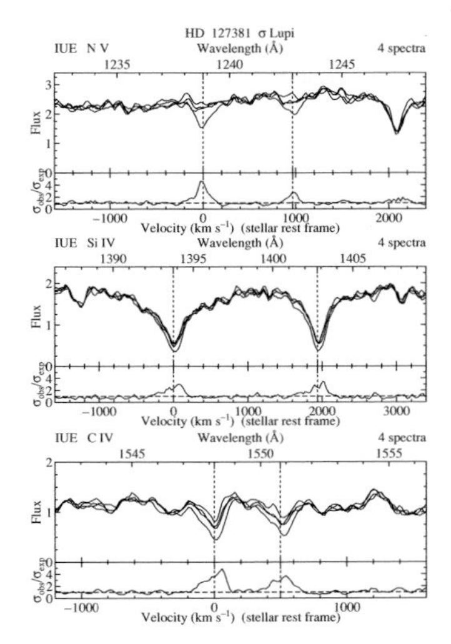

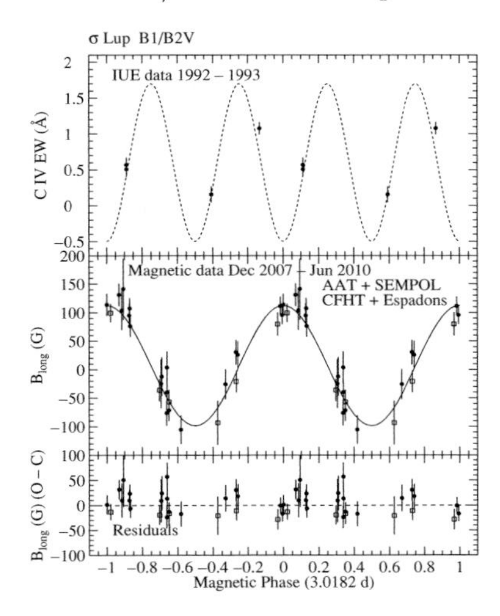

Figure 1. *(a) Left:* Variable UV wind lines of σ Lup. Such variability is only found in magnetic B stars. *(b) Right, middle:* phase plot of SEMPOL (□) + ESPaDOnS (•) magnetic data of σ Lup with the weighted-fit period of 3.0182 d. The best cosine model fit is overplotted. *Lower panel:* Residuals of magnetic data. *Upper panel:* Equivalent width of C IV wind lines. The expected double sine curve (not a fit), as observed in other magnetic B stars is overplotted.

From the LSD spectra we computed the mean longitudinal field (B_l). The smallest error bars are about 16 G. The best fit of the function $B_l(t) = B_0 + B_{\max}\cos(2\pi(t-t_0)/P)$ gives: $B_0 = 7\pm5$ G, $B_{\max} = 106\pm9$ G, $P = 3.01819\pm0.00033$ d, and $t_0 =$ JD 2455103.12 ± 0.56 with a reduced $\chi^2 = 1.0$. This function together with the data as a function of phase are plotted in Fig. 1b (middle). We identify the photometric period with the rotation period. From $v\sin i$ and the estimated stellar radius follows $i > 50°$. The magnetic tilt angle β is then constrained by the observed ratio $B_{\max}/B_{\min} = \cos(i+\beta)/\cos(i-\beta) = -1.14^{+0.27}_{-0.36}$, implying β close to 90°.

The photometric period is 3.0186 ± 0.0004 d (Jerzykiewicz and Sterken 1992), determined 3147 cycles earlier. The extrapolated epoch of maximum light coincides within the uncertainties with the epoch of maximum (positive) magnetic field. The assumption that these epochs are equal allows a more accurate determination of the period: $P = 3.01858\pm0.00014$ d, i.e. within 12 s. New photometry would be able to confirm this.

The three previously discovered magnetic B stars (β Cep, ζ Cas and V2052 Oph) showed a double sine wave in the equivalent width of the UV wind lines, with the maximum emission (minimum EW) coinciding with maximum field strength, i.e. at the phase when a magnetic pole is pointed towards the observer. Because of the particular spacing of the 4 datapoints a fit of a double sine wave is not meaningful, but the expected curve (with arbitrary scaling, see Fig. 1b, top) suggests similar behavior for σ Lup as well. This supports a model with the emitting material in the magnetic equatorial plane.

References

Henrichs, H. F., de Jong, J. A., Donati, J.-F., Catala, C. *et al.* 2000, in: M. A. Smith, H. F. Henrichs, & J. Fabregat (eds.), *IAU Colloq. 175: The Be Phenomenon in Early-Type Stars*, ASP-CS 214, p. 324

Jerzykiewicz, M. & Sterken, C. 1992, *A&A* 261, 477

Levenhagen, R. S. & Leister, N. V. 2006, *MNRAS* 371, 252

Neiner, C., Geers, V. C., Henrichs, H. F., Floquet, M. *et al.* 2003a, *A&A* 406, 1019

Neiner, C., Henrichs, H. F., Floquet, M., Frémat, Y. *et al.* 2003b, *A&A* 411, 565

Active OB stars: structure, evolution, mass loss, and critical limits
Proceedings IAU Symposium No. 272, 2010
C. Neiner, G. Wade, G. Meynet & G. Peters, eds.

doi:10.1017/S1743921311010301

Rigid Field Hydrodynamic simulations of the magnetosphere of σ Orionis E

Nicholas R. Hill[1], Richard H. D. Townsend[1], David H. Cohen[2] and Marc Gagné[3]

[1]Dept. of Astronomy, University of Wisconsin-Madison, Madison, WI 53706, USA
[2]Dept. of Physics and Astronomy, Swarthmore College, Swarthmore, PA 19081, USA
[3]Dept. of Geology and Astronomy, West Chester University, West Chester, PA 19383, USA

Abstract. We present Rigid Field Hydrodynamic simulations of the magnetosphere of σ Ori E. We find that the X-ray emission from the star's magnetically confined wind shocks is very sensitive to the assumed mass-loss rate. To compare the simulations against the measured X-ray emission, we first disentangle the star from its recently discovered late-type companion using *Chandra* HRC-I observations. This then allows us to place an upper limit on the mass-loss rate of the primary, which we find to be significantly smaller than previously imagined.

Keywords. hydrodynamics, stars: magnetic fields, stars: mass loss, X-rays: stars

1. Introduction

The B2Vpe star σ Ori E has long been known to possess a strong ($\sim 10\,\mathrm{kG}$), dipolar magnetic field (e.g., Landstreet & Borra 1978). It is widely believed that the star's strong, relatively hard X-ray emission (e.g., Sanz-Forcada et al. 2004; Skinner *et al.* 2008) arises when wind streams, channeled and confined by the strong field, collide with each other and shock-heat to millions of Kelvin. To test this hypothesis, Townsend *et al.* (2007) formulated a new Rigid Field Hydrodynamic (RFHD) approach for simulating the time-dependent wind flow along field lines, which are assumed rigid in accordance with the star's very large magnetic confinement parameter, $\eta_* \sim 10^7$ (see ud-Doula & Owocki 2002).

2. RFHD Analysis

We have modified the RFHD code described in Townsend et al. (2007) to incorporate energy transport by field-parallel electron thermal conduction. We have also introduced an algorithm that limits the time-step to the smallest characteristic time scale of the differing processes (hydrodynamic and energetic) in the simulation; this is to improve coupling between these processes.

The most notable result from these modifications is an overall cooling of the magnetosphere, relative to simulations based on previous versions of the RFHD code. This is due to thermal conduction, which transfers heat from the hot, low-density post-shock regions to the cool, high-density equatorial accumulation disk, where it can be radiated away efficiently. As a consequence, typical magnetospheric temperatures do not reach the levels reported by Townsend et al. (2007).

A further significant finding, illustrated in Fig. 1, is that the X-ray differential emission measure (DEM) is very sensitive to changes in the mass-loss rate, as parametrized via the $\bar{Q}$ introduced by Gayley (1995) to characterize the overall opacity available for line driving in the Castor, Abbott & Klein (1975) wind formalism.

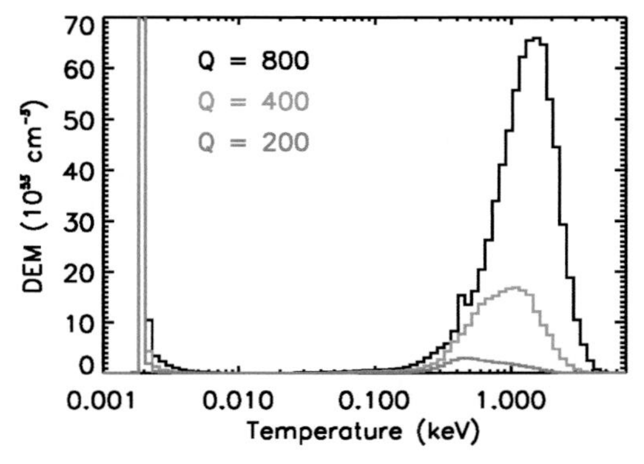

Figure 1. The time-averaged DEM from three 20 Msec RFHD simulations of σ Ori E, each differing only in the choice of opacity parameter $\bar{Q}$. Increasing $\bar{Q}$ (bottom curve to top curve) leads to disproportionately stronger X-ray emission, and also hardens the spectrum.

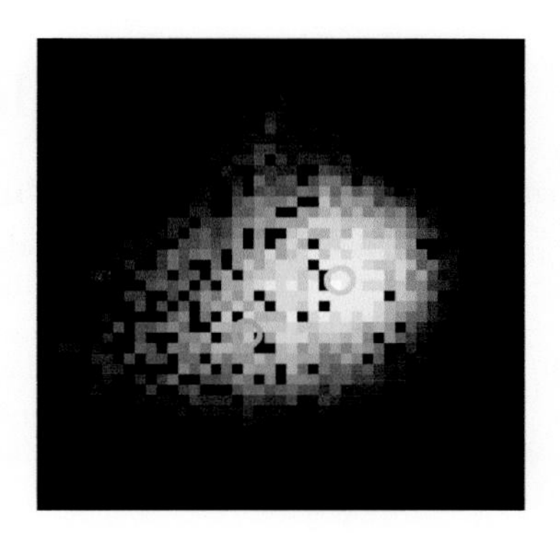

Figure 2. Lucy-Richardson deconvolution of the *Chandra* HRC-I image of σ Ori E. The lower-left circle marks the position of the primary, and the upper-right circle the position of the companion as measured by Bouy *et al.* (2009).

3. Observational Comparison

When comparing our simulated DEMs against X-ray observations of σ Ori E, we must contend with the possibility of emission from the late-type companion claimed by Bouy et al. (2009). Using a Lucy-Richardson deconvolution of the *Chandra* HRC-I observations of the star (PI: S. Wolk), we find that approximately two-thirds of the observed X-rays (during quiescence) come from a source at the same offset and position angle as the proposed companion (see Fig. 2). This both confirms the companion's existence, and indicates that only one-third of the observed X-rays originate from the primary.

Complementary *Chandra* ACIS-I observations (Skinner et al. 2008) indicate an overall X-ray emission measure ($0.2 - 3\,\mathrm{keV}$) of $\sim 2 \times 10^{53}\,\mathrm{cm}^{-3}$, so the emission measure of the primary should be on the order $7 \times 10^{52}\,\mathrm{cm}^{-3}$. This is approximately half that predicted by our $\bar{Q} = 200$ simulation, indicating that the mass loss rate of the primary must be *less* than the $\sim 2.4 \times 10^{-11}\,M_{\odot}\,\mathrm{yr}^{-1}$ derived from this simulation. Standard CAK theory predicts significantly higher mass-loss rates for a B2V star, on the order of $10^{-9}\,M_{\odot}\,\mathrm{yr}^{-1}$; thus, we conclude that there is something unusual about the wind of the σ Ori E primary.

Acknowledgements

We acknowledge support from NASA grant *LTSA*/NNG05GC36G.

References

Bouy, H., Huélamo, N., Martín, E. L., Marchis, F. *et al.* 2009, *A&A*, 493, 931
Castor, J. I., Abbott, D. C., & Klein, R. I. 1975, *ApJ*, 195, 157
Gayley, K. G. 1995, *ApJ*, 454, 410
Landstreet, J. D. & Borra, E. F. 1978, *ApJ* (Letters), 224, L5
Sanz-Forcada, J., Franciosini, E., & Pallavicini, R. 2004, *A&A*, 421, 715
Skinner, S. L., Sokal, K. R., Cohen, D. H., Gagné, M. *et al.* 2008, *ApJ*, 683, 796
Townsend, R. H. D., Owocki, S. P., & Ud-Doula, A. 2007, *MNRAS*, 382, 139
ud-Doula, A. & Owocki, S. P. 2002, *ApJ*, 576, 413

Active OB stars: structure, evolution, mass loss, and critical limits
Proceedings IAU Symposium No. 272, 2010
C. Neiner, G. Wade, G. Meynet & G. Peters, eds.

doi:10.1017/S1743921311010313

Magnetic fields of massive stars

Swetlana Hubrig[1], Michel Curé[2], Ilya Ilyin[1], and Markus Schöller[3]

[1]Astrophysikalisches Institut Potsdam, An der Sternwarte 16, Potsdam, Germany
email: shubrig@aip.de

[2]Departamento de Fisica y Astronomía, Facultad de Ciencias,
Universidad de Valparaíso, Chile

[3]European Southern Observatory, Karl-Schwarzschild-Str. 2,
85748 Garching bei München, Germany

Abstract. We recently carried out a spectropolarimetric study of a sample of massive O-type stars and pulsating β Cephei stars using the SOFIN echelle spectrograph at the 2.56 m Nordic Optical Telescope and the low-resolution FORS 2 spectrograph at the VLT in spectropolarimetric mode. The sample consists of massive stars already detected as magnetic in the course of our previous low-resolution polarimetric observations with FORS 1 and a few O-type stars with magnetic field detections reported in the literature.

Keywords. stars: magnetic fields, techniques: polarimetric, stars: early-type, stars: rotation, stars: oscillations, stars: kinematics, stars: individual (HD 108, HD 191612, V1449 Aql, ξ^1 CMa)

---

1. Observations and analysis

To date, only a small number of O, early B-type, and Wolf-Rayet stars have been investigated for magnetic fields, and as a result, only about two dozen magnetic massive early type stars are known. On the other hand, a lot of effort has been put into the research of massive stars in recent years in order to properly model the effects of rotation, stellar winds, surface chemical composition, and evolution. Polarimetric spectra of the massive O-type stars HD 108, HD 36879, 15 Mon, 9 Sgr, and HD 191612 were obtained with the SOFIN spectrograph, which is equipped with three optical cameras and mounted at the Cassegrain focus of the NOT. The new magnetic field measurements confirm the presence of longitudinal magnetic fields in these stars (e.g. Hubrig *et al.* 2008, 2009a). The presence of a longitudinal magnetic field in the Of?p star HD 108, $\langle B_z \rangle = -150 \pm 10$ G, was recently reported by Martins *et al.* (2010), using NARVAL and ESPaDOnS observations. For this star, Nazé *et al.* (2006) suggested a period between 50 and 60 years. Applying the moment technique, we obtain a mean longitudinal magnetic field $\langle B_z \rangle = -168 \pm 35$ G. A mean longitudinal magnetic field $\langle B_z \rangle = -220 \pm 38$ G in the Of?p star HD 191612 was measured by Donati *et al.* (2006), combining spectropolarimetric observations from four nights. The rotation period of 537.6 d was determined by Howarth *et al.* (2007). Using ephemeris from this work, the magnetic field measurements of Donati *et al.* (2006) have been carried out at a rotation phase of $\sim$0.24. Our measurements using eight spectral lines result in $\langle B_z \rangle = 450 \pm 153$ G at rotation phase 0.43. The difference in phase between the measurement of Donati *et al.* (2006) and our measurement is about 0.19. Thus, we observe a change of polarity over $\sim$100 days.

Over several years, we undertook a magnetic field survey for main-sequence pulsating B-type stars, namely the slowly pulsating B (SPB) stars and β Cephei stars, with FORS 1/2 in spectropolarimetric mode at the VLT, allowing us to detect in four β Cephei stars and 16 SPB stars, for the first time, longitudinal magnetic fields of the order of a few hundred Gauss. The SOFIN measurements of ξ^1 CMa, separated by one year,

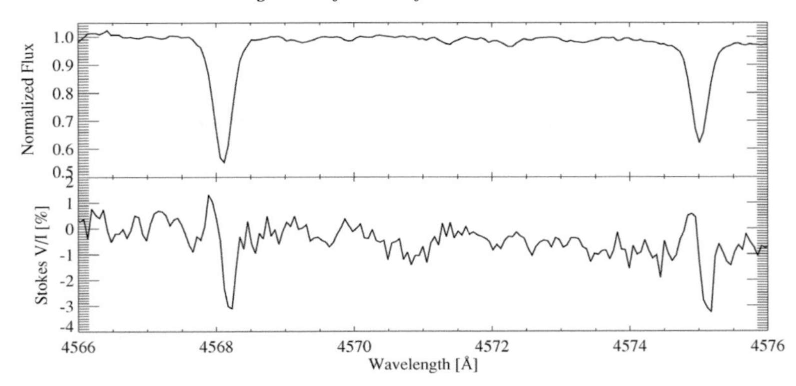

Figure 1. SOFIN I and V spectra of ξ^1 CMa in the spectral region around the Si III (Mult. 2) lines.

$\langle B_z \rangle = 386 \pm 139$ G and $\langle B_z \rangle = 297 \pm 126$ G, confirmed our previous detection of a slightly variable magnetic field in this star (Hubrig *et al.* 2006). In Fig. 1 we present SOFIN I and V spectra of this star in the spectral region around the Si III (Mult. 2) lines. Another β Cephei star, V1449 Aql, previously reported by Hubrig *et al.* (2009b) as magnetic, was found to show the strongest mean longitudinal magnetic field among the magnetic β Cephei stars, of the order of -800 G. The detection of such a strong magnetic field in this massive star is of special interest due to the recently discovered solar-like pulsations using CoRoT observations (Belkacem *et al.* 2009).

2. Summary

Although it was possible to recognize a few hot magnetic stars as being peculiar on the basis of their spectral morphology prior to their field detection (Walborn 2006), the presence of a magnetic field can also be expected in stars of other classification categories. Our measurements of several massive stars indicate that magnetic fields are possibly present in stars with very different observed properties in visual, X-ray, and radio domains. Future magnetic field measurements of massive stars in field and cluster stars will constrain the conditions controlling the presence of magnetic fields, and the implications of these fields on their mass-loss rate and evolution.

References

Belkacem, K., Samadi, R., Goupil, M.-J., Lefèvre, L. *et al.* 2009, *Science* 324, 1540

Donati, J.-F., Howarth, I. D., Bouret, J.-C., Petit, P. *et al.* 2006, *MNRAS* 365, L6

Howarth, I. D., Walborn, N. R., Lennon, D. J., Puls, J. *et al.* 2007, *MNRAS* 381, 433

Hubrig, S., Briquet, M., Schöller, M., De Cat, P. *et al.* 2006, *MNRAS* 369, L61

Hubrig, S., Schöller, M., Schnerr, R. S., González, J. F. *et al.* 2008, *A&A* 490, 793

Hubrig, S., Stelzer, B., Schöller, M., Grady, C. *et al.* 2009a, *A&A* 502, 283

Hubrig, S., Briquet, M., De Cat, P., Schöller, M. *et al.* 2009b, *AN* 330, 317

Martins, F., Donati, J.-F., Marcolino, W. L. F., Bouret, J.-C. *et al.* 2010, *MNRAS* 407, 1423

Nazé, Y., Barbieri, C., Segafredo, A., Rauw, G. *et al.* 2006, *IBVS* 5693

Walborn, N. R. 2006, in: *The Ultraviolet Universe: Stars from Birth to Death*, Proc. IAU Joint Discussion 4, #19, Prague, Czech Republic

Active OB stars: structure, evolution, mass loss, and critical limits
Proceedings IAU Symposium No. 272, 2010
C. Neiner, G. Wade, G. Meynet & G. Peters, eds.

doi:10.1017/S1743921311010325

Magnetic fluxes of massive stars: statistics and evolution

Alexander F. Kholtygin[1], **Sergei N. Fabrika**[2], **Natalia A. Drake**[1] **and Andrei P. Igoshev**[1]

[1]Astronomical Instituut, Saint-Petersburg University,
198504, ul Shaxmatova, Petrodvoretz, Rissia
email: afkholtygin@gmail.com

[2]Special Astrophysical Observatoty, Nizhnii Arxyz, Russia

Abstract. The statistical properties of magnetic fields and magnetic fluxes of OB stars were investigated. The mean magnetic fluxes of massive OB stars appear to be 3 order larger than those for neutron stars.

Keywords. stars: early-type, stars: magnetic fields, stars: neutron

---

1. Magnetic Fields

We collect the recent measurements of the magnetic fields of OB stars from the catalogue of Bychkov *et al.* (2009) and the newest data from the literature. Basing on these data we have investigated the statistical properties of an ensemble of magnetic fields of OB stars. As a statistical measure of the stellar magnetic field we used the *rms* longitudinal magnetic field $\mathcal{B}$. The mean magnetic fields for OB stars averaged over the spectral subclasses are plotted in Fig. 1 (left panel). There is an unexpectedly large jump between O and B star mean fields.

We have calculated the normalized differential magnetic field function (MFF) $F(\mathcal{B})$ for OB stars, which is determined accordingly Kholtygin *et al.* (2010a) in such a way that a value of $F(\mathcal{B})d\mathcal{B}$ indicates the probability that a *rms* field $\mathcal{B}$ is in an interval $(\mathcal{B}, \mathcal{B}+d\mathcal{B})$ (right panel in Fig. 1). We found that the MFF for $\mathcal{B} > 400\,\mathrm{G}$ can be approximated by a power function: $F(B) = 0.33(B/1kG)^{-1.82}$. For smaller values of $\mathcal{B}$ the values of the MFF are lower than those obtained from the power fit by more than an order of magnitude. We suppose that this deviation can be connected with observational bias.

To study this effect we suppose that the real MFF is described by the power law for all values of $\mathcal{B} < M_{\min} = 20\,\mathrm{G}$. We calculated with our Monte-Carlo code the field detection probability $P(n, \sigma, \mathcal{B})$ for a star with the *rms* field $\mathcal{B}$, a fixed error σ of field measurement and n polarimetric observations (for details see Kholtygin *et al.* 2010a). It appears that a value of $P(n, \sigma, \mathcal{B})$ weakly depends on the n for $n \geqslant 3$. We use this probability to restore the quasi-observed MFF which could be obtained in a case of observation of all stars using the polarimeter with a fixed value of σ. The restored values were calculated by Kholtygin *et al.* (2010b). It appeared that the observed MFF cannot be made to agree with the restored one for any possible values of σ. We connect this fact with an additional factor of decreasing MFF, probably with destroying the weak fields by field instabilities, as proposed by Aurière *et al.* (2007).

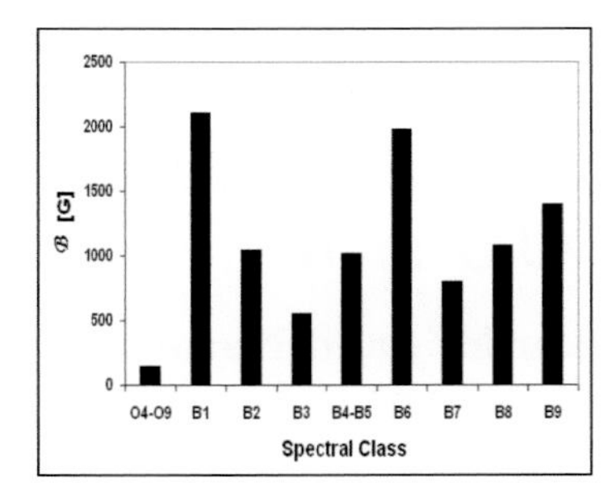

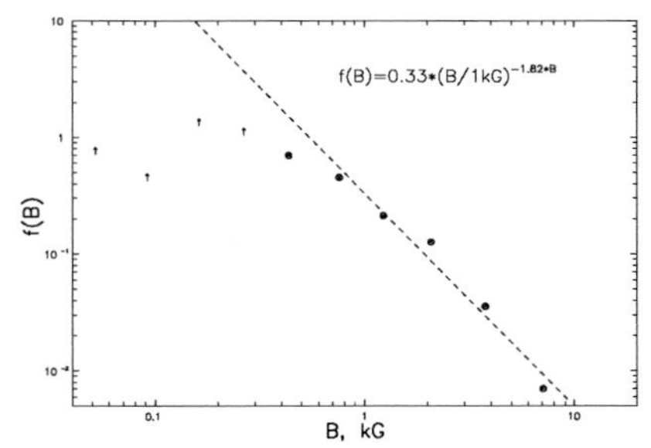

Figure 1. Left Panel: Mean Magnetic Fields of OB stars for different spectral subclasses. **Right Panel**: MFF for OB stars obtained on the measured magnetic fields for $\mathcal{B} \geqslant 400\,\mathrm{G}$ (points) and for $\mathcal{B} < 400\,\mathrm{G}$ (arrows). Power fit for MFF is shown with a dashed line.

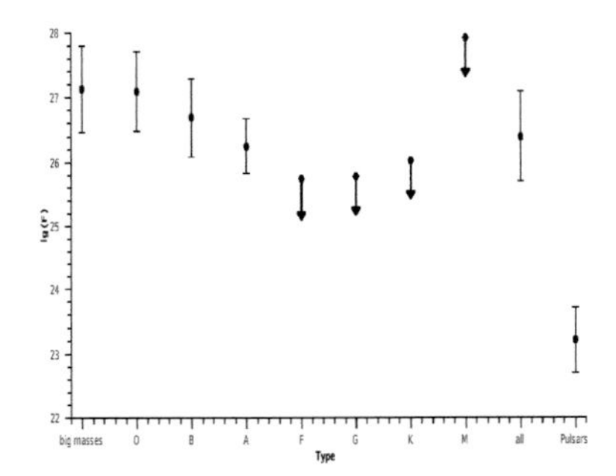

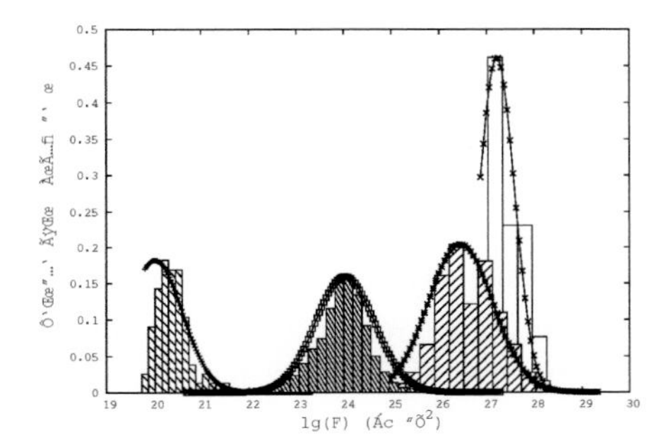

Figure 2. Left Panel: Mean magnetic fluxes for normal stars, massive stars and pulsars vs. their spectral class. **Right Panel**: Distributions of the magnetic fluxes for millesecond pulsars, normal pulsars, OB stars and magnetars (from left to right) and their log.-normal fit.

2. Magnetic Fluxes

For all studied stars we have estimated the magnetic fluxes using a relation $\mathcal{F} = 4\pi R_*^2 \mathcal{B}$, where R_* is a photospheric radius of the star. We established that the magnetic fluxes for stars of all spectral classes have a log-normal distribution. Mean magnetic fluxes as a function of spectral class are given in Fig. 2 (left panel). For stars of F and later spectral classes only upper limits of magnetic fluxes can be found (arrows). We also calculated the magnetic fluxes for neutron stars (NSs) using their magnetic fields from ATNF Pulsar Catalogue (http://www.atnf.csiro.au/research/pulsar/psrcat) and McGill SGR/AXP Online Catalog (http://www.physics.mcgill.ca/~pulsar/magnetar/main.html) and accepting for all NS the typical radius $R_* = 10\,\mathrm{km}$. Mean magnetic fluxes for most NSs appear to be up to 3 orders of magnitude lower than those for massive OB stars. This means that massive stars lose the lion's share of their magnetic flux. The magnetic flux distributions for NSs and OB stars are plotted in Fig. 2 (right panel). The distribution of magnetic fluxes for magnetars is narrow and it is shifted by 0.4 dex relative to that of OB stars.

References

Bychkov, V. D., Bychkova, L. V., & Madej, J. 2009, *MNRAS*, 394, 1338

Aurière, M., Wade, G. A., Silvester, J., Lignières, F. *et al.* 2007, *A&A*, 475, 1053

Kholtygin, A. F., Fabrika, S. N., Drake, N. A., Bychkov, V. D. *et al.* 2010a, *Astron. Lett.*, 36, 370

Kholtygin, A. F., Fabrika, S. N., Drake, N. A., Bychkov, V. D. *et al.* 2010b, *Kinematics and Physics of Celestial Bodies*, 26, 181

Active OB stars: structure, evolution, mass loss, and critical limits
Proceedings IAU Symposium No. 272, 2010
C. Neiner, G. Wade, G. Meynet & G. Peters, eds.

doi:10.1017/S1743921311010337

Line Profile microvariability and wind structure for OB stars

Alexander F. Kholtygin[1], Sergei N. Fabrika[2], and Natalia P. Sudnik[1]

[1]Astronomical Institute, Saint-Petersburg University, 198504, Petrodvoretz, Russia
email: afkholtygin@gmail.com

[2]Special Astrophysical Observatory, Nizhnii Arxyz, Russia

Abstract. We report the results of a search for line profile variability (LPV) in the spectra of OB stars. The wavelets were used for looking for the irregular LPV in spectra of program stars. We connect the appearance of irregular details in the LPV with the formation and dissipation of the small-scale substructures (clumps) in the wind.

Keywords. stars: early-type, stars: variables: other, line: profiles, stars: winds, outflows

---

1. Line profile variability

Our program of searching for small amplitude microvariations of the line profiles (microLPV) in spectra of OB stars had started in 2001 (Kholtygin *et al.* 2003). The observations were made with the 1.8 m telescope of the Korean Bohyunsan Optical Astronomical Observatory and 1-m and 6-m telescopes of the Special Astrophysical Observatory, Russia. More than 1000 spectra of 12 OB stars were obtained. Both regular and stochastic microLPV in spectra of all program stars are found.

For the detection of microLPV of various nature we use the smooth Time Variation Spectra (smTVS, see Kholtygin *et al.* 2006 for details):

$$smTVS(\lambda, S) = \left[\frac{1}{N-1}\left(\sum_{i=1}^{n}[F_i(\lambda, S) - \overline{F(\lambda, S)}]^2\right)\right]^{1/2}, \qquad (1.1)$$

where $F_i(\lambda, S)$ is the flux in a line at wavelength λ, smoothed with a Gaussian filter with width S, i the number of the spectrum, N the total number of spectra, and $\overline{F(\lambda, S)}$ the mean of all smoothed line fluxes. For $S = 0$ the $smTVS(\lambda, 0)$ spectra corresponds to the *Time Variation Spectra* introduced by Fullerton *et al.* (1996). The $smTVS$ spectrum for the HeII 4200 line in a spectrum of λ Cep is plotted in Fig. 1 (left panel). The microLPV for these lines is seen at all widths. For values of $S > 1$Å we can see a very weak LPV of the line CIII 4187, which cannot be detected by other methods. In the Fourier spectrum of LPV in spectra of λ Cep there is a wide peak at $\nu = 3.46 \pm 0.3d^{-1}$. This can be explained by non-radial photospheric pulsations with a period $P = 6.9 \pm 0.6h$.

2. Wind structure

As a tool to detect the microdetails in a line profile connected with small clumps in the wind we use the *dynamical wavelet spectra* of residuals of the line profile (see Kholtygin *et al.* 2006 for details):

$$\mathcal{W}_t(s, V, t) = \frac{1}{\sqrt{S}}\int_{-\infty}^{\infty} f(V, t)\psi\left(\frac{x - V}{S}\right)dx\,, \qquad (2.1)$$

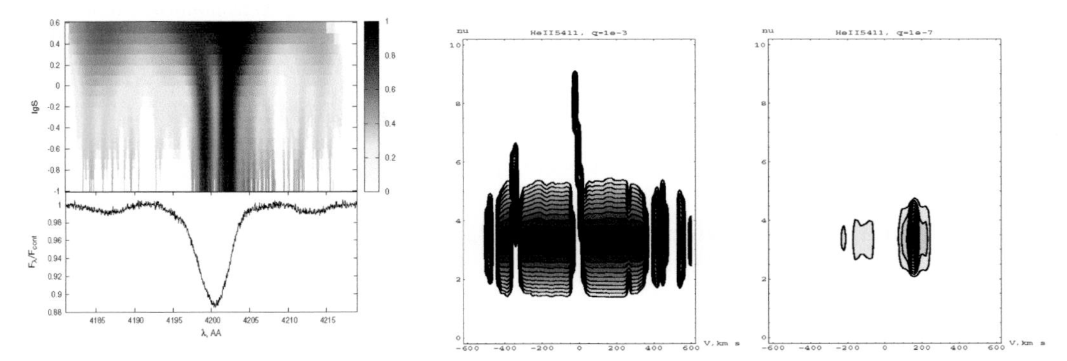

Figure 1. Left Panel: *smTVS* spectra for HeII λ 4200 line in spectra of λ Cep (top) and mean line profile (bottom). **Right Panel**: Fourier-spectra of the line HeII λ 5411 LPV in spectra of λ Cep for FAP level $q = 10^{-3}$ (left) and $q = 10^{-7}$ (right).

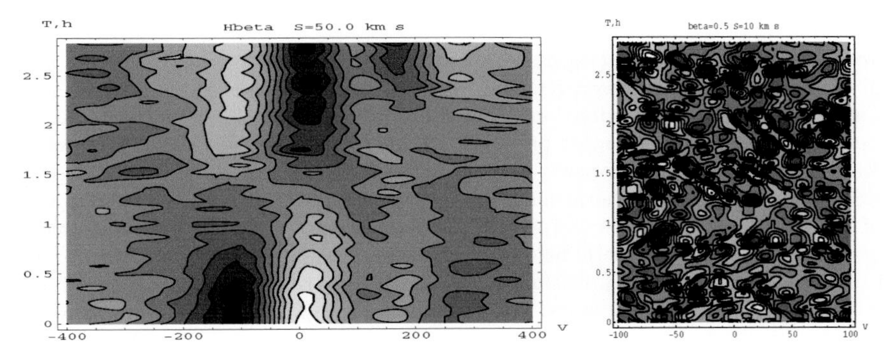

Figure 2. Left Panel: *Dynamical wavelet spectra* for line H_β in spectra of δ Ori A for the velocity interval -400 km/s $\leqslant V \leqslant 400$ km/s and a scale $S = 50$ km/s. The time is counted in hours from beginning of observations. **Right Panel**: The same as in the Left Panel, but for $S = 10$ km/s and -100 km/s $\leqslant V \leqslant 100$ km/s.

where $f(V,t)$ is an analysed function. In our case $f(V,t) = F(V,t) - \overline{F(V,t)}$. Here $F(V,t)$ is an individual line profile measured at the moment t, $\overline{F(V,t)}$ is the mean profile averaged over all observations. For the sake of convenience we use the Doppler shift V from the line centre instead of wavelength λ. In this case the scale S of the wavelet transform is expressed in km/s. As a mother wavelet we exploit the MHAT-wavelet $\psi(x) = (1 - x^2)\exp(-x^2/2)$.

In Fig. 2 we present the *dynamical wavelet spectra* of LPV in the line H_β in spectra of δ Ori A for scales $S = 50$ km/s (left panel) and $S = 10$ km/s (right panel). For the larger scale $S = 50$ km/s we see in the *dynamical wavelet spectrum* mainly regular LPV, connected with non-radial pulsations with a period $P \approx 4^{\rm h}$ (see Kholtygin *et al.* 2006). At the same time, the numerous irregular details in the dynamical wavelet spectra for $S = 10$ km/s can be connected with small clumps in the wind of the main component Aa^1 of the triple system δ Ori A, having the velocity dispersion $\sigma_V = 10 - 20$ km/s.

References

Kholtygin, A. F., Monin, D. N., Surkov, A. E., & Fabrika, S. N. 2003, *Astron. Lett.*, 29, 175

Kholtygin, A. F., Burlakova, T. E., Fabrika, S. N., Valyavin, G. G. *et al.* 2006, *Astron. Rep.*, 50, 887

Fullerton, A. W., Gies, D. R., & Bolton, C. T. 1996, *ApJS*, 103, 475

Active OB stars: structure, evolution, mass loss, and critical limits
Proceedings IAU Symposium No. 272, 2010
C. Neiner, G. Wade, G. Meynet & G. Peters, eds.

doi:10.1017/S1743921311010349

The search for magnetic fields in mercury-manganese stars

Vitalii Makaganiuk[1], Oleg Kochukhov[1], Nikolai Piskunov[1], Sandra V. Jeffers[2], Christopher M. Johns-Krull[4], Christoph U. Keller[2], Michiel Rodenhuis[2], Frans Snik[2], Henricus C. Stempels[1] and Jeff A. Valenti[3]

[1]Department Physics and Astronomy, Uppsala University, Uppsala, Sweden
[2]Sterrekundig Instituut, Universiteit Utrecht, Utrecht, The Netherlands
[3]Space Telescope Science Institute, San Martin Dr, Baltimore, MD, USA
[4]Department of Physics and Astronomy, Rice University, Houston, TX, USA

Abstract. Mercury-manganese (HgMn) stars were considered to be non-magnetic, showing no evidence of surface spots. However, recent investigations revealed that some stars in this class possess an inhomogeneous distribution of chemical elements on their surfaces. According to our current understanding, the most probable mechanism of spot formation involves magnetic fields. Taking the advantage of a newly-built polarimeter attached to the HARPS spectrometer at the ESO 3.6m-telescope, we performed a high-precision spectropolarimetric survey of a large group of HgMn stars. The main purpose of this study was to find out how typical it is for HgMn stars to have weak magnetic fields. We report no magnetic field detection for any of the studied objects, with a typical precision of the longitudinal field measurements of 10 G and down to 1 Gauss for some of the stars. We conclude that HgMn stars lack large-scale magnetic fields typical of spotted magnetic Ap stars and probably lack any fields capable of creating and sustaining chemical spots. Our study confirms that alongside the magnetically altered atomic diffusion, there must be other structure formation mechanism operating in the atmospheres of late-B main sequence stars.

Keywords. instrumentation: polarimeters, stars: magnetic fields, stars: chemically peculiar

---

1. Introduction

Mercury-manganese (HgMn) stars form a subclass of the upper main sequence chemically peculiar (CP) stars, showing notable overabundance of Hg, Mn, Y, Sr and other, mostly heavy, chemical elements with respect to the solar chemical composition. HgMn stars are frequently found in binaries and lie on the H-R diagram between the early-A and late-B spectral types, which corresponds to $T_{\rm eff} = 9500 - 16000$ K. A number of studies during the last decade reported variability of some spectral lines, indicating the presence of chemical spots in HgMn stars (Adelman *et al.* 2002, Kochukhov *et al.* 2005, Hubrig *et al.* 2006). It is believed that the presence of magnetic field is necessary for creating inhomogeneities in stellar atmospheres. In order to clarify the magnetic status of HgMn stars we conducted a major spectropolarimetric survey of these stars. We investigated 47 HgMn stars with $T_{\rm eff} = 10500 - 14500$ K and $v_{\rm e} \sin i = 70$ km s^{-1}. Most of these objects were never studied before with high-resolution spectropolarimetry. Taking advantage of a new polarimeter HARPSpol, attached to the HARPS instrument at the ESO 3.6-m telescope in La Silla, Chile, we were able to push the limits of magnetic field detection in early-type stars down to levels.

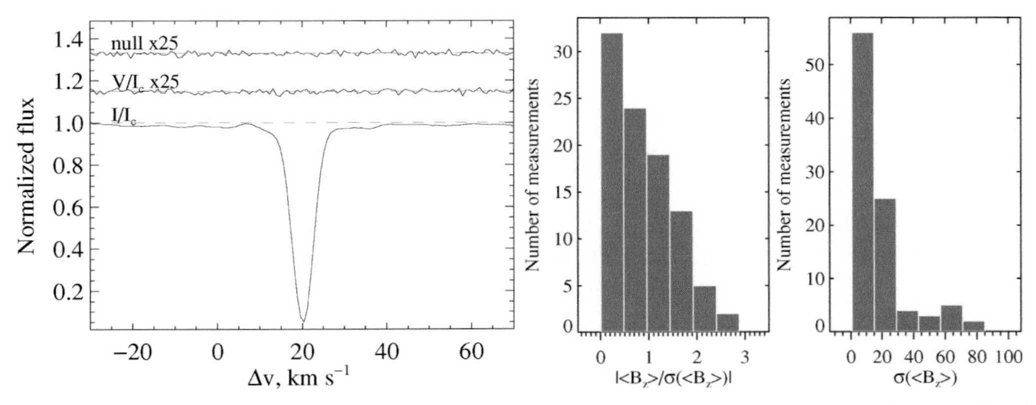

Figure 1. LSD profiles for HD 71066.

Figure 2. Left: the distribution of longitudinal field normalized by its error bar. Right: the distribution of error bars of $\langle B_z \rangle$ in Gauss.

2. Magnetic field search

It is impossible to draw any conclusion about the presence of weak polarization signatures in individual spectral lines because of noise. To circumvent this problem, we use a multi-line addition technique called Least-Squares Deconvolution (LSD, Donati *et al.* 1997). With the help of this method we were able to decrease noise by a factor of 7-16. The LSD profiles obtained for HD 71066 are shown in Fig. 1. For the assessment of magnetic field detection from Stokes V line profiles we employed χ^2 probability statistics. The analysis of LSD profiles of all HgMn stars yielded no magnetic field detection.

We did not detect any longitudinal magnetic field $\langle B_z \rangle$. For many stars the precision of our $\langle B_z \rangle$ measurements is better than 10 G. The best precision, ± 0.81 G, was achieved for HD 71066. The overall distribution of measurements and error bars is presented in Fig. 2.

3. Conclusions

We confirm that HgMn stars are non-magnetic. Based on the results of our study we can conclude that spot formation in HgMn stars cannot be caused by the global magnetic field. We can also rule out a complex configuration of the magnetic field as it would reveal itself in Stokes V profiles. Hereby, this class of CP stars remains a puzzling example of the chemical spots formation in massive stars, which is possibly related to hydrodynamic instabilities in chemically stratified stellar atmospheres (Kochukhov *et al.* 2007).

References

Adelman, S. J., Gulliver, A. F., Kochukhov, O. P., & Ryabchikova, T. A. 2002, *ApJ*, 575, 449
Kochukhov, O., Piskunov, N., Sachkov, M., & Kudryavtsev, D. 2005, *A&A*, 439, 1093
Hubrig, S., González, J. F., Savanov, I., Schöller, M. *et al.* 2006, *MNRAS*, 371, 1953
Donati, J.-F., Semel, M., Carter, B. D., Rees, D. E. *et al.* 1997, *MNRAS*, 291, 658
Kochukhov, O., Adelman, S. J., Gulliver, A. F., & Piskunov, N. 2007, *Nature Physics*, 3, 526

Active OB stars: structure, evolution, mass loss, and critical limits
Proceedings IAU Symposium No. 272, 2010
C. Neiner, G. Wade, G. Meynet & G. Peters, eds.

doi:10.1017/S1743921311010350

Discovery of a strong magnetic field in the rapidly rotating B2Vn star HR 7355

Mary E. Oksala[1,2], Gregg A. Wade[2], Wagner L. F. Marcolino[3,4], Jason H. Grunhut[2,5], David Bohlender[6], Nadine Manset[7], Richard H. D. Townsend[8], and the MiMeS Collaboration

[1]Department of Physics and Astronomy, University of Delaware, Newark, DE, USA
[2]Department of Physics, Royal Military College of Canada, Kingston, Ontario, Canada
[3]LAM-UMR 6110, CNRS & Univ. de Provence, Marseille , France
[4]Observatòrio Nacional, Rio de Janeiro, Brazil
[5]Department of Physics, Queen's University, Kingston, Ontario, Canada
[6]National Research Council of Canada, Herzberg Institute of Astrophysics,Victoria, Canada
[7]Canada-France-Hawaii Telescope Corporation, Kamuela, HI, USA
[8]Department of Astronomy, University of Wisconsin-Madison, Madison, WI, USA

Abstract. We report on the detection of a strong, organized magnetic field in the helium-variable early B-type star HR 7355 using spectropolarimetric data obtained with ESPaDOnS on CFHT by the MiMeS large program. We also present results from new V-band differential photometry obtained with the CTIO 0.9m telescope. We investigate the longitudinal field, using a technique called Least-Squares Deconvolution (LSD), and the rotational period of HR 7355. These new observations strongly support the proposal that HR 7355 harbors a structured magnetosphere similar to that in the prototypical helium-strong star, σ Ori E.

Keywords. stars: magnetic fields, stars: rotation, stars: early-type, stars: circumstellar matter, stars: individual (HR 7355), techniques: polarimetric

---

1. Introduction

HR 7355 (HD 182180) is a bright B2Vn helium-strong star originally classified as a Be star due to Hα emission present in its spectrum (Abt & Cardona 1984). Previous studies of this star show a $v \sin i \sim 300$ km s^{-1} (Abt *et al.* 2002) with a $P_{rot} \sim 0.52$ d (Koen & Eyer 2002), as well as variation in helium, Hα, and brightness, suggesting the presence of a magnetosphere (Rivinius *et al.* 2008). HR 7355 is the most rapidly rotating helium-strong star, rotating near its critical velocity, providing an excellent testbed for magnetospheres under the effects of rapid rotation.

2. Method

Least-Squares Deconvolution (LSD) describes the stellar spectrum as the convolution of a mean Stokes I or V profile, representative of the average shape of the line profile, and a line mask, describing the position, strength and magnetic sensitivity of all lines in the spectrum. From the LSD mean Stokes I and V profiles, we calculate the longitudinal magnetic field, B_ℓ:

$$B_\ell = -2.14 \times 10^{11} \frac{\int v V(v) dv}{\lambda g c \int [1 - I(v)] dv} \tag{2.1}$$

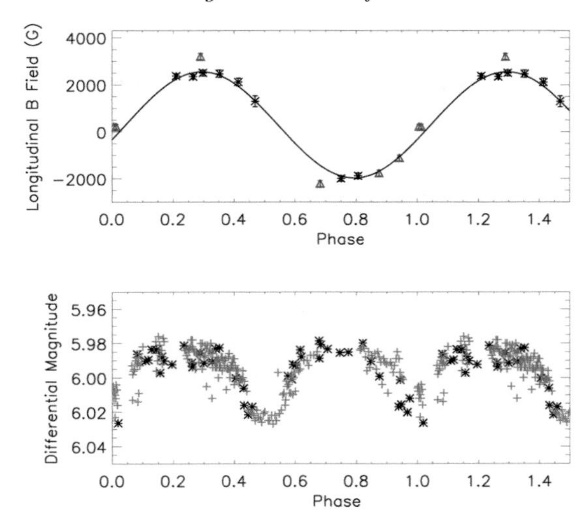

Figure 1. Top: Longitudinal magnetic field measurements for HR 7355 and the best-fit first order sine curve. Oksala *et al.* (2010) (asterisks) and Rivinius *et al.* (2010) (diamonds) with 1 σ error bars. Bottom: The V-band photometric light curve for HR 7355 including both HIPPARCOS photometry (asterisks) and new CTIO data (plus signs).

(Wade *et al.* 2000), where λ is the average wavelength and g is the average Landé factor in the mask. I_c is the continuum value of the intensity profile. The integral is evaluated over the full velocity range of the mean profile.

3. Results

We detect a strong magnetic field on HR 7355, the most rapidly rotating helium-strong star discovered thus far. A simultaneous independent confirmation of the field detection has been obtained with FORS at the VLT by Rivinius *et al.* 2010. The longitudinal magnetic field varies sinusoidally with the rotation period, with extrema -2 to 2.5 kG. Assuming a dipole magnetic field, the polar value of the magnetic field is $\sim$ 13-17 kG. The photometric (brightness) light curve constructed from HIPPARCOS archival data and new CTIO measurements shows two minima separated by 0.5 in rotational phase and occurring 0.25 cycles before/after the magnetic extrema. Using the Scargle periodogram, eclipse-like photometric variations give a highly precise $P_{rot} = 0.5214404(6)$ days. We confirm spectral variability of helium and metal lines, as well as variability of Hα emission. Hα emission indicates circumstellar material extending out to 5 $R_\star$ from the star, rotating rigidly with the stellar surface. We conclude that HR 7355 is a magnetic oblique rotator with a magnetosphere, mirroring the physical picture for σ Ori E (Townsend *et al.* 2005).

References

Abt, H. A. & Cardona, O. 1984, *ApJ*, 285, 190
Abt, H. A., Levato, H., & Grosso, M. 2002, *ApJ*, 573, 359
Koen, C. & Eyer, L. 2002, *MNRAS*, 331, 45
Oksala, M. E., Wade, G. A., Marcolino, W. L. F., Grunhut, J. *et al.* 2010, *MNRAS*, 405, 51
Rivinius, T., Štefl, S., Townsend, R. H. D., & Baade, D. 2008, *A&A*, 482, 255
Rivinius, T., Szeifert, T., Barrera, L., Townsend, R. H. D. *et al.* 2010, *MNRAS*, 405, 46
Townsend, R. H. D., Owocki, S. P., & Groote, D. 2005, *ApJ* (Letters), 630, 81
Wade, G. A., Donati, J.-F., Landstreet, J. D., & Shorlin, S. L. S. 2000, *MNRAS*, 313, 851

Active OB stars: structure, evolution, mass loss, and critical limits
Proceedings IAU Symposium No. 272, 2010
C. Neiner, G. Wade, G. Meynet & G. Peters, eds.

doi:10.1017/S1743921311010362

τ Sco: the discovery of the clones

Véronique Petit[1], Derck L. Massa[2], Wagner L. F. Marcolino[3], Gregg A. Wade[4], Richard Ignace[5] and the MiMeS Collaboration

[1]Dept. of Geology & Astronomy, West Chester University, West Chester, PA 19383, USA
email: VPetit@wcupa.edu

[2]Space Telescope Science Institute, 3700 N. San Martin Drive, Baltimore, MD 21218, USA

[3]Observatório Nacional-MCT, CEP 20921-400, São Cristóvão, Rio de Janeiro, Brasil

[4]Dept. of Physics, Royal Military College of Canada, Kingston, Canada, K7K 4B4

[5]Dept. of Physics & Astronomy, East Tennessee State University, Johnson City, TN 37614, USA

Abstract. The B0.2 V magnetic star τ Sco stands out from the larger population of massive magnetic OB stars due to its remarkable, superionized wind, apparently related to its peculiar magnetic field – a field which is far more complex than the mostly-dipolar fields usually observed in magnetic OB stars. τ Sco is therefore a puzzling outlier in the larger picture of stellar magnetism – a star that still defies interpretation in terms of a physically coherent model.

Recently, two early B-type stars were discovered as τ Sco analogues, identified by the striking similarity of their UV spectra to that of τ Sco, which was – until now – unique among OB stars. We present the recent detection of their magnetic fields by the MiMeS collaboration, reinforcing the connection between the presence of a magnetic field and a superionized wind. We will also present ongoing observational efforts undertaken to establish the precise magnetic topology, in order to provide additional constrains for existing models attempting to reproduce the unique wind structure of τ Sco-like stars.

Keywords. stars: early-type, stars: magnetic fields, ultraviolet: stars, stars: individual (HD 66665, HD 63425), techniques: polarimetric

---

1. The young magnetic B-type star τ Sco

The magnetic field of τ Sco is unique because it is structurally far more complex than the mostly-dipolar fields ($l = 1$) usually observed in magnetic OB stars, with significant power in spherical-harmonic modes up to $l = 5$ and a mean surface field strength of ~ 300 G (Donati *et al.* 2006). τ Sco also stands out from the crowd of early-B stars because of its stellar wind anomalies, as diagnosed through its odd UV spectrum. These anomalies, unique to this star, are indicative of a highly ionised outflow.

Interestingly, the wind lines of τ Sco vary periodically with the star's 41 d rotation period (Donati *et al.* 2006). Clearly the magnetic field exerts an important influence on the wind dynamics. What is not clear is whether the wind-line anomalies described above are a consequence of the unusual complexity of τ Sco's magnetic field, a general consequence of wind confinement in this class of star, or perhaps even unrelated to the presence of a magnetic field.

2. The τ Sco Clones

We present two early B-type stars – HD 66665 and HD 63425 – that we identified to be the first τ Sco analogues. These stars were first discovered by their UV spectra, which are strikingly similar to the UV spectrum of τ Sco. Spectropolarimetric observations

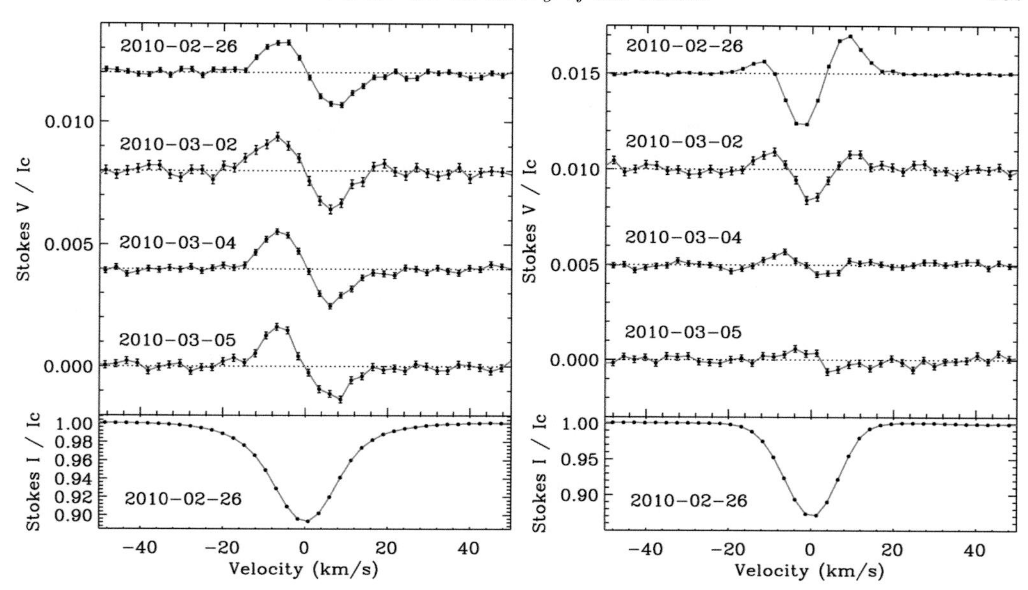

Figure 1. Mean Stokes I absorption line profiles (bottom) and circular polarisation Stokes V profiles (top) of HD 63425 (left) and HD 66665 (right) obtained with ESPaDOnS.

of HD 66665 and HD 63425 were taken with ESPaSOnS at the Canada-France-Hawaii Telescope, in the context of the Magnetism in Massive Stars Large Program. We acquired 4 high-resolution, broad-band, intensity (Stokes I) and circular polarisation (Stokes V) for each star.

In order to determine the stellar and wind parameters of HD 66665 and HD 63425 we used non-LTE model atmospheres from the CMFGEN code (Hillier & Miller 1998). The physical parameters are similar to τ Sco's. In order to increase the magnetic sensitivity of our data, we applied the Least-Squares Deconvolution (LSD) procedure (Donati *et al.* 1997), which enables the simultaneous use of many spectral lines to detect a magnetic field Stokes V signature. All observations led to a significant detection of a (time-variable) magnetic signal (see Figure 1). The same analysis was performed on the diagnostic null profiles, and no signal was detected.

As the exact rotation phases of our observations are not known, we used the method described by Petit et al. (2008, in prep), which compares the observed Stokes V profiles to a rotation independent, dipolar oblique rotator model, in a Bayesian statistic framework. We can obtain a conservative estimate of the surface field strength, in a dipolar approximation. The average surface field strengths are of the same scale as the surface field of τ Sco. However, more phase-resolved observations are required in order to assess the potential complexity of their magnetic field, and verify if it is linked to the wind anomalies.

References

Donati, J.-F., Semel, M., Carter, B. D., Rees, D. E. *et al.* 1997, *MNRAS*, 291, 658
Donati, J.-F., Howarth, I. D., Jardine, M. M., Petit, P. *et al.* 2006, *MNRAS*, 370, 629
Hillier, D. J. & Miller, D. L. 1998, *ApJ* 496, 407
Petit, V., Wade, G. A., Drissen, L., Montmerle, T. *et al.* 2008, *MNRAS* 387, L23

Active OB stars: structure, evolution, mass loss, and critical limits
Proceedings IAU Symposium No. 272, 2010
C. Neiner, G. Wade, G. Meynet & G. Peters, eds.

doi:10.1017/S1743921311010374

Magnetic fields, winds and X-rays of massive stars in the Orion nebula cluster

Véronique Petit[1], Gregg A. Wade[2], Evelyne Alecian[3], Laurent Drissen[4], Thierry Montmerle[3] and Asif ud-Doula[5]

[1]Dept. of Geology & Astronomy, West Chester University, West Chester, PA 19383, USA
email: VPetit@wcupa.edu
[2]Dept. of Physics, Royal Military College of Canada, Kingston, K7K 7B4, Canada
[3]Laboratoire d'Astrophysique de Grenoble, F-38041 Grenoble Cdex 9, France
[4]Dépt. de Physique, Université Laval, Québec, Canada
[5]Penn State Worthington Scranton, Dunmore PA 18512, USA

Abstract. In some massive stars, magnetic fields are thought to confine the outflowing radiatively-driven wind. Although theoretical models and MHD simulations are able to illustrate the dynamics of such a magnetized wind, the impact of this wind-field interaction on the observable properties of a magnetic star - X-ray emission, photometric and spectral variability - is still unclear. The aim of this study is to examine the relationship between magnetism, stellar winds and X-ray emission of OB stars, by providing empirical observations and confronting theory. In conjunction with the COUP survey of the Orion Nebula Cluster, we carried out spectropolarimatric ESPaDOnS observations to determine the magnetic properties of massive OB stars of this cluster.

Keywords. stars: early-type, stars: magnetic fields, X-rays: stars, techniques: polarimetric

---

1. Introduction

The Chandra Orion Ultradeep Project (COUP) was dedicated to observe the Orion Nebula Cluster (ONC) in X-rays. The OBA sample (20 stars) was studied with the goal of disentangling the respective roles of winds and magnetic fields in producing X-rays (Stelzer *et al.* 2005). The production of X-rays by radiative shocks (Lucy & White 1980, Owocki & Cohen 1999) should be the dominant mechanism for the subsample of 9 O to early-B stars which have strong winds. However, aside from 2 of those stars, all targets showed X-ray intensity and/or variability which were inconsistent with the small shock model predictions. We have undertaken a study with ESPaDOnS to explore the role of magnetic fields in producing this diversity of X-ray behaviours.

Eight stars of the COUP strong winds OB subsample were observed with the echelle spectropolarimeter ESPaDOnS at CFHT. The mean Stokes I and V profiles were extracted with the Least Squares Deconvolution technique (LSD) of Donati *et al.* (1997), which allows the use of many lines to increase the level of detection of a magnetic Stokes V signature. Formal signal detection was achieved for 3 stars: θ^1 Ori C (for which a field has already been detected by Donati *et al.* 2002), LP Ori (HD 36982) and NU Ori (HD 37061).

With foreknowledge of the shape of an expected deviation, we can add some information to the formal χ^2 statistics. It has been shown that with multiple noisy observations, it is possible to pick up an underlying signal by computing the odds ratios of the no magnetic field model (M_0) to a magnetic model (M_1), in a Bayesian framework (Petit *et al.* in prep). As the exact rotation phases of our observations are not known, we used the method described by Petit *et al.* (2008), which compares the observed Stokes V profiles to a rotation independent, dipolar oblique rotator model. As can be seen from the

Table 1. Odds ratios, surface dipolar strength and magnetic wind confinement of ONC stars.

ID	HD	Spec Type	$\log(M_0/M_1)$	$B_{\rm pole}$ [1][G]	$B_{\rm pole}(\eta_\star = 1.0)$ [G]
θ^1 Ori C	37022	O7 V †	**-110**	$1785\ ^{+494}_{-652}$	268
θ^2 Ori A	37041	O9.5 V †	0.38	< 118	118
θ^1 Ori A	37020	B0.5 V †	0.51	< 57	63
θ^1 Ori B	37023	B0.5 V †	0.38	< 79	77
NU Ori	37061	B0.5 V †	**-15**	$465\ ^{+116}_{-179}$	66
θ^2 Ori B	37042	B0.5 V	0.25	< 103	45
LP Ori	36982	B1-2 V	**-37**	$1\,020\ ^{+199}_{-302}$	15
JW 660		B3 V †	0.14	$< 1\,287$	16

[1] Median of the posterior probability density marginalised for $B_{\rm pole}$ and 68.3% credible region for magnetic stars for detected stars, upper limit of the 95.4% credible region for the non-detections.
† Confirmed or suspected binaries.

computed odds (Table 1) in the case of the detected stars, the magnetic oblique rotator model is favoured by many orders of magnitude ($\log(M_0/M_1) < 0$). For the non-detected stars, any improvement of the fit to the data achieved by assuming a magnetic field is not sufficient to justify employing this more complex magnetic model.

By performing a Bayesian parameter estimation for the dipole model, we can obtain the probability density function marginalised for the dipole strength, and put constraints on the values admissible by our observations ($B_{\rm pole}$ in Table 1).

2. Wind confinement

According to our observations, the 3 magnetic massive stars of the ONC have fields that should be strong enough to dynamically influence their stellar winds at a significant level (see the minimum field required for confinement in Table 1). However, this field-wind interaction is not reflected in any systematic way in the X-ray properties of these stars. Furthermore, no fields strong enough to dynamically influence the wind are found in other ONC massive stars that Stelzer *et al.* (2005) considered to be "prime candidates" for magnetism. From this we conclude that X-ray variability, intensity and hardness enhancement are not systematically correlated with the presence of a magnetic field.

More detailed studies of the field geometries of these magnetic stars will serve as inputs to new models (Townsend *et al.* 2007) and 3D MHD simulations of magnetic wind confinement (e.g. ud-Doula *et al.* 2008), to better understand the mechanisms that lead to this variety of X-ray properties.

References

Donati, J.-F., Babel, J., Harries, T. J., Howarth, I. D. *et al.* 2002, *MNRAS* 333, 55
Donati, J.-F., Semel, M., Carter, B. D., Rees, D. E. *et al.* 1997, *MNRAS*, 291, 658
Lucy, L. B. & White, R. L. 1980, *ApJ*, 241, 300
Owocki, S. P. & Cohen, D. H. 1999, *ApJ*, 520, 833
Petit, V., Wade, G. A., Drissen, L., Montmerle, T. *et al.* 2008, *MNRAS* 387, L23
Stelzer, B., Flaccomio, E., Montmerle, T., Micela, G. *et al.* 2005, *ApJS*, 160, 557
Townsend, R. H. D., Owocki, S. P., & Ud-Doula, A. 2007, *MNRAS* 382, 139
ud-Doula, A., Owocki, S. P., & Townsend, R. H. D. 2008, *MNRAS* 385, 97

Active OB stars: structure, evolution, mass loss, and critical limits
Proceedings IAU Symposium No. 272, 2010
C. Neiner, G. Wade, G. Meynet & G. Peters, eds.

doi:10.1017/S1743921311010386

Magnetism of the He-weak star HR 2949†

Thomas Rivinius[1], Gregg A. Wade[2], Richard H. D. Townsend[3], Matthew Shultz[2], Jason H. Grunhut[2], Otmar Stahl[4] and the MiMeS collaboration

[1]ESO, Chile; [2]RMC, Canada; [3]UW Madison, USA; [4]LSW/ZAH Heidelberg, Germany

Abstract. A magnetic field and rotational line profile variability (*lpv*) is found in the He-weak star HR 2949. The field measured from metallic lines varies in a clearly non-sinusoidal way, and shows a phase lag relative to the morphologically similar He I equivalent width variations. The surface abundance patterns are strong and complex, and visible even in the hydrogen lines.

Keywords. stars: binaries, stars: early-type, stars: magnetic fields

---

1. Introduction

The mid-B star HR 2949 was for many years considered to be non-variable, and indeed was listed in standard star catalogues. However, Rivinius *et al.* (2003) noted that the star is actually variable both photometrically as well as spectroscopically. They classified it as a He-weak star. Here we report the discovery of a magnetic field and describe the line profile variations (*lpv*) due to surface abundance inhomogeneities. The existence of these inhomogeneities was already proposed by Rivinius *et al.* (2003), but could not be investigated further on the basis of solely four spectra.

2. Observations

Eight archival spectra have been obtained with the *FEROS* instrument at La Silla, providing a resolving power of 48 000 over a spectral range of 375 to 890 nm. Spectropolarimetric (Stokes V) data were taken with ESPaDOnS at the 3.6 m CFHT in the 2009/2010 season, in total 16 measurements. The observations were obtained as part of the Magnetism in Massive Stars (MiMeS) Large Program.

Koen & Eyer (2002) quote a period of $P = 1.9093$ d from *HIPPARCOS* photometric data. Re-analysing these data using the Lomb-Scargle statistic, we obtain $P = 1.9083 \pm 0.00025$ d. In the spectra, the helium lines show the strongest variability. The time base covered by the spectra is long, 1999 to 2009, but there is severe aliasing. However, the aliases are narrow and there is only one peak in reasonable agreement with the Hipparcos data, at $P = 1.90871 \pm 0.00007$ d. As the epoch we chose the date of the ESPaDOnS measurement showing the most negative field, so that our adopted ephemeris is:

$$T_{\rm minB}(\rm HJD) = 2\,455\,223.987 + 1.90871(7) \times E$$

3. Spectral Variations

The magnetic field as measured by ESPaDOnS (using Least-Squares Deconvolution applied to metallic lines) confirms the photometric and spectroscopic periods. Figure 1 shows the respective phased variations. The field varies between -760 and $+200$ G. There

† Based on observations under ESO programs 073.C-0337, 076.C-0164 and the MiMeS large program at the CFHT.

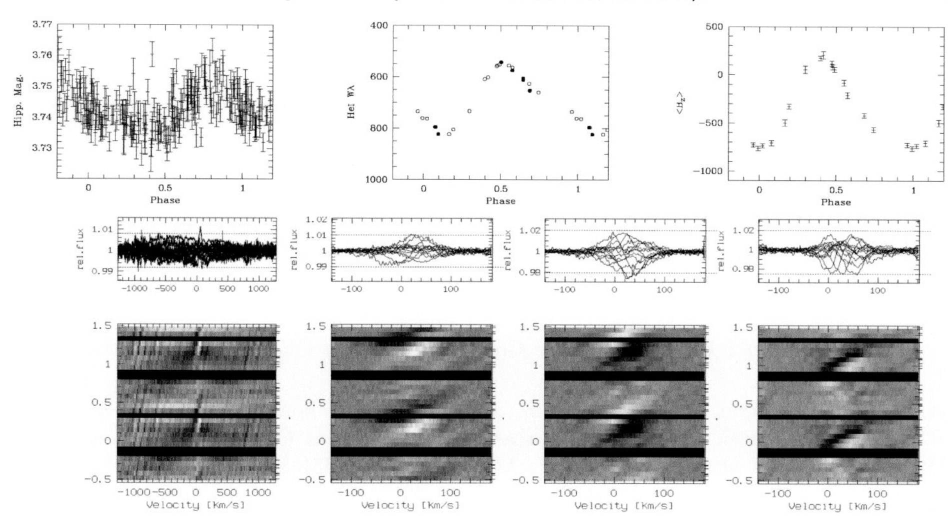

Figure 1. Upper row, from left to right: Hipparcos photometry, equivalent width variability of the combined He I 4009 & 4026 lines (ESPaDOnS, open symbols; FEROS, filled symbols), and magnetic data. Lower row, left to right: Exemplary types of line profile variation seen in HR 2949: Balmer wings, Balmer cores (both Hβ), helium (He I 4388), metal (Fe I 4949).

is clear rotational line profile variability (*lpv*), with no line in the spectrum unaffected. Taking the behavior of the He I lines as a starting point, the other lines can be sorted into three variability morphology groups: **Lines similar to helium** are those of Mg I 8807, Mg II 4481, O I 8446. They show a clear modulation of EW. Lines behaving **inversely to He I**, are Si II 4128 and Si III 4553, Cr II 4588, P II 6043, P III 4222, and the Balmer lines. Also these vary in EW with rotation. The majority of lines show a third pattern, e.g. S II 5640, Fe II 4949, C II 4267, Ca II 8662, Sr II 4215, Ti II 4564 and thus shall be called **metal group** behavior. While these lines show clear *lpv*, their EW is rather non-variable. Finally, some lines show different types of **hybrid** variation patterns.

It is noteworthy that the Balmer lines show a clear pattern. The line core shows an abundance-like pattern well described as an "anti-helium"-group behavior. We attribute the pattern to an actual variation of the H-abundance, rather than to a change in local surface parameters. However, there are also changes in the Balmer line wings probably caused by local parameter variations. Although the effect is subtle, less than 0.5 %, and in spite of the notoriously difficult normalization of the Balmer wings in echelle data the pattern is identical in all lines.

4. Conclusions

HR 2949 is a magnetic star of the He-weak/Si-strong type. Its surface shows the clear signature of the typically complex abundance patterns characteristic to such stars. In addition to the abundance variation, the variability of the Balmer line wings indicates some other local parameter changes, such as pressure or temperature, linked to the magnetic field.

References

Koen, C. & Eyer, L. 2002, *MNRAS*, 331, 45

Rivinius, T., Stahl, O., Baade, D., & Kaufer, A. 2003, *IBVS* 5397

Active OB stars: structure, evolution, mass loss, and critical limits
Proceedings IAU Symposium No. 272, 2010
C. Neiner, G. Wade, G. Meynet & G. Peters, eds.

doi:10.1017/S1743921311010398

Searching for weak or complex magnetic fields in polarized spectra of Rigel

Matthew Shultz[1,2], Gregg A. Wade[2], Coralie Neiner[3], Nadine Manset[4], Véronique Petit[2,5], Jason H. Grunhut[1,2], Edward Guinan[6], David A. Hanes[1] and the MiMeS collaboration

[1]Queen's University, Canada [2]Royal Military College, Canada [3]Paris-Meudon Observatory
[4]Canada-France-Hawaii Telescope Corporation [5]West Chester U [6]Villanova U

Abstract. Seventy-eight high-resolution Stokes V, Q and U spectra of the B8Iae supergiant Rigel were obtained with the ESPaDOnS spectropolarimeter at CFHT and its clone NARVAL at TBL in the context of the Magnetism in Massive Stars (MiMeS) Large Program, in order to scrutinize this core-collapse supernova progenitor for evidence of weak and/or complex magnetic fields. In this paper we describe the reduction and analysis of the data, the constraints obtained on any photospheric magnetic field, and the variability of photospheric and wind lines.

Keywords. stars: magnetic fields, stars: variables: other, stars: activity, stars: winds, outflows

1. Introduction: Physical Parameters and Observations

Rigel: a blue supergiant, the closest and most readily studied Type II supernova progenitor, and a known α Cygni variable, is the subject of a global monitoring campaign known as the 'Rigel-thon', involving long-term spectroscopic monitoring, Microvariability and Oscillations in STars (MOST) space photometry, and spectropolarimetry.

Like most OB stars, Rigel (β Ori A) shows no sign of an easily detected magnetic field, however apparent brightness and sharp spectral lines make it practical to ask if the star possesses a weak or complex field geometry which might be revealed within a high resolution data set. Thus, over the epoch 09/2009–02/2010, 65 Stokes V (circular polarization) and 13 Stokes Q and U (linear polarization) spectra spanning 370-1000 nm with a mean resolving power R $\sim$ 65000 at 500 nm, were taken with the ESPaDOnS spectropolarimeter at CFHT and its clone Narval at TBL. Integration times were typically of a few seconds duration. The densest spectropolarimetric sampling was concurrent with the collection of MOST data.

The physical radius was determined from the interferometric angular diameter, $\theta_D = 2.76 \pm 0.01$ mas (Aufdenberg *et al.*, 2008) together with a distance of 240 $\pm$ 50 pc calculated using the Hipparcos parallax of 4.22 $\pm$ 0.81 mas: thus R = 70 $\pm$ 14 $R_\odot$. The star appears to be a slow rotator, with $v\sin i$ = 36 $\pm$ 5 km/s (Przybilla *et al.* 2006), giving an upper bound on the rotation period P_{rot} of $\sim$ 98 d; calculation of the breakup velocity ($\sim$ 250 km/s) provides a lower limit of $\sim$ 14 d.

2. Results and Analysis

Least Squares Deconvolution (LSD) was employed to extract high S/N ratio mean Stokes I, V, and diagnostic N profiles from the circular polarization spectra. The LSD line mask was cleaned to eliminate contamination from telluric, emission and Balmer lines, and to remove weak or apparently absent lines. Ultimately $\sim$ 90 lines remained, and their weights were empirically adjusted to reflect observed line depths. The typical

S/N ratio in Stokes V mean profiles was $\sim$ 20,000. No significant signal was detected in either Stokes V or diagnostic N. Each LSD profile was then analyzed to determine the longitudinal magnetic field B_l. No significant longitudinal field was detected, with a median 1-σ uncertainty in individual measurements of 13 G. The distribution of B_l values inferred from Stokes V is statistically identical to that inferred from diagnostic N. The measured B_l was then compared to a grid of synthetic longitudinal field curves corresponding to dipoles with $0° \leqslant \beta \leqslant 90°$, $0° \leqslant i \leqslant 90°$ (with the data folded according to the theoretical maximum P_{rot} = 93 d at i = 90° and progressively shorter periods at smalled i, with ten different phase offsets tested at each period), and polar field strengths B_d from 0 to $\sim$ 3 kG. For (i = 90°, β = 90°) the maximum B_d compatible with the data at 3-σ confidence is $\sim$ 20 G, while B_d is constrained below $\sim$ 50 G for intermediate values of i and β. Fields at this level, if present within the photosphere, remain capable of strongly influencing the wind (ud-Doula & Owocki, 2002), with a wind magnetic confinement parameter $\eta_* \sim 2 - 90$, assuming $\dot{M} \sim 10^{-7}$–10^{-6} $M_\odot$/y (Barlow & Cohen 1977, Abbot *et al.* 1980, Puls *et al.* 2008) and $v_\infty \sim$ 400–600 km/s (Bates *et al.* 1980).

Rigel is a long-known α Cygni variable (Sanford, 1947), with significant line profile variability (LPV) in Hα as well as various metal lines, which may be associated with any or all of: mass loss events, photospheric spots, corotating interacting structures, and/or g- or p-mode pulsations (Kaufer *et al.* 1996a, 1997). Distinct LPV is seen in Hα as compared to metal lines: Hα is in strong emission, variable over a broad velocity range and apparently aperiodic (consistent with earlier spectroscopic monitoring, Kaufer *et al.* 1996a, 1996b, Israelian *et al.* 1997); metal lines showed little apparent emission excess, but their variability was suggestive of periodic behaviour. Amongst the most complexly variable of the metal lines is the O triplet at 777 nm.

3. Conclusions & Future Work

No evidence of a magnetic field is obtained in 65 high precision Stokes V observations of Rigel. Significant variability is observed in numerous spectral lines, with some suggestion of periodicity on the order of $\sim$ 1 month in metallic lines. Further modeling of Stokes V profiles must be performed to obtain quantitative constraints on various potential field topologies, e.g. the dynamo-generated field proposed by Cantiello *et al.* (2009); a more rigorous analysis of LPV may help to identify periodic behaviour.

References

Abbott, D. C., Bieging, J. H., Churchwell, E., & Cassinelli, J. P. 1980, *ApJ*, 238, 196
Aufdenberg, J. P., Ludwig, H.-G., Kervella, P., Mérand, A. *et al.* 2008, in: A. Richichi, F. Delplancke, F. Paresce, & A. Chelli (eds.), *The Power of Optical/IR Interferometry: Recent Scientific Results and 2nd Generation*, ESO Astrophysics Symposia, p. 71
Barlow, M. J. & Cohen, M. 1977, *ApJ*, 213, 737
Bates, B., Giaretta, D. L., McCartney, D. J., McQuoid, J. A. *et al.* 1980, *MNRAS*, 190, 611
Cantiello, M., Langer, N., Brott, I., de Koter, A. *et al.* 2009, *A&A*, 499, 279
Israelian, G., Chentsov, E., & Musaev, F. 1997, *MNRAS*, 290, 521
Kaufer, A., Stahl, O., Wolf, B., Gaeng, T. *et al.* 1996a, *A&A*, 305, 887
Kaufer, A., Stahl, O., Wolf, B., Gaeng, T. *et al.* 1996b, *A&A*, 314, 599
Kaufer, A., Stahl, O., Wolf, B., Fullerton, A. W. *et al.* 1997, *A&A*, 320, 273
Puls, J., Vink, J. S., & Najarro, F. 2008, *A&AR*, 16, 209
Przybilla, N., Butler, K., Becker, S. R., & Kudritzki, R. P. 2006, *A&A*, 445, 1099
Sanford, R. F. 1947, *ApJ*, 105, 222
ud-Doula, A. & Owocki, S. P. 2002, *ApJ*, 576, 413

Active OB stars: structure, evolution, mass loss, and critical limits
Proceedings IAU Symposium No. 272, 2010
C. Neiner, G. Wade, G. Meynet & G. Peters, eds.

doi:10.1017/S1743921311010404

Line profiles of OB star winds using a Monte Carlo method

Brankica Šurlan[1,2] and Jiří Kubát[1]

[1]Astronomický ústav, Akademie věd České republiky, CZ-251 65 Ondřejov
email: `surlan, kubat@sunstel.asu.cas.cz`
[2]Matematički Institut SANU, Kneza Mihaila 36, 11001 Beograd, Serbia

Abstract. The solution of the radiative transfer in an expanding atmospheres using the Monte Carlo method is presented. We applied our method to winds of several OB stars. In our calculation, the velocity and density structure is assumed to be given. Selected line profiles are shown.

Keywords. radiative transfer, line: formation, stars: winds, outflows

1. Introduction

We started to develop a code for the solution of the radiative transfer problem which could be used for the solution of line formation in a 3-D inhomogeneous medium. The code is based on a Monte Carlo method. As a first step, we developed a spherically symmetric code for the *formal* solution of the radiative transfer equation, i.e. for a solution of this equation for given velocity, temperature, and density stratification. This task has already been done by many authors for different astrophysical applications, e.g. Abbott & Lucy (1985), Whitney (1991), Lucy & Abbott (1993), de Koter *et al.* (1997), Vink *et al.* (1999), Carciofi & Bjorkman (2008), Wood *et al.* (2004), and Kromer & Sim (2009), to name at least some of them. Nevertheless, in order to understand all processes more deeply and to have better control of the code's behaviour, we decided to develop an independent code. Its detailed description will be published elsewhere.

2. Wind model and results

The velocity structure $v(r)$ of the wind is calculated using the β-velocity law. Knowing the velocity structure and the mass-loss rate $\dot{M}$, the density structure $\rho(r)$ follows from the continuity equation. The temperature $T(r)$ was taken as independent of r and equals the approximate value of the radiation temperature $T_{\rm rad} \approx 3/4\,T_{\rm eff}$. This value was used for calculation of the LTE ionization and excitation equilibrium in the wind, while the electron density was calculated consistently. We further assume that all electrons in the wind come from hydrogen ionization, and that the opacity of the medium consists of only line scattering under the Sobolev approximation. For depth discretization we used equidistant spacing in the logarithmic scale. Depth points split the wind into zones. The density $\rho(r)$ was taken to be constant within a zone equal to the value at the lower radius of the zone. The radial velocity $v(r)$ is linearly interpolated inside the zone. The flux at the lower boundary of the wind was computed as an emergent flux from the static spherically symmetric NLTE model atmosphere calculated using a code of Kubát (2003). Parameters for selected stars were taken from Crowther *et al.* (2006) and are listed in Table 1.

Table 1. Stellar and wind parameters taken from Crowther *et al.* (2006).

Name	$T_{\rm eff}$ [K]	$\log g$	R_* $[R_\odot]$	v_∞ $[km/s]$	$\dot{M}$ $[10^{-6} M_\odot/yr]$	β
HD 14818	18500	2.40	46.1	565	0.55	2.0
HD 13854	21500	2.55	37.4	920	0.85	2.0
HD 115842	25500	2.85	34.2	1180	2.00	1.5
HD 30614	29000	3.00	26.0	1560	5.00	1.5

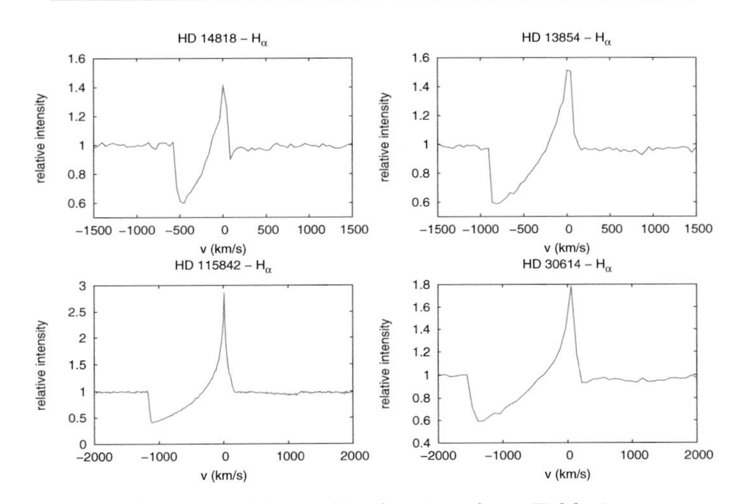

Figure 1. Hα profiles for stars from Table 1.

We plot the profiles of the Hα line of four OB stars obtained as a result of our calculation using Monte Carlo radiative transfer. Stellar and wind parameters are given in Table 1. The flux is expressed as relative intensity with respect to the local continuum. The terminal velocities, which follow from the theoretical profiles, correspond well to measured terminal velocities of the wind (v_∞ in Table 1).

Our code is able to solve the line formation problem in an expanding medium for the spherically symmetric case. In our future work we aim at generalizing the code to 3D and including inhomogeneities.

Acknowledgements

This research was supported by GA ČR grants 205/08/0003 and 205/08/H005.

References

Abbott, D. C. & Lucy, L. B. 1985, *ApJ* 288, 679
Carciofi, A. C. & Bjorkman, J. E. 2008, *ApJ* 684, 1374
Crowther, P. A., Lennon, D. J., & Walborn, N. R. 2006, *A&A* 446, 279
de Koter, A., Heap, S. R., & Hubeny, I. 1997, *ApJ* 477, 792
Kromer, M. & Sim, S. A. 2009, *MNRAS* 398, 1809
Kubát, J. 2003, in: N. Piskunov, W. W. Weiss, & D. F. Gray (eds.), *Modelling of Stellar Atmospheres*, IAU Symposium 210, p. 6P
Lucy, L. B. & Abbott, D. C. 1993, *ApJ* 405, 738
Whitney, B. A. 1991, *ApJS* 75, 1293
Wood, K., Mathis, J. S., & Ercolano, B. 2004, *MNRAS* 348, 1337
Vink, J. S., de Koter, A., & Lamers, H. J. G. L. M. 1999, *A&A* 350, 181

Active OB stars: structure, evolution, mass loss, and critical limits
Proceedings IAU Symposium No. 272, 2010
C. Neiner, G. Wade, G. Meynet & G. Peters, eds.

doi:10.1017/S1743921311010416

Monte-Carlo simulations of linear polarization in clumpy OB-star winds

Richard H. D. Townsend and Nick Mast

Department of Astronomy, University of Wisconsin-Madison, Madison, WI 53706, USA

Abstract. We present results from Monte-Carlo simulations of linear polarization in clumped OB-star winds. We find that previous single-scattering models of clumped winds have overestimated the degree of polarization, even in cases where individual clumps are optically thin. An application to P Cygni suggests the star's wind is more fragmented than previously thought.

Keywords. polarization, radiative transfer, scattering, methods: numerical, techniques: polarimetric, stars: mass loss, stars: early-type, stars: winds, outflows, stars: individual (P Cygni)

1. Introduction

A wide variety of observational diagnostics indicate that the radiation-driven winds of OB stars are clumpy. However, the properties of the clumps (small or large? optically thick or thin?), and their origins (wind instabilities? photospheric perturbations?), remain unclear. One promising approach to resolving these uncertainties is analysis of the continuum linear polarization arising from electron scattering in the clumped wind. By comparing the measured (time-varying) Stokes vector against predictions from polarization models, constraints can be placed on the nature of the clumping.

2. Simulations

We model the Stokes vector for a clumpy wind outflow using YARG, a Monte-Carlo radiative transfer code that tracks polarized photon propagation through an ensemble of spherical electron-scattering clumps. As these clumps advect outward in accordance with the standard β velocity law $v(r) = v_\infty(1 - R_*/r)^\beta$, their radius ℓ grows in proportion to

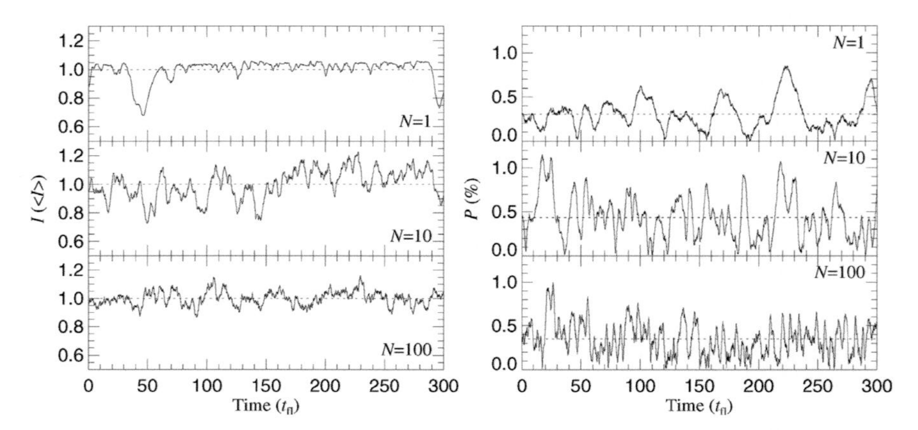

Figure 1. Time-series intensity (left) and fractional linear polarization (right) for the $N = 1, 10$ and 100 simulations. The abscissa is in units of the wind flow time $t_{fl} = 3.3$ d, and the dotted lines indicate mean levels.

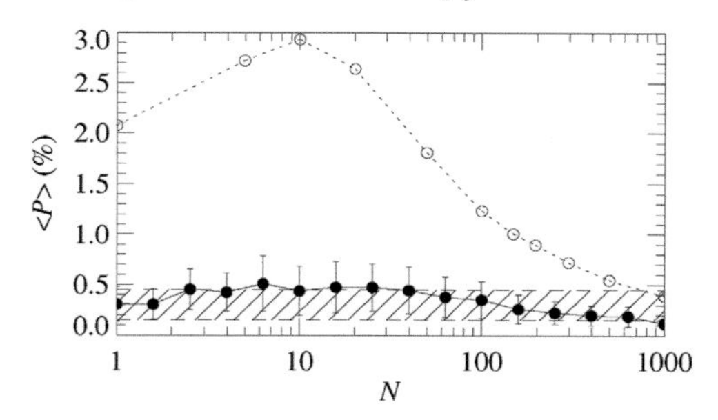

Figure 2. The simulated mean linear polarization $\langle P \rangle$ and 1-σ variability limits of P Cygni, plotted as a function of N (lower curve: Monte-Carlo, present work; upper curve: single-scattering, from D07). The hatched region shows the observed $0.3\% \pm 0.15\%$ polarization.

r. Two parameters establish the overall wind clumping properties: the mean number of clumps N emitted at the stellar surface per unit wind flow time $t_{\rm fl} \equiv R_*/v_\infty$, and the initial clump radius ℓ_*. In the porosity formalism of Cohen *et al.* (2008), these parameters correspond to a terminal porosity length $h_\infty = 3R_*^3/N\ell_*^2$.

As an initial application of YARG, we simulate time-series Stokes vectors for the LBV P Cygni. Parameters follow those of Davies *et al.* (2007; hereafter D07): $\beta = 1$, $\pi\ell_*^2 = 0.04R_*^2$, and the overall wind optical depth scale is $\tau_* \equiv \kappa\dot{M}/4\pi R_* v_\infty = 0.42$, corresponding to an optical depth from the sonic point to infinity of $\tau_{\rm s} = 1.3$. Time series spanning $300\,t_{\rm fl}$ are calculated for $\log N = 1.0, 1.2, 1.4, \ldots, 2.8, 3.0$.

3. Results

Fig. 1 presents the time-series intensity I and fractional linear polarization $P = \sqrt{Q^2 + U^2}/I$, for the $N = 1, 10$ and 100 YARG simulations. Fig. 2 combines the data from all simulations to plot the mean polarization $\langle P \rangle$, and its 1-σ variability limits, as a function of N. Also shown are data from the single-scattering P Cygni models of D07, which are seen to be larger by a factor of 3–6. This difference arises due to multiple scatterings: although individual clumps are optically thin for $N \gtrsim 20$, the wind of P Cygni *as a whole* remains optically thick ($\tau_{\rm s} > 1$). Thus, multiple, different-clump scatterings are inevitable, reducing net polarization levels to below the single-scattering limit.

D07 quote a mean *observed* polarization of 0.3% for P Cygni, with variability at the ±0.15% level. These two values are consistent with our $N \sim 100$ simulations (Fig. 2). Percy *et al.* (2001) likewise report light variations with a typical semi-amplitude $\Delta V \sim 0.1$, again consistent with the $N \sim 100$ case (see lower-left panel of Fig. 1). Thus, we conclude that P Cygni has a moderately clumped wind ($N \sim 100$, $h_\infty \sim 2.4\,R_*$), in contrast to the smoother wind ($N \gtrsim 1,000$) proposed by D07.

Acknowledgements

We acknowledge support from NASA grant *LTSA*/NNG05GC36G.

References

Cohen, D. H., Leutenegger, M. A., & Townsend, R. H. D. 2008, in: W.-R. Hamann, A. Feldmeier, & L. M. Oskinova (eds.), *Clumping in Hot-Star Winds*, p. 209

Davies, B., Vink, J. S., & Oudmaijer, R. D. 2007, *A&A*, 469, 1045

Percy, J. R., Evans, T. D. K., Henry, G. W., & Mattei, J. A. 2001, in: M. de Groot & C. Sterken (eds.), *P Cygni 2000: 400 Years of Progress*, ASP-CS 233, p. 31

Active OB stars: structure, evolution, mass loss, and critical limits
Proceedings IAU Symposium No. 272, 2010
C. Neiner, G. Wade, G. Meynet & G. Peters, eds.

doi:10.1017/S1743921311010428

Influence of decoupling effect on stellar wind variability

Viktor Votruba[1,2], Klára Šejnová[2], Pavel Koubský[1] and Daniela Korčáková[1]

[1]Stellar Department, Astronomical Institute AV ČR, v.v.i., Fričova 298, Ondřejov 25165, Czech Republic; email: votruba@physics.muni.cz

[2]Institute of Theoretical Physics and Astrophysics, Masaryk University, Faculty of Natural Science, Kotlářská 2, Brno 61137, Czech Republic; email: klarka@physics.muni.cz

Abstract. A detailed investigation of momentum transfer in radiatively driven stellar winds shows that for the thin wind case ion decoupling may occur. The decoupling of absorbing ions significantly affects dynamics of the wind. We analysed such effects using our hydrodynamic code and predicted spectral changes with the help of the SHELLSPEC code.

Keywords. stars: winds, outflows, stars: oscillations (including pulsations)

1. Introduction

The assumption of a one-component flow used by the CAK theory is acceptable for most cases of stellar winds from O and hot B stars. But for stars with a low-density wind it is necessary to use a detailed description of the momentum transfer. As was shown first by Springmann & Pauldrach (1992), the gas and radiation field can decouple in the flow generated by a low-density radiatively driven wind. This is because in the low-density wind, Coulombic collisions are not effective and cannot support momentum transfer from absorbing ions to the passive plasma. As a result, the wind decouples at a certain point.

A very interesting result from the analysis of the decoupling effect is the generation of pulsating shells, which was first suggested by Porter & Skouza (1999). In the case when winds decouple at the point where the local velocity of the flow is still smaller than the escape velocity, the passive plasma is still gravitationally bound to the star. This implies that matter is decelerated and falls back down to the star. The interaction with an outflow from the star leads to the pulsating shells.

2. Numerical Simulation

We use the hydrodynamic code developed by Feldmeier (1995) for the simulation of pulsating shells, which uses an Euler scheme with Van Leer flux splitting. To model the decoupling effect we terminated the radiative force at a distance r_d, where the flow was still gravitationally bound to the star. It was also necessary to change the outflow boundary condition and suppress material flow from the outer boundary. The chosen stellar parameters can be found in Table 1. The time of the simulation was roughly ~2d.

Table 1. Stellar parameters of the model

M [M$_{\rm sol}$]	R_s [R$_\odot$]	T [K]	$\dot{M}$ [M$_\odot$.y^{-1}]	$r_{\rm d}$ [R$_s$]
13.0	7.0	19300	1.2 10^{-9}	1.25

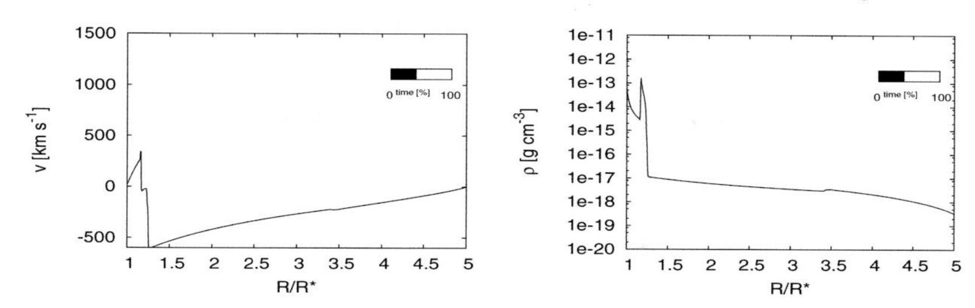

Figure 1. Snapshot of the pulsating shell for a model star. *Left panel:* Velocity distribution in the flow. *Right panel:* Density distribution in the flow. Spherical shells are clearly visible as enhanced density regions in the plot.

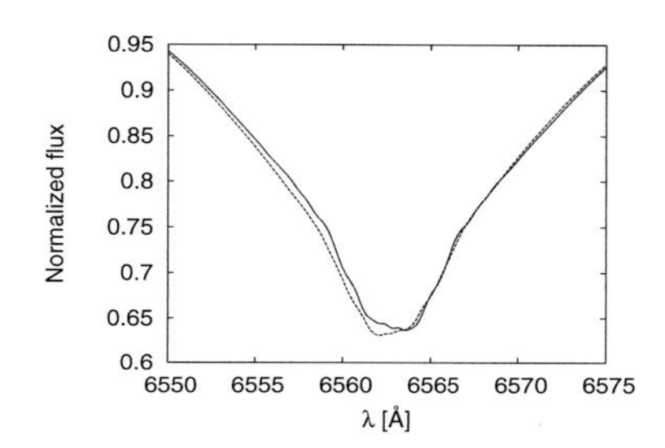

Figure 2. H_α line profiles. Two lines mark two different groups of input parameters. The solid line corresponds to the infalling shell with velocity $\sim$$-500$ km.s^{-1}, density $\sim$1. 10^{-14} g.cm^{-3} and thickness $\sim$7 R_{sol}. The dashed line corresponds to outflow velocity $\sim$500 km.s^{-1}, density $\sim$1.$^{-15}$ g.cm^{-3} and thickness $\sim$12 R_{sol}.

3. Synthetic spectral changes

The key question is if there is an observable signature of the decoupling effect. The most obvious effect of decoupling is that the observed terminal velocity will be much lower than predicted from CAK theory. This effect can be also detected directly from the stellar spectra. As was first predicted by Porter & Skouza (1999), pulsating shells may provide a varying component to the hydrogen lines. Under the assumption of an optically thin shell we used the SHELLSPEC code for the estimation of spectral line variations.

SHELLSPEC, developed by Budaj & Richards (2004), is designed to solve simple radiative transfer along the line of sight in moving media. The output of the code is: synthetic spectra, light curve or trailing spectrogram. Since generated shells pulsate roughly periodically, we analysed 2 different cases: when the outflowing shell is moving away from the star and when the shell is falling back toward the star. For both stages we derived the geometry and dynamical properties of the problem from our numerical simulations and applied the SHELLSPEC code. The resulting spectral variations are shown in Fig. 2.

Acknowledgements

This research was supported by grants 205/09/P476 (GA ČR) and 205/08/H005 (GA ČR). The Astronomical Institute Ondřejov is supported by project AV0Z10030501.

References

Budaj, J. & Richards, M. T. 2004, *Contributions of the Astronomical Observatory Skalnate Pleso* 34, 167

Feldmeier, A. 1995, *A&A*, 299, 523

Porter, J. M. & Skouza, B. A. 1999, *A&A*, 344, 205

Springmann, U. W. E. & Pauldrach, A. W. A. 1992, *A&A*, 262, 515

Active OB stars: structure, evolution, mass loss, and critical limits
Proceedings IAU Symposium No. 272, 2010
C. Neiner, G. Wade, G. Meynet & G. Peters, eds.

doi:10.1017/S174392131101043X

Of?p stars: a class of slowly rotating magnetic massive stars

Gregg A. Wade[1], Jason H. Grunhut[1], Wagner L. F. Marcolino[2], Fabrice Martins[3], Ian D. Howarth[4], Yael Nazé[5], Nolan R. Walborn[6] and the MiMeS Collaboration

[1]RMC, Canada; [2]ONB, Brazil; [3]GRAAL, France; [4]UCL, UK; [5]Liège, Belgium; [6]STScI, USA

Abstract. Only 5 Of?p stars have been identified in the Galaxy. Of these, 3 have been studied in detail, and within the past 5 years magnetic fields have been detected in each of them. The observed magnetic and spectral characteristics are indicative of organised magnetic fields, likely of fossil origin, confining their supersonic stellar winds into dense, structured magnetospheres. The systematic detection of magnetic fields in these stars strongly suggests that the Of?p stars represent a general class of magnetic O-type stars.

Keywords. techniques: spectroscopic, stars: magnetic fields, stars: individual (HD 191612, HD 108, HD 148937)

---

1. Introduction

The enigmatic Of?p stars are identified by a number of peculiar and outstanding observational properties. The classification was first introduced by Walborn (1972) according to the presence of C III λ4650 emission with a strength comparable to the neighbouring N III lines. Well-studied Of?p stars are now known to exhibit recurrent, and apparently periodic, spectral variations (in Balmer, He I, C III and Si III lines) with periods ranging from days to decades, strong C III λ4650 in emission, narrow P Cygni or emission components in the Balmer lines and He I lines, and UV wind lines weaker than those of typical Of supergiants (see Nazé *et al.* 2010 and references therein).

Only 5 Galactic Of?p stars are known (Walborn *et al.* 2010): HD 108, HD 148937, HD 191612, NGC 1624-2 and CPD-28°2561. Three of these stars - HD 108, HD 148937 and HD 191612 - have been studied in detail. In recent years, HD 191612 was carefully examined for the presence of magnetic fields (Donati *et al.* 2006), and was clearly detected. Recent observations, obtained chiefly within the context of the Magnetism in Massive Stars (MiMeS) Project (Martins *et al.* 2010; Wade *et al.*, in prep) have furthermore detected magnetic fields in HD 108 and HD 148937, thereby confirming the view of Of?p stars as a class of slowly rotating, magnetic massive stars.

2. HD 191612

HD 191612 was the first Of?p star in which a magnetic field was detected (Donati *et al.* 2006). Subsequent MiMeS observations with ESPaDOnS@CFHT (Wade *et al.*, in prep) confirm the existence of the field, and demonstrate the sinusoidal variability of the longitudinal field with the Hα and photometric period of 537.6 d. As shown in Fig. 1, the longitudinal field, Hα and photometric extrema occur simultaneously when folded according to the 537.6 d period. This implies a clear relationship between the magnetic field and the circumstellar envelope. We interpret these observations in the context of the oblique rotator model, in which the stellar wind couples to the kilogauss dipolar magnetic

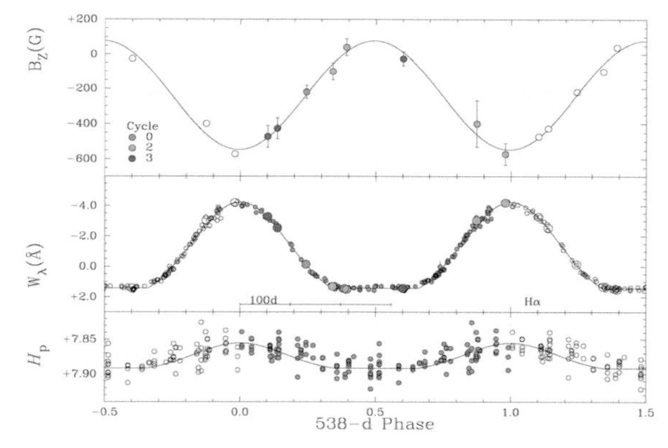

Figure 1. Longitudinal field (top), Hα EW (middle) and Hipparcos mag (bottom) of the Of?p star HD 191612, all phased according to the 537.6 d period. From Wade *et al.*, in preparation.

field, generating a dense, structured magnetosphere, resulting in all observables varying according to the stellar rotation period.

3. HD 108

HD 108 was the second Of?p star in which a magnetic field was detected (Martins *et al.* 2010). Based on long-term photometric and spectroscopic monitoring, HD 108 is suspected to vary on a timescale of 50-60 y (Nazé *et al.* 2001). The magnetic observations acquired by Martins *et al.* from 2007-2009 show at most a marginal increase of the longitudinal field during more than 2 years of observation. This supports the proposal that the variation timescale is in fact the stellar rotational period, and that HD 108 is a magnetic oblique rotator that has undergone extreme magnetic braking.

4. HD 148937

HD 148937 was recently observed intensely by the MiMeS Collaboration, resulting in the detection of circular polarisation within line profiles indicative of the presence of an organised magnetic field of kilogauss strength (Wade *et al.*, in prep). Although the field is consistently detected in the observations, no variability is observed, in particular according to the 7.03 d spectral period. This result supports the proposal by Nazé *et al.* (2010) that HD 148937 is observed with our line-of-sight near the stellar rotational pole.

References

Donati, J.-F., Howarth, I. D., Bouret, J.-C., Petit, P. *et al.* 2006, *MNRAS*, 365, L6
Martins, F., Donati, J.-F., Marcolino, W. L. F., Bouret, J.-C. *et al.* 2010, *MNRAS* 407, 1423
Nazé, Y., Ud-Doula, A., Spano, M., Rauw, G. *et al.* 2010, *A&A* 520A, 59
Nazé, Y., Vreux, J.-M., & Rauw, G. 2001, *A&A*, 372, 195
Walborn, N. R. 1972, *JRASC* 66, 71
Walborn, N. R., Sota, A., Maíz Apellániz, J., Alfaro, E. J. *et al.* 2010, *ApJ* (Letters), 711, L143

Active OB stars: structure, evolution, mass loss, and critical limits
Proceedings IAU Symposium No. 272, 2010
C. Neiner, G. Wade, G. Meynet & G. Peters, eds.

doi:10.1017/S1743921311010441

Magnetic fields in classical Be stars: results of our long-term program with FORS1 at the VLT

Ruslan V. Yudin[1], **Swetlana Hubrig**[2], **Michail A. Pogodin**[1] **and Markus Schoeller**[3]

[1]Central Astronomical Observatory of the Russian Academy of Sciences at Pulkovo, Saint-Petersburg, Russia, email: ruslan61@gao.spb.ru

[2]Astrophysical Institute Potsdam, Germany

[3]European Southern Observatory, Garching, Germany

Abstract. We report the results of our search for magnetic fields in a representative sample of classical Be stars carried out during 2006-2008 using low-resolution spectropolarimetry with FORS1 at the VLT. Among the 28 classical Be stars studied, detections of a magnetic field were achieved in seven stars (i.e. ~25%). The detected magnetic fields are rather weak, not stronger than ~150G. Among the Be stars studied with time series, one Be star, λ Eri, displays cyclic variability of the magnetic field with a period of 21.12 min.

Keywords. stars: emission-line, Be, magnetic fields

1. Introduction

A number of physical processes in classical Be stars (e.g., angular momentum transfer to a CS disk, channelling stellar wind matter, accumulation of material in an equatorial disk, etc.) are more easily explainable if magnetic fields are invoked (e.g. Brown *et al.* 2004; Cassinelli *et al.* 2002). Maheswaran (2003, 2005) developed the Magnetic Rotator Wind Disk model, in which Keplerian disks may be formed by magnetic fields of the order of a few tens of Gauss. Very recently, Maheswaran & Cassinelli (2009) obtained solutions for the structure and evolution of a protodisk region, i.e. the disk region that is initially formed when wind material is channelled by dipole-type magnetic fields towards the equatorial plane, showing that magnetorotational instability may assist in the formation of a quasisteady disk. According to their calculations, magnetic fields of the order of a few tens of Gauss will be able to channel wind flow into a protodisk region. Due to the high rotation of Be stars and the presence of strong Balmer emission lines, magnetic field measurements are difficult. We report here the results of our search for magnetic fields in a representative sample of classical Be stars carried out during four observing runs in the years 2005–2008 using low-resolution spectropolarimetry with FORS 1 at the VLT.

2. Results and Discussion

During our runs with FORS1 using low-resolution (R 2000 and 4000) spectropolarimetric data we studied 28 Be stars in the range of spectral classes from B9 to O8. Among this sample the detections of a magnetic field were achieved in 8 stars, most of them just occasionally on single nights (Hubrig *et al.* 2007; 2009a; 2009b). The magnetic fields of Be stars appear to be very weak, generally of the order of 100G and less. Significant polarization features have been revealed in CS components of the Ca II K&H lines in the spectra of 5 Be stars. No doubt that these lines are of circumstellar (CS) origin. These features are likely to be formed in the equatorial gaseous disks surrounding the stars. Besides, a clear Stokes V signature is observed in a redshifted absorption component of the H_β line profile in the spectrum of HD58011. Profiles of such a type are indicators of the presence of matter infall from the CS disk onto the star. Therefore, we can assume that the infalling flow in the CS envelope of HD58011 is magnetized. Note that the magnetic field detection in Be stars is independent of their spectral classes and was achieved for stars of spectral types from B9 to O8. In a framework of our observing programs we also carried out time resolved magnetic field measurements of nine classical Be stars. We were able to obtain during one hour 20-30 consecutive measurements for the stars depending on its brightness.

In the obtained amplitude spectra a 2.4σ peak corresponding to a period of 21.12 min was detected in the data set of measurements carried out using hydrogen lines in the star λ Eri in August 2006. This peak appears at a 2.2σ in the data set of measurements carried out using the whole spectrum. Note, however, that our observations obtained 16 months later on two consecutive nights on 2007 November 27 and 28 over few hours did not reveal any significant periodicity in any of the data sets. Other Be stars: QY Car, δ Cen, a Ara, and ϵ Tuc show weak signals in the Fourier transforms of our data sets, corresponding to periods of 21.86 min, 27.74 min, 9.37 min, and 4.27 min, respectively.

Actually, we found that five stars out of nine from our program with time series show periodic phase variations of magnetic fields with short periods (minutes, tens minutes). Since the topology of the magnetic field is not known, it is difficult to estimate the impact of non-radial pulsations causing strong line asymmetries on our measurements. It is quite possible that lines of different elements behave differently with respect to their pulsation amplitudes and shapes of the line profiles. Nevertheless, non-radial pulsations can be responsible for a periodic spectral and photometric variability observed on short-term scales (less than 1 day).

3. Conclusions

Our search for magnetic fields in Be stars revealed that while their magnetic fields are rather weak, fields of the order of 100G and less are not rare (7 Be stars out of 28 or $\sim$1/4). Since a large fraction of stars in our sample was observed only once, a non-detection of their magnetic field may be explained by temporal variability of their magnetic fields. A cyclical variability with short periods (minutes, tens of minutes) was detected in 5 stars out of nine with time series. Observations of Be stars at different stages of their phase transition "Be→Be-shell→normal B" would be very useful for understanding of the role of magnetic fields to Be-phenomenon. It is beyond doubt that magnetic fields are present in Be stars, but they are not strong !

References

Brown, J. C., Telfer, D., Li, Q., Hanuschik, R. *et al.* 2004, *MNRAS* 352, 1061
Cassinelli, J. P., Brown, J. C., Maheswaran, M., Miller, N. A. *et al.* 2002, *ApJ* 578, 951
Hubrig, S., Yudin, R. V., Pogodin, M., Schöller, M. *et al.* 2007, *AN* 328, 1133
Hubrig, S., Schöller, M., Savanov, I., Yudin, R. V. *et al.* 2009a, *AN* 330, 708
Hubrig, S., Schöller, M., Briquet, M., De Cat, P. *et al.* 2009b, *The Messenger* 135, 21
Maheswaran, M. 2003, *ApJ* 592, 1156
Maheswaran, M. 2005, in: R. Ignace & K. G. Gayley (eds.), *The Nature and Evolution of Disks Around Hot Stars*, ASP-CS 337, p. 259
Maheswaran, M. & Cassinelli, J. P. 2009, *MNRAS* 394, 415

Active OB stars: structure, evolution, mass loss, and critical limits
Proceedings IAU Symposium No. 272, 2010
C. Neiner, G. Wade, G. Meynet & G. Peters, eds.

doi:10.1017/S1743921311010453

Measurements of magnetic fields in Herbig Ae/Be stars and stars with debris disks at the VLT 8-m telescope: statistical results of our long-term program

Ruslan V. Yudin[1], Swetlana Hubrig[2], Michail A. Pogodin[1], Markus Schoeller[3] and Ilya Ilyin[2]

[1]Central Astronomical Observatory of the Russian Academy of Sciences at Pulkovo, Saint-Petersburg, Russia, email: ruslan61@gao.spb.ru

[2]Astrophysical Institute Potsdam, Germany

[3]European Southern Observatory, Garching, Germany

Abstract. We present the results of magnetic field measurements for a sample of 23 young Herbig Ae/Be (HAEBEs) stars and 12 stars with debris disks. The spectropolarimetric data were obtained during four observing runs in 2003-2008 at the European Southern Observatory with the multi-mode instrument FORS 1 installed at the 8 m Kueyen telescope. Among the 23 HAEBEs studied, stellar magnetic fields of about 100-150G have been detected in 11 stars (i.e. ~50%). The presence of circumstellar polarization signatures formed in the stellar wind supports the assumption that the magnetic centrifuge is one of the main mechanisms of the wind acceleration. No field detection at a significance level of 3σ was achieved in stars with debris disks.

Keywords. stars: magnetic fields, stars: early-type, stars: emission-line

1. Introduction

Numerous theoretical works predict the existence of a global magnetic field around HAEBEs. Nevertheless over an extended period of years all attempts to obtain reliable direct measurements of magnetic fields of HAEBEs have been rather unsuccessful. In the last years definite evidence for the presence of magnetic fields of the order of about 100G has been presented for several HAEBEs by Hubrig *et al.* (2004, 2006). After that time direct spectropolarimetric observations of several HAEBEs showed that magnetic fields are indeed present in intermediate mass pre-main sequence stars (e.g., Wade *et al.* 2007; Hubrig *et al.* 2009) indicating that magnetic fields are important ingredients of the star formation process (McKee & Ostriker 2007). Here we summarized our results of magnetic field measurements for HAEBEs obtained in a framework of our long-term program. For the measurement of the magnetic fields we used spectropolarimetric observations obtained with the 8m VLT+FORS1 (using low-resolution R=2000) during four observing runs in the years 2003-2008.

2. Results and Discussion

During four runs we were able to obtain circular polarization data for 23 HAEBEs and 12 debris disc stars. Among this sample the detections of a magnetic field were achieved in twelve HAEBEs, in most of them just occasionally on single nights (Hubrig *et al.* 2004; 2006; 2009). For a few stars we investigated the photospheric and circumstellar

(CS) magnetic field components separately, showing that the spectropolarimetric results strongly depend on the level of the CS contribution to the stellar spectra. Strong distinct Zeeman features at the position of the Ca II H&K lines were detected in four HAEBEs. These lines are very likely formed at the base of the stellar wind, as well as in the accretion gaseous flow and frequently display multi-component complex structures in both the Stokes V and the Stokes I spectra. In two HAEBEs, HD31648 and HD190073, such a structure was especially noticeable, and from their study we concluded that a magnetic field is present in both stars, but is most likely of circumstellar origin. For HD31648, we detected a magnetic field $B_z = 87 \pm 22$G. One of the HAEBEs, HD101412, showed the largest magnetic field strength ever measured in intermediate mass pre-main-sequence stars with $B_z = -454 \pm 42$G, confirming the previous FORS1 detection by Wade *et al.* 2007. Quite recently high-resolution spectropolarimetric observations with UVES at Kueyen/UT2 at the VLT and HARPS at the 3.6m telescope on La Silla (Hubrig *et al.* 2010) revealed that HD101412 possesses the strongest magnetic field ever measured in any Herbig Ae star, with a surface magnetic field $< B >$ up to 3.5kG. The evidence for the presence of Zeeman features in circumstellar Ca doublet lines in HD31648 was also confirmed by our recent observations at high spectral resolution, R = 30000, with SOFIN at the 2.56m Nordic Optical Telescope. The magnetic field in 12 Vega-like stars, if present at all, is less than 100G, and is almost below the detection limit of spectropolarimetric measurements with FORS 1.

3. Conclusions

Among the 23 HAEBEs studied, the detections of a magnetic field were achieved in 11 stars (or ~50%). Apart from HD101412, we measured weak fields (~100-300G) in other HAEBEs, but the high resolution spectropolarimetric observations show that fields can be much stronger. These strong (~kG) fields are certainly present on the surface of few HAEBEs, but are difficult to detect due to very broad spectral lines. No definite detection was achieved for 12 stars with debris discs. Careful analysis of polarimetric spectra shows that previous (and sometimes recent) unsuccessful attempts to detect magnetic fields in HAEBEs and previous discrepancies in estimations of magnetic fields for a few stars observed on different dates, can possibly be explained by: a) weak level of their magnetic fields; b) possible variability of these fields; and c) variable contribution of circumstellar emission and absorption to the observed spectra. The measured magnetic field in some HAEBEs using low-resolution spectropolarimetry are frequently related to CS spectral lines and not to photospheric lines. The most sensitive indicator of the CS magnetic field in HAEBEs is the CaII doublet. Circular polarization features corresponding to this doublet are observed in most program targets. The magnetic field diagnosed in the CaII lines is generated in the CS matter in the vicinity of the stellar surface where the base of the stellar wind as well as gaseous flows infalling onto the star are likely located.

References

Hubrig, S., Schöller, M., & Yudin, R. V. 2004, *A&A* 428, L1
Hubrig, S., Yudin, R. V., Schöller, M., & Pogodin, M. A. 2006, *A&A* 446, 1089
Hubrig, S., Stelzer, B., Schöller, M., Grady, C. *et al.* 2009, *A&A*, 502, 283
Hubrig, S., Schöller, M., Savanov, I., González, J. F. *et al.* 2010, *AN* 331, 361
McKee, C. F. & Ostriker, E. C. 2007, *ARAA* 45, 565
Wade, G. A., Bagnulo, S., Drouin, D., Landstreet, J. D. *et al.* 2007, *MNRAS*, 376, 1145

Discussion during the welcome cocktail and coffee breaks. From left to right and top to bottom: Doug Gies and Pavel Koubsky; Jiri Krticka, ? and Stan Owocki; Philippe Stee and Christopher Russell; Catherine Lovekin, Adrian Potter and Nathan Smith; Eduardo Janot-Pacheco, Jean Zorec, ?, and Antono Pereyra; Gloria Koenigsberger, Mary Oksala, Richard Townsend and Thomas Rivinius; Myron Smith and Hideyuki Saio; Jean-Paul Zahn, Stéphane Mathis and André Maeder.

Active OB stars: structure, evolution, mass loss, and critical limits
Proceedings IAU Symposium No. 272, 2010
C. Neiner, G. Wade, G. Meynet & G. Peters, eds.

doi:10.1017/S1743921311010465

Massive stars in globular clusters: drivers of chemical and dynamical evolution

Thibault Decressin†

Argelander Institute for Astronomy (AIfA), Auf dem Hügel 71, D-53121 Bonn, Germany
email: Thibaut.Decressin@unige.ch

Abstract. Massive stars have a strong impact on globular cluster evolution. First providing they rotate initially fast enough they can reach the break-up velocity during the main sequence and a mechanical mass-loss will eject matter from the equator at low velocity. Rotation-induced mixing will also bring matter from the convective core to the surface. From this ejected matter loaded in H-burning material a second generation of stars will born. The chemical pattern of these second generation stars are similar to the one observed for stars in globular cluster with abundance anomalies in light elements. Then during the explosion as supernovae the massive stars will also clear the cluster of the remaining gas. If this gas expulsion process acts on short timescale it can strongly modify the dynamical properties of clusters by ejecting preferentially first generation stars.

Keywords. stars: abundances, stars: evolution, stars: mass loss, globular clusters: general, stellar dynamics, methods: n-body simulations

1. Introduction

Globular clusters are self-gravitating aggregates of tens of thousands to millions of stars that have survived over a Hubble time. Many observations show that these objects are composed of multiple stellar populations. The first evidence rests on the chemical analysis that reveals large star-to-star abundance variations in light elements in all individual clusters studied so far, while the iron abundance stays constant (for a review see Gratton *et al.* 2004). These variations include the well-documented anticorrelations between C-N, O-Na, Mg-Al, Li-Na and F-Na (Kraft 1994; Carretta *et al.* 2007; Gratton *et al.* 2007; Pasquini *et al.* 2007; Carretta *et al.* 2010; Lind *et al.* 2009). H-burning at high temperature around 75×10^6 K is required to explain this global chemical pattern (Arnould *et al.* 1999; Prantzos *et al.* 2007). As the observed chemical pattern is present in low-mass stars both on the red giant branch (RGB) and at the turn-off that do not reach such high internal temperatures, the abundance anomalies must have been inherited at the time of formation of these stars.

Besides, deep photometric studies provide another indications for multiple populations in individual GCs with the discoveries of multiple giant and sub-giant branches or main sequences. In ω Cen a blue main sequence has been discovered (Bedin *et al.* 2004) that is presumably related to a high content in He (Piotto *et al.* 2005; Villanova *et al.* 2007). A triple main sequence has been discovered in NGC 2808 (Piotto *et al.* 2007). The additional blue sequences are explainable by a higher He content of the corresponding stars which shifts the effective temperatures towards hotter values. He-rich stars have also been proposed to explain the morphology of extended horizontal branch (hereafter HB) seen in many globular clusters (see e.g., Caloi & D'Antona 2005). Whereas no direct

† Present address: Geneva Observatory, University of Geneva, 51 ch. des Maillettes, 1290 Versoix, Switzerland

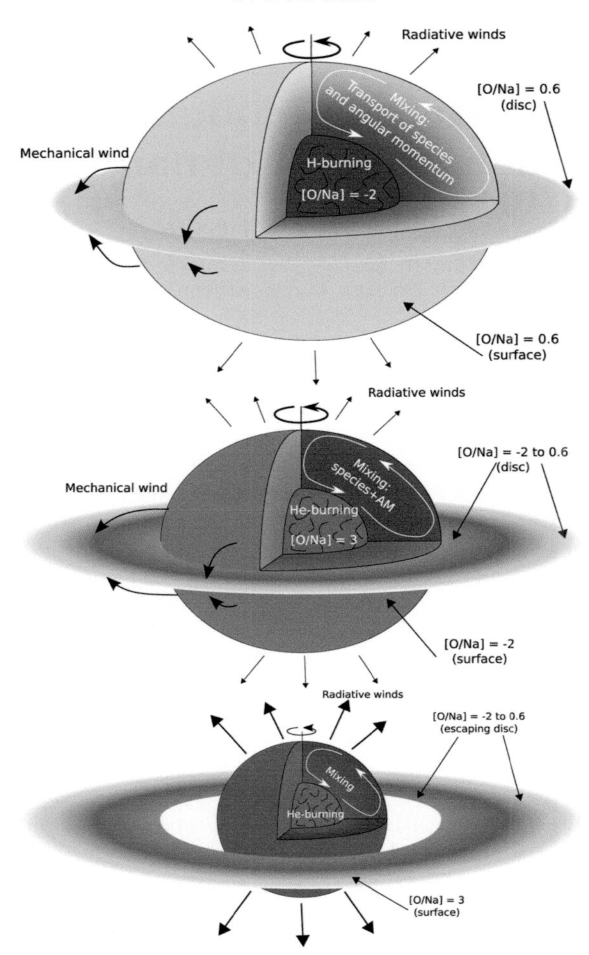

Figure 1. Schematic view of the evolution of fast rotating massive stars. The colours reflect the chemical composition of the various stellar regions and of the disc. (top) During the main sequence, a slow outflowing equatorial disc forms and dominates matter ejection with respect to radiative winds. (middle) At the beginning of central He-burning, the composition of the disc material spans the range in [O/Na] observed today in low-mass cluster stars. The star has already lost an important fraction of its initial mass. (bottom) Due to heavy mass loss, the star moves away from critical velocity and does not supply its disc anymore; radiatively-driven fast wind takes over before the products of He-burning reach the stellar surface.

observational link between abundance anomalies and He-rich sequences has been found, this link is easily understood theoretically as abundance anomalies are the main result of H burning to He.

These observed properties lead to the conclusion that globular clusters born from giant gas clouds first form a generation of stars with the same abundance pattern as field stars. Then a polluting source enriches the intracluster-medium with H-burning material out of which a chemically-different second stellar generation forms. This scheme can explain at the same time the abundance anomalies in light elements and He-enrichment.

Two main candidates that reach the right temperature for H-burning have been proposed to be at the origin of the abundance anomalies (Prantzos & Charbonnel 2006): (a) intermediate mass stars evolving on the thermal pulses along the asymptotic giant branch (hereafter TP-AGB), and (b) main sequence massive stars. After being first proposed by Cottrell & Da Costa (1981) the AGB scenario has been extensively studied (see e.g., Ventura & D'Antona 2008a,b, 2009, see Ventura, this volume) and has been seriously challenged by rotating AGB models that predict unobserved CNO enrichment in low-metallicity globular clusters (Decressin *et al.* 2009).

On the other hand, as being suggested by Brown & Wallerstein (1993) and Wallerstein *et al.* (1987), massive stars can also pollute the interstellar medium (ISM) of a forming cluster (see Smith 2006; Prantzos & Charbonnel 2006). In particular Decressin *et al.* (2007b) show that fast rotating massive stars (with a mass higher than $\sim 25\ M_{\odot}$) are good candidates for the self-enrichment of globular clusters (see § 2). An alternative suggestion has been proposed by de Mink *et al.* (2009) to consider non-conservative mass-transfer from binaries stars. In the following we will only consider the consequences of the pollution by rotating massive stars.

2. Chemical population of globular clusters

In the wind of fast rotating massive stars (WFRMS) scenario sketched in Fig. 1, rotationally-induced mixing transports H-burning products (and hence matter with correct abundance signatures) from the convective core to the stellar surface. Providing initial rotation is high enough, the stars reach the break-up on the main sequence evolution. As a result a mechanical wind is launched from the equator that generates a disk around the star similar to that of Be stars (e.g. Townsend *et al.* 2004). Later, when He-burning products are brought to the surface, the star has already lost a high fraction of its initial mass and angular momentum, so that it no longer rotates at the break-up velocity. Matter is then ejected through a classical fast isotropic radiative wind. From the matter ejected in the disk, a second generation of stars may be created with chemical pattern in agreement with observations.

The slow winds ejected by massive stars have no Li has this fragile elements burn at 3.5×10^6 K. The dilution of the slow winds (Li-free) with the pristine ISM (Li-rich) can explain the presence of Li in stars with abundance anomalies as observed in several clusters (see e.g., Pasquini *et al.* 2005; Lind *et al.* 2009; D'Orazi & Marino 2010; D'Orazi *et al.* 2010). By taking this dilution into account we can reproduce the trends of the Li-Na anticorrelation observed in NGC 6752 as observed by (Pasquini *et al.* 2005) as well as the anticorrelation between O and Na (see Decressin *et al.* 2007a).

3. Dynamical consequences for globular clusters

Based on the determination of the composition of giant stars in 19 globular clusters by Carretta *et al.* (2009) the stars with abundance anomalies populate between 50 to 80% of the cluster stars.

How to produce such a high fraction of chemically peculiar stars? The main problem is that assuming a Salpeter (1955) IMF for the polluters, the accumulated mass of the slow winds ejected by the fast rotating massive stars would only provide 10% of the total number of low-mass stars. To match the observations thus requires either (a) a flat IMF with a slope of 0.55 instead of 1.35 (Salpeter's value), or (b) that 95% of the first generation stars have escaped the cluster (Decressin *et al.* 2007a). Here we first verify whether such a high loss of stars is possible, and which are the main processes that could drive it.

3.1. *Dynamical evolution of globular clusters*

First we assume that the globular clusters display primordial mass segregation so that the massive stars are located at their center. Since we expect that the formation of the second generation of low-mass stars happens locally around individual massive stars (see Decressin *et al.* 2007a for more details), the second generation of stars will also be initially more centrally concentrated than the first generation. In such a situation, two competitive processes act in the clusters: the loss of stars in the outer cluster parts will first reduce the number of bound first generation stars; and the dynamical spread of the initially more concentrated second generation stars will stop this differential loss when the two populations are dynamically mixed.

Our analysis, based on the N-body models computed by Baumgardt & Makino (2003) with the collisional Aarseth N-body code NBODY4 (Aarseth 1999), is presented in detail in Decressin *et al.* (2008).

As these models have been computed for a single stellar population, we apply the following process to mimic the formation of a cluster with two dynamically distinct populations: we sort all the low-mass stars ($M \leqslant 0.9$ M$_\odot$) according to their specific energy (i.e., their energy per unit mass). We define the second stellar generation as the stars with lowest specific energies, (i.e., those which are most tightly bound to the cluster due to their small central distance and low velocity). The number of second generation stars is given by having their total number representing 10% of the total number of low-mass stars. Initially, first generation stars show an extended distribution up to 40 pc whereas the second generation stars (with low specific energy) are concentrated within 6 pc around the centre.

Initially, as only stars of the first generation populate the outer part of the cluster owing to their high specific energy, only these first generation stars are lost in the early times. This lasts until the second generation stars migrate towards the outer part of the cluster. Depending on the cluster mass, it takes between 1 to 4 Gyr (about 2–3 relaxation times) to start losing second generation stars. Due to the time-delay to lose second generation stars, their fraction of second generation stars relative to first generation ones increases first. Then it tends to stay nearly constant as soon as the two distributions are similar. This fraction only increases by a factor of 2.5 over the cluster history. Therefore, these second generations stars can account for 25% of the low-mass stars present in the clusters. Compared to the observed ratios (50 to 80% in 19 globular clusters) the internal dynamical evolution and the dissolution due to the tidal forces of the host Galaxy are not efficient enough. An additional mechanism is thus needed to expel the first generation stars more effectively.

3.2. *Gas expulsion*

As it operates early in the cluster history (a few million years after cluster formation at the latest), initial gas expulsion by supernovae is an ideal candidate for such a process. As the gas still present after the star formation is quickly removed, it ensues a strong lowering of the potential well of the cluster so that the outer parts of the cluster can become unbound.

Baumgardt & Kroupa (2007) computed a grid of N-body models to study this process and its influence on cluster evolution by varying the free parameters: star formation efficiency, SFE, ratio between the half-mass and tidal radius, r_h/r_t, and the ratio between the timescale for gas expulsion to the crossing time, $\tau_M/t_{\rm Cross}$. They show in particular that, in some extreme cases, the complete disruption of the cluster can be induced by gas expulsion. This process has also been used successfully by Marks *et al.* (2008) to explain

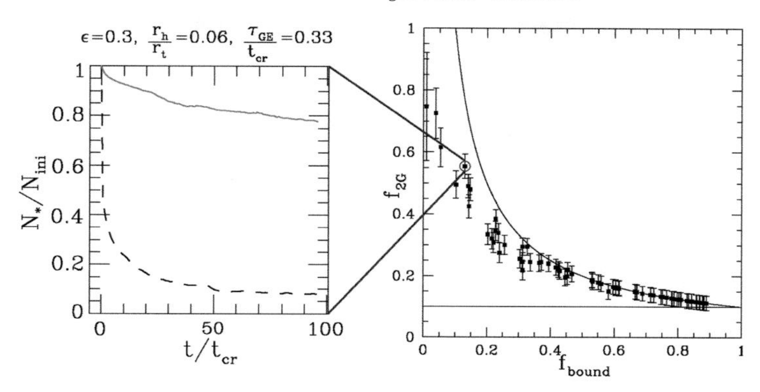

Figure 2. *Left:* Number of first (dotted line) and second (full line) generation stars relative to their initial number as a function of time for a cluster with the following initial properties: $\epsilon = 0.3$, $r_{\rm h}/r_{\rm t} = 0.06$ and $\tau_{GE}/t_{\rm cr} = 0.33$. *Right:* Fraction of second-generation stars as a function of the final fraction of bound stars at the end of the computations of Baumgardt & Kroupa (2007), i.e., after about 100 initial crossing times. Dashed lines indicate limiting cases where no second-generation stars are lost (upper) and no preferential loss of first-generation stars occurs (lower). Estimates of the statistical errors are also included based on the number of first, N_1, and second, N_2, generation stars bound to the cluster.

the challenging correlation between the central concentration and the mass function of globular clusters as found by De Marchi *et al.* (2007).

We have applied the same method as the one we used in § 2.2 to the models of Baumgardt & Kroupa (2007). Fig. 2 (left panel) shows that in the case of a cluster which loses around 90% of its stars, the ejection of stars from the cluster mostly concerns first generation ones. At the end of the computation only 7% of first generation stars remain bound to the cluster along with most of second generation stars. Therefore the number ratio between the second to first generation stars increases by a factor of 10: half of the population of low-mass stars still populating the cluster are second generation stars. The final fraction of second generation stars is strongly dependent of the initial properties of the cluster as shown in Fig. 2 (right panel) where this number is shown as a function of the fraction of star still bound to the cluster for the whole set of computation done by Baumgardt & Kroupa (2007) (see Decressin *et al.* 2010 for more details).

Thus if globular clusters are born mass segregated, dynamical processes (gas expulsion, tidal stripping and two-body relaxation) can explain the number fraction of second generation stars with abundance anomalies. Similar conclusions have been reached by D'Ercole *et al.* (2008).

Acknowledgements

TD acknowledges support from the Swiss National Science Foundation (FNS) and from the "Programme National de Physique Stellaire" of CNRS/INSU, France.

References

Aarseth, S. J. 1999, *PASP*, 111, 1333
Arnould, M., Goriely, S. & Jorissen, A. 1999, *A&A*, 347, 572
Baumgardt, H. & Kroupa, P. 2007, *MNRAS*, 380, 1589
Baumgardt, H. & Makino, J. 2003, *MNRAS*, 340, 227
Bedin, L. R., Piotto, G., Anderson, J., Cassisi, S. *et al.* 2004, *ApJ* (Letters), 605, L125
Brown, J. A. & Wallerstein, G. 1993, *AJ*, 106, 133
Caloi, V. & D'Antona, F. 2005, *A&A*, 435, 987

Carretta, E., Bragaglia, A., Gratton, R. G., Recio-Blanco, A. *et al.* 2010, *A&A*, 516, A55+
Carretta, E., Bragaglia, A., Gratton, R. G., Lucatello, S. *et al.* 2009, *A&A*, 505, 117
Carretta, E., Bragaglia, A., Gratton, R. G., Lucatello, S. *et al.* 2007, *A&A*, 464, 927
Cottrell, P. L. & Da Costa, G. S. 1981, *ApJ* (Letters), 245, L79
De Marchi, G., Paresce, F., & Pulone, L. 2007, *ApJ* (Letters), 656, L65
de Mink, S. E., Pols, O. R., Langer, N., & Izzard, R. G. 2009, *A&A*, 507, L1
Decressin, T., Baumgardt, H., Charbonnel, C., & Kroupa, P. 2010, *A&A*, 516, A73+
Decressin, T., Baumgardt, H., & Kroupa, P. 2008, *A&A*, 492, 101
Decressin, T., Charbonnel, C., & Meynet, G. 2007a, *A&A*, 475, 859
Decressin, T., Charbonnel, C., Siess, L., Palacios, A. *et al.* 2009, *A&A*, 505, 727
Decressin, T., Meynet, G., Charbonnel, C., Prantzos, N. *et al.* 2007b, *A&A*, 464, 1029
D'Ercole, A., Vesperini, E., D'Antona, F., McMillan, S. L. W., & Recchi, S. 2008, *MNRAS* 391, 825
D'Orazi, V., Lucatello, S., Gratton, R., Bragaglia, A. *et al.* 2010, *ApJ* (Letters), 713, L1
D'Orazi, V. & Marino, A. F. 2010, *ApJ* (Letters), 716, L166
Gratton, R., Sneden, C. & Carretta, E. 2004, *ARAA*, 42, 385
Gratton, R. G., Lucatello, S., Bragaglia, A., Carretta, E. *et al.* 2007, *A&A*, 464, 953
Kraft, R. P. 1994, *PASP*, 106, 553
Lind, K., Primas, F., Charbonnel, C., Grundahl, F. *et al.* 2009, *A&A*, 503, 545
Marks, M., Kroupa, P. & Baumgardt, H. 2008, *MNRAS*, 386, 2047
Pasquini, L., Bonifacio, P., Molaro, P., Francois, P. *et al.* 2005, *A&A*, 441, 549
Pasquini, L., Bonifacio, P., Randich, S., Galli, D. *et al.* 2007, *A&A*, 464, 601
Piotto, G., Bedin, L. R., Anderson, J., King, I. R. *et al.* 2007, *ApJ* (Letters), 661, L53
Piotto, G., Villanova, S., Bedin, L. R., Gratton, R. *et al.* 2005, *ApJ*, 621, 777
Prantzos, N. & Charbonnel, C. 2006, *A&A*, 458, 135
Prantzos, N., Charbonnel, C., & Iliadis, C. 2007, *A&A*, 470, 179
Salpeter, E. E. 1955, *ApJ*, 121, 161
Smith, G. H. 2006, *PASP*, 118, 1225
Townsend, R. H. D., Owocki, S. P., & Howarth, I. D. 2004, *MNRAS*, 350, 189
Ventura, P. & D'Antona, F. 2008a, *A&A*, 479, 805
Ventura, P. & D'Antona, F. 2008b, *MNRAS*, 385, 2034
Ventura, P. & D'Antona, F. 2009, *A&A*, 499, 835
Villanova, S., Piotto, G., King, I. R., Anderson, J. *et al.* 2007, *ApJ*, 663, 296
Wallerstein, G., Leep, E. M., & Oke, J. B. 1987, *AJ*, 93, 1137

Discussion

D. BAADE: Why does the evolution ends with the second generation, i.e. why isn't there a third generation

T. DECRESSIN: In my models I consider that second generation stars consist only of low-mass stars. If this assumption is relaxed and the formation of massive second generation stars are allowed, you will run out of matter to form the third generation: a 60 $M_\odot$ star (first generation) can only produce a 30 $M_\odot$ star (second generation)

D. BOMANS: Since we observe proto-globular clusters which have cleaned out their gas and have ages of $\sim 5-10\times10^6$ yr, what is the timescale of the formation of your second generation?

T. DECRESSIN: The timescale for the formation of second generation stars need to be short to avoid that supernovae cleared out the slow winds of massive stars during the primordial gas expulsion process. It should be similar to the lifetime of massive polluters, i.e., a few Myr.

Active OB stars: structure, evolution, mass loss, and critical limits
Proceedings IAU Symposium No. 272, 2010
C. Neiner, G. Wade, G. Meynet & G. Peters, eds.

doi:10.1017/S1743921311010477

Populations of OB-type stars in galaxies

Christopher J. Evans

UK Astronomy Technology Centre, Blackford Hill, Edinburgh, EH9 3HJ, UK
email: chris.evans@stfc.ac.uk

Abstract. One of the challenges for stellar astrophysics is to reach the point at which we can undertake reliable spectral synthesis of unresolved populations in young, star-forming galaxies at high redshift. Here I summarise recent studies of massive stars in the Galaxy and Magellanic Clouds, which span a range of metallicities commensurate with those in high-redshift systems, thus providing an excellent laboratory in which to study the role of environment on stellar evolution. I also give an overview of observations of luminous supergiants in external galaxies out to a remarkable 6.7 Mpc, in which we can exploit our understanding of stellar evolution to study the chemistry and dynamics of the host systems.

Keywords. Galaxy: stellar content, Magellanic Clouds, stars: early-type, stars: fundamental parameters

1. Introduction

One of the prime motivations to study stellar evolution is to use that knowledge to develop tools to explain integrated-light observations of distant star clusters and galaxies. Consider the recent multi-wavelength study of a gravitationally-lensed galaxy at a redshift of $z = 2.3$ by Swinbank *et al.* (2010). By virtue of the lens, individual regions are resolved in sub-millimetre imaging, each ~100 pc in scale. These are intense regions of star formation on a comparable spatial scale to that of 30 Doradus, viewed at a time when the universe was significantly younger. Such observations are only possible at present due to the magnification of the lens but, with future facilities such as the Atacama Large Millimetre Array (ALMA) and Extremely Large Telescopes (ELTs), we can expect comparable observations in unlensed systems in the coming years. One of the real tests of stellar astrophysics is to reach the point at which we are confident that we can interpret integrated-light spectroscopy of such distant systems accurately, exploiting our understanding of massive stars to obtain new insights into the processes at work during one of the most critical epochs of galaxy evolution.

Population synthesis codes such as Starburst99 (Leitherer *et al.* 1999) are the 'bridge' from studies of individual stars to analysis of entire populations on galaxy scales. The rest-frame ultraviolet (UV) is rich with the signatures of stellar winds and, for high-redshift galaxies, is redshifted into the optical (e.g. Pettini *et al.* 2002) and, ultimately, into the near-infrared for the most distant systems. To model the rest-frame UV a new spectral library for Starburst99 has been calculated by Leitherer *et al.* (2010) using the WM-Basic model atmosphere code (Pauldrach, Hoffmann & Lennon, 2001). Fig. 1 shows a comparison of library spectra from the *International Ultraviolet Explorer (IUE)* with spectra calculated using the new WM-Basic library. In general there is excellent agreement, a worthy testament to theoretical developments in recent years! Many of the apparent differences are related to observational issues such as narrow interstellar features and the significant wings of the Lyman-α absorption from Galactic H I. The most discrepant stellar feature is O V λ1371Å, known to be sensitive to wind inhomogeneities ('clumping'), the effects of which are not incorporated in WM-Basic at present.

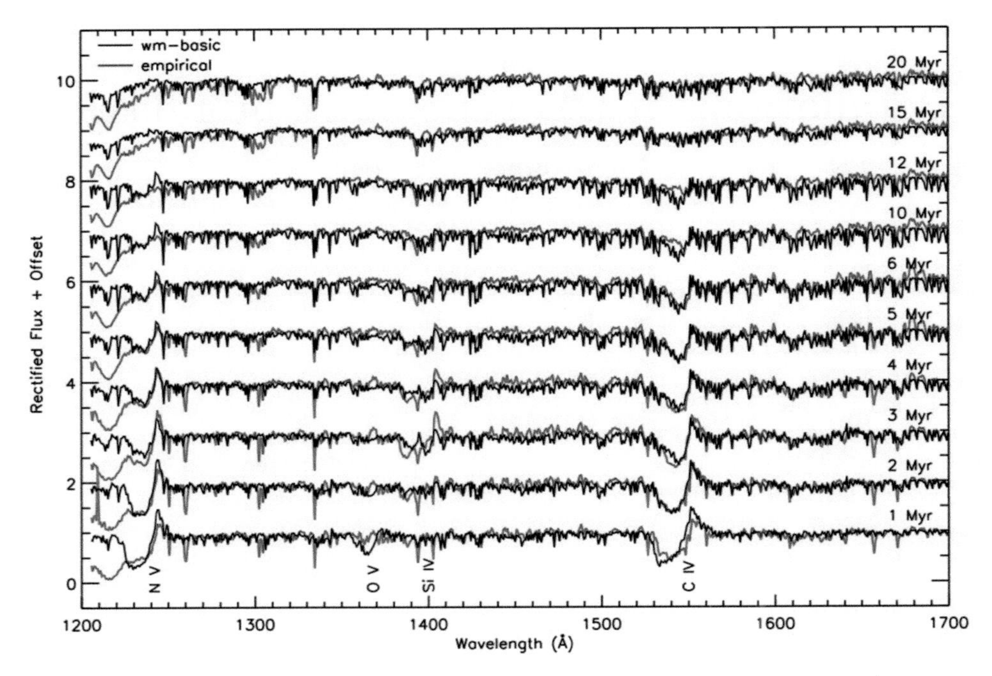

Figure 1. Single-stellar populations from the new Starburst99 WM-Basic library (solar metallicity, black lines) compared to empirical *IUE* templates (red lines), from Leitherer *et al.* (2010).

The comparison in Fig. 1 reminds me of comment that a high-redshift astronomer once said to me at a conference: *'... but stars are done aren't they?'* At the time I argued strongly to the contrary – while we now have a significant command of stellar astrophysics, a broad range of fundamental questions still eludes us for massive stars, including basic issues such as their formation (Zinnecker & Yorke, 2007) and end-points (Smith *et al.* 2010).

We can use observations of populations of massive stars to improve our understanding of the physics and evolution of the stars themselves and, once we are confident of a decent grasp of their behaviour, we can use them as tracers of the properties of their host galaxies. In the following sections, I summarise recent developments in terms of the role of environment on the evolution of massive stars (Sec. 2), observational studies of luminous supergiants in external galaxies (Sec. 3), and the potential of the next generation of ground-based telescopes in studies of massive stars (Sec. 4).

2. Massive Stars & Metallicity

The Large and Small Magellanic Clouds (LMC and SMC) are metal deficient when compared to the solar neighbourhood, with metallicities (Z) of approximately 50% and 20-25% solar (e.g. Trundle *et al.* 2007). This presents us with the chance to study the properties of massive stars in regimes which span a range in metallicity comparable to those found in galaxies in the early universe (e.g. Erb *et al.* 2006; Nesvadba *et al.* 2008).

In the course of investigating near-IR photometry from the *Spitzer Space Telescope*, Bonanos *et al.* (2009, 2010, these proceedings) have compiled catalogues of published spectral classifications for over 7,000 massive stars in the Clouds. This illustrates the significant progress we have made in determining the stellar content of two of our nearest neighbours. Over the past decade the focus has been to exploit the latest generation of

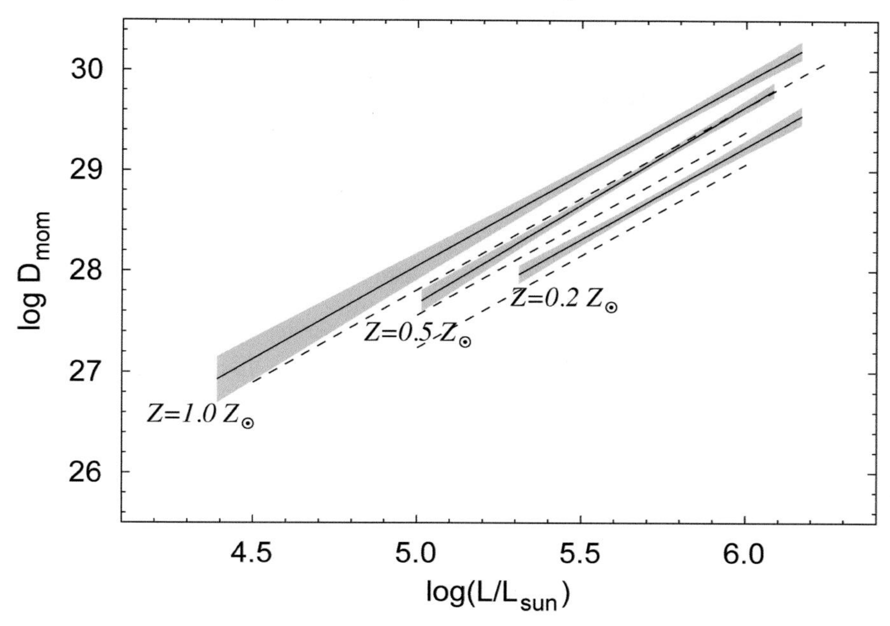

Figure 2. The observed wind-momentum–luminosity relations (solid lines) for O-type stars from the FLAMES survey (Mokiem *et al.* 2007b) compared with theoretical predictions (dashed lines). The upper, middle and lower relations are the Galactic, LMC and SMC results, respectively.

ground-based telescopes to deliver sufficient spectral resolution and signal-to-noise for quantitative atmospheric analysis of large samples of massive stars in the Clouds. This includes the VLT-FLAMES Survey of Massive Stars (Evans *et al.* 2005, 2006), and more targeted studies of O-type stars (Massey *et al.* 2004, 2005, 2009) and B-/Be-type stars (Martayan *et al.* 2006, 2007). Analysis of these samples has been used to investigate a broad range of parameters, including the Z-dependence of stellar wind intensities, effective temperatures, and rotational velocities, each of which is now briefly discussed.

2.1. *Metallicity-dependent stellar winds*

A consequence of the theory of radiatively-driven stellar winds is that their intensities should be dependent on Z, with weaker winds at lower metallicities (Kudritzki *et al.* 1987; Vink *et al.* 2001). Indications for such an offset were seen from analysis of 22 stars in the Clouds (Massey *et al.* 2005), with more comprehensive evidence provided by the larger sample from the FLAMES survey (Mokiem *et al.* 2007b). The FLAMES results are shown in Fig. 2, with fits to the stellar wind-momenta, D_{mom} (which is a function of the mass-loss rate, terminal velocity and stellar radius). Fig. 2 also shows the theoretical predictions using the prescription from Vink *et al.* (2001) – the relative separations of the observed fits are in good agreement, finding a Z-dependence with exponents in the range 0.72-0.83 (depending on assumptions regarding clumping in the winds), as compared to $Z^{0.69 \pm 0.10}$ from theory. Analysis of early B-type supergiants in the SMC (Trundle *et al.* 2004; Trundle & Lennon 2005) also reveals weaker winds compared to their Galactic counterparts (Crowther, Lennon & Walborn, 2006).

2.2. *Stellar Effective Temperatures*

The effects of line-blanketing are reduced at lower metallicities due to the diminished cumulative opacity from the metal lines. This leads to less 'back warming' by trapped radiation, thus requiring a hotter model to reproduce the observed line ratios.

The consequences of this are clearly seen in the temperatures obtained for O-type stars in the SMC, which are hotter than those found for Galactic stars with the same spectral type (Massey *et al.* 2005; Mokiem *et al.* 2006). The temperatures for stars in the LMC are seen to fall neatly between the SMC and Galactic results (Mokiem *et al.* 2007a). A similar Z-dependence was also found in the effective temperatures derived for B-type stars observed with FLAMES (Trundle *et al.* 2007). Typical differences from this effect are 5-10% when moving from solar to SMC metallicity. Note that there are also temperature differences for cooler supergiants (e.g. Evans & Howarth, 2003; Levesque *et al.* 2006), but in the opposite sense and for different reasons (see Evans, 2009).

2.3. *Stellar rotational velocities*

Stellar rotation strongly affects the evolution of all O- and B-type stars via, for example, changes in their main-sequence lifetimes (Meynet & Maeder, 2000). The effects of rotation also manifest themselves via mixing of chemically-processed material, leading to changes in the surface abundances of elements such as nitrogen (e.g. Brott *et al.*, these proceedings).

The increased fraction of Be- to normal B-type stars at lower metallicity (Maeder, Grebel & Mermilliod, 1999) pointed to a Z-dependence in stellar rotation rates. Investigations into the effects of Z on stellar rotational rates have largely focussed on studies of B-type stars, to avoid the potential complications of angular momentum loss due to strong stellar winds in more massive stars. In the Magellanic Clouds it can be difficult to define an appropriate 'field' or 'cluster' sample for comparison, given that the rotation rates for stars in clusters appear faster than for those in the field population (Keller, 2004; Strom *et al.* 2005; Wolff *et al.* 2008). However, a trend of faster velocities at lower Z appears to be borne out by comparisons between rotation rates in the Galaxy and the LMC (Keller, 2004) and, more recently, including the SMC (Martayan *et al.* 2007; Hunter *et al.* 2008).

Mokiem *et al.* (2006) found tentative evidence for different rotational velocity distributions for O-type stars in the SMC compared to the Galaxy, but with some reservations given the potential effects of mass-loss and macroturbulence. New results suggest that macroturbulence is ubiquitous in O-type stars (e.g. Simón-Díaz *et al.*, these proceedings), with Penny & Gies (2009) suggesting that its magnitude is also Z-dependent. This perhaps points to an origin similar to the convective effects argued by Cantiello *et al.* (2009) to account for *microturbulence* in massive stars (also seen to have a Z-dependence).

2.4. *What next?*

The VLT-FLAMES Tarantula Survey is a new ESO Large Programme which has obtained multi-epoch spectroscopy of over 1,000 stars in the 30 Doradus region of the LMC (Evans *et al.* 2010a). 30 Dor is the largest H II region in the Local Group, providing us with an excellent stellar nursery to build-up a large observational sample of the most massive stars; the top-level motivations for the survey are summarised by Lennon *et al.* (these proceedings).

Multi-epoch, radial velocity studies of OB-type stars in Galactic clusters have found binary fractions in excess of 50% (see Sana & Evans, these proceedings), and the results of Bosch, Terlevich & Terlevich (2009) suggest a similarly large binary fraction for the O-type stars in 30 Dor. The Tarantula Survey was designed to combine high-quality spectroscopy for quantitative atmospheric analysis, with repeat observations for detection of massive binaries (over a longest baseline of one year).

This strategy has already paid dividends in terms of putting strong constraints on the nature of 30 Dor #016, a massive O2-type star on the western fringes of 30 Dor

Table 1. Summary of published spectroscopic observations and analyses of luminous blue supergiants in external galaxies from the Araucaria project.

Galaxy	d [Mpc]	12+log(O/H)	References
IC 1613	0.7	7.90 ± 0.08	Bresolin *et al.* (2007)
WLM	0.9	7.83 ± 0.12	Bresolin *et al.* (2006), Urbaneja *et al.* (2008)
NGC 3109	1.3	7.76 ± 0.07	Evans *et al.* (2007)
NGC 300	1.9	–	Bresolin *et al.* (2002); Urbaneja *et al.* (2003); Urbaneja *et al.* (2005a)
NGC 55	1.9	–	Castro *et al.* (2008)

(Evans *et al.* 2010b). Previous spectroscopy revealed a peculiar radial velocity, but the new FLAMES spectra enabled a massive companion to be ruled out to a high level of confidence, suggesting the star as an ejected runaway.

A further example from the new survey is the observation of two separate components in some of the spectra of R139 (Taylor *et al.*, in preparation). R139 is just over 1′ to the north of R136, the dense cluster at the core of 30 Dor, and was classified by Walborn & Blades (1997) as O7 Iafp. Previously reported as a single-lined binary with a period of 52.7 d (Moffat, 1989), recent efforts by Schnurr *et al.* (2008) found no evidence for binarity in R139 (within a range of periods of up to 200 d), although they noted it as having a slightly variable radial velocity. Chené *et al.* (these proceedings) also report new observations of radial velocity variations in R139 from a separate monitoring campaign.

3. Beyond the Magellanic Clouds

The BA-type supergiants are the most intrinsically luminous ('normal') stars at optical wavelengths. The 8-10 m class telescopes have given us the means by which we can obtain spectroscopy for quantitative analysis of individual blue supergiants in galaxies well beyond the Magellanic clouds, providing us with estimates of their physical parameters, chemical abundances, and also providing an alternative distance diagnostic in the form of the flux-weighted gravity–luminosity relationship (Kudritzki, Bresolin & Przybilla, 2003; Kudritzki *et al.* 2008).

Analysis of high-resolution spectroscopy of luminous A-type supergiants in each of NGC 6822 (Venn *et al.* 2001), M31 (McCarthy *et al.* 1997, Venn *et al.* 2000), WLM (Venn *et al.* 2003) and Sextans A (Kaufer *et al.* 2004) provided some of the first examples of this type of work in external galaxies, with analysis of larger samples of B-type supergiants in M31 by Trundle *et al.* (2002) and in M33 by Urbaneja *et al.* (2005b)

More recently, the Araucaria project has combined optical and near-IR imaging of Cepheids with spectroscopy of luminous blue supergiants to refine the distance determinations to nearby galaxies (Gieren *et al.* 2005). The Araucaria project has obtained low resolution (~5 Å) optical spectroscopy with the VLT of BA-type supergiants in a number of external galaxies, as summarised in Table 1.

This has provided us with estimates of oxygen abundances/metallicities in the irregular dwarf galaxies IC 1613 and WLM, and in the main body of NGC 3109. In the larger spirals, the stellar abundances allow study of the radial abundance trends (e.g. Urbaneja *et al.* 2005a), providing a useful comparison for nebular abundance determinations. In the case of NGC 300, Bresolin *et al.* (2009) find that 'strong-line' nebular diagnostics (such as the R_{23} ratio) overestimate the true abundance by a factor of two (or more) compared to estimates from auroral nebular lines and the stellar analyses, highlighting the caution that should be employed in metallicity estimates in distant star-forming galaxies (e.g. Kewley & Ellison, 2008).

4. Future Prospects in the Era of ELTs

Observations from the Araucaria project highlight the power of using individual stars to investigate the properties of their host galaxies. Indeed, one of the most remarkable observations in this context is the VLT spectroscopy of two luminous supergiants in NGC 3621, at a distance of 6.7 Mpc (Bresolin *et al.* 2001). However, with the exception of the very brightest stars, our potential of exploring the massive star content of external galaxies is limited by the sensitivity of current facilities. With primary apertures in excess of 20 m, the ELTs, the next generation of optical-IR telescopes, will revolutionise our ground-based capabilities, particularly when coupled with adaptive optics (AO) to correct for atmospheric turbulence, thus delivering huge gains in both sensitivity and angular resolution.

Spectroscopy of stellar populations in external galaxies is one of the key elements of the science cases toward the ELTs. This includes using stars as tracers of properties of the host galaxies, and to extend our studies of environmental effects on stellar evolution in systems such as starbursts and very metal-poor galaxies. Efforts toward building the ELTs are increasingly global, with three projects now in the advanced stages of their design, fund-raising and planning for instrumentation – the Giant Magellan Telescope (GMT), the Thirty Meter Telescope (TMT), and the European Extremely Large Telescope (E-ELT).

4.1. *MAD: An AO pathfinder for ELTs*

An integral part of the ELTs is the use of AO to deliver improved image quality in the near-IR, and there is an understandable desire to maximise the field-of-view over which one can obtain both good and uniform correction. A key technical development in plans toward the E-ELT was an on-sky demonstration of multi-conjugate adaptive optics (MCAO), which uses multiple deformable mirrors to correct for different layers of turbulence in the atmosphere. This was realised as the Multi-conjugate Adaptive optics Demonstrator (Marchetti *et al.* 2007), delivering AO-corrected near-IR imaging over a 1×1 arcmin field. MAD was commissioned at the VLT in early 2007.

As part of a science demonstration programme, MAD was used to obtain H- and K_s-band imaging of R136 (Campbell *et al.* 2010). The exquisite resolution achieved in these data is shown by Fig. 3, in which the two central WN5h stars in the core of R136 ('a1' and 'a2', separated by $0\farcs1$) are spatially resolved. This provides near-comparable angular resolution to optical imaging with the *HST* (e.g. Hunter *et al.* 1995), at wavelengths less affected by the significant and variable extinction toward 30 Dor.

When combined with AO-corrected IFU spectroscopy from Schnurr *et al.* (2009), these data have been used by Crowther *et al.* (2010) to argue that three of the central stars in R136 have current masses in excess of 150 $M_\odot$. From cluster IMF simulations they find the massive stars in R136 are consistent with an upper initial mass-limit of $\sim$300 $M_\odot$. This contrasts with the claim by Figer (2005) of an upper limit of 150 $M_\odot$ from analysis of the Arches cluster. To address this difference, Crowther *et al.* returned to the most massive members of the Arches with new photometry and revised distance/extinction estimates, finding greater current masses than those from Martins *et al.* (2008), and consistent with evolutionary tracks for stars with initial masses in excess of 150 $M_\odot$.

5. Summary

Work continues apace in characterising the massive star populations in external galaxies (e.g. Herrero *et al.* 2010) and to improve our understanding of stellar evolution in massive O-type stars (e.g. the Tarantula Survey). We now have a handle on the role of metallicity on some key aspects of stellar evolution, but there remain a number of

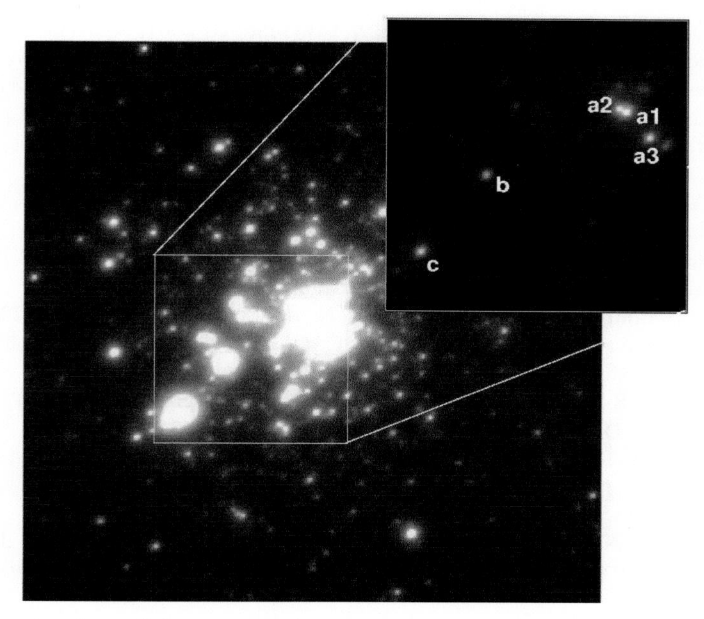

Figure 3. Central 12″ × 12″ (eqv. 3 × 3 pc) of the K_s-band MAD image, with an expanded view of the central 4″ × 4″ showing the four WN5h very massive stars analysed by Crowther *et al.* (2010; components a1, a2, a3, and c).

issues to be addressed before we can claim to really understand 'populations' in distant galaxies, such as the effects of binary evolution and revision to the accepted upper mass limit for single stars.

Inclusion for the effects of binary evolution has been argued to account for some of the observed properties of low-redshift galaxies (e.g. Han, Podsiadlowski & Lynas-Gray, 2007; Brinchmann, Kunth & Durret, 2008). The high incidence of massive binaries (Sana & Evans, these proceedings) suggests further work is required to characterise their typical properties and assess their impact on our interpretation of unresolved populations.

Lastly, the new results of Crowther *et al.* (2010) urge for investigation of the effects of very massive stars in population synthesis codes. Although very rare, and almost certainly short-lived, Crowther *et al.* show that the four WN5h stars at the core account for ∼45% of the Lyman-continuum ioinizing flux and ∼35% of the mechanical power of the respective totals from stars in the central 5 pc of R136. The inferred star-formation rates for the components of high-redshift galaxies from Swinbank *et al.* (2010) and Jones *et al.* (2010) are an order of magnitude larger than that from 30 Dor but one could, very naively, image scaling-up the density and intensity of R136 to larger scales – in which case the initial mass function would suggest a considerable input of radiation/energy from very massive stars.

Acknowledgements

I gratefully acknowledge financial support from the IAU.

References

Bonanos, A. Z., Massa, D. L., Sewilo, M., Lennon, D. J. *et al.* 2009, *AJ*, 138, 1003
Bonanos, A. Z., Lennon, D. J., Köhlinger, F., van Loon, J. T. *et al.* 2010, *AJ*, 140, 416
Bosch, G., Terlevich, E., & Terlevich, R. 2009, *AJ*, 137, 3437

Bresolin, F., Kudritzki, R.-P., Mendez, R. H., & Przybilla, N. 2001, *ApJ* (Letters), 548, L159
Bresolin, F., Gieren, W., Kudritzki, R.-P., Pietrzyński, G. *et al.* 2002, *ApJ*, 567, 277
Bresolin, F., Pietrzyński, G., Urbaneja, M. A., Gieren, W. *et al.* 2006, *ApJ*, 648, 1007
Bresolin, F., Gieren, W., Kudritzki, R.-P., Pietrzyński, G. *et al.* 2009, *ApJ*, 700, 309
Bresolin, F., Urbaneja, M. A., Gieren, W., Pietrzyński, G. *et al.* 2007, *ApJ*, 671, 2028
Brinchmann, J., Kunth, D., & Durret, F. 2008, *A&A*, 485, 657
Campbell, M. A., Evans, C. J., Mackey, A. D., Gieles, M. *et al.* 2010, *MNRAS*, 405, 421
Cantiello, M., Langer, N., Brott, I., de Koter, A. *et al.* 2009, *A&A*, 499, 279
Castro, N., Herrero, A., Garcia, M., Trundle, C. *et al.* 2008, *A&A*, 485, 41
Crowther, P. A., Lennon, D. J. & Walborn, N. R. 2006, *A&A*, 446, 279
Crowther, P. A., Schnurr, O., Hirschi, R., Yusof, N. *et al.* 2010, *MNRAS* 408, 731
Erb, D. K., Shapley, A. E., Pettini, M., Steidel, C. C. *et al.* 2006, *ApJ*, 644, 813
Evans, C. J. & Howarth, I. D. 2003, *MNRAS*, 345, 1223
Evans, C. J., Smartt, S. J., Lee, J.-K., Lennon, D. J. *et al.* 2005, *A&A*, 437, 467
Evans, C. J., Lennon, D. J., Smartt, S. J. & Trundle, C. 2006, *A&A*, 456, 623
Evans, C. J., Bresolin, F., Urbaneja, M. A., Pietrzyński, G. *et al.* 2007, *ApJ*, 659, 1198
Evans, C. J. 2009, in: J. T. van Loon & J. M. Oliveira (eds.), *The Magellanic System: Stars, Gas, and Galaxies*, IAU Symposium 256, p. 325
Evans, C. J., Bastian, N., Beletsky, Y., Brott, I. *et al.* 2010a, in: R. de Grijs & J. R. D. Lépine (eds.), *Star clusters: basic galactic building blocks throughout time and space*, IAU Symposium 266, p. 35
Evans, C. J., Walborn, N. R., Crowther, P. A., Hénault-Brunet, V. *et al.* 2010b, *ApJ* (Letters), 715, L74
Figer, D. F. 2005, *Nature*, 434, 192
Gieren, W., Pietrzynski, G., Bresolin, F., Kudritzki, R.-P. *et al.* 2005, *The Messenger*, 121, 23
Han, Z., Podsiadlowski, P. & Lynas-Gray, A. E. 2007, *MNRAS*, 380, 1098
Herrero, A., Garcia, M., Uytterhoeven, K., Najarro, F. *et al.* 2010, *A&A*, 513, A70
Hunter, D. A., Shaya, E. J., Holtzman, J. A., Light, R. M. *et al.* 1995, *ApJ*, 448, 179
Hunter, I., Lennon, D. J., Dufton, P. L., Trundle, C. *et al.* 2008, *A&A*, 479, 541
Jones, T. A., Swinbank, A. M., Ellis, R. S., Richard, J. *et al.* 2010, *MNRAS*, 404, 1247
Kaufer, A., Venn, K. A., Tolstoy, E., Pinte, C. *et al.* 2004, *AJ*, 127, 2723
Keller, S. C. 2004, *PASA*, 21, 310
Kewley, L. J. & Ellison, S. L. 2008, *ApJ*, 681, 1183
Kudritzki, R. P., Pauldrach, A. & Puls, J. 1987, *A&A*, 173, 293
Kudritzki, R. P., Bresolin, F. & Przybilla, N. 2003, *ApJ* (Letters), 582, L83
Kudritzki, R.-P., Urbaneja, M. A., Bresolin, F., Przybilla, N. *et al.* 2008, *ApJ*, 681, 269
Leitherer, C., Schaerer, D., Goldader, J. D., González Delgado, R. M. *et al.* 1999, *ApJS*, 123, 3
Leitherer, C., Ortiz Otálvaro, P. A., Bresolin, F., Kudritzki, R.-P. *et al.* 2010, *ApJS*, 189, 309
Levesque, E. M., Massey, P., Olsen, K. A. G., Plez, B. *et al.* 2006, *ApJ*, 645, 1102
Maeder, A., Grebel, E. K. & Mermilliod, J.-C. 1999, *A&A*, 346, 459
Marchetti, E., Brast, R., Delabre, B., Donaldson, R. *et al.* 2007, *The Messenger*, 129, 8
Martayan, C., Frémat, Y., Hubert, A.-M., Floquet, M. *et al.* 2006, *A&A*, 452, 273
Martayan, C., Frémat, Y., Hubert, A.-M., Floquet, M. *et al.* 2007, *A&A*, 462, 683
Martins, F., Hillier, D. J., Paumard, T., Eisenhauer, F. *et al.* 2008, *A&A*, 478, 219
Massey, P., Zangari, A. M., Morrell, N. I., Puls, J. *et al.* 2009, *ApJ*, 692, 618
Massey, P., Puls, J., Pauldrach, A. W. A., Bresolin, F. *et al.* 2005, *ApJ*, 627, 477
Massey, P., Bresolin, F., Kudritzki, R. P., Puls, J. *et al.* 2004, *ApJ*, 608, 1001
McCarthy, J. K., Kudritzki, R.-P., Lennon, D. J., Venn, K. A. *et al.* 1997, *ApJ*, 482, 757
Meynet, G. & Maeder, A. 2000, *A&A*, 361, 101
Moffat, A. F. J. 1989, *ApJ*, 347, 373
Mokiem, M. R., de Koter, A., Evans, C. J., Puls, J. *et al.* 2006, *A&A*, 456, 1131
Mokiem, M. R., de Koter, A., Evans, C. J., Puls, J. *et al.* 2007a, *A&A*, 465, 1003
Mokiem, M. R., de Koter, A., Vink, J. S., Puls, J. *et al.* 2007b, *A&A*, 473, 603
Nesvadba, N. P. H., Lehnert, M. D., Davies, R. I., Verma, A. *et al.* 2008, *A&A*, 479, 67

Pauldrach, A. W. A., Hoffmann, T. L. & Lennon, M. 2001, *A&A*, 375, 161
Penny, L. R. & Gies, D. R. 2009, *ApJ*, 700, 844
Pettini, M., Rix, S. A., Steidel, C. C., Adelberger, K. L. *et al.* 2002, *ApJ*, 569, 742
Schnurr, O., Moffat, A. F. J., St-Louis, N., Morrell, N. I. *et al.* 2008, *MNRAS*, 389, 806
Schnurr, O., Chené, A.-N., Casoli, J., Moffat, A. F. J. *et al.* 2009, *MNRAS*, 397, 2049
Smith, N., Li, W., Filippenko, A. V. & Chornock, R. 2010, *ArXiv e-prints* 1006, .
Strom, S. E., Wolff, S. C. & Dror, D. H. A. 2005, *AJ*, 129, 809
Swinbank, A. M., Smail, I., Longmore, S., Harris, A. I. *et al.* 2010, *Nature*, 464, 733
Trundle, C., Dufton, P. L., Lennon, D. J., Smartt, S. J. *et al.* 2002, *A&A*, 395, 519
Trundle, C., Lennon, D. J., Puls, J., & Dufton, P. L. 2004, *A&A*, 417, 217
Trundle, C. & Lennon, D. J. 2005, *A&A*, 434, 677
Trundle, C., Dufton, P. L., Hunter, I., Evans, C. J. *et al.* 2007, *A&A*, 471, 625
Urbaneja, M. A., Herrero, A., Bresolin, F., Kudritzki, R.-P. *et al.* 2003, *ApJ* (Letters), 584, L73
Urbaneja, M. A., Herrero, A., Bresolin, F., Kudritzki, R.-P. *et al.* 2005a, *ApJ*, 622, 862
Urbaneja, M. A., Herrero, A., Kudritzki, R.-P., Najarro, F. *et al.* 2005b, *ApJ*, 635, 311
Urbaneja, M. A., Kudritzki, R.-P., Bresolin, F., Przybilla, N. *et al.* 2008, *ApJ*, 684, 118
Venn, K. A., McCarthy, J. K., Lennon, D. J., Przybilla, N. *et al.* 2000, *ApJ*, 541, 610
Venn, K. A., Lennon, D. J., Kaufer, A., McCarthy, J. K. *et al.* 2001, *ApJ*, 547, 765
Venn, K. A., Tolstoy, E., Kaufer, A., Skillman, E. D. *et al.* 2003, *AJ*, 126, 1326
Vink, J. S., de Koter, A., & Lamers, H. J. G. L. M. 2001, *A&A*, 369, 574
Walborn, N. R. & Blades, J. C. 1997, *ApJS*, 112, 457
Wolff, S. C., Strom, S. E., Cunha, K., Daflon, S. *et al.* 2008, *AJ*, 136, 1049
Zinnecker, H. & Yorke, H. W. 2007, *ARAA*, 45, 481

Active OB stars: structure, evolution, mass loss, and critical limits
Proceedings IAU Symposium No. 272, 2010
C. Neiner, G. Wade, G. Meynet & G. Peters, eds.

doi:10.1017/S1743921311010489

Populations of Be stars: stellar evolution of extreme stars

Christophe Martayan[1,2], Thomas Rivinius[1], Dietrich Baade[3], Anne-Marie Hubert[2] and Jean Zorec[4]

[1]ESO Chile; email: cmartaya@eso.org
[2]GEPI-Observatoire de Meudon, France
[3]ESO Germany
[4]Institut d'Astrophysique de Paris, France

Abstract. Among the emission-line stars, the classical Be stars known for their extreme properties are remarkable. The Be stars are B-type main sequence stars that have displayed at least once in their life emission lines in their spectrum. Beyond this phenomenological approach some progresses were made on the understanding of this class of stars. With high-technology techniques (interferometry, adaptive optics, multi-objects spectroscopy, spectropolarimetry, high-resolution photometry, etc) from different instruments and space mission such as the VLTI, CHARA, FLAMES, ESPADONS-NARVAL, COROT, MOST, SPITZER, etc, some discoveries were performed allowing to constrain the modeling of the Be stars stellar evolution but also their circumstellar decretion disks. In particular, the confrontation between theory and observations about the effects of the stellar formation and evolution on the main sequence, the metallicity, the magnetic fields, the stellar pulsations, the rotational velocity, and the binarity (including the X-rays binaries) on the Be phenomenon appearance is discussed. The disks observations and the efforts made on their modeling is mentioned. As the life of a star does not finish at the end of the main sequence, we also mention their stellar evolution post main sequence including the gamma-ray bursts. Finally, the different new results and remaining questions about the main physical properties of the Be stars are summarized and possible ways of investigations proposed. The recent and future facilities (XSHOOTER, ALMA, E-ELT, TMT, GMT, JWST, GAIA, etc) and their instruments that may help to improve the knowledge of Be stars are also briefly introduced.

Keywords. stars: emission-line, Be, stars: formation, stars: evolution, stars: rotation, stars: abundances, stars: magnetic fields, stars: oscillations (including pulsations), binaries: general, surveys, gamma rays: bursts

---

1. Introduction

The emission-line stars are spread over the entire HR digram. They concern young, not evolved stars or at the opposite evolved stars and of course Main Sequence stars. They are or not massive stars. For instance, among others, one can found T Tauri stars, UV Ceti stars, flare stars, Mira stars, HBe/Ae stars, B[e] stars, Be/Oe stars, Of stars, Supergiant stars, LBV, WR stars. This document concerns mainly OeBeAe stars. The OeBeAe stars have properties similar to OBA type star (see Evans contribution, this volume) and specific properties due to their extreme nature. The first Be star, γ Cas was discovered by the father Secchi in 1866 (Secchi 1867).

Collins (1987) wrote the first definition of Be stars: "they are non-supergiant OBA-type stars that have displayed at least once in their spectrum emission lines (Hα)." As example, see Fig. 1. Actually the emission-lines come from the circumstellar decretion disk (Struve 1931) formed by episodic matter ejections from the central star. Are the

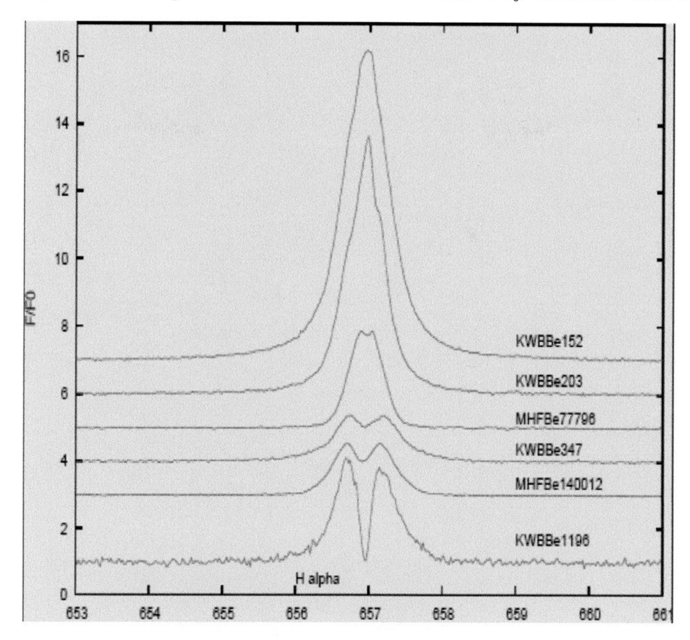

Figure 1. Examples of Hα emission-line in different LMC Be stars. The shape depends among other parameter of the inclination angle.

matter ejections related to the rotation? Be stars are very fast rotators, rotating close to the breakup velocity but often an additional mechanism is needed for the star being able to eject the matter. In the next sections, the rotation and other mechanisms (binarity, stellar evolution, magnetic fields, pulsations) are examined as well as the effects of the metallicity, the stellar evolution, etc.

Porter & Rivinius (2003) presented different kind of emission-line stars with their known properties and how they differ to the "classical Be stars". Since 2003, with the improvements of the technology (interferometry, MOS, etc) and the efforts made on the modeling and theory, the knowledge of Be stars was improved. The next sections will provide some very summarized informations (since 2004, more than 240 refeered articles were published dealing with Be stars) about these recent developments (see also contribution by Baade, this volume).

1.1. *Techniques for detecting a Be star*

Before studying Be stars, it is necessary to find/recognize them. There are several ways to detect them. The first possibility is to use the photometric techniques. Combining different colour/colour or colour/magnitude diagrams would help to pre-classify the stars and detect potential Be star candidates. Keller *et al.* (1999) provide examples of CMD with given thresholds above which the Be stars could fall. They also show that Be stars tend to form a redder sequence than normal B stars.

Dachs *et al.* (1988) have shown that the infrared excess is related to the circumstellar disk of Be stars. It is also linked to the Hα equivalent width. More recent studies found similar results in other wavelength domains such as in the infrared with the AKARI survey (Ita *et al.* 2010) or with SPITZER (Bonanos *et al.* 2010). However, this kind of study has some limits due to the intrinsic properties of Be stars (variability, change of phases Be-B). McSwain & Gies (2005) in their photometric survey of Galactic open clusters found 63% of the known Be stars due to the transience of the Be phenomenon.

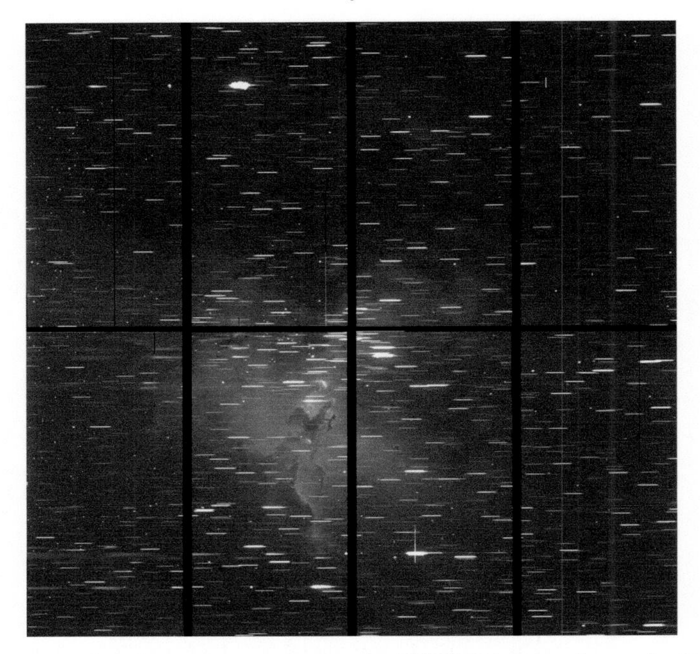

Figure 2. Example of slitless spectroscopy in the NGC6611 open cluster by Martayan *et al.* (2008). The stars appear as spectra.

Thus with the photometry the coupling of indices is necessary. It is also necessary to disentangle the local reddening and the infrared excess due to the circumstellar disk of Be stars.

Another way to find the emission-line stars, and the Be stars, is to do slitless spectroscopy and combine the spectroscopic diagnostic with the photometry as presented in Martayan *et al.* (2010a). The slitless spectroscopy gives the advantage not to be sensitive to the diffuse nebulosity and the spectra are not contaminated by other nebular lines.

Another possibility for finding Be stars is to study the behaviour and the variation of lightcurves. The Be phenomenon is transient, as a consequence the star could appear as a Be star or as a B star when the disk was blown up. Moreover, Be stars show often outbursts. These characteristics help to find this kind of stars using the lightcurves variations as published by Mennickent *et al.* (2002) and Sabogal *et al.* (2005).

1.2. *The Surveys*

Be stars with the techniques mentionned above can be found mainly in open clusters and in fields. In the Galaxy, there are different surveys, let us mention, the photometric survey by McSwain & Gies (2005) and the slitless spectroscopic survey by Mathew *et al.* (2008).

In other galaxies such as the Magellanic Clouds but also in the Milky Way, there are the following surveys among others: a slitless spectroscopic survey of the SMC by Meyssonnier & Azzopardi (1993), another in the SMC/LMC/MW by Martayan *et al.* (2010a, 2008) (see Fig. 2).

And recently, with the VLT-FLAMES, some spectroscopic surveys were and are performed in different galaxies (Evans *et al.* 2005, Martayan *et al.* 2006, 2007). With the polarimetry, some Be stars were found in the SMC/LMC (Wisniewski *et al.* 2007, Wisniewski & Bjorkman 2006), and with FORS2 in IC1613 (Bresolin *et al.* 2007), which correspond to the farthest Be stars detected.

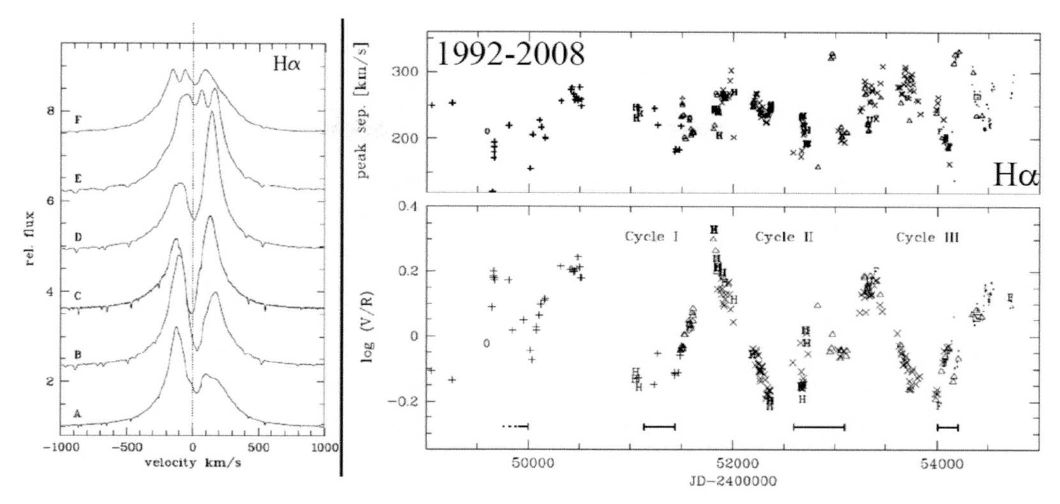

Figure 3. Example of V/R variations in the Be star ζ Tau. Figure adapted from Stefl *et al.* (2009).

2. Mid/long term monitoring of Be stars, V/R variations

2.1. *Monitoring of Be stars*

Why long-term monitorings are needed? Due to the transience of the Be phenomenon for having a chance to detect a Be star, a mid or long term monitoring of B/Be stars is needed. Indeed, for the follow-up of the variability of the star related to outbursts and/or pulsations, or binarity, or modifications of the CS disk, etc, a monitoring is needed.

Surveys such as the photometric ones of MACHO or OGLE are useful in that sense but also the spectroscopic surveys. Here the amateur astronomers have a role to play. With the new efficient, good quality, low cost spectrograph that can be mounted in amateur telescopes, see contribution by Blanchard *et al.* (this volume), some useful good quality observations can be performed.

Moreover, without long-term monitoring (more than 10 years of spectroscopy) the results concerning the Be star μ Cen and the predictions of the outbursts dates could not have been obtained. For all details about this pioneer study, see Rivinius *et al.* (1998a), Rivinius *et al.* (1998b).

This kind of study could thanks to the databases be performed more easily now, let us cite the Be Star Spectra database http://basebe.obspm.fr/basebe/ (Neiner *et al.* 2007, contribution by De Batz *et al.*, this volume).

2.2. *The V/R variation of the emission lines*

A long-term spectroscopic monitoring of Be stars can show in certain cases variations of the V and R peaks of the emission-lines. This variation is now understood and is related to global one-armed oscillations in the circumstellar disk. The case of the Be star ζ Tau (see Stefl *et al.* 2009) is representative of that phenomenon with a V/R cycle of about 1400 days without correlation with the period of 133 days of the companion. Fig. 3 from Stefl *et al.* (2009) illustrates this phenomenon. The (polarimetric, spectroscopic, interferometric) observations are properly reproduced by a modeling of a one-armed spiral viscous disk as described in Carciofi *et al.* (2009), see also contributions by Carciofi and Stee, this volume.

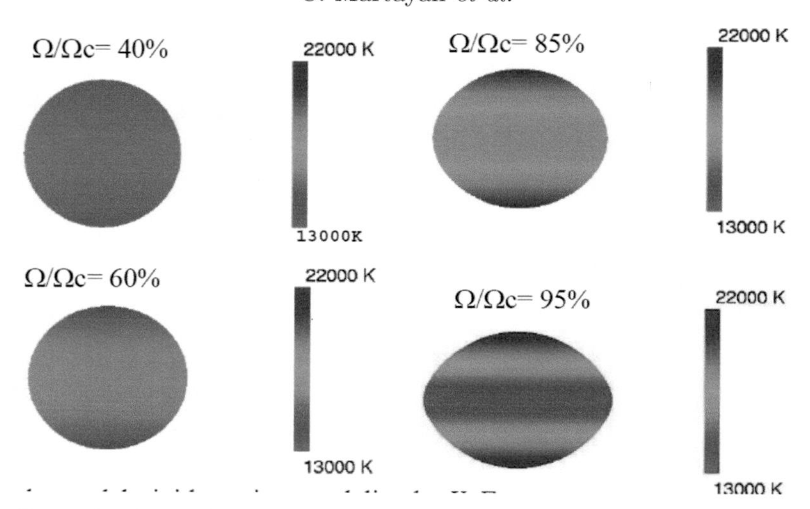

Figure 4. Example of star shape depending on the Ω/Ω_c ratio in the Roche model and rigid rotation approach. Due to the Von Zeipel effect there is a temperature gradient from the poles to the equator. Courtesy by Y. Frémat, see also Frémat *et al.* (2005).

3. Stellar winds

The diagnostic lines are mostly present in the UV and optical but also several of them are in the infrared. Let us mention the lines of HeI, HeII, Si IV, C IV, N V, Brγ, etc. For more details, see Martins (2011), Mokiem *et al.* (2006), and Henrichs *et al.*, this volume.

The structure of the wind differs among B stars. For the normal B stars, the wind is spherical, while for Be stars, we can expect a radiative bi-polar wind mainly due to the Von Zeipel effect. The poles in case of fast rotation becomes brighter than the equator, it explains this asymetry. In case of Be stars, one can also expect an additional mechanical wind at the equator that creates the CS disk. In such case the needed value of the Ω/Ω_c ratio (angular velocity to critical angular velocity ratio) for ejecting matter has to be determined (see Sect. 4).

Fortunately, recent interferometric facilities (such as the VLTI-AMBER), allowed to measure the shape of the CS environment of Be stars and both winds were found. It is for example the case of the Be stars α Arae, Achernar (Meilland *et al.* 2007, Kervella *et al.* 2006).

4. Rotational velocities

Be stars are known to be very fast rotator but how fast are they rotating? Fig. 4 shows effect of the rotation on the shape of the stars for different Ω/Ω_c in the Roche model, rigid rotation approach. Next sections provide some clues about this point.

4.1. *From the interferometry*

The interferometric observations show that all observed Be stars are found flattened revealing high Ω/Ω_cratios. The flattening of Achernar was found equal to 1.56 ±0.05 by Domiciano de Souza *et al.* (2003). However, this value could also be due to remaining small disk indicating that Achernar is not rotating at the breakup velocity but close to it.

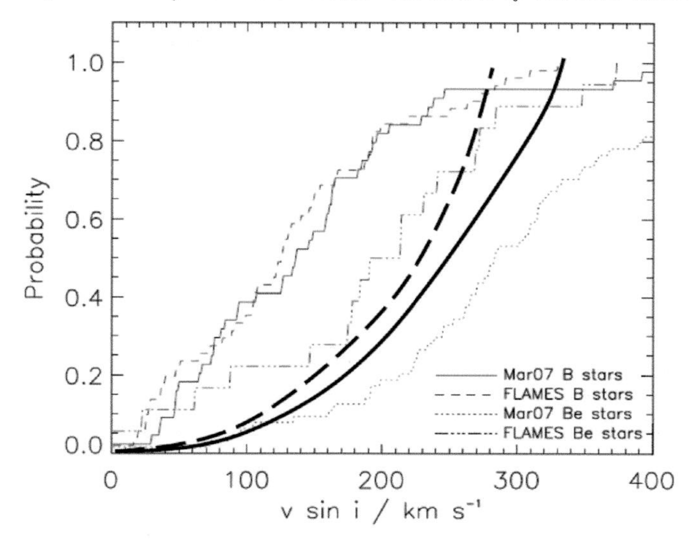

Figure 5. Vsini measurements of B and Be stars by Hunter *et al.* (2008) compared to those of Martayan *et al.* 2007. The dotted curve corresponds to Ω/Ω_c=70% and the solid curve to Ω/Ω_c=80% (line saturation effect with 1 He line measurement). The curves come from McSwain *et al.* (2008).

4.2. *Line saturation effect at high Ω/Ω_c*

Townsend *et al.* (2004) showed that the HeI 4471 Å line has a saturation effect at high Ω/Ω_c, typically above 80%. This implies that the Vsini measurements based on the use of 1 He line are limited in case of fast rotating stars, typically Be stars and possibly Bn stars. Using several He lines as well as metallic lines, the saturation effect occurs again but at higher Ω/Ω_c (~90-95%), see Frémat *et al.* (2005). This explains the discrepancy between the measurements of Be stars Vsini between Hunter *et al.* (2008) and Martayan *et al.* (2007). Fig. 5 shows the measurements differences from Hunter *et al.* (2008) and the saturation curves at Ω/Ω_c=70 and 80 % from McSwain *et al.* (2008). It clearly shows that the Be star Vsini measurements done by Hunter *et al.* (2008) are saturated and the measurements by Martayan *et al.* (2007) saturate at ~90%.

4.3. *Fast rotation effects correction*

Accordingly to the previous section, in case of fast rotators, one has to take into account the fast rotation effects (flattening, gravitational darkening). These effects introduce underestimates of the Vsini, and it implies too that the spectral type/luminosity classes are modified from the late to earlier spectral type and from evolved to less evolved luminosity classes. The codes BRUCE by Townsend *et al.* (2004) and FASTROT by Frémat *et al.* (2005) are able to properly correct the fundamental parameters of the fast rotating stars.

4.4. *The veiling effect*

Another important effect in case of Be stars is the veiling effect due to the circumstellar disk that affects the continuum level and thus the lines depth. Not taking this effect into account will imply bad measurements of the fundamental parameters but also of the chemical abundances. For taking this effect into account, Ballereau *et al.* (1995) proposed a method based on the He4471 emission-line EW measurements. Other possibility is to prefer the bluest lines when spectrum fitting is performed but also to determine the disk influence by fitting the Spectral Energy Distribution. The new spectrocopic facility, the VLT-XSHOOTER that covers the near UV to the K-band simultaneously should help in that approach. Alternatively, there is the method consisting to fit the Balmer discontinuity as described by Zorec *et al.* (2009).

4.5. *Metallicity effects*

Keller (2004), Martayan *et al.* (2006, 2007), and Hunter *et al.* (2008) found that SMC OB stars rotate faster than LMC OB stars, which rotate faster than their Galactic counterparts. In the SMC, the mass-loss of OB stars by radiatively driven winds was found lower than in the Galaxy by Bouret *et al.* (2003) and Vink (2007). Consequently, the stars should loose less angular momentum and could rotate faster (Maeder & Meynet 2001). The comparisons between SMC Be stars Vsini fairly agree with the theoretical models by Ekström *et al.* (2008). The SMC Be stars are found to be rotating very close to the critical velocity.

4.6. *ZAMS rotational velocities*

The ZAMS rotational velocities of SMC, LMC, and Milky Way (MW) intermediate-mass Be stars were determined by Martayan *et al.* (2007). They found a metallicity effect on the ZAMS rotational velocities. At lower metallicity Be stars rotate faster since their birth. This could be related to an opacity effect, for an identical Ω/Ω_c, at lower metallicity, the radii are smaller thus the stars can rotate faster. The comparison of the SMC ZAMS rotational velocities of intermediate-mass and massive Be stars with the theoretical tracks by Ekström *et al.* (2008) are also in a fair agreement.

4.7. *Number of Be stars vs. the metallicity and redshift*

As a consequence of faster rotational velocities at lower metallicity, one can expect to find more Be-type stars in low-metallicity environments. Using the ESO-WFI in its slitless mode (Baade *et al.* 1999), Martayan *et al.* (2010a) found 3 to 5 times more Be stars in SMC open clusters than in the Galactic ones (McSwain & Gies 2005), see Fig. 6. This quantified the results by Maeder *et al.* (1999) and Wisniewski *et al.* (2008).

Another consequence of this increase of the Be stars number with decreasing metallicity is when a survey of OBA-type populations is performed, one should find an increased number/ratio of OeBeAe stars in the sample. This was exactly the case of the survey by Bresolin *et al.* (2007) who found 6 Be stars in the 6 MS B stars observed in the low metallicity galaxy IC1613. Actually at lower metallicity than the SMC, and probably at higher redshift, the stars could rotate again faster, then one can expect to have more Be stars but also that the Be phenomenon is extended to other categories of stars.

4.8. *Chemical abundances*

The measurement of chemical abundances is very important for testing the stellar evolution models with fast rotation, including the rotational mixing due to the rotation. Moreover, it is also important for distinguishing the stars following a chemical homogeneous evolution than the usual evolution (Maeder 1987, Yoon *et al.* 2006). It is also of interest to know them for the pulsating stars of different metallicity environments.

Lennon *et al.* (2005) found 2 Be stars without N enhancement, while the theoretical models using the rotational mixing expect a N enhancement and a C depletion. Dunstall *et al.* (this volume) and Peters (this volume) also found no N enrichment in Be stars. Does it mean that the rotational mixing theory is not able to explain the measurements? Not necessarily, Porter (1999) found that due to the temperature gradient in Be stars, the ions are displaced from the poles to the equator. This could occur to the N, implying that the corresponding lines become weaker and the measurements biased.

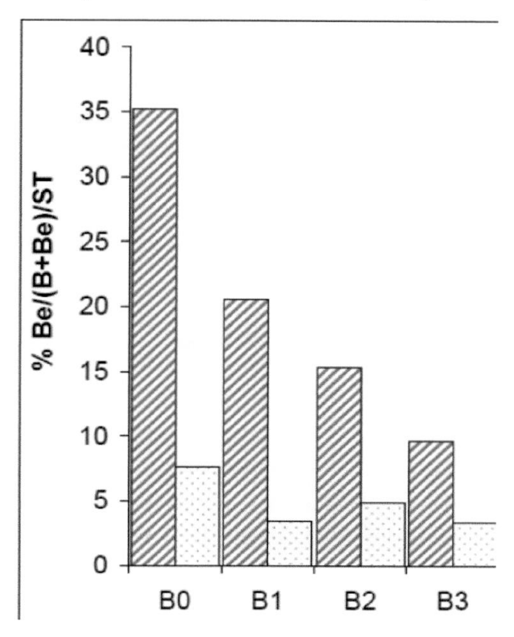

Figure 6. Ratio of Be to B stars by spectral categories in the SMC and MW. Figure adapted from Martayan *et al.* (2010a).Red dashed bars are for the SMC, white bars for the MW.

5. Stellar pulsations

Be stars lie in β Cep and SPB regions, thus p and g modes are expected to be found. The κ mechanism is at the origin of the observed pulsations due to the iron bump. A beating of non radial pulsations combined to the fast rotation of Be stars could be at the origin of the matter ejection. According to Townsend *et al.* (2004) an $\Omega/\Omega_c = 95\%$ is needed for launching the matter in orbit, while Cranmer (2009) shows that depending on the Ω/Ω_c value the ejected matter will be part of the wind or will form a disk. Rivinius *et al.* (1998a) have shown that the combination fast rotation + beating of non radial pulsations is able to reproduce and predict the outbursts in the Be star μ Cen. More recently, with the new photometric space missions MOST, COROT, KEPLER, a large number of pulsational frequencies (g and p modes) were found in several Be stars. Huat *et al.* (2009), in the Be star HD49330, have found that the pulsational frequencies but also their amplitude change between the quiescence phase, the pre burst phase, and during the burst phase. Is it related to the subsurface convection zones in OB stars? For more details, see Cantiello contribution (this volume).

6. Spectropolarimetry: magnetic fields and disks

The spectropolarimetry is a useful tool for finding the Be stars due to their disks but also for detecting the magnetic fields. Recent spectropolarimetric facilities such as ESPADONS, NARVAL, allow the search of magnetic fields in Be stars but up to now, there is only 1 Be star in which a magnetic field was found: ω Ori (Neiner *et al.* 2003). The presence of magnetic field (Ω transport from the core to the surface or magnetic reconnection) combined to the fast rotation could also be an additional mechanism for ejecting matter and create the Be star CS disk.

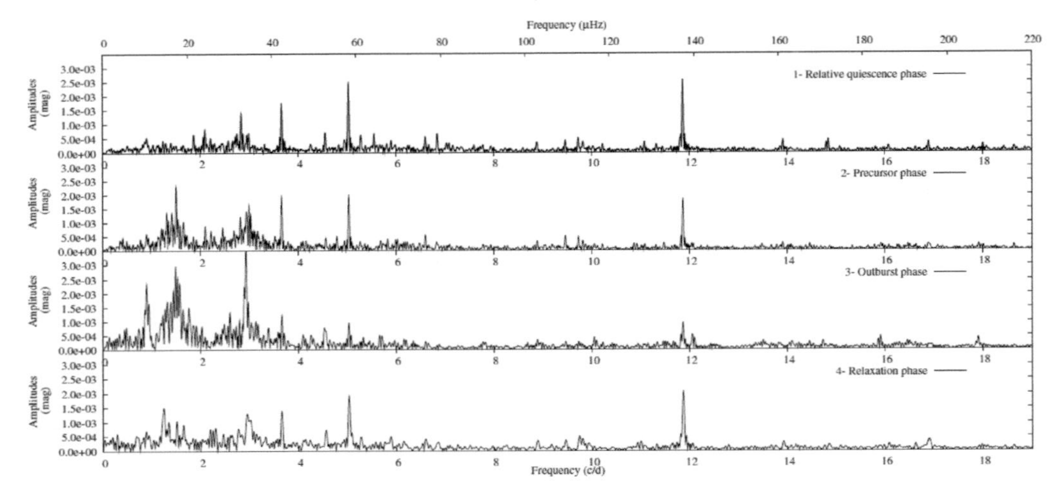

Figure 7. Evolution of pulsational frequencies in the Be star HD49330 from COROT data during the different pre, during, and after an outburst phases. The figure comes from Huat *et al.* (2009).

7. Circumstellar disks

Huge progresses were performed thanks to the interferometry (VLTI, CHARA, NPOI) and the multi-wavelengths studies. The main result is that the disk is rotating in a Keplerian way around its central star, see for example Meilland *et al.* (2007). In addition, the thermal structure of the CS disk is also better known (see Tycner contribution, this volume).

8. Stellar formation and evolution, Gamma Ray Bursts

What kind of objects is at the origin of Be stars and their fast rotation? Could they be related to the Herbig Ae/Be objects? Are the fast B rotator Bn stars at the Be star origin, while their number is roughly equal to the number of Be stars in the MW? What is the IMF of the Be stars? Does it differ from the normal B stars? It seems that the open cluster density has no effect on the appearance of Be stars (McSwain & Gies 2005, Martayan *et al.* 2010a).

Taking into account the fast rotation effects, Zorec *et al.* (2005) derived the evolutionary status of Be stars in the MW (see Fig. 8). Their location/life in the MS strongly depends on their mass and on the evolution of the Ω/Ω_c ratio. In the LMC, the diagram of Be stars appearance is similar than in the MW but in the SMC, at the opposite of the MW, massive Be stars and Oe stars appear or are still existing in the second part of the MS (Martayan *et al.* 2007). These differences are also consistent with the Ω/Ω_c evolution at different metallicities.

However, the life of a Be star is not finished at the end of the MS. It seems that the more massive SMC Be and Oe stars could play a role in the explanation of the type 2 GRBs (also called Long Gamma Ray Bursts). Martayan *et al.* (2010b) were able to reproduce the number and ratio of LGRBs using the populations of SMC massive Be and Oe stars accross their stellar evolution. The type 3 GRBs (without SN counterpart) could be explained in a binary scenario involving Be-binary stars (Tutukov & Fedorova 2007).

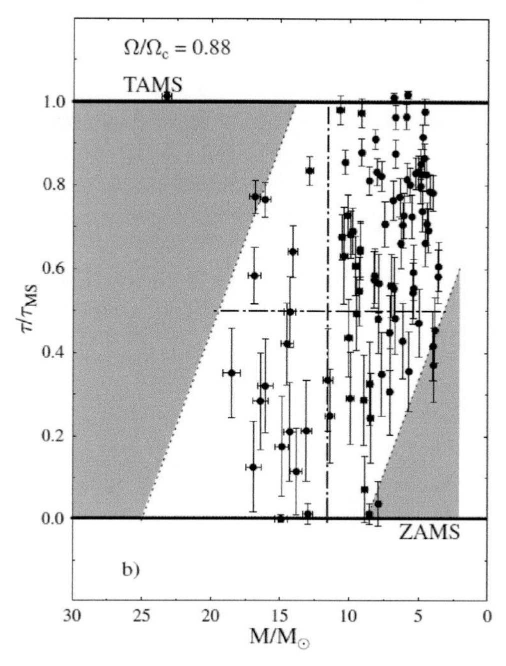

Figure 8. Evolutionary diagram in the MS of the Be stars in the MW by Zorec *et al.* (2005).

9. Binarity

While 60 to 70% of O stars are known to be binaries (see Sana *et al.* 2009 and his contribution in this volume), the question remains opened for the B and Be stars. The last study by Oudmaijer & Parr (2010) using the VLT-NACO AO facility found that: 29 to 35 $\pm$ 8% of the nearby B stars are binaries and that 30 to 33 $\pm$ 8% of the nearby Be stars are binaries. They scanned the separations larger than 20 AUs with a flux contrast ratio of 10 corresponding to M companions. Scanning smaller scales needs the interferometry. About 40% of all Be stars observed with the VLTI and CHARA were found to have a companion.

However, certain type of Be stars are known to be binaries such as the Be-X rays (see for example Coe *et al.* 2008). The γ Cas-like stars have possible companions and have magnetic field activity in their CS environment (Smith *et al.* 2006).

About the Be-binary disruptions scenario when the companion has exploded in SN, one should find that Be stars are runaway stars or see SN remnants. In both cases, this is not observed for the large majority of Be stars. The chemical abundances could help to find a potential previous interaction with a companion. In that case, one can expect abundances anomalies.

10. Conclusion

To better understand the properties of these stars, it is necessary to do multiwavelength studies and multi-techniques studies with the use for example of: the VLT-XSHOOTER, IR and X-rays satellites, MOS such as the VLT-FLAMES, the interferometry, the AO facilities, and if possible simultaneous observations.

In the future, it will be possible to do similar studies than today but in very far galaxies, to scan other metallicity ranges, etc, thanks to the ELTs and their instruments but also to ALMA, SKA, etc.

As shown, the Be phenomenon could be not restricted to the late O, B, and early A stars and could also not be restricted to the MS, specially in lower metallicity environments. Some of the questions about them and their CS environment are now solved but some others are still opened and the definition of Be stars should be revised.

Finally, one can propose this new definition of the Be phenomenon:

This is a star with innate or acquired very fast rotation, which combined to other mechanism such as non-radial pulsations beating leads to episodic matter ejections creating a CS decretion disk or envelope.

This definition implies that the stars are not restricted to the MS, not restricted to B-type, and from the CS material the emission lines come.

References

Baade, D., Meisenheimer, K., Iwert, O., Alonso, J. *et al.* 1999, *The Messenger*, 95, 15

Ballereau, D., Chauville, J., & Zorec, J. 1995, *A&AS*, 111, 457

Bonanos, A. Z., Lennon, D. J., Köhlinger, F., van Loon, J. T. *et al.* 2010, *AJ*, 140, 416

Bouret, J.-C., Lanz, T., Hillier, D. J., Heap, S. R. *et al.* 2003, *ApJ*, 595, 1182

Bresolin, F., Urbaneja, M. A., Gieren, W., Pietrzyński, G. *et al.* 2007, *ApJ*, 671, 2028

Carciofi, A. C., Okazaki, A. T., Le Bouquin, J.-B., Štefl, S. *et al.* 2009, *A&A*, 504, 915

Coe, M. J., Schurch, M., Corbet, R. H. D., Galache, J. *et al.* 2008, *MNRAS*, 387, 724

Collins, II, G. W. 1987, in: A. Slettebak & T. P. Snow (eds.), *Physics of Be Stars*, IAU Colloquium 92, p. 3

Cranmer, S. R. 2009, *ApJ*, 701, 396

Dachs, J., Kiehling, R. & Engels, D. 1988, *A&A*, 194, 167

Domiciano de Souza, A., Kervella, P., Jankov, S., Abe, L. *et al.* 2003, *A&A*, 407, L47

Ekström, S., Meynet, G., Maeder, A., & Barblan, F. 2008, *A&A*, 478, 467

Evans, C. J., Smartt, S. J., Lee, J.-K., Lennon, D. J. *et al.* 2005, *A&A*, 437, 467

Frémat, Y., Zorec, J., Hubert, A.-M., & Floquet, M. 2005, *A&A*, 440, 305

Huat, A.-L., Hubert, A.-M., Baudin, F., Floquet, M. *et al.* 2009, *A&A*, 506, 95

Hunter, I., Lennon, D. J., Dufton, P. L., Trundle, C. *et al.* 2008, *A&A*, 479, 541

Ita, Y., Matsuura, M., Ishihara, D., Oyabu, S. *et al.* 2010, *A&A*, 514A, 2

Keller, S. C., Wood, P. R. & Bessell, M. S. 1999, *A&AS*, 134, 489

Keller, S. C. 2004, *PASA*, 21, 310

Kervella, P. & Domiciano de Souza, A. 2006, *A&A*, 453, 1059

Lennon, D. J., Lee, J.-K., Dufton, P. L., & Ryans, R. S. I. 2005, *A&A*, 438, 265

Maeder, A. 1987, *A&A*, 173, 247

Maeder, A., Grebel, E. K. & Mermilliod, J.-C. 1999, *A&A*, 346, 459

Maeder, A. & Meynet, G. 2001, *A&A*, 373, 555

Martayan, C., Frémat, Y., Hubert, A.-M., Floquet, M. *et al.* 2006, *A&A*, 452, 273

Martayan, C., Frémat, Y., Hubert, A.-M., Floquet, M. *et al.* 2007, *A&A*, 462, 683

Martayan, C., Floquet, M., Hubert, A. M., Neiner, C. *et al.* 2008, *A&A*, 489, 459

Martayan, C., Baade, D. & Fabregat, J. 2010a, *A&A*, 509A, 11

Martayan, C., Zorec, J., Frémat, Y. & Ekström, S. 2010b, *A&A*, 516A, 103

Martins, F. 2011, *Bulletin de la Societe Royale des Sciences de Liege* 80, 29

Mathew, B., Subramaniam, A. & Bhatt, B. C. 2008, *MNRAS*, 388, 1879

McSwain, M. V. & Gies, D. R. 2005, *ApJS*, 161, 118

McSwain, M. V., Huang, W., Gies, D. R., Grundstrom, E. D. *et al.* 2008, *ApJ*, 672, 590

Meilland, A., Stee, P., Vannier, M., Millour, F. *et al.* 2007, *A&A*, 464, 59

Mennickent, R. E., Pietrzyński, G., Gieren, W., & Szewczyk, O. 2002, *A&A*, 393, 887

Meyssonnier, N. & Azzopardi, M. 1993, *A&AS*, 102, 451

Mokiem, M. R., de Koter, A., Evans, C. J., Puls, J. *et al.* 2006, *A&A*, 456, 1131

Neiner, C., Hubert, A.-M., Frémat, Y., Floquet, M. *et al.* 2003, *A&A*, 409, 275

Neiner, C., de Batz, B., Mekkas, A., Cochard, F., & Martayan, C. 2007, in: J. Bouvier, A. Chalabaev, & C. Charbonnel (eds.), *SF2A-2007: Proceedings of the Annual meeting of the French Society of Astronomy and Astrophysics*, p. 538
Oudmaijer, R. D. & Parr, A. M. 2010, *MNRAS*, 405, 2439
Porter, J. M. 1999, *A&A*, 341, 560
Porter, J. M. & Rivinius, T. 2003, *PASP*, 115, 1153
Rivinius, T., Baade, D., Stefl, S., Stahl, O. *et al.* 1998a, *A&A*, 333, 125
Rivinius, T., Baade, D., Stefl, S., Stahl, O. *et al.* 1998b, *A&A*, 336, 177
Sabogal, B. E., Mennickent, R. E., Pietrzyński, G., & Gieren, W. 2005, *MNRAS*, 361, 1055
Sana, H., Gosset, E. & Evans, C. J. 2009, *MNRAS*, 400, 1479
Secchi, A. 1867, *Astronomical register*, 5, 40
Smith, M. A., Henry, G. W., & Vishniac, E. 2006, *ApJ*, 647, 1375
Štefl, S., Rivinius, T., Carciofi, A. C., Le Bouquin, J.-B. *et al.* 2009, *A&A*, 504, 929
Struve, O. 1931, *ApJ*, 73, 94
Townsend, R. H. D., Owocki, S. P. & Howarth, I. D. 2004, *MNRAS*, 350, 189
Tutukov, A. V. & Fedorova, A. V. 2007, *Astron. Rep.*, 51, 847
Vink, J. S. 2007, in: R. J. Stancliffe, G. Houdek, R. G. Martin, & C. A. Tout (eds.), *Unsolved Problems in Stellar Physics: A Conference in Honor of Douglas Gough*, AIP-CP 948, p. 389
Wisniewski, J. P. & Bjorkman, K. S. 2006, *ApJ*, 652, 458
Wisniewski, J. P., Bjorkman, K. S., Magalhães, A. M., Bjorkman, J. E. *et al.* 2007, *ApJ*, 671, 2040
Wisniewski, J. P., Clampin, M., Grady, C. A., Ardila, D. R. *et al.* 2008, *ApJ*, 682, 548
Yoon, S.-C., Langer, N., & Norman, C. 2006, *A&A*, 460, 199
Zorec, J., Frémat, Y., & Cidale, L. 2005, *A&A*, 441, 235
Zorec, J., Cidale, L., Arias, M. L., Frémat, Y. *et al.* 2009, *A&A*, 501, 297

Discussion

T. Rivinius: Comment on the remark by A. Miroshnichenko about the Be binary fraction. The binary fraction could well be 50 %. What is really important is that in studies where B and Be stars are investigated, after having made sure that the same selection effects apply to both, the binary fraction is found the same for B and Be stars. At least for binary separations wide enough not to destroy the disk by interaction.

R. Prinja: Are the fast, high-ionization polar winds of Be star clumps (and perhaps structures) in a manner akin to massive OB stars? Fundamental differences in wind characteristics of these classes of hot stars may constrain the pivotal role of for example Fe-bump driven convection and the role of disk-wind driving in Be stars.

C. Martayan: The clumping and the wind structure of Be stars are not yet known. Maybe the current interferometric programs on Be stars could bring some preliminary informations.

A. Maeder: During the cycle Be-B, are luminosity variations observed?

C. Martayan: Yes, during the life cycle of the CS disk of Be stars, some luminosity variations were reported and they follow a sequence that is shown by De Wit *et al.* (2006).

Active OB stars: structure, evolution, mass loss, and critical limits
Proceedings IAU Symposium No. 272, 2010
C. Neiner, G. Wade, G. Meynet & G. Peters, eds.

doi:10.1017/S1743921311010490

Infrared properties of active OB stars in the Magellanic Clouds from the Spitzer SAGE survey

Alceste Z. Bonanos[1], Danny J. Lennon[2], Derck L. Massa[2], Marta Sewilo[2], Fabian Köhlinger[2], Nino Panagia[2], Jacco Th. van Loon[3], Chris J. Evans[4], Margaret Meixner[2], Karl D. Gordon[2] and the SAGE teams

[1]Institute of Astronomy & Astrophysics, National Observatory of Athens, I. Metaxa & Vas. Pavlou St., P. Penteli, 15236 Athens, Greece
bonanos@astro.noa.gr

[2]Space Telescope Science Institute, 3700 San Martin Drive, Baltimore, MD, 21218, USA

[3]Astrophysics Group, Lennard-Jones Laboratories, Keele University, Staffordshire ST5 5BG, UK

[4]UK Astronomy Technology Centre, Royal Observatory Edinburgh, Blackford Hill, Edinburgh, EH9 3HJ, UK

Abstract. We present a study of the infrared properties of 4922 spectroscopically confirmed massive stars in the Large and Small Magellanic Clouds, focusing on the active OB star population. Besides OB stars, our sample includes yellow and red supergiants, Wolf-Rayet stars, Luminous Blue Variables (LBVs) and supergiant B[e] stars. We detect a distinct Be star sequence, displaced to the red, and find a higher fraction of Oe and Be stars among O and early-B stars in the SMC, respectively, when compared to the LMC, and that the SMC Be stars occur at higher luminosities. We also find photometric variability among the active OB population and evidence for transitions of Be stars to B stars and vice versa. We furthermore confirm the presence of dust around all the supergiant B[e] stars in our sample, finding the shape of their spectral energy distributions (SEDs) to be very similar, in contrast to the variety of SED shapes among the spectrally variable LBVs.

Keywords. catalogs, galaxies: individual (LMC, SMC), infrared: stars, stars: early-type, stars: emission-line, Be

---

1. Introduction

The *Spitzer Space Telescope* Legacy Surveys SAGE ("Surveying the Agents of a Galaxy's Evolution", Meixner *et al.* 2006) and SAGE-SMC (Gordon *et al.* 2010) have for the first time made possible a comparative study of the infrared properties of massive stars at a range of metallicities, by imaging both the Large and Small Magellanic Clouds (LMC and SMC). In Bonanos *et al.* (2009, Paper I) and Bonanos *et al.* (2010, Paper II), we presented infrared properties of massive stars in the LMC and SMC, which we summarize below. The motivation was threefold: (a) to use the infrared excesses of massive stars to probe their winds, circumstellar gas and dust, (b) to provide a template for studies of other, more distant, galaxies, and (c) to investigate the dependence of the infrared properties on metallicity. Papers I and II were the first major compilations of accurate spectral types and multi-band photometry from 0.3–24 μm for massive stars in any galaxy, increasing by an order of magnitude the number of massive stars for which mid-infrared photometry was available.

Infrared excess in hot massive stars is primarily due to free-free emission from their ionized, line driven, stellar winds. Panagia & Felli (1975) and Wright & Barlow (1975) first computed the free-free emission from ionized envelopes of hot massive stars, as a function of the mass-loss rate ($\dot{M}$) and the terminal velocity of the wind (v_∞). The properties of massive stars, and in particular their stellar winds (which affect their evolution) are expected to depend on metallicity (Z). For example, Mokiem *et al.* (2007) found empirically that mass-loss rates scale as $\dot{M} \sim Z^{0.83\pm0.16}$, in good agreement with theoretical predictions (Vink *et al.* 2001). The expectation, therefore, is that the infrared excesses of OB stars in the SMC should be lower than in the LMC, given that $\dot{M}$ is lower in the SMC. Furthermore, there is strong evidence that the fraction of classical Be stars among B-type stars is higher at lower metallicity (Martayan *et al.* 2007b). Grebel *et al.* (1992) were the first to find evidence for this, by showing that the cluster NGC 330 in the SMC has the largest fraction of Be stars of any known cluster in the Galaxy, LMC or SMC. More recent spectroscopic surveys (Martayan *et al.* 2010) have reinforced this result. We are also interested in quantifying the global dependence of the Be star fraction on metallicity. The incidence of Be/X-ray binaries is also much higher in the SMC than in the LMC (Liu *et al.* 2005), while the incidence of Wolf-Rayet (WR) stars is much lower; therefore, a comparison of infrared excesses for these objects is also of interest.

2. Spectral type and Photometric Catalogs

We have compiled catalogs of massive stars with known spectral types in both the LMC and SMC from the literature. We then cross-matched the stars in the SAGE and SAGE-SMC databases, after incorporating optical and near-infrared photometry from recent surveys of the Magellanic Clouds. The resulting photometric catalogs were used to study the infrared properties of the stars. The LMC spectral type catalog contains 1750 massive stars. A subset of 1268 of these are included in the photometric catalog, for which uniform photometry from $0.3 - 24$ μm in the $UBVIJHK_s$+IRAC+MIPS24 bands is presented in Paper I. The SMC spectral type catalog contains 5324 massive stars; 3654 of these are included in the photometric catalog, for which uniform photometry from $0.3 - 24$ μm is presented in Paper II. All catalogs are available electronically.

3. Active OB stars

3.1. *O/Oe and early-B/Be stars*

In Figure 1, we plot J_{IRSF} vs. $J_{IRSF} - [3.6]$, $J_{IRSF} - [5.8]$ and $J_{IRSF} - [8.0]$ colors for the 1967 early-B stars from our SMC catalog, respectively, denoting their luminosity classes, binarity and emission line classification properties by different symbols. We compare the observed colors with colors of plane-parallel non-LTE TLUSTY stellar atmosphere models (Lanz & Hubeny 2003, 2007) of appropriate metallicity and effective temperatures. For reference, reddening vectors and TLUSTY models reddened by $E(B-V) = 0.2$ mag are also shown. We clearly detect infrared excesses from free-free emission despite not having dereddened the stars, as in the LMC. At longer wavelengths, the excess is larger because the flux due to free–free emission for optically thin winds remains essentially constant with wavelength. Fewer stars are detected at longer wavelengths because of the decreasing sensitivity of *Spitzer* and the overall decline of their SEDs. We find that the majority of early-B supergiants in the SMC exhibit lower infrared excesses, when compared to their counterparts in the LMC, due to their lower mass-loss rates, although certain exceptions exist and deserve further study.

The CMDs allow us to study the frequency of Oe and Be stars, given the low foreground and internal reddening for the SMC. Our SMC catalog contains 4 Oe stars among 208 O stars, of which one is bluer than the rest. There are 16 additional stars with $J_{IRSF} - [3.6] > 0.5$ mag and $J_{IRSF} < 15$ mag (including all luminosity classes), whose spectra appear normal (although the Hα spectral region in most cases was not observed). We refer to these as "photometric Oe" stars and attribute their infrared excesses to free-free emission from a short-lived, possibly recurrent circumstellar region, whose Hα emission line was not detected during the spectroscopic observations either because the gas had dispersed or because the region was optically thick to Hα radiation or the observation spectral range just did not extend to Hα. Given the expectation of lower $\dot{M}$ at SMC metallicity, we argue that such a region is more likely to be a transient disk rather than a wind. Assuming these are all Oe stars, we find a $10 \pm 2\%$ fraction of Oe stars among the O stars in the SMC. The error in the fraction is dominated by small number statistics. In contrast, there are 4 Oe and 14 "photometric Oe" stars (with $J_{IRSF} - [3.6] > 0.5$ mag and $J_{IRSF} < 14.5$ mag) out of 354 O stars in the LMC (despite the higher $\dot{M}$ at LMC metallicity), which yields a $5 \pm 1\%$ fraction of Oe stars among O stars in the LMC.

Turning to the early-B stars, the most striking feature in Figure 1 is a distinct sequence displaced by ~ 0.8 mag to the red. A large fraction of the stars falling on this redder sequence have Be star classifications, although not all Be stars reside there. Given that the circumstellar gas disks responsible for the emission in Be stars are known to completely vanish and reappear between spectra taken even 1 year apart (see review by Porter & Rivinius 2003, and references therein), the double sequence reported here provides further evidence for the transient nature of the Be phenomenon. A bimodal distribution at the $L-$band was previously suggested by the study of Dougherty *et al.* (1994), which included a sample of 144 Galactic Be stars. Our larger Be sample, which is essentially unaffected by reddening, and the inclusion of all early-B stars, clearly confirms the bimodal distribution. It is due to the much larger number of Be stars classified in the SMC, in comparison to the LMC, as well as the higher fraction of Be stars among early-B stars in the SMC, which is $19 \pm 1\%$ vs. $4 \pm 1\%$ in the LMC when considering only the spectroscopically confirmed Be stars (cf. $\sim 17\%$ for < 10 Myr B0–5 stars; Wisniewski *et al.* 2006). Excluding the targeted sample of Martayan *et al.* (2007a, 2007b) does not significantly bias the statistics, since the fraction only decreases to $15 \pm 1\%$. We caution that incompleteness in our catalogs could also affect the determined fractions, if our sample turns out not to be representative of the whole population of OB stars.

We proceed to define "photometric Be" stars as early-B type stars with an intrinsic color $J_{IRSF} - [3.6] > 0.5$ mag, given that a circumstellar disk or envelope is required to explain such large excesses. Including these "photometric Be" stars and using the same color and magnitude cuts as for the "photometric Oe" stars above, yields fractions of Be stars among early-B stars of $27 \pm 2\%$ for the SMC and $16 \pm 2\%$ for the LMC (cf. 32% from young SMC clusters; Wisniewski *et al.* 2006). We compare our results with the fractions determined by Maeder *et al.* (1999) from young clusters, i.e. 39% for the SMC and 23% for the LMC, finding ours to be lower, although the sample selections were very different.

These preliminary statistics (available for the first time for Oe stars) indicate that both Oe and Be stars are twice as common in the SMC than in the LMC. We emphasize the importance of including the "photometric Be" stars, which significantly increase the frequencies of Oe/O and Be/early-B stars determined and are crucial when comparing such stars in different galaxies. This novel method of confirming Oe and Be star candidates from their infrared colors or a combination of their optical and infrared colors, as recently suggested by Ita *et al.* (2010) is complementary to the detailed spectroscopic analyses by

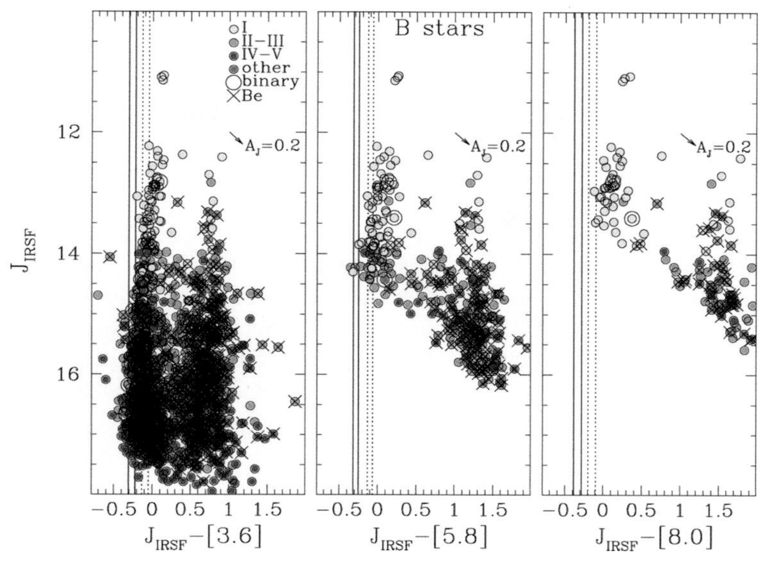

Figure 1. Infrared excesses (J_{IRSF} vs. $J_{IRSF} - [3.6]$, $J_{IRSF} - [5.8]$ and $J_{IRSF} - [8.0]$) for 1967 early-B stars in the SMC. Supergiants are shown in yellow, giants in green, main-sequence stars in blue, stars with uncertain classifications ("other") in red, binaries with a large circle and Oe stars with an ×. The solid lines correspond to 30kK and 50kK TLUSTY models with $\log g = 4.0$. A reddening vector for $E(B - V) = 0.2$ mag is shown, as well as reddened TLUSTY models by this same amount (dotted lines). The more luminous stars exhibit larger infrared excesses, which increase with λ.

e.g. Negueruela *et al.* (2004) on individual Oe stars to understand their nature, although it is limited to galaxies with low internal reddening. We finally note that the spectral types of Oe stars in the SMC (O7.5Ve, O7Ve, O4-7Ve and O9-B0III-Ve) and the LMC (O9Ve (Fe II), O7:Ve, O8-9IIIne, O3e) are earlier than those of known Galactic Oe stars, which are all found in the O9-B0 range (Negueruela *et al.* 2004).

Finally, we note that the brightest Be stars in the SMC ($J_{IRSF} \sim 13.2$ mag) are brighter than the brightest Be stars in the LMC ($J_{IRSF} \sim 13.4$ mag), i.e. there is a 0.7 mag difference in absolute magnitude, given the 0.5 mag difference in the distance moduli.

3.2. *Supergiant B[e] stars*

In the SMC photometric catalog, we have detected 7 luminous sources with colors typical of sgB[e] stars (see Paper I for an introduction), i.e. $M_{3.6} < -8$, $[3.6] - [4.5] > 0.7$, $J - [3.6] > 2$ mag. Five of these are previously known sgB[e] stars (with R50; B2-3[e] being the brightest in all IRAC and MIPS bands), while R4 (AzV 16) is classified as an LBV with a sgB[e] spectral type. In addition to these, we find that 2dFS1804 (AFA3kF0/B[e]) has a very similar SED (and therefore infrared colors) to the known sgB[e] 2dFS2837 (AFA5kF0/B[e]). Evans *et al.* (2004) also remarked on the similarity of their spectra. We therefore confirm the supergiant nature of 2dFS1804. The similarity of the SEDs of these sgB[e] stars, despite the various optical spectral classifications, implies that all are the same class of object. The cooler, composite spectral types indicate a lower mass and perhaps a transitional stage to or from the sgB[e] phenomenon. The only difference we find between the sgB[e] stars in the SMC vs. the LMC is that on average they are ~1-2 mag fainter (in absolute terms).

3.3. *Luminous Blue Variables*

All 3 known LBVs in the SMC: R4 (AzV 16, B0[e]LBV), R40 (AzV 415, A2Ia: LBV) and HD 5980 (WN6h;LBV binary), were detected at infrared wavelengths. R4 is the more reddened LBV, whereas the colors of HD 5980 (a well known eccentric eclipsing binary, see e.g. Foellmi *et al.* 2008) are similar to those of the LBVs in the LMC. We find their SEDs to differ, given their very different spectral types. Moreover, we find evidence for variability, which can be confirmed from existing light curves in the All Sky Automated Survey (ASAS) (Pojmanski 2002), as pointed out by Szczygiel *et al.* (2010), who studied the variability of the massive stars presented in Paper I in the LMC. The various SED shapes and spectral types observed depend on the time since the last outburst event and the amount of dust formed.

References

Bonanos, A. Z., Massa, D. L., Sewilo, M., Lennon, D. J. *et al.* 2009, *AJ*, 138, 1003
Bonanos, A. Z., Lennon, D. J., Köhlinger, F., van Loon, J. T. *et al.* 2010, *AJ*, 140, 416
Dougherty, S. M., Waters, L. B. F. M., Burki, G., Cote, J. *et al.* 1994, *A&A*, 290, 609
Evans, C. J., Lennon, D. J., Trundle, C., Heap, S. R. *et al.* 2004, *ApJ*, 607, 451
Foellmi, C., Koenigsberger, G., Georgiev, L., Toledano, O. *et al.* 2008, *Rev. Mexicana AyA* 44, 3
Gordon, K. D., Meixner, M., Blum, R., *et al.* 2010, *AJ*, in preparation
Grebel, E. K., Richtler, T., & de Boer, K. S. 1992, *A&A*, 254, L5
Ita, Y., Matsuura, M., Ishihara, D., Oyabu, S. *et al.* 2010, *A&A*, 514A, 2
Lanz, T. & Hubeny, I. 2003, *ApJS*, 146, 417
Lanz, T. & Hubeny, I. 2007, *ApJS*, 169, 83
Liu, Q. Z., van Paradijs, J., & van den Heuvel, E. P. J. 2005, *A&A*, 442, 1135
Maeder, A., Grebel, E. K., & Mermilliod, J.-C. 1999, *A&A*, 346, 459
Martayan, C., Floquet, M., Hubert, A. M., Gutiérrez-Soto, J. *et al.* 2007a, *A&A*, 472, 577
Martayan, C., Frémat, Y., Hubert, A.-M., Floquet, M. *et al.* 2007b, *A&A*, 462, 683
Martayan, C., Baade, D. & Fabregat, J. 2010, *A&A*, 509, A11
Meixner, M., Gordon, K. D., Indebetouw, R., Hora, J. L. *et al.* 2006, *AJ*, 132, 2268
Mokiem, M. R., de Koter, A., Vink, J. S., Puls, J. *et al.* 2007, *A&A*, 473, 603
Negueruela, I., Steele, I. A., & Bernabeu, G. 2004, *AN*, 325, 749
Panagia, N. & Felli, M. 1975, *A&A*, 39, 1
Pojmanski, G. 2002, *Acta Astronomica*, 52, 397
Porter, J. M. & Rivinius, T. 2003, *PASP*, 115, 1153
Szczygieł, D. M., Stanek, K. Z., Bonanos, A. Z., Pojmański, G. *et al.* 2010, *AJ*, 140, 14
Vink, J. S., de Koter, A., & Lamers, H. J. G. L. M. 2001, *A&A*, 369, 574
Wisniewski, J. P. & Bjorkman, K. S. 2006, *ApJ*, 652, 458
Wright, A. E. & Barlow, M. J. 1975, *MNRAS*, 170, 41

Discussion

MIROSHNICHENKO: What kind of positions do you have in your catalog of OB stars in the LMC (Spitzer, 2MASS, optical)?

BONANOS: We have used the best coordinates available, e.g. from Brian Skiff's updated lists for the Sanduleak catalog, which are generally accurate to <1".

WISNIEWSKI: How do you exclude or differentiate Herbig Be stars from classical Be stars in your data? Herbigs' transitional disks can show similar optical spectroscopic features:

IR colors (especially given the dust content of the SMC/LMC) and candidate Herbigs have already been identified in the SMC/LMC, see e.g. Lamers *et al.* 1999; de Wit *et al.* 2002, 2003, 2005; Bjorkman *et al.* 2005.

BONANOS: We have not differentiated between them, as our sample was selected from the literature by mainly targeting OB stars in clusters. None of the stars in our catalog have HBe classifications, however some could be HBe stars.

Active OB stars: structure, evolution, mass loss, and critical limits
Proceedings IAU Symposium No. 272, 2010
C. Neiner, G. Wade, G. Meynet & G. Peters, eds.

doi:10.1017/S1743921311010507

The B[e] phenomenon in the Milky Way and Magellanic Clouds

Anatoly S. Miroshnichenko[1], Nadine Manset[2], Francesco Polcaro[3], Corinne Rossi[4], and Sergey Zharikov[5]

[1]Dept. of Physics & Astronomy, University of North Carolina at Greensboro, P.O. Box 26170, Greensboro, NC 27402–6170, USA
email: a_mirosh@uncg.edu
[2]CFHT Corporation, 65–1238 Mamalahoa Highway, Kamuela, HI 96743
[3]Istituto di Astrofisica Spaziale e Fisica Cosmica, INAF, Via del Fosso del Cavaliere 100, 00133, Roma, Italy
[4]Universitá La Sapienza Roma - Pza A Moro 5, I–00162 Roma, Italy
[5]Instituto de Astronomía, Universidad Nacional Autónoma de México, Apartado Postal 877, 22800, Ensenada, BC, Mexico

Abstract. Discovered over 30 years ago, the B[e] phenomenon has not yet revealed all its puzzles. New objects that exhibit it are being discovered in the Milky Way, and properties of known objects are being constrained. We review recent findings about objects of this class and their subgroups as well as discuss new results from studies of the objects with yet unknown nature. In the Magellanic Clouds, the population of such objects has been restricted to supergiants. We present new candidates with apparently lower luminosities found in the LMC.

Keywords. stars: early-type, infrared: stars, circumstellar matter

1. Introduction

The B[e] phenomenon was observationally defined by Allen & Swings (1976) on the basis of optical spectroscopic and near-infrared photometric data. It refers to the simultaneous presence of forbidden line emission (e.g., [O I], [Fe II], [N II], and sometimes [O III] lines) in addition to permitted line emission (e.g., Balmer and Fe II lines) and large IR excesses in the spectra of B-type stars. These observational features make it different from the Be phenomenon. The presence of forbidden lines indicates that the gaseous component of the circumstellar (CS) envelopes is more extended than in Be stars. The large IR excess is a manifestation of CS dust which is not present in Be stars. Also, the B[e] phenomenon occurs in a wider variety of objects than the Be phenomenon (see Miroshnichenko 2006 for a recent review).

Allen & Swings have already noticed the variety of objects with the B[e] phenomenon. Some twenty years later Lamers *et al.* (1998) summarized available data and concluded that the B[e] phenomenon occurs in four stellar groups with well-understood nature and evolutionary status (Herbig Ae/Be stars, symbiotic binaries, supergiants, and compact Planetary Nebulae). At the same time, nearly half of the original list of 65 objects with the B[e] phenomenon remained unclassified. A few of them were well-studied (e.g., FS CMa = HD 45677 and V742 Mon = HD 50138), but their derived properties did not allow to fit them within any of the above groups. Other unclassified objects have not been studied enough to classify them until recently.

A few years ago Miroshnichenko (2007) and Miroshnichenko *et al.* (2007) analyzed both historic and their own data on the unclassified objects and concluded that most of them can be separated into a new group. The group was called FS CMa type objects after the prototype object with the B[e] phenomenon (Swings 2006). These authors also showed that FS CMa objects are neither pre-main-sequence Herbig Ae/Be stars nor symbiotic binaries, their luminosity range (log $L/L_{\odot} \sim 2.5$–4.5) is below that of supergiants, and they are unlikely to be at the post-AGB evolutionary stage. The main observational features of this group are described in the other paper by Miroshnichenko *et al.* in these proceedings. One of these features is a very strong emission-line spectrum which is hard to explain by the evolutionary mass loss from a single star of the described luminosity range (Vink, de Koter, & Lamers 2001). Therefore, it was suggested that FS CMa objects are binary systems observed after a rapid mass-exchange phase (Miroshnichenko 2007). No direct mass transfer seems to be currently observed in any of these objects.

Currently, only a few objects with the B[e] phenomenon can be called unclassified due to insufficient data. All Herbig Ae/Be stars, not only seven from the original list, exhibit the B[e] phenomenon. Therefore, the original list of 65 objects has been expanded significantly. Among the variety of objects with the B[e] phenomenon, only supergiants and FS CMa type objects seem to form dust while having a B-type star in their content. Here we will concentrate on the expansion of the list of objects which belong to the FS CMa group in the Milky Way and the Large Magellanic Cloud (LMC).

2. Finding new candidates

Another feature of the FS CMa group is a sharp decrease of the IR flux at $\lambda > 10\mu$m. It allowed Miroshnichenko *et al.* (2007) to expand the group by finding nine IRAS sources (whose fluxes were accurately measured in three IRAS photometric bands at 12, 25, and 60 μm) with such a flux behavior that positionally coincide with early-type emission-line stars. However, this procedure only works for relatively bright IR sources, because the sensitivity of IRAS decreases with wavelength. Also, the lowest-mass post-AGB objects, RV Tau type stars, turned out to have IRAS colors within the same range.

Nevertheless, a new photometric criterion for FS CMa objects was established in the same paper. It was found that the observed $J - K$ color-indices of FS CMa objects exceed $\sim$1.3 mag, while RV Tau stars show bluer colors. We decided to use this criterion in combination with not very large optical colors and search for new candidates with the B[e] phenomenon in a catalog of emission-line stars by Kohoutek & Wehmeyer (1997), which we cross-correlated with the NOMAD catalog by Zacharias *et al.* (2005).

As a result, we found sixteen new candidates. Five of them, randomly picked, were observed and the presence of forbidden lines was found in all of them. Additionally, a very faint object with the B[e] phenomenon ($V \sim 17$ mag) was accidentally found. It turned out to positionally coincide with the 2MASS source 03094640+6418429. Fragments of our spectra of some of these objects are shown in Fig. 1. They are most likely FS CMa objects, because luminosity sensitive absorption lines (e.g., Si III 5739 Å, Miroshnichenko *et al.* 2004) typical for supergiants, were not found in their spectra. Other newly discovered and spectroscopically confirmed objects with the B[e] phenomenon include IRAS 02110+6212 = VES 723, IRAS 21263+4927, IRAS 20090+3809, and MWC 485.

We continue searching the NOMAD catalog and have already found another dozen of new candidates. The photometric search is not exhaustive, because we only look for objects with not very different magnitudes in the blue and red band. Redder objects may be just very cool stars. Nevertheless, currently the number of FS CMa objects and candidates approaches 70.

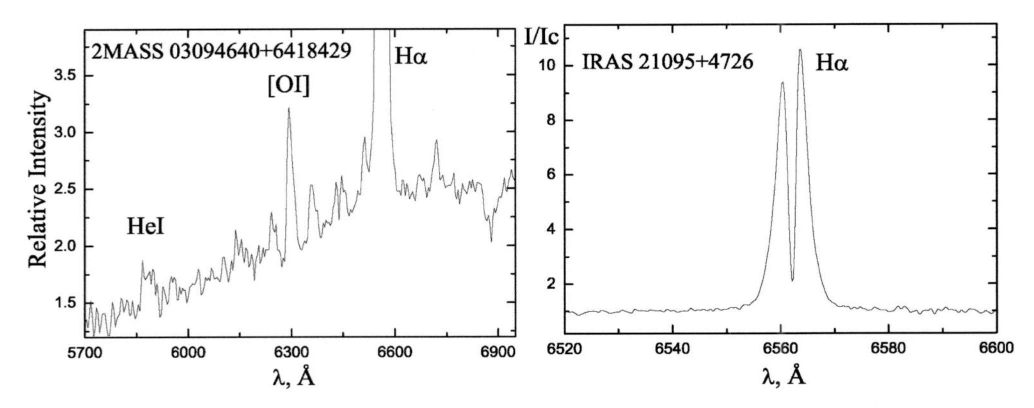

Figure 1. Left panel: Low-resolution spectrum of 2MASS 03094640+6418429 obtained at the 1.82-m telescope of the Padova Observatory ($R \sim 1500$). Right panel: High-resolution spectrum of IRAS 21095+4726 obtained at the 2.12-m telescope of the San Pedro Martir Observatory ($R \sim 15000$).

3. New candidates in the Large Magellanic Cloud

The first object (R 126) with the B[e] phenomenon in the Magellanic Clouds was reported by Zickgraf *et al.* (1985). Another seven were added by Zickgraf *et al.* (1986) and four more by Gummersbach, Zickgraf, & Wolf (1995). All these objects are located beyond the main-sequence. Ten of them are supergiants with luminosities $\log L/L_{\odot} = 4.7 - 6.1$, while the remaining two (Hen S59 and Hen S137) are late B-type stars with a luminosity of $\log L/L_{\odot} \sim 4$.

It is easier to separate supergiants from less luminous stars in the Clouds with their low interstellar extinction than in the Milky Way. All the supergiants with the B[e] phenomenon in the LMC have optical brightnesses of $V = 11 - 13$ mag objects, while Hen S59 and Hen S137 have $V \sim 14$ mag. In an attempt to find more candidates to the list of objects with the B[e] phenomenon in the LMC, we positionally cross-correlated a catalog of optical photometry by Zaritsky *et al.* (2004) and the 2MASS catalog (Cutri *et al.* 2003).

In total, we found nearly 100 positionally close objects that have slightly reddened colors of B-type stars and large IR color-indices, but for only nine of them the optical-IR position offset does not exceed 1 arcminute. These objects are listed in Table 1. One of

Table 1. Candidates for objects with the B[e] phenomenon in LMC. The objects names are from Bohanan & Epps (1974) - BE74, Henize (1956) - LHA120, Andrews & Lindsay (1964) - AL, Ardeberg *et al.* (1972) - ARDB, and the rest is from the USNO–B1.0 survey (Monet *et al.* 2003). The coordinates are given for the epoch 2000.0.

Name	R.A.	Dec.	V	K	$J-K$
ARDB 54	4:54:43.5	−70:21:27.8	12.77	11.55	1.15
0218-0100858	5:45:29.5	−68:11:45.7	14.02	11.48	1.62
0203-0138943	5:41:43.7	−69:37:38.3	14.11	11.14	2.00
0181-0125572	5:27:47.6	−71:48:52.6	14.24	11.77	1.72
BE74 540	5:12:09.1	−71:06:49.7	14.27	12.11	1.61
BE74 580	5:24:17.4	−71:31:50.0	14.56	12.24	1.55
0225-0105286	5:24:57.9	−67:24:57.9	14.67	12.51	1.40
LHA120-N 148B	5:31:42.2	−68:34:53.9	15.42	13.41	1.50
AL 190	5:26:30.7	−67:40:36.5	15.69	12.48	2.12

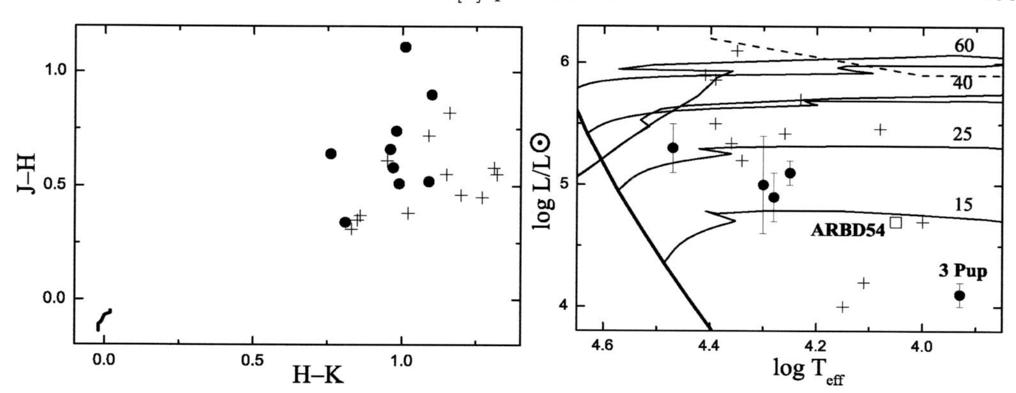

Figure 2. Left panel: A $(J-H) \sim (H-K)$ diagram for the known Clouds objects with the B[e] phenomenon (pluses) and the objects from Table 1 (circles). The solid line in the lower left corner represents intrinsic colors of B-type supergiants (Wegner 1994). Right panel: A Hertzsprung-Russell diagram for supergiants with the B[e] phenomenon and well-constrained luminosities in the Milky Way (circles) and in the Clouds (pluses). The newly found candidate ARDB 54 is shown by an open square. The solid lines are the zero-age main-sequence and evolutionary tracks for single stars (Schaller *et al.* 1992) labeled with initial masses in solar units. The dashed line is the Humphreys-Davidson stability limit.

these objects, ARDB 54, is a supergiant (see right panel of Fig. 2). All the others have $V = 14.0 - 15.7$ mag and represent lower luminosity objects.

A near-IR color-color diagram for the recognized objects with the B[e] phenomenon in the Magellanic Clouds and the objects from Table 1 is shown in the left panel of Fig. 2. If forbidden line emission is found in their spectra, then they probably belong to the FS CMa group. We plan to obtain both photometric and spectroscopic observations of these objects in the nearest future.

4. Conclusions

Using our photometric criteria for separation of objects with the B[e] phenomenon, we found nearly 20 new candidates in the Milky Way and eight in the LMC. One new supergiant candidate is also found in LMC. The presence of the B[e] phenomenon in five Galactic objects was confirmed. Thus, we expanded the Galactic FS CMa group to nearly 50 members and 20 candidates. The number of objects with the B[e] phenomenon in the LMC may be doubled, if its presence in the spectra of the reported candidates is confirmed.

Acknowledgements

This research has made use of the SIMBAD database operated at CDS, Strasbourg, France, and of data products from the Two Micron All Sky Survey, which is a joint project of the University of Massachusetts and the Infrared Processing and Analysis Center/California Institute of Technology, funded by the National Aeronautics and Space Administration and the National Science Foundation.

References

Allen, D. A. & Swings, J. P. 1976, *A&A*, 47, 293
Andrews, A. D. & Lindsay, E. M. 1964, *Irish Astronomical Journal*, 6, 241
Ardeberg, A., Brunet, J. P., Maurice, E., & Prevot, L. 1972, *A&AS*, 6, 249

Bohannan, B. & Epps, H. W. 1974, *A&AS*, 18, 47

Cutri, R.M., Skrutskie, M. F., van Dyk, S., Beichman, C. A. *et al.* 2003, *The IRSA 2MASS All-Sky Point Source Catalog* (NASA/IPAC Infrared Science Archive)

Gummersbach, C. A., Zickgraf, F.-J. & Wolf, B. 1995, *A&A*, 302, 409

Henize, K. G. 1956, *ApJS*, 2, 315

Kohoutek, L., & Wehmeyer, R. 1997, *Astron. Abh. Hamburg. Sternw.*, 11

Lamers, H. J. G. L. M., Zickgraf, F.-J., de Winter, D., Houziaux, L. *et al.* 1998, *A&A*, 340, 117

Miroshnichenko, A. S., Levato, H., Bjorkman, K. S., Grosso, M. *et al.* 2004, *A&A*, 417, 731

Miroshnichenko, A. S. 2006, in: M. Kraus & A. S. Miroshnichenko (eds.), *Stars with the B[e] Phenomenon*, ASP-CS 355, p. 13

Miroshnichenko, A. S. 2007, *ApJ*, 667, 497

Miroshnichenko, A. S., Manset, N., Kusakin, A. V., Chentsov, E. L. *et al.* 2007, *ApJ*, 671, 828

Monet, D. G., Levine, S. E., Canzian, B., Ables, H. D. *et al.* 2003, *AJ*, 125, 984

Schaller, G., Schaerer, D., Meynet, G., & Maeder, A. 1992, *A&AS*, 96, 269

Swings, J.-P. 2006, in: M. Kraus & A. S. Miroshnichenko (eds.), *Stars with the B[e] Phenomenon*, ASP-CS 355, p. 3

Vink, J. S., de Koter, A., & Lamers, H. J. G. L. M. 2001, *A&A*, 369, 574

Wegner, W. 1994, *MNRAS*, 270, 229

Zacharias N., Monet D. G., Levine S. E., Urban S. E., Gaume R., & Wycoff G. L. 2005, *American Astronomical Society Meeting 205*, BAAS 36, p. 1418

Zaritsky, D., Harris, J., Thompson, I. B., & Grebel, E. K. 2004, *AJ*, 128, 1606

Zickgraf, F.-J., Wolf, B., Stahl, O., Leitherer, C. *et al.* 1985, *A&A*, 143, 421

Zickgraf, F.-J., Wolf, B., Leitherer, C., Appenzeller, I. *et al.* 1986, *A&A*, 163, 119

Active OB stars: structure, evolution, mass loss, and critical limits
Proceedings IAU Symposium No. 272, 2010
C. Neiner, G. Wade, G. Meynet & G. Peters, eds.

doi:10.1017/S1743921311010519

Massive variable stars at very low metallicity?

Dominik J. Bomans and Kerstin Weis

Astronomical Institute, Ruhr-University Bochum,
Universitätsstr. 150, 44801, Bochum, Germany
email: bomans@astro.rub.de, kweis@astro.rub.de

Abstract. Observational contraints on the evolution and instabilities of massive stars at very low metallicities are limited. Most of the information come from HST observations of one target, I Zw 18. Recent distance estimates of I Zw 18 put it at 17 Mpc, moving detailed studies of single stars clearly beyond the range of current ground based telescopes. Since massive stars with metallcities of 1/10 of solar and below are our best proxies for massive stars in (proto-) galaxies around the time of reionization, finding them and studying their evolution and instabilities is of premium importance for our understanding of galaxy formation, feedback, and the IGM reionization. Here we present pilot study results of variable stars in two more nearby extremely low metallicity galaxies, UGC 5340 and UGCA 292, and comment on the possibilities of more detailed studies of variable massive stars with new ground-based instrumentation.

Keywords. stars: variables: other, stars: evolution, stars: early-type, (stars:) supergiants, stars: winds, outflows, galaxies: stellar content, galaxies: irregular, instrumentation: spectrographs

1. Why is very low metallicity exciting?

It is long known from theory of stellar evolution and stellar atmospheres that there are significant changes when going to stars at low metallicity. These effects are already apparent in the Magellanic Clouds, which provide nice laboratories for massive stars at 0.4 and 0.2 of solar metallicity. Still, there is a significant gap between the metallicity of the Magellanic Clouds and the metallicities observed at high redshifts, despite a significant spread of metallicities observed in DLA and Lyman break galaxies. This is especially true for the epoch of galaxy formation to the end of reionization ($z > 6$) (e.g., Schaerer & de Barros 2010). Observational studies of massive stars at metallicities significantly below 1/10 of solar are therefore important for our understanding not only of galaxy formation but also stellar feedback at high redshift. Massive stars and subsequent supernovae provide the energy input driving galactic outflows and winds (Leitherer *et al.* 1992), which are critical ingredients of galaxy formation and evolution. At very low metallicity, stellar evolution, stellar atmospheres, and the instabilities in stars should be very different. This affects stellar winds, late evolutionary phases, variability, (evolution of) rotation, convection, SN types, binary star evolution, and progenitors of long GRBs.

2. Nearby very low metallicity galaxies

To get a better handle on the stellar population and their time dependent energy input to their host galaxies in the early universe, observations of the best possible local proxies are mandatory. Unfortunately, such extreme metal-poor galaxies (metallicities of less than $\sim$1/10 of solar) are quite rare objects (e.g., Kunth & Östlin 2000). The prototype of these galaxies is I Zw 18. Not surprisingly, it was repeatably observed with HST, but it turned out to be quite difficult to study due to high crowding, high and variable background and

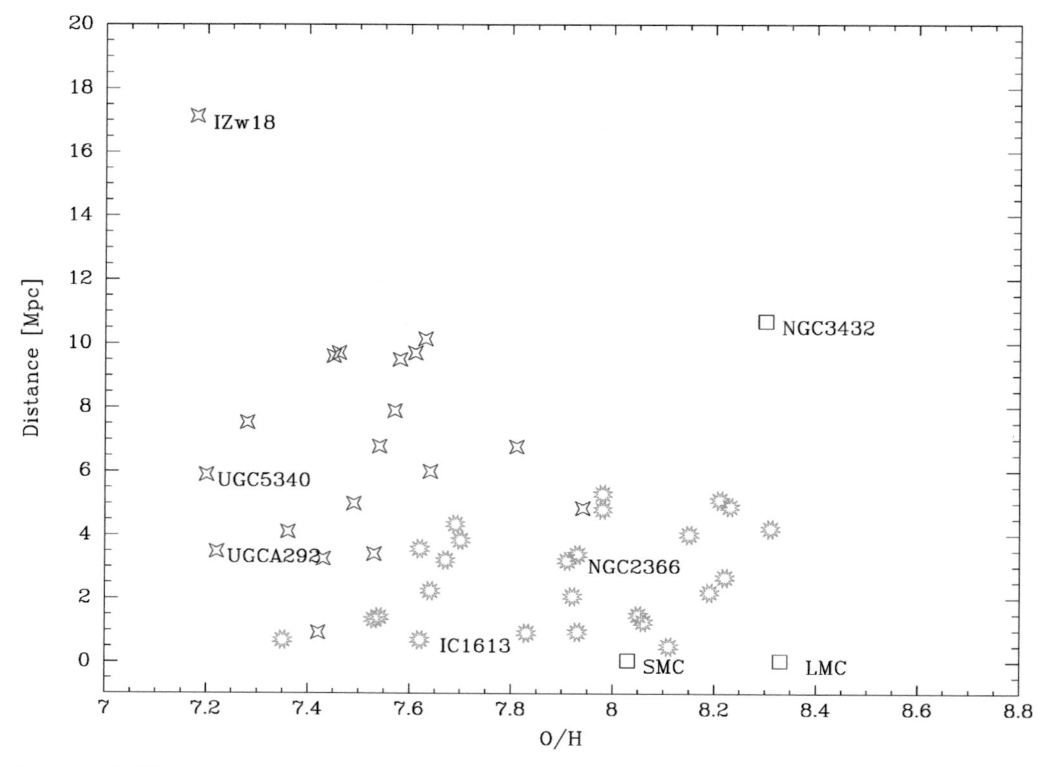

Figure 1. Distance versus metallicity of a compilation of the most metal poor galaxies in the local universe. Green stars are from the compilation of Lee *et al.* (2006), black squares some important other objects, blue stars are all currently known nearby very low metallicity galaxies.

faintness of its stars. I Zw 18 has now a securely determined distance of 18 Mpc (Aloisi *et al.* 2007; Fiorentini et al. 2010), which places it beyond the reach for ground based studies of its stars. The analysis of the HST data material showed that I Zw 18 does contain five periodic variables, 3 of them inside the Cepheid instability strip (2 of the Cepheids have periods $>$ 100 days), and 34 candidat/non-periodic variables (Fiorentino *et al.* 2010). Since the Meynet & Maeder (2005) rotation models imply an extension of the mass range of stars with LBV-like instabilities downward (e.g., Weis 2010, these proceedings), the location of the brightest and bluest of these candidate variables in I Zw 18 is consistent with the CMD location of rapidly rotating LBV-like stars.

Clearly, for detailed studies of massive stars at these metallicities, more targets at lower distance are important. The basic problem to reach this aim is that such low metallicities are found locally only in low mass galaxies (e.g. Lee *et al.* 2006). Therefore a significant recent star formation rate in these galaxies is needed to ensure useful sample sizes of (very) massive and therefore short lived stars. The short evolutionary time scales therefore implies that not every low metallicity dwarf galaxy will provide examples of all short lived, transitional phases.

3. Very low metallicity galaxies in the local volume

Which are the galaxies with significant recent star formation at low metallcity and how many do exist in the Local Volume? As Local Volume we define here the sphere with 11 Mpc radius of the Milky Way. It is roughly the maximal volume in which massive single stars are accessible for detailed spectral analysis with 10 m class ground based telescopes, e.g. Bresolin *et al.* (2001). The problem for the study of single stars out to these distances is not only the faintness of the targets but also the spatial resolution. Dense groups and clusters of massive stars become unresolved, or just barely resolved, producing a significant light contamination to targets in or near them. In the case of luminous transients, like LBV outbursts, these problems are significantly reduced, as the star outshines its environment. In such a case the Local Volume may be even a conservative limit, see for example the LBV transient in NGC 3432 (Pastorello *et al.* 2010). In the case of very bright transients, time enters the problem. Due to their transient nature, they will unpredictively be bright enough for study. The other problem is that studying only these transient introduces a classical Malmquist bias to a lot of the analyses. We started to compile a data base of very low metallicity local galaxies, the current state of it is plotted in Fig.1. While I Zw 18 is clearly not the best target anymore, it is doable with the HST, and several more potentially useful targets within 8 Mpc are apparent.

4. Apparently very massive variable stars a low metallicity

UGC 5340 is an extremely low metallicity galaxy with an abundance of ∼1/40 solar. The LBV candidate in UGC 5340 was a serendipitous spectroscopic discovery. When taking a spectrum of a HII knot in 2008, Pustilnik *et al.* (2008) noted that it is significantly different from a spectrum taken 3 years earlier. The difference spectrum clearly showed P Cygni profiles in the Balmer lines and a blue continuum. They interpreted this finding as the brightening of an LBV, an interpretation later supported by Izotov & Thuan (2009).

Still, one should still be skeptical, since slight misalignment between the slit position of the observations in such a relatively distant (D ∼ 8 Mpc) and a complex background may lead to spurious results. To check the presence of a variable source, we compiled a ground based light curve of the object using own and archival imaging data. The result is plotted in Fig.2. Clearly, the knot is variable by more that a 1 mag over the last 50 years. If the measurement from 1988 defines the quiescent state of the most luminous star in the knot, than the star showed brightening by $> 2^{mag}$ since then. For the interpretation one has to keep in mind that these measurements are integrated values for the unresolved (or barely resolved) ionizing cluster of an HII region. Therefore the age of the cluster should be below 3×10^6 yr. Our preliminary STARBURST99 simulations even imply $\sim 1 \times 10^6$ yr (Bomans *et al.* 2010).

All these pieces of evidence seem to be consistent with a very massive, highly variable star, dominating the luminosity and color of the cluster in its bright phases. Before jumping to the interpretation of an LBV, there are a few odd aspects: the brightening appears to be at constant or bluer color, which excludes a classical S Dor variability (e.g. van Genderen 2001). The wind terminal velocity estimated from the 2008 difference spectrum is $\sim$800 km s^{-1}, which looks more more like a wind than mass ejection event (a giant eruption).

NGC 2366 is a strongly star forming (Lee *et al.* 2009) dwarf irregular galaxy at a distance of 3 Mpc and a metallicity of ∼ 1/10 solar. The massive stars are clearly driving material out into the halo of this galaxy (Martin 1998; van Eymeren *et al.* 2009), making it a good laboratory for stellar feedback studies. In NGC 2366, Drissen *et al.* (2001) noted the sudden appearance of a stellar source inside its brightest giant HII region. They found an increase by 3.1^{mag} to V $\sim -10.2^{mag}$ and a rise in brightness while getting hotter (bluer colors). The spectra changed with time (Petit *et al.* 2006) and the early spectra are similar

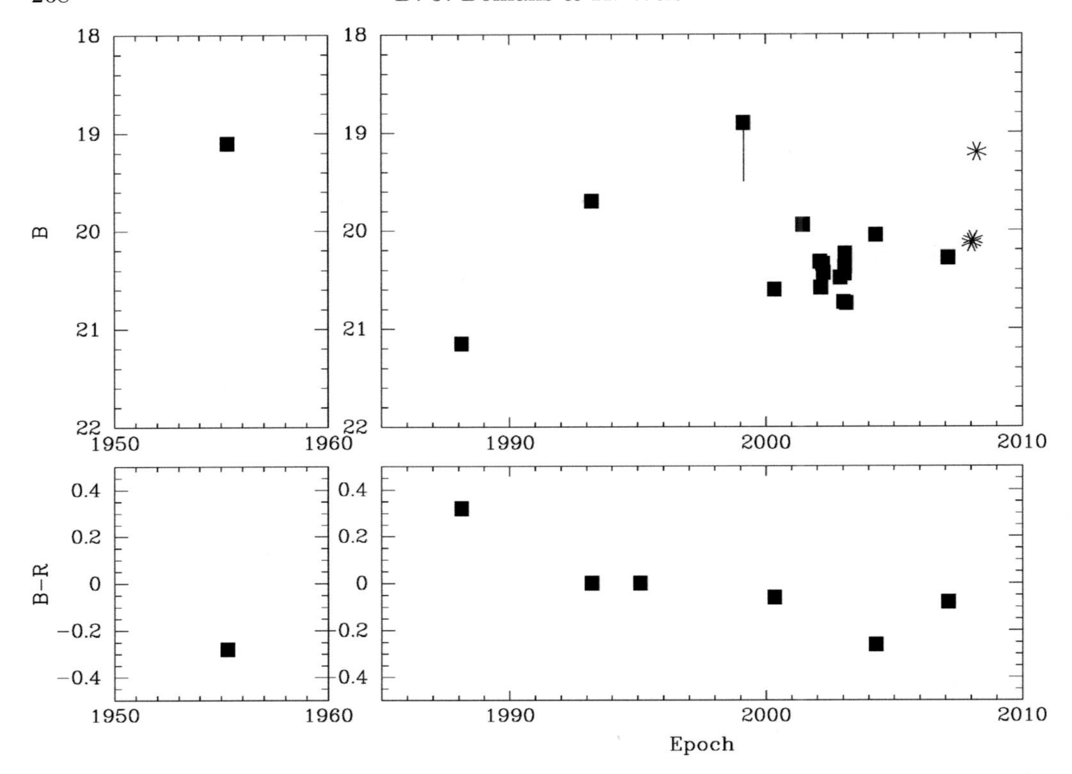

Figure 2. Lightcurve of the cluster plus transient source in UGC 5340. Black squares are archival and our photometric data (mostly from CCD), asterisks denote spectroscopic fluxes converted to R band magnitudes.

to the one of the UGC 5340 transient. Again the spectra are more like wind instead of sudden mass ejection (giant eruption) and the wind terminal velocity $\sim 250\,\mathrm{km\,s^{-1}}$.

There appear to be distinct similarities between the UGC 5340 transient and NGC 2366 V1. One is tempted to speculate that we see a LBV-like variability, but somewhat different from classical S Dor cycle. Within that cycle the star should encounter the bistability jump, which is function of metallicity. This might lead to a less pronounced change in the star's spectrum (and the S Dor variability) and may yield a different spectroscopic behaviour for LBVs like UGC 5340 and NGC 2366 V1 (see e.g., Weis 2010, these proceedings).

5. Very low metallicity variables at lower mass?

Observational results on the evolution of massive (but not extremely massive) stars and the variability at very low metallicities are very sparse, too. The I Zw 18 results of Fiorentino *et al.* (2010) show a rich population of possible variables especially in the blue loop area of lower mass supergiants. Differences to higher metallicity samples are not yet clear, since the apparent statistics needs to take the complex star formation history in I Zw 18 into account.

V39 in IC 1613, a very metallicity dwarf irregular galaxy in the Local Group (1/14 of solar), is another example of strange variable stars at the lower end of the mass region expected for LBV-like stars (Herrero *et al.* 2010).

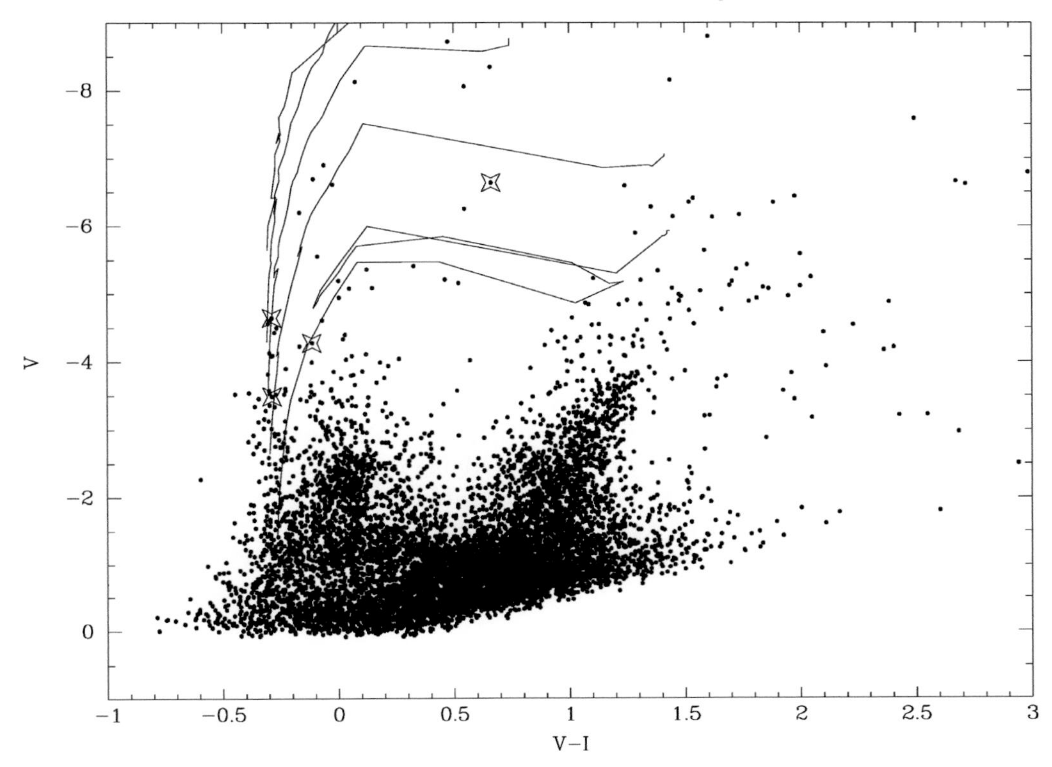

Figure 3. Preliminary HST CMD of UGCA 292. The 4 objects marked with stars are high probability variable stars. Geneva tracks (without rotation for 60, 40, 25, 15, and 9 $M_{\odot}$ are overplotted.

Using our database of very low metallicity galaxies, we startet a pilot project to search for massive variable stars in these galaxies using HST archival data. Fig.3 shows as an example the color-magnitude diagram (CMD) of UGCA 290 ($\sim 1/40$ of solar metallicity). The HST data provided a high quality CMD with clear structure in the region of upper main sequence and blue supergiant plume. We detected four high probability variables, one of which is located in the Hertzsprung gap. This pilot analysis based on only two epochs can can obviously provide just a snapshot of the variable massive star population. Photometric monitoring will be the next step, where the HST data provide useful priors for the photometry from ground-based data.

6. Instrumentation aspects

Imaging and photometric monitoring are important for the identification and characterization of the massive (variable) stars, but for many parameters, at least intermediate dispersion spectroscopy is needed. Together with our need for high spatial resolution this defines a mayor challenge, which at least two new instruments at 8-10m class telescope will help to overcome. At the LBT, running since 2010, the LUCIFER NIR multi-object spectrograph, is fully utilizing the very good median seeing in the NIR at Mt. Graham. At the VLT, starting 2012, the multi-IFU spectrograph MUSE with adaptive optics will be big progress for solving the crowding problem. Both instruments provide good spectral resolution and high throughput for the spectral analysis of stellar parameters in Local Volume galaxies. Clearly, the future ELTs will allow the study of massive stars in galaxies well beyond this distance (e.g. Evans 2010, these proceedings), but there are several objects with can be studied with current instrumentation.

7. Conclusions

Stars are very different at ~ 1/40 solar metallicity than SMC stars, as recently shown by comparing their new HST-COS spectrum of the I Zw 18 cluster with UV spectra of SMC (1/5 solar metallicity) stars (Heap *et al.* 2010). The metal lines is the COS spectrum were extremely weak, even compared to SMC stars. Still, 'normal' stellar types exist, even Cepheids and WR stars. Very low metallicity galaxies are rare in the Local Volume, but there are a few good targets. Unfortunately, I Zw 18 is not necessarily one of them.

In UGC 5340 a LBV-like transient appeared in 2008. It appears to have similar properties as NGC 2366 V1. Both do not behave like normal, high metallicity LBVs, therefore we may see the effects of metallicity on the underlying instability. IC 1613 V39, the variable stars detected in UGCA 290, and the bright, blue non-periodic variable in I Zw 18 point at roads to the physics of very metal-poor massive stars.

New instruments are providing access to spectroscopy of the very metal poor active OB stars NOW!

References

Aloisi, A., Clementini, G., Tosi, M., Annibali, F. *et al.* 2007, *ApJ* (Letters), 667, L151
Bomans, D.J., *et al.* 2010, in: P. Williams & G. Rauw (eds.), *The multi-wavelength view of Hot, Massive Stars*, 39th Liège International Astrophysical Colloquium, in press
Bresolin, F., Kudritzki, R.-P., Mendez, R. H. & Przybilla, N. 2001, *ApJ* (Letters), 548, L159
Drissen, L., Crowther, P. A., Smith, L. J., Robert, C. *et al.* 2001, *ApJ*, 546, 484
van Eymeren, J., Marcelin, M., Koribalski, B., Dettmar, R.-J. *et al.* 2009, *A&A*, 493, 511
Fiorentino, G., Contreras Ramos, R., Clementini, G., Marconi, M. *et al.* 2010, *ApJ*, 711, 808
van Genderen, A. M. 2001, *A&A*, 366, 508
Heap, S. *et al.* 2010, in: P. Williams & G. Rauw (eds.), *The multi-wavelength view of Hot, Massive Stars*, 39th Liège International Astrophysical Colloquium, in press
Herrero, A., Garcia, M., Uytterhoeven, K., Najarro, F. *et al.* 2010, *A&A*, 513, A70
Izotov, Y. I. & Thuan, T. X. 2009, *ApJ*, 690, 1797
Kunth, D. & Östlin, G. 2000, *A&AR*, 10, 1
Lee, H., Skillman, E. D., Cannon, J. M., Jackson, D. C. *et al.* 2006, *ApJ*, 647, 970
Lee, J. C., Kennicutt, R. C., José G. Funes, S. J., Sakai, S. *et al.* 2009, *ApJ*, 692, 1305
Leitherer, C., Robert, C. & Drissen, L. 1992, *ApJ*, 401, 596
Martin, C. L. 1998, *ApJ*, 506, 222
Meynet, G. & Maeder, A. 2005, *A&A*, 429, 581
Pastorello, A., Botticella, M. T., Trundle, C., Taubenberger, S. *et al.* 2010, *MNRAS* 408, 181
Petit, V., Drissen, L., & Crowther, P. A. 2006, *AJ*, 132, 1756
Pustilnik, S. A., Tepliakova, A. L., Kniazev, A. Y., & Burenkov, A. N. 2008, *MNRAS*, 388, L24
Schaerer, D. & de Barros, S. 2010, *A&A*, 515, A73

Active OB stars: structure, evolution, mass loss, and critical limits
Proceedings IAU Symposium No. 272, 2010
C. Neiner, G. Wade, G. Meynet & G. Peters, eds.

doi:10.1017/S1743921311010520

Discussion – Populations of OB stars in galaxies

Gregg A. Wade[1] **and Hideyuki Saio**[2]

[1]RMC, Kingston, Canada
[2]Astronomical Institute, Graduate School of Science, Tohoku University, Sendai, Miyagi 980-8578, Japan

Abstract. This article summarises the discussion following Session 3 (Populations of OB stars in galaxies) of IAUS 272.

This discussion followed talks by N. Langer (The elusive massive main sequence stars), T. Decressin (Massive stars in globular clusters: drivers of chemical and dynamical evolution), C. Evans (Populations of OB stars in galaxies), C. Martayan (Populations of Be stars: stellar evolution of extreme stars), A. Bonanos (Infrared properties of Active OB stars in the Magellanic Clouds from the Spitzer SAGE Survey), A. Mironichenko (The B[e] phenomenon in the Milky Way and Magellanic Clouds) and D. Bomans (Massive variable stars at very low metallicity?).

J. Zorec: Comment related to presentations by Martayan / Maeder / Miroschnichenko. The problems of abundance determination in fast rotators (Be stars) is challenging. First, massive Be stars are required to get lines of CNO in the optical. In addition, these stars should be viewed pole-on to reduce $v \sin i$ and produce sharp lines. One must use models including rotational effects, otherwise one will tend to overestimate the N abundance. Veiling effects are potentially a problem, and need to be corrected, in particular for pole-on stars because it is for these objects that we have the largest contribution from the (face-on) disc. The calibration presented by Martayan is only one of several possible. Uncertainties result from lack of knowledge of intrinsic magnitude of "naked" star (i.e. there is a contribution form the disc). Also, it is not clear what the structure of envelope does to the scatter. So positions of stars on that diagram are uncertain.

G. Peters: Comments concerning abundances/bipolar structures of Be stars. UV avoids veiling problems, and various C and O lines available. But emission lines are present as well, usually in doubly-ionised lines (e.g. Si II and IV can be fit well, but not Si III). This emission consitutes evidence for external plasma at temperatures higher than the disc (20 kK), probably from a bipolar flow. Concerning $v \sin i$: has personally found vsini's from UV features to be slightly smaller than in the optical, possibly due to temperature variation of photosphere from pole to equator. Discussion of binary and binary fraction for Be stars. One issue not discussed in the talks is that very close binaries have not been examined. With Gies, has found subdwarf companions around 2 objects. Also Be counterparts of Algol systems, etc.

J. Puls: A completely different topic: what is the present state of understanding B[e] supergiants? Question directed to the evolutionary people, there are always two scenarios discussed. Do these result from binary merger or post-supergiant evolution?

A. Maeder: if such a star rotates quite fast, then it should have a fast low density polar wind, and a slow, high density equatorial wind where dust can form, so this would be consistent with the "peanut" shape. The presence of companion may also be possible, but

why would a companion make the polar ejection? Is the frequency of mergers compatible with the observed number of such stars? Perhaps this is not sufficiently frequent?
Inaudible question.
A. Mironichenko: From the observational point of view, about 10 B[e] supergiants confirmed in the galaxy. Most are binaries not yet merged. For example, in my talk I showed B1 supergiant + F supergiant, with a period of about 1/2 year. So most are actually in binary systems.
N. Langer: A few comments in this respect. One B[e] supergiant for which strong evidence exists that merger was the source, R4 in SMC. Still is a binary, has an main sequence A star companion (5 $M_{\odot}$), so B star should have long been gone, so must be from merger. Evidence for a merger in this case is quite evident. From binary population synthesis statistics, 10% of all MS stars undergo mergers during their MS evolution - a rather high fraction of stars can be expected to merge. Concerning Be stars: the binary incidence is very important. In Be X-ray binaries, all the B-type main sequence stars are Be stars, so binary is related to Be phenomenon. If a neutron star is formed in a supernova there will be a very strong kick given to the neutron star and ejection from the binary system, so most Be X-ray binaries should have been disrupted. It is expected that 95% of such systems get disrupted. Therefore for each Be X-ray binary that we see with a neutron star companion, there should be many single Be stars (that were originally in binary systems, but from which the neutron star has been ejected).
O. Chesneau: Regarding B[e] stars: From an observer's point of view, the formation of disk is a consequence of binarity, perhaps without a merger. In the case of a merger, perhaps the disc was formed before merger, and is not a consequence of merger. It is difficult to imagine disc as a consequence of merger - how to form long-lived Keplerian dense, dusty disc? No explanation has been offered except binary. This conclusion is unavoidable in the planetary nebula community - you are obliged to invoke angular momentum input from a companion, even if it is just a 10 Jupiter-mass planet.
M. Smith: Topic that could not be fit into one of the talks, but appropriate in this context: environment of central black hole of Milky Way. IR studies show that there are several B stars orbiting within a few AU of Milky Way's central black hole. Where do these stars come from? How did they form?
D. Baade: Recent workshop in Garching about nuclear star clusters. The clusters form in gas flow around black hole (stars are so young they must have formed in situ). A relationship exists between the mass of the cluster and velocity dispersion of parent galaxy (and hence the mass of the black hole). Related to feedback in the formation of galaxies on various scales.
(Inaudible discussion)
G. Meynet: A point that deserves further explanation and needs to be explained by models, is non-continuous formation of Be star discs - produced in discrete mass loss events. To know the timescales is important to understand physics behind the phenomenon. If it is due to critical velocity, we would expect this kind of back-and-forth evolution as it reaches $v_{\rm crit}$, loses mass, slows down, etc. Would a binary be able to produce this kind of episodic phenomenon in the injecton of matter into the disc?
C. Martayan: Two comments: First about Langer's remark. In the case of binary disruption, shouldn't we find Be stars as runaways, or associated with supernova remnants, if this were true? But most are found in open clusters. Consider δ Sco which undergoes mass ejection when companion passes periastron. So this shows at least one connection to binarity, but may not be the case for all Be stars.
N. Langer: In the binary scenario, you spin up a B star by accretion (from the companion?) and left with a rapidly-rotating star. How is this different from a single star that

is born rapidly rotating? Regarding Be stars / Be X-ray binaries: supernova remnants might still be around, but lifetime is just $10^4 - 10^5$ years, so only small fraction of stars would be expected to show that. Need statistics to test this. What about them being runaway stars from the supernova explosion? In principle right, but many systems are wide binaries. Perhaps through conservative mass transfer they may evolve to very wide orbits which are relatively immune to generating runaways.
D. Baade: You are right, there is no a priori reason to believe that single star and member of binary system would evolve differently after spin-up has occurred. However, this does not necessarily make them both Be stars: what is the 2nd (or 3rd or 4th) ingredient required? e.g. if pulsation is important, then internal structure must also be the same.
A. Maeder: It is sure that a supernova remnant has limited lifetime. A critical parameter for merging is for how long do we see polar ejection and disc? Also question of lifetime of the merging phenomenon. Not sure it is much longer than remnant lifetime. N. Langer, do you have an idea of the lifetime of the events associated with a merging event, e.g. the peculiarities associated with that?
N. Langer: Well, once the stars merge, they live on as a rapidly rotating single star for a long time. Whether or not it is a Be star or not depends on D. Baade's remark, that there may be factors involved that we do not know.
W. Huang: Recent result of B star rotation survey. Low vs. high mass B stars behave differently.
G. Peters: A phenomenon that is known more in the binary community is a class of Algol binaries that show long-term cycles of activity. Orbits are of order 10 days, but cycles are 6 months-2 years in the apparent mass transfer activity. Mass transfer in these systems seems to vary cyclically on long timescales. Perhaps relevant to discussion of binarity?
S. Owocki: To follow up on George's and Dietrich's comments: we should distinguish between ω/ω_{crit} and v/v_{crit}. To launch material with just thermal energy from stellar surface you need v/v_{crit} 95%, but this is ω/ω_{crit} of around 98-99%. So must be very close to critical to use just thermal energy. Pulsation could allow ejection at lower values of v/v_{crit}. If there is no process that forces the star to continually spin up to approach critical (e.g. internal evolution) the ejection process would only occur once. So such internal evolution much occur to continually push star toward critical, as we see episodic ejections.

Active OB stars: structure, evolution, mass loss, and critical limits
Proceedings IAU Symposium No. 272, 2010
C. Neiner, G. Wade, G. Meynet & G. Peters, eds.

doi:10.1017/S1743921311010532

Evolutionary study of the Be star 28 Tau

Nazhatulshima Ahmad[1], Mohd. Zambri Zainuddin[1], Mohd. Sahar Yahya[2], Peter P. Eggleton[3] and Hakim L. Malasan[4]

[1]Physics Dept., Faculty of Science, University of Malaya, 50603 Kuala Lumpur, Malaysia
[2]Math. Div., Center of Fundamental Sciences, University of Malaya
[3]Lawrence Livermore National Laboratory, US Livermore CA 94551 0808 United State
[4]Astronomy Division, FMIPA, ITB, Jl. Ganesa 10, Bandung 40132 Indonesia

Abstract. We present an evolutionary study of 28 Tau, a Be star, in connection with its rapid rotation. The photometric data during the absence of its envelope in 1921 have been used to determine the effective temperature and luminosity of the star at the main sequence of the HR diagram. From an evolutionary model, we found that the mass and radius of the star are about 3.2 $M_\odot$ and 3.2 $R_\odot$ respectively. The equatorial rotation velocity of the star,ν_e found to be close to its critical velocity,ν_{cr} where $\nu_e/\nu_{cr} \simeq 0.87$.

Keywords. stars: fundamental parameters, stars: evolution, stars: rotation

1. Introduction

28 Tau (Pleione, BU Tau, HD 23862, HR 1180), a B8Vpe star (Hoffleit, 1995), is a well-known typical example of a Be star observed since the $18^{\rm th}$ century. The variation of Be phases of 28 Tau has been reported since the early $19^{\rm th}$ century until 2007 (Goraya *et al.*, 1990, Katahira *et al.*, 1996, Tanaka *et al.*, 2007). The hydrogen emission lines were absent for about 32 years and have returned in late 1938 as a noticeable emission line with a fine central reversal (McLaughlin, 1938). The absence of hydrogen emission lines – which indicate the destruction of the circumstellar matter - influences the measurement of the surface brightness of the star. In this study, photometric data during the absence of the envelope have been used to determine the effective temperature and luminosity of the star on the HR diagram. The evolutionary tracks were created using Eggleton code (Eggleton & Kiseleva-Eggleton 2002) with consideration of the star's rotation.

2. The evolutionary model

The evolutionary track has been created at specific mass and rotation, taking into account the dynamo activity, mass loss by dynamo-driven stellar wind and magnetic braking, and metallicity Z = 0.02. 28 Tau has been found to be a spectroscopic binary by Katahira *et al.* (1996) with the secondary star expected to be a low-mass helium star or white dwarf. He concluded that the circumstellar gas around the primary star is supplied by the primary itself, based on the mass-loss rate of its companion. Hence, in this study we only evolve 28 Tau as a single star. With an initial mass, $M = 3.2 M_\odot$ and rotational period, $P_{\rm rot} = 0.334$ days, we found the best approximated values with $\log T_{\rm eff} = 4.0359$ and $\log L = 2.1077$, based on photometric data during the absence of its envelope using the bolometric flux method (Gray, 1992) and the correlation of $T_{\rm eff}$ and BC (Flower, 1996). Table 2 shows the age in Myr, rotational velocity in km/s, critical velocity in km/s and the ratio of rotational and critical velocities of each model in Fig 1. Model 3 is an estimated position of 28 Tau on the HR diagram observed in 1921 where the ratio of

Table 1. The photometric data taken from Binnendijk (1949) only using photovisual and Pv method. The bold data were chosen for this study in obtaining the $T_{\rm eff}$ and L of 28 Tau.

Year	$\mathbf{m}_v$	**Met**	**Year**	$\mathbf{m}_v$	**Met**	**Year**	$\mathbf{m}_v$	**Met**
1911	4.87±0.004	Pv	1914-1915	4.95±0.04	v	1928	5.00±0.06	Pv
1913	4.99±0.05	Pv	1916-1917	4.94±0.04	Pv	1931	4.96±0.06	Pv
1914-1915	4.92±0.03	Pv	**1921**	**4.85±0.03**	**Pv**	1936	5.04±0.04	Pg

Notes: Method : Pv-photovisual, v-visual and Pg-photographic

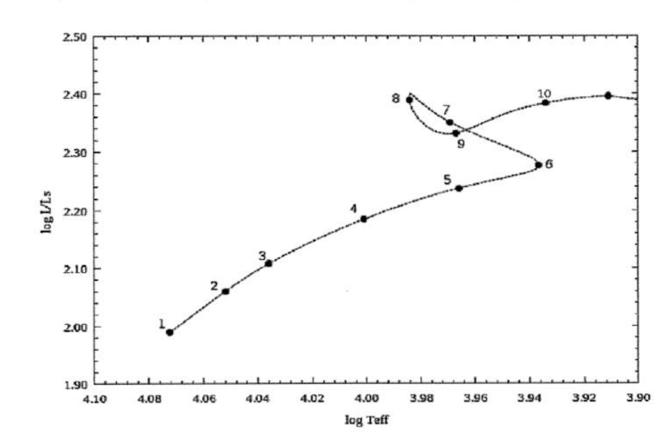

Figure 1. The evolutionary model of 28 Tau with initial M = 3.2M$_\odot$ and P$_{rot}$ = 0.334days.

Table 2. The values of age, P$_{rot}$(d), rotational velocity, ν, and critical velocity, ν_{cr}, at each model of the evolutionary track in Fig 1.

Model	**Age**	ν_e	ν_{cr}	ν_e/ν_{cr}	**Model**	**Age**	ν_e	ν_{cr}	ν_e/ν_{cr}
1	0.01	347.02	458.55	0.76	6	316.24	202.23	283.04	0.71
2	99.0	343.93	420.44	0.82	7	321.75	231.48	292.38	0.79
3	**154.73**	**341.26**	**394.01**	**0.87**	8	322.01	256.01	296.21	0.86
4	232.20	320.28	347.74	0.92	9	322.09	230.66	294.06	0.78
5	283.85	263.46	310.30	0.85	10	322.26	218.32	264.69	0.82

rotational to critical velocities is 0.87. The critical velocity is given by $\nu_{cr} = \sqrt{GM_\star/R_e}$ $= \sqrt{2GM_\star/3R_p}$ where R_e and R_p are the radius at equator and pole respectively.

3. Result and discussion

The ratio of rotational velocity of 28 Tau to its critical velocity during the main sequence band has been calculated in the range of 0.71-0.92. The upper limit is close to the ratio by Townsend *et al.* (2004) i.e 0.95. From the models we suggest that the current mass of 28 Tau is 3.19 M$_\odot$, radius of 3.2 R$_\odot$ and rotational period is 0.475 days.

References

Binnendijk, L. 1949, *AJ*, 54, 117B
Eggleton, P. P. & Kiseleva-Eggleton, L. 2002, *ApJ*, 575, 461
Flower, P. J. 1996, *ApJ*, 469, 355
Gray, D. F. 1992, *The observation and analysis of stellar photospheres*, Cambridge Astrophys. Ser., Vol. 20, p. 337
Goraya, P. S., Sharma, S. D., Malhi, J. S., & Tur, N. S. 1990, *Ap&SS*, 174, 1-11
Hoffleit, D. & Warren, Jr., W. H. 1995, *Bright Star Catalogue, 5th Revised Ed*
Katahira, J.-I., Hirata, R., Ito, M., Katoh, M. *et al.* 1996, *PASJ*, 48, 317
McLaughlin, D. B. 1938, *ApJ*, 88, 622
Tanaka, K., Sadakane, K., Narusawa, S.-Y., Naito, H. *et al.* 2007, *PASJ*, 59, L35
Townsend, R. H. D., Owocki, S. P., & Howarth, I. D. 2004, *MNRAS*, 350, 189-195

Active OB stars: structure, evolution, mass loss, and critical limits
Proceedings IAU Symposium No. 272, 2010
C. Neiner, G. Wade, G. Meynet & G. Peters, eds.

doi:10.1017/S1743921311010544

B stars in open clusters: fundamental parameters

Yael Aidelman[1], Lydia S. Cidale[1,2], Juan Zorec[3] and María L. Arias[1,2]

[1]Facultad de Ciencias Astronómicas y Geofísicas, Universidad Nacional de La Plata, Paseo del Bosque S/N, 1900 La Plata, Argentina; email: aidelman@fcaglp.unlp.edu.ar

[2]Instituto de Astrofísica de La Plata (CONICET-UNLP), Argentina

[3]Institut d'Astrophysique de Paris, UMR 7095 CNRS-Université Pierre & Marie Curie 98bis bd. Arago, 75014 Paris, France

Abstract. We use the BCD spectrophotometric classification system to derive fundamental parameters of B stars in NGC 2439, NGC 3766 and NGC 6087. We are able to perform a complete study of each open cluster by deriving spectral classification of its members, distance modulus and age as well.

Keywords. stars: distances, stars: emission-line, Be, stars: fundamental parameters

1. Introduction

Color-magnitude and color-color diagrams of open clusters are important tools to derive distances and ages useful to study the structure of the Galaxy. In addition, photometric studies provide information on the interstellar extinction and stellar evolution. Nevertheless, some cluster's age and distance remain somewhat uncertain and so are the properties of the stars belonging to these systems. The uncertainties could be related to the overlapping of different stellar groups in the line of sight, and to the presence of a generally inhomogeneous interstellar medium, and circumstellar envelopes around early-type stars.

In this work, we aim at obtaining distances, ages and fundamental parameters of B stars in galactic clusters based on the BCD spectrophotometric classification system (Barbier & Chalonge, 1941; Chalonge & Divan, 1952).

2. Methodology, Observations and Results

The BCD system is based on measurable quantities in the stellar continuum spectrum around the Balmer's discontinuity (BD). In particular the height of the BD is a strong function of $T_{\rm eff}$ while the spectral average position of the BD measured by λ_1 is related to the star's surface gravity [for details see Zorec *et al.* (2009)]. One of the advantages of the BCD system is that D and λ_1 are free from interstellar extinction and absorption/emission from the circumstellar envelope (Zorec & Briot, 1991). Furthermore, D and λ_1 allow us to determine not only the fundamental parameters $T_{\rm eff}$, log g, the spectral type and the luminosity class of a star but also M_v and M_{bol}, making use of the calibrations given by Zorec (1986) and Zorec *et al.* (2009).

Low resolution spectra in the range 3500-4600 Å were taken during multiple observing runs in 2002 March and 2003 February, using the B&C spectrograph attached to the 2.15m telescope in CASLEO, Argentina. We observed 11 stars of NGC 2439, 32 of NGC 3766, and 15 of NGC 6087.

Individual distance moduli for the stars of each cluster were derived using apparent magnitudes and color excesses from photometric data available in the literature, together

Table 1. Be stars with second component in the BD. Stars nomenclature is taken from White (1975) for NGC 2439, Ahmed (1962) for NGC 3766, and Fernie (1961) and Breger (1966) for NGC 6087. m_v values were taken from SIMBAD database.

ID	ID	D [dex]	λ_1 [Å]	ST & LC	$T_{\rm eff}$ [K]	log g [dex]	M_v [mag]	M_{bol} [mag]	m_v [mag]	$(m_v - M_v)_0$ [mag]
NGC 2439 070	CD-31 4897b	0.162	32.23	B3 II	17500	2.89	-4.80	-5.90	12.11	15.763±0.5
...	HD 62033	0.358	61.88	B8 V	12200	4.26	0.22	-0.80	8.32	6.86±0.3
NCG 3766 232	HD 100943	0.123	21.02	B5 Ib	16000	2.25	-8.00	-8.63	7.15	14.55±0.5
NGC 3766 240	ALS 2401	0.224	30.95	B4 III	15500	2.75	-4.00	-4.75	9.61	13.01±0.5
NGC 3766 264	HD 306657	0.198	84.58	B3	19000	$>$ 4.30	-1.10	-3.10	10.49	10.99±0.5
...	HD 308852	0.318	53.71	B6 V	13700	4.10	-4.43	-1.35	10.10	13.93±0.3
NGC 6087 007	HD 146483	0.300	30.67	B6 III	12500	2.83	-2.80	-3.50	8.29	10.46±0.5
NGC 6087 007	HD 146483	0.240	60.02	B4 V	16700	4.21	-0.88	-2.50	8.29	8.54±0.3
NGC 6087 009	HD 146484	0.285	79.42	B8 VI	14000	$>$ 4.30	-0.65	-1.70	9.48	9.50±0.3
NGC 6087 009	HD 146484	0.350	52.58	B7 V	12500	4.00	-0.45	-1.00	9.48	9.30±0.3
NGC 6087 010	HD 146324	0.290	41.00	B6 III	15000	3.35	-1.75	-2.50	7.92	9.04±0.5
NGC 6087 011	HD 146294	0.290	67.95	B7 VI	14100	$\sim$ 4.40	-0.28	-1.57	9.43	9.08±0.3
NGC 6087 014	CPD-57 7791	0.370	67.03	B9 V	11300	$\sim$ 4.39	0.52	-0.50	9.70	8.55±0.3
NGC 6087 156	CD-57 6346	0.230	73.79	B6 VI	15000	$\sim$ 4.43	-0.63	-2.00	9.20	9.20±0.3

with the BCD absolute magnitudes. The distance modulus for each cluster is an average of the individual determinations. Our values are 12.48 ± 0.44 mag, 10.07 ± 0.31 mag and 9.32 ± 0.33 mag for NGC 2439, NGC 3766 and NGC 6087, respectively. Cluster ages were derived by fitting the isochrones computed by Bressan *et al.* (1993). We obtained 12.6 Myr $<$ t $<$ 20 Myr for NGC 2439, 16 Myr $<$ t $<$ 24 Myr for NGC 3766 and 40 Myr $<$ t $<$ 79 Myr for NGC 6087. The results show excellent agreement with previous photometric determinations. Moreover, the BCD system has allowed us to detect 12 stars with the Be phenomenon, since they display a second BD which is an indicative of the presence of an extended envelope (see Table 1). Seven of these objects have been reported as Be stars for the first time. Likewise, the observation of the BD in two different epochs revealed that the stars 007 and 009 of NGC 6087 are variable.

3. Conclusions

The BCD method has allowed us to perform a complete study of the members of open clusters. We derive not only the spectral classification but the cluster's distance modulus and age as well. It is worth mentioning that the BCD spectrophotometric system is a powerful tool to study far galactic and extragalactic clusters with the large telescope generation since BCD parameters are free of interstellar and circumstellar extinction. Furthermore, the method is appropriate for the study and detection of Be stars.

References

Ahmed, F. 1962, *Publications of the Royal Observatory of Edinburgh*, 3, 60
Barbier, D. & Chalonge, D. 1941, *Annales d'Astrophysique* 4, 30
Breger, M. 1966, *PASP*, 78, 293
Bressan, A., Fagotto, F., Bertelli, G., & Chiosi, C. 1993, *A&AS*, 100, 647
Chalonge, D. & Divan, L. 1952, *Annales d'Astrophysique*, 15, 201
Fernie, J. D. 1961, *ApJ*, 133, 64
White, S. D. M. 1975, *ApJ*, 197, 67
Zorec, J. 1986, Thèse d'État (Université Paris VII)
Zorec, J. & Briot, D. 1991, *A&A*, 245, 150
Zorec, J., Cidale, L., Arias, M. L., Frémat, Y. *et al.* 2009, *A&A*, 501, 297

Active OB stars: structure, evolution, mass loss, and critical limits
Proceedings IAU Symposium No. 272, 2010
C. Neiner, G. Wade, G. Meynet & G. Peters, eds.

doi:10.1017/S1743921311010556

Analyzing the δ Sco binary in anticipation of a disk-star collision

Ashley Ames[1], **Christopher Tycner**[1] **and Robert Zavala**[2]

[1]Dept. of Physics, Central Michigan University
[2]USNO, Flagstaff Station

Abstract. A current investigation is underway into the possible collision between a circumstellar disk and the secondary star in the δ Scorpii binary system. δ Scorpii is a prime candidate for a disk-star collision since the primary star has a circumstellar disk and the secondary star has a highly elliptical orbit with a period of approximately 10.5 years making the periastron passage very close to the primary star. The Navy Prototype Optical Interferometer (NPOI) was used to spatially resolve the two stars as well as the cirumstellar disk around the primary. A revised orbit with new orbital parameters has been calculated using observations obtained between 2005 and 2010. Our results indicate periastron passage will occur on 2011-07-03.

Keywords. instrumentation: high angular resolution, astrometry, binaries: general

1. Introduction

δ Scorpii (HD143275, HR5953, FK5 0594) is a well known binary system with a highly eccentric orbit and a period of 10.5 years. The primary star is classified as a Be star with a gaseous circumstellar disk and the secondary is a B2-type star (Tango *et al.* 2009). The goal for this project is to refine the orbit of the secondary with respect to the primary and test for the possibility of a disk-star collision. The δ Sco binary system was observed with the NPOI for a total of 108 nights. Data on two nights were obtained in July 2000 and the rest were obtained from June 2005 to August 2010.

2. Experimental Data

All the raw observational data, based on 108 nights, have been reduced using a standard NPOI reduction pipeline. The astrometric information is extracted from the reduced data giving the position of the secondary with respect to the primary. The NPOI observations were compared to orbits based on parameters from Mason *et al.* (2009), Tango *et al.* (2009), and Miroshnichenko *et al.* (2001). The new astrometric data, combined with radial velocities from Miroshnichenko *et al.* (2001), were used to calculate a refined orbit (shown in Figure 1) with the revised binary parameters listed in Table 1. Since the instrument used has a wide range of baselines, the resulting narrow-angle astrometry has high precision. In combination with extensive orbital coverage, including observations close to the previous periastron passage reults in the best orbital parameters to date.

3. Summary

The binary orbit has been refined using astrometric data obtained from NPOI and radial velocities from Miroshnichenko *et al.* (2001). The orbit that is obtained gives a better fit to the data than previous findings. The next periastron passage date has been

Table 1. The orbital elements for δ Sco.

Element	Mason *et al.* (2009)	Miroshnichenko *et al.* (2001)	Tango *et al.* (2009)	This Work
Period(y)	10.68 ± 0.05	10.58[1]	10.74 ± 0.02	10.81 ± 0.002
Semimajor axis (mas)	104 ± 6	107[1]	98.3 ± 1.2	99.0 ± 0.05
inclination, i (deg)	39 ± 8	38 ± 5	38 ± 6	30.3 ± 0.24
Long. of asc. node, Ω (deg)	153 ± 9	175[1]	175.2 ± 0.6	172.3 ± 0.6
eccentricity, e	0.94[2]	0.94 ± 0.01	0.9401 ± 0.0002	0.941 ± 0.0007
Long. periastron, ω	29 ± 12	-1 ± 5	1.9 ± 0.1	2.6 ± 0.7
T (Epoch of Periastron)	*J*2000.693[2]	*J*2000.693± 0.008	*J*2000.69389± 0.00007	*J*2000.6942± 0.0011

Notes:
[1] Parameter adopted from Hartkopf *et al.* (1996) solution.
[2] Parameter adopted from Miroshnichenko *et al.* (2001) solution.

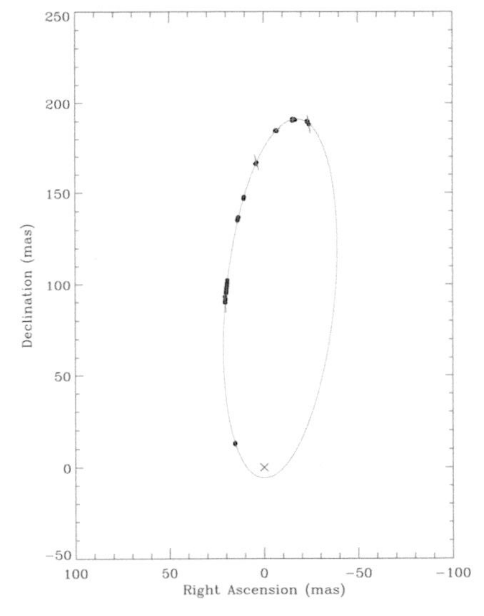

Figure 1. Binary orbit based on orbital parameters obtained from NPOI data. The black circles are the data points and the locations of the primary is marked with an X.

calculated to be 2011-07-02, $22h \pm 1.1d$. Our results indicate the periastron passage will occur 24 days later than the prediction of Tango *et al.* (2009).

Acknowledgements

The Navy Prototype Optical Interferometer is a joint project of the Naval Research Laboratory and the US Naval Observatory, in cooperation with Lowell Observatory, and is funded by the Office of Naval Research and the Oceanographer of the Navy. A.A. and C.T. would also like to thank the Physics Department at Central Michigan University for support.

References

Mason, B. D., Hartkopf, W. I., Gies, D. R., Henry, T. J. *et al.* 2009, *AJ*, 137, 3358
Hartkopf, W. I., Mason, B. D., & McAlister, H. A. 1996, *AJ*, 111, 370
Miroshnichenko, A. S., Fabregat, J., Bjorkman, K. S., Knauth, D. C. *et al.* 2001, *A&A*, 377, 485
Tango, W. J., Davis, J., Jacob, A. P., Mendez, A. *et al.* 2009, *MNRAS*, 396, 842

Active OB stars: structure, evolution, mass loss, and critical limits
Proceedings IAU Symposium No. 272, 2010
C. Neiner, G. Wade, G. Meynet & G. Peters, eds.

doi:10.1017/S1743921311010568

ArasBeam: when amateurs contribute to Be star research

François Cochard[1,2], Valérie Desnoux[2] and Christian Buil[2]

[1]Shelyak Instruments, France - http://www.shelyak.com
email: francois.cochard@shelyak.com

[2]Aras, Amateur Ring for Astronomical Spectroscopy, http://astrosurf.com/aras
email: valerie.desnoux@free.fr, christian.buil@wanadoo.fr

Abstract. Since 2003, the amateur astronomical community has decided, in collaboration with the Paris-Meudon Observatory, to coordinate their observations to get the best spectral survey of Be stars as possible. A database for amateur and professional Be star spectra, BeSS, has been created. Spectrographs (up to R=20000) and software tools have been developed for amateurs. Among them, *ArasBeam* is a web-based tool designed to organize amateur Be spectral observations. A very simple color coding indicates to any observer which stars must be observed on the following night to get the best possible survey of Be stars. So far, more than 11000 amateur spectra have been collected in BeSS. About all bright Be stars (up to magnitude 8) listed in BeSS and visible from the Northern hemisphere have been observed at least one time. In addition, 6 outbursts have been detected by amateurs in the last 2 years.

Keywords. stars: emission-line, Be, techniques: spectroscopic, stars: activity, surveys

1. Amateurs, spectroscopy and Be stars

Be stars are easy targets for small instruments (300 stars up to magnitude 8, Northern hempisphere), their spectra evolve on different timescales (from days to years), periodically or with episodic outbursts. These stars require a continuous observation, to better understand the Be phenomenon.

Amateur spectroscopists are numerous, dispatched over the world, and available very quickly. They can provide observations that are complementary to professional observations. The amateur community has developed several tools to contribute to Be spectroscopic observations :

- spectrographs, such as Lhires III or eShel (see Thizy *et al.*, these proceedings),
- software for spectra processing (Iris, Vspec, Audela,...),
- a mailing list: Spectro-l (http://groups.yahoo.com/group/spectro-l),
- *and* ArasBeam, a web-based tool to coordinate Be observations.

Our spectra are uploaded in the BeSS database (see de Batz *et al.*, these proceedings). The observing program is simple: we focus on Hα, observe all Be stars up to V magnitude 8 at least once a year, detect as many outbursts as possible, and track them intensively.

2. The tool

Arasbeam is available at *http://arasbeam.free.fr/?lang=en*. ArasBeam is a website developed and managed by amateurs. It allows to show the list of Be stars from BeSS catalog with dynamic sorting keys and display all existing BeSS spectra for a given Be star. For each Be star an ideal observing period is defined depending on its activity. By default, this period is one year. In this way, we provide priorities for observations,

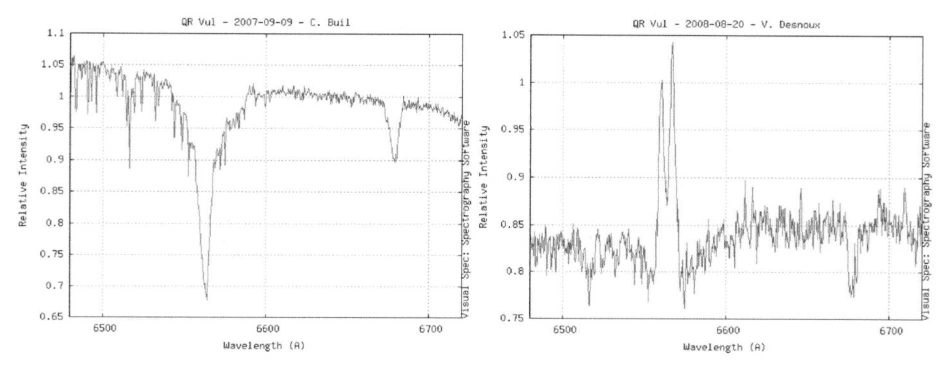

Figure 1. QR Vul, before and after the outburst discovery in Aug. 2008.

based on ideal periods and the latest observations. This is given with a visual color coding. ArasBeam also allows to distribute news on Be observations via a multi-authors platform. Finally, ArasBeam is meant to coordinate amateur observations: it is when amateurs work together that they become really complementary to professionnals.

3. Current results

- 31 amateur observers recorded in BeSS (mainly in the Northern hemisphere);
- More than 11000 amateur spectra have been uploaded in BeSS so far. 3000 of them are at Hα;
- 295 Be stars up to V magnitude 8 have at least one spectra. This covers 97% of the BeSS catalog up to V=8 for the Northern hemisphere;
- 6 outbursts have been detected in 3 years of observations: QR Vul (see Fig. 1), HD 22780, λ Cyg, λ Eri, HD 37149, and HD 34959;
- 5 other Be stars, known as higly active, have a high observation rate: θ CrB, 66 Oph, V1040 Sco, and HD 57682;
- 17 stars have more than 10 spectra obtained during the last year.

4. Next steps

In the future, we plan to increase the number of observers and observations, by communication, training, and improving tools. We also plan to analyze the current data to classify stars depending on their activity. Based on these data, optimize the ideal observing period for each Be star. The goal is to increase the number of outburst detections.

5. Conclusion

Spectrography is quite new to amateur astronomers. However, this activity is growing fast and is now mature. The data we have collected during the last years are very complementary to professionnal data: we are limited in magnitude and signal-to-noise ratio, but we have a much better time coverage, for a high number of stars. Knowing that about 80% of professionnal observations are obtained in spectroscopy, we can easily understand that when amateurs start working in this area, this opens a lot of opportunity for new Professionals/Amateurs collaborations. The Be star observing program is a perfect example of such a succesful collaboration. This program is possible thanks to an active work on both sides, amateur and professionnal.

Active OB stars: structure, evolution, mass loss, and critical limits
Proceedings IAU Symposium No. 272, 2010
C. Neiner, G. Wade, G. Meynet & G. Peters, eds.

doi:10.1017/S174392131101057X

Spectrographs for small telescopes

Olivier Thizy and François Cochard

Shelyak Instruments, Les Roussets, 38420 Revel, France
email: olivier.thizy@shelyak.com, francois.cochard@shelyak.com

Abstract. Shelyak Instruments is a company founded in 2006 offering a full range of spectrographs designed for Astronomy, shipping World Wide. Current users are (1) Public and private observatories who want to setup small telescope for scientific programs or training; (2) Universities for education and demonstrations; (3) Experienced amateurs observers who work in Pro/Amateur collaborations. Lhires III Littrow high resolution spectrograph and eShel optical fibre fed echelle solution are presented with some scientific results including OB stars in which those instruments had a significant contribution.

Keywords. instrumentation: spectrographs, stars: emission-line, Be

1. Introduction

In 2003, following a CNRS pro/am Astrophysics school in Oleron, several amateurs designed and industrialized a Littrow spectrograph (LHIRES) to reach power of resolution R>10000. Shelyak Instruments now continue the distribution of the Lhires III.

Lhires III is a F/10 Littrow (F = 200mm) spectrograph with a 15/19/23/35μm slit. Power of resolution R is around 18000. Calibration is done with an internal neon lamp.

In 2008, the first commercial optical fiber fed echelle spectrograph - eShel - was available to universities, professional observatories, and amateur astronomers willing to contribute to pro/am collaboration.

Eshel system includes a F/6 Fiber Injection and Guiding Unit, a 50μm optical fiber, a ThAr Calibration Unit and an echelle spectrograph (125mm F/5 collimator, R2 echelle grating, cross-dispersing prism, 85mm F/1.8 objective) with a power of resolution R above 10000.

Shelyak Instruments also introduced in 2010 a Low Resolution spectrograph LISA: F/5, 23μm slit, R = 1000, near infra-red mode.

Figure 1 shows a picture of the Lhires III and eShel spectrographs.

2. Scientifical Results

Lhires III main target was Be stars monitoring. An overall pro/am collaboration infrastructure has been put in place: Spectro-L discussion group, ARAS and ARASBeAm web sites, BeSS spectra database, practical workshops... Complemented with the eShel, Lhires III instrument is allowing dozens of amateur around the world to contribute in the Be star monitoring with a continuuous spectroscopic coverage of the bright targets and, as a result, height outbursts discovered in two years and 11000 spectra have been put in the database by amateur astronomers (see ARASBeAm poster, those proceedings).

Other hot stars have been good targets for the Lhires III such as Wolf-Rayet WR140 during 2009 periastron campaign. Hundreds spectra have been taken and the excess emission on top of the CIII 5696 flat line was particularly studied (Fahed *et al.*, these proceedings).

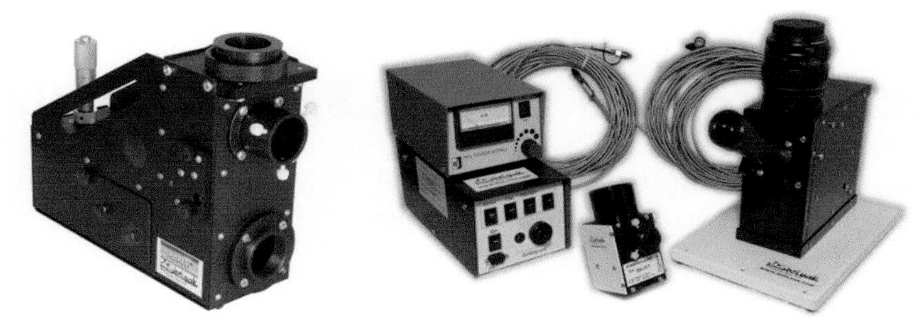

Figure 1. Lhires III and eShel spectrographs

In parallel to WR140 observations, a "Mons campaign" has been put in place to study other OB stars: HD 14134, HD 42087, HD 43384 and HD 52382 supergiants; HD 45314 and HD 60848 Oe stars (Morel *et al.*, these proceedings).

Some pulsating stars have been studied with eShel spectrograph but results have not been published yet: β Cep (spectra in BeSS) and BW Vul showing shock waves impact on spectral line shifts.

Luminous Blue Variable P cygni is a target of choice for the Lhires III with studied by Ernst Pollmann and other amateur spectroscopists.

During 2009/2010, an intensive observing campaign has been launched on ϵ Aur. The transit of the secondary object and its surrounding material is being followed by several amateur astronomers. Robin Leadbeater is focusing on KI 7699 line in particular with great results (Leadbeater & Stencel 2010). At mid eclipse, more than 100 spectra on KI 7699 line were recorded and around 400 amateur spectra on Hα or sodium D lines.

Several spectroscopic binaries have been studied with Lhires III. But eShel opened the door for much higher radial velocity accuracy studies. It has allowed a group of amateurs (Buil *et al.*, see http://astrosurf.com/buil/extrasolar/obs.htm) to be the first to record four exoplanets with commercially available spectrographs: τ Boo, HD 189733, HD 195019 and 51 Peg (Buil 2009). An opportunity exists for amateurs to search for new exoplanets around hotter stars or stars with higher rotational speed where the instrument resolution will not limit the radial velocity measurements.

3. Conclusion

With off-the-shelf spectrogaphs now widely available, amateur spectroscopy has entered a new era. Professional observatories and universities have now access to affordable standard equipment. Amateur made also some significant contribution to professional programs.

In addition, there is an educational aspect of those instruments with the introduction for several students to high resolution spectroscopy and echelle design.

In the future, more emphasis should be put on spectra archiving and access (extension of BeSS model to other star type) and more structured targeted campaigns (like for WR140 and ϵ Aur; for exemple δ Sco periastron in 2011). But it appears now obvious that small telescopes, with more total observing time and flexibility than large telescopes, will contribute to more pro/am collaboration initiatives.

References

Leadbeater, R. & Stencel, R. 2010, *ArXiv e-prints* 1003.3617

Active OB stars: structure, evolution, mass loss, and critical limits
Proceedings IAU Symposium No. 272, 2010
C. Neiner, G. Wade, G. Meynet & G. Peters, eds.

doi:10.1017/S1743921311010581

BeSS, the official Be Star Spectra database

Bertrand de Batz[1], Coralie Neiner[1], Michèle Floquet[2] and François Cochard[3]

[1]LESIA, Observatoire de Paris-Meudon, France
email: B.deBatz@obspm.fr

[2]GEPI, Observatoire de Paris-Meudon, France

[3]Shelyak Instruments, Les Roussets, 38420 Revel, France

Abstract. We present the status of the BeSS database, which contains a catalogue of all known classical Be stars and a large collection of their spectra obtained at any wavelength, any epoch, and from various sources, from amateur astronomer spectra to professional high-resolution high signal-to-noise echelle spectra. Efficient data retrieval in such a heterogeneous data collection is possible with a wide range of selection criteria thanks to their storage in the fits format and via a web interface (http://basebe.obspm.fr) as well as via the Virtual Observatory. BeSS already contains over 49000 spectra and has allowed the detection of several outbursts.

Keywords. astronomical data bases: miscellaneous, catalogs, stars: activity, stars: emission-line, Be, stars: variables: other

1. The BeSS database

BeSS contains the catalog of all known (2026) Be stars, continuously kept up-to-date, with available fundamental parameters and their references. Anyone willing to share his/her spectra is encouraged to submit those spectra via the dedicated interface. The spectra are carefully checked before becoming public in BeSS. Over 49000 spectra of about 600 Be stars and 63 different sources are already available. The stellar information and spectra stored in BeSS can be used, e.g., for the study of the history of H_α emission in an individual Be star, statistical studies of Be star properties or outbursts, the preparation of Be star observations,...

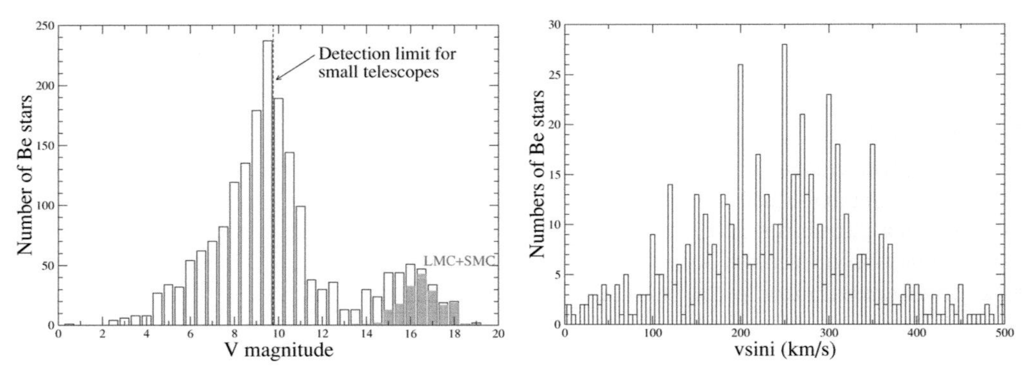

Figure 1. Left panel: distribution in V magnitude of all Be stars with known V magnitude. Right panel: distribution of projected rotational velocity for all Be stars with derived vsini.

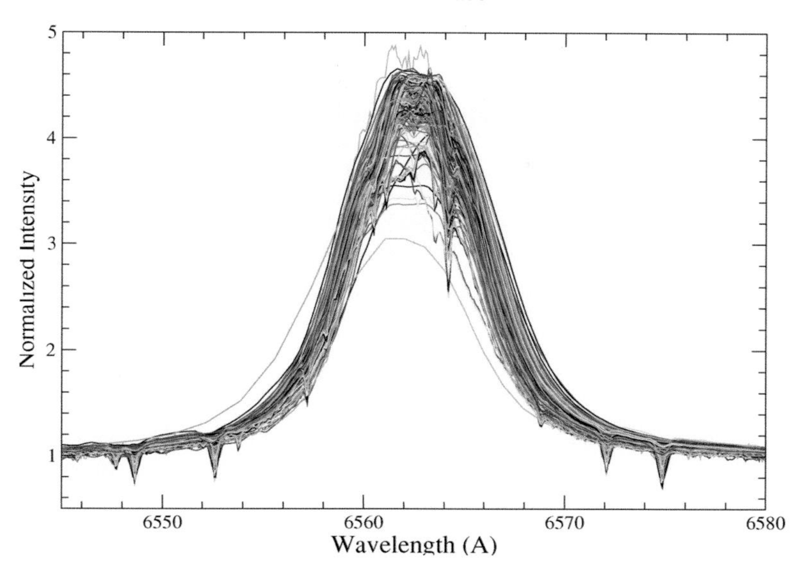

Figure 2. The 187 H_α spectra available in BeSS for the star γ Cas.

2. Statistics of Be stars

Figure 1 shows examples of a statistical use of BeSS. The left histogram shows the distribution in magnitude of all Be stars in the BeSS catalog with known V magnitude. A clear observational bias is seen at the detection limit of small telescopes. Be stars from the LMC and SMC VLT surveys are also visible on the right side of the histogram (in shaded bars). The right histogram shows the distribution of projected rotational velocity (vsini in km.s^{-1}) for all Be stars in the BeSS catalog with derived vsini. The distribution peaks around vsini=250 km.s^{-1}.

3. Study of individual Be stars

Figure 2 shows an example of all 187 H_α spectra available in the BeSS database for the star γ Cas. These spectra have been obtained by 20 different observers, including 15 amateur observers, during a 19-year period (1991-2010). More spectra are available in Bess for this star at other wavelengths. Variability in the strong emission at H_α is clearly observed in γ Cas and corresponds to the change of structure of the dense circumstellar disk of the Be star. In other Be stars, several outbursts have been detected thanks to BeSS spectra (see Cochard *et al.*, these proceedings).

4. Conclusions

The BeSS database is a unique tool for the long-term study of individual Be stars at various wavelength and the detection of outbursts. It also allows the statistical study of Be stars as a population. Finally, its interactive catalog is very useful for the preparation of observations or the selection of adequate targets for a research program. Scientists wishing to contribute to BeSS are welcome to upload their archival spectra into BeSS.

Active OB stars: structure, evolution, mass loss, and critical limits
Proceedings IAU Symposium No. 272, 2010
C. Neiner, G. Wade, G. Meynet & G. Peters, eds.

doi:10.1017/S1743921311010593

Multicolour studies of β Cephei stars in the LMC

Chris A. Engelbrecht[1], Fabio A. M. Frescura[2] and Sashin L. Moonsamy[2]

[1]Department of Physics, University of Johannesburg, P.O. Box 524, Auckland Park, Johannesburg 2006, South Africa; email: chrise@uj.ac.za

[2]School of Physics, University of the Witwatersrand, Private Bag 3, WITS, Johannesburg 2050, South Africa; email: fabio.frescura@wits.ac.za

Abstract. Preliminary results of a four-week multi-colour photometric campaign on previously identified β Cephei stars as well as newly-discovered variable stars in two respective LMC fields are presented. Besides the two targeted β Cephei stars, at least six further presumed B variables are detected. The strongest identified periods appear to lie on the longward end of the galactic β Cep instability strip, as predicted by model calculations.

Keywords. stars: early-type, techniques: photometric, stars: variables: other

1. Introduction

Pigulski & Kolaczkowski (2002) announced the first discovery of β Cephei pulsators in the LMC. Theoretical analyses of pulsational stability had previously predicted that early B main-sequence stars with metallicities lower than Z = 0.01 should not pulsate at all (e.g. Pamyatnykh 1999). Following this announcement, and announcements of 92 β Cep candidates in the LMC by Kolaczkowski & Pigulski (2006) - also see Diago *et al.* (2009) - more detailed studies adopting a variety of opacity calculations and metal mixtures indicated that β Cep pulsations could be explained in low-metallicity environments after all (Miglio *et al.* 2007a; Miglio *et al.* 2007b; Zdravkov & Pamyatnykh 2008), but with the longer-period modes favoured over the shorter-period ones. It has thus become very relevant to study the exact nature of the detected β Cep pulsations seen in low-metallicity environments, in order to subject the new models to more precise constraints. As noted by Handler (2008), observations in at least three different photometric filters are advised to ensure effective mode identification. We ran a 4-week UBVRI campaign on two of the LMC β Cep stars considered by Pigulski & Kolaczkowski (2002) to attempt a mode identification analysis, and to search for other pulsating stars in the surrounding fields. We labelled the main targets LMC1 and LMC2 respectively. Pigulski and Kolaczkowski clearly identified 5 regular periodicities in LMC1, and two in LMC2.

2. Observations and Analysis

We observed the two fields surrounding LMC1 and LMC2, respectively, for four weeks spread over December 2009 and January 2010, using the 1.0 m telescope at the Sutherland station of the South African Astronomical Observatory (SAAO). Table 1 displays the strongest provisional pulsation frequencies (and their corresponding amplitudes in B) for a small selection of stars observed in the two chosen fields. We confirm most of the frequencies reported by Pigulski & Kolaczkowski (2002). Three of these four stars only display strong signals below 4 cycles per day, which is on the low end of the β

Table 1. Firmly detected periodicities in four stars in the LMC.

Star ID	Frequency(c/d)	Amplitude (mmag)
LMC1	2.00	42
	3.76	28
	1.70	20
LMC2	3.50	17
LMC1 V2	4.81	4
	7.13	4
	3.09	3
	7.29	2
LMC2 V2	2.96	28
	3.10	17
	3.75	16

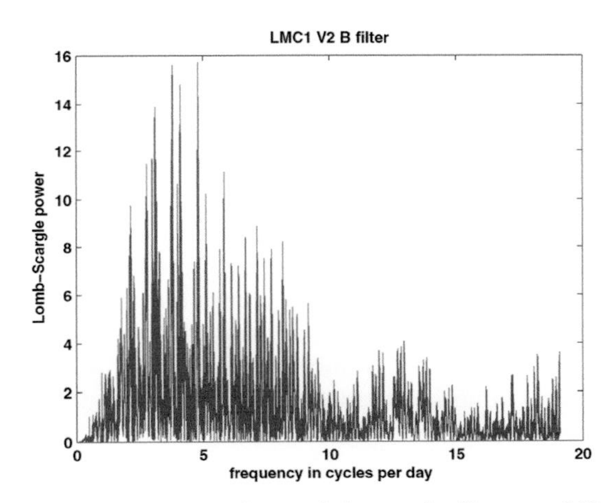

Figure 1. Lomb-Scargle periodogram of one of the newly-discovered B variables, LMC1 V2.

Cep spectrum for stars in our galaxy. This is in agreement with expectations for low-metallicity β Cep stars. The Lomb-Scargle periodogram for one of the newly-discovered B variables (LMC1 V2) is shown in Figure 1.

Acknowledgements

We thank the SAAO for generous amounts of observing time and the NRF for funding. CAE thanks Luis Balona for valuable discussions.

References

Diago, P. D., Gutiérrez-Soto, J., Fabregat, J., & Martayan, C. 2009, *Communications in Asteroseismology* 158, 184

Handler, G. 2008, *Communications in Asteroseismology* 157, 106

Kołaczkowski, Z., Pigulski, A., Soszyński, I., Udalski, A. *et al.* 2006, *MemSAI*, 77, 336

Miglio, A., Montalbán, J., & Dupret, M.-A. 2007a, *MNRAS*, 375, L21

Miglio, A., Montalbán, J., & Dupret, M.-A. 2007b, *Communications in Asteroseismology* 151, 48

Pamyatnykh, A. A. 1999, *AcA*, 49, 119

Pigulski, A. & Kołaczkowski, Z. 2002, *A&A*, 388, 88

Zdravkov, T. & Pamyatnykh, A. A. 2008, *Journal of Physics Conference Series*, 118, 012079

Active OB stars: structure, evolution, mass loss, and critical limits
Proceedings IAU Symposium No. 272, 2010
C. Neiner, G. Wade, G. Meynet & G. Peters, eds.

doi:10.1017/S174392131101060X

Statistical search for Be stars candidates in the Small Magellanic Cloud

Alejandro García-Varela[1], Beatriz Sabogal[1] and Ronald E. Mennickent[2]

[1]Departamento de Física, Universidad de los Andes, Bogotá, Colombia
email: josegarc@uniandes.edu.co

[2]Departamento de Astronomía, Facultad de Ciencias Físicas y Matemáticas, Universidad de Concepción, Chile

Abstract. A report of a systematic search for Be star candidates in the Small Magellanic Cloud using statistical selection criteria is presented. The results are compared with those obtained with a standard photometric method to search for Be star candidates.

Keywords. methods: statistical, stars: variables: other, astronomical data bases: miscellaneous

1. Introduction

The standard method to search for variable stars inside photometric data bases like OGLE or MACHO implies that light curves are extracted and the mean magnitude and average standard deviation for each star are computed. Then, the stars are grouped and the average magnitude and average standard deviation are calculated per bin. The resulting pairs are plotted and fit with a polynomial, leading to a function f representing the standard deviation as a function of mean magnitude. Stars with standard deviation values larger than a threshold value (αf) are picked out as variable star candidates. Visual inspection is needed to find the variable stars. The disadvantage of this method is that it consumes a lot of time inspecting light curves, while the discovery of new variable stars in a secure way is its main advantage.

2. The statistical method

During the OGLE II project (Udalski *et al.* 1997, Udalski *et al.* 2002), VI photometry maps of the Galactic Bulge (GB) and the Small Magellanic Cloud (SMC) were obtained. Within the OGLE II catalog of variable stars of the SMC we looked for stars with absolute V magnitudes in the typical range of the Galactic Be stars, i.e. $-6 < M_V < 0$ (Wegner 2000, Garmany & Humphreys 1985), and with V-I colours in the range for classical Be stars. In order to search and find Be star candidates on the SMC, we established the following process:

1. To avoid noisy detections, we used a filter that rejected measurements larger than 2.5σ times the mean I-band magnitude in all light curves.
2. We selected the stars with amplitudes A_I between 0.05 mag (the limit of noise in OGLE II data) and 0.3 mag. Almost the 70% of the Be stars have amplitudes less than this upper limit (Sabogal *et al.* 2005).
3. We applied the statistical properties found for the GB Be stars to identify the Be star candidates in the SMC. The GB Be star candidates found by Sabogal *et al.* (2008) show a parabolic correlation between kurtosis excess (KE) and skewness (S) of the I magnitudes (Fig. 1). The skewness is a measure of the asymmetry of a distribution; for a

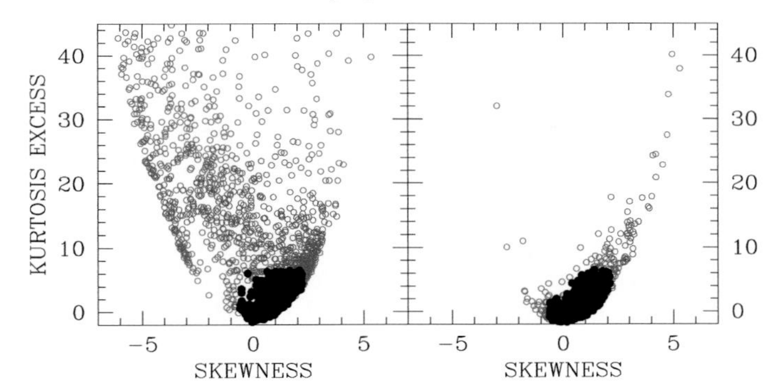

Figure 1. Statistical properties of GB (left) and SMC (right) stars. Open circles represent the variable stars candidates. Filled circles represent the Be star candidates.

Gaussian distribution S is zero. Negative values of S of a sample of data indicate that the left tail of the distribution is long relative to the right tail. Kurtosis (K) is a measure of whether the distribution is sharp or flattened relative to a normal distribution. Datasets with high K value tend to have a distinct peak relative to the mean and decline rather rapidly. Since (K) of a normal distribution is 3, we adopted $KE = K - 3$ as the statistical variable to examine whether the distributions of I magnitudes are sharp.

Analyzing the parabolic behavior of the KE vs S diagram for GB variable star candidates, we found that the Be star candidates are located in the range $KE < 6.5$ and S between $[-0.6, 2.2]$. KE and S of the I magnitudes of SMC variable stars also presented a parabolic behavior (Fig. 1), and the SMC stars falling in the range previously mentioned corresponded to the Be star candidates selected by standard photometric methods.

3. Discussion and Conclusion

5007 variable stars were selected of the OGLE II catalog of the SMC galaxy. Applying the statistical method described in previous section we found that the variable stars of the SMC show the same parabolic behavior in the KE vs S diagram than the variable stars of the GB. The locus of the GB Be stars in that correlation allowed us to find the Be stars of the SMC. This could imply that the range of KE and S found is valid for Be star candidates in different galaxies, and in that case this statistical method would be a useful method to search for Be stars candidates. We found 1623 candidates to Be stars in the SMC, most of them previously reported by Mennickent *et al.* (2002) in a search that used the traditional method. Those authors visually inspected 5007 light curves (LC) searching for Be stars, while, in our case, we only needed to inspect 1623 (only 32% of the LC inspected in the traditional method). We conclude that searching for Be star candidates is faster and more efficient using a statistical method than a traditional one.

References

Garmany, C. D. & Humphreys, R. M. 1985, *AJ*, 90, 2009
Mennickent, R. E., Pietrzyński, G., Gieren, W., & Szewczyk, O. 2002, *A&A*, 393, 887
Sabogal, B. E., Mennickent, R. E., Pietrzyński, G. & Gieren, W. 2005, *MNRAS*, 361, 1055
Sabogal, B. E., Mennickent, R. E., Pietrzyński, G., García, J. A. *et al.* 2008, *A&A*, 478, 659
Udalski, A., Kubiak, M., & Szymanski, M. 1997, *AcA*, 47, 319
Udalski, A., Szymanski, M., Kubiak, M., Pietrzynski, G. *et al.* 2002, *AcA*, 52, 217
Wegner, W. 2000, *MNRAS*, 319, 771

Active OB stars: structure, evolution, mass loss, and critical limits
Proceedings IAU Symposium No. 272, 2010
C. Neiner, G. Wade, G. Meynet & G. Peters, eds.

doi:10.1017/S1743921311010611

Spectroscopic Hα and Hγ survey of field Be stars: 2004–2009

Erika D. Grundstrom[1,2], Douglas R. Gies[3], Christina Aragona[4], Tabetha S. Boyajian[3], E. Victor Garcia[2], Amber N. Marsh[4], M. Virginia McSwain[4], Rachael M. Roettenbacher[4], Stephen J. Williams[3] & David W. Wingert[3]

[1]Dept. of Physics & Astronomy, Vanderbilt University, Nashville, TN 37235 USA
email: `erika.grundstrom@vanderbilt.edu`

[2]Dept. of Physics, Fisk University, Nashville, TN 37208, USA

[3]CHARA, Georgia State University, Atlanta, GA 30302, USA

[4]Dept. of Physics, Lehigh University, Bethlahem, PA 18015, USA

Abstract. Massive O- and B-type stars are "cosmic engines" in the Universe and can be the dominant source of luminosity in a galaxy. The class of Be stars are rapidly rotating B-type stars that lose mass in an equatorial, circumstellar disk (Porter & Rivinius 2003) and cause Balmer and other line emission. Currently, we are unsure as to why these stars rotate so quickly but three scenarios are possible: they may have been born as rapid rotators, spun up by binary mass transfer, or spun up during the main-sequence evolution of B stars. In order to investigate these scenarios for this population of massive stars, we have been spectroscopically observing a set of 115 field Be stars with the Kitt Peak Coudè Feed telescope in both the Hα and Hγ wavelength regimes since 2004. This time baseline allows for examination of variability properties of the circumstellar disks as well as determine candidates for closer examination for binarity. We find that 90% of the observed stars show some variability with 8% showing significant variability over the 5-year baseline. Such values may be compared with the significant variability seen in some clusters such as NGC 3766 (McSwain 2008). Also, while ~20% of the sample consists of known binaries, we find that another 15-30% of the sample shows indications of binarity.

Keywords. stars: emission-line, Be, stars: winds, outflows, circumstellar matter, techniques: spectroscopic, surveys

---

Since fall 2004, we have spectroscopically investigated 115 field Be stars taken from the Yudin (2001) catalog of bright Be stars - our sample covers magnitudes 2 to 8. We chose the Hα region to investigate disk properties (i.e., size, morphology) and also the Hγ region to investigate stellar properties (i.e., rotation speed, $\log g$) but in some cases we can also examine disk properties. Our spectral resolution $R = \lambda/\Delta\lambda$ for the Hα region was ~5000 (2004, 2006) and ~11,500 (2008). For the Hγ region, it was ~11,000.

For each of the 115 Be stars we studied, we generated a "quadruple plot" such as the example shown in Fig. 1. Also, while examining the Hγ region spectra, we noticed that one can see changes in spectral features (for instance, presence of Fe II emission and absorption features) and the most notable case was Pleione (HD 23862) - a star that has long been known to change between B, Be, and Be-shell states, but many others exhibited this behavior as well (see Fig. 2). The term "shell" star describes a star that has many absorption or emission lines associated with Fe II transitions (which indicates one is looking through or at circumstellar material). We found that fully 50% of our sample shows non-temperature-related Fe II emission or absorption features with a number showing variable Fe II features.

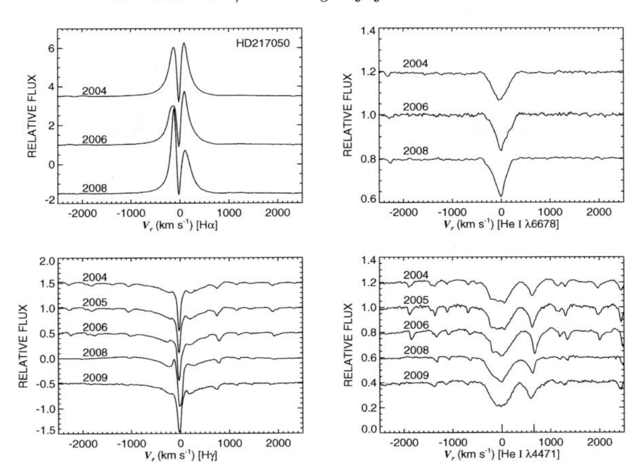

Figure 1. Example of a "quadruple plot" (this of HD 217050). In each quadrant are plots of the normalized fluxes of the seasonal averages of the spectra, offset for clarity, and in radial velocity space. Each is centered on the rest wavelength of a different spectral feature: *top left* - Hα, *top right* - He I λ6678, *bottom left* - Hγ, and *bottom right* - He I λ4471 (with a mark at $\sim$670 km s^{-1} delineating Mg II λ4481).

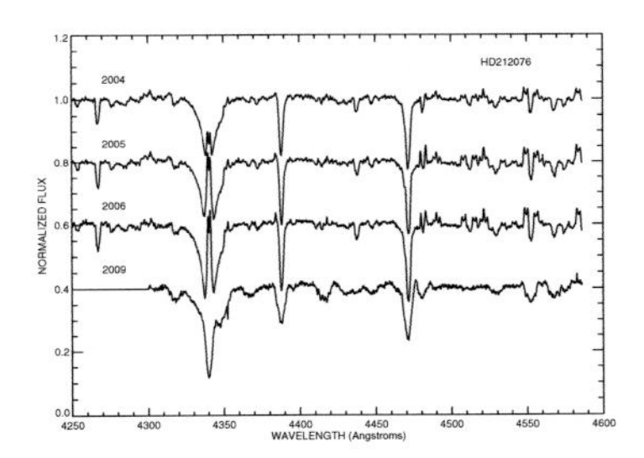

Figure 2. A plot of seasonal averages of the normalized flux versus wavelength for the blue spectrum of 31 Peg (HD 212076) - notice how all emission features diminish greatly in 2009. These spectra are offset for clarity.

Especially from the Hγ region spectra, one can also see radial velocity and line profile variations which can be indicative of binarity and/or nonradial pulsation. Through this survey, we have identified $\sim$40 new nonradial pulsation and/or binarity candidates.

Results from this survey and results concerning particular interesting stars are forthcoming and the reduced spectra will be made public in hopes the astronomical community will utilize them. We will continue to monitor these Be stars whenever possible as well as investigate nonradial pulsation and binary candidates more fully and at higher resolution. Another goal is to compare these field Be stars to Be stars in clusters.

Acknowledgements

This work has been supported in part by a grant from NASA grant # NNX08AV70G, NSF Career grant AST-0349075, and the Vanderbilt University Learning Sciences Institute.

References

McSwain, M. V. 2008, *ApJ*, 686, 1269
Porter, J. M. & Rivinius, T. 2003, *PASP*, 115, 1153
Yudin, R. V. 2001, *A&A*, 368, 912

Active OB stars: structure, evolution, mass loss, and critical limits
Proceedings IAU Symposium No. 272, 2010
C. Neiner, G. Wade, G. Meynet & G. Peters, eds.

doi:10.1017/S1743921311010623

Are the stellar winds in IC 1613 stronger than expected?

Artemio Herrero[1,2], Maria Garcia[1,2], Katrien Uytterhoeven[3], Francisco Najarro[4], Daniel J. Lennon[5], Sergio Simón-Díaz[1,2], Norberto Castro[1,2], Joachim Puls[6], Jorick S. Vink[7], Miguel A. Urbaneja[8], and Alex de Koter[9,10]

[1]Instituto de Astrofísica de Canarias, E38200 La Laguna, Spain; [2]Departamento de Astrofísica, Universidad de La Laguna, E38205, La Laguna, Spain; [3]CEA, Saclay, France; [4]CAB (CSIC-INTA), Torrejón de Ardoz, Spain; [5]STScI, Baltimore, USA; [6] Univ.-Sternwarte Munich, Germany; [7] Armagh Observatory, Armagh, Northern Ireland; [8] IfA, Hawaii, USA; [9] Astronomical Institut Anton Pannekoek, Amsterdam, The Netherlands;[10]Astronomical Institut, Utrecht, The Netherlands

Abstract. In this poster we present the results of our analyses of three early massive stars in IC 1613, whose spectra have been observed with VIMOS and analyzed with CMFGEN and FASTWIND. One of the targets resulted a possible LBV and the other two are Of stars with unexpectedly strong winds. The Of stars seem to be strongly contaminated by CNO products. Our preliminary results may represent a challenge for the theory of stellar atmospheres, but they still have to be confirmed by the analysis of more objects and a more complete coverage of the parameter space.

Keywords. stars: atmospheres, stars: early-type, stars: fundamental parameters, stars: mass loss, galaxies: individual (IC 1613)

---

1. Why and how we observed stars in IC1613

IC 1613 is a Local Group dwarf irregular galaxy with a very low metallicity and ongoing massive star formation. Therefore, it is an ideal place to test theories of stellar atmospheres, winds and evolution, and compare their predictions with the actual behaviour of massive stars. For this reason, we have obtained multiwavelength (UBVRI) photometry of IC 1613 with the WFC attached to the INT@ORM (Garcia *et al.* 2009) and have secured VIMOS@VLT spectra of IC 1613 stars selected using this photometry. We used the HR-Blue and HR-Orange gratings, yielding R $\approx$ 2100, λ = 3900-7000 Å and SNR $\approx$ 100 after combining 19 blue and 10 red exposures (see Herrero *et al.* 2010, for more details).

2. The nature of V39

The first object we analyzed was variable V39 in IC1613. The nature of V39 has been debated since its discovery. Sandage (1971), in his analysis of IC1613 based on unpublished previous work by Baade, points out that it is the only peculiar variable in the field, due to its inverted beta-Lyrae light curve. Baade never considered it a Cepheid, because it is too bright to fit the P-L relation compared to other Cepheids with the same period. In our observations, V39 displays P-Cygni profiles in the Balmer and Fe II lines at $\lambda\lambda$4924, 5018 and 5168 Å, as well as He I 5876 Å (the only He I line seen in the spectrum; its presence is confirmed because it can be independently seen in the blue and red spectrograms).

Table 1. Results of the quantitative analysis of IC1613 Of stars. Mass loss rates are in units of solar masses per year, and terminal velocities (adopted from V_∞-Z relations) are in km s^{-1}

Star ID	Sp Type	Teff	log g	$R/R_\odot$	$\log(L/L_\odot)$	$\dot{M}$	V_∞	Y(He)	log(N/H)+12
62024	O5 If	38500	3.50	10.5	5.34	1.60×10^{-6}	1490	0.18	8.00
67559	O7 I(f)	34000	3.40	12.3	5.25	1.23×10^{-6}	1450	0.18	7.80

The V39 spectrum changes from early A in the blue to late G in the red, but without significant spectral variability (the lack of important spectral variability is particularly remarkable in H_α). Mantegazza *et al.* (2002) suggested that V39 is actually the casual superposition of a Galactic W Vir star and an IC1613 Red Supergiant. However, this is not consistent with our observations, because the spectrum has a unique radial velocity, consistent with the IC1613 systemic velocity and the (small) H_α profile variations are not consistent with that of an W Vir object. Moreover, a binary system is not consistent with the observed spectroscopic lack of variability in short timescales.

Our analysis of V39, performed by means of CMFGEN, reveals that its stellar parameters and location on the Hertzsprung-Russell Diagram are consistent with a low-luminosity LBVc or SgB[e] star (see Herrero *et al.* 2010 for details), which raises the question of how to produce such an object at low metallicities. We propose V39 to be an LBVc surrounded by a hot thick disk: the LBV would be responsible for the P-Cygni profiles observed in the spectrum and the disk would precess and produce the photometric variability and additional absorption in the Balmer lines. The model is consistent with the large reddening of V39 (>5 times the average IC1613 internal reddening; unfortunately, no Spitzer images are available) but has to be confirmed or rejected with new observations.

3. Of stars in IC 1613

As a second step, we have selected two Of stars from our sample of IC1613 massive stars. The analyses have been carried out using the latest FASTWIND version, that includes N III dielectronic recombination (Rivero González *et al.*, 2011, in prep.). Although results are still preliminary, there are two points of great interest (to be confirmed): (a) the stars have higher mass-loss rates and wind momenta than expected for the metallicity of IC 1613; (b) the N and He abundances suggest strong CNO contamination at the surface (see Tab. 1). While the second one can be expected from theory, that predicts increased mixing at lower metallicities, the first one is not predicted by the theory of radiatively driven winds. If confirmed, it would indicate an unexpected behaviour of stellar winds at low metallicities. However, before adopting such a conclusion, more objects have to be analyzed and a more complete coverage of the parameter space (wind clumping, beta parameter, microturbulence) is mandatory.

References

Garcia, M., Herrero, A., Vicente, B., Castro, N. *et al.* 2009, *A&A*, 502, 1015
Herrero, A., Garcia, M., Uytterhoeven, K., Najarro, F. *et al.* 2010, *A&A*, 513A, 70
Mantegazza, L., Antonello, E., Fugazza, D., Covino, S. *et al.* 2002, *A&A*, 388, 861
Sandage, A. 1971, *ApJ*, 166, 13

Active OB stars: structure, evolution, mass loss, and critical limits
Proceedings IAU Symposium No. 272, 2010
C. Neiner, G. Wade, G. Meynet & G. Peters, eds.

doi:10.1017/S1743921311010635

Photometric and spectroscopic study of candidate Be stars in the Magellanic Clouds

Paul KT[1], Annapurni Subramaniam[2], Blesson Mathew[2], Ronald E. Mennickent[3], and Beatriz Sabogal[4]

[1]Christ University,Bangalore , [2]Indian institute of Astrophysics,Bangalore, [3]University of Concepcion, Chile, [4]Universidad de los Andes, Colombia

1. Introduction

Mennickent *et al.* (2002) presented a catalogue of 1056 Be star candidates in the Small Magellanic cloud (SMC) by studying light curve variation using OGLE II data base. They classified these Be star candidates of the SMC in four categories: Type 1 stars showing outbursts (139 stars); Type 2 stars showing sudden luminosity jumps (154 stars); Type 3 stars showing periodic or near periodic variations (78 stars); Type 4 stars showing light curves similar to Galactic Be stars (658 stars). They suggested that Type 4 could be Be stars. On the other hand, they suggested that Type-3 stars may not be linked to the Be star phenomenon at all.

Based on a similar inspection of OGLE II data, Sabogal *et al.* (2005) classified Be candidates in the Large Magellanic Cloud (LMC) also as Type 1 (581 stars) , Type 2 (150 stars), Type 3 (149 stars) and Type 4 stars (1468 stars). However most of the type 4 stars in LMC are found to be reddened and located parallel to the main sequence, this feature was not found in the same diagrams of the SMC. The photometric properties of Type 1 and Type 3 stars on the LMC are very different from those of the SMC. Thus, the various types of stars identified based on variability seem to differ between the LMC and the SMC.We have studied the near IR properties of various types of Be star candidates in the LMC and SMC by cross matching IRSF and OGLE II catalogs.We aim to study the properties of various types and compare them in the LMC and SMC, in particular, the type 4 stars. We also correlated the LMC and SMC Be star candidates with galactic counter parts through NIR colour-colour diagram. We also present results from a spectroscopic study of a 49 stars from Types 1, 2 and 3 in the SMC and 12 stars belonging to type 1 and type 3 in the LMC. The spectral features are used to identify their spectral class and to identify the nature of these stars, including the Be properties.

2. Results

In order to obtain their near-IR properties, we used the near-IR IRSF catalog (Kato *et al.* 2007; http://pasj.asj.or.jp/v59/n3/590315). The optically identified stars are cross-matched with near-IR IRSF calaloge to confirm its candidature in IRSF. The near-IR photometric magnitude in J, H and Ks bands for the identified candidate stars were taken from IRSF database. These stars were used to study the near-IR properties of the various Types. Optical spectra were obtained for 61 candidate stars in the LMC and SMC. Observations were made during October 2002 using the 1.5m CTIO Telescope. 56 Blue(3700-5500)Å and 51 red (5700-7000)Å spectra were obtained with a resolution of 3.7Å.

We cross correlated 1640 stars identified by Sabogal *et al.* (2005). Of these, 399 are Type 1 stars, 92 are Type 2 stars, 91 are Type 3 stars and 989 stars are Type 4. The

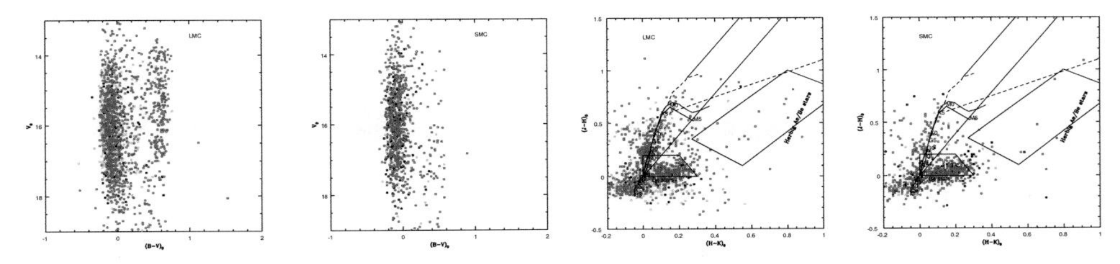

Figure 1. CMDs ((B$-V)_0$ vs V_0)and CCDs($(H-K)_0$ Vs $(J-H)_0$) for the cross matched stars in the LMC and SMC

SMC sample contains 841 stars, of which 89 are Type 1 stars, 131 are type 2 stars, 65 are Type 3 stars and 554 are Type 4 stars.

Adopting $R = A_v/E(B-V) = 3.1$ on the basis of previously published values, colour magnitude diagrams with (B$-V)_0$ vs V_0 were plotted for the LMC and the SMC(type 1 to type 4 shown in different colours) We notice the presence of Type -4 stars parallel to the main sequence in the range of B-V =0.4 to 0.7 mag, in the LMC, which is absent in the SMC. These figures suggest that the cross-identified stars have sampled all the types and this subset can be used to study the properties of various types in the near-IR. We estimated the dereddened $(J-H)_0$ and $(H-K)_0$ values of the cross matched sources. Reddening corrections were made using formulae given by (Bessell & Brett 1988) by taking E(B$-$V) = 0.1 mag. Color-Color diagrams (CCD), were plotted with $(H-K)_0$ Vs $(J-H)_0$ for all the types in the LMC and SMC. The CCD for the LMC shows that most of the stars are populated near the $(J-H)_0 \sim 0.0$, with a small range in $(H-K)_0$, as a more or less horizontal band. We do also notice that some stars occupy a location above this band, on the MS, giving rise to a clumpy appearance. The stars in this clump are mostly Type 4 stars. The CCD for the SMC also shows the horizontal band like distribution, but the clumpy population is not found in the SMC. We identify a few sources with large near-IR excess in the LMC, belonging to Type 3 and Type 4. SMC shows fewer sources with near-IR excess.

We compared the spectra of the Be candidate stars with those of the standard stars and the Be stars were found to be in the range from O5 to A3. We find that type 1, type 2 and type 3 stars have similar NIR properties in the LMC and in the SMC. Spectral analysis of selected stars from type 1, type 2 and type 3 shows that most of them belong to early spectral type (early A and B type). Spectra of most of the stars show H_α emission. Hence we conclude that type 1, 2, 3 stars could be Be stars. Among the sample studied type 3 stars in the LMC have relatively less H_α EW when compared to those in SMC. Type 2 stars are likely to be Be-stars and unlikely to be pre-MS stars, since they lack near IR excess. Type 4 stars in the LMC fall in two groups, one with nearIR properties similar to those of the Galactic Be stars, and a NEW GROUP with different nearIR properties. The type 4 stars in the SMC have nearIR properties similar to the Galactic Be stars.

References

Mennickent, R. E., Pietrzyński, G., Gieren, W., & Szewczyk, O. 2002, *A&A* 393, 887

Sabogal, B. E., Mennickent, R. E., Pietrzyński, G., & Gieren, W. 2005, *MNRAS* 361, 1055

Active OB stars: structure, evolution, mass loss, and critical limits
Proceedings IAU Symposium No. 272, 2010
C. Neiner, G. Wade, G. Meynet & G. Peters, eds.

doi:10.1017/S1743921311010647

The VLT-FLAMES Tarantula survey

Daniel J. Lennon[1], Christopher J. Evans[2], Nate Bastian[3], Yuri Beletsky[4], Ines Brott[5], Matteo Cantiello[6], Giovanni Carraro[4], J. Simon Clark[7], Paul A. Crowther[8], Alex de Koter[9], Selma E. de Mink[1], Philip L. Dufton[1]0, Paul Dunstall[10], Mark Gieles[11], Goetz Gräfener[12], Vincent Hénault-Brunet[2], Artemio Herrero[13], Ian D. Howarth[14], Norbert Langer[6], Jesus Maíz Apellániz[15], Nevena Markova[16], F. Paco Najarro[17], Joachim Puls[18], Hugues Sana[9], Sergio Simón-Díaz[13], Stephen J. Smartt[10], Vanessa E. Stroud[19], William D. Taylor[2], Jacco T. van Loon[20], Jorick S. Vink[12] and Nolan R. Walborn[1]

[1]ESA/STScI, 3700 San Martin Drive, Baltimore, MD 21218, USA, email: lennon@stsci.edu;
[2] ATC, Royal Observatory Edinburgh, Blackford Hill, Edinburgh, EH9 3HJ, UK; [3] University of Exeter; [4] ESO; [5] Utrecht University; [6] University of Bonn; [7] The Open University; [8] University of Sheffield; [9] University of Amsterdam; [10] The Queens University of Belfast; [11] University of Cambridge; [12] Armagh Observatory; [13] Instituto de Astrofisica de Canarias; [14] University College London; [15] Instituto de Astrofsica de Andaluca, Granada; [16] NAO, Bulgaria; [17] IEM-CSIC, Madrid; [18] University of Munich; [19] Faulkes, Cardiff University; [20] Keele University;

Abstract. The Tarantula Survey is an ESO Large Programme which has obtained multi-epoch spectroscopy of over 1,000 massive stars in the 30 Doradus region of the Large Magellanic Cloud. The assembled consortium will exploit these data to address a range of fundamental questions in both stellar and cluster evolution.

Keywords. stars: early-type, stars: fundamental parameters, binaries: spectroscopic, open clusters and associations: individual (30 Doradus)

---

1. Overview

The Tarantula Nebula (30 Doradus, NGC 2070) in the Large Magellanic Cloud (LMC) is the brightest and most massive H II region in the Local Group. With its rich stellar populations, 30 Dor is an ideal laboratory in which to investigate a number of important outstanding questions regarding the physics, evolution, binary fraction, and chemical enrichment of the most massive stars. Building on the successes of the VLT-FLAMES Survey of Massive Stars (Evans *et al.* 2005, 2006), we introduce the Tarantula Survey, a new time-sampled, spectroscopic survey of over 1,000 massive stars in the 30 Dor region. Our scientific motivations are discussed in more depth by Evans *et al.* (2010), and include:

- Effects of stellar rotation on surface abundances,
- Feedback to the interstellar medium from stellar winds and ionizing radiation,
- Physical properties and evolution of massive O-type stars,
- Star cluster dynamics in the context of infanty mortality,
- Binary fraction analysis,
- Providing a census of the nearest 'starburst'.

In terms of shaping the observational strategy, the most pertintent issue is binarity. Recent results (see Sana & Evans, these proceedings) find large binary fractions from multi-epoch spectroscopic studies in Galactic star clusters, as well as an apparently rich binary population in 30 Dor (Bosch, Terlevich & Terlevich, 2009). To gain a true

Figure 1. $20' \times 20'$ V-band WFI image showing the FLAMES-GIRAFFE targets in and around 30 Dor (north to the top, east to the left).

understanding of the evolution of massive-star populations, the effects of binarity need to be fully included in our theoretical models of stellar evolution.

The majority of the data have been collected using the Medusa mode of the VLT-FLAMES instrument, which feeds 132 optical fibres into the Giraffe spectrograph. Nine field configurations were used (each with the same central pointing) to survey ~1000 stars, covering 3980–5050 Å at a resolution of ~40 km s^{-1}, and 6480-6790 Å at a greater resolution of ~20 km s^{-1}. The distribution of targets is shown in Fig. 1.

At the core of 30 Dor is the dense cluster R136, thought to contain the most massive stars in the local universe (Crowther *et al.* 2010). Five areas of this inner region were observed with the ARGUS integral-field unit, to probe the dynamics of this important cluster.

2. Status

All the observations have been completed, with the data reduced and released to the consortium. Work has now begun in earnest towards the first papers, including classification of the different spectra, analysis of the stellar radial velocities/binarity, quantitative analysis of the O-type spectra with contemporary model atmospheres, and analysis of the nebular gas profiles.

References

Bosch, G., Terlevich, E., & Terlevich, R. 2009, *AJ*, 137, 3437

Crowther, P. A., Schnurr, O., Hirschi, R., Yusof, N. *et al.* 2010, *MNRAS* 408, 731

Evans, C. J., Smartt, S. J., Lee, J.-K., Lennon, D. J. *et al.* 2005, *A&A*, 437, 467

Evans, C. J., Lennon, D. J., Smartt, S. J., & Trundle, C. 2006, *A&A*, 456, 623

Evans, C. J., Bastian, N., Beletsky, Y., Brott, I. *et al.* 2010, in: R. de Grijs & J. R. D. Lépine (eds.), *Star clusters: basic galactic building blocks throughout time and space*, IAU Symposium 266, p. 35

Active OB stars: structure, evolution, mass loss, and critical limits
Proceedings IAU Symposium No. 272, 2010
C. Neiner, G. Wade, G. Meynet & G. Peters, eds.

doi:10.1017/S1743921311010659

Variability of young massive stars in the Arches cluster

Kostas Markakis[1], Alceste Z. Bonanos[1], Grzegorz Pietrzynski[2], Lucas Macri[3], Kris Z. Stanek[4]

[1]National Observatory of Athens, Institute of Astronomy & Astrophysics, I. Metaxa & Vas. Pavlou St., P. Penteli 15236, Athens, Greece; email: markakis@astro.noa.gr

[2]Warsaw University Observatory, Al. Ujazdowskie 4, 00-478 Warszawa, Poland

[3]Department of Physics, Texas A&M University, College Station, TX 77842-4242, USA

[4]The Ohio State University, 140 West 18th Avenue, Columbus, OH 43210, USA

Abstract. We present preliminary results of the first near-infrared variability study of the Arches cluster, using adaptive optics data from NIRI/Gemini and NACO/VLT. The goal is to discover eclipsing binaries in this young (2.5 ± 0.5 Myr), dense, massive cluster for which we will determine accurate fundamental parameters with subsequent spectroscopy. Given that the Arches cluster contains more than 200 Wolf-Rayet and O-type stars, it provides a rare opportunity to determine parameters for some of the most massive stars in the Galaxy.

Keywords. Galaxy: center, infrared: stars, open clusters and associations: individual (Arches cluster), binaries: eclipsing, stars: variables: other, stars: Wolf-Rayet

1. Introduction

One of the most important questions is how massive can the most massive stars in the Universe be today. In other words what is the upper limit of the Initial Mass Function in the Universe. The Arches Cluster provides us with a unique opportunity to address this question because it has all the criteria of the ideal place to look at for massive eclipsing binary systems. It lies near the Galactic Center which is a very dense region that benefits the formation of massive stars and the cluster itself is very young which can guarantee that its stars will not have evolved significantly.

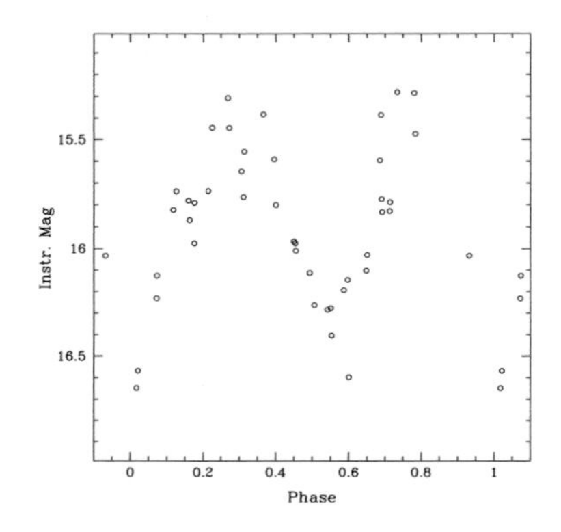

Figure 1: K_s−band light curve of the candidate eclipsing binary in the Arches cluster.

2. Datasets & Reduction

We used two datasets in the K_s band. The first dataset was obtained with Gemini's NIRI infrared camera which consisted of 16 observations of 30 and 1 second exposures respectively covering 8 nights from April to July of 2006. The NIRI data are pending a linearity correction. The second dataset was obtained with the VLT's NACO infrared camera and consisted of 46 observations of 20 seconds exposures each covering 31 nights from June of 2008 to March of 2009. The reduction of the NIRI images was performed with the IRAF† Gemini v1.9 package while the reduction of the NACO images was performed via the NACO reduction pipeline, based on ESO's Common Pipeline Library.

3. Image Subtraction & Photometry

We used the image subtraction package ISIS (Alard & Lupton 1998, Alard 2000), which is optimal for detecting variables in crowded fields, together with IRAF's DAOPHOT (Stetson 1987) package on the reference image from ISIS. Although ISIS allows for a spatially variable PSF it does not provide us with an accurate PSF model due to the anisoplanatic effects introduced by the imperfect correction of atmospheric turbulence by the adaptive optics. The light curve presented in Fig. 1 has been obtained with ISIS and its overall shape strongly suggests that this may be a contact eclipsing binary. In order to obtain more accurate photometry and confirm our result, we are currently using the StarFinder code (Diolaiti *et al.* 2000). which was designed to extract an empirical PSF from the image, that also takes into account the anisoplanatic effects caused by the adaptive optics.

4. Preliminary results

We present the light curve of an eclipsing binary candidate in the Arches cluster from the NACO data (Fig. 1), which corresponds to the second star in the catalog of Figer *et al.* (2002). It has a 10.49-day period and is likely in a contact configuration. It has a spectral type of WN7 (Blum *et al.* 2001) and an initial mass greater than 120 $M_\odot$ (Figer *et al.* 2002). It has also been identified as radio source AR10 with a mass-loss rate of $\sim$1.9$\times$ 10$M_\odot$ yr^{-1} (Lang *et al.* 2001b) and as X-Ray source A6, probably associated with the close pair of radio sources AR6 and AR10 (Wang *et al.* 2006).

Acknowledgements

K.M. & A.Z.B. acknowledge support from the IAU and the European Commission for an FP7 Marie Curie International Reintegration Grant.

References

Alard, C. & Lupton, R. H. 1998, *ApJ*, 503, 325
Alard, C. 2000, *A&AS*, 144, 363
Blum, R. D., Schaerer, D., Pasquali, A., Heydari-Malayeri, M. *et al.* 2001, *AJ*, 122, 1875
Diolaiti, E., Bendinelli, O., Bonaccini, D., Close, L. *et al.* 2000, *A&AS*, 147, 335
Figer, D. F., Najarro, F., Gilmore, D., Morris, M. *et al.* 2002, *ApJ*, 581, 258
Lang, C. C., Goss, W. M. & Rodríguez, L. F. 2001, *ApJ* (Letters), 551, L143
Stetson, P. B. 1987, *PASP*, 99, 191
Wang, Q. D., Dong, H., & Lang, C. 2006, *MNRAS*, 371, 38

† IRAF is distributed by the NOAO, which are operated by the Association of Universities for Research in Astronomy, Inc., under cooperative agreement with the NSF.

Active OB stars: structure, evolution, mass loss, and critical limits
Proceedings IAU Symposium No. 272, 2010
C. Neiner, G. Wade, G. Meynet & G. Peters, eds.

doi:10.1017/S1743921311010660

Massive Oe/Be stars at low metallicity: candidate progenitors of long GRBs?

Christophe Martayan[1,2], Dietrich Baade[3], Juan Zorec[4], Yves Frémat[5], Juan Fabregat[6] and Sylvia Ekström[7]

[1]ESO Chile; [2]GEPI-Observatoire de Meudon, France; [3]ESO Germany;
[4]Instritut d'Astrophysique de Paris, France; [5]Royal Observatory of Belgium, Brussels, Belgium;
[6]Valencia University, Spain; [7]Geneva Observatory, Switzerland

Abstract. At low metallicity B-type stars rotate faster than at higher metallicity, typically in the SMC. As a consequence, a larger number of fast rotators is expected in the SMC than in the Galaxy, in particular more Be/Oe stars. With the ESO-WFI in its slitless mode, we examined the SMC open clusters and found an occurence of Be stars 3 to 5 times larger than in the Galaxy. The evolution of the angular rotational velocity seems to be the main key on the understanding of the specific behaviour and stellar evolution of such stars at different metallicities. With the results of this WFI study and using observational clues on the SMC WR stars and massive stars, as well as the theoretical indications of long gamma-ray burst progenitors, we identify the low metallicity massive Be and Oe stars as potential LGRB progenitors. Therefore the expected rates and numbers of LGRB are calculated and compared to the observed ones, leading to a good probability that low metallicity Be/Oe stars are actually LGRB progenitors.

Keywords. gamma rays: observations, stars: supernovae: general, stars: rotation, Magellanic Clouds

---

1. Long GRBs and the collapsar model

In the collapsar model for Ib/c supernovae (Woosley 1993), matter with high specific angular momentum is accreted by the already-formed black hole with a short delay. The intermediate disk structure gives rise to the formation of relativistic jets emitting the so-called long gamma-ray bursts (GRBs). The gamma-ray radiation is strongly relativistically beamed with opening angles of order 10 degrees.

In order for a rapidly rotating core to be present, quasi-homogeneous chemical evolution is essential. Stellar evolution models show that, in stars with very high initial rotation rates, mixing is faster as chemical gradients are built up by nuclear burning (for more details see Yoon & Langer 2005).

The angular momentum still available at this late phase of stellar evolution is the larger, the less mass, and therefore, angular momentum, the progenitor star has lost during its lifetime. Since radiatively driven mass loss from massive stars decreases with metallicity, this has led to the notion that low-metallicity star-forming regions should be preferred hosts of GRBs. In fact, observations are beginning to support this picture (e.g., Modjaz *et al.* 2008). Moreover, the GRB rate per unit mass increases with redshift z. But it is not clear whether this requires any explanation other than the general increase in star formation with redshift (up to $z \sim 2$).

2. Oe/Be stars

In the Milky Way, the near-main sequence stars with the highest rotation velocities are Oe/Be stars. Moreover, many of them rotate at ~90% of the critical rate. Two arguments predict a larger fraction of nearly critically rotating massive stars in low-metallicity environments: (i) The mass (and angular momentum) loss rates are lower and (ii), all else being equal, the radii are smaller. Various comparisons of open clusters in the Magellanic Clouds and the Galaxy have concluded (cf. Martayan, Baade & Fabregat 2010) that the abundance of Oe/Be stars increases with decreasing metallicity Z. If one keeps in mind that rapid rotation only is a necessary, but not also sufficient, condition for a rapidly rotating O/B star to become an Oe/Be star, this can pass as a confirmation of the expectation. In any event, because of their emission lines, Oe/Be stars probably are more reliable tracers of a population of rapidly rotating OB stars than broad photospheric lines would be. Therefore, Oe/Be stars more massive than $\sim 18\, M_{\odot}$ (metallicity dependent; Yoon, Langer & Norman 2006) qualify as candidates, also at low Z. Note that, perhaps, Oe/Be stars do not quite reach the theoretical upper mass limit for the progenitors of GRBs (Yoon, Langer & Norman 2006).

3. GRBs and Oe/Be stars in a synopsis

The strong relativistic beaming makes GRBs observable out to extreme redshifts. But it also implies that much fewer nearby GRBs are detected. Therefore, while it is still relatively straightforward to determine the number of Oe/Be stars per SMC mass and also the space density of such environments as a function of redshift can be estimated, statistics with just a few dozen entries (GRBs) pose serious problems. Nevertheless, using a distribution of beam angles determined from observations, 3–6 long GRBs are predicted at $z \leqslant 0.2$ per 11-year period while the observed number is 8 (cf. Martayan *et al.* 2010). Since the agreement is better than 2 σ, there is no need to fine tune adjustment parameters such as the range in spectral type, from which GRB progenitors are drawn, or the number, type, and star-formation level of the assumed host galaxies. But the excellent agreement could still be coincidental.

4. Discussion

In addition to the challenges already mentioned, a number of other questions still warrant further examination:

• Do Oe/Be stars really preserve their large intial angular momentum throughout their evolution? After all, they regularly eject rotationally supported disks. This behavior is not observed in On/Bn stars, i.e. rapidly rotating O/B stars without emission lines.

• Are there other differences in the evolution of Be and Bn stars?

• Is the SMC a good proxy for GRB-forming environments?

• How incomplete are the observations, especially of GRBs?

• Can candidate progenitors other than Oe/Be stars reach the GRB state along different evolutionary paths?

References

Martayan, C., Baade, D., & Fabregat, J. 2010, *A&A*, 509, A11

Martayan, C., Zorec, J., Frémat, Y., & Ekström, S. 2010, *A&A*, 516, A103

Modjaz, M., Kewley, L., Kirshner, R. P., Stanek, K. Z. *et al.* 2008, *AJ*, 135, 1136

Woosley, S. E. 1993, *ApJ*, 405, 273

Yoon, S.-C. & Langer, N. 2005, *A&A*, 443, 643

Yoon, S.-C., Langer, N., & Norman, C. 2006, *A&A*, 460, 199

Active OB stars: structure, evolution, mass loss, and critical limits
Proceedings IAU Symposium No. 272, 2010
C. Neiner, G. Wade, G. Meynet & G. Peters, eds.

doi:10.1017/S1743921311010672

IR mass loss rates of LMC and SMC O stars

Derck L. Massa[1], Alex W. Fullerton[1], Danny J. Lennon[1] and Raman K. Prinja[2]

[1]STScI, Baltimore, MD, USA

[2]Dept. of Astronomy, UCL, London, UK

Abstract. We use a combination of VJHK and *Spitzer* [3.6], [5.8] and [8.0] photometry, to determine IR excesses in a sample of LMC and SMC O stars. This sample is ideal for determining excesses because: 1) the distances to the stars, and hence their luminosities, are well-determined, and; 2) the very small line of sight reddenings minimize the uncertainties introduced by extinction corrections. We find IR excesses much larger than expected from Vink *et al.* (2001) mass loss rates. This is in contrast to previous wind line analyses for many of the LMC stars which suggest mass loss rates much less than the Vink *et al.* predictions. Together, these results indicate that the winds of the LMC and SMC O stars are strongly structured (clumped).

Keywords. stars: early-type, stars: mass loss, galaxies: Magellanic Clouds

1. Introduction

The effects of structure on wind diagnostics is now widely accepted. It causes Hα and radio and IR continuum measurements (n_e^2 diagnostics) to *over*-estimate mass loss rates, $\dot{M}$. Structure, and its accompanying porosity, can also cause wind line diagnostics to *under*-estimate $\dot{M}$ (Prinja & Massa 2010). We examine this effect in lightly reddened Magellanic Cloud O stars by deriving $\dot{M}$s from *Spitzer* IR excesses, and comparing the results to theoretical expectations from Vink *et al.* (2001). If clumping is significant, we expect the IR excesses to over-estimate the mass loss rates. Furthermore, comparing the LMC and SMC results allows us to examine metallicity effects on clumping.

2. Extinction Corrections

To determine IR excesses, the continua must be corrected for reddening, and this requires the assumption that a portion of the continuum is free of excess and has a known slope, typically $(B - V)$. Applying an extinction curve with an inappropriate $A(V)$ can introduce enormous errors in the inferred IR excess. In this regard, LMC and SMC O stars have a substantial advantage compared to Galactic O stars.

3. The Model

We use a generalization of the Lamers & Waters (1984) model which agrees with results from the Puls *et al.* (1996) Fastwind model when similar parameters are used. We ignore the effects of disks or non-standard, slowly accelerating velocity laws at this time. For the stellar parameters the following were adopted: For the LMC we used the Martins *et al.* (2002) calibrations for the stellar parameters, except for luminosities, where a distance modulus of 18.52 mag was used. We assumed $Z(\mathrm{LMC})/\mathrm{Z}_\odot = 0.6$. For the SMC we used the Massey *et al.* (2009) Spectral Type $\rightarrow T_{eff}$ calibration, a $DM = 18.91$ mag to determine $\log L(\mathrm{SMC})/\mathrm{L}_\odot$, and the Leitherer et al. (2010) $Z(\mathrm{SMC})/\mathrm{Z}_\odot = 0.2$ grid to determine masses. When measured terminal velocities, v_∞, were not available, we

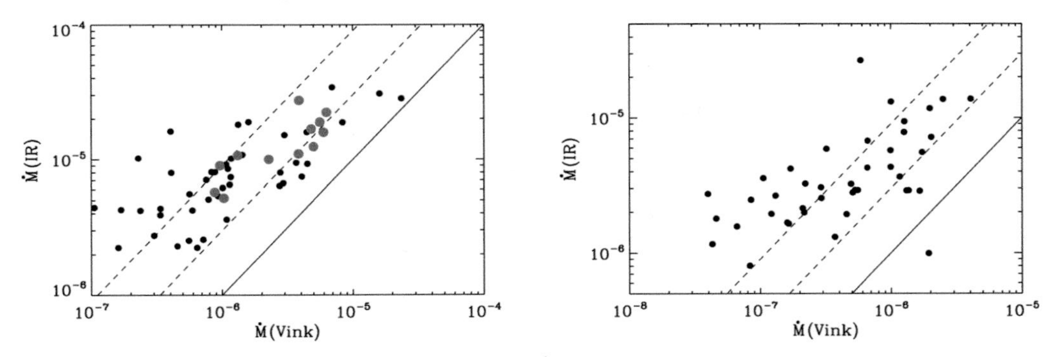

Figure 1. Mass loss rates from *Spitzer* excesses, $\dot{M}$(IR), as a function of theoretical expectations, $\dot{M}$(Vink), for the LMC (left) and SMC (right). For the LMC, P V results, infer $\dot{M}$s much smaller than $\dot{M}$(Vink). The red points depict stars which will be observed by *HST*. The solid line indicates $\dot{M}(\mathrm{IR}) = \dot{M}(\mathrm{Vink})$ and the dashed lines give $\dot{M}(\mathrm{IR}) = 3$ and $9\times\dot{M}(\mathrm{Vink})$.

used the Vink *et al.* (2001) formulae relating escape velocity and v_∞. TLUSTY (Lanz & Hubeny 2003) models of the appropriate temperatures, surface gravities and metallicities were used for the bare photospheres, and an $R(V) = 3.1$ extinction curve was assumed throughout (Fitzpatrick & Massa 2009).

4. Results

All of the IR mass loss rates (see, Fig. 1) are larger than expected and, as is well known, LMC mass loss rates are larger than SMC rates for similar stellar parameters. Furthermore, the relative disagreement between the IR, $\dot{M}$(IR), and Vink *et al.* (2001) mass loss rates, $\dot{M}$(Vink), is similar for the LMC and the SMC.

5. Conclusions

• IR excesses for LMC and SMC O stars are larger than those expected from theory by factors of 3 – 10.
• For LMC stars, the mass loss rates are vastly larger than those expected from UV wind lines (Massa *et al.* 2003, Fullerton *et al.* 2006).
• The relative disagreement between theory and observation is similar for both galaxies.
• If, as expected, clumping causes of the disagreement, then its effect appears to be weakly dependent on metallicity.

References

Fitzpatrick, E. L. & Massa, D. 2009, *ApJ*, 699, 1209
Fullerton, A. W., Massa, D. L., & Prinja, R. K. 2006, *ApJ*, 637, 1025
Lamers, H. J. G. L. M. & Waters, L. B. F. M. 1984, *A&A*, 136, 37
Lanz, T. & Hubeny, I. 2003, *ApJS*, 146, 417
Martins, F., Schaerer, D., & Hillier, D. J. 2002, *A&A*, 382, 999
Massa, D., Fullerton, A. W., Sonneborn, G. & Hutchings, J. B. 2003, *ApJ*, 586, 996
Massey, P., Zangari, A. M., Morrell, N. I., Puls, J. *et al.* 2009, *ApJ*, 692, 618
Prinja, R. K. & Massa, D. L. 2010, *A&A* 521, L55
Puls, J., Kudritzki, R.-P., Herrero, A., Pauldrach, A. W. A. *et al.* 1996, *A&A*, 305, 171
Vink, J. S., de Koter, A., & Lamers, H. J. G. L. M. 2001, *A&A*, 369, 574

Active OB stars: structure, evolution, mass loss, and critical limits
Proceedings IAU Symposium No. 272, 2010
C. Neiner, G. Wade, G. Meynet & G. Peters, eds.

doi:10.1017/S1743921311010684

A statistical study of galactic bright Be stars

Anatoly S. Miroshnichenko

Dept. of Physics & Astronomy, University of North Carolina at Greensboro,
P.O. Box 26170, Greensboro, NC 27402–6170, USA; email: a_mirosh@uncg.edu

Abstract. Creation of a modern database with a wide variety of data is an important step toward a better understanding of the Be phenomenon. In an effort to do that, I refined the existing catalog of Galactic Be stars and collected available observational data for 340 brightest objects and present some results of these data analysis. New candidates for Be binaries, which seem to represent a large fraction of Be stars, are suggested. Importance of the circumstellar optical continuum for modeling of the Be star envelopes is illustrated.

Keywords. stars: emission-line, Be, circumstellar matter

Introduction

The Be phenomenon is observationally defined as the presence of emission lines in the spectra of rapidly-rotating B-type stars. Stars with the Be phenomenon show continuum excess radiation (due to free-free and bound-free transitions in the circumstellar gas), variations in brightness and spectral lines, and periods of complete loss of the emission spectrum. They have disk-like envelopes with no dust and are though to be at the main-sequence evolutionary stage.

Among many catalogs of emission-line stars, only one intended to list Be stars alone (Jaschek & Egret 1982). It contains very limited information, but can be used for creating a modern database. I collected IR fluxes (IRAS, 2MASS) for all catalog stars and other available data for 340 brightest ones, including 19 found since 1982 ($V \leqslant 7.5$ mag): binary status (detection, orbital period), polarization, parallax, rotational velocity, fundamental parameters, variability data (multicolor optical and near-IR photometry, Hα line).

The main results of my study of this material are as follows. Objects accessible from the northern hemisphere ($\delta \geqslant -20°$) have been better observed than more southern objects. Observations of $V \geqslant 8$ mag stars are needed for further constraining the list of Galactic

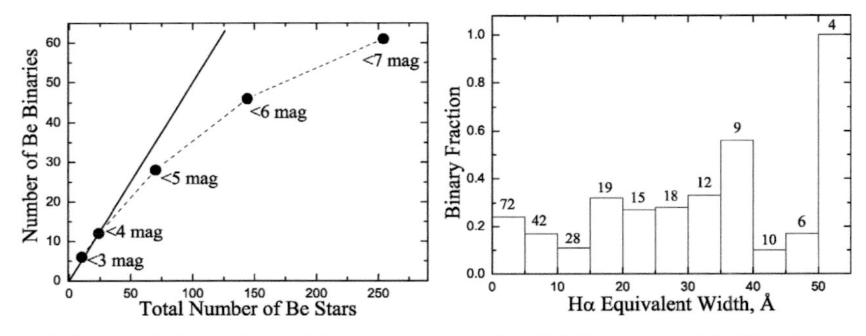

Figure 1. Left panel: Cumulative binary statistics for 254 Be stars with V < 7 mag. Circles show the numbers of known binaries brighter than the indicated level. The solid line marks a 50% binary fraction. The number of recognized binaries used here is 61. Right panel: Binary fractions are shown as a function of the Hα equivalent width (EW). The maximum Hα strength was used whenever possible. The total number of Be stars used in each bin is shown above it.

Table 1. Statistics of the original catalog

Category	Number
Total number of objects (duplicates excluded)	1149
Herbig Ae/Be and Vega-type	12
Objects with the B[e] phenomenon and LBV	10
Supergiants and WR stars	48
Bright giants (luminosity type II)	25
Luminosity type II/III	24
Oe type stars	4
No line emission confirmed	3
Other unrelated objects	2
Objects with no luminosity type	103
Objects with no good line profiles	>500

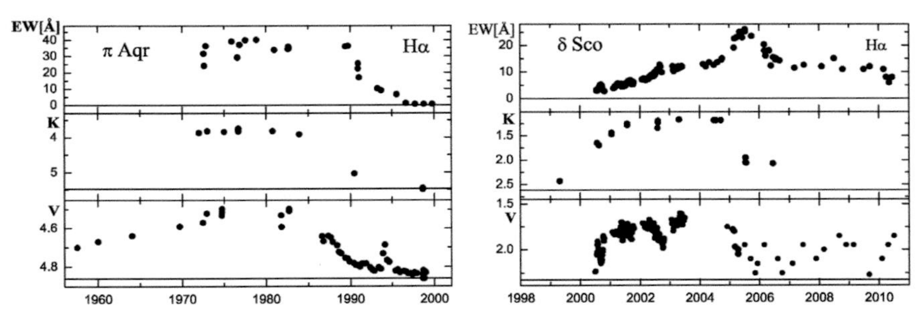

Figure 2. Brightness and Hα variations of π Aqr (left) and δ Sco (right). Solid lines show the no-disk level of the parameters. Data for π Aqr are taken from Bjorkman *et al.* (2002), for δ Sco from Carciofi *et al.* (2006), Halonen *et al.* (2008), Pollmann (2010), and Otero (see http://varsao.com.ar/delta_Sco.htm).

Be stars. The progressively lower observed binary fraction for fainter objects (Fig. 1, left) is a selection effect: fainter stars are less studied. Therefore, over 50% of Be stars can be binary systems. The right panel of Fig. 1 may be misleading, as not many stars have been observed extensively. However, many bright Be stars with strong Hα emission (e.g., γ Cas, π Aqr, χ Oph, ø Cas, 66 Oph) were recognized as binaries. Good candidates for the secondary companion search include: ψ Per, HD 7636, HD 50083, EW Lac, V777 Cas, V659 Mon, NV Pup, MWC 19, HD 206773.

Fig. 2 shows brightening of two Be binaries during a strong line-emission phase. These data show that both mass loss and the disk structure are inhomogeneous. Modeling of emission-line profiles in stars with no edge-on disk that does not account for the optical excess radiation gives wrong results for the disk parameters. This effect also leads to an overestimated star's luminosity by up to 70%. π Aqr has a circular orbit with a 84.3–day period. Currently it shows weak signs of emission. The simultaneous optical fading and Hα EW increase of δ Sco (orbital period 10.8 years – see Ames *et al.* in these proceedings, eccentricity $e = 0.94$) in 2005 are due to dissipation of the disk inner part.

References

Jaschek, M. & Egret, D. 1982, in: M. Jaschek & H.-G. Groth (eds.), *Be Stars*, IAU Symposium 98, p. 261

Bjorkman, K. S., Miroshnichenko, A. S., McDavid, D., & Pogrosheva, T. M. 2002, *ApJ*, 573, 812

Carciofi, A. C., Miroshnichenko, A. S., Kusakin, A. V., Bjorkman, J. E. *et al.* 2006, *ApJ*, 652, 1617

Halonen, R. J., Jones, C. E., Sigut, T. A. A., Zavala, R. T. *et al.* 2008, *PASP*, 120, 498

Pollmann, E. 2010, *B*e Star Newsletter, No. 40, in press

Active OB stars: structure, evolution, mass loss, and critical limits
Proceedings IAU Symposium No. 272, 2010
C. Neiner, G. Wade, G. Meynet & G. Peters, eds.

doi:10.1017/S1743921311010696

The e-MERLIN Cyg OB2 radio survey (COBRaS†)

Raman K. Prinja and Danielle Fenech

Dept. of Physics & Astronomy, University College London,
Gower Street, London WC1E 6BT, UK

Abstract. The e-MERLIN Cyg OB2 Radio Survey (COBRaS) is designed to exploit e-MERLIN's enhanced capabilities to conduct uniquely probing, targeted deep-field mapping of the massive Cyg OB2 association in our Galaxy. The project aims to deliver (between 2010 to 2013) the most detailed radio census for the most massive OB association in the northern hemisphere, offering direct comparison to not only massive clusters in general, but also young globular clusters and super star clusters. With the COBRaS Legacy project we will assemble a uniform dataset of lasting value that is critical for advancing our understanding of current astrophysical problems in the inter-related core themes of (i) mass loss and evolution of massive stars, (ii) the formation, dynamics and content of massive OB associations, and (iii) the frequency of massive binaries and the incidence of non-thermal radiation.

Keywords. stars: evolution, stars: mass loss, radio continuum: stars

1. Introduction

We introduce here a e-MERLIN Legacy project is designed to conduct uniquely deep radio mapping of the tremendously rich Cyg OB2 stellar association in our Galaxy. The Cyg OB2 radio survey (COBRaS) has been awarded Legacy status, with a substantial allocation of 252 hrs for C-band (5 Ghz) and 42 hrs for L-band (1.6 Ghz) e-MERLIN observations (www.merlin.ac.uk/legacy/projects/cobras.html). Our goal is to conduct the most detailed radio census for the most massive OB star association in the northern hemisphere, offering direct comparisons to not only massive stellar clusters in general, but also young globular clusters and super star clusters. This programme will advance our understanding of the role of evolution in stellar mass-loss, constrain stellar cluster formation scenarios, and provide new perspectives of the formation of massive stars in high binary fractions. The substantial COBRaS dataset will be assembled between 2010 and 2013, and will ultimately also be combined with other multi-waveband international surveys of the Cygnus X region, from both current (Spitzer and Chandra)

Several factors combine to make the Cyg OB2 association a uniquely important laboratory for studying the collective and individual properties of massive stars. The COBRaS project will exploit the e-MERLIN datasets to obtain solutions to current fundamental problems in three key areas of massive star astrophysics: (i) the mass loss, energy feedback and evolution process, (ii) the kinematics and formation of massive stellar clusters, and (iii) the incidence of non-thermal radiation and the frequency of massive binaries.

2. Planned e-MERLIN pointings

The COBRaS project will map the core of Cyg OB2 at 5 GHz, going to a depth of $\sim 3\ \mu$Jy (1-sigma), plus additional pointings at 1.6 GHz (see Fig. 1). The investment of

† COBRaS: http://www.homepages.ucl.ac.uk/~ucapdwi/cobras/team_members.html

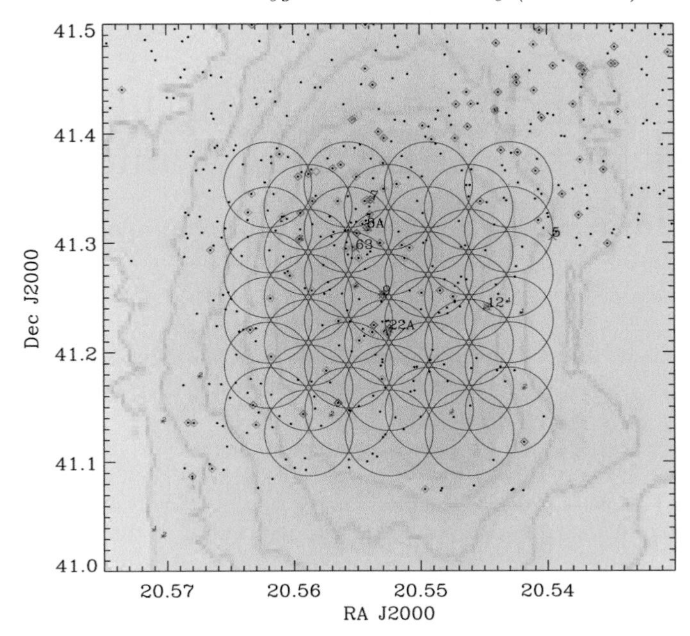

Figure 1. COBRaS Legacy project planned C-band mosaicing of the core of Cyg OB2, 42 (6-hr) pointings are plotted and a further 7 will be executed for the L-band. The background figure is the outline of the Cyg OB2 association (based on counts from the 2MASS survey).

a substantial amount of e-MERLIN observing time ($\sim$ 300 hrs awarded) to this young massive star cluster is important as it is one of only a few examples known in our Galaxy. As Cyg OB2 is a smaller version of the super star clusters (SSCs) seen in e.g. M82, it can therefore serve as a Rosetta Stone to help interpret the information from these much more distant clusters.

The 5 GHz frequency is the primary band for our purposes: its 2 GHz broad bandwidth allows us to determine not only the flux, but (generally) the spectral index as well. This will allow us to distinguish between thermal and non-thermal radiation. We estimate that at least 10^3 sources will be detected in our survey. For the O stars, based on estimated thermal radio fluxes, all supergiants will be detectable, most giants, but only the brightest main-sequence stars. Another key goal is to use the high spatial resolution offered by e-MERLIN to obtain milliarcsec accuracy astrometric observations of the radio stars within Cyg OB2 at multiple epochs, in order to determine their proper motions. As part of our consortium (see below) we have an ongoing program of high resolution spectroscopic observations of the massive stellar populations within Cyg OB2 to identify and characterise massive stars, which provide radial velocity measurements of comparable accuracy; taken together these complementary datasets will allow a full 3 dimensional picture of the kinematics of Cyg OB2 to be constructed.

The COBRaS e-MERLIN Legacy dataset will be secured between 2010 to 2013. It will allow us to deliver new results on mass-loss via clumped outflows, high energy phenomena associated with massive binary stars, and the dynamics of clusters of stars.

Active OB stars: structure, evolution, mass loss, and critical limits
Proceedings IAU Symposium No. 272, 2010
C. Neiner, G. Wade, G. Meynet & G. Peters, eds.

doi:10.1017/S1743921311010702

A spectroscopic study of Be-like variable stars in the Small Magellanic Cloud

Beatriz Sabogal[1], **Alejandro García-Varela**[1] **and Ronald E. Mennickent**[2]

[1]Departamento de Física, Universidad de los Andes, Bogotá, Colombia
email: bsabogal@uniandes.edu.co

[2]Departamento de Astronomía, Facultad de Ciencias Físicas y Matemáticas, Universidad de Concepción, Chile

Abstract. Photometric searches for Be stars in environments with different metallicities have led to the discovery of many Be-like star variables. The knowledge of these types of variables is still fragmentary. This work presents the preliminary results of analyzing FLAMES+GIRAFFE spectra of a sample of these Be-like stars that we have found in the Small Magellanic Cloud (SMC).

Keywords. stars: early-type, stars: emission-line, Be, stars: variables: other

1. Introduction

Mennickent *et al.* (2002) performed a photometric search for Be star candidates in the OGLE II database of SMC discovering new kind of variable stars in the same range of colours and magnitudes than classical Be stars. These objects were called Type-1 stars (showing outbursts), Type-2 stars (showing jumps), Type-3 stars (showing periodic or quasi-periodic variations) and Type-4 stars (showing irregular variations). The same kind of variable stars were found in the Large Magellanic Cloud (Sabogal *et al.* 2005) and in the direction of the Galactic Bulge (Sabogal *et al.* 2008). Samples of Type-1 stars were studied by de Wit *et al.* (2006) and by Mennickent *et al.* (2009). These studies showed that Type-1 stars could be stars with an optically thick circumstellar envelope that evolves in an optically thin one before to dissipate into the interstellar medium. In order to follow up the study of these new variables, we are analyzing medium resolution spectra of 111 blue variable stars of the SMC in order to obtain clues to understand the mechanisms causing the different variations and their relationships with the Be phenomenon. The sample consists of 7 Type-1 stars, 19 Type-2 stars, 43 Type-4, 18 Double Periodic Variables (DPVs) (Mennickent *et al.* 2003). The 24 remaining stars are periodic or multiperiodic Be stars, non-periodic Be stars, candidates to SPBs and non variable stars.

2. Data

A total of 3712 spectra were obtained in service mode with GIRAFFE/FLAMES spectrograph mounted at UT2/VLT/ESO telescope in Paranal Observatory (Chile). The LR MEDUSA mode was used, with a spectral range of 6437 Å- 7182 Å, and a central wavelength of 6822 Å. The resolving power was 8600, and the average seeing was 0.8 arcsec. We have data for 5 epochs: October, November and December 2007, January and February 2008. The spectra were reduced and calibrated with GIRAFFE pipeline and normalized to the continuum using IRAF routine "continuum".

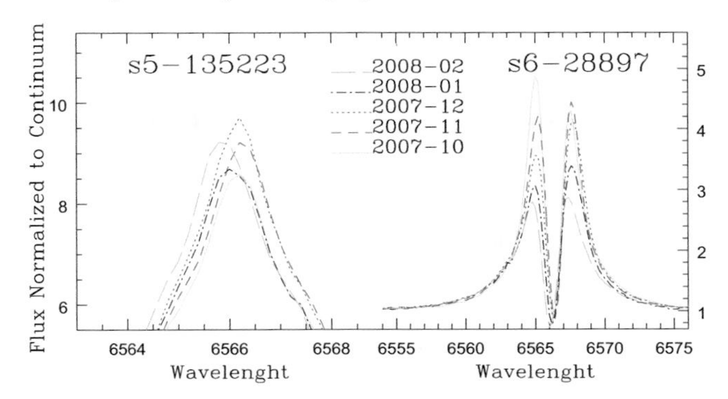

Figure 1. Variability of the H_α profile of a Type-1 star (Left: intensity increases during the first 3 months and proportionally decreases during the 2 following months) and a Type-4 star (Right: intensity of the 2 components decreases during the five observation periods).

3. Preliminary Results

From visual inspection of the spectra of the total sample mentioned in previous section, it is evident that Type-1 stars show strong and mostly double peaked H_α emission lines, although single and wine-bottle sharp profiles are also present (Fig. 1). Type-2 stars show double and single peaked H_α emission lines at the same proportion. The aspect of H_α lines of Type-4 stars is similar to that of Type-2 stars, although some of these Type-4 stars have more complicated H_α profiles. DPVs also show both double and single peaked H_α emission lines. We found also one Type-4 star (smc-s5-196382) with an absorption H_α line. These preliminary results show that there is not a clear separation between the 4 classes of stars regarding their H_α profiles. Some of the double-peaked profiles show nearly equal V and R emission components but other stars have variable differences in the intensities of these emission components and only very slow variations in the equivalent width of the line (Fig. 1). It seems that a tendency between the time-scale of Type-1 star outbursts and the time-scale of changes in the strength of the H_α line exists, in agreement with the model of de Wit *et al.* (2006). This fact needs to be confirmed for all Type-1 stars with a follow up program obtaining simultaneosly photometric and spectroscopic data for such stars. It is also possible that a correspondence exists between the constant state of Type-2 stars during a long time-scale and the small changes in the intensity of their H_α lines. This needs to be confirmed too.

Double-peaked H_α profiles resemble those of Be stars and indicate the presence of a flattened, equatorial Keplerian disk that is seen by an observer viewing the star at an intermediate or edge-on angle. The H_α profiles shown in Fig. 1 (right panel), presenting changes in the intensities of the blue and red emission components, are commonly explained by the appearance and slow drift of a disc-like region of enhanced gas density. These aspects provide important evidence that the 4 classes of stars studied actually are Be stars and that their variability is associated with the Be phenomenon.

References

de Wit, W. J., Lamers, H. J. G. L. M., Marquette, J. B., & Beaulieu, J. P. 2006, *A&A*, 456, 1027

Mennickent, R. E., Pietrzyński, G., Gieren, W., & Szewczyk, O. 2002, *A&A*, 393, 887

Mennickent, R. E., Pietrzyński, G., Diaz, M., & Gieren, W. 2003, *A&A*, 399, L47

Mennickent, R. E., Sabogal, B., Granada, A., & Cidale, L. 2009, *PASP*, 121, 125

Sabogal, B. E., Mennickent, R. E., Pietrzyński, G., & Gieren, W. 2005, *MNRAS*, 361, 1055

Sabogal, B. E., Mennickent, R. E., Pietrzyński, G., García, J. A. *et al.* 2008, *A&A*, 478, 659

Active OB stars: structure, evolution, mass loss, and critical limits
Proceedings IAU Symposium No. 272, 2010
C. Neiner, G. Wade, G. Meynet & G. Peters, eds.

doi:10.1017/S1743921311010714

The IACOB spectroscopic database of galactic OB stars

Sergio Simón-Díaz[1,2], Norberto Castro[1,2], Miriam Garcia[1,2] and Artemio Herrero[1,2]

[1]Instituto de Astrofísica de Canarias, E-38200 La Laguna, Tenerife, Spain
[2]Departamento de Astrofísica, Universidad de La Laguna, E-38205 La Laguna, Tenerife, Spain

Abstract. We present the IACOB spectroscopic database, the largest homogeneous database of high-resolution, high signal-to-noise ratio spectra of Northern Galactic OB-type stars compiled up to date. The spectra were obtained with the FIES spectrograph attached to the Nordic Optical Telescope. We briefly summarize the main chareceristics and present status of the IACOB, first scientific results, and some future plans for its extension and scientific exploitation.

Keywords. stars: early-type, techniques: spectroscopic, catalogs, astronomical data bases: miscellaneous

---

1. Introduction

In an epoch in which we count with a new powerful generation of stellar atmopshere codes including all the important physics for the modelling of massive OB stars, with (clusters of) high efficiency computers allowing the computation of large grids of stellar models in more than reasonable computational times, and with the possibility to obtain good quality, medium resolution spectra of hundreds O and B-type stars in clusters outside the Milky way in just one snapshot (see e.g. the *FLAMES I & II Surveys of Massive Stars*, Evans *et al.* 2008, 2010; see also Lennon *et al.*, these proceedings), the compilation of medium and high-resolution spectroscopic databases of OB stars in our Galaxy is becoming more and more important. With this idea in mind, two years ago we began to compile the IACOB spectroscopic database, aiming at constructing the largest database of multiepoch, high resolution, high signal-to-noise ratio (S/N) spectra of Galactic Northern OB-type stars. The IACOB perfectly complements the efforts also devoted in the last years by the GOSSS (P.I. Maíz-Apellaniz) and the OWN (P.I's Barbá & Gamen, leading a multi-epoch, high-resolution spectroscopic survey of Galactic O and WR stars in the Southern hemisphere; see Barbá *et al.* 2010) teams.

2. Characteristics of the IACOB and present status

We are using the FIES spectrograph at the 2.56 m Nordic Optical Telescope (NOT) in the Roque de los Muchachos observatory (La Palma, Spain) to compile spectra for the IACOB (see http://www.not.iac.es). A summary of the instrumental configuration and observing dates (before Sept. 2010) is presented in Table 1. 720 spectra of 105 stars with spectral types O4-B2 and luminosity classes ranging from I (Supergiants) to V (Dwarfs) have already been compiled. The O-type targets were selected among those stars with V $\leqslant$ 8 included in the GOS catalogue (GOSC, Maíz-Apellániz *et al.* 2004). The main part of the B-type stars sample correspond to the works presented in Simón-Díaz (2010) and Simón-Díaz *et al.* (2010). The final spectra normally have S/N $\geqslant$ 200.

Table 1. General characteristics of the IACOB v1.0 spectroscopic database.

Instrumental configuration		Observing run & Dates	
Telescope: NOT2.56 m	Spect. range: 3800 - 7000 Å	08 A-D: 2008/11/05-08,	10 D: 2010/06/22
Instrument: FIES	Res. power: 46000	09 A-D: 2009/11/09-12,	10 E: 2010/07/15
Mode: med-res	Sampling: 0.03 Å/pix	10 A-C: 2010/06/05-07,	10 F: 2010/08/07

3. Some scientific results using spectra from the IACOB

There are already 2 published papers using data from the IACOB (and more in prep.). In Simón-Díaz (2010), we used the stellar atmosphere code FASTWIND (Puls *et al.* 2005) to perform a thorough self-consistent spectroscopic analysis of 13 early B-type stars from the various subgroups comprising the Orion OB1 association; this study showed that the dispersion of O and Si abundances between stars in the various subgroups found in previous analyses (e.g. Cunha & Lambert 1992) was a spurious result, being the consequence of a bad characterization of the abundance errors propagated from the uncertainties in the stellar parameter determination. In Simón-Díaz *et al.* (2010) we showed the first observational evidence for a correlation between macroturbulent broadening and line-profile variations in OB Supergiants using spectroscopic timeseries for a sample of 13 OB Sgs; this may support the hypothesis that macroturbulent broadening in this type of stars is likely a result of the collective effect of stellar pulsations. A subsample of IACOB spectra has been used within the *FLAMES-II Survey of Massive Stars: the Tarantula Survey* consortium to construct an atlas of medium resolution spectra of Galactic OB-type stars (Sana *et al.* in prep.). We plan to use this atlas for the spectral classification of the massive stars in 30 Dor. Finally, the scientific exploitation of the IACOB spectra concerning the quantitative spectroscopic analysis of the stars has already began and a series of papers with results will be published soon (see details in Simón-Díaz *et al.*, in prep.).

4. Future plans for the IACOB

In the next semesters, we will continue with the compilation of spectra for the IACOB, observing stars with V⩽8 in at least three epochs (more in the case of known or newly detected binaries). Our idea is to make public the database via the Virtual Observatory in the next year. In the meantime, interested people can have access to the database under request to the author (`ssimon@iac.es`). The complete list of stars will be published in Simón-Díaz *et al.*, in prep.. We will acknowledge any observer who having obtained FIES spectra will like to add the spectra to the IACOB database after scientific exploitation.

References

Barbá, R. H., Gamen, R., Arias, J. I., Morrell, N. *et al.* 2010, in: T. Rivinius & M. Curé (eds.), *The Interferometric View on Hot Stars*, Rev. Mexicana AyA Conference Series 38, p. 30

Cunha, K. & Lambert, D. L. 1992, *ApJ*, 399, 586

Evans, C., Hunter, I., Smartt, S., Lennon, D. *et al.* 2008, *The Messenger*, 131, 25

Evans, C. J., Bastian, N., Beletsky, Y., Brott, I. *et al.* 2010, in: R. de Grijs & J. R. D. Lépine (eds.), *Star clusters: basic galactic building blocks throughout time and space*, IAU Symposium 266, p. 35

Maíz-Apellániz, J., Walborn, N. R., Galué, H. Á. & Wei, L. H. 2004, *ApJS*, 151, 103

Puls, J., Urbaneja, M. A., Venero, R., Repolust, T. *et al.* 2005, *A&A*, 435, 669

Simón-Díaz, S. 2010, *A&A* 510, 22

Simón-Díaz, S., Herrero, A., Uytterhoeven, K., Castro, N. *et al.* 2010, *ApJ* (Letters), 720, L174

Speakers in action. From left to right and top to bottom: Alex Fulerton, Jose Groh, Sylvia Ekstrom, Thomas Rivinius, Stéphane Mathis, Hideyuki Saio, Maria-Fernanda Nieva, and Véronique Petit,

Active OB stars: structure, evolution, mass loss, and critical limits
Proceedings IAU Symposium No. 272, 2010
C. Neiner, G. Wade, G. Meynet & G. Peters, eds.

doi:10.1017/S1743921311010726

Observations of circumstellar disks

Philippe Stee

UMR 6525 H. Fizeau, Université de Nice Sophia Antipolis, Centre National de la Recherche Scientifique, Observatoire de la Côte d'Azur,
Avenue Copernic, F-06130 Grasse, France
email: philippe.stee@oca.eu

Abstract. In this review I will present recent results obtained between 2005, which was the last "Be-stars" meeting in Sapporo, and 2010, on the observations of circumstellar disks around active hot stars.

Keywords. stars: emission-line, Be, stars: early-type, stars: winds, outflows, stars: rotation, stars: mass loss, stars: imaging, techniques: interferometric, techniques: photometric, techniques: polarimetric, techniques: spectroscopic

---

1. Introduction

The observation of circumstellar disks around active OB stars is a very active and attractive research field and more than 65 papers with the keywords "observations + disks + Be stars" were found between 2005 and 2010 using the Astrophysics Data System (ADS). These papers were "classified" by myself into various (arbitrary) themes, namely photometry (3 papers), spectroscopy (13), polarimetry (3), interferometry (19), binarity (4), model constraints (8), variability (6), i.e. Non Radial Pulsations, outbursts, one-arm oscillations, etc...) and x-rays (6). Note that interferometric techniques have been really fruitful and seem now to be at a mature age with more results on circumstellar disks coming out from these techniques rather than from spectroscopic observations. I have not included in this review model constraints since this is done by Alex Carciofi in his review. Neither have I presented results on x-rays since this is presented in the contribution by David Cohen, also in these proceedings.

Since the field covered by observations of circumstellar disks is very large, we will try to concentrate on some well identified topics and try to answer only five main questions:

- What is the disk geometry?
- Do we have a disk, a wind or both?
- What are the density and temperature distributions within the disk?
- What are the disk kinematics?
- How does the disk form and dissipate?

We must also stress that disk observations are strongly wavelength and time dependent, with the important role played by amateur astronomer in the spectroscopic and photometric long term following or "alert" mode of some interesting targets, see for instance the BeSS database†. There is also a large spread of measurements, which may sometimes appear to be inconsistent but since active hot stars are by definition variable it is always difficult to conclude if these differences are linked to an intrinsic source variability or due to a different physical interpretation or simply something that went wrong in the interpretation/simulation. Nevertheless, there is also, and hopefully, some general trends that we will try to describe in the following sections. This paper has the following

† http://basebe.obspm.fr

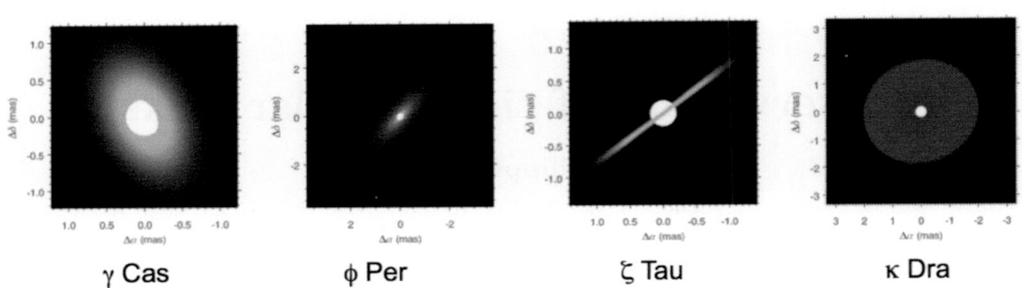

Figure 1. Modeled images using CHARA data by Gies *et al.* (2007)

structure: in Sect. 2 we present results regarding the extensions of the circumstellar disks using various techniques. In Sect. 3 their shapes and the azimuthal morphology are described. Sect. 4 is dedicated to the determination of their radial density distribution. In Sect. 5 we discuss the recent observations of winds around some Be stars and debate the link between the disk and the wind. Sect. 6 is dedicated to the disk kinematics determination whereas in Sect. 7 we present some possible scenarios regarding the disk formation/dissipation with a special focus on the outburst scenario. The roles of metallicity and binarity are discussed in Sect. 8 and the conclusions, where we will try to answer the initial five questions, are drawn in Sect. 9.

2. Disk geometry: extension

Using spectro-astrometry measurements at 1.28 μm Oudjmaijer *et al.* (2008) found an extension of a few dozen stellar radii for α Col and ζ Tau whereas Arias *et al.* (2006) obtained 2.0 ± 0.8 $R_\star$ in Fe II optically thick lines between 4230-7712 Å using the SAC method. The size of the hot CO (2-0 and 3-1 bands of the CO overtone) around the (young) Be star 51 Oph was resolved with VLTI/AMBER at 2.3 μm with a size of $\sim$ 0.15 AU and a more distant nearby continuum of $\sim$ 0.25 AU by Tatulli *et al.* (2008) confirming that the CO band-heads originate in a dust free hot gaseous disk and the continuum emitting region is closer to the star than the dust sublimation radius by at least a factor two. Meilland *et al.* (2009) have observed 7 classical Be stars with the VLTI/MIDI between 8 and 12 μm, namely P Car, ζ Tau, κ CMa, α Col, δ Cen, β CMi, α Arae. They found that the sizes of the disks do not vary strongly with wavelength within this spectral domain which is a very different conclusion compared to B[e] stars, with increasing sizes as a function of wavelength. Moreover the size of α Arae's disk was found to be identical at 2, 8 and 12 μm, which might be due to disk truncation by a companion. Finally it seems from their studies that envelopes of late type Be stars might be smaller than for early type.

γ Cas and ψ Per were observed by Tycner *et al.* (2006) with the NPOI interferometer in the Hα domain. They found that a uniform disk or a ring-like model were inconsistent with their data and that a Gaussian model was fitting pretty well the measurements. The γ Cas disk was also consistent with the orbital parameters already published. Nevertheless, higher precision binary solutions were mandatory to test for a possible disk truncation by the secondary. The disk of ψ Per was found to be truncated by the presence of a companion as already predicted by Waters (1986). γ Cas, ψ Per, ζ Tau and κ Dra were observed with the CHARA interferometer in the K-band by Gies *et al.* (2007) (see Fig. 1). They found, using Gaussian elliptical fits of visibilities, that the disk size

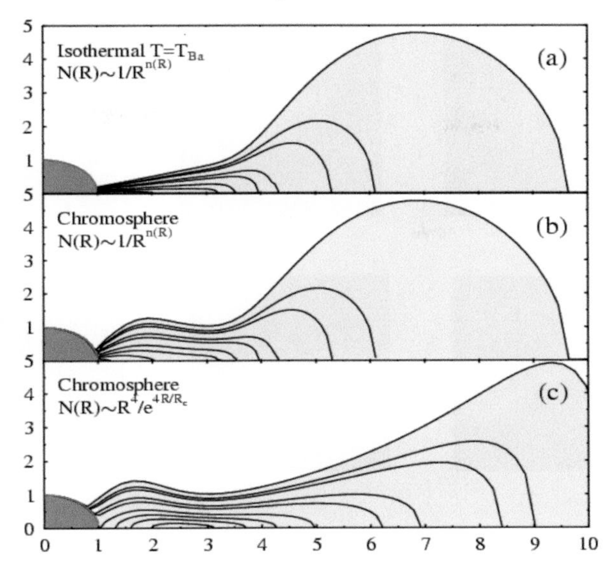

Figure 2. Shape of discs with Keplerian scale heights from Zorec *et al.* (2007)

in the K-band was smaller than in Hα due to a larger Hα opacity and relatively larger neutral hydrogen fraction with increasing disk radius. All these Be stars are known binaries and this binarity effect was found to be more important for ψ Per and κ Dra. Tycner *et al.* (2008) using the NPOI interferometer observed χ Oph and obtained a good fit of the Hα emitting disk with a circularly symmetric Gaussian with a FWHM diameter of 3.46 ± 0.07 mas. The extension of 48 Per and ψ Per in the Hα domain was determined by Delaa *et al.* (2010) using a Gaussian disk model and found to be respectively 2.1 ± 0.2 and 4.0 ± 0.2 with a flattening ratio of 1.3 and 2.9.

3. Disk geometry: shape and the azimuthal morphology

The shape of the disk around Be stars is still a very active and controversial topic, i.e. a very flattened disk of 1-3° of opening angle versus more "spheroidal" geometry. Arias *et al.* (2006) have observed in the Balmer lines 18 southern B stars and found a flattened geometry but without a quantitative value, whereas Zorec *et al.* (2007) obtained a semi-height scales perpendicular to the equatorial plane h $\leqslant 0.5$ R$_\star$ from Fe II optically thick line observations of 17 Be stars using the SAC method.

ζ Tau was found to exhibit asymmetrical IR emission lines (He I, O I, Fe II) and Brackett, Paschen, Pfund series lines with opposite V/R (V $>$ R) compared to Hα Balmer lines (V $<$ R). This was interpreted by Wisniewski *et al.* (2007) as a direct evidence of a density wave with different average azimuthal morphology in the inner versus outer disk region. This asymmetry was also measured by Schaefer *et al.* (2010) with the CHARA/MIRC interferometer in the H band. They have evidenced this asymmetry along the minor axis of the disk with a 10° change in the Position Angle (PA) of the disk between 2007, Nov and 2008, Sep. The cyclic variability of the disk of ζ Tau was successfully explained with a 2D global disk oscillation model by Stefl *et al.* 2009 and Carciofi *et al.* 2009 (see Fig. 3) where they have used the NLTE 3D code HDUST to fit simultaneously VLTI/AMBER data and V/R variations in the Hα and Brγ lines. An asymmetry was also detected in the disk of κ CMa with the VLTI/AMBER interferometer in the K band by Meilland *et al.* (2007b) but this asymmetry was found to be poorly explained within the "one-armed"

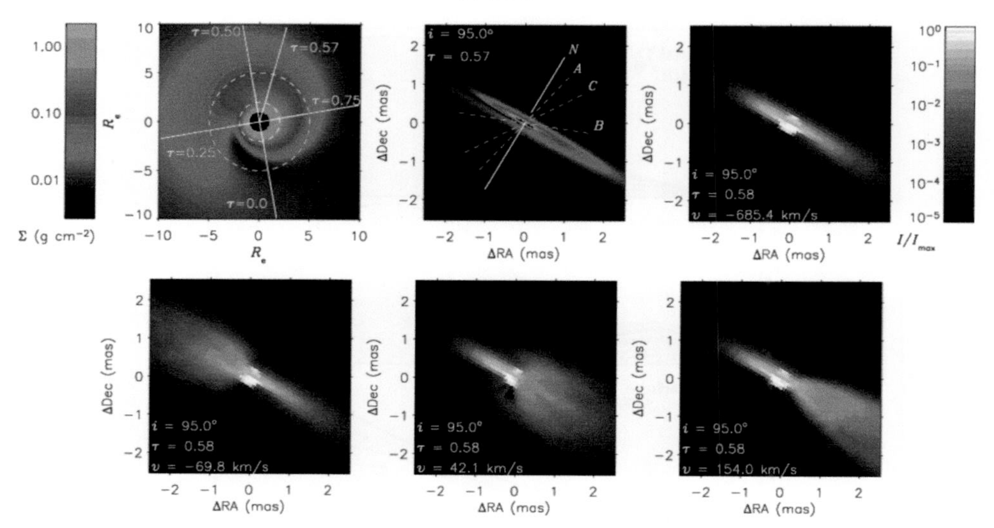

Figure 3. ζ Tau density perturbation pattern from Carciofi's model (Carciofi *et al.* 2009)

viscous disk framework Okazaki (1997) since the "pseudo period" they obtained was too long compared to predictions for a disk with a radius of 23 $R_\star$.

4. Disk geometry: radial density distribution

The value of the "n" index of the density distribution within the circumstellar disk is also highly debated in the community. We recall that the density distribution is often described as:

$$\rho(r) = \rho_0 \left(\frac{R_\star}{r}\right)^n \tag{4.1}$$

where $R_\star$ is the stellar radius and ρ_0 the density at the base of the disk.

Zorec *et al.* (2007) found that $n \leqslant 1$ for r $\leqslant$ 3 $R_\star$, using the SAC method and and a non-LTE interpretation of the FeII line formation in the disc, n $\sim$ 0.5 near the star from near-UV spectropolarimetry measurements (see Fig. 2) whereas Gies *et al.* (2007) obtained values for n between 0.16 and 3.19 and emphasized that n is smaller in binaries with smaller semi-major axis, namely ψ Per and κ Dra, which is a confirmation that binary companions do influence the disk properties. Interestingly, Jones *et al.* (2008) have fitted observed Hα visibilities from the NPOI interferometer and observed Hα line profiles using the NLTE code BEDISK. They obtained n values of 4.2, 2.1, 4.0 and ρ_0 = 1.5 10^{-10}, 3.0 10^{-12} and 8.0 10^{-10} g cm^{-3} respectively for κ Dra, β Psc, υ Cyg. Note that there is a large discrepancy between the value found by Gies *et al.* (2007), i.e. 0.16 and the one from Jones *et al.* (2008), i.e. 4.2. This discrepancy is deeply discussed in the paper by Jones *et al.* (2008). χ Oph was observed by Tycner *et al.* (2008) with the NPOI interferometer and modeled with BEDISK. They obtained n = 2.5 and ρ_0 = 2.0 10^{-11} g cm^{-3}.

The Be star δ Sco was also intensively observed since this binary star is supposed to have its companion at periastron around July 2011(some American colleagues would guess July 4th, I bet on July 14 !). Thus, Carciofi *et al.* (2006) have studied its disk properties from continuum modeling, photometric and spectro-polarimetric observations. They found that in order to explain their observations (between 2000 and 2005) they have to advocate for a change in the mass loss rate AND in the disk geometry. The disk was

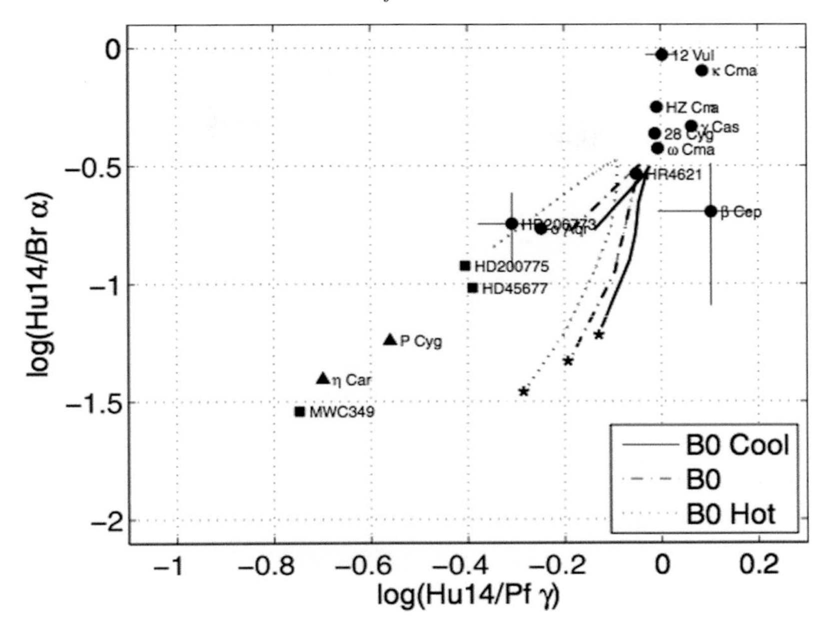

Figure 4. Predicted line flux ratios computed from theoretical disc models by Jones *et al.* (2009) for values of $n = 3.0$, 3.5 and 4.2 with a range of ρ_0

found to be non isothermal and fully ionized. The inner disk is supposed to be unflared in contrast to large flaring expected for isothermal models. They also found that $n \neq 3.5$ as this is the case for isothermal disks and they obtained a total mass loss of 1.5 10^{-9} $M_\odot$ yr^{-1} and a disk radius of 7 $R_\star$. New photometric observations of δ Sco were carried out in 2006 (and taking into account the 10 months observations earlier by Carciofi) in the J, H and K bands by Halonen *et al.* (2008). From the modeling of the Hα emiision line they obtained $\rho_0 = 5.0\ 10^{-10}$ g cm^{-3} and $n = 4.0$ even if they mention that a range for n from 3.5 to 4.5 and ρ_0 from 5.0 10^{-11} to 5.0 10^{-10} is acceptable. They found a "global" agreement between Carciofi's code and BEDISK and a detailed comparison between both codes is foreseen. On the other hand, we recall that Zorec *et al.* (2007) obtain $n < 1$ for $r < 3\ R_\star$ and that the density distribution near the star is constant or increasing. This $n <$ value is also important in order to reproduce the "flat" distribution of the linear polarization observed in the far-UV. Finally Carciofi *et al.* (2006) explain that in order to reproduce the variations in brightness of the system, some material must be ejected at higher stellar latitudes.

From the Spectral Energy Distribution (SED) study, using spectro-photometric data from 0.4 to 4.2 μm of northern sky Be stars Touhami *et al.* (2010), found that stars with a strong Hα excess have also an IR exces. This IR excess correlates with the Hα Equivalent Width (EW), with the largest scatter from the data for densest and largest disks. They obtain a better correlation with high excitation transitions, for instance Hu14. The IR continuum and these high excitation transitions are supposed to occured in the inner and dense part of the disk. Thus, the less marked correlation between the IR excess and the Hα EW might be linked with changes in the density distribution in the outer part of the disk, maybe due to a temporal evolution of the disk or the influence of a binary companion. The IR domain is a useful spectral domain to constrain the disk geometry. For instance Jones *et al.* (2009) used the observed flux ratio Hu14/Brγ and Hu14/Pfγ defined by Lenorzer *et al.* (2002) to constrain basic properties of Be disks with

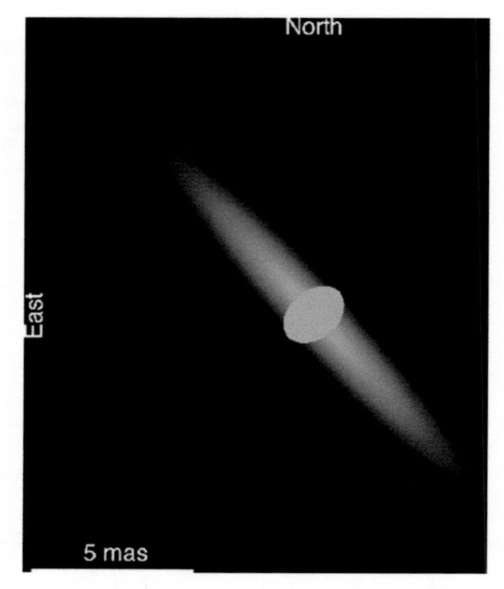

Figure 5. Graphical representation of the best-fit model intensity distribution of Achernar by Kervella & Domiciano (2006)

the BEDISK code (see Fig. 4). They observed a continuum turn-down at mm and radio wavelengths which was explained by a geometrical disk change, i.e. an increase of n at large radial distances from the star, already predicted by Waters *et al.* (1991).

5. Disk geometry: disk and/or wind ?

Recently polar winds along the rotational axis of Be disks were evidenced using interferometric techniques. For instance Kervella & Domiciano (2006) have measured a polar extension on the fast rotating Be star Achernar using the VLTI/VINCI instrument at 1.631 and 2.175 μm (Fig. 5). This wind have an extension of 17.6 $\pm$ 4.9 mas and a relative (envelope to star) near-IR flux contribution of 4.7 $\pm$ 0.3 %. Kanaan *et al.* (2008) used a wind+disk model to interpret the whole set of VLTI/VINCI data of Achernar in the K band and found the these data cannot be explained by a rotationally distorted Be star with a companion alone. A polar wind contribution is mandatory to fit the data. This polar wind contribution, interpreted as a polar wind, does not appear to be linked to the presence of a disk or a ring. Thus, it seems possible to have a wind without a disk in the circumstellar environment of a Be star. Meilland *et al.* (2007a) have also evidenced a dense equatorial (Keplerian) disk with a polar wind along the rotational axis of α Arae using the VLTI/AMBER instrument in the K band (Fig. 6).

Another interesting possibility is to test if the disk formation/dissipation is creating rings around the central star or not. Using theoretical SED, Brγ line profiles and visibilities for two scenarios explaining the disk dissipation: an outburst scenario forming rings or a slowly decreasing mass loss, Meilland *et al.* (2006) have shown that a clear signature of the ring dissipation scenario would be the disappearance of the high velocity tails in the emission lines and a nearly constant peak separation, as well as a clear interferometric signature with an increasing second lobe and a displacement of the first zero of the visibility. Note that a ring like structure was already evidenced by Millour *et al.* (2011)

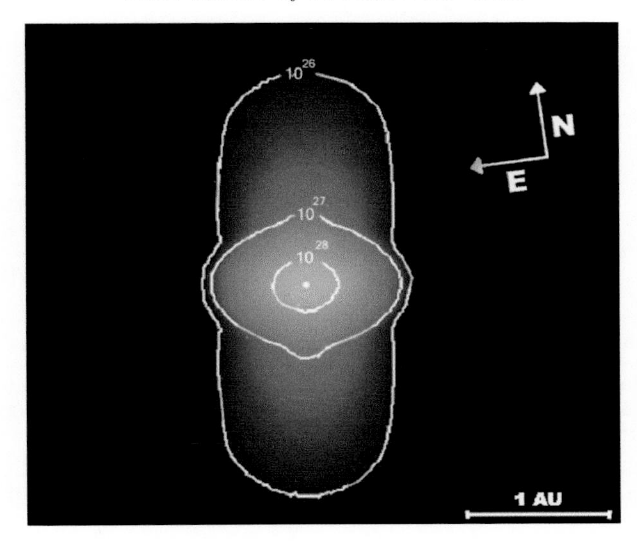

Figure 6. Intensity map in the continuum at 2.15 m obtained with SIMECA from Meilland *et al.* (2007a) best model parameters

around the B[e] star HD 62623 from a direct image reconstruction method without any a priori model from VLTI/AMBER spectrally resolved data.

6. Circumstellar disks: kinematics

Thanks to spectrally resolved interferometric observations, direct measurement of the kinematics with the disk of Be stars was possible. Meilland *et al.* (2007a) was the first to directly determine the Keplerian rotation of the disk around α Arae with the VLTI/ AMBER instrument in the K band. On the other side, same type of measurements were applied for κ CMa by Meilland *et al.* (2007b) who obtained that the rotation within the disk was not Keplerian and $\beta = 0.32$ was found for this star assuming that the rotation law follows $V_\phi(r,\theta) = \sin\theta\ (\frac{R}{r})^\beta$. The kinematics within the disk of the B[e] star HD 62623 was also found to be Keplerian by Millour *et al.* (2011) and the first spectrally resolved reconstructed images were obtained, again without any a priori model from VLTI/AMBER measurements. A comparison between α Arae, κ CMa and Achernar kinematics was presented by Stee & Meilland (2009) and summarized in Table 1.

48 Per and ψ Per were also observed by Delaa *et al.* (2010) with the CHARA/VEGA instrument at 0.65 μm and the rotation within the disk of 48 Per was found to be Keplerian, i.e. $\beta = 0.5$ whereas for ψ Per a $\beta = 0.35$ value was obtained which is very similar to the one obtained for κ CMa, i.e. $\beta = 0.3$. Thus, the departure of the Keplerian rotation might be linked to a density perturbation such as a "one-armed" oscillation and/or a perturbation due to a close enough companion. Note that Keplerian rotation was also found for β CMi's disk by Oudmaijer *et al.* using very high resolution spectroscopic data (see his poster contribution in these proceedings).

Another important parameter to obtain (also needed to constrain the disk kinematics) is the inclination angle of the rotational axis with respect to the line of sight. For that purpose, Mackay *et al.* (2009) have computed theoretical normalized stokes visibilities of Be disks and has shown that if we are able to obtain an accuracy on the interferometric polarization observations better than 10^{-3}–10^{-4}, Stokes Q visibilities can remove model degeneracies and can estimate the disk inclination angle.

Equatorial Disk			
	α Arae	κ CMa	Achernar
Stellar Rotation	97 % V_c	52 % V_c	$\sim V_c$
Radius (K band)	32 $R_\star$	23 $R_\star$ independent of λ	Intermittent $R_{max} \sim 4.8\ R_\star$ if Keplerian
Flux (K band)	40%	50%	< 5% (2002)
Expansion	negligible	negligible	0.2 kms^{-1}
Disk rotation (β parameter)	0.48 (quasi Keplerian)	0.3 (sub Keplerian)	no disk detected in 2000
Polar Wind			
Extension	>10 $R_\star$	not detected	> 10 $R_\star$
Flux (K band)	1-5%	X	3-4% (2002) (H band)
Opening angle	50 $\pm 10^o$	X	5-40°

Table 1. Comparison of the stellar rotation, geometry and kinematics of the circumstellar disk of the 3 studied Be stars α Arae, κ CMa from Stee & Meilland (2009).

7. Disk formation/dissipation: Non Radial Pulsations (NRP) and Outbursts

Observational proof for a direct link between NRP or Outbursts and disk formation is growing. For instance, using multiple epochs Hα spectroscopy of 47 stars in the open cluster NGC 3766 Mc Swain *et al.* (2008) have confirmed that 16 objects were Be stars and discovered one new Be star. No particular stage of main sequence evolution were found but they are all rapid rotators (with rotation at least between 70-80 % of the critical velocity). The have also observed disk size changes which are consistent with NRP ($l = 2$, $m = \pm 2$) for the formation of these highly variable disks. Note that, in this very interesting paper, they were also able to estimate the disk formation rates for many targets. Mennickent *et al.* (2009) have done L-band observations of 13 outbursting Be stars with the ISAAC instrument. They found different groups of stars regarding the Brα, Pfγ and Humphreys lines optical depth. This was consistent with the description by de Wit (2006) for Be stars outbursts in termes of ejection of an optically thick disk that expands and becomes optically thin before dissipation. A large broadening was also observed in the IR lines which was attributed to vertical velocity fields near the central star. The outburst scenario was also investigate by Kanaan *et al.* (2008) and was found to be compatible to interpret spectroscopic observations of Achernar and the formation of an outflowing ring with a radial velocity ~ 0.2 kms^{-1}. Rapid polarization variability of Achernar was also interpreted as evidence of photospheric and circumstellar activity by Carciofi *et al.* (2007). Thanks to a detailed modeling of the polarization they suggest that short term variations originate from discrete mass ejection events which produce transient inhomogeneities in the inner disk. They also explain the observed long term variations as due to the formation of an inner ring following one or several mass ejection events. The same kind of study was done by Wisniewski *et al.* (2010) using spectro-polarimetric observations to study the formation/dissipation of the disk around π Aquarii and 60

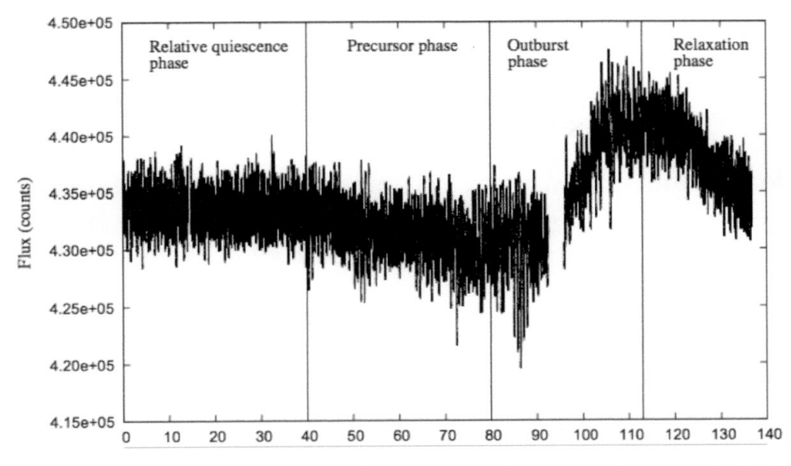

Figure 7. CoRoT light curve of HD 49330 in flux from Huat *et al.* (2009)

Cygni. They observed a minor outburst during a quiescent phase, with similar lifetimes as for μ Cen, which was interpreted as injection and viscous dissipation of individual blobs into the inner disk. Observed deviations from the mean intrinsic polarization angle during polarization outburst may be due to deviations from axisymmetry. These deviations are certainly indicative of injection and circularization of new blobs into the inner disk, in the plane of the disk or in a slightly inclined (non coplanar) orbit.

Nevertheless, the first clear and direct evidence of an ejection of matter that coincides with the constructive beating of two main frequencies during an outburst phase was obtained by Huat *et al.* (2009) and Floquet *et al.* (2009) from the observation of the CoRoT light curve of HD 49330 (see Fig. 7). What is still not clear is the fact that the variations in pulsation modes are producing the outburst, as already predicted by Rivinius *et al.* (1998) or if this is the occurence of the outburst that is producing the excitation of modes, e.g. a change in the stellar structure. Monitoring the precursor phase of the outburst of 28 CMa (October 2008) in photometry, high resolution optical and near IR spectroscopy and interferometry, Stefl *et al.* (2010) have observed cyclic smaller outbursts. These observations seem to be in better agreement with the dynamical model by Okazaki (2007) where the inner disk has (partially) reacreted to the central star.

Regarding possible disk formation/dissipation scenarios, Granada *et al.* (2010) have observed 8 Be stars in the K and L bands with the GEMINI/NIRI instrument. Two distinct groups were found: the first one with optically thin IR lines (28 Tau, 66 Oph, V923 Aql, 28 Cyg) and the second one with optically thin IR lines (88 Her, BK Cam) and one Be star in between: EW Lac. For 28 Tau, 66 Oph and 28 Cyg the Hα weakening observed is consistent with a dissipating disk. For 88 Her and BK Cam the disk seems to be growing certainly due to recent mass ejection episodes. Observed variations in Hα and polarization were interpreted as direct evidence of the spatial motion of the disk and the presence of a precessing disk outside the equatorial plane. The innermost regions of the envelope were found to rotate faster and/or scattering effects must be relevant in these dense regions.

Since many Be stars seem to have permanent disks whereas some others have very variable disks we may wonder if this is due to different physical properties. For that purpose, Mc Swain *et al.* (2009) have studied the disk formation/dissipation in 4 southern open clusters. 296 stars were observed at multiple epochs in Hα. They have detected 12 new transients whereas 17 additional Be stars with relatively stable disks were confirmed.

Thus, they have investigated physical differences between these transient Be stars, stable Be and normal B type stars. Be stars were found to be all faster rotators but no significant physical differences between transient and stable Be stars were evidenced.

8. Disk formation: metallicity, critical rotation and binarity

Metallicity has certainly a great effect on the evolution and the formation of Be star disks as demonstrated by Martayan *et al.* (2010) using Hα slitless spectroscopy of 4437 stars with a survey complete to B2/B3 spectral types on the main sequence. They obtained that in the Small Magellanic Cloud (SMC) Be stars are more frequentby a factor 3-5 and that the fractional critical angular rotation rate Ω/Ω_c is the main parameter governing the occurence of the Be phenomenon. This Be phenomenon seems to cover preferentially the second half of the main sequence evolution. Moreover, considering the evolution, the initial mass and the metallicity effect, Ω/Ω_c is still the main dominant parameter.

Binarity was also very often advocate as THE clue for the Be phenomenon, i.e. the formation of a circumstellar disk in a close binary system. Even if the number of binary Be stars is rapidly growing, especially with recent interferometric detections, and if binarity is not the main physical process to form a Be disk, it has certainly very important effects on the disk size and evolution as we will see in the following examples. For instance Catanzaro (2008) has observed β Cephei, even if we may wonder if β Cephei is really a typical Be star, nevertheless its Hα emission line seems to be typical for Be stars. In fact, this emission is attributed to the secondary component since β Cephei is a spectroscopic binary system. From archived spectra he has recovered the evolution of the Hα line profile od the secondary as a function of time. He found that the typical double-peak emission was strongly variable and was able to estimate the vsini of the secondary to be 230 kms^{-1} and estimate the outer radius of the disk to be 3.28 $R_\star$ (assuming a Keplerian rotation). He also found thaht the Radial Velocity (RV) and V/R variations observed were in agreement with Okazaki (1996) model. Achernar was also classified as "binary" since the detection of a companion at a few tens of AU from VLTI/VISIR measurements by Kervella & Domiciano (2007). The secondary was found to be compatible with a late A type main sequence star with a Position Angle (PA) almost aligned with the equatorial plane of Achernar. The projected linear separation they obtain was 12.3 AU. Last but not least, δ Cen was also detected as a new binary by Meilland *et al.* (2008) using VLTI/AMBER spectro interferometry in the H and K bands. Assuming a radius of 5.9 $R_\odot$, a distance of 121 pc, the measured separation of 68.7 mas they obtained corresponds to 300 $\pm$ 50 $R_\star$. The companion spectral class ranges between B4V and A0III, its mass between 4-7 $M_\odot$ and the Period lower limit estimates was equal to 4.6 years.

9. Conclusion

Finally we may wonder if it is possible to answer to the 5 previous questions ?

• What is the disk geometry ?

A few mas (a few tens of $R_\star$), thin close to the central star but flaring after a few stellar radii. Truncated when the Be star is embedded within a binary system. The opening angle of the disk is still poorly constrained due mainly to the disk inclination uncertainty. Nevertheless, it is certainly NOT as thin as it was previously supposed, i.e. with a 1-2° of opening angle. The disk geometry, as seen during this review, is strongly wavelength (or chemical species) dependent.

• Do we have a disk, a wind or both ?
Certainly both with the possibility to have a wind but no disk. They were strong debates during this IAU meeting regarding the wind detection by interferometric measurements and the fact that this detection was model dependent or not. There is clearly a polar extension in the interferometric data. Is it a wind or something else, the answer is not trivial and the solution may not be unique. On the other side, these interferometric observations are compatible with a polar wind model, which include a NLTE radiative transfer simulation, free-free and free-bound continuum emission. Whether this wind as to be optically thick or not is not, from my point of view, the right question: what we only need is some photons coming out of this wind and reaching the detector regardless the optical thickness of the wind. Moreover, Achernar's wind was observed in the near-IR continuum (K-band) and, if this continuum was optically thick, it would infer that the Balmer lines are also (and even more) optically thick. Since Achernar is seen close to edge-on we would have seen a strong "single-peak" Hα emission line which is not the case since the Hα line at this epoch was very faint and "double-peaked". Does it means that the wind opacity is sufficient to be seen in the near-IR continuum and that the Hα source function is too faint due to a very strong Hydrogen ionization caused by the hotter polar caps ? If the wind is optically thick in the K band it must be very closed to the star, i.e. $\leqslant 1\ R_\star$ and optically thin and extended further away. A nice confirmation of the stellar wind hypothesis would be to observer other Bn stars and try to see with VLTI/AMBER or VEGA/CHARA if there is also an extended emission detected along their polar axis, mainly due to the Von Zeipel effect.

• How is the density and temperature distribution within the disk? Well, this is not very easy to answer. Clearly the n index is $\neq 3.5$ but can change as a function of the stellar radius. The Temperature seems to be cooler in the disk mid-plane and not isothermal. In many cases, this density distribution seems not to be axisymmetric with nice confirmations of the 2D global disk oscillations scenario.

• What is the disk kinematics?
Clearly Keplerian but with some departure. Are these departure linked to these global disk oscillations in some cases or related to a putative companion? The situation for these non Keplerian detections are not very clear and have certainly to be confirmed. Whatever this rotation is, the disk around Be star is clearly dominated by the rotation with only a very small radial velocity (< 1 kms^{-1}).

• How do the disk form and dissipate?
The critical rotation and the critical angular rotation rate Ω/Ω_c together with the outbursts scenario are clearly the most favorable physical processes to create a circumstellar disk.

Acknowledgements

This work has been supported by the French Program National en Physique Stellaire (PNPS) and by the Centre National de la Recherche Scientifique (CNRS). This research has made use of SIMBAD database, operated at CDS, Strasbourg, France and of NASA's Astrophysics Data System.

References

Arias, M. L., Zorec, J., Cidale, L., Ringuelet, A. E. *et al.* 2006, *A&A*, 460, 821

Catanzaro, G. 2008, *MNRAS*, 387, 759

Carciofi, A. C., Miroshnichenko, A. S., Kusakin, A. V., Bjorkman, J. E. *et al.* 2006, *ApJ*, 652, 1617

Carciofi, A. C., Magalhães, A. M., Leister, N. V., Bjorkman, J. E. *et al.* 2007, *ApJ* (Letters), 671, L49
Carciofi, A. C., Okazaki, A. T., Le Bouquin, J.-B., Štefl, S. *et al.* 2009, *A&A*, 504, 915
Delaa, O., Stee, P., Meilland, A. *et al.* 2010, *A&A*, in preparation
de Wit, W. J., Lamers, H. J. G. L. M., Marquette, J. B. & Beaulieu, J. P. 2006, *A&A*, 456, 1027
Floquet, M., Hubert, A.-M., Huat, A.-L., Frémat, Y. *et al.* 2009, *A&A*, 506, 103
Gies, D. R., Bagnuolo, Jr., W. G., Baines, E. K., ten Brummelaar, T. A. *et al.* 2007, *ApJ*, 654, 527
Granada, A., Arias, M. L. & Cidale, L. S. 2010, *AJ*, 139, 1983
Halonen, R. J., Jones, C. E., Sigut, T. A. A., Zavala, R. T. *et al.* 2008, *PASP*, 120, 498
Huat, A.-L., Hubert, A.-M., Baudin, F., Floquet, M. *et al.* 2009, *A&A*, 506, 95
Jones, C. E., Molak, A., Sigut, T. A. A., de Koter, A. *et al.* 2009, *MNRAS*, 392, 383
Jones, C. E., Tycner, C., Sigut, T. A. A., Benson, J.A. *et al.* 2008, *ApJ*, 687, 598
Kanaan, S., Meilland, A., Stee, P., Zorec, J. *et al.* 2008, *A&A*, 486, 785
Kervella, P. & Domiciano de Souza, A. 2006, *A&A*, 453, 1059
Kervella, P. & Domiciano de Souza, A. 2007, *A&A*, 474, L49
Lenorzer, A., de Koter, A. & Waters, L. B. F. M. 2002, *A&A*, 386, L5
Mackay, F. E., Elias, N. M., Jones, C. E. & Sigut, T. A. A. 2009, *ApJ*, 704, 591
Martayan, C., Baade, D. & Fabregat, J. 2010, *A&A*, 509, A11
McSwain, M. V., Huang, W., Gies, D. R., Grundstrom, E. D. *et al.* 2008, *ApJ*, 672, 590
McSwain, M. V., Huang, W. & Gies, D. R. 2009, *ApJ*, 700, 1216
Meilland, A., Stee, P., Zorec, J. & Kanaan, S. 2006, *A&A*, 455, 953
Meilland, A., Stee, P., Vannier, M., Millour, F. *et al.* 2007a, *A&A*, 464, 59
Meilland, A., Millour, F., Stee, P., Domiciano de Souza, A. *et al.* 2007b, *A&A*, 464, 73
Meilland, A., Millour, F., Stee, P., Spang, A. *et al.* 2008, *A&A*, 488, L67
Meilland, A., Stee, P., Chesneau, O. & Jones, C. 2009, *A&A*, 505, 687
Mennickent, R. E., Sabogal, B., Granada, A. & Cidale, L. 2009, *PASP*, 121, 125
Millour, F., Meilland, A., Chesneau, O. *et al.* 2011, *A&A*, 526A, 107
Okazaki, A. T. 1996, *PASJ*, 48, 305
Okazaki, A. T. 1997, *A&A*, 318, 548
Okazaki, A. T. 2007, in: A. T. Okazaki, S. P. Owocki, & S. Stefl (eds.), *Active OB-Stars: Laboratories for Stellar and Circumstellar Physics*, ASP-CS 361, p. 230
Oudmaijer, R. D., Parr, A. M., Baines, D. & Porter, J. M. 2008, *A&A*, 489, 627
Rivinius, T., Baade, D., Stefl, S., Stahl, O. *et al.* 1998, *A&A*, 333, 125
Schaefer, G. H., Gies, D. R., Monnier, J. D., Richardson, N. *et al.* 2010, in: T. Rivinius & M. Curé (eds.), *The Interferometric View on Hot Stars*, Rev. Mexicana AyA Conference Series 38, p. 107
Stee, P. & Meilland, A. 2009, in: J.-P. Rozelot & C. Neiner (eds.), *The Rotation of Sun and Stars*, Lecture Notes in Physics 765 (Berlin Springer Verlag), p. 195
Štefl, S., Rivinius, T., Carciofi, A. C., Le Bouquin, J.-B. *et al.* 2009, *A&A*, 504, 929
Stefl, S., Rivinius, T., Le Bouquin, J.-B., Carciofi, A. *et al.* 2010, in: T. Rivinius & M. Curé (eds.), *The Interferometric View on Hot Stars*, Rev. Mexicana AyA Conference Series 38, p. 89
Tatulli, E., Malbet, F., Ménard, F., Gil, C. *et al.* 2008, *A&A*, 489, 1151
Touhami, Y., Richardson, N. D., Gies, D. R., Schaefer, G. H. *et al.* 2010, *PASP*, 122, 379
Tycner, C., Gilbreath, G. C., Zavala, R. T., Armstrong, J. T. *et al.* 2006, *AJ*, 131, 2710
Tycner, C., Jones, C. E., Sigut, T. A. A., Schmitt, H. R. *et al.* 2008, *ApJ*, 689, 461
Waters, L. B. F. M. 1986, *A&A*, 162, 121
Waters, L. B. F., Marlborough, J. M., van der Veen, W. E. C., Taylor, A. R. *et al.* 1991, *A&A*, 244, 120
Wisniewski, J. P., Kowalski, A. F., Bjorkman, K. S., Bjorkman, J. E. *et al.* 2007, *ApJ* (Letters), 656, L21
Wisniewski, J. P., Draper, Z. H., Bjorkman, K. S., Meade, M. R. *et al.* 2010, *ApJ*, 709, 1306
Zorec, J., Arias, M. L., Cidale, L. & Ringuelet, A. E. 2007, *A&A*, 470, 239

Active OB Stars: structure, evolution, mass loss, and critical limits
Proceedings IAU Symposium No. 272, 2010
C. Neiner, G. Wade, G. Meynet & G. Peters, eds.

doi:10.1017/S1743921311010738

The circumstellar discs of Be stars

Alex C. Carciofi

Instituto de Astronomia, Geofísica e Ciências Atmosféricas,
Universidade de São Paulo, Rua do Matão 1226, Cidade Universitária,
05508-900, São Paulo, SP, BRAZIL email: carciofi@usp.br

Abstract. Circumstellar discs of Be stars are thought to be formed from material ejected from a fast-spinning central star. This material possesses large amounts of angular momentum and settles in a quasi-Keplerian orbit around the star. This simple description outlines the basic issues that a successful disc theory must address: 1) What is the mechanism responsible for the mass ejection? 2) What is the final configuration of the material? 3) How the disc grows? With the very high angular resolution that can be achieved with modern interferometers operating in the optical and infrared we can now resolve the photosphere and immediate vicinity of nearby Be stars. Those observations are able to provide very stringent tests for our ideas about the physical processes operating in those objects. This paper discusses the basic hydrodynamics of viscous decretion discs around Be stars. The model predictions are quantitatively compared to observations, demonstrating that the viscous decretion scenario is currently the most viable theory to explain the discs around Be stars.

1. Introduction

Recent years witnessed an important progress in our understanding of the circumstellar discs of Be stars, largely due to interferometric observations capable of angularly resolve those objects at the milliarcsecond (mas) level (see Stee, these proceedings, for a review).

In the late nineties, discs around Be stars were considered to be equatorially enhanced outflowing winds, and several models and mechanisms to drive the outflow were proposed (see Bjorkman 2000, for a review). The much stronger observational constraints available today allow us to rule out several theoretical scenarios that were proposed in the past (e.g. the wind compressed disc models of Bjorkman & Cassinelli 1993). As an example, spectrointerfometry and spectroastrometry have directly probed the disc kinematics (Meilland *et al.* 2007a, Štefl *et al.*, these proceedings, Oudmaijer *et al.*, these proceedings), revealing that Be discs rotate very close to Keplerian. As a result, the viscous decretion scenario, proposed originally by Lee *et al.* (1991) and further developed by Porter (1999), Okazaki (2001), Bjorkman & Carciofi (2005), among others, has emerged as the most viable scenario to explain the observed properties of Be discs.

2. Discs Diagnostics

Before reviewing the basic aspects of the theory of circumstellar discs, it is useful to put in perspective what observations tells us about the structure and kinematics of those discs.

2.1. *The formation loci of different observables*

One important issue to consider when analyzing observations is that different observables probe different regions of the disc; it is therefore useful to be able to make a correspondence between a given observable (say, the continuum flux level at a given wavelength) and the part of the disc whence it comes.

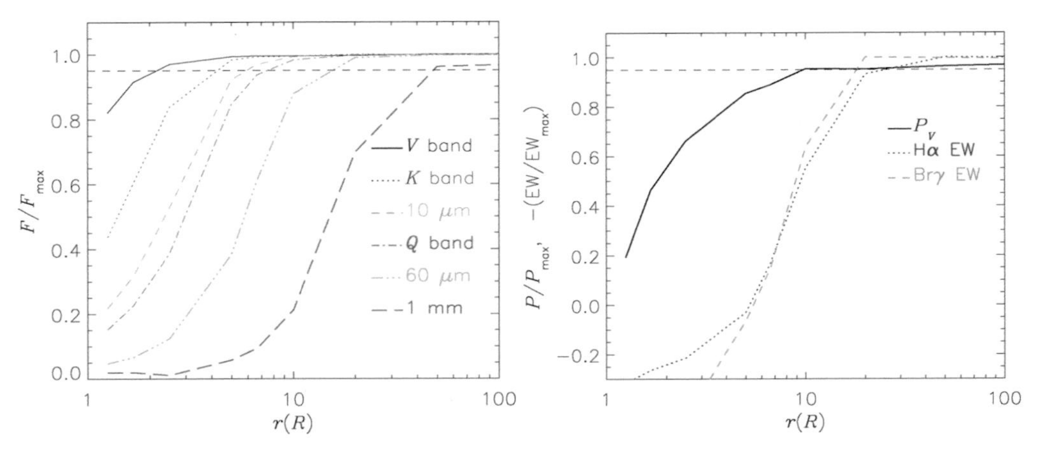

Figure 1. The formation loci of different observables. The calculations assume a rapidly rotating B1Ve star ($T_{\rm eff}^{\rm pole} = 25\,000$ K, $\Omega/\Omega_{\rm crit} = 0.92$) surrounded by viscous decretion discs with different sizes. The results correspond to a viewing angle of 30°. *Left: continuum emission.* Plotted is the ratio between the observed flux to the maximum flux, $F_{\rm max}$, as a function of the disc outer radius. $F_{\rm max}$ corresponds to the flux of a model with a disc outer radius of 1000 R. *Right: polarization and line equivalent width.* Calculations were carried out with the HDUST code (Carciofi & Bjorkman 2006).

To make such a correspondence, we consider a typical Be star of spectral type B1Ve and calculate the emergent spectrum arising from viscous decretion discs of different outer radii (see Sect. 3 for a detailed description of the structure of viscous discs). In Fig. 1 we present results for continuum emission, continuum polarization and the equivalent width of emission lines.

Let us discuss first the continuum and line emission. The reason the spectra of Be stars show continuum excess or emission lines is that the dense disc material acts as a pseudo-photosphere that is much larger than the stellar photosphere. For the pole-on case (inclination $i = 0$), the effective radius of the pseudo-photosphere is the location where the vertical optical depth $\tau_\lambda(R_{\rm eff}) = 1$. For continuum Hydrogen absorption, the opacity depends on the wavelength as $\kappa_\lambda \propto \lambda^2$ (if one neglects the wavelength dependence of the absorption gaunt factors) and thus the size of the pseudo-photosphere increases with wavelength (see Carciofi & Bjorkman 2006, appendix A, for a derivation of $R_{\rm eff}(\lambda)$).

In Fig. 1, left panel, the intersection of the different lines (each corresponding to a given wavelength) with the horizontal dashed line marks the position in the disc whence about 95% of the continuum excess comes. For instance, from the Figure we see that the V band excess is formed very close to the star, within about 2 R, whereas the excess at 1 mm originates from a much larger volume of the disc. As we shall see below, the fact that the continuum excess flux at visible wavelengths forms so close to the star makes such observations indispensable to study systems with complex temporal evolution because the V band continuum emission has the fastest response to disc changes as a result of photospheric activity.

In the right panel of Fig. 1 we show how the line emission and continuum polarization grows with radius. For the optically thick Hα and Brγ lines the disc emission only fills in the photospheric absoprtion profile when the disc size is about 5 R. Both lines have pseudo-photospheres that extend up to about 20 R. For polarization, the results are at odds with the common belief that polarization is formed close to the star; it is seen that 95% of the maximum polarization is only reached when the disc size is about 10 R.

2.2. *Disc Thickness*

In a nice example of the diagnostics potential of interferometric and polarimetric observations combined, Wood, Bjorkman, & Bjorkman (1997) and Quirrenbach *et al.* (1997) showed that the circumstellar disc of ζ Tau is geometrically thin (opening angle of 2.5°). This result was confirmed by Carciofi *et al.* (2009) from a more detailed analysis. Other studies based on the fraction of Be-shell stars vs. Be stars typically find larger values for the opening angle (e.g. 13° from Hanuschik 1996) but this discrepancy is accommodated by the fact that the geometrical thickness of Be discs increases with radius (flared disc) and different observables probe different disc regions (Fig. 1). In any case, there is no doubt that the Be discs are flat, geometrically thin structures.

2.3. *Disc density distribution*

Models derived both empirically (e.g., Waters 1986) and theoretically (e.g., Okazaki 2001 and Bjorkman & Carciofi 2005) usually predict (or assume) a power-law fall off of the disc density [$\rho(r) \propto r^{-n}$]. Values for n vary widely in the literature, typically in the range $2 < n < 4$. One question that arises is whether this quoted range for the density slope is real or not.

Below we see that the viscous decretion disc model, in its simplest form of an isothermal and isolated disc, predicts a slope of $n = 3.5$ for the disc density. On one hand, the inclusion of other physical effects can change the value of the above slope. For instance, non-isothermal viscous diffusion results in much more complex density distribution that cannot be well represented by a simple power-law, and tidal effects by a close binary may make the density slope shallower (Okazaki *et al.* 2002). On the other hand, it is also true that many of quoted values for n in the literature are heavily influenced by the assumptions and methods used in the analysis and should be viewed with caution.

2.4. *Disc dynamics*

Since the earliest detections of Be stars, it became clear that a rotating circumstellar disc-like material was the most natural explanation for the observed double-peaked profile of emission lines, but the problem of how the disc rotates has been an open issue until recently.

The disc rotation law is profoundly linked with the disc formation mechanism; therefore, determining observationally the disc dynamics is of great interest. Typically, one can envisage three limiting cases for the radial dependence of the azimuthal component of the velocity, v_ϕ, depending on the forces acting on the disc material

$$v_\phi = \begin{cases} V_{\rm rot}(r/R)^{-1} & \text{radiatively driven outflow,} \\ V_\star(r/R) & \text{magnetically dominated disc,} \\ V_{\rm crit}(R/r)^{1/2} & \text{disc driven by viscosity.} \end{cases} \tag{2.1}$$

In the first case (radiatively driven outflow) the dominant force on the material is the radially directed radiation pressure that does not exert torques and thus conserves angular momentum ($V_{\rm rot}$ is thus the rotation velocity of the material when it left the stellar surface). The second case corresponds to a rigidly rotating magnestosphere a la σ Ori E (Townsend *et al.* 2005), in which the plasma is forced to rotate at the same speed as the magnetic field lines ($V_\star$). The last case corresponds to Keplerian orbital rotation, written in terms of the critical velocity, $V_{\rm crit} \equiv (GM/R)^{1/2}$, which is the Keplerian orbital speed at the stellar surface. This case requires a fine-tunning mechanism such that *the centrifugal force, v_ϕ^2/r, exactly balances gravity at all radii.* As we shall see below, viscosity does provide the fine-tunning mechanism capable of producing a Keplerian disc.

Porter & Rivinius (2003) reviewed the then existing observational constraints on the disc kinematics and concluded that "all of the kinematic evidence seems to point to a disc velocity field dominated by rotation, with little or no radial flow, at least in the regions where the kinematic signatures of emission and absorption are significant". Today, spectrointerferometry and spectroastrometry provides clear-cut evidence that, in most systems observed and analysed so far (κ CMa being the only possible exception, Meilland *et al.* 2007b), the discs rotate in a Keplerian fashion (Meilland *et al.* 2007b, Štefl *et al.*, these proceedings, Oudmaijer *et al.*, these proceedings). This is an important result, seeing that it indicates that viscosity is the driving mechanism of the outflow.

2.5. *Cyclic V/R variations*

About 2/3 of the Be stars present the so-called V/R variations, a phenomenon characterised by the quasi-cyclic variation in the ratio between the violet and red emission peaks of the H I emission lines. These variations are generally explained by global oscillations in the circumstellar disc forming a one-armed spiral density pattern that precesses around the star with a period of a few years (Kato 1983, Okazaki 1991, Okazaki 1997).

Recently, Carciofi *et al.* (2009) provided a quantitative verification of the global disc oscillation theory from a detailed modelling of high-angular resolution AMBER data of the Be star ζ Tau. From a theoretical perspective, the existence of density waves in rotating discs, as suggested by Kato (1983), imposes most stringent constraints on the rotation velocity, since mode confinement requires that the rotation law be Keplerian within about 1% and the radial flow be hightly subsonic ($\lesssim 0.01\ c_s$, Okazaki 2007).

2.6. *Long-term variations*

One of the most intriguing types of variability observed in the Be stars is the aperiodic transition between a normal B phase (discless phase) and a Be phase whereby the disc is lost and rebuilt in timescales of months to years (e.g., Clark *et al.* 2003, Štefl *et al.*, these proceedings). The varying amount of circumstellar gas manifests itself as changes in line profiles (Clark *et al.* 2003), continuum brightness and colors (Harmanec 1983) and polarization (Draper *et al.*, these proceedings). A fine example of the secular process of disc formation and dissipation, and its effects on the continuum brightness and colors, is shown in Fig. 2: the outburst phase responsible for the disc build-up lasted about 300 hundred days (phases I and II, during which the star got brighter and redder) and was followed by a quiescent phase of about 500 days (phases III and IV, during which the star slowly went back to its original appearance as the previously built disc dissipated). As discussed below, the timescales involved in the disc formation and dissipation are generally consistent with the timescales of viscous diffusion.

2.7. *Short-term variations*

Small-scale, short-term variations are quite common in Be stars, and possess a complex phenomenology (see Rivinius 2007, for a review). On one hand, some observed variations (e.g. line profile variations due to non-radial pulsations) are associated with the photosphere proper; others, on the other hand, are thought to originate from the very base of the disc and are, therefore, the manifestation of the physical process(es) that is (are) feeding the disc (e.g. short-term V/R variations of emission lines). To date, μ Cen remains the only system in which the ejection mechanism have been unambiguously identified (in this case non-radial pulsations, Rivinius *et al.* 1998). In the viscous diffusion theory outlined below, it is assumed that matter is ejected by the star and deposited at the inner boundary of the disc with Keplerian or super-Keplerian speeds; clearly, the current

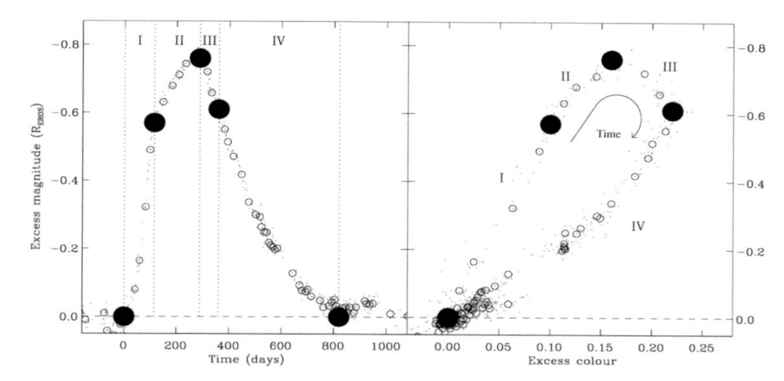

Figure 2. Light curve (*left panel*) and CMD (*right panel*) of the variation of the star OGLE 005209.92-731820.4 (de Wit 2006, reproduced with permission).

status of the theory is still unsatisfactory inasmuch as the fundamental link between the disc and the photosphere proper (i.e., the feeding mechanism) is still unknown.

Studies of short-term variations associated with the disc (e.g., Rivinius *et al.* 1998) are generally consistent with the following scenario: mass injection by some mechanism that causes a transient asymmetry in the inner disc that manifests itself by, e.g., short-term V/R variations, followed by circularization and dissipation in a viscous scale (see below). Here, again, viscosity seems to be playing a major role in shaping the evolution of ejecta.

3. Viscous Decretion Disc Models

The observational properties outlined above give strong hints as to the ingredients a successful theory for the structure of the Be discs must have. The theory must explain the keplerian rotation, slow outflow speeds, small geometrical thickness and must also account for the timescales of disc build-up and dissipation.

The only theory to date that satisfies those requirements is the viscous decretion disc model (Lee *et al.* 1991). This model is essentially the same as that employed for protostellar discs (Pringle 1981), the primary difference being that Be discs are outowing, while pre- main-sequence discs are inflowing. In this model, it is supposed that some yet unknown mechanism injects material at the Keplerian orbital speed into the base of the disc. Eddy/turbulent viscosity then transports angular momentum outward from the inner boundary of the disc (note that this requires a continual injection of angular momentum into the base of the disc). If the radial density gradient is steep enough, angular momentum is added to the individual fluid elements and they slowly move outward. To critically test such decretion disc models of Be stars against observations, we must determine the structure of the disc from hydrodynamical considerations.

3.1. *Viscous discs fed by constant decretion rates*

There are many solutions available for the case of a viscous disc fed by a constant decretion rate, all agreeing in their essentials. Below we outline the mains steps of one possible derivation (Bjorkman 1997) to the goal of discussing the properties of the solution and confronting them to the observations. More detailed analyses as well as different approaches to the problem can be found in Lee *et al.* (1991), Okazaki (2001), Bjorkman & Carciofi (2005), Okazaki (2007), Jones *et al.* (2008) and Carciofi & Bjorkman (2008).

3.2. *Hydrostatic Structure*

Our goal is to write and solve the Navier Stokes fluid equations in cylindrical coordinates (ϖ, ϕ, z) which in the steady-state case have the following general form

$$\frac{1}{\varpi}\frac{\partial}{\partial\varpi}(\varpi\rho v_\varpi) + \frac{1}{\varpi}\frac{\partial}{\partial\phi}(\rho v_\phi) + \frac{\partial}{\partial z}(\rho v_z) = 0, \tag{3.1}$$

$$v_\varpi\frac{\partial v_\varpi}{\partial\varpi} + \frac{v_\phi}{\varpi}\frac{\partial v_\varpi}{\partial\phi} + v_z\frac{\partial v_\varpi}{\partial z} - \frac{v_\phi^2}{\varpi} = -\frac{1}{\rho}\frac{\partial P}{\partial\varpi} + f_\varpi, \tag{3.2}$$

$$v_\varpi\frac{\partial v_\phi}{\partial\varpi} + \frac{v_\phi}{\varpi}\frac{\partial v_\phi}{\partial\phi} + v_z\frac{\partial v_\phi}{\partial z} + \frac{v_\varpi v_\phi}{\varpi} = -\frac{1}{\rho\varpi}\frac{\partial P}{\partial\phi} + f_\phi, \tag{3.3}$$

$$v_\varpi\frac{\partial v_z}{\partial\varpi} + \frac{v_\phi}{\varpi}\frac{\partial v_z}{\partial\phi} + v_z\frac{\partial v_z}{\partial z} = -\frac{1}{\rho}\frac{\partial P}{\partial z} + f_z, \tag{3.4}$$

where ρ is the gas mass density, P is the pressure and f_ϖ, f_ϕ and f_z are the components of the external forces acting on the gas.

Let us initially ignore any viscous effects (inviscid disc) and assume that the only force acting on the gas is gravity. If we further assume circular orbits — $v_\varpi = 0$, $v_\phi \neq 0$, and $v_z = 0$, an assumption that will be droped later on when viscosity is included — the only non-trivial fluid equations are the ϖ- and z-momentum equations (Eqs. 3.2 and 3.4), which take the form

$$\frac{1}{\rho}\frac{\partial P}{\partial\varpi} = \frac{v_\phi^2}{\varpi} + f_\varpi, \tag{3.5}$$

$$\frac{1}{\rho}\frac{\partial P}{\partial z} = f_z, \tag{3.6}$$

where the external force components are given by the gravity of the spherical central star

$$f_\varpi = -\frac{GM\varpi}{(\varpi^2 + z^2)^{3/2}}, \tag{3.7}$$

$$f_z = -\frac{GMz}{(\varpi^2 + z^2)^{3/2}}. \tag{3.8}$$

To specify the pressure, we introduce the equation of state, $P = c_s^2\rho$, where $c_s = (kT)^{1/2}(\mu m_\mathrm{H})^{-1/2}$. In this last expression, k is the Boltzmann constant, μ is the gas molecular weight and m_H is the mass of the hydrogen atom.

In the thin-disc limit ($z \ll \varpi$), we obtain

$$v_\phi = V_\mathrm{crit}\,(R/\varpi)^{1/2}, \tag{3.9}$$

$$\frac{\partial\ln(c_s^2\rho)}{\partial z} = -\frac{V_\mathrm{crit}^2 Rz}{c_s^2\varpi^3}. \tag{3.10}$$

The above equations mean that the disc rotates at the Keplerian orbital speed and is hydrostatically supported in the vertical direction. To determine the vertical disc structure we must solve Eq. (3.10). Assuming an isothermal disc one obtains

$$\rho(\varpi, z) = \rho_0(\varpi)\exp\left[-0.5(z/H)^2\right], \tag{3.11}$$

where ρ_0 the disc density at the mid-plane ($z = 0$), and the disc scale height is given by

$$H(\varpi) = (c_s/v_\phi)\varpi. \tag{3.12}$$

Since $v_\phi \propto \varpi^{-0.5}$, we obtain the familiar result that for an isothermal disc the scaleheight grows with distance from the star as $H \propto \varpi^{1.5}$.

As we shall see below, it is useful to express $\rho(\varpi, z)$ in terms of the disc surface density, Σ, written as

$$\Sigma(\varpi) = \int_{-\infty}^{\infty} \rho(\varpi, z)\, dz = \sqrt{2\pi} H \rho_0. \tag{3.13}$$

Thus

$$\rho(\varpi, z) = \frac{\Sigma(\varpi)}{\sqrt{2\pi} H(\varpi)} \exp\left[-0.5(z/H)^2\right]. \tag{3.14}$$

3.3. *Viscous Outflow*

Clearly, in Eq. (3.14) the disc density scale, $\Sigma(\varpi)$, is completely undetermined, because for a inviscid Keplerian disc we can choose to put an arbitrary amount of material at a given radius. To set the density scale, we must include a mechanism — viscous diffusion — to transport material from the star outwards.

Viscous flows grow in a viscous diffusion timescale

$$\tau_{\rm diff} = \varpi^2/\nu, \tag{3.15}$$

where ν is the kinematic viscosity. One problem that has already been noted long ago (Shakura & Sunyaev 1973) is that for molecular viscosity the diffusion timescale is much too long. Shakura & Sunyaev (1973) appealed instead to the so-called eddy (or turbulent) viscosity, which they parameterized as

$$\nu = \alpha c_s H, \tag{3.16}$$

where $0 < \alpha < 1$. The α parameter describes the ratio of the product of the turbulent eddy size and speed to the product of disc scaleheight and sound speed. In other words, it is assumed that the largest eddies can be at most about the size of the disc scaleheight and that the turnover velocity of the eddies cannot be arger than the sound speed (otherwise, the turbulence would be supersonic and the eddies would fragment into a series of shocks). With this value of the viscosity, the viscous diffusion timescale becomes

$$\tau_{\rm diff} = \frac{V_{\rm crit}}{\alpha c_s^2} \sqrt{\varpi R} \tag{3.17}$$

$$\approx 20{\rm yr} \left(\frac{0.01}{\alpha}\right) \sqrt{\frac{\varpi}{R}}. \tag{3.18}$$

Studies of the formation and dissipation of the discs around Be stars find typical time scales of months to a few years (e.g., Wisniewski *et al.* 2010); therefore, an α of the order of 0.1 or larger is required to match the observed timescales (see Sect. 4).

If we add viscosity to the fluid equations, we can still assume that the disc is axisymmetric and that the vertical structure is hydrostatic ($v_z = 0$). However, the presence of an outflow implies that $v_\varpi \neq 0$. The ϖ- and z-momentum equations are the same as before, so v_ϕ and ρ are the same as in the pure Keplerian case [eqs. (3.9) and (3.14)].

Two fluid equations remain to be solved, the continuity equation, Eq. (3.1), and the ϕ-momentum equation, Eq. (3.3). The continuity equation, written in terms of the surface density,

$$\frac{\partial}{\partial \varpi}(2\pi\varpi\Sigma v_\varpi) = 0 \tag{3.19}$$

means that the mass decretion rate, $\dot{M} \equiv 2\pi\varpi\Sigma v_\varpi$, is a constant (independent of ϖ).

The viscous outflow speed is given by

$$v_{\varpi} = \frac{\dot{M}}{2\pi\varpi\Sigma}. \tag{3.20}$$

This ϕ-momentum equation now is more complicated because viscosity exerts a torque, which is described by the viscous shear stress tensor, $\pi_{\varpi\phi}$. Including this shear stress, the ϕ-momentum equation becomes

$$v_{\varpi}\frac{\partial v_{\phi}}{\partial\varpi} + \frac{v_{\varpi}v_{\phi}}{\varpi} = \frac{1}{\rho\varpi^2}\frac{\partial}{\partial\varpi}(\varpi^2\pi_{\varpi\phi}), \tag{3.21}$$

where

$$\pi_{\varpi\phi} = \nu\rho\varpi\frac{\partial(v_{\phi}/\varpi)}{\partial\varpi} = -\frac{3}{2}\alpha c_s^2\rho. \tag{3.22}$$

Multiplying Eq. (3.21) by $\rho\varpi^2$ and integrating over ϕ and z, we find

$$\dot{M}\frac{\partial}{\partial\varpi}(\varpi v_{\phi}) = \frac{\partial}{\partial\varpi}(\mathcal{T}), \tag{3.23}$$

where

$$\mathcal{T} = \int_{-\infty}^{\infty}\varpi\pi_{\varpi\phi}2\pi\varpi dz = -3\pi\alpha c_s^2\varpi^2\Sigma \tag{3.24}$$

is the viscous torque. The ϕ-momentum equation, Eq. (3.21), expresses the fact that the change in the angular momentum flux — ϖv_{ϕ} being the specific angular momentum — is given by the gradient of the viscous torque. Since the continuity equation implies that $\dot{M}$ is constant, we integrate Eq. (3.21) over ϖ to obtain

$$\mathcal{T}(\varpi) = \dot{M}V_{\text{crit}}\sqrt{\varpi R} + \text{constant}. \tag{3.25}$$

Substituting Eq. (3.24) into Eq. (3.25) and solving for the surface density we find

$$\Sigma(\varpi) = \frac{\dot{M}}{3\pi\alpha c_s^2}\left(\frac{GM}{\varpi^3}\right)^{1/2}\left[(R_0/\varpi)^{1/2} - 1\right]. \tag{3.26}$$

Eq. (3.26) describes the surface density of an unbounded disc (i.e. a disc that is allowed to grow indefinitely). The integration constant R_0 is a parameter that depends on the integration constant of Eq. (3.25) and is related to the physical size of the disc; for time-dependent models, such as those of Okazaki (2007), R_0 grows with time and thus R_0 is related with the age of the disc.

3.4. *Properties of the Solution*

We have now completed the hydrodynamic description of an isothermal, unbounded viscous decretion disc. To fully determine the problem one must specify the decretion rate, $\dot{M}$, the value of α, and the disc age (or size). Assuming that the disc is sufficiently old, $R_0 \gg R$, in which case Eq. (3.26) becomes a simple power-law with radius, $\Sigma(\varpi) \propto \varpi^{-2}$. From Eqs. (3.14) and (3.12) we obtain that the isothermal disc density profile is quite steep, $\rho \propto \varpi^{-3.5}$.

Another important property of isothermal viscous discs can be readily derived from Eq. (3.12). Since v_{ϕ} is much larger than the sound speed (the former is of the order of several hundreds of km/s whereas the later is a few tens of km/s), the disc scaleheight is small compared to the stellar radius, i.e., the disc is geometrically thin. Finally, from Eq. (3.20) we find that, for large discs, the radial velocity is a linear function of the radial distance, $v_{\varpi} \propto \varpi$.

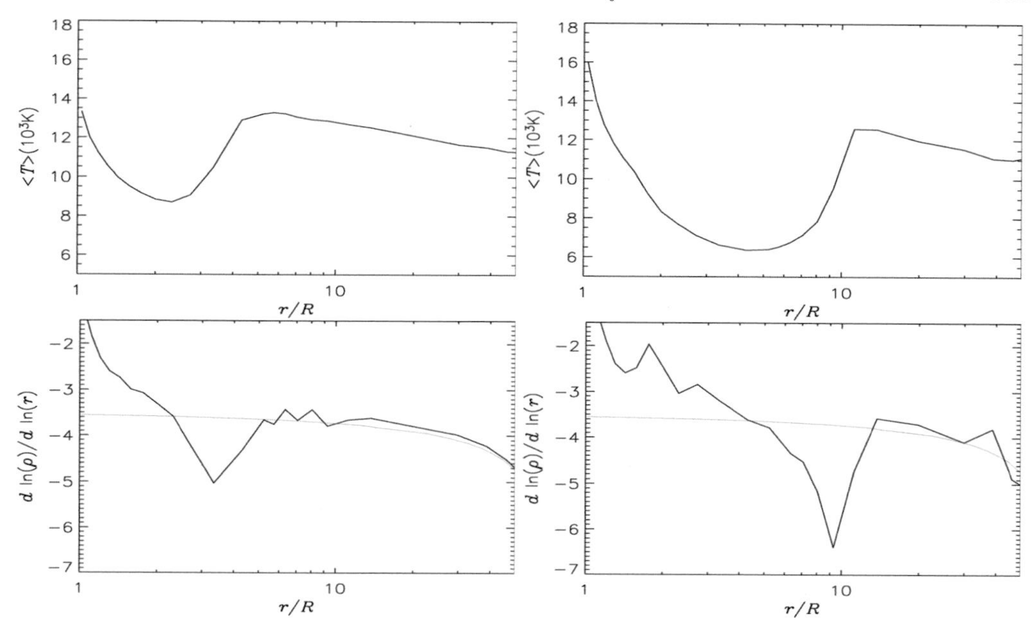

Figure 3. *Top*: Vertically averaged disc temperature. *Bottom*: Local power-law index of the density profile. *Left:* Results for a low-density model ($\rho_0 = 4.2 \times 10^{-12}$ g cm^{-3}). *Right:* Results for a high-density model ($\rho_0 = 3.0 \times 10^{-11}$ g cm^{-3}). Adapted from Carciofi & Bjorkman (2008).

3.5. *Temperature structure*

The temperature structure of viscous discs were investigated by Carciofi & Bjorkman (2006) and Sigut & Jones (2007). They found that those discs are highly nonisothermal, mainly on their denser inner parts. An example of the temperature structure is shown in Fig. 3. The temperature initially drops very quickly close to the stellar photosphere; when the disc becomes optically thin vertically the temperature rises back to the optically thin radiative equilibrium temperature, which is approximately constant, as in the winds of hot stars. The density is the most important factor that controls the temperature structure: for high density models the amplitude of the temperature variations is much larger and the nonisothermal region extends much farther out into the disc.

3.6. *Non-isothermal effects on the disc structure*

From Eqs. (3.14) and (3.26) we see that both the viscous diffusion and the vertical hydrostatic solutions depend on the gas temperature; one can expect, therefore, that the complex temperature structure of the disc might have effects on the disc density structure. Carciofi & Bjorkman (2008) and Sigut *et al.* (2009) calculated the vertical density structure of a disc in consistent vertical hydrostatic equilibrium. They found that the temperature decrease causes the disc to collapse, becoming much thinner in the inner regions. This collapse redistributes the disc material toward the equator, increasing the midplane density by a factor of up to 3 relative to an equivalent isothermal model.

Carciofi & Bjorkman (2008) investigated, in addition, how the temperature affects the viscous diffusion. The combination of the radial temperature structure, disc scaleheight, and viscous transport produces a complex radial dependence for the disc density that departs very much from the simple $n = 3.5$ power-law. As shown in Fig. 3, the equivalent radial density exponent varies between $n = 2$ in the inner disc to $n = 5$ near the temperature minimum, eventually rising back to the isothermal value $n = 3.5$ in the outer disc.

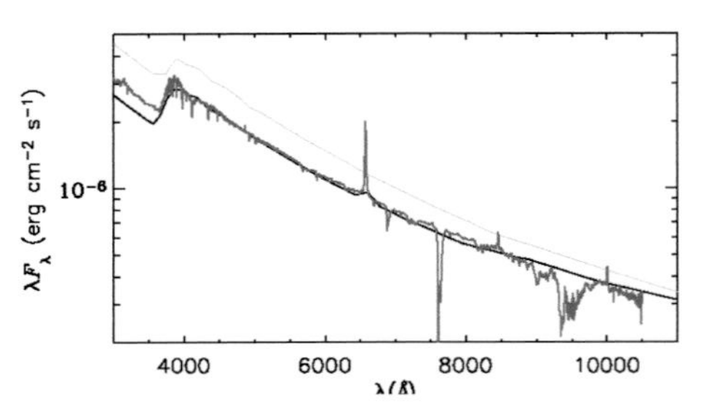

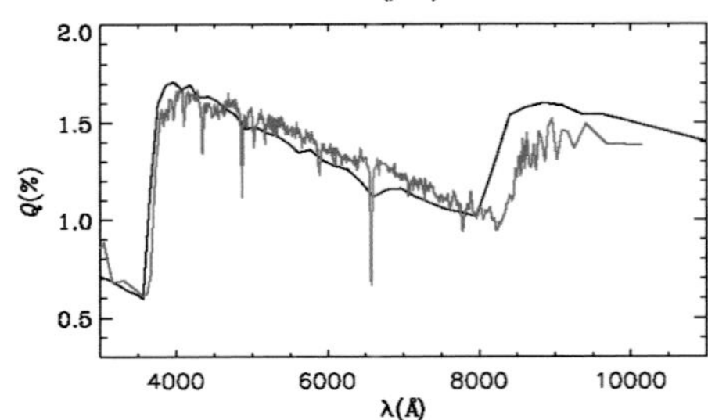

Figure 4. *Left:* Fit to the SED of ζ Tau. The dark grey lines are the observations and the black lines the model results. The light grey line corresponds to the unattenuated stellar SED and gives a measure of how the disc affects the emergent flux. Adapted from Carciofi *et al.* (2009).

We conclude that non-isothermal effects on the viscous diffusion may account for the at least part of the large scatter of the index n reported in the literature (Sect. 2.3).

3.7. *Two test cases: ζ Tau and χ Oph*

The model described above makes several predictions about the disc structure that are in qualitative agreement with the observations (Sect. 2): viscous discs are geometrically thin, rotate in a near keplerian fashion ($v_\phi \propto \varpi^{-1/2}$ and $v_\varpi \ll v_\phi$) and have a steep density fall-off ($n = -3.5$ in the isothermal case). This model and its predictions must now be quantitatively compared to observations.

A successful verification of the viscous decretion disc model have been obtained in the case of the Be star ζ Tau (Carciofi *et al.* 2009). This star is particularly suitable for such study because it had shown little or no secular evolution in the past 18 years or so (Štefl *et al.* 2009); therefore, a constant mass decretion rate, as assumed above when deriving the disc structure, is a good approximation for this system. Some of the results obtained for ζ Tau are shown in Fig. 4. The radial dependence of the disc density, temperature, and opening angle all affect the slope of the visible and IR SED (Waters 1986, Carciofi & Bjorkman 2006), as well as the shape of the intrinsic polarization. Therefore, the fact that the model reproduces the detailed shapes of SED and spectropolarimetry represents a non-trivial test of the Keplerian decretion disc model.

Using a somewhat different model, Tycner *et al.* (2008) successfully fitted several observations of the Be star χ Oph, including high angular resolution interferometry. In their modeling the index n of the density power-law is a free parameter. Interestingly, their best fitting model was for a much flatter density law ($n = 2.5$). Whether this flatter density profile can be accommodated by including some other physical process in addition to viscosity (e.g., tidal effects by a binary) or is an indication that the viscous decretion disc model is not a good model for this system remains to be verified by further analysis.

4. Viscous Discs Fed by Non-constant Decretion Rates

All the discussion so far have been focused on the problem of a disc fed by a constant decretion rate that grows to a given size in a viscous timescale. Clearly, those models can only be applied to objects, such as ζ Tau, that went through a sufficiently long and stable decretion phase. A different approach is needed if one wants to investigate dynamically active systems such as the one shown in Fig. 2.

Okazaki (2007), Haubois *et al.* (these proceedings), and Jones *et al.* (2008) described the solution of the viscous diffusion problem for systems with non-constant mass decretion rates. Fig. 5 shows a series of models of the lightcurve of the Be star 28 CMa

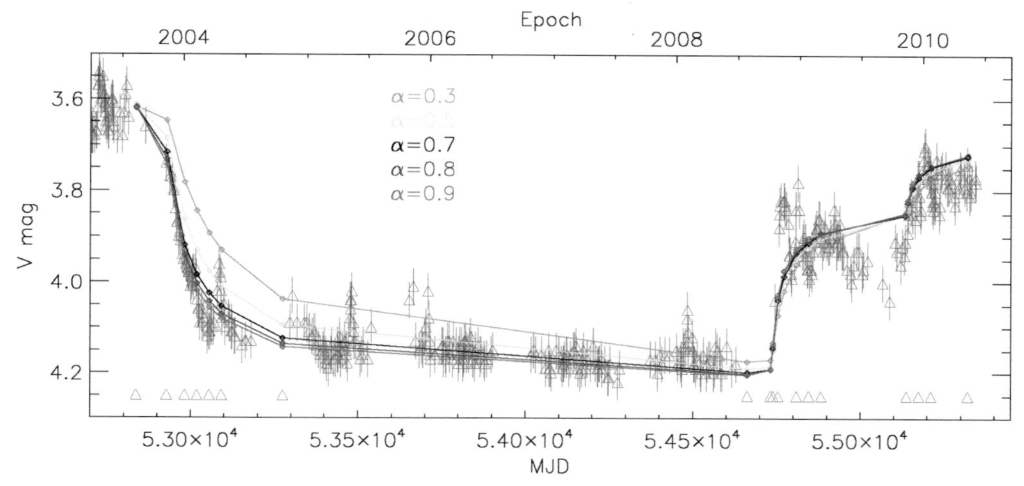

Figure 5. The visual light curve of 28 CMa fitted with a dynamical viscous decretion disc model with different values of the viscosity parameter, α (Štefl *et al.*, these proceedings)

(Carciofi et al., in prep.), that uses the computer code SINGLEBE by Atsuo Okazaki to solve the 1D viscous diffusion problem and the HDUST code to calculate the emergent spectrum (Haubois *et al.*, these proceedings). The simulation begins after a few years of disc evolution to account for the previous disc build-up. At $t = 2003.6$ mass decretion is turned off and the system evolves passively until $t = 2008.8$, when the recent outburst started (Štefl *et al.*, these proceedings). The model with $\alpha = 0.9$ is the one that reproduces best the lightcurve at all phases. This is, to our knowledge, the first time the viscosity parameter is determined for a Be star disc.

5. Conclusions

This paper discussed the basic hydrodynamics that determine the structure of viscous decretion disc models. Those are, to date, the most satisfactory models for Be discs because they can account quantitatively for most of their observational properties, namely the fact that they are geometrically thin, have Keplerian rotation and steep radial density profiles. It has been recently shown that the viscous decretion disc model can also explain the temporal evolution of dynamically active systems such as 28 CMa. Studies of such system allows for the determination of the viscosity parameter α and the mass decretion rate, $\dot{M}$, two quantities that are very difficult to determine otherwise.

Acknowledgements

The author acknowledges support from FAPESP grant 2010/08809-5 and CNPq grant 308985/2009-5.

References

Bjorkman, J. E. & Cassinelli, J. P. 1993, *ApJ*, 409, 429

Bjorkman, J. E. 1997, in: J. P. de Greve, R. Blomme & H. Hensberge (eds.), *Stellar Atmospheres: Theory and Observations*, LNP (New York: Springer) 497, 239

Bjorkman, J. E. 2000, in: *IAU Colloq. 175: The Be Phenomenon in Early-Type Stars*, ASP-CS 214, 435

Bjorkman, J. E. & Carciofi, A. C. 2005, in: R. Ignace & K. G. Gayley (eds.), *The Nature and Evolution of Disks Around Hot Stars*, ASP-CS 337, p. 75
Carciofi, A. C. & Bjorkman, J. E. 2006, *ApJ*, 639, 1081
Carciofi, A. C. & Bjorkman, J. E. 2008, *ApJ*, 684, 1374
Carciofi, A. C., Okazaki, A. T., Le Bouquin, J.-B., Štefl, S. *et al.* 2009, *A&A*, 504, 915
Clark, J. S., Tarasov, A. E. & Panko, E. A. 2003, *A&A*, 403, 239
Jones, C. E., Sigut, T. A. A. & Porter, J. M. 2008, *MNRAS*, 386, 1922
Hanuschik, R. W. 1996, *A&A*, 308, 170
Harmanec, P. 1983, *Hvar Observatory Bulletin* 7, 55
Kato, S. 1983, *PASJ*, 35, 249
Lee, U., Osaki, Y. & Saio, H. 1991, *MNRAS*, 250, 432
Meilland, A., Stee, P., Vannier, M., Millour, F. *et al.* 2007a, *A&A*, 464, 59
Meilland, A., Millour, F., Stee, P., Domiciano de Souza, A. *et al.* 2007b, *A&A*, 464, 73
Okazaki, A. T. 1991, *PASJ*, 43, 75
Okazaki, A. T. 1997, *A&A*, 318, 548
Okazaki, A. T. 2001, *PASJ*, 53, 119
Okazaki, A. T., Bate, M. R., Ogilvie, G. I. & Pringle, J. E. 2002, *MNRAS*, 337, 967
Okazaki, A. T. 2007, in: A. T. Okazaki, S. P. Owocki, & S. Stefl (eds.), *Active OB-Stars: Laboratories for Stellar and Circumstellar Physics*, ASP-CS 361, p. 230
Porter, J. M. 1999, *A&A*, 348, 512
Porter, J. M. & Rivinius, T. 2003, *PASP*, 115, 1153
Pringle, J. E. 1981, *ARAA*, 19, 137
Quirrenbach, A., Bjorkman, K. S., Bjorkman, J. E., Hummel, C. A. *et al.* 1997, *ApJ*, 479, 477
Rivinius, T., Baade, D., Stefl, S., Stahl, O. *et al.* 1998, *A&A*, 333, 125
Rivinius, T. 2007, in: A. T. Okazaki, S. P. Owocki, & S. Stefl (eds.), *Active OB-Stars: Laboratories for Stellar and Circumstellar Physics*, ASP-CS 361, p. 219
Shakura, N. I. & Sunyaev, R. A. 1973, *A&A*, 24, 337
Sigut, T. A. A. & Jones, C. E. 2007, *ApJ*, 668, 481
Sigut, T. A. A., McGill, M. A. & Jones, C. E. 2009, *ApJ*, 699, 1973
Štefl, S., Rivinius, T., Carciofi, A. C., Le Bouquin, J.-B. *et al.* 2009, *A&A*, 504, 929
Townsend, R. H. D., Owocki, S. P. & Groote, D. 2005, *ApJ* (Letters), 630, L81
Tycner, C., Jones, C. E., Sigut, T. A. A., Schmitt, H. R. *et al.* 2008, *ApJ*, 689, 461
Waters, L. B. F. M. 1986, *A&A*, 162, 121
Wisniewski, J. P., Draper, Z. H., Bjorkman, K. S., Meade, M. R. *et al.* 2010, *ApJ*, 709, 1306
de Wit, W. J., Lamers, H. J. G. L. M., Marquette, J. B. & Beaulieu, J. P. 2006, *A&A*, 456, 1027
Wood, K., Bjorkman, K. S. & Bjorkman, J. E. 1997, *ApJ*, 477, 926

Discussion

PULS: Excellent talk! How are the occupation numbers required for the profile synthesis calculated?

CARCIOFI: The computer code HDUST calculates the occupation numbers by solving the coupled problem of the radiative transfer, radiative equilibrium and statistical equilibrium in the NLTE regime. Because the code uses the Monte Carlo method, it can handle arbitrary 3D density and velocity distributions.

LEE: What is the timescale of the V/R variability?

CARCIOFI: If the question refers to cyclic V/R variability, such as observed in ζ Tau, the timescale is of a few years

Active OB stars: structure, evolution, mass loss, and critical limits
Proceedings IAU Symposium No. 272, 2010
C. Neiner, G. Wade, G. Meynet & G. Peters, eds.

doi:10.1017/S174392131101074X

Spatially resolving the wind and disk structures around active B-type stars

Christopher Tycner

Department of Physics, Central Michigan University, Mount Pleasant, MI 48859, USA

Abstract. Long-baseline optical and IR interferometers now routinely resolve the wind and disk-like structures around early-type stars. The typical angular scales resolved by current generation of instruments are well bellow the milli-arcsecond level. These type of observations allow, in some cases for the first time, placing very tight constraints on current theories and models of the circumstellar structures around these type of stars. Specific examples of observations obtained at the Navy Prototype Optical interferometer of the spatially resolved regions around a luminous blue variable star P Cyg and a B-type star with circumstellar disk are presented. The need for connection between interferometric observables and physical parameters predicted by theory and numerical models are emphasized.

Keywords. techniques: interferometric, stars: emission-line, Be

1. Introduction

Interferometric observations of line-emitting early-type stars have shown to be invaluable in constraining the circumstellar regions associated with these type of stars. Most notably the Hα emission line, which happens to be one of the strongest emission lines, provides some of the strongest interferometric signals of the line-emitting regions (see, for example, Stee *et al.* (1995), Quirrenbach *et al.* (1997), Tycner *et al.* (2006)). Although some of the first studies concentrated on spatially resolving the line-emitting regions and have successfully constrained their geometries, it is now possible to directly compare the interferometric observables to detailed numerical disk models.

2. Methodology

To go beyond simple geometrical models representing the line-emitting regions around early-type stars, one can utilize numerical disk modeling routines that not only can solve for self-consistent temperature structure by balancing all the heating and cooling mechanisms within the disk, but can provide model interferometric observables that can be directly compared to observations.

To illustrate this strategy we utilize the BEDISK code of Sigut & Jones (2007). BEDISK computes a non-LTE radiative equilibrium model with solar chemical composition for the circumstellar gas. The numerical model provides monochromatic images of the circumstellar disk over a specific range of wavelengths. This includes the regions associated with the Hα emission line, which is of particular interest if models are to be compared to interferometric observations obtained over that line. Figure 1 illustrates the typical density structure that is used as a starting point and is described by a simple two-component parameterization of the form:

$$\rho(R, Z) = \rho_0 \left(\frac{R_\star}{R}\right)^n e^{-\left(\frac{Z}{H}\right)^2}, \tag{2.1}$$

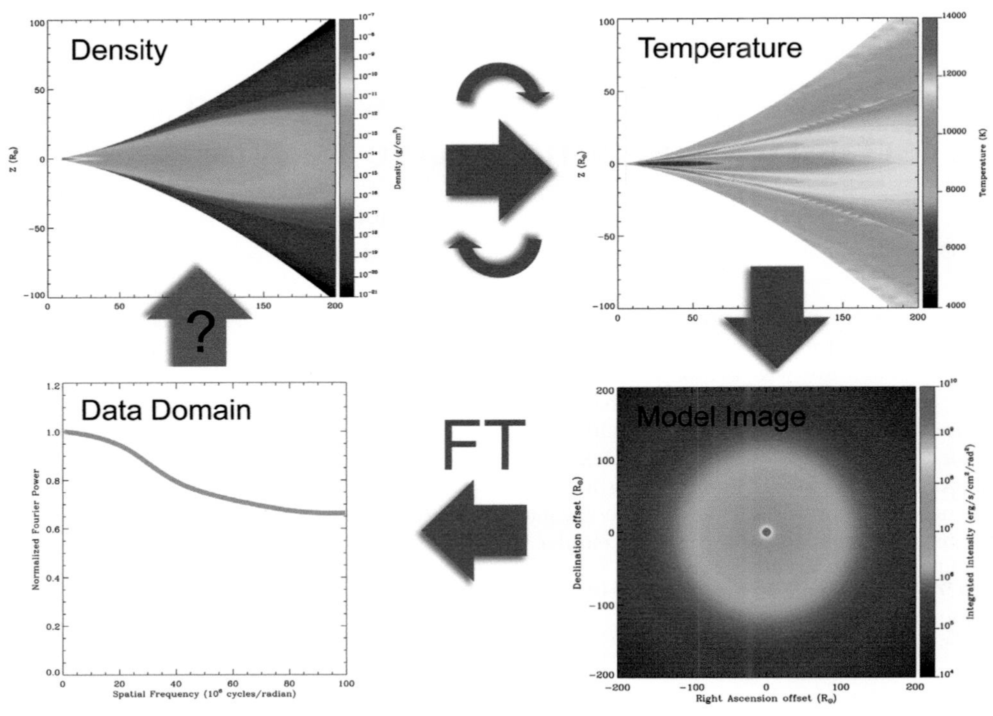

Figure 1. Schematic representation of a typical data flow involved in a comparison of numerical disk model to interferometric observations. A density structure (*upper-left*) is selected, which in turn is used to calculate (iteratively) a self-consistent temperature structure (*upper-right*). Once a final density and temperature structure is obtained, a synthetic image in the desired spectral bandpass can be obtained (*lower-right*). Finally, a Fourier transform of the synthetic image provides model values that can be compared directly to the interferometric observables (*lower-left*). The entire process can (if desired) be repeated in an iterative sequence.

where ρ_0 is the density at the inner edge of the disk in the equatorial plane, n is the index in the radial power-law, H is the scale height in the Z-direction perpendicular to the plane of the disk, and $R_\star$ is the stellar radius (see Sigut & Jones (2007) for more details). The vertical density structure and the temperature distribution throughout the disk are iterated until all the heating and cooling mechanisms are balanced. The final density and temperature structure is then used to calculate a two-dimensional model image over a wavelength range of interest. Because long-baseline interferometric observations represent the Fourier transform of the source structure on the sky, the final step involves the Fourier transform of the synthetic image. Only then the model can be compared directly to the interferometric observables.

In practice the entire data flow (as illustrated in Figure 1) is either, repeated in an iterative sequence so that a best-fit numerical disk model is found, or as it was demonstrated by Tycner *et al.* (2008) an extensive grid of models is compared to interferometric observations. The latter approach allows not only a selection of 'best' model, but also testing for any inherent correlations between the free model parameters, such as ρ_0 and n.

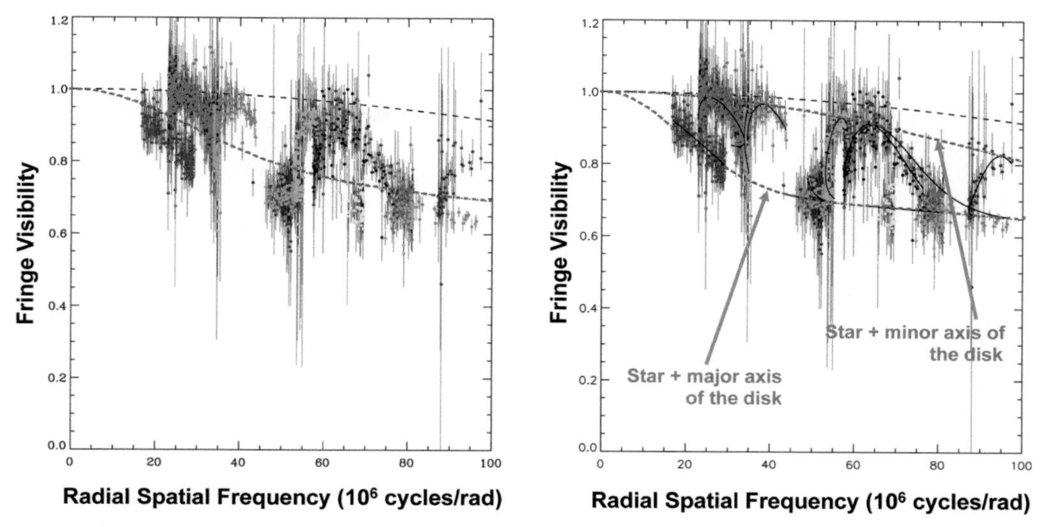

Figure 2. Square visibility data acquired in the Hα channel for the Be star ζ Tau using 15 unique baselines (shown in different colors). Both, circularly symmetric (*left panel*) and elliptical (*right panel*) Gaussian models are shown. In the case of the elliptical model the extreme model values at minor and major axes are shown (*dashed-dotted lines*). The mostly unresolved central star represented by uniform disk model is also shown (*dashed line*).

3. Be Star

The Navy Prototype Optical Interferometer (NPOI) operates over a spectral range between 560 and 870 nm. The interferometric signal (which consists of squared visibility measures and sometimes closure phases) is dispersed over 15 spectral channels. Therefore, interferometric observations of Hα-emitting sources contain all the line emission in one spectral channel. Figure 2 illustrates a typical signature seen for Be stars that do not possess circularly symmetric disk emission as seen on the sky. Most of this type of asymmetry can be attributed to a simple geometrical projection of how the circumstellar disk is viewed by the observer (i.e., the inclination angle with respect to the plane of the disk).

It is evident from Figure 2 that not only the Hα region is fully resolved, but that it cannot be represented by models that have circular symmetry. Because most, it not all, of such asymmetry can be attributed to a simple projection angle, that means that such data can be compared to circularly symmetric numerical disk models as long as the projection angle is properly accounted for. In the case of the BEDISK numerical routine, this is possible with a ray tracing routine known as BERAY (see contribution by Sigut, this volume).

4. LBV Star

In addition to resolving Hα-emitting circumstellar structures around B-type stars, one can utilize the same technique to resolve the wind structure around an LBV star. Figure 3 shows the interferometric data from Balan *et al.* (2010) acquired in the spectral channel containing the Hα emission line of the LBV star P Cyg. The wind structure is fully resolved and data at long enough baselines (i.e., high enough Fourier frequencies)

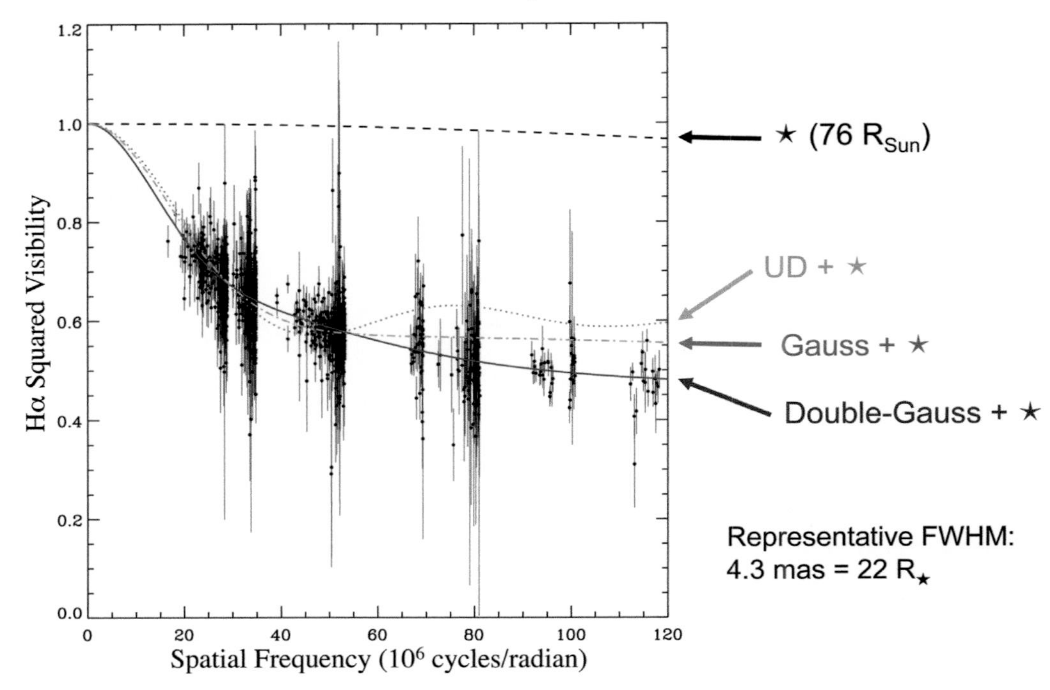

Figure 3. Squared visibility data acquired in the Hα channel of P Cyg. The model representing the central star (*dashed-line*) is shown along with three possible models representing the radial fall-off of the Hα emission from the wind structure.

has been acquired that uniform disk (UD) or Gaussian radial intensity distributions can be excluded. The data presented in Figure 3 has been modeled with a two-component Gaussian model, however this only represents a simple geometric parameterization.

Describing the radial intensity distribution around P Cyg with only a few parameters, such as provided by a double-Gaussian model, can be useful when simplicity is desired. However, a much more complete analysis could be accomplished if such interferometric data would be compared directly to model predicted observables. Furthermore, if various competing models describing the wind structure would publish observables that could be either directly, or after an appropriate transform, be compared to interferometric observables, such observations would provide yet another critical test and constraint for such models.

5. Conclusions

We have demonstrated how interferometric data for circumstellar disks and wind structures can be directly used to constrain the functional form of the radial intensity distribution on the sky. Although this can provide angular measurements of such regions with unprecedented accuracy, what is needed now is a systematic way of comparing such observations to theoretical models. This could be easily accomplished if analytical and numerical models published in the literature would accompany their predictions with both one- and two-dimensional observables. For example, publishing synthetic images (preferably in digital format) as illustrated in Figure 4 would allow for a direct comparison to data acquired via interferometry and possibly a combination of observational methods.

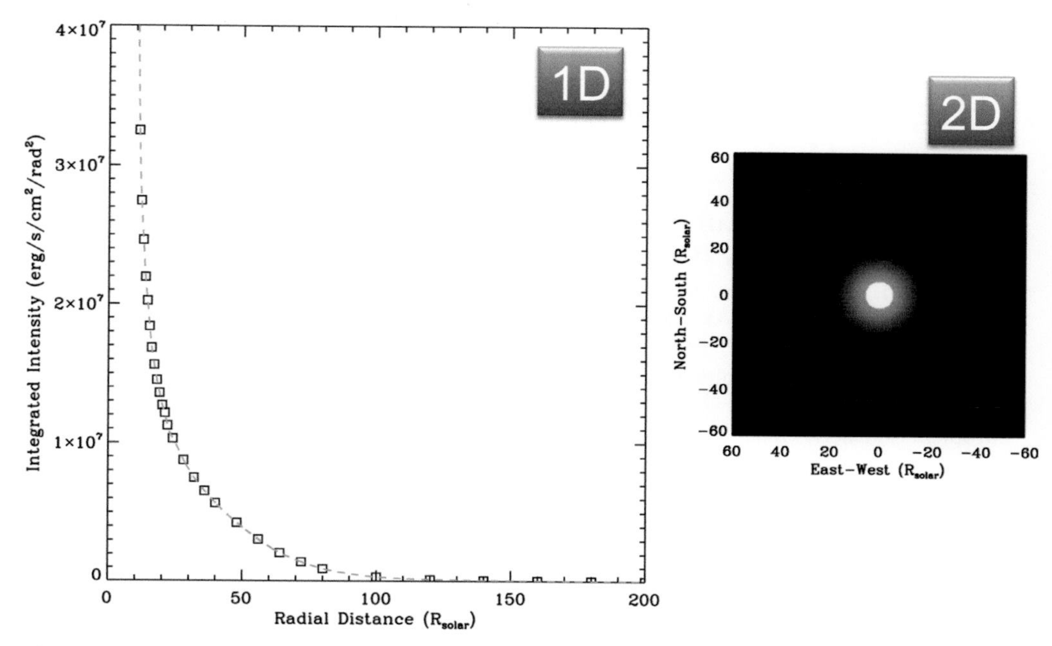

Figure 4. Integrated intensity of a synthetic model as a function of radial distance from the central star over a spectral channel containing an Hα emission line. Both, radial profile of the intensity distribution (*left panel*) and a 2-D contour plot of the intensity (*right panel*) are shown.

Acknowledgements

The Navy Prototype Optical Interferometer is a joint project of the Naval Research Laboratory and the US Naval Observatory, in cooperation with Lowell Observatory, and is funded by the Office of Naval Research and the Oceanographer of the Navy. C.T. would also like to acknowledge, with thanks, financial support from the Central Michigan University.

References

Balan, A., Tycner, C., Zavala, R. T., Benson, J. A. *et al.* 2010, *AJ*, 139, 2269

Quirrenbach, A., Bjorkman, K. S., Bjorkman, J. E., Hummel, C. A. *et al.* 1997, *ApJ*, 479, 477

Sigut, T. A. A. & Jones, C. E. 2007, *ApJ*, 668, 481

Stee, P., de Araujo, F. X., Vakili, F., Mourard, D. *et al.* 1995, *A&A*, 300, 219

Tycner, C., Gilbreath, G. C., Zavala, R. T., Armstrong, J. T. *et al.* 2006, *AJ*, 131, 2710

Tycner, C., Jones, C. E., Sigut, T. A. A., Schmitt, H. R. *et al.* 2008, *ApJ*, 689, 461

Active OB stars: structure, evolution, mass loss, and critical limits
Proceedings IAU Symposium No. 272, 2010
C. Neiner, G. Wade, G. Meynet & G. Peters, eds.

doi:10.1017/S1743921311010751

High spatial resolution monitoring of the activity of BA supergiant winds

Olivier Chesneau[1], Luc Dessart[2], Andreas Kaufer[3], Denis Mourard[1], Otmar Stahl[4], Raman K. Prinja[5], and Stan P. Owocki[6]

[1]UMR 6525 H. Fizeau, Univ. Nice Sophia Antipolis, CNRS, Observatoire de la Côte d'Azur, Av. Copernic, F-06130 Grasse, France
email: olivier.chesneau@oca.eu

[2]Laboratoire d'Astrophysique de Marseille, Université de Provence, CNRS, 38 rue Frédéric Joliot-Curie, F-13388 Marseille Cedex 13, France

[3]European Southern Observatory, Alonso de Cordova 3107, Casilla 19001, Santiago 19, Chile

[4]Zentrum für Astronomie der Universität Heidelberg, Landessternwarte, Königstuhl 12, 69117 Heidelberg, Germany

[5]Department of Physics & Astronomy, University College London, Gower Street, London, WC1E 6BT, UK

[6]Bartol Research Institute, Dept. of Physics & Astronomy, Univ. of Delaware Newark, DE 19716 USA

Abstract. There are currently two optical interferometry recombiners that can provide spectral resolutions better than 10000, AMBER/VLTI operating in the H-K bands, and VEGA/CHARA, recently commissioned, operating in the visible. These instruments are well suited to study the wind activity of the brightest AB supergiants in our vicinity, in lines such as Hα or Brγ. We present here the first observations of this kind, performed on Rigel (B8Ia) and Deneb (A2Ia). Rigel was monitored by AMBER in two campaigns, in 2006-2007 and 2009-2010, and observed in 2009 by VEGA; whereas Deneb was monitored in 2008-2009 by VEGA. The extension of the Hα and Brγ line forming regions were accurately measured and compared with CMFGEN models of both stars. Moreover, clear signs of activity were observed in the differential visibility and phases. These pioneer observations are still limited, but show the path for a better understanding of the spatial structure and temporal evolution of localized ejections using optical interferometry.

Keywords. techniques: high angular resolution, stars: activity, stars: mass loss, stars: individual (HD 34085, HD 197345)

---

1. Introduction

Supergiants of spectral types B and A (BA-type supergiants) are evolved massive stars of typical initial mass of 25-40 $M_\odot$ and high luminosity ($\gtrsim 10^5\ L_\odot$). Their luminosity and temperature place them among the visually brightest massive stars, a particularly interesting aspect for extragalactic astronomy. Nearby BA supergiants have been analyzed with sophisticated radiative-transfer tools, and among them the closest ones Deneb (α Cygni, HD 197345, A2 Ia) and Rigel (β Orionis, HD 34085, B8Ia).

Most BA supergiants rotate slowly ($v \sin i$ of about 25-40km s^{-1}), at least relative to their terminal wind-velocity of $\sim$200-400 km s^{-1} (Fraser *et al.* 2010, Lefever *et al.* 2007). Intensive spectroscopic monitoring of the activity of some wind sensitive lines, such as Hα has lead to the conclusion that variability in the stellar winds of luminous hot stars is localized and structured and rotates around the stars. Obviously, the winds are modulated by patches on the stellar surfaces produced either by non-radial pulsation (NRP) patterns or magnetic surface structures.

2. VEGA/CHARA observations in the visible

The VEGA recombiner of the CHARA array (Mt Wilson, CA) is a recently commissioned facility that provides spectrally dispersed interferometric observables, with a spectral resolution reaching R =30 000, and a spatial resolution of less than one *mas* Mourard *et al.* (2009). The instrument recombines currently the light from two telescopes, but 3-4 telescope recombination modes are foreseen in a near future. As an example, the extension of the Hα line-forming region of the prototypical Herbig star AB Aur was resolved for the first time Rousselet-Perraut *et al.* (2010).

The Hα line of bright, slow rotators such as Deneb or Rigel can be isolated from the continuum, and the spatial properties of the line-forming region can thus be studied with unprecedented resolution. Using the smallest baseline of the CHARA array (baseline of 34m), we conducted a pioneering temporal monitoring of Deneb uncovering a high level of activity in the Hα line-forming region. Rigel was also observed a few times. The information in a line was extracted differentially by comparing the properties of the fringe between a reference channel centered on the continuum of the source, and a sliding science narrow channel.

For the most accurate estimates to date of the diameters of Rigel and Deneb, we refer to Aufdenberg *et al.* (2008) in which CHARA/FLUOR observations in the K band with baselines reaching 300 m are described. These observations infer a UD angular diameter of 2.76 ± 0.01 mas, and 2.363 ± 0.002 mas, for Rigel and Deneb, respectively.

3. New perspectives for estimating the mass-loss rate

Many useful diagnostics of mass loss are used from the radio to the X-rays, based on the determination of the free-free continuum emission at radio wavelengths, the fitting of some line emission (in particular Hα, ultraviolet (UV) resonance-line absorption or or more recently the fitting of X-ray spectra (Cohen *et al.* these proceedings). These estimators suffer from different biases depending on whether their physical mechanisms is linearly or quadratically dependent on the local density. They also suffer different degrees of contamination and require different ancillary knowledge of, e.g., the excitation or ionization conditions in the wind, the velocity structure of the wind, or the distance to the star.

We discuss here some new ways to constrain the consistency of the mass-loss rate or the wind velocity by measuring accurately the limb-darkening of massive close-by BA supergiants photosphere, and the extension of the line forming regions of some wind-sensitive lines such as Hα. Of course, these estimations are not free from their own biases, although they are based from observables that are very different from those that are commonly used.

Radiative-transfer calculations were carried out with the line-blanketed non-LTE model-atmosphere code CMFGEN. In this study, we used the stellar parameters obtained by Przybilla et al. (2006) and Schiller & Przybilla (2008) for Rigel and Deneb, respectively. An important conclusion is that the second lobe of the visibility curve is far more sensitive to any fluctuation of the mass-loss rate in the visible than the infrared. The reasons are twofold: a higher sensitivity to limb-darkening effects in the visible, and a more extended continuum-formation region, despite the very limited amount of flux involved (the wind remains in any case optically thin). Balmer bound-free cross-sections increase from 400 to 800nm. This causes the continuum photosphere to shift weakly in radius across this wavelength range and also alters the limb-darkening properties of the star.

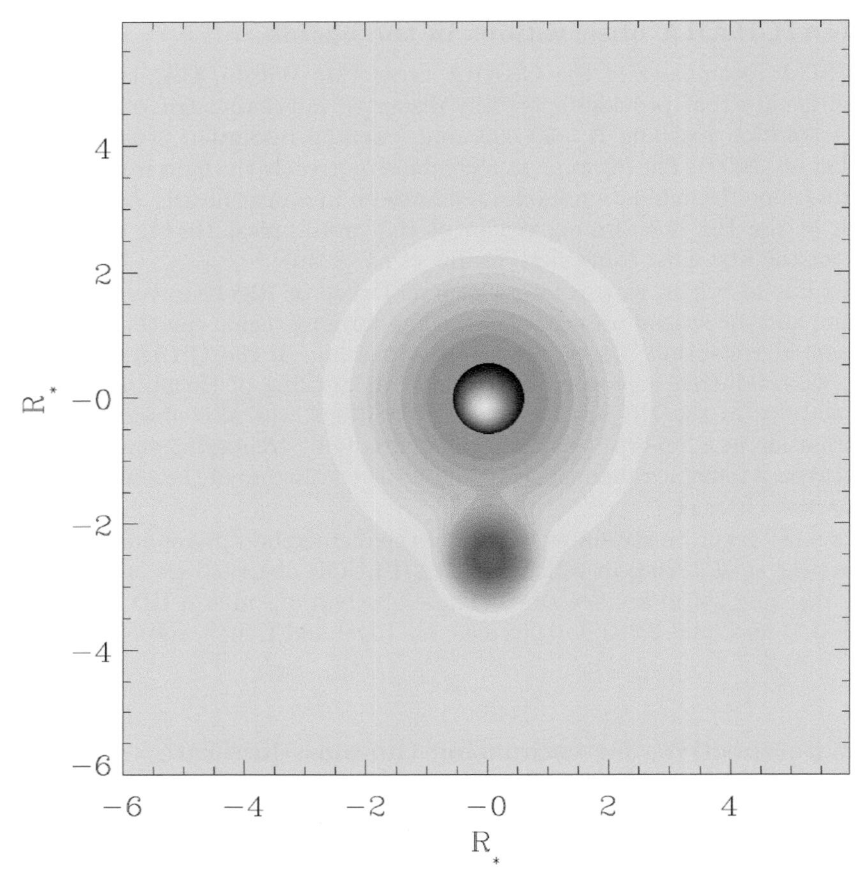

Figure 1. Illustrative picture summarizing the observations from FLUOR/CHARA (Aufdenberg *et al.* 2006), providing the low inclination fast-rotator photospheric model and the VEGA/CHARA observation of July 2008, showing a the Hα emission around the star. The picture was not consistently generated. The Hα envelope was created in the frame of a non-rotating CMFGEN, and the perturbation artificially created by locally multiplying the flux by a coefficient following a Gaussian distribution. The FWHM of the Gaussian is about half the radius of the star, and the flux increase at peak is 10 times the local flux. The position, FWHM and flux contrast of the perturbation were chosen to fit the interferometric data.

4. AMBER/VLTI observations in the near-IR

An interferometric monitoring campaign of Rigel has been undertaken at the ESO/ Paranal observatory with the the near-infrared instrument of the VLTI using the highest spectra resolution (R=12000) and baselines reaching 125m. Two periods of monitoring were performed in 2006-2007 and in 2009-2010. These observations were completed by intensive spectroscopic observations using the FEROS and BESE spectrographs†. The data are under analysis and only preliminary results are discussed here. The first test was to apply the model of Rigel used to account for the two observations of Rigel in Hα with VEGA/CHARA. The Brγ visibilities are much deeper than predicted by the model, and the mass-loss rate must be significantly increased to match the observations. This can hardly be explain by the intrinsic variability of the stars and the use of non-contemporary

† This is a FEROS copy running at the 1.5m Bochum telescope on Armazones

observations, and might be an indication that the model is not consistently accounting for the visible AND near-IR observations.

With VEGA/CHARA, a significant S-shape differential phase signal was observed once with the North-South baseline. This S-shape signal can be interpreted as a rotation of the Hα line forming region. The latest dynamical spectrum of Hα from the BESO data set shows that a large scale event is seen in the Hα around JD2455240 (12-02-2010). A clear change in the profile of the Brγ line as seen by AMBER is visible, corresponding to a marked photocenter offset in the red part of the line indicating that the ejected material was located in a particular direction at a defined radial velocity.

Both the rotation and the kinematics of the ejection event are not currently accounted for by the CMFGEN models developed so far, that do not include any kinematics apart from the wind velocity law. This is not an issue in the case of Deneb which is probably seen at large inclination, and it was possible to account for the large phase signal affecting the full Hα line by a perturbation of the model. However, it is not possible with such a non-rotating model to generate a perturbation that would affect only part of the line. This is why a refined rotating model for Rigel is necessary.

5. On the fast rotation of BA supergiants

Aufdenberg *et al.* presented evidence that Deneb is a fast rotator, based on 24 high accuracy FLUOR/CHARA visibilities in the K'-band with projected baselines ranging from 106 m to 310 m, sampling the first and second lobes of the visibility curve. They detected slight departures from a purely spherical model at a level of about 2% in the near-IR, a discovery that had not been anticipated for an AB supergiant. They tentatively fitted the data with a rotating model atmosphere, and demonstrated that despite the low $v \sin i$ of the star, a model at half critical speed, seen nearly pole-on may account for the interferometric observations. That no evidence of rotation was detected in the differential phases is an additional argument for a low $v \sin i$, and therefore a nearly pole-on configuration for Deneb. Such a signal was observed on Rigel by VEGA in Hα and by AMBER in Brγ which suggests that Rigel is seen at significantly higher inclination, with an axis direction approximately oriented north-south. Note that the *vsini* of Deneb and Rigel is about 20km s^{-1}and 35km s^{-1}, respectively.

Rapidly rotating evolved supergiants appear to be relatively rare and one might wonder whether they might share some relationship with B[e] star. As discussed in Fraser *et al.* 2010, there are at least three channels by which these stars could have reached their current evolutionary state: (1) they had extremely high rotational velocities on the main sequence, and have spun down normally as they evolved; (2) they have not spun down as much as most other stars as they became supergiant stars; (3) they had a moderate velocity on the main sequence but have experienced an additional source of angular momentum that has allowed them to maintain their rotational velocity, and this source might be a close companion.

The example of the A[e] supergiant HD 62623 is very informative in this context (see Millour *et al.*, Meilland *et al.* these proceedings). This is the coolest member of the so-called B[e] stars, surrounded by dense, dusty disks. HD62623 exhibits a rich emission line spectrum and only few photospheric lines are observable, alike many B[e] stars. Spectroscopically and spatially resolved observations of AMBER/VLTI have shown that the supergiant lies in a cavity, and is surrounded by a dense disk of plasma. The Brγ line from the supergiant is *in absorption* showing that the star is not much different from a normal member of its class such as Deneb, albeit with a significantly large (but not overwhelming) *vsini* of about 50km s^{-1}. By contrast, the Balmer and Bracket lines are much larger (*vsini* of about 120km s^{-1}), and the AMBER observations demonstrated that they originate from a disk of plasma, most probably in Keplerian rotation (Millour, Meilland *et al.* submitted). It is difficult to understand how HD 62623 would exhibit a much larger mass-loss rate compared to its class, and how such a dense equatorial disk could have been generated. However, high quality spectral monitoring performed more than 15yrs ago showed that a companion star rotates close to the supergiant with a period of about 136 days Plets et al.(1995). The spectroscopic signal is weak, hence the mass ratio is very large, and the companion is probably a solar mass star (given that the recent interferometric observations have provided invaluable constraints on the inclination of the system). This tightens further the connection between B[e] stars and binarity

The recent Spitzer observations of the mid-IR excess observed in LMC B[e] stars provide further evidence of the analogy to the class of post-asymptotic giant branch stars with binary companions and dusty, circumbinary disks. The SEDs are all similar, within the LMC group of B[e] stars and also compared with galactic onces, leading Kastner *et al.* (2010) to quote this sentence in their abstract:'we speculate that B[e] supergiant stars may be post-red supergiants in binary systems with orbiting, circumbinary disks that are derived from post-main-sequence mass loss' .

6. Prospects

The observations reported here are a pioneering effort in order to improve upon the spectral and spatial resolution of the interferometric facilities. An extension of the VEGA observations is possible for sources with an apparent angular diameter between 0.5 and 1.5 mas, that are large enough, but also bright enough to make use of the highest spectral dispersion of the VEGA instrument (limiting magnitude of about 3). This concerns the AB supergiants closer than 1.5 kpc, and the supergiants in the Orion complex ($d\sim$500 pc) are in this context of particular interest. Simultaneous 3 telescope recombination would provide much better *uv* coverage than the one presented in this paper. The number of sources accessible to AMBER observations are more restricted as the spatial resolution in the near-IR is decreased by about 4, and no baselines longer than 130m are to date accessible. Moreover, the Brγ line is formed closer to the photosphere. Paβ is the H band is better suited, but the best line by far would be the HeI1080 line in the J band. Long baselines FLUOR/CHARA observations of Rigel and Deneb should be repeated to confirm that these measures are independent of the star activity (Miroshnichenko 2007). The B[e] star with the largest infrared excess of the class, HD 87643 appeared also, in view of the AMBER/VLTI observations, as a unique and quite extreme large separation binary system. The large nebula that surrounds this system shows periodic arc that suggest a possible link between extreme period of mass ejection and the periastron passages of the companion with a highly eccentric orbit (Millour *et al.* 2009).

References

Aufdenberg, J. P., Ludwig, H.-G., Kervella, P., Mérand, A. *et al.* 2008, in: A. Richichi, F. Delplancke, F. Paresce, & A. Chelli (eds.), *The Power of Optical/IR Interferometry: Recent Scientific Results and 2nd Generation*, ESO Astrophysics Symposia, p. 71

Aufdenberg, J. P., Mérand, A., Ridgway, S. T., Coudé du Foresto, V. *et al.* 2006, in: *Bulletin of the American Astronomical Society*, 38, p. 84

Fraser, M., Dufton, P. L., Hunter, I., & Ryans, R. S. I. 2010, *MNRAS*, 404, 1306

Kastner, J. H., Buchanan, C., Sahai, R., Forrest, W. J. *et al.* 2010, *AJ*, 139, 1993

Lefever, K., Puls, J., & Aerts, C. 2007, *A&A*, 463, 1093

Millour, F., Chesneau, O., Borges Fernandes, M., Meilland, A. *et al.* 2009, *A&A*, 507, 317
Miroshnichenko, A. S. 2007, *ApJ*, 667, 497
Mourard, D., Clausse, J. M., Marcotto, A., Perraut, K. *et al.* 2009, *A&A*, 508, 1073
Plets, H., Waelkens, C., & Trams, N. R. 1995, *A&A*, 293, 363
Przybilla, N., Butler, K., Becker, S. R., & Kudritzki, R. P. 2006, *A&A*, 445, 1099
Rousselet-Perraut, K., Benisty, M., Mourard, D., Rajabi, S. *et al.* 2010, *A&A*, 516, L1
Schiller, F. & Przybilla, N. 2008, *A&A*, 479, 849

Active OB stars: structure, evolution, mass loss, and critical limits
Proceedings IAU Symposium No. 272, 2010
C. Neiner, G. Wade, G. Meynet & G. Peters, eds.

doi:10.1017/S1743921311010763

X-ray spectral diagnostics of activity in massive stars

David H. Cohen[1], **Emma E. Wollman**[2] **and Maurice A. Leutenegger**[3]

[1]Department of Physics and Astronomy, Swarthmore College,
500 College Ave., Swarthmore, Pennsylvania, 19081, USA
email: cohen@astro.swarthmore.edu

[2]Department of Physics, California Institute of Technology,
Pasadena, California, 91125, USA
email: ewollman@caltech.edu

[3]NASA/Goddard Spaceflight Center, Code 662, Greenbelt, Maryland, 20771, USA
email: maurice.a.leutenegger@nasa.gov

Abstract. X-rays give direct evidence of instabilities, time-variable structure, and shock heating in the winds of O stars. The observed broad X-ray emission lines provide information about the kinematics of shock-heated wind plasma, enabling us to test wind-shock models. And their shapes provide information about wind absorption, and thus about the wind mass-loss rates. Mass-loss rates determined from X-ray line profiles are not sensitive to density-squared clumping effects, and indicate mass-loss rate reductions of factors of 3 to 6 over traditional diagnostics that suffer from density-squared effects. Broad-band X-ray spectral energy distributions also provide mass-loss rate information via soft X-ray absorption signatures. In some cases, the degree of wind absorption is so high, that the hardening of the X-ray SED can be quite significant. We discuss these results as applied to the early O stars ζ Pup (O4 If), 9 Sgr (O4 V((f))), and HD 93129A (O2 If*).

Keywords. line: formation, shock waves, stars: winds, outflows, X-rays: stars

---

1. Introduction

Soft X-ray emission is ubiquitous in O stars, and it is generally accepted that it arises in numerous shock-heated regions embedded in these stars' powerful and dense radiation-driven winds. The broad emission lines seen in high-resolution X-ray spectra of O stars confirm this scenario qualitatively. In this paper, we present quantitative analysis of resolved X-ray emission lines observed in three very early O stars, from which we are able to place constraints on the kinematics and spatial distribution of the shock-heated plasma and thereby test predictions of numerical simulations of wind shocks. We also show how the degree of attenuation by the bulk wind in which the shocked plasma is embedded can be measured both from resolved emission lines and from lower resolution broadband X-ray spectra in order to estimate the mass-loss rates of O star winds.

We restrict our discussion to "normal" massive stars, where binarity and the associated colliding wind shock (CWS) X-ray emission and magnetically channeled wind shock (MCWS) X-ray emission is absent or negligible. The dominant paradigm for X-ray production in normal O and early-B stars is the embedded wind shock (EWS) scenario, and the specific mechanism for EWSs is usually assumed to involve the line-driving instability (LDI), either in a self-excited mode (Owocki, Castor, & Rybicki 1988) or in a mode where the instability is seeded by perturbations at the base of the wind (Feldmeier, Puls, & Pauldrach 1997).

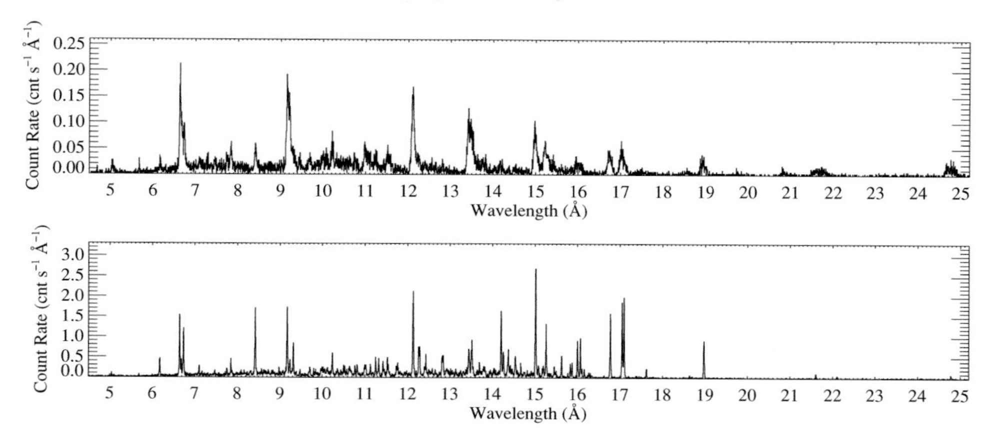

Figure 1. *Chandra* medium energy grating (MEG) spectra of the O4 If star, ζ Pup (top) and for comparison, of the G star, Capella (bottom). The spectrum of the O star is harder (strongest lines at shortest wavelengths) but by comparing H-like and He-like line strengths (e.g. of Mg at 8.42 Å and 9.2 Å, respectively) it is evident that the higher temperature plasma is found on the G star. Finally, note that the emission lines are unresolved in the Capella spectrum and are significantly broadened in the ζ Pup spectrum.

The morphology of high-resolution X-ray spectra of normal massive stars reveals some important qualitative properties of O star X-rays. In Fig. 1 we compare the O supergiant ζ Pup to the coronal G star, Capella, to highlight some of these properties. The O star's spectrum is harder, overall, than the G star's, however this is due not to higher plasma temperatures, but rather to the effects of wind absorption, consistent with the X-rays arising in the dense stellar wind of the O star. This broadband view of the X-ray spectra also shows quite obviously that the emission lines in the O star are much broader than the (unresolved) lines in the G star, as the EWS scenario predicts. In the next section, we show how quantitative information can be derived from the Doppler broadened X-ray emission lines.

2. The X-ray line profile model applied to ζ Pup

To extract information from individual resolved line profiles, we fit a simple wind-shock model informed by the LDI simulations, in which numerous shock-heated regions are distributed throughout the wind above some shock onset radius, $R_{\rm o}$, with local emission measure assumed to scale with the local ambient wind density squared (Owocki & Cohen 2001). The kinematic profile of the X-ray plasma is assumed to trace the same beta-velocity law that describes the bulk wind. This assumption is based on the results of numerical simulations of EWSs that show accelerated pre-shock wind streams being decelerated back down the local ambient wind velocity (Runacres & Owocki 2002). The attenuation due to continuum opacity in the bulk wind in which the shock-heated plasma is embedded is described by the characteristic optical depth parameter, $\tau_* \equiv \kappa \dot{M}/4\pi R_* v_\infty$, where κ is the (wavelength dependent) opacity of the bulk wind, $\dot{M}$ is the wind mass-loss rate, R_* is the stellar radius, and v_∞ is the wind terminal velocity. The absorption of X-rays imparts a characteristic blue-shifted and asymmetric shape to the emission line profiles due to the preferential attenuation of red-shifted line photons emitted in the far hemisphere of the wind, while leaving blue-shifted line photons emitted from the near hemisphere much less attenuated.

For each emission line in the *Chandra* spectrum of an O star, we can fit this empirical profile model and derive best-fit values of $R_{\rm o}$ and τ_* by minimizing the C statistic, and

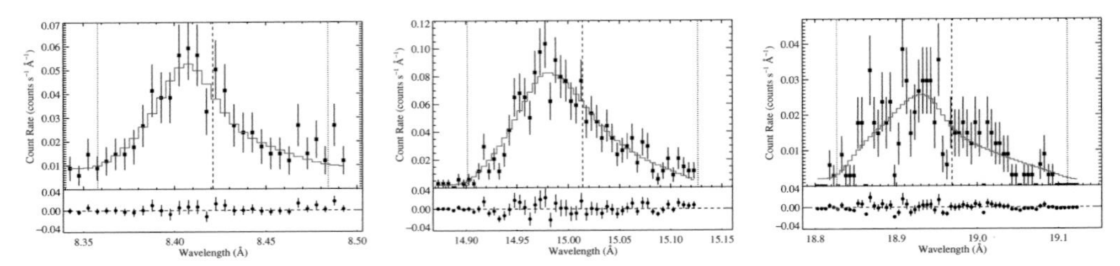

Figure 2. Fits to three lines in the *Chandra* spectrum of ζ Pup. From left to right: the Lyα line of Mg XII at 8.42 Å, the Fe XVII line at 15.01 Å, and the Lyα line of O VIII at 18.97 Å. The vertical dashed lines in each panel represent the laboratory rest wavelengths of each transition, while the flanking dotted lines represent the Doppler shifts associated with the wind terminal velocity. The characteristic broad, blue shifted, and asymmetric profile shapes are evident, as is an increase in the shift and asymmetry with wavelength, as is expected from the form of the continuum opacity of the bulk wind, which generally increases with wavelength. The characteristic optical depths of these three lines are roughly $\tau_* = 1, 2,$ and 3, respectively.

place confidence limits on them via the $\Delta\chi^2$ formalism applied to the C statistic. For ζ Pup (summarizing the results published in Cohen *et al.* 2010), we find – for 16 lines and line complexes in the *Chandra* grating spectrum (three representative lines and their best-fit profile models are shown in Fig. 2) – a universal value for the shock-onset radius of $R_o \approx 1.5$ R$_*$, which is consistent with numerical simulations of the LDI (Feldmeier, Puls, & Pauldrach 1997, Runacres & Owocki 2002). We also find a range of characteristic optical depths, τ_*, for the 16 emission lines, consistent with the expected wavelength trend in the atomic opacity. By calculating a detailed opacity model, and assuming standard values for the stellar radius and wind terminal velocity, we fit the ensemble of characteristic optical depths to find a best-fit mass-loss rate, via $\dot{M} = 4\pi R_* v_\infty \tau_*(\lambda)/\kappa(\lambda)$. The values of the onset radius, R_o, and of the characteristic optical depths, τ_*, are shown in Fig. 3. The panel with the τ_* values also shows the best-fit model of the wavelength-dependent optical depths, from which we derive a mass-loss rate of $3.5 \pm 0.3 \times 10^{-6}$ M$_\odot$ yr^{-1}.

We emphasize that the mass-loss rate determination from the X-ray profiles represents a factor of roughly three reduction from the traditional Hα-derived mass-loss rate that ignores clumping (Markova *et al.* 2004). And that this modest reduction in the mass-loss rate is consistent with newer determinations using Hα and radio and IR free-free excesses that *do* account for clumping (Puls *et al.* 2006). Also, we note that for a high signal-to-noise *Chandra* spectrum with many emission lines, like that of ζ Pup, the statistical error on the derived mass-loss rate is small (about 10%), but that the actual uncertainty is dominated by uncertainty in the wind opacity model, which in turn is dominated by uncertainty in the elemental abundances. Individual elemental abundances do not have a large effect, but the overall metallicity does. The model we use in this paper (and which was used in Cohen *et al.* (2010)) uses subsolar metallicity and C, N, and O abundances altered by CNO processing. If future abundance determinations are made which supersede the current ones, the mass-loss rate should be rescaled in inverse proportion to the metallicity adjustment (more metals cause higher opacity which would then require lower wind column densities and so lower mass-loss rates).

3. Other O stars

We can apply the same type of line profile analysis to other O stars observed with the *Chandra* grating spectrometer. Here we present preliminary analysis of the early O

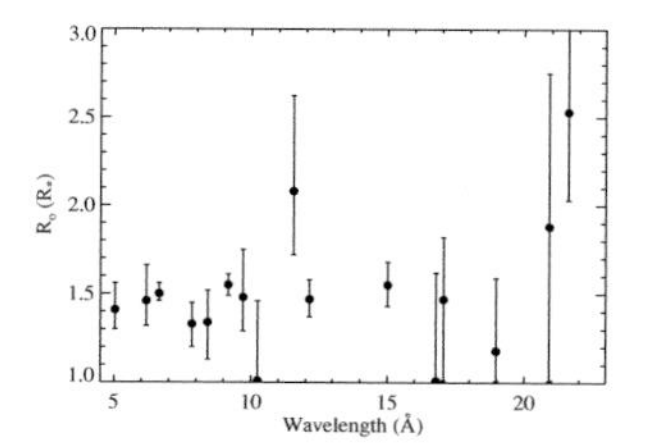

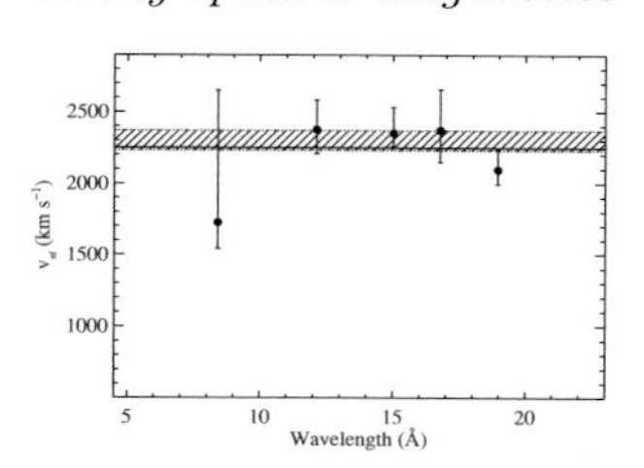

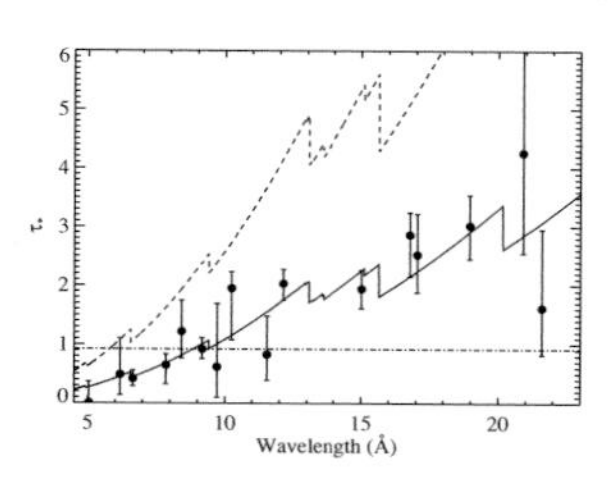

Figure 3. Results from fitting the wind-profile model to the emission lines in the *Chandra* spectrum of ζ Pup. From left to right: the shock onset radii, R_o, which are consistent with a universal value of 1.5 R_*; the wind terminal velocities for the five strongest, unblended lines in the spectrum, which are consistent with the value for the bulk wind, of $v_\infty = 2250$ km s^{-1}, derived from UV spectra (this value is represented by the horizontal line, while the cross hatched region is the 68% confidence limit on the mean value of the five fitted terminal velocities shown as points with error bars); and the τ_* values from each of the 16 fitted line profiles. This last panel shows that a constant value of the characteristic optical depth provides a poor fit, as does a model that incorporates the continuum opacity of the bulk wind but assumes a high value for the mass-loss rate (8.3×10^{-6} M$_\odot$ yr^{-1}), while a model with the mass-loss rate as a free parameter provides a good fit, with a mass-loss rate of 3.5×10^{-6} M$_\odot$ yr^{-1}.

main sequence star, 9 Sgr, at the center of the Lagoon Nebula, and the very early O supergiant, HD 93129A, in Tr 14 in Carina. Both stars have binary companions, but in neither case are the emission lines in the grating spectrum significantly contaminated by the harder emission associated with CWB X-rays.

There are nine lines and line complexes in the 9 Sgr *Chandra* grating spectrum with high enough signal-to-noise for line profile modeling to provide meaningful constraints. We assume a wind velocity parameter $\beta = 0.7$ and find a mean shock onset radius of $R_o = 1.4$ R$_*$, consistent with the EWS scenario. The ensemble of τ_* values can be fit, given a model of the bulk wind opacity (which we calculate assuming solar abundances), to derive a mass-loss rate. We find a mass-loss rate of $\dot{M} = 3.4 \times 10^{-7}$ M$_\odot$ yr^{-1}, which represents a factor of six reduction over the traditional mass-loss rate derived from Hα and radio free-free emission, assuming a smooth wind (Lamers & Leitherer 1993, Puls *et al.* 1996). In Fig. 4 we show the R_o and τ_* results.

The O2 If* star, HD 93129A, is the earliest O star in the Galaxy and, according to Taresch *et al.* (1997), has the highest mass-loss rate of any O star, with $\dot{M} = 1.8 \times 10^{-5}$ M$_\odot$ yr^{-1}. More recent modeling (though also ignoring clumping effects) gives $\dot{M} = 2.6 \times 10^{-5}$ M$_\odot$ yr^{-1} and a wind terminal velocity of $v_\infty = 3200$ km s^{-1} (Repolust, Puls, & Herrero 2004). This extremely strong and dense stellar wind provides an interesting test of the EWS scenario for X-ray production in O stars. Indeed, the *Chandra* spectrum is quite hard, which if assumed to be due to high temperature would make an EWS interpretation implausible. However, the H-like Si line strength is very weak, compared to the He-like Si line strength, indicating a plasma emission temperature of no more than 8 million K, which is consistent with LDI simulations of wind shocks. The hardness of the X-ray spectrum appears instead to be due to severe attenuation of the soft X-ray emission by both the interstellar medium and the star's own wind.

Because of the absent soft X-rays, there are only four lines and line complexes in the *Chandra* grating spectrum available for fitting. We show the τ_* results in Fig. 4 with the mass-loss rate fit superimposed. For this star, too, we find a modest mass-loss rate reduction of roughly a factor of four over the value derived from Hα fitting assuming no clumping.

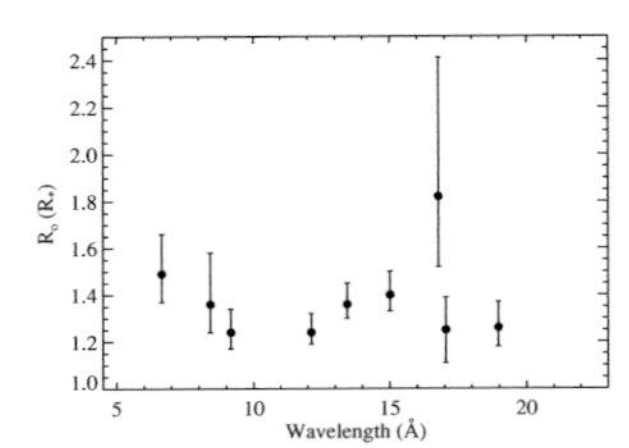

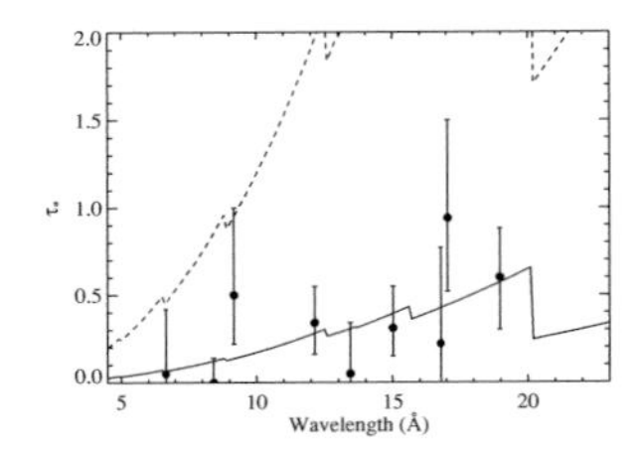

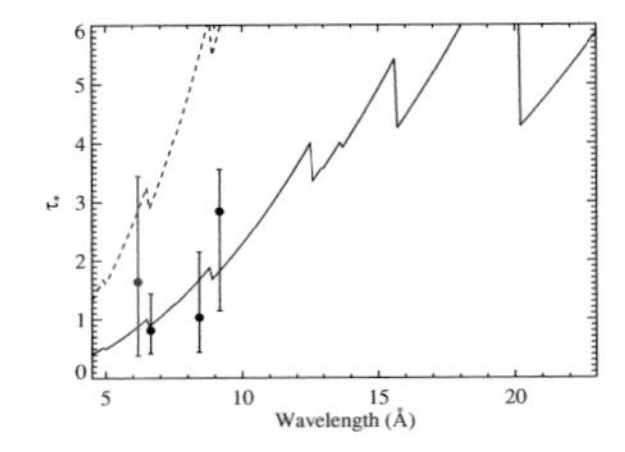

Figure 4. Results from fitting the wind-profile model to the emission lines in the *Chandra* spectrum of 9 Sgr and HD 93129A. The shock onset radii for 9 Sgr (left) are consistent with a value of $R_o = 1.4$ R_*, while the characteristic optical depths (center) are well fit by a model that has a mass-loss rate of $\dot{M} = 3.4 \times 10^{-7}$ $M_\odot$ yr^{-1} (solid line), a factor of six below the unclumped Hα mass-loss rate (dotted line). For HD 93129A we show the τ_* values (right) along with the best-fit mass-loss rate model (solid line; $\dot{M} = 6.8 \times 10^{-6}$ $M_\odot$ yr^{-1}) and for comparison, the traditional unclumped Hα mass-loss rate model. The opacity model used for HD 93129A assumes altered CNO abundances (Taresch *et al.* 1997).

4. Broadband X-ray properties

Given the strong wind absorption in the X-ray spectra of early O stars, we have modeled the broadband spectral energy distributions using simple one- and two-temperature thermal emission spectral models (e.g. APEC Smith *et al.* 2001) along with wind attenuation, using the newly published radiation transport model, *windtabs* (Leutenegger *et al.* 2010). This model accounts for the spatial distribution of the emitting plasma within the absorbing wind, and thus has a much more gradual decrease of transmission vs. fiducial optical depth than is seen in the exponential absorption model that describes an absorbing medium in between the background emitter and the observer, such as those employed in interstellar absorption models. The *windtabs* absorption model also uses a realistic photoionization opacity model that includes partially ionized metals and fully ionized H and He. We fit the *Chandra* zeroth-order spectrum (a CCD low-resolution spectrum) of HD 93129A with this APEC and *windtabs* model and find a low plasma temperature of 0.6 keV – fully consistent with the LDI simulation results – and a significant wind column density, corresponding to a mass-loss rate of $\dot{M} = 8 \times 10^{-6}$ $M_\odot$ yr^{-1}, which is consistent with the value we find from fitting the individual line profiles (shown in the third panel of Fig. 4). We show the HD 93129A zeroth-order spectrum and best-fit APEC and *windtabs* model in Fig. 5, along with the grating spectrum of the star, in which the low H-like/He-like line ratios, indicative of low plasma temperatures, can be seen.

5. Conclusions

The X-ray emission from normal massive stars can be understood in the context of embedded wind shocks due to the line-driving instability. Specifically, the X-ray emitting plasma shares the same kinematic profile as the bulk wind. It is spatially distributed throughout the wind above an onset radius of roughly $R_o = 1.5$ R_* and – from broadband modeling – the plasma temperatures are less than 10 million K, in accord with the predictions of LDI simulations of EWSs. These relatively low temperatures can be reconciled with the relatively hard observed spectra by taking wind attenuation of the soft X-rays into account. When we model the effect of wind attenuation on individual emission lines, we find that their modestly blue-shifted and asymmetric profiles can be reproduced using mass-loss rates that are lower by a factor of 3 to 6 compared to traditional mass-loss rates that ignore clumping (and are consistent with newer determinations that

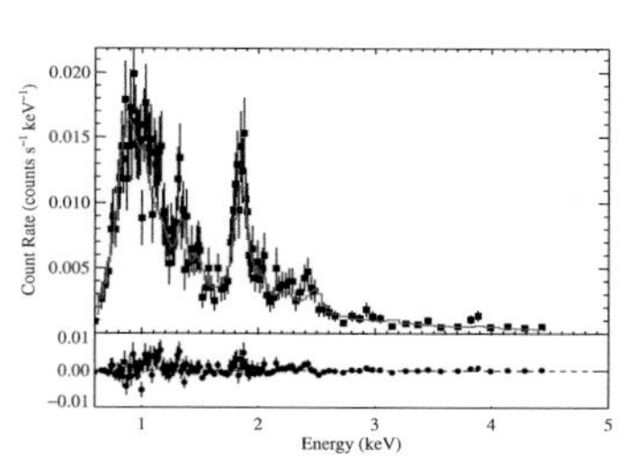

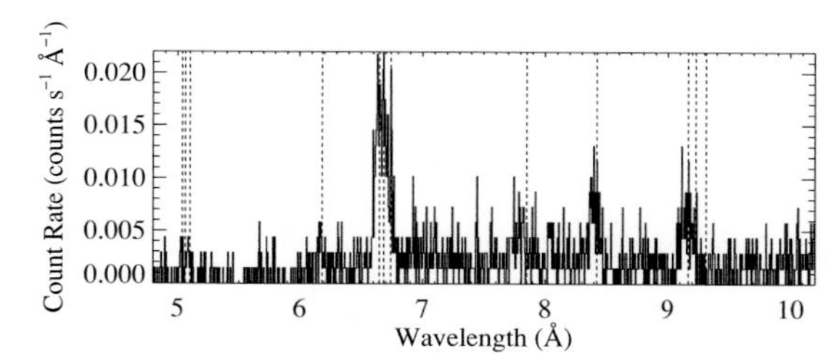

Figure 5. The zeroth-order, low-resolution *Chandra* spectrum of HD 93129A, with the best-fit thermal emission model attenuated by stellar wind absorption (left). This modeling shows that the emission temperature of the plasma is relatively low – kT = 0.6 keV – while the effects of wind attenuation are significant; explaining the observed hardness. The grating spectrum of this star is shown in the right-hand panel (dashed lines indicate rest wavelengths of important emission lines). Note that while the spectrum is quite hard (no significant emission longward of 10 Å), the Si XIV Lyα line at 6.18 Å is very weak compared to the Si XIII complex near 6.7 Å. This very low H-like/He-like line ratio is indicative of plasma with a temperature of no more than 8 million K (0.7 keV).

account for the clumping). And furthermore, we find that when we model the broadband spectral properties and account for the effects of wind attenuation using a realistic radiation transport model in conjunction with a realistic opacity model, we derive similar mass-loss rate values.

Finally, we note that in this short paper we do not have the space to discuss in detail the possible role of *porosity* in generating the only modestly blue-shifted and asymmetric profiles. Porosity arises from clumping on very large scales, where individual clumps are optically thick to X-ray photoelectric absorption (Oskinova, Feldmeier, & Hamann 2006). Our modeling suggests that porosity does not need to be invoked in order to explain the observed X-ray properties of the early O stars we discuss here. Their properties are well explained by modest mass-loss rate reductions. Furthermore, porosity requires clumping, by definition (but not the other way around). So, once clumping is invoked, and the density-squared diagnostics are adjusted accordingly, there is no longer any need to invoke porosity to explain the data. This and other aspects of porosity are addressed in the end-of-session discussion, later in these proceedings.

References

Cohen, D. H., Leutenegger, M. A., Wollman, E. E., Zsargó, J. *et al.* 2010, *MNRAS*, 405, 2391
Feldmeier, A., Puls, J. & Pauldrach, A. W. A. 1997, *A&A*, 322, 878
Lamers, H. J. G. L. M. & Leitherer, C. 1993, *ApJ*, 412, 771
Leutenegger, M. A., Cohen, D. H., Zsargó, J., Martell, E. M. *et al.* 2010, *ApJ*, 719, 1767
Markova, N., Puls, J., Repolust, T., & Markov, H. 2004, *A&A*, 413, 693
Oskinova, L. M., Feldmeier, A., & Hamann, W.-R. 2006, *MNRAS*, 372, 313
Owocki, S. P., Castor, J. I., & Rybicki, G. B. 1988, *ApJ*, 335, 914
Owocki, S. P. & Cohen, D. H. 2001, *ApJ*, 559, 1108
Puls, J., Kudritzki, R.-P., Herrero, A., Pauldrach, A. W. A. *et al.* 1996, *A&A*, 305, 171
Puls, J., Markova, N., Scuderi, S., Stanghellini, C. *et al.* 2006, *A&A*, 454, 625
Repolust, T., Puls, J., & Herrero, A. 2004, *A&A*, 415, 349
Runacres, M. C. & Owocki, S. P. 2002, *A&A*, 381, 1015
Smith, R. K., Brickhouse, N. S., Liedahl, D. A., & Raymond, J. C. 2001, *ApJ* (Letters), 556, L91
Taresch, G., Kudritzki, R. P., Hurwitz, M., & Bowyer, S. *et al.* 1997, *A&A*, 321, 531

Active OB stars: structure, evolution, mass loss, and critical limits
Proceedings IAU Symposium No. 272, 2010
C. Neiner, G. Wade, G. Meynet & G. Peters, eds.

doi:10.1017/S1743921311010775

Activity of Herbig Be stars and their environment

Evelyne Alecian

Laboratoire d'Astrophysique, Observatoire de Grenoble,
BP 53, F-38041 Grenoble Cedex 9, France
email: evelyne.alecian@obs.ujf-grenoble.fr

Abstract. The Herbig Ae/Be stars are the high-mass counterparts of the T Tauri stars, and are therefore considered as the pre-main sequence progenitors of the A/B stars. These stars are still contracting towards the main sequence, and are surrounded by dust and gas, remnants of their parental molecular cloud. In order to understand the formation processes at high mass, as well as the magnetic and rotation properties of the MS A/B stars, it is fundamental to understand the structure of the circumstellar matter of the Herbig Ae/Be stars, as well as the interaction of these PMS stars with their close environment. In this talk I will review our current knowledge about the properties of the circumstellar environment of the Herbig Ae/Be stars as well as the possible physical processes at the origin of their observed activities.

Keywords. stars: pre–main-sequence, stars: activity, stars: circumstellar matter, stars: early-type, stars: emission-line, stars: formation

---

1. Introduction

In this paper I will review the current knowledge of the environment and activity of Herbig Ae/Be (HAeBe) stars by focusing only on the observed common properties and phenomena among Herbig Ae (HAe) and Herbig Be (HBe) stars. More information on the activity and environment of HAe stars can be found in various reviews on HAeBe stars (e.g. Waters & Waelkens 1998, Dullemond & Monnier 2010). This review is certainly not exhaustive and might reflect the interest of the author.

The following sections introduce first the reader to the objects called the Herbig Ae/Be stars, then discuss the pre-main sequence evolution in the HR diagram for intermediate- and high-mass stars, in order to give the definition of the HAe and HBe stars that I will use all along the paper. Section 2 and 3 will describe the environments of these stars, while a summary is given in Section 4.

1.1. *The Herbig Ae/Be stars*

In order to find the higher mass counterparts to the T Tauri stars, pre-main sequence (PMS) stars of low-mass, Herbig (1960) defined a sub-class of objects with the following criteria: (*a*) the spectral-type is A or earlier, with emission lines, (*b*) the star lies in an obscured region, and (*c*) the star illuminates bright luminosity in its immediate vicinity. While the emission lines condition has been chosen by analogy with T Tauris stars, condition (*b*) has been defined to select young stars in close proximity with their birth-places, which excludes more evolved objects (e.g. Wolf Rayet or Be stars), and condition (*c*) guarantees the physical association of the star with its surrounding dark cloud.

Since Herbig (1960), the selection criteria evolved, and the number of members of HAeBe stars increased. Finkenzeller & Mundt (1984) remarked that the HAeBe stars are characterised with a stronger IR excess. The most complete, and most recent, catalog of HAeBe stars has been compiled by Thé *et al.* (1994, TWP hereinafter), that listed

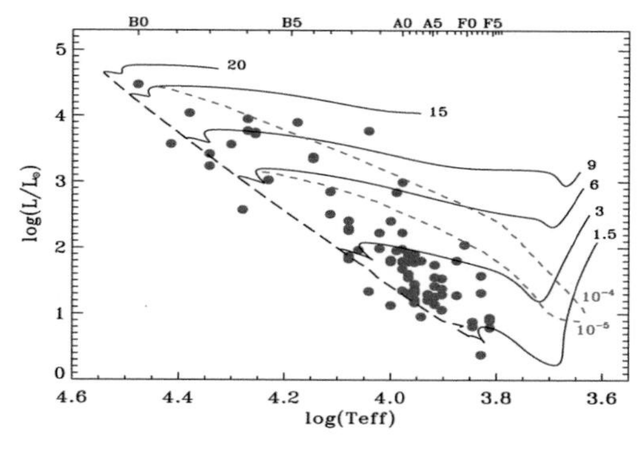

Figure 1. A sample of 80 HAeBe stars plotted in an HR diagram (Alecian *et al.* in prep.). The PMS evolutionary tracks has been computed using the code CESAM (Morel 1997).

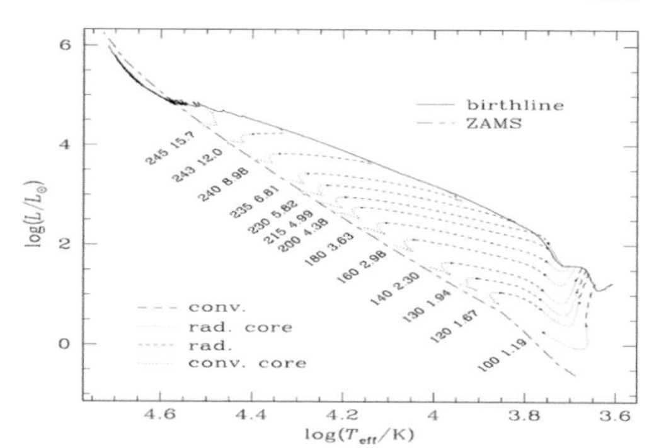

Figure 2. PMS evolutionary tracks calculated by Behrend & Maeder (2001). Couples of numbers represent the age of the stars on the ZAMS (in unit of 10^3 yr), and its mass.

108 HAeBe members, all of spectral type Ae and Be, as well as 23 probable members with spectral type Fe. All these stars respect one of the HAeBe criteria chosen by TWP: a strong near- or far-IR excess. In addition, the members and probable members can fulfill one or more Herbig (1960) criteria, as well as many others such as: an anomalous extinction law, or the Mg II (2800 Å) doublet in emission.

Since the work of Thé *et al.* (1994), the selection of Herbig Ae/Be members has been mainly done with the presence of IR excess and/or emission lines (e.g. Hernández *et al.* 2004), and includes spectral types of F5 and higher. Not all HAeBe stars can be associated with an obscured region or a bright nebulosity, but most of them are (see Vieira *et al.* 2003). New HAeBe candidates have been identified by many authors, often associated with star forming regions, or young clusters or associations (e.g. Pauzen *et al.* 2007). Today the exact number of HAeBe stars is unknown, but from the works of Thé *et al.* (1994) and other authors previously cited, we can estimate the number of field HAeBe stars around 140.

1.2. *Evolutionary status of the Herbig Ae/Be stars*

The pre-main sequence nature of the HAeBe stars has been confirmed by Strom *et al.* (1972) who demonstrates that all the stars of their sample have surface gravities appropriate to pre-main sequence or zero-age main sequence stars. Alecian *et al.* (in prep.) made a spectropolarimetric suvey of about 80 stars, and plotted all the stars in the HR diagram (Fig. 1)superimposed with PMS evolutionary tracks at different masses and the zero-age main sequence (ZAMS), computed with CESAM 2K (Morel 1997), as well as the birthlines computed by Palla & Stahler (1992) using two mass accretion rates during the proto-stellar phase: 10^{-5} and 10^{-4} $M_{\odot}.yr^{-1}$. The birthlines are defined by the loci in the HR diagram where newly born stars become visible in the optical wavelength.

Palla & Stahler (1993, PS93 hereinafter) argue that the mass accretion rate during the proto-stellar phase should be identical at all masses and equal to 10^{-5} $M_{\odot}.yr^{-1}$, due to the lack of HBe stars more massive than about 6 $M_{\odot}$. Since their work, many massive HAeBe stars have nonetheless been identified (Fig. 1), and the question about the appropriate mass accretion rate is still unanswered. It seems reasonable to assume that the protostellar accretion rate could vary within a reasonable range depending on the mass of the core. Behrend & Maeder (2001, BM01 hereinafter) proposes that the mass accretion rate depends of the mass of the growing star. As a result, the mass accretion rate should therefore be larger for larger mass of the formed star. In Fig. 2 are plotted the birthline and the PMS evolutionary tracks calculated by BM01. While, with a unique accretion rate of 10^{-5} $M_\odot$.yr^{-1} as proposed by PS93, the mass maximum of a PMS star is around 6 $M_\odot$, a modulated mass accretion rate, as proposed by BM01, allow the birthline to reach the ZAMS at much higher masses (around 20 $M_\odot$).

Massive Herbig Ae/Be stars are therefore theoretically predicted, and are observed. I will call the Herbig Be (HBe) stars, stars with masses larger than 5 $M_\odot$, and Herbig Ae (HBe), stars with masses comprised between 1.5 and 5 $M_\odot$. In Fig. 1, most of the Herbig Be stars have spectral types earlier than B4. For simplicity, I will consider in the following that B4 traces a reasonable limit between HAe and HBe stars. The following of this paper will justify the choice of this limit.

In Fig. 1, we observe easily that the number of HBe stars is much smaller than the number of HAe stars. Two main reasons can explain this: (i) the PMS evolution at high-mass is much faster than at intermediate-mass, and (ii) the confusion can very often be made between classical Be stars and Herbig Be stars. Furthermore, for long time it was assumed that massive Herbig Be stars couldn't exist. For all these reasons our knowledge of the environment of the Herbig Be stars is less developed than for the Herbig Ae stars, as will be shown in the following of this paper.

2. Properties of the environment of the Herbig Ae/Be stars

2.1. *The gaseous environment*

2.1.1. *Large-scale molecular surroundings*

The physical association of Herbig Ae/Be stars with their associated dark clouds has been demonstrated by Finkenzeller & Jankovics (1984), by measuring no significant motion of a sample of 27 HAeBes relative to the clouds. The first radio surveys of these stars seem to show that they are located near the edges of their parent cores. The characteristic sizes were evaluated between 0.1 and 0.8 pc, and their masses between 150 and 2000 $M_\odot$. Large column density gradients, near the positions of the central stars, suggest a gas clearing, either by the interaction between the stars and the cores, or by internal molecular dissociation (e.g. Hillenbrand 1995).

Molecular outflows (Fig. 3) have also been detected in many Herbig Ae/Be stars (e.g. Levreault 1988, Matthews *et al.* 2007). Many of them appear bipolar. Their sizes are ranging from 0.07 to 5 pc, and the expansion velocities are estimated between 6 and 60 km.s^{-1}. While outflows are commonly observed in young stellar objects, the physical processes at the origin of these outflows are still highly debated, but seem to be strongly correlated with jets driven by the central young stars (e.g. Bachiller 1996).

2.1.2. *Gaseous structure of the environment at lower scales*

The interferometric millimeter and submillimeter observations of the environment of many HAeBe stars allowed to explore the gas and dust closer to the star, and revealed disk of cold gas in Keplerian rotation (Fig. 4), sometimes with a central gap (e.g. Mannings & Sargent 1997, 2000). Panić *et al.* (2010) have even measured a difference in temperature between one side and the other of the disk. The origin of this asymmetry is still not understood. The sizes of the disks have been estimated between $\sim$ 85 and $\sim$ 450 AU, while their masses range from $\sim$ 0.005 to $\sim$ 0.03 $M_\odot$. The work of Dent *et al.* (2005) revealed a decrease of the disk mass with time.

The inner gas structure ($<$ 100 AU) of the environment of the Herbig Ae/Be stars can be probed using the IR and UV signatures of molecules. The first studies of the inner gas have used the infra-red (IR) spectral lines of CO and H^2O, which are consistent

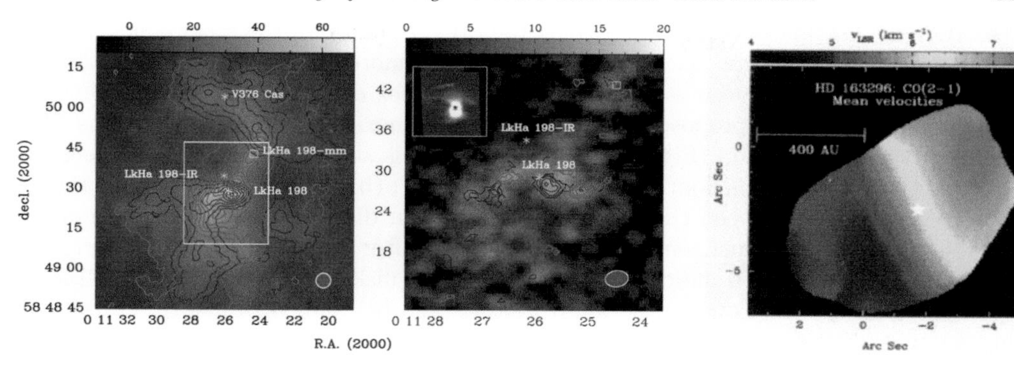

Figure 3. Gray scale showing the zeroth moment over the entire range of CO emission toward the region surrounding LkHα 198 (*left*), and centered on LkHα 198 (*right*). Optically visible HAeBe stars are marked as stars, while the embedded millimeter source is indicated by the square. Contours illustrate the emission toward us (blue) and opposite to us (red) (Matthews *et al.* 2007).

Figure 4. Intensity weighted-mean velocities of the CO radio emission observed around HD 163296. The star symbol represents the star position (Mannings & Sargent 1997).

with hot gas distributed in a disk (e.g. Najita *et al.* 2007). However the properties of the gaseous disk derived from these minor tracers must be considered with caution, as many assumptions used in the models are poorly known (see Deleuil *et al.* 2010).

H^2, the main gaseous constituent of the close environment of HAeBe stars, is difficult to detect in the IR, and can be easily contaminated with background extended emission (e.g. Thi *et al.* 2001, Sheret *et al.* 2003). While many attempts of H^2 detection have been made for HAeBe stars (Carmona *et al.* 2008), only two positive detections have been reported. The inner gas seem to be located in a disk within $\sim$ 35 AU or less, and its mass is estimated in a range of 10^{-2} to 1 M_{Jup} (e.g. Martin-Zaïdi *et al.* 2007).

Contrary to the IR-domain, thousands of H^2 emission or absorption lines can be observed in ultraviolet (UV), and have been detected in many HAeBe stars (e.g. Bouret *et al.* 2003). The few studies of the UV H^2 lines, that can trace warm gas within 1 AU, revealed the presence of a flared-disk and warm and/or hot excited media very close to the HAe stars, while the observations of the HBe stars are more consistent with a photodissociation region (PDR), and therefore a large circumstellar envelope (e.g. Martin-Zaïdi *et al.* 2008). Martin-Zaïdi and collegues argue that the difference observed between HAe and HBe stars comes from the fastest evolution of the most massive stars, around which it is more likely to find molecular remnants.

Vink *et al.* (2002) propose to probe the structure of the innermost environment of HAeBe stars by studying the linear polarisation accross the emission profile Hα. Their work show that the regions emitting Hα are flattened on small scales. Based on differences in the signature of the linear polarisation between HAe and HBe stars, they argue that while in HBe the scenario of a classical accretion from the disk to the star is favoured, magnetospheric accretion is more likely to happen around the Herbig Ae stars.

2.2. *The dusty environment*

2.2.1. *The spectral energy distribution*

The IR-excesses of the HAeBe stars, are assumed to come from the dust in the close environment of the star, reprocessing the stellar light. Hillenbrand *et al.* (1992) analysed the spectral energy distribution (SED) of 47 HAeBe stars, and showed that 30 HAeBe stars of their sample display SED consistent with a model of disk with a central hole.

They derived disk masses in the range 0.01 to 6 $M_{\odot}$, the radii of the inner edges from 15 to 175 AU, as well as mass accretion rates, for which more realistic values (from 3.10^{-9} to 10^{-6} $M_{\odot}.yr^{-1}$) have been estimated by Garcia Lopez *et al.* (2006).

Malfait *et al.* (1998) proposed an evolutionary scenario, based on the various SED shapes of the HAeBe stars (Fig. 5): (a) first the star is still embedded in its molecular cloud, in which the SED is dominated by the IR excess ; (b) the environment is flattened, with a central hole between the star and its accretion disk, and the SED can now be reproduced by a single-dust temperature model ; (c) after some time a two-temperature model fit better the SED, indicating a gap inside the disk, that could be due to planet formation ; the two last steps describe β-Pictoris like (d) and Vega-like (e) stars, with only far-IR excess, that decreases with time as the dust disk vanishes.

Other authors (e.g. Meeus *et al.* 1998) propose that the differences observed in the SED can also be due to the line of sight inclination, the inner disk holes, and other physical parameters. Miroshnichenko *et al.* (1997) even proposes that the SED shapes can be reproduces with spherical envelopes. The modeling of the SED to probe the circumstellar dust of HAeBe stars should therefore be done with other restraining observations, such as interferometric data (e.g. Eisner *et al.* 2004).

2.2.2. *Direct detection*

Direct evidence of dust disks have first been obtained with continuum radio observations, showing disks of smaller size and much less massive ($\sim 10^{-4}$ $M_{\odot}$) than the gas disk (Mannings & Sargent 2000). Coronography and IR imagery allowed to detect the disk of few stars with sometimes the inner hole or a gap inside the disk (e.g. Augerau *et al.* 2001, Wisniewski *et al.* 2008). Adaptive optic have also been able to image the disc around few Herbig Ae/Be stars (e.g. Chen *et al.* 2006). The inner cavities appear optically thin in the HAe stars, while there is evidence that optically thick inner cavities are more often observed in HBe stars (Monnier *et al.* 2005, Kraus *et al.* 2008).

IR interferometric data of various HAeBe stars have been interpreted as flared disk with an inner puffed-up rim (e.g. Doucet *et al.* 2007, Eisner *et al.* 2007). The flared disk scenario have also found strong support with the work of Lagage *et al.* (2006), by observing an offset between the peak flux of the emission of the circumstellar material, and the geometrical center of the image of the emission (see also Okamoto *et al.* 2009). IR imaging polarimetry (e.g. Hales *et al.* 2006) and milimeter interferometry (Raman *et al.* 2006) bring also other evidence of flaring disks.

2.2.3. *Disk properties*

The inferred disk scales from the various data vary from 0.3 to 1000 AU (e.g. Eisner *et al.* 2003, Okamoto *et al.* 2009). The work of Eisner *et al.* (2007) revealed a radial temperature gradient systematically steeper in HBe stars, than in HAe stars.

IR spectroscopy of many HAe/Be stars have revealed the presence of PAH (Polycyclic Aromatic Hydrocarbon) as well as silicate (e.g. Acke & van den Ancker 2004). No correlation is observed between PAH and silicate emission, however a clear correlation between PAH and the mid-IR excess is observed. Meeus *et al.* (2001) proposed a model to reproduce these observations (Fig.6): the disk around HAeBe stars is composed of 3 components: (I) an optically thick, geometrically thin disk causing the near and far-IR excess (observed in all the stars), (II) an optically thin/thick inner puffed-up rim causing the silicate emission (observed in all the stars), and (III) a geometrically thick, optically thin flared dust layer below and above the midplane of the optically thick, geometrically thin disk, responsible of the PAH emission and the strong mid-IR excess (only observed in stars displaying PAH emission). All stars possess components (I) and (II), while only

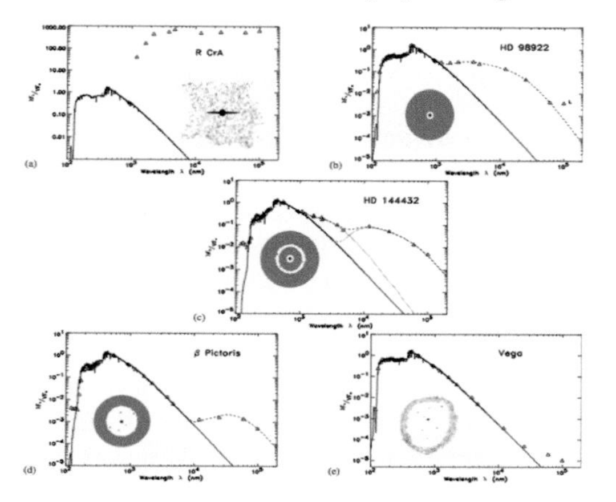

Figure 5. The evolutionary scenario proposed by Malfait *et al.* (1998).

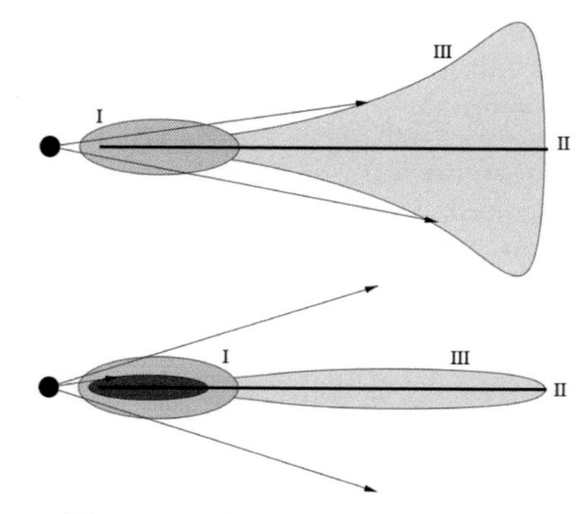

Figure 6. The model of HAeBe disks proposed by Meeus *et al.* (2001).

few of them possess the flared components. In the other stars an optically thin dust layer is still present but cannot be flared due to the optically thick puffed-up inner rim that block the UV radiation from the stars. This qualitative model have been developed quantitatively by Dullemond *et al.* (2001), and tested with success on many HAeBe star (Eisner *et al.* 2003, Acke & van den Ancker 2004).

According to Dullemond *et al.* (2001), the disk inner rim is heated due to the direct exposition to the stellar flux, and therefore is puffed-up. The assumption of a puffed-up inner rim is required to reproduce the near-IR and interferometric properties of HAeBe stars (e.g. Natta et al. 2001, Tatulli *et al.* 2007). Furthermore, a puffed-up rim can only appear if the inner part of the disk is optically thin. The observed inner holes in many HAeBe stars (e.g. Grady *et al.* 2005), as well as the low mass accretion rate derived by Garcia Lopez *et al.* (2006) are therefore additional arguments in favour of this model.

The origin of optically thin cavities observed between the stars and the innermost edge of the disk is still not well understood, and different theories exist. The presence of a magnetosphere, and the action of magnetospheric accretion could be one of the reasons of the clearance. Indication of magnetospheric accretion have ben observed in HAe stars (e.g. Muzerolle *et al.* 2004, Mottram *et al.* 2007). Another theory is the planet formation. Many clues of ongoing planet formation can be found in the literature, such as discret absorption component in the UV and optical spectra (e.g.Grady *et al.* 2006), or long-term photometric cyclicity (e.g. Shevchenko & Ezhkova, 2001).

In the case of HBe stars, their UV radiation could be sufficient for photodissociating the innermost part of the disk. Many clues of photodissociation around HBe stars have been obtained recently: Okamoto *et al.* (2009) have detected photoevaporation tracers in the IR spectrum of HD 200775 ; the mass of the disks around HBe stars are usually 5 to 10 times smaller than those around lower mass stars (Alonso-Albi *et al.* 2009) ; and Berné *et al.* (2009) showed HBe IR spectra typical of photodissociated regions.

All these recent results show that the impact of the stellar radiation on the gas and matter surrounding HBe stars is relatively important in determining the structure of the close environment of these stars.

2.2.4. *Accretion diagnostics*

While many clues of magnetospheric accretion onto the T Tauri stars exist, it is not clear if accretion exists and how it operates onto HAeBe stars. Contrary to the T Tauri stars, there isn't any clear evidence of magnetospheric accretion, such as veiling, or the presence of blueshifted forbidden emission lines (e.g. [OI] 6300 Å). We observe, however, in some stars, Hα profiles similar to the T Tauri ones, which are therefore assumed to have the same origin. We also observe inverse P-Cygni profiles, or redshifted circumstellar absorption features. However these features can be well understood in terms of clumpy accretion, which is enforced by typical spectroscopic and photometric variability (e.g. Mora *et al.* 2004). While most of these clumpy accretion diagnostics are observed in HAe stars, we can observe them also in few HBe stars (e.g. Boley *et al.* 2009).

2.3. *Winds and jets*

Finkenzeller & Mundt (1984) have detected P Cygni profiles in the Hα and Mg II h & k lines, in a large number of HAeBe stars. The presence of P Cygni profiles definitely confirm that some Herbig Ae/Be stars possess a stellar wind. Böhm & Catala (1994) argue that the presence, in about half of their sample, of forbidden emission lines, such as [OI] (6300 Å), centered on the stellar radial velocity, indicates the presence of a stellar wind. Radio emission detected in HAeBe stars by Skinner *et al.* (1993) is predominantly thermal, and in many cases wind-related. Finally, spectro-interferometric observations, around the Brγ line, of few HAe/Be stars are better interpreted with the presence of an optically thick disk, and a stellar wind whose the apparent size is much larger than the disk extension (e.g. Malbet *et al.* 2007).

Beside the presence of large-scale molecular outflows driven by HAeBe stars, mentioned in Sec. 2.1, optical outflows, such as Herbig-Haro (HH) objects or jets, have been associated with many Herbig Ae/Be stars (e.g. Ray *et al.* 1990, Gomez *et al.* 1997). According to Mundt & Ray (1994) these outflows are 2 to 3 times faster than those driven by low-mass objects, and 70% of them are highly collimated. Recently, collimated bipolar microjets driven by HD 163296 and HD 104237, have been detected very close to the stars, at distances down to 7 AU (e.g. Grady *et al.* 2004).

The origin of the winds and jets associated with HAeBe stars is poorly understood. Corcoran & Ray (1997) propose the theory of accretion driven winds. They argue that the positive correlations observed between the forbidden emission lines [OI] and Hα, or the IR excess, imply a strong link between outflows and disk. However, in many HAeBe stars, the absence of high-mass disks (especially in HBe stars), as well as the evidence of low-mass accretion rates (e.g. Garcia Lopez *et al.* 2006), are indication of passive disk. In the HBe stars, that emit a strong radiation field, theories like radiation driven winds might be more appropriate (e.g. Babel & Montmerle 1997).

3. The activity of the Herbig Ae/Be stars

I will only expose below our knowledge on magnetic related activity of Herbig Ae/Be. The reason is that other activity types (photometric variability or pulsations) have only been reported in HAe stars (e.g. Herbst & Shevchenko 1999 ; Böhm *et al.* 2004).

3.1. *Evidence of magnetic fields in Herbig Ae/Be stars*

Many indirect evidence of magnetic fields are observed in Herbig Ae/Be stars. Highly-ionised species, such as N V or O VI, are observed in emission in the spectra of HAeBe stars (e.g. Roberge *et al.* 2001), and X-rays have also been reported in HAeBe stars (e.g. Hamaguchi *et al.* 2005). These emissions are believed to come from very high-temperature regions close to the stellar surface, such as hot corona or chromosphere (e.g. Bouret *et al.* 1997). Non-thermal radio observations of few HAeBe stars have also been reported by Skinner *et al.* (1993), suggesting a magnetic origin. Finally, rotational modulations of wind lines have been observed in few HAeBe stars, and are interpreted as being formed in winds structured by magnetic fields (e.g. Catala *et al.* 1991).

For all these reasons, Herbig Ae/Be stars have been assumed to host magnetic fields. Many attempt of direct magnetic detection have been made without much success (e.g. Catala *et al.* 1999, Wade *et al.* 2007). Until recently only one marginal detection in HD 104237 have been reported by Donati *et al.* (1997). It is only with the emergence of the new generation of spectropolarimeters ESPaDOnS and Narval, installed at the Canada-France-Hawaii Telescope (CFHT, HAwaii) and at the Telescope Bernard Lyot (TBL, France), respectively, that a large survey of HAeBe stars could be performed, leading to the detection of 8 magnetic stars, and an incidence of $\sim 6\%$ (e.g. Alecian *et al.* 2009).

The origin of the magnetic fields in Herbig Ae/Be stars and in their decedents, the main sequence A/B stars, is highly debated. These stars possess a small convective core surrounded by a large radiative envelope. Some very young HAeBe stars are even totally radiative. They therefore lack the convective envelope favourable to the generation of magnetic fields, as in the low-mass stars. It has been known for long time that about 5% of the main sequence A/B stars possess strong magnetic fields, organised on large-scales, and stable over many years (e.g. Donati & Landstreet 2009). None of the theories including a core dynamo have been able to reproduce the magnetic characteristics of these stars (Moss 2001).

Today, the favoured hypothesis is a fossil origin. This theory implies that the magnetic fields observed in the main sequence A/B stars are remnants of fields either present in molecular clouds from which these stars formed, or that they were generated by a dynamo during the early stages of star formation. The spectropolarimetric survey of HAeBe stars have brought very strong argument in favour of this hypothesis, by discovering that the progenitor of the magnetic A/B stars possess also strong magnetic fields organised on large scales, and that are stable on many years (Wade *et al.* in prep.). A fossil link has therefore been established between the PMS and the MS phases of intermediate- and high-mass stars.

3.2. *X-rays from Herbig Ae/Be stars*

The first large studies of X-ray emissions from HAeBe stars, using the satellites *Einstein* and *ROSAT* lead to an X-ray incidence close to 50% (e.g. Damiani *et al.* 1994). Thorough studies of binarity of HAeBe stars showed that most of the X-ray emission in the direction of HAeBe stars can be attributed to one or more close late-type companions (Stelzer *et al.* 2009). However in few cases, it is very unlikely that a late-type companion is responsible of the X-ray emission, and in 2 peculiar cases (AB Aur and HD 104237), based on detailed analyses of X-ray data, the companion theory has been totally rejected (e.g. Telleschi *et al.* 2007). Most interesting are the X-ray periodic variability of AB Aur and HD 104237 coinciding with the rotational periods of the stars (Telleschi *et al.* 2007, Testa *et al.* 2008), but also with the period of the modulations of non-photospheric lines formed in the winds (e.g. Catala *et al.* 1999). In addition, AB Aur and HD 104237 show the presence of emission lines of highly ionised species, confirming the existence of high-energy phenomena in their surroundings. Besides, Damiani *et al.* (1994) observed that the HBe stars are X-ray brighter than HAe stars. Such a correlation is not expected in the low-mass companion theory. Some HAeBe stars are therefore very likely X-ray emitters. Among them we find 2 HBe stars HD 259431 and HD 200775, that share some of the high-energy properties of AB Aur and HD 104237. These stars are therefore not unique and seem to describe a specific sub-class.

The most promising theories, capable of explaining the observed X-ray properties of this class of stars, involve magnetic fields: the Corotating Interaction Regions (CIR,

Bouret *et al.* 1997) and the Magnetically Confined Winds Shocks (MCWS, Babel & Montmerle 1997). However, no correlation is observed between the presence of magnetic fields and the emission of X-rays in HAeBe stars : while HD 104237 and HD 200775 have been detected as magnetic, AB Aur and HD 259431 do not possess strong magnetic fields. On the other side, Bouret *et al.* (1997) argue that a surface magnetic field of only 100 G in the CIR theory would be sufficient to reproduce the UV and X-ray properties of AB Aur, which is below our limit of magnetic detection ($\sim$ 500 G, Wade *et al.* in prep.) Magnetic fields, too faint to be detected, could therefore exist in more HAeBe stars, and could be at the origin of the X-rays in HAeBe stars.

4. Summary

At large-scale the molecular gas surrounding the Herbig Be stars seem to be concentrated in disks. Molecular outflows, such as jets and winds are also observed at very large-scales. Molecular gas has been detected down to few AU, and is very likely distributed in flared disks, but also in hot media very close to the star. Spectropolarimetric and UV observations seem to describe differences between HAe and HBe stars, the latter being surrounded with classical accretion disk and molecular remnants, while the former could experience magnetospheric accretion, and lack of molecular remnants.

Combined with other type of data (e.g. UV, optical, or IR spectroscopy), the analysis of the SED of HAeBe stars reached the conclusion that HBe stars are also surrounded with dusty disk, which have been directly detected using various instrumentation. The current favoured scenario consists of flared-disk with a puffed-up inner rim, heated by the radiation field of the star. The photometric and spectroscopic differences observed between HAe and HBe stars could be explain by the absence of flared-disk in HBe stars, either because the disk would have been photodissociated, or because the puffed-up inner rim is too opaque for allowing the external part of the disk to flare, or a combination of both. Many other clues have been recently obtained indicating photodissociation around HBe stars. This could therefore explain the presence of much less massive disk around HBe than around HAe stars. The strong radiation field of the HBe stars seem to have a largest impact on the structure of its environment, than in HAe stars.

While it sounds evident that PMS HBe stars are still accreting from their disk, there is no clear indication of accreting matter onto the stars. The nature and physical origin of this potential accretion is still an open question.

Strong clues of stellar winds have been reported from spectroscopic and interferometric observations. Optical jets and Herbig-Haro objects have also been associated with many HAeHBe stars. The physical processes at the origin of these outflows are not known. While accretion driven mechanisms could be active in HAe stars, the evidence of passive disks around HBe stars does not favoured this hypothesis.

While indirect magnetic proofs have been observed for long time in HAeBe stars, it is only recently that direct detections of magnetic fields in HAeBe stars have been obtained. Their characteristic strongly support a fossil link between PMS and MS A/B stars. However the detailed processes of this theory during the star formation are not known, and should be investigated in the future in order to validate it.

There are convincing arguments that a small number of HAeBe stars emit X-rays. While the favoured hypotheses for their origin include all magnetic fields, no clear observational link has been drawn between X-rays and magnetic fields. More investigation of these theories and of the X-ray and environment properties of these stars are required in order to understand the high-energy phenomena observed around them.

To conclude, many work has been done in order to understand the properties of the environment of the HAeBe stars. However most of them concern only HAe stars, which is due to a very small number of known HBe stars. In order to progress in this domain, it is crucial to increase this number. A very efficient way is to observe very young clusters and associations. It would increase the HBe number by factor of at least 10. However these clusters are far away and magnitude limited for the current ground instrumentation. In the future it would therefore be very helpful to increase the sensibility of our instrumentation and to be able to use it on bigger telescope (the 8m and 40m classes).

References

Acke, B. & van den Ancker, M. E. 2004, *A&A*, 426, 151

Akeson, R. L., Ciardi, D. R., van Belle, G. T., Creech-Eakman, M. J. *et al.* 2000, *ApJ*, 543, 313

Alecian, E., Wade, G. A., Catala, C., Bagnulo, S. *et al.* 2009, *MNRAS*, 400, 354

Alonso-Albi, T., Fuente, A., Bachiller, R., Neri, R. *et al.* 2009, *A&A*, 497, 117

Augereau, J. C., Lagrange, A. M., Mouillet, D., & Ménard, F. 2001, *A&A*, 365, 78

Babel, J. & Montmerle, T. 1997, *A&A*, 323, 121

Bachiller, R. 1996, *ARAA*, 34, 111

Behrend, R. & Maeder, A. 2001, *A&A*, 373, 190

Berné, O., Joblin, C., Fuente, A., & Ménard, F. 2009, *A&A*, 495, 827

Böhm, T. & Catala, C. 1994, *A&A*, 290, 167

Böhm, T., Catala, C., Balona, L., & Carter, B. 2004, *A&A*, 427, 907

Boley, P. A., Sobolev, A. M., Krushinsky, V. V., van Boekel, R. *et al.* 2009, *MNRAS*, 399, 778

Bouret, J.-C., Catala, C., & Simon, T. 1997, *A&A*, 328, 606

Bouret, J.-C., Martin, C., Deleuil, M., Simon, T. *et al.* 2003, *A&A*, 410, 175

Cantó, J., Rodríguez, L. F., Calvet, N., & Levreault, R. M., 1984, *ApJ*, 282, 631

Carmona, A., van den Ancker, M. E., Henning, T., Pavlyuchenkov, Y. *et al.* 2008, *A&A*, 477, 839

Catala, C., Czarny, J., Felenbok, P., Talavera, A. *et al.* 1991, *A&A*, 244, 166

Catala, C., Donati, J. F., Böhm, T., Landstreet, J. *et al.* 1999, *A&A*, 345, 884

Chen, X. P., Henning, T., van Boekel, R., & Grady, C. A. 2006, *A&A*, 445, 331

Corcoran, M. & Ray, T. P. 1997, *A&A*, 321, 189

Damiani, F., Micela, G., Sciortino, S., & Harnden, Jr., F. R. 1994, *ApJ*, 436, 807

Deleuil, M., Bouret, J. C., Feldman, P., Lecavelier Des Etangs, A. *et al.* 2010, in: T. Montmerle, D. Ehrenreich, & A.-M. Lagrange (eds.), *EAS Publications Series* 41, p. 155

Dent, W. R. F., Greaves, J. S. & Coulson, I. M. 2005, *MNRAS*, 359, 663

Donati, J.-F., Semel, M., Carter, B. D., Rees, D. E. *et al.* 1997, *MNRAS*, 291, 658

Donati, J.-F. & Landstreet, J. D. 2009, *ARAA*, 47, 333

Doucet, C., Habart, E., Pantin, E., Dullemond, C. *et al.* 2007, *A&A*, 470, 625

Dullemond, C. P., Dominik, C., & Natta, A. 2001, *ApJ*, 560, 957

Dullemond, C. P. & Monnier, J. D. 2010, *ARAA*, 48, 205

Eisner, J. A., Lane, B. F., Akeson, R. L., Hillenbrand, L. A. *et al.* 2003, *ApJ*, 588, 360

Eisner, J. A., Lane, B. F., Hillenbrand, L. A., Akeson, R. L. *et al.* 2004, *ApJ*, 613, 1049

Eisner, J. A., Chiang, E. I., Lane, B. F., & Akeson, R. L. 2007, *ApJ*, 657, 347

Finkenzeller, U. & Mundt, R. 1984, *A&AS*, 55, 109

Finkenzeller, U. & Jankovics, I. 1984, *A&AS*, 57, 285

Garcia Lopez, R., Natta, A., Testi, L., & Habart, E. 2006, *A&A*, 459, 837

Geisel, S. L. 1970, *ApJ* (Letters), 161, L105

Gomez, M., Kenyon, S. J., & Whitney, B. A. 1997, *AJ*, 114, 265

Grady, C. A., Woodgate, B., Torres, C. A. O., Henning, T. *et al.* 2004, *ApJ*, 608, 809

Grady, C. A., Woodgate, B., Heap, S. R., Bowers, C. *et al.* 2005, *ApJ*, 620, 470

Grady, C. A., Williger, G. M., Bouret, J.-C., Roberge, A. *et al.* 2006, in: G. Sonneborn, H. W. Moos, & B.-G. Andersson (eds.), *Astrophysics in the Far Ultraviolet: Five Years of Discovery with FUSE*, ASP-CS 348, p. 281

Hales, A. S., Gledhill, T. M., Barlow, M. J., & Lowe, K. T. E. 2006, *MNRAS*, 365, 1348

Hamaguchi, K., Yamauchi, S., & Koyama, K. 2005, *ApJ*, 618, 360
Herbig, G. H. 1960, *ApJS*, 4, 337
Herbst, W. & Shevchenko, V. S. 1999, *AJ*, 118, 1043
Hernández, J., Calvet, N., Briceño, C., Hartmann, L. *et al.* 2004, *AJ*, 127, 1682
Hillenbrand, L. A., Strom, S. E., Vrba, F. J., & Keene, J. 1992, *ApJ*, 397, 613
Hillenbrand, L.A., 1995, *PhD thesis*, University of California
Imhoff, C. L. 1994, in: P. S. The, M. R. Perez, & E. P. J. van den Heuvel (eds.), *The Nature and Evolutionary Status of Herbig Ae/Be Stars*, ASP-CS 62, p. 107
Kenyon, S. J. & Hartmann, L. 1995, *ApJS*, 101, 117
Kraus, S., Preibisch, T., & Ohnaka, K. 2008, *ApJ*, 676, 490
Lagage, P.-O., Doucet, C., Pantin, E., Habart, E. *et al.* 2006, *Science*, 314, 621
Levreault, R. M. 1988, *ApJS*, 67, 283
Malbet, F., Benisty, M., de Wit, W.-J., Kraus, S. *et al.* 2007, *A&A*, 464, 43
Malfait, K., Bogaert, E., & Waelkens, C. 1998, *A&A*, 331, 211
Mannings, V. & Sargent, A. I. 1997, *ApJ*, 490, 792
Mannings, V. & Sargent, A. I. 2000, *ApJ*, 529, 391
Marconi, M. & Palla, F. 1998, *ApJ* (Letters), 507, L141
Martin-Zaïdi, C., Lagage, P.-O., Pantin, E., & Habart, E. 2007, *ApJ* (Letters), 666, L117
Martin-Zaïdi, C., Deleuil, M., Le Bourlot, J., Bouret, J.-C. *et al.* 2008, *A&A*, 484, 225
Matthews, B. C., Graham, J. R., Perrin, M. D., & Kalas, P. 2007, *ApJ*, 671, 483
Meeus, G., Waelkens, C. & Malfait, K. 1998, *A&A*, 329, 131
Meeus, G., Waters, L. B. F. M., Bouwman, J., van den Ancker, M. E. *et al.* 2001, *A&A*, 365, 476
Miroshnichenko, A., Ivezic, Z. & Elitzur, M. 1997, *ApJ* (Letters), 475, L41
Monnier, J. D., Millan-Gabet, R., Billmeier, R., Akeson, R. L. *et al.* 2005, *ApJ*, 624, 832
Mora, A., Eiroa, C., Natta, A., Grady, C. A. *et al.* 2004, *A&A*, 419, 225
Morel, P. 1997, *A&AS*, 124, 597
Moss, D. 2001, in: G. Mathys, S. K. Solanki, & D. T. Wickramasinghe (eds.), *Magnetic Fields Across the Hertzsprung-Russell Diagram*, ASP-CS 248, p. 305
Mottram, J. C., Vink, J. S., Oudmaijer, R. D., & Patel, M. 2007, *MNRAS*, 377, 1363
Mundt, R. & Ray, T. P. 1994, in: P. S. The, M. R. Perez, & E. P. J. van den Heuvel (eds.), *The Nature and Evolutionary Status of Herbig Ae/Be Stars*, ASP-CS 62, p. 237
Muzerolle, J., D'Alessio, P., Calvet, N., & Hartmann, L. 2004, *ApJ*, 617, 406
Najita, J.R., Carr, J.S., Glassgold, A.E., & Valenti, J.A., 2007, in: B. Reipurth, D. Jewitt & K. Keil (eds.), *Protostars and Planets V* (University of Arizona Press), p. 507
Natta, A., Prusti, T., Neri, R., Wooden, D. *et al.* 2001, *A&A*, 371, 186
Okamoto, Y. K., Kataza, H., Honda, M., Fujiwara, H. *et al.* 2009, *ApJ*, 706, 665
Palla, F. & Stahler, S. W. 1992, *ApJ*, 392, 667
Palla, F. & Stahler, S. W. 1993, *ApJ*, 418, 414
Panić, O., van Dishoeck, E. F., Hogerheijde, M. R., Belloche, A. *et al.* 2010, *A&A*, 519, A110
Paunzen, E., Netopil, M., & Zwintz, K. 2007, *A&A*, 462, 157
Raman, A., Lisanti, M., Wilner, D. J., Qi, C. *et al.* 2006, *AJ*, 131, 2290
Ray, T. P., Poetzel, R., Solf, J., & Mundt, R. 1990, *ApJ* (Letters), 357, L45
Roberge, A., Lecavelier des Etangs, A., Grady, C. A., Vidal-Madjar, A. *et al.* 2001, *ApJ* (Letters), 551, L97
Sako, S., Yamashita, T., Kataza, H., Miyata, T. *et al.* 2005, *ApJ*, 620, 347
Sheret, I., Ramsay Howat, S. K., & Dent, W. R. F. 2003, *MNRAS*, 343, L65
Shevchenko, V. S. & Ezhkova, O. V. 2001, *Astron. Lett.*, 27, 39
Skinner, S. L., Brown, A., & Stewart, R. T. 1993, *ApJS*, 87, 217
Stelzer, B., Robrade, J., Schmitt, J. H. M. M., & Bouvier, J. 2009, *A&A*, 493, 1109
Strom, S. E., Strom, K. M., Yost, J., Carrasco, L. *et al.* 1972, *ApJ*, 173, 353
Tatulli, E., Isella, A., Natta, A., Testi, L. *et al.* 2007, *A&A*, 464, 55
Telleschi, A., Güdel, M., Briggs, K. R., Skinner, S. L. *et al.* 2007, *A&A*, 468, 541
Testa, P., Huenemoerder, D. P., Schulz, N. S., & Ishibashi, K. 2008, *ApJ*, 687, 579

Thé, P. S., de Winter, D., & Pérez, M. R., 1994, *A&AS*, 104, 315
Thi, W. F., van Dishoeck, E. F., Blake, G. A., van Zadelhoff, G. J. *et al.* 2001, *ApJ*, 561, 1074
Vieira, S. L. A., Corradi, W. J. B., Alencar, S. H. P., Mendes, L. T. S. *et al.* 2003, *AJ*, 126, 2971
Vink, J. S., Drew, J. E., Harries, T. J., & Oudmaijer, R. D. 2002, *MNRAS*, 337, 356
Wade, G. A., Bagnulo, S., Drouin, D., Landstreet, J. D. *et al.* 2007, *MNRAS*, 376, 1145
Waters, L. B. F. M. & Waelkens, C. 1998, *ARAA*, 36, 233
Wisniewski, J. P., Clampin, M., Grady, C. A., Ardila, D. R. *et al.* 2008, *ApJ*, 682, 548

Discussion

WISNIEWSKI: Recent coronographic imaging surveys which examine the radial surface brightness profiles show a wide range of power-laws for Group I vs Group II Meeus sources (see e.g. Grady *et al.*, Wisniewski *et al.*), which raises very serious doubts that there is an evolutionary sequence between Group I and Group II. Also, some Group II show jets (e.g. HD 163296) which shouldn't be present if they were older systems.

GAGNE: In the magnetically detected HAeBe stars, you have magnetic dipole geometries and field strengths that can be input into modified versions of existing magnetic/wind models (RRM, RFHD, MCWS, cf. the work of Towsend *et al.* and Owocki *et al.*).

MIROSHNICHENKO: We have just finished modelling of our results of high-resolution spectro-interferometry of the Herbig Be star MWC 297 in the Brackett-gamma line region. An accretion-powered disk-wind model was used. We were able to successfully reproduce the Br-gamma line profile as well as the line and nearby continuum visibilities. This may lead to a better understanding of the mentioned problem of the hydrogen emission origin in Herbig Ae/Be stars.

Active OB stars: structure, evolution, mass loss, and critical limits
Proceedings IAU Symposium No. 272, 2010
C. Neiner, G. Wade, G. Meynet & G. Peters, eds.

doi:10.1017/S1743921311010787

Dust formation of Be stars with large infrared excess

Chien-De Lee[1] **and Wen-Ping Chen**[1,2]

[1]Graduate Institute of Astronomy, National Central University,
Jhongli 32001, Taiwan
email: `m959009@astro.ncu.edu.tw`

[2]Department of Physics, National Central University,
Jhongli 32001, Taiwan
email: `wchen@astro.nuc.edu.tw`

Abstract. Classical Be stars, in addition to their emission-line spectra, are associated with infrared excess which is attributable to free-free emission from ionized gas. However, a few with exceptionally large near-infrared excess, namely with J−H, and $H-K_s$ both greater than 0.6 mag—and excess emission extending to mid- and far-infrared wavelengths—must be accounted for by thermal emission from circumstellar dust. Evolved Be stars on the verge of turning off the main sequence may condense dust in their expanding cooling envelopes. The dust particles should be very small in size, hence reprocess starlight efficiently. This is in contrast to Herbig Ae/Be stars for which the copious infrared excess arises from relatively large grains as part of the surplus star-forming materials.

Keywords. stars: early-type, stars: evolution, stars: emission-line, Be, stars: pre–main-sequence, infrared: stars

1. Introduction

When the first emission-line star was detected by Padre Angelo Secchi in 1866 (Porter & Rivinius 2003), the term "Be stars" referred to a general class of early-type stars exhibiting emission lines in their spectra. After about a hundred years, early-type pre-main sequence stars, also with emission-line spectra, were identified by Herbig (1960) and later called Herbig Ae/Be (HABe) stars. To distinguish from HABe stars and giant emission-line stars, the early-type main-sequence stars with emission lines are called classical Be (CBe) stars.

CBe stars are fast rotators, with a near-critical equatorial speed. They also show infrared (IR) emission in excess of photospheric radiation, often attributed to free-free emission from ionized gas in an extended circumstellar envelope. This is thought to differ from the large IR excess seen in HABe stars, which arises from reprocessing of starlight by circumstellar dust left over from star formation. A few CBe stars show abnormally large IR excess, yet appear not to belong to a young stellar population. We present their IR properties and discuss the possible dust formation mechanism.

2. Infrared excess

The left panel of Fig. 1 shows where typical HABe, B[e], and CBe stars occupy in a 2MASS J-H versus H-Ks diagram. Compared with HABe or B[e] stars, the majority of CBe stars appear only slightly redder, with J-H $\lesssim 0.1$ and $H-K_s \lesssim 0.2$, than the early-type main sequence locus. A few CBe stars, however, exhibit near-IR excess as large as that of HABe stars, having (J−H) and $(H-K_s) \sim 0.6$–0.7 mag.

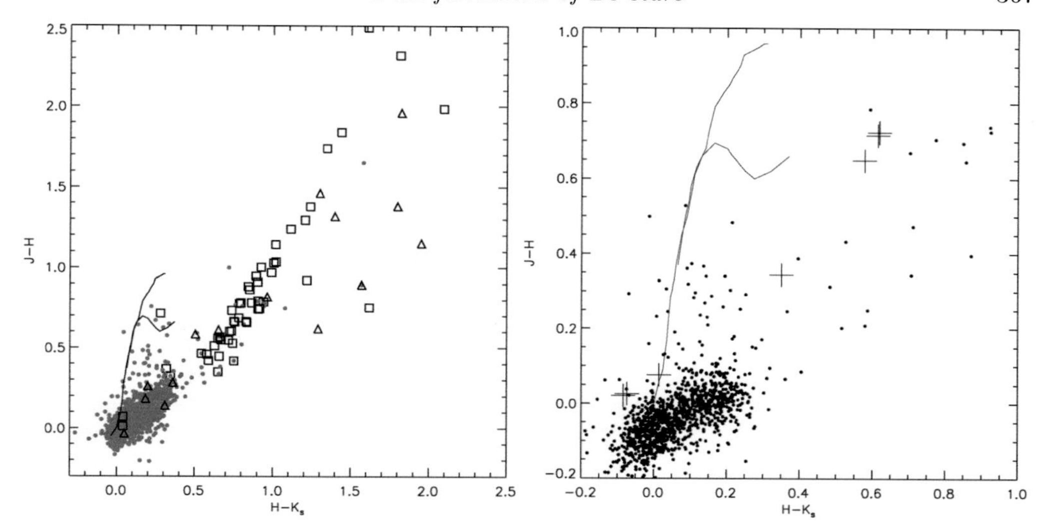

Figure 1. (*Left*): In the 2MASS J−H versus H−K$_s$ diagram, most CBe stars (grey dots) are distinctly separated from Herbig Ae/Be (squares) and B[e] stars (trangles). The sample of CBe stars is taken from Zhang, Chen & Yang (2005), and data of both Herbig Ae/Be and B[e] stars are taken from de Winter *et al.* (2001). (*Right*): The same color-color diagram but only CBe stars are shown. The crosses indicate how free-free emission with different levels of electron number density and emitting size affects the observed near-IR colors of stars. From bottom left, each cross represents an increase of an order, from 10^0 to 10^6 times in emissivity (see text) surrounging a B0 star with a gas temperature of 20,000 K (Gehrz & Hackwell 1974).

The intensity of free-free emission from an ionized gas is proportional to the square of the electron density (n_e^2) and the volume of the emitting gas (R_g^3). For CBe stars, typical values $R_g = 10^{12}$–10^{13} cm and $n_e = 10^{11}$–10^{12} cm^{-3} (Gehrz & Hackwell 1974). The right panel of Fig. 1 illustrates the extent to which free-free emission from a hot plasma alters the position of a star, here assuming a B0 photosphere, in the diagram. Setting $R_0 = 10^{12}$ cm and $n_0 = 10^{11}$ cm^{-3}, each cross represents an order increase, from bottom left of $10^0\, n_0^2 R_0^3$ moving upwards and to the right until $10^6\, n_0^2 R_0^3$. The last two crosses, for 10^5 and 10^6 times of $n_0^2 R_0^3$—unphysically large for a CBe star—are already close together; in fact any increment in density or volume sees nothing but an asymptotic approach to J-H ≈ 0.7, and H-K$_s$ ≈ 0.6.

The fact that free-free emission from a stellar envelope has this behavior is understood as follows. When the contribution of free-free emission is weak, the observed spectral energy distribution (SED), i.e., the addition of the radiation from the photosphere and from the envelope, is only slightly modified from the photospheric (blackbody) radiation. This is seen in most CBe stars. As the free-free emission, which has a fixed frequency dependence, becomes progressively prominent, it dominates the spectral running between J and H, and between H and K, thus the fixed J−H and H−K$_s$ colors. Inclusion of free-free absorption only exacerbates the case. A cool companion may also contribute to near-IR excess. But in no cases would the excess emission extend to mid- or far-IR, or beyond. The only plausible explanation for a large IR excess remain thermal emission from circumstellar dust.

Here we discuss the 6 stars in Fig. 1 which show prominent near-IR excess, namely HD 98922, HD 50138, HD 85567, CD−49 3441, HD 259431 and HD 181615/6. Fig. 2 shows their SEDs, each along with a blackbody fit with a photospheric temperature appropriate for the star, plus a single (maximum) temperature, $T_d^{\max}$ fit to the circumstellar dust. The inner radius at which T_d^{max} occurs is also labeled in each panel. The excess of HD 98922, HD 50138, HD 85567, and CD−49 3441 extends to mid-IR but they not associated with any obvious star-forming region or nebulocity, as diagnosed by DSS images, each of 1 degree field around each target. Therefore they should not be HABe stars. Whether a pre-main sequence star can be seen in isolation of any nebulocity or other young stars is an open issue. An HABe star, unlike a T Tauri star, lacks the youth signature such the lithium absorption line in the spectrum, so difficult to distinguish from an early-type main-sequence star. The short pre-main sequence lifetime also makes it difficult even for a run-away HABe star to transverse afar from its parental cloud.

HD 259431 could be an HAeBe star, not only because the IR excess has an increasing trend toward far-IR but it is seen against a dark cloud. The IR excess however is smaller than that of a typical HABe star, so it is still possible to be a main-sequence Be star.

HD 181615/6 is a binary which the SED (see Fig. 2) consistent with what Samus *et al.* (2004) suggested, i.e., with an O9 V star plus a supergiant B8 I with the companion suffering a significant extinction ($A_V \sim 15$ mag). The supergiant component is believed to be embedded in an envelope (Dudley & Jeffery 1990), suffering a significant extinction ($A_V \sim 15$ mag). Our SED analysis should be taken with caution if the observations at different wavelengths have been collected at separate epochs, so a particular component in the binary might dominate a certain wavelength range in a particular orbital phase. It is nevertheless obvious that while a companion may contribute, at least partly, to the near-IR excess, it cannot account for the increasing excess extending to mid- and far-IR wavelengths. There must ample of dust grains in their atmosphere(s).

3. Dust formation of evolved Be stars

The large IR excess of the 6 Be stars must come mainly from circumstellar dust emission. One of them may be HABe stars, for which the dust is part of the surplus material from star formation. These dust grains have grown in dense molecular clouds to micron to sub-micron sizes. At least some of these stars are CBe stars. Schild & Romanishin (1976) suggested that the fast rotation is the result of conservation of angular momentum, as a CBe star uses up its hydrogen in the contracting core. Singh & Chaubey (1987) studied some 30 CBe stars and concluded that they are indeed near the turn-off points of the main sequence, in support of the core contraction scheme. As such, a CBe star puffs off mass in a nonspherical fashion, likely in a toroid or disk configuration, because of the fast rotation. The expanding material then cools off and condenses to form dust grains, which reprocess starlight to produce the observed excessive IR emission. The freshly condensed dust grains — in contrast to those in HABe stars — could be tiny, e.g., nano-particles, so are efficient emitters. In particular, the disk geometry allows a CBe star to suffer relatively little optical extinction, as is the case for many CBe stars. The star-disk boundary layer produces the characteristic emission lines, just like in an HABe star.

To delineate the evolutionary status, CBe stars in star clusters are utilized. Slettebak (1985) found CBe stars in two open clusters, NGC 3766 and NGC 4755, to be at any stage on the main sequence. In a sample of some 100 CBe stars in about 50 star clusters, Fabregat & Torrejon (2000) concluded that star clusters younger than 10 Myr tend to lack of CBe stars, with the maximum incidence occurring at ages of 13–25 Myr. This implies that CBe stars already evolve to a late phase on the main sequence. The double clusters h and χ with an age of 13.5–14 Myr (Currie *et al.* 2010), is at the stage for an early-B type star to turn off, and most of the CBe stars in the double clusters are, indeed, very close to the end of main sequence. Furthermore, Currie *et al.* (2008) presented a JHK_s two-color diagram of h and χ Per in which a few CBe stars show very large near-IR

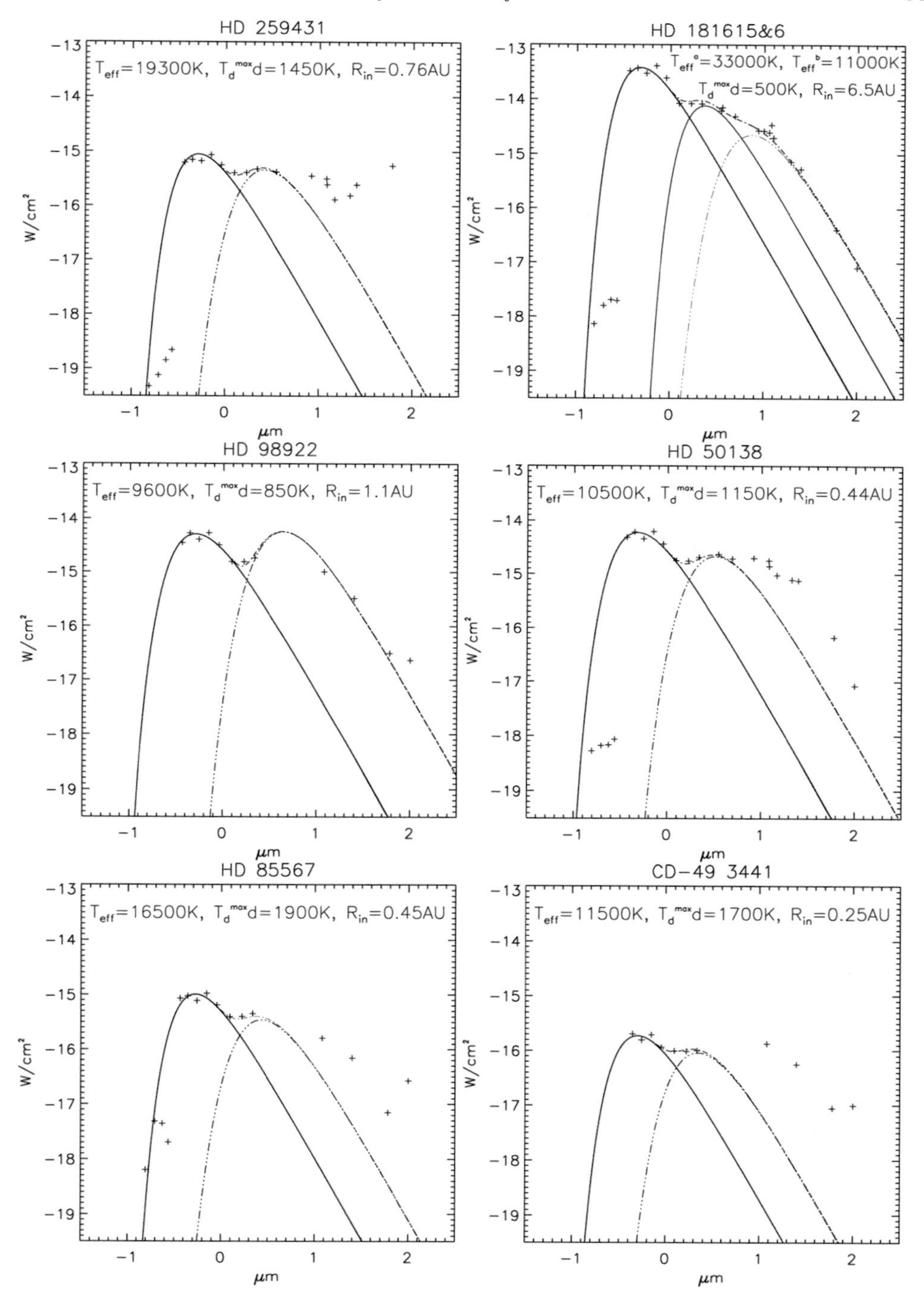

Figure 2. The sample of Be stars show excess emission not only in near-IR, but extends to mid- and far-IR wavelengths.

excess, implying that these CBe stars are just in the beginning of the dust formation. Recently, Mathew *et al.* (2008) claimed to have identified a CBe star in IC 1590 which is only $\sim$ 4 Myr old, marking the nature of CBe stars even more puzzling ever.

4. Conclusion

Most CBe stars show moderate near-IR excess which can be explained by hot circumstellar plasma. Those with IR excess larger than 0.6 mag in both J−H and H−K_s, however, cannot be accounted for by free-free emission, but require thermal emission from circumstellar dust. These dust grains are produced by condensation in the expanding stellar atmosphere, and may have drastically different physical properties from those found around HABe stars.

References

Currie, T., Hernandez, J., Irwin, J., Kenyon, S. J. *et al.* 2010, *ApJS*, 186, 191
Currie, T., Kenyon, S. J., Balog, Z., Rieke, G. et al. 2008, *ApJ*, 672, 558
Dudley, R. E. & Jeffery, C. S. 1990, *MNRAS*, 247, 400
Fabregat, J. & Torrejón, J. M. 2000, *A&A*, 357, 451
Gehrz, R. D., Hackwell, J. A. & Jones, T. W. 1974, *ApJ*, 191, 675
Herbig, G. H. 1960, *ApJS*, 4, 337
Mathew, B., Subramaniam, A. & Bhatt, B. C. 2008, *MNRAS*, 388, 1879
Porter, J. M. & Rivinius, T. 2003, *PASP*, 115, 1153
Samus, N. N, Durlevich, O. V. *et al.* 2004, *Combined General Catalogue of Variable Stars*, VizieR On-line Data Catalog: II/250
Schild, R. & Romanishin, W. 1976, *ApJ*, 204, 493
Singh, M. & Chaubey, U. S. 1987, *Ap&SS*, 129, 251
Slettebak, A. 1985, *ApJS*, 59, 769
de Winter, D., van den Ancker, M. E., Maira, A., Thé, P. S. *et al.* 2001, *A&A*, 380, 609
Zhang, P., Chen, P. S. & Yang, H. T. 2005, *New Astron.*, 10, 325

Discussion

RIVINIUS: Comment: In the Hα of your objects, three have classical P Cygni profiles. This is rather the signature of an expanding wind than that of a rotating disk. Only in the latter case it could be a CBe star.

LEE: Be stars may have stellar winds that produce P Cygni profiles, and at the same time a non-spherical envelope.

WISNIEWSKI: Comment: Just because a star is "isolated" and not associated with a star-forming region does not mean it cannot be a pre-main sequence star. HD 163296, for example is a proto-type of the class of objects known as "isolated HABe stars".

LEE: The existence of many "isolated" HABe stars is not conclusive. In the case of HD 163296, it is actually not isolated, but instead associated with dark clouds, as seen in Fig. 3. It is part of a prominent H II region and a cloud complex.

MIROSHNICHENKO: Comment: The six objects with the large IR excesses presented here do not belong to the class of Be stars. Three of the objects with the B[e] phenomenon are members of the FS CMa group, two other are pre-main-sequence HABe stars, and one binary system with a peculiar supergiant. Miroshnichenko *et al.* (2010, these proceedings) show that circumstellar dust in the FS CMa objects is not fresh (recently formed). The envelopes of Be stars do not contain dust, so these objects need to be excluded from lists of Be stars.

LEE: The term of "classical Be stars" is confusing, to say the least. Any Be stars with circumstellar dust are interesting. Here we present evidence of such dust in at least some

Figure 3. The DSS blue image with HD 163296 at the center. The field is 1 deg square with east to the left and north to the top. Some dark clouds are seen, which is part of prominent emission nebula and dark cloud filaments to the west.

of the CBe stars. The formation, evolution, and properties of these dust grains deserve further investigation.

Active OB stars: structure, evolution, mass loss, and critical limits
Proceedings IAU Symposium No. 272, 2010
C. Neiner, G. Wade, G. Meynet & G. Peters, eds.

doi:10.1017/S1743921311010799

Nebulae around Luminous Blue Variables – large bipolar variety

Kerstin Weis

Astronomisches Institut, Ruhr-Universität Bochum,
Universitätsstr. 150, 44781, Bochum, Germany
email: kweis@astro.rub.de

Abstract. During the LBV phase—a short transitional phase of only the very massive stars—strong stellar winds sometimes accompanied by eruptions form small circumstellar LBV nebulae around a substantial number of LBV stars. Analyzing the morphology and and kinematics of the LBV nebulae even weakly bipolar components can be detected, leading to the conclusion that about 50% have—to some degree—a bipolar structure. A global overview of our current observational knowledge of LBV nebulae is summarized, including their morphology, sizes and kinematic parameters.

Keywords. stars: variables: other, stars: circumstellar matter

1. Stars in the LBV phase

In recent years the number of *Luminous Blue Variables (LBVs)* listed in the literature has increased substantially. What is a LBVs? In 1984 Conti introduced the name LBV to unite the Hubble-Sandage, S Dor variables with the P Cyg and η Car type stars into one class, see Conti (1984). Are all LBVs luminous, bright and variable? Yes and no. LBVs do show photometric variations that are irregular and with different timescales and amplitudes. The variability intrinsic to LBVs, is the S Dor variability (for a detailed description see e.g. van Genderen 2001). This variability has its origin in the change of the stellar spectrum, and therefore is accompanied by a change in color. In addition some LBVs show *giant eruptions*, in which their brightness increases significantly for only a few years. Are LBVs blue? The answer is sometimes. If the star is in the hot phase within the S Dor cycle it is a blue supergiant. In the cool phase the LBV however can have a spectrum as cool as type A or F. Are LBVs luminous? The LBV phase is encounter by massive stars, the most luminous. The current list of LBVs however includes stars which show a luminosity as low as log $L/L_\odot \sim 5.5$ or equivalent to an initial mass of $25\,M_\odot$. This is lower as original proposed for LBVs (mass limit $50\,M_\odot$), but well within the limit of about $22\,M_\odot$ set by stellar evolution models that include an initial rotation rate of $v_{rot} = 300\,\mathrm{km\,s^{-1}}$ (Meynet & Maeder 2005). Therefore, LBVs are luminous but not necessarily all are very luminous.
LBVs are stars in a short, a few 10^4 years, instable phase. Most likely their proximity to the Humpheys-Davidson limit—or from the theoretical point of view the Ω- and/or Γ-limit—causes their photometric and spectroscopic variability as well as increases their mass loss rate (up to $10^{-4}\,M_\odot \mathrm{yr}^{-1}$), and maybe to the formation of nebulae. For further details on the LBV phase the reader is referred to Humphreys & Davidson (1994) and the more recent proceedings of conferences on the topic. Note however, that currently a clear definition of a LBV is missing! The stars show no really unique spectral feature, not for all a S Dor cycle has been observed and not all do have a nebula.

2. LBV nebulae

The LBVs larger mass loss and sometimes eruptions form LBV nebulae which show strong [N II] emission (CNO processed material), which have the following parameters.

Morphology:

Among the resolved LBV nebulae (Fig. 2 & 3), several are spherical, an example is S 61 (Weis 2003). Some nebulae do show an additional outflow or convexity, e.g. Sk $-69°$ 279 has an outflow to the north (Fig. 1, Weis & Duschl (2002)). A complete irregular structure is rare, with R 143 the best and only example (Weis 2003). Situated within the 30 Dor region R 143 lies in an area of higher density and interstellar turbulence, making it likely that the ambient medium had a larger impact on the nebula's structure. A large fraction of nebulae are bipolar. Bipolarity is seen either as an hourglass shape like in η Car (Weis 2001) and HR Car (Weis *et al.* 1997), or bipolar attachments—*caps*—as in WRA 751 (Weis 2000) or R 127 (Weis 2003). At least η Car and P Cyg show several distinct nebula parts, the *outer ejecta*, *Homunculus*, and *Little Homunculus* for η Car and the inner and outer shell of P Cyg. A current statistic of the morphology of the nebulae (Tab. 1) yields that about 50% are bipolar, 40% spherical and only 10% are of irregular shape. Taking only the Galactic objects into account bipolarity increases to 75% !

Size:

The smallest LBV nebula, diameter of ~ 0.2 pc, surrounds HD 168625 (Fig. 1). The Homunculus and inner nebula of P Cyg are comparable in size. Sk $-69°$ 279 (Fig. 1) has with diameter of 4.5 pc or 4.5 pc×6.2 pc including its outflow the largest nebula. Typical sizes of the nebulae are about 1-2 pc (Tab. 1). A collage of the LBV nebulae drawn to scale is given in Fig. 3. It visualize that all LMC nebulae are larger as the Galactic. This may be due a detection problem caused by the lower resolution (1 pc $\sim 4''$).

Kinematics:

So far, with the exception of η Car, expansion velocities for LBV nebulae have been determined through radial velocities (as in Tab. 1). The slowest expansion velocity, 14 km s^{-1}, is detected in Sk $-69°$ 279 the physically largest nebula. Including η Car, the largest expansion velocity is detected in the outer ejecta with 3200 km s^{-1} (Smith & Morse 2004). Otherwise P Cyg (140 and 185 km s^{-1}, inner and outer shell respectively) and HR Car

Table 1. Parameters of LBV nebulae in the Milky Way and LMC. Slashes separate values for nebula that consists of two distinct parts. Maximum size are either the largest extent as diameter or major and minor axes. For hourglass shaped bipolar nebulae, the radius and expansion velocities (marked with *) is given for one lobe. Table adapted from Weis (2001), Weis (2003).

LBV	host galaxy	maximum size [pc]	radius [pc]	v_{exp} [km/s]	kinematic age [10^3 yrs]	morphology
η Carinae	Milky Way	0.2/0.67	0.05/0.335	300*/10 − 3200		bipolar
AG Carinae	Milky Way	1.4 × 2	0.4	$\sim$ 25*	$\sim$ 30	bipolar
HD 168625	Milky Way	0.13 × 0.17	0.075	40	1.8	bipolar ?
He 3-519	Milky Way	2.1	1.05	61	16.8	spherical/elliptical
HR Carinae	Milky Way	0.65 × 1.3	0.325	75*	4.2	bipolar
P Cygni	Milky Way	0.2/0.84	0.1/0.42	110 − 140/185	0.7/2.1	spherical
Pistol Star	Milky Way	0.8 × 1.2	0.5	60	8.2	spherical
Sher 25	Milky Way	0.4×1	0.2×0.5	30 − 70	6.5 − 6.9	bipolar
WRA 751	Milky Way	0.5	0.25	26	9.4	bipolar
R 71	LMC	< 0.1?	< 0.05?	20	2.5 ?	?
R 84	LMC	< 0.3 ?	< 0.15?	24 (split)	6 ?	?
R 127	LMC	1.3	0.77	32	23.5	bipolar
R 143	LMC	1.2	0.6	24 (split)	49	irregular
S Dor	LMC	< 0.25?	<0.13?	< 40 (FWHM)	3.2 ?	?
S 61	LMC	0.82	0.41	27	15	spherical
S 119	LMC	1.8	0.9	26	33.9	spherical/outflow
Sk −69° 279	LMC	4.5 × 6.2	2.25	14	157	spherical/outflow

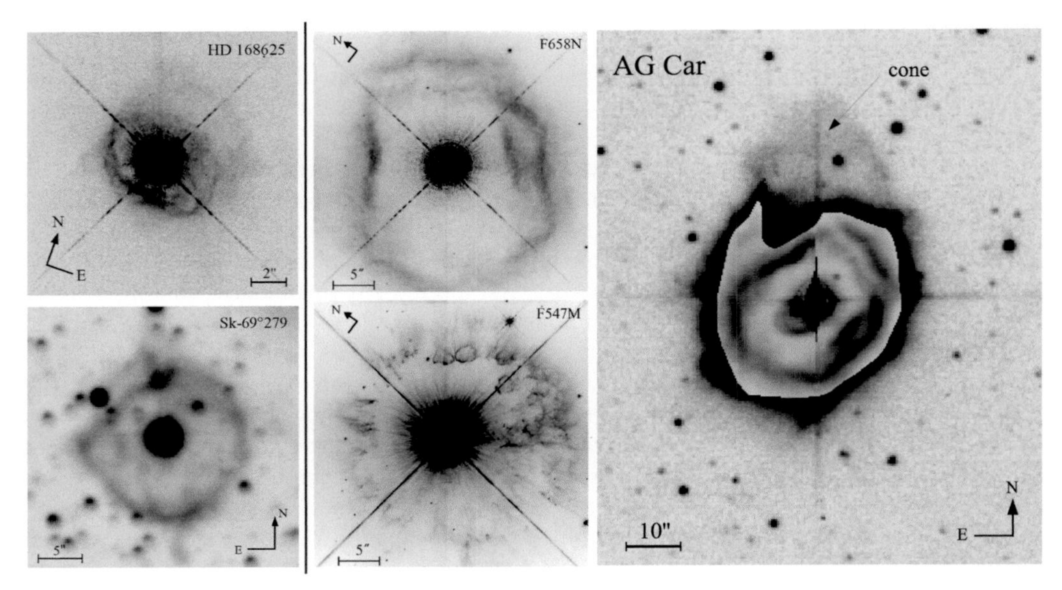

Figure 1. *Left section:* On top a HST image of HD 168625 the smallest and below a ground based frame of Sk $-69°$ 279 the largest nebula. *Right section:* AG Car HST images in the F658N filter showing [N II] emission and the F547 frame with scattered light from dust. A deep [N II] ground based image shows the fainter (including the cone) and superimposed the brighter emission.

(75 km s^{-1}) hold the record. More typical or average values lie around 50 km s^{-1}. LBV nebulae in the LMC, compared to the Galactic, have on average a slower expansion velocity. Some nebulae do show outflows or regions that move faster than the main body. S 119 has an outflow which moves with 140 km s^{-1} but the expansion velocity is only about 25 km s^{-1} (Weis, Duschl & Bomans 2003). Bipolarity of the nebulae is also detected kinematically. The hourglass shaped nebulae reveal two expansion ellipses one for each of the two lobes (e.g. HR Car). Nebulae that own their bipolar appearance to attached caps (e.g. WRA 751) have a redshifted and an antipodal blueshifted cap.

3. AG Carinae

Fig. 1 shows HST- and ground based images of the nebula of AG Car. The new deep image at the right reveals a total extend of 1.4×2 pc. Compared to earlier measurements it nearly doubled in size! Our analysis of long-slit echelle observations shows the presence of two expansion ellipses, manifesting a blue and a redshifted shell which are spatially superimposed (with the redshifted shell shifted to the north-east), proving a bipolar nebula. A large, cone shaped structure is part of the redshifted shell. The nebula is consequently larger and bipolar, with two distinct shells (Weis & Duschl in prep.).

4. The case of bipolarity

A significant fraction, about 50%, of the nebulae are bipolar—both in morphology and kinematics. Bipolarity appears as hourglass shaped structures or is seen in attached caps. Both types of bipolarity are confirmed kinematically. It is worth noting that also Planetary Nebula do show the same types of bipolar morphologies! The PN Hubble 5 (alias Hb 5, see e.g. Corradi & Schwarz (1993)) looks indeed like a tiny twin to the Homunculus, while the Cat's Eye nebula (NGC 6543, e.g. Balick & Preston 1987) is similar to R 127. In

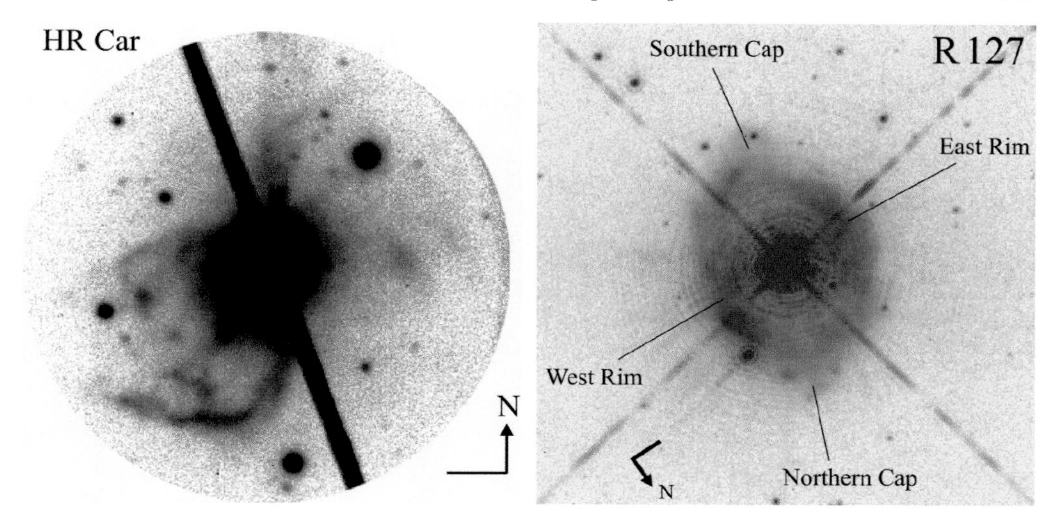

Figure 2. Two of a kind. The bipolarity in LBV nebulae is manifested either as in classical hourglass shapes like seen here in HR Car (left, Weis (2001)) or in bipolar caps that are attached to the main body of the nebula as in the case of R 127 (right, Weis 2003) .

general there is a large similarity of the morphologies and kinematics of LBV nebulae and PNs, implying, at least to some degree, the same physical—hydrodynamical—scenario for their formation, as for example discussed in Frank (1999).

Which physical processes would help to make bipolar LBV nebulae? Among the most natural mechanism is rotation and/or the formation of an equatorial density enhancement (maybe disk) or gradient either by the stellar wind or the ISM. Different mass loss phases in which the wind changes from equatorial to polar like during the passage of the bistability jump could do it, as it should be the case as the LBV passes through an S Dor cycle. Binary evolution may provide a scenario for forming a preferentially bipolar structure. What may prevent a bipolar structure? A very dense and/or turbulent ISM might suppress or destroy a bipolar shape originally inhibited. Does metalicity play a role? Connected to that: Why is the percentage of bipolar LBVs higher in the Milky Way as in the LMC? Metalicity decreases the mass loss rate for line driven winds, it may also have an effect on the bistability jump (change of mass loss from polar to equatorial)! With a lower metalicity, the effect could be weaker and the change of the orientation of the wind be not as strong and yield fewer/weaker bipolar structures in the LMC. Any hints on which scenario works? For at least two objects, AG Car (see previous chapter) and HR Car (Weis *et al.* (1997)) there is a good indication that the bipolar nebulae might be the result of rotation, as both stars have been identified as fast rotators (Groh, Hillier & Damineli (2006) and Groh *et al.* (2009)).

5. Summary and Conclusions

Now, what is already known, what has to be done and what is striking?

To know list:

LBV nebulae show a large range of shapes and sizes. Regarding their morphology about 50% are bipolar. Bipolar is popular ! Several are spherical, some do show an outflow structure and only one so far is irregular. The average size of a LBV nebula is 1-2 pc, and at least for now none is larger than about 5 pc. Typical expansion velocities of LBV nebulae are of the order 50 km s^{-1} but can be as low as a few km/s and as high as several

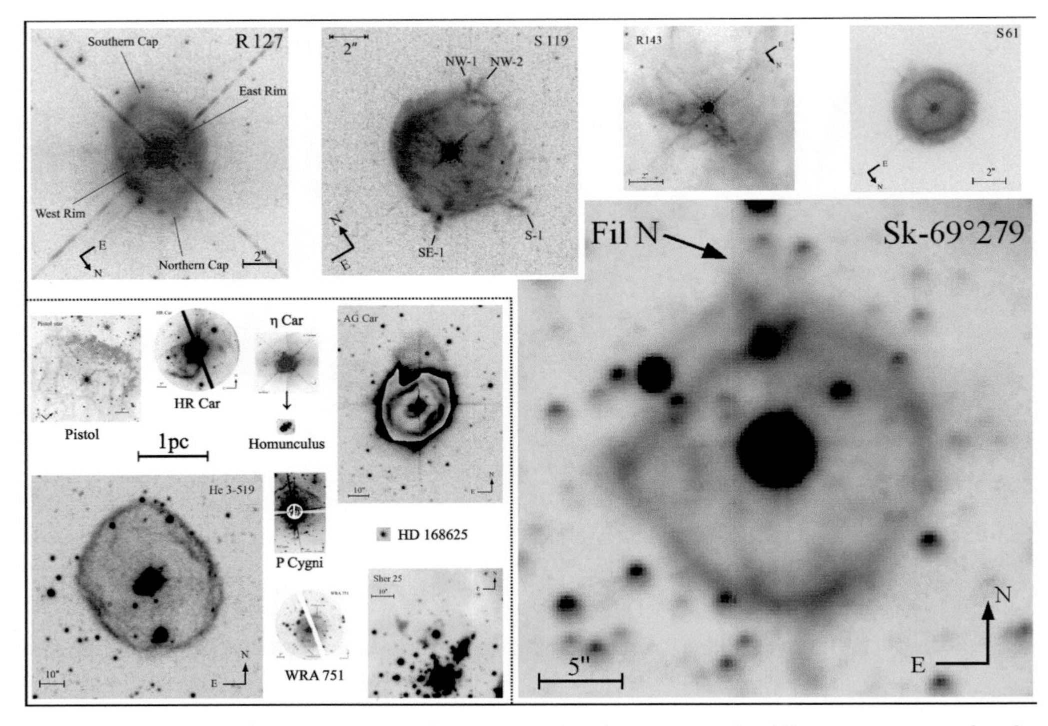

Figure 3. Resolved Galactic and LMC LBV nebulae drawn to scale. All images, except for the Pistol star, are taken either with an H_α or [N II] filter. Galactic nebulae are concentrated in the lower left section, within the dashed line.

thousand (η Car). The kinematic ages are for most of the nebulae several thousand years and therefore well within the expected duration of the LBV phase. The LBV nebulae in the LMC are larger and expand slower, compared to those in the Milky Way.

To do list:

Are all nebulae created equal, or are some the result of wind-wind interaction while other have been created by eruptions? With only two LBVs with nebulae known to have had a giant eruption (η Car, P Cyg), that question is hard to answer. Given the large kinematic ages, other LBV nebulae could have been formed in an eruption that has not been observed. Finding many bipolar nebulae we need to ask: How far does fast rotation play role? At least for AG Car and HR Car this seems to be the case. Therefore it is of interest to check other LBVs with bipolar nebulae, also to figure if the two types of bipolarity (hourglass and caps), results from stars with a higher and slower rotation rate.

To wonder list:

Finally why do not all LBVs do have a nebula? Do LBVs need to pass at least one cycle of the S Dor variability, switching from a hot to cool to hot phases, to form a nebula? Are different wind phase (keyword: bistability jump) responsible for the nebulae to form? Is this effect weaker in the more metal poor galaxies (lower Z) and might prevent the formation of LBV nebulae at all? And/or is it the proximity of the stars to the instability limits (Ω- or Γ-limit)? Are the nebulae formed by instability and ejection, rather than a continuous hydrodynamic processes? At least for η Car this seems a likely scenario. What forms LBV nebulae? Wind-wind interaction, instability and eruptions or are all scenarios possible and if differentiable?

Luminous **B**lue **V**ariables are stars that show a **L**arge **B**ipolar **V**ariety in their nebula. Rotation, binarity, eruptions and the bistability jump are likely origins for the formation of these geometrically distinct structures. Which scenario holds?

...the answer my friend is blowing in the wind...

References

Balick, B. & Preston, H. L. 1987, *AJ*, 94, 958

Conti, P. S. 1984, in: A. Maeder & A. Renzini (eds.), *Observational Tests of the Stellar Evolution Theory*, IAU Symposium 105, p. 233

Corradi, R. L. M. & Schwarz, H. E. 1993, *A&A*, 269, 462

Frank, A. 1999, *New Astron. Revs*, 43, 31

van Genderen, A. M. 2001, *A&A*, 366, 508

Groh, J. H., Hillier, D. J. & Damineli, A. 2006, *ApJ* (Letters), 638, L33

Groh, J. H., Damineli, A., Hillier, D. J., Barbá, R. *et al.* 2009, *ApJ* (Letters), 705, L25

Humphreys, R. M. & Davidson, K. 1994, *PASP*, 106, 1025

Meynet, G. & Maeder, A. 2005, *A&A*, 429, 581

Smith, N. & Morse, J. A. 2004, *ApJ*, 605, 854

Weis, K. 2000, *A&A*, 357, 938

Weis, K. 2001, in: R. E. Schielicke (eds.), *Reviews in Modern Astronomy*, 14, p. 261

Weis, K. 2003, *A&A*, 408, 205

Weis, K. & Duschl, W. J. 2002, *A&A*, 393, 503

Weis, K., Duschl, W. J., Bomans, D. J., Chu, Y.-H. *et al.* 1997, *A&A*, 320, 568

Weis, K., Duschl, W. J. & Bomans, D. J. 2003, *A&A*, 398, 1041

Discussion

MEYNET: LBV phenomena can arise from a relative broad range of initial mass stars. Is there any trend of the frequency of bipolarity with mass/luminosity?

WEIS: So far I do not see a clear trend. Bipolar nebula are found around very massive/luminous LBVs (e.g η Car and AG Car) as well as around lower mass and less bright LBVs (e.g. HR Car). There is also not a difference concerning hourglass shaped and bipolarity due to caps (R 127 is massive/luminous and WRA 751 fainter and less massive).

PRINJA: You presented an interesting analogy between LBV nebulae and PN—in PN, binaries may be very important in sculpting the nebulae (during the commen envelope phase)—so to what degree do you think binarity may dominate in the formation of bipolar nebulae in LBVs?

WEIS: Currently I know only two LBVs for which a binary scenario has been invoked, that is η Car and HD 5980. All others and therefore all other bipolar nebulae result from apparently single stars.

SMITH: You showed a faint outer shell around AG Car. What is the kinematic age of this outer shell, and what is the age of the previously known inner shell?

WEIS: There is not an inner and an outer shell. The fainter emission detected further out is only an extension of the nebula. It is not an additional outer structure or new nebula but an attached feature. Therefore there is only one kinematic age indicated in my table.

Active OB stars: structure, evolution, mass loss, and critical limits
Proceedings IAU Symposium No. 272, 2010
C. Neiner, G. Wade, G. Meynet & G. Peters, eds.

doi:10.1017/S1743921311010805

Discussion – Circumstellar environment of active OB stars

summarized by Douglas R. Gies[1] and Richard H. D. Townsend[2]

[1]Center for High Angular Resolution Astronomy and Department of Physics and Astronomy, Georgia State University, P. O. Box 4106, Atlanta, GA 30302-4106, U.S.A.

[2]Department of Astronomy, University of Wisconsin-Madison, Madison, WI 53706, USA

This session dealt with the circumstellar gas surrounding active OB stars, and the discussion broadly focused on disks, wind morphology and structures, X-ray emission, and mass loss rates.

Alex Lobel noted that the GAIA mission will record the Paschen 14 λ8598 line of hydrogen, and he asked about the state of models of line formation in Be star disks with cooler, equatorial zones. Alex Carciofi responded that two-dimensional models for optically thick disks solved for the radiation field and temperature distribution self-consistently. He suggested that even high level transitions of hydrogen, like Paschen 14, were probably optically thick out to regions well above and below the cooler equatorial zone in Be stars with dense disks. Vladimir Strelnitski called attention to MWC349, an object hosting a disk with possible solid-body instead of Keplerian rotation.

Thomas Rivinius commented on the claims from interferometric observations of polar winds in Be stars. He doubted that significantly dense polar winds exist because: (1) the winds from the polar regions of Be stars should be similar to those of normal B-stars, i.e., relatively weak; (2) if optically thick polar winds are present, they should appear in other kinds of observations (e.g., emission lines); (3) absolute visibility measurements with VLTI/Amber are difficult, and the detection of polar winds from other interferometers (NPOI, CHARA) has not been forthcoming. Jo Puls added that wind models indicate that the IR continuum emission might extend to 1.2 stellar radii but not to 10 stellar radii. Philippe Stee responded that while the interferometric results are interpreted using a polar wind model, the emission may result instead from a disk wind. Anthony Meilland commented that some image reconstruction work on VLTI/VINCI observations of the Be star Achernar yields evidence of polar winds without relying on a wind model framework.

The discussion then turned to the issue of the hour-glass shaped outflows from LBVs and other luminous stars. Stan Owocki noted that the Homunculus Nebula surrounding η Carinae probably resulted from faster wind flows and higher mass loss from the polar regions. He asked if there is any evidence that more mass is ejected from the polar regions of LBVs in general. Kerstin Weis responded that many LBVs show bipolar outflows, but the mass loss rates as a function of latitude were not well known. Nathan Smith added that the answer requires data on both gas density and solid angle of emission, and such work has only been done for η Carinae and P Cygni. Dietrich Baade pointed out that a similar hour-glass morphology exists around the site of SN1987A (whose progenitor was a blue-supergiant). He advocated a search for circumstellar gas envelopes around other early-type supergiants, especially in the Magellanic Clouds where the extinction is low. He wondered if there was a consensus about the origin of bipolar outflows in rotation and/or magnetic fields. Baade also reminded us that supernovae often show evidence of asymmetries (detected through polarimetry), and he speculated about a general connection between the asymmetries pre- and post-SN. Nathan Smith responded that in the case of SN1987A, high resolution observations of the expanding cloud from the explosion displayed an asymmetry orthogonal to that of the shells from the pre-SN wind loss, so he doubted that the SN asymmetries were related to the pre-SN rotation axis.

Coralie Neiner redirected the discussion to X-rays from active OB stars. She noted that many of the newly detected magnetic OB stars from the MiMeS program belong to the *weak-wind* category, and she asked about the X-ray properties of the stars with measured magnetic fields. David Cohen responded that there is a large diversity of X-ray properties among the magnetic stars, ranging from strong X-ray emitters like θ^1 Ori C to rather weakly X-ray emitting stars.

Nathan Smith brought up the all-important issue of mass loss rates. He noted that the mass loss rates derived from X-ray studies are lower than those from standard UV line analyses and are closer to those for clumpy winds. He asked whether X-ray diagnostics are based upon the wind properties in environments with similar densities to those for the UV line studies. David Cohen answered affirmatively that the X-rays form throughout the wind and act to back-light the wind material seen along the line of sight. He noted that the K-shell edges seen in Chandra X-ray spectroscopy of O-stars arise from similar ions to those that form P Cygni wind lines in the UV spectrum.

Rich Townsend raised the issue of porosity in the wind, i.e., the idea that the wind is spatially porous and structured into optically thick clumps (see Owocki & Cohen 2006, ApJ, 648, 565). David Cohen emphasized that creating such optically thick clumps would require sweeping up wind gas over a large length scale. Rayleigh-Taylor instabilities would likely develop and make smaller clumps. Thus, if porosity is invoked to revise upwards mass loss rates from X-ray diagnostics, then clumping must inevitably follow, and hence mass loss rate indicators based upon density squared diagnostics would need to be lowered. Jorick Vink discussed the similarity of the mass loss rates from theory with those from Hα that include the downward revision from clumping, while for wind models invoking porosity, the theoretical and observationally implied mass loss rates differ. David Cohen concluded the discussion with a plea for more research on the mass loss rates and all their observational diagnostics (X-ray, Hα, P V lines, IR and radio continuum excess, etc.). He noted the development of a new code by Rich Townsend that characterizes the wind clumping properties and derives estimates of observable quantities like the net polarization.

Active OB stars: structure, evolution, mass loss, and critical limits
Proceedings IAU Symposium No. 272, 2010
C. Neiner, G. Wade, G. Meynet & G. Peters, eds.

doi:10.1017/S1743921311010817

Spectral variation of BU Tau

Kalju Annuk

Tartu Observatory,
61602 Tõravere, Estonia
email: annuk@aai.ee

Abstract. We present the preliminary result of our spectroscopic observations covering a period over more than ten years, from 1999 up to 2010. We analyze line profile variations of several lines before the shell phase and during the shell phase. The measured radial velocity, equivalent width and line intensity variations are presented.

Keywords. stars: emission-line, Be, stars: individual (BU Tau), line: profiles

1. Introduction

BU Tau (Pleione) is well–known B/Be star located in the Pleiades cluster. It is accepted that this star has periodic (P$\sim$34 years) shell formation. So far, there has been clearly observed two shell phases, in 1938–1954 and 1972–1988. The duration of shell phase lasts about 16 years. The third shell phase started in 2005–2006 and it has to be continued up to 2022. BU Tau is also known as a photometric variable. Although this object has been under observation and investigation more than 100 years, there are still several unsolved questions.

2. Observations and reduction

Observations were carried out with the 1.5 m telescope of the Tartu Observatory, using a Cassegrain spectrograph ASP-32. All observations were obtained with the 1800 l/mm grating. The nitrogen cooling CCD camera Orbis-1 (512x512 pixels) was used in 1999 up to March 2006. Starting from April 2006, the Peltier cooling CCD Andor Newton DU-970N (1600x200 pixels) has been used. The spectral resolution was $\sim$0.26 Å/pixel (with Orbis-1) or $\sim$0.17 Å/pixel (with Andor Newton) at the H_α. Three spectral regions were observed: around the H_α, H_β and λ <4000 Å. The MIDAS package has been used for the reduction. All spectra were calibrated with a ThAr lamp and normalized to the continuum level. Radial velocities of the H_α emission lines were measured by the bisector method and radial velocities of the other lines by fitting a gaussian profiles.

3. Discussion

The variation of the equivalent width (EW) of the H_α line is presented in the left part of Figure 1. We can see that the EW has continuously decreased from the value $\sim$35Å (November 1999) up to $\sim$25Å (March 2006). At that time the star was in a Be phase. During the next $\sim$450 days the EW very quickly decreased reaching to its minimum value $\sim$5.5Å. This period corresponds to the shell formation phase.

The central intensities of the H_α and H_β absorption components (right part of Figure 1) behave in the same way as the equivalent width of the H_α line. We have to mention that central intensities of the H_α and H_β have a very good correlation. Starting the March 2006, the central intensity of the H_β line is smaller than 1.0.

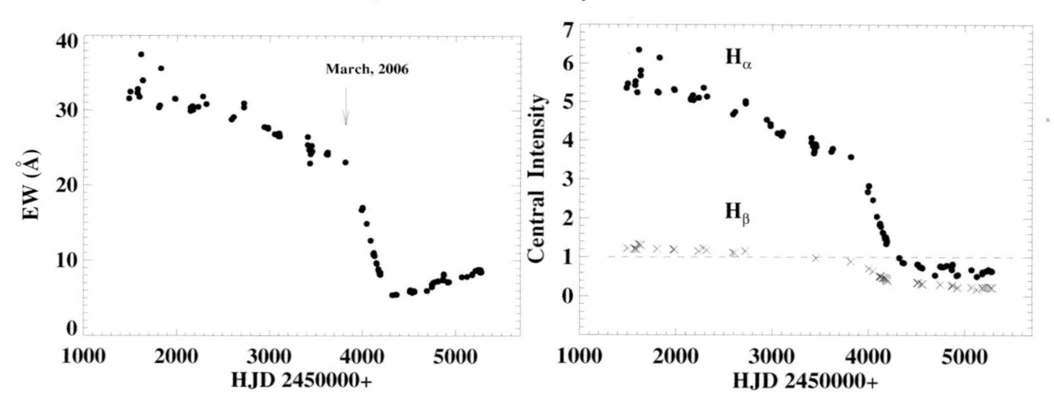

Figure 1. *Left:* The equivalent width variation of the H_α line. *Right:* The central intensities of the H_α and H_β absorption component.

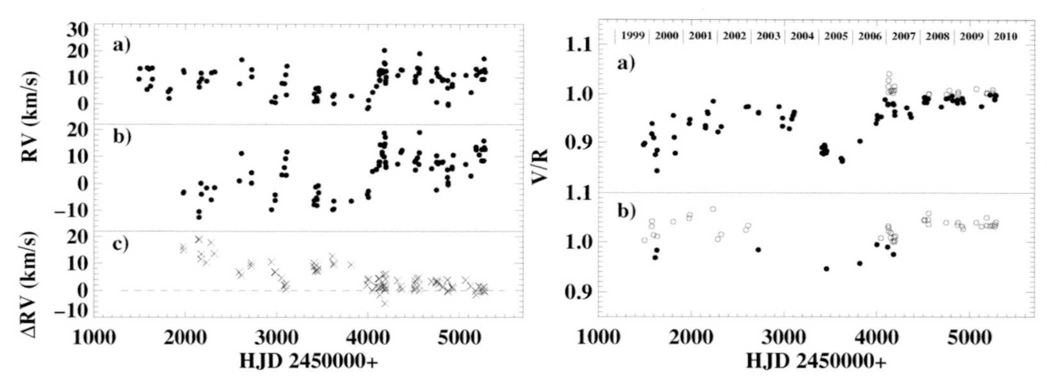

Figure 2. *Left:* The radial velocities of the H_α: a) emission component (RV_{em}); b) absorption component (RV_{abs}); c) emission minus absorption. *Right:* The V/R variations: a) of the H_α line; b) of the H_β line.

The radial velocity variation of the H_α is shown in the left part of Figure 2. As the line profile is somewhat asymmetric then we used for the calculation of radial velocities of emission component the levels which correspond to the more symmetric part of the profile. There is very strong correlation between radial velocities of emission and absorption components. At mean the radial velocity of emission is somewhat higher than the radial velocity of absorption (see Figure 2c).

The measured V/R relations of the H_α and H_β lines are plotted in the right part of Figure 2. We can see that in the most cases the V/R<1 (filled circles) for H_α and V/R>1 (open circles) for H_β. The violet (V) peak of the H_α line was first time higher than the red (R) peak (V/R>1) in the middle of January 2007. For both lines the deep minimum took place in 2005, directly before the starting of shell phase formation. During the years of 1999–2002 there has been a remarkable trend that V/R was increased. At the same time the fluctuation was quite large. Our observations show that there is a good correlation between the variations of V/R relations of the H_α and H_β lines.

Acknowledgements

This work was supported by the target-financed project SF0060030s08 financed by the Ministry of Education and Research of Estonia.

Active OB stars: structure, evolution, mass loss, and critical limits
Proceedings IAU Symposium No. 272, 2010
C. Neiner, G. Wade, G. Meynet & G. Peters, eds.

doi:10.1017/S1743921311010829

FRACS: modelling of the dust disc of the B[e] CPD-57° 2874 from VLTI/MIDI data

Philippe Bendjoya, Armando Domiciano de Souza and Gilles Niccolini

Université de Nice Sophia Antipolis, Observatoire de la Côte d'Azur, CNRS UMR 6525
email: bendjoya@unice.fr

Abstract. The physical interpretation of spectro-interferometric data is strongly model dependent. On one hand, models involving elaborate radiative transfer solvers are in general too time consuming to perform an automatic fitting procedure and derive astrophysical quantities and their related errors. On the other hand, using simple geometrical models does not give sufficient insights into the physics of the object. We developed a numerical tool optimised for mid-infrared (mid-IR) interferometry, the Fast Ray-tracing Algorithm for Circumstellar Structures (FRACS).Thanks to the short computing time required by FRACS, best-fit parameters and uncertainties for several physical quantities were obtained, such as inner dust radius, relative flux contribution of the central source and of the dusty CSE, dust temperature profile, disc inclination.

Keywords. circumstellar matter, stars: emission-line, Be, techniques: interferometric

---

1. Introduction

B[e] supergiants are luminous, massive post-main sequence stars presenting non spherical wind, forbidden lines, and hot dust on a disc-like structure. We use mid-IR spectro-interferometric observations from VLTI/MIDI to resolve and study the CSE of the Galactic B[e] supergiant CPD-57°2874 (Domiciano de Souza et al. 2011). For a physical interpretation of the observables (visibilities and spectrum) we used our ray-tracing radiative transfer technique (FRACS). FRACS is based on the ray-tracing technique, without scattering, supplemented with the use of quadtree meshes and the full symmetries of the axisymmetrical problem to signicantly decrease the computing time necessary to obtain e.g.monochromatic images and visibilities.

2. Parametrized Model

FRACS is fully described in Niccolini *et al.* (2011). We assume the specific intensity from the central regions of the star to be a power-law : $I^{\rm s}_{\lambda} = I^{\rm s}_{\lambda_0}\, \left(\frac{\lambda_0}{\lambda}\right)^{\alpha}$ A radius $R_{\rm s}$=54 $R_{\rm sun}$ was adopted for the central region as a scaling factor. The number density of dust grains is given by: $n(r,\theta) = n_{\rm in}\, \left(\frac{R_{\rm in}}{r}\right)^2 \left(\frac{1+A_2}{1+A_1}\right) \frac{1+A_1\,(\sin\theta)^m}{1+A_2\,(\sin\theta)^m}$ where (r,θ) are the radial coordinate and co-latitude, and $n_{\rm in}$ is the dust grain number density at $\theta = 90°$ and at $r = R_{\rm in}$, which is the inner dust radius where dust starts to survive. A_1 controls the ratio between the equatorial and polar mass loss rates. A_2 indicates how much faster is the polar wind compared to the slow equatorial wind. Parameter m controls how fast the mass loss drops from the equator to the pole. The dust grain opacity was calculated in the Mie theory for silicate dust and for a dust size distribution following a power-law $\propto a^{-3.5}$, where a is the dust grain radius. The temperature structure of the dusty region is given by: $T(r) = T_{\rm in}\, \left(\frac{R_{\rm in}}{r}\right)^{\gamma}$ where $T_{\rm in}$ is the dust temperature at the disc inner radius

$R_{\rm in}$. The inclination of the disc plane towards the observer is i, and the position angle (from North to East) of the maximum elongation of the sky-projected disc is $\rm PA_d$. Thus, the 10 free parameters of the model are: $I^{\rm s}_{\lambda_0}$, α, $T_{\rm in}$, γ, $R_{\rm in}$, i, $\rm PA_d$, A_2, $n_{\rm in}$, and m.

3. FRACS philosophy

The radiative transfer equation (RTE hereafter) is integrated along a set of rays making use of the symmetries of the problem. We seek to produce intensity maps within seconds and we want our numerical method to deal with a large range of density and temperature structures. Regarding the above mentioned constraints, the numerical integration of RTE is more efficiently computed using a mesh based on a tree data structure (quadtrees/octrees). The mesh also distribute the integration points along the rays according to the variations of the medium emissivity. From a computed intensity map it is then possible to derive flux and a set of visibilities using the observationnal bases. A fit of the observed data is then possible by means of a χ^2 minimization algorithm.

4. Experimental DATA

The interferometric observations of CPD-57 2874 were performed with MIDI, the mid-infrared 2-telescope beam-combiner instrument of ESOs VLTI. All four 8.2 m unit telescopes (UTs) were used. The N-band spectrum as well as spectrally dispersed fringes have been recorded between $\lambda \simeq 7.5\mu$m and $\lambda \simeq 13.5\mu$m with a spectral resolution of R $\simeq 30$ using a prism. In total, $n_B = 10$ data sets have been obtained with projected baselines ranging from 40 m to 130 m, and baseline position angles between $\simeq 8°$ and $\simeq 105°$ (from North to East). VLTI/MIDI also provides spectral uxes of CPD-57°2874 in the N band.

5. Results

We give the best-fit model (visibilities and fluxes) parameters and uncertainties derived from a χ^2 minimization (minimum reduced $\chi^2 = 0.54$). The uncertainties were estimated from the $\chi^2_{\rm r}$ maps. Values are computed for an estimated distance (from FEROS spectroscopic observations) of 1.7 kpc: $I^{\rm s}_{\lambda_0}$ (10^5 W m^{-2} μm^{-1} str^{-1}) $= 2.2\,^{+0.7}_{-0.7}$, $\alpha = 2.4\,^{+1.3}_{-1.2}$, $T_{\rm in}(K) = 1500$, $R_{\rm in}(AU) = 12.7^{+3.6}_{-2.9}$, i (°) $= 61.3\,^{+10.8}_{-18.2}$, $\rm PA_d$ (°) $= 140.3^{+12.3}_{-14.0}$, $A_2 = -0.98$, $n_{\rm in}$ (m^{-3})$= 0.30$,$m = 332$

6. Conclusion

FRACS is a method to fit inteferometric data, fast enough to make an automatic search of glogal minimum in a multi dimensional physical parameter space. It is just between pure geometrical methods which are very fast but avoid to reach physical parameters and Monte Carlo methods which give a deep physical caracterization but which are extremely computational time consumers.

References

Domiciano de Souza, A., Bendjoya, P., Niccolini, G., Chesneau O., Borges Fernandes, M. *et al.* 2011, *A&A*, 525, A22

Niccolini, G., Bendjoya, P., Domiciano de Souza, A. 2011, *A&A*, 525, A21

Active OB stars: structure, evolution, mass loss, and critical limits
Proceedings IAU Symposium No. 272, 2010
C. Neiner, G. Wade, G. Meynet & G. Peters, eds.

doi:10.1017/S1743921311010830

The circumstellar environment of the FS CMa star IRAS 00470+6429

Alex C. Carciofi[1], **Anatoly S. Miroshnichenko**[2] **and Jon E. Bjorkman**[3]

[1] Instituto de Astronomia, Geofísica e Ciências Atmosféricas, Universidade de São Paulo, Rua do Matão 1226, Cidade Universitária, 05508-900, São Paulo, SP, BRAZIL email: carciofi@usp.br

[2] Department of Physics and Astronomy, University of North Carolina at Greensboro, Greensboro, NC 27402, USA

[3] Ritter Observatory, M.S. 113, Dept. of Physics and Astronomy, University of Toledo, Toledo, OH 43606-3390, USA

Abstract. FS CMa type stars are a recently described group of objects with the B[e] phenomenon that exhibit strong emission-line spectra and strong IR excesses. In this paper we report the first attempt for a detailed modeling of IRAS 00470+6429, for which we have the best set of observations. Our modeling is based on two key assumptions: the star has a main-sequence luminosity for its spectral type (B2) and that the circumstellar (CS) envelope is bimodal, composed of a slowly outflowing disk-like wind and a fast polar wind. Both outflows are assumed to be purely radial. We adopt a novel approach to describe the dust formation site in the wind that employs timescale arguments for grain condensation and a self-consistent solution for the dust destruction surface. With the above assumptions we were able to reproduce satisfactorily many observational properties of IRAS 00470+6429, including the HI line profiles and the overall shape of the spectral energy distribution.

Keywords. circumstellar matter, stars: emission-line, Be, stars: winds, outflows, stars: activity

---

1. Introduction

This series of papers is devoted to studying objects from the Galactic FS CMa type group. The group comprises about 40 objects with the B[e] phenomenon, which refers to the simultaneous presence of forbidden lines in the spectra and strong excesses of IR radiation (Miroshnichenko 2007, Lamers *et al.* 1998). Its members show properties of B-type stars with typical main-sequence luminosities $[\log(L/L_{\rm sol}) \sim 2.5 - 4.5]$. On the other hand, they exhibit extremely strong emission-line spectra that are on average more than an order of magnitude stronger than those of Be stars of similar spectral types.

One of the main problems in understanding those objects has to do with the presence of significant amounts of hot dust in their circumstellar region that cannot be left from the time of the stars' formation. It is also hard to explain the dust formation using typical mass loss rates from single stars of an appropriate mass range ($\sim 5 - 20$ $M_{\rm sol}$). Our observations show that over 30% of the group objects show signs of secondary companions in the spectra (e.g., Li I 6708 Å absorption line). Therefore, one might assume that the large amounts of circumstellar matter near FS CMa type objects may be a consequence of the binary evolution, although direct evidence of the mass transfer between the stellar companions is not observed.

The observed properties of the first object for which we collected a good wealth of data, IRAS 00470+6429, were presented by Miroshnichenko *et al.* (2009). In the present paper we make use of the Monte Carlo code HDUST (Carciofi & Bjorkman 2006) to determine the physical parameters of this intricate system.

2. Results

In this paper we adopted an ad hoc model for IRAS 00470+6429 which is physically motivated by the current knowledge of the CS envelopes of sgB[e]. The model consists of a 6 R_{sol} central star with $T_{eff} = 20\,000$ K and $L = 5\,100\ L_{sol}$, surrounded by a bimodal CS envelope composed of a dense, slowly outflowing disk-like wind and a fast polar wind. In the model the underlying physical reason for the bimodal envelope is that the mass loss is, somehow, enhanced around the equator. From the analysis of the observed SED and H I line profiles, it was found that the equatorial outflow is at least 200 times denser than the fast polar wind and about 10 times slower. In addition, the opening angle of this slowly outflowing disk (defined as the co-latitude for which the mass loss rate drops to half of its value at the equator) is about 7°.

We determined that the integrated stellar mass loss rate is $\dot{M} = 2.5 — 2.9{\times}10^{-7}\ M_{sol}\ yr^{-1}$. The two extremes correspond to fits to observations made in December 2006, but separated by an interval of two weeks. This gives a quantitative measure of how the mass loss varies in short timescales. The above value of $\dot{M}$, while much smaller than that of sgB[e], which is of the order of 10^{-6} — 10^{-5} $M_{sol}\ yr^{-1}$, is at least 100 times larger than that of main-sequence stars of spectral type B. We adopted a prescription for dust formation based on two complementary criteria. In order to form dust at a given point in the CS envelope, the equilibrium temperature of the dust grains must be smaller than a given grain destruction temperature, $T_{destruction}$, and the gas density must be larger than a critical value. Several important radiative transfer effects are considered in our calculations, such as the shielding of the dust by the optically thick inner CS material and the fact that differently sized grains have different equilibrium temperatures.

To investigate the properties of the dusty content of IRAS 00470+6429 we studied dust models with three different grain size distributions, one with the standard MRN distribution (a = 0.05—0.25 μm), one with both small and large grains (a = 0.05—10 μm), and one with only large grains (a = 1—50 μm). The dust was assumed to be oxygen-based (silicates), an assumption supported by evolutionary arguments and our Spitzer observations of FS CMa stars. The only model capable of reproducing the observed IR excess in the entire 2 — 13 μm range was the model with only large grains ($a_{min} = 1\ \mu$m), because the presence of small grains always results in a strong 9.7 μm silicate emission, which is not observed. One consequence of the prevalence of large grains around IRAS 00470+6429 is that the bulk density of the grain material must be very small ($\rho_{dust} \sim 0.1$ g cm^{-3}). Therefore, the observed shape of the IR excess seems to firmly indicate that the CS dust grains of IRAS 00470+6429 are both very large and fluffy. This work is described in detail in a recently accepted paper to the Astrophysical Journal (Carciofi, Miroshnichenko & Bjorkman 2010).

References

Carciofi, A. C. & Bjorkman, J. E. 2006, *ApJ*, 639, 1081

Carciofi, A. C., Miroshnichenko, A. S. & Bjorkman, J. E. 2010, *ApJ*, 721, 1079

Lamers, H. J. G. L. M., Zickgraf, F.-J., de Winter, D., Houziaux, L. *et al.* 1998, *A&A*, 340, 117

Miroshnichenko, A. S. 2007, *ApJ*, 667, 497

Miroshnichenko, A. S., Chentsov, E. L., Klochkova, V. G., Zharikov, S. V. *et al.* 2009, *ApJ*, 700, 209

Active OB stars: structure, evolution, mass loss, and critical limits
Proceedings IAU Symposium No. 272, 2010
C. Neiner, G. Wade, G. Meynet & G. Peters, eds.

doi:10.1017/S1743921311010842

Analysis of the Balmer discontinuity behavior of Be stars by the Monte Carlo method

Alicia Cruzado[1]

[1],Facultad de Ciencias Astrónomicas y Geofísicas, Universidad Nacional de La Plata, Paseo de Bosque s/n, La Plata, Buenos Aires, Argentina
email: acruzado@fcaglp.unlp.edu.ar

Abstract. For a given photospheric model, we study the behavior of the BD as different density and temperature distributions in the circumstellar envelope are assumed. For non spherically symmetric envelopes, we analyze the variation of the BD when the angle of observation varies. The radiation transfer through the medium is handled by means of the Monte Carlo method. We calculate the flux emitted by the star+envelope system in a small wavelength range around the BD. The calculations are made under LTE conditions.

Keywords. stars: early-type, circumstellar matter

1. Methodology

The Monte Carlo technique we used to solve the radiation transfer through the medium is illustrated en Fig. 1

The ratios of the number of photons emitted at each frequency (N_ν) to the total number (N) have been set deterministically taking into account that $N_\nu/N=L_\nu V\nu/cL$, where V is the highest value of the velocity in the medium, L and L_ν are the ν and total luminosity, respectively.

2. Test

The photospheric jump of normal stars could be reproduced with our code. Also we have been able to reproduce some results obtained by conventional methods (Crivellari & Simonneau 1994, Gros *et al.* 1997) for spherically symmetric envelopes. These results constitute the test of our code.

3. Envelope Model

To describe the density distribution in the envelope we have considered monotonic decreasing functions with r, as well as distributions that take into account an accumulation of material near the star (Zorec *et al.* 2007). To take into account non spherically simetric matter distribution, decay from equator to pole have been considered. Regards the temperature, it is assumed constant through the envelope, taking values in the range 5000K-30000k.

The opacity sources we have considered are: hydrogen bound-free and free-free transitions, electron scattering and H Rayleigh scattering.

The calculations have been performed under LTE conditions.

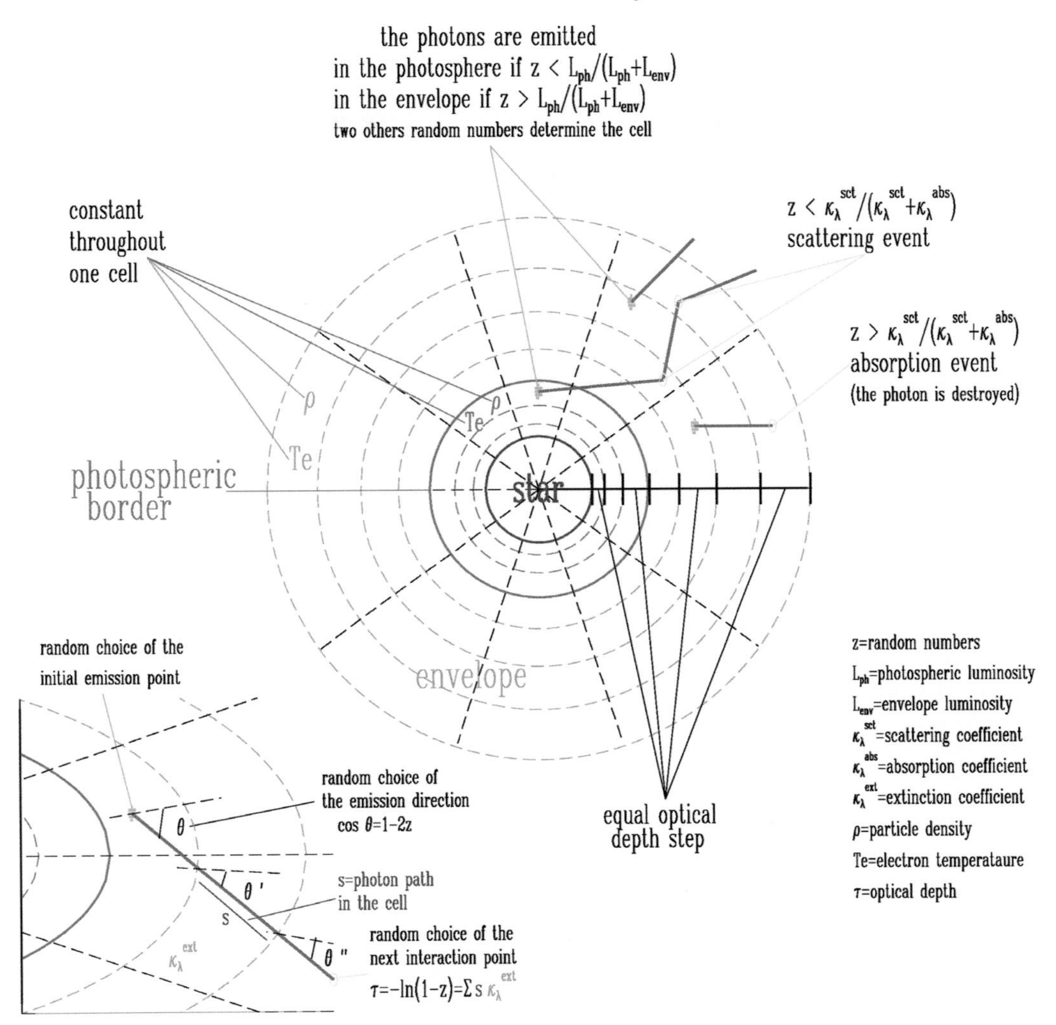

Figure 1. Monte Carlo Method illustration

4. Results and Conclusions

From our preliminary results some general trends could be observed. For a given spectral type, the appearance of the BD depends on the density distribution in the envelope, the temperature in the envelope, and the angle of observation. For very low densities and/or geometrically thin envelopes no second jump is observed. For a given spatial distribution of the material around the star, the size of a second jump in emission depends on the envelope temperature. In spatially not too extended envelopes, the second jump can be observed in absorption for low temperatures in the envelope.

References

Crivellari, L. & Simonneau, E. 1994, *ApJ* 429, 331

Gros, M., Crivellari, L. & Simonneau, E. 1997, *ApJ* 489, 331

Zorec, J., Arias, M. L., Cidale, L., & Ringuelet, A. E. 2007, *A&A* 470, 239

Active OB stars: structure, evolution, mass loss, and critical limits
Proceedings IAU Symposium No. 272, 2010
C. Neiner, G. Wade, G. Meynet & G. Peters, eds.

doi:10.1017/S1743921311010854

Disk-loss and disk-renewal phases in classical Be stars – II. Detailed analysis of spectropolarimetric data

Zachary H. Draper[1], John P. Wisniewski[1], Karen S. Bjorkman[2], Jon E. Bjorkman[2], Xavier Haubois[3], Alex C. Carciofi[3], Marilyn R. Meade[4]

[1]Department of Astronomy, University of Washington,
email: zhd@u.washington.edu jwisnie@u.washington.edu

[2]Ritter Observatory, Department of Physics & Astronomy, University of Toledo

[3]IAG, Universidade de Sao Paulo

[4]Space Astronomy Lab, University of Wisconsin-Madison

Abstract. In Wisniewski *et al.* (2010), paper I, we analyzed 15 years of spectroscopic and spectropolarimetric data from the Ritter and Pine Bluff Observatories of 2 Be stars, 60 Cygni and π Aquarii, when a transition from Be to B star occurred. Here we analyze the intrinsic polarization, where we observe loop-like structures caused by the rise and fall of the polarization Balmer Jump and continuum V-band polarization being mismatched temporaly with polarimetric outbursts. We also see polarization angle deviations from the mean, reported in paper I, which may be indicative of warps in the disk, blobs injected at an inclined orbit, or spiral density waves. We show our ongoing efforts to model time dependent behavior of the disk to constrain the phenomena, using 3D Monte Carlo radiative transfer codes.

Keywords. circumstellar matter, stars: individual (π Aquarii, 60 Cygni)

---

1. Balmer Jump vs Continuum Polarization

The time evolution of the intrinsic continuum V-band polarization (V-pol) of π Aqr is shown in the left panel of Figure 1. We find evidence of clockwise loop-like structures (Fig. 1, middle panel) when comparing the evolution of the polarization across the Balmer Jump (BJ) vs V-pol, particularly during polarimetric outburst events (red in Fig. 1). 60 Cyg also displays this behavior.

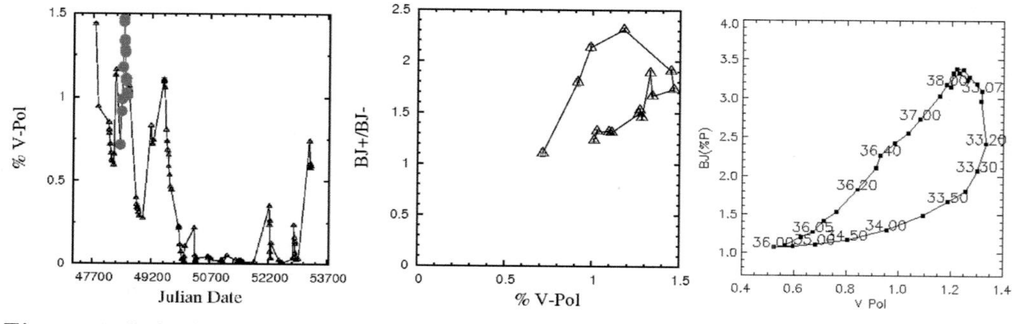

Figure 1. Left: Time evolution of the intrinsic continuum V-band polarization (V-pol) of π Aqr; Middle: Loop-like structure of the evolution of the polarization across the Balmer Jump vs V-pol; Right: Modeled clockwise loop structures.

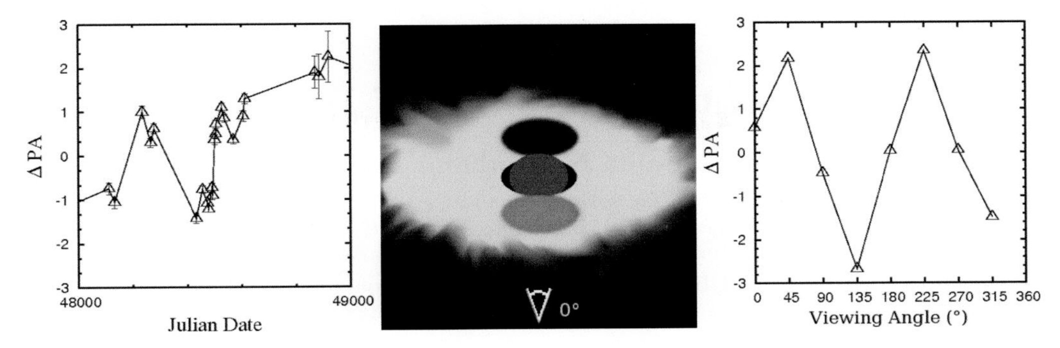

Figure 2. Left: Deviations in the mean PA during polarimetric outbursts of π Aqr; Middle: Modelled density enhancement (orange) and decrement (black) on opposite sides of the disk; Right: Computed polarization from different viewing angles in a counter-clockwise motion.

2. Modeling Polarization Loops

We use the non-LTE 3D Monte Carlo code developed by Carciofi & Bjorkman (2006), HDUST, to investigate the origin of the loop-like behavior of the BJ vs V-pol during polarimetric outbursts. Interestingly, we find that the clockwise loop structures can be reproduced when the mass-loss from the central star which feeds the disk is turned on (6 to 12 o′clock, right panel of Fig. 1) then off (12 to 6 o′clock, right panel of Fig. 1). We therefore suggest that this diagnostic can provide insight into the time dependence of the density of the innermost disk region. Counter-clockwise loops are also observed in π Aqr and 60 Cyg and will require further modeling to ascertain thier origin. We note that de Wit *et al.* (2006) found similar loop structures in CMD diagrams of Be stars.

3. PA Variations

In paper I, we detected variations in the PA of disks during polarimetric outbursts, and speculated that these variations could be indicative of warps, non-equatorial blob injections, or spiral density waves in the inner disk. The left panel of Fig. 2 depicts deviations in the mean PA during polarimetric outbursts of π Aqr. As shown in the middle panel, we model a density enhancement on one side (orange) of the disk and a decrement on the opposing side (black) using HDUST. The result of computing the polarization from different viewing angles in a counter-clockwise motion are shown in the right panel of Fig. 2. This model suggests that asymeteric material being ejected from the star can potentially match observed deviations in the PA.

References

Carciofi, A. C. & Bjorkman, J. E. 2006, *ApJ*, 639, 1081

de Wit, W. J., Lamers, H. J. G. L. M., Marquette, J. B., & Beaulieu, J. P. 2006, *A&A*, 456, 1027

Wisniewski, J. P., Draper, Z. H., Bjorkman, K. S., Meade, M. R. *et al.* 2010, *ApJ*, 709, 1306 (paper I)

Active OB stars: structure, evolution, mass loss, and critical limits
Proceedings IAU Symposium No. 272, 2010
C. Neiner, G. Wade, G. Meynet & G. Peters, eds.

doi:10.1017/S1743921311010866

Infrared continuum sizes of Be star disks

Douglas R. Gies[1], Yamina N. Touhami[1] and Gail H. Schaefer[2]

[1]Center for High Angular Resolution Astronomy and Department of Physics and Astronomy, Georgia State University, P. O. Box 4106, Atlanta, GA 30302-4106, U.S.A.
email: gies@chara.gsu.edu, yamina@chara.gsu.edu

[2]Georgia State University, CHARA Array, P. O. Box 48, Mount Wilson, CA 91023, U.S.A.
email: schaefer@chara-array.org

Abstract. Long baseline interferometry now offers us the opportunity to measure the dimensions of Be star circumstellar disks across the spectrum. This includes the near-infrared continuum where the emission is dominated by bound-free and free-free emission from the ionized disk gas. Here we present the results of calculations of the disk sizes and continuum flux excesses for a simple version of the viscous decretion model of the disk. We compare these results to recent 18 μm flux measurements from the AKARI infrared satellite all-sky survey.

Keywords. techniques: interferometric, circumstellar matter, stars: emission-line, Be

---

Be stars are bright in the infrared due to emission from the ionized gas in their circumstellar disks (Waters *et al.* 1987; Dougherty *et al.* 1994; Zhang *et al.* 2005). It is now possible to angularly resolve the disks of relatively nearby Be stars with long baseline interferometers like the CHARA Array and VLTI (Gies *et al.* 2007; Meilland *et al.* 2007). Since the IR flux excess depends on both the source function (dependent on temperature) and projected disk size, a combined interferometric and flux excess analysis can yield valuable information on the disk gas temperature. Gies *et al.* (2007) presented a realization of the viscous decretion model for isothermal disks of Be stars that they used to create model images, interferometric visibilities, and IR fluxes. The model is based upon four main parameters: the gas base density ρ_0, the radial density exponent n, the disk inclination i, and the disk-to-star temperature ratio $T_d/T_{\rm eff}$. Here we use the model to calculate the disk IR fluxes from 1.7 to 18μm in wavelength.

We begin with a default model with a central star with $T_{\rm eff} = 30$ kK, $R/R_\odot = 10$, and $M/M_\odot = 15.5$. The adopted disk parameters are $n = 3$, $i = 45^\circ$, $T_d/T_{\rm eff} = 2/3$, and an outer boundary disk radius $R_{\rm out}/R_\star = 21.4$. We show in Figure 1 the predicted 18μm flux excess $E^\star(V^\star - m_\lambda) = 2.5 \log(1 + F_d/F_\star)$ and ratio of the disk radius (HWHM) to stellar radius along the projected major axis of the disk. These are plotted for a range in base density of $\log \rho_0 = -12.0$ to -9.7 (units of g cm^{-3}). Also shown are several other model sequences made by varying one of the disk parameters. The position in the diagram is most sensitive to disk density and inclination, but there is a systematic displacement between models with different assumed $T_d/T_{\rm eff}$, which indicates that the near-IR fluxes and interferometric diameters can be used to estimate disk temperature (especially for dense, large disks).

We can test the model predictions through a comparison of the the 2MASS and AKARI fluxes (Ita *et al.* 2010) for a sample of Be stars from Dougherty *et al.* (1994) . We determined for each star an observational excess pegged to the V-band flux

$$E^\star(V^\star - m_\lambda) = V - m_\lambda - E(B-V) \times (3.10 - R_\lambda) - (V - m_\lambda)(\text{Kurucz})$$

where the ratio of interstellar extinction to reddening is $R_\lambda = A_\lambda/E(B-V)$ and $(V - m_\lambda)$(Kurucz) is the intrinsic stellar colour derived from monochromatic sampling of flux

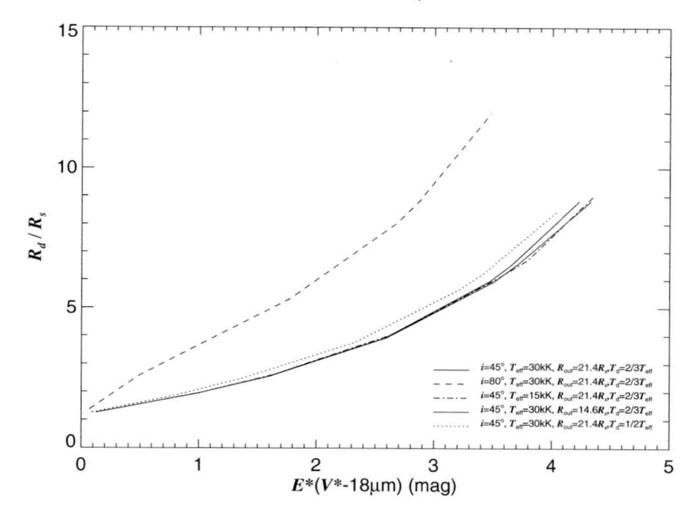

Figure 1. A plot of the disk HWHM versus the 18μm flux excess for several models.

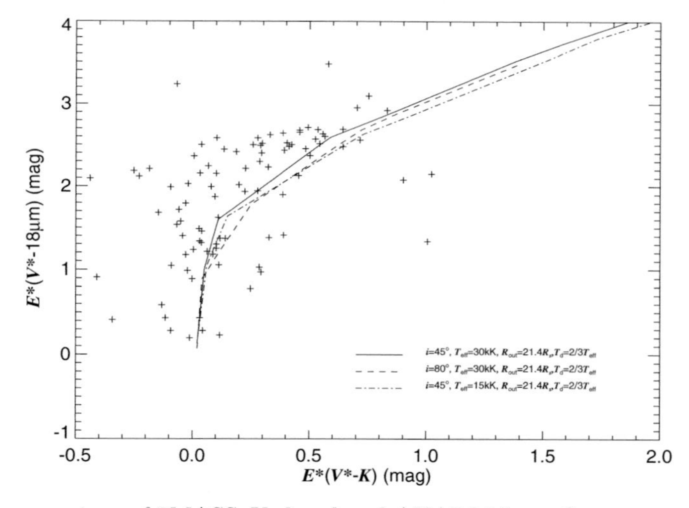

Figure 2. A comparison of 2MASS K_s-band and AKARI 18 μm flux excesses with models.

ratios of model spectra with a Vega spectrum from R. L. Kurucz. The equivalence of the model and observed colour excesses is based upon the assumption that the disk contributes no flux in the V-band, and this neglect may lead to somewhat negative colour excesses for the dense disk cases. A comparison of the K_s and 18 μm flux excesses is shown in Figure 2. The overall agreement suggests that the viscous decretion disk model provides a satisfactory description of the near-IR flux excesses.

References

Dougherty, S. M., Waters, L. B. F. M., Burki, G., Cote, J. *et al.* 1994, *A&A*, 290, 609

Gies, D. R., Bagnuolo, Jr., W. G., Baines, E. K., ten Brummelaar, T. A. *et al.* 2007, *ApJ*, 654, 527

Ita, Y., Matsuura, M., Ishihara, D., Oyabu, S. *et al.* 2010, *A&A*, 514, A2

Meilland, A., Stee, P., Vannier, M., Millour, F. *et al.* 2007, *A&A*, 464, 59

Waters, L. B. F. M., Cote, J. & Lamers, H. J. G. L. M. 1987, *A&A*, 185, 206

Zhang, P., Chen, P. S. & Yang, H. T. 2005, *New Astron.*, 10, 325

Active OB stars: structure, evolution, mass loss, and critical limits
Proceedings IAU Symposium No. 272, 2010
C. Neiner, G. Wade, G. Meynet & G. Peters, eds.

doi:10.1017/S1743921311010878

K- and L-band spectroscopy of Be stars

Anahí Granada[1,2,4], María L. Arias[1,2], Lydia S. Cidale[1,2] and Ronald E. Mennickent[3]

[1]Facultad de Ciencias Astronómicas y Geofísicas, Universidad Nacional de La Plata, Argentina
[2]Instituto de Astrofísica La Plata, CCT La Plata-CONICET-UNLP, Argentina
[3]Departamento de Astronomía, Universidad de Concepción, Chile
[4]Observatoire de Genève. Université de Genève, Suisse

email: granada@fcaglp.unlp.edu.ar

Abstract. We describe the behaviour of IR hydrogen emission lines of a sample of Be stars and discuss the physical properties of the circumstellar envelopes of Be stars classified in Groups I and II (Mennickent *et al.* 2009). We find that while Humphreys and Pfund lines of Group I stars form in an optically thick envelope/disk, Group II stars show Pfund lines that form in an optically thick medium and Humphreys lines originating in optically thinner regions. The transition between Groups I and II could be understood in terms of the evolution of the circumstellar disk of the star and might bring clues on the mechanism originating the Be phenomenon.

Keywords. stars: emission-line, Be, circumstellar matter

---

1. Introduction

Be stars have a circumstellar envelope revealed in the infrared through moderate flux excesses and the appearance of hydrogen recombination lines. K- and L- band spectra of Be stars present numerous lines of Pfund (Pf), Humphreys (Hu) and Brackett (Br) series. Their profiles are sensitive to the physical properties and dynamical structure of the line-forming regions and thus become useful probes of circumstellar environments. An inspection of the L-band hydrogen spectra of Be stars allows a classification in three groups following Mennickent *et al.* (2009): Group I contains the stars with Brα and Pfγ equally intense as Hu lines, Group II consists of those with Brα and Pfγ more intense than Hu lines, and Group III is made up of those with no detected emission. This classification scheme reflects the optical depth conditions in the Be star envelope. Group I stars have a more compact Hu line-forming region than Group II stars, whereas Group III stars might have lost their envelopes (Mennickent *et al.* 2009, Granada *et al.* 2010). In this work we analize K- and L- band spectra of a sample of 26 B-type emission line stars (see Table 1) and relate them to the optical depths of their line-forming regions.

2. Results

We show the line fluxes of Hu and Pf lines relative to a reference transition (Fig. 1), for the stars in Table 1. The continous lines indicate the optically thin limit predicted by Menzel Case B recombination theory while the dashed line represents the thick case.

Hu and Pf lines of Group I stars come from an optically thick envelope (Figs. 1.1a and 1.1b). Fig. 1.2a shows the range of Hu line flux ratios covered by Be stars of Group II. In some cases these ratios are close to Menzel Case B, whereas some objects depart from this case. Pf line ratios (Fig. 1.2b) correspond to an optically thick envelope. B[e] stars have Hu and Pf line ratios (Fig. 1.3), close to those of Menzel Case B, likely to form in an isothermal stellar wind (Lenorzer *et al.* 2002a).

Table 1. Sample. a)Lenorzer *et al.*(2002b); b)Mennickent *et al.*(2009); c)Granada *et al.*(2010)

Star (HD)	Name	S.T.	Group	Ref	Star (HD)	Name	S.T.	Group	Ref
	MWC349A	O9III[e]	II	a	120991	V767 Cen	B3IIIe	I	b
5394	γ Cas	B0IVesh	II	a	148259	OZ Nor	B2II	I	b
20336	BK Cam	B2.5 Ve	I	c	162732	88 Her	Bpshe	I	c
23862	28 Tau	B8 IVev	II	c	164284	66 Oph	B2Ve	II	c
29441	V1150 Tau	B2.5Vne	I	b	178175	V4024 Sgr	B2V	I	b
45677	MWC 142	B2V[e]	II	a	183656	V923 Aql	Bpshe	II	c
50013	κ CMa	B1.5IVne	I	a	186272	V341 Sge	B2.5V	II	b
56139	ω CMa	B2IV-Ve	II	a	187811	12 Vul	B2.5Ve	I,II	a,b
93308	η Car	Bpe	II	a	191610	28 Cyg	B2.5Ve	II	a,c
94910	AG Car	B2pe	II	a	193237	P Cyg	B2pe	II	a
105521	V817 Cen	B3 IVe	I	b	200775	MWC 361	B2V[e]	II	a
105435	δ Cen	B2IVne	II	b	209409	omi Aqr	B7IVe	II	a
120324	μ Cen	B2IV-Ve	II	b	217050	EW Lac	B3 IVshe	II	c

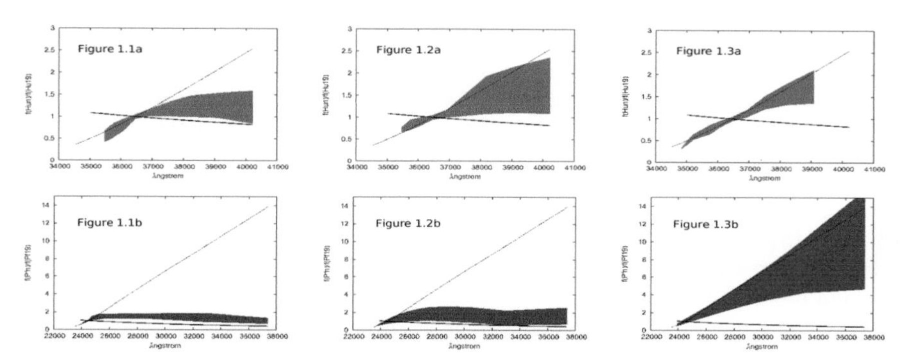

Figure 1. Line Flux Ratios: 1) Group I; 2) Group II; 3) Other emission lines

3. Discussion

Most of the objects in Table 1 are B2-3 type stars. Thus, if we consider similar central stars, the differences observed in the IR spectra evidence differences in the line-forming regions. We obtain a rough estimate of the column density of atoms in the lower excitation level (that is N_6 for Hu lines and N_5 for Pf lines) as well as the relative extension, Δr of the forming region of Hu (R_{Hun}/R_{Hu16}) and Pf (R_{Pfn}/R_{Pf16}) lines, using line flux ratios (Granada *et al.* 2010). We find that for Group I stars the Hu line-forming regions seem to be more dense and compact than those for Group II stars; whereas the mean column density obtained for Group I stars ranges from 1.2 to 3.7×10^{14} cm^{-1} and Δr from 0.38 to 1.28, for Hu_{18} to Hu_{14}, Group II stars have column densities from 0.22 to 1.39×10^{14} cm^{-1} with Δr from 0.28 to 1.90 for the same lines. Moreover, higher members of Pf series for B2 stars of both Groups also form in compact and dense regions.

Many Be stars of our sample show strong spectroscopic variability. For 12 Vul and 28 Cyg, we reported changes from one Group to another, which are possibly indicating structural changes in the circumstellar environment (Mennickent *et al.* 2009, Granada *et al.* 2010). Time resolved near-IR spectroscopy of Be stars would allow us to study the origin and evolution of the envelopes as well as to set constraints to different models.

References

Granada, A., Arias, M. L. & Cidale, L. S. 2010, *AJ*, 139, 1983
Lenorzer, A., de Koter, A. & Waters, L. B. F. M. 2002a, *A&A*, 386, L5
Lenorzer, A., Vandenbussche, B., Morris, P., de Koter, A. *et al.* 2002b, *A&A*, 384, 473
Mennickent, R. E., Sabogal, B., Granada, A., & Cidale, L. 2009, *PASP*, 121, 125

Active OB stars: structure, evolution, mass loss, and critical limits
Proceedings IAU Symposium No. 272, 2010
C. Neiner, G. Wade, G. Meynet & G. Peters, eds.

doi:10.1017/S174392131101088X

Investigating the continuum linear polarization of Be stars

Robbie J. Halonen[1], Frances E. Mackay[1], Carol E. Jones[1], and T. A. Aaron Sigut[1]

[1]Dept. of Physics and Astronomy
The University of Western Ontario
London, ON, Canada
email: rhalonen@uwo.ca

Abstract. In order to understand the mechanisms that govern the development of circumstellar disks surrounding classical Be stars, we use computational codes to create theoretical models of these particular objects with their gaseous environments and we compare the predicted observables to astronomical observations. In this study, we present the use of the non-LTE radiative transfer code of Sigut & Jones (2007) to examine the effect of a self-consistent thermal structure and realistic chemical composition on the polarization of the classical Be star γ Cassiopeia. Primarily, we investigate the effect of several improvements on the pioneering work of Poeckert & Marlborough (1978) in calculating the polarization levels of γ Cas. We establish best-fit models for the same observations and analyze the implications of the differences between our results and those obtained by Poeckert & Marlborough.

Keywords. stars: emission-line, Be, methods: numerical, radiative transfer, polarization

1. Introduction

Observations of Be stars, through techniques such as photometry or spectroscopy, can be used to place important constraints on the geometry and density structure of their circumstellar disks. Non-zero polarization levels constitute a key observable for investigating the nature of Be stars, particularly for probing the geometric properties of the circumstellar environment. Along with observations, detailed modeling can be used to theoretically ascertain fundamental disk parameters. Computational modeling efforts began with the work of Poeckert & Marlborough (1978), hereafter PM. PM used a radiative transfer code to predict the polarization of γ Cas. Although their models were more or less ad hoc due to the complexity of the problem and the computational limitations at the time, PM's model was remarkably successful in predicting the polarization and other observables for the star.

2. Overview

Our calculations use the output level populations, temperature structure, and radiation field of the non-LTE radiative transfer code BEDISK developed by Sigut & Jones (2007) for the underlying model of the Be star disk. This computational code formally solves the coupled problems of radiative transfer, statistical equilibrium and radiative equilibrium to provide a self-consistent calculation of the thermal structure of the disk. The constructed disk is assumed to be axisymmetric about the stellar rotation axis and symmetric about the midplane and to have an initial density structure of the form $\rho = \rho_0 (R/R_{star})^{-n}$ in

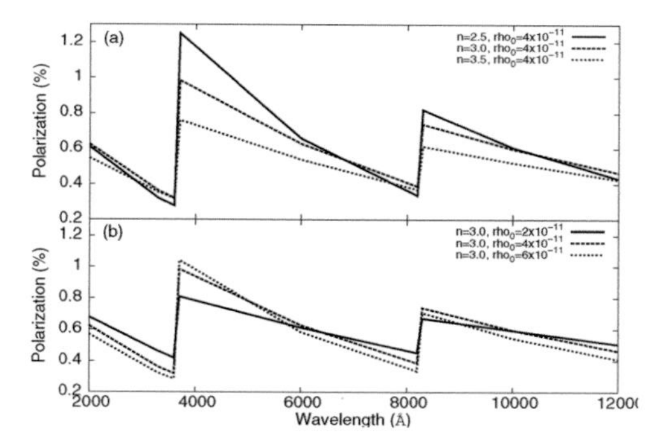

Figure 1. The effect of varying n and ρ_0 from the density equation are illustrated in the top and bottom graphs, respectively. These models include a self-consistent calculation of the thermal structure of the disk and the use of a realistic chemical composition.

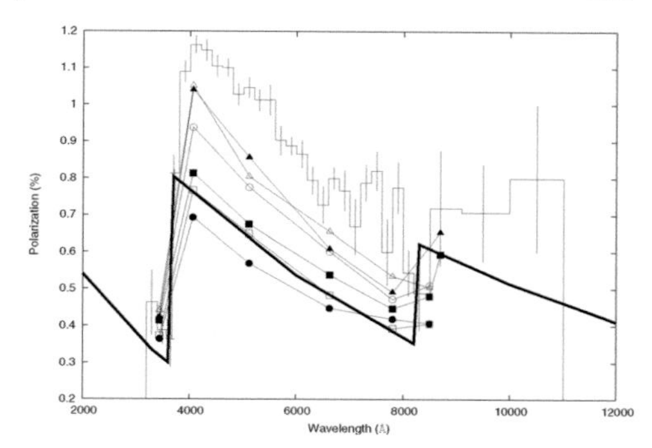

Figure 2. The thick line represents the best fit to the 1976 observations (open squares) modeled by PM. The fit is an improvement over that of PM whose model predicted consistently higher levels than measured. See Mackay *et al.* (2010) for further discussion.

the equatorial plane. The gas is taken to be in vertical, hydrostatic, isothermal equilibrium perpendicular to the plane of the disk. The net polarization of the disk is determined by integrating the Stokes parameters at individual grid points along lines of sight through the disk, with the results summed over the projected area of the disk. The use of the BEDISK code in providing the necessary inputs into the polarization computation allows us to significantly improve PM's method with regards to both the state of the gas and the extent of computational sampling used throughout the disk.

3. Results and Future Work

The most notable improvements of the BEDISK code are the inclusion of self-consistent calculation of the thermal structure of the disk and the use of a realistic chemical composition for the enveloping gas. The profound effect that these improvements have on the thermal structure of the disk significantly affect the level populations and the number of scatterers present in the disk and, therefore, influence the predicted polarization. Furthermore, the BEDISK code uses an increased number of grid points, both in the disk and on the star, in order to improve the accuracy of the solution. Our access to PM's code provides us with the unique opportunity to investigate the effect that various computational parameters have on the polarization result published by PM. For a full discussion of the results of our study, see Mackay *et al.* (2010).

We are currently investigating the influence of line effects and line blanketing on the continuum polarization. Furthermore, we hope to include the process of multiple scattering into our calculations. Lastly, we are preparing an extensive study of the correlations between the continuum linear polarization and other principal observational features of Be stars.

References

Mackay, F. E., Halonen, R. J., Jones, C. E., Bjorkman, K. S., Sigut, T.A.A. & Meade, M. R. 2010, *ApJ*, submitted

Poeckert, R. & Marlborough, J. M. 1978, *ApJ*, 220, 940

Sigut, T. A. A. & Jones, C. E. 2007, *ApJ*, 668, 481

Active OB stars: structure, evolution, mass loss, and critical limits
Proceedings IAU Symposium No. 272, 2010
C. Neiner, G. Wade, G. Meynet & G. Peters, eds.

doi:10.1017/S1743921311010891

The dynamical evolution of Be star disks

Xavier Haubois[1], Alex C. Carciofi[1], Atsuo T. Okazaki[2], and Jon E. Bjorkman[3]

[1]Instituto de Astronomia, Geofísica e Ciências Atmosféricas, Universidade de São Paulo, Rua do Matão 1226, Cidade Universitária, São Paulo, SP 05508-900, Brazil
email: xhaubois@astro.iag.usp.br

[2]Faculty of Engineering, Hokkai-Gakuen University, Toyohira-ku, Sapporo 062-8605, Japan

[3]University of Toledo, Department of Physics & Astronomy, MS111 2801 W. Bancroft Street Toledo, OH 43606, USA

Abstract. We present a novel theoretical tool to analyze the dynamical behaviour of a Be disk fed by non-constant decretion rates. It is mainly based on the computer code HDUST, a fully three-dimensional radiative transfer code that has been successfully applied to study several Be systems so far, and the SINGLEBE code that solves the 1D viscous diffusion problem. We have computed models of the temporal evolution of different types of Be star disks for different dynamical scenarios. By showing the behaviour of a large number of observables (interferometry, polarization, photometry and spectral line profiles), we show how it is possible to infer from observations some key dynamical parameters of the disk.

Keywords. circumstellar matter, radiative transfer, stars: emission-line, Be

1. Presentation of the simulations

In order to analyze the observational signatures of different dynamical scenarios, we use the SINGLEBE code (Okazaki, 2007) that computes the temporal evolution of the surface density for a given stellar mass loss history and an α viscosity parameter (Shakura & Sunyaev, 1973) of the disk. We then use the surface density at a given time as an input for a three-dimensional non-LTE Monte Carlo code called HDUST (Carciofi & Bjorkman, 2006). By repeating this for different epochs of the disk evolution, we can follow the evolution of several observables (SED, polarization, images, etc).

A wide range of observables can be derived from the HDUST simulations. By means of a comparative work, they can unveil the observational features of a specific dynamical scenario occurring in a viscous disk. The final objective of these quantities is to be analyzed and compared to real polarimetric, photometric, spectroscopic or interferometric observations. This method has the potential to allow one observer to unambiguously infer the size, the α parameter and the mass loss history of the observed system.

2. Some results

As preliminary results, we present some observables and their correlation which clearly show different temporal behaviour depending on the α parameter (Fig. 1). We present results for a dynamical scenario in which 3 year long outbursts are separated by 3 year long quiescent phases. The influence of α is understood by considering that the larger α, the faster the viscous diffusion occurs.

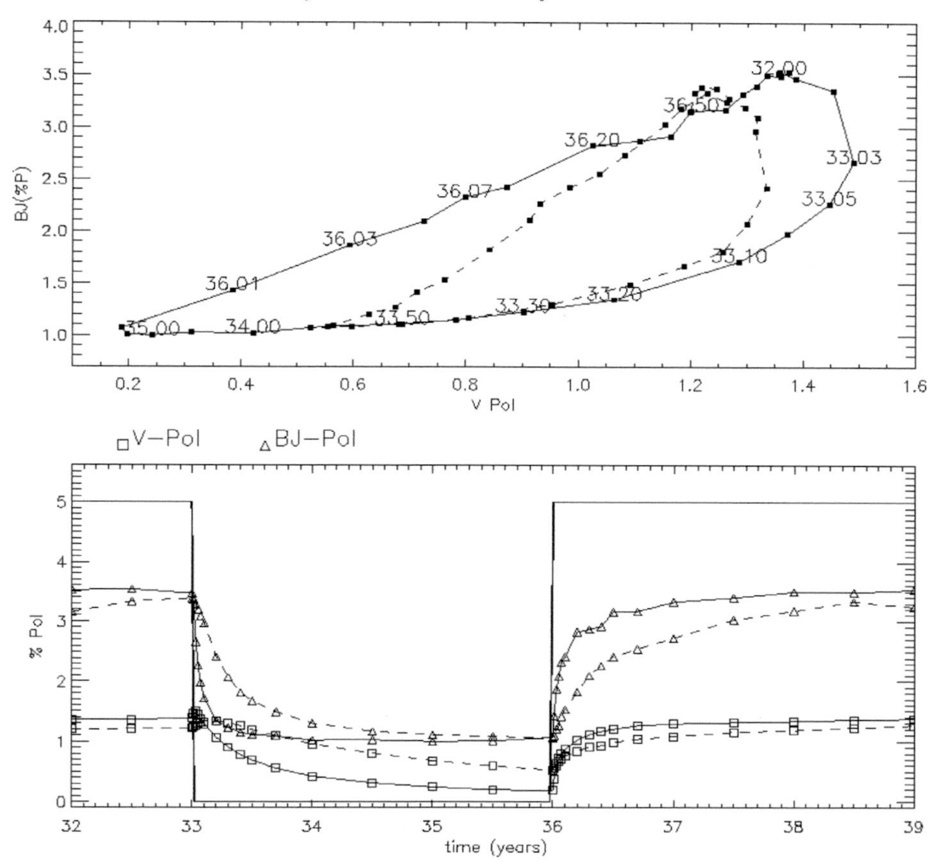

Figure 1. Upper panel: polarization change across the Balmer Jump (BJ) vs V band polarization for two α parameter values (0.1 in dashed line, 0.7 in full line) and at an inclination angle of 70 degrees. Epochs are marked and counted in years, the correlation loops move clockwise. Lower panel: temporal evolution of these two quantities. The thickest line shows the mass loss history arbitrarily scaled to the range of the graphic. It has been turned on and off every 3 years from 0 to 39 years. That figure depicts the 32-39 years period.

3. Conclusion and Perspectives

These simulations can predict the observational signatures of dynamical scenarios by generating a wide range of observables. Some of these observables are thus powerful tools to estimate some key parameters of observed Be systems such as the α viscosity parameter. This work is currently ongoing and more complex dynamical scenarios have to be investigated (Haubois *et al.*, in preparation). Moreover, the results of these studies being more easily understandable with videos, we plan to make them accessible to the community in a near future through a dedicated website (http://www.astro.iag.usp.br/beacon/).

References

Carciofi, A. C. & Bjorkman, J. E. 2006, *ApJ*, 639, 1081

Haubois, X., Carciofi, A. C., Bjorkman, J. E., & Okazaki, A. T., in preparation

Okazaki, A. T. 2007, in: A. T. Okazaki, S. P. Owocki, & S. Stefl (eds.), *Active OB-Stars: Laboratories for Stellar and Circumstellar Physics*, ASP-CS 361, p. 230

Shakura, N. I. & Sunyaev, R. A. 1973, *A&A*, 24, 337

Active OB stars: structure, evolution, mass loss, and critical limits
Proceedings IAU Symposium No. 272, 2010
C. Neiner, G. Wade, G. Meynet & G. Peters, eds.

doi:10.1017/S1743921311010908

Probing Be star disks: new insights from Hα spectroscopy and detailed numerical models

Carol E. Jones[1], Christopher Tycner[2], Jessie Silaj[1], Ashly Smith[2] and T. A. Aaron Sigut[1]

[1]Dept. of Physics and Astronomy, *The* University of Western Ontario
London, Ontario, Canada N6A 3K7; email: cejones@uwo.ca

[2]Dept. of Physics, Central Michigan University, Mt. Pleasant, MI 48859, USA

Abstract. Hα high resolution spectroscopy combined with detailed numerical models is used to probe the physical conditions, such as density, temperature, and velocity of Be star disks. Models have been constructed for Be stars over a range in spectral types and inclination angles. We find that a variety of line shapes can be obtained by keeping the inclination fixed and changing density alone. This is due to the fact that our models account for disk temperature distributions self-consistently from the requirement of radiative equilibrium. A new analytical tool, called the variability ratio, was developed to identify emission-line stars at particular stages of variability. It is used in this work to quantify changes in the Hα equivalent widths for our observed spectra.

Keywords. stars: emission-line, Be, line: profiles, radiative transfer, circumstellar matter, stars: activity

---

1. Introduction

B-emission or Be stars are characterized by Balmer emission lines in their spectra due to the presence of a disk-like distribution of circumstellar gas. Often the Hα Balmer line is the most prominent feature. Consequently, the study of the Hα spectral line, including its shape, equivalent width (EW), and variability, offers a valuable probe of Be star disks.

Here we present the main results from a thorough study of Hα line profiles for Be stars built from models with a wide range of input parameters, on the basis of non-LTE calculations of Be disk systems. Our models were compared to 69 Be star/disk systems (Silaj *et al.* 2010). We also investigate the variability of many of these systems by monitoring the change in Hα EW calculated from observed spectra acquired over a period of about 4 years. We found a simple method to quantify the changes in the spectra, called the variability ratio. This technique requires relatively little observational data, and can be used to find these stars in particular stages of variability. Alternatively, it can be used to determine when a star/disk system is changing and needs to be monitored closely.

2. Results

A common misconception in the literature is that the shape of certain spectral lines can be used to infer the inclination angle of a disk system. We showed that this was generally not true in all cases and that computing a self-consistent disk thermal structure is critical to interpreting observations. Fig. 1 shows B0 Hα line profiles (far left) and their corresponding temperature structures (right) for four different models with inclinations of 20° and 45°. (The power law index, n, and the assumed base density at the stellar surface, ρ_o, are parameters related to the density distribution (see Sigut & Jones (2007)

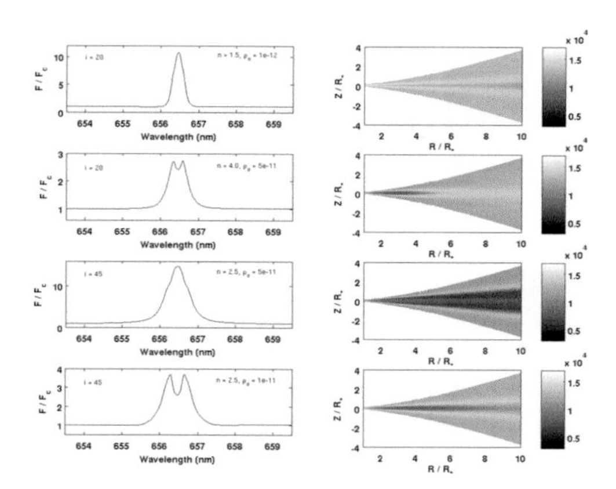

Figure 1. Model B0 Hα line profiles (left) and their corresponding temperature structures (right) for four different models with inclinations of 20° and 45°. KEY: Line profile shapes can vary due to changes in disk density alone when the temperature is accounted for self-consistently (Silaj *et al.* 2010).

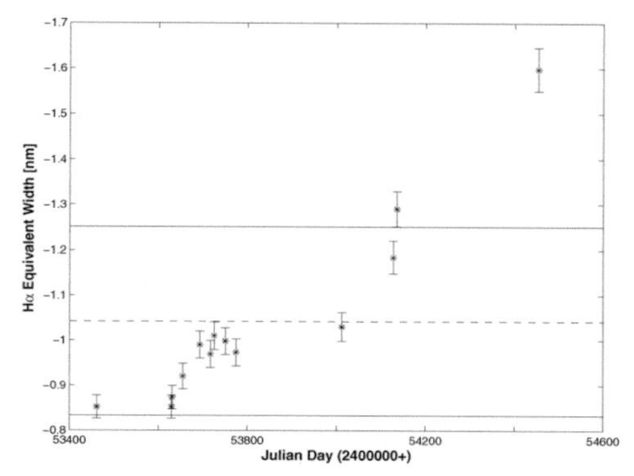

Figure 2. A plot showing the change in Hα equivalent width as a function of time for the star BK Cam (HR 985). The estimated error for each observation is shown by the errors bars. The middle dashed horizontal line corresponds to the mean of the EW. The upper and lower solid horizontal lines correspond to $\overline{EW} \pm$ one standard deviation (Jones *et al.* 2011).

for greater detail). Notice that for a given inclination the line shifts from singly peaked to doubly peaked with a change in disk density alone.

A necessary prerequisite to the development of successful dynamical models will be timely observations that adequately sample the disk loss and disk growth events for Be stars. Many Be stars are known to be variable and the study of this variability has had a long history. The variability occurs over a wide range of time scales from periods much less than a day to periods as long as decades. A successful model must account for this observed variability. Therefore, it is crucial to monitor the system at particular stages of variability, for example, during a disk loss or growth event, if we hope to improve our understanding of these systems. We have developed a new tool that offers a simple but powerful method to place bounds on the degree of variability of particular systems based on statistical analysis of changes in Hα EW over time.

Fig. 2 shows an example of the change in Hα EW for the star, BK Cam (HR 985) during our investigation. We calculate a variability ratio, R from the ratio of standard deviation to the mean uncertainty. Based on our preliminary work, we calculate $R = 6.6$ which falls into our variable category. We are currently refining our statistics for our program stars.

3. Future Work

We are currently completing a thorough study of Hα profiles for Be shell stars. As a continuation of our work on the variability of Be stars, we are using our detailed models, in steps, to simulate dynamic models.

References

Jones, C. E., Tycner, C., & Smith, A. D. 2011, *AJ*, 141, 150
Silaj, J., Jones, C. E., Tycner, C., Sigut, T. A. A. *et al.* 2010, *ApJS*, 187, 228
Sigut, T. A. A. & Jones, C. E. 2007, *ApJ*, 668, 481

Active OB stars: structure, evolution, mass loss, and critical limits
Proceedings IAU Symposium No. 272, 2010
C. Neiner, G. Wade, G. Meynet & G. Peters, eds.

doi:10.1017/S174392131101091X

The inhomogeneous wind of the LBV candidate Cyg OB2 No.12

Valentina G. Klochkova[1], Eugene L. Chentsov[1] and Anatoly S. Miroshnichenko[2]

[1]Special Astrophysical Observatory RAS, Nizhnij Arkhyz, 369167, Russia
email: valenta@sao.ru, echen@sao.ru

[2]Dept. of Physics & Astronomy, University of North Carolina at Greensboro,
P.O. Box 26170, Greensboro, NC 27402–6170, USA; email: a_mirosh@uncg.edu

Abstract. We present the results of high-resolution spectroscopy of the extremely luminous star Cyg OB2 No. 12. We identified about 200 spectral features in the range 4552–7939 Å, including the interstellar Na I, K I lines and numerous very strong DIBs, along with the He I, C II, and Si II lines. An MK spectral type we derived for the object is B4.5±0.5 Ia^+. Our analysis of the radial velocity data shows the presence of a gradient in the stellar atmosphere, caused by both atmospheric expansion and matter infall onto the star. The Hα emission displays broad Thompson wings, a slightly blue-shifted P Cyg type absorption component and a time-variable core absorption. We conclude that the wind is variable in time.

Keywords. stars: supergiants, stars: circumstellar matter

---

1. Introduction

For stars in clusters the evolutionary stage, age, and luminosity can be determined more reliably, whereas they are rather uncertain for field stars. It is especially important to study group members that are rare, such as LBV–stars. From this point of view, young Cyg OB2 association is of special interest. Many unevolved O/Of–stars have been identified there as well as an LBV candidate – the variable star No. 12. Its luminosity is $\log L/L_\odot = 6.26$ (de Jager 1998) at the association distance of 1.7 kpc.

2. Observations and results

Optical spectra of Cyg OB2 No. 12 were taken using the échelle spectrographs of the 6-meter telescope of the Special Astrophysical Observatory. On June 12, 2001, we used the PFES spectrograph (Panchuk *et al.* 1997) with a 1040×1170-pixel CCD at the prime focus and got a spectrum with a resolution of $R = \lambda/\Delta\lambda \sim 15000$ (20 km s^{-1}). Later we used the NES spectrograph (Panchuk *et al.* 2009) equipped with a 2048×2048-pixel CCD and an image slicer and obtained spectra with $R \sim 60000$ (5 km s^{-1}) on April 12, 2003 and on December 8, 2006.

The spectral types we derived for three dates were the same within the errors: B5.0±0.5, B4.8±0.5 and B4.0±0.5. The luminosity type is Ia^+. The high luminosity is supported by the strong O I 7773 Å IR–triplet whose equivalent width of 1.14 Å corresponds to an absolute visual magnitude of $M_V < 8$ mag.

The radial velocities (V_r) measured from the absorption line cores vary with time and with the line intensity. The weakest lines give V_r lower than $V_{sys} = -11$ km s^{-1} (Klochkova & Chentsov 2004) by 5, 14 and 15 km s^{-1} in 2001, 2003, 2006, respectively, suggesting a variable expansion rate of the layers where they form. The left panel of

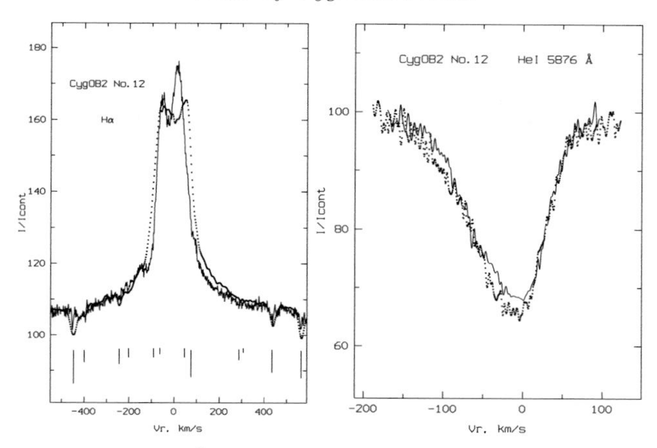

Figure 1. Hα (left) and He I 5876 Å (right) profiles in the spectra of Cyg OB2 No. 12 obtained on June 12, 2001 (dotted line) and April 12, 2003 (solid line). The absorption features seen within the Hα core correspond to transitions in the line rather than to the telluric spectrum, whose contribution was carefully removed. The vertical dashes show the positions of telluric lines, with the dash lengths proportional to the line strength.

Fig. 1 shows that the Hα profile varies with time, but its principal features are preserved: a strong bell-shaped emission, with a dip at the short-wavelength slope, a sheared peak, and extended Thompson wings. The blue-shifted absorption is barely visible in June 2001 and is more pronounced in April 2003, but can be traced at least to $V_r = -160\,\mathrm{km\,s^{-1}}$ in both cases; i.e., to the same limit that is reached by the blue wings of the absorption lines of Si II and He I (the latter shown in the right panel of Fig. 1).

The wind terminal velocity is $\sim 150\,\mathrm{km\,s^{-1}}$. The intensity inversions in the upper part of the Hα profile indicate that the wind is not uniform. In addition to the high velocity material mentioned above, it contains a fair amount of material that is nearly stationary relative to the star or is even falling onto the stellar surface. Coexistence of lines with direct and inverse P Cygni profiles in the same spectrum, and even combinations of such features in the profile of the same line leads us to reject a spherical symmetry wind. It is possible that the slow part of the wind also contributes to the absorption profiles. So far, this possibility is supported by the coincident velocities for the central dips of the Hα line and the well-formed cores of strong absorption lines (He I 5876 Å in 2001 and Si II 6347 Å in 2003), as well as by the fact that the blue shift of all the absorption lines in the 2003 spectrum relative to their positions in 2001 was accompanied by a similar shift of the central dip in Hα. At any rate, both the hydrogen lines and the strongest absorption lines in the visual spectrum Cyg OB2 No. 12 are partially formed in the wind.

Acknowledgements

This research was supported by the Russian Foundation for Basic Research (project no. 08–02–00072 a).

References

de Jager, C. 1998, *A&AR*, 8, 145

Klochkova, V. G. & Chentsov, E. L. 2004, *Astron. Rep.*, 48, 1005

Panchuk, V. E., Najdenov, I. D., Klochkova, V. G., Ivanchik, A. B. *et al.* 1997, *Bull. Special Astrophys. Obs.*, 44, 127

Panchuk, V. E., Klochkova, V. G., Yushkin, M.V., & Naidenov, I. D. 2009, *Opticheskii Zhurn.*, 76, 42

Active OB stars: structure, evolution, mass loss, and critical limits
Proceedings IAU Symposium No. 272, 2010
C. Neiner, G. Wade, G. Meynet & G. Peters, eds.

doi:10.1017/S1743921311010921

Hydrodynamical simulations of Pinwheel nebula WR 104

Astrid Lamberts[1], Sebastien Fromang[2] and Guillaume Dubus[1]

[1]Laboratoire d'Astrophysique de Grenoble, UMR 5571 CNRS, Université Joseph Fourier, BP 53, 38041 Grenoble, France, email: astrid.lamberts@obs.ujf-grenoble.fr

[2]Laboratoire AIM, CEA/DSM-CNRS-Université Paris Diderot, IRFU/Service d'Astrophysique CEA-Saclay F-91191 Gif-sur-Yvette, France

Abstract. Pinwheel Nebulae are colliding wind binaries (CWB) composed of a Wolf-Rayet star and an early-type star. We first compare our simulations to analytic solutions for CWB. Then we perform large scale 2D simulations of the particular system WR 104. We determine the properties of the gas in the winds and confirm the flow in the spiral has a ballistic motion.

Keywords. hydrodynamics, methods: numerical, stars: individual (WR 104), binaries, stars: winds, outflows

---

1. Introduction

Massive stars possess highly supersonic line-driven winds. In binary systems, the interaction of two winds creates a double-shocked structure. For each wind there is a free unshocked component upstream of the shock and a dense, hot shocked wind downstream. The winds are separated by a contact discontinuity (CD). Their momentum flux ratio is given by $\eta = \frac{\dot{M}_1 v_{\infty 1}}{\dot{M}_2 v_{\infty 2}}$. The subscripts 1 and 2 respectively stand for the stronger and weaker wind. For $\eta \gg 1$ the weaker wind is collimated. The whole structure looks like a cometary tail and is turned into a spiral due to orbital motion. As a prototype of CWB, we consider WR 104, composed of an O-B star and a WC9 star. The O-B wind is collimated into a very narrow spiral and shows a clear example of Pinwheel Nebula in infrared (Tuthill *et al.* 2008). The infrared emission is due to the presence of dust in the WR wind, whose origin is still poorly constrained. Up to now simple models of the dust emission in WR 104 have assumed ballistic motion of the shocked winds along an Archimedian spiral (Harries *et al.* 2004).

2. Comparison to analytic models

We use the code hydro RAMSES (Teyssier 2002) with Adaptive Mesh Refinement which enables us to locally increase the spatial resolution. We properly resolve the shock formation and determine the large scale structure at reasonable computational cost. We compared our 2D and 3D adiabatic and isothermal simulations to the solutions of Canto *et al.* (1996) and Antokhin *et al.* (2004) with no orbital motion. The 2D results are shown in figure 1. We show the density map and overplot the analytic solutions.

In the adiabatic case one can clearly distinguish the two shocks and the position of the CD is approximately matched by the analytic solutions. In the isothermal case, the structure is dominated by the 'thin shell instabilities' (Pittard 2009). This cannot be modelled with the analytic solutions. Moreover, the analytic solutions do not account for the orbital motion and for the effect of thermal pressure. These effects can be properly modelled using hydrodynamical simulations.

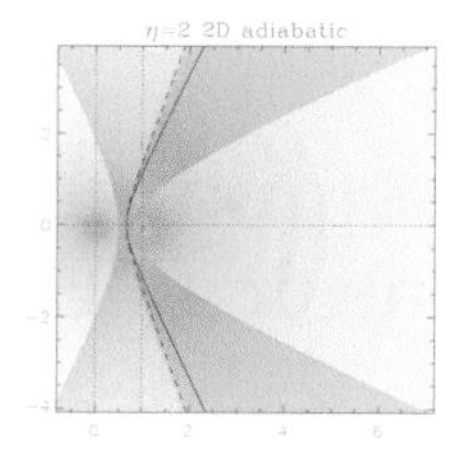

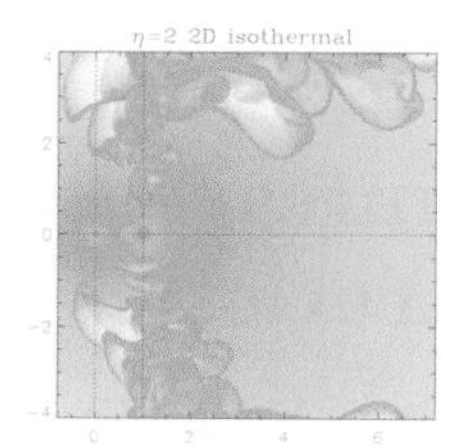

Figure 1. 2D density maps. The stars are located at the intersections of the dotted lines. The dashed line represents the solution from Canto *et al.* (1996), the solid line the solution from Antokhin *et al.* (2004). The length scale is given the binary separation.

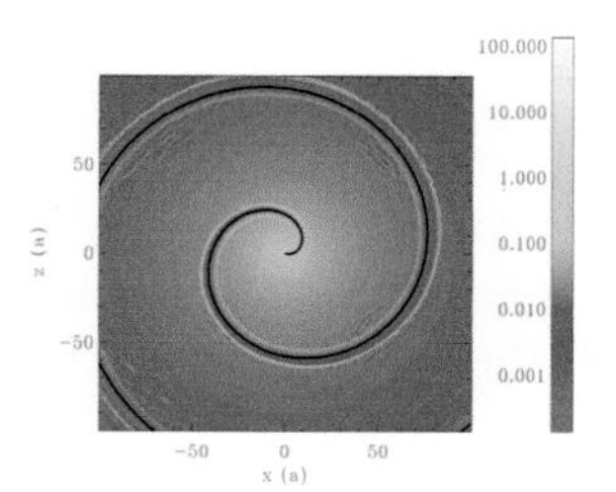

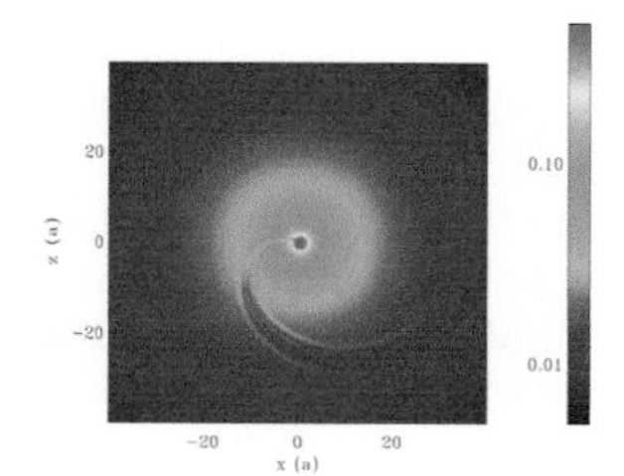

Figure 2. Density in WR 104, adiabatic (left panel) and isothermal (right panel) case.

3. The case of WR 104

We made a 2D simulation of WR 104 with adiabatic winds. As $\eta = 305$ in WR 104, very high resolution is required to resolve the shocks. The density map is shown in figure 2 on the left panel. The theoretical Archimedean spiral (solid line) matches the results of the simulation. Different components of the wind can be seen. Most of the gas is the free WR wind. The densest zone is the shocked WR wind at the edges of the spiral. The density in the shocked O-B wind in the spiral appears to be steady. The velocity is constant in the spiral, confirming ballistic motion.

Cooling is important in the WR wind, suggesting an isothermal equation of state is more appropriate. We thus expect the presence of thin shell instabilities, which might help mixing. We made a 2D simulation with an isothermal equation of state, as is shown on the right panel of fig. 2. No instabilities can be seen. The extreme confinement of the O-B star wind could prevent their developpement. This could also be due to insuficient numerical resolution

4. Summary and conclusions

We performed 2D large scale simulations of WR 104, completely modelling one step of the spiral structure. This work confirms ballistic motion along Archimedean spiral. More analysis is necessary to put stronger constrains on dust formation. Further work will focus on the phenomena triggering or preventing the thin shell instabilities.

References

Antokhin, I. I., Owocki, S. P. & Brown, J. C. 2004, *ApJ*, 611, 434
Canto, J., Raga, A. C. & Wilkin, F. P. 1996, *ApJ*, 469, 729
Harries, T. J., Monnier, J. D., Symington, N. H., & Kurosawa, R. 2004, *MNRAS*, 350, 565
Pittard, J. M. 2009, *MNRAS*, 396, 1743
Teyssier, R. 2002, *A&A*, 385, 337
Tuthill, P. G., Monnier, J. D., Lawrance, N., Danchi, W. C. *et al.* 2008, *ApJ*, 675, 698

Active OB stars: structure, evolution, mass loss, and critical limits
Proceedings IAU Symposium No. 272, 2010
C. Neiner, G. Wade, G. Meynet & G. Peters, eds.

doi:10.1017/S1743921311010933

Near-infrared excess and emission characteristics of classical Be stars

Chien-De Lee[1], Wen-Ping Chen[1,2] and Daisuke Kinoshita[1]

[1]Graduate Institute of Astronomy, National Central University,
Jhongli 32001, Taiwan,
email: m959009@astro.ncu.edu.tw

[2]Department of Physics, National Central University,
Jhongli 32001, Taiwan,
email: wchen@astro.nuc.edu.tw

Abstract. Classical Be (CBe) stars are fast-rotating emission-line stars associated with infrared excess often attributed to plasma free-free emission. A few with exceptionally large near-infrared excess, namely with (J−H) and (H−K_s) both greater than 0.6 mag, however, must be accounted for by thermal emission from circumstellar dust. From 2007 to 2009, spectra of more than 100 CBe stars have been collected. We present some of these spectra and discuss how temporal correlation (or lack of) among spectral features would provide possible diagnosis of the origin of the CBe phenomena.

Keywords. stars: early-type, stars: evolution, stars: emission-line, Be, stars: pre–main-sequence, infrared: stars

---

1. Be stars with large near-infrared excess

Near-infrared excess of most CBe stars is due to free-free emission. A few show significantly large excess in the 2MASS colors (Zhang, Chen & Yang 2005) which cannot be explained by gas only, but requires thermal emission from circumstellar dust. Lee & Chen (2009) suggested that CBe stars are turning-off the main sequence. Dust properties in such evolved main-sequence stars should be markedly different from those in pre-main sequence or in post-main sequence stars. To diagnose the possible relation between gas activity and dust formation in the circumstellar environments, we obtained optical spectra of more than 100 Be stars with the Lulin One-meter Telescope in Taiwan, the SMARTS 1.5 m in Chile, and the 2.16 m telescope in Xinglong, China. Additional infrared photometry has been acquired — in a few cases simultaneously with optical spectra — with the SMARTS 1.5 m, and the 1.88 m telescope of the Okayama Observatory in Japan.

2. Emission lines and near-infrared excess

Fig. 1 shows the near-infrared excess and Balmer activity of CBe stars in a 2MASS/JHK_S color-color diagram. Stars with both Hα and Hβ in absorption — signifying low gas activity — show little near-infrared excess due to lack of free-free emission. Those with Hα in emission, but Hβ and higher Balmer lines in absorption, have moderate near-infrared excess. CBe stars with Hα and Hβ both in emission are highly active, and exhibit large near-infrared excess as well. Some CBe stars have excess emission extending to far-infrared or longer, so must be accounted for by dust emission (Lee & Chen, these proceedings). CBe star are known to vary with timescales from days to years, photometrically and spectroscopically. The 2MASS data shown in Fig. 1 were taken at the same epoch, but the spectra were not. Simultaneous observations of active CBe stars

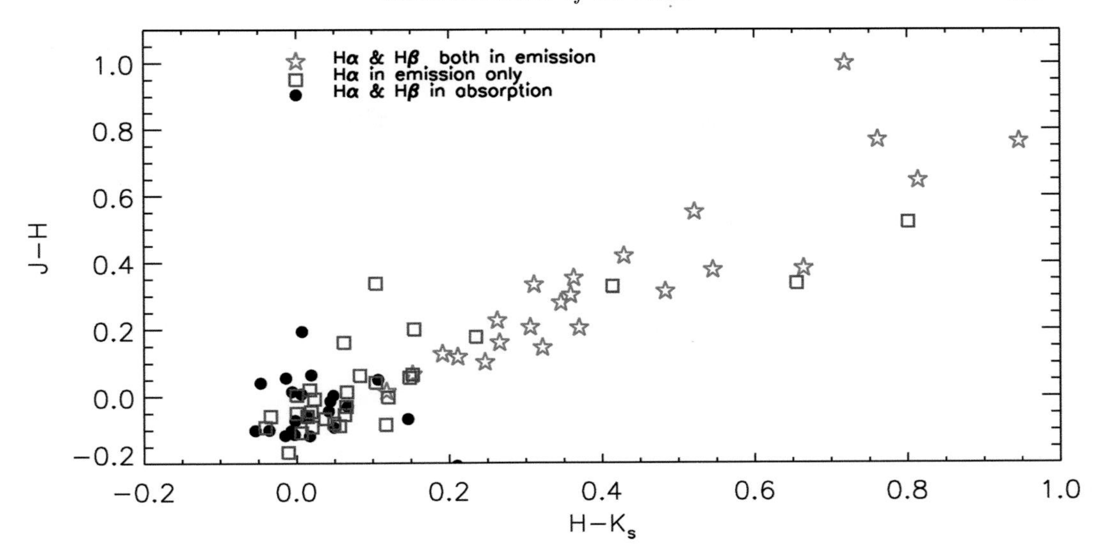

Figure 1. The emission lines and near-infrared excess of CBe stars.

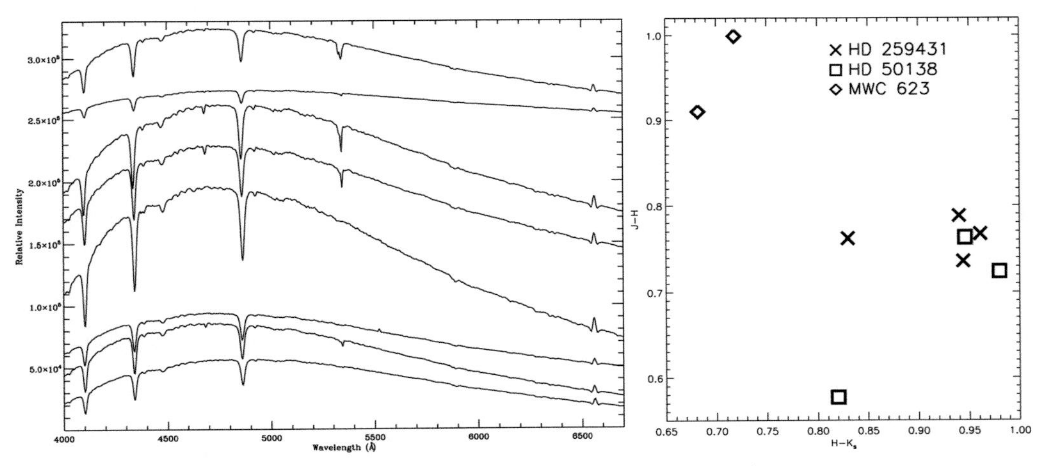

Figure 2. *Left*: HD 142926 shows rapid variations of its He II 5412 Å line within half an hour. *Right*: HD 259431 (crosses), HD 50138(squares) and MWC 623 (diamonds) show variable infrared colors on time scales of hours.

hence are desirable in order to probe the gas activity and dust formation mechanism. So far, variability of near-infrared excess (in hours) and emission lines (within half an hour) have been detected (Fig. 2). We are currently processing data taken simultaneously of the same CBe stars.

References

Lee, C. D. & Chen, W. P. 2009, in: B. Soonthornthum, S. Komonjinda, K. S. Cheng, & K. C. Leung (eds.), *Stellar pulsation: challenges for theory and observation*, ASP-CS 404, p. 302

Zhang, P., Chen, P. S. & Yang, H. T. 2005, *New Astron.* 10, 325

Active OB stars: structure, evolution, mass loss, and critical limits
Proceedings IAU Symposium No. 272, 2010
C. Neiner, G. Wade, G. Meynet & G. Peters, eds.

doi:10.1017/S1743921311010945

Resolving the dusty circumstellar environment of the A[e] supergiant HD 62623 with the VLTI/MIDI

Antony Meilland[1], Sameer Kanaan[2], Marcelo Borges Fernandes[3], Olivier Chesneau[4], Florentin Millour[1,4], Philippe Stee[4], and Bruno Lopez[4]

[1]Max-Planck Institut fr Radioastronomie, Bonn, Germany;
[2]Universidad de Valparaso, Chile; [3]Observatorio Nacional, Rio de Janeiro, Brazil;
[4]UMR Fizeau , UNS-OCA-CNRS, Nice, France

Abstract. HD 62623 is one of the very few A-type supergiants showing the B[e] phenomenon. We studied the geometry of its circumstellar envelope in the mid-infrared using the VLTI/MIDI instrument. Using the radiative transfer code MC3D, we managed to model it as a dusty disk with an inner radius of 3.85 AU, an inclination angle of 60°, and a mass of $2 \times 10^{-7} M_\odot$. It is the first time that the dusty disk inner rim of a supergiant star exhibiting the B[e] phenomenon is significantly constrained. The inner gaseous envelope likely contributes up to 20% to the total N band flux and acts like a reprocessing disk. Finally, the hypothesis of a stellar wind deceleration by the companion gravitational effect remains the most probable case since the bi-stability mechanism is not efficient for this star.

Keywords. techniques: high angular resolution, stars: emission-line, Be, stars: winds, outflows

---

The B[e] phenomenon is defined by strong Balmer lines in emission, low-excitation and forbidden emission lines, and strong infrared excess due to hot circumstellar dust. However, these objects are not a homogeneous group of stars in terms of stellar evolution, as such features have been detected for both pre-main sequence and evolved stars. Thus, this group of stars has been divided into four subclasses: B[e] supergiants, Herbig AeB[e] stars, compact planetary nebulae B[e]-type stars, and symbiotic B[e]-type stars.

HD 62623 is one of the very few A-type supergiants showing the B[e] phenomenon. The formation mechanisms responsible for supergiant B[e] stars' circumstellar environment is still an open issue. Zickgraf *et al.* (1985) proposed a general scheme for these stars consisting of a hot and fast line-driven polar wind responsible for the presence of forbidden lines and a slowly expanding dense equatorial region where permitted-emission lines and dust can form. The deviation from the spherical geometry for this model can be due to rotation or binarity.

The VLTI/MIDI (Leinert et al 2004) observations were carried out at Paranal Observatory between 2006 and 2008 with the 1.8 m Auxiliary Telescopes (ATs). We obtained nine visibility measurements with projected baselines ranging from 13.4 to 71.4m and with various orientations on the skyplane. All observations were made using the SCI-PHOT mode, which enables a better visibility calibration since the photometry and the interferometric fringes are recorded simultaneously. Thanks to the PRISM low spectral dispersion mode, we obtained spectrally resolved visibility and spectra with R=30 in the N band (7.513.5 μm). Standard reduction packages MIA and EWS were used to reduce these data.

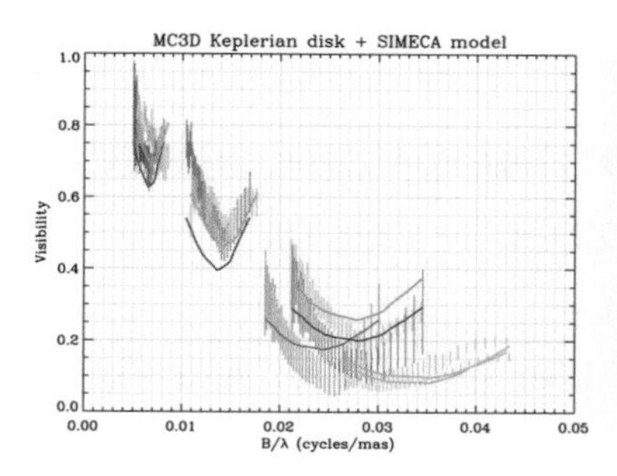

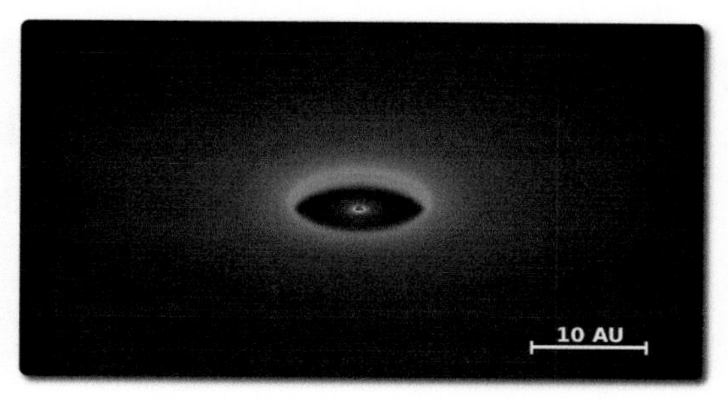

Figure 1. Left: Visibility plotted as a function of the spatial frequency for the nine baselines. The best SIMECA + MC3D model curves are overplotted as solid lines. Right: Corresponding SIMECA (in blue) + MC3D (in orange) composite image for the best-fit model.

We tried to model our interferometric measurements using MC3D (Wolf *et al.* 1999), a Monte-Carlo radiative transfer code for dusty circumstellar envelopes. We probed different disk structures including equatorial wind and Keplerian viscous disk. No model managed to fit all the visibilities simultaneously. In the best-fit models, the longest baselines visibilities are underestimated This suggests the presence of structures not fully resolved with the 70 m baselines, which are not taken into account in our MC3D model.

The residual visibility might be due to inhomogeneities or clumping in the disk. However, the most probable hypothesis is that it originates from a non-resolved inner gaseous environment surrounding the central star. We decided to take into account the emission of the ionized gas and modeled it using SIMECA (Stee 1996), a code developed to model gaseous environment of hot stars. As the extension of the gaseous emission is much smaller than the dust inner radius, we could compose the two models in an ad-hoc way by adding directly the complex visibilities extracted from each model.

The fit of the visibilities and a composite image of the MC3D + SIMECA best-fit model are presented in Fig 1. We were able to constrain the total mass of the disk, i.e. about 2×10^{-7} $M_{\odot}$, and the dust sublimation radius and temperature, i.e. 3.9 ± 0.6 AU and 1250 K, respectively. It is the first time that the dusty disk inner rim of a supergiant star exhibiting the B[e] phenomenon is significantly constrained. HD 62623 is seen under an intermediate inclination angle of $60 \pm 10\,^{o}$ and with a position angle of its major-axis on the sky plane roughly perpendicular to polarization measurement, i.e. $15 \pm 10\,^{o}$. As expected from the existence of the residual visibility, 10 to 20% of the total N-band flux are likely to originate from the free-free emission of the circumstellar ionized gas.

Using vsini measurement, stellar parameters, and our estimation of the object inclination angle, we were able to constrain HD 62623 rotational velocity, i.e. between 30 and 60% of its critical velocity. Consequently, rotation seems not to be efficient enough to explain the break of the spherical symmetry of the mass-loss and the formation of the equatorial disk. The putative presence of low mass companion could explain such geometry. However, considering the expected brightness ratio and the accuracy on our measurements, it cannot be detected in our dataset. These VLTI/MIDI observations and modelling are presented in details in Meilland *et al.* (2010).

References

Leinert, C., van Boekel, R., Waters, L. B. F. M., Chesneau, O. *et al.* 2004, *A&A*, 423, 537
Meilland, A., Kanaan, S., Borges Fernandes, M., Chesneau, O. *et al.* 2010, *A&A*, 512A, 73
Stee, P. 1996, *A&A*, 311, 945
Wolf, S., Henning, T., & Stecklum, B. 1999, *A&A*, 349, 839
Zickgraf, F.-J., Wolf, B., Stahl, O., Leitherer, C. *et al.* 1985, *A&A*, 143, 421

Active OB stars: structure, evolution, mass loss, and critical limits
Proceedings IAU Symposium No. 272, 2010
C. Neiner, G. Wade, G. Meynet & G. Peters, eds.

doi:10.1017/S1743921311010957

Imaging "Pinwheel" nebulae with optical long-baseline interferometry

Florentin Millour[1,2], Thomas Driebe[1,3], Jose H. Groh[1], Olivier Chesneau[2], Gerd Weigelt[1], Adriane Liermann[1] and Anthony Meilland[1]

[1]Max-Planck Institute for Radioastronomy, auf dem Hügel 69, 53121 Bonn, Germany
email: fmillour@oca.eu

[2]Observatoire de la côte d'Azur, Bd de l'Observatoire, 06304 Nice, France

[3]German Aerospace Center (DLR), Königswinterer Str. 522-524, 53227 Bonn, Germany

Abstract. Dusty Wolf-Rayet stars are few but remarkable in terms of dust production rates (up to $\dot{M} = 10^{-6} M_{\odot}/yr$). Infrared excesses associated to mass-loss are found in the sub-types WC8 and WC9. Few WC9d stars are hosting a "pinwheel" nebula, indirect evidence of a companion star around the primary. While few other WC9d stars have a dust shell which has been barely resolved so far, the available angular resolution offered by single telescopes is insufficient to confirm if they also host "pinwheel" nebulae or not. In this article, we present the possible detection of such nebula around the star WR 118. We discuss about the potential of interferometry to image more "pinwheel" nebulae around other WC9d stars.

Keywords. techniques: high angular resolution, techniques: interferometric, binaries (including multiple): close, stars: individual (WR 118), stars: winds, outflows, stars: Wolf-Rayet

---

1. WR 118

In 2008, we observed the dusty Wolf-Rayet (WR) star WR 118 using the Astronomical Beam Recombiner (AMBER, Petrov *et al.* 2007), at the focus of the Very Large Telescope Interferometer (VLTI). WR 118 had already been observed using speckle interferometry, on the BTA 6m telescope, in Russia (Yudin *et al.* 2001). The AMBER visibilities and closure phases were acquired at spatial frequencies up to 5 times larger than the previous speckle observations.

We clearly resolved the system with AMBER, with visibilities decreasing up to ≈ 55 cycles/arc-second, and increasing again above. Such a visibility behavior is typical of an object containing a sharp edge in its intensity distribution. In addition, the closure phase, measured at three different moments during the night, is clearly non-equal to zero, meaning that WR 118's dusty nebula is asymmetric.

We modelled WR 118's dusty nebula with several geometrical models, including a clumpy spherical wind and a "pinwheel" nebula. This last model provides a physical description of the system and best match the observed data. Therefore, we concluded that WR 118 probably hosts a "pinwheel" nebula, detected for the first time using long-baseline interferometry (Millour *et al.* 2009a).

2. Preparing future observations

Repeating the AMBER observations on WR 118 to confirm its "pinwheel" nature is the next step in this research program. One step forward is to assess the feasibility of imaging such targets with the current and future capabilities of the VLTI. For that,

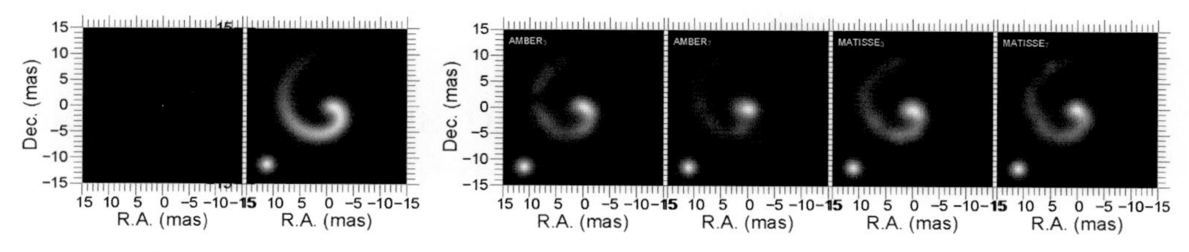

Figure 1. Simulations of aperture synthesis of "pinwheel" nebulae, using the parameters from WR 118. From left to right: the model used convolved with a beam equivalent to a 130m diameter telescope, synthesised images for AMBER observations during 3 and 7 nights, and synthesised images for MATISSE observations during 3 and 7 nights.

we perform simulations of the current VLTI instrument AMBER (K-band, 2.2 microns) and of the second generation planned VLTI instrument MATISSE (L-band, 3.5 microns) observations. We start from the model of WR 118 found previously and simulate a realistic observation (typical V^2 errors of 5%) with different baselines and one measurement per hour. Then, we use the MIRA software (Thiébaut 2008) to reconstruct an image of the target from these simulations. Four cases are simulated here:

- AMBER observations during 3 nights with 3 telescopes configurations,
- AMBER observations during 7 nights with 7 telescopes configurations,
- MATISSE observations during 3 nights with 3 telescopes configurations,
- MATISSE observations during 7 nights with 7 telescopes configurations.

As expected, we find that AMBER provides some imaging capabilities, even using as few as 3 nights of observation. One can recognize in the reconstructed image all the features from the original image (Fig. 1). However, as experienced on real datasets (Millour *et al.* 2009b), the relative flux of the different features is not well constrained, because of the lack of 66% of phase information. Increasing, by more than a factor 2, the number of nights used, only marginally improves the quality of the image reconstruction.

On the other hand, MATISSE simulations show a better agreement on the fluxes of the different features, even using "only" three nights of observation. We also note that the loss in angular resolution from the K-band and L-band do not apparently affect the quality of the image reconstruction.

3. Conclusion

We showed here that long-baseline interferometry is now clearly mature to confirm the brightest suspected "pinwheel" nebulae, using the current abilities of AMBER on the VLTI. In addition, we also showed that AMBER will be able to synthethise an image, where the "pinwheel" nebula would be qualitatively recognized, using at least three nights of observing time. The use of the second generation MATISSE instrument will permit to have more realistic flux measurements in the different parts of the image, with the same amount of observing time.

References

Millour, F., Driebe, T., Chesneau, O., Groh, J. H. *et al.* 2009a, *A&A*, 506, L49

Millour, F., Chesneau, O., Borges Fernandes, M., Meilland, A. *et al.* 2009b, *A&A*, 507, 317

Petrov, R. G., Malbet, F., Weigelt, G., Antonelli, P. *et al.* 2007, *A&A*, 464, 1-12

Thiébaut, E. 2008, in: M. Schöller, W. C. Danchi, & F. Delplancke (eds.), *Optical and Infrared Interferometry*, SPIE Conference Series 7013, p. 43

Yudin, B., Balega, Y., Blöcker, T., Hofmann, K.-H. *et al.* 2001, *A&A*, 379, 229

Active OB stars: structure, evolution, mass loss, and critical limits
Proceedings IAU Symposium No. 272, 2010
C. Neiner, G. Wade, G. Meynet & G. Peters, eds.

doi:10.1017/S1743921311010969

Images of unclassified and supergiant B[e] stars disks with interferometry

Florentin Millour[1,2], Anthony Meilland[1], Olivier Chesneau[2], Marcelo Borges Fernandes[3], Jose H. Groh[1], Thomas Driebe[1,4], Adrianne Liermann[1] and Gerd Weigelt[1]

[1]Max-Planck Institute for Radioastronomy, auf dem Hügel 69, 53121 Bonn, Germany
email: fmillour@oca.eu

[2]Observatoire de la côte d'Azur, Bd. de l'Observatoire, 06304 Nice, France

[3]Observatorio Nacional, Rua General José Cristino, 77, Rio de Janeiro, Brazil

[4]German Aerospace Center (DLR), Königswinterer Str. 522-524, 53227 Bonn, Germany

Abstract. B[e] stars are among the most peculiar objects in the sky. This spectral type, characterised by allowed and forbidden emission lines, and a large infrared excess, does not represent an homogenous class of objects, but instead, a mix of stellar bodies seen in all evolutionary status. Among them, one can find Herbig stars, planetary nebulae central stars, interacting binaries, supermassive stars, and even "unclassified" B[e] stars: systems sharing properties of several of the above. Interferometry, by resolving the innermost regions of these stellar systems, enables us to reveal the true nature of these peculiar stars among the peculiar B[e] stars.

Keywords. techniques: high angular resolution, techniques: interferometric, binaries (including multiple): close, stars: individual (HD 87643, HD 62623), stars: mass loss, stars: winds, outflows

1. Introduction

We started an observing program covering the brightest unclassified and (candidate) supergiant B[e] stars, by using the Very Large Telescope Interferometer. For now, about 10 targets have been observed using a combination of AMBER (near-IR) and MIDI (mid-IR) observations. Here we focus on imaging and therefore stick to AMBER observations of two of the targets: the unclassified B[e] star HD87643 and the supergiant A[e] star HD 62623.

2. HD 87643

We used an extensive dataset spanning several orders of magnitude of spatial resolution to partly unveil the nature of this stunning system.

Wide-field images: Many new details are seen in our wide field image of the nebula around HD87643: blown up structures in the large scale nebula, which could be the result of an eruption much like in LBVs, and a series of arcs, unseen before, which could be the result of periodic ejections of matter.

Interferometric images: Our interferometric images, made using the AMBER/VLTI instrument, show a companion star to the bright star. In addition, the primary star exhibits an extended shell (4 mas), and background emission is detected.

Interpretation: Using a model involving both stellar components plus a resolved background, we were able to separate individual spectra in H, K and N bands. We find that the primary star is enshrouded in a dust shell heated at the dust-sublimation temperature: we resolved the inner region of the HD87643 disk. The secondary component

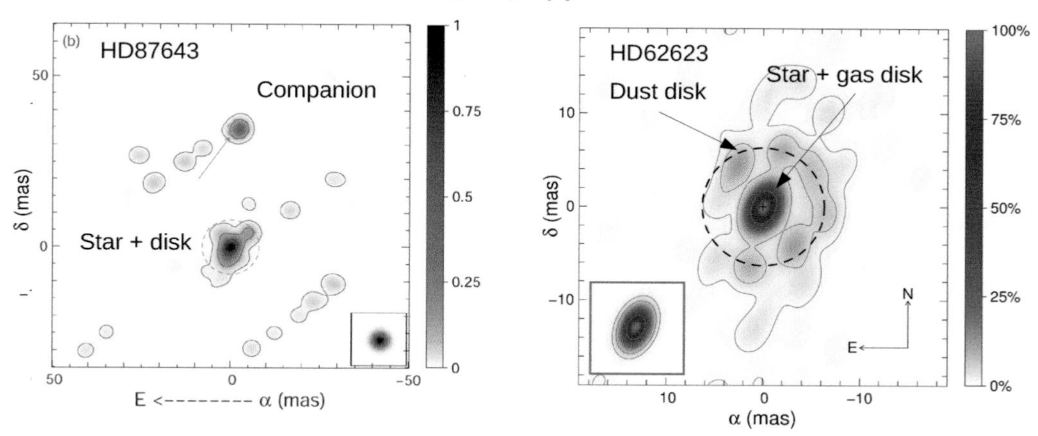

Figure 1. Left: AMBER/VLTI image of the star HD87643, revealing a companion star and a disk around the primary star. Right: AMBER/VLTI image of the previously-putative disk around the star HD62623. A spectrally-resolved image-cube could be synthesised, evidencing Keplerien rotation in this disk.

is also enshrouded in dust, at temperatures ranging from 500 to 50K. The resolved background holds most of the silicate emission, hence being likely composed of colder condensed dust. As a conclusion, and since the distance to the system is still poorly constrained, HD87643 could be a YSO instead of an evolved object. See Millour *et al.* 2009 for more details.

3. HD 62623

This star was already extensively studied by spectroscopy Plets *et al.* 1995. Hence it was already known that it hosted a circumstellar disk.

The continuum: We clearly resolve the gas and the dust disk in the continuum, imaging for the first time the disk of an evolved star. The image shows an outer ring corresponding to the inner rim where dust sublimation occurs, while we also resolve a central region where free-free emission from the gas takes place. The dust emission is asymmetric, which we attribute to an inclination effect. This direct picture strikingly confirms what was indirectly inferred from spectroscopic data previously.

The Brγ line: Our high-spectral resolution image cube in the Brγ line directly shows the motion of the gas in different velocity bins. The combination of a model of the dust, which gives the on-sky orientation and inclination of the disk, and of a model of the rotating gas, allows us to claim that we detected Keplerian rotation in that disk. We can now clearly rule out expanding wind models for HD 62623. Instead, we have a Keplerian rotating disk, where expansion is negligible. Such a disk, unexpected for massive stars, is common in young stellar objects like Herbig stars. In the case of HD 62623, the presence of a very close unseen companion, previously detected by radial velocities, might be the key to the formation of such a rotating disk.

References

Millour, F., Chesneau, O., Borges Fernandes, M., Meilland, A. *et al.* 2009, *A&A*, 507, 317
Plets, H., Waelkens, C. & Trams, N. R. 1995, *A&A*, 293, 363

Active OB stars: structure, evolution, mass loss, and critical limits
Proceedings IAU Symposium No. 272, 2010
C. Neiner, G. Wade, G. Meynet & G. Peters, eds.

doi:10.1017/S1743921311010970

Properties of the circumstellar dust in galactic FS CMa objects

Anatoly S. Miroshnichenko[1], Richard O. Gray[2], Karen S. Bjorkman[3], Richard J. Rudy[4], David K. Lynch[5] & Alex C. Carciofi[6]

[1]Dept. of Physics & Astronomy, University of North Carolina at Greensboro, P.O. Box 26170, Greensboro, NC 27402–6170, USA; email: a_mirosh@uncg.edu; [2]Dept. of Physics & Astronomy, Appalachian State University, Boone, NC 28608, USA; [3]Ritter Observatory, University of Toledo, Toledo, OH 43606, USA; [4]The Aerospace Corp., M2/266, P.O.Box 92957, Los Angeles, CA 90009, USA; [5]Thule Scientific, Topanga, CA 90290, USA; [6]Instituto de Astronomia, Geofisica e Ciencias Atmosfericas, Universidade de São Paulo, Rua do Matao 1226, Cidade Universitaria, São Paulo, SP 05508–900, Brazil

Abstract. FS CMa objects are a group of hot stars that exhibit the B[e] phenomenon. The group was defined a few years ago on the basis of the formerly known unclassified B[e] stars and newly discovered objects. One of their main features is the presence of hot circumstellar dust whose properties were unknown. We present IR spectra of nearly 20 FS CMa objects obtained with the Spitzer Space Telescope. Dusty features, such as broad silicate bands in emission and narrow bands that are usually explained by PAHs, are detected. The IR fluxes are compared to those detected by IRAS and MSX. Main results of the data analysis are briefly discussed.

Keywords. stars: early-type, infrared: stars, circumstellar matter

---

1. Introduction

The B[e] phenomenon is the simultaneous presence of line emission (forbidden: [O I], [Fe II], [N II], sometimes [O III] and permitted: H, He, and Fe II) and large IR excesses due to hot circumstellar (CS) dust in the spectra of B-type stars (Allen & Swings 1976). It is found in five groups of stars (Lamers *et al.* 1998): pre-main-sequence stars, symbiotic binaries (a cool giant and a white dwarf or a neutron star), proto-planetary/planetary nebulae, supergiants, and FS CMa objects (formerly known as unclassified B[e] stars, Miroshnichenko 2007). The presence of CS dust near hot stars may be due to various reasons. In pre-main-sequence stars, it is inherited from protostellar clouds. In proto-planetary nebulae, it is left from the previous, asymptotic giant branch (AGB) evolutionary stage. In symbiotic binaries, CS dust is formed in ejecta of cool giants, as in AGB stars. In B[e] supergiants with very dense radiation-driven winds, dust can be formed due to self-shielding of parts of the winds (e.g., due to clumping) from dust-destroying UV radiation. Only B[e] supergiants and FS CMa objects seem to form dust, when they have a B-type star in their content. FS CMa objects are the least studied. Our Spitzer program was aimed at taking IR spectra of a sample of 25 Galactic FS CMa objects with IRS in the range 5–37 μm for the first time.

2. The Galactic FS CMa group

The group comprises ∼50 members and ∼20 candidates (Miroshnichenko *et al.* 2010), ∼30% of which are detected binary systems. The main defining features of FS CMa objects include: 1) A B to early-A type star with an extremely strong emission-line spectrum; 2) A sharp decrease of the IR flux at $\lambda > 10\mu$m indicating that the dusty

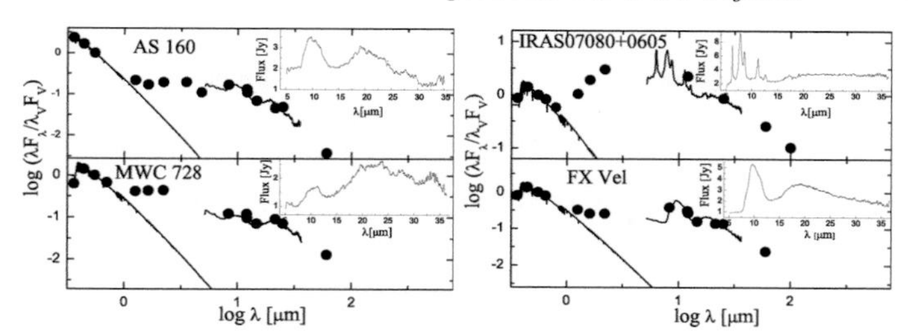

Figure 1. Spectral energy distributions (SED) of some program objects (various photometric data are shown by the dots) and their Spitzer spectra (shown by the solid lines in logarithmic units on the main plots and in flux units on the insets). Solid lines through the dereddened optical parts of the SEDs are model atmospheres (Kurucz 1994) for $T_{\rm eff}$ = 10,000 K (FX Vel and IRAS 07080+0605), 12,000 K (MWC 728), and 19,000 K (AS 160).

envelopes are compact and either stable for a long time or recently created; 3) Luminosity ($\log L/L_\odot$ = 2.5–4.5) is typical for a 3–10 $M_\odot$ single star; 4) The Balmer line emission is ~10 times stronger than that in supergiants, Be, and Herbig Ae/Be stars. It is too strong to be explained by the evolutionary mass loss from a single star of similar luminosity.

3. Spitzer data

All 25 objects were observed, most spectra have good quality, data problems were found in the spectra of four objects, two objects were not detected. Silicate emission features were detected in all observed objects except for IRAS 07080+0605 (Fig. 1). The structure of the 10–μm feature and presence of forsterite bands at 23.3, 27.8, and 33.7 μm suggest that the dust was formed some time ago and already processed by stellar radiation. Objects with late-B/early-A spectral type stars also exhibit emission features which are attributed to Polycyclic Aromatic Hydrocarbons (Leger & Puget 1984) or to Small Carbonaceous Molecules (Bernstein & Lynch 2009). The IR flux level of all the objects is comparable with that of earlier observations (MSX, IRAS).

4. Conclusions

The Spitzer IRS spectra revealed the chemical composition of the Galactic FS CMa objects for the first time. The data constrain the amount of CS dust. This will allow us to estimate their role in the Galactic dust production. Previously B-type stars of intermediate and low luminosity were not considered as dust producers. Simultaneous modeling of the gaseous and dusty envelopes in FS CMa objects together with their orbital parameters will enable us to understand what kind of binary systems can pass through this evolutionary stage and solve the long-standing problem of the existence of the B[e] phenomenon in objects whose evolutionary state was unknown for over 30 years.

References

Allen, D. A. & Swings, J. P. 1976, *A&A*, 47, 293

Bernstein, L. S. & Lynch, D. K. 2009, *ApJ*, 704, 226

Kurucz, R. 1994, *Smithsonian Astrophys. Obs.*, CD-ROM No. 19

Lamers, H. J. G. L. M., Zickgraf, F.-J., de Winter, D., Houziaux, L. *et al.* 1998, *A&A*, 340, 117

Leger, A. & Puget, J. L. 1984, *A&A*, 137, L5

Miroshnichenko, A. S. 2007, *ApJ*, 667, 497

Miroshnichenko, A. S., Polcaro, V. F., Rossi, C., Zharikov, S. V., & Gray, R. O. 2010, *American Astronomical Society 215*, BASS 42, p. 340

Active OB stars: structure, evolution, mass loss, and critical limits
Proceedings IAU Symposium No. 272, 2010
C. Neiner, G. Wade, G. Meynet & G. Peters, eds.

doi:10.1017/S1743921311010982

Variability monitoring of OB stars during the Mons campaign

Thierry Morel[1], Gregor Rauw[1], Thomas Eversberg[2], Filipe Alves[3], Wolfgang Arnold[4], Thomas Bergmann[5], Nelson G. Correia Viegas[6], Rémi Fahed[7], Alberto Fernando[8], Luis F. Gouveia Carreira[9], Thomas Hunger[10], Johan H. Knapen[11], Robin Leadbeater[12], Filipe Marques Dias[13], Anthony F. J. Moffat[7], Norbert Reinecke[14], José Ribeiro[15], Nando Romeo[16], José Sánchez Gallego[11], Eva M. dos Santos[6], Lothar Schanne[17], Otmar Stahl[18], Barbara Stober[19], Berthold Stober[19], Klaus Vollmann[2], Mike F. Corcoran[20], Sean M. Dougherty[21], Kenji Hamaguchi[20], Julian M. Pittard[22], Andy M. T. Pollock[23], and Peredur M. Williams[24]

[1] Institut d'Astrophysique et de Géophysique, Université de Liège, 4000 Liège, Belgium; [2] Schnörringen Telescope Science Institute (STScI), Am Kielshof 21a, 51105 Köln, Germany; [3] Av. Portugal 616C - 2765–272 Estoril, Portugal; [4] Burggraben 3, 61206 Wöllstadt, Germany; [5] Eichendorffstrasse 8, 63538 Grosskrotzenburg, Germany; [6] Rua Nuno Ataide Mascarenhas, No. 47, 2.Esq. 8100–610 Loule, Portugal; [7] Département de Physique, Université de Montréal, Montréal (Québec) H3C 3J7, Canada; [8] Alto Ajuda, Rua 27 - No. 215 1300–581, Lisbon, Portugal; [9] R. Rego de Agua LT 24 RC Esq., Marrazes, 2400 Leiria, Portugal; [10] Normannenweg 39, 59519 Möhnesee-Körbecke, Germany; [11] Instituto de Astrofísica de Canarias, E-38200 La Laguna, Tenerife, Spain; [12] Three Hills Observatory, The Birches, Torpenhow CA7 1JF, UK; [13] Rua Almirante Campos Rodrigues, Edf. Girassol, 5F, 1500–036 Lisbon, Portugal; [14] Fontainegraben 150, 53123, Bonn, Germany; [15] R. Venezuela 29 3 Esq. - 1500–618 Lisbon, Portugal; [16] Virulylaan 30, 2267 BS Leidschendam, The Netherlands; [17] Hohlstrasse 19, 66333 Völklingen, Germany; [18] ZAH, Landessternwarte Königstuhl, 69117 Heidelberg, Germany; [19] Nelkenweg 14, 66907 Glan-Münchweiler, Germany; [20] Laboratory for High Energy Astrophysics, Goddard Space Flight Center, Greenbelt, USA; [21] Herzberg Institute for Astrophysics, Penticton, British Columbia V2A 6J9, Canada ; [22] School of Physics and Astronomy, The University of Leeds, Leeds LS2 9JT, UK; [23] ESA XMM-Newton Science Operations Centre, Villafranca del Castillo, Spain; [24] Institute for Astronomy, University of Edinburgh, Royal Observatory, Edinburgh, UK

Abstract. We present preliminary results of a 4-month campaign carried out in the framework of the Mons project, where time-resolved Hα observations are used to study the wind and circumstellar properties of a number of OB stars.

Keywords. line: profiles, stars: early-type, stars: winds, outflows, stars: individual (HD 14134)

1. Context

The Mons project is a collaboration between professional and amateur astronomers, which was primarily set up to monitor the periastron passage of the colliding-wind binary system WR 140 centred on January 12, 2009 (Fahed *et al.*, this volume).† A dedicated spectroscopic campaign was organised from December 2008 to March 2009 using the 50-cm Mons telescope at Teide Observatory. Time-resolved observations of the Hα line (6360–6950 Å, 0.34 Å pix^{-1}) were also obtained for a small sample of early B-type supergiants and Oe stars to investigate the properties of their large-scale wind structures

† See also `http://www.stsci.de/wr140/index_e.htm`

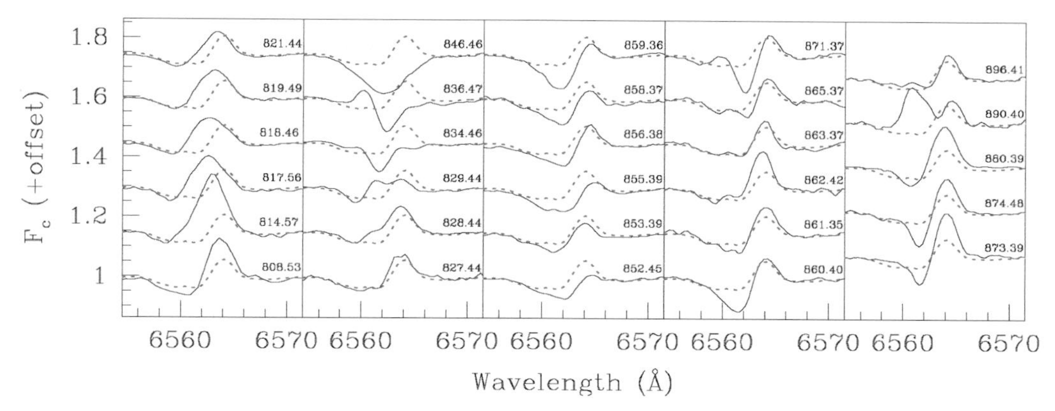

Figure 1. Variations of the Hα line in the B3 supergiant HD 14134. The mean profile is overplotted as a dashed line. The date of the observations (HJD–2,454,000) is indicated.

and circumstellar material, respectively. The B1–B3 supergiants were selected from Morel *et al.* (2004) based on previous indication of cyclical changes (HD 14134 and HD 42087) or strong variations (HD 43384 and HD 52382). Here we present an overview of the variations exhibited by these objects (the data for the Oe stars HD 45314 and HD 60848 are still being reduced) and briefly discuss forthcoming developments in the data analysis.

2. Preliminary results and perspectives

Variability studies in the UV domain have shown that the winds of OB stars are likely made up of large-scale streams (the 'co-rotating interaction regions'; CIRs) whose formation may be triggered by the existence of non-uniform physical conditions at the stellar surface (due, e.g., to magnetic structures or pulsations; Cranmer & Owocki 1996). Optical wind lines can also be used to probe the physical properties of these structures. In particular, revealing rotational modulation in these features would provide evidence that the CIRs extend relatively close to the star and are possibly directly emerging from the photosphere. Strong, daily line-profile variations are observed in all the targets, as illustrated in the case of HD 14134 in Fig.1 by the great variety of profiles observed (strong emission/absorption, double peaked, classical or even inverse P-Cygni profile). This star is of particular interest because of the previous detection of a 12.8-d periodic signal both in photometry and in spectroscopy (Morel *et al.* 2004).

Our efforts will now be directed towards the detection of a periodic behaviour that could allow us to identify the physical processes that drive the variations. For instance, a dipole magnetic field tilted with respect to the rotational axis in the Oe stars is expected to induce changes modulated by the rotational period, whereas the variations should take place on much longer timescales if they arise from some kind of disk instability. On the other hand, high-resolution spectroscopic observations of the B3 supergiant HD 14134 are scheduled in November 2010 at OHP (France) to examine the existence of pulsations and to eventually link the variations taking place in the photosphere to those in the wind.

References

Cranmer, S. R. & Owocki, S. P. 1996, *ApJ*, 462, 469

Morel, T., Marchenko, S. V., Pati, A. K., Kuppuswamy, K. *et al.* 2004, *MNRAS*, 351, 552

Active OB stars: structure, evolution, mass loss, and critical limits
Proceedings IAU Symposium No. 272, 2010
C. Neiner, G. Wade, G. Meynet & G. Peters, eds.

doi:10.1017/S1743921311010994

Effect of Be-disk evolution on the global one-armed oscillations

Finny Oktariani[1] **and Atsuo T. Okazaki**[2]

[1]Department of Cosmoscience, Graduate School of Science, Hokkaido University, Kita-ku, Sapporo 060-0810, Japan
email: finny@astro1.sci.hokudai.ac.jp

[2]Faculty of Engineering, Hokkai-Gakuen University, Toyohira-ku, Sapporo 062-8605, Japan
email: okazaki@elsa.hokkai-s-u.ac.jp

Abstract. We consider the effect of density distribution evolution on the global one-armed oscillation modes in disks around Be stars. Previous studies of global oscillations in Be disks assumed a power-law density distribution of the disk. However, observational results show that some Be stars exhibit evidence of formation and dissipation of the equatorial disk. This causes the disk density distribution can be far from a power-law form. Performing calculations for several times in the disk formation and dissipation stages, we find one-armed modes confined to the inner part of the disk in both stages. In the disk formation stage, the oscillation frequency stays approximately constant after the disk is fully developed. In the dissipation stage of the Be disk, the local precession frequency is, in general, higher than in the disk formation stage. Thus, we expect that V/R periods become shorter as the innermost part of the disk starts to accrete.

Keywords. stars: emission-line, Be, stars: oscillations

---

1. Introduction

Be stars are non-supergiant B-type stars that show emission lines in their Balmer lines of hydrogen. The emission lines are attributed to a cool circumstellar disk around the star. Many Be stars show long-term V/R variations which are variations of Violet (V) to Red (R) peak ratio in double peaked emission line profiles. This variation is attributed to the precession of global one-armed oscillation in Be disk (Okazaki 1991; Papaloizou *et al.* 1992). Some Be stars show evidences of formation and dissipation of disk, during which the density distribution in the disk is likely far from the usually-assumed, power-law distribution. We study the effect of the Be disk evolution on the global one-armed oscillation.

2. Density distribution

We calculated the density evolution in an isothermal, decretion disk around a Be star, solving a 1-D, diffusion-type equation of the surface density (e.g., equation (12) of Okazaki *et al.* 2002) for the formation and dissipation stages. The obtained results can be seen in the upper panels of Fig. 1. In the early times of the formation stage, we have a density distribution with steep slope. The mass is concentrated in the area near the star. Then, by the means of viscosity the mass flows to the outer part, making the density distribution closer to that with a power-law. In the dissipation stage, when no mass is ejected from the star, the disk is lost from the innermost part, causing a gap between the star and the disk. Then, the density distribution has a peak near the inner radius. We then calculate one-armed eigenmodes using the obtained density distribution at a particular time.

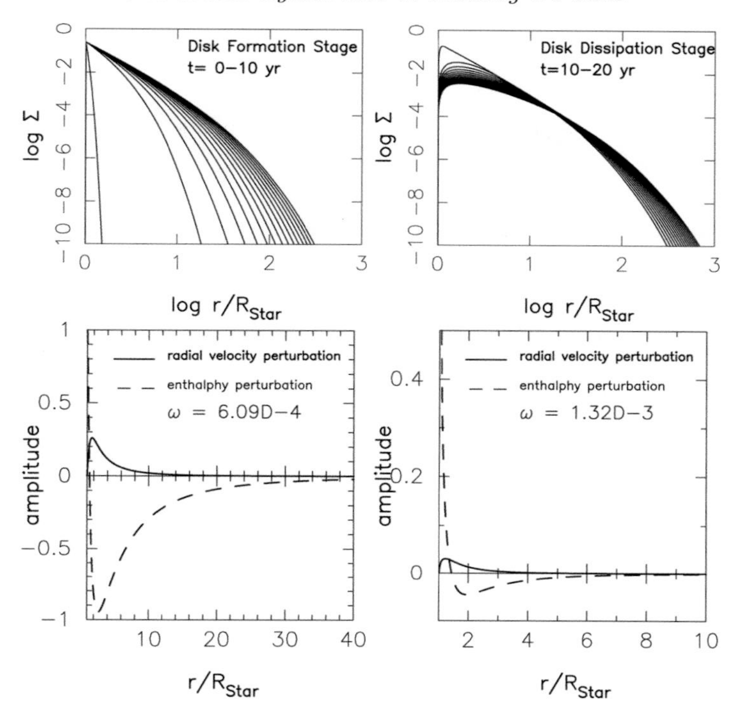

Figure 1. Surface density evolution in the disk around a single Be star in the formation stage (upper left panel) and dissipation stage (upper right panel). The two adjacent lines correspond to $t \approx 10$ months. We used viscosity parameter $\alpha = 0.1$, and mass ejection rate $\dot{M} \sim 10^{-9} M_\odot\,\mathrm{yr}^{-1}$ in the disk formation stage. The lower panels show the eigenfunctions of the $m = 1$ modes for a single Be star in the disk formation stage (lower left panel) and dissipation stage (lower right panel).

3. Effect of disk density evolution on global one-armed modes

We superposed linear global one-armed modes $\propto \exp[i(\omega t - \varphi)]$ in the unperturbed disk model. We found one armed modes confined to the inner part of the disk both in formation and dissipation stages. In the disk formation stage, ω stays approximately constant after the disk is fully developed. No eigenfrequency found before the disk is developed. In the disk dissipation stage, we found that the eigenmodes are more strongly confined than those in the formation stage, and ω also stays approximately constant throughout the stage. In the dissipation stage, the local precession frequency is generally higher than in the formation stage. Hence, we have higher ω, i.e., a shorter period of V/R variability. Comparing our results with those calculated using a power-law density distribution (Oktariani & Okazaki 2009), we found that several years after the disk formation started, the eigenfrequency becomes close to the previous results. This is because the density distribution becomes close to the power-law form when the disk is fully developed.

References

Okazaki, A. T. 1991, *PASJ*, 43, 75
Okazaki, A. T., Bate, M. R., Ogilvie, G. I., & Pringle, J. E. 2002, *MNRAS*, 337, 967
Oktariani, F. & Okazaki, A. T. 2009, *PASJ*, 61, 57
Papaloizou, J. C., Savonije, G. J., & Henrichs, H. F. 1992, *A&A*, 265, L45

Active OB stars: structure, evolution, mass loss, and critical limits
Proceedings IAU Symposium No. 272, 2010
C. Neiner, G. Wade, G. Meynet & G. Peters, eds.

doi:10.1017/S1743921311011008

Spectrally and spatially resolved Hα emission from Be stars: their disks rotate Keplerian

René D. Oudmaijer[1], Hugh E. Wheelwright[1], Alex C. Carciofi[2], Jon E. Bjorkman[3] and Karen S. Bjorkman[3]

[1] School of Physics & Astronomy, University of Leeds, Woodhouse Lane, Leeds LS2 9JT, UK

[2]Instituto de Astronomia, Geofísica e Ciências Atmosféricas, Universidade de São Paulo, Rua o Matão 1226, Cidade Universitária, São Paulo, SP 05508-900, Brazil

[3]Department of Physics & Astronomy, University of Toledo, MS111 2801 W. Bancroft Street, Toledo, OH 43606, USA

Abstract. We test whether Be star disks rotate in a Keplerian or an Angular Momentum Conserving fashion. This is done by employing sub-milli arcsecond spectroastrometry around Hα. We spatially resolve the disks, and are the first to do so at such a high spectral resolution. We fit the emission line profiles with parametric models. The Keplerian models reproduce the spectro-astrometry, whereas the AMC models do not, thereby supporting the viscous disk model for Be stars.

Keywords. line: profiles, techniques: high angular resolution, stars: emission-line, Be

---

Knowing the kinematics of Be star disks will inform us about the formation of these disks. We aim to directly determine the kinematics of Be disks: Keplerian as expected for viscous excretion disks or angular momentum conserving, requiring different models? Such a study has only been done previously using barely spectrally resolved interferometric data (Meilland *et al.* 2007). We embarked on a high angular resolution survey of half a dozen objects, and present here our results on β CMi

We approach the problem using spectro-astrometry. This technique allows the detection of extended structures at sub-milli-arcsecond (mas) precision in longslit spectra (see also e.g. Wheelwright *et al.* 2010; Oudmaijer et al. 2008). As an example, the left graph in Fig.1 shows several velocity slices of a rotating disk. The disk is only a few mas large, and note that the higher the velocity, the smaller the disk. The right-hand

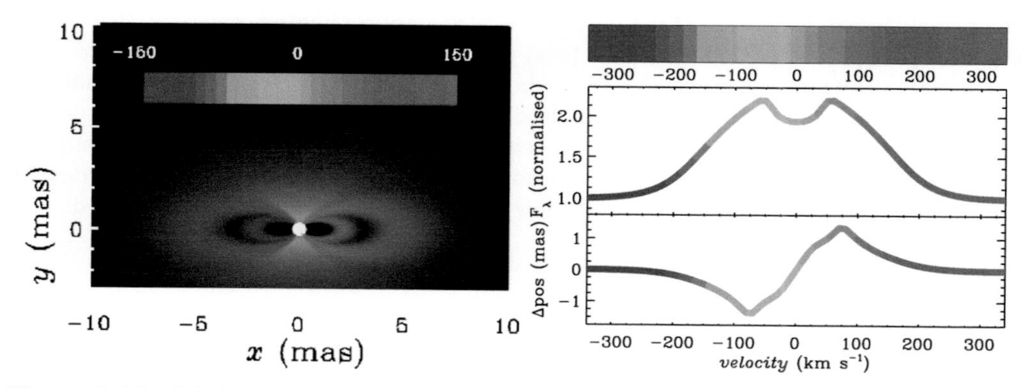

Figure 1. The left-hand image shows the spatially extended emission from a rotating disk for different velocities. Right: The top panel shows the resulting Hα emission line profile, and the bottom panel the expected excursion in the spectro-astrometry.

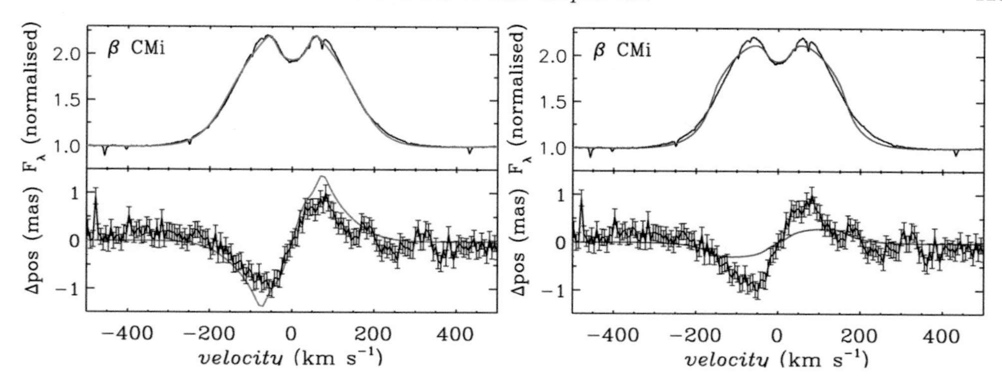

Figure 2. The top panels show the Hα line profile of β CMi, and the bottom panels its spectro-astrometric signature. Overplotted are a Keplerian fit (left) and an AMC fit (right) to the emission line. The Keplerian model reproduces the spectro-astrometry very well. The AMC model not at all.

panel shows how the total intensity spectrum of the line would look like. Normally this emission is not spatially resolved due to the small disks. However, if we take a longslit spectrum in the optical we can measure the position of the peak emission (essentially the photo-centre) as function of velocity with sub-milli-arcsecond precision, even in arcsecond seeing. The bottom panel of the right-hand figure shows how such a disk looks like in spectro-astrometry. The blue-shifted emission is located on one side of the star and the red-shifted emission on the other side, as expected. The spectro-astrometry measures the position of the photo-centre of the line emission and this shifts up to 1 mas.

We show our UVES/VLT data of β CMi (B8Ve) in Fig. 2. The rotating disk is clearly detected in the spectro-astrometry (bottom-panel), and indeed, has the same shape as expected from Fig. 1. This is the first time that this line is spatially resolved at such a high spectral resolution ($\sim$4 km/s). We interpret the data with a model to determine the disk's kinematical structure.

We fit the spectrum and predict the astrometry using both a Keplerian rotating and angular momentum conserving (AMC) disk using the prescription of Grundstrom & Gies (2006). Both models can fit the Hα emission line profile without any problems. However, while the Keplerian disk (left) also comfortably reproduces the spatial profile, the AMC does not at all (right). This is because of its steeper drop in rotation speed, the AMC disk is typically 10 times smaller (with a photo-centre several times smaller) than the Keplerian disk when matching the observed rotation velocities.

To summarize, both models easily fit the emission line profiles, but the additional constraint of spatially resolved data allows us to distinguish between the two. We conclude that the disks around Be stars rotate in a Keplerian fashion, and are not Angular Momentum Conserving. We have data of more Be stars in hand and are fitting the data simultaneously with the infrared excess and polarization using the HDUST model by Carciofi & Bjorkman (2006).

References

Carciofi, A. C. & Bjorkman, J. E. 2006, *ApJ* 639, 1081

Grundstrom, E. D. & Gies, D. R. 2006, *ApJ* (Letters), 651, L53

Meilland, A., Stee, P., Vannier, M., Millour, F. *et al.* 2007, *A&A*, 464, 59

Oudmaijer, R. D., Parr, A. M., Baines, D., & Porter, J. M. 2008, *A&A*, 489, 627

Wheelwright, H. E., Oudmaijer, R. D., de Wit, W. J., Hoare, M. G. *et al.* 2010, *MNRAS*, 408, 1840

Active OB stars: structure, evolution, mass loss, and critical limits
Proceedings IAU Symposium No. 272, 2010
C. Neiner, G. Wade, G. Meynet & G. Peters, eds.

doi:10.1017/S174392131101101X

Hα spectropolarimetry of GG Car

Antonio Pereyra[1], Francisco X. de Araújo[1]†, Antonio M. Magalhães[2], Marcelo Borges Fernandes[1] and Armando Domiciano de Souza[3]

[1]Observatório Nacional, Rua General José Cristino 77, São Cristovão, Rio de Janeiro, Brazil
[2]Dept. de Astronomia, IAG, Univ. de São Paulo, Rua do Matão 1226, São Paulo, Brazil
[3]UMR 6525 H. Fizeau, Univ. Nice Sophia Antipolis, CNRS, Obs. de la Côte d'Azur, France

Abstract. We present spectropolarimetric observations of the B[e] supergiant star GG Car at two epochs. Polarization line effects along Hα are analysed using the Q–U diagram. In particular, the polarization position angle (PA) obtained using the line effect allows to constrain the symmetry axis of the disk/envelope. The depolarization line effect around Hα is evident in the Q–U diagram for both epochs, confirming that light from the system is intrinsically polarized. A rotation of the PA along Hα is also observed, indicating a counter-clockwise rotating disk. The intrinsic PA calculated using the line effect (∼85°) is consistent between our two epochs, suggesting a clearly defined symmetry axis of the disk.

Keywords. polarization, stars: individual (GG Car), circumstellar matter

1. Introduction

Early studies of GG Car (Pickering 1896, Pickering & Fleming 1896) showed that it displayed a peculiar spectrum with strong emission lines. McGregor *et al.* (1988) identified CO absorption bands along with characteristics of the B[e] phenomenon, which include strong Balmer lines in emission, low excitation permitted emission lines (e.g. FeII), forbidden emission lines (e.g. [FeII] and [OI]), and strong infrared excess (Zickgraf 1998). Lamers et al. (1998) classified this object as a Galactic B[e] supergiant (sgB[e]) based on estimates of its effective temperature and luminosity. Previous optical broadband long-term polarization variability (e.g. Gnedin et al. 1992) indicates that GG Car does have intrinsic polarization, which probably originated in light scattering off one of the binary components or a variable circumstellar disk/wind. Spectropolarimetry is a powerful tool because it provides insight into stellar envelopes where scattering opacities exist without the need to resolve the envelope (Magalhães *et al.* 2006). In this sense, it yields additional information about the envelope geometry and the structure of the line formation region.

2. Observations and results

The observations were performed during 2 runs in 2006 April and May using the 1.6m telescope at the Observatório do Pico dos Dias (OPD-LNA), Brazil. We used IAGPOL, the IAG imaging polarimeter (Magalhães *et al.* 1996), installed in the Eucalyptus-IFU spectrograph. This setup provides a spectral range of ∼600 Å around Hα and a resolution R=4000, or ∼0.3 Å/pixel.

Our Hα spectropolarimetry is shown in Fig. 1 for our 2 epochs. Significant changes in the polarization level and its PA, indicative of a detected line effect, are observed across the Hα emission for both dates, which suggests that intrinsic polarization in GG Car is present. The $Q-U$ diagrams show the line effect evident around the emission line. In 2006 Apr., the line effect is consistent with depolarization showing a linear excursion that points approximately to the (Q, U) coordinates origin. On the other hand, the 2006 May

† In memoriam

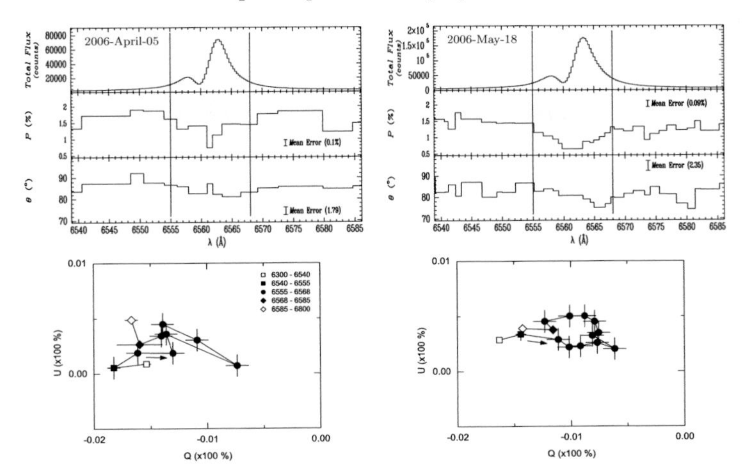

Figure 1. Spectropolarimetry of GG Car around Hα at 2 epochs. *Top*: The vertical lines indicate the region where the line effect is evident. The spectra show the total flux (top), polarization (middle), and polarization PA (bottom) binned using a variable bin size with a constant polarization error (per bin) of 0.1%. *Bottom.*: $Q - U$ diagram showing the line effect (black dots). Arrows indicate the direction in which the wavelength increases. Mean values before and after the line effect are also shown (see symbols).

data also exhibit depolarization but with a clearly defined PA rotation, which resembles a mix between a loop and a linear excursion. The non-detection of the loop in the 2006 Apr. data may be due to the lower signal-to-noise ratio in this measurement. Interestingly, the direction in which the wavelength increases around the line effect has the same sense in our 2 epochs. This direction is such that if a disk exists around GG Car it must rotate counter-clockwise as seen by an observer on Earth (Poeckert & Marlborough 1977).

To determine the intrinsic PA (PA_{int}), we completed a linear fit to the points that display the line effect in the $Q-U$ diagram. The PA_{int} values are practically identical ($\sim 85°$) during our 2 epochs considering the individual errors and represents a clearly defined symmetry axis for the disk. If this is true, the PA_{int} will be perpendicular or parallel to the disk plane depending on whether we have an optically thin or optically thick disk, respectively (Vink *et al.* 2005). Consistent with this issue, VLTI/MIDI interferometry (Domiciano de Souza 2009, priv. comm.) also detects a disk plane close to being perpendicular to our PA_{int} indicative of an optically thin disk.

References

Gnedin, Y. N., Kiselev, N. N., Pogodin, M. A., Rosenbush, A. E. *et al.* 1992, *Soviet Astronomy Letters* 18, 182

Lamers, H. J. G. L. M., Zickgraf, F.-J., de Winter, D., Houziaux, L. *et al.* 1998, *A&A*, 340, 117

Magalhães, A. M., Rodrigues, C. V., Margoniner, V. E., Pereyra, A. *et al.* 1996, in: W. G. Roberge & D. C. B. Whittet (eds.), *Polarimetry of the Interstellar Medium*, ASP-CS 97, p. 118

Magalhães, A. M., Melgarejo, R., Pereyra, A., & Carciofi, A. C. 2006, in: M. Kraus & A. S. Miroshnichenko (eds.), *Stars with the B[e] Phenomenon*, ASP-CS 355, p. 147

McGregor, P. J., Hyland, A. R. & Hillier, D. J. 1988, *ApJ*, 324, 1071

Pickering, E. C. 1896, *AN* 141, 169

Pickering, E. C. & Fleming, W. P. 1896, *ApJ*, 4, 142 (see also Harvard Circ. No. 9)

Poeckert, R. & Marlborough, J. M. 1977, *ApJ*, 218, 220

Vink, J. S., Drew, J. E., Harries, T. J., Oudmaijer, R. D. *et al.* 2005, *MNRAS*, 359, 1049.

Zickgraf, F.-J. 1998, in: A. M. Hubert & C. Jaschek (eds.), *B[e] stars*, Astrophysics and Space Science Library 233, p. 1

Active OB stars: structure, evolution, mass loss, and critical limits
Proceedings IAU Symposium No. 272, 2010
C. Neiner, G. Wade, G. Meynet & G. Peters, eds.

doi:10.1017/S1743921311011021

The spectral variations of MWC 314

Corinne Rossi[1], Antonio Frasca[2], Ettore Marilli[2], Michael Friedjung[3], Gérard Muratorio[4]

[1]Dipartimento di Fisica, Universita La Sapienza, Roma, Italy;
email:`corinne.rossi@uniroma1.it`

[2] INAF – Osservatorio Astrofisico di Catania, Italy;

[3] Institut d'Astrophysique CNRS+Paris 6, Paris, France; [4] OMP/LAM, Marseille, France

Abstract. New spectra of MWC314 are presented; they indicate that the V/R emission line flux ratios show signs of varying in an opposite way to the absorption line radial velocities. The latter appear to be due to apparently non-periodic pulsations, perhaps in strange modes.

Keywords. stars: emission-line, Be, stars: individual (MWC 314)

1. Introduction

There has been confusion about whether MWC 314 is a B[e] supergiant or a luminous blue variable. However, it does not show the typical time variations of the latter class. In addition, it appears to be one of the most luminous stars in the Galaxy (Miroshnichenko *et al.* 1998). Permitted emission line profiles show two peaks, suggesting formation in a rotating disk and we considered it to belong rather to the B[e] class (Muratorio *et al.* 2008). Rapid variations in the displacements of the absorption lines, previouly suggested by Wisniewski *et al.* (2006), could indicate binarity (Muratorio *et al.* 2008) with a period of ~ 31 days. In order to confirm the suggested periodicity, we obtained additional spectra at the Catania Observatory (OAC) during 2008 and 2009 in the range 4300–6850 Å with resolution $R \simeq 21\,000$. In some cases we could follow the star for several consecutive nights.

2. Data analysis

The new observations confirmed that: a) The heliocentric radial velocity of the absorption lines is strongly variable, while that of the barycentre of the emission lines remains constant (see Figure 1, from top to bottom: June 15, 20, 24, July 3, 30, August 31, September 4, 2009); b) The intensities of the emission line peaks and the position of the central dip vary (see Figure 2); c) The ratios V/R, obtained from double Gaussian fits of the emission line fluxes, vary in an opposite way to the mean of the absorption line radial velocities (see Figure 3 top and bottom, respectively). This effect does not appear to be explicable by blending of the absorption lines and the emission line peaks, suggesting a more profound physical effect. There is no direct connection between the absorption lines and the minimum between the two peaks of each emission line. In any case more analysis must be undertaken. The periodicity of the radial velocity variation is not confirmed. Though binarity is still possible, all the radial velocity variations cannot be so explained. Strange modes of pulsation (see the review by Glatzel & Chernigovski, 2001) which can be chaotic, have been invoked to explain the variations of luminous blue variables and especially massive stars for which the luminosity/mass ratio in solar units is more than 10^3. That appears to be the case for our star.

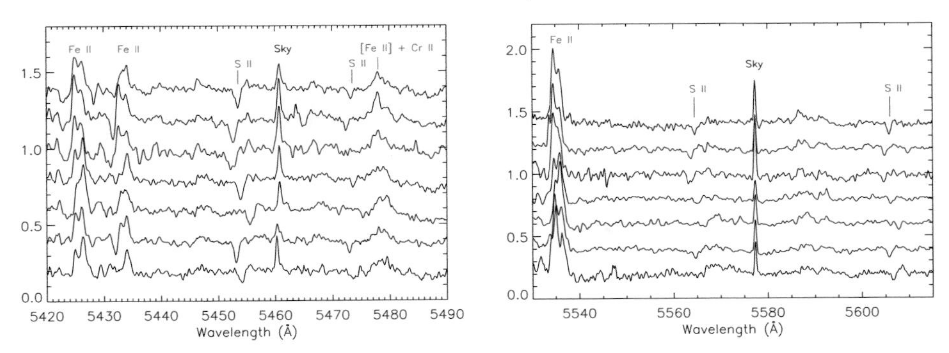

Figure 1. Selected regions from spectra taken at OAC in Summer 2009.

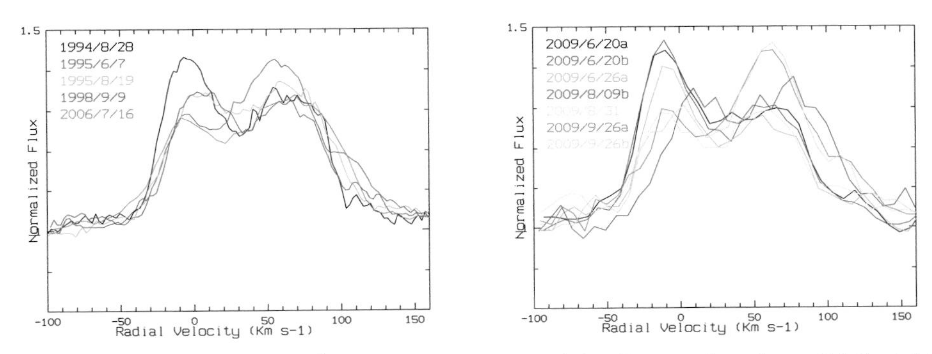

Figure 2. Fe II line 6416.90 Å: Older observations, left. Recent data from OAC, right.

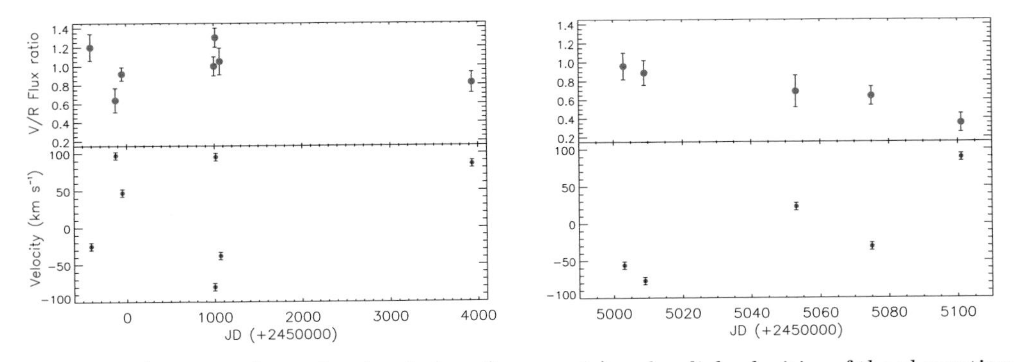

Figure 3. V/R ratios of permitted emissions (top panels) and radial velocities of the absorptions as a function of JD (lower panels). Older observations, left. Recent data from OAC, right.

References

Miroshnichenko, A. S., Fremat, Y., Houziaux, L., Andrillat, Y. *et al.* 1998, *A&AS*, 131, 469

Glatzel, W. & Chernigovski, S. 2001, in: M. de Groot & C. Sterken (eds.), *P Cygni 2000: 400 Years of Progress*, ASP-CS, 233, p. 227

Muratorio, G., Rossi, C. & Friedjung, M. 2008, *A&A* 487, 637

Wisniewski, J. P., Babler, B. L., Bjorkman, K. S., Kurchakov, A. V. *et al.* 2006, *PASP*, 118, 820

Active OB stars: structure, evolution, mass loss, and critical limits
Proceedings IAU Symposium No. 272, 2010
C. Neiner, G. Wade, G. Meynet & G. Peters, eds.

doi:10.1017/S1743921311011033

Multi-epoch interferometric observations of the Be star ζ Tau

Gail H. Schaefer[1], Douglas R. Gies[2], John D. Monnier[3], Noel D. Richardson[2], Yamina N. Touhami[2] and Ming Zhao[4]

[1] The CHARA Array of Georgia State University, Mount Wilson Observatory, Mount Wilson, CA, 91023, U.S.A., email schaefer@chara-array.org
[2] Center for High Angular Resolution Astronomy and Department of Physics and Astronomy, Georgia State University, P.O. Box 4106, Atlanta, GA 30302, U.S.A
[3] Department of Astronomy, University of Michigan, Ann Arbor, Michigan 48109, USA
[4] Jet Propulsion Laboratory, Pasadena, California 91101, USA

Abstract. We present interferometric observations of the Be star ζ Tau obtained using the MIRC beam combiner at the CHARA Array during four epochs in 2007–2009. Fitting a geometric model to the data reveals a nearly edge-on disk with a FWHM of $\sim$ 1.8 milli-arcsec in the *H*-band. The non-zero closure phases indicate an asymmetry in the brightness distribution. Interestingly, when combining our results with previously published interferometric observations of ζ Tau, we find a correlation between the position angle of the disk and the spectroscopic V/R ratio, suggesting that the tilt of the disk might be precessing. This work is part of a multi-year monitoring campaign to investigate the development and outward motion of asymmetric structures in the disks of Be stars.

Keywords. techniques: interferometric, circumstellar matter, stars: emission-line, Be

---

The bright Be star ζ Tau is an ideal target for optical/infrared interferometry. We used the MIRC beam combiner at the CHARA Array to measure interferometrically the size and orientation of the disk of ζ Tau in the *H*-band over four epochs in 2007–2009. The visibilities provide information on the size, orientation, and inclination of the disk. The non-zero closure phases indicate an asymmetry in the brightness distribution.

We fit a two component geometric model to the MIRC data obtained for ζ Tau. The model is composed of a uniform disk with an angular diameter 0.40 mas ($R = 5.5\ R_{\odot}$) to fit the central star and an elliptical Gaussian surface brightness distribution to model the circumstellar disk. To account for the asymmetry, we modulated the elliptical Gaussian disk by a sinusoid as a function of azimuth. This creates a skewed disk model where the sinusoid causes the brightness distribution to peak on one side of the disk and places a depression in the brightness on the other side. The models are shown in Figure 1.

Over the four epochs, the FWHM of the major axis of the disk ranges between 1.6 – 2.1 mas in the *H*-band. This is similar to the K'-band size of 1.8 mas computed by Gies *et al.* (2007) and smaller than the 3.1 – 4.5 mas FWHM measured in Hα by Quirrenbach *et al.* (1997) and Tycner *et al.* (2004). We find that the star contributes on average about 55% of the light in the *H*-band. In comparison, the star contributes 41% of the flux in the K'-band (Gies *et al.* 2007). These ratios are consistent with near-IR excess fluxes observed by Touhami *et al.* (2010).

Combining our results with previous interferometric observations, we find that the position angle of the major axis of the ζ Tau disk varies as a function of the spectroscopic V/R phase (see Fig. 2). The disks of Be stars may develop a global, one-armed spiral instability. The oscillation mode forms a spiral density enhancement that precesses prograde with the disk rotation with a cycle time of a few years. We suspect that the

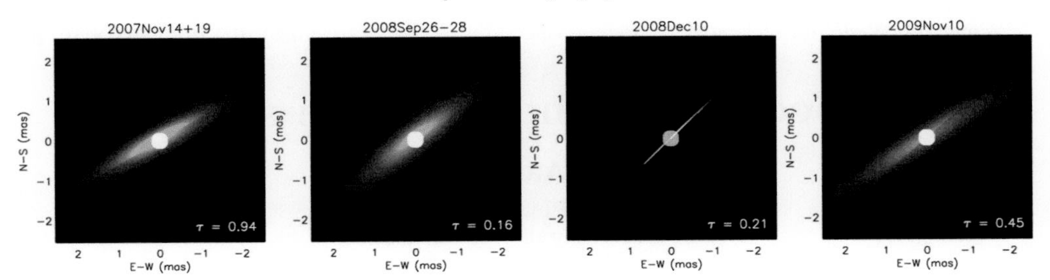

Figure 1. Best-fit geometric models for ζ Tau during the epochs of the MIRC observations. The spectroscopic V/R phase τ is indicated in the bottom right of each panel.

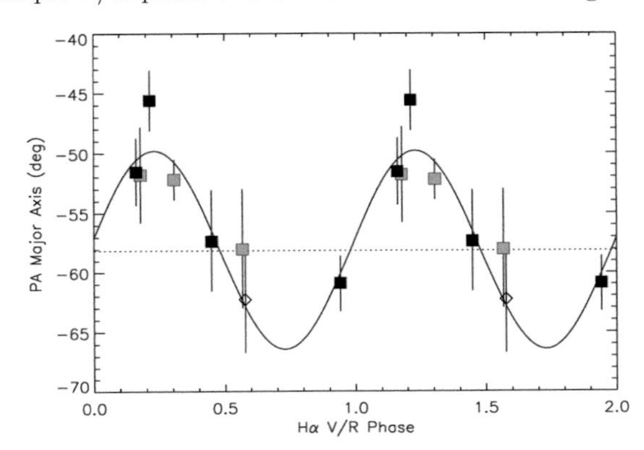

Figure 2. Position angle of the disk major axis for ζ Tau derived from interferometry plotted against the V/R phase. The measurements are repeated over two cycles to emphasize phase continuity. The filled black squares represent our MIRC observations, the gray squares are previously published near-IR measurements (Gies *et al.* 2007, Štefl *et al.* 2009, Carciofi *et al.* 2009), and the open diamond indicates the Hα result from Tycner *et al.* (2004). The solid line shows a sinusoidal weighted fit of the variation (mean PA -58.0°±1.4°, semiamplitude 8.1°±1.7°). The dotted line indicates the mean position angle of -58.1°±1.2° determined from linear polarization observations (McDavid 1999, Štefl *et al.* 2009).

position angle variations result from a tilt of the disk that could be generated by vertical motions of the gas caused by the spiral density enhancement (Ogilvie 2008; Oktariani & Okazaki 2009) as it moves through the disk. Additionally, we find that the asymmetry in the light distribution of the disk roughly corresponds to the expected location of the density enhancement in the spiral oscillation model.

We plan to continue monitoring changes in the structure and orientation of the disk of ζ Tau with future observations at the CHARA Array.

References

Carciofi, A. C., Okazaki, A. T., Le Bouquin, J.-B., Štefl, S. *et al.* 2009, *A&A*, 504, 915

Gies, D. R., Bagnuolo, Jr., W. G., Baines, E. K., ten Brummelaar, T. A. *et al.* 2007, *ApJ*, 654, 527

McDavid, D. 1999, *PASP*, 111, 494

Ogilvie, G. I. 2008, *MNRAS*, 388, 1372

Oktariani, F. & Okazaki, A. T. 2009,*PASJ*, 61, 57

Quirrenbach, A., Bjorkman, K. S., Bjorkman, J. E., Hummel, C. A. *et al.* 1997, *ApJ*, 479, 477

Štefl, S., Rivinius, T., Carciofi, A. C., Le Bouquin, J.-B. *et al.* 2009, *A&A*, 504, 929

Touhami, Y., Richardson, N. D., Gies, D. R., Schaefer, G. H. *et al.* 2010, *PASP*, 122, 379

Tycner, C., Hajian, A. R., Armstrong, J. T., Benson, J. A. *et al.* 2004, *AJ*, 127, 1194

Active OB stars: structure, evolution, mass loss, and critical limits
Proceedings IAU Symposium No. 272, 2010
C. Neiner, G. Wade, G. Meynet & G. Peters, eds.

doi:10.1017/S1743921311011045

Spectral synthesis for Be stars

T. A. Aaron Sigut

Department of Physics and Astronomy
The University of Western Ontario
London, Ontario, Canada N6A 3K7
email: asigut@uwo.ca

Abstract. A new monochromatic imaging and spectral synthesis package for Be stars, based on the BEDISK code, is introduced. Example images and spectra are given for for H I and Fe II. Predicted Fe II equivalents widths are also compared to recent observations by Arias *et al.* (2006) and show good agreement, although only for very dense disks.

Keywords. stars: emission-line, Be, circumstellar matter, radiative transfer

1. Introduction & Motivation

Be stars are main sequence B stars that are surrounded by an equatorial disk. The disk is thought to form via outflow from the stellar photosphere, but the mechanism that creates the disk is uncertain (Porter & Rivinius, 2003). There is good evidence that the disks are Keplerian and rotationally supported (Hummel & Vrancken, 2000), but the central B stars are not critically rotating (Cranmer, 2005), and the source of additional angular momentum at the inner edge of the disk is unknown. Be stars account for $\approx 17\%$ of all B stars and are interesting because they couple fundamental themes of stellar pulsation, rotation and angular momentum, and mass loss.

The mass loss mechanism can be constrained by determining the physical conditions in the inner disk, such as the kinetic temperature (T) and density scale height (H). The BEDISK code (Sigut & Jones, 2007; Sigut, Jones & McGill, 2009) can self-consistently predict both T and H given a density model for the equatorial plane of the disk. There have been suggestions that the inner density scale heights of Be star disks are enhanced over the prediction of simple vertical hydrostatic equilibrium (Chauville *et al.* 2001; Zorec *et al.* 2007). These results are based on the comparison of (unresolved) H I and Fe II spectra between nearly pole-on ($i \approx 0^o$) and nearly edge-on ($i \approx 90^o$) systems.

Figure 1. A B1Ve star seen at $i = 65^o$ in a ± 20 Å wavelength bin centred on Hα.

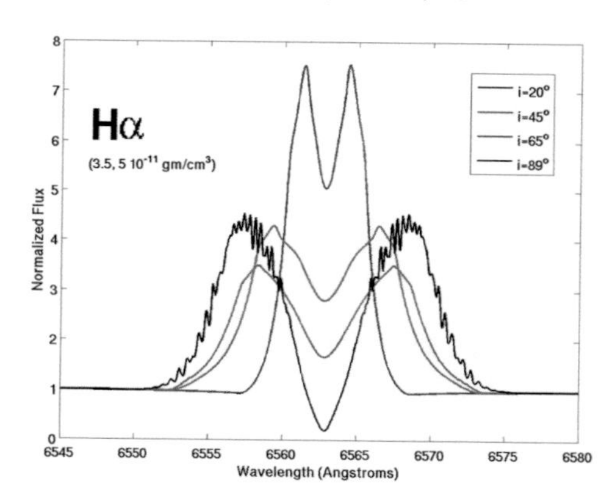

Figure 2. Example Hα line profiles for the same disk density model as in Figure 1.

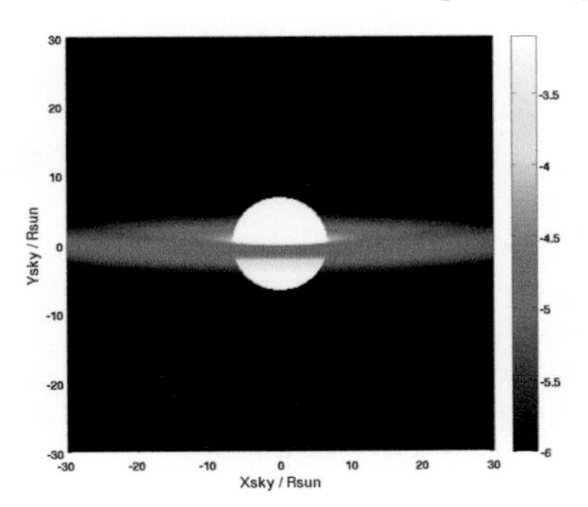

Figure 3. A B2Ve star seen at $i = 75^o$ in a ± 2 Å bin centred on Fe II 4233Å.

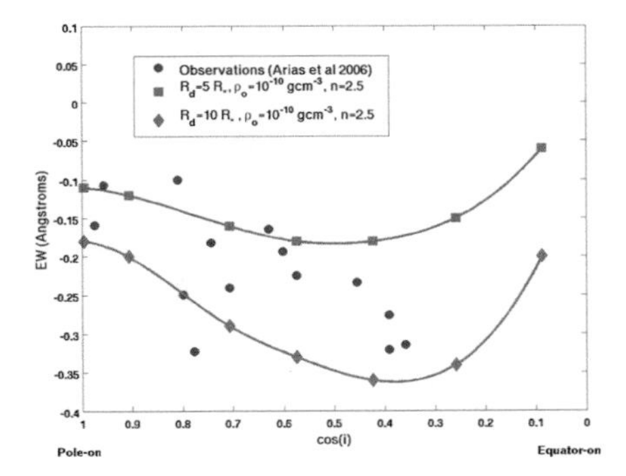

Figure 4. Predicted and observed equivalent widths of Fe II 4233Å as a function of $\cos i$.

2. Beray

To investigate the inner scale heights of Be star disks, a new radiative transfer code, BERAY, has been developed which can predict images on the sky and unresolved spectra given a BEDISK density and temperature structure. The equation of radiative transfer is solved along a series of rays ($\approx 10^5$) through the star-plus-disk system. An example monochromatic Hα image is shown in Figure 1 for a disk surrounding a B1 main sequence star seen at an inclination of $i = 65^o$ with an equatorial density model of $\rho(R) = 510^{-11}(R_*/R)^{3.5}$ g cm^{-3}. While the density model is axisymmetric, the monochromatic image on the sky is not due to increased photon escape for rays with larger $|dv/dz|$. Figure 2 shows the Hα line profile of this model seen for various inclinations.

3. Fe II in Be Stars

BERAY/BEDISK can also be used to predict detailed spectra of metals. An image in a bin of ± 2 Å around Fe II 4233Å (multiplet 27, $z\,^4D^o - b\,^4P^e$) of the disk $\rho(R) = 10^{-10}(R_*/R)^{2.5}$ g cm^{-3} surrounding a B2V star is shown in Figure 3. Zorec *et al* (2007) suggest that the optical depths in Fe II emission lines as compared between pole-on and equator-on Be star systems require enhanced scale heights in the inner disk, above the prediction of gravitational equilibrium. Figure 4 compares observed Fe II equivalent widths for a sample of B2V stars (Arias *et al.* 2006) with BEDISK/BERAY calculations for a dense disk, $\rho(R) = 10^{-10}(R_*/R)^{2.5}$ g cm^{-3} in gravitational equilibrium, seen at several inclinations. The observed equivalent widths are well-bracketed by models with R_{disk} between 5 and 10 R_*. However, Zorec *et al.* (2007) note that such dense disks may by physically unrealistic. BERAY/BEDISK is currently being used to model all H I and Fe II observations of Arias *et al.* (2006) to search for realistic sets of ρ_o, n, and R_{disk} to see if increased inner disk scale heights are required by the data.

References

Arias, M. L., Zorec, J., Cidale, L., Ringuelet, A. E. *et al.* 2006, *A&A* 460, 821
Chauville, J., Zorec, J., Ballereau, D., Morrell, N. *et al.* 2001, *A&A* 378, 861
Cranmer, S. R. 2005, *ApJ* 634, 585
Hummel, W. & Vrancken, M. 2000, *A&A*, 359, 1075
Porter, J. M. & Rivinius, T. 2003, *PASP* 115, 1153
Sigut, T. A. A., McGill, M. A. & Jones, C. E. 2009, *ApJ* 699, 1973
Sigut, T. A. A. & Jones, C. E. 2007, *ApJ* 668, 481
Zorec, J., Arias, M. L., Cidale, L., & Ringuelet, A. E. 2007, *A&A* 470, 239

Active OB stars: structure, evolution, mass loss, and critical limits
Proceedings IAU Symposium No. 272, 2010
C. Neiner, G. Wade, G. Meynet & G. Peters, eds.

doi:10.1017/S1743921311011057

Do the γ Cas X-rays come from the Be Star?

Myron A. Smith[1] and R. Lopes de Oliveira[2]

[1]Dept. of Physics & Astronomy, Catholic University of America,
Washington, D.C. 20064, USA
email: msmith@stsci.edu

[2]Universidade de São Paulo, Instituto de Física de São Carlos,
Caixa Postal 369, 13560-970, São Carlos, SP, Brazil

Abstract. We discuss the origin of the hard X-rays in γ Cas and its analogs. Of great importance are their temporal correlations with optical/UV signatures, suggesting an origin near the Be star.

Keywords. stars: emission-line, Be, stars: winds, outflows, stars: magnetic fields, X-rays: stars

1. Introduction

γ Cas (B0.5IVe) is the prototype of a group of some 8 known Galactic B0.5-1e stars with hot thermal ($kT \geqslant 10$ keV) X-ray spectra indicating a complex geometry and strong Hα emission. γ Cas itself exhibits continuous flaring on timescales of a few seconds (requiring densities of $\rho > 10^{14}$ cm^{-3}; see "scatter" in Fig. 1), a few hours, and 2-3 month cycles. The origin of the X-rays is a mystery. Multiband correlations suggest that they could arise from physical interactions near the Be star, e.g. from magnetic stresses arising from differential rotation between solid-body rotation near the surface and the Keplerian disk.

2. Lines of observational evidence

Simultaneous X-ray/UV spectral campaign in January 1996 established strong correlations between flux variations of X-rays, the inverted UV continuum and in lines formed above the Be star from species below and above the dominant ion stage(see Fig. 1). X-ray flux seems to influence the ionization of the local circumstellar plasma.

At times of high X-ray flux, light dips are too brief to be due to surface spots and must be due to passages of translucent clouds $\sim$0.3R$_*$ above the surface (Smith *et al.* 1998), consisting of both warm and cool plasma (Smith & Robinson 1999). Spectral lines in the optical and UV spectrum of γ Cas frequently exhibit "migrating subfeatures" in absorption. These *msf* move blue to red across the profile with an acceleration of +95 km s^{-2} and arise in small cloudlets in front of the star at mainly the same times as the UV flux dips. The *msf* are also present in optical lines of the γ Cas analog star HD 110432 (Smith & Balona 2006) and AB Dor, which is a well known active K dwarf with a complex magnetic field configuration. All of these activities find a common explanation in CS clouds forced into corotation near X-ray active centers on the Be star. This inference is supported by the finding of overbroadening in the weakest hydrogen Brackett emission lines of the γ Cas spectrum, indicating that the inner CS region rotates more rapidly than the Be star (Hony *et al.* 2000), where v*sin i* $\approx$ 400 km s^{-1}.

The light curve of γ Cas shows periodic, gray nonsinusoidal signatures of amplitude 0.003 mags. that are stable over 10 years. The period is 1.21581$\pm$0.00004 days (Smith *et al.* 2006). This is undoubtedly a signature of rooted surface chemical inhomogeneities. There are no periodic UV resonance line variations. Thus the inferred presence of surface fields must take the form of a complex topology.

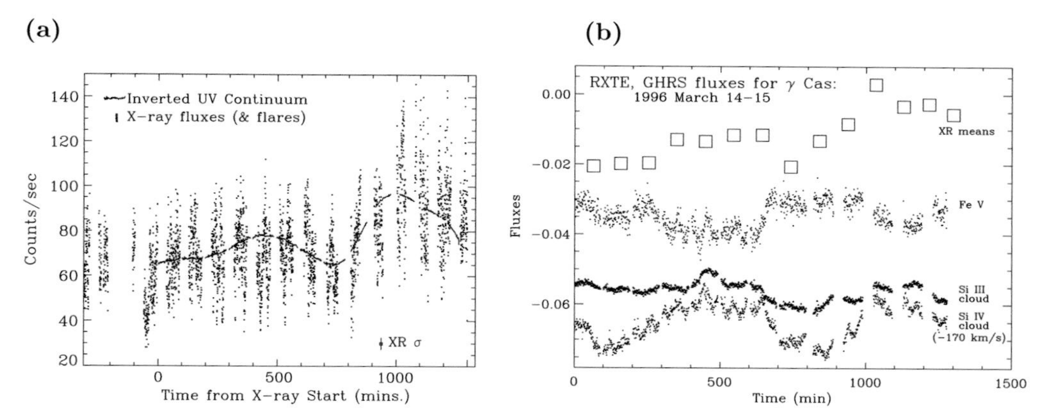

Figure 1. **(a)** X-ray and inverted UV continuum fluxes; **(b)** X-ray with Si III, Si IV and Fe V line flux.

The same optical light curve shows 2-3 month-long optical cycles with larger amplitudes in the red. In 2000 and 2001 these cycles were found to be well correlated with X-ray cycles (Robinson *et al.* 2002). (At different times a similar X-ray and optical cycle has been found in the γ Cas analog HD 110432.) Robinson & Smith (2000) suggested that variable physical conditions in the disk, e.g. a magnetic dynamo, mediate the X-ray emission.

Our 1996 HST/GHRS spectra disclose the presence of high velocity absorptions, up to $+2000$ km s^{-1} (Smith & Robinson 1999), implying that CS blobs accelerate toward the star. These velocities are sufficient to excite the X-ray flares observed in γ Cas.

Putting this all together, we believe correlations between X-ray and UV/optical diagnostics show the X-rays are produced near or on the surface of the Be star. We speculate that shear in angular rotation between the solid-body CS complexes near the Be star and its Keplerian disk cause magnetic lines of force to stretch, break, and reconnect, thereby catapulting ambient debris towards the star and causing copious hard X-ray emission.

The building blocks for this idea to work are a rapidly rotating Be star, possibly a complex surface field, and a robust disk. γ Cas itself is a 204 day binary in a nearly circular orbit (Miroshnichenko et al. 2002). This condition and a case for two other γ Cas analogs being blue stragglers suggest the possibility that prior prior angular momentum could have spun up the Be star. When occurring together these conditions would satisfy the requirement that the γ Cas stars occur infrequently but are not unique.

In our view these observations seem to point to Be star-disk interactions. Yet, it cannot be corroborated to date that any specific mechanism that mediates X-ray variability, such as a magnetorotational instability, is self-sustaining or has predictive power.

References

Hony, S., Waters, L. B. F. M., Zaal, P. A., & de Koter, A. *et al.* 2000, *A&A*, 355, 187
Miroshnichenko, A. S., Bjorkman, K. S., & Krugov, V. D. 2002, *PASP*, 114, 1226
Robinson, R. D., Smith, M. A., & Henry, G. W. 2002, *ApJ*, 575, 435
Robinson, R. D. & Smith, M. A. 2000, *ApJ*, 540, 474
Smith, M. A., Robinson, R. D. & Hatzes, A. P. 1998, *ApJ*, 507, 945
Smith, M. A. & Balona, L. 2006, *ApJ*, 640, 491
Smith, M. A. & Robinson, R. D. 1999, *ApJ*, 517, 866
Smith, M. A., Henry, G. W., & Vishniac, E. 2006, *ApJ*, 647, 1375

Active OB stars: structure, evolution, mass loss, and critical limits
Proceedings IAU Symposium No. 272, 2010
C. Neiner, G. Wade, G. Meynet & G. Peters, eds.

doi:10.1017/S1743921311011069

The 2008+ outburst of the Be star 28 CMa - a multi-instrument study†

Stan Štefl[1], Alex C. Carciofi[2], Dietrich Baade[4], Thomas Rivinius[1], Sebastian Otero[3], Jean-Baptiste Le Bouquin[5], Juan Fabregat[6], Atsuo T. Okazaki[7] and Fredrik Rantakyrö[8]

[1]ESO Chile; [2]São Paulo, Brazil; [3]Buenos Aires, Argentina; [4]ESO, Germany; [5]Grenoble, France; [6]Valencia, Spain; [7]Sapporo, Japan; [8]Gemini Chile

Abstract. Optical and IR spectra, optical to sub-mm photometry, visual imaging polarimetry, and IR high-resolution spectro-interferometry are being used to monitor the new outburst of 28 CMa, which started in 2008 and so far closely resembles previous ones. First modeling based on viscous decretion and focused on constraining the disk viscosity parameter, α, is presented.

Keywords. stars: circumstellar matter, stars: emission-line, Be, stars: mass loss

---

1. Introducing 28 CMa and its outburts

28 CMa is an early-type (B2 IV), pole-on ($v \sin i = 80$ km/s) Be star. Like most Galactic early-type Be stars, its line-profile variability can be modeled as $\ell = 2, m = +2$ nonradial pulsation (Maintz *et al.* 2003) and it undergoes discrete mass loss events (outbursts). Spectroscopic long-term monitoring (Štefl *et al.* 2003) only found a single 1.371-d period so that the outbursts cannot probably be attributed to multi-mode beating.

During outbursts, mass is transferred to a circumstellar disk. Because of the pole-on perspective of 28 CMa, this star's outbursts manifest themselves as brightenings by ~ 0.5 mag in visual light. They are spaced by 6-8 years and last about 2-4 years. Fluctuations by $\sim$0.1 mag within $\sim$10 days suggest that such major outbursts actually are phases of enhanced activity with numerous minor mass-loss events. This paper gives a first brief account of the most recent outburst detected through SO's naked-eye monitoring in 2008.

2. Observations and their modeling

Following the first alert, a broad suite of observations was undertaken. In addition to the dense visual photometry, *JHKL* photometry was obtained with the Mk II photometer of SAAO (4 epochs) and CAIN-II Tenerife/TCS camera (1 epoch); *Q*1 and *Q*3 measurements were added with *VISIR* on the VLT (1 epoch), and observations at 0.87 mm / 345 GHz were secured with *LABOCA* on *APEX* (1 epoch). Optical echelle spectra are available from *UVES/VLT* (Oct 2008-Mar 2009), *FEROS* / La Silla (2 epochs) and the 1-m telescope at Pico dos Dias Observatory (Jan 2009 - May 2010). *BVRI* imaging polarimetry was made with the 0.6-m telescope at Pico dos Dias (4 epochs). Finally, time was obtained with *AMBER* in its high-spectral resolution mode for interferometry with three *VLTI* Auxiliary Telescopes (3 epochs). Efforts are being undertaken to continue the observations through the decline to the visually faint phase, which is assumed to mark the end of the present active period. Following the very successful modeling of ζ Tau by

† Based partly on observations collected at ESO; props. 282.D-5014, 284.D-5043.

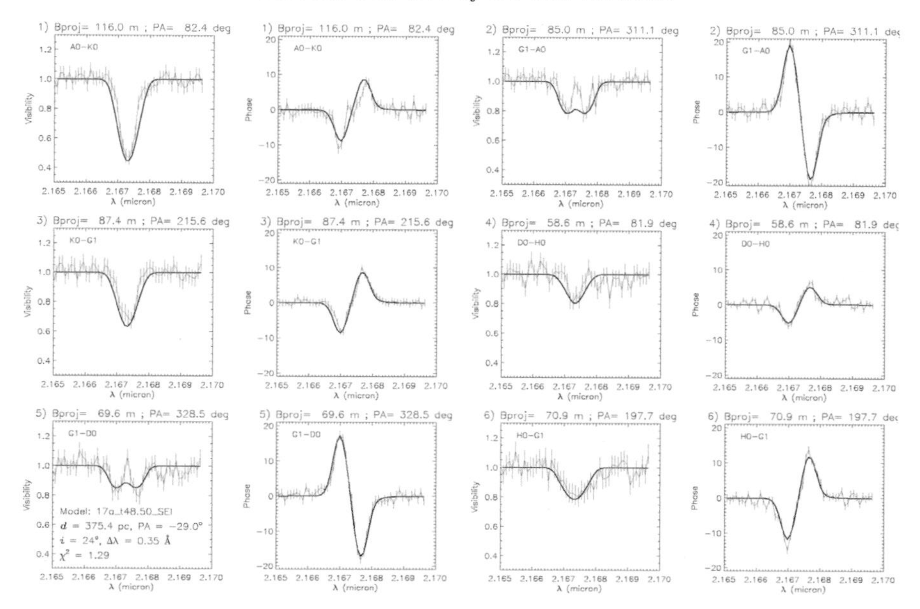

Figure 1. Differential visibilities and phases obtained at the six used baselines in April 2009. The model (black line) fits data well except the "central peak" in visibilities at PA = 310 - 330°.

the HDUST code and viscous disk decretion (Carciofi *et al.* 2009), the analysis of the observations of 28 CMa is proceeding along the same lines.

3. First preliminary results

To the extent that comparable observations exist, the present outburst closely follows the two previous ones. Of all indicators studied, Brγ was the last to echo the disk build-up and a large phase amplitude is seen with *AMBER*. This confirms model calculations as well as other observations, e.g. of ζ Tau, that Brγ line emission does not form close to the central star. As for ζ Tau, the 28 CMa data are in agreement with Keplerian rotation, but the closure phase indicates less asymmetry in the disk. The differential visibilities and phases are shown in Fig. 1.

Because of the high level of repetivity of the visual light curve, an attempt was made to reproduce its shape between the decline from one outburst, when, in the model, decretion was switched off through the rise of the next one, at which time decretion was switched on. The disk was assumed to evolve completely passively between the two mass-loss phases. Since an outburst reaches its maximum very quickly, the inner disk fills and saturates quickly and so does not usefully constrain the disk viscosity parameter, α. This is different during the phases right after termination of the mass supply to the disk and the subsequent evolution. In both cases, $\alpha \sim 0.9$ well reproduces the observations (see Fig. 5 in Carciofi, these proceedings). This is unexpected as it would imply that the viscous torque depends very little on disk density, which varies a lot during this time interval.

References

Carciofi, A. C., Okazaki, A. T., Le Bouquin, J.-B., & Štefl, S. *et al.* 2009, *A&A*, 504, 915
Maintz, M., Rivinius, T., Štefl, S., & Baade, D. *et al.* 2003, *A&A*, 411, 181
Štefl, S., Baade, D., Rivinius, T., & Stahl, O. *et al.* 2003, *A&A*, 411, 167

Portraits of young researchers in front of their posters. From left to right and top to bottom: André-Nicolas Chene, Rémi Fahed, Chloé Fourtune-Ravard, Nick Hill, Evgenia Koumpia, Meghan McGill, Christopher Russell, and Matthew Shultz.

Active OB stars: structure, evolution, mass loss, and critical limits
Proceedings IAU Symposium No. 272, 2010
C. Neiner, G. Wade, G. Meynet & G. Peters, eds.

doi:10.1017/S1743921311011070

Asteroseismic observations of OB stars

Peter De Cat[1], Katrien Uytterhoeven[2], Juan Gutiérrez-Soto[3], Pieter Degroote[4], and Sergio Simón-Díaz[5]

[1]Royal observatory of Belgium, Ringlaan 3, B-1180 Brussel, Belgium
email: Peter.DeCat@oma.be

[2]Lab. AIM, CEA/DSM-CNRS-Université Paris Diderot, CEA, IRFU, SAp, Saclay, 91191, Gif-sur-Yvette, France
email: katrien.uytterhoeven@cea.fr

[3]Instituto de Astrofísica de Andalucía (CSIC), Apartado 3004, 18080 Granada, Spain
email: jgs@iaa.es

[4]Instituut voor Sterrenkunde, KU Leuven, Celestijnenlaan 200D, B-3001 Leuven, Belgium
email: Pieter.Degroote@ster.kuleuven.be

[5]Instituto de Astrofísica de Canarias, E-38200 La Laguna, Tenerife, Spain
email: ssimon@iac.es

Abstract. The region of the hot end of the main-sequence is hosting pulsating stars of different types and flavours. Pulsations are not only observed for Slowly pulsating B stars (mid to late B-type stars; high order g-modes) and β Cephei stars (early B-type stars; low order p/g-modes) but are also causing variability in Be stars and OB-supergiants. In this review we give an overview of the asteroseismic observations that are currently available for these types of stars. The first asteroseismic results were solely based on ground-based observations. Recently, the arrival of space-based data gathered by space missions like MOST, COROT and KEPLER has led to important discoveries for massive stars, highlighting their excellent asteroseismic potential. We show that, despite the unprecedented precision of the space-based data, there is still a clear need for ground-based follow-up observations.

Keywords. stars: early-type, stars: emission-line, Be, stars: variables: other, stars: oscillations (including pulsations)

---

1. Introduction

The title of this review consists of three parts. The adjective *asteroseismic* is derived from "asteroseismology", which refers to the science in which stellar (aster) oscillations (seismo) are studied (logy) to gain information of stars in general, and about their interior in specific. By matching the observed and theoretically predicted frequency spectrum, severe constraints can be obtained on important parameters like the mass, the internal rotation law (rigid or not), the metallicity and the effects of convection (extent of convective overshooting; size of convective layers). For massive stars, the internal structure globally consists of a convective core and a radiative envelope. Stellar oscillations are described by four numbers: (1) the pulsation frequency f or period P (the timescale of reoccurring patterns), (2) the degree ℓ (number of node-lines on the surface), (3) the azimuthal number m (number of node-lines on the surface perpendicular on the equator; the sign is linked to the direction of the propagation) and (4) the radial order n (approximation of the number of node-surfaces between the core and the surface). Different steps in the analysis of the data are needed to determine these numbers: frequency analysis for f, mode identification for ℓ and m and theoretical modelling for n. Based on their surface behaviour, a distinction is made between radial modes ($l=m=0$; only radial motions; symmetry maintained during pulsation cycle) and non-radial modes ($\ell \neq 0$; both radial

Table 1. Overview of the specifications of the space missions MOST, COROT and KEPLER.

Space mission	Size telescope	Photometric Accuracy	Length of time-series
MOST	15/17.3-cm	few μmag	typically 30 days
COROT	30-cm	10 μmag	30-150 days
KEPLER	1.4-m	1.7 μmag	3.5 years

and transverse motions occur; nodelines on surface). In the presence of stellar rotation and/or the presence of a magnetic field, the frequency degeneracy in m is lifted: a nonradial mode with degree ℓ will be splitted into its $2\ell+1$ components, resulting into a so called frequency multiplet. Based on their physical behaviour, a distinction is made between acoustic modes (p-modes, for which the restoring force is pressure) and gravity modes (g-modes, for which the restoring force is buoyancy). For p-modes, the radial motions are dominant (highest amplitudes in core of spectral lines) while for the g-modes the tangential motions are more pronounced (higher amplitudes in the wings of the spectral lines). While p-modes carry information from the surface layers, the g-modes probe the deepest layers of the star, making them very interesting from an asteroseismic point of view. In general, the observed periods for p-modes are shorter then those of g-modes (see below).

At least four different types of *observations* can be used to assemble the ingredients needed for in-depth asteroseismic studies (a number of well-identified modes and restrictions on stellar parameters), i.e. photometry, spectroscopy, spectropolarimetry and interferometry. Each type has its own advantages and disadvantages. Time-series of photometric observations are in general easy to gather and are particularly useful to detect and study low-ℓ modes (the amplitude of high-ℓ modes are generally too small due to cancellation effects). In the case of multi-colour photometry, the calibration of the photometric system can be used to determine stellar parameters ($T_{\rm eff}$, $\log g$, metallicity) and reddening, but it only allows a partial mode identification because the observables used in the photometric mode identification technique (photometric amplitude ratios/phase differences, Dupret *et al.* 2003) only depend on ℓ (e.g. Handler *et al.* 2004; Handler *et al.* 2006; De Cat *et al.* 2007). Time-series of spectroscopic observations are complementary to those in photometry in several ways. Low to medium resolution spectra (R$<$30 000) provide information on extra stellar parameters (radial velocity $v_{\rm rad}$, projected rotational velocity $v \sin i$, abundances) while time-series of high-quality (SNR$\geqslant$200), high-resolution (R$\geqslant$30 000) spectra are used for a detailed study of line profile variations (LPVs; e.g. De Cat *et al.* 2005; Briquet *et al.* 2005; Briquet *et al.* 2009). They allow a determination of both ℓ and m of the observed modes and of constraints on the inclination i and rotational parameters through the application of spectroscopic mode identification techniques (Moment Method: Briquet & Aerts 2003; Intensity Period Search: Telting & Schrijvers 1997; Fourier Parameter Fit method: Zima 2006). Moreover, spectroscopic observations are more sensitive to modes with a higher ℓ and a lower amplitude because we are not dealing with an integrated quantity. For magnetic stars, spectropolarimetry allows to study magnetic field variations for a determination of the rotation period and the magnetic geometry, which can shrink the free parameter space of the mode identification significantly (Neiner *et al.* 2003a,b,c) Recently, magnetic field detections have been reported for several OB stars (Hubrig *et al.* 2006; Hubrig *et al.* 2007; Hubrig *et al.* 2009). Also interferometric observations can be useful, in particular for the measurement of the radii of bright nearby stars and for a full orbit determination of binaries (e.g. β Cen: Davis *et al.* 2005; λ Sco: Tango *et al.* 2006). Such observations can either be obtained from observatories on Earth (ground-based) or with space missions (space-based). Because ground-based observations can only be done during the night, time-series of single-site observations are

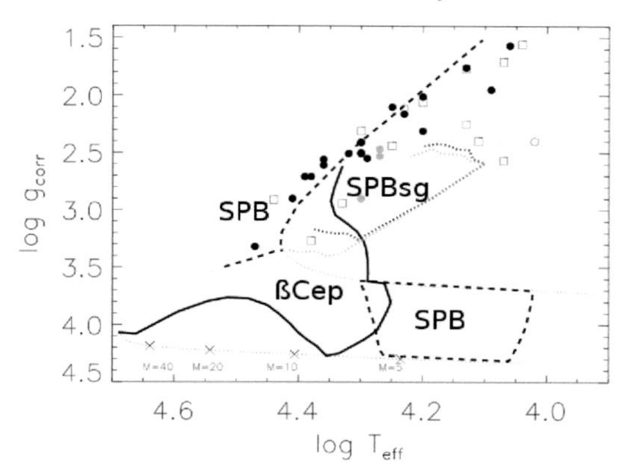

Figure 1. The $\log T_{\rm eff}$ - $\log g$ diagram for the OB-stars. It shows the theoretical instability strip for low-n p/g-modes ("β Cep"; thick solid line) and for high-n g-modes ("SPB"; thick dashed line) calculated by Pamyatnykh (1999), and for g-modes in post-TAMS models for $\ell = 1$ (grey dotted line) and $\ell = 2$ (black dotted line) as calculated by Saio *et al.* (2006) ("SPBsg"). The circles and squares indicate the objects studied by Lefever *et al.* (2007). (altered version of figure taken from Lefever *et al.* 2007)

generally heavily affected by aliasing effects (spurious frequencies introduced by the time sampling) which hampers the frequency analysis (e.g. Uytterhoeven *et al.* 2007). These aliasing can partly be solved by performing multi-site campaigns (e.g. ν Eri: Handler *et al.* 2004; Aerts *et al.* 2004a). Asteroseismic space missions like MOST, COROT and KEPLER generally allow continuous photometric monitoring of stars for long periods (see Table 1). Moreover, the quality of the data is not degraded by the Earth's atmosphere which leads to time-series of white-light photometry with a stunning accuracy down to micromagnitude level (Michel *et al.* 2008; Jenkins *et al.* 2010; Gilliland *et al.* 2010) and free from the typical alias frequencies encountered in ground-based observations.

This raises the question: *Are space-based observations sufficient for a state-of-the-art asteroseismic study?* Or, in other words: *Are ground-based follow-up observations really worth the effort?* In what follows, we will give an answer by focussing on the main asteroseismic results obtained recently for types of variable *OB stars*, i.e. the hot main-sequence pulsators (β Cephei stars, slowly pulsating B stars and hybrid pulsators; Section 2), the Be stars (Section 3) and OB-supergiants (Section 4).

2. Hot main-sequence pulsators

The hot end of the main-sequence hosts two well-established classes of pulsating stars. β Cephei (β Cep) stars are early B-type (B 0-B 3) main-sequence stars that oscillate in low-radial-order n p/g-modes for which the observed periods are short (2-12 hours) and the amplitudes low (in general < 40 mmag in light and < 20 $km\,s^{-1}$ in radial velocity) (see Stankov & Handler 2005 for a review). Slowly pulsating B (SPB) stars are mid to late B-type (B 2-B 9) main-sequence stars pulsating in high-radial-order n g-modes with longer periods (0.3-5 days) and lower amplitudes (in general < 20 mmag in light and < 10 $km\,s^{-1}$ in radial velocity) (see De Cat 2007 for a review). The pulsations in both classes of pulsating stars are driven by the κ mechanism acting on the iron opacity bump at around 200 000 K (e.g. Dziembowski *et al.* 1993, Gautschy & Saio 1993). Their theoretical instability strips as derived by Pamyatnykh (1999) are given in Fig. 1. Theoretical frequency spectra of β Cep stars are rather sparse making asteroseismic modeling possible with a small number of well-identified modes. For SPB stars, much more well-identified modes are needed because their theoretical frequency spectra are much denser.

Table 2. Overview of the results of in-depth asteroseismic results obtained for β Cep stars so far. In the colums, we give for each object the HD number, the variable star name (if available), the spectral type, the numbers of modes used for the asteroseismic modeling, the identification of the radial mode, the overshoot, the velocity of the rotation core, the detection of g-modes, and the references, respectively.

HD number	Name	SpT	#fit	radial mode	$\alpha_{\rm ov}$	rotation core	g	Ref
HD 129929	V836 Cen	B3V	3	fundamental	0.10(5)	4x faster than surf.	no	[1],[2]
HD 29248	ν Eri	B2III	4	fundamental	0.05(5)	3x faster than surf.	yes	[3],[4]
HD 16582	δ Ceti	B2IV	2	1st overtone	0.20(5)		no	[5]
HD 44743	β CMa	B2III	2	1st overtone	0.20(5)		no	[6]
HD 157056	θ Oph	B2IV	3	fundamental	0.44(7)	(not conclusive)	no	[7]
HD 214993	12 Lac	B2III	2+	1st overtone	<0.4	(faster than surf.?)	yes	[8]

[1] Aerts *et al.* (2003), [2] Dupret *et al.* (2003), [3] Ausseloos *et al.* (2004), [4] Pamyatnykh *et al.* (2004), [5] Aerts *et al.* (2006b), [6] Mazumdar *et al.* (2006), [7] Briquet *et al.* (2007), [8] Desmet *et al.* (2009)

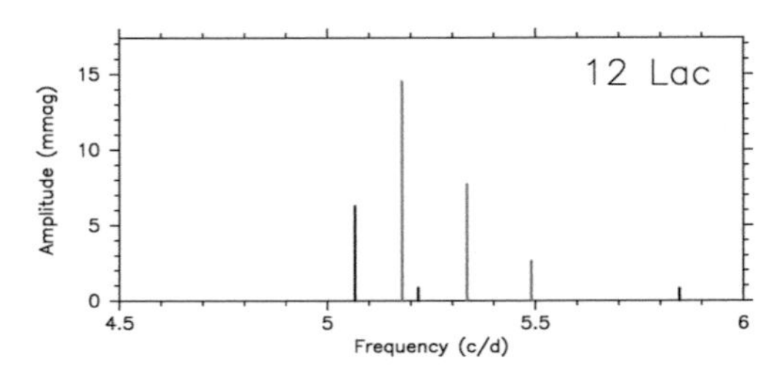

Figure 2. The observed frequencies of the β Cep star 12 Lac. The frequencies indicated in red (colours in electronic version only) give the false impression that we are dealing with a rotationally splitted $\ell=1$ mode.

The first in-depth asteroseismic study of a *β Cephei (β Cep) star* was solely based on single-site ground-based photometry. In a time-span of 21 years (!), a total of about 1500 observations in the 7-colour photometric system of Geneva were collected for V836 Cen (HD 129929), resulting in the detection of six frequencies (Aerts *et al.* 2004b). These were identified with the radial fundamental ($\ell=0,\mathrm{p}_1$) mode, two components of the $\ell=1,\mathrm{p}_1$ mode and two components of the $\ell=2,\mathrm{g}_1$ mode (Dupret *et al.* 2004). This study gave the first evidence for *convective overshooting* at the core and, thanks to the observations of two rotationally splitted modes, of *non-rigid rotation* (Aerts *et al.* 2003).

The best studied β Cep star solely based on ground-based observations is ν Eri, for which a large scale photometric and spectroscopic multi-site campaign was organised in 2002-2003, with an extension of the photometric multi-site campaign to 2004. The frequency analysis of the data revealed 34 frequencies in photometry (14 independent and 20 combination frequencies; Handler *et al.* 2004; Jerzykiewicz *et al.* 2005) and 20 frequencies in spectroscopy (8 independent and 12 combination frequencies; Aerts *et al.* 2004a; De Ridder *et al.* 2004), including two g-modes, making ν Eri a so-called β Cep/SPB *hybrid pulsator* (see below). For the modeling, the frequencies of the four identified modes can be used: $\ell=0,\mathrm{p}_1$, $\ell=1,\mathrm{g}_1$, $\ell=1,\mathrm{p}_1$, and $\ell=1,\mathrm{p}_2$. By doing so, four of the five free parameters (X, Z, $\alpha_{\rm ov}$, M, age) can be easily determined with standard models, but a *problem with mode excitation* occurs: the $\ell=1,\mathrm{p}_2$ is not excited in the resulting model. Different solutions have been proposed to solve this problem. Ausseloos *et al.* (2004) assumed a global metal enhancement while Pamyatnykh *et al.* (2004) suggested a local iron enrichment in the driving zone due to diffusion.

In Table 2 we give an overview of the in-depth asteroseismic studies of β Cep stars, from the era before the COROT satellite, enforcing the main conclusions of the studies

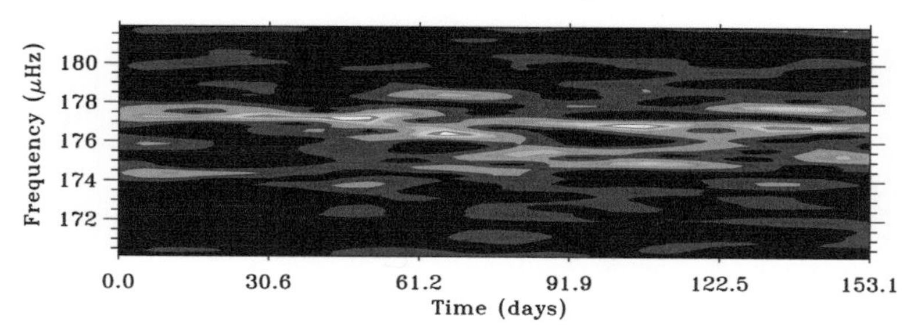

Figure 3. Time-frequency diagram, using a Morlet wavelet with a 20-day width, showing the solar-like oscillations of V1449 Aql after prewhitening the β Cep and SPB modes. (figure taken from Belkacem *et al.* 2009).

highlighted above. Fig.2 shows the observed photometric frequencies of 12 Lac (Handler *et al.* 2006). From the nearly equidistant frequencies given in red, one might conclude that 12 Lac exhibits a rotationally split $\ell = 1$ mode and use this information for asteroseismic modeling. However, spectroscopic mode identification revealed that these frequencies originate from an (ℓ,m)=(1,1) mode (left), a radial mode (middle) and an (2,1) mode (right) (Desmet *et al.* 2009). This example shows that *asteroseismic modeling without mode identification should be avoided*, and illustrates the need for multi-colour and spectroscopic information in seismic studies of B-type stars. Indeed, space missions, such as COROT and KEPLER only provide white-light photometry, which does not allow direct mode-identification. The optimal scientific exploitation of the seismic space data requires complementary ground-based data (e.g. Uytterhoeven 2009; Uytterhoeven *et al.* 2010a,b).

However, COROT has lead to important new results for β Cep stars as can be illustrated by the stories of V1449 Aql (HD 180642) and HD 46149. Before COROT, the B1.5II-III star V1449 Aql was known as a monoperiodic β Cep star with a radial mode (5.4871 d^{-1}) of very high amplitude showing strong non-linear effects (Aerts 2000). It was observed for 156 days by COROT and an intensive ground-based follow-up campaign in multi-colour photometry and high-resolution spectroscopy was organised. The new datasets show that V1449 Aql is clearly multiperiodic and exhibits both β Cep and SPB-like frequencies. Two of the newly found frequencies are identified with $\ell = 3$ modes (Briquet *et al.* 2009). The 11 independent and 22 low-order combination frequencies discovered in the COROT data were modeled in detail by comparing three different models of which *nonlinear resonant mode coupling* seems to be the statistically preferred one (Degroote *et al.* 2009b). Moreover, V1449 Aql is the first massive star for which hints for damped acoustic modes excited by turbulent convection associated with the Fe-opacity bump in the upper layers are found (Fig. 3; Belkacem *et al.* 2009). Evidence for *solar-like oscillations* for which the observed frequency range and spacings are compatible with theoretical predictions is also found for the late O-type star HD 46149 based on observations of the COROT satellite (Degroote *et al.* 2010b).

Compared to these successes, asteroseismic studies of *slowly pulsating B (SPB) stars* are still in their infancy. This is mainly caused by observational constraints: typical pulsation periods are of the order of a day, what makes it extremely difficult to monitor the periodic variations from the ground, and amplitudes are fairly small (below 20 mmag; 10 $km\,s^{-1}$). Moreover, the abovementioned techniques currently used for spectroscopic mode identification have been developed and extensively tested for p-mode pulsations. While these methods are often successful for p-mode pulsators, their application to g-modes seems not to be straightforward (De Cat *et al.* 2005; Zima *et al.* 2007).

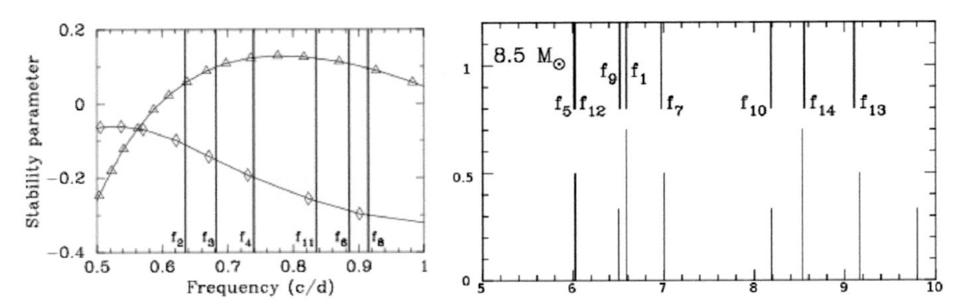

Figure 4. Comparison of theoretically predicted pulsation frequencies of an 8.5 M$_\odot$ model of γ Peg fitting f_1 ($\ell=0$) and f_5 ($\ell=1$) with the observed pulsation frequencies indicated with $f_1, ..., f_{14}$. **Left:** Stability of the theoretical $\ell=1$ (diamonds) and $\ell=2$ (triangles) modes. If the stability parameter is above (below) zero, the corresponding mode is excited (stable). **Right:** The model frequencies are plotted at the bottom with arbitrary amplitudes of 0.7 for $\ell=0$, 0.5 for $\ell=1$, and 0.33 for $\ell=2$. (figures taken from Belkacem *et al.* 2009).

Consequently, not many well-studied cases exist. Dedicated ground-based spectroscopic multi-site campaigns for main-sequence g-mode pulsators are currently ongoing in an attempt to fill this void (De Cat *et al.* 2009).

Since the discovery of SPB stars by Waelkens (1991), HD 160124 has remained the record holder for the maximum number of frequencies for 15 years (8 frequencies between 0.3 and 1.1 d^{-1}). Thanks to 37 days of continuous monitoring with the satellite MOST, up to 20 significant frequencies were detected in the resulting time-series of white-light photometry of the B5 II/III star HD 163830 (Aerts *et al.* 2006a). The two lowest frequencies were tentatively attributed to stellar rotation while the remaining 18 frequencies are consistent with low-ℓ, high-n nonradial g-modes of seismic models of an evolved 4.5 M$_\odot$ star. This study nicely illustrates that space-based data are able to increase the number of observed pulsation frequencies significantly (see also Poretti *et al.* 2009; Chapellier *et al.* 2010), but the lack of a formal mode identification prevents us to move on to asteroseismic modeling.

The theoretical instability strips of β Cep and SPB stars have a common part that should host stars in which low-n p/g-modes and high-n g-modes are simultaneously excited (Fig. 1). The existence of the so called *hybrid pulsators* was already illustrated above. They are theoretically predicted (e.g. Miglio *et al.* 2007) and are of particular asteroseismic interest because different internal regions can be probed at the same time.

γ Peg (HD 886) was one of the first suspected hybrid β Cep/SPB stars (Chapellier *et al.* 2006). It was recently subjected to a detailed study based on 10 weeks of radial velocity monitoring and 30 days of white-light photometry from the MOST satellite, which lead to the detection of eight β Cep-modes and six SPB-modes (Handler *et al.* 2009). The observed p-modes of γ Peg can be fitted extremely well with an 8.5 M$_\odot$ model (right panel of Fig. 4). Note that *all the expected p-modes are observed*! Only a minor frequency difference is found between f_{13} and its theoretical counterpart, i.e. an $\ell=1$,p$_2$ mode (= the p-mode for which the excitation problems were encountered in the analysis of ν Eri; see above). The low $v \sin i$ value of γ Peg ($\sim$3 $km\,s^{-1}$) is consistent with the assumption that f_5 and f_{12} are components of a rotationally splitted $\ell=1$ mode. Also the frequency spacings of the observed g-modes agree well with those of the theoretically predicted $\ell=1$ modes, but these dipole modes are all *stable* (left panel of Fig. 4). The systematic shift to lower frequencies might be indicative that all the observed g-modes are prograde dipole modes, which seems to be common amongst main-sequence g-mode pulsators (Wright et al., in preparation). As the authors say themselves, this study is “a demonstration of the value of combining space based photometry with ground-based spectroscopy”.

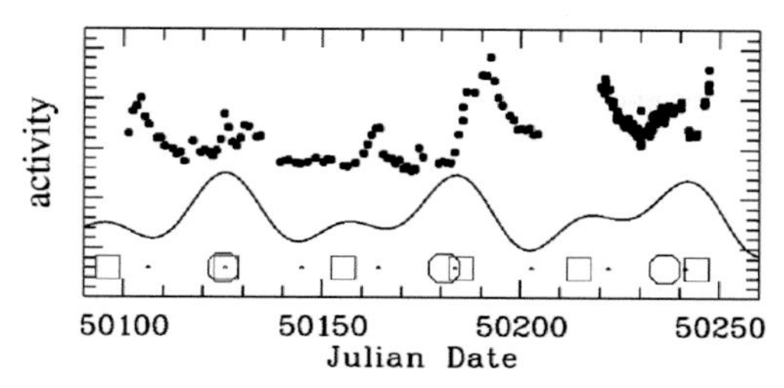

Figure 5. Circumstellar activity (points) of μ Cen in 1996 compared to the combined amplitude of the multiperidicity (solid line). The symbols mark times when two strong modes have no phase difference. (figure taken from Rivinius *et al.* 1998c).

The most exciting hybrid discovery was done for the young B3V star HD 50230, a star that has been observed for 137 days with the COROT satellite. The data revealed its hybrid nature as hundreds of g-modes with mmag amplitudes and tens of p-modes with μmag amplitudes were detected. More importantly, there is strong evidence for a *quasi uniform period spacing in the observed g-modes* (Degroote *et al.* 2010a). Such a period spacing is predicted for high-n g-modes in the asymptotic regime but was never observed before. The deviations from a uniform period spacing could be nicely fitted with a sinusoid with a period of 2450 sec and a decreasing amplitude with a maximum of 240 sec. These results imply that already 60% of the initial hydrogen is consumed and that there is a smooth gradient of chemical composition outside the convective core.

The examination of the light curves of 358 candidate B pulsators observed in the exoplanet field of the initial run of COROT (55 days) has lead to a wealth of objects with different types and flavours (Degroote *et al.* 2009a). A subset of these variable stars fills the gap between the theoretical instability strip of the SPB and the δ Scuti stars but their true nature can only be unraveled if (preferably high-resoltion) spectra would become available (Uytterhoeven 2009; cf. ongoing large program on the ESO/Flames multi-fiber spectrograph; P.I. C. Neiner). There is a similar *need for classification spectra* for the candidate B-type pulsators that are currently being observered with the space mission KEPLER (Uytterhoeven *et al.* 2010a,b; Lehmann *et al.* 2010).

3. Be stars

Be stars are non-supergiant B-type stars for which emission lines are observed at least once (see Porter & Rivinius 2003 for a review). The objects are flattened because of their very fast rotation. Be stars show long term variations that are linked to episodes of mass ejections (outbursts), leading to a decretion disk. It is still not clear what causes these outbursts. Be stars also show short term variations that could be a reflection of stellar pulsations (β Cep-type modes, SPB-type modes) and/or of rotational modulation (spots, clouds). So the key question for Be star research is: What causes the Be phenomenon?

The B2IV-Ve star μ Cen (HD 120324) has been observed intensively with ground-based instruments in the nineties (Rivinius *et al.* 1998a,b; Rivinius *et al.* 2001). A total of six frequencies were found that seemed to be phase coherent over five years. By comparing the circumstellar activity with the combined amplitude of the multiperiodic variations, there is clear evidence that the times of the combined maxima coincide with the times of the onset of the outbursts (Fig. 5; Rivinius *et al.* 1998c). This seems to suggest that *pulsation triggers the mass transfer from the star to the disk.*

The satellite MOST has provided data for several studies of massive emission-line stars. The 24 days of MOST observations of the O9.5Ve star ζ Oph (HD 149757) have revealed 16 significant frequencies, of which six are in common with the eight significant frequencies

Table 3. Overview of the SPBe stars that have been observed by the MOST satellite. In the colums, we give for each object the HD number, the variable star name (if available), the spectral type, the total time base of the MOST observations (in days), the rotation frequency (in d^{-1}), the number of detected frequencies, the number of frequency groups, and the references, respectively.

HD number	**Name**	**SpT**	MOST	$f_{\rm rot}$	**#freq**	**#groups**	**Ref**
HD 127756		B1/B2Vne	31 d	0.86 d^{-1}	30	3	[1]
HD 163868		B5Ve	37 d	1.38 d^{-1}	60	3	[2]
HD 217543		B3Vpe	26 d	1.73 d^{-1}	40	3	[1]
HD 58715	β CMi	B8Ve	41 d	2.76 d^{-1}	20	1	[3]

[1] Cameron *et al.* (2008), [2] Walker *et al.* (2005b), [3] Saio *et al.* (2007).

detected in the varations of ground-based high-resolution spectra (Walker *et al.* 2005a). The dominant mode with frequency 5.1806 d^{-1} is interpreted as the radial first overtone. The observed frequencies are compatible with low-n radial and nonradial p/g-modes modified by rotation, i.e. *β Cep-pulsations excited by the κ mechanism.* Therefore, we could call ζ Oph a *β Cephei Be (β Cepe) star.* Unfortunately, no unambiguous rotational frequency could be identified for ζ Oph.

For HD 163868, up to 60 significant frequencies are detected in the MOST lightcurve spanning 37 days (Walker *et al.* 2005b). The observed groups around 14 and 7 hours are compatible with prograde $|m|=1$ and $|m|=2$ even g-modes, respectively (ℓ-m is even (odd) for "even (odd) modes"). The tail of the observed group around 8 days coincides with a group of unstable retrograde $|m|=1$ odd r-modes (Rossby modes) predicted by representative seismic models. Hence, the observed frequencies can be interpreted as *nonradial g/r-modes distorted by fast rotation and excited by the κ mechanism*, making HD 163868 the prototype of the *slowly pulsating Be (SPBe) stars.* Three other SPBe stars have been detected thanks to the MOST satellite (Saio *et al.* 2007; Cameron *et al.* 2008). A short overview of their results is given in Table 3. Cameron *et al.* (2008) derived from the resulting rotation frequencies that all these SPBe stars are *rotating close to the critical value.* Their rotation frequency decreases with increasing temperature which is a reflection the increase in radius.

The satellite COROT has made an important contribution in Be star research. In Table 4, a short summary of the results for the observed stars is given (for details: see overview of Gutiérrez-Soto et al., these proceedings). HD 49330 showed a moderate outburst of 30 mmag during the COROT observations (Huat *et al.* 2009; Floquet *et al.* 2009). The outburst could be subdivided in four phases: quiescent, precursor, outburst and relaxation. The amplitudes of the observed g(p)-modes in(de)creased during the precursor and outburst phase, while they de(in)creased during the relaxation phase. Like for μ Cen, this indicates that the *pulsations might cause the outburst. Or* is it the *outburst* that *causes the excitation of pulsation modes* - a possible scenario suggested by Owocki (2005)? The COROT data of HD 50209 has provided *proof that g-mode pulsations occur in late Be stars* (Diago *et al.* 2009). Indeed, the rotational frequency $f_{\rm rot} = 0.679\ d^{-1}$ has been observed together with a rotationally split g-mode with $m=0$,-1,-2,-3 (0.10811 d^{-1}+ n$f_{\rm rot}$ with n=0,1,2,3). The observed variations of HD 175869 can be fully explained by *rotational modulation*, but nonradial pulsations can not be excluded (Gutiérrez-Soto *et al.* 2009). If this star exhibits a dipolar magnetic field, then the longitudinal component must be below 400 Gauss. For HD 181231, the frequency 0.624 d^{-1} is compatible with the *rotation frequency* (as derived from the fundamental parameters) *and* 0.695 d^{-1} is interpreted as a *nonradial g-mode* with ℓ~3 (Neiner *et al.* 2009). The observed 14-day modulation is due to beating between $f_{\rm rot}$ and the g-mode, while an additional modulation with a period of

Table 4. Overview of the Be stars that have been observed by the COROT satellite. In the columns, we give for each object the HD number, the spectral type, the total time base of the COROT observations (in days), the rotation frequency (in d^{-1}), the detection of g-modes, the detection of p-modes, the variability of the photometric amplitudes of the modes, the variability of the spectral lines, the detection of a magnetic field, and the references, respectively.

HD number	SpT	COROT	$f_{\rm rot}$	g	p	amplitude	LPVs	$\overline{B}$	Ref
HD 49330	B0.5IVe	136 d (LRa1)	0.87 d^{-1}	yes	yes	yes	yes		[1],[2]
HD 50209	B8IVe	136 d (LRa1)	0.679 d^{-1}	yes	no	yes	no		[3]
HD 175869	B8IIIe	27.3 d (SRc1)	0.639 d^{-1}	(no)	(no)	yes	no	no	[4]
HD 181231	B5IVe	157 d (LRc1)	0.624 d^{-1}	yes	no	yes	yes	no	[5]

[1] Huat *et al.* (2009), [2] Floquet *et al.* (2009), [3] Diago *et al.* (2009), [4] Gutiérrez-Soto *et al.* (2009), [5] Neiner *et al.* (2009).

116 days can be due either to beating or a zonal mode. This star can not host a dipolar magnetic field with a longitudinal component above 650 Gauss. It is clear that pulsations are present in early, mid and late Be stars, indicating that Be stars have an excellent asteroseismic potential. No LPVs were detected for HD 50209 and HD 175869 while the periods present in the variations of the photospheric lines of HD 49330 and HD 181231 could be linked with those of the photometry. For the latter stars, also variations in the emission lines were found which means that their disk is either being generated or is changing/moving. Note that complementary ground-based follow-up observations were needed for the detection of the rotation frequency, for the study of the variability of the spectral lines, and for the search for magnetic fields!

4. OB-Supergiants

OB-Supergiants are massive, evolved stars. The presence of non-radial pulsations in OB-supergiants was first suggested in the context of an explanation for the semi-regular light and radial variations observed in *α Cygni variables* (periods of 5 - 100 days; amplitudes of 0.01 - 0.1 mmag; e.g. van Genderen 2001).

Recently, more firm evidence for non-radial pulsations in a B supergiant was found, thanks to 37 days of observations with the satellite MOST. A total of 48 frequencies were detected for the B2Ib/II star HD 163899 (Saio *et al.* 2006). These frequencies range from 0 to 3 d^{-1}, embracing both g- and p-mode pulsations. Indeed, according to the seismic models for post-TAMS stars of Saio *et al.* (2006), both p- and g-modes can be unstable in OB-supergiants (Fig. 6). These g-modes have a great asteroseismic potential and can be excited because they are reflected at the convective zone associated to the hydrogen-burning shell. The location of HD 163899 in the H-R diagram is compatible with the picture of simultaneous p- and g-modes (Fig. 6). This star is considered as the first member of the *slowly pulsating B supergiant (SPBsg) stars*, which are distinct from the α Cygni variables. The theoretical instability strips for $\ell = 1$ and $\ell = 2$ g-modes in SPBsg stars based on the models of Saio *et al.* (2006) are given in Fig. 1.

More observational evidence for g-mode instabilities in OB-supergiants was found by Lefever *et al.* (2007). They carefully derived the atmospheric and wind parameters of a sample of 28 periodically variable and 12 comparison B supergiants. These variable B supergiants were classified as such by Waelkens *et al.* (1998) and show variations with periods ranging from 1 to 25 days. Several of them show evidence for multi-periodicity. Their resulting location in the H-R diagram is consistent with the suggestion that the observed variations are due to *opacity driven g-modes* (circles and squares in Fig. 1).

More evidence for g-mode pulsations is found from an unexpected angle. For OB-supergiants, the total line broadening of spectral lines can only be explained after

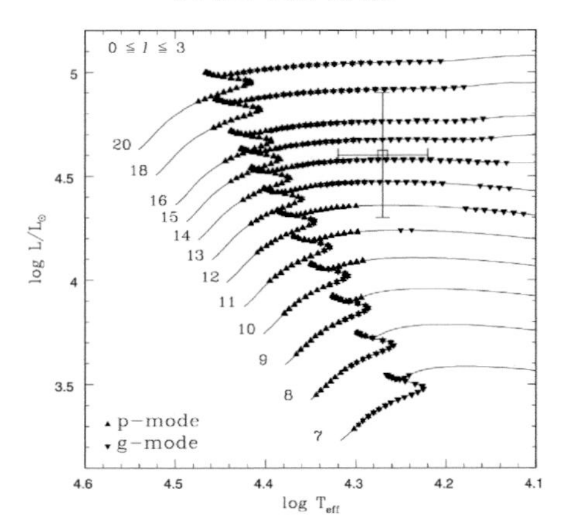

Figure 6. On the evolutionary tracks of massive stars, the seismic models with unstable p-modes are indicated with triangles and those with unstable g-modes with inverted triangles (only modes up to $\ell = 3$ are considered). The square with error bars indicates the estimated position of HD 163899 which is compatible with the picture of the simultaneous existence of p- and g-modes. (figure taken from Saio *et al.* 2006).

introduction of an important extra line-broadening called macroturbulence. It was believed to originate from large-scale turbulent motions but this interpretation has recently been questioned because it would imply highly supersonic velocity fields in many cases. Aerts *et al.* (2009) suggested the occurrence of hundreds of low-amplitude non-radial g-modes as an alternative physical explanation for macroturbulence. Simón-Díaz *et al.* (2010) examined spectroscopic observations for a sample of 12 O9.5-B2 supergiant stars. They found firm observational evidence for a strong correlation between the average size of the macroturbulence and the peak-to-peak amplitudes in the first (centroid velocity) and third (skewness) normalised velocity moment of the photospheric LPVs. These observed observations agree extremely well with those predicted by the simulations of Aerts *et al.* (2009), making g-mode pulsations a valid explanation for macrotubulence. A dedicated ground-based spectroscopic follow-up campaign for selected OB-supergiants is currently ongoing (Simón-Díaz *et al.* 2009), promising new exciting prospects for the seismic analysis of the not well-studied class of OB-supergiants.

The luminous blue variable supergiant HD 50064 has been observed for 137 days by the satellite COROT. Afterwards, 14 ground-based high-resolution échelle spectra were gathered at three epochs in a time span of 169 days by Aerts *et al.* (2010). They detected a period of 37 days in the time series of COROT photometry and discovered a sudden amplitude increase with a factor of 1.6 that occurred once. The photometric period was also observed in the spectroscopic variations. The mass loss rate, as derived from the Balmer lines, seems to undergo variations on a similar timescale. The observed period is tentatively interpreted as a *radial strange mode oscillation* that is triggering periodic mass-loss episodes in which a circumstellar envelope is being generated.

5. Conclusions

A better knowledge of the internal structure and the evolution of massive stars is of great importance because they are the progenitors of type II, Ib or Ic supernovae. These stars enrich their immediate environment with heavy elements at the end of their lives. We have discussed the results of some representative examples to prove that the β Cephei, slowly pulsating B, Be and OB-supergiant stars all have an excellent asteroseismic potential. Recently, progress has already been made thanks to the space missions MOST and

COROT, and even more exciting results are being expected in the near future from the KEPLER mission too. However, as we have shown, complementary ground-based support has been essential to get these results and will remain mandatory to make even more progress in the future!

References

Aerts, C. 2000, *A&A*, 361, 245

Aerts, C., De Cat, P., Handler, G., & Heiter, U. *et al.* 2004a, *MNRAS*, 347, 463

Aerts, C., De Cat, P., Kuschnig, R., & Matthews, J. M. *et al.* 2006a, *ApJ* (Letters), 642, L165

Aerts, C., Lefever, K., Baglin, A., & Degroote, P. *et al.* 2010, *A&A*, 513, L11

Aerts, C., Marchenko, S. V., Matthews, J. M., & Kuschnig, R. *et al.* 2006b, *ApJ*, 642, 470

Aerts, C., Puls, J., Godart, M., & Dupret, M.-A. 2009, *A&A*, 508, 409

Aerts, C., Thoul, A., Daszyńska, J., & Scuflaire, R. *et al.* 2003, *Science*, 300, 1926

Aerts, C., Waelkens, C., Daszyńska-Daszkiewicz, J., & Dupret, M.-A. *et al.* 2004b, *A&A*, 415, 241

Ausseloos, M., Scuflaire, R., Thoul, A., & Aerts, C. 2004, *MNRAS*, 355, 352

Belkacem, K., Samadi, R., Goupil, M.-J., & Lefèvre, L. *et al.* 2009, *Science*, 324, 1540

Briquet, M. & Aerts, C. 2003, *A&A*, 398, 687

Briquet, M., Lefever, K., Uytterhoeven, K., & Aerts, C. 2005, *MNRAS*, 362, 619

Briquet, M., Morel, T., Thoul, A., & Scuflaire, R. *et al.* 2007, *MNRAS*, 381, 1482

Briquet, M., Uytterhoeven, K., Morel, T., & Aerts, C. *et al.* 2009, *A&A*, 506, 269

Cameron, C., Saio, H., Kuschnig, R., & Walker, G. A. H. *et al.* 2008, *ApJ*, 685, 489

Chapellier, E., Le Contel, D., Le Contel, J. M., & Mathias, P. *et al.* 2006, *A&A*, 448, 697

Chapellier, E., Rodríguez, E., Auvergne, M., Uytterhoeven K., *et al.* 2010, *A&A*, 525, A23

Davis, J., Mendez, A., Seneta, E. B., & Tango, W. J. *et al.* 2005, *MNRAS*, 356, 1362

De Cat, P. 2007, *Communications in Asteroseismology* 150, 167

De Cat, P., Briquet, M., Aerts, C., & Goossens, K. *et al.* 2007, *A&A*, 463, 243

De Cat, P., Briquet, M., Daszyńska-Daszkiewicz, J., & Dupret, M. A. *et al.* 2005, *A&A*, 432, 1013

De Cat, P., Wright, D. J., Pollard, K. R., & Maisonneuve, F. *et al.* 2009, in: J. A. Guzik & P. A. Bradley (eds.), *American Institute of Physics Conference Series*, AIP-CP 1170, p. 480

Degroote, P., Aerts, C., Baglin, A., & Miglio, A. *et al.* 2010a, *Nature*, 464, 259

Degroote, P., Aerts, C., Ollivier, M., & Miglio, A. *et al.* 2009a, *A&A*, 506, 471

Degroote, P., Briquet, M., Auvergne, M., & Simón-Díaz, S. *et al.* 2010b, *A&A*, 519, A38

Degroote, P., Briquet, M., Catala, C., & Uytterhoeven, K. *et al.* 2009b, *A&A*, 506, 111

De Ridder, J., Telting, J. H., Balona, L. A., & Handler, G. *et al.* 2004, *MNRAS*, 351, 324

Desmet, M., Briquet, M., Thoul, A., & Zima, W. *et al.* 2009, *MNRAS*, 396, 1460

Diago, P. D., Gutiérrez-Soto, J., Auvergne, M., & Fabregat, J. *et al.* 2009, *A&A*, 506, 125

Dupret, M.-A., De Ridder, J., De Cat, P., & Aerts, C. *et al.* 2003, *A&A*, 398, 677

Dupret, M.-A., Thoul, A., Scuflaire, R., & Daszyńska-Daszkiewicz, J. *et al.* 2004, *A&A*, 415, 251

Dziembowski, W. A., Moskalik, P., & Pamyatnykh, A. A. 1993, *MNRAS*, 265, 588

Floquet, M., Hubert, A.-M., Huat, A.-L., & Frémat, Y. *et al.* 2009, *A&A*, 506, 103

Gautschy, A. & Saio, H. 1993, *MNRAS*, 262, 213

Gilliland, R. L., Jenkins, J. M., Borucki, W. J., & Bryson, S. T. *et al.* 2010, *ApJ* (Letters), 713, L160

Gutiérrez-Soto, J., Floquet, M., Samadi, R., & Neiner, C. *et al.* 2009, *A&A*, 506, 133

Handler, G., Jerzykiewicz, M., Rodríguez, E., & Uytterhoeven, K. *et al.* 2006, *MNRAS*, 365, 327

Handler, G., Matthews, J. M., Eaton, J. A., & Daszyńska-Daszkiewicz, J. *et al.* 2009, *ApJ* (Letters), 698, L56

Handler, G., Shobbrook, R. R., Jerzykiewicz, M., & Krisciunas, K. *et al.* 2004, *MNRAS*, 347, 454
Huat, A.-L., Hubert, A.-M., Baudin, F., & Floquet, M. *et al.* 2009, *A&A*, 506, 95
Hubrig, S., Briquet, M., De Cat, P., & Schöller, M. *et al.* 2009, *AN*, 330, 317
Hubrig, S., Briquet, M., Schöller, M., & De Cat, P. *et al.* 2006, *MNRAS*, 369, L61
Hubrig, S., Briquet, M., Schöller, M., & De Cat, P. *et al.* 2007, in: A. T. Okazaki, S. P. Owocki, & S. Stefl (eds.), *Active OB-Stars: Laboratories for Stellar and Circumstellar Physics*, ASP-CS 361, p. 434
Jenkins, J. M., Caldwell, D. A., Chandrasekaran, H., & Twicken, J. D. *et al.* 2010, *ApJ* (Letters), 713, L120
Jerzykiewicz, M., Handler, G., Shobbrook, R. R., & Pigulski, A. *et al.* 2005, *MNRAS*, 360, 619
Lehmann, H, Tkachenko, A., Semaan, T., Gutiérrez-Soto, J. *et al.* 2010, *A&A*, 526, A124
Lefever, K., Puls, J., & Aerts, C. 2007, *A&A*, 463, 1093
Mazumdar, A., Briquet, M., Desmet, M., & Aerts, C. 2006, *A&A*, 459, 589
Michel, E., Baglin, A., Auvergne, M., & Catala, C. *et al.* 2008, *Science*, 322, 558
Miglio, A., Montalbán, J., & Dupret, M.-A. 2007, *MNRAS*, 375, L21
Neiner, C., Geers, V. C., Henrichs, H. F., & Floquet, M. *et al.* 2003a, *A&A*, 406, 1019
Neiner, C., Gutiérrez-Soto, J., Baudin, F., & de Batz, B. *et al.* 2009, *A&A*, 506, 143
Neiner, C., Henrichs, H. F., Floquet, M., & Frémat, Y. *et al.* 2003c, *A&A*, 411, 565
Neiner, C., Hubert, A.-M., Frémat, Y., & Floquet, M. *et al.* 2003b, *A&A*, 409, 275
Owocki, S. 2005, in: R. Ignace & K. G. Gayley (eds.), *The Nature and Evolution of Disks Around Hot Stars*, ASP-CS 337, p. 101
Pamyatnykh, A. A. 1999, *AcA*, 49, 119
Pamyatnykh, A. A., Handler, G. & Dziembowski, W. A. 2004, *MNRAS*, 350, 1022
Poretti, E., Michel, E., Garrido, R., & Lefèvre, L. *et al.* 2009, *A&A*, 506, 85
Porter, J. M. & Rivinius, T. 2003, *PASP* 115, 1153
Rivinius, T., Baade, D., Stefl, S., & Stahl, O. *et al.* 1998a, *A&A*, 333, 125
Rivinius, T., Baade, D., Stefl, S., & Stahl, O. *et al.* 1998b, *A&A*, 336, 177
Rivinius, T., Baade, D., Stefl, S., & *et al.* 1998c, in: L. Kaper & A. W. Fullerton (eds.), *Cyclical Variability in Stellar Winds*, p. 207
Rivinius, T., Baade, D., Štefl, S., & Townsend, R. H. D. *et al.* 2001, *A&A*, 369, 1058
Saio, H., Cameron, C., Kuschnig, R., & Walker, G. A. H. *et al.* 2007, *ApJ*, 654, 544
Saio, H., Kuschnig, R., Gautschy, A., & Cameron, C. *et al.* 2006, *ApJ*, 650, 1111
Simón-Díaz, S., Herrero, A., Uytterhoeven, K., & Castro, N. *et al.* 2010, *ApJ* (Letters), 720, L174
Simón-Díaz, S., Uytterhoeven, K., Herrero, A., & Castro, N. 2009, in: J.A. Guzik & P.A. Bradley (eds.), *Stellar pulsation: Challenges for theory and observation*, AIP Conf. Ser., 1170, p. 397
Stankov, A. & Handler, G. 2005, *ApJS*, 158, 193
Tango, W. J., Davis, J., Ireland, M. J., & Aerts, C. *et al.* 2006, *MNRAS*, 370, 884
Telting, J. H. & Schrijvers, C. 1997, *A&A*, 317, 723
Uytterhoeven, K. 2009, *Communications in Asteroseismology* 158, 156
Uytterhoeven, K., Briquet M., Bruntt H., & De Cat, P. *et al.* 2010a, *AN*, in press, arXiv1003.6093
Uytterhoeven, K., Poretti, E., Rodríguez, E., & De Cat, P. *et al.* 2007, *A&A*, 470, 1051
Uytterhoeven, K., Szabo, R., Southworth, J., & Randall, S. *et al.* 2010b, *AN*, in press, arXiv1003.6089
van Genderen, A. M. 2001, *A&A*, 366, 508
Waelkens, C. 1991, *A&A*, 246, 453
Waelkens, C., Aerts, C., Kestens, E., & Grenon, M. *et al.* 1998, *A&A*, 330, 215
Walker, G. A. H., Kuschnig, R., Matthews, J. M., & Reegen, P. *et al.* 2005a, *ApJ* (Letters), 623, L145
Walker, G. A. H., Kuschnig, R., Matthews, J. M., & Cameron, C. *et al.* 2005b, *ApJ* (Letters), 635, L77
Zima, W. 2006, *A&A*, 455, 227
Zima, W., De Cat, P. & Aerts, C. 2007, *Communications in Asteroseismology* 150, 189

Active OB stars: structure, evolution, mass loss, and critical limits
Proceedings IAU Symposium No. 272, 2010
C. Neiner, G. Wade, G. Meynet & G. Peters, eds.

doi:10.1017/S1743921311011082

Pulsations in Wolf-Rayet stars: observations with MOST

André-Nicolas Chené[1,2] and Anthony F. J. Moffat[3,4]

[1]Departamento de Física, Universidad de Concepción, Casilla 160-C, Concepción, Chile
email: achene@astro-udec.cl

[2]National Research Council of Canada, Herzberg Institute of Astrophysics,
5071, West Saanich Road,Victoria (BC), V9E 2E7

[3]Département de Physique, Université de Montréal,
C. P. 6128, succ. centre-ville, Montréal (Qc) H3C 3J7

[4]Centre de Recherche en Astrophysique du Québec, Canada
email: moffat@astro.umontreal.ca

Abstract. Photometry of Wolf-Rayet (WR) stars obtained with the first Canadian space telescope *MOST* (Microvariability and Oscillations of STars) has revealed multimode oscillations mainly in continuum light that suggest stellar pulsations could be a significant contributing factor to the mass-loss rates. Since the first clear detection of a pulsation period of P = 9.8h in WR123, two other stars have also shown periods of a few days, which must be related to stellar pulsations.

Keywords. stars: Wolf-Rayet, stars: oscillations (including pulsations), stars: winds, outflows

1. Introduction

The wind momentum (dM/dtv) of the Wolf-Rayet (WR) stars is a factor 10 times higher than the radiative momentum outflow rate (L/c); radiation pressure doesnt seem to be sufficient to initiate the strong winds. Hence, another driving mechanism should be present near, or at the surface of the star. Theoretical work suggest that strange-mode pulsations (SMPs) are present in the envelope of hot and luminous stars with a large luminosity-to-mass ratio, where the thermal timescale is short compared to the dynamical timescale, and where radiation pressure dominates (Glatzel *et al.* 1993). Hence, the most violent SMPs are expected in classical WR stars, where SMPs manifest themselves in cyclic photometric variability with periods ranging from minutes to hours (Glatzel *et al.* 1999). However, these variability are expected to be epoch-dependent, with small amplitude, therefore, very difficult to detect from the ground.

The space telescope *MOST* (Microvariability and Oscillations of STars, Matthews *et al.* 1999, Walker *et al.* 2003) contains a 15-cm Rumak-Maksutov telescope imaging onto a CCD detector via a custom optical broadband (350-750 nm) filter. From its polar Sun-synchronous orbit of altitude 820 km and period 101 min, *MOST* has a continuous viewing zone (CVZ) about 54° wide within which it can monitor target fields for up to two months without interruption. Targets brighter than V∼6 mag are observed in Fabry imaging mode, while fainter targets (like most of the WR stars in the CVZ) are observed in direct imaging mode, similar to standard CCD photometry with a groundbased instrument. The photometry is non-differential, but given the orbit, thermal and design characteristics of *MOST*, experience has shown that it is a very photometrically stable platform even over long timescales (with repeatability of the mean instrumental flux from a non-variable target with V∼11-12 to within about 1 mmag over a month).

2. Three First Stars Observed With *MOST*

Since its launch, *MOST* has observed one WR star per year. The first one is the famous WR 123 of the WN8 type (Lefèvre *et al.* 2005). Its light curve has a total amplitude of 10 mmag and shows variations on timescales of ∼hours to days. It displays dips which seem to occur every 14–15 days, but these features are not always identical and they were not sufficiently well monitored during the 38 days of observation. Also, the periodogram shows a significant peak at ~ 2.45 c d^{-1} (i.e. P∼9.8 hours). Two quite different scenarios have been proposed to explain these observations. Townsend & MacDonald (2006) find that a deep, hot Fe opacity bump can lead to g-mode pulsations, while Dorfi et al. (2006) find that a cooler Fe opacity bump can produce SMPs. The latter is in contrast with Glatzel *et al.* (1993, 1999) work, where SMP periods of order 10 minutes were predicted to prevail in WR stars. The longer timescales and period found in WR 123 are most likely a result of the puffed-up nature of WN8 stars ($R_* \sim 15 R_\odot$, according to Crowther *et al.* 1995), compared to their much more compact early-type WN brethren. In these two theoretical studies, the former group assumed a traditionally small stellar radius ($R_* \sim 2 R_\odot$), while the latter group assumed significant hydrogen (X_H=0.35, compared to the observed value of 0.00. Both of these assumptions contradict what we know about this star, thus casting some doubt on their applicability.

On the second year was observed WR 103, a dust-making WR star of the Carbon sequence. No clear period could be found although the light curve shows clear variations with an amplitude of ∼5 mmag (Marchenko *et al.* 2006). Simultaneous spectroscopy has shown that the changes are coming from the continuum light, and not from the expanding wind, where the spectral emission are formed.

Finally, WR 111 (WC5) was observed in 2006. Moffat *et al.* (2008) have found no coherent Fourier components above the 50 part per million level over the whole interval for frequencies f<10 c d^{-1}. Simultaneous spectroscopic observations reveal a normal level of stochastic clumps propagating in the wind, which has no effect on period detection.

3. WR 124

WR 124 = 209BAC = QR Sge = Merrill's star is a moderately bright ($v = 11.08$, $b-v = 1.07$), reddened ($A_v = 4.43$ mag), northern [RA(2000) = 19:11:30.88, DEC(2000) = +16:51:38.2] Galactic (l = 50.198°, b = +3.310°) WR star in the cool-nitrogen sequence with type WN8h (Smith *et al.* 1996). Located at a distance of 3.35 kpc (Marchenko, Moffat & Crowther, in prep.), it is a runaway star (e.g. Moffat *et al.* 1998).

Fig. 1 shows the entire *MOST* light curve of WR 124 (*upper*) and a control star (*lower*), binned for each *MOST* orbit. Although significant variations are seen on timescales longer than ∼a day, no such variations are seen on shorter timescales and binning allows one to increase the photometric point-to-point *rms* precision from ∼3 mmag before binning to ∼0.5 mmag after binning. Fig. 1 shows a time-frequency plot with straddled 8-day (double the best period) sampling. There is a family of three plausible frequencies for WR 124, but first examination of the data shows that the most probable frequency is unique, and is located at ∼0.19 c d^{-1} (P∼5.3 d). Simultaneous spectral monitoring gives a period of P∼4.45 d. The spectral line-profile variability of WR 124 is very similar to the one observed for WR 123, although with a smaller amplitude. Since both these stars are of similar spectral types, the same type of pulsation may be present in each. Also,

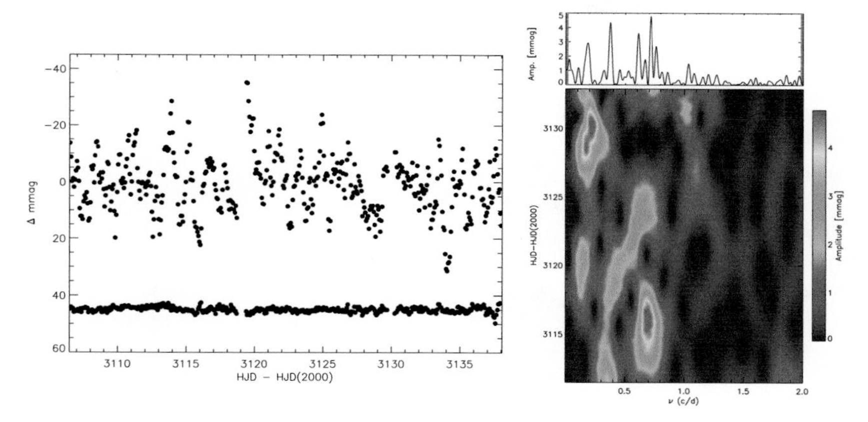

Figure 1. *Right*: *MOST* light curve of WR 124 in 101-minute MOST-orbit bins. *Left*: Fourier amplitude spectrum of the binned *MOST* light curve and Time-frequency Fourier plot with 8-day running windows in time.

from previous ground-based observations, all the ~10 relatively well-observed WN8 stars show a similar high degree of intrinsic variability (Antokhin *et al.* 1995, Marchenko *et al.* 1998), implying that all cool WR stars may show similar behavior.

4. WR 110

WR 110 is a moderately bright (v = 10.30, $b - v$ = 0.75), reddened (A_v = 3.83 mag), southern [RA(2000) = 18:07:56.96, DEC(2000) = -19:23:56.8] Galactic (l = 10.80^o, b = +0.39^o) WR star in the mid-nitrogen sequence with type WN5-6b (Smith *et al.* 1996). Although to date it lacks any obvious binary signature, a comprehensive radial velocity search has not yet been carried out. However, WR 110 does have relatively high X-ray flux, with a significant hard component above 3 keV, compared to its lower-energy 0.5 keV emission, as seen in most WN stars (Skinner *et al.* 2002). The lower-energy component likely arises due to small-scale turbulent shocks in the wind, with velocity dispersion ~100 km s^{-1}, as seen in virtually all WR-star winds (Lépine & Moffat 1999) and other luminous hot stars (e.g. Eversberg *et al.* 1998, Lépine & Moffat 2008. The former hard component implies velocities of at least 10^3 km s^{-1}, which are more difficult to explain, without invoking an additional mechanism to provide such high speeds (e.g. accretion onto a low-mass companion or colliding winds with a massive companion). However, WR 110 emits normal thermal radio emission from its wind (Skinner *et al.* 2002), which lends little support for its binary nature.

Recently St-Louis *et al.* (2009) have examined a northern sample of 25 WR stars for the presence of Corotating-Interaction Regions (CIRs) via large-scale variations exhibited on broad, strong emission lines. Some 20% of these stars show clear signatures of such effects, with WR 110 situated at the limit between those stars that show CIRs and those that do not. In addition, some (and possibly all, given the limited number of spectra) stars in the sample unsurprisingly show small-scale wind-clump variations. As we show in this investigation, it is possible and even likely, that WR 110's hard X-rays arise in the shocks produced by its CIRs rotating at velocities of ~ 10^3 km s^{-1} relative to the ambient wind.

Fig. 2 shows the entire *MOST* light curve of WR 110 (*upper*) and a control star (*lower*), binned for each *MOST* orbit. Fig 2 also shows the Fourier spectrum for the *MOST* photometry. Here we see that the most significant peak occurs at a frequency of ν_0 = 0.245 cd^{-1} (P = 4.08 d), with harmonics of this frequency at, or close to, $n \times \nu_0$, with

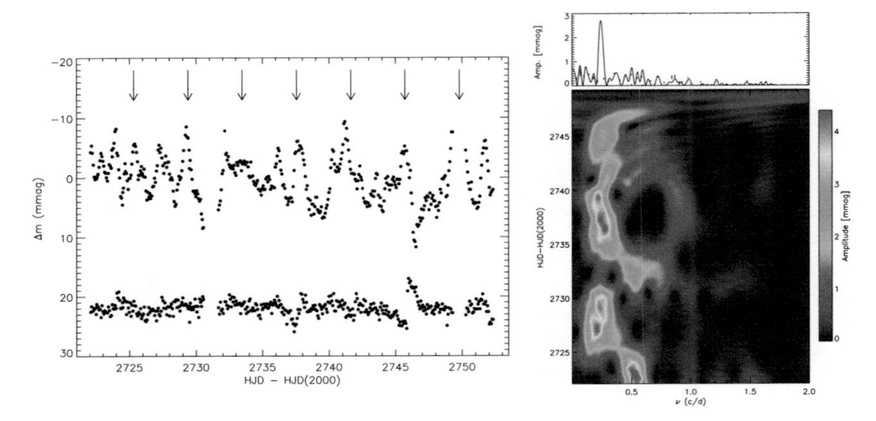

Figure 2. *Right*: *MOST* light curve of WR 110 in 101-minute MOST-orbit bins. The vertical lines indicate intervals of period 4.08 d starting at t_0 = HJD 2454270.31. *Left-Top*: Fourier amplitude spectrum of the binned *MOST* light curve. The highest peak refers to the adopted fundamental frequency corresponding to a period of P = 4.08 d. Harmonics are indicated at frequencies at multiples of 2 through 6 times fundamental. *Left-Bottom* : Time-frequency Fourier plot with 8-day running windows in time.

$n = 2$, 3, 4, 5 and 6. A few other peaks do occur at lower relative amplitudes, notably at periods longer than ~a day, but with no obvious order or interrelationship. No significant peaks occur at longer than a day. We assume that all this implies one dominating (non-sinusoidal) periodicity with P = 4.08d, plus mainly stochastic noise (probably related to wind clump activity) on timescales longer than a day. Examination of the light curve in Fig. 2 shows relatively sharp cusps occurring on this time scale (i.e. P = 4.08 d, starting at t_0 = HJD 2 454 270.27) in all seven cases but one (the third case), although even then there is a rise in brightness, though not cusp-like. The cusps have typical widths at the base of ~ a day. Such cusps are clearly non sinusoidal, requiring a series of successive harmonics to reproduce them, as observed. Fig. 2 also shows a time-frequency plot with straddled 8-day (double the best period) sampling. This plot reveals that the 4.08-day period stands out at all times except just before the middle of the run and at the ends (where edge effects come into play). The former is compatible with the apparently suppressed third cusp, as noted above.

The most likely scenario to explain the low-amplitude (~1 %) 4.08 d periodicity and cusp-like behavior in the *MOST* light curve, along with mostly unrelated spectral variations, appears to be large-scale, emitting over-density structure rotating with the wind. An attractive scenario for this are the CIRs (or their source at the base of the wind) proposed for O star winds by Cranmer & Owocki (1996). We now apply this idea to WR 110 in the following way.

First, we assume that the CIRs are continuum-emitting thermal sources, heated by the associated shock action created by a hot spot on the underlying rotating star with the ambient wind. Most of their emission must arise close to the star, where the ambient wind density is highest, as inspired by the simulations of Cranmer & Owocki (1996). Alternatively, the emission could arise in a hot spot at or close to the stellar surface, that produces the CIR. To simplify our calculations, we assume a point source located at some radius $R_s = \gamma R_*$ from the center of the star, with $\gamma > 1$ and R_* being the stellar core radius. For simplicity (and following Cranmer & Owocki 1996) we assume the CIR to initiate at the stellar rotation equator.

In this scenario, the changes in the light-curve are caused by the assumed CIR-associated point source attenuated by the wind and seen at different angles with the line of sight as it rotates on the near side of the WR star. In this study, we use a simple WR wind model derived by Lamontagne *et al.* (1996) in the context of wind eclipses for WR + O systems. Hence, the change in magnitudes is defined as :

$$\Delta m = constant - 2.5 \log_{10} \left(I_{WR} + I_s \mathrm{e}^{-\tau} \right), \tag{4.1}$$

with arbitrary constant and where I_{WR} is the intensity of the WR star and I_s, the intensity of the hot spot. Here the total opacity τ between the source and the observer, passing through the WR wind is :

$$\tau = k \int_{z_0}^{\infty} d\left(z/R_s\right) \left[r^2 \left(1 - R_*/r\right)^{\beta} \right]^{-1} \tag{4.2}$$

where $z_0 = -(R_s \sin i) \cos 2\pi\phi$, with R_s, the radial position of the hot spot (or the region of light emission) and i, the inclination of the rotation axis relative to the observer. We allow for a WR wind with a β law, in which the actual β value will be fitted. Since the CIR source is much closer to the inner WR wind than the O star is in the (non photo-spheric eclipsing) binary case, we expect that $\beta = 0$ will not be a good choice, as it was in the case of WR + O binaries. Following Skinner *et al.* (2002), a radio-based thermal mass-loss rate for the WR star of $\dot{M} = 1.6\ 10^{-5} M_{\odot}\, yr^{-1}$, after dividing by a factor three to allow for wind clumping; taking wind terminal speed of v_{∞} =2100 km s^{-1}; and assuming a WR core radius $R_* = 4R_{\odot}$, we find an opacity constant as in Lamontagne *et al.* (1996) $k = \alpha\sigma_e/[4\pi m_p v_{\infty} R_s] = 0.0028/\gamma$. In our case, we are primarily interested in the phases centered at phase 0.50 (i.e. when the source is between the star and the observer) since, normally, the CIR point source will not be visible as it rotates on the back side of the WR star.

We calculated the Equation 4.1 numerically, centered at phase 0.5 for half a rotation period, as a function of the parameters β, I_s/I_{WR}, i, and γ. We consider two cases: (a) an optically thin point source, whose emission is isotropic in direction, and (b) and an optically thick source whose emission is maximum when seen perpendicularly to the stellar surface, dropping to zero when seen parallel to the surface (i.e. like pancakes on or near the stellar surface). In case (a) we have the light curve excess from the hot spot and in case (b), we have to multiply the net spot intensity (I_s) from Equation 4.1 by the projection factor $(\sin i) cos[2\pi(\phi - 0.5)]$.

Taking the second cusp in Fig. 2 as the cleanest, most representative form of the hotspot light curve, we fit the above parameters to match the data best, for i = 90° (to give the maximum effect). In case (b), the values of γ and β, which are always strongly coupled (since an increase in γ making the hotspot source further away from the star, has to be compensated for by an increase in absorbing material at larger distances, i.e. an increased β value, in order to reproduce best the observed light-curve cusps), tend to be slightly smaller than in case (a), since the projection factor in case (b) already provides a start (but only a start) to the correct form for the cuspy light curve, that requires less extended wind opacity to bring about the observed cuspy shape. Indeed, it is the varying opacity through the WR wind as the WR star rotates, which leads to the cuspy shape; the closer the source is to the stellar surface (i.e. γ closer to unity), the smaller the β value (and less extended the wind) necessary to give the cuspy shape. All the best solutions for the thin and the thick cases are plotted in Fig. 3.

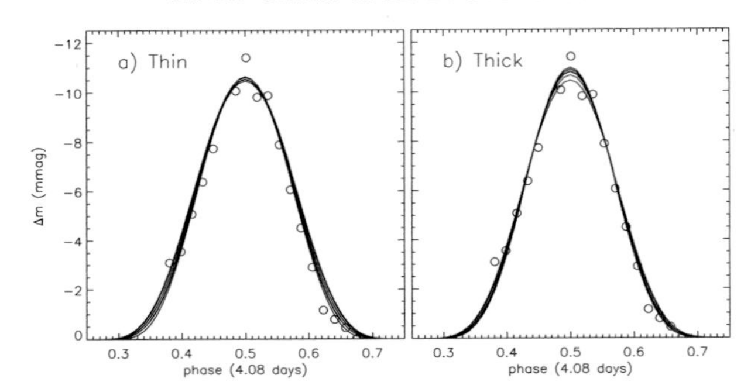

Figure 3. The data points refer to the second cusp indicated by a vertical line in Fig. 2. The curves indicate the best fits for (a) an optically thin point source and (b) an optically thick source (see text).

References

Antokhin, I., Bertrand, J.-F., Lamontagne, R., & Moffat, A. F. J. *et al.* 1995, *AJ*, 109, 817
Cranmer, S. R. & Owocki, S. P. 1996, *ApJ*, 462, 469
Crowther, P. A., Hillier, D. J., & Smith, L. J. 1995, *A&A*, 293, 403
Dorfi, E. A., Gautschy, A., & Saio, H. 2006, *A&A*, 453, L35
Eversberg, T., Lepine, S., & Moffat, A. F. J. 1998, *ApJ*, 494, 799
Glatzel, W., Kiriakidis, M., Chernigovskij, S., & Fricke, K. J. 1999, *MNRAS*, 303, 116
Glatzel, W., Kiriakidis, M., & Fricke, K. J. 1993, *MNRAS*, 262, L7
Lamontagne, R., Moffat, A. F. J., Drissen, L., & Robert, C. *et al.* 1996, *AJ*, 112, 2227
Lefèvre, L., Marchenko, S. V., Moffat, A. F. J., Chené, A. N. *et al.* 2005, *ApJ* (Letters), 634L, 109
Lépine, S. & Moffat, A. F. J. 1999, *ApJ*, 514, 909
Lépine, S. & Moffat, A. F. J. 2008, *AJ*, 136, 548
Matthews, J., Kuschnig, R., Walker, G., & Johnson, R. *et al.* 1999, *JRASC*, 93, 183
Marchenko, S. V., Moffat, A. F. J., Eversberg, T., & Morel, T. *et al.* 1998, *MNRAS*, 294, 642
Marchenko, S., Lefevre, L., Moffat, A. F., & Zhilyaev, B. E. *et al.* 2006, in: *Bulletin of the American Astronomical Society*, 38, p. 121
Moffat, A. F. J., Marchenko, S. V., Bartzakos, P., & Niemela, V. S. *et al.* 1998, *ApJ*, 497, 896
Moffat, A. F. J., Marchenko, S. V., Zhilyaev, B. E., & Rowe, J. F. *et al.* 2008, *ApJ* (Letters), 679L, 45
Skinner, S. L., Zhekov, S. A., Güdel, M., & Schmutz, W. 2002, *ApJ*, 572, 477
Smith, L. F., Shara, M. M., & Moffat, A. F. J. 1996, *MNRAS*, 281, 163
St-Louis, N., Chené, A.-N., & Schnurr, O. & Nicol, M.-H. 2009, *ApJ*, 698, 1951
Townsend, R. H. D. & MacDonald, J. 2006, *MNRAS*, 368, L57
Walker, G., Matthews, J., Kuschnig, R., & Johnson, R. *et al.* 2003, *PASP*, 115, 1023

Discussion

PULSE: You have mentioned that the two theoretical studies on WR123 have assumed wether stellar radius too small (Tonwsend & MacDonald 2006), or an hydrogen abundance too high (Dorfi *et al.* 2006). Since Rich is in the auditory, he might be able to comment on this?

TOWNSEND: The stellar radius depends quite critically on the presumed hydrogen abundance. We don't see any evidence for hydrogen in WR 123. Thus, I have assumed an X_H=0 for the star, and my stellar evolution models then predict a small hydrostatic radius for the star. With this small radius, the observed MOST variability must be due to g modes rather than strange modes.

Active OB stars: structure, evolution, mass loss, and critical limits
Proceedings IAU Symposium No. 272, 2010
C. Neiner, G. Wade, G. Meynet & G. Peters, eds.

doi:10.1017/S1743921311011094

Short-term variations in Be stars observed by the CoRoT and Kepler space missions

Juan Gutiérrez-Soto[1,2], Coralie Neiner[2], Juan Fabregat[3], Antonino Francesco Lanza[4], Thierry Semaan[5], Monica Rainer[6] and Ennio Poretti[6]

[1]Instituto de Astrofísica de Andalucía (CSIC), Glorieta de la Astronomía s/n 18008, Granada, Spain; email: jgs@iaa.es

[2]LESIA, Observatoire de Paris, CNRS, Université Paris Diderot, 5 place Jules Janssen, 92190 Meudon, France

[3]Observatorio Astronómico de la Universidad de Valencia, Calle Catedrático Agustín Escardino 7, 46980 Paterna, Valencia, Spain

[4]INAF - Osservatorio Astrofisico di Catania, via S. Sofia, 78, 95123 Catania, Italy

[5]GEPI, Observatoire de Paris, CNRS, Université Paris Diderot, 5 place Jules Janssen, 92190 Meudon, France

[6]INAF - Osservatorio Astronomico di Brera, via E. Bianchi 46, 23807 Merate (LC), Italy

Abstract. The COROT and KEPLER space missions are collecting very high-precision long-duration photometric data of many Be stars, allowing us to better understand the origin of their short-term variability and the link between these variations and the Be phenomenon. In this paper, we present a brief summary of the results obtained in the analysis of several Be stars observed with COROT in terms of pulsations. In addition, we show that variations of the Be star HD 175869 can be explained as two active regions separated by 150 degrees or as unstable pulsating modes in a star with an extensive mixing in radiative layers corresponding to a core overshooting of 0.35Hp. A preliminary study of the photometric and spectroscopic variability seen in the B1.5IVe star HD 51193 is performed. Currently the KEPLER satellite is observing the only confirmed Be star in its field of view, namely KIC 6954726. From low-resolution spectra we derived a spectral type of B2.5Ve for this star and we studied the long-term variation of the emission in the Hα line. The 3.5-year KEPLER light curve will allow us to detect even more close frequencies than with COROT and to perform a detailed analysis of the amplitude variations in a Be star.

Keywords. stars: oscillations, stars: emission-line, Be

1. Introduction

Be stars are main sequence or slightly evolved B-type stars with emission in the Balmer lines arising from a circumstellar disk. It is generally agreed that the disk is fed by episodic mass ejections (outbursts) but the mechanism at work is yet uncertain. Be stars are very fast rotating stars, with an average angular velocity of 90% of the critical breakup velocity (Frémat *et al.* 2005). Therefore, an additional mechanism is needed to provide the angular momentum to eject material into the circumstellar disk. Osaki in 1986 proposed non-radial pulsations as the mechanism to produce these outbursts. More recently, Rivinius *et al.* (1998) found spectroscopically that the mass-loss episodes in the star μ Cen could be explained by the beating of pulsating modes. A correlation between outbursts and pulsations was also found photometrically in two stars observed with COROT (Huat *et al.* 2009, see below; Gutiérrez-Soto *et al.* 2010). Alternatively, Balona (2009) suggested that the eruptions could be of magnetic origin and proposed a simple model for Be stars based

Table 1. Overview of the first Be stars that were observed with the COROT satellite. In the colums, we give for each object: the HD number, spectral type, inclination angle in ° (with an error bar of 10° for all stars), $v \sin i$ (in $km\, s^{-1}$), rotation frequency, the main low and high frequencies (in d^{-1}) detected with COROT, the frequencies detected in line-profile variations, and the references.

Star	SpT	i	$v \sin i$	f_{rot}	low freq.	high freq.	LPV	Ref
HD 49330	B0.5IVe	40	280±10	0.96±0.35	1.47, 2.94, 0.87	11.86, 16.89	11.86, 16.89, 1.51	[1],[2]
HD 51193	B1.5IV	37	220±25	0.89±0.3	1.40, 0.72, 2.62, 1.78	no	1.78, 1.39, 0.71	[6]
HD 181231	B5IVe	45	190±25	0.76±0.32	0.62,0.69,1.24	no	0.69	[5]
HD 175869	B8IIIe	50	170±10	0.64±0.1	0.64, 1.28	no	no	[4]
HD 50209	B8IVe	60	200±20	0.62±0.2	1.48, 2.16,0.79, 0.67	no	no	[3]

[1] Huat *et al.* (2009), [2] Floquet *et al.* (2009), [3] Diago *et al.* (2009), [4] Gutiérrez-Soto *et al.* (2009), [5] Neiner *et al.* (2009), [6] Frémat *et al.* (2006) and this work.

on magnetic fields. However, no clear detection of magnetic fields in Be stars has been found so far.

Be stars show different timescales of variability in their light curves and spectra: from weeks to decades, caused by changes in the envelope; and from hours to days, produced by non-radial pulsations or/and rotational modulation. A long discussion was held during the last decade about the origin of these short-term variations (Baade & Balona 1994).

From an exhaustive spectroscopic campaign of 27 early-type Be stars, Rivinius *et al.* (2003) showed that the short-term periodic line profile variability (LPV) of these objects is due to non-radial pulsations. The LPV of the majority of stars in the sample were identified with the mode $\ell = m = 2$. More recently, the MOST satellite observed 4 Oe/Be stars during 30-40 consecutive days and suggested that non-radial pulsations are involved in all rapidly rotating Be stars (see Cameron *et al.* 2008 and references therein). COROT has shown that most of the observed Be stars are multiperiodic and that pulsations are present in early to late Be stars (see below for more details).

Alternatively, Balona (1995) found a correlation between the photometric periods and their projected rotational velocities and proposed that short-term variability in Be stars is caused by stellar spots or co-rotating clouds attached to the star by a magnetic field. Again this model assumes the presence of a stellar magnetic field and no detection has been obtained so far.

2. Overview of the COROT results

COROT consists of a space telescope dedicated to stellar seismology and the search for extrasolar planets. It is observing up to 150 consecutive days of the same fields with a precision of a few μmag and a data sampling of 32s (Auvergne *et al.* 2009). For a summary of the first results, see Michel *et al.* (2008).

2.1. COROT *observations*

In Table 1 a short summary of the results obtained for the first Be stars observed with COROT is given. $f_{\rm rot}$ was taken from the Tables in each reference, assuming that the rotational velocity is at 90% of the critical velocity.

The earliest Be star HD 49330 was found to be a hybrid Be star showing p- and g-modes (Huat *et al.* 2009). The amplitudes of the p- and g-modes varies significantly along the COROT observations. An outburst occured just after the amplitude of the two main g-modes were at maximum and while the amplitude of the p-modes had drastically decreased. After the outburst the amplitude of the g-modes decreased and the p-modes recovered a larger amplitude. This showed, for the first time in photometry, a clear

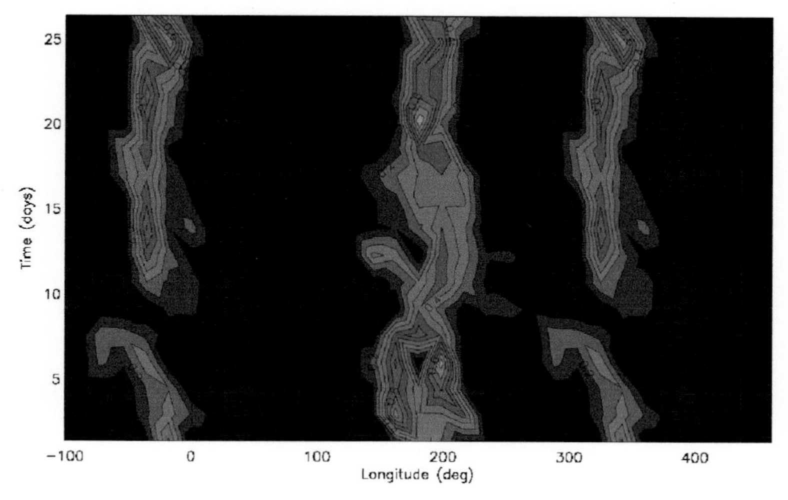

Figure 1. Isocontour plot of the filling factor versus longitude and time for the star HD 175869. Two regions are always active and separated by 150 degrees.

correlation between pulsations and outbursts. In addition, thanks to the study of spectroscopic time series we could identify the degree of the p-modes ($\ell \sim 4$ and 6).

The other early-type Be star, HD 51193, only shows g-modes in its light-curve and several of them could be identified in spectroscopy as well. See details below.

Multiple g-mode frequencies were detected in the COROT light curve of the mid-type Be stars HD 181231 (Neiner *et al.* 2009). One of the frequency is also detected in spectroscopy.

The 26-day light curve of the late-type star HD 175869 is dominated by a small-amplitude frequency and its harmonics (Gutiérrez-Soto *et al.* 2009). These variations are interpreted as inhomogeneities at the surface of the star, although non-radial pulsations are not excluded (see below in Sect.2.2).

Multiple periods due to g-modes have been detected in the late-type Be star HD 50209 (Diago *et al.* 2009). In addition, for the first time, the frequencies showed a clear pattern produced by the rotational splitting, which is not expected by models for such a rapidly rotating star (it rotates at least at 90% of the critical angular velocity). This star does not show a double wave phase diagram.

2.2. *Modelling the star HD 175869*

Non-radial pulsations could explain the variations seen in the COROT light curve of the star HD 175869. Applying pulsating models, as in Saio *et al.* (2007), we found unstable g-modes that are close to the observed frequencies. However, the star should be in the core-hydrogen burning stage elongated by an extensive mixing in the radiative layers corresponding to a core overshooting of 0.35 Hp. See Neiner *et al.* (2010), Lovekin *et al.* (2010).

As the light curve of the star HD 175869 is a double-wave with a period compatible with the rotational period, these variations could also be explained as spots or inhomogeneities in or close to the photosphere that rotate with the star. In order to confirm/reject this hypothesis, we applied the Maximum Entropy method (Lanza *et al.* 2009) to model the light curve as spots. The light curve was divided into 17 subsets of duration of the assumed rotational period (1.56 days, 0.64 d^{-1}) and each one was modelled independently from the others, giving us the possibility to trace spot evolution one rotation after the other.

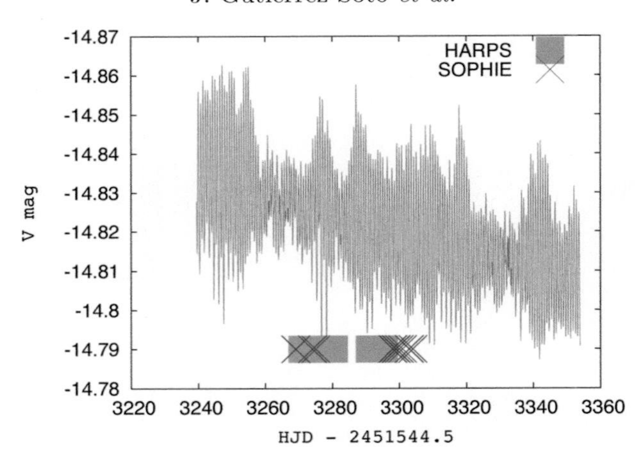

Figure 2. COROT light curve of the star HD 51193. The time of observations with HARPS and SOPHIE are indicated.

The residuals of this modelling are of the order of the residuals of the fitting given in Gutiérrez-Soto *et al.* (2009).

The first result is that there are two main active longitudes, where individual spots form and evolve. They are separated by about 150 degrees and have comparable levels of activity as measured by the spot filling factors (see Fig. 1). We would expect a separation of 180 degrees between the active regions if the spots are produced by a dipole magnetic field. However, if the dipole magnetic field is offcentered, i.e. if there is a quadrupole component, the regions where the spots would be attached may be separated by less than 180 degrees. The results will be published in a forthcoming paper (Gutiérrez-Soto, Lanza *et al.* 2010).

3. New results

3.1. *HD 51193*

The star HD 51193 was observed in the seismology fields of COROT during 114 days (LRa2 run). The light curve, depicted in Fig. 2, shows a clear beating, also seen in most of the Be stars observed with COROT. A dominant peak is found in the periodogram at 1.399 d^{-1} with an amplitude of 15 mmag. Many frequencies close to this one are detected with amplitudes ranging from 5 to 0.7 mmag, which could be produced by amplitude changes. Other peaks were found at frequencies 0.721 d^{-1} (~3 mmag), 0.156 d^{-1} (~2 mmag), 2.617 d^{-1} (~1 mmag), and 2.798 d^{-1} (~1mmag). No clear combination is found among these frequencies, except for the frequency 2.798 d^{-1} which is twice 1.399 d^{-1}. No frequencies higher than 5 d^{-1} are found in the periodogram of this early-type Be star.

High-resolution spectra were obtained with HARPS at the 3.6-m telescope in La Silla (PI E. Poretti, ESO LP 182.D-0356) and with SOPHIE at the 1.93m telescope at the OHP in December 2008/January 2009 (PI P. Mathias). They were collected simultaneously with the COROT data, as can be seen in Fig. 2.

A preliminary study of the line-profile variations reveals several frequencies at 1.78, 1.40 and 0.71 d^{-1} with coherent phase distributions across the line-profile. In Fig. 3 we display the power and phase distributions across the He I 4388 line for the frequencies 1.78 and 0.71 d^{-1}. Note the different slope of the phase for each frequencies, suggesting a different degree or azimuthal order of the mode.

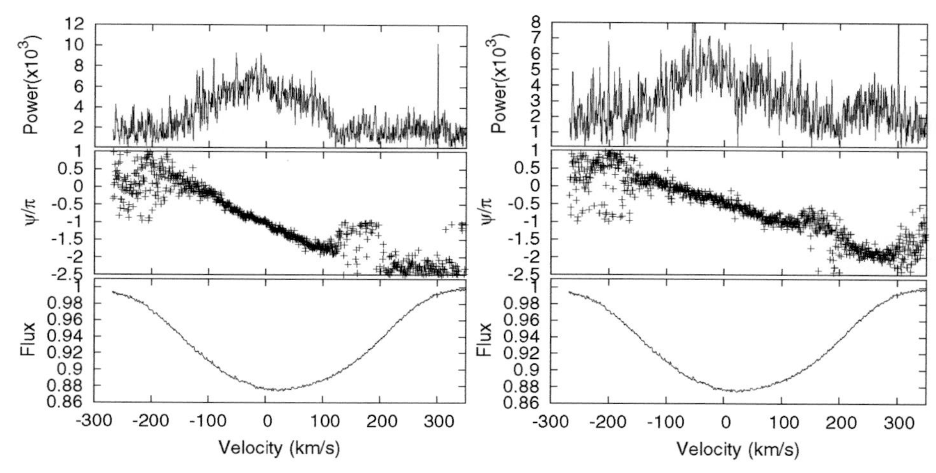

Figure 3. Power distribution (top) and phase distribution (middle) across the He I 4388 line profile (bottom) for f = 1.78 (left) and 0.71 d^{-1} (right), for the star HD 51193.

The multiple frequencies detected in the COROT light curve and the spectroscopic time series are attributed to non-radial pulsations.

3.2. KEPLER: *a promising future*

The KEPLER mission (Koch *et al.* 2010) is designed to detect Earth-like planets around solar-type stars by the transit method. KEPLER is continuously monitoring the brightness of over 100 000 stars for at least 3.5 yr in a 105-square degree fixed field of view.

The only confirmed Be star being observed by KEPLER is the star KIC 6954726. The low-resolution spectra obtained during the last 2 years suggest that this star is a B2.5Ve. In addition, we have followed the variation of its Hα line: it is always in emission and its equivalent width has been increasing since 2008.

The longer duration of the KEPLER observations with respect to COROT runs (typically 150 days) will give us the opportunity to distinguish between intrinsic amplitude variations produced by transient frequencies or amplitude variations due to the beating of very close frequencies. Hopefully, this long duration will also allow us to observe an outburst with KEPLER.

4. Conclusions

Some questions have arised from these studies. Can we explain the multiple frequencies found in most of Be stars observed with COROT and MOST with stellar spots or co-rotating clouds? The observed frequencies are coherent and independent and are often numerous, which would require a high number of independant spots i.e. a complex magnetic field. In addition, note that no clear detection of a magnetic field in a Be star has been obtained so far. Non-radial pulsations thus seem a simpler explanation and pulsation models are very successful in reproducing observations. On the other hand, can we explain the amplitude variations and double-wave shape seen in many of the COROT Be stars with non-radial pulsations?

All these questions will hopefully be answered with the help of the space missions COROT, KEPLER and later on PLATO. In addition, non-perturbative methods (eg. Ballot *et al.* 2010) are needed to model the non-radial pulsations in these rapidly rotating Be stars: they will give us the keys to understand these observations.

Acknowledgements

J. Gutiérrez-Soto acknowledges PNPS and the IAU for awarding him a grant to participate to the symposium. We wish to thank the COROT team for the acquisition and reduction of the COROT data. The HARPS data was obtained as part of the ESO Large Programme LP182.D-0356 (PI: E. Poretti). This research is also based on spectroscopic data from the SOPHIE spectrograph at OHP (France).

References

Auvergne, M., Bodin, P., Boisnard, L., & Buey, J.-T. *et al.* 2009, *A&A*, 506, 411

Baade, D., Balona, L. A. 1994, in: L. A. Balona, H. F. Henrichs, & J. M. Le Contel (eds.), *Pulsation; Rotation; and Mass Loss in Early-Type Stars*, IAU Symposium 162, p. 311

Balona, L. A. 1995, *MNRAS*, 277, 1547

Balona, L. 2009, in: J. A. Guzik, & P. A. Bradley (eds.), *Stellar Pulsation: Challenges for theory and observation*, AIP Conf. Ser., 1170, p. 339

Ballot, J., Lignières, F., Reese, D. R., & Rieutord, M. 2010, *A&A*, 518A, 30

Cameron, C., Saio, H., Kuschnig, R., Walker, & G. A. H. *et al.* 2008, *ApJ*, 685, 489

Diago, P. D., Gutiérrez-Soto, J., Auvergne, M., & Fabregat, J. *et al.* 2009, *A&A*, 506, 125

Floquet, M., Hubert, A.-M., Huat, A.-L., & Frémat, Y. *et al.* 2009, *A&A*, 506, 103

Frémat, Y., Zorec, J., Hubert, A.-M., & Floquet, M. 2005, *A&A*, 440, 305

Frémat, Y., Neiner, C., Hubert, A.-M., & Floquet, M. *et al.* 2006, *A&A*, 451, 1053

Gutiérrez-Soto, J., Floquet, M., Samadi, R., & Neiner, C. *et al.* 2009, *A&A*, 506, 133

Gutiérrez-Soto, J., Semaan, T., Garrido, R., & Baudin, F. *et al.* 2010, *AN*, in press, arXiv:1010.1910

Gutiérrez-Soto, J., Lanza, A. F. *et al.* 2010, *A&A* in preparation

Huat, A.-L., Hubert, A.-M., Baudin, F., & Floquet, M. *et al.* 2009, *A&A*, 506, 95

Koch, D. G., Borucki, W. J., Basri, G., & Batalha, N. M. *et al.* 2010, *ApJ* (Letters), 713, L79

Lanza, A. F., Pagano, I., Leto, G., & Messina, S. *et al.* 2009, *A&A*, 493, 193

Lovekin, C.C., Neiner, C., Saio, H., & Mathis, S. *et al.* 2010, *AN*, in press

Michel, E., Baglin, A., Weiss, W. W., & Auvergne, M. *et al.* 2008, *Communications in Asteroseismology* 156, 73

Neiner, C., Gutiérrez-Soto, J., Baudin, F., & de Batz, B. *et al.* 2009, *A&A*, 506, 143

Neiner, C., Mathis, S., Saio, H., & Lovekin, C. *et al.* 2010, *A&A* in preparation

Osaki, Y. 1986, *PASP*, 98, 30

Rivinius, T., Baade, D., Stefl, S., *et al.* 1998, in: L. Kaper & A.W. Fullerton (eds.), *Cyclical Variability in Stellar Winds*, ESO Astrophysics Symposia (Berlin, New York: Springer-Verlag), p. 207

Rivinius, T., Baade, D., & Štefl, S. 2003, *A&A* 411, 229

Saio, H., Cameron, C., Kuschnig, R., & Walker, G. A. H. *et al.* 2007, *ApJ*, 654, 544

Discussion

M. SMITH: Unfortunately, it appears that there are some difficulties about observing Be stars with KEPLER. First, the difficulty to find B stars with the KEPLER photometric catalogue: it was dispensed with their "u" filter of the original filter set from the sloan survey and therefore, the KEPLER project's ground-based photometry does not allow to distinguish between a mid O star and an O star. Second, the fixed KEPLER field of view is centered several degrees from the Galactic plane. Furthermore, KEPLER cannot easily observe bright stars at all because of constraints in downlinking and transmitting data to the Earth. For these reasons the observable OB stars that one will likely find will be intrinsically luminous or "runaway" stars from the Galactic plane. One can hope to be lucky and perhaps find a good variable or Be candidates.

Active OB stars: structure, evolution, mass loss, and critical limits
Proceedings IAU Symposium No. 272, 2010
C. Neiner, G. Wade, G. Meynet & G. Peters, eds.

doi:10.1017/S1743921311011100

Seismic modelling of OB stars

Marc-Antoine Dupret[1], **Mélanie Godart**[1], **Kevin Belkacem**[1] **and Arlette Noels**[1]

[1]Institut d'Astrophysique, Géophysique et Océanographie de l'Université de Liège, Allée du 6 Août 17, 4000 Liège, Belgium
email: MA.Dupret@ulg.ac.be

Abstract. A review of the ability of asteroseismology to probe the internal physics of OB stars is presented. The main constraints that can be obtained from the frequency spectrum in p- and g-modes pulsators are discussed. Next, we consider energetic aspects of the pulsations in OB stars and show how such study also allows to constrain their internal physics. The cases of p-mixed modes (β Cep stars), g-modes (SPB stars), strange modes and stochastically excited modes are considered.

Keywords. stars: evolution, stars: interiors, stars: oscillations, stars: rotation, stars: variables: other, turbulence, convection

---

1. Uncertainties in stellar physics

It is useful to begin with recalling some important problems in stellar physics for which observational constraints can be obtained from asteroseimology.

1.1. *Convective overshoot and penetration*

In all main-sequence massive stars, a convective core is present as a consequence of the high temperature dependence of the CNO cycle. The region just above this core is important and poorly known in these stars. A key ability of asteroseismology is to probe it (Noels *et al.* 2010). Because of their inertia, convective elements penetrate slightly in the subadiabatic layers just above the convective core. As the Peclet number is very high there, we may assume that these plumes keep their heat content as they penetrate in these regions. Hence, the buoyancy braking is large and the plumes cool down the gas around them. (Zahn 1991) proposes to call this process "convective penetration", in contrast with the small Peclet numbers cases called "convective overshoot". We can expect that this cooling increases the temperature gradient until it becomes nearly adiabatic. This decreases significantly the braking and the plumes can penetrate further. The extension and description of this extra-mixed overshooting region is thus very uncertain. Its size is generally parametrized as a fraction of the pressure scale height : $\Delta r_{ov} = \alpha_{ov} |dr/d\ln P|$, where α_{ov} is the so-called overshooting parameter. It must be stressed that stellar evolution strongly depends on the extension of this extra-mixing which continuously transports fresh material into the core where it is burned.

1.2. *Semi-convection*

One of the following two criteria is generally used for the determination of the convective core boundary. On the one hand, the Schwartzschild criterion $\nabla_{rad} = \nabla_{ad}$ assumes thermal neutrality (null convective heat transfer) at the boundary ($\nabla_{rad} \equiv 3\kappa PL/(16\pi acGmT^4)$, $\nabla_{ad} \equiv \partial \ln T/\partial \ln P|_s$). On the other hand, the Ledoux criterion $\nabla_L = \nabla_{ad}$ assumes dynamical neutrality (null force) at the boundary($\nabla_L \equiv \nabla_{rad} - \frac{\phi}{\delta}\nabla_\mu$

with $\nabla_\mu \equiv dln\mu/dlnP$). In massive stars, the convective core mass sometimes grows with time or decreases slowly. In principle, this should lead to a discontinuity of chemical composition at the boundary of the convective core. However, with the assumption of full mixing below and no mixing above, it appears impossible to define consistently this boundary. The only solution is to assume partial mixing in a so-called semi-convective zone above the convective core. In this semi-convective region, $\nabla_{ad} \leqslant \nabla \leqslant \nabla_L$ ($\nabla \equiv dlnT/dlnP$ is the "real gradient" corresponding to the temperature stratification in the star), but the exact temperature and chemical stratifications remain quite uncertain. Key constraints on this region could be obtained from asteroseismology (Noels *et al.* 2010).

1.3. *Microscopic diffusion*

Microscopic chemical transport is certainly present in massive stars under the action of two forces: the gravitation and the radiative forces (impulsion of photons transmitted to the gas). The gravitation force is proportional to the mass so that the heaviest particles are pushed down. The absorption of photons varies strongly from one type of particle to another. Particles with high cross section in the frequency range around the maximum of the Planck function are strongly pushed up. This is important in massive stars where radiation pressure is significant. The action of these forces implies a chemical transport in stellar interiors under the form of a microscopic diffusive process. However, significant uncertainties remain associated to these processes (collision integrals, shielding, ...).

1.4. *Macroscopic chemical and angular momentum transfer*

As detailed by Zahn (these proceedings), rotation is at the origin of two macroscopic transfer processes of very distinct natures in radiative zones. First, the Von Zeipel theorem shows that rotation must generate a meridional circulation. After a transient phase, this circulation settles in a quasi-stationary regime allowing to advect angular momentum (and at the same time chemicals) towards the surface where it is lost for example by winds. Second, the horizontal shear and other instabilities due to differential rotation generate turbulence. This turbulence is expected to be much more vigorous in the horizontal plane because of the stabilizing effect of buoyancy, justifying the so-called shellular approximation where differential rotation in latitude is assumed negligible. Turbulence in radiative zones acts as a *diffusive* transfer mechanism, which is qualitatively very different from the *advection* by meridional circulation. The main source of uncertainties lies in the turbulent diffusion coefficients that must be used to model this process (Mathis *et al.* 2004).

Another transfer mechanism that was shown to be important these last years comes from gravity waves and modes. Progressive waves emitted at the boundary of convective zones transfer angular momentum in the radiative zone where they are dissipated. Works of Charbonnel & Talon (2005) show that this mechanism is able to explain the solid body rotation in the solar radiation zone. It should not be forgotten that another possibly important mechanism in massive stars is the transfer by stationary gravity modes. This transfer process is expected to weaken the differential rotation caused by evolutionary core contraction (Lee & Saio 1993). Finally, we mention that magnetic field and differential rotations certainly influence each other, we refer to Zahn (these proceedings) for more details. For a general review of all these processes, see e.g. Talon (2008).

1.5. *Opacities*

Many stellar physics problems of the 20th century were solved thanks to improvements in the computations of opacities. The last important one was the explanation of the driving of β Cep stars thanks to the OPAL opacity tables. It is clear that the determination of precise opacities is not yet a solved problem. Asteroseismology and more specifically non-adiabatic asteroseismology (see Sect. 3.2) give important constraints on the opacities, indicating that models with the current best opacities: OP (Seaton 2005) and OPAL (Rogers & Iglesias 1992) are not yet able to explain some observations.

2. Interpreting the frequencies

We first recall that there are mainly two types of pulsation modes: the pressure (or p-) modes similar to acoustic standing waves and the gravity (or g-) modes in which the restoring force is the buoyancy. These two families are not exclusive, many times in intermediate and high mass stars, we also encounter the so-called mixed modes that have at the same time a gravity mode nature in the core and a pressure mode nature in the envelope of the star.

2.1. *Seismic probe by pressure and mixed modes*

The frequency spectrum of pressure modes mainly depends on the sound speed profile $c(r)$ in the envelope of the star. Its square is given by:

$$c^2 = \Gamma_1 P/\rho \,. \tag{2.1}$$

As in an ideal gas, $P/\rho \propto T/\mu$, we see that probing the sound speed means essentially probing the temperature and the internal chemical composition throughout μ (mean molecular weight) and $\Gamma_1 \equiv \partial \ln P / \partial \ln \rho|_s$. Probing this sound speed was already achieved with high precision for the Sun by helioseismology. Asteroseismology aims to do the same for other stars, but with less precision since the number of observed modes is by far lower.

In the context of OB stars, p-mixed modes pulsations are observed in β Cep stars. It is important to emphasize that such oscillations are very different from the p-modes oscillations observed in the Sun and other solar-like stars. Solar-like oscillations correspond to high order p-modes near the asymptotic regime. As a consequence, their frequency patterns are regular, mode identification is "easy", but the number of individual constraints is reduced by the asymptotic degeneracy. On the contrary, each p-mixed mode in a β Cep star probes a very different region of the star (see for example the kernels illustrated in Fig. 5 of Pamyatnykh *et al.* 2004). That means that each frequency gives a completely independent constraint on the stellar interior model. Even with few modes, key informations can be obtained, if the modes are identified. Several examples illustrate the power of asteroseismology applied to β Cep stars: for example ν Eri (Pamyatnykh *et al.* 2004; Ausseloos *et al.* 2004; Dziembowski & Pamyatnykh 2008) and HD 129929 (Aerts *et al.* 2003; Dupret *et al.* 2004). The main unknown global parameters of a massive star are its mass, age, initial hydrogen fraction X, heavy elements fraction Z and the overshooting parameter. Hence, with five identified modes of different (ℓ, n), it is possible to determine them with precision. With constraints on effective temperature, gravity and metallicity from spectroscopy or photometry, this minimum required number of identified modes can be reduced. Non-adiabatic analysis also gives additional constraints (see Sect. 3.2). Best constrained cases such as ν Eri show that standard models are not able to fully reproduce the observations.

For the same reason (mixed nature of the modes), the kernels associated to rotation show that each multiplet of a β Cep star probes the rotation rate in a distinct region (see Fig. 6 in Pamyatnykh *et al.* 2004 and Fig. 7 in Dupret *et al.* 2004). With two multiplets, it is already possible to test if solid body rotation is compatible with seismic observations. The results differ from a particular star to another: in ν Eri and HD 129929, asteroseismology shows that the core rotates about three times faster than the envelope. But in θ Oph (Briquet *et al.* 2007), solid rotation is compatible with seismic observations. Hence the question arises: which physical processes act differently in stars at the same location in the HR diagram, producing solid rotation in one case and strong differential rotation in another. Maybe, the answer could come from the magnetic field ? Indeed, the freezing effect of the magnetic field is well known and, if strong enough, this could produce solid body rotation. Unfortunately, no magnetic field has been detected in θ Op (Hubrig *et al.* 2006). We mentioned above the possibility of angular momentum transfer by gravity modes. Depending on the pulsational history of a star, angular momentum transfer could be significant in some of them, leading to solid body rotation; while this transfer could be weaker in others, letting evolutionary core contraction produce the differential rotation. However, quantitative models and simulations are needed to confirm this possibility.

Macroscopic transfer of angular momentum and chemicals by differential rotation could also be constrained from the frequencies of identified p-mixed modes. Indeed, Montalbàn *et al.* (2008) showed that the frequencies of p-mixed modes in β Cep models are significantly affected by turbulent mixing. Seismic constraints on these processes for particular stars are expected to be obtained in the near future.

2.2. *Seismic probe by high order gravity modes*

The period spectrum of gravity modes mainly depends on the so-called Brunt-Vaïsälä (BV) frequency, also called Buoyancy frequency. Its square is defined by:

$$N^2 = g\left(\frac{1}{\Gamma_1}\frac{d\ln P}{dr} - \frac{d\ln\rho}{dr}\right). \tag{2.2}$$

We have approximately (fully ionized ideal gaz) :

$$N^2 \simeq \frac{g^2\rho}{P}(\nabla_{ad} - \nabla + \nabla_\mu), \tag{2.3}$$

where $\nabla_\mu \equiv dln\mu/dlnP$ is the "μ-gradient" corresponding to the molecular weight variation throughout the star. Fine details of the internal structure of stars appear throughout ∇ and ∇_μ and can be probed by gravity modes (if observed).

In the asymptotic limit, the periods of g-modes are approximately given by the following law:

$$P_k = \frac{\pi^2(2k+n_e)}{\sqrt{\ell(\ell+1)}\int_{r_0}^{R}|N|/r dr}, \tag{2.4}$$

where r_0 is the radius of the convective core. The integral of the BV frequency appearing in the denominator is thus the main physical quantity that can be deduced from the g-modes spectrum, simply by considering the period spacing between consecutive modes. Typically, as the star evolves, the density contrast between the core and the envelope increases, therefore the BV frequency increases in the core and the period spacing decreases. Of course, information more useful than the mean period spacing can be potentially extracted from the g-modes spectrum. Let's consider the typical case of a main-sequence star with a receding convective core letting behind it a region of variable molecular weight. According to Eq. (2.3), the resulting strong molecular weight gradient produces a peak in the BV frequency. In particular, a strong discontinuity of N is predicted at the top of this μ-gradient region. The effect of such discontinuity on the g-mode period spacing can be determined. As shown e.g. in Miglio et al. (2008), instead of being perfectly constant as in Eq. (2.4), this period spacing $\Delta P(k) = P_k - P_{k-1}$ oscillates with a "wavelength":

$$\Delta k \simeq \frac{\int_{r_0}^{R} |N|/r dr}{\int_{r_0}^{r_1} |N|/r dr} \tag{2.5}$$

where r_1 corresponds to the location of the discontinuity. For a ZAMS model, there is not yet a μ-gradient region, thus $\Delta k \to \infty$ and the period spacing is independent of k. As the star evolves, the μ-gradient region increases in size and thus the denominator of Eq. (2.5) increases. Hence, an oscillation of $\Delta P(k)$ appears with a "wavelength" Δk which decreases progressively. This is illustrated for example in Fig. 16 of Miglio et al. (2008). Hence, if such oscillation of the period spacing is observed, it allows to determine accurately the evolutionary state of a star. More quantitatively, *Δk is a precise indicator of the central hydrogen abundance X_c of a MS star.*

What about overshooting now ? Miglio et al. (2008) showed that, for massive stars with fixed X_c, $\int_{r_0}^{r_1} (N/r)dr$ and thus Δk do not depend significantly on the amount of overshooting (assuming instantaneous mixing). This is different for intermediate mass stars such as γ Dor stars where the nuclear burning region is slightly larger than the convective core. But this does not mean that overshooting cannot be constrained from the g-modes periods in massive stars. Stellar evolution tracks strongly depend on overshooting. If, for example, the effective temperature, gravity and metallicity of the star are determined with enough precision, combined seismic and non-seismic constraints could allow us to constrain overshooting.

Finally, a very important ability of g-modes is to probe the μ-gradient region and thus chemical transfer mechanisms due for example to differential rotation. We have seen above that the discontinuity of ∇_μ at the top of this region leads to an oscillation of $\Delta P(k)$. Macroscopic chemical transfer mechanisms such as turbulent diffusion smooth this discontinuity (see Fig. 22 of Miglio *et al.* 2008). As a consequence, the amplitude of the oscillation of $\Delta P(k)$ is smaller and decreases with k. With high turbulent diffusivity, ΔP becomes independent of k. A clear signature of diffusive mixing can thus be detected in the period spacing.

A nice application of g-modes asteroseismology was recently possible thanks to the detection of hundreds of modes in the SPB HD 50230 by CoRoT (Degroote *et al.* 2010a). The very important discovery was the detection of 8 consecutive g-modes showing a clear oscillation in the function $\Delta P(P)$. Based on these observations, seismic modelling was possible. The main results are that 60% of the central hydrogen has already been consumed, the overshooting parameter is larger than 0.2 and the small amplitude of the deviation to constant spacing indicates a smooth gradient of chemical composition above the convective core incompatible with instantaneous mixing.

2.3. *Mode identification*

An important problem in asteroseismology is the mode identification. In the asymptotic limit, the spectrum of frequencies is expected to be regular. We already mentioned the quasi-constant period spacing predicted for high order gravity modes. In high order pressure modes, quasi-equidistant frequency separation is expected, according to the following approximate relation:

$$\nu_{n,\ell} \simeq \left(n + \frac{\ell}{2} + \frac{1}{4} + \alpha\right) \Delta\nu \; - \; (A\ell(\ell+1) - \delta)\frac{\Delta\nu^2}{\nu_{n,\ell}} \; - \; m(1 - C_{n,\ell}) f_{\rm rot} \tag{2.6}$$

where $\Delta\nu$ is called the large separation. Such regularities are easy to detect in solar-type oscillations. They allow us to identify the modes, which makes seismic modelling possible without need of additional informations.

However, such regularities are not present in low-order modes such as those observed in β Cep and δ Sct stars, because of their mixed character. Mode identification is thus impossible on the basis of the frequencies alone. Two types of additional observables can be used for mode identification. First, we have the spectroscopic line-profile variations. They result from the pulsational surface velocity field (Doppler effect) and allow us to constrain it. In particular, their interpretation allows us to determine the degree ℓ and azimuthal order m of the modes, see e.g. the review of Telting (2008) and references therein. Secondly, we have the photometric amplitudes and phase differences of magnitude variations in different passbands. They depend on the degree ℓ of the modes, so that it can be determined by comparing theoretical predictions with observations. The basics of this method are detailed in Watson (1988). Several improvements have been proposed to take non-adiabatic and rotational effects into account, see Handler (2008) for a review.

2.4. *Fast rotation*

All previous discussions were valid for slow rotators. In fast rotators such as Be stars, seismic modelling is much more difficult. Fast rotation has two important effects. Firstly, centrifugal force breaks down the spherical symetry when $\Omega^2 R^3/(GM)$ is significant. Secondly, the Coriolis force plays a large role in the mode dynamics when Ω/σ is significant (Ω is the rotation rate and σ the angular pulsation frequency). Usual modelling tools for slow rotators are evidently not adequate for rapid rotators.

Concerning the internal structure, a new generation of full 2D stellar models are needed. The most rigorous approach would be to develop 2D stellar evolution codes fully including the deformation due to centrifugal force and the transfer processes associated with differential rotation. Works in this direction are being done with spectral (Espinosa & Rieutord 2007) and finite differences (Deupree 1995) numerical approaches; but the task remains hard. A simpler and efficient approach is to use the Self Consistent Field (SCF) method (see e.g. Jackson *et al.* 2005; Roxburgh 2004 for the most recent papers), which is valid for conservative rotation laws (solid body or cylindrical). It is also useful to mention the "characteristics" method proposed by Roxburgh (2006) which allows us to compute 2D structure models for arbitrary differential rotation laws. The input of the code are here a 1D structure model and the rotation law $\Omega(r, \theta)$; the output is a 2D deformed model with the same mean structure. Computation times are much shorter with such an approach, which would make possible to compute numerous models as required for seismic modelling. Finally, a widely used but approximate approach is to model the structural effect of rotation in a perturbative way: spherical symetric component plus perturbation proportional to $P_2(\cos\theta)$.

Let's consider now the modelling of oscillations in rapid rotators. Here also, a new generation of oscillation codes is required. The perturbative approach does not apply when the centrifugal deformation and/or Ω/σ are significant. In these cases, the coupling between oscillations and rotation must be modelled in a non-perturbative way. Several adiabatic oscillation codes have been developed for this purpose: with a finite difference approach (Clement 1998), a full spectral approach (Reese *et al.* 2006) and a mixed finite difference-spectral approach (Reese *et al.* 2009, Ouazzani et al. in preparation). Concerning non-adiabatic oscillation codes, the best present models are obtained with the method of Lee & Baraffe (1995) in which a perturbative approach is adopted for the structure models and a spectral non-perturbative approach is used for the oscillations.

As shown by several studies, the mode geometry is completely different from spherical harmonics in fast rotators (see e.g. Fig. 3 in Reese *et al.* 2009). The p-modes of fast rotators fall into three categories: island, chaotic and whispering gallery modes (Lignières

& Georgeot 2009). The structure of the oscillation frequency spectrum is also completely different from the slow rotation case: in the asymptotic limit, old regularities such as Eq. (2.6) are no longer valid, but new ones appear. Reese et al. (2009) proposed new asymptotic laws which better agree with the frequency patterns in fast rotators.

It is clear that the main price for accurate modelling of fast rotators pulsations is the computation time. Asteroseismology by the direct method generally requires the computation of many structure models and their oscillations in order to find the best fit solution. Such an approach is currently not possible because of computation time. New strategies feasible even with a small number of models are needed (Lovekin & Goupil 2010).

3. Energetic aspects of stellar oscillations

3.1. *Driving mechanism*

The first part of this review considered adiabatic modelling of stellar oscillations. Such an approach is adequate for the determination of the frequencies because the modes mainly propagate in deep enough regions where the oscillations are quasi-adiabatic. But such an approach does not allow us to study the energetic aspects of oscillations such as the driving mechanisms at the origin of the oscillations. Non-adiabatic modelling is required to study these aspects. The main uncertainties in such models concern the time-dependent interaction between convection and oscillations. Fortunately, this difficult aspect does not play a significant role in massive stars. In OB stars, the driving of the modes comes from a standard κ-mechanism occurring in the iron M-shell opacity bump at $T \approx 2\times10^5$ K (Dziembowski *et al.* 1993a; 1993b). The main driving or damping of the modes always occurs in the transition zone where the thermal relaxation time is of the order of the pulsation period. In β Cep stars, the transition region for low order pressure modes coincides with the iron opacity bump, which explains their efficient driving. In the cooler SPB stars, the iron bump is deeper, coinciding with the transition zone for long period high order gravity modes.

3.2. *Non-adiabatic asteroseismology*

Asteroseismology is generally seen as a method for constraining internal structure of stars by using the information given by the observed pulsation frequencies. However, this is often not possible without mode identification. We call *non-adiabatic asteroseismology* a method of constraining stellar interiors based on observables which are theoretically determined by non-adiabatic computations. What are these observables ? Firstly, non-adiabatic computations determine which modes of a given model are excited and thus able to grow up to observable amplitudes. The constraint is here obtained by comparing the range of observed frequencies to the predicted range of excited modes. Secondly, we mentioned in Sect. 2.3 the photometric mode identification method. Theoretical determination of the multi-color photometric amplitude ratios and phase differences requires the knowledge of the normalized amplitude and phase of effective temperature variations, which are obtained by non-adiabatic computations. The constraints come here from the comparison between theoretical and observed amplitude ratios and phase differences.

Non-adiabatic predictions in massive stars mainly depend on the opacity κ and its derivatives $\partial\ln\kappa/\partial\ln T|_\rho$ and $\partial\ln\kappa/\partial\ln\rho|_T$ in the driving region. Non-adiabatic asteroseismology in massive stars gives thus the ability to constrain the opacity. The current most precise opacities are given in the OPAL (Rogers & Iglesias 1992) and OP (Opacity project) tables. The last improvements in OP and a comparison with OPAL are presented in Badnell *et al.* (2005). The opacities also strongly depend on the adopted mixture. Miglio *et al.* (2007a) computed instability strips (IS) of β Cep and SPB stars with OPAL and OP opacity tables and two different mixtures: Grevesse & Noels (1993) and the new solar mixture of Asplund *et al.* (2005). They found significant differences between OP and OPAL IS: mainly the IS of SPB stars obtained with OP extends to hotter stars compared to OPAL. These differences indicate that current uncertainties in opacity computations have a significant observable effect. They also found a significant impact of the metal mixture on the IS: higher overtones are excited with the mixture of Asplund *et al.* (2005).

How do theory and observations compare ? Hybrid pulsators in which both β Cep type p-mixed modes and SPB type g-modes are observed are particularly constraining. An important example is ν Eri in which the range of observed modes is larger than what theory predicts. More modes are excited with OP than with OPAL, but the excitation of the highest overtone is still not obtained (Dziembowski & Pamyatnykh 2008). Another interesting case is γ Peg (Handler *et al.* 2009).

β Cep and SPB candidates were recently discovered in the SMC (Karoff *et al.* 2008; Diago *et al.* 2008), which challenges theory. Indeed, the metallicity of SMC is very low ($Z \approx 0.001 - 0.004$) so that the iron opacity bump is not strong enough to allow efficient κ-driving of the modes (Salmon *et al.* 2009). A significant increase of the opacity in the iron bump would be required to excite β Cep and SPB modes in the SMC (Salmon et al. in preparation).

Finally, as above-mentioned, comparison between theoretical and observed multi-color photometric amplitude ratios and phase differences can also be used to constrain the opacity and metallicity (see Daszyńska & Walczak 2009, Daszyńska & Walczak 2010 for the most recent studies of this type).

3.3. *Energetic aspects in fast rotators*

We discussed in Sect. 2.4 the effect of fast rotation on the frequencies and mode geometry under the adiabatic approximation. Non-adiabatic non-perturbative pulsation codes would be required to study the excitation and damping mechanisms in fast rotators. Present models include approximations. For example, the traditional approximation is often used for the study of the driving mechanisms (Townsend 2005) and in the non-adiabatic models required for photometric mode identification (Townsend 2003). We recall however that this approximation assumes spherical symetry of the equilibrium model. It is justified for g-modes in slow or moderate rotators but not in fast rotators.

Non-adiabatic models of fast rotators are complex but there is at least one simple effect with significant observational impact. In the energy equation, what matters is the pulsation frequency from the point of view of the corotating frame. The frequency range of excited modes expressed in this frame does not depend strongly on m. We have

$$\sigma_I = \sigma_c - m\Omega \tag{3.1}$$

where σ_I is the frequency in the observer inertial frame and σ_c is the frequency in the corotating frame. As a simple consequence of this relation, the excited frequencies in the observer frame strongly depend on m. For example, the retrograde ($m > 0$) modes can have very long observed periods. A very interesting extreme case is when $\sigma_c < \Omega$, as can be the case for the g-modes of Slowly Pulsating Be (SPBe) stars. From Eq. (3.1), modes with different m are therefore separated in distinct groups. This is excellent news, as it solves the problem of mode identification ! Such an effect was indeed observed in several SPBe stars (Cameron *et al.* 2008).

3.4. *Driving non-radial pulsations in post MS B stars*

Post MS stars are characterized by a high density contrast between the core and the envelope. As a consequence, the BV frequency is very large in the radiative core and all non-radial modes behave like g-modes in these deep regions. Theory predicts that the radiative damping of these modes is very large because of their high wavenumber ($k_r = \sqrt{\ell(\ell+1)}N/(\sigma r)$). Therefore, non-radial modes are not expected to be excited in post MS B stars. The detection of many non-radial modes in a B supergiant by MOST was thus a big surprise, but not for a long time since it could be explained in the paper reporting their discovery by Saio *et al.* (2007). As a consequence of the evolution of the convective core during the MS phase, the temperature gradient ∇ becomes flat and slightly subadiabatic above it. When the H-burning shell appears during stellar evolution, L/m increases there, the temperature gradient easily becomes superadiabatic, which leads to the appearence of an Intermediate Convection Zone (ICZ) above the H-burning shell. Saio *et al.* (2007) showed that this ICZ can act as a propagation barrier for the non-radial modes: most are trapped in the radiative core below it, but some others are trapped in the envelope. These last modes have negligible amplitudes in the core and do not undergo significant radiative damping. They are driven by a κ-mechanism in the Fe opacity bump like in MS SPB stars.

The presence of the ICZ at the origin of these modes depends on several processes. As shown by Godart *et al.* (2009), mass loss inhibits the appearance of the ICZ; the observation of these mode gives thus an upper limit on the mass loss rate. Overshooting also has such an inhibiting effect, as well as rotational mixing. Finally, depending on the convection criterion (Schwartzschild versus Ledoux), the resulting ICZ is of very different nature (Lebreton *et al.* 2009). These processes and their consequences can thus be constrained by the observation of non-radial modes in B supergiants.

3.5. *Strange modes*

As a consequence of the combined actions of the L- and M-shell iron opacity bumps and the high radiative pressure, a sound speed inversion can be present in the envelope of massive stars. This produces a cavity where modes can be trapped: the strange modes (Saio *et al.* 1998). Strange modes are of different kinds: some have an adiabatic counterpart, others are purely thermal. There are also the convective strange modes (Saio, these proceedings): g^- modes which are shown to oscillate when non-adiabatic aspects are included. Mode trapping can also occur in the stellar atmosphere when a temperature inversion is present (Godart et al. these proceedings), producing another kind of strange modes. As strange modes propagate in very superficial layers, they hold informations about these regions (sound speed, ...). However, non-linear effects are strong because of their very low inertia. Strange modes could also play a role in mass ejection such as in LBV stars (Glatzel 2009). Finally, we mention the recent and interesting detection by CoRoT (and in spectroscopy) of a mode which could be a strange mode (Aerts *et al.* 2010).

3.6. *Solar-like modes in massive stars*

High order p-modes have recently been dicovered by CoRoT in a β Cep star (Belkacem *et al.* 2009) and in a O-star (Degroote *et al.* 2010b). A time-frequency (or wavelet) analysis clearly shows their stochastic nature, which makes them similar to the solar modes. Stochastic excitation models (Belkacem *et al.* 2010) show that these modes could be excited by the turbulent motions in the iron subsurface convection zone (Cantiello, these proceedings) or near the top of the convective core. The eigenfunction shape in the excitation region significantly affects the observed amplitudes. In the first case (subsurface excitation), a modulation of the amplitudes with frequency is predicted. In the second, a decrease of the amplitudes with frequency is predicted, simply because high frequency modes have small normalized amplitudes in the core. Current large uncertainties in the description of these region (convective velocities, eddy-time correlation, ...) also have a significant impact on the predicted amplitudes. The comparison with observations gives thus a unique opportunity to improve the knowledge of these regions. Of course, the frequencies of these modes can also be used to constrain the envelope structure.

References

Aerts, C., Thoul, A., Daszyńska-Daszkiewicz, J., & Scuflaire, R. *et al.* 2003, *Science* 300, 1926

Aerts, C., Lefever, K., Baglin, A., & Degroote, P. *et al.* 2010, *A&A*, 513, L11

Asplund, M., Grevesse, N., Sauval, A. J., & Allende Prieto, C. *et al.* 2005, *A&A*, 431, 693

Ausseloos, M., Scuflaire, R., Thoul, A., & Aerts, C. 2004, *MNRAS*, 355, 352

Badnell, N. R., Bautista, M. A., Butler, K., & Delahaye, F. *et al.* 2005, *MNRAS*, 360, 458

Belkacem, K., Samadi, R., Goupil, M.-J., & Lefèvre, L. *et al.* 2009, *Science*, 324, 1540

Belkacem, K., Dupret, M. A., & Noels, A. 2010, *A&A*, 510, A6

Briquet, M., Morel, T., Thoul, A., & Scuflaire, R. *et al.* 2007, *MNRAS*, 381, 1482

Cameron, C., Saio, H., Kuschnig, R., & Walker, G. A. H. *et al.* 2008, *ApJ*, 685, 489

Charbonnel, C. & Talon, S. 2005, *Science*, 309, 2189

Clement, M. J. 1998, *ApJS*, 116, 57

Daszyńska-Daszkiewicz, J. & Walczak, P. 2009, *MNRAS*, 398, 1961

Daszyńska-Daszkiewicz, J. & Walczak, P. 2010, *MNRAS*, 403, 496

Degroote, P., Aerts, C., Baglin, A., & Miglio, A. *et al.* 2010a, *Nature*, 464, 259

Degroote, P., Briquet, M., Auvergne, M., & Simón-Díaz, S. *et al.* 2010b, *A&A*, 519A, 38

Deupree, R. G. 1995, *ApJ*, 439, 357

Diago, P. D., Gutiérrez-Soto, J., Fabregat, J., & Martayan, C. 2008, *A&A*, 480, 179

Dupret, M.-A., Thoul, A., Scuflaire, R., & Daszyńska-Daszkiewicz, J. *et al.* 2004, *A&A*, 415, 251

Dziembowski, W. A. & Pamiatnykh, A. A. 1993a, *MNRAS*, 262, 204

Dziembowski, W. A., Moskalik, P. & Pamyatnykh, A. A. 1993b, *MNRAS*, 265, 588

Dziembowski, W. A. & Pamyatnykh, A. A. 2008, *MNRAS*, 385, 2061

Espinosa Lara, F. & Rieutord, M. 2007, *A&A*, 470, 1013

Glatzel, W. 2009, *Communications in Asteroseismology* 158, 252

Godart, M., Noels, A., Dupret, M.-A., & Lebreton, Y. 2009, *MNRAS*, 396, 1833

Grevesse, N. & Noels, A. 1993, in: Hauck B. & Paltani S. R. D. (eds.), *La Formation des Eléments Chimiques*, AVCP (Lausanne), p. 205

Handler, G. 2008, *Communications in Asteroseismology* 157, 106

Handler, G., Matthews, J. M., Eaton, J. A., Daszyńska-Daszkiewicz, J. *et al.* 2009, *ApJ* (Letters), 698, L56

Hubrig, S., Briquet, M., Schöller, M., & De Cat, P. *et al.* 2006, *MNRAS*, 369, L61

Jackson, S., MacGregor, K. B., & Skumanich, A. 2005, *ApJS*, 156, 245

Karoff, C., Arentoft, T., Glowienka, L., & Coutures, C. *et al.* 2008, *MNRAS*, 386, 1085

Lebreton, Y., Montalbán, J., Godart, M., & Morel, P. *et al.* 2009, *Communications in Asteroseismology* 158, 277

Lee, U. & Saio, H. 1993, *MNRAS*, 261, 415

Lee, U. & Baraffe, I. 1995, *A&A*, 301, 419

Lignières, F. & Georgeot, B. 2009, *A&A*, 500, 1173

Lovekin, C. C. & Goupil, M.-J. 2010, *A&A*, 515A, 58

Mathis, S., Palacios, A., & Zahn, J.-P. 2004, *A&A*, 425, 243

Miglio, A., Montalbán, J., & Dupret, M.-A. 2007a, *MNRAS*, 375, L21

Miglio, A., Montalbán, J., Noels, A., & Eggenberger, P. 2008, *MNRAS*, 386, 1487

Montalbán, J., Miglio, A., Eggenberger, P., & Noels, A. 2008, *AN*, 329, 535

Noels, A., Montalban, J., Miglio, A., & Godart, M. *et al.* 2010, *Ap&SS*, 328, 227

Pamyatnykh, A. A., Handler, G., & Dziembowski, W. A. 2004, *MNRAS*, 350, 1022
Reese, D., Lignières, F., & Rieutord, M. 2006, *A&A*, 455, 621
Reese, D. R., MacGregor, K. B., Jackson, S., & Skumanich, A. *et al.* 2009, *A&A*, 506, 189
Rogers, F. J. & Iglesias, C. A. 1992, *ApJS*, 79, 507
Roxburgh, I. W. 2004, *A&A*, 428, 171
Roxburgh, I. W. 2006, *A&A* 454, 883
Saio, H., Baker, N. H., & Gautschy, A. 1998, *MNRAS*, 294, 622
Saio, H., Cameron, C., Kuschnig, R., & Walker, G. A. H. *et al.* 2007, *ApJ*, 654, 544
Salmon, S., Montalbán, J., Miglio, A., & Morel, T. *et al.* 2009, in: J. A. Guzik & P. A. Bradley (eds.), *Stellar pulsation: challenges for theory and observation*, AIP-CP 1170, p. 385
Seaton, M. J. 2005, *MNRAS*, 362, L1
Telting, J. H. 2008, *Communications in Asteroseismology* 157, 112
Talon, S. 2008, in: C. Charbonnel & J.-P. Zahn (eds.), *EAS Publications Series* 32, p. 81
Townsend, R. H. D. 2003, *MNRAS*, 343, 125
Townsend, R. H. D. 2005, *MNRAS*, 360, 465
Watson, R. D. 1988, *Ap&SS*, 140, 255
Zahn, J.-P. 1991, *A&A*, 252, 179

Active OB stars: structure, evolution, mass loss, and critical limits
Proceedings IAU Symposium No. 272, 2010
C. Neiner, G. Wade, G. Meynet & G. Peters, eds.

doi:10.1017/S1743921311011112

Radial and nonradial oscillations of massive supergiants

Hideyuki Saio

Astronomical Institute, Graduate School of Science, Tohoku University,
Sendai, Miyagi 980-8578, Japan
email: saio@astr.tohoku.ac.jp

Abstract. Stability of radial and nonradial oscillations of massive supergiants is discussed. The kappa-mechanism and strange-mode instability excite oscillations having various periods in wide ranges of the upper part of the HR diagram. In addition, in very luminous ($\log L/L_\odot \gtrsim 5.9$) models, monotonously unstable modes exist, which probably indicates the occurrence of optically thick winds. The instability boundary is not far from the Humphreys-Davidson limit. Furthermore, it is found that there exist low-degree ($\ell = 1, 2$) oscillatory convection modes associated with the Fe-opacity peak convection zone, and they can emerge to the stellar surface so that they are very likely observable in a considerable range in the HR diagram. The convection modes have periods similar to g-modes, and their growth-times are comparable to the periods. Theoretical predictions are compared with some of the supergiant variables.

Keywords. stars: oscillations (including pulsations), supergiants, instabilities, convection

1. Introduction

Light and velocity variations on various time-scales are common in very luminous stars (e.g., van Genderen 1989, van Leeuwen *et al.* 1998). Those variations are caused by various kinds of instabilities. Here we discuss mainly the cause of microvariations of massive supergiants based on linear stability analyses applied to evolutionary models of massive stars. The evolution models were calculated by a Henyey-type code using OPAL opacity tables (Iglesias & Rogers 1996). Wind mass-loss is included for the evolutionary models for $M_i \gtrsim 30 M_\odot$ ($M_i =$ initial mass) based on the mass-loss rates of Vink *et al.* (2001). Linear stability analyses were performed using the methods given in Saio, Winget & Robinson (1983) for radial modes and Saio & Cox (1980) for nonradial modes, where the outer mechanical boundary condition was modified to $\delta P_{\rm gas} \to 0$ ($\delta P_{\rm gas}$ means the Lagrangean perturbation of gas pressure) taking into account the fact that radiation pressure is dominant near the outer boundary. It should be noted, however, that the effect of winds on radial and nonradial oscillations is not included because the effect is not well understood yet.

2. Stability of radial modes

Figure 1 shows instability boundaries of spherical symmetric modes and selected evolutionary tracks. Short-dashed line indicates the instability boundary for low-order modes. The nearly vertical "finger" ($4.4 \gtrsim \log T_{\rm eff} \gtrsim 4.3$) is the well-known β Cep instability strip, in which low-order radial modes and nonradial p-modes are excited by the kappa-mechanism at the Fe-peak of opacity around $T \sim 2 \times 10^5$K. At the luminous part of the instability strip, the cool-side boundary bents to become horizontal around $\log(L/L_\odot) \sim 5.8$. This is due to the strange-mode instability which occurs in models with sufficiently high luminosity to mass ratios as $L/M \gtrsim 10^4 L_\odot/M_\odot$. The properties of

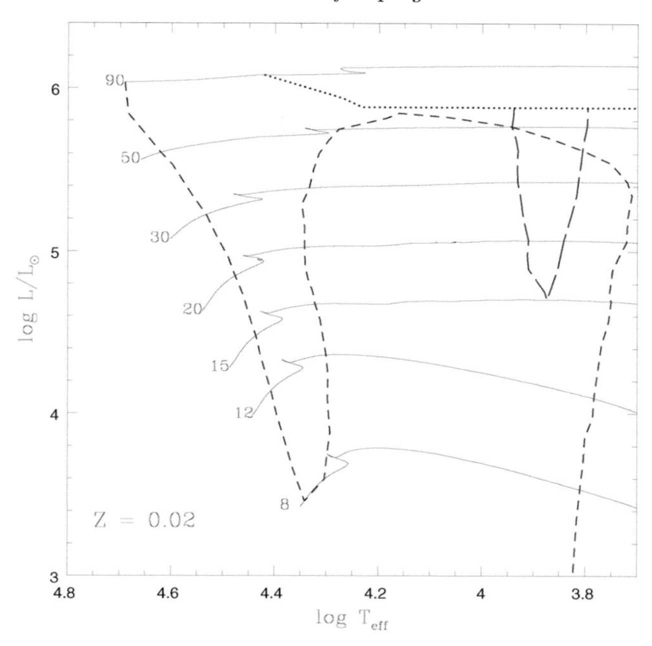

Figure 1. Instability boundaries of radial modes and selected evolutionary tracks. Short- and long-dashed lines indicate stability boundaries for low-order and relatively high-order radial modes, respectively. Dotted line indicates the boundary for monotonously unstable radial modes. Numbers along evolutionary tracks indicate initial masses in solar units. The instability boundaries for radial modes also represent approximately the boundaries for low-degree ($\ell \lesssim 2$) nonradial p-modes.

the strange modes have been investigated by e.g., Glatzel & Kiriakidis (1993), Glatzel (1994), Saio, Baker & Gautschy (1998) (see also Saio 2009).

Nearly vertical part of the boundary around $\log T_{\rm eff} \sim 3.8$ indicates the well-known blue edge of the Cepheid instability strip. (Red edge is not obtained because the perturbation of convective flux is neglected in this analysis.)

In the vertical narrow region indicated by long-dashed line around $\log T_{\rm eff} \sim 3.9 - 3.8$, relatively high-order radial modes are excited around hydrogen ionization zone. (Low-degree nonradial modes with similar frequencies are also excited.) The amplitude of these modes are extremely confined to the outermost layers above the hydrogen ionization zone. Since these modes exist even under the NAR approximation where thermal-time is set to be zero (Gautschy & Glatzel 1990), they may be classified as strange modes. These modes have got little attention so far (cf. Gautschy 2009).

Dotted line in Fig. 1 shows the boundary above which monotonously unstable modes exist. The growth times of these modes are much shorter than the timescale of evolution. The presence of such monotonously unstable modes have not been recognized before. Such a mode probably corresponds to the presence of an optically thick wind as investigated by Kato & Iben (1992) for WR stars. It is interesting to note that the boundary is not far from the Humphreys-Davidson limit (Humphreys & Davidson 1979).

3. Stability of nonradial modes

The instability ranges of low-degree high-order g-modes are shown in Fig. 2 by solid and dotted lines for $\ell = 2$ and $\ell = 1$ modes, respectively. These modes are excited by the kappa-

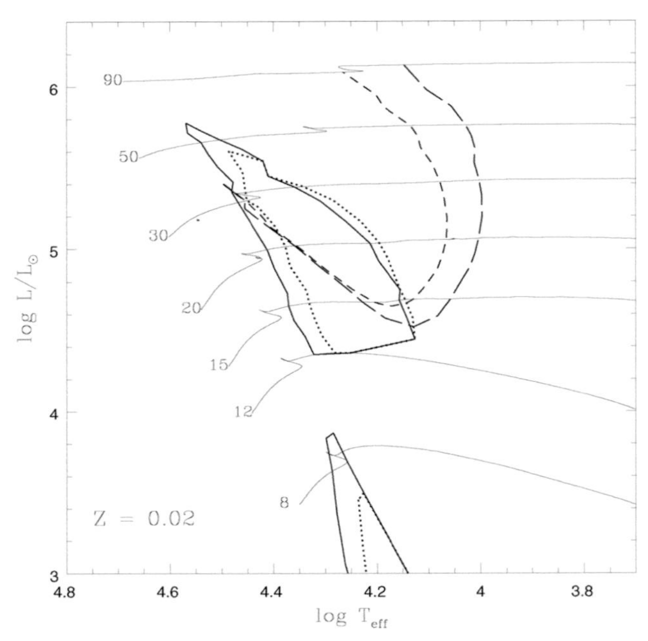

Figure 2. Ranges where nonradial oscillations are expected to be detected. The instability boundaries for SPB-type low-degree high-order g-modes are shown by solid ($\ell = 2$) and dotted ($\ell = 1$) lines. The ranges where oscillatory convection modes should be observable are indicated by long- and short-dashed lines for $\ell = 1$ and $\ell = 2$, respectively.

mechanism at the Fe opacity peak. They are responsible for the long-period variations in slowly pulsating B (SPB) stars. Such g-modes can be excited even in supergiants (SPBsg) because a shell convection zone associated with hydrogen burning reflects some g-modes and hence suppresses dissipation otherwise expected in the core (Saio *et al.* 2006).

The long dashed and short dashed lines in Fig. 2 indicate ranges where oscillatory convection modes of $\ell = 1$ and $\ell = 2$, respectively, are expected to be observable. It is well known that linear convection modes are monotonously (dynamically) unstable in *adiabatic analyses*. Shibahashi & Osaki (1981) found, however, that the high-degree ($\ell \geqslant 10$) convection modes become overstable (oscillatory) when the nonadiabatic effect is included in luminous ($L/L_\odot = 10^5$) models hotter than the cepheid instability strip.

In our massive star models, it is found that low-degree ($\ell \leqslant 2$) oscillatory convection modes exist associated with the Fe-opacity peak convection zones, and some of these modes are expected to be observable. Since the growth time of a convection mode is short (comparable to the period), the mode is expected to have a large amplitude in the convective zone. Therefore, the visibility of the oscillatory convective modes can be measured by the ratio of the photospheric amplitude to the maximum amplitude in the interior (mostly in the convection zone). Assuming that an oscillatory convection mode is observable when the ratio is larger than 0.2, the boundaries of the visible ranges are shown in Fig. 2 by dashed lines. Oscillatory convection modes are visible in sufficiently luminous ($\log L/L_\odot \gtrsim 4.6$) B-type stars. Although there are many oscillatory convection modes in a star, only one or two modes for a given ℓ are visible because the other modes are well confined to the convection zone. Periods of these modes are comparable to g-modes much longer than those of radial modes. These oscillatory convection modes might be responsible for long-period variations in supergiant stars.

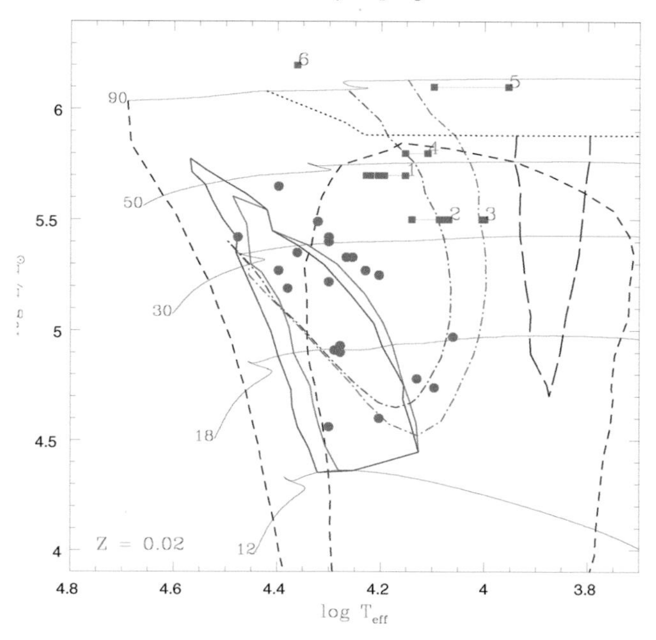

Figure 3. Ranges where (quasi-)periodic variable stars are expected on the HR diagram are compared with supergiant variables analyzed by Lefever *et al.* (2007) (big filled circles) and LBVs by Lamers *et al.* (1998) (filled squares). Positions at different LBV (or S Dor) phases for each star are connected with thin lines; 1 = R 712, 2 = HR Car, 3 = 164 G Sco, 4 = S Dor, 5 = R 1273, 6 = AG Car.

4. Comparison with supergiant variables

Observed positions of variable supergiants on the HR diagram are compared with the instability and visibility boundaries in Fig. 3. Their periods-$T_{\rm eff}$ relations are compared with theoretical ones in Figs. 4 and 5. Fig. 3 indicates that all the hotter ($\log T_{\rm eff} \gtrsim 4$) and luminous ($\log L/L_\odot \gtrsim 4.5$) stars are expected to show (quasi-)periodic variations, which is consistent with the observational fact that no or at most a very little number of stable supergiants exist in a spectral range of O9 – A0 as found by van Genderen (1989).

Big dots in Fig. 3 are relatively less luminous supergiant variables analyzed by Lefever *et al.* (2007). This figure indicates most of them to have masses ranging from $\sim 15M_\odot$ to $\sim 30M_\odot$. They are located on the HR diagram in the g-mode instability regions or visible range of oscillatory convection modes. Fig. 4 compares the periods of these stars as function of the effective temperature with theoretical ones of $15M_\odot$ and $30M_\odot$ models. This figure indicates that for most of these stars periods seem consistent to low-degree high-order g-modes (SPBsg) or oscillatory convection modes. We note, however, that for the coolest three stars the periods are shorter than any of the excited modes. Although these three stars are located on the HR diagram in the region where oscillatory convection modes should be visible (Fig. 3), the periods are much shorter than those of the oscillatory convection modes. If these effective temperatures are accurate, an unknown excitation mechanism might be working in these stars.

Fig. 5 compares theoretical periods of very massive models ($M_i = 50M_\odot$ and $70M_\odot$) with periods of microvariations of some LVB stars analyzed by Lamers *et al.* (1998), each of which has different periods and effective temperature depending on the LBV (S Dor) phases. Figs. 5 and 3 indicate that the microvariations of R 712 (1), S Dor (4) and AG Car (6) are consistent to the properties of oscillatory convection modes.

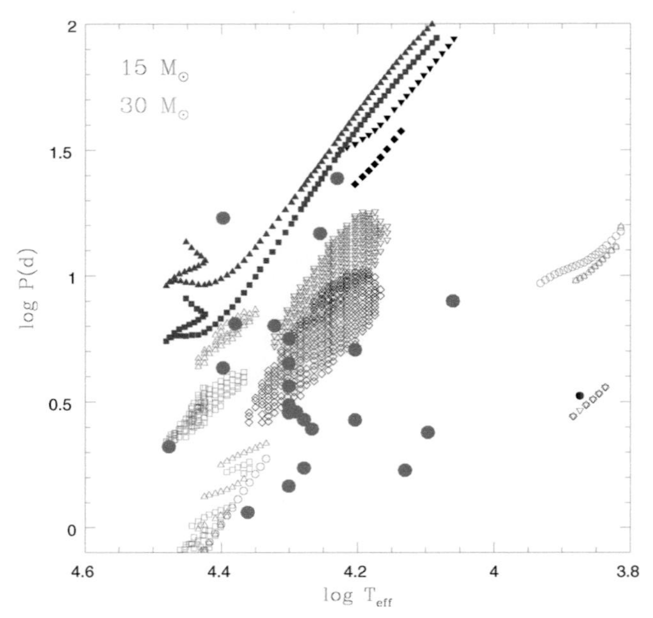

Figure 4. Periods of excited modes and of observable oscillatory convection modes versus effective temperature, where open ($30M_\odot$) and filled ($15M_\odot$) circles are radial modes; triangles ($30M_\odot$) and inverted triangles ($15M_\odot$) are $\ell = 1$ modes; squares ($30M_\odot$) and diamonds ($15M_\odot$) are $\ell = 2$ modes. Convection modes are shown by filled symbols for nonradial modes. Big dots show observed periods-$T_{\rm eff}$ relations of supergiants from Lefever *et al.* (2007).

The periods of microvariations of HR Car (2), 164 G Sco (3), and AG Car (5) are consistent with periods of strange modes. However, luminosities of HR Car and 164 G Sco are too low for the strange modes to exist (Fig. 3). Further investigations are needed to resolve the discrepancy.

5. Summary

We have discussed various instabilities that occur in massive supergiants. Radial modes and nonradial p- and g-modes are excited by the kappa-mechanism at the Fe opacity bump at $T \sim 2 \times 10^5$.

In a star with a very high luminosity to mass ratio of $L/M \gtrsim 10^4 L_\odot/M_\odot$, the strange mode instability works for radial and nonradial modes. Strange modes seem to be responsible for quasi-periodic variations in some of the luminous supergiants.

In addition, it is found that in very luminous models ($\log L/L_\odot \gtrsim 5.9$) a monotonously unstable radial mode exists, which is probably related to the occurrence of an optically thick wind. It is interesting to note that the instability boundary roughly coincides with the Humphreys-Davidson limit.

Furthermore, we found that low-degree ($\ell = 1, 2$) oscillatory convection modes exist in the convection zones caused by the Fe opacity peak, and that some of them can emerge to the stellar surface and hence be observable. The oscillatory convection modes have periods of $10 \sim 10^2$days depending of the effective temperature, which are longer than those of strange modes. The growth-times are comparable to the periods. They seem to be consistent with the properties of long-period microvariations in LVB stars (see Saio 2010 for further discussions).

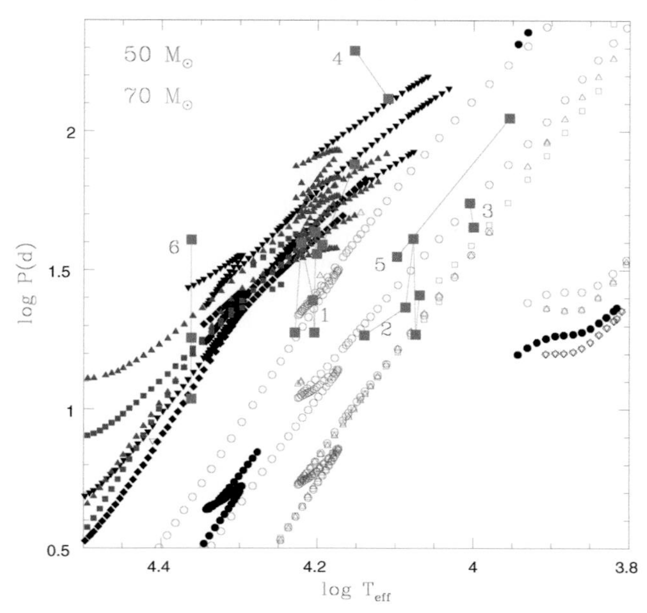

Figure 5. Theoretical periods of excited modes and of observable oscillatory convection modes versus effective temperature, where open ($70M_\odot$) and filled ($50M_\odot$) circles are radial modes; triangles ($70M_\odot$) and inverted triangles ($50M_\odot$) are $\ell = 1$ modes; and squares ($70M_\odot$) and diamonds ($50M_\odot$) are $\ell = 2$ modes. Convection modes are shown by filled symbols for nonradial modes. Big filled squares connected with thin lines show observed periods-$T_{\rm eff}$ relations of LBVs obtained by Lamers *et al.* (1998). Periods at different LBV (or S Dor) phases for each star are connected with thin lines; 1 = R 712, 2 = HR Car, 3 = 164 G Sco, 4 = S Dor, 5 = R 1273, 6 = AG Car.

References

Gautschy, A. 2009, *A&A*, 498, 273
Gautschy, A. & Glatzel, W. 1990, *MNRAS*, 245, 597
Glatzel, W. 1994, *MNRAS*, 271, 66
Glatzel, W., & Kiriakidis, M. 1993, *MNRAS*, 263, 375
Humphreys, R. M. & Davidson, K. 1979, *ApJ*, 232, 409
Iglesias, C. A. & Rogers, F. J. 1996, *ApJ*, 464, 943
Kato, M. & Iben, Jr., I. 1992, *ApJ*, 394, 305
Lamers, H. J. G. L. M., Bastiaanse, M. V., Aerts, C., & Spoon, H. W. W. 1998, *A&A*, 335, 605
Lefever, K., Puls, J., & Aerts, C. 2007, *A&A*, 463, 1093
Saio, H. 2009, *Communications in Asteroseismology*, 158, 245
Saio, H., 2010, *MNRAS*, accepted, arXiv:1011.4729
Saio, H., Baker, N. H., & Gautschy, A. 1998, *MNRAS*, 294, 622
Saio, H. & Cox, J. P. 1980, *ApJ*, 236, 549
Saio, H., Kuschnig, R., Gautschy, A., & Cameron, C. *et al.* 2006, *ApJ*, 650, 1111
Saio, H., Winget, D. E., & Robinson, E. L. 1983, *ApJ*, 265, 982
Shibahashi, H. & Osaki, Y. 1981, *PASJ*, 33, 427
van Genderen, A. M. 1989, *A&A*, 208, 135
van Leeuwen, F., van Genderen, A. M., & Zegelaar, I. 1998, *A&AS*, 128, 117
Vink, J. S., de Koter, A., & Lamers, H. J. G. L. M. 2001, *A&A*, 369, 574

Active OB stars: structure, evolution, mass loss, and critical limits
Proceedings IAU Symposium No. 272, 2010
C. Neiner, G. Wade, G. Meynet & G. Peters, eds.

doi:10.1017/S1743921311011124

The multiplicity of massive stars

Hugues Sana[1] and Christopher J. Evans[2]

[1]Sterrenkundig Instituut Anton Pannekoek, University of Amsterdam,
Postbus 94249, NL-1090 GE Amsterdam, The Netherlands
email: h.sana@uva.nl

[2]UK Astronomy Technology Centre, Royal Observatory Edinburgh
Blackford Hill, Edinburgh, EH9 3HJ, UK
email: chris.evans@stfc.ac.uk

Abstract. Binaries are excellent astrophysical laboratories that provide us with direct measurements of fundamental stellar parameters. Compared to single isolated stars, multiplicity induces new processes, offering the opportunity to confront our understanding of a broad range of physics under the extreme conditions found in, and close to, astrophysical objects.

In this contribution, we will discuss the parameter space occupied by massive binaries, and the observational means to investigate it. We will review the multiplicity fraction of OB stars within each regime, and in different astrophysical environments. In particular we will compare the O star spectroscopic binary fraction in nearby open clusters and we will show that the current data are adequately described by an homogeneous fraction of $f \approx 0.44$.

We will also summarize our current understanding of the observed parameter distributions of O + OB spectroscopic binaries. We will show that the period distribution is overabundant in short period binaries and that it can be described by a bi-modal Öpik law with a break point around $P \approx 10$ d. The distribution of the mass-ratios shows no indication for a twin population of equal mass binaries and seems rather uniform in the range $0.2 \leqslant q = M_2/M_1 \leqslant 1.0$.

Keywords. binaries (including multiple): close, binaries: general, binaries: spectroscopic, binaries: visual, stars: early-type, open clusters and associations: individual (Col 228, IC1 805, IC 1848, IC 2944, NGC 330, NGC 346, NGC 2004, NGC 2244, NGC 6231, NGC 6611, N 11, Tr 14, Tr 16, West 1, 30 Dor)

---

1. Introduction

Massive stars have many fascinating aspects, which extend well beyond stellar physics alone. One of their most striking properties is conceptually very simple: their high-degree of multiplicity. Most O- and early B-type stars are found in binaries and multiple systems. Even single field stars are often believed to have been part of a multiple system in the past, then ejected by a supernova kick or by dynamical interaction. To ignore the multiplicity of early-type stars is equivalent to neglecting one of their most defining characteristics.

In this review we concern ourselves with the multiplicity of stars more massive than 8 $M_\odot$ on the zero-age main sequence, which have spectral types earlier than B3 V. Our approach is to focus on their observational properties, with the emphasis on O-type binaries, although early B-type binaries feature in some of the quoted works. Despite the importance of detailed studies of individual objects, our prime motivation here is to consider the broader results from the literature, in an attempt to lift the veil on some of the general properties of the binary population of early-type stars.

The distributions of the orbital parameters of massive binaries, as a population, are of fundamental importance to stellar evolution, yet remain poorly constrained. These distributions trace the products of star formation and the early dynamical evolution of the host systems, and are necessary ingredients to population synthesis studies. Only with an understanding of these distributions can we hope to recover accurate predictions for some of the exotic late stages of binary evolution.

This contribution is structured as follows. Section 2 describes some of the physical processes and observational biases that are present in multiple systems compared to single stars. Section 3 introduces the different parts of parameter space occupied by massive binaries, and the observational means to investigate them; Section 4 then reviews the multiplicity fraction of OB stars within each regime, and in different astrophysical environments. Section 5 attempts to summarize our current understanding of the parameter distributions of O + OB spectroscopic binaries. Finally, Section 6 provides a summary.

2. Physical processes and observational biases

Binaries are excellent astrophysical laboratories that provide us with direct measurements of fundamental parameters such as stellar masses and radii. Multiplicity induces new processes compared to isolated single stars, offering the opportunity to confront our understanding of a broad range of physics under the extreme conditions found in, and close to, astrophysical objects. Moreover, if one fails to take multiplicity into account, observations (and their analysis) can be significantly biased or misleading. Most critically, early-type binaries with orbital periods of up to 10 years follow significantly different evolutionary paths, an aspect that can also impact the outputs of population synthesis models (e.g., Vanbeveren 2009). By way of additional motivation to understand multiplicity in massive stars, some of the observational and evolutionary impacts include:

Different evolutionary paths: Binarity significantly affects the evolutionary path of the components of the systems compared to single stars. Tidal effects in close binaries modifies the evolution of stellar rotation rates, thus also the induced rotational-mixing of enriched material into their photospheres de Mink *et al.* (2009). Roche-lobe overflow will result in mass and angular momentum transfer, spinning up the secondary to its critical rotation rate Packet (1981); Langer *et al.* (2008). While the gaining star might be rejuvenated by the increase in mass Braun & Langer (1995), the primary will see a reduction in the life-time of its red supergiant phase Eldridge *et al.* (2008). A common-envelope phase and/or stellar mergers are other possible outcomes of binary evolution. The impacts on observed stellar populations are numerous, including modified surface abundances, modified enrichment of the interstellar medium, the rate of supernova and γ-ray burst explosions, and on the number of evolved systems such as Wolf-Rayet stars and high-mass X-ray binaries (e.g., Izzard *et al.* 2006; Brinchmann *et al.* 2008).

Wind collisions: In binaries, the powerful stellar wind from the stars may interact with one another or with the surface of the star with the weaker wind Usov (1992). The supersonic collision heats the gas to temperature up to several 10^7 K Stevens *et al.* (1992). In several cases, the wind-wind interaction is also to accelerate particles up to relativistic energies. The signature of the wind collision can be observed throughout the electromagnetic spectrum, through non-thermal radio (and possibly X- and γ-ray) emission De Becker (2007), through X-ray thermal emission Parkin & Pittard (2010) and via a contribution to the recombination lines in the optical and infrared Sana *et al.* (2001). In massive binaries containing evolved stars with very dense winds, the wind interaction region can act as a nucleation site for dust particles, creating structures such as the pinwheel nebulae Tuthill *et al.* (2008). These effects can provide indirect indiciations of multiplicity. However, if multiplicity is not considered, wind collision can lead to erroneous estimates of fundamental properties such as intrinsic X-ray luminosities Sana *et al.* (2006), spectral classifications, and stellar mass-loss rates (as measured from the strength of, e.g., the Hα line).

Struve-Sahade effect: In its most generalized form, the Struve-Sahade (S-S) effect can be described as the variation in the apparent strength of the spectrum of one or both components when the star is approaching/receding (for an example, see e.g. Sana *et al.* 2001). Various physical effects can induce a S-S signature: gaseous streams in the systems, ellipsoidal variations, surface streams, and changes in the local surface temperature due to, e.g., mutual illumination or heating from a wind-wind collision (e.g., Bagnuolo *et al.* 1999; Linder *et al.* 2007).

Cluster dynamical mass: Ignoring the contribution of binaries to the stellar velocity dispersion in clusters (in both integrated-light observations of distant systems and studies of resolved clusters), can lead to a significant overestimate of their dynamical mass Bosch *et al.* (2009); Gieles *et al.* (2010). For example, some of the disagreement in the mass-to-light ratio of young extragalactic clusters might arise from the binary properties of their red supergiant populations Gieles *et al.* (2010).

Supermassive stars: Unresolved multiple systems have often been confused with very high mass stars due to their large luminosity. Numerous objects have indeed seen their masses revised at the light of improvements of the observing facilities (e.g. the case of R136: Cassinelli *et al.* 1981; Weigelt & Baier 1985; Crowther *et al.* 2010).

3. The parameter space

Before discussing the multiplicity properties of populations of massive stars, we attempt to give the reader a feel for the typical parameter space that needs to be investigated. Our aim is to provide a qualitative overview of the orders of magnitude involved; the values and sketches should only be considered as indicative!

While many more parameters are involved, it is useful to restrain our discussion to a two-dimensional space. Indeed the detection efficiency of most of the observing techniques can be discussed in terms of the orbital separation (or, equivalently, of the orbital period) and of the mass- or flux-ratio of the components. For a given evolutionary stage, the mass-ratio can directly be related the flux ratio and we will therefore assume a direct equivalence between these two values. This simplified approach assumes that observations with sufficient time-sampling are available, and knowingly neglects the second-order effects of eccentricity and orbital inclination on the detection probabilities.

Mass-ratio ($q = M_2/M_1$): In principle, the range of possible mass-ratios spans equal-mass binaries ($q = 1.0$) to a system with a massive star with a light companion ($q << 1$). For example, an O5 + M8 system would have a mass ratio of only $q \sim 0.002$. Of course, a companion with such a low mass would be very hard to detect, but the absence of observational clues does not preclude their existence. There are other observational issues, such as the likelihood that low-mass companions are still in the pre-main sequence phase – observations at longer wavelengths could provide crucial information in this scenario. The range of flux-ratios that require scrutiny can reach up to 10^5, providing a significant observational challenge.

Separations (d): An estimate of the minimal separation can be adopted as the distance at which two main-sequence stars would enter a contact phase. For typical O- and early B-type primaries, this corresponds to rough separations of 20 $R_\odot$ or 0.1 AU, equating to periods of 1-2 days depending of the system mass. The outer separation boundary is more of a grey zone that depends on both the system environment and on the timescale involved. In this context, we consider two arguments. The first makes the distinction between *hard* and *soft* binary systems, i.e., between systems that have a large likelihood of surviving a three-body interaction, versus systems that will be easily disrupted. Heggie (1975) defined *hard* binaries as systems in which the binding energy (E_b) is larger than the kinetic energy (E_k) brought about by an encounter :

$$|E_b| > E_k(encounter) = \frac{<m><v^2>}{2}, \tag{3.1}$$

where $<m>$ and $<v^2>$ are the typical mass and velocity dispersions of stars in a given cluster. Following Portegies Zwart *et al.* (2010) and adopting an effective cluster radius of 1 pc and cluster masses in the range 2.5×10^3 to 10^5 $M_\odot$, one estimates the maximum separation of *hard* binaries to be in the range of 10^3 to several 10^4 A.U.

A second more qualitative argument emphasized by Maíz Apellániz (2010) points out that massive stars have short life-times. One could therefore limit the parameter space to orbital periods of 10^5 to 10^6 yr as only these systems would accomplish a significant number of orbits during their life-time. Following the third Kepler law, this also corresponds to typical separations of several 10^4 AU. Interestingly, this means that most of the massive binaries are hard binaries, that will be difficult to disrupt over their life-time. The observed maximum range of separations considered here is in line with the statement of Abt (1988) that the more massive stars can sustain companions up to several 10^4 AU or more.

Observational techniques: Investigating such a large parameter space requires a combination of techniques (Fig. 1), each characterized by their own sensitivities and observational biases. Short-period close binaries are probed efficiently through spectroscopy, while very wide binaries, with angular separations larger than a couple of arcseconds can be detected by classical, high-contrast imaging. Enhanced imaging techniques such as adaptive optics (AO) and lucky imaging can provide about an order of magnitude in terms of closer separation and can also reach large flux contrasts. In principle, the gap between the spectroscopic and imaging regimes can be bridged with speckle interferometry, and ground-based and space interferometry. Speckle interferometry has the potential for large surveys but, to date, its applications have been limited to flux ratios of about ten Mason *et al.* (2009). Space and ground-based interferometry can reach separations of milliarcsecond scales, at flux ratios of up to 100, but are much more costly to operate and no large survey has yet been attempted.

Combining these various methods allows us in principle to explore the full range of separations for massive binaries out to a distance of ≈ 5 kpc. In practise, these techniques are not equally sensitive and do not offer the same detection probability in their respective regions of parameter space. For example, spectroscopy is very efficient for short-period binaries, with periods of up to a couple of years. The detection probability however decreases dramatically for long-period systems (see, e.g., Fig. 2 of Evans *et al.* 2010), in part due to the reduced radial velocity (RV) signal and also due to the longer timescales involved. Moreover, eccentric systems are harder to detect due the narrower window (sometimes less than a tenth of the orbital cycle) during which the RV variations are concentrated. Imaging techniques (classical, lucky, or AO-corrected) share a common bias in which the achievable contrast varies as a function of the separation (see e.g., Fig. 2 of Maíz Apellániz 2010).

Detailed comprehension of the limitations of each technique and of their observational bias is of prime importance in order to retrieve the global multiplicity properties of massive star populations.

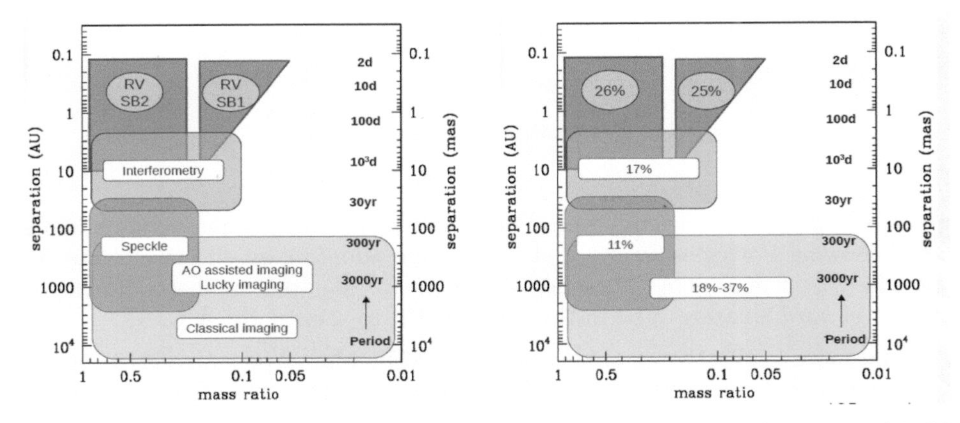

Figure 1. Left-hand panel: typical parameter space for massive binaries. A primary of 40 $M_\odot$ at a distance of 1 kpc has been assumed to construct this sketch. The relevant regions for various detection techniques have been overlaid. Right-hand panel: measured multiplicity in those parts of parameter space (see text for details).

4. The multiplicity fraction of O-type stars

4.1. *Spectroscopic binary fraction in various separation regimes*

The right-hand panel of Fig. 1 gives an overview of the results from recent surveys, including the minimum multiplicity fraction obtained in each part of parameter space from the relevant technique:

Spectroscopy: The most comprehensive overview of the spectroscopic binary (SB) fraction is provided by Mason *et al.* (2009). Based on a review of the literature covering more than 300 O-type objects, these authors found over half of the sample to be part of a SB system. The systems are separated, almost equally, into single- (SB1) and double-lined (SB2) systems.

Speckle interferometry: In the same paper, Mason *et al.* (2009) provide speckle observations of 385 O-type stars, thus covering almost all of the targets in the Galactic O star catalog Maíz-Apellániz *et al.* (2004). 11% of the objects in the Mason *et al.* (2009) sample are found to have speckle companions.

Enhanced imaging techniques: At larger separations, AO-corrected and lucky imaging surveys (respectively, Turner *et al.* (2008) – 138 O stars – and Maíz Apellániz (2010) – 128 O stars) found that 37% of the O stars are part of wide multiple systems. These two studies are mostly limited to the northern hemisphere and are thus missing some of the richer massive star clusters and associations in the southern sky. Part of this gap is filled by the AO campaigns of Duchêne *et al.* (2001) and Sana *et al.* (2010b) on, respectively, NGC 6611 and Tr 14. Both studies revealed a lower multiplicity fraction of 18% for their sample of OB stars. Yet, (part of) this difference results from the fact that these two regions are dense clusters. In these environments, disentangling the true pairs from chance alignment with stars in the same clusters becomes more challenging and only a smaller separation range can be investigated reliably. Interestingly, both Duchêne *et al.* (2001) and Sana *et al.* (2010b) concluded that OB stars have more companions than lower mass-stars.

Interferometry: As mentioned earlier, interferometry is less suitable for surveys. To the best of our knowledge, only one homogeneous survey has been attempted so far. Nelan *et al.* (2004) targeted a limited sample of 23 O-type stars in the Carina region with the *Hubble Space Telescope* fine guidance sensor, resolving close-by companions for four stars.

Table 1. Overview of the spectroscopic binary fraction in clusters.

Object	# O stars	Binary fraction[a]	Ref	Object	# O stars	Binary fraction[a]	Ref.
	Nearby clusters				Distant/extragalactic clusters		
NGC 6611	9	0.44	1	West 1	20	0.30	9
NGC 6231	16	0.63	2	30 Dor	54	0.45	10
IC 2944	14	0.53	3	NGC346	19	0.21	11
Tr 16	24	0.48	4	N11	44	0.43	11
IC 1805	8	0.38	5	NGC2004	4	0.25	11
IC 1848	5	0.40	5	NGC 330	6	0.00	11
NGC 2244	6	0.17	6				
Tr 14	6	0.00	7		Milky Way O star population		
Col 228	15	0.33	8	Clusters & OB associations	305	0.57	12

Notes: [a] The quoted binary fraction is a lower limit as each new detection will increase it.
References: 1. Sana *et al.* (2009), 2. Sana *et al.* (2008), 3. Sana *et al.* (2010a), 4. Literature review, 5. Hillwig *et al.* (2006), 6. Mahy *et al.* (2009), 7. Penny *et al.* (1993), García *et al.* (1998), 8. Sana *et al.* (in prep.), 9. Ritchie *et al.* (2009), 10. Bosch *et al.* (2009), 11. Evans *et al.* (2006), 12. Mason *et al.* (2009)

Combining information from these various ranges, a minimum multiplicity fraction close to 70% for the population of Galactic O-type stars is reached Mason *et al.* (2009). Given the detection limits of these campaigns, there is ample scope for the true multiplicity fraction to be even larger.

Despite the quality of the observations collected so far, improvements are still needed in each of the ranges covered by the various observing techniques described above:

- Homogeneous AO and lucky imaging campaigns have been mostly limited to the northern sky. Extending such work to the rich and dense clusters and star-formation regions of the southern hemisphere is highly desirable,
- Higher flux contrasts are needed in the 10-100 mas separation regime. Techniques such as sparse-aperture masking coupled with AO could, in principle, bring some improvements,
- The separation range 5-100 AU remains almost unexplored,
- About half the known and suspected SBs lack an orbital solution. As a consequence, the distribution of the the orbital parameters remains largely uncertain (see also Section 5).

4.2. *Spectroscopic binary fraction in clusters*

Mason *et al.* (2009) investigated the dependence of the SB fraction on environment by comparing stars from clusters and associations with runaway and field stars, finding that the first category harbours many more binaries and multiple systems. This picture is mostly consistent with an ejection scenario for the field/runaway stars in which most of the multiple systems would be disrupted. In this section, we take a different approach and look for differences in the multiplicity fraction of various clusters. Several authors have indeed proposed the SB fraction to be related to the cluster density (e.g., Penny *et al.* 1993; García & Mermilliod 2001).

To support our discussion, Table 1 summarizes the SB fraction of O-star rich clusters (i.e., clusters with at least five O-type stars), with Fig. 2 giving a graphical comparison of the SB fractions in the various samples. Focusing on the qualitatively homogeneous sample formed by the nearby clusters, we calculate an average binary fraction of $f = 0.44 \pm 0.05$. While some deviations are observed around this average value, each can be explained by statistical fluctuations. Even the extreme case of Tr 14, with no known spectroscopic companions to its six O-type stars, is not statistically significant. For instance, the probability to have six single stars, drawn from an underlying binomial

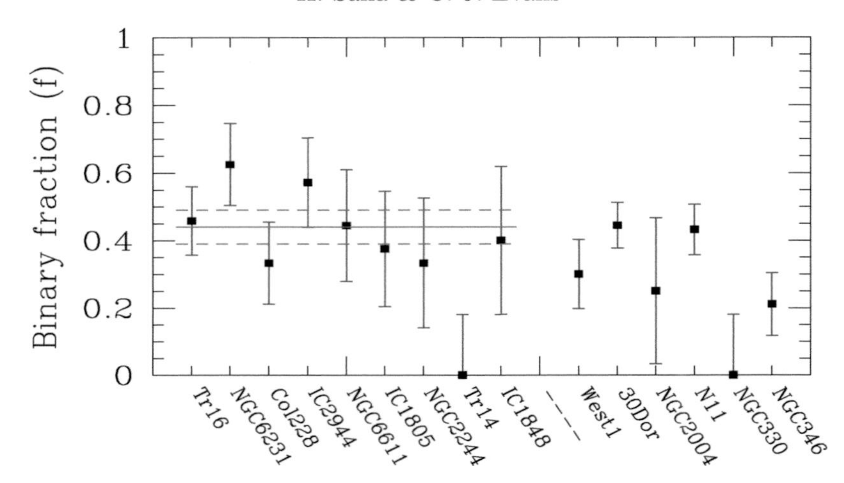

Figure 2. Spectroscopic binary fraction of nearby (left) and distant/extragalactic (right) clusters. The plain line and dashed lines indicate the average fraction and 1σ dispersion computed from the nearby cluster sample.

distribution with a multiplicity fraction of $f = 0.44$ is 3%. Assuming that parent population is the same, the chance of obtaining zero binaries in any one of our nine clusters (given the size of their respective O star population) is 13%, such that we cannot reject the null hypothesis. Of course, the fact that Tr 14 is the densest and possibly the youngest of the nine clusters in our sample is intriguing.

The multiplicity properties from distant and extragalactic clusters are less constrained and should be considered as lower limits, in part because some of these works have a limited baseline and/or a limited number of epochs. Aside from the case of NGC 330, there is again no fundamental disagreement with the results from the nearby cluster sample. With no companions seen for six O-type stars, the NGC 330 sample is similar, in terms of size and binary fraction, to Tr 14. Sample size effects could be invoked (as for Tr 14), but the fact that the much larger population of B-type stars in NGC 330 also show a depleted binary population Evans *et al.* (2006) is appealing. Interestingly, NGC 330 is an older region with a very low surface density, in strong contrast with the properties of Tr 14.

In summary, while some variations of the binary fraction might occur in peculiar situations, the null hypothesis of a common parent distribution cannot be rejected given the current data set. Adopting a uniform binary fraction of $f \approx 0.44$ is thus the most relevant description of the current data. As a direct consequence of this result, one can however reject with a very high confidence the null hypothesis that all O stars are spectroscopic binaries.

5. Distributions of the orbital parameters of spectroscopic binaries

This section provides an overview of our current knowledge of the orbital parameter distributions for O-type spectroscopic binaries. In doing so, it is useful to define two samples (Table 2):

- *The Galactic O-star sample:* mostly based on the sample of Mason *et al.* (2009). While Mason *et al.* (2009) only concentrate on the multiplicity aspect, we perform our own literature review to search for estimates of periods, mass-ratios and eccentricities. When no orbital solution was available, we estimated the mass-ratios of SB2 systems by

Table 2. Overview of the two O star samples used to derive the distributions of the orbital parameters. The first part of the table indicates the number of O stars, the number of O-type binaries and the binary fraction of the two samples. The second part of the table provides the number and the fraction of systems with constraints on their periods, mass-ratios and eccentricities.

#	Galactic O stars	Nearby rich clusters
# O stars	305	82
# binaries	173	38
Binary fraction	0.57	0.46
# periods	102 (59%)	33 (87%)
# mass-ratios	76 (44%)	29 (76%)
# eccentricities	86 (50%)	30 (79%)

Note: The sample of nearby clusters is formed by IC 1805, IC1848, IC 2944, NGC6231, NGC 6611 and Tr16.

adopting typical masses for the components as a function of their spectral classification Martins *et al.* (2005). Compared to the review of Mason *et al.* (2009), we also include information that became available in the last two years, as well as preliminary results from our work.

- *The nearby O-star rich clusters:* a subsample of the Galactic O-star sample, focusing on the O-star rich clusters within $\approx$ 3 kpc. These clusters have been more thoroughly studied so that the scope for observational biases is more limited.

The binary fraction of the two samples appear to be different, with the Galactic O-star sample displaying more binaries. A possible explanation for this is provided by García & Mermilliod (2001), who noted that the O stars in poor clusters (i.e., clusters with only one or two O-type stars) were almost all multiple. These clusters are not included in our second sample, which may pull the binary fraction to lower values.

While the Galactic O-star sample is the most comprehensive, only about 50% of the binaries have constraints on their orbital solution (Fig. 3), leaving a lot of room for observational biases. For example, the orbital solutions are more difficult to obtain for long-period high eccentricity systems. There might thus be an uneven representation of various parameter ranges in the observed distribution functions. The situation is much improved for the cluster sample, as almost 80% of the systems have proper orbital solutions and 87% have estimates of the orbital period. We therefore argue that the distributions derived from the cluster sample are much less affected by observational biases. In the following, we will compare the parameter distributions built from the two samples to one another and to analytical distributions commonly used to represent the properties of the massive star binary population.

Period: Fig. 3 provides an overview of the respective samples with the cumulative number distributions of the orbital periods. It shows that the period distribution function obtained from the cluster sample is almost fully constrained, but that uncertainties could still affect the Galactic sample. However, the cumulative distribution functions (CDFs) are mostly in agreement (Fig. 4, left-hand panel). Both CDFs show an overabundance of short periods, with 50 to 60% of the systems having a period shorter than 10 days. Consequently, the CDF of observed periods in the spectroscopic regime can not be represented by the traditional Öpik Law†. As already suggested by Sana *et al.* (2008), a much better representation of the period CDF is provided by a bi-uniform distribution

† Öpik's Law states that the distribution of separations is flat in logarithmic space. The corresponding period distribution should be flat in $\log P$ as well.

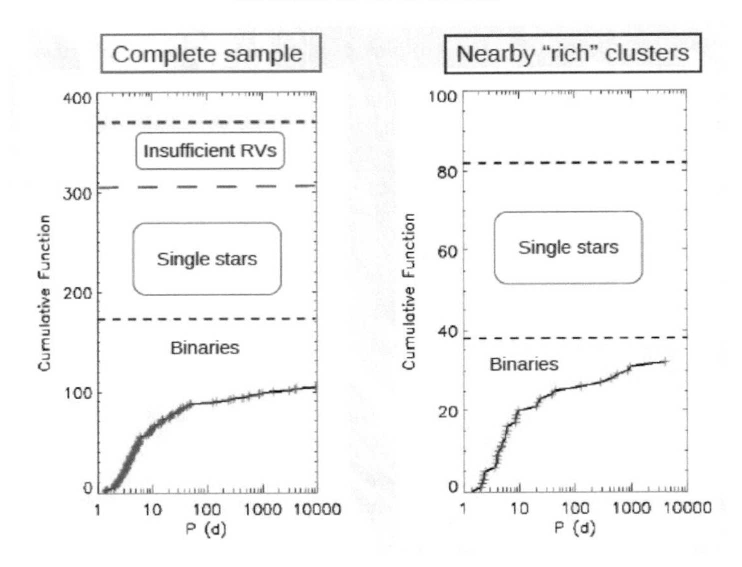

Figure 3. Cumulative number function of orbital periods for the complete sample (left) and for the nearby cluster sample (right). This plot aims to give a graphical impression of the potential biases affecting the two samples. Normalised cumulative distribution functions for systems with solutions are given in Fig. 4.

in $\log P$ (which one could consider a 'broken' Öpik Law) such that:

$$CDF(P) = \begin{cases} \frac{F_{\rm break}\left(\log P - \log P_{\rm min}\right)}{\log P_{\rm break} - \log P_{\rm min}}, & \text{for } \log P_{\rm min} \leqslant \log P \leqslant P_{\rm break} \\ F_{\rm break} + \frac{\left(1-F_{\rm break}\right)\left(\log P - \log P_{\rm break}\right)}{\log P_{\rm max} - \log P_{\rm break}}, & \text{for } P_{\rm break} < \log P \leqslant \log P_{\rm max} \end{cases} \tag{5.1}$$

where P is expressed in days. Adopting a break-point at $P_{\rm break} \approx 10$ d, with upper and lower limits of $\log P/{=}0.3$ and 3.5 d and considering that the binaries are evenly spread in the short and long period regimes (i.e., $F_{\rm break} \approx 0.5$), Eq. 5.1 becomes:

$$CDF(P) = \begin{cases} \frac{5}{7}\log P - \frac{10.5}{7}, & \text{for } 0.3 \leqslant \log P \leqslant 1.0 \\ \frac{1}{5}\log P - \frac{3}{10}, & \text{for } 1.0 < \log P \leqslant 3.5 \end{cases} \tag{5.2}$$

Eqs. 5.1 and 5.2 give an empirical description of the CDF of the observed periods. The latter should still be corrected for the detection probability (mostly affecting longer periods) and for the systems lacking orbital solutions (also more likely to affect the longer-period regime). The exact location of the lower and upper limits and of the 'break' still needs to be more tightly constrained. That said, the general behaviour and the overabundance of short-period spectroscopic binaries appear clear.

Mass-ratio: The CDFs of the mass-ratios (Fig. 4, middle panel) are well reproduced by a uniform distribution in the range $0.2 < q < 1.0$. The Galactic O-star sample shows slightly fewer systems with $q < 0.6$; this can be (partly) explained by observational biases as the detection of the secondary signature for systems with large mass differences (i.e., large flux contrasts) requires very high-quality data that are not always available for the Galactic sample. SB1 binaries represent about 20-25% of the cluster sample. For these stars, one cannot directly estimate the mass-ratio. However, we note that the fraction

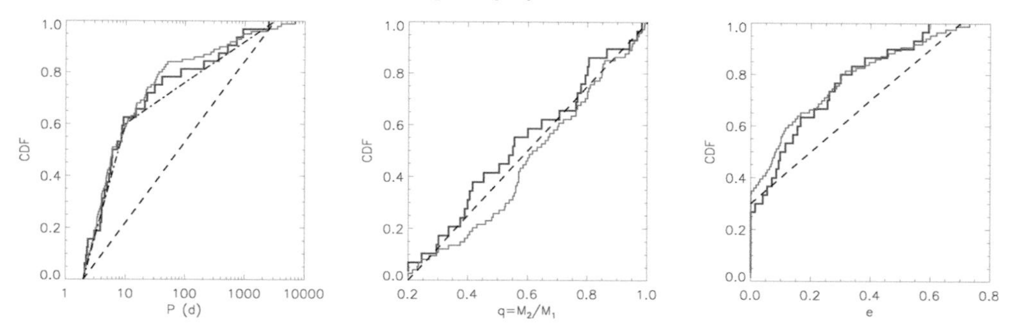

Figure 4. Cumulative distribution functions (CDFs) of the periods (P), mass-ratios (q) and eccentricities (e). The plain thin/magenta and thick/blue lines indicate the CDFs of, the Galactic O-star sample and the nearby cluster sample, respectively. Left-hand panel: the dashed line shows an Öpik Law over this range of periods, while the dot-dashed line indicates the alternative law given by Eq. 5.2. Middle panel: the dashed line indicates a uniform distribution in the considered range. Right-hand panel: the dashed line indicates a uniform distribution for $e > 0$.

of SB1 is roughly compatible with an extension of the uniform CDF towards $q < 0.2$; testing this statement will require detailed simulations.

As a direct consequence of a uniform mass-ratio CDF, the presence of a twin population with $q > 0.95$ proposed by Pinsonneault & Stanek (2006) can be rejected. Another implication resides in the fact that massive binaries cannot be formed by random pairing from a Salpeter/Kroupa IMF. Our results rather suggest the presence of a mechanism that favors the creation of O + OB binaries. Such a mechnaism could find part of its origin in the early dynamical evolution, where companion exchanges favor the capture of more and more massive secondaries. It could also trace a particular formation mechanism Zinnecker & Yorke (2007).

Eccentricity: The CDF of the eccentricities (Fig.4, right-hand panel) is characterized by an overabundance of circular and low eccentricity systems. Indeed, 25-30% of the systems displayed a circular orbit, while another 30% have $e < 0.2$. This behaviour contradicts the expected properties of a purely thermal binary population, which can be qualitatively explained by the large fraction of short-period systems for which tidal dissipation will tend to circularize the orbit.

An analytical description of the observed CDF for eccentric systems can be provided through $CDF(e > 0) \propto e^{0.5}$ in the range $0.0 < e < 0.8$. However, as 20% of the cluster sample and 50% of the Galactic sample are lacking robust eccentricities and as biases are most likely to affect larger eccentricities, we cannot consider this relation as definite. That said, one would expect that $CDF(e)$ will remain overabundant towards low eccentricity systems.

6. Summary

We have attempted to provide an overview of our current knowledge of the important multiplicity properties of massive stars. We described some of the physical processes and observational biases that lead to binaries behaving differently compared to single stars. We then briefly described the observational parameter space that one needs to explore to investigate massive binaries, and we discussed the challenges of probing it homogeneously. Despite these difficulties, it is now well established that the vast majority of O-type stars are part of a multiple system. The typical separation between the multiple components covers at least 4 order of magnitudes.

At least 45-55% of the O star population in clusters and OB associations is comprised by spectroscopic binaries, with a lower fraction found for field and runaway stars Mason *et al.* (2009). Here we have investigated possible variations of the multiplicity fraction among clusters with a rich O star population. While room for small variations remains due to our limited sample and due to the small O star population of some clusters, the binary fraction can mostly be considered as uniform with a value close to 44%. Given the current data set, one can hardly argue that the multiplicity fraction is significantly correlated with the cluster density (at least not in the range covered in our sample) . While density can still play a role, for example, to explain the difference observed between O-star rich and O-star poor clusters, its impact among rich clusters remain questionable in light of the current data. It is well accepted that most O-type stars are part of a multiple systems, but a similar statement does not hold when limiting ourselves to spectroscopic companions. Given the observed SB binary fraction and the sample sizes, it is unlikely that the underlying fraction of SBs is larger than 70-75%.

Finally, we have constructed CDFs for the periods, mass-ratios and eccentricities for two samples of massive binaries. The Galactic O-star sample is more extensive but has been studied less homogeneously. The second sample, based on the O star binary population in six rich nearby open clusters, is more homogeneous and is less susceptible to detection biases. There are some differences in the CDFs of the two samples (see Fig. 4), but two-sided Kolmogorov-Smirnoff tests do not reveal statistically significant deviations. These differences can be qualitatively understood in terms of different observational effects. Currently, the observed CDFs for P, q and e of spectroscopic O-type binaries can be analytically described by the following functions:

- *Periods:* a broken Öpik Law with a break point at $P \sim 10$ days,
- *Mass-ratios:* a uniform distribution down to $q = 0.2$, potentially extending in the SB1 domain (i.e., for $q < 0.2$),
- *Eccentricities:* 25-30% of the characterised systems have circular orbits. $CDF(e > 0)$ shows a square-root dependance with e, but detailed considerations of bias are lacking at present.

A quantitative analysis of the effects of the detection limit and of other observational biases would be highly desirable (although not trivial) in order to: (i) assess the completness and the exactness of the observed CDFs; (ii) retrieve the underlying distributions.

In conclusion, significant progress has been made in the past two decades but uncertainties on the exact multiplicity properties of massive stars remain numerous. In particular, an homogeneous exploration of the parameter space, the distribution function of the orbital parameters and the impact of the environment on the multiplicity properties are likely the areas in which observational progresses are the most crucially needed. Fortunately, numerous projects are currently underway which aim at improving our knowledge of these aspects. It is our hope to have drawn attention to the importance of a proper understanding of the detection limits and of the observational biases that affect each survey. These are necessary information to consider in order to glue all the pieces together toward a global view of the massive star properties across the full reach of parameter space and in different environements.

Acknowledgements

The authors warmly thank the organizers for their invitation and for their flexibility. The authors also wished to express their thanks to M. De Becker, A. de Koter, S. de Mink, M. Gieles, E. Gosset, P. Massey and S. Portegies Zwart for useful discussion in the preparation and redaction of this review.

References

Abt, H. A. 1988, *ApJ*, 331, 922

Bagnuolo, Jr., W. G., Gies, D. R., Riddle, R., & Penny, L. R. 1999, *ApJ*, 527, 353

Bosch, G., Terlevich, E., & Terlevich, R. 2009, *AJ*, 137, 3437

Braun, H. & Langer, N. 1995, *A&A*, 297, 483

Brinchmann, J., Kunth, D., & Durret, F. 2008, *A&A*, 485, 657

Cassinelli, J. P., Mathis, J. S., & Savage, B. D. 1981, *Science*, 212, 1497

Crowther, P. A., Schnurr, O., Hirschi, R., Yusof, N. *et al.* 2010, *MNRAS*, 408, 731

De Becker, M. 2007, *A&AR*, 14, 171

de Mink, S. E., Cantiello, M., Langer, N., & Pols, O. R. *et al.* 2009, *A&A*, 497, 243

Duchêne, G., Simon, T., Eislöffel, J., & Bouvier, J. 2001, *A&A*, 379, 147

Eldridge, J. J., Izzard, R. G., & Tout, C. A. 2008, *MNRAS*, 384, 1109

Evans, C. J., Bastian, N., Beletsky, Y., & Brott, I. *et al.* 2010, in: R. de Grijs & J. R. D. Lépine (eds.), *Star clusters: basic galactic building blocks throughout time and space*, IAU Symposium 266, p. 35

Evans, C. J., Lennon, D. J., Smartt, S. J., & Trundle, C. 2006, *A&A*, 456, 623

García, B., Malaroda, S., Levato, H., & Morrell, N. *et al.* 1998, *PASP*, 110, 53

García, B. & Mermilliod, J. C. 2001, *A&A*, 368, 122

Gieles, M., Sana, H. & Portegies Zwart, S. F. 2010, *MNRAS*, 402, 1750

Heggie, D. C. 1975, *MNRAS*, 173, 729

Hillwig, T. C., Gies, D. R., Bagnuolo, & Jr., W. G., Huang, W. *et al.* 2006, *ApJ*, 639, 1069

Izzard, R. G., Dray, L. M., Karakas, A. I., & Lugaro, M. *et al.* 2006, *A&A*, 460, 565

Langer, N., Cantiello, M., Yoon, S.-C., & Hunter, I. *et al.* 2008, in: F. Bresolin, P. A. Crowther, & J. Puls (eds.), *Massive Stars as Cosmic Engines*, IAU Symposium 250, p. 167

Linder, N., Rauw, G., Sana, H., & De Becker, M. *et al.* 2007, *A&A*, 474, 193

Mahy, L., Nazé, Y., Rauw, G., & Gosset, E. *et al.* 2009, *A&A*, 502, 937

Maíz-Apellániz, J., Walborn, N. R., Galué, H. Á., & Wei, L. H. 2004, *ApJS*, 151, 103

Maíz Apellániz, J. 2010, *A&A*, 518, A1

Martins, F., Schaerer, D., & Hillier, D. J. 2005, *A&A*, 436, 1049

Mason, B. D., Hartkopf, W. I., Gies, D. R., & Henry, T. J. *et al.* 2009, *AJ*, 137, 3358

Nelan, E. P., Walborn, N. R., Wallace, D. J., & Moffat, A. F. J. *et al.* 2004, *AJ*, 128, 323

Packet, W. 1981, *A&A*, 102, 17

Parkin, E. R. & Pittard, J. M. 2010, *MNRAS*, 406, 2373

Penny, L. R., Gies, D. R., Hartkopf, W. I., & Mason, B. D. *et al.* 1993, *PASP*, 105, 588

Pinsonneault, M. H. & Stanek, K. Z. 2006, *ApJ* (Letters), 639, L67

Portegies Zwart, S. F., McMillan, S. L. W., & Gieles, M. 2010, *ARAA*, 48, 431

Ritchie, B. W., Clark, J. S., Negueruela, I., & Crowther, P. A. 2009, *A&A*, 507, 1585

Sana, H., Gosset, E., & Evans, C. J. 2009, *MNRAS*, 400, 1479

Sana, H., Gosset, E., Nazé, Y., & Rauw, G. *et al.* 2008, *MNRAS*, 386, 447

Sana, H., James, G., & Gosset, E. 2010a, *MNRAS*, submitted

Sana, H., Momany, Y., Gieles, M., Carraro, G. *et al.* 2010b, *A&A*, 515, A26

Sana, H., Rauw, G., & Gosset, E. 2001, *A&A*, 370, 121

Sana, H., Rauw, G., Nazé, Y., & Gosset, E. *et al.* 2006, *MNRAS*, 372, 661

Stevens, I. R., Blondin, J. M. & Pollock, A. M. T. 1992, *ApJ*, 386, 265

Turner, N. H., ten Brummelaar, T. A., Roberts, L. C., & Mason, B. D. *et al.* 2008, *AJ*, 136, 554

Tuthill, P. G., Monnier, J. D., Lawrance, N., & Danchi, W. C. *et al.* 2008, *ApJ*, 675, 698

Usov, V. V. 1992, *ApJ*, 389, 635

Vanbeveren, D. 2009, *New Astron. Revs*, 53, 27

Weigelt, G. & Baier, G. 1985, *A&A*, 150, L18

Zinnecker, H. & Yorke, H. W. 2007, *ARAA*, 45, 481

Active OB stars: structure, evolution, mass loss, and critical limits
Proceedings IAU Symposium No. 272, 2010
C. Neiner, G. Wade, G. Meynet & G. Peters, eds.

doi:10.1017/S1743921311011136

Evolutionary models of binaries

Walter van Rensbergen, Nicki Mennekens, Jean-Pierre de Greve, Kim Jansen and Bert de Loore

Astrophysical Institute, Vrije Universiteit Brussel
Pleinlaan 2, B-1050 Brussels, Belgium
email: wvanrens@vub.ac.be, nmenneke@vub.ac.be, jpdgreve@vub.ac.be, kim.jansen@telenet.be, cdeloore@pandora.be

Abstract. We have put on CDS a catalog containing 561 evolutionary models of binaries: J/A+A/487/1129 (Van Rensbergen+, 2008). The catalog covers a grid of binaries with a B-type primary at birth, different values for the initial mass ratio and a wide range of initial orbital periods. The evolution was calculated with the Brussels code in which we introduced the spinning up and the creation of a hot spot on the gainer or its accretion disk, caused by impacting mass coming from the donor. When the kinetic energy of fast rotation added to the radiative energy of the hot spot exceeds the binding energy, a fraction of the transferred matter leaves the system: the evolution is *liberal* during a short lasting era of rapid mass transfer. The spin-up of the gainer was modulated using both strong and weak tides. The catalog shows the results for both types. For comparison, we included the evolutionary tracks calculated with the conservative assumption. Binaries with an initial primary below 6 $M_{\odot}$ show hardly any mass loss from the system and thus evolve conservatively. Above this limit differences between liberal and conservative evolution grow with increasing initial mass of the primary star.

Keywords. binaries: general, stars: evolution, stars: mass loss, stars: variables: other, stars: statistics, stars: fundamental parameters, catalogs

1. Introduction

Codes calculating conservative evolution of binaries have been developed by e.g. Paczyński (1967a,b), Kippenhahn, & Weigert (1967), Kippenhahn *et al.* (1967) and Nelson & Eggleton (2001). Mass loss during liberal evolution is defined by a quantity β giving the fraction of mass lost by the donor (subscript d) that is accreted by the gainer (subscript g):

$$\dot{M}_g = -\beta \, \dot{M}_d \quad with\, 0 \leqslant \beta \leqslant 1 \tag{1.1}$$

Conservative evolution is thus characterized by $\beta = 1$. Liberal evolution further depends on the amount of angular momentum which is taken away by the matter that leaves the system. This is characterized by Podsiadlowski *et al.* (1992) as a quantity α, determined by the location where the mass leaves the system. Our liberal code assumes that matter is lost from the hot spot on the gainer (or accretion disk around it) so that the escaping matter removes only the angular momentum of the gainer's orbit. In that case one obtains a time dependent value of α:

$$\alpha = \left(\frac{M_d}{M_g + M_d}\right)^2 \tag{1.2}$$

A typical value of α during the fast and liberal era of Roche lobe overflow during H core burning of the donor (RLOF A) is then $\alpha \approx 0.25$ (q $\approx$ 1). When this liberal era is succeeded by a fast and liberal era of RLOF B during H shell burning, the value of α turns out to be much smaller and can easily be calculated from relation (1.2).

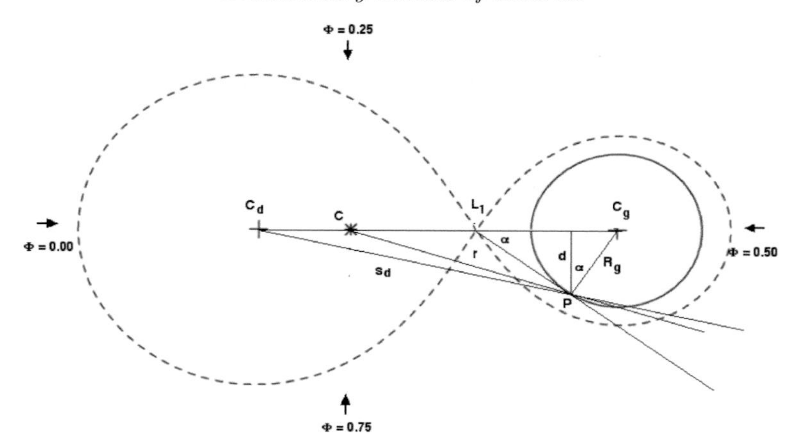

Figure 1. Geometry of a semi-detached binary, showing the impact parameter d and the impact point P at the equator of the gainer or at the edge of its accretion disk. The lines of sight at different orbital phases are indicated by arrows.

Values of $\beta < 1$ were proposed by Meurs & Van den Heuvel (1989) in order to theoretically reproduce the observed numbers of persistent strong massive binary X-ray sources and the number of observed Wolf-Rayet binaries. Binary evolutionary calculations with constant values of β (e.g. $\beta = 0.5$) were then reproduced by various authors, e.g. De Loore & De Greve (1992).

Although mass loss from binaries is needed in evolutionary theory, it would be astonishing that β does not depend on the mass transfer rate. It is more plausible that evolution of semi-detached binaries remains conservative ($\beta = 1$) during the long lasting quiet eras of slow RLOF, and that mass can only be lost from the system ($\beta < 1$) during short lasting violent eras of rapid mass transfer.

Calculations yielding time dependent behavior of β for massive binaries have been published by Wellstein *et al.* (2001), showing binary evolution which is conservative most of the time but severely liberal during epochs of fast mass transfer. A scenario for liberal evolution of binaries with an intermediate mass primary at birth was discussed by Van Rensbergen *et al.* (2008, 2010a,b). Spin-up and hot spots created on the gainer by mass transferred from the donor can drive mass out of a binary. The scenario is discussed in detail in this contribution.

2. Geometry of the system

Spin-up of the gainer and the creation of a hot spot on the gainer's equator are caused by the impact of RLOF-material starting from the first Lagrangian point L_1 and impacting at P as is shown in Fig. 1. This figure also illustrates that the hot spot is turned towards the observer near phase $\Phi = 0.75$ only, as already stated by Peters (2001) for six Algols with a main sequence B-type gainer. Moreover, when the criterion of Lubow & Shu (1975) shows that the gainer is surrounded by an accretion disk, the point P is located at its edge.

3. The spin-up of the gainer

Conservation of angular momentum spins the gainer up due to the impact of RLOF-material coming from the donor. Mass located near the gainer's equator gets loosely

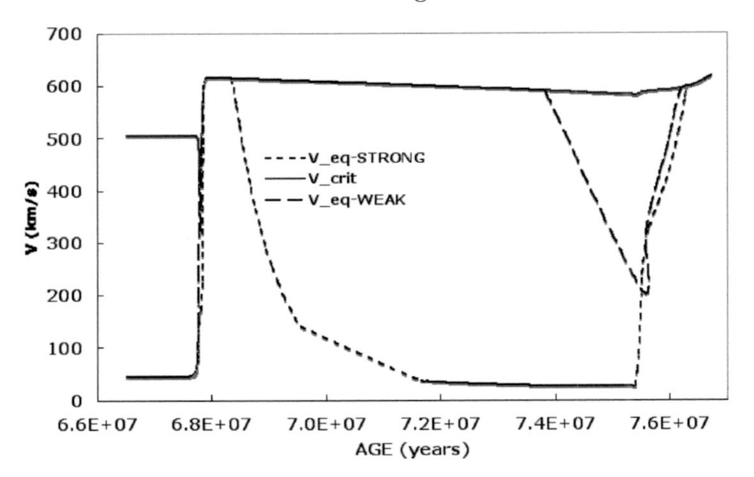

Figure 2. Spin-up of the gainer of a $(6{+}3.6)M_\odot$ binary with an initial period of 3 days.

bound when the gainer rotates rapidly. The spinning-up of the gainer is characterized by an enhancement of its rotational angular momentum ΔJ_g^+ which is given in cgs-units by Packet (1981), corrected with the impact-parameter d shown in Fig. 1:

$$\Delta J_g^+ = 6.05 \times 10^{51} \times \left[\frac{R_g}{R_\odot}\right]^{\frac{1}{2}} \times \left[\frac{M_g}{M_\odot} + \frac{\Delta M_g}{2\ M_\odot}\right]^{\frac{1}{2}} \times \frac{\Delta M_g}{M_\odot} \times \frac{d}{R_g} \tag{3.1}$$

This spin-up is, however, counteracted by tidal interactions which were first studied by Darwin (1879). The formalism for tidal downspinning is taken from Zahn (1977), who gives a suitable approximation for the synchronisation time-scale:

$$\tau_{sync}\ (yr) = q^{-2} \times \left[\frac{a}{R_g}\right]^6 \tag{3.2}$$

This expression uses the semi-major axis a of the binary and a mass-ratio q, in which the star that has to be synchronized is in the denominator. This is the *gainer* in our case, so that $q = \frac{M_d}{M_g}$. Tidal interactions modulate the angular velocity of the gainer ω_g with the angular velocity ω_{orb} of the system. According to Tassoul (2000) one can write:

$$\frac{1}{\omega_g - \omega_{orb}} \times \frac{d\omega_{orb}}{dt} = -\frac{1}{\tau_{sync} \times f_{sync}} = -\frac{1}{t_{sync}} \tag{3.3}$$

Using the moment of inertia I_g of the gainer we find the expression which was used by Detmers *et al.* (2008) in their scenario for *liberal* evolution of massive close binaries:

$$\Delta J_g^- = I_g \times (\omega_{orb} - \omega_g) \times \left[1 - e^{\left(\frac{-\Delta t}{\tau_{sync} \times f_{sync}}\right)}\right] \tag{3.4}$$

Tidal interactions spin the gainer down when $\omega_g > \omega_{orb}$. Tides spin the gainer up when $\omega_g < \omega_{orb}$. Weak tidal interactions are represented with $f_{sync} = 1$ whereas $f_{sync} = 0.1$ implies strong tides. When the upspinning stops at the end of RLOF, tidal interactions settle the system into a situation with $\omega_g = \omega_{orb}$. Expression (3.4) then implies that $\Delta J_g^- = \Delta J_g^+ = 0$. Synchronisation is achieved and angular momentum remains conserved. One sees in Fig. 2 that the gainer is spun up to critical velocity quickly during the rapid era of mass transfer when core H-burning of the donor is at work. This is the case for slow tidal interaction as well as for a binary undergoing strong tides. During the slow phase of mass transfer the rotation of the gainer synchronizes rapidly in the case of strong tidal interaction. In the case of weak tidal interaction, synchronization is never achieved before the onset of RLOF B. The rotation of the gainer spins up again towards critical velocity when a second era of fast mass transfer occurs during H-shell burning of the donor. Rapid rotation favors the gainer to lose mass into interstellar space. The example shown in Fig. 2 turns out to evolve marginally *liberal*.

4. The gainer's hot spot

The temperature of the hot spot is conditioned by that part of the accretion luminosity that can be transformed into radiation. It is customary to define the accretion luminosity by its value that it would have if L_1 is located at infinity: $L_{acc}^{\infty} = G \times \frac{M_g \times \dot{M}_d}{R_g}$. The position of L_1 in a semi-detached binary is however not located at infinity but determined by the geometry shown in Fig. 1. Using the values of the potential energy U in the co-rotating system, the real accretion luminosity for a semi-detached binary is thus given by:

$$L_{acc} = U(L_1) - U(P) = D \times L_{acc}^{\infty} \quad with \ \ 0 < D < 1 \tag{4.1}$$

The quantity D is zero for a contact system with a gainer which is not spun up and does not show a hot spot. Values of D are shown in the last column of Table 1 for the 13 semi-detached binaries that were used for the calibration of the hot spot characteristics.

The efficiency of the liberal scenario is further defined by the value of the factor of radiative efficiency $\tilde{K}$ which defines the radiation pressure caused by L_{acc} in the small surface area of the hot spot. van Rensbergen *et al.* (2010a) showed that in the case of direct impact on the gainer's equator, the quantity $\tilde{K}$ can be calculated with:

$$\tilde{K} \;=\; \frac{\left[\frac{R_g}{R_\odot}\right]^2 \times \left[T_{spot}^4 - T_{eff,g}^4\right]}{\frac{L_{acc}}{L_\odot} \times (5770)^4} \tag{4.2}$$

There are only 13 systems found in the literature with sufficiently reliable observed data to evaluate the quantity $\tilde{K}$ from relation 4.2. According to the criterion of Lubow & Shu (1975) ten systems are direct impact systems, while three others have a transient accretion disk. In the case of the formation of a hot spot on the edge of an accretion disk we have to replace in Eq. (4.2) R_g by R_{disk}. We further replaced $T_{eff,g}$ by T_{disk} for β *Lyr* with an opaque and optically thick accretion disk. Two other systems have a transparent and optically thin accretion disk so that their hot spot is observed as a region with a higher temperature as the underlying stellar surface. Table 1 contains the data for 13 interacting binaries that enable us to find a tentative evaluation of $\tilde{K}$, with a (rather poor) best fit as a function of the total mass of the system. The values mentioned in Table 1 are taken from van Rensbergen *et al.* (2010b) and references therein.

$$\tilde{K} = 3.9188 \times \left[\frac{M_d}{M_\odot} + \frac{M_g}{M_\odot}\right]^{1.645} \tag{4.3}$$

5. The catalog

The tentative calibration of $\tilde{K}$ shown in Eq. (4.3) luckily differs only slightly from the expression found by van Rensbergen *et al.* (2010a) with data for 11 interacting binaries. The latter expression was used to construct the catalog. The catalog covers a grid of binaries between 3 and 15 $M_\odot$ for the initially most massive component, different values

Table 1. Data used to determine $\tilde{K}$ for 13 semi-detached binaries. The first ten binaries are direct impact systems, the last three lines contain disk systems.

System	$M_d + M_g$	$R_d + R_g$	$T_{eff,d}$	$T_{eff,g}$	dM/dt	T_{Spot}	$\tilde{K}$	D
VW Cep	0.90 + 0.25	0.93 + 0.50	5000	5200	8.769E-08	7076	6.528	0.0445
AR Boo	0.35 + 0.90	0.65 + 1.00	5398	5100	1.484E-07	5539	1.064	0.0535
CN And	1.30 + 0.51	1.43 + 0.92	6500	5911	1.215E-07	6485	8.251	0.0261
KZ Pav	0.80 + 1.20	1.66 + 1.50	5000	6500	2.629E-08	7357	12.544	0.280
V361 Lyr	1.26 + 0.87	1.02 + 0.72	6200	4500	2.178E-07	11021	5.185	0.156
RT Scl	1.63 + 0.72	1.67 + 1.02	7000	4800	9.500E-08	9300	36.363	0.0923
CL Aur	1.35 + 2.24	2.51 + 2.58	6323	9420	1.302E-07	10598	45.080	0.178
U Cep	1.86 + 3.57	4.40 + 2.41	4975	11215	5.092E-07	30000	305.631	0.577
U Sge	1.99 + 5.45	5.64 + 4.11	5500	12250	2.036E-06	20000	49.185	0.510
SV Cen	8.56 + 6.05	5.90 + 5.00	14000	23000	1.626E-04	37580	211.238	0.0297
System	$M_d + M_g$	$R_d + R_{disk}$	$T_{eff,d}+T_{eff,g}$	$T_{disk} + T_{edge,disk}$	dM/dt	T_{Spot}	$\tilde{K}$	D
SW Cyg	0.50 + 2.50	4.30 + 3.44	4891 + 9000	6308 + 4968	2.130E-07	13060	106.032	0.587
V356 Sgr	3.00 + 11.00	13.20 + 9.07	8600 + 16500	6174 + 4299	4.442E-07	17050	659.326	0.561
β Lyr	4.25 + 14.1	16.70 + 15.88	13000 + 28000	18279 + 8919	3.440E-05	22590	135.355	0.446

for the initial mass ratio and a wide range of initial orbital periods so that cases A and B are well represented. The evolution of every binary was calculated so as to allow mass to leave the system when the added energy of rapid rotation and radiation from a hot spot exceeds the binding energy of matter located in the hot spot. This situation can only occur during epochs of rapid RLOF, when the mass transfer rate exceeds a well defined critical value. The quantity β, defined in Eq. (1.1), is thus time dependent. It equals unity most of the time, but can become small during eras of fast mass transfer.

The updated catalog contains 561 conservative and liberal evolutionary tracks and is available at the Centre de Données Stellaires (CDS). Binaries with an initial primary mass $\in$ [3-5] $M_\odot$ are calculated in one mode only since they evolve conservatively. Binaries with an initial primary mass $\in$ [6-15] $M_\odot$ are calculated in the liberal mode. Results for evolution with weak and strong tidal interaction are given separately. Conservative tracks are always added so that the reader is able to compare results of liberal (in two different tidal modes) and conservative evolution.

References

Darwin, G. 1879, *Philosophical Transactions of the Royal Society of London*, 170, 447

De Loore, C. & De Greve, J. P. 1992, *A&AS*, 94, 453

Detmers, R. G., Langer, N., Podsiadlowski, P., & Izzard, R. G. 2008, *A&A*, 484, 831

Kippenhahn, R. & Weigert, A. 1967, *ZfA*, 65, 251

Kippenhahn, R., Kohl, K., & Weigert, A. 1967, *ZfA*, 66, 58

Lubow, S. H. & Shu, F. H. 1975, *ApJ*, 198, 383

Mennekens, N., Vanbeveren, D., De Greve, J. P., & De Donder, E. 2010, *A&A*, 515, A89

Meurs, E. J. A. & van den Heuvel, E. P. J. 1989, *A&A*, 226, 88

Nelson, C. A. & Eggleton, P. P. 2001, *ApJ*, 552, 664

Packet, W. 1981, *A&A*, 102, 17

Paczyński, B. 1967a, *AcA*, 17, 193

Paczyński, B. 1967b, *AcA*, 17, 355

Peters, G. J. 2001, in: D. Vanbeveren (eds.), *The Influence of Binaries on Stellar Population Studies*, Astrophysics and Space Science Library 264, p. 79

Podsiadlowski, P., Joss, P. C., & Hsu, J. J. L. 1992, *ApJ*, 391, 246

Tassoul, J.-L. 2000, *Stellar Rotation*, Cambridge University Press (New York)

van Rensbergen, W., De Greve, J. P., De Loore, C., & Mennekens, N. 2008, *A&A*, 487, 1129 van Rensbergen, W., De Greve, J. P., Mennekens, N., Jansen, K., & De Loore, C. 2010a, *A&A*, 510, A13

van Rensbergen, W., De Greve, J.-P., Mennekens, N., Jansen, K., & De Loore, C. 2010b, *A&A* 528, A16

Wellstein, S., Langer, N., & Braun, H. 2001, *A&A*, 369, 939

Zahn, J.-P. 1977, *A&A*, 57, 383

Discussion

KOENINGSBERGER: Are both the mass-donor and mass-gainer rotating synchronously? If not, how did you deal with rotation?

VAN RENSBERGEN: The binary is born with well defined values of masses, radii and orbital periods. Before RLOF starts, there is time to synchronize the system. But after the onset of RLOF, the rotation of the gainer is modulated by spin-up through mass impacting from the donor and spin-down due to tidal interaction. We did not (although we should) take the enhanced rotational velocity of the gainer into account to calculate modifications of its internal structure. Neither did we follow up the rotation of the donor.

MYRON SMITH: I would like to raise your attention to the small group of γ Cas stars. This is an important group of Be stars known for their hard X-ray emissions. γ Cas itself is a widely spaced (P = 504 days, e $\approx$ 0) binary, and two others are arguably blue stragglers. It would be of great interest to know what the products are in terms of the secondary stars. Are these degenerate products white dwarfs, neutron stars or something else? A grid of evolutionary models of binary systems would provide important checks as to how these strange systems have come to be.

VAN RENSBERGEN: Binaries with short initial orbital periods are followed up until over-contact when both stars are still on the main sequence. The evolution of systems with longer initial periods is only calculated until exhaustion of He in the core of the donor. From that moment on further evolution can be predicted. Implementing the data of our catalog into the population code used by Mennekens *et al.* (2010), we find that binaries with an initial mass of the primary below $7M_\odot$ eventually yield many WD+WD systems. Binaries with one or two neutron stars can be expected from the evolution of the most massive systems in our catalog.

MATHIS: You speak about two tidal interactions (the weak and the strong tidal interactions). Could you give more details about this?

VAN RENSBERGEN: For the spin-up of the gainer we used the formalism of Packet (1981), including the fact that the effect diminishes strongly when the orbit of the semi-detached binary is narrow. The spin-down is settled by tidal interaction as formulated in principle by Darwin (1879), treated rigorously in his book on stellar rotation by Tassoul (2000) and modulated by Zahn (1977) who makes a difference between strong and weak tidal interaction. We calculated the evolution of binaries using both types of tides. Although we found some significant individual differences when the mass loss from the system is large, we also found that the global result (e.g. distribution of mass ratios and orbital periods of Algols) remains very much the same.

Active OB stars: structure, evolution, mass loss, and critical limits
Proceedings IAU Symposium No. 272, 2010
C. Neiner, G. Wade, G. Meynet & G. Peters, eds.

doi:10.1017/S1743921311011148

Discussion – Periodic variations and asteroseismology of active OB stars

Thomas Rivinius[1] **and Juan Fabregat**[2]

[1]ESO, Santiago, Chile

[2]Observatorio Astronómico de la Universidad de Valencia, Calle Catedrático Agustín Escardino 7, 46980 Paterna, Valencia, Spain

Abstract. We summarize the discussion held after the session on periodic variations and asteroseismology. The session not only included seven talks, but as well thirty excellent posters were shown. It was impossible to summarize all these in the available frame of a discussion, and so this work focuses on very few sub-topics only mentioned in the actual discussion session. These topics were the relation of pulsation and turbulence, pulsation in close binaries, the observed photometric variability, the connection of pulsations and outburst, and bipolar flows.

Keywords. stars: early-type, stars: emission-line, Be, stars: mass loss, stars: oscillations (including pulsations), line: profiles

---

1. Pulsation and turbulence

In this proceedings are several contributions on the relation between pulsation, in particular the pulsational pattern considering contributions from a set of periods varying in strength over time, and a line broadening parameter that was characterized as a type of turbulence (de Cat, Simón-Díaz *et al.*, Godart *et al.*, all this volume).

In order to set the discussion, first a reminder on the respective terms:

Microturbulence is a parameter, expressed as velocity, that mainly adds equivalent width to the intrinsic profile. The origin is more or less a still unknown, but possibly pulsation-related, velocity distribution related to the subsurface convection. Observationally, it finds its expression in e.g. the requirement to include such a parameter to obtain agreement on the temperature/gravity obtained from ionization balances of various elements, and of the same element over more than two ionization stages. As it is a *local* phenomenon, it does not change the global broadening function of the line profile.

Macroturbulence, in turn, are a sort of still unknown velocity fields over the *total stellar surface*. In the conventional treatment macroturbulence does not change the total equivalent width of a line, whereas micro turbulence of course does. The effect of macroturbulence is, however, to modify the shape of the line as observed in integrated light. This is most obvious in hot and luminous stars, for which already long ago a lack of low $v \sin i$ objects was found. Macroturbulence in this sense is an additional line broadening function.

In this context, it is important to remember that the **line width parameter** called $v \sin i$ is *not* directly the product of the equatorial rotational velocity with the sine of the inclination, but in fact a parameterization for the measured line-width, and various techniques to determine this parameters have to be carefully cross-calibrated in what they actually measure.

Despite the above description of local vs. global, an important technical point was raised 35 years ago by Gray (1975), who actually pointed out that the broadening mechanism for sharp line late type stars can be best described by what he called radial tangential microturbulence models. The physical basis was a picture in which the convection cells were envisioned as rectangular, and Gray described in that way how the broadening mechanism alters the line profiles: it put a lot of power in the wings and very little at small velocities, i.e. in the core. Indeed, if you take a look at the Fourier transform of the resulting line profile, it turns out that that gives you a very different signature from rotation compared to the one you would expect from microturbulence.

Now the interesting question is whether this also applies for hot type stars. Smith & Gray (1976) actually worked on such a test, and came to the conclusion that indeed it seems to be that the same type of microturbulent description provides some basis to explain the observations. Nowadays, 35 years later not only there are much better observing techniques to probe (relatively) sharp lined hot stars much better, but the same is true for the analysis tools. It would be worthwhile to repeat the task done then, because this provides a powerful tool to see whether this type of microturbulence is important in relatively sharp lined hot stars.

Such studies have, in fact, been carried out on high resolution spectral time series of B supergiants to study the effect of microturbulence (Símon-Díaz and Herrero, 2007; Símon-Díaz *et al.*, 2010). Applying the Fourier transform technique it was possible to disentangle rotation and the extra broadening. Very tight correlations exist between the size of the extra broadening, the line profile variations and the asymmetry of the lines, which support the claim that the additional broadening is caused by pulsations, oscillation or some similar kind of time variable phenomena.

For microturbulence, there is as of yet no physical explanation for the velocity field, nevertheless its order of magnitude has been measured with various techniques. When microturbulence is measured in the visual spectral domain, many people who use the O II lines to do so obtain supersonic values. On the other hand, this might be somewhat suspicious in particular for the more luminous objects. For instance, for luminosity class III stars, microturbulence of 20 km/s is obtained sometimes, which is difficult to believe. On the other hand, it seems to be an often overlooked fact that, if one uses ultraviolet diagnostics, much lower microturbulence are usually inferred than those from the visual domain. In the past this has caused some debate until one was mutually convinced the respective other side just had a wrong measurement. However, after close inspection, this effect of the microturbulent velocity, scaling somehow with wavelength, is found to be a real one by Peters (this volume).

In the context of supersonic velocities, S. Owocki reminds that the first order assumption of the sonic velocity as an upper limit for pulsational or turbulent velocities might be misleading: Only when supersonic turbulence runs in small scale, it does quickly produce shocks and gets dissipated, as is supposed in the classical picture. However, on a large spatial scale velocities can well be supersonic and not produce shocks at all. For strictly ordered motions, like rotation, this is intuitively clear, but it is as well true for large scale turbulent (or pulsational) motions. What really counts is the divergence of the velocity field. Thus, if you have a large enough scale, there is no reason why even a turbulent velocity field cannot be supersonic.

2. Pulsations in close binaries

About 20 years ago, there was a thorough spectroscopic study on Spica (α Vir) by Smith (1985a,1985b). Based on the moving bumps apparent in the spectral line profiles it was concluded that Spica shows non-radial pulsation, in a mode that consists of almost exclusive horizontal motions. In case of p-modes, the only way to understand this is in the frame of a Rossby wave, but for g-modes such motions are as well be possible for waves with long intrinsic periods (typically retrograde ones).

However, more recent studies, using modern techniques to confirm non-radial pulsations and mode identification, are lacking, and it seems in fact Spica has been ignored somewhat in terms of pulsation research, either because it was considered as "too bright" to be interesting, or because of its binary nature with a somewhat awkward 4-day period, making it hard to get a complete phase diagram from a single observatory in less than a few years. Consequently, the most interesting question, namely whether and to what extent those bumps and the binarity are connected to each other is still unanswered.

3. Photometric variability

With all the new photometric data in hand, both from long term lensing projects as well as from the new satellite databases, an interesting exercise is to re-visit and compare these with what used to be the state-of-the art of our knowledge before we had such data-strings.

One question that is hoped to be answered with satellite data is that of the photometric variability of O-type stars. O stars display absorption components in the UV lines. Theoretically these can for instance be explained by the presence of bright spots in the surface of the star. If this is correct, however, O stars should present the photometric signature of rotational modulation, and this way the photometric data may give us the observational tools to advance on this question.

However, O stars are a too rare a species to have been caught by chance in number in the fields observed so far, and while one of the CoRoT fields focuses on a cluster with many O type stars, providing a good database for this kind of study, at the moment the results from this observation are still pending. Nevertheless, some first evidence from three stars point to rotational modulation an at least one of them, and a lot of incoherent variability, in the shape of red frequency noise, in the others. On the other hand, de Groot (this volume) does find evidence for solar-like oscillations in O type stars.

Another type of stars for which these data mean a breakthrough are the Be stars. In particular for those stars, 20 to 30 years ago, often claims of double- and even triple-wave variability were made, i.e. variations that produce a better and less scattered phase diagram when they were sorted with twice or three times the period returned from a search for sinusoidal variations. As this terminology has somewhat disappeared from the discussion, one may wonder what might have happened to them, or are they all understood as multiperiodic variables nowadays? This is a particularly relevant question as such light curves were used as an argument to interpret short term variability as rotational modulation.

As it turned out, the non-observation of double wave periods with MOST is probably a selection effect, while they are clearly such light curves present in the CoRoT data. However, with the 150 d long time strings observed so far by CoRoT the answer to the above question is still difficult, as this is a relatively short time-base to resolve closely spaced periods. At least from the point of view of a formal analysis, one can say almost all stars indeed have a dominant frequency and the double of this frequency, i.e. its first harmonic. The amplitude of the first harmonic sometimes is higher, and sometimes weaker, than the first relevant frequency.

4. The pulsation-outburst connection

Be stars have revealed interesting results as well in another respect: Some show a nice correlation between the excitation of the modes, or amplitude of the modes, and the beginning of an outburst. For some Be stars these outbursts have been confirmed spectroscopically, for some there is only photometric data, but the pattern is becoming increasingly clear: There is a connection between the spectroscopic outbursts, i.e. the ejection of material into the close circumstellar environment from the photosphere, and the short-period variability. Examples are HD 49 330 (Huat *et al.*, 2009), Corot 102 719 279 (Gutiérrez-Soto *et al.*, 2010) and Achernar (α Eri, Goss *et al.*, 2010).

What is typically seen is that the overall photometric amplitude is high in times of outburst, and often additional modes become detectable as well. However, at this moment we only have a clear correlation, but the direction of the causality is unclear, i.e. we do not know whether it is the short periodic variability triggering, or even driving, the outburst, or whether it is rather the additional material in the close circumstellar environment altering the properties of the periodic variation, for instance through modification of the outer boundary condition.

Re-observing the stars for which such a correlation was found is crucial to answer the question of regularity for such outbursts and possibly to investigate the causality relation.

5. Bipolar flows

The conference logo itself, showing a star surrounded by a disk and having a bi-polar outflow, poses a question: Do these two properties co-exist in Be stars?

Observational data, obtained with interferometric techniques, supports such flows (e.g. on Achernar, Kervella & Domiciano de Souza, 2006, or α Ara, Meilland *et al.*, 2007). In more massive stars with a higher mass loss than main-sequence B stars not only winds are common but their concentration above the poles for rapid rotators are both theoretically understood and observationally confirmed: A wind gets concentrated towards the pole by very rapid rotation. However, these objects, O-type stars, do not have disks. Some B-type stars do, and while such a phenomenon as polar winds would certainly not be expected for normal B stars, this is less clear for Be stars.

The observations pointing to a polar wind were so far obtained with VINCI (which was a very reliable instrument) and AMBER (which at least at the time of the observations had its problems with absolute visibility calibration). Both instruments then worked in broad-band mode, however, in K or H+K, respectively, meaning they are sensitive only to changes of size in the continuum light. Unfortunately, other interferometric instruments could so far not confirm polar winds.

The optical thickness of such a polar wind to be seen in broad band interferometry, in any case, would have to be non-negligible. However, even if Be stars rotate at critical velocity, a theoretical, radiation driven wind based on a luminosity of the star would not be sufficiently dense, or in other words: the stellar poles even in a critically rotating star are not bright enough. Mass loss rates of at least equivalent to $10^{-5}\dot{M}_\odot$/yr, spread over a small area, would be necessary, and it is hardly feasible that a Be star can do anything close to that locally. Indeed, the known mass loss rates for main sequence B stars are much smaller, and main sequence normal mid-B stars, have almost no detectable wind at all.

In this context, it might be worthwhile to re-visit the IUE spectral database. For instance, in normal Be stars, the observed spectrum has quite weak Si III lines. Modern spectral synthesis codes reproduce the observed Si III lines quite nicely, so this line can be used as a diagnostic for the presence of polar winds in Be stars. One signature of the polar winds in Be stars should be emission in the ultraviolet Si III line. On the other hand, Grady, Bjorkman, & Snow (1987) did not detect polar winds, but rather report enhanced winds above Be stars in intermediate latitudes.

References

Goss, K. J. F., Karoff, C., Chaplin, W. J., & Elsworth, Y. *et al.* 2010, *MNRAS* tmp, 1668
Grady, C. A., Bjorkman, K. S., & Snow, T. P. 1987, *ApJ*, 320, 376
Gray, D. F. 1975, *ApJ*, 202, 148
Gutiérrez-Soto J., Semaan T., Garrido R., & Baudin F. *et al.* 2010, *AN* 331, P51, arXiv:1010.1910
Huat, A.-L., Hubert, A.-M., Baudin, F., & Floquet, M. *et al.* 2009, *A&A*, 506, 95
Kervella, P. & Domiciano de Souza, A. 2006, *A&A*, 453, 1059
Meilland, A., Stee, P., Vannier, M., & Millour, F. *et al.* 2007, *A&A*, 464, 59
Simón-Díaz, S., Herrero, A., Uytterhoeven, K., & Castro, N. *et al.* 2010, *ApJ* (Letters), 720, L174
Simón-Díaz, S. & Herrero, A. 2007, *A&A*, 468, 1063
Smith, M. A. 1985a, *ApJ*, 297, 206
Smith, M. A. 1985b, *ApJ*, 297, 224
Smith, M. A. & Gray, D. F. 1976, *PASP*, 88, 809

Active OB stars: structure, evolution, mass loss, and critical limits
Proceedings IAU Symposium No. 272, 2010
C. Neiner, G. Wade, G. Meynet & G. Peters, eds.

doi:10.1017/S174392131101115X

Very massive binaries in R 136

André-Nicolas Chené[1,2], Olivier Schnurr[3,4], Paul A. Crowther[3], Eduardo F. Lajus[5] and Anthony F. J. Moffat[6]

[1]Departamento de Física, Universidad de Concepción
email: achene@astro-udec.cl

[2]National Research Council of Canada, Herzberg Institute of Astrophysics

[3]Department of Physics and Astronomy, University of Sheffield

[4]Astrophysikalisches Institut Potsdam

[5]Facultad de Ciencias Astronómicas y Geofísicas, Universidad Nacional de La Plata

[6] Département de Physique, Université de Montréal

Abstract. As recent observations have shown, luminous, hydrogen-rich WN5-7h stars (and their somewhat less extreme cousins, O3f/WN6 stars) are the most massive main-sequence stars known. However, not nearly enough very massive stars have been reliably weighed to yield a clear picture of the upper initial-mass function (IMF). We therefore have carried out repeated high-quality spectroscopy of four new O3f/WN6 and WN5-7h binaries in R136 in the LMC with GMOS at Gemini-South, to derive Keplerian orbits for both components, respectively, and thus to directly determine their masses. We also monitored binary candidates and other, previously unsurveyed stars, to increase the number of very massive stars that can be directly weighed.

Keywords. stars: fundamental parameters (masses), stars: Wolf-Rayet, (stars:) binaries: eclipsing, (stars:) binaries: spectroscopic

---

1. Introduction

How massive can the most massive stars get? So far, there is no clear answer to this question since very little is known about the top end of the initial-mass function (IMF) when it comes to the highest masses, mainly because very few of these stars have been observed yet. The most massive star directly weighed so far is NGC3603-A1 with 116 ± 33 $M_{\odot}$ Schnurr *et al.* (2008a). However, there is a severe lack of confirmed, dynamical masses from model-independent, Keplerian orbits of double-eclipsing binaries when it comes to masses above 40 $M_{\odot}$. One of the best locations to find very massive binaries is the greater 30 Doradus region, the famous Tarantula nebula in the LMC. The most massive stars do not have O-type, absorption-line spectra but resemble nitrogen-rich Wolf-Rayet stars (subtype WN5-7ha), i.e. they have an emission-line spectrum. Schnurr *et al.* (2008b) identified five WN binaries around R136, the ionizing cluster of 30 Dor, but were unable to detect the secondaries. Later, Schnurr *et al.* (2009a) reported that one of them, R145, could indeed be a very massive (>120 $M_{\odot}$) star, but had difficulties detecting the secondary, so the masses remain ill-constrained. Clearly, though, R145 is a key system for the study of very massive stars. Schnurr *et al.* (2009b) surveyed the WN5h stars in the very core of R136, but did only find one binary candidate, R136c (BAT99-112). However, the single central WN5h stars are suspected to have initial masses up to 320 $M_{\odot}$ Crowther *et al.* (2010), thereby exceeding the canonical upper-mass limit of 150 $M_{\odot}$ by a factor of two! Skalkowski *et al.* (unpublished) also reported many binaries and binary candidates among the "normal" O-star population around R136. Follow-up

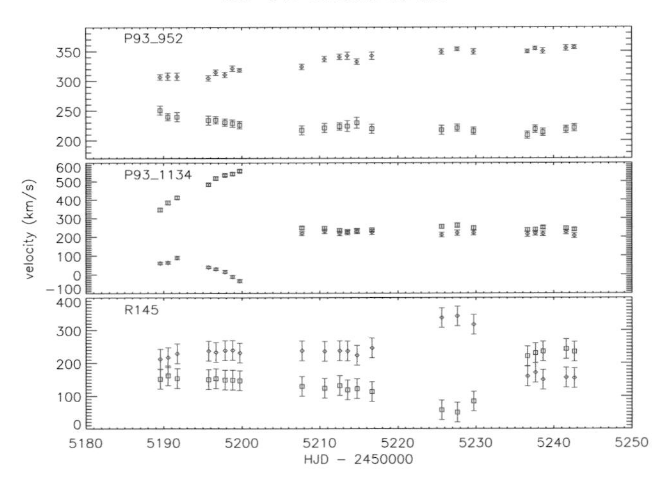

Figure 1. *Top*: RV curve of the primary (red diamonds) and the secondary (blue squares) of the binary system P93-952 (BAT99-107). *Middle*: Same for P93-1134 (BAT99-116). *Bottom*: Same for P93-1788 (R145).

observations are required to confirm binaries, to weigh them, and thus to "fill the gaps" of our knowledge of very massive stars.

2. Preliminary results

Among the O star of our sample, several binaries have been identified, viz. three SB1 (P93-930, P93-974 and P93-1140) and three SB2 systems (P93-805, P93-830 and P93-871). Also, two new binaries were found among the WN stars, i.e. P93-952 (BAT99-107) and P93-1134 (BAT99-116). P93-952 is a "weird" Of/WN star of uncertain spectral type. Moreover, atmosphere models do not yield satisfactory fits. It is moderately bright in X-rays. Schnurr *et al.* 2008b) reporte marginally variable radial velocities (RVs), but the RV curve (Fig. 1) shows that it is in fact a long-period binary consisting of two extreme O3f/WN6 stars! As for P93-1134 (BAT99-116), it is one of the X-ray brightest WN stars known (Portegies Zwart *et al.* 2002) and was suspected of having a compact companion (Schnurr *et al.* 2008b). Our new observations now show that it seems to be a highly eccentric binary containing two extreme emission-line O3f/WN6 stars (see Fig. 1)! Rapid photometry during one night does not find the tentative 1.5h periodicity that was reported in X-rays Guerrero & Chu (2008).
Also, new observations of R145 (BAT99-119) were obtained (Fig. 1). This star was previously found to be a 159-day binary with orbital inclination of $i = 40°$ (Schnurr *et al.* 2009a). Preliminary analysis of the new data suggest it's an equal-mass system consisting of two $\sim 120 M_\odot$ WN6ha stars, so potentially it's the most massive binary known so far!

References

Crowther, P. A., Schnurr, O., Hirschi, R., Yusof, N. *et al.* 2010, *MNRAS*, 408, 731
Guerrero, M. A. & Chu, Y.-H. 2008, *ApJS*, 177, 216
Schnurr, O., Casoli, J., Chené, A.-N., & Moffat, A. F. J. *et al.* 2008a, *MNRAS*, 389, L38
Schnurr, O., Moffat, A. F. J., St-Louis, N., & Morrell, N. I. *et al.* 2008b, *MNRAS*, 389, 806
Schnurr, O., Moffat, A. F. J., Villar-Sbaffi, A., & St-Louis, N. *et al.* 2009a, *MNRAS*, 395, 823
Schnurr, O., Chené, A.-N., Casoli, J., & Moffat, A. F. J. *et al.* 2009b, *MNRAS*, 397, 2049
Portegies Zwart, S. F., Pooley, D., & Lewin, W. H. G. 2002, *ApJ*, 574, 762

Active OB stars: structure, evolution, mass loss, and critical limits
Proceedings IAU Symposium No. 272, 2010
C. Neiner, G. Wade, G. Meynet & G. Peters, eds.

doi:10.1017/S1743921311011161

Using the orbiting companion to trace WR wind structures in the 29d WC8d + O8-9IV binary CV Ser

Alexandre David-Uraz[1] **and Anthony F. J. Moffat**[1]

[1]Département de Physique, Université de Montréal
C.P. 6128, Succursale Centre-Ville, Montréal, QC H3C 3J7, Canada
emails: alexandre@astro.umontreal.ca, moffat@astro.umontreal.ca

Abstract. We have used continuous, high-precision, broadband visible photometry from the MOST satellite to trace wind structures in the WR component of CV Ser over more than a full orbit. Most of the small-scale light-curve variations are likely due to extinction by clumps along the line of sight to the O companion as it orbits and shines through varying columns of the WR wind. Parallel optical spectroscopy from the Mont Megantic Observatory is used to refine the orbital and wind-collision parameters, as well as to reveal line emission from clumps.

Keywords. stars: Wolf-Rayet, stars: winds, outflows

1. Introduction

The primary aim of our project was to probe the structures in the wind of the WR component in the CV Ser WR + O binary by using high-precision photometry. The basic idea is shown in fig. 1. At different phases, clumps with varying sizes will go through the O star's line of sight, thus producing random dips in the light curve. We can then analyze these dips to find constraints on the sizes and shapes of the clumps.

However, in early studies (Hjellming & Hiltner (1963), Stepień (1970), Kuhi & Schweizer (1970), Cowley *et al.* (1971)), CV Ser has proved to be a misbehaving binary system, with the depth of its eclipse varying with time. It was even reported to have stopped eclipsing. The most plausible explanation is that the wind of the dust-forming Wolf-Rayet component changed its structure between observations. However, for the first time, we show evidence for two consecutive eclipses with different depths, which might suggest a rapidly varying mass-loss rate.

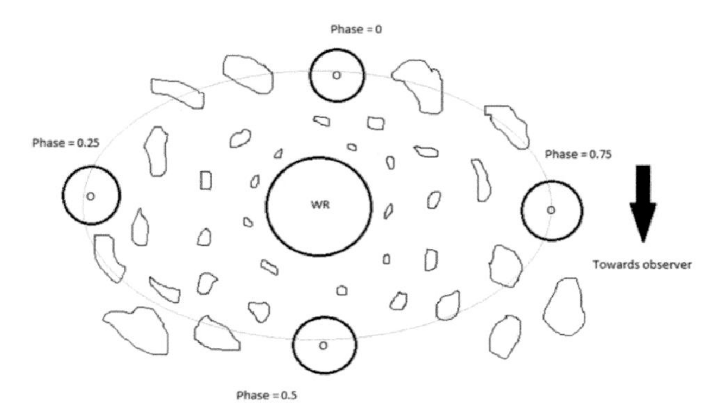

Figure 1. Intuitive model of the effects of clumps on the light curve.

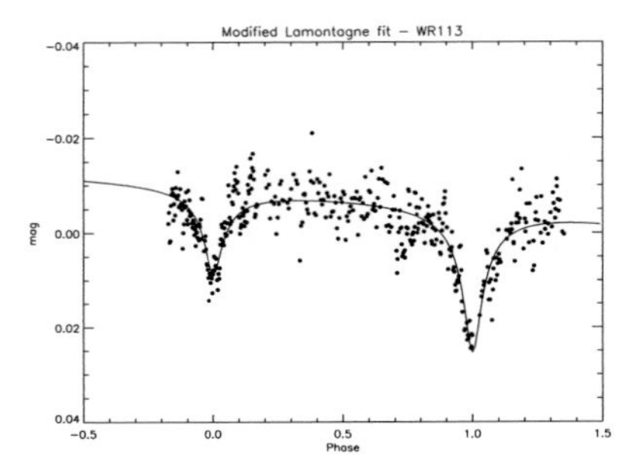

Figure 2. Best fit to the modified Lamontagne *et al.* model.

2. Preliminary results

We tried to fit our light curve using a first order model for atmospheric eclipses, as described in Lamontagne *et al.* (1996):

$$\Delta m = \Delta m_0 + A\left(\frac{\pi/2 + \arcsin\epsilon}{\sqrt{1-\epsilon^2}}\right) \tag{2.1}$$

(with $\epsilon = (\sin i)\cos 2\pi\phi$, $A = \frac{(2.5\log e)k}{(I_{WR}/I_O)}$ and $k = \frac{\alpha\sigma_e \dot{M}}{4\pi m_p v_\infty a}$.)

However, since our two eclipses do not have the same depth, it was necessary to make a small adjustment. We let k vary linearly with time, which could indicate a change in the mass loss parameters (most probably $\dot{M}$ and/or v_∞). We also had to allow Δm_0 to vary linearly with time, most likely due to detector drift. The best fit is shown in fig. 2.

Assuming the values given in Lamontagne *et al.* (1996) for most parameters for CV Ser except $\dot{M}$, the fit yields a mass-loss rate of about $2.9 \cdot 10^{-5} M_\odot/year$ for the first eclipse. If we assume that the variation of the depth of the eclipse is entirely due to the variation of the mass-loss rate, we then find that the Wolf-Rayet mass-loss rate increases by 70% from the first eclipse to the second eclipse!

3. What's next?

There is still much to be done with CV Ser. First, Fourier and wavelet analyses will help detect coherent or random variations in the light curve (which can hopefully be linked to clumps). The spectra of both components of the binary system can be separated using the "shift and add" method. We will also check for changes in the spectra which could be linked to variations in the properties of the WR wind (mass-loss rate, temperature, terminal velocity, ionisation, etc.). Lührs' model can be applied to the excess emission to model the shock cone and compare the value of i obtained to the one given by Lamontagne's model. Ultimately, the goal of this project is to characterize the clumping phenomenon. New data (MOST and ground-based spectra) taken during summer 2010 will help us complete our analysis.

References

Cowley, A. P., Hiltner, W. A., & Berry, C. 1971, *A&A*, 11, 407
Hjellming, R. M. & Hiltner, W. A. 1963, *ApJ*, 137, 1080
Kuhi, L. V. & Schweizer, F. 1970, *ApJ* (Letters), 160, L185
Lamontagne, R., Moffat, A. F. J., Drissen, L., & Robert, C. *et al.* 1996, *AJ*, 112, 2227
Stępień, K. 1970, *AcA*, 20, 13

Active OB stars: structure, evolution, mass loss, and critical limits
Proceedings IAU Symposium No. 272, 2010
C. Neiner, G. Wade, G. Meynet & G. Peters, eds.

doi:10.1017/S1743921311011173

Spectroscopic follow-up of the colliding-wind binary WR 140 during the 2009 January periastron passage

Rémi Fahed[1], Anthony F. J. Moffat[1], Juan Zorec[2], Thomas Eversberg[3], André-Nicolas Chené[4], Filipe Alves*, Wolfgang Arnold*, Thomas Bergmann*, Luis F. Gouveia Carreira*, Filipe Marques Dias*, Alberto Fernando*, José Sánchez Gallego*, Thomas Hunger*, Johan H. Knapen*, Robin Leadbeater*, Thierry Morel*, Grégor Rauw*, Norbert Reinecke*, José Ribeiro*, Nando Romeo*, Eva M. dos Santos*, Lothar Schanne*, Otmar Stahl*, Barbara Stober*, Berthold Stober*, Nelson G. Correia Viegas*, Klaus Vollmann*, Michael F. Corcoran[5], Sean M. Dougherty[6], Julian M. Pittard[7], Andy M. T. Pollock[8], and Peredur M. Williams[9]

[1]Université de Montréal, Montréal, Canada; [2]Institut d'astrophysique de Paris, Paris, France; [3]Schnörringen Telescope Science Institute, Köln, Germany; [4]Herzberg Institute of Astrophysics, Victoria, Canada; [5]Laboratory for High Energy Astrophysics, Goddard Space Flight Center, Greenbelt, USA; [6]Herzberg Institute for Astrophysics, Penticton, Canada; [7]University of Leeds, Leeds, UK; [8]ESA XMM-Newton Science Operations Centre, Villafranca del Castillo, Spain; [9]University of Edinburgh, Royal Observatory, Edinburgh, UK; * MONS pro-am collaboration

Abstract. We present the results from the spectroscopic follow-up of WR140 (WC7 + O4-5) during its last periastron passage in january 2009. This object is known as the archetype of colliding wind binaries and has a relatively large period ($\simeq$8 years) and eccentricity ($\simeq$0.89). We provide updated values for the orbital parameters, new estimates for the WR and O star masses and new constraints on the mass-loss rates.

Keywords. binaries: general, stars: fundamental parameters, stars: Wolf-Rayet, stars: winds, outflows

1. Introduction

WR140 is a very eccentric WC7+O5 colliding-wind binary (CWB) system with an eccentricity of 0.89 and a long period of 7.94 years. It is also the brightest Wolf-Rayet star in the northern hemisphere and is considered as the archetype of CWB. We present here the results from a spectroscopic follow-up, unique in time coverage and resolution. The observation campaign was a worldwide collaboration involving professional and amateur astronomers and took place during a period of 4 months around periastron passage in January 2009.

2. Radial velocities

The WR star radial velocities were measured by cross correlation with a reference spectrum and the O star radial velocities by measuring the centroid of the photospheric absorption lines (see Fig. 1). We notably find a higher eccentricity than previously

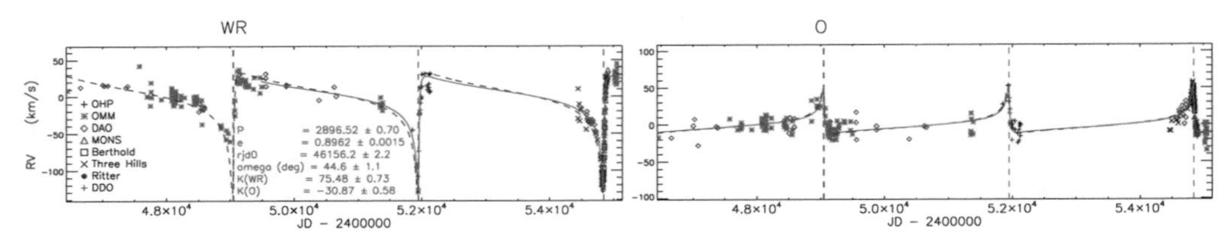

Figure 1. Measured radial velocities of the WR star and of the O star together with the fit for the orbital solution (full line). We included data from the last periastron campaign in 2001 (M03). The black dashed line is the orbital solution from M03. The dashed vertical lines show the position of the periastron passage. The best fit parameters are indicated in grey.

published ($e = 0.896 \pm 0.002$ cf. 0.881 ± 0.005 from Marchenko *et al.* (2003) = M03) and an updated value for the period (2896.5 ± 0.7 d instead of 2899.0 ± 1.3 d).

3. Excess emission

The presence of a shock cone around the O star induces an excess emission that we measured on the CIII 5696 flat top line. This excess emission appears first, just before periastron passage, on the blue side of the line, and then moves quickly to the red side, just after periastron passage, before it disappears. We fitted the radial velocity and the width of this excess as a function of orbital phase using a simple geometric model (Luehrs 1997). We find a value for the inclination of $52 \pm 8°$ (cf. $58 \pm 5°$ from Dougherty *et al.* 2005) which gives the following estimation for the stellar masses : $M_{WR} = 18.4 \pm 1.8\ M_\odot$ and $M_O = 45.1 \pm 4.4\ M_\odot$ (cf. 19 $M_\odot$ and 50 $M_\odot$ from M03). From the half opening angle of the shock cone (Canto *et al.* 1996), we also find a wind momentum ratio $\eta = 0.028 \pm 0.009$.

4. Conclusion

The 2009 periastron campaign on WR140 provided updated values for the orbital parameters, new estimates for the WR and O star masses and new constraints on the mass-loss rates. However, our capability to measure the shock cone parameters with confidence and to understand its underlying physics is limited by the over simplistic approach of our model. A more sophisticated theoretical investigation should be done. Meanwhile, the measured d^{-2} dependency of the excess (where d is the orbital separation) strongly suggests that some kind of isothermal process is involved here. Links with observations in other spectral domains (X,IR and radio) will certainly provide valuable clues about the physics. Finally, we will attempt to isolate the WR spectrum from the O-star spectrum from our data in order to identify the spectral type of the latter more precisely.

References

Canto, J., Raga, A. C., & Wilkin, F. P. 1996, *ApJ*, 469, 729

Dougherty, S. M., Beasley, A. J., Claussen, M. J., & Zauderer, B. A. *et al.* 2005, *ApJ*, 623, 447

Luehrs, S. 1997, *PASP*, 109, 504

Marchenko, S. V., Moffat, A. F. J., Ballereau, D., & Chauville, J. *et al.* 2003, *ApJ*, 596, 1295

Active OB stars: structure, evolution, mass loss, and critical limits
Proceedings IAU Symposium No. 272, 2010
C. Neiner, G. Wade, G. Meynet & G. Peters, eds.

doi:10.1017/S1743921311011185

Pulsations in massive stars: effect of the atmosphere on the strange mode pulsations

Mélanie Godart[1], Marc-Antoine Dupret[1], Arlette Noels[1], Conny Aerts[2], Sergio Simón-Díaz[3,4], Karolien Lefever[5], Joachim Puls[6], Josefina Montalban[1] and Paolo Ventura[7]

[1]Astrophysics, Geophysics, and Oceanography Department, University of Liége, Belgium
email: godart@astro.ulg.ac.be;
[2]Instituut voor Sterrenkunde, KULeuven, Belgium; [3]Instituto de Astrofísica de Canarias, E-38200, La Laguna, Tenerife, Spain; [4]Departamento de Astrofísica, Universidad de La Laguna, E-38205 La Laguna, Tenerife, Spain; [5]Belgisch Instituut voor Ruimte Aeronomie (BIRA), Brussels, Belgium; [6]Universitäts-Sternwarte, Scheinerstrasse 1, 81679 München, Germany; [7]INAF Osservatorio Astronomico di Roma, Monteporzio, Italy

Abstract. Recent space observations with CoRoT and ground-based spectroscopy have shown the presence of different types of pulsations in OB stars. These oscillations could be due to acoustic and gravity modes, solar-like oscillations or even other pulsations of large growth rates. We present a first attempt at interpreting the latter as strange modes.

Keywords. stars: atmospheres, stars: mass loss, stars: oscillations (including pulsations)

1. Introduction: strange modes properties

Massive stars are expected to present β Cephei type modes (low order p and g-modes with periods of the order of hours) and Slowly Pulsating B star (SPB) type modes (high order g-modes with periods around 1 day). Moreover, theoretical models predict, for highly luminous stars ($L/M \sim 10^4$), so called strange mode pulsations (e.g. Saio *et al.* 1998). In this preliminary study, some strange mode properties are presented together with the effect of the atmosphere model on these pulsations. Strange mode oscillations are trapped into a cavity caused by a density inversion located near the iron opacity bump. This inversion appears in the external region of high L/M ratio stars, at an optical depth τ of about 5.3 in log, where the radiation pressure plays an important role. The confinement of the pulsation in a small propagation region alters the behaviour of these modes compared to acoustic or gravity modes: the growth rate is very large and the dimensionless frequencies decrease along the evolution on the main sequence (MS), i.e. with a decreasing effective temperature $T_{\rm e}$ (Fig. 1). As the MS evolution proceeds, the cavity widens and the trapping becomes less effective (Godart *et al.*, 2010).

2. Effect of the atmosphere: FASTWIND v.s. Eddington

When considering a dynamic atmosphere model, FASTWIND (Puls *et al.* 2005), rather than a static grey Eddington atmosphere, a temperature (T) inversion appears at $\log\tau \sim -2.0$ and, therefore, a new trapping region occurs. The T inversion is more pronounced when located at smaller τ, hence the "atmospheric" trapping cavity is more efficient. To characterize the wind, we use the wind strength parameter, $\log Q = \log[\dot{M}/(R v_\infty)^{1.5}]$, whose value is $\log Q \simeq -15$ ($\log Q \simeq -12$) for a negligible (resp. strong) wind. Fig. 2 illustrates the cavity dimension in function of $\log Q$ and $\log T_{\rm e}$; the point size is proportional to its dimension expressed in τ. The location and extent in τ of the T inversion is

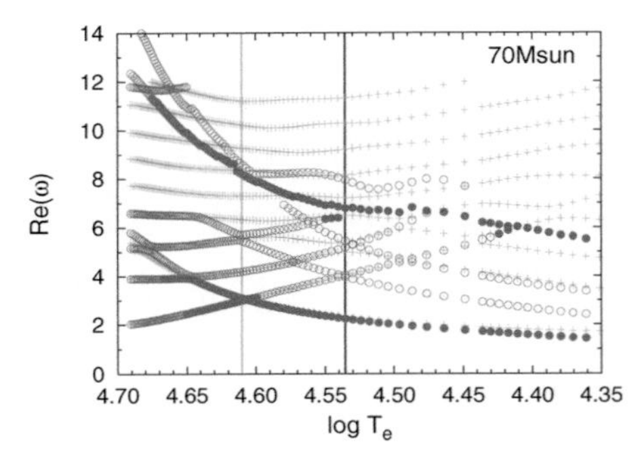

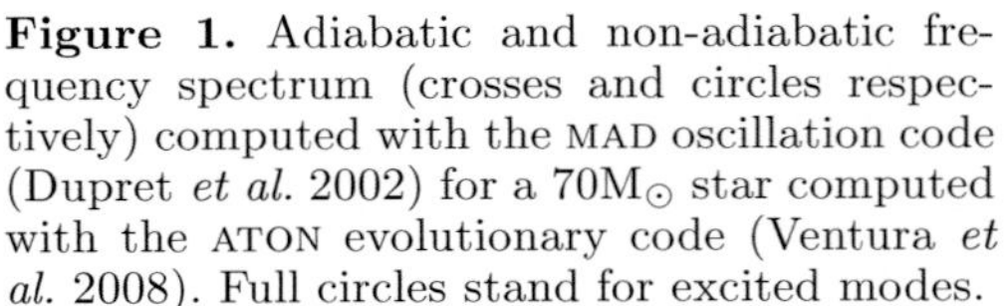

Figure 1. Adiabatic and non-adiabatic frequency spectrum (crosses and circles respectively) computed with the MAD oscillation code (Dupret *et al.* 2002) for a $70\mathrm{M}_{\odot}$ star computed with the ATON evolutionary code (Ventura *et al.* 2008). Full circles stand for excited modes.

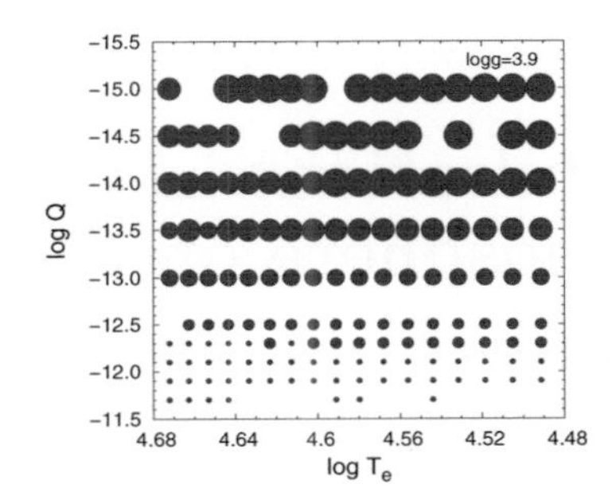

Figure 2. Effect of the wind strength parameter (Q) on the temperature inversion location. The point size is proportional to the cavity dimension expressed in optical depth. The smallest dots stand for a cavity located at $\log\tau > -1$, and the largest ones stand for $\log\tau < -2.5$.

mostly sensitive to $\log Q$. The cavity is thus more effective for weaker winds. We check the effect of the atmosphere model for $40\mathrm{M}_{\odot}$ and $70\mathrm{M}_{\odot}$ at the beginning (ZAMS), the middle (MMS) and the end of the MS (TAMS). No effect from the atmosphere model is noticeable on the strange mode eigenfunctions and frequencies for the $40\mathrm{M}_{\odot}$ models. For the $70\mathrm{M}_{\odot}$ star, no strange mode is found at the ZAMS and the spectrum is not altered by the different atmospheres. Excited strange modes are found in the $70\mathrm{M}_{\odot}$ MMS model, independently of the atmosphere model and the frequencies and eigenfunctions are similar. The "iron bump" cavity of the strange mode propagation then widens and in the $70\mathrm{M}_{\odot}$ TAMS model, the associated strange modes have lost their strange mode behaviour. However, thanks to the new "atmospheric" cavity, an excited strange mode is found, whereas it is obviously not present in the model with an Eddington atmosphere.

3. Conclusion

During most of the MS phase, for stars from 40 to $70\mathrm{M}_{\odot}$, we find excited strange modes, indenpendently of the considered atmosphere model, due to the presence of a trapping cavity near the iron opacity bump. These models are located in a HR diagram region corresponding to O stars, which have been observed by the CoRoT space mission. However no such modes have yet been detected. On the contrary, from CoRoT photometry and from spectroscopy, Aerts *et al.* (2010) detected a strange mode candidate in the B6Ia HD 50064: our $70\mathrm{M}_{\odot}$ TAMS model is located in the same region of the HR diagram. However this model can only excite strange mode with an "atmospheric" cavity, i.e. with a FASTWIND atmosphere model. In that case, we find an excited strange mode with a period of 15 days (for indication only). Such a strange mode could be susceptible of inducing mass loss.

References

Aerts, C., Lefever, K., Baglin, A., & Degroote, P. *et al.* 2010, *A&A*, 513, L11
Dupret, M.-A., De Ridder, J., Neuforge, C., & Aerts, C. *et al.* 2002, *A&A*, 385, 563
Godart, M., Dupret, M.-A., & Noels, A., *et al.* 2010, *AN*, in press
Puls, J., Urbaneja, M. A., Venero, R., Repolust, T. *et al.* 2005, *A&A*, 435, 669
Ventura, P., D'Antona, F., & Mazzitelli, I. 2008, *Ap&SS*, 316, 93
Saio, H., Baker, N. H., & Gautschy, A. 1998, *MNRAS*, 294, 622

Active OB stars: structure, evolution, mass loss, and critical limits
Proceedings IAU Symposium No. 272, 2010
C. Neiner, G. Wade, G. Meynet & G. Peters, eds.

doi:10.1017/S1743921311011197

Non-radial pulsations in the Be/X binaries 4U 0115+63 and SAX J2103.5+4545

Juan Gutiérrez-Soto[1,2], Pablo Reig[3,4], Juan Fabregat[5] and Lester Fox-Machado[6]

[1]Instituto de Astrofísica de Andalucía (CSIC), Glorieta de la Astronomía s/n 18008, Granada, Spain, email: jgs@iaa.es
[2]GEPI, Observatoire de Paris, CNRS, Université Paris Diderot, 5 place Jules Janssen, 92190 Meudon, France
[3]IESL, Foundation for Research and Technology, 71110 Heraklion, Crete, Greece
[4]University of Crete, Physics Department, PO Box 2208, 710 03 Heraklion, Crete, Greece
[5]Observatorio Astronómico de la Universidad de Valencia, Calle Catedrático Agustín Escardino 7, 46980 Paterna, Valencia, Spain
[6]Observatorio Astronómico Nacional, Instituto de Astronomía Universidad Nacional Autónoma de México, Ap. P. 877, Ensenada, BC 22860, Mexico

Abstract. The discovery of non-radial pulsations (NRP) in the Be/X binaries of the Magellanic Clouds (MC, eg. Fabrycky 2005, Coe *et al.* 2005, Schmidtke & Cowley 2005) provided a new approach to understand these complex systems, and, at the same time, favoured the synergy between two different fields: stellar pulsations and X-ray binaries. This breakthrough was possible thanks to the MACHO and OGLE surveys. However, in our Galaxy, only two Be/X have been reported to show NRP: GRO J2058+42 (Kiziloglu *et al.* 2007) and LSI+61 235 (Sarty *et al.* 2009). Our objective is to study the short-term variability of Galactic Be/X binaries, compare them to the Be/X of the MC and to the isolated Galactic Be observed with COROT and KEPLER. We present preliminary results of two Be/X stars, namely 4U 0115+63 and SAX J2103.5+4545 showing multiperiodicity and periodicity respectively, most probable produced by NRP.

Keywords. stars: oscillations (including pulsations), stars: emission-line, Be, X-rays: binaries, stars: individual (4U 0115+63, SAX J2103.5+4545)

1. Introduction

Be/X-ray binaries consist of a neutron star orbiting a Be star. The early-type companion is believed to have the same physical properties as an isolated Be star. However, the structure and evolution of the equatorial disk is affected by the presence of the compact companion Reig (2007). While there are numerous detections of NRP in isolated Galactic Be stars, especially after the launch of the COROT mission (Gutierrez-Soto *et al.*, these proceedings), very little work on the search of variability associated with NRP in Galactic BeX exists. NRP can manifest themselves through multi-period photometric variability by modulating the stellar surface temperature. Typical periods associated with NRP in these stars range from few hours to 2 days.

We have set up a project to investigate the short-term optical photometric variability of BeX in the Milky Way. The main goal of this project is to detect NRP and compare their frequencies and amplitudes with those of isolated Galactic Be stars and with BeX in other galaxies with different metallicity content.

The Galactic Be/X studied in this project are 4U 0115+63, SAX J2103.5+4545, AO 0535+26 and 4U 2206+54. In this paper we only show the results for the two first targets.

Optical differential CCD-photometry was obtained in 3 sites in two observing campaigns between 2008 and 2009. We used the 1.3m telescope at the Skinakas observatory

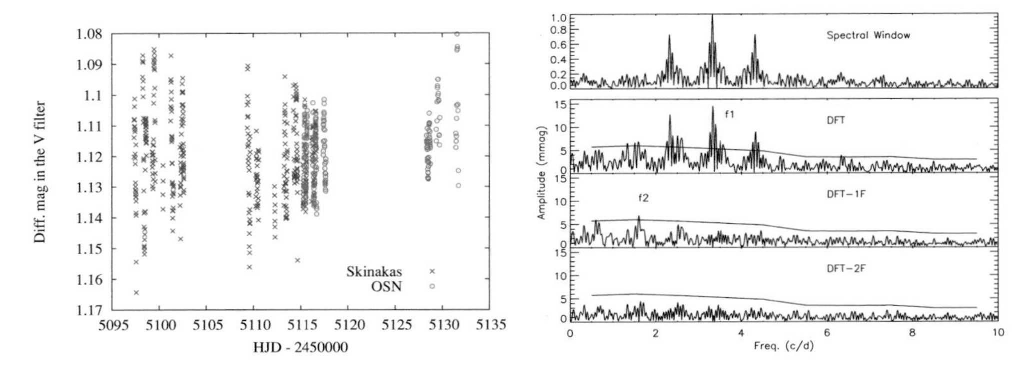

Figure 1. Left: The Skinakas (blue crosses) and the OSN data (purple circles) for the star 4U 0115+63. **Right**: The subsequent periodograms of the Skinakas data for the star 4U 0115+63. The horizontal line represents the 3.5 signal-to-noise level

(Crete), the 0.84 m telescope at the OAN (San Pedro Martir, Mexico) and the 1.5 m telescope at the Observatorio de Sierra Nevada (OSN) located in Granada, Spain. A total of 945 datapoints in 29 nights and 497 datapoints in 22 nights were collected for the stars 4U 0115+63 and SAX J2103.5+4545 respectively. The data were obtained primarily in the Johnson V.

2. Results

The target 4U 0115+63 showed variations in 2008, with a frequency of 3.3 d^{-1} and an amplitude of 18 mmag. In 2009, 2 significant frequencies at 3.33 and 1.60 d^{-1}, with amplitudes of 14 mmag and 7 mmag respectively were detected in the Skinakas data. As seen in the Be stars observed with the COROT mission (eg. Gutiérrez-Soto *et al.* 2009), the relationship between the two largest frequencies is very close to 2:1. The main frequency 3.33 d^{-1} is too high to be the rotational frequency. If we assume that the frequency 1.60 $d^{-1}$is the rotational frequency, the Be star should rotate at critical velocity. Therefore, these variations are most probable due to NRP.

A frequency at 2.23 d^{-1} with an amplitude of 4 mmag was found in the light curve of SAX J2103.5+4545. Using the radius and mass of a star with spectral type B0 derived by Vacca *et al.* (1996), and assuming that the star is rotating at break-up velocity, we calculate a rotational frequency of 1.6 d^{-1}. Therefore, the frequency 2.23 d^{-1} is attributed to non-radial pulsations.

References

Coe, M. J., Negueruela, I., & McBride, V. A. 2005, *MNRAS*, 362, 952
Fabrycky, D. 2005, *MNRAS*, 359, 117
Gutiérrez-Soto, J., Floquet, M., Samadi, R., & Neiner, C. *et al.* 2009, *A&A*, 506, 133
Reig, P. 2007, *MNRAS*, 377, 867
Kızıloğlu, U., Kızıloğlu, N., Baykal, A., & Yerli, S. K. *et al.* 2007, *A&A*, 470, 1023
Sarty, G. E., Kiss, L. L., Huziak, R., & Catalan, L. J. J. *et al.* 2009, *MNRAS*, 392, 1242
Schmidtke, P. C. & Cowley, A. P. 2005, *AJ*, 130, 2220
Vacca, W. D., Garmany, C. D., & Shull, J. M. 1996, *ApJ*, 460, 914

Active OB stars: structure, evolution, mass loss, and critical limits
Proceedings IAU Symposium No. 272, 2010
C. Neiner, G. Wade, G. Meynet & G. Peters, eds.

doi:10.1017/S1743921311011203

Non-radial pulsations in the CoRoT Be Star 102761769

Eduardo Janot Pacheco[1], Laerte B.P. de Andrade[1], Marcelo Emilio[2], Juan Carlos Suárez[3] and Andressa Jendreieck[1]

[1]Instituto de Astronomia, Geofísica e Ciências Atmosféricas, Universidade de São Paulo, Rua do Matão 1226, 05508-090, São Paulo/SP, Brazil

[2]Observatorio Astronomico/DEGEO, Universidade Estadual de Ponta Grossa, Av. Carlos Cavalcanti, 4748 Ponta Grossa, / PR, Brazil - 84030-900

[3]Instituto de Astrofísica de Andalucía (CSIC), Glorieta de la Astronomía, E-18008 Granada, Spain

Abstract. We investigate non-radial pulsations of the CoRoT IR1 Be Star 102761769, with a projected stellar rotation estimated to be 120±15 km/s. If the star is a typical galactic Be star it rotates near the critical velocity. We propose an alternative scenario, where the star could be seen nearly equator-on rotating at a relatively moderate velocity say, $\approx$ 120 km/s and therefore the nonradial oscillations could be modeled. In order to identify the pulsation modes of the observed frequencies, we computed a set of models representative of CoRoT 102761769 by means of the adiabatic pulsation package FILOU. Results indicate that the two frequencies are compatible with a high-g mode as predicted by pulsation models of Be stars.

Keywords. stars: oscillations (including pulsations), stars: rotation, stars: emission-line, Be

1. Photometric and spectroscopic observations

In this work we investigate the variability of the CoRoT IR1 Be Star 102761769 observed in the exoplanet field during 54.6 days. Time series analysis of the light curve was made using both Cleanest and Singular Spectrum Analysis algorithms. We found two close frequencies related to the star at $f_1 = 2.465$ c/d (28.5 μHz) and $f_2 = 2.441$ c/d (28.2 μHz). The precision to which those frequencies were found is 0.018 c/d (0.2 μHz). One low-resolution spectrum of the star was obtained at Observatorio del Roque de los Muchachos (La Palma). The projected stellar rotation is estimated to be 120 km/s $\pm$ 15 km/s from the Fourier transform of spectral lines. If CoRoT 102761769 is a typical galactic Be star it rotates near the critical velocity. The critical rotation frequency of a typical B5-6 star is about 3.5 c/d (40.5 μHz), suggesting that the above frequencies are really due to stellar pulsations rather than linked to the star's rotation.

2. Representative models

In order to identify the pulsation modes of the observed frequencies in the star, we computed a set of asteroseismic models representative of CoRoT 102761769 by varying its main physical parameters (Emilio *et al.* 2010): its mass (M $= 3.8 \pm 0.2\ M_{\odot}$) and equatorial rotation velocity ($v_{rot,s} = 90$ or 120 km/s), assuming a solar abundance of elements (Figure 1). The structure models were computed using the evolutionary code CESAM (Morel, 1997), assuming, as would be plausible for such stars, an uniform rotation for the core and differential rotation for the envelope. For this latter, local conservation of the angular momentum during the evolution was assumed (Suárez *et al.* 2009), resulting in a radial differential rotation profile of the type $\Omega(r) = \Omega_s[1 - \eta_0(r)]$.

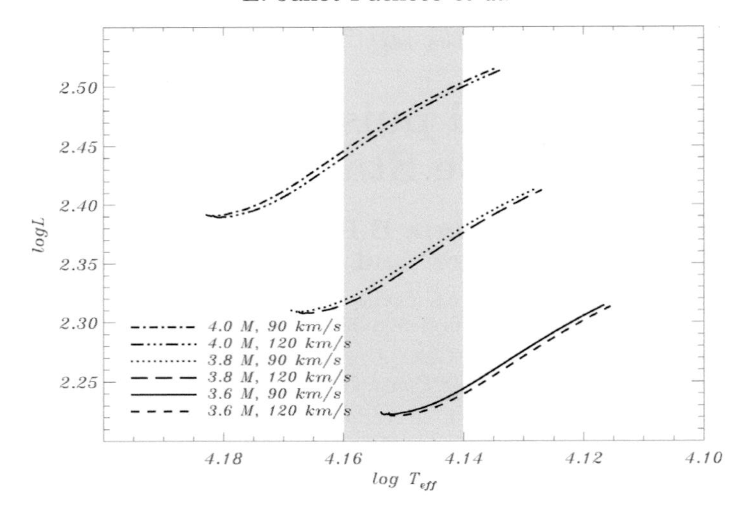

Figure 1. HR diagram showing the tracks of evolutionary paths calculated for CoRoT 102761769, for three different masses (3.6, 3.8 and 4.0 $M_{\odot}$) and two different surface equatorial velocities (90 and 120 km/s). The shaded area corresponds to the uncertainty in effective temperature.

3. Splittings and asymmetries for low l multiplets

Oscillation frequencies were then derived using the adiabatic pulsation package FILOU (Tran Minh & Léon 1995; Suárez & Goupil 2008) presuming the effects of rotation up to second order, and assuming mode coupling as would be expected of a star with high rotational velocity. The code compensates for the equilibrium models without non-spherical components by means of a linear perturbation analysis. A rotation of 120 km/s for an intermediate-mass star can be considered within the limit of validity of the perturbation theory (Suárez *et al.* 2006). Caution must be taken in any case until similar computations with a non-perturbative oscillation theory can be done for realistic stellar models. The resulting frequency spectra were then scanned for those modes that better approach the observed f_1 and f_2, inside an effective temperature uncertainty derived from Huang & Gies (2006) as $\log(T_{eff}) = 4.16 \pm 0.01$ for a B5-6V star. The central value considered for the effective temperature was that of a 3.8 solar mass star in the middle of the main sequence. The results indicate that for all the models considered the frequencies can only be described by a high-g mode, g_5 ($l = 1$, $m = 0$).

Acknowledgements

The authors would like to acknowledge the financial support of IAG-USP, CNPq and FAPESP.

References

Emilio, M., Andrade, L., Janot-Pacheco, E., & Baglin, A. *et al.* 2010, *A&A*, 522A, 43
Huang, W. & Gies, D. R. 2006, *ApJ*, 648, 591
Morel, P. 1997, *A&AS*, 124, 597
Suárez, J. C., Goupil, M. J., & Morel, P. 2006, *A&A*, 449, 673
Suárez, J. C. & Goupil, M. J. 2008, *Ap&SS*, 316, 155
Suárez, J. C., Moya, A., Amado, P. J., & Martín-Ruiz, S. *et al.* 2009, *ApJ*, 690, 1401
Tran Minh, F., & Léon, L. 1995, in: I. W. Roxburgh & J.-L. Masnou (eds.) *Physical Processes in Astrophysics*, p. 219

Active OB stars: structure, evolution, mass loss, and critical limits
Proceedings IAU Symposium No. 272, 2010
C. Neiner, G. Wade, G. Meynet & G. Peters, eds.

doi:10.1017/S1743921311011215

Asteroseismology and rotation in the main sequence

Andressa Jendreieck[1], **Eduardo Janot Pacheco**[1], **Laerte B. P. de Andrade**[1] **and Juan Carlos Suárez**[2]

[1]Instituto de Astronomia, Geofísica e Ciências Atmosféricas, Universidade de São Paulo, Rua do Matão 1226, 05508-090, São Paulo/SP, Brazil

[2]Instituto de Astrofísica de Andalucía, Glorieta de la Astronomía, E-18008 Granada, Spain

Abstract. In this project, we study the effects of stellar rotation on the pulsation predictions for stars in the Main Sequence following the series δ Scu, γ Dor, SPB, Be and β Cep. The objects' rotation in this series span from a few km/s to a few hundreds of km/s. We will compare theoretical predictions yielded by the codes CESAM/FILOU with published data from the MOST and CoRoT satellites. A better diagnostic of the rotation effects on stellar pulsations will help to improve the oscillatory models.

Keywords. stars: oscillations (including pulsations), stars: rotation

1. Introduction

The behavior of the oscillations is determined by the cavity where they are formed, giving direct information of the internal structure of the star (Christensen-Dalsgaard, 2003). Non-radial oscillation has been detected in almost all types of stars in many stages of stellar evolution (e.g., Aerts *et al.*, 2008).

The physical nature of the oscillations are either of the nature of standing acoustic waves (commonly referred to as pressure modes or p modes) or internal gravity waves (g modes). The oscillations modes are described by spherical harmonics with numbers l and m in a radial field with number n (Unno *et al.*, 1989).

2. Rotation

A star with uniforme angular velocity has a coordinate system in the frame rotating with the star (r', θ', ϕ') related with the coordinates (r, θ, ϕ) in an inertial frame through (e.g., Christensen-Dalsgaard, 2003)

$$(r', \theta', \phi') = (r, \theta, \phi - \Omega t) \tag{2.1}$$

Thus the frequency is split uniformly according to m by

$$\omega_m = \omega_0 + m\Omega \tag{2.2}$$

3. Purpose

The project aim to study the effect of rotation in the oscillation spectrum predicted theoretically for stars in the main sequence.

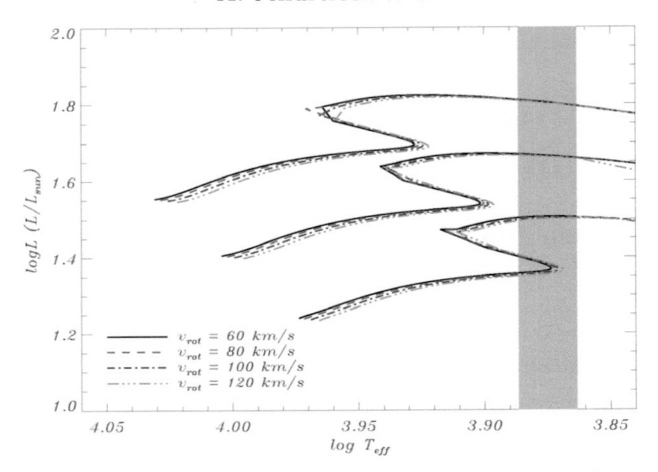

Figure 1. Evolutionary tracks of stars models with masses of 2.0 $M_{\odot}$ (bottom lines), 2.2 $M_{\odot}$ and 2.4 $M_{\odot}$(upper lines). The grey box represents the uncertainty in effective temperature by Poretti *et al.* (2009).

4. Methods

The theoretical predictions are yielded by the codes

- CESAM (Morel, 1997) that calculate the stellar internal structure and evolution;
- FILOU (Suárez & Goupil, 2008) that calculate the oscillations frequencies.

5. Preliminary Results

The preliminary results are for the δ Scu star HD 50844 observed with CoRoT and published by Poretti *et al.* (2009). The evolutionary tracks for a star with masses of 2.0 - 2.4$M_{\odot}$ and rotational velocity 60 - 120 km/s were calculated with CESAM. Figure 1 shows the models that falls inside the uncertainty box. The oscillations frequencies were already calculated with FILOU and are been analyzed.

6. Future Prospects

Compare the frequencies calculated for the star HD 50844 with the observations with CoRoT. Do the same analysis with other stars (different masses and rotation velocity) in the main sequence.

Acknowledgements

I would like to thank IAG-USP, CNPq and FAPESP for financial support in this project.

References

Aerts, C., Christensen-Dalsgaard, J., Cunha, M., & Kurtz, D. W. 2008, *Solar Phys.*, 251, 3

Christensen-Dalsgaard, J. 2003, *Lectures Notes on Stellar Oscillations* (Denmark, Aarhus Universitet), 5th Ed.

Morel, P. 1997, *A&AS*, 124, 597

Poretti, E., Michel, E., Garrido, R., & Lefèvre, L. *et al.* 2009, *A&A*, 506, 85

Suárez, J. C. & Goupil, M. J. 2008, *Ap&SS*, 316, 155

Unno, W., Osaki, Y., Ando, H., Saio, H., & Shibahashi, H. 1989, *Nonradial oscillations of stars* (Tokyo: University of Tokyo Press), 2nd Ed.

Active OB stars: structure, evolution, mass loss, and critical limits
Proceedings IAU Symposium No. 272, 2010
C. Neiner, G. Wade, G. Meynet & G. Peters, eds.

doi:10.1017/S1743921311011227

The WR/LBV system HD 5980: wind-velocity – brightness correlations

Gloria Koenigsberger[1], Leonid Georgiev[2], D. John Hillier[3], Nidia Morrell[4], Rodolfo Barbá[5] and Roberto Gamen[6]

[1]Instituto de Ciencias Físicas, Universidad Nacional Autónoma de México, Cuernavaca, Morelos, 62210, Mexico, email: gloria@astro.unam.mx

[2]Instituto de Astronomía, Universidad Nacional Autónoma de México, Apdo. Postal 70-264, México D.F. 04510, Mexico, email: georgiev@astro.unam.mx

[3]Department of Astronomy, 3941 O'Hara Street, University of Pittsburg, Pittsburg, PA 15260, USA, email: djh@rosella.phyast.pitt.edu

[4]Las Campanas Observatory, The Carnegie Observatories, Colina El Pino s/n, Casillas 601, La Serena, Chile, email: nmorrell@lco.cl

[5]Departamento de Física, Universidad de la Serena, Benavente 980, La Serena, Chile; ICATE-CONICET, San Juan Argentina, email: rbarba@dfuls.cl

[6]Facultad de Ciencias Astronómicas y Geofísicas, Universidad Nacional de La Plata, Box and Instituto de Astrofísica de La Plata (CCT La Plata-CONICET), Paseo del Bosque S/N, B1900FWA, La Plata, Argentina, email: rgamen@gmail.com

Abstract. The massive eclipsing system HD 5980 in the Small Magellanic Cloud presented sudden ∼1–3 mag eruptive events in 1993-1994, the nature of which is still unexplained. We recently showed that these brief eruptions occurred at the beginning of an extended high state of activity which is characterized by large emission-line intensities and that this high state is currently ending (Koenigsberger *et al.* 2010). *Star A*, the more massive member of the 19-day binary, is responsible for the spectacular spectral variations observed over the past 3 decades (see Figure 1). It has a He-enriched stellar wind and is over-luminous for its mass, implying an advanced evolutionary state (Koenigsberger *et al.* 1998). Data obtained over the past 3 decades show that *Star A's* wind speed slowed down as the system brightened. Also present in these data is a correlated increase in emission-line strength, visual and UV brigthness. The latter suggests that the high activity state in HD 5980 may be attributed to a bolometric luminosity increase, consistent with the results of Drissen *et al.* (2001). Hence, HD 5980 may be providing the important clues needed for understanding the behavior of other luminous blue variables and for understainding the evolutionary transition between massive O-type stars and Wolf-Rayet stars.

Keywords. binaries: eclipsing, stars: Wolf-Rayet, stars: individual (HD 5980)

References

Drissen, L., Crowther, P. A., Smith, L. J., Robert, C. *et al.* 2001, *ApJ*, 546, 484
Koenigsberger, G., Pena, M., Schmutz, W., & Ayala, S. 1998, *ApJ*, 499, 889
Koenigsberger, G., Georgiev, L., Hillier, D. J., Morrell, N. *et al.* 2010, *AJ*, 139, 2600

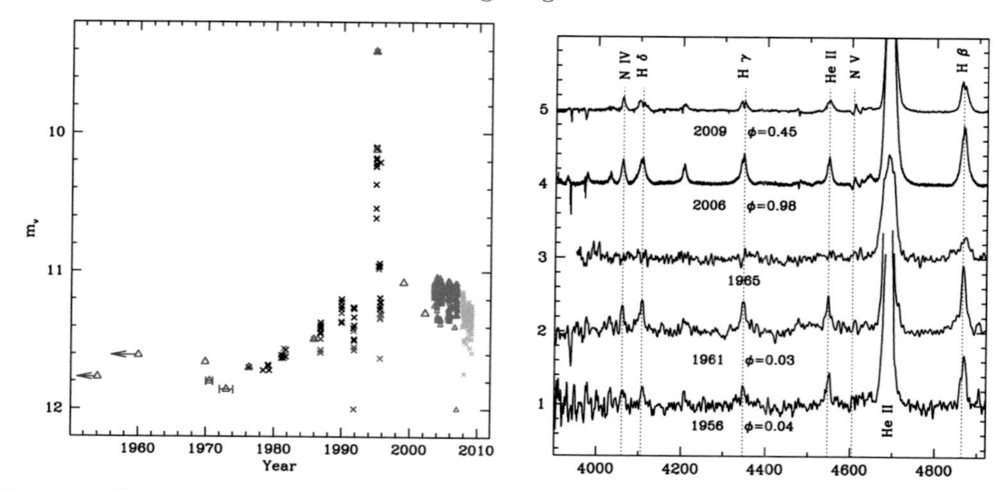

Figure 1. Visual magnitude measurements of HD 5980 from *IUE* (crosses), *SWOPE* and *ASAS* and other sources (left) and evolution of the visual spectrum between 1956 and 2009 illustrating the varying strength in emission-line intensities (right).

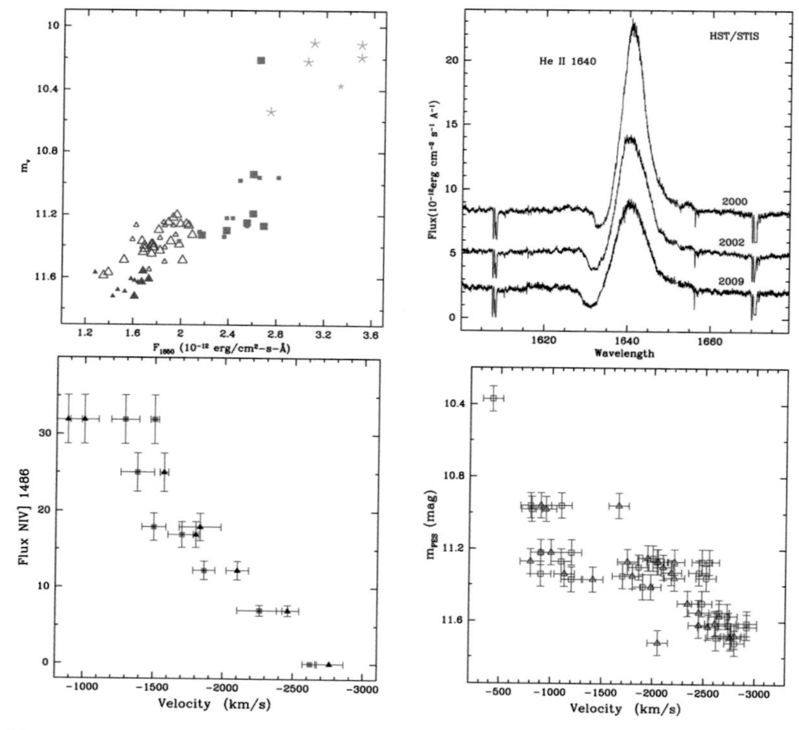

Figure 2. Top: m_v values plotted against the corresponding UV continuum flux level at λ1800 Å, showing a correlated increase in continuum brightness over a broad wavelength region (left). The right-hand plot shows the HST/STIS spectra obtained at orbital phase ~0.0 (when *Star A* eclipses its companion), at 3 different epochs, showing how emission line intensity correlates with extent of P Cygni absorption components. **Bottom:** The emission-line intensity *vs.* wind-velocity correlation (left) and the visual-magnitude *vs.* wind-velocity correlation (right), from *IUE* and *HST/STIS* spectra obtained between 1979–2009.

Active OB stars: structure, evolution, mass loss, and critical limits
Proceedings IAU Symposium No. 272, 2010
C. Neiner, G. Wade, G. Meynet & G. Peters, eds.

doi:10.1017/S1743921311011239

Line profile variability and tidal flows in eccentric binaries

Gloria Koenigsberger[1], Edmundo Moreno[2] and David M. Harrington[3]

[1]Instituto de Ciencias Físicas, Universidad Nacional Autónoma de México,
Cuernavaca, Morelos, 62210, Mexico
email: gloria@astro.unam.mx

[2]Instituto de Astronomía, Universidad Nacional Autónoma de México,
Apdo. Postal 70-264, México D.F. 04510, Mexico
email: edmundo@astro.unam.mx

[3]Institute for Astronomy, University of Hawaii,
Box 2680 Woodlawn Drive, Honolulu, HI, 96822, USA
email: dmh@ifa.hawaii.edu

Abstract. A number of binary systems display enhanced activity around periastron passage which may be caused by the tidal interactions. We have developed a time-marching numerical calculation from first principles that computes the surface deformations, the perturbed velocity field, the energy dissipation rates and the photospheric line-profiles in a rotating star with a binary companion in an eccentric orbit. The method consists of solving the equations of motion for a grid of elements covering the surface of star m_1, subjected to gravitational, centrifugal, Coriolis, gas pressure and viscous shear forces (Moreno *et al.* 1999, Toledano *et al.* 2007, Moreno *et al.* 2011). At selected times during the orbital cycle, the velocities of surface elements on the visible hemisphere of the star are projected along the observer's line of sight and the photospheric line-profile calculation is performed (Moreno *et al.* 2005). Direct comparison with observational photospheric line profile variability is then possible, showing that the general features are reproduced (Harrington *et al.* 2009). In this poster we show the example of a highly eccentric system ($e = 0.8$, $P = 15$ d). The surface deformation changes rapidly from that of an "equilibrium tide" at periastron to one with smaller-scale structure shortly thereafter. The computed line profiles display the presence of large blue-to-red migrating "bumps" around periastron, with smaller scale structure appearing later in the orbital cycle. Because the growth rate of the surface perturbations increases very abruptly at periastron, instabilities are expected to arise which may cause the observed activity and mass-ejection events around this orbital phase.

Keywords. binaries: spectroscopic, stars: oscillations (including pulsations), stars: activity, line: profiles

---

References

Harrington, D., Koenigsberger, G., Moreno, E., & Kuhn, J. 2009, *ApJ*, 704, 813
Moreno, E. & Koenigsberger, G. 1999, *Rev. Mexicana AyA*, 35, 157
Moreno, E., Koenigsberger, G., & Toledano, O. 2005, *A&A*, 437, 641
Moreno, E., Koenigsberger, G. & Harrington, D. M. 2011, *in preparation*
Toledano, O., Moreno, E., Koenigsberger, G., & Detmers, R. *et al.* 2007, *A&A*, 461, 1057

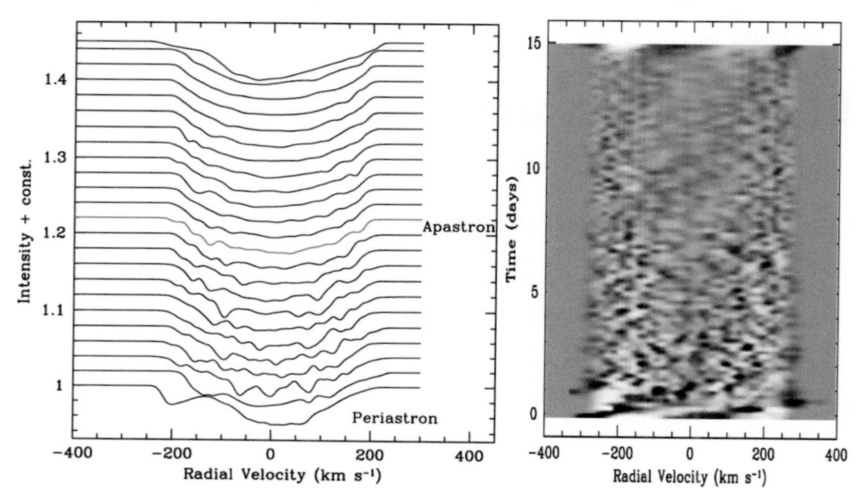

Figure 1. Example of computed photospheric line-profile variability arising in the primary star ($m_1 = 5$ M$_\odot$, $R_1 = 3.2$ R$_\odot$) of a highly eccentric ($e = 0.8$), 15-day binary system. The initial equatorial rotation velocity is 192 km s^{-1}, which corresponds to a rotation rate which is 1.2 faster than the relative orbital motion of the companion at periastron. The companion's mass is $m_2 = 4$ M$_\odot$. The left panel contains a sample of the computed line profiles, distributed evenly over the orbital cycle, starting with periastron passage at the bottom of the plot. Although "bumps" are visible throughout the orbital cycle, their amplitude is clearly largest around periastron. In the grey-scale representation of the line-profile residuals (right) the transition from broad blue-to-red migrating "bumps" at periastron to smaller-scale structure a little later in the orbital cycle is evident.

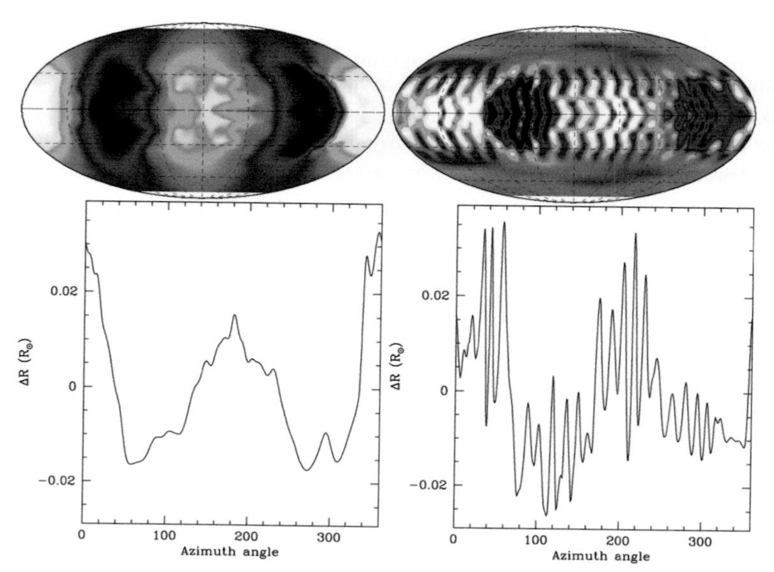

Figure 2. Example of the surface distortion at two orbital phases, periastron (left) and 0.833 days later (right) for the same model binary system described in Fig. 1. These figures illustrate the rapid change from an "equilibrium tide" configuration at periastron to one with structures of a much smaller scale. The maps (top) are in Mollweide projection, with azimuth angle $\varphi = 0^\circ$ at left and 360° at right. The companion, m_2, is located at $\varphi = 0^\circ$. White corresponds to maximum positive deformation. The plots (bottom) give the amplitude in $R_\odot$ of the deformation at the equatorial latitude.

Active OB stars: structure, evolution, mass loss, and critical limits
Proceedings IAU Symposium No. 272, 2010
C. Neiner, G. Wade, G. Meynet & G. Peters, eds.

doi:10.1017/S1743921311011240

Fundamental parameters of 4 massive eclipsing binaries in Westerlund 1

Eugenia Koumpia and Alceste Z. Bonanos

National Observatory of Athens, Institute of Astronomy & Astrophysics,
I. Metaxa & Vas. Pavlou St., Palaia Penteli GR-15236 Athens, Greece
email: koumpia@astro.noa.gr, bonanos@astro.noa.gr

Abstract. Westerlund 1 is one of the most massive young clusters known in the Local Group, with an age of 3-5 Myr. It contains an assortment of rare evolved massive stars, such as blue, yellow and red supergiants, Wolf-Rayet stars, a luminous blue variable, and a magnetar, as well as 4 massive eclipsing binary systems (Wddeb, Wd13, Wd36, WR77o, see Bonanos 2007). The eclipsing binaries present a rare opportunity to constrain evolutionary models of massive stars, the distance to the cluster and furthermore, to determine a dynamical lower limit for the mass of a magnetar progenitor. Wddeb, being a detached system, is of great interest as it allows determination of the masses of 2 of the most massive unevolved stars in the cluster. We have analyzed spectra of all 4 eclipsing binaries, taken in 2007-2008 with the 6.5 meter Magellan telescope at Las Campanas Observatory, Chile, and present fundamental parameters (masses, radii) for their component stars.

Keywords. open clusters and associations: individual (Westerlund 1), stars: fundamental parameters, stars: early-type, binaries: eclipsing, stars: Wolf-Rayet

---

Westerlund 1 (Wd1) is one of the most massive compact young star clusters known in the Local Group. It was discovered by Westerlund (1961), but remained largely unstudied for many years due to high interstellar extinction in its direction. The cluster contains a large number of rare, evolved high-mass stars including 6 yellow hypergiants, 4 red supergiants, 24 Wolf-Rayet stars, a luminous blue variable and many OB supergiants. In addition, X-ray observations have revealed the presence of the magnetar CXO J164710.2-455216, a slow X-ray pulsar that must have formed from a high-mass progenitor star (Muno *et al.* 2006). Wd1 is believed to have formed in a single burst of star formation, implying the constituent stars are of the same age and composition.

The study of eclipsing binaries in this cluster is important for several reasons: (1) the determination of fundamental parameters of the component stars (mass, radii, etc.), in order to test the evolutionary models of massive stars, (2) EBs provide an independent measurement of the distance, based on the expected absolute magnitude, and (3) the determination of a dynamical lower limit for the mass of the magnetar progenitor.

We have analyzed spectra of 4 eclipsing binaries taken in 2007-2008 with the 6.5 meter Magellan telescope at Las Campanas Observatory, Chile. The spectra were reduced and extracted using IRAF‡. We used two methods to determine the radial velocities: Gaussian fitting of the line cores with the IRAF *noao.rv.rvidlines* package and χ^2 minimization, which finds the least χ^2 from the observed spectrum and fixed synthetic TLUSTY models (Lanz & Hubeny 2003). In both cases, we chose to use the narrow Helium lines ($\lambda\lambda 6678, 7065$), as they are less sensitive to systematics, rather than the broader

‡ IRAF is distributed by the NOAO, which are operated by the Association of Universities for Research in Astronomy, Inc., under cooperative agreement with the NSF.

Table 1. Fundamental Parameters of Eclipsing Binaries in Wd1

Binary	**P(days)**	**$M_1(M_\odot)$**	**$M_2(M_\odot)$**	**$R_1(R_\odot)$**	**$R_2(R_\odot)$**	**$\log g_1$**	**$\log g_2$**	**eccentr**	**Incl(^{o})**
Wddeb	4.447	9.62	13.84	5.47	6.22	3.94	3.99	0.177	84.46
Wd36	3.182	10.75	12.51	9.45	10.15	3.52	3.52	0(fixed)	72.72
WR77o	3.520	49.67	19.87	18.06	11.81	3.62	3.59	0(fixed)	58.53
Wd13	9.267	29.91	23.86	26.88	24.08	3.05	3.05	0(fixed)	55.08

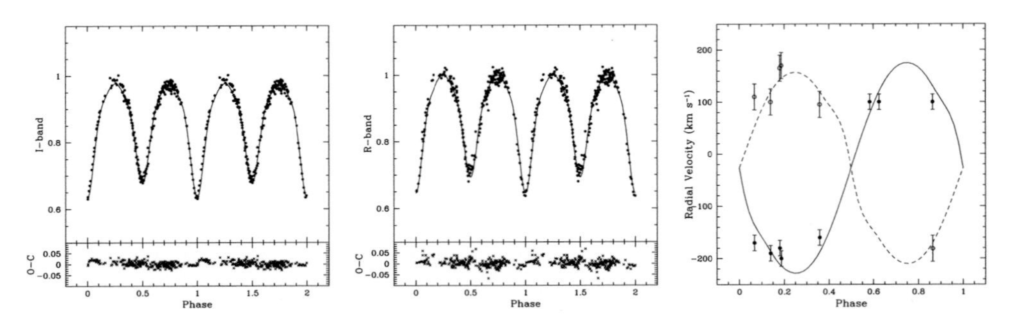

Figure 1. Phased I, R light curves and radial velocity curve of Wd36, respectively.

hydrogen lines. We adopted the velocities resulting from the second method as they seemed to be more robust.

Our preliminary results for the parameters of the members of the 4 systems (period, masses, radii, surface gravities of the components, eccentricity and inclination), as modelled using PHOEBE software (Prsa & Zwitter 2005), are presented in Table 1.
Wd13 is a semi-detached, double-lined spectroscopic binary (B0.5Ia$^+$ and OB types, Negueruela *et al.* 2010) in a circular orbit. We can see both absorption and emission lines in its spectra. **Wd36** is an overcontact, double-lined spectroscopic binary system (OB-type) in a circular orbit (see Fig. 1). **WR77o** is probably a double contact binary system in an almost circular orbit. It is a single lined spectroscopic binary and thus it is difficult to determine the parameters of its component stars. The spectroscopic visible star is a Wolf-Rayet star of spectral type WN6-7 (Negueruela & Clark 2005), given the large line widths of 2000 km s^{-1}. **Wddeb** is a detached, double-lined spectroscopic binary system (OB-type) with an eccentricity of almost 0.2.

Acknowledgements

EK & AZB acknowledge support from the IAU and the European Commission for an FP7 Marie Curie International Reintegration Grant.

References

Bonanos, A. Z. 2007, *AJ*, 133, 2696
Lanz, T. & Hubeny, I. 2003, *ApJS*, 146, 417
Muno, M. P., Clark, J. S., Crowther, P. A., & Dougherty, S. M. *et al.* 2006, *ApJ* (Letters), 636, 41
Negueruela, I. & Clark, J. S. 2005, *A&A*, 436, 541
Negueruela, I., Clark, J. S., & Ritchie, B. W. 2010, *A&A*, 516A, 78
Prša, A. & Zwitter, T. 2005, *ApJ*, 628, 426
Westerlund, B. 1961, *AJ*, 66T, 57

Active OB stars: structure, evolution, mass loss, and critical limits
Proceedings IAU Symposium No. 272, 2010
C. Neiner, G. Wade, G. Meynet & G. Peters, eds.

doi:10.1017/S1743921311011252

The nature of the light variations of chemically peculiar stars CU Vir and HD 64740

Jiri Krtička[1], Hana Marková[1], Zdenek Mikulášek[1], Theresa Lüftinger[2], David Bohlender[3], Juraj Zverko[4], and Jozef Žižňovský[4]

[1]Masaryk University, Kotlářská 2, CZ-611 37 Brno, Czech Republic
email: krticka@physics.muni.cz, 175960@mail.muni.cz, mikulas@physics.muni.cz

[2]Institute for Astronomy of the University of Vienna, Vienna, Austria
email: lueftinger@astro.univie.ac.at

[3]National Research Council of Canada, Herzberg Institute of Astrophysics, Victoria, Canada
email: david.bohlender@nrc-cnrc.gc.ca

[4]Astronomical Institute, Slovak Academy of Sciences, Tatranská Lomnica, Slovakia
email: zve@ta3.sk, ziga@ta3.sk

Abstract. The nature of the light variations of chemically peculiar stars was studied in detail only in a very few cases. To better understand the mechanisms of light variability of these stars, we study the light variations of the well-known magnetic chemically peculiar star CU Vir and one of the least amplitude variable stars HD 64740. We show that the light variability of these stars is induced by flux redistribution in spots of enhanced abundance of chemical elements (e.g., helium, silicon, iron or chromium), and by the stellar rotation. We conclude that this is a promising model for the explanation of the light variability of most chemically peculiar stars.

Keywords. stars: atmospheres, stars: chemically peculiar, stars: spots, stars: variables: others

1. Modelling of light variations of magnetic chemically peculiar stars

The light variations of magnetic chemically peculiar stars are still not very well understood. These stars show inhomogeneous surface distribution of different elements, including helium, silicon or iron. Krtička *et al.* (2007, 2009) showed that the redistribution of the emergent flux due to the bound-free (continuum) and bound-bound (line) transitions in the stellar atmosphere can explain the light variability of HD 37776, and HR 7224. They concluded that the comparison of observed and simulated light variations can provide an important new test of modern model stellar atmospheres.

The model stellar atmospheres are calculated using the code TLUSTY (Hubeny & Lanz 1995, Lanz & Hubeny 2007) assuming a fixed stellar effective temperature and surface gravity. The abundance of chemical elements concerned is set according to the maps of surface elemental distribution derived from spectroscopy.

The emergent fluxes from individual surface elements are calculated using the code SYNSPEC. From these fluxes the magnitudes of the star in individual photometric filters are derived. Finally, these simulated light variations are compared with the observed ones. We do not use any free parameter to fit the observed light curves.

2. Results

The observed light variations of CU Vir in *vby* filters of Strömgren photometric system can be explained as a result of inhomogeneous distribution of silicon, chromium, and

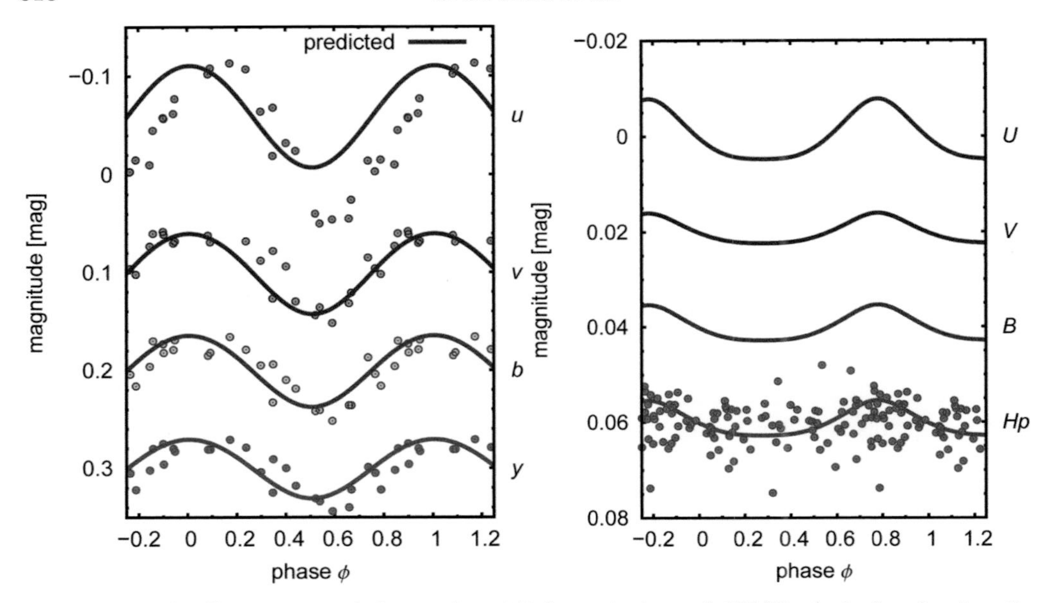

Figure 1. Left: Comparison of the predicted light variations of CU Vir (calculated using the helium, silicon, chromium, and iron surface distribution according to Kuschnig *et al.* (1999)) and the observed one (Pyper & Adelman 1985) in different colors. Right: Comparison of the predicted light variations of HD 64740 (calculated using the helium and silicon surface distribution according to Bohlender & Landstreet 1990), and the observed one (Perryman 1997).

iron on the stellar surface (see Fig. 1). This causes the redistribution of the flux from the ultraviolet to the visible part of the spectrum, and, due to the stellar rotation, the light variations. However, there is still some disagreement in the *u* filter, which will be the subject of the further research.

The reason why some stars show only very small amplitude of the light variations is that their surface chemical composition does not significantly differ from the solar one. This was shown for one of the least amplitude variable stars, HD 64740. The low-amplitude light variations of this star observed by the Hipparcos satellite can be explained by the inhomogeneous surface distribution of helium and silicon (Fig. 1).

Comparison of the observed and simulated light variations serves as an independent test of modern model stellar atmospheres.

Acknowledgements

This work was supported by grants GAAV IAA301630901, MEB 060807/WTZ CZ-11/2008, VEGA 2/0074/09, and MUNI/A/0968/2009.

References

Bohlender, D. A. & Landstreet, J. D. 1990, *ApJ*, 358, 274

Hubeny, I. & Lanz, T. 1995, *ApJ*, 439, 875

Kuschnig, R., Ryabchikova, T. A., Piskunov, N. E., & Weiss, W. W. *et al.* 1999, *A&A*, 348, 924

Krtička, J., Mikuláček, Z., Zverko, J., & Žižňovský, J. 2007, *A&A*, 470, 1089

Krtička, J., Mikulášek, Z., Henry, G. W., & Zverko, J. *et al.* 2009, *A&A*, 499, 567

Lanz, T. & Hubeny, I. 2007, *ApJS*, 169, 83

Perryman M.A.C. & ESA 1997, *The Hipparcos and Tycho Catalogues*, ESA-SP 1200 (Noordwijk, Netherlands: ESA Publications Division)

Pyper, D. M. & Adelman, S. J. 1985, *A&AS*, 59, 369

Active OB stars: structure, evolution, mass loss, and critical limits
Proceedings IAU Symposium No. 272, 2010
C. Neiner, G. Wade, G. Meynet & G. Peters, eds.

doi:10.1017/S1743921311011264

Long-term spectroscopic monitoring of LBVs and LBV candidates

Alex Lobel[1], Jose H. Groh[2], Kelly Torres[1] and Nadya Gorlova[3]

[1]Royal Observatory of Belgium
Ringlaan 3, B-1180, Brussels, Belgium
email: Alex.Lobel@oma.be; alobel@sdf.lonestar.org; Kelly.Torres@oma.be

[2]Max Planck Institute for Radio Astronomy,
Auf dem Hügel 69, 53121 Bonn, Germany
email: jgroh@mpifr-bonn.mpg.de

[3]Institute of Astronomy, Katholieke Universiteit Leuven,
Celestijnenlaan 200D BUS 2401, 3001 Leuven, Belgium
email: nadya@ster.kuleuven.be

Abstract. We present results of a long-term spectroscopic monitoring program (since mid 2009) of Luminous Blue Variables with the new HERMES echelle spectrograph on the 1.2m Mercator telescope at La Palma (Spain). We investigate high-resolution (R = 80,000) optical spectra of two LBVs, P Cyg and HD 168607, the LBV candidates MWC 930 and HD 168625, and the LBV binary MWC 314. In P Cyg we observe flux changes in the violet wings of the Balmer Hα, Hβ, and He I lines between May and Sep 2009. The changes around 200 to 300 km s^{-1} are caused by variable opacity at the base of the supersonic wind from the blue supergiant.

We observe in MWC 314 broad double-peaked metal emission lines with invariable radial velocities over time. On the other hand, we measure in the photospheric S II λ5647 absorption line, with lower excitation energy of ~14 eV, an increase of the heliocentric radial velocity centroid from 37 km s^{-1} to 70 km s^{-1} between 5 and 10 Sep 2009 (and 43 km s^{-1} on 6 Apr 2010). The increase of radial velocity of ~33 km s^{-1} in only 5 days can confirm the binary nature of this LBV close to the Eddington luminosity limit.

A comparison with VLT-UVES and Keck-Hires spectra observed over the past 13 years reveals strong flux variability in the violet wing of the Hα emission line of HD 168625 and in the absorption portion of the Hβ line of HD 168607. In HD 168625 we observe Hα wind absorption at velocities exceeding 200 km s^{-1} which develops between Apr and June 2010.

Keywords. stars: emission-line, stars: individual (P Cyg, MWC 314, MWC 930, HD 168607, HD 168625), stars: variables: other

1. Introduction

The HERMES instrument on the 1.2m Mercator telescope at La Palma (Raskin & Van Winckel 2008) is a new high-efficiency fiber-fed bench-mounted cross-dispersed echelle spectrograph that observes the complete wavelength range from 420 nm to 900 nm in a single exposure with R=80,000. TAC reviewed HERMES observation programs of the contributing research institutions started mid 2009. We present first results of a long-term high-resolution spectroscopic monitoring program of 3 LBVs and 2 LBV candidates (up to V=11$^{\rm m}$.0). The HERMES monitoring program will provide invaluable new clues about the structure and dynamics of LBV atmospheres, the physics of their extended winds, and the strong line broadening mechanisms in these rare massive hot stars near the Eddington luminosity limit. The monitoring program will be crucial for documenting the enigmatic LBV outburst events, for detecting new long-period LBV binaries and the reliable determination of the orbital parameters.

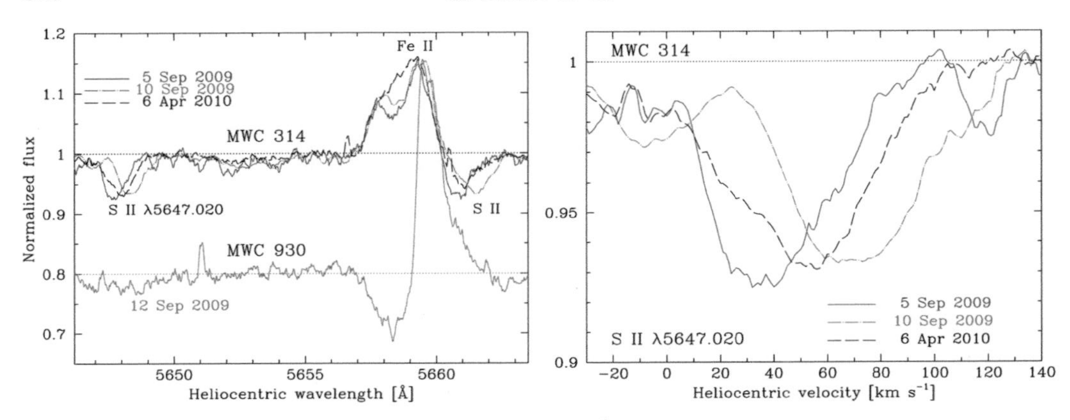

Figure 1. Large radial velocity changes of ~33 km s^{-1} observed in S II lines of MWC 314 over 5 days with Mercator-HERMES confirm its LBV binary nature (*see text*).

2. The LBV Binary MWC 314 and LBV Candidate MWC 930

MWC 314 was observed with HERMES on 5 and 10 Sep 2009, and on 6 April 2010. Muratorio *et al.* (2008) found evidence for a ~30 day orbital period and proposed that its double-peaked permitted and forbidden ionic emission lines (Fe II, Cr II, Ti II) are formed in a warm disk rotating around the star. Its orbital parameters are however still unknown at present. We observe with HERMES many broad and double-peaked optical emission lines (i.e., Fe II λ5657) reminiscent of a Be-star spectrum. The emission line flux maxima are strongly variable due to wind opacity changes in the emission line formation region. The radial velocities of these optical emission lines are however invariable, signaling a stable circumstellar or circumbinary (disc) envelope. We observe large changes in the radial velocity centroid of photospheric S II absorption lines from 37 km s^{-1} (*blue curve in the right-hand panel of Fig. 1*) to 70 km s^{-1} (*red curve*) in only 5 days. Our short-term spectroscopic observations with HERMES hence confirm the binarity of MWC 314. We observe many strong optical P Cygni profiles in the LBV candidate MWC 930 in Sep 2009 (*magenta curve in Fig. 1*) signaling the wind properties of a massive hot star.

3. Summary

Long-term monitoring with Mercator-HERMES of the optical spectrum of the prototypical LBV P Cyg reveals variability at the base of its supersonic wind that can be linked to moderate V-changes (of $0^{\mathrm{m}}.1$ to $0^{\mathrm{m}}.2$) over a period of ~4 months. We find strong indications for the binary nature of MWC 314 from large radial velocity changes observed in photospheric absorption lines during less than one week. We observe prominent P Cygni profiles in the optical spectrum of LBV candidate MWC 930, signaling the presence of a central massive hot star. The optical spectral lines of LBV candidate HD 168625 are less variable, although we also observe clear signatures of expanding Hα wind variability on short time-scales of ~1 month. The optical spectrum of HD 168607 reveals large line profile changes over the past 12 years confirming its LBV designation.

References

Raskin, G. & Van Winckel, H. 2008, in: I. S. McLean & M. M. Casali (eds.), *Optical and Infrared Interferometry*, SPIE Conference Series 7014, p. 178

Muratorio, G., Rossi, C., & Friedjung, M. 2008, *A&A*, 487, 637

Active OB stars: structure, evolution, mass loss, and critical limits
Proceedings IAU Symposium No. 272, 2010
C. Neiner, G. Wade, G. Meynet & G. Peters, eds.

doi:10.1017/S1743921311011276

HD 150136: towards one of the most massive systems?

Laurent Mahy[1], **Eric Gosset**[1], **Hugues Sana**[2], **Gregor Rauw**[1], **Thomas Fauchez**[1] **and Christian Nitschelm**[3]

[1]Institut d'Astrophysique et de Géophysique, University of Liège,
Allée du 6 Août, 17, Bât. B5C, B-4000, Liège, Belgium
email: mahy@astro.ulg.ac.be

[2]Sterrenkundig Instituut 'Anton Pannekoek', Universiteit van Amsterdam,
Postbus 94249, NL-1090 GE Amsterdam, The Netherlands

[3]Instituto de Astronomía, Universidad Católica del Norte,
Avenida Angamos 0610, Antofagasta, Chile

Abstract. We present the preliminary results of an intensive monitoring devoted to HD 150136. Already quoted as an O3+O6 binary, we detected a third O-type component physically linked to the system, making it one of the nearest (1.3 kpc) most massive systems known until now ($\sim 134 M_{\odot}$). To determine the physical parameters of this system, we applied a disentangling program to study individually the three components. It allows us to constrain their spectral types and to derive a new orbital solution for the short-period system.

Keywords. stars: individual (HD 150136), binaries: spectroscopic, stars: fundamental parameters

1. Overview on HD 150136

HD 150136 was classified by Niemela & Gamen (2005) as an O3V+O6V binary system. This system has an orbital period of about 2.66 days and presents variability in the X-ray domain on a one-day time scale (Skinner *et al.* 2005).

Sixty-four spectra of HD 150136 were collected from 1999 to 2006 and 14 others in 2009 with the 1.5m and 2.2m telescopes, at La Silla, equipped with FEROS. These high-resolution spectra allowed us to detect a third component and to derive spectral types of O3, O6 and O6.5–O7 for the primary, the secondary and the third star, respectively.

2. Orbital solution of the short-term binary and properties of the third star

We used a disentangling program, based on the method of González & Levato (2006) and adapted to triple systems, which also measures the radial velocities (RVs) by cross-correlation even at phases where the spectra are heavily blended. We applied a Fourier method (Heck *et al.* 1985) to the differences of these RVs to refine the orbital period of the short-term system. This yields a period of 2.67±0.01 days, i.e., similar to the previous one (Niemela & Gamen 2005). The RV curve is given in Fig. 1 (left panel) whilst the orbital parameters are listed in Table 1 (T_0 refers the time of the primary conjunction). We fitted the primary and the secondary by using the CMFGEN atmosphere code (Hillier & Miller 1998). The stellar parameters were constrained as in Mahy *et al.* (2010) but we were not able to estimate the wind parameters because the wind diagnostic lines, in the optical domain, present variations impossible to disentangle. We derived $T_{\rm eff}$ of about

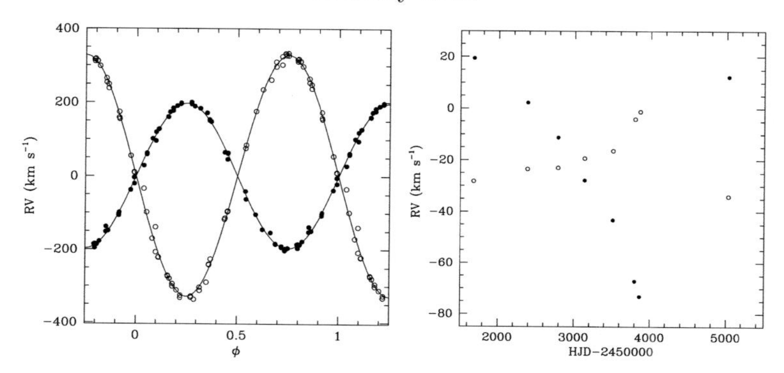

Figure 1. *Left:* RV curve of the short-term binary system. Full circles indicate the primary, the open ones represent the secondary star. *Right:* Evolution of the RVs of the third component and the mean systemic velocity of the short-period binary (full/open circles) as a function of time.

Table 1. Orbital parameters of the short-term binary system.

Parameters	Primary	Secondary
P(days)	2.67 ± 0.01	
e	0.0 (fixed)	
T_0 (HJD)	$2\,451\,318.518 \pm 0.002$	
K (km s^{-1})	196.9 ± 1.0	328.3 ± 1.7
$a \sin i$ ($R_\odot$)	10.4 ± 0.1	17.3 ± 0.1
$M \sin^3 i$ ($M_\odot$)	25.1 ± 0.3	15.0 ± 0.2
Q (M_1/M_2)	1.667 ± 0.009	
rms (km s^{-1})	9.75	

45.7 and 39.8 kK, $\log(L/L_\odot)$ of 5.85 and 5.43 and $\log g$ of 4.0, for the primary and the secondary, respectively, suggesting respective masses of 67 and 40 $M_\odot$.

The evolution with time of the RVs of the third component and of the mean systemic velocity of the short-period binary (Fig. 1, right panel) reveals, for the first time, that the three components are physically linked. The expected period of this long-term system is certainly larger than 10 years and the orbit is clearly eccentric.

3. Future works and conclusions

The high-resolution of our data allowed us to show the existence of a 3rd component in the HD 150136 system. However, we are not yet able to constrain with high accuracy the parameters of this 3rd star. The short-term binary system is composed of O3V and O6V stars, probably in contact (Skinner *et al.* 2005), with an inclination close to 46°. This system also likely features a wind interaction zone, as revealed by the complex profile variations of the He II 4686 and H_α lines. This will be investigated in a future paper.

References

González, J. F. & Levato, H. 2006, *A&A*, 448, 283
Heck, A., Manfroid, J., & Mersch, G. 1985, *A&AS*, 59, 63
Hillier, D. J. & Miller, D. L. 1998, *ApJ*, 496, 407
Mahy, L., Rauw, G., Martins, F., Nazé, Y. *et al.* 2010, *ApJ*, 708, 1537 *ApJ*, 708, 1537
Niemela, V. S. & Gamen, R. C. 2005, *MNRAS*, 356, 974
Skinner, S. L., Zhekov, S. A., Palla, F., & Barbosa, C. L. D. R. 2005, *MNRAS*, 361, 191

Active OB stars: structure, evolution, mass loss, and critical limits
Proceedings IAU Symposium No. 272, 2010
C. Neiner, G. Wade, G. Meynet & G. Peters, eds.

doi:10.1017/S1743921311011288

The spectroscopic orbits and physical parameters of GG Car

Paula E. Marchiano[1], Elisande Brandi[1,2,3], Maria F. Muratore[1], Claudio Quiroga[1,2], Osvaldo Ferrer[1,4] and Lia García[1,2]

[1]Facultad de Ciencias Astronómicas y Geofísicas, Universidad Nacional de La Plata, Paseo del Bosque, S/N, 1900 La Plata, Argentina
email: pmarchiano@fcaglp.unlp.edu.ar

[2]Instituto de Astrofísica La Plata, CCT La Plata, CONICET

[3]Comisión de Investigaciones Científicas de la Prov. de Buenos Aires (CIC)

[4]Consejo Nacional de Investigaciones Científicas y Técnicas (CONICET)

Abstract. GG Car is a peculiar B type star with emission lines classified as a B[e] supergiant star. In this work we present a spectral analysis of this system based on spectra obtained at Casleo. We fit the spectral energy distribution adopting a model for the gas and dust circumstellar components and thus we obtain the physical parameters of the star and its environment.

Keywords. stars: individual (GG Car), stars: emission-line, stars: fundamental parameters

---

1. Introduction

Hernandez *et al.* (1981) were the first to investigate the possible periodic variability of the spectrum of GG Car. Gosset *et al.* (1985) used photoelectric photometry in both the standard UBV and Stromgren uvby systems and they determined two possibles periods: 31,020 and 62,039 days, then the first value was definitely confimed from the radial velocities with spectroscopic data along the light curve. Finally, it is known that GG Car has an infrared excess of the type that appears to be associated with circumstellar dust. From IUE low resolution data, Brandi *et al.* (1987) propose a value of the B-V excess of $E_{B-V} = 0,52 \pm 0,04$.

2. The spectroscopy orbits and the spectral distribution

The observational material consists of 55 spectra which were obtained with the 2.15 m "Jorge Sahade" telescope at CASLEO (San Juan, Argentina). An echelle spectrograph REOSC and a teck CCD detector were used, resulting a resolution $R \sim 12000$ and an spectral range of 4000 to 7100 Å and 5000 to 8700 Å.

The He I ($\lambda\lambda$ 4471, 5875, 6678, 7065) line profiles show a large variability and several absorption components were detected. One of these components corresponds to the blue-shifted absorption of a P-Cyni profile. Our task was to calculate the radial velocities of the other components and the result of this analysis led us to find that they could be associated with the two stars of the binary system.

Table 1 shows the orbital elements obtained for combined solutions. Fig. 1 *Left* shows the radial velocity curves obtained for both components of the system. Continuous curves give the result of the best fit with $P = 31$ days.

In order to reproduce the continuum energy distribution of GG Car (Fig. 1 *Right*), we adopt B-type flux models (Kurucz, 1979) surrounded by a gaseous circumstellar envelope and outermost dust layers. Taking into account the classification criterion of Lopes *et al.*

Table 1. Orbital solutions obtained from the HeI absorption lines 4471, 5875, 6678 and 7065 Å

Comp.	P [days]	K [kms^{-1}]	γ_0 [kms^{-1}]	e	ω [deg]	$T_{\circ}^{(1)}$ [JD24...]	$T_{conj}^{(2)}$ [JD24...]	$\Delta T^{(3)}$ [days]	$a \sin i$ [AU]	$M \sin^3$i [M$_{\odot}$]
Prim.	31.033± 0.008	65.8±7.3	-162.1± 4.3	0.28± 0.06	272± 12	52020.96	52051.93	2.44	0.18±0.02	18±3
Sec.		143.3±8.7							0.39±0.02	8±2

(1) $T_{\circ}$: time of periastron passage.
(2) T_{conj} : time of spectroscopic conjunction.
(3) $\Delta T = T_{conj} - T_{phot-min}$.
$T_{phot-min}$=2444260.21 is the mean epoch of the minimum (Min I) given by Gosset *et al.* (1985).
The orbital phases were calculated with the ephemeris: $T_{conj} = 2452051.93 + 31.033E$.

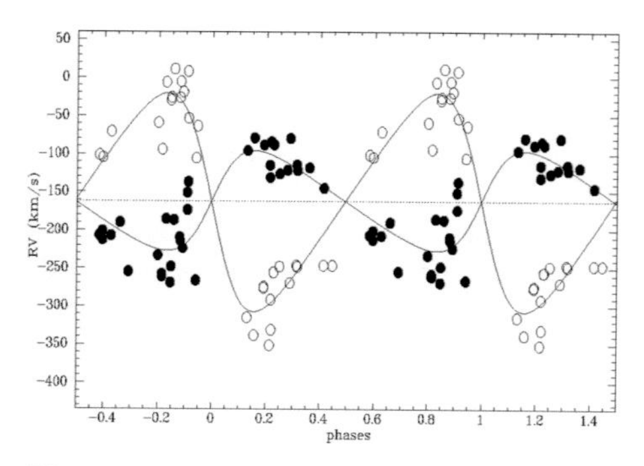

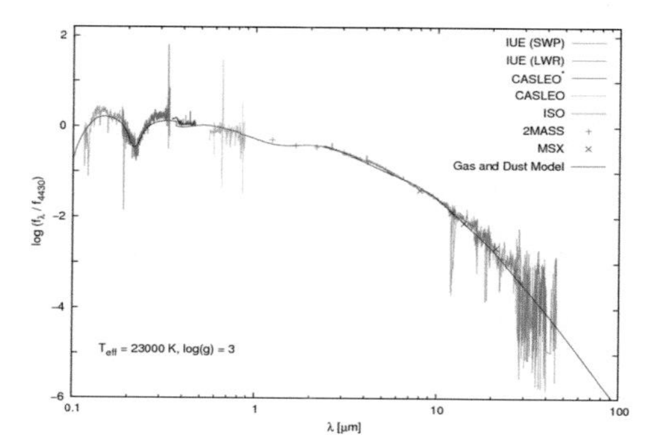

Figure 1. Left: *Fill circle*: primary component. *Empty circle*: secondary component. *Continuous curve*: the best fit of the orbital solutions. **Right:** The observed energy distribution using available photometric and spectroscopic data and the theoretical model (see Muratore *et al.*, these proceedings.)

(1992) and Clark *et al.* (2000) together with the calibration temperature scale derived by Zorec *et al.* (2009), we conclude that the star is consistent with a B0-B2 spectral type.

3. Conclusions

We have determined the orbit for both components of the binary system GG Car through a detailed study of the He I lines and we have confirmed the photometric period. The orbital parameters are indicated in Table 1 and a mass ratio $q \cong 2,179$ was obtained.

We have modeled GG Car as a central star with $T_{eff} = 23,000K$ and $log(g) = 3$ surrounded by a spherical envelope consisting of two regions: a layer close to the central star of 3, 5 stellar radius composed of ionized gas and other outermost layers composed of dust with $E_{B-V} = 0,3$. For the interstellar medium, an $E_{B-V} = 0,18$ was added.

References

Brandi, E., Gosset, E., & Swings, J.-P. 1987, *A&A* 175, 151
Clark, J. S. & Steele, I. A. 2000, *A&AS* 141, 65
Gosset, E., Hutsemekers, D., Swings, J. P., & Surdej, J. 1985, *A&A* 153, 71
Hernández, C. A., Lopez, L., Sahade, J., Thackeray, A. D. 1981, *PASP* 93, 747
Kurucz, R. L. 1979, *ApJS* 40, 1
Lopes, D. F., Damineli Neto, A., & de Freitas Pacheco, J. A. 1992, *A&A* 261, 482
Zorec, J., Cidale, L., Arias, M. L., & Frémat, Y. *et al.* 2009, *A&A* 501, 297

Active OB stars: structure, evolution, mass loss, and critical limits
Proceedings IAU Symposium No. 272, 2010
C. Neiner, G. Wade, G. Meynet & G. Peters, eds.

doi:10.1017/S174392131101129X

Hα emission variability in the γ-ray binary LS I +61 303

M. Virginia McSwain[1], Erika D. Grundstrom[2], Douglas R. Gies[3] and Paul S. Ray[4]

[1]Dept. of Physics, Lehigh University, Bethlehem, PA 18015
email: mcswain@lehigh.edu

[2]Physics and Astronomy Department, Vanderbilt University, Nashville, TN 37235
email: erika.grundstrom@vanderbilt.edu

[3]Center for High Angular Resolution Astronomy and Department of Physics and Astronomy, Georgia State University, Atlanta, GA 30303
email: gies@chara.gsu.edu

[4]Space Science Division, Naval Research Laboratory, Washington, DC 20375
email: paul.ray@nrl.navy.mil

Abstract. LS I +61 303 is an exceptionally rare example of a Be/X-ray binary that also exhibits MeV–TeV emission, making it one of only a handful of "γ-ray binaries". Here we present Hα spectra that show strong variability during the 26.5 day orbital period and over decadal time scales. The Hα line profile exhibits a dramatic emission burst shortly before apastron, observed as a redshifted shoulder in the line profile, as the compact source moves almost directly away from the observer. Here we investigate several possible origins for this red shoulder, including an accretion disk, tidal mass transfer stream, turbulent gas in the wake of the neutron star, and a compact pulsar wind nebula in the system.

Keywords. accretion, accretion disks, stars: winds, outflows, stars: emission-line, Be, stars: individual (LS I +61 303)

---

1. Introduction

LS I +61 303 is a high mass X-ray binary (HMXB) that is also a confirmed source of very high energy γ-ray emission. The system consists of a B0 Ve optical star and an unknown compact companion in a highly eccentric, 26.5 day orbit (Aragona *et al.* 2009). We recently obtained an extensive collection of red optical spectra of LS I +61 303 to determine an updated orbital ephemeris for the spectroscopic binary. These observations also recorded the evolution of the Hα emission during a full orbital cycle (Fig. 1, left). We subtracted the mean emission line profile, shifted to the rest velocity of the optical star, to investigate the emission residuals carefully. A prominent emission feature stands out as a "red shoulder" in the Hα line profile near $\phi(\mathrm{TG}) \sim 0.6$ as the compact companion is receding from the observer. A more complete discussion of the short-term and long-term variability of the Hα emission, and its origin, is available in McSwain *et al.* (2010).

2. Discussion

To investigate the origin of the red shoulder, we measured the equivalent width of Hα, $W_{\mathrm{H}\alpha}$, by directly integrating over each line profile. $W_{\mathrm{H}\alpha}$ rises dramatically with the onset of the red shoulder. We also used Gaussian fits of the peak residual emission to measure its radial velocity, V_r, and full width half maximum, FWHM. The low FWHM of our difference spectra is generally consistent with a spiral density wave within the Be

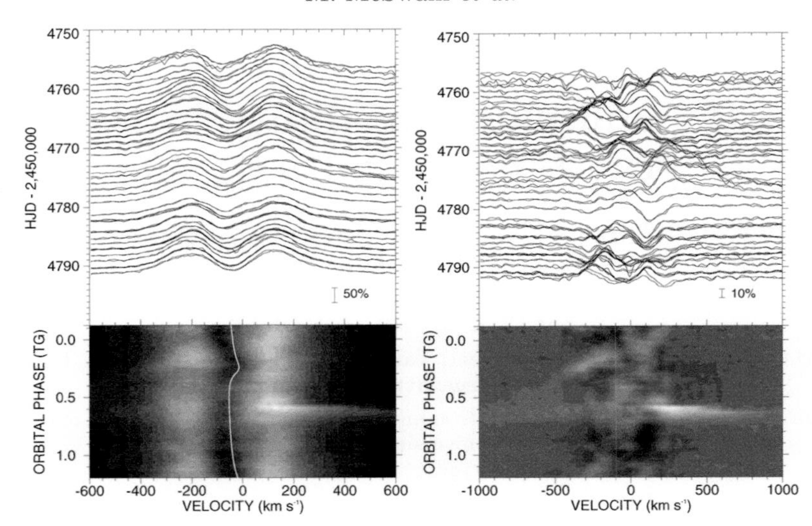

Figure 1. *Left:* The upper plot shows the Hα line profile of LS I +61 303 over our continuous 35 nights of observation, sorted by HJD, and the lower plot shows a gray-scale image of the same line. *Right:* Emission residuals, or difference spectra, in the same format.

circumstellar disk. The FWHM of the red shoulder is significantly higher, suggesting a more turbulent region of gas and an origin outside the circumstellar disk.

Paredes *et al.* (1994) first proposed that the increase in Hα emission in the red shoulder originates in an accretion disk. However, the observed V_r of the red shoulder does not correspond to the orbital motion of a compact companion. Therefore an accretion disk is an unlikely source of the red shoulder.

Neither can the red shoulder be due to turbulent gas trailing the neutron star. The orbital velocity of the neutron star is faster than the circumstellar disk only very close to periastron. If the Keplerian disk extended out to the neutron stars location, the disk would actually stream past the slow moving neutron star near apastron.

Grundstrom *et al.* (2007) propose the development of a tidal stream within the Be disk near periastron, extending beyond the truncation radius of the Be star disk and into the vicinity of the compact companion. The red shoulder could be formed as the induced tidal stream falls back onto the Be circumstellar disk. The resulting infall velocity of the tidal stream is somewhat comparable to the observed V_r of the red shoulder.

Alternatively, LS I +61 303 may contain a shrouded pulsar whose relativistic wind interacts with the optical stars wind to form a cone-shaped wind shock region (eg. Dubus 2006). The unusually broad FWHM of the red shoulder emission is consistent with a Balmer-dominated shock (BDS) sometimes observed in pulsar wind nebulae (Heng 2010). The temporary red shoulder may suggest that the BDS only forms when the high density tidal stream interacts with the neutron star.

References

Aragona, C., McSwain, M. V., Grundstrom, E. D., & Marsh, A. N. *et al.* 2009, *ApJ*, 698, 514
Dubus, G. 2006, *A&A*, 456, 801
Grundstrom, E. D., Caballero-Nieves, S. M., Gies, D. R., & Huang, W. *et al.* 2007, *ApJ*, 656, 437
Heng, K. 2010, *PASA*, 27, 23
McSwain, M. V., Grundstrom, E. D., Gies, D. R., & Ray, P. S. 2010, *ApJ*, 724, 379
Paredes, J. M., Marziani, P., Marti, J., & Fabregat, J. *et al.* 1994, *A&A*, 288, 519

Active OB stars: structure, evolution, mass loss, and critical limits
Proceedings IAU Symposium No. 272, 2010
C. Neiner, G. Wade, G. Meynet & G. Peters, eds.

doi:10.1017/S1743921311011306

Optical spectroscopy of DPVs and the case of LP Ara

Ronald E. Mennickent[1], Darek Graczyk[1], Zbigniew Kołaczkowski[2], Gabriela Michalska[2], Daniela Barría[1] and Ewa Niemczura[2]

[1]Dpto. de Astronomía, Universidad de Concepción, Chile, [2]Instytut Astronomiczny Uniwersytetu Wrocławskiego, Wrocław, Poland

Abstract. We present preliminary results of our spectroscopic campaign of a group of intermediate mass interacting binaries dubbed "Double Periodic Variables" (DPVs), characterized by orbital light curves and additional long photometric cycles recurring roughly after 33 orbital periods (Mennickent *et al.* 2003, 2005). They have been interpreted as interacting, semi-detached binaries showing cycles of mass loss into the interstellar medium (Mennickent *et al.* 2008, Mennickent & Kołaczkowski 2009). High resolution Balmer and helium line profiles of DPVs can be interpreted in terms of mass flows in these systems. A system solution is given for LP Ara, based on modeling of the ASAS V-band orbital light curve and the radial velocity of the donor star.

Keywords. stars: binaries, stars: early-type

1. Spectra of Galactic DPVs and report on the analysis of LP Ara

During recent years we have monitored a sample of Galactic DPVs with high resolution optical spectrographs. From the inspection of the spectral region around Hα and He I 5875 we find in all cases evidence for blended emission or absorption profiles of complex morphology (Fig. 1). The He I 5875 profiles are usually broad and shallow, being the AU Mon He I 5875 profile exceptionally deep among Galactic DPVs.

LP Ara (HD 328568, 2MASS J16400178-4639348, B = 10.48, B-V = 0.28) is classified as an eclipsing binary of β Lyr type in SIMBAD (simbad.u-strasbg.fr/simbad/). Spectral types B8+[A8] and mass ratio $q = 0.090$ were given by Svechnikov & Kuznetsova (1990). From modeling of photometric observations made with the INTEGRAL/OMC camera, Zasche (2010) found a semidetached system with orbital period $P_o = 8.53282038$ d, $i = 77.1^o$, $q = 0.2$, ratio between stellar temperatures and radii $T_1/T_2 = 1.143$ and $R_1/R_2 = 1.135$ and no third light. The above authors did not correct their observations for the additional long photometric cycle $P_l = 273$ days reported by Michalska *et al.* (2009).

We compared the spectrum taken near the long cycle maximum at $\Phi_o = 0.96$ with a grid of synthetic model spectra in a region deployed of H I and He I lines. We find the best fit for the secondary star with the model $T_{eff} = 9500$ K, log g = 3.0 and v_2 sini = 65 km/s. We modeled the ASAS-3 light curve and radial velocity of LP Ara with a Wilson-Devinney code obtaining $P_o = 8.53295$ d, $T_1 = 16400$ K, $q = 0.30$, $i = 83.9^o$, orbital separation $a = 41.1\ R_\odot$, mass funtion $f(m) = 5.70 \pm 0.36$, $M_1 = 9.84\ M_\odot$, $M_2 = 2.98\ M_\odot$, $R_1 = 5.3\ R_\odot$, $R_2 = 11.6\ R_\odot$, $\log g_1 = 4.0$, $\log g_2 = 2.8$ and V-band luminosity ratio $k = L_1/L_2 = 1.50$. Typical errors of derived physical parameters are $\approx$ 10–20%. LP Ara is a double lined spectroscopic binary, however according to the present state of our analysis, only lines from the secondary component strictly follow the orbital motion. The use of ASAS-3 photometry corrected for long period changes yields a model free from systematic effects within 5% accuracy. If third light or additional structures do exist they contribute below 5% to the total orbital light.

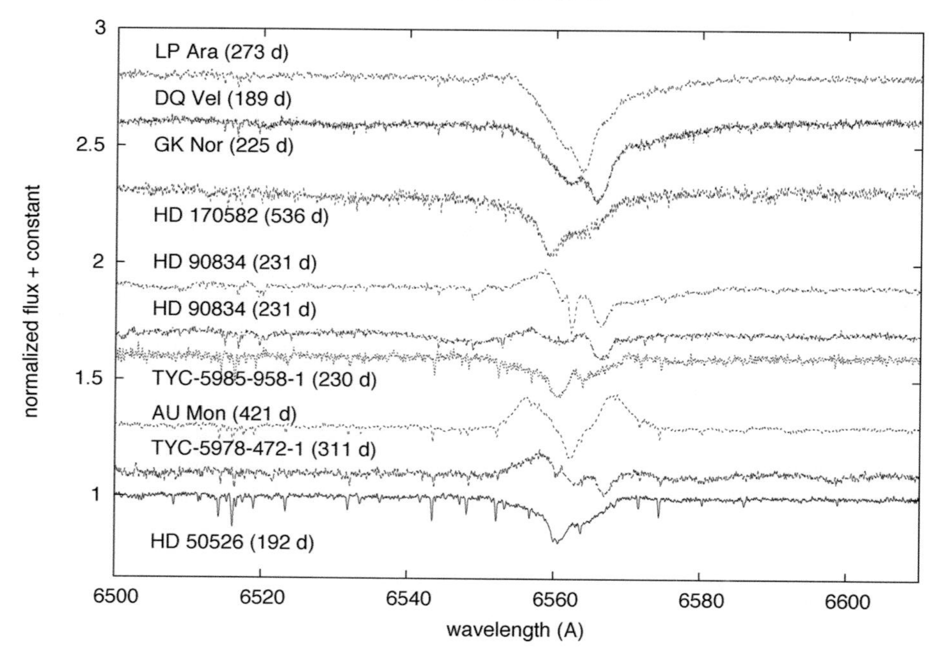

Figure 1. Spectra of Galactic DPVs around Hα at randomly selected orbital phases. The long period, found by us from a study of ASAS-3 ligth curves, is given in parenthesis. Two spectra of HD 90834 illustrate line profile variability. Sharp absorption features are telluric lines.

2. Conclusions

DPV Hα profiles are complex and usually show asymmetric absorption/emission features varying with the orbital period as well as with the long cycle. This fact suggests that *often* the line emission region is not disc-like, but more as an irregular structure, a fact already noted for V 393 Sco (Mennickent *et al.* 2010). The photometric regularity of DPVs (Michalska *et al.* 2009) place them apart from active Algols (W Serpentids). The rotational velocities of emitting material in some DPVs are much larger than expected for Keplerian orbits around B-type primaries. The system parameters for LP Ara fit the global scheme of low mass ratios found in other DPVs, e.g. OGLE LMC-SC8-125836 and V 393 Scorpii (Mennickent *et al.* 2008, 2010) and AU Mon (Desmet *et al.* 2010).

References

Desmet, M., Frémat, Y., Baudin, F., & Harmanec, P. *et al.* 2010, *MNRAS* 401, 418

Mennickent, R. E., Cidale, L., Díaz, M., & Pietrzyński, G. *et al.* 2005, *MNRAS* 357, 1219

Mennickent, R. E. & Kołaczkowski, Z. 2009, *Binaries: Key to Comprehension of the Universe*, ASP-CS, in press, arXiv0908:3900

Mennickent, R. E., Kołaczkowski, Z., Graczyk, D., & Ojeda, J. 2010, *MNRAS* 405, 1947

Mennickent, R. E., Kołaczkowski, Z., Michalska, G., & Pietrzyński, G. *et al.* 2008, *MNRAS* 389, 1605

Mennickent, R. E., Pietrzyński, G., Diaz, M., & Gieren, W. 2003, *A&A* 399, L47

Michalska, G., Mennickent, R. E., Kołaczkowski, Z. & Djurašević, G. 2009, *ASP-CS*, arXiv:0910.4359

Svechnikov, M. A. & Kuznetsova, E. F. 1990, *Katalog priblizhennykh fotometricheskikh i absoliutnykh elementov zatmennykh peremennykh zvezd* (Sverdlovsk: Izd-vo Ural'skogo universiteta)

Zasche, P. 2010, *New Astron.* 15, 150

Active OB stars: structure, evolution, mass loss, and critical limits
Proceedings IAU Symposium No. 272, 2010
C. Neiner, G. Wade, G. Meynet & G. Peters, eds.

doi:10.1017/S1743921311011318

An OB-type eclipsing binary system ALS 1135

Gabriela Michalska[1], **Ewa Niemczura**[1], **Marek Steslicki**[1], **and Andrew Williams**[2]

[1]Instytut Astronomiczny Uniwersytetu Wrocławskiego, Poland
emails: (michalska, eniem, steslicki)@astro.uni.wroc.pl

[2]Perth Observatory, Walnut Road, Bickley, Perth 6076, Australia
email: andrew@physics.uwa.edu.au

Abstract. We present new physical and orbital parameters of an early-type double-lined eclipsing binary system ALS 1135. The $UBVI_C$ light curves and radial velocity curves were modeled simultaneously by means of the Wilson-Devinney code. As a result, we obtained inclination and size of the orbit, as well as masses, radii and effective temperatures of the components.

Keywords. stars: early-type, binaries: eclipsing

1. Introduction

ALS 1135 is a member of OB association Bochum 7 of the age of about 6 Myr located at the distance of 4.8 kpc (Sung *et al.* 1999). Corti *et al.* (2003) discovered that ALS 1135 is a single-lined spectroscopic binary with a period of 2.7532 d, and classified the main component as O6.5 V((f)). Fernández Lajús & Niemela (2006) found the presence of faint lines of secondary component and from the radial velocities of both stars and the ASAS photometry obtained the orbital solution and physical parameters of both components. Based on the ASAS photometry for ALS 1135, Pigulski and Michalska (2007) found additional periodic variations with frequency of 2.31095 d^{-1}. If these pulsations originate in the primary component, it would be the first O-type star among β Cephei class variables and an excellent object for asteroseismology. To confirm this result the photometric and spectroscopic campaign for this star was conducted.

2. Observations

The $UBVI_C$ observations of ALS 1135 were carried out with the 1-m telescope at the South African Astronomical Observatory (SAAO) during 13 nights between January 23 and February 5, 2008. Between November 2007 and January 2009, the BVI_C observations with 0.6-m Perth-Lowell Automated Telescope at Perth Observatory were taken.

The spectroscopic observations of ALS 1135 were carried out between 25 and 27 January 2008 with the ESO New Technology Telescope (NTT; La Silla, Chile) and the ESO Multi-Mode Instrument (EMMI) in the cross-dispersed échelle mode. Additionally, on 5th January 2008, a single spectrum of ALS 1135 was obtained with the MIKE spectrograph, attached to the 6.5-m Magellan-Clay Telescope at Las Campanas Observatory.

3. Analysis and Results

The orbital period found from our analysis is equal to 2.753189 d and the time of the primary minimum is equal to HJD 24552070.144(13). The search for variable stars around

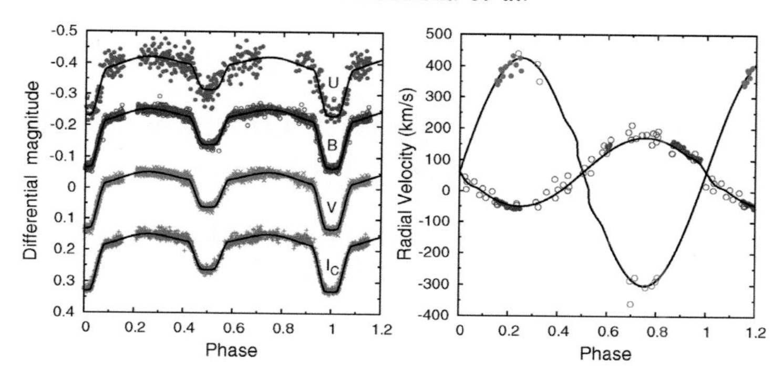

Figure 1. Light and radial-velocity curves of ALS 1135. Solid line represent the best fit.

ALS 1135 reveal that additional variations found by Pigulski & Michalska (2007) were originated in neighboring EW-type system which was not resolved in the ASAS data.

To obtain atmospheric parameters of the primary component we used a grid of fluxes calculated for TLUSTY model atmospheres for O-type star (Lanz & Hubeny 2003). The solar helium abundance, metallicity and a microturbulence velocity equal to $10\,\mathrm{km\,s^{-1}}$ were assumed. The temperature of the primary component was established to 37 500 K, $\log g = 3.8\,\mathrm{cm\,s^{-2}}$ and $v \sin i = 200 \pm 20\,\mathrm{km\,s^{-1}}$. The EMMI/NTT data were used to derive heliocentric radial velocities for both ALS 1135 components.

Our $UBVI_\mathrm{C}$ light curves and radial-velocity curves (Fig. 1) were modeled simultaneously by means of the latest version of the Wilson-Devinney (W-D) code (Wilson & Devinney 1971). To the analysis we included also the radial velocities derived by Corti *et al.* (2003) and Fernández Lajús & Niemela (2006) (open circles in Fig. 1). The W-D code was run with geometry used for detached binaries. The albedos and gravity darkening coefficients was set to 1.0. We assumed circular orbit, synchronization and allowed no spots. The following parameters were adjusted during the fits: mass ratio (q), radial velocity of the mass center (V_γ), phase shift, surface potentials, effective temperature of the secondary component ($T_{\mathrm{eff},2}$), inclination (i) and monochromatic luminosity of the primary component. As a result of the modeling, we obtained new, more accurate orbital and physical parameters of the system: $q = 0.306(4)$, $a = 26.9(2)\,\mathrm{R}_\odot$, $i = 79.21(6)^\circ$, $V_\gamma = 63.4(9)\,\mathrm{km\,s^{-1}}$, $T_{\mathrm{eff},2} = 25\,900(70)\,\mathrm{K}$, $M_1 = 26.3(7)\,\mathrm{M}_\odot$, $M_2 = 8.1(6)\,\mathrm{M}_\odot$, $R_1 = 10.31(8)\,\mathrm{R}_\odot$ and $R_2 = 3.60(3)\,\mathrm{R}_\odot$.

The analysis of new photometric observation of ALS 1135 reveals that periodic variations with frequency $2.31095\ \mathrm{d}^{-1}$ found in this star from ASAS-3 data (Pigulski and Michalska 2007) are caused by contamination with neighboring EW-type system.

Acknowledgements

This work was supported by Polish MNiSzW grant N N203 302635 and the Proyecto FONDECYT No 3085010.

References

Corti, M., Niemela, V., & Morrell, N. 2003, *A&A*, 405, 571

Fernández Lajús, E. & Niemela, V. S. 2006, *MNRAS*, 367, 1709

Lanz, T. & Hubeny, I. 2003, *ApJS*, 146, 417

Pigulski, A. & Michalska, G. 2007, *AcA*, 57, 61

Sung, H., Bessell, M. S., Park, B.-G., & Kang, Y. H. 1999, *Journal of Korean Astronomical Society*, 32, 109

Wilson, R. E. & Devinney, E. J. 1971, *ApJ*, 166, 605

Active OB stars: structure, evolution, mass loss, and critical limits
Proceedings IAU Symposium No. 272, 2010
C. Neiner, G. Wade, G. Meynet & G. Peters, eds.

doi:10.1017/S174392131101132X

Fast rotating stars resulting from binary evolution will often appear to be single

Selma E. de Mink, Norbert Langer and Robert G. Izzard

Argelander-Institut für Astronomie der Universität Bonn, Germany
email: S.E.deMink@gmail.com

Abstract. Rapidly rotating stars are readily produced in binary systems. An accreting star in a binary system can be spun up by mass accretion and quickly approach the break-up limit. Mergers between two stars in a binary are expected to result in massive, fast rotating stars. These rapid rotators may appear as Be or Oe stars or at low metallicity they may be progenitors of long gamma-ray bursts.

Given the high frequency of massive stars in close binaries it seems likely that a large fraction of rapidly rotating stars result from binary interaction. It is not straightforward to distinguish a a fast rotator that was born as a rapidly rotating single star from a fast rotator that resulted from some kind of binary interaction. Rapidly rotating stars resulting from binary interaction will often appear to be single because the companion tends to be a low mass, low luminosity star in a wide orbit. Alternatively, they became single stars after a merger or disruption of the binary system during the supernova explosion of the primary.

The absence of evidence for a companion does not guarantee that the system did not experience binary interaction in the past. If binary interaction is one of the main causes of high stellar rotation rates, the binary fraction is expected to be smaller among fast rotators. How this prediction depend on uncertainties in the physics of the binary interactions requires further investigation.

Keywords. stars: evolution, stars: rotation, binaries, stars: emission-line, Be, gamma rays: theory

---

Rotation gives rise to various phenomena in stars, for example the deformation of fast-rotators to an oblate shape (e.g. Zhao, these proceedings) and mixing processes in stellar interiors (e.g. Zahn and Ekström *et al.*, these proceedings). Rotation also seems to be an essential ingredient in the production of disks around classical Be and Oe stars and it may lead to the formation of long GRB progenitors at lower metallicity (Martayan, these proceedings).

The close binary fraction among massive stars is high. About 45% of the massive stars in OB associations show radial velocity variations in their spectra (e.g. Sana et al, these proceedings) indicating the presence of a nearby companion. Binary interaction can lead to rapid rotation rates in various ways, see Fig. 1. Given the relative ease with which binary systems produce fast rotating stars and the high fraction of massive stars that are found in binary systems, one may consider the following question: *"What fraction of rapid rotators result from binary interaction?"*. This question is relevant for various other open questions. Are Be stars formed in binaries (e.g. Pols *et al.* 1991, Ekström *et al.* 2008)? Are the surface abundances in fast rotating stars signatures of rotational mixing or from binary mass transfer (e.g. Hunter *et al.* 2008)? Do binaries produce the progenitors of long gamma-ray bursts (Cantiello *et al.* 2007)?

Identifying the role of binary interaction in the production of rapid rotators is not straightforward. The low-luminosity companion star in a post-interaction binary is typically very hard to detect, see Fig 1. **(I)** Mass transfer will widen the orbit and reduce

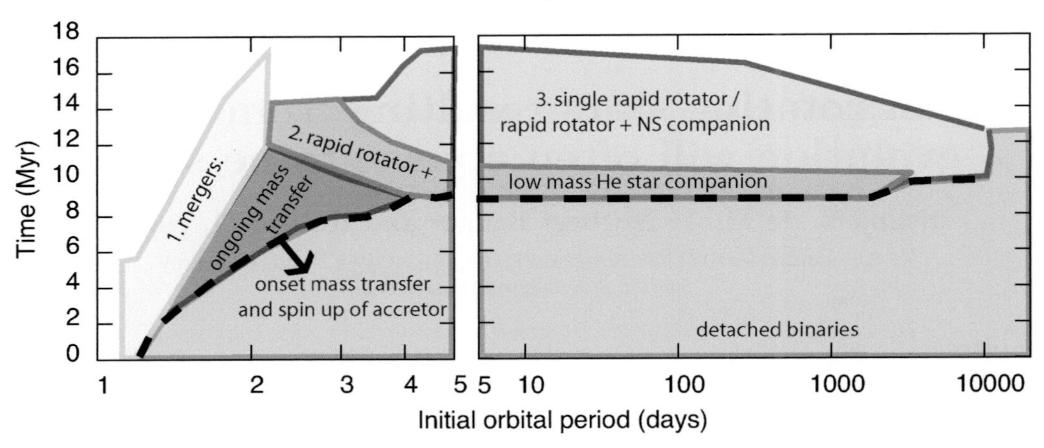

Figure 1. Schematic depiction of the evolutionary stages of a 20+15$M_{\odot}$ binary as a function of the initial orbital period (x-axis) and age of the system (y-axis). The thick dashed line indicates the onset of mass transfer, which occurs later for wide binaries or is avoided in the widest systems. This diagram indicates that rapid rotators resulting from binary interaction are often single or appear to be single. (1) Mergers in close binaries result in massive, rapidly rotating, main-sequence stars that are single. (2) Mass transfer drives the accreting star to critical rotation. Initially the companion will be a low mass, low luminosity helium star which will be hard to detect. (3) If the helium star is massive enough it will explode and is likely to disrupt the binary system, leaving its companion behind as a fast spinning single star. This diagram is based on a grid of models computed with an upgraded version of the binary population synthesis code described by Izzard *et al.* (2006) based on Hurley *et al.* (2002), De Mink *et al.* (in prep.).

the mass and luminosity of the originally most massive star. **(II)** When the companion explodes, the binary system is likely to be disrupted. The accretor is left behind as a rapidly rotating single star. **(III)** Finally, mergers also produce rapidly rotating single stars.

We come to the somewhat counter-intuitive conclusion: fast rotating stars that were produced during binary interactions will, in many or even most cases, appear to be single stars. In fact, if binary interaction is one of the main causes of high stellar rotation rates, a smaller binary fraction for fast rotators is expected. How this prediction depends on uncertainties in the modeling of the mass transfer phase requires further investigation.

References

Cantiello, M., Yoon, S.-C., Langer, N., & Livio, M. 2007, *A&A*, 465, L29
Ekström, S., Meynet, G., Maeder, A., & Barblan, F. 2008, *A&A*, 478, 467
Hunter, I., Brott, I., Lennon, D. J., & Langer, N. *et al.* 2008, *ApJ* (Letters), 676, 29
Hurley, J. R., Tout, C. A., & Pols, O. R. 2002, *MNRAS*, 329, 897
Izzard, R. G., Dray, L. M., Karakas, A. I., & Lugaro, M. *et al.* 2006, *A&A*, 460, 565
Pols, O. R., Coté, J., & Waters, L. B. F. M., Heise, J. 1991, *A&A*, 241, 419

Active OB stars: structure, evolution, mass loss, and critical limits
Proceedings IAU Symposium No. 272, 2010
C. Neiner, G. Wade, G. Meynet & G. Peters, eds.

doi:10.1017/S1743921311011331

Emission features in a B[e] binary system V2028 Cyg

Jan Polster[1,2], Daniela Korčáková[2], Viktor Votruba[1,2], Petr Škoda[2], Miroslav Šlechta[2] and Blanka Kučerová[1]

[1]Faculty of Science, Masaryk University in Brno, Kotlářská 2, 611 37 Brno, Czech Republic
[2]Astronomical Institute AV ČR, Fričova 298, 251 65 Ondřejov, Czech Republic
email: polster@physics.muni.cz, kor@sunstel.asu.cas.cz

Abstract. We present a preliminary analysis of our six-year observation campaign of the B[e] stellar system V2028 Cyg (MWC 623). The time variability of spectral features is described.

Keywords. stars: emission-line, binaries: general, stars: individual (V2028 Cyg)

1. Introduction

A variable star V2028 Cyg (MWC 623) shows a B[e] phenomenon. The nature of the central object is very difficult to determine due to the extended circumstellar medium. The spectrum of V2028 Cyg is dominated by strong Balmer emissions, especially Hα. It also contains permitted and forbidden emission lines of low-ionized or neutral atoms (Fe II, [Fe II], [O I], ...).

We present measurements of some emission features in spectra of our six-year observation campaign.

84 spectra in the Hα region were taken with the Ondřejov 2m telescope using a coudé slit spectrograph with resolving power $R_{\mathrm{H}\alpha} \sim 12500$. Additional four spectra were obtained from the ELODIE archive ($R = 42000$ in the wavelength interval 3906 – 6811 Å).

2. Radial velocities and bisectors

Radial velocities of the metallic emissions were measured by Gaussian fitting. We used the double Gaussian method (Schneider & Young, 1980) for the wings and the polynomial fitting in the case of the peak to measure the Hα radial velocities. The overall behaviour of the Hα line profile is described by bisectors (Fig. 2).

3. Conclusions

Radial velocities of the metallic emissions (Fig. 1) show certain scatter, but no period or trend was found. There is a clear minimum of radial velocity in JD = 2454155 ± 17 in the case of Hα wings (Fig. 3). The peak velocity values are more scattered, but there can be resolved a minimum in JD = 2454683 ± 33.

The bisector measurements (Fig. 2) reveal, that the whole lower part of the Hα profile (under the hump) behaves the same way as the wings. The peak above the hump varies differently. It indicates existence of at least two different kinematical environments in the envelope, where Hα emission arises.

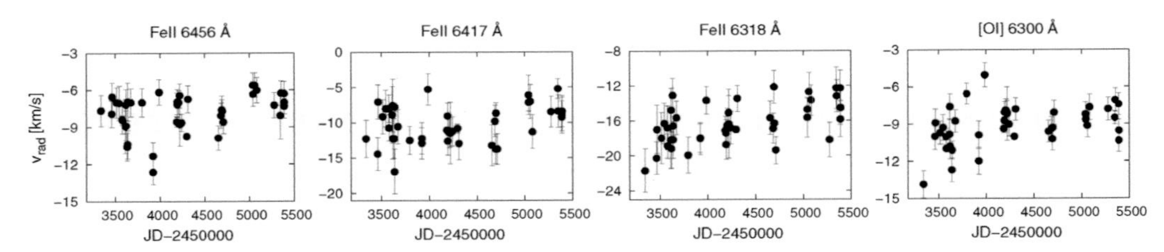

Figure 1. Radial velocities of [O I] and Fe II emission lines.

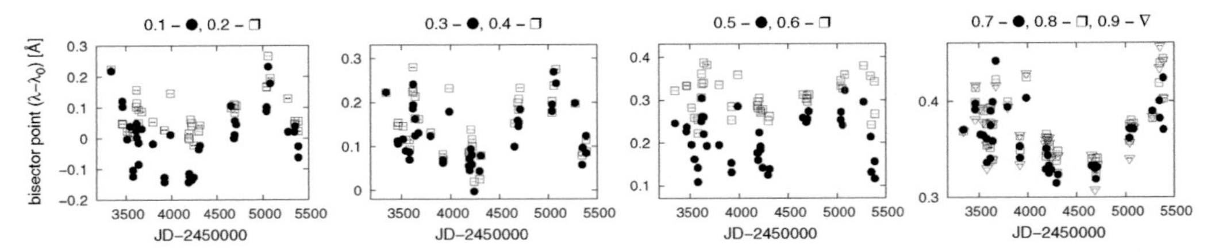

Figure 2. Bisector points of Hα line in relative heights from 0.1 (wings) to 0.9 (peak).

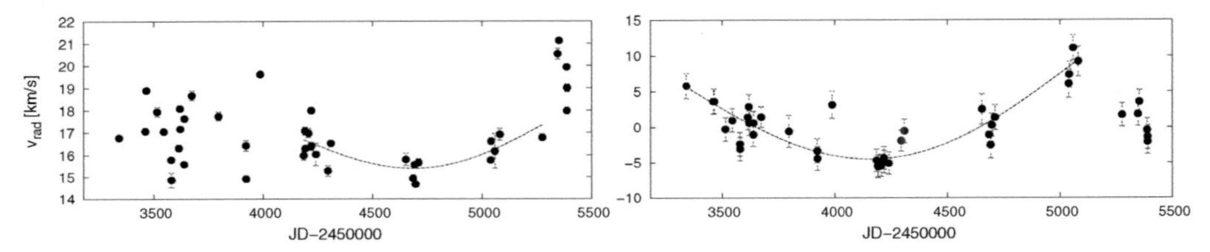

Figure 3. Radial velocities of the Hα peak (left) and wings (right). The fitting function is plotted in the graph.

The half maximum widths of metallic lines are ~ 1 Å, which restricts the maximum value of the velocity gradient in the region of their formation to ~ 50 km/s. No observed line shows P Cyg profile and the emissions of metals have a blue-shifted maximum.

Currently, we are working on a numerical modeling of the bisector time variability in order to explain these observed phenomena.

Acknowledgements

This research is supported by grants 205/09/P476 (GA ČR), 205/08/H005(GA ČR) LC06041 (MŠMT ČR) and MUNI/A/0968/2009. The Astronomical Institute Ondřejov is supported by a project AV0Z10030501.

Reference

Schneider, D. P. & Young, P. 1980, *ApJ*, 238, 946

Active OB stars: structure, evolution, mass loss, and critical limits
Proceedings IAU Symposium No. 272, 2010
C. Neiner, G. Wade, G. Meynet & G. Peters, eds.

doi:10.1017/S1743921311011343

The effects of μ gradients on pulsations of rapidly rotating stars

Daniel R. Reese[1], **Francisco Espinosa Lara**[2] **and Michel Rieutord**[3]

[1] LESIA, CNRS UMR 8109, Observatoire de Paris, 92195 Meudon, France,
email: daniel.reese@obspm.fr

[2] GEPI, CNRS UMR 8111, Observatoire de Paris, 92195 Meudon, France,
email: francisco.espinosa@obspm.fr

[3] LATT, CNRS UMR 5572, Université de Toulouse, 31400 Toulouse, France
email: rieutord@ast.obs-mip.fr

Abstract. Recently, Reese *et al.* (2008), Lignières & Georgeot (2008) and Lignières & Georgeot (2009) showed that the frequencies of low-degree acoustic modes in rapidly rotating stars, also known as "island modes", follow an asymptotic formula, the coefficients of which can be deduced from ray dynamics. We investigate how this asymptotic behaviour is affected by μ gradients by comparing pulsation spectra from models with and without such a discontinuity.

Keywords. stars: oscillations, stars: rotation, stars: abundances

---

Rapid rotation strongly modifies both the geometry of pulsation modes and the organisation of the associated frequency spectrum. Recently, using both 2D calculations and ray dynamics, Reese *et al.* (2008), Lignières & Georgeot (2008) and Lignières & Georgeot (2009) showed that the frequencies of low-degree acoustic modes in rapidly rotating stars, also called "island modes", obey an asymptotic formula, for which two of the coefficients can be given in terms of travel time integrals. An open question is how such a formula will be affected by a strong compositional gradient.

In order to answer this question, we study the pulsation modes of rapidly rotating 3 $M_\odot$ stellar models with the following compositions:

- *Model 1*: uniform composition ($X = 0.70$, $Z = 0.02$)
- *Model 2*: H poor interior ($X = 0.35$), H rich envelope ($X = 0.70$)
- *Model 3*: H rich interior ($X = 0.70$), H poor envelope ($X = 0.35$)

Model 3 is unrealistic, since nuclear reactions deplete hydrogen starting from the interior, but remains useful for the purposes of comparison. Furthermore, the density jumps in models 2 and 3 were chosen to be close to the surface in order to investigate the effects of μ gradients on acoustic modes. In reality, such jumps are likely to be located near the core and to have a stronger effect on gravito-inertial modes.

The models were produced using the ESTER code (Espinosa Lara & Rieutord 2007, Rieutord & Espinosa Lara 2009), which uses a multi-domain spectral approach, and a surface-fitting spheroidal coordinate system based on Bonazzola *et al.* (1998). In the present study, the rotation profiles have a constant value on the surface and were r and θ dependant in the interior, as a result of baroclinic effects. The pulsation modes were calculated using the TOP code (Reese *et al.* 2009), fully including the effects of spheroidal distortion and differential rotation.

Figure 1 shows the meridional cross section of the same pulsation mode in models 2 and 3. The sound speed decreases across the jump when going towards the interior of model 2, thereby causing the pulsation mode to be slightly deflected towards the centre, whereas the opposite occurs in model 3.

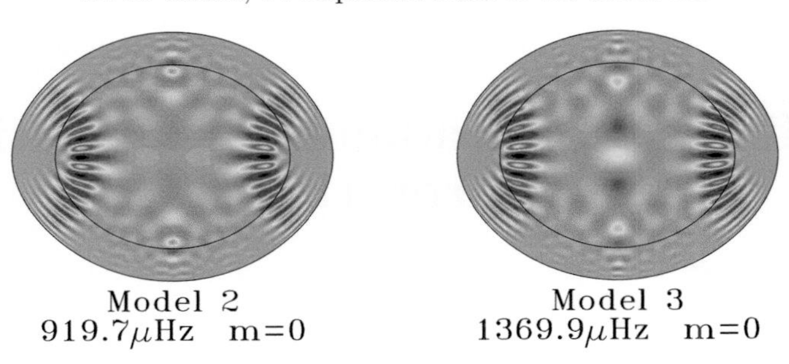

Figure 1. Meridional cross-section of the same pulsation mode in models 2 and 3. The jump in sound speed, represented by the black internal line, causes a slight deflection of the mode.

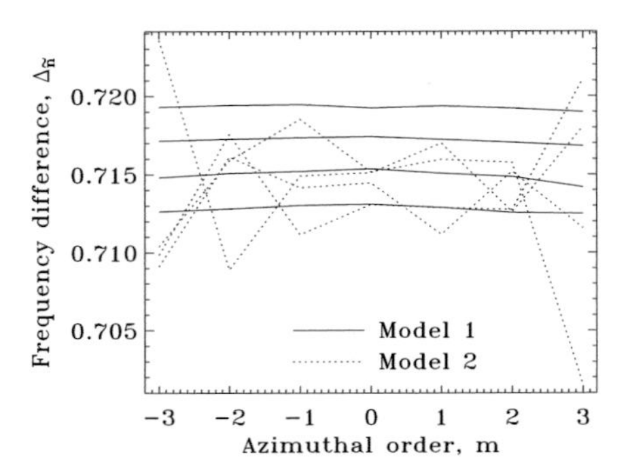

Figure 2. Frequency separations, as a function of m, between high order modes with $\tilde{\ell} = 0$ and consecutive $\tilde{n}$ values.

Figure 2 shows the frequency separations, $\Delta_{\tilde{n}} = \omega_{\tilde{n}+1,\tilde{\ell},m} - \omega_{\tilde{n},\tilde{\ell},m}$ of high order modes as a function of the azimuthal order, m. The pseudo-radial orders, $\tilde{n}$, go from 36 to 39. As expected, model 1 displays a smooth behaviour, since there are no μ gradients, whereas model 2 has an oscillatory behaviour. Nonetheless, these results must be taken with caution since the uncertainties on the frequencies of model 2, possibly as a result of numerical irregularities in the derivatives of Ω, are of the same order of magnitude as the oscillations present in the figure.

Acknowledgements

DRR is currently supported by the CNES ("Centre National d'Etudes Spatiales") through a postdoctoral fellowship. The numerical calculations were carried out on the Vargas supercomputer (IBM SP Power 6) from the "Institut du Développement et des Ressources en Informatique Scientifique" (IDRIS). FEL thanks the ANR SIROCO.

References

Bonazzola, S., Gourgoulhon, E., & Marck, J.-A. 1998, *Phys. Rev. D*, 58, 104020
Espinosa Lara, F. & Rieutord, M. 2007, *A&A*, 470, 1013
Lignières, F. & Georgeot, B. 2008, *Phys. Rev. E*, 78, 016215
Lignières, F. & Georgeot, B. 2009, *A&A*, 500, 1173
Reese, D., Lignières, F., & Rieutord, M. 2008, *A&A*, 481, 449
Reese, D. R., MacGregor, K. B., Jackson, S., Skumanich, A. *et al.* 2009, *A&A*, 506, 189
Rieutord, M. & Espinosa Lara, F. 2009, *Communications in Asteroseismology*, 158, 99

Active OB stars: structure, evolution, mass loss, and critical limits
Proceedings IAU Symposium No. 272, 2010
C. Neiner, G. Wade, G. Meynet & G. Peters, eds.

doi:10.1017/S1743921311011355

Multiplicity in 5 $M_\odot$ stars

Nancy Remage Evans

Smithsonian Astrophysical Observatory
60 Garden St., MS 4, Cambridge MA 02138 USA
email: nevans@cfa.harvard.edu

Abstract. Binary/multiple status can affect stars at all stages of their lifetimes: evolution onto the main sequence, properties on the main sequence, and subsequent evolution. 5 $M_\odot$ stars have provided a wealth of information about the binary properties fairly massive stars. The combination of cool evolved primaries and hot secondaries in Cepheids (geriatric B stars) have yielded detailed information about the distribution of mass ratios. and have also provided a surprisingly high fraction of triple systems. Ground-based radial velocity orbits combined with satellite data from Hubble, FUSE, IUE, and Chandra are needed to provide full information about the systems, including the masses. As a recent example, X-ray observations can identify low mass companions which are young enough to be physical companions. Typically binary status and properties (separation, eccentricity, mass ratio) determine whether any stage of evolution takes an exotic form.

Keywords. binaries: general, stars: variables: Cepheids

---

1. Introduction

Understanding the configurations involving massive stars, the processes which shape them, and the objects they evolve into is formidable–though being undertaken with equally formidable observing and computing resources. In the interplay of rotation, winds, and magnetic fields binary/multiple status sometimes take on a leading role, sometimes plays a relatively passive role. It is, however, frequently a significant factor and progress is being made in determining binary/multiple properties. This contribution discusses two aspects for $5M_\odot$ stars, the frequency of higher multiplicity systems and the identification of low mass companions.

2. Multiplicity

Cepheids (post-main sequence He burning stars of typically 5 $M_\odot$) have provided some new insights into system multiplicity. Frequently they have hot companions which can be studied uncontaminated by the light of the primary. This has lead to the identification of triple systems in a number of ways. For the best studied sample (18 stars with orbits and ultraviolet spectra of the companions), 44% (possibly 50%) are actually triple systems (Evans *et al.* 2005).

3. Low Mass Companions

A second important area where we are obtaining new information about $5M_\odot$ systems is in the identification of low mass companions. These are the most difficult companions to detect, either in photometric or spectroscopic (radial velocity) studies. We are exploring the use of X-ray data to determine the fraction of late B (B3-A0) stars which have low

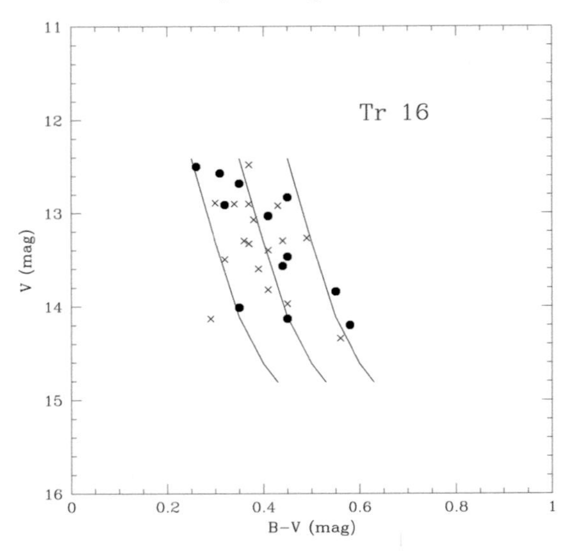

Figure 1. The sample of late B stars in Tr 16. Lines show the ZAMS (center) with a range of E(B-V) = ± 0.1 mag. Dots are X-ray sources; x's are not detected.

mass companions. Late B do not in general produce X-rays themselves. However, low mass stars (mid-F through K spectral types) at the same age ($\simeq$ 50 Myr) are strong X-ray producers, much stronger than field stars of the same temperature. Fig 1 shows data from the Chandra ACIS image of Tr 16 (Evans *et al.* 2011; Townsley *et al.* 2011, Albacete-Colombo *et al.* 2008). With many exciting more massive stars in Tr 16, little attention has been paid to late B stars, so the first step was to establish a sample. A list was compiled of stars within 3' of η Car (the center of Tr 16), with an appropriate combination of V and B-V for the ZAMS (distance of 2.3 kpc, E(B-V) = 0.55 mag ±0.1 mag), and proper motions consistent with cluster membership (Cudworth *et al.* 1993). 39% of the stars are detected, spread through the luminosity range as would be expected for a random occurrence. These are identified as having low mass companions. This approach is complementary to photometric or interferometric surveys such as the O/B survey of Mason *et al.* (2009) and the IUE survey of Cepheids (Evans 1992) and probes new parameter space.

Acknowledgements

Funding for this work was provided by Chandra X-ray Center NASA Contract NAS8-39073.

References

Albacete-Colombo, J. F., Damiani, F., Micela, G., Sciortino, S. *et al.* 2008, *A&A*, 490, 1055
Cudworth, K. M., Martin, S. C., & Degioia-Eastwood, K. 1993, *AJ*, 105, 1822
Evans, N. R. 1992, *ApJ*, 384, 220
Evans, N. R., Carpenter, K. G., Robinson, R., Kienzle, F. *et al.* 2005, *AJ*, 130, 789
Evans, N. R., DeGioia-Eastwood, K, Gagne, M., Townsley, L. *et al.* 2011, preprint, (Chandra Carina Complex Project)
Mason, B. D., Hartkopf, W. I., Gies, D. R., Henry, T. J. *et al.* 2009, *AJ*, 137, 3358
Townsley, L. *et al.* 2011, preprint, (Chandra Carina Complex Project)

Active OB stars: structure, evolution, mass loss, and critical limits
Proceedings IAU Symposium No. 272, 2010
C. Neiner, G. Wade, G. Meynet & G. Peters, eds.

doi:10.1017/S1743921311011367

Masses of the astrometric SB2 ζ Ori A†

Thomas Rivinius[1], **Christian A. Hummel**[2] **and Otmar Stahl**[3]

[1]ESO Chile; [2]ESO, Germany; [3]ZAH/LSW Heidelberg, Germany

Abstract. We report the first dynamic mass for an O-type supergiant, the interferometrically resolved SB2 system ζ Ori A (O9.5Ib+B0/1). The separation of the system excludes any previous mass-transfer, ensuring that the derived masses can be compared to single star evolutionary tracks.

Keywords. stars: binaries, stars: early-type, stars: fundamental parameters (masses)

---

1. Introduction

Stellar masses in the upper HRD are notoriously hard to constrain. Very few masses are known independently from stellar evolution or wind models, mostly using eclipsing SB2 systems. For stars that have already evolved away from the main sequence, this is a problematic technique, though: These are usually quite narrow short period systems, which means that the possibility of mass transfer via overflow having altered the evolutionary paths is high. Interferometry is in principle able to overcome this problem in cases where the orbit can be measured for an SB2.

Hummel *et al.* (2000) found the O9.5 Ib primary of ζ Ori A to be a multiple star. In addition to the well known B-component of spectral type B0 at V = 3.77mag, several arcseconds away, they found another companion (Ab) 40 mas away, about 2 mag fainter than the primary (Aa), i.e. the magnitude of Ab is similar to that of component B.

2. Observations

Additional Interferometric observations to complete the orbit coverage (Fig. 1, right). were obtained at the NPOI in Flagstaff, Arizona. Spectroscopic observations were taken with the echelle instruments HEROS and FEROS.

In the spectra all He I lines as well as He II 4686 have a relatively narrow core with varying RV in one sense, while the line wings are shifted in anti-phase wrt. the cores. This is the signature of an SB2 binary. For some lines (almost) exclusive formation in the O9.5 component can be assumed. The best candidates are the He II lines, typically not seen in B-type stars, except He II 4686. The radial velocities of this line were measured with Gaussian fits to the line center (Fig. 1, left). There are as well very weak and rather narrow lines that are not expected in the O9 Ib star. These narrow lines are RV variable in the same sense as the cores of stronger lines, i.e. they belong to the companion and are indicative for an early type B star.

In addition to Gaussian fits, the RV curve was also measured by spectral disentangling with VO-KOREL, a virtual observatory tool based on the KOREL code by Hadrava (1995).

† Based on observations under ESO programs 076.C-0431, 080.A-9021, and 083.D-0589.

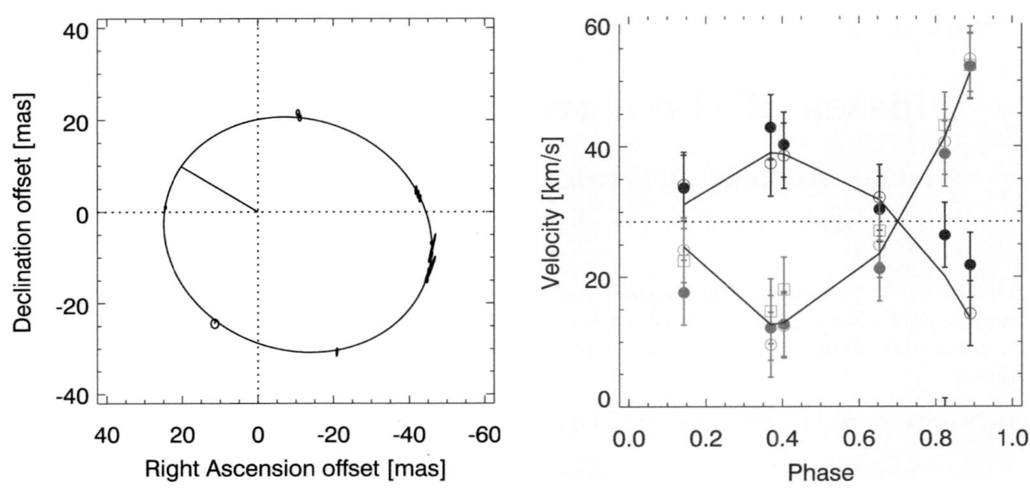

Figure 1. *Left panel*: The interferometric orbit. The periastron is marked by the line from the origin to the orbit. The secondary progresses clockwise. *Right panel*: he measured RVs of both components. The filled disks are Gaussian centers (He II and O II. The other RVs are from spectral disentangling of O II 4940 (rectangles) and He I 6678 (circles)

3. Spectro-interferometric parameters

The orbital parameters were derived from a simultaneous fit to the **interferometric** and spectroscopic data:

$P_{\mathrm{orb.}}$	2677.5 ± 6.9 d	inclination	$137.9 \pm 1.1°$
Primary periastron date	JD=2 452 747.9 $\pm$ 5.2	semi-major axis	36.1 ± 0.3 mas
Eccentricity	0.34 ± 0.01	q	0.66
Periastron long.	$208.4 \pm 1.1°$	γ	28.6 km/s
Ascending Node	$86 \pm 0.8°$		
M_{Aa}	$24.8 \pm 5.6\,\mathrm{M}_\odot$	M_{Ab}	$16.4 \pm 4.9\,\mathrm{M}_\odot$

This is the first directly measured (preliminary) mass for an O-type supergiant unaffected by mass transfer, as well being the first application of the interferometric/SB2 technique to an O-star binary. The 09.5 Ib star ζ Ori Aa was found to have a mass of:

$$\boldsymbol{M_{\zeta\,\mathrm{Ori\ Aa}} = 24.8 \pm 5.6\,\mathrm{M}_\odot}$$

We as well give a mass for its early B-type companion Ab, which is $16.4_\odot$. This value is in reasonable agreement with the estimate of the companion to be a little evolved very early B-type star. The mass of ζ Ori Aa is also in good agreement with the theoretically expected value for its spectral type, i.e. its track mass.

We stress that this is still a preliminary determination, based on only a few spectral lines and not yet using all available spectroscopic data. Taking advantage of the full wavelength range of the available echelle data, it will be possible to determine the radial velocity curve of both components with much higher confidence.

References

Hadrava, P. 1995, *A&AS*, 114, 393

Hummel, C. A., White, N. M., Elias, II, N. M., Hajian, A. R. *et al.* 2000, *ApJ* (Letters), 540, L91

Active OB stars: structure, evolution, mass loss, and critical limits
Proceedings IAU Symposium No. 272, 2010
C. Neiner, G. Wade, G. Meynet & G. Peters, eds.

doi:10.1017/S1743921311011379

The resonant B1II + B1II binary BI 108†

Thomas Rivinius[1], **Ronald E. Mennickent**[2], **Zbigniew Kołaczkowski**[3]
[1]ESO, Chile; [2]U. de Concepcíon, Chile; [3]U. Wroclawskiego, Poland

Abstract. BI 108 is a luminous variable star in the Large Magellanic Cloud classified B1 II. The variability consists of two resonant periods (3:2), of which only one is orbital, however. We discuss possible mechanisms responsible for the second period and its resonant locking.

Keywords. stars: binaries, stars: early-type

1. Introduction

The observed period of BI 108 (OGLE-ID: LMC SC9-125719, MACHO-ID 79.5378.25) of 10.73 d has six equidistant but distinct minima and some symmetry around the deepest minimum. At close inspection, it became clear that the lightcurve can actually be disentangled into two periods with a resonant ratio of 3:2 (see figures in Kołaczkowski *et al.* in press). From October 2008 to January 2009 20 spectra were taken at ten epochs with the echelle spectrograph UVES, mounted at the 8.2 m telescope UT2 on Cerro Paranal.

2. Orbital parameters and spectral disentangling

The obtained spectra are of an SB2 composite nature, however, only one single period is present in the radial velocities, namely $P_{\rm orb.} = 5.37$ d. Star A and B have an almost circular orbit, similar mass, and are in a similar evolutionary stage, being both very early B supergiants, B0 II+B0 II. No circumstellar gas, i.e. no mass transfer, was detected. The spectral disentangling was done with VO-KOREL based on KOREL, Hadrava (1995):

$P_{\rm sup.}$	10.73309 d	Periastron long.	93°
$P_{\rm res.}$	3.57793 d	K_1	170 km/s
$P_{\rm orb.}$	5.36654 d	K_2	225 km/s
Epoch (min. light)	MJD=51 163.3915	q	0.76
Periastron date	MJD=54 742.8345	$a \sin i$	41.5 $R_\odot$
Eccentricity	0.08	$M \sin^3 i$	33.6 $M_\odot$

The resonant period is not completely absent in the spectra, however, but modulates the line strengths. In particular, the total equivalent width (EW) over both components is about constant, while each component varies strongly, i.e. it seems as if a certain fraction of equivalent width is exchanged between the stars with a period of $P_{\rm res.} = 3.58$ d, although admittedly our data is too scarce to claim so with much certainty. For brevity, we call this behavior "EW-shuffling" (see Fig. 1, right) in the following.

3. Discussion

A good hypothesis for the explanation of the system has to fulfill a number of criteria: **1.)** Explain why the superperiod is disentangled into two resonant periods. **2.)** Provide a strong and stable locking mechanism between the two periods. **3.)** Explain the phase

† Based on observations under ESO program 382.D-0311.

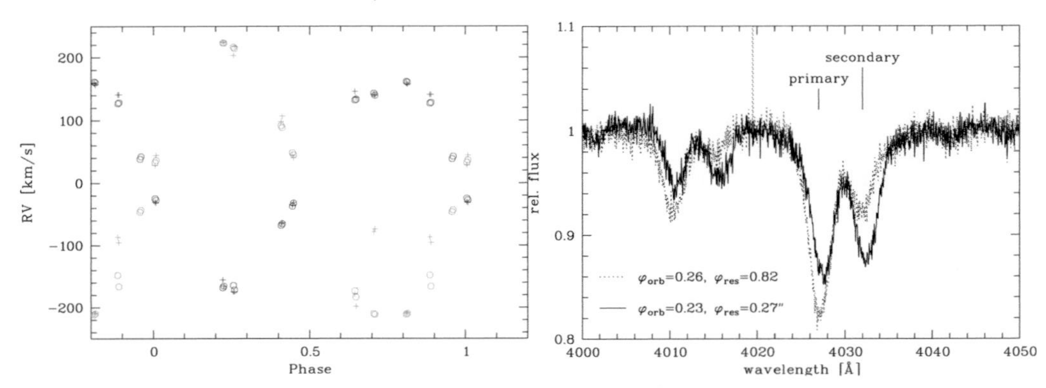

Figure 1. Left panel: Orbital velocities determined by VO-KOREL disentangling. Crosses are values derived from SiIII 4553, circles from Hβ. Sort with $P_{\rm orb.}$=5.36654 d.
Right panel: UVES spectra taken at similar orbital, but opposite resonant quadrature phase. The "EW shuffling" between both lines is obvious.

relation, i.e. **a.)** $T_{\rm min,res.} = T_{\rm min,orb.}$, and **b.)** the RV vs. light curve relations, not as a line-of-sight coincidence, but for all aspects. **4.)** Provide an evolution scenario, in which the current system, hosting (at least) two B0 II stars, could have evolved from a plausible off-the-shelf system. **5.)** Provide at least a toy model for the EW shuffling.

In the following we discuss several hypothesis: **Lagrangian triple:** A Lagrangian triple is a system with three stars A+B+C on a common orbit, trailing each other with $\Delta\phi = 120°$. In a more general scheme, concentric orbits for stars with unequal masses are possible as well. EW shuffling might be due to the unrecognized component switching between A and B. However, such a system quite strongly violates criteria 2, 3b, and 4. **Hierarchical system:** A system of type (Aa+Ab)+(Ba+Bb) was the original suggestion when the system was first published Ofir, 2008. However, this suggestion falls short in criteria 1, 2, 3a, and 3b. **Two-star magnetic resonance:** This hypothesis assumes only two stars A+B, with the spin-orbit relation locked by magnetic fields. Criteria 1 and 3 are naturally fulfilled and 4 is not a major problem a priori, but the 3:2 locking ratio might be an issue for criterion 2 (at least for dipoles). Yet neither the EW shuffling nor the photometric curve look like anything a magnetic star usually has to offer. **Two-star tidal resonance:** Also only two stars A+B, but their spin-orbit relation are locked due to tidal interaction. Again criteria 1 and 3 are naturally fulfilled, while according to Witte & Savonije (2001) 2 and 4 are plausible. Geometrical distortion and light modulation by a strong tidal wave is at least a toy model for the EW behaviour.

As Witte & Savonije (2001) point out in a study of a $10\,{\rm M}_\odot$+$10\,{\rm M}_\odot$ main sequence binary, such a tidal resonant locking might actually be quite common and stable during extended phases of the orbital circularization. Since we deal with massive stars, it is as well plausible that the stars have evolved away from the main sequence before being fully circularized. Although the tidally locked scenario leaves uncomfortably many open questions, it is the one requiring the least extreme assumptions, satisfying Occam's razor.

References

Kołaczkowski, Z., Mennickent, R. E., & Rivinius, T., in: *Binaries: Key to Comprehension of the Universe*, ASP-CS, in press, arXiv:1004.5464

Hadrava, P. 1995, *A&AS* 114, 393

Ofir A. 2008, *IBVS* 5868

Witte, M. G. & Savonije, G. J. 2001, *A&A* 366, 840

Active OB stars: structure, evolution, mass loss, and critical limits
Proceedings IAU Symposium No. 272, 2010
C. Neiner, G. Wade, G. Meynet & G. Peters, eds.

doi:10.1017/S1743921311011380

The (B0+?)+O6 system FN CMa†: a case for tidal-pulsational interaction?

Thomas Rivinius[1], Otmar Stahl[2], Stanislas Štefl[1], Dietrich Baade[3], Richard H.D. Townsend[4] and Luis Barrera[5]

[1]ESO Chile; [2]LSW/ZAH Heidelberg, Germany; [3]ESO Germany; [4]UW Madison, USA; [5]UMCE Santiago, Chile

Abstract. FN CMa is visually double with a separation of ~0.6 arcsec. Sixty high-cadence VLT/*UVES* spectra permit the A and B components to be disentangled, as the relative contribution of each star to the total light entering the spectrograph fluctuates between exposures due to changes in seeing. Component A exhibits rapid line-profile variations, leading us to attribute the photometric variability seen by HIPPARCOS (with a derived $P = 0.08866$ d) to this component. From a total of 122 archival and new echelle spectra it is shown that component A is an SB1 binary with an orbital period of 117.55 days. The eccentricity of 0.6 may result in tidal modulation of the pulsation(s) of component Aa.

Keywords. stars: binaries: general, stars: oscillations (including pulsations), stars: early-type

1. Introducing FN CMa

FN CMa (HD 53 974) is a bright ($V = 5.4$ mag) B0.5 III star and visually double. Within about a century, the relative position of components A and B, which are separated by ~0.6 arcsec, has changed marginally at most. A is brighter than B by about 1.2 mag.

2. Observations and data reduction

The ESO Science Archive contains 60 VLT/*UVES* echelle spectra of FN CMa obtained within 1.4 hours for a study of interstellar medium, and three more spectra from *FEROS* at the 2.2-m ESO/MPG telescope, La Silla. In 2009 and 2010, an additional 59 echelle spectra were secured with the *BESO* spectrograph, a clone of *FEROS* mounted on the Bochum 1.5-m Hexapod Telescope on Cerro Armazones.

As a result of variable seeing and imperfect guiding, some UVES spectra contain a significantly higher fraction of light from component B than others. Since the light combination is geometric it has no spectral dependency, and thus, under the assumption that certain spectral features are due to either A (e.g., Si III 4553) or B (e.g., He II 4540) alone, a simple linear set of equations can be used for the disentangling of the spectra from the two stars over the entire wavelength range. The inferred spectral light ratios, between 0.75 and 0.85, are in good agreement with the known magnitude difference.

3. Results

FN CMa B: This component has a spectrum typical of mid-O main-sequence stars. Compared to the B0.5III primary, it would be considerably underluminous if the pair were physical. However, assuming an O subdwarf companion does not help because,

† Based on data from ESO programs 076.C-0431, 076.C-0164, and the BESO spectrograph.

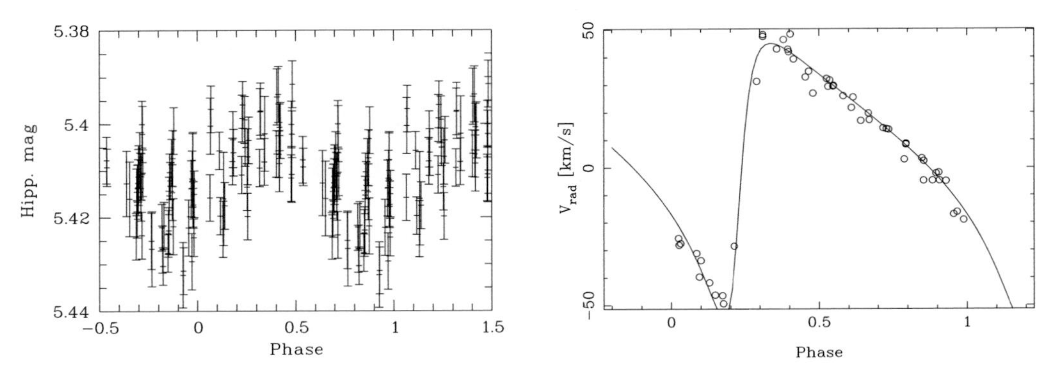

Figure 1. Left panel: *HIPPARCOS* photometry of FN CMa phased with a period of 2.13 h. Right panel: The radial velocity curve of FN CMa Aa (P = 117.55 d).

then, component B would be about 2 mag *over*luminous. Components A and B display the same set of interstellar Ca sc ii K lines except that the redmost one is significantly stronger in B. Considering also incipient emission in N III and Hα, we conclude that component B is best described as an O6V((f)) background star.

FN CMa A: In the literature, FN CMa has a record of low-amplitude photometric variability and modulated spectral line profiles. But there is no consensus about its nature. Our analysis of the *HIPPARCOS* photometry yields a period of 0.08866 d (2.13 h; see left panel of Fig. 1) with ~0.02 mag amplitude. The combination of spectral type, period, and amplitude makes FN CMa a β Cephei star candidate, as already suggested by other observers. This is further supported by the rapid spectral line-profile variability of component A (however, at just 1.4 h, the *UVES* data string is too short and the *BESO* spectra are not sufficiently densely sampled to attempt an independent period determination). In any case, given the spectral variations seen in component A, we attribute the photometric variability to this component as well.

Much larger-amplitude long-term radial-velocity variability is apparent from the *BESO* data: FN CMa A is itself an SB1 binary with the following properties:

Period [d]	117.55 ± 0.33
Periastron epoch [JD]	2 453 779.5 ± 4
Periastron longitude [deg]	247 ± 7
e	0.60 ± 0.05
K_1 [km/s]	49.8 ± 3.5
γ [km/s]	5.9 ± 1.5

The radial-velocity curve of FN CMa Aa is shown in Fig. 1 (right panel). Its relatively large amplitude suggests that the so-far (directly) undetected component FN CM Ab is a fairly massive star. However, it appears too faint to be the carrier of the rapid variability.

4. Discussion

The high eccentricity and moderate orbital period of the subsystem FN CMa Aa+Ab may enable searches for a tidal modulation of the pulsation of component Aa. Since FN CMa is bright and situated in a region with numerous other pulsating OB stars, it might be worthwhile including it in the target lists of wide-angle asteroseismology satellites such as BRITE.

Active OB stars: structure, evolution, mass loss, and critical limits
Proceedings IAU Symposium No. 272, 2010
C. Neiner, G. Wade, G. Meynet & G. Peters, eds.

doi:10.1017/S1743921311011392

Light curves of the Be stars of NGC 3766

Rachael M. Roettenbacher[1] and M. Virginia McSwain[1]

[1]Lehigh University
Department of Physics, 16 Memorial Drive E, Bethlehem, PA, 18015, USA
email: rmr207@lehigh.edu, mcswain@lehigh.edu

Abstract. Nonradial pulsations (NRPs) are a possible formation mechanism for the equatorial disks surrounding Be stars. The open cluster NGC 3766 has a high fraction of transient Be stars, Be stars that have been observed with both emission due to a circumstellar disk and a non-emitting B-type spectrum. Because of the large fraction of transient Be stars, this cluster is a prime location for studying the formation mechanisms of Be star disks. We observed NGC 3766 for more than 25 nights over three years to generate Strömgren *uvby* light curves of the Be population. We present the results of a period search to investigate the presence of NRPs.

Keywords. open clusters and associations: individual (NGC 3766), stars: emission-line, Be, stars: oscillations (including pulsations)

1. Introduction

NGC 3766 is an open cluster rich with transient Be stars (McSwain *et al.* 2008). Be stars are non-supergiant B-type stars that have exhibited emission features in the Balmer or other spectral lines at some point (Porter & Rivinius 2003).

NRPs are spherical harmonic waves that move across stellar surfaces driven from below by an opacity mechanism produced by ionized iron (Gutierrez-Soto 2007). These pulsations, in conjunction with rapid rotation, are possibly the cause of the equatorial disks surrounding Be stars (Porter & Rivinius 2003). For more discussion on the mechanisms behind NRPs, see Buta & Smith (1979).

2. Analysis and Results

We obtained Strömgren *uvby* photometric data in 2008 March and June, 2009 February and May, and 2010 March with the CTIO 0.9-m telescope. We applied standard reduction routines in IRAF using the *quadred* package. The photometric calibrations used non-variable B stars with known magnitudes from Shobbrook (1985, 1987). Balona & Engelbrecht (1986) and McSwain *et al.* (2008) list the B and Be stars we used for potential nonradial pulsation candidates. Sample b- and y-band light curves for one cluster member, No. 1 (using identifiers from the WEBDA database) over the course of one week (2010 March) are shown in Figure 1, left. With the oscillation frequencies determined by PDM (see Figure 1, right), we folded the light curves in plots of magnitude versus phase (see Figure 2).

3. Conclusions and Future Work

We find that some of the Be stars have multiple modes of NRPs, for example No. 1 (see Figures 1 and 2). The light curves and periods presented here, as well as those of many other members of NGC 3766, will be used to model NRPs. We will assume a spherically

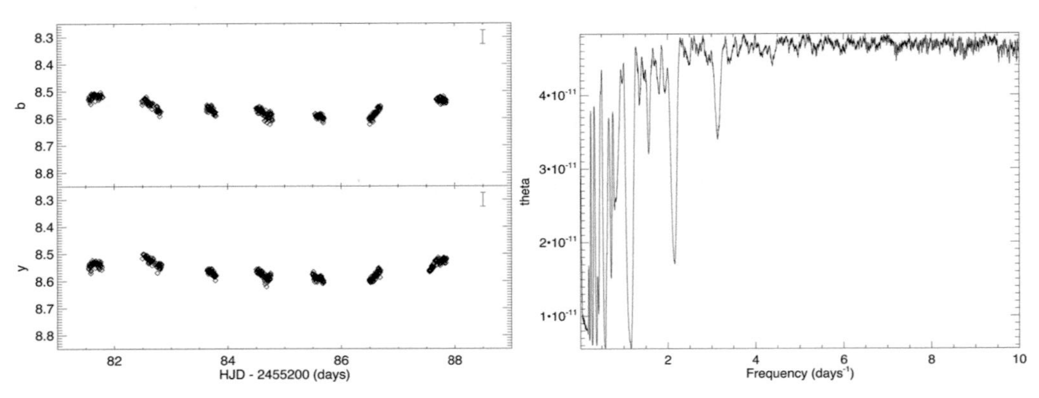

Figure 1. *Left:* Strömgren b- and y-band light curves for No. 1 from 2010 March. Representative error bars are in the upper right corner. *Right:* A sample plot of frequency versus θ, the PDM statistic for significance, for No. 1. The minima are possible frequencies of NRPs coupled with the rotation period of the star.

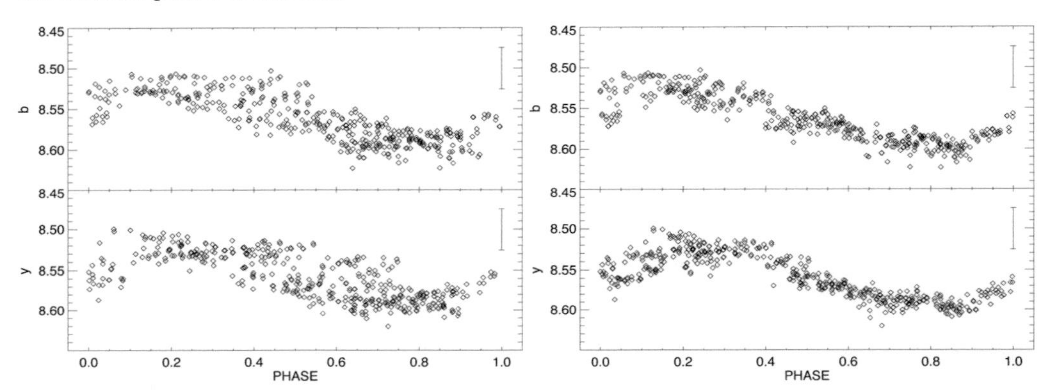

Figure 2. *Left:* Folded Strömgren b- and y-band light curves for No. 1 from 2010 March with a period of 11.16 hours. *Right:* The same light curves are folded with a period of 20.39 hours.

symmetric model of NRPs and model the observed light curves, following an element of the stellar surface as it pulsates and rotates.

Acknowledgements

We gratefully acknowledge travel support to CTIO from the Sigma Xi Grants-in-Aid of Research Program. We also thank Charles Bailyn and the SMARTS Consortium for their help in scheduling these observations. We are grateful for an institutional grant and travel support from Lehigh University.

References

Balona, L. A. & Engelbrecht, C. A. 1986, *MNRAS*, 219, 131
Buta, R. J. & Smith, M. A. 1979, *ApJ*, 232, 213
Gutiérrez-Soto, J., Fabregat, J., Suso, J., Lanzara, M. *et al.* 2007, *A&A*, 476, 927
McSwain, M. V., Huang, W., Gies, D. R., Grundstrom, E. D. *et al.* 2008, *ApJ*, 672, 590
Porter, J. M. & Rivinius, T. 2003, *PASP*, 115, 1153
Shobbrook, R. R. 1985, *MNRAS*, 212, 591
Shobbrook, R. R. 1987, *MNRAS*, 225, 999

Active OB stars: structure, evolution, mass loss, and critical limits
Proceedings IAU Symposium No. 272, 2010
C. Neiner, G. Wade, G. Meynet & G. Peters, eds.

doi:10.1017/S1743921311011409

Spectral and photometric study of Be Stars in the first exoplanet fields of CoRoT

Thierry Semaan[1], Christophe Martayan[2,1], Yves Frémat[3,1], Anne-Marie Hubert[1], Juan Gutiérrez Soto[4,1], Coralie Neiner[5] and Juan Zorec[6]

[1] GEPI, Observatoire de Paris, CNRS, Université Paris Diderot, France
[2] ESO, Alonso de Cordova, 3107, Vitacura, Santiago, Chile
[3] Royal observatory of Belgium, 3 avenue circulaire, 1180 Brussels, Belgium
[4] Instituto de Astrofsica de Andalucía , Apartado 3004, 18080 Granada, Spain
[5] LESIA, Observatoire de Paris, CNRS, Université Paris Diderot, UPMC, France
[6] Institut d'Astrophysique de Paris (IAP), CNRS, Université Pierre et Marie Curie, France

Abstract. First we investigate the spectral and photometric properties (colours, magnitudes) of a sample of faint Be stars observed in the first exoplanet fields of CoRoT (IR1, LRA1 and LRC1). We determine the fundamental parameters by fitting ESO-FLAMES/GIRAFFE spectra with synthetic models taking account for non-LTE effects. After that we correct these parameters from fast rotation effects. We also study the location of each star in the (logL vs logT) HR diagram. Second we start to analyse the CoRoT light curves to investigate further the possible correlation between the pulsating properties and the fundamental parameters of the stars.

Keywords. stars: emission-line, Be, stars: fundamental parameters

1. Introduction

A spectroscopic survey has been initiated to characterize the variable faint stars discovered in exoplanetary fields of CoRoT (PI: C. Neiner). Thanks to this programme we obtained spectra for a fraction (about 10%) of the stars observed by CoRoT in the first exoplanetary fields (IR1 and LRA1 toward the Galactic anticentre direction and LRC1 toward the Galactic centre direction). In these 3 fields CoRoT observed 31286 variable stars and we obtained a spectral coverage for 4005 ones. The spectra were recorded at medium resolution with the multi-object spectrograph FLAMES/GIRAFFE mounted at the ESO-VLT/UT2. In the majority of the cases we acquired one spectrum in the blue domain (R=6400; $\lambda\lambda$= 3964-4567 Å) and two spectra in the red one (R=8600; $\lambda\lambda$= 6438-7184 Å). Our goal is the detection of emission line stars (ELS) by observation of Balmer lines, the identification of Be stars and the determination of their fundamental parameters. We discovered 17 OBAe in the IR1 and LRA1 and only one in the LRC1. We have been able to determine the fundamental parameters for only 10 stars due to the lack of spectra in the blue domain.

2. Determination of the parameters

To determine the fundamental parameters of Be stars we use the procedure developed by Frémat *et al.* (2005), which takes account for veiling and fast rotation effects. The method is decomposed in 3 parts :

Apparent fundamental parameters determination: We use the GIRFIT programme to derive the effective temperature, surface gravity, projected rotational velocity and radial velocity. This programme adjusts the observations between 4000-4500 Å with theoretical

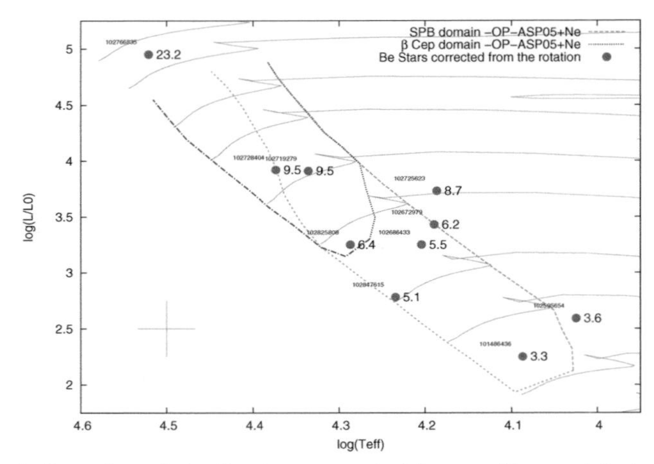

Figure 1. Location of the Be stars corrected from veiling and fast rotation.

spectra interpolated in a grid of stellar fluxes computed with the SYNSPEC programme and from model atmospheres calculated with TLUSTY Hubeny & Lanz (1995) or / and ATLAS9 Kurucz (1993).

Correction of veiling: The spectra of the Be stars are affected by emission lines. When the emission is strong in lines, the continuum is also affected. We use an empirical approach developed by Ballereau *et al.* (1995).

Correction of fast rotation effect: Be stars are very fast rotators. This rotation induces a modification of the shape of the star (flattening) and different physical processes (gravitational darkening). We correct parameters by adopting a rate of 90 % of the breakup velocity.

3. Results

The 18 ELS detected in the first exoplanetary fields of CoRoT are probably classical Be stars. We determine the apparent fundamental parameters and correct them from veiling and fast rotation ($\omega/\omega_c = 0.9$). The mass, luminosity, radius of each target have been derived from evolutionary tracks from Schaller *et al.* (1992). In the HR diagramme 6 Be stars are located in the first part in the main sequence and 4 Be stars are close to the end of the main sequence (TAMS) (Fig. 1). The instability strips for β Cephei and SPB stars taken from Miglio *et al.* (2007) are depicted in Fig. 1. We have begun a global analysis of the CoRoT light curve of the Be stars. Generally the frequency spectrum shows a forest of frequencies around one or two main frequencies as well as several isolated frequencies. However we note that two Be stars, which are at the frontier of the SPB domain, show a very poor frequency spectrum. In a next step, we will study the global characteristics of the frequency spectra derived from the CoRoT light curves for these objects in relation to their location in the HR diagramme.

References

Ballereau, D., Chauville, J., & Zorec, J. 1995, *A&AS* 111, 423
Frémat, Y., Zorec, J., Hubert, A.-M., & Floquet, M. 2005, *A&A* 440, 305
Hubeny, I. & Lanz, T. 1995, *ApJ* 439, 875
Kurucz, R. 1993, *Smithsonian Astrophys. Obs.*, CD-ROM No. 13
Miglio, A., Montalbán, J., & Dupret, M.-A. 2007, *Communications in Asteroseismology*, 151, 48
Schaller, G., Schaerer, D., Meynet, G., & Maeder, A. 1992, *A&AS*, 96, 269

Active OB stars: structure, evolution, mass loss, and critical limits
Proceedings IAU Symposium No. 272, 2010
C. Neiner, G. Wade, G. Meynet & G. Peters, eds.

doi:10.1017/S1743921311011410

Is macroturbulence in OB Sgs related to pulsations?

Sergio Simón-Díaz[1,2], Artemio Herrero[1,2], Katrien Uytterhoeven[3], Norberto Castro[1,2], Conny Aerts[4,5] and Joachim Puls[6]

[1]Instituto de Astrofísica de Canarias, E-38200 La Laguna, Tenerife, Spain; email: ssimon@iac.es; [2]Departamento de Astrofísica, Universidad de La Laguna, E-38205 La Laguna, Tenerife, Spain; [3]Laboratoire AIM, CEA/DSM-CNRS-Université Paris Diderot; CEA, IRFU, SAp, centre de Saclay, 91191, Gif-sur-Yvette, France; [4]Instituut voor Sterrenkunde, Katholieke Universiteit Leuven, Celestijnenlaan 200D, 3001 Leuven, Belgium; [5]IMAPP, Department of Astrophysics, Radboud University Nijmegen, PO Box 9010, 6500 GL Nijmegen, the Netherlands; [6]Universitätssternwarte München, Scheinerstr. 1, 81679 München, Germany

Abstract. As part of a long term observational project, we are investigating the macroturbulent broadening in O and B supergiants (Sgs) and its possible connection with spectroscopic variability phenomena and stellar oscillations. We present the first results of our project, namely firm observational evidence for a strong correlation between the extra broadening and photospheric line-profile variations in a sample of 13 Sgs with spectral types ranging from O9.5 to B8.

Keywords. stars: early-type, stars: atmospheres, stars: oscillations, stars: rotation, supergiants

1. Introduction

The presence of an important extra line-broadening mechanism (in addition to the rotational broadening and usually called macroturbulence) affecting the spectra of O and B Sgs is well established observationally (see Simón-Díaz *et al.* 2010, and references therein). Lucy (1976) postulated that this extra broadening may be identified with surface motions generated by the superposition of numerous non-radial oscillations. More recently, Aerts *et al.* (2009) computed time series of line profiles for evolved massive stars broadened by rotation and hundreds of low amplitude non-radial gravity mode oscillations and showed that the resulting profiles could mimic the observed ones. Stellar oscillations are a plausible explanation for the extra broadening in O and B Sgs, but this hyphotesis needs to be observationally confirmed.

2. The macroturbulence – LPV connection

As a first step, in Simón-Díaz *et al.* (2010), we investigated the possible connection between the macroturbulent broadening and the presence and temporal behaviour of line-profile variations (LPVs) in a sample of 11 late-O and early-B Sgs, 2 late B-Sgs, and 2 late-O, early-B dwarfs. To this aim, we obtained and analyzed time series of high resolution (R$\sim$46000), high S/N spectra obtained with FIES@NOT in two observing runs. We applied the Fourier transform (Gray 1976) and the goodness-of-fit techniques to disentangle and measure the contributions from rotational ($v\sin i$) and macroturbulent ($\Theta_{\rm RT}$) broadening to the Si III 4567 and/or the O III 5592 line profiles. We quantified the LPVs in these lines by means of the first, $\langle v \rangle$, and third, $\langle v^3 \rangle$, normalized velocity moments of the line. These moments are related to the centroid velocity and the skewness of the line profile, respectively, and are well suited to investigate whether an observed

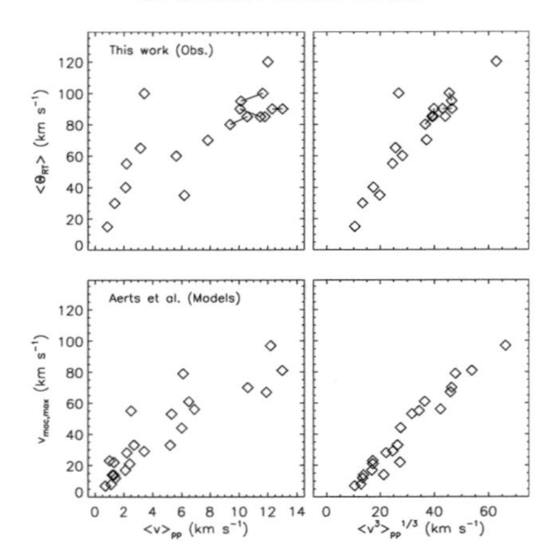

Figure 1. (Top) Empirical relations between the average size of the macroturbulent broadening ($\langle\Theta_{\rm RT}\rangle$) and the peak-to-peak amplitude of the first and third moments of the line profile. Solid lines connect results from four stars observed in both campaigns. (Bottom) similar plots with data from Table 1 in Aerts *et al.* (2009), based on simulations of line profiles broadened by rotation and by hundreds of low amplitude non-radial gravity mode pulsations. The simulations lead to clear trends which are compatible with spectroscopic observations.

line profile is subject to time-dependent line asymmetry, as expected in the case of a pulsating star.

We found a clear positive correlation between the average size of the macroturbulent broadening, $\langle\Theta_{\rm RT}\rangle$, and the peak-to-peak amplitude of $\langle v\rangle$ and $\langle v^3\rangle$ variations (see Fig. 1). To our knowledge, this is the *first clear observational evidence for a connection between extra broadening and LPVs in early B and late O Sgs.*

3. Is macroturbulent broadening in OB-Sgs caused by pulsations?

Non-radial oscillations have been often suggested as the origin of LPVs and photosperic lines in OB Sgs; however, a firm confirmation (by means of a rigorous seismic analysis) has not been achieved yet. From a theoretical point of view, Saio *et al.* (2006) showed that g-modes can be excited in massive post-main sequence stars, as the g-modes are reflected at the convective zone associated with the H-burning shell. Lefever *et al.* (2007) presented observational evidence of g-mode instabilities in a sample of photometrically variable B Sgs from the location of the stars in the ($\log T_{\rm eff}$, $\log g$)-diagram.

These results, along with our observational confirmation of a tight connection between macroturbulent broadening and parameters describing observed LPVs render stellar oscillations the most probable physical origin of macroturbulent broadening in B Sgs; however, it is too premature to consider them as the only physical phenomenon to explain the unknown broadening.

References

Aerts, C., Puls, J., Godart, M., & Dupret, M.-A. 2009, *A&A*, 508, 409

Gray, D. F. 1976, *The observation and analysis of stellar photospheres* (New York: Wiley-Interscience), 1st Ed.

Lefever, K., Puls, J., & Aerts, C. 2007, *A&A*, 463, 1093

Lucy, L. B. 1976, *ApJ*, 206, 499

Saio, H., Kuschnig, R., Gautschy, A., Cameron, C. *et al.* 2006, *ApJ*, 650, 1111

Simón-Díaz, S., Herrero, A., Uytterhoeven, K., Castro, N. *et al.* 2010, *ApJ* (Letters), 720, L174

Active OB stars: structure, evolution, mass loss, and critical limits
Proceedings IAU Symposium No. 272, 2010
C. Neiner, G. Wade, G. Meynet & G. Peters, eds.

doi:10.1017/S1743921311011422

SC3-63371 and SC4-67145: two wind-interacting A+B binaries

Myron A. Smith[1] and Ronald E. Mennickent[2]

[1]Dept. of Physics & Astronomy, Catholic University of America, Washington, D.C. 20064, USA; email: msmith@stsci.edu

[2]Dept. de Astronomia, Universidad de Concepción, Casilla 160-C, Concepcíon, Chile

Abstract. We report on two SMC objects with remarkable light curves and spectra. Both are in fact A I + B III binaries that exhibit interacting winds. We show a sketch of the geometry.

Keywords. stars: emission-line, Be, stars: winds, outflows, stars: variables: other

---

1. Introduction

SC3-63371 ("SC3") and SC4-67145 ("SC4") are 2 of 8 Small Magellanic Cloud "Type 3" variables (Mennickent *et al.* 2002: Be candidates defined by nearly periodic I-band light curves) brighter than $M_v = -4.2$. Previous low-dispersion spectra exhibited strong H/K and D lines and Hα emissions. These features indicated conflicting A and Be type classifications. These unusual properties motivated us to obtain 3 high dispersion spectra of both objects in 2004, 2007, and 2009 and a far-UV (*FUSE* satellite) observation. These results are published fully by Mennickent & Smith (2010a). We also undertook a several year photometric campaign to characterize flux variability. The photometry has been published by Mennickent *et al.* (2010b). Light curves of both objectss show short periods, possibly due to nonradial pulsations. SC3's light curve also exhibits a regular 238.1 day period, complicated by an ellipsoidal variation and a reflection effect that suggests the star is in a highly eccentric binary. SC4 exhibits a periodic 184.26 day eclipse, the duration of which is much too long to arise from the transit of a stellar companion, suggesting the presence of a long column density of circumstellar matter. Our radial velocity measures (Fig. 1, inset) are consistent with the binary interpretation for SC4.

2. Spectroscopic Analysis

We have used SYNSPEC code to model the Ca II K line and derived spectral types of A6 I and A5 II with the optical spectrum. These types and classes agree well with matchings of comparison stars in the UVES Paranal Spectral Atlas as well as a comparison of optical magnitudes and the SMC's distance modulus of 19.0. Modeling strengths of several Fe III lines and the C III λ1176 complex likewise suggests spectral types of B3 III for both UV spectra. These comparisons indicate that both objects are A + B binaries. The $v\sin i$ values of all components are moderate (20-75 km s^{-1}). Remarkably, the 2 objects out of the 8 brightest Mennickent Type 3 stars are evolutionarily very similar.

The far blue spectra of SC3 and SC4 disclose sharp Balmer cores up to ~H30. In the optical metallic lines of both SC3 and SC4 exhibit components of the photosphere and a circumbinary (CB) disk. However, red He I lines present in spectra of (only) SC3 suggest a hot medium with a relative blueshift of some 74 km s^{-1}. This indicates that the hot region responsible for them is not cospatial with the metallic line forming regions.

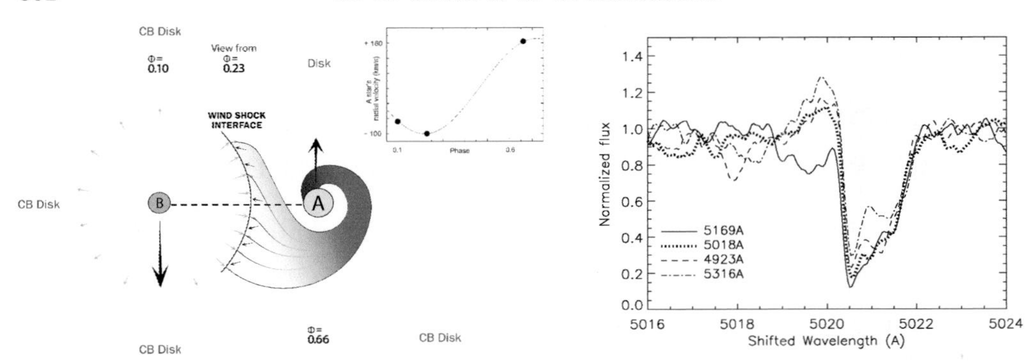

Figure 1. Left: Cartoon of wind-wind interaction and gas streaming into the ≈A5 II star of SMC4. Right: Difference in the blue and red wings of common-multiplet Fe II lines in a single spectrum of SC4 in 2009 (photometric phase 0.66). The weakest lines (e.g. λ5316) exhibit blue wing emissions and the strongest line of the series, λ5169, shows absorption.

The optical line profiles of SC4 at all times show stationary "Blue Absorption Components, a phenomenon reported to date apparently for only one other star, N8 (putatively sgB[e], Heydari-Malayeri 1990). In addition, weak emissions are present in medium-excitation that can mask the visibility of the stellar/CB components. In some phase, secondary blue absorptions or emissions can be observed, even for example in optically thin lines of a common Fe II multiplet. Fig. 2 shows overplotting of such Fe II lines from a single spectrum, showing changes from absorption (weak multiplet member) to emission (strong member). Such variations among multiplet members speak to formation in a inhomogeneous medium. We suggest that these optically thin components form across a narrow wind-wind shocking zone, where physical conditions can change over relatively small scale lengths. We believe also that in the SC3 system the He I line emissions arise in a i more energetic wind-wind shock, distinct from other line formation regions.

The strong Hα emission in the spectra of both objects suggests an extended equatorial disk. However, the small velocity separations of the V and R emission peaks, when assessed against the long periods of these systems, strongly suggests the disks are circumbinary and not circumstellar. The radial velocity variations of SC4 suggest the characteristic disk size is three times the binary separation. However, the He II emissions and the qualitatively different blue components - absorption to emission such as in Fig. 2 - suggest that the formations of these features in shocks. Fig. 1 is a sketch of the geometry we envision, as seen from the pole of the binary. Here observers see the effects of wind-wind shocks between the star in which wind detritus spirals into a column toward the A supergiant. At phase $\phi \sim 0.0$ the observer looks through this column, which is thick enough to exhibit the 184 day light eclipses in the photometry.

Post-ZAMS Ae stars have been known to exist mainly among supergiants, and their spectra exhibit weak ephemeral Hα emission components. But we now see that "Be-like" Balmer emissions also occur in spectra of a new class of wind-interacting A I + B binaries.

References

Heydari-Malayeri, M. 1990, *A&A*, 234, 233

Mennickent, R. E., Pietrzyński, G., Gieren, W., & Szewczyk, O. 2002, *A&A*, 393, 887

Mennickent, R. E. & Smith, M. A., *et al.* 2010a, *MNRAS*, 407, 734

Mennickent, R. E., & Smith, M. A., Kołaczkowski, Z., Pietrzyński, G., Soszyński, I. 2010b, *PASP*, 122, 662

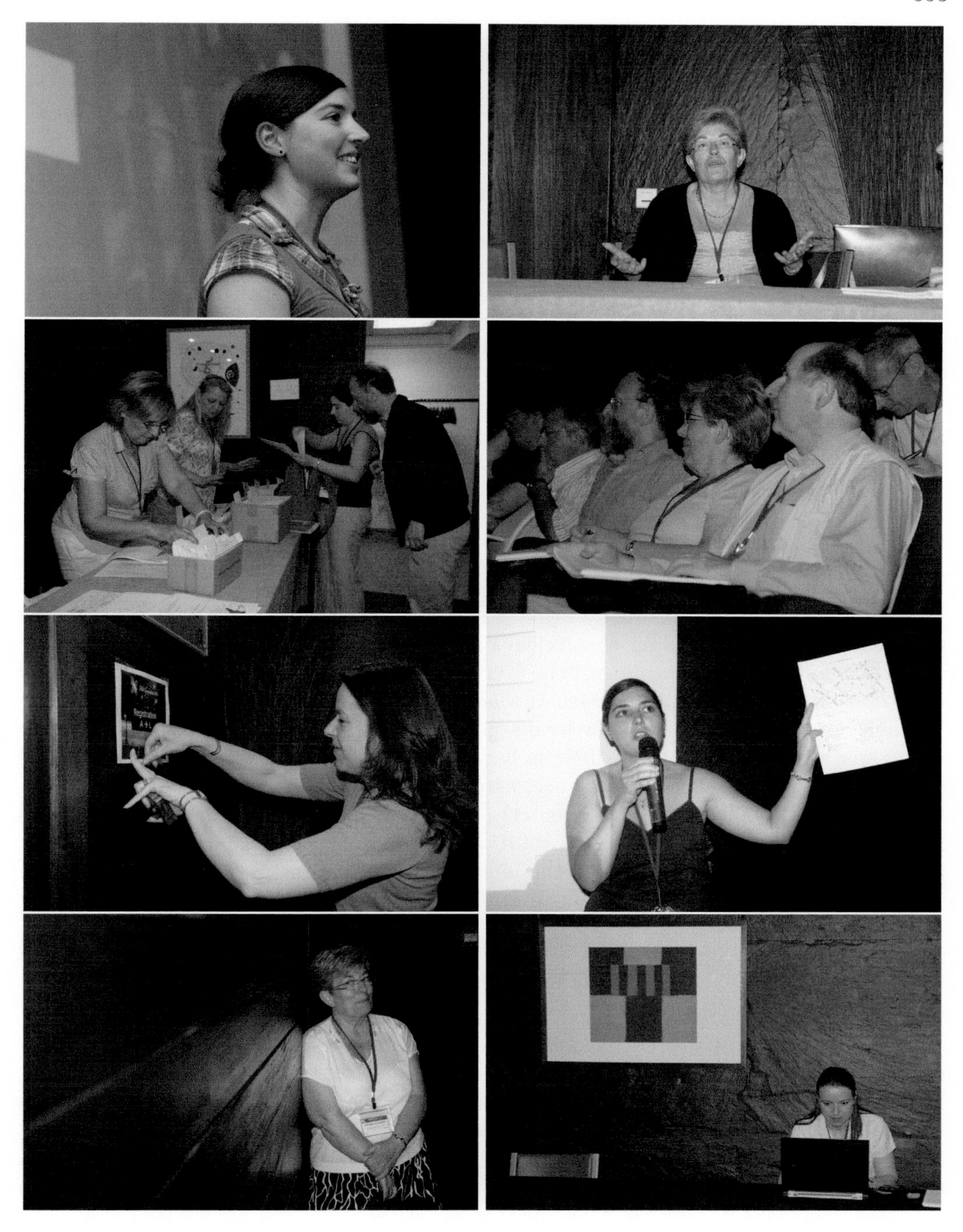

The core of the LOC. From left to right and top to bottom: Evelyne Alecian; Michèle Floquet; Olga Martions, Annick Oger, Evelyne Alecian, and Bertrand de Batz; Bertrand de Batz, Michèle Floquet and Bernard Leroy; Coralie Neiner; Evelyne Alecian; Michèle Floquet; Coralie Neiner.

Active OB stars: structure, evolution, mass loss, and critical limits
Proceedings IAU Symposium No. 272, 2010
C. Neiner, G. Wade, G. Meynet & G. Peters, eds.

doi:10.1017/S1743921311011434

OB-stars as extreme condition test beds

Joachim Puls[1], **Jon O. Sundqvist**[1] **and Jorge G. Rivero González**[1]

[1] Universitätssternwarte der Ludwig-Maximilians-Universität München
Scheinerstr. 1, D-81679 München, Germany
email: uh101aw@usm.uni-muenchen.de (J.P.)

Abstract. Massive stars are inherently extreme objects, in terms of radiation, mass loss, rotation, and sometimes also magnetic fields. Concentrating on a (personally biased) subset of processes related to pulsations, rapid rotation and its interplay with mass-loss, and the bi-stability mechanism, we will discuss how active (and normal) OB stars can serve as appropriate laboratories to provide further clues.

Keywords. hydrodynamics, instabilities, line: formation, stars: abundances, stars: early-type, stars: evolution, stars: mass loss, stars: oscillations, stars: rotation, stars: winds, outflows

---

1. Introduction

Massive stars are inherently extreme objects, in terms of radiation, mass loss, rotation and sometimes also magnetic fields. Thus, they can serve as test beds for extreme conditions and corresponding theoretical predictions. Such tests are, e.g., particularly important for our understanding of the (very massive) First Stars and for the physics of fast rotation in massive stars, which is a key ingredient in the collapsar model of long Gamma Ray Bursts. In this review we discuss how a variety of physical processes present in massive stars can affect both their stellar photospheres and/or winds, and how active (and normal) OB stars can be, and are, used as appropriate laboratories to provide further clues. In the following, we concentrate on a (personally biased) subset of processes related to pulsations (Sect. 2), rapid rotation and its interplay with mass-loss (Sect. 3), and mass-loss itself, particularly on the bi-stability mechanism (Sect. 4).

2. Pulsations

2.1. *Pulsating B-supergiants*

Well outside the instability strips of β Cep and slowly pulsating B-stars (SPB), Waelkens *et al.* (1998) via HIPPARCOS detected 29 periodically variable B-*supergiants*. A corresponding instability region had not been predicted at that time. Meanwhile, however, Pamyatnykh (1999) and Saio *et al.* (2006, see also this volume) identified such regions for pre-TAMS and post-TAMs objects, respectively, with SPB-type of oscillations (high order g-modes). These regions are indicated in Fig. 1, together with results from quantitative spectroscopy by Lefever *et al.* (2007), for those of the above 29 supergiants with sufficient spectral information. Obviously, most of these objects are located very close to the high gravity limit of the predicted pre-TAMS or within the predicted post-TAMS instability strips for evolved stars. Together with their multi-periodic behaviour, this strongly suggests that these objects are opacity-driven non-radial pulsators (NRPs), and thus are ideal **test beds** for asteroseismologic studies of evolved massive stars. Note that Lefever *et al.* (2007) found additional periodically variable objects not known to be pulsators so far, and suggested, from their pulsational behaviour and their positions, that these objects are g-mode pulsators as well. Two of them, HD 64760 (B0.5 Ib) and

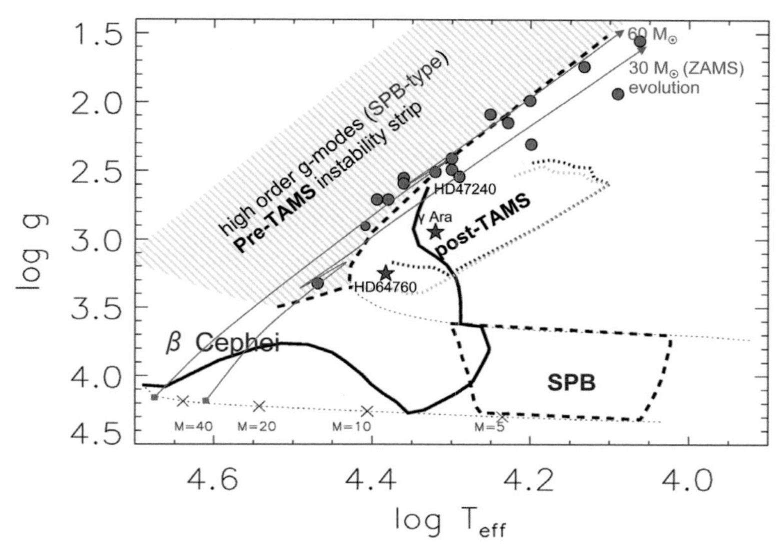

Figure 1. log $T_{\rm eff}$-log g diagram for hot massive stars. Indicated are the instability regions for β Cep stars (solid bold), SPBs close to the ZAMS (dashed bold) and SPBs of evolved type as predicted by Pamyatnykh (1999) (pre-TAMS, blue hatched) and Saio *et al.* (2006) (post-TAMS, dotted). Red dots correspond to the positions of slowly pulsating B-supergiants from the sample by Waelkens *et al.* (1998) as derived by Lefever *et al.* (2007), and blue asterisks are two newly suggested g-mode pulsators from the same work. See text for further details. Adapted from Lefever *et al.* (2007).

γ Ara (= HD 157246, B1 Ib), are explicitly indicated in Fig. 1, together with HD 47240 (B1 Ib) from the original sample by Waelkens *et al.*, and will be referred to later on.

2.2. *Macroturbulence*

A present key problem in atmospheric diagnostics by high resolution spectroscopy is the finding that the line-profiles from (at least) late O- and B-supergiants display substantial extra-broadening (in addition to the well-known effects from rotation etc.), which has been termed 'macro-turbulence' (for details and references, see Simón-Díaz *et al.* 2010). This extra-broadening can be simulated by allowing for a *supersonic* Gaussian or quasi-Gaussian velocity distribution in photospheric regions, which is difficult to justify physically. Recently, however, Aerts *et al.* (2009) showed that such extra-broadening can be reproduced from the *collective* effect of low-amplitude g-mode oscillations. First hints that this scenario might be realistic have been found by Simón-Díaz *et al.* (2010), from a tight observed correlation between the peak-to-peak amplitudes of velocity moments measured from (variable) photospheric profiles of B-supergiants and the derived macro-turbulent broadening. Given the ubiquity of macro-turbulence in hot massive stars, this tentatively suggests that a large fraction of OB-stars are non-radial pulsators (see also Fullerton, this volume).

2.3. *Triggering of structure/clump formation*

With respect to their stellar winds, pulsations in massive stars might be responsible for inducing large-scale structures, such as co-rotating interaction regions (CIRs, see Fullerton, this volume), and, particularly, for triggering the formation of clumps: To reproduce the observed X-ray emission from hot stellar winds, $L_{\rm x} \approx 10^{-7} L_{\rm bol}$, the line-driven (or deshadowing) instability related to radiative line driving needs to be excited by deep-seated photospheric disturbances of a multitude of frequencies (NRPs?), giving rise to strong clump-clump collisions and consequently strong shocks (Feldmeier *et al.* 1997a,b). For models with self-excited instability alone, the predicted X-ray emission is much too weak. Moreover, such perturbations might be responsible for triggering the on-set of *deep-seated* wind-clumping (Sect. 4), as implied from various diagnostics (e.g., Bouret *et al.* 2005; Puls *et al.* 2006; Sundqvist *et al.* 2011).

2.4. *Strange mode oscillations*

In addition to 'conventional' pulsations, another class of quasi-periodic, *dynamical* instabilities are predicted to occur in the envelopes of luminous stars with large $L/M > 10^3$. These are the so-called strange-mode oscillations (for details and references, see Saio and Chené, this volume), which should be particularly strong in WR-stars and might even help to initiate their winds (e.g., Wende *et al.* 2008). So far, there is no direct evidence of these predictions, though the strongest amplitudes of optical lpv in O-stars are located within the region of predicted strange mode oscillations (Fullerton *et al.* 1996), and at least for one WR star such oscillations might actually have been observed (see Chené, this volume). Alternative **test beds** to check the reality of strange mode oscillations might be late B-/early A-supergiants (as suggested by Puls, Glatzel, & Aerts as targets for the micro-satellite BRITE), since, in comparison to WRs, these objects have less dense winds and 'convenient' frequencies (on the order of a few to tens of days), with predicted amplitudes of 0.1 mag (W. Glatzel, priv. comm.). Indeed, the COROT observations of the late B-supergiant HD 50064 (Aerts *et al.* 2010) showed a period of 37 days, with a sudden amplitude change by a factor of 1.6. Together with other evidence (variable $\dot{M}$ etc.), Aerts *et al.* tentatively interpreted this finding as the result of a strange mode oscillation.

3. Rapid rotation

3.1. *Photospheric deformation and gravity darkening*

Rapid rotation affects the stellar photosphere in (at least) two ways. First, it becomes deformed, with $R_{\rm eq}/R_{\rm pole} = 1.5$ at critical rotation (using a Roche model with point mass distribution, see Zhao, this volume, and Cranmer & Owocki 1995 for details and references). The first observational **test bed** which confirmed the basic effect was the brightest Be star known, Achernar = α Eri (VLTI observations by Domiciano de Souza *et al.* 2003).

The second effect is gravity darkening, first suggested by von Zeipel (1924), who assumed rotational laws that can be derived from a potential, e.g., uniform or cylindrical. An important extension was provided by Maeder (1999), who considered the more realistic case of shellular rotation in radiative envelopes, where the angular velocity is assumed to be constant on horizontal surfaces (Zahn 1992). In result, the photospheric flux is proportional to the *effective* gravity, $\vec{F} \propto \vec{g}_{\rm eff}(1+\zeta(\vartheta))$, with $|\zeta(\vartheta)| < 0.1$ in most cases and $\zeta = 0$ in the original von Zeipel case. The effective gravity is the vector sum of gravitational and centrifugal acceleration, $\vec{g}_{\rm eff} = \vec{g}_{\rm grav} + \vec{g}_{\rm cent}$, i.e., lower at the equator than at the pole, with $\vec{g}_{\rm eff}(\text{pole}) = \vec{g}_{\rm grav}$. Note that here $\vec{g}_{\rm eff}$ is *independent of the radiative acceleration*! Neglecting $\zeta(\vartheta)$, for radiative envelopes we obtain $T_{\rm eff}(\vartheta) \propto g_{{\rm eff},\perp}^{1/4}$, i.e., $T_{\rm eff}$ decreases towards the equator, in dependence of the normal component of $\vec{g}_{\rm eff}$.

Both effects are demonstrated in Fig. 2, for a typical O-supergiant rotating close and very close to critical rotation. Deformation and gravity darkening become significant only for rotational speeds higher than roughly 70% of the critical one! **Test beds** to check

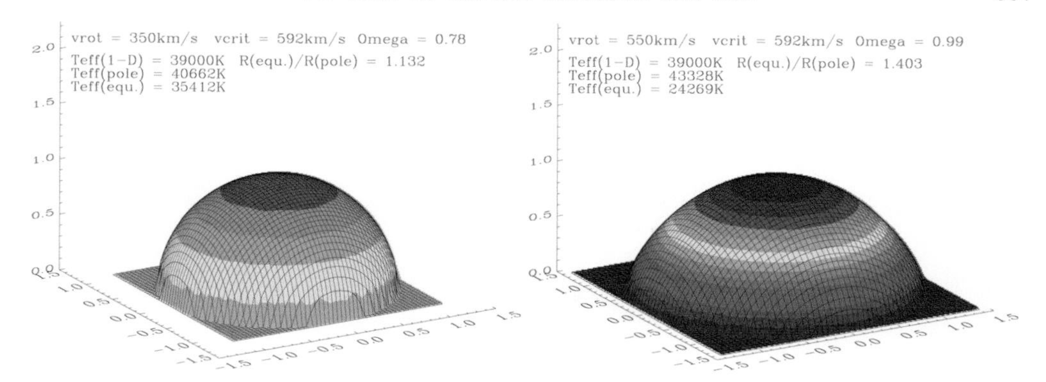

Figure 2. Predicted photospheric deformation and gravity darkening for a star similar to ζ Pup (1-D values: $T_{\rm eff}$=39 kK, R_*=19$R_\odot$, $\log g$=3.6) but rotating with 78% (left) and 99% (right) of its critical angular velocity. Both figures are on the same scale, with identical color coding for $T_{\rm eff}(\vartheta)$. The ratios $R_{\rm eq}/R_{\rm pole}$ and the effective temperatures at the hotter pole and cooler equator are indicated within the figures.

both effects are discussed by Zhao (this volume; see also the summary provided by van Belle 2010).

3.2. *Rapid rotation and winds*

Standard line-driven wind theory (for a recent review and references, see Puls *et al.* 2008) predicts that the mass-loss rate of a non (slowly) rotating star scales as

$$\dot{M} \propto (N_{\rm eff} L)^{1/\alpha} \left(g_{\rm grav} R_\star^2 (1-\Gamma)\right)^{1-1/\alpha},$$

with $N_{\rm eff}$ the effective number of driving lines (proportional to the force-multiplier parameter k, Castor *et al.* 1975, CAK), CAK parameter α (corresponding to the steepness of the line-strength distribution function), and Eddington factor Γ. Accounting for rotation (and Γ not too large), we find that the mass-loss rate depends on co-latitude θ,

$$\begin{aligned} \dot{M}(\theta) \quad &\propto \quad \left(N_{\rm eff}(\theta) F(\theta) R_\star^2(\theta)\right)^{1/\alpha(\theta)} \left(g_{\rm eff}(\theta) R_\star^2(\theta)(1-\Gamma)\right)^{1-1/\alpha(\theta)} \\ &\overset{\rm von\ Zeipel}{\propto} \quad \left(N_{\rm eff}(\theta)\right)^{1/\alpha(\theta)} g_{\rm eff}(\theta) R_\star^2(\theta) \end{aligned} \tag{3.1}$$

(cf. Owocki *et al.* 1998). This expression renders two possibilities. i) If the ionization equilibrium is rather constant w.r.t. θ (as it is the case for O-stars), we obtain a *prolate* wind structure, since $g_{\rm eff}(\theta)$ is largest at the pole. This is the $g_{\rm eff}$-effect, see Owocki *et al.* (1998); Maeder (1999); Maeder & Meynet (2000). ii) *If*, on the other hand, the ionization equilibrium were strongly dependent on θ, this would imply an *oblate* wind structure if the increase of $N_{\rm eff}$ and the decrease of α towards the equator (as a consequence of decreasing ionization) could overcompensate the decrease of $g_{\rm eff}$. Such a situation (the κ-effect, see Maeder 1999; Maeder & Meynet 2000) *might* occur in B-supergiants (but see Sect. 4). Note, however, that *no* thin disk can be formed by this process alone. Note also that self-consistent 2-D hydro/NLTE calculations (though somewhat simplified) for rapidly rotating B-stars around $T_{\rm eff}$= 20 kK (i.e, just in the region where the κ effect might be expected) by Petrenz & Puls (2000) still resulted in a prolate wind structure, since the ionization effects turned out to be only moderate.

Of course, these predictions need to be checked observationally, particularly when considering their importance regarding stellar evolution (e.g., a pronounced polar mass loss would lead to less loss of angular momentum), and with respect to mass-loss

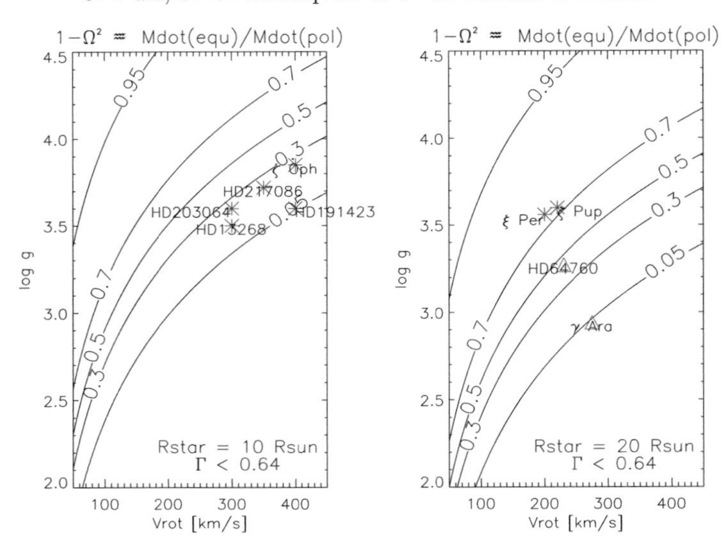

Figure 3. Iso-contours of $1 - \Omega^2 \approx \dot{M}(\mathrm{eq})/\dot{M}(\mathrm{pole})$ as a function of v_{rot} and $\log g$, for typical dwarfs (left) and supergiants (right). Overplotted are the positions of some rapidly rotating Galactic O-stars (asterisks) and B-supergiants (triangles), assuming a minimum $v_{\mathrm{rot}} = v \sin i$.

diagnostics (when do we need 2-D models?). To our knowledge, clear observational evidence for aspherical winds is still missing.† So, what are potential **test beds**?

In Fig. 3 we have plotted theoretical iso-contours of $1 - \Omega^2 \approx \dot{M}(\mathrm{eq})/\dot{M}(\mathrm{pole})$ (from Eq. 3.1 and assuming N_{eff} and α to be constant; $g_{\mathrm{eff}} \approx g_{\mathrm{grav}}(1 - \Omega^2)$ and $\Omega = \omega/\omega_{\mathrm{crit}}$) as a function of v_{rot} and $\log g$, for typical dwarfs (left) and supergiants (right). Overplotted are the locations of well-known Galactic rapid rotators, for a minimum value of $v_{\mathrm{rot}} = v \sin i$ (data from Repolust *et al.* 2004 and Lefever *et al.* 2007). Asterisks denote O-stars, triangles B-supergiants. All O-dwarfs/giants (left) are predicted to have a significant mass-loss contrast, below 0.3. Unfortunately, their (average) mass-loss rate is too low to lead to substantial effects in the optical wind-lines and the IR continuum, though UV-spectra should be affected by deviations from spherical symmetry. For the fast rotating O-supergiants, on the other hand, the predicted effect is rather small, so nothing might be visible. Vink *et al.* (2009), using linear H_α spectro-polarimetry, conclude that most winds from rapidly rotating O-stars are spherically symmetric. For the two rapidly rotating B-supergiants, HD 64760 and particularly γ Ara, the situation is more promising, and they might be used as **test beds** to check the impact of rotation on the global wind topology. Remember that HD 64760 (see also Sect. 2) is one of the best studied objects in the UV (thanks to the IUE mega-campaign, Massa *et al.* 1995) - with the detection of CIRs and 'PAMS' (see Fullerton, this volume), both presumably related to its non-radial pulsations –, and has also been studied in the optical to clarify the interaction between NRPs and CIRs (Kaufer *et al.* 2006).

Fig. 4 (left) displays the corresponding H_α line profiles, for the above two rapidly rotating B-supergiants and for HD 47240 (see also Sect. 2), with a somewhat lower $v \sin i$. At first glance, these profiles might indicate the presence of a disk or an oblate wind (e.g.,

† The polar wind structures claimed for the Be-stars Achernar (Kervella & Domiciano de Souza 2006) and α Ara (Meilland *et al.* 2007) from NIR interferometry still need to be confirmed, given that - as discussed during this conference - for such low mass-loss rates the IR-photosphere is very close to the optical one (in other words, the IR-excess from the wind is very low).

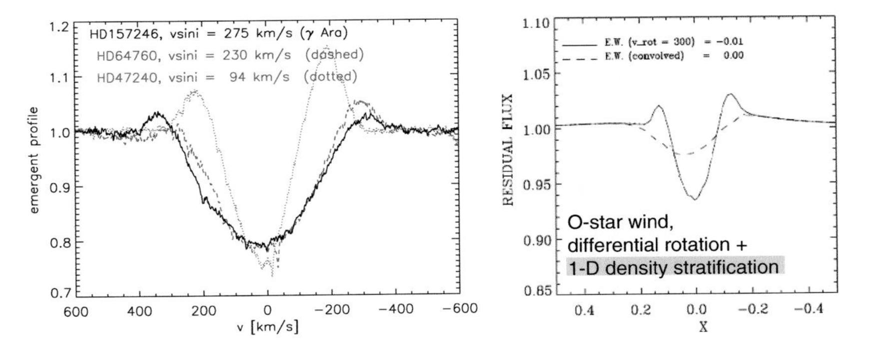

Figure 4. Left: H_α line profiles for three rapidly rotating B-supergiants (spectra and data from Lefever *et al.* 2007). Right: Theoretical H_α line profiles, for an O-star wind with *spherical* density stratification. Dashed: Profile convolved with a rotation profile of width 300 km s^{-1}. Solid: 2-D line transfer allowing for differential rotation, $v_{\rm rot} \propto 1/r$. Adapted from Petrenz & Puls (1996).

the κ-effect from above), but this is not necessarily the case. As shown in Fig. 4 (right), even a *spherical* wind can give rise to double-peaked profiles, when accounting for the wind's differential rotation (due to the so-called resonance-zone effect, Petrenz & Puls 1996). Note that in this case the profile only depends on the product $v \sin i$ and not on the individual factors. For a 2-D density stratification, however, the profiles will look different for a prolate or oblate topology, and will depend on the individual values of $v_{\rm rot}$ and $\sin i$ as well, which might induce a certain dichotomy. Interestingly, UV spectroscopy (via IUE) of γ Ara by Prinja *et al.* (1997) gave indications for a *prolate* geometry, mainly because of missing or weak emission peaks in the P Cygni profiles.

3.3. *The* $\Omega\Gamma$ *limit*

An interesting question is what happens if a star is rapidly rotating *and* close to the Eddington limit. After a controversial discussion (Langer *et al.* 1997; Glatzel 1998), Maeder & Meynet (2000) were able to solve this problem in an elegant way. For the following discussion, it is only important to note that the *total* acceleration due to gravity, centrifugal forces, and radiation pressure gradients can be expressed as $\vec{g_{\rm tot}} = \vec{g_{\rm eff}}(1-\Gamma_\Omega)$, where the effective gravity remains defined as previously, and $\Gamma_\Omega/\Gamma > 1$ is a function of $v_{\rm rot}/v_{\rm crit}$. Consequently, the total acceleration can become zero before the nominal Eddington limit is reached, and this new limit is called the $\Omega\Gamma$ limit. As shown by Maeder & Meynet (2000), the combination of rapid rotation and large Γ can affect the total (polar-angle integrated) mass-loss rate from a radiation driven wind considerably,

$$\frac{\dot{M}(\text{rotating})}{\dot{M}(\text{non-rotating})} \approx \left(\frac{1-\Gamma}{\Gamma/\Gamma_\Omega - \Gamma}\right)^{\frac{1}{\alpha}-1} \begin{cases} = O(1) & \text{for not too fast rotation and low } \Gamma \\ \gg 1 & \text{for fast rotation and considerable } \Gamma \\ \text{but:} & \text{max. } \dot{M} \text{ limited because } L \text{ limited} \end{cases}$$

since α is on the order of 0.4 ... 0.6. To identify potential **test beds** to check this important prediction, in Fig. 5 we have plotted the iso-contours of the $v_{\rm rot}$ required for a significantly increased mass-loss rate, as a function of $T_{\rm eff}$ and $\log g$. (A factor of four compared to the non-rotating case was chosen to allow for an easy observational check.) The red shaded region comprises the approximate locations of Galactic OB-supergiants. Overplotted are the positions of some rapidly rotating supergiants, O-types (asterisks) and B-types (triangles). The numbers in brackets are the observed $v \sin i$. Again, O-supergiants are not suited as test beds, since they would need to rotate much faster than

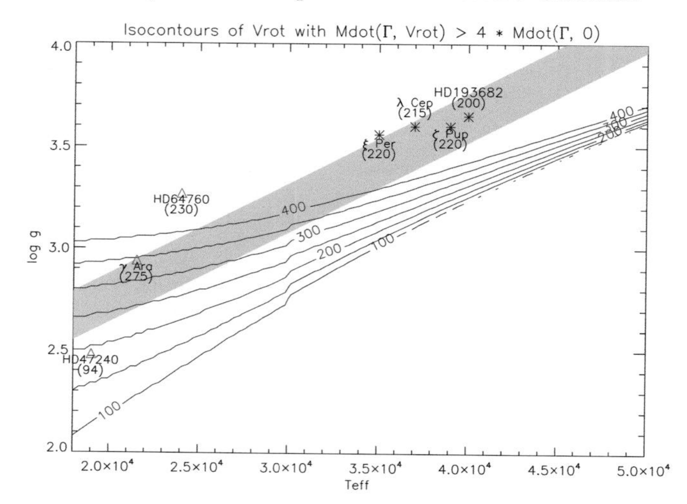

Figure 5. Iso-contours of $v_{\rm rot}$ for which the total mass-loss rate is predicted to become increased by a factor of four, compared to the non-rotating case, as a function of $T_{\rm eff}$ and $\log g$, together with the positions and $v \sin i$-values of some rapidly rotating O- and B-supergiants (asterisks and triangles, respectively). See text.

400 km s^{-1} to show the required increase in $\dot{M}$. Interestingly, however, at least γ Ara (and maybe HD 47240 - if its $\sin i$ were 0.5) are located at the 'right' position and might be worth being investigated in detail, e.g., by means of interferometry (see Chesneau, this volume) in combination with 2-D NLTE modeling (Georgiev *et al.* 2006). The outcome of such investigations will be of particular relevance for stellar evolution with rapid rotation, especially in the early Universe (see Ekström, this volume).

4. Mass loss

As we have seen, rapidly rotating B-supergiants are ideal **test beds** to check a number of theoretical predictions. Unfortunately, however, there exists only few such objects, since there is a rapid drop of rotation below $T_{\rm eff} \approx 20$ kK, as is obvious from the distribution of $v \sin i$ (e.g., Howarth *et al.* 1997). In a recent letter, Vink *et al.* (2010) tried to explain this finding based on two alternative scenarios (see also Langer, this volume). In *scenario I*, the low rotation rates of B-supergiants are suggested to be caused by braking due to an increased mass loss for $T_{\rm eff} < 25$ kK, where this increased mass loss should be due to the so-called bi-stability jump. Vink *et al.* termed this process 'bi-stability braking'.

4.1. *The bi-stability jump*

The bi-stability jump itself has often been discussed and referred to in the literature, and goes back to findings by Pauldrach & Puls (1990) when modeling the wind of P Cygni. These findings were generalized by Vink *et al.* (2000, 2001) in their work on stellar wind models for OB-stars: In the intermediate/late O-star and early B-star regime, the major contribution of driving lines in the lower wind (which are responsible for initiating the mass-loss rate) is from Fe IV. Below $T_{\rm eff} = 23$ kK, however, Fe IV recombines more or less abruptly to Fe III. Since Fe IIIhas more effective lines than Fe IV, $N_{\rm eff}$ increases (in parallel with a decrease of α, see also Puls *et al.* 2000), which leads to an increase in $\dot{M}$ and a decrease in the terminal velocity, v_∞. This is the theoretical basis for the κ-effect discussed in Sect. 2. Vink *et al.* (2000) predict a typical increase in $\dot{M}$ by a factor of

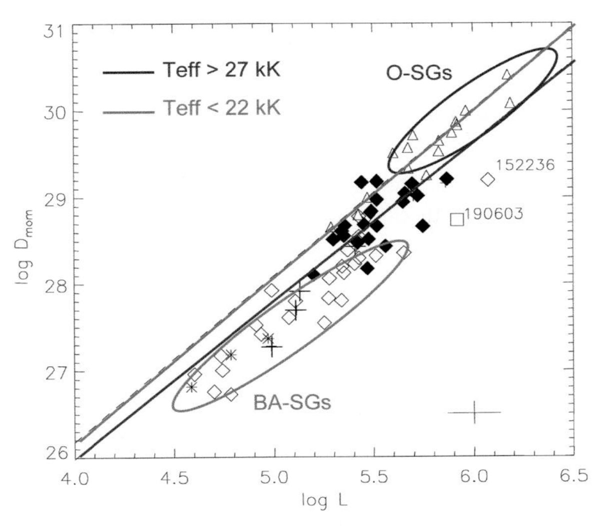

Figure 6. Modified wind-momenta for Galactic O- and B-supergiants. Dashed (red) and solid (blue) lines represent the predictions by Vink *et al.* (2000) for objects below and above 23 kK, respectively. Triangles are O-supergiants ($T_{\rm eff}$ > 27 kK), filled diamonds early B-sgs with $T_{\rm eff}$ > 22 kK and open triangles B-supergiants below this value. The cross in the lower right displays the typical error bars. Adapted from Markova & Puls (2008).

five, and a decrease in v_∞ by a factor of two. Consequently, the wind-momenta from massive stars with $T_{\rm eff}$< 23 kK should be *higher* than those from stars of earlier spectral types (see dashed/solid lines in Fig. 6.) This important prediction needs to be checked observationally, not least because present day evolutionary codes often incorporate the corresponding 'mass-loss recipe'.

4.2. *Test beds for the bi-stability jump (I): B[e]-supergiants?*

The hybrid spectrum of B[e] supergiants can be explained by a two-component wind, with an outflowing 'disk' (equatorial wind) of low velocity, high density and low ionization, and a high velocity, low density and highly ionized polar wind (Zickgraf *et al.* 1986, 1989). A first explanation of this wind structure was given by Lamers & Pauldrach (1991), who combined the effects of fast rotation and bi-stability jump, following the calculations from Pauldrach & Puls (1990) for the latter: The high values of $g_{\rm eff}$ together with the high ionization at the pole ($T_{\rm eff}$ calculated via von Zeipel) then give rise to a fast and thin polar wind, whereas the low $g_{\rm eff}/T_{\rm eff}$ values at the equator induce a slow and dense wind. (Note that v_∞ scales with the photospheric escape velocity and thus with $g_{\rm eff}$, see below).

Until now, the B[e]-sg mechanism is heavily debated. Owocki *et al.* (1998) pointed out that Lamers & Pauldrach (1991), though accounting for gravity darkening when calculating the ionization, did not include its impact on $\dot{M}$ (Eq. 3.1). When gravity darkening is included into the accelerating flux, $F(\theta) \propto g_{\rm eff}(\theta)$, a 'disk' formation becomes almost impossible, due to the counteracting effects of bi-stability and increased polar flux (see also Puls *et al.* 2008 and references therein). Simulations by Pelupessy *et al.* (2000), on the other hand, indicated that the bi-stability mechanism can work even when consistently accounting for gravity darkening, at least for a density contrast up until ten (observed values are on the order of hundred). Curé *et al.* (2005) showed that near critical rotation enables the wind to 'switch' from the standard, fast-accelerating solution to a slow, shallow-accelerating velocity law. This, in combination with the bi-stability effect, can lead to the formation of a slow and dense equatorial wind. Madura *et al.* (2007), finally, confirmed and explained the 'Curé-effect', but argued that gravity darkening is still a problem when aiming at a significant density contrast.

4.3. *Test beds for the bi-stability jump (II): 'normal' B-supergiants*

Thus, it is still unclear whether B[e] supergiants can be used to verify the bi-stability effect. Consequently, we now consider 'normal' B-supergiants. One of the predictions by Vink *et al.* (2000) is an abrupt decrease of v_∞(see also Lamers *et al.* 1995) around the bi-stability jump (around 23 kK). Let us first consider this effect. Standard line-driven wind theory predicts that $v_\infty \approx 2.24\alpha/(1-\alpha)v_{\rm esc}$ (e.g., Puls *et al.* 1996), and a compilation of different measurements/analyses (mostly based on Evans *et al.* 2004 and Crowther *et al.* 2006) by Markova & Puls (2008) shows that the average ratio $v_\infty/v_{\rm esc} \approx 3.3$ for $T_{\rm eff}$ > 23 kK and $v_\infty/v_{\rm esc} \approx 1.3$ for $T_{\rm eff} < 18$ kK, with a *gradual* decrease in between (see also the original work by Evans *et al.* 2004; Crowther *et al.* 2006). Thus, there *is* an effect on v_∞, but we also have to check the behaviour of the mass-loss rates. Conventionally, this is done by plotting the modified wind-momentum rate, $D_{\rm mom}$, as a function of the stellar luminosity, since one of the major predictions from radiation driven wind theory is the well-known wind-momentum luminosity relation (WLR, Kudritzki *et al.* 1995),

$$\log D_{\rm mom} = \log\left(\dot{M} v_\infty (R_\star/R_\odot)^{1/2}\right) \approx x \log(L/L_\odot) + \text{offset(spect. type, metallicity)},$$

where x has a similar dependence as the offset. (Theoretically, $x = (\alpha - \delta)^{-1}$, where $\delta \approx 0.1$ accounts for ionization effects.)

Fig. 6 compares observationally inferred modified wind-momentum rates for OB-supergiants with the predictions from Vink *et al.* (2000) (for details, see Markova & Puls 2008). As pointed out above, the predicted WLR for B-stars lies *above* the one for O-stars (more increase in $\dot{M}$ than decrease in v_∞), whereas the observations show the opposite. The observed O-star rates (triangles, encircled in blue) lie above the predictions, which can be explained by clumping effects (see below), whereas the observed B-star rates for $T_{\rm eff}$ < 22 kK lie well below the predictions and those for $T_{\rm eff} > 22$ kK just connect the O-star regime and the cooler B-stars. With respect to $\dot{M}$ itself, a careful analysis shows that $\dot{M}$ either decreases in concert with v_∞(more likely), or at least remains unaffected (less likely). Globally, however, we do not see the predicted increase in $\dot{M}$, though a certain maximum around the location of the jump might be present (Benaglia *et al.* 2007). Thus, at least below the bi-stability jump there is a severe problem. Either the predicted $\dot{M}$ for cooler objects are too high, or the 'observed' (i.e., derived) ones are too low. Accounting for the observed O-star rates, the latter seems unlikely (and the inclusion of clumping would even increase the discrepancy for the B-stars). A way out of the dilemma might be the potential impact of the 'slow' wind solution (see above) on BA-supergiants, as suggested by Granada *et al.* (this volume).

4.4. *A separate population?*

Returning to the problem of the low rotation rates of B-supergiants and accounting for the above dilemma, one has to admit that *if* indeed the mass-loss rates were not increasing at the bi-stability jump, then there would be no bi-stability braking, and the rapid drop of rotation below $T_{\rm eff} = 20$ kK still needs to be explained. To this end, Vink *et al.* (2010) discuss an alternative *scenario II* (see also Langer, this volume): The cooler, slowly rotating supergiants might form an entirely separate, non core hydrogen-burning population, e.g., they might be products of binary evolution (though this is not generally expected to lead to slowly rotating stars), or they might be post-RSG or blue-loop stars.

Support of this second scenario is the finding that the majority of the cooler objects

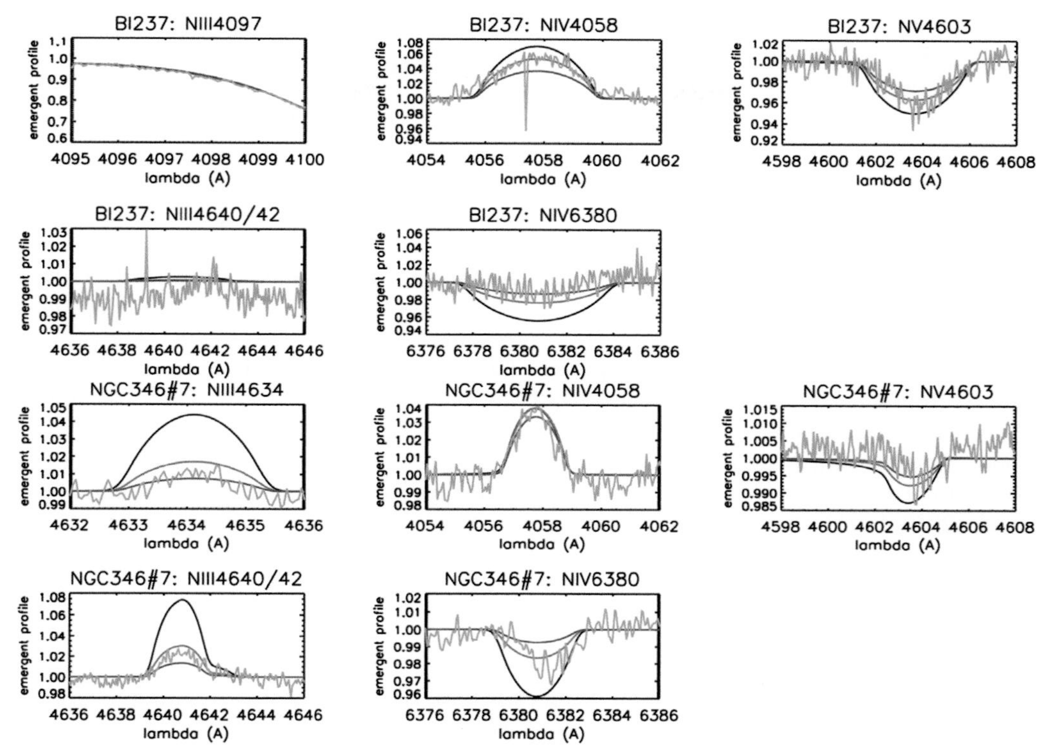

Figure 7. Strategic Nitrogen line profiles in the optical from ionization stages III to V for two early O-type stars. Observations in green, solid lines are synthetic profiles (calculated by the atomospheric code FASTWIND, Puls *et al.* 2005) for different abundances: $N/N_\odot$=0.2 (blue), 0.4 (red) and 1.0 (black). Upper panels: BI237 (O2V((f*)) in the LMC), with $T_{\rm eff}$ = 52 kK and $\log g$ = 4.0. The derived Nitrogen abundance is $N/N_\odot$ = 0.4 or [N/H] = 7.38. Lower panels: NGC#7 (O4V((f$^+$)) in the SMC), with $T_{\rm eff}$ = 45 kK and $\log g$ = 4.0. The derived Nitrogen abundance is $N/N_\odot \approx 0.2\ldots\ 0.4$ or [N/H] $\approx 7.08\ldots\ 7.38$.

(here: in the LMC) is *strongly Nitrogen-enriched*, which was one of the outcomes of the VLT-FLAMES survey of massive stars (Brott, this volume; see also Evans *et al.* 2008 for a brief summary of the project). Vink *et al.* argue that "although rotating models can in principle account for large N abundances, the fact that such a large number of the cooler objects is found to be N enriched suggests an evolved nature for these stars."

4.5. *Nitrogen abundances from O-stars*

So far, Nitrogen abundances could be derived only for a subset of the VLT-FLAMES sample stars, and corresponding data are missing particularly for the most massive and hottest stars. Indeed, when inspecting the available literature for massive stars, one realizes that metallic abundances, in particular of Nitrogen, which is *the* key element to check evolutionary predictions, are scarcely found for O-type stars. The simple reason is that they are difficult to determine, since the formation of NIII/IV lines (and lines from similar ions of C and O) is problematic due to the impact of various processes that are absent or negligible at cooler spectral types, e.g., dielectronic recombination, mass-loss, and clumping. Within the VLT-FLAMES project, progress is under way (Rivero-González 2010), and in Fig. 7 we show two examples of N-abundance determinations for two early type O-stars in the LMC and SMC. Though no detailed comparison with evolutionary models has been made yet, the derived abundances for both objects are consistent, within the error-bars, with the average abundances from corresponding B-type stars of early evolutionary stages, which are [N/H] = 7.13 ±0.29 for the LMC and [N/H] = 7.24 ±0.31 for the SMC, respectively (Hunter *et al.* 2009).

4.6. *Wind clumping*

Mass loss is pivotal for the evolution/fate of massive stars (e.g., the formation of GRBs critically depends on the loss of angular momentum due to mass loss, see Ekström and Langer, this volume), their energy release, and their stellar yields. Thus, reliable mass-loss rates are urgently required (ideally better than a factor of two, Meynet *et al.* 1994). O-star mass-loss rates derived from the optical/radio have been found to be higher than predicted by the widely used mass-loss recipe from Vink *et al.* (2000) (see Fig. 6). The present hypothesis assumes that this discrepancy is due to neglected wind-clumping (small scale density inhomogeneities), originating from the line-driven instability, which results in overestimated mass-loss rates when using recombination-based diagnostics (Puls *et al.* 2008, and references therein). To check and infer the effects due to optically thin and thick clumps, and due to porosity in velocity space, on the various diagnostics, Sundqvist *et al.* (2010, 2011) have used the well observed star λ Cep (O6I) as a **test bed** to derive a mass-loss rate of $1.5{\cdot}10^{-6}\,M_\odot$/yr. This is a factor of four lower than corresponding 'unclumped' values and a factor of two lower than the predictions by Vink *et al.* (2000).

5. Very brief summary and conclusions

We discussed OB-stars as extreme condition **test beds**, regarding effects due to pulsations, rapid rotation, and mass-loss. Rapidly rotating B-supergiants (though scarce) are particularly well suited to check a number of theoretical predictions, and the B1 supergiant γ Ara may be a prime candidate for future diagnostics.

Acknowledgements

The authors gratefully acknowledge a travel grant from the IAU and the local organizers of this conference (J.P.), a grant from the IMPRS, Garching (J.O.S.), and a research grant from the German DFG (J.G.R.G.).

References

Aerts, C., Puls, J., Godart, M., & Dupret, M.-A. 2009, *A&A* 508, 409
Aerts, C., Lefever, K., Baglin, A., Degroote, P. *et al.* 2010, *A&A* (Letters), 513, L11
Benaglia, P., Vink, J. S., Martí, J., Maíz Apellániz, J. *et al.* 2007, *A&A* 467, 1265
Bouret, J.-C., Lanz, T., & Hillier, D. J. 2005, *A&A* 438, 301
Castor, J. I., Abbott, D. C., & Klein, R. I. 1975, *ApJ*, 195, 157
Cranmer, S. R. & Owocki, S. P. 1995, *ApJ*, 440, 308
Crowther, P. A., Lennon, D. J., & Walborn, N. R. 2006, *A&A*, 446, 279
Curé, M., Rial, D. F., & Cidale, L. 2005, *A&A*, 437, 929
Domiciano de Souza, A., Kervella, P., Jankov, S., Abe, L. *et al.* 2003, *A&A* (Letters), 407, L47
Evans, C. J., Lennon, D. J., Trundle, C., Heap, S. R. *et al.* 2004, *ApJ*, 607, 451
Evans, C., Hunter, I., Smartt, S., Lennon, D. *et al.* 2008, *The Messenger*, 131, 25
Feldmeier, A., Kudritzki, R.-P., Palsa, R., Pauldrach, A. W. A. *et al.* 1997, *A&A*, 320, 899
Feldmeier, A., Puls, J., & Pauldrach, A. W. A. 1997, *A&A*, 322, 878
Fullerton, A. W., Gies, D. R., & Bolton, C. T. 1996, *ApJS*, 103, 475
Georgiev, L. N., Hillier, D. J., & Zsargó, J. 2006, *A&A*, 458, 597
Glatzel, W. 1998, *A&A* (Letters), 339, L5
Howarth, I. D., Siebert, K. W., Hussain, G. A. J., & Prinja, R. K. 1997, *MNRAS*, 284, 265

Hunter, I., Brott, I., Langer, N., Lennon, D. J. *et al.* 2009, *A&A*, 496, 841
Kaufer, A., Stahl, O., Prinja, R. K., & Witherick, D. 2006, *A&A*, 447, 325
Kudritzki, R.-P., Lennon, D. J., & Puls, J. 1995, in: J.R. Walsh & I.J. Danziger (eds.), *Science with the VLT*, Proc. ESO Workshop (Berlin: Springer), p. 246
Kervella, P. & Domiciano de Souza, A. 2006, *A&A*, 453, 1059
Lamers, H. J. G. & Pauldrach, A. W. A. 1991, *A&A* (Letters), 244, L5
Lamers, H. J. G. L. M., Snow, T. P., & Lindholm, D. M. 1995, *ApJ*, 455, 269
Langer, N., Heger, A., & Fliegner, J. 1997, in: T. R. Bedding, A. J. Booth, & J. Davis (eds.), *Fundamental stellar properties: The interaction between observation and theory*, IAU Symposium 189, p. 343
Lefever, K., Puls, J., & Aerts, C. 2007, *A&A*, 463, 1093
Maeder, A. 1999, *A&A*, 347, 185
Maeder, A. & Meynet, G. 2000, *A&A*, 361, 159
Madura, T. I., Owocki, S. P., & Feldmeier, A. 2007, *ApJ*, 660, 687
Markova, N. & Puls, J. 2008, *A&A*, 478, 823
Massa, D., Fullerton, A. W., Nichols, J. S., Owocki, S. P. *et al.* 1995, *ApJ* (Letters) 452, L53
Meilland, A., Stee, P., Vannier, M., Millour, F. *et al.* 2007, *A&A* 464, 59
Meynet, G., Maeder, A., Schaller, G., Schaerer, D. *et al.* 1994, *A&AS*, 103, 97
Owocki, S. P., Cranmer, S. R., & Gayley, K. G. 1998, in: A. M. Hubert & C. Jaschek (eds.), *B[e] stars*, Astrophysics and Space Science Library 233, p. 205
Pamyatnykh, A. A. 1999, *AcA*, 49, 119
Pauldrach, A. W. A. & Puls, J. 1990, *A&A*, 237, 409
Pelupessy, I., Lamers, H. J. G. L. M., & Vink, J. S. 2000, *A&A*, 359, 695
Petrenz, P. & Puls, J. 1996, *A&A*, 312, 195
Petrenz, P. & Puls, J. 2000, *A&A*, 358, 956
Prinja, R. K., Massa, D., Fullerton, A. W., Howarth, I. D. *et al.* 1997, *A&A*, 318, 157
Puls, J., Kudritzki, R.-P., Herrero, A., Pauldrach, A. W. A. *et al.* 1996, *A&A*, 305, 171
Puls, J., Springmann, U. & Lennon, M. 2000, *A&AS*, 141, 23
Puls, J., Urbaneja, M. A., Venero, R., Repolust, T. *et al.* 2005, *A&A*, 435, 669
Puls, J., Markova, N., Scuderi, S., Stanghellini, C. *et al.* 2006, *A&A*, 454, 625
Puls, J., Vink, J. S., & Najarro, F. 2008, *A&AR*, 16, 209
Repolust, T., Puls, J., & Herrero, A. 2004, *A&A*, 415, 349
Rivero-González, J. 2010, *Diploma-Thesis* (LMU München)
Saio, H., Kuschnig, R., Gautschy, A., Cameron, C. *et al.* 2006, *ApJ*, 650, 1111
Simón-Díaz, S., Herrero, A., Uytterhoeven, K., Castro, N. *et al.* 2010, *ApJ* (Letters) 720, 174
Sundqvist, J. O., Puls, J., & Feldmeier, A. 2010, *A&A*, 510A, 11
Sundqvist, J. O., Puls, J., Owocki, S., *et al.* 2011, in: P. Williams & G. Rauw (eds.), *The multi-wavelength view of Hot, Massive Stars*, 39th Liège International Astrophysical Colloquium, in press
van Belle, G. T. 2010, in: T. Rivinius & M. Curé (eds.), *The Interferometric View on Hot Stars*, Rev. Mexicana AyA Conference Series 38, p. 119
von Zeipel, H. 1924, *MNRAS*, 84, 665
Vink, J. S., de Koter, A., & Lamers, H. J. G. L. M. 2000, *A&A*, 362, 295
Vink, J. S., de Koter, A., & Lamers, H. J. G. L. M. 2001, *A&A*, 369, 574
Vink, J. S., Davies, B., Harries, T. J., Oudmaijer, R. D. *et al.* 2009, *A&A*, 505, 743
Vink, J. S., Brott, I., Gräfener, G., Langer, N. *et al.* 2010, *A&A* (Letters), 512, L7
Waelkens, C., Aerts, C., Kestens, E., Grenon, M. *et al.* 1998, *A&A*, 330, 215
Wende, S., Glatzel, W., & Schuh, S. 2008, in: A. Werner & T. Rauch (eds.), *Hydrogen-Deficient Stars*, ASP-CS 391, p. 319
Zahn, J.-P. 1992, *A&A*, 265, 115
Zickgraf, F.-J., Wolf, B., Leitherer, C., Appenzeller, I. *et al.* 1986, *A&A*, 163, 119
Zickgraf, F.-J., Wolf, B., Stahl, O., & Humphreys, R. M. 1989, *A&A*, 220, 206

Active OB stars: structure, evolution, mass loss, and critical limits
Proceedings IAU Symposium No. 272, 2010
C. Neiner, G. Wade, G. Meynet & G. Peters, eds.

doi:10.1017/S1743921311011446

Fundamental parameters of "normal" B stars in the solar neighborhood

Maria-Fernanda Nieva[1] and Norbert Przybilla[2]

[1]Max-Planck-Institut für Astrophysik, Postfach 1317, D-85741 Garching, Germany
email: fnieva@mpa-garching.mpg.de

[2]Dr. Karl Remeis-Observatory Bamberg & ECAP, University Erlangen-Nuremberg, Sternwartstr. 7, D-96049 Bamberg, Germany
email: przybilla@sternwarte.uni-erlangen.de

Abstract. Understanding phenomena of activity in stars, like pulsations or magnetism, benefits from systematic comparisons of some key physical parameters of active with those of "normal" stars. Here we concentrate on a careful derivation of fundamental parameters of a well selected sample of 27 "normal" B stars in nearby OB associations and in the field. A quantitative spectral analysis methodology based on hybrid non-LTE techniques is applied to high-resolution and high-S/N spectra. Results derived from the pure spectroscopic analysis are compared to other data/indicators of stellar parameters in order to prove the reliability of the method. Very good agreement is obtained among all of them. Besides the fundamental parameters, the chemical composition of the stars is also determined at high precision, turning out to be highly homogeneous. A comparative study of the present results with those of well known active massive stars will help to improve our understanding of the driving mechanisms of activity.

Keywords. stars: abundances, stars: atmospheres, stars: early-type, stars: fundamental parameters

---

1. Introduction

In order to improve our understanding of the driving mechanisms of activity in stars, systematic comparisons of fundamental parameters like temperature, mass, age, radius, luminosity, chemical composition of "normal" and active – pulsating or magnetic – stars have to be performed. Once the parameters are determined at high precision, further comparisons of the observed characteristics of the stars surface with predictions of stellar evolutionary models may help us to understand the nature of the stars.

There are different methods to derive fundamental parameters of massive OB stars. Eclipsing binaries offer a direct method to determine masses, radii and distances based on the measurement of Doppler shifts in the spectrum and the geometrical configuration of the system. For single stars, interferometry is the most direct method to derive the star's radius. However, the nearest OB main sequence stars have an angular diameter below the measurement threshold of the current interferometers. Indirect methods are based on e.g. photometry, spectrophotometry, spectroscopy or asteroseismology. Photometric calibrations may provide only rough estimates of parameters that need to be confirmed with other, more accurate techniques. Spectrophotometry is also used to determine temperatures e.g. via the Balmer jump (see e.g. Adelman *et al.* 2002). The method can be used for early B-type stars, however it is more precise for stars of spectral class late-B or A, where the jump is larger and more sensitive to temperature changes. Photometric and spectrophotometric analyses depend on details of model atmospheres. On the other hand, asteroseismology also offers the possibility to derive fundamental parameters like

Table 1. Spectroscopic indicators for for $T_{\rm eff}$ and $\log g$ determination.

HD	H	He I	He II	C II	C III	C IV	O I	O II	Ne I	Ne II	Si III	Si IV	Fe II	Fe III
36512	x	x	x	x	x	x		x		x	x	x		x
149438	x	x	x	x	x	x		x	x	x	x	x		x
63922	x	x	x	x	x	x		x		x	x	x		x
34816	x	x	x	x	x			x	x	x	x	x		x
36822	x	x	x	x	x	x		x	x	x	x	x		x
36960	x	x	x	x	x			x	x	x	x	x		x
36591	x	x	x	x	x			x	x	x	x	x		x
205021	x	x	x	x	x			x	x	x	x	x		x
61068	x	x	x	x	x			x	x	x	x	x		x
35299	x	x		x	x		x	x	x	x	x	x	x	x
216916	x	x		x	x		x	x	x	x	x	x	x	x
74575	x	x		x	x		x	x	x	x	x	x	x	x
886	x	x		x	x		x	x	x		x	x	x	x
29248	x	x		x	x		x	x	x		x	x	x	x
16582	x	x		x	x		x	x	x		x	x	x	x
122980	x	x		x	x		x	x	x		x		x	x
35708	x	x		x	x		x	x	x		x	x	x	x
3360	x	x		x	x		x	x	x		x	x	x	x
160762	x	x		x			x	x	x		x		x	x
209008	x	x		x			x	x	x		x		x	x

mass, age, effective temperature and surface gravity. The interpretation of the observed frequency spectrum relies on stellar model assumptions, e.g. chemical composition, convection and rotation (see Briquet *et al.* 2010 for an example of an asteroseimic study of a O9V star).

Among the indirect methods, quantitative spectroscopy has become a very powerful technique to derive reliable fundamental parameters of massive stars. In the past years we have improved the spectral modeling and analysis of unevolved early B-type stars, being able to derive unprecedentedly accurate stellar atmospheric parameters from the spectrum only. The method can be applied to objects that are not deformed due to high rotation rates. Here, we go one step further to show that also the fundamental parameters derived by means of our techniques are highly reliable and can be used as reference for further studies of active stars. Moreover, a first systematic comparison of chemical patterns of "normal" and active stars using our techniques has already been undertaken as discussed in Przybilla & Nieva (these proceedings), showing promising results.

2. The sample: "normal" stars

An original sample of 27 well studied "normal" early B-type stars in OB associations and the field of the solar neighbourhood was selected from previous analyses of chemical abundances (references in Nieva & Przybilla, in prep.). The observational material comprises high-resolution (~40 000–48 000) and high S/N (~250-800) spectra with broad wavelength coverage obtained with FEROS on the ESO 2.2m telescope in La Silla, with FOCES on the 2.2m telescope at Calar Alto/Spain and also collected from the ELODIE archive (1.93 m telescope of the Observatoire de Haute-Provence). This material

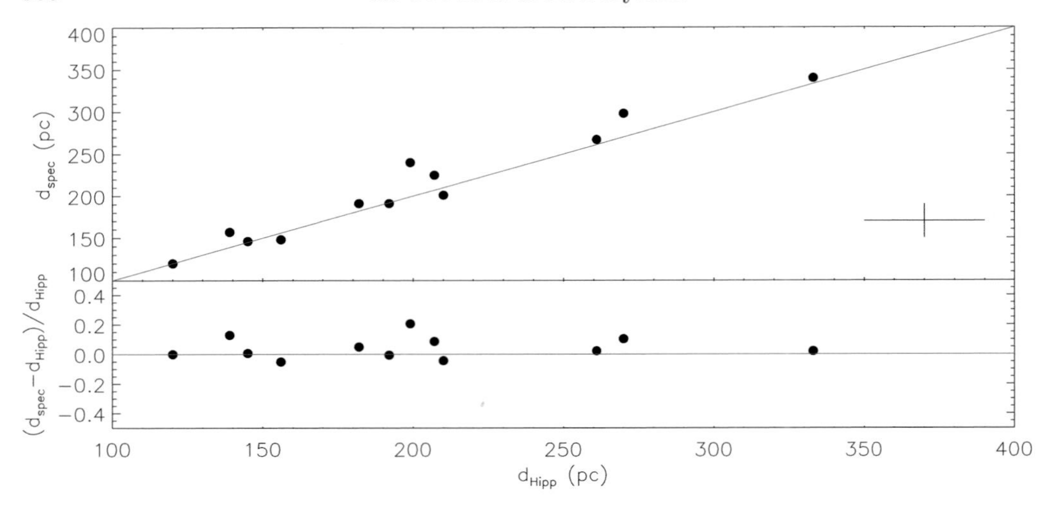

Figure 1. Spectroscopic vs. Hipparcos distances.

constitutes one of the highest quality for the quantitative spectral analysis of early B-type stars in the solar neighbourhood so far. After careful visual inspection of the spectra, 7 stars were identified as spectroscopic binaries or chemically peculiar stars and were removed from the sample. The final star sample containing 20 objects is listed in Table 1. Despite the stars were considered as "normal" in the past decades, half of them turned out to be of the β-Cephei type and some are magnetic, e.g. τ Sco (HD 149438).

3. Spectral Models and Analysis

Among the OB stars, early B-type stars on the Main Sequence are the best-constrained objects in terms of quantitative spectral analysis. In contrast to their hotter (early O) and luminous (OB supergiant) siblings, their atmospheric structure is much less sensitive to non-local thermodynamic equilibrium effects (non-LTE) and stellar winds. Classical atmospheric models assuming plane-parallel geometry, homogeneity and LTE are proven to represent their atmospheric structure very well (see e.g. Nieva & Przybilla 2007). However, most of their spectral lines are still subject to pronounced non-LTE effects nowadays well understood and constrained in the optical (see Przybilla 2008).

The model atmospheres are computed here with the ATLAS9 code (Kurucz 1993b) and line blanketing is realised by means of opacity distribution functions ODFs (Kurucz 1993a). Non-LTE level populations are computed with DETAIL (Giddings 1981), that can treat even complex ions in a realistic way. The synthetic spectra are calculated with SURFACE (Butler & Giddings 1985), using refined line-broadening theories. Non-LTE spectra of all elements are computed using our most recent model atoms (see Przybilla, Nieva & Butler 2008 for references).

The analysis method is based on the use of different spectroscopic indicators, i.e. several Balmer lines and multiple ionization equilibria. The aim is to reproduce simultaneously all indicators via an iterative line-fitting procedure in order to derive atmospheric parameters and chemical abundances self-consistently. The stellar parameters primarily derived with this technique are the effective temperature T_{eff}, surface gravity $\log g$, microturbulence ξ, macroturbulence ζ, projected rotational velocity $v \sin i$ and elemental abundances $\varepsilon(\text{X})$. Table 1 lists the star sample with the respective spectroscopic indicators, denoting with boxes established ionization equilibria. Within this method, the most relevant sources

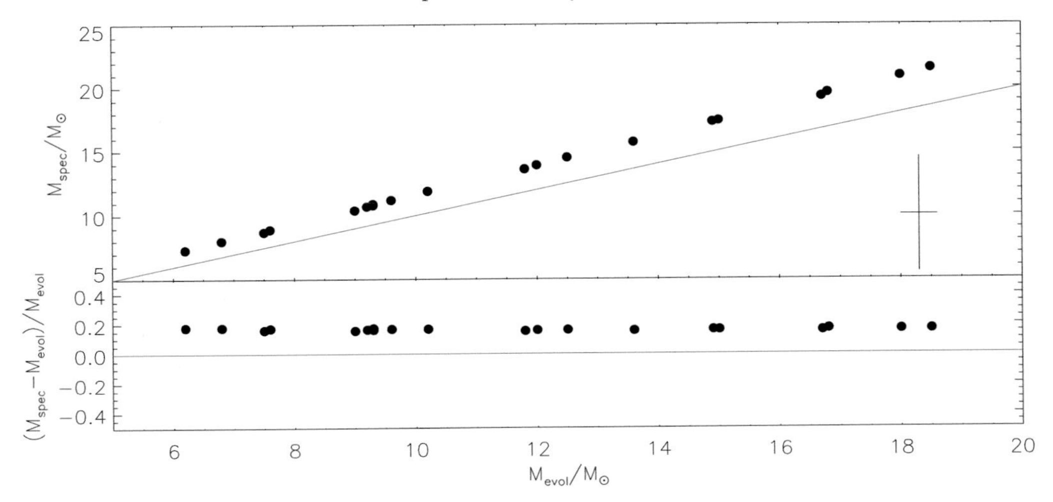

Figure 2. Spectroscopic vs. evolutionary masses.

of systematic errors were identified and consequently minimized (see Nieva & Przybilla 2008, 2010 for a discussion on this). Furthermore, the atmospheric parameters serve to compute the whole emergent stellar flux and to derive photometric quantities, like bolometric correction, colour excess and absolute magnitude, in an intermediate step to compute luminosity L, mass M, distance d and radius R.

4. Testing spectroscopic fundamental parameters

Stellar parameters derived spectroscopically can be tested against results obtained by other means. Here, we discuss tests for surface gravities, masses, effective temperatures, bolometric corrections, luminosities and chemical composition.

The determination of spectroscopic distances depends strongly on the surface gravity, more than on other quantities like evolutionary mass, effective temperature (through the model atmosphere flux at the stellar surface) and dereddened apparent magnitude (Ramspeck *et al.* 2001). Therefore, surface gravities can be tested by comparing the spectroscopic distances with distances derived from Hipparcos parallaxes (new reduction by van Leeuwen 2007), as shown in Fig. 1. The spectroscopic and *Hipparcos* distances agree very well within the uncertainties.

The spectroscopic masses depend on the gravity, the distance (therefore, also indirectly on the evolutionary mass) and the effective temperature. The distance allow to determine the luminosity, and radii are derived from from L and $T_{\rm eff}$. The spectroscopic masses can be compared with masses derived from stellar evolutionary tracks (e.g. Meynet & Maeder 2003), as displayed in Fig. 2. They show a constant offset of $\sim$ 15%, that may be due to the different metallicity of the evolutionary models ($Z=0.020$) and the star sample ($Z=0.014$), among other reasons that need to be investigated. However, the offset is relatively small when compared to a variable factor of typically 0–2 for stars up to 25 $M_{\odot}$ (e.g. Repolust *et al.* 2004).

Other tests – not shown here – prove the reliability of the results concerning the effective temperatures, bolometric corrections, luminosities and chemical composition. Effective temperatures derived from multiple ionization equilibria coincide with those determined via spectral energy distributions from the UV to the IR, but are by far more precise. Bolometric corrections for the hotter stars agree with values from Vacca *et al.*

(1996) and Martins *et al.* (2005). The mass-luminosity relation of the sample agrees very well with theory (Maeder 2009, p. 625ff.). Furthermore, the chemical composition of the star sample is highly homogeneous, in agreement with recent findings by Przybilla, Nieva & Butler (2008). Details on the remaining comparisons not shown here can be found in Nieva (in prep.) and Nieva & Przybilla (in prep.).

5. Conclusions

The determination of fundamental parameters can be performed following different, direct or indirect methods. No matter which technique we use, it is necessary to cross-check the results with independent methods or measurements in order to test their reliability. Further comparisons between observational constraints and stellar evolutionary model predictions are only meaningful when the derived stellar parameters are reliable and describe the physical properties of the stars consistently. Here, we presented a very well studied sample of "normal" early B-type stars representative for the solar neighbourhood. The sample has very high-quality spectra and has been carefully inspected before the analysis. Several stellar – atmospheric and fundamental – parameters derived by means of our self-consistent spectroscopic method have been successfully tested against independent measurements, indicators and model predictions. This study offers a solid basis for further investigations by comparing some key properties of "normal" and active stars (see Przybilla & Nieva, these proceedings).

References

Adelman, S. J., Pintado, O. I., Nieva, M. F., Rayle, K. E. *et al.* 2002, *A&A*, 392, 1031

Butler, K., & Giddings, J. R. 1985, *Newsletter of Analysis of Astronomical Spectra*, No. 9 (Univ. London)

Briquet, M., Aerts, C., Baglin, A., Nieva, M.F. *et al.* 2010, *A&A*, submitted

Giddings, J. R. 1981, Ph.D. Thesis (Univ. London)

Kurucz, R. L. 1993a, *Smithsonian Astrophys. Obs.*, CD-ROM No. 2–12

Kurucz, R. L. 1993b, *Smithsonian Astrophys. Obs.*, CD-ROM No. 13

Maeder, A. 2009, *Physics, Formation and Evolution of Rotating Stars*, Astronomy and Astrophysics Library (Springer Berlin Heidelberg)

Martins, F., Schaerer, D., & Hillier, D. J. 2005, *A&A*, 436, 1049

Meynet, G. & Maeder, A. 2003, *A&A*, 411, 543

Nieva, M. F. & Przybilla, N. 2007, *A&A*, 467, 295

Nieva, M. F. & Przybilla, N. 2008, *A&A*, 481, 199

Nieva, M.-F. & Przybilla, N. 2010, in: C. Leitherer, P. Bennett, P. Morris, & J. van Loon (eds.), *Hot and Cool: Bridging Gaps in Massive Star Evolution*, ASP-CS 425, p. 146

Przybilla, N. 2008, in: S. Röser (eds.), *Reviews in Modern Astronomy*, 20, p. 323

Przybilla, N., Nieva, M.-F., & Butler, K. 2008, *ApJ* (Letters), 688, L103

Vacca, W. D., Garmany, C. D., & Shull, J. M. 1996, *ApJ*, 460, 914

Van Leeuwen, F. 2007, *Hipparcos, the New Reduction of the Raw Data*, Astrophysics and Space Science Library 350

Ramspeck, M., Heber, U., & Edelmann, H. 2001, *A&A*, 379, 235

Repolust, T., Puls, J., & Herrero, A. 2004, *A&A*, 415, 349

Active OB stars: structure, evolution, mass loss, and critical limits
Proceedings IAU Symposium No. 272, 2010
C. Neiner, G. Wade, G. Meynet & G. Peters, eds.

doi:10.1017/S1743921311011458

Eruptive outflow phases of massive stars

Nathan Smith

Steward Observatory, University of Arizona, 933 N. Cherry Ave., Tucson, AZ 85721, USA
email: nathans@as.arizona.edu

Abstract. I review recent progress on understanding eruptions of unstable massive stars, with particular attention to the diversity of observed behavior in extragalatic optical transient sources that are generally associated with giant eruptions of luminous blue variables (LBVs). These eruptions are thought to represent key mass loss episodes in the lives of massive stars. I discuss the possibility of dormant LBVs and implications for the duration of the greater LBV phase and its role in stellar evolution. These eruptive variables show a wide range of peak luminosity, decay time, expansion speeds, and progenitor luminosity, and in some cases they have been observed to suffer multiple eruptions. This broadens our view of massive star eruptions compared to prototypical sources like Eta Carinae, and provides important clues for the nature of the outbursts. I will also review and discuss some implications about the possible physical mechanisms involved, although the cause of the eruptions is not yet understood.

Keywords. instabilities, circumstellar matter, stars: evolution, stars: mass loss, supernovae: general, stars: winds, outflows

1. Introduction and Background

Almost sixty years have passed since, as a result of attempts to produce standard candles for cosmology, Hubble & Sandage (1953) discovered the class of luminous, blue, irregular variables in M31 and M33 that we now collectively refer to as luminous blue variables (LBVs) in any galaxy. A few key conferences in the late 1980s and 1990s established some paradigms for LBVs and the evolution of massive stars in general, some of which may be in need of revision.

In the context of this conference on "active" OB stars, the LBVs are perhaps a hideous extreme example of stellar activity. However, they can be viewed as cases where the effects of rotation, pulsation, binaries, and perhaps even magnetic fields may have rather extreme consequences when the a star is near the Eddington limit. In that sense, there is hopefully some synergy between LBVs and the various other types of stars discussed at this meeting.

LBVs exhibit so-called "microvariability" in their photometry and also undergo well-known S-Doradus excursions when the star changes color at relatively constant bolometric luminosity (although see the talk by J. Groh in these proceedings). However, they are most notable and mysterious for their giant eruptions, when the stars are thought to increase their bolometric luminosity to be above the classical Eddington limit, during which time they may eject large amounts of mass — anywhere form 0.1 – 10 $M_{\odot}$. Smith & Owocki (2006) have argued that when this is combined with the facts that LBV eruptions repeat, and that mass-loss rates for O-type stars are lower than we used to think, that LBVs probably dominate the shedding of the H envelope in massive single stars. This may have significant implications, since LBV eruptions do not necessarily depend on metallicity.

However, we still have no clear idea what causes the giant eruptions of LBVs, and we have no good formulation for how the eruptive behavior scales with initial mass and metallicity, or if it depends on binarity. Since the dominant mass-loss machanism in stellar evolution is so poorly understood, we cannot have very much faith in the predictions of the fates of massive stars in stellar evolution models, or how this scales with metallicity.

We do, however, know that giant LBV eruptions certainly occur because we observe them, and advances can be made in constraining their properties. Giant LBV eruptions are bright and can be seen in other galaxies. They are detected by accident in systematic supernova searches that are conducted — like Hubble & Sandage's early work — in the pursuit of standard candles for cosmology, and so they are sometimes called "SN impostors". Several dozen SN impostor giant eruptions have now been seen in nearby galaxies, but LSST will vastly increase the number of these transients. Hopefully this will allow us to improve our knowledge of the statistics of LBVs. For now, we must be content with studying the few examples we have and gleaning as many clues about their physics as we can. In this paper, I briefly review some of the observed properties of LBV stars, and I emphasize some new results including the distribution of observed properties in giant LBV eruptions and their connection to Type IIn supernovae. Much of what I discussed in my talk at IAU Symposium 272 is presented in more detail in two recent papers (Smith *et al.* 2010a; Smith & Frew 2010), and the reader is reffered to these for more information.

2. Lifetime of the LBV Phase, and Ducks that Don't Quack

"If it looks like a duck, and quacks like a duck, we have at least to consider the possibility that we have a small aquatic bird of the family anatidae on our hands."

...Douglas Adams

This is a slightly different formulation of the more familiar "If it looks like a duck..." phrase, which was often used in connection with LBV eruptions in the 1980s and 1990s (e.g., Conti 1995, 1997; Bohannan 1997), suggesting that you can't really be sure that a duck is a duck unless you hear it quack. The point was that although pretty much everything in the upper left part of the HR diagram is luminous, blue, and at least somewhat variable if you look closely enough, the classification "LBV" was to be reserved for a specific class of stars that are observed to undergo more violent eruptions (i.e. they "quack" rather loudly), and that this therefore indicated some particular inherent instability in the star, which is not present in all supergiants. In the same breath, however, it was sometimes admitted that a star which had erupted in the past (or will erupt in the near future) might not necessarily be exhibiting signs of that instability *right now.* This problem is illustrated in Figure 1.

LBVs are extremely rare — there are only a handful known in the Milky Way or in any nearby galaxy. However, there is a larger number of stars that closely resemble LBVs in their observed spectral properties and location on the HR diagram. Some of these have massive circumstellar shells that resemble LBV shells, as in several recent examples detected by Spitzer (Gvaramdze *et al.* 2010; Wachter *et al.* 2010). These are typically called "LBV candidates" until they are actually seen to have an eruption. The Ofpe/WN9 stars are a good example of a class of stars which resemble LBVs and often have circumstellar shells; there are documented examples of confirmed LBVs that are Ofpe/WN9 stars in their quiescent hot states (e.g., R127, AG Car). If Ofpe/WN9 stars are really dormant LBVs, it would imply that LBVs may go through relatively long periods of time when they are not erupting. In other words, they may have extended "dormant" phases in between major eruptions, like volcanoes or geysers.

Figure 1. The reader can deduce that the object on the left is obviously a Duck, even though this conference proceedings volume is not accompanied by an audio CD containing a recording of it quacking. On a related note, the panel at right shows several hot massive stars surrounded by LBV-like dust shells detected in the mid-IR (Gvaramadze *et al.* 2010).

Another sign of long dormant phases of LBVs is illustrated by the example of P Cygni, which is the nearest and the first LBV. It underwent a giant LBV eruption in 1600 AD, with a second eruption 55 years later. After being faint for another 50 years, it brightened at the beginning of the 18th century. Since then, however, *P Cygni hasn't done much of anything to suggest that it is an unstable star*. Had we started observing it around 1700 AD, instead of 1600 AD, we would never know that it is an LBV. We would see a relatively tame blue supergiant with a strong wind and a shell nebula, and we would simply call it an LBV candidate. (Note that P Cygni is not as hot as an Ofpe/WN9 star, suggesting that there may be many other blue supergiants that are dormant LBVs as well.) Similarly, many stars discovered in the Galactic center resemble LBVs in their luminosity and spectral properties, but because they are only seen in the IR, we do not have the benfit of many decades or centuries of continuous photometric monitoring, so we have not necessarily observed eruptive behavior in all the Galactic center LBVs.

The rarity of LBV has led to important conjectures about their evolutionary phase and the lifetime of the LBV phase itself. Many authors have discussed this, but the argument usually goes something like this, as discussed in the review by Bohannan (1997): There were 5 LBVs and 115 WR stars known in the LMC at that time. If we assume that all WR stars are descended from LBVs, and that the WR lifetime is the core-He burning lifetime of about 5×10^5 yr, the rarity of LBVs then suggests that the LBV phase only lasts only a few 10^4 yr. It was widely concluded, therefore, that the LBV phase represents an extremely brief and fleeting *transitional* phase between the core-H burning main sequence of O-type stars and the core-He burning phase of WR stars.

This line of reasoning suffers from some fallacies, and the derived age is probably wrong. It ignores the possibility of dormant phases of LBVs, as noted above, and offers no good explanation for the large number of LBV candidates and other blue supergiants that are necessarily evolved massive stars as well. It also ignores the fact that about 1/3 of stars counted as "WR stars" are actually WNH stars (see Smith & Conti 2008) and are probably not in core-He burning yet.

Returning to the analogy with volcanoes, one could reproduce a similar fallacy: there are typically something like 1 or 2 major volcanic eruptions on Earth each year (where "major" means more than 0.1 km^3 of tephra). Some of us experienced the unfortunate consequences of this for international travel earlier this year. One could then say that since there are several thousand major mountains on Earth, each of which has an average geological age of around 10^8 yr, that the lifetime of a typical volcano is only a few 10^4 yr. This is, of course, a severe underestimate for the lifetime of a volcano because volcanoes spend most of their time in dormant phases. We know this because mountains with a crater or with evidence for a history of eruptions are counted as real volcanoes, and may erupt again in the future. Similarly, one could take inventory of the number of ducks quaking at any instant and vastly underestimate the true number of aquatic birds of the family anatidae.

Deriving the correct lifetime for LBVs depends on the "duty cycle" of the unstable LBV phase. In other words, we need to know what fraction of the time an LBV might be dormant by our observational standards, and correct for that. We have no theoretical prediction of this time, since there is no theoretical prediction of LBVs. There is, however, an expectation that LBVs recover from major eruptions and go through a relatively quiescent period where they re-establish thermal equilibrium. Both P Cygni and η Car have mulitple shell nebulae that suggest time periods of order 10^3 yr in between major eruptions. Moreover, Massey *et al.* (2007) counted only 6 LBVs in M31 and M33, but they counted over 100 LBV candidates. This suggests a factor of 10-20 more LBVs than are counted by active LBVs at any time, implying a duty cycle of 5–10% for the manifestation of LBV instability during the greater evolutionary phase in which we find LBVs. If we re-do the calculation above (now including the fact that 1/3 of WR stars are WNH), then we find that the lifetime over which a massive star could be an LBV is more like $(2\text{–}5)\times10^5$ yr.

This paints a very different picture for the evolutionary state of LBVs, where they spend a substantial fraction (or all) of their core-He burning lifetime as an LBV (or candidate LBV), punctuated by intermittent episodes of eruptive instability. If some of these LBVs make it all the way to core collapse before shedding their H envelopes, it may explain the observed connection between LBVs and Type IIn supernovae (Smith *et al.* 2007, 2008, 2010a, 2010c; Gal-Yam & Leonard 2009; etc.).

Of course, the comments above are predicated on the notion that all WR stars are descended from LBVs, allowing us to calculate the LBV lifetime by comparison to the assumed WR lifetime. This hypothesis may be wrong if, for example, a substantial fraction of WR stars have shed their H envelopes via Roche lobe overflow in binary systems (see, e.g., Smith *et al.* 2010c for implications from Type Ibc supernovae). In that case, the fraction of LBVs+candidates to WR stars depends on both the relative lifetimes and the fraction of massive stars in close binaries. One gets the impression that our paradigms of massive star evolution need to be taken back to the drawing board.

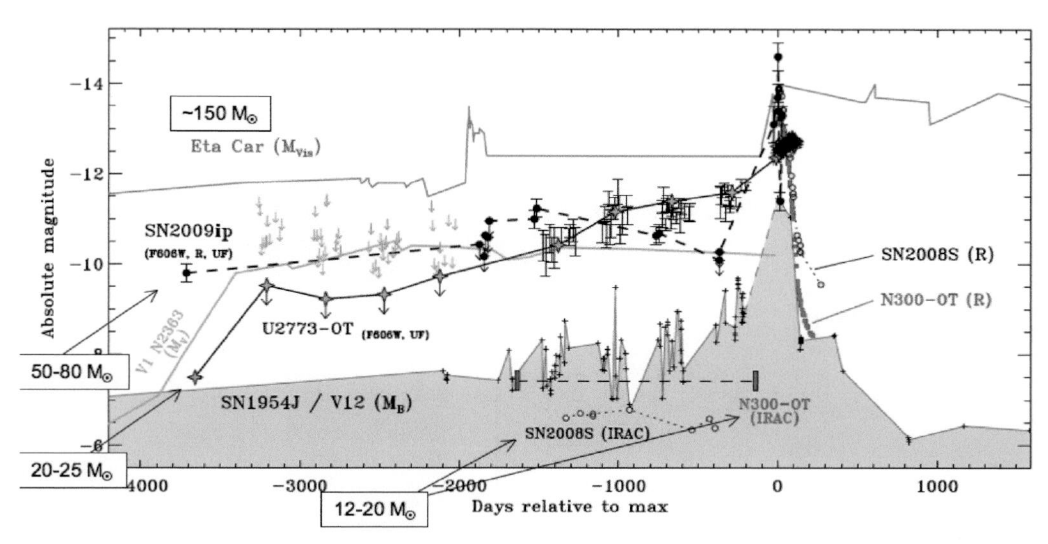

Figure 2. Light curves of various LBV-like transients (from Smith *et al.* 2010a), with a few cases where approximate initial masses for the progenitor stars have been estimated. References for individual sources of photometry can be found in that paper.

3. A Diverse Range of Observed Properties

LBVs are by definition associated with the most luminous and most massive stars in any galaxy, but their initial mass range is actually rather wide. LBVs are thought to arise from stars with initial masses ranging from 20 or 25 $M_\odot$ up to the most massive stars known (Smith, Vink & de Koter 2004). (The lower-luminosity LBVs with initial masses of 20 or 25 $M_\odot$ up to about 40 $_\odot$ are thought to reach their unstable state only after they have been through substantial mass loss in a previous RSG phase, thereby increasing their L/M ratio.) This mass range is perhaps a result of how we identify them: we define the LBV variability as a brightening at visual wavelengths that corresponds roughly to the star's bolometric correction (e.g., Humphreys & Davidson 1994). The S Doradus instability strip is slanted on the HR diagram, so that more luminous LBVs are hotter in their quiescent state. As a result, these hotter and more luminous LBVs have a larger bolometric correction, and consequently, brighten more at visual wavelengths when they undergo an S Dor eruption.† These are the classical LBVs. LBVs at the bottom end of the initial mass range have cooler quiescent temperatures and, consequently, smaller bolometric correction and less pronounced brightenings in a normal S Dor stage. In fact, if we were to extrapolate the S Dor instability strip to lower luminosities and cooler temperatures, it would cross the temperture for cool 7000–8000 K eruptive states of LBVs at luminosities that correspond to initial masses of $\sim$20 $M_\odot$. In other words, whatever instability causes the LBV phase might manifest itself somewhat differently below 20–25 $M_\odot$, and we might not recognize these stars as LBVs because of how we define the observed LBV variability.

In fact, recent studies of extragalactic transient sources have revealed some transients that closely resemble LBV giant eruptions, but which — unexpectedly — seem to have progenitor stars of lower masses around 10–20 $M_\odot$ or even lower. These transients and other LBVs are reviewed recently by Smith *et al.* (2010a; see also Smith *et al.* 2009,

† Note that it was originally hoped that calibrating this would allow LBVs to be used as standard candles.

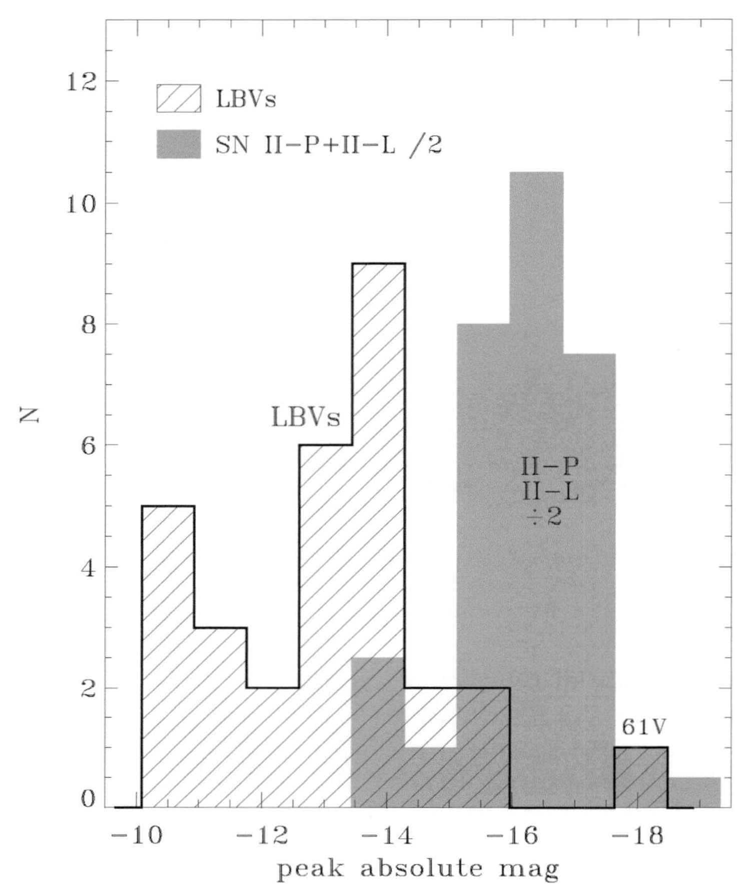

Figure 3. A histogram of the distribution of peak absolute magnitudes for LBV-like eruptions (hatched) compared to normal Type II-P and II-L supernovae (gray; numbers divided by 2) from the Berkeley SN search (figure from Smith *et al.* 2010a; see that paper for details). The LBV farthest to the right is SN 1961V, which Smith *et al.* (2010a) have argued is not really an LBV, but rather, a true core-collapse SN IIn.

2010b; Prieto *et al.* 2008, 2009; Thompson *et al.* 2009; Gogarten *et al.* 2009). The true nature of these sources is still debated, however; it is not clear if they are manifestations of LBV-like instability extending to stars with lower-initial masses, or if they are something altogether different originating from intermediate-mass stars. The light curves for some LBV-like transients are shown in Fig. 2 (from Smith *et al.* 2010b), concentrating on some sources that show detections of their progenitor stars before a giant LBV-like eruption. This is meant to demonstrate the range of initial luminosities and masses for stars that undergo giant LBV-like eruptions. Some of the stars even show precursor variability before the eruption begins, like SN 2009ip and UGC2773-OT (Smith *et al.* 2010b).

Smith *et al.* (2010a) has also discussed the diversity in the observed properties of the eruptions themselves. Figures 3 and 4 show histograms of the distributions of peak absolute visual magnitude (a combination of V and R magnitudes) and the distribution of expansion speeds (measured from Hα). LBV-like eruptions span a range in absolute peak magnitude from around −10 to −16 mag, peaking at −14 mag. The luminous end of the distribution overlaps with the faintest core-collapse SNe, but one can usually distinguish the two based on spectra (see Smith *et al.* 2009, 2010b). The low-luminosity end of the

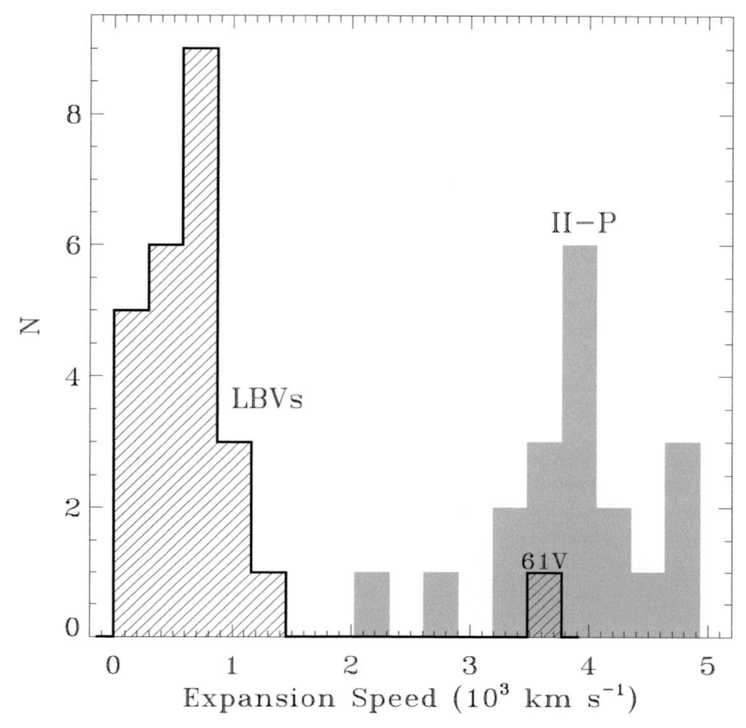

Figure 4. A histogram of the expansion speeds for LBV-like eruptions (hatched) compared to SNe II-P (figure from Smith *et al.* 2010a; see that paper for details).

distribution of LBV-like eruptions is muddy; there may be a mix of LBV giant eruptions and S Doradus outbursts (see Smith *et al.* 2010a for more details). One problematic case is the prototypical SN impostor SN 1961V, which is much brighter than any other LBV eruption. SN 1961V also stands out in its observed expansion speed (Fig. 4) which is much faster than other LBVs and more in line with core-collapse SNe. Based on these points and other information, Smith *et al.* (2010a) have argued that SN 1961V was in fact not an LBV giant eruption, but a true core-collapse SN IIn. The other LBVs have expansion speeds that range from around 100 to 1000 km s^{-1}, much slower than speeds for core-collapse SNe, indicating less energetic explosions.

4. Some Detailed Examples

There has been a recent increase in studies of extragalactic transients that seem analogous to LBV giant eruptions, perhaps due in part to the increased community-wide interest in transient sources of all types, and perhaps also because a substantial fraction of the SN community seems to finally be getting bored of Type Ia SN cosmology. Whatever the reason, extragalacitc LBV-like eruptions are receiving more attention and we have more examples of them, with the result that the increased number do not support some long-help paradigms about LBVs.

In particular, LBV eruptions were thought to always have cool ∼8000 K pseudo photospheres (Humphreys & Davidson 1994), but this is apparently wrong. Some do indeed exhibit apparent temperatures in this range and have F-supergiant like spectra; the recent transient UGC2773-OT is a good example, and its spectrum is shown in Fig. 5. However, several LBV giant eruptions exhibit hotter temperatures with smooth blue continua and

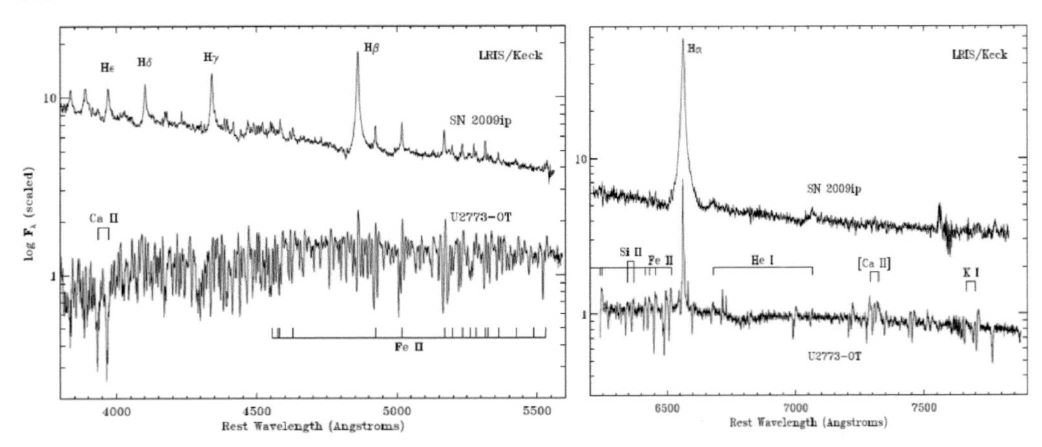

Figure 5. Visual-wavelength spectra of two recent transients, demonstrating the range of spectral properties in LBV outbursts. SN 2009ip is an example of a hotter LBV, with blue continuum and strong and relatively broad Balmer emission lines, plus evidence for a blast wave. UGC 2773-OT is more characteristic of cooler LBV wind-dominated spectra in their F supergiant state. Both are from Smith *et al.* (2010b).

strong Balmer emission lines. Some even show evidence for fast blast waves of 5,000 km s^{-1} out in front of the bulk of ejecta moving at around 600 km s^{-1}. SN 2009ip is an example of this (Figure 5; Smith *et al.* 2010b), as is the more familiar case of η Carinae (Smith 2008). The spectral diversity of these LBV giant eruptions is discussed in more detail by Smith *et al.* (2010a).

An exciting recent development is that some LBVs exhibit multiple brief eruptions, partly as a consequence of continued monitoring. It was already known that both η Car and P Cygni had secondary eruptions about 50 yr after their initial giant eruptions (Humphreys & Davidson 1994; Humphreys *et al.* 1999). However, a recent study by Smith & Frew (2010) shows an even more complicated situation for η Car, with two brief precursor eruptions in 1838 and 1843 that preceded the peak of the Great Eruption in 1844, and which appear to have occurred near times of periastron. Moreover, the LBV-like eruption SN 2000ch (see Wagner *et al.* 2004) was later discovered to have multiple subsequent eruptions in 2008 and 2009 (Pastorello *et al.* 2010). Very recently, Drake *et al.* (2010) reported the discovery of another subsequent eruption of SN 2009ip, which was discussed above. In all cases, the repeated eruptions appear to be very brief (few to 10^2 days), and not the multi-year affairs as seen in more conventional LBV eruptions. The physical cause of these repeated brief outbursts is not yet known, but Smith *et al.* (2010a) and Pastorello *et al.* 2010) have mentioned the possibility of binary interactions, among other potential causes. Smith (2010) discusses a particular way that such a model might work.

Lastly, there is continually mounting evidence that LBV-like eruptions seem to precede the particular class of supernovae known as Type IIn, where the narrow (n) lines of H arise when the SN blast wave encounters extremely dense circumstellar material ejected immediately before the outburst. This has been discussed elsewhere by multiple authors at previous conferences. The point I would like to emphasize here is that the range of properties inferred for the precursor eruptions of SNe IIn seems to roughly match the diversity in properties exhibited by LBV eruptions themselves (expansion speed, mass-loss rates, composition), but there are no other known stars with sufficient mass-loss rates to match SN IIn progenitors. The SN IIn/LBV connection will likely become clearer with

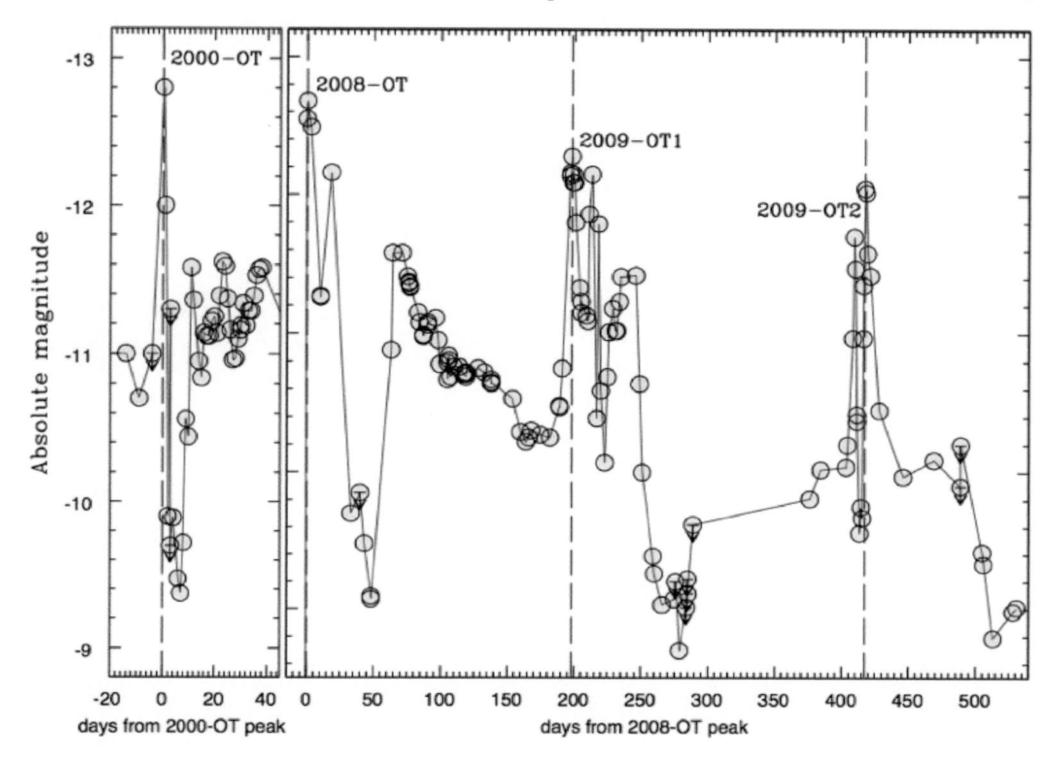

Figure 6. The R-band light curve of the multiple eruptions of SN 2000ch (in 2000, and then again in 2008-2009), from Pastorello *et al.* (2010).

more studies of LBV eruptions and of SNe IIn, and especially cases where LBV progenitor stars are seen to explode as SNe IIn (e.g., Gal-Yam & Leonard 2009).

References

Bohannan, B. 1997, in: A. Nota & H. Lamers (eds.), *Luminous Blue Variables: Massive Stars in Transition*, ASP-CS 120, p. 3

Conti, P. S. 1995, in: K. A. van der Hucht & P. M. Williams (eds.), *Wolf-Rayet Stars: Binaries; Colliding Winds; Evolution*, IAU Symposium 163, p. 565

Conti, P. S. 1997, in: A. Nota & H. Lamers (eds.), *Luminous Blue Variables: Massive Stars in Transition*, ASP-CS 120, p. 387

Drake, A. J., Prieto, J. L., Djorgovski, S. G., Mahabal, A. A. *et al.* 2010, *The Astronomer's Telegram*, 2897

Gal-Yam, A. & Leonard, D. C. 2009, *Nature*, 458, 865

Gogarten, S. M., Dalcanton, J. J., Murphy, J. W., Williams, B. F. *et al.* 2009, *ApJ*, 703, 300

Gvaramadze, V. V., Kniazev, A. Y., & Fabrika, S. 2010, *MNRAS*, 405, 1047

Hubble, E. & Sandage, A. 1953, *ApJ*, 118, 353

Humphreys, R. M. & Davidson, K. 1994, *PASP*, 106, 1025

Humphreys, R. M., Davidson, K., & Smith, N. 1999, *PASP*, 111, 1124

Massey, P., McNeill, R. T., Olsen, K. A. G., Hodge, P. W. *et al.* 2007, *AJ*, 134, 2474

Pastorello, A., Botticella, M. T., Trundle, C., Taubenberger, S. *et al.* 2010, *MNRAS* 408, 181

Prieto, J. L., Kistler, M. D., Thompson, T. A., Yüksel, H. *et al.* 2008, *ApJ* (Letters), 681, L9

Prieto, J. L., Sellgren, K., Thompson, T. A., & Kochanek, C. S. 2009, *ApJ*, 705, 1425

Smith, N. 2008, *Nature*, 455, 201

Smith, N. 2010, *MNRAS*, submitted, arXiv:1010.3770

Smith, N. & Conti, P. S. 2008, *ApJ*, 679, 1467
Smith, N. & Frew, D. J. 2010, *ArXiv e-prints* 1010.3719
Smith, N., Hinkle, K. H., & Ryde, N. 2009, *AJ*, 137, 3558
Smith, N. & Owocki, S. P. 2006, *ApJ* (Letters), 645, L45
Smith, N., Vink, J. S., & de Koter, A. 2004, *ApJ*, 615, 475
Smith, N., Li, W., Foley, R. J., Wheeler, J. C. *et al.* 2007, *ApJ*, 666, 1116
Smith, N., Chornock, R., Li, W., Ganeshalingam, M. *et al.* 2008, *ApJ*, 686, 467
Smith, N., Li, W., Silverman, J., Ganeshalingam, M., & Filippenko, A. 2010a, *MNRAS*, submitted, arXiv:1010.3718
Smith, N., Miller, A., Li, W., Filippenko, A. V. *et al.* 2010b, *AJ*, 139, 1451
Smith, N., Li, W., Filippenko, A. V., & Chornock, R. 2010c, *MNRAS*, in press, arXiv:1006:3899
Thompson, T. A., Prieto, J. L., Stanek, K. Z., Kistler, M. D. *et al.* 2009, *ApJ*, 705, 1364
Wachter, S., Mauerhan, J. C., Van Dyk, S. D., Hoard, D. W. *et al.* 2010, *AJ*, 139, 2330
Wagner, R. M., Vrba, F. J.; Henden, A. A., Canzian, B. *et al.* 2004, *PASP*, 116, 326

Active OB stars: structure, evolution, mass loss, and critical limits
Proceedings IAU Symposium No. 272, 2010
C. Neiner, G. Wade, G. Meynet & G. Peters, eds.

doi:10.1017/S174392131101146X

Massive stars at (very) high energies: γ-ray binaries

Guillaume Dubus and Benoît Cerutti

Laboratoire d'Astrophysique de Grenoble, UMR 5571 Université J. Fourier & CNRS, France
email: gdubus@obs.ujf-grenoble.fr

Abstract. γ-ray binaries are systems that emit most of their radiative power above 1 MeV. They are associated with O or Be stars in orbit with a compact object, possibly a young pulsar. Much like colliding wind binaries, the pulsar generates a relativistic wind that interacts with the stellar wind. The result is non-thermal emission from radio to very high energy γ-rays. The wind, radiation and magnetic field of the massive star play a major role in the dynamics and radiative output of the system. They are particularly important to understand the high energy physics at work. Inversely, γ-ray binaries offer novel probes of stellar winds and insights into the fate of O/B binaries.

Keywords. radiation mechanisms: nonthermal, pulsars: general, stars: winds, outflows, gamma rays: observations, gamma rays: theory, X-rays: binaries

---

High energy (HE) γ-rays have energies above 100 MeV, very high energy γ-rays (VHE) are above 100 GeV. This definition has an instrumental origin: both HE and VHE γ-rays are stopped in the upper atmosphere but the Cherenkov light from the electromagnetic showers initiated by VHE γ-rays can be detected from the ground using large telescopes with fast detectors. Our knowledge of the HE and VHE sky is increasing rapidly thanks to a successful collection of new instruments. In HE γ-rays the *Fermi γ-ray space telescope* has been operating for two years. Its first catalog, published earlier this year, contains 1500 sources when 250 were previously known. In VHE γ-rays, more than 80 sources have been discovered by the arrays of Cherenkov telescopes that have now been operating for 5-7 years (HESS, VERITAS, MAGIC): only a handful of confirmed sources were known in 2003. This bonanza has started to impact many fields of astrophysics, including massive stars, and will hopefully continue in the future with projects like the *Cherenkov Telescope Array* (CTA). Although this paper will be concerned with γ-ray binaries, the recent discoveries of HE emission from stellar clusters like 30 Dor (Abdo *et al.* 2010b) or from the direction of η Car are also susceptible to be of interest to the massive star community.

1. Binaries at high energies

There is, at present (7/2010), four confirmed binaries emitting HE or VHE γ-rays: PSR B1259-63, LS 5039, LS I+61 303 and Cyg X-3 (Holder 2009). Three other candidates can be added: HESS J0632+057 (confirmed, variable, VHE and radio source coincident with the Be star MWC148 but without evidence yet for a compact companion), Cygnus X-1 (well-known black hole candidate orbiting an O star but the VHE γ-ray detection during a brief flare was marginal and has not been independently confirmed; AGILE has also reported marginal HE detections not confirmed by *Fermi*) and the latest addition, V407 Cyg, a symbiotic system (white dwarf orbiting a red giant) where HE γ-rays were detected during a nova outburst (Abdo *et al.* 2010a). V407 Cyg is a nice illustration of the surprises the HE sky has in store for us.

V407 Cyg aside, the other systems are all high mass X-ray binaries where a compact object, neutron star or black hole, orbits a Be or an O star. The four confirmed binaries all show modulations at the orbital period in γ-rays. In LS 5039, the orbital period can be determined independently from the VHE modulation with a precision comparable to that obtained with radial velocities. Optical spectroscopy gives access to the orbital and stellar parameters (see McSwain, these proceedings). It remains the essential first step from which stems all understanding of these systems.

Microquasars. The binaries can be set in two groups. The Cygnus sources are both well-known accreting sources where material from the massive star falls onto the compact object, releasing gravitational energy. The compact object in Cyg X-3 orbits a Wolf-Rayet star every 4.8 hours. Accretion is accompanied by the occasional ejection. Radio flares and X-ray state changes are then seen. The radio emission has been resolved into relativistic jets. γ-ray emission from Cyg X-3 has been detected only during soft X-rays states and high flux radio states, that is when the relativistic jet is known to be present (The Fermi LAT Collaboration, Abdo *et al.* 2009). This implies γ-ray emission is related in some way to the accretion-ejection physics. This holds great promise to further our understanding of the link between non-thermal emission and accretion/ejection, processes that are also at work in AGNs and γ-ray bursts. The detected γ-ray spectral luminosity represents $<$10 % of the X-ray luminosity.

γ-ray binaries. Emission above 1-10 MeV dominates the radiative output in those systems, hence their name (Dubus 2006). Their X-ray luminosities are moderate ($10^{33}-10^{34}$ erg s^{-1}), their optical spectra are dominated by the massive star. All are radio sources and that is rare amongst high-mass X-ray binaries. However, it is really in γ-rays that the unusual character of these objects is revealed and sensitive Galactic Plane surveys carried out by CTA in the future should discover more examples.

2. γ-ray binaries as binary pulsar wind nebulae

Similarities in the spectra and behaviour of γ-ray binaries have led to the suggestion that all may belong to a class of systems different from the accreting microquasars. The key is PSR B1259-63, a 48 ms radio pulsar in a 3.5 year orbit around a Be star. Timing of the radio pulses enables a very good determination of the orbit, but also to measure the spindown rate of the pulsar. The corresponding rotational power $\dot{E}$ that is lost is 8 10^{35} erg s^{-1}, the inferred magnetic field is $\approx 3\ 10^{11}$ G and the spindown timescale is $\approx 3\ 10^{5}$ years. Radio flaring is associated with the crossing of the Be disc. Radio flares are seen in LS I+61 303 (the orbital ephemeris is derived from them) and radio variability has been detected in HESS J0632+057 (Skilton *et al.* 2009); both have Be companions.

Pulsar winds. The magnetic field dominates the immediate environment of the neutron star. Particles accelerated in the magnetosphere are thought to be responsible for the pulsed γ-ray emission seen now from $\approx$ 60 pulsars by *Fermi* (pulsars are the dominant HE population in the Galactic Plane). Particles on the field lines cannot co-rotate with the neutron star beyond the light cylinder radius ($cP_{\rm ns}/2\pi$), where their angular speed attains c. Magnetic field lines open up and a relativistic wind is launched, carrying away the rotational energy of the neutron star. Pulsar winds interacting with the surrounding material (supernova remnant, ISM) create pulsar wind nebulae: the canonical PWN is the Crab nebula but dozens have now been detected in VHE γ-rays (here also PWN are the dominant VHE population in the Galactic Plane). In PSR B1259-63 the interaction occurs with the stellar wind or the disc of the Be star. The other established pulsars with B/Be companions are PSR J1740-3052 and PSR J0045-7319. Unfortunately, they are too far to be detectable in γ rays ($\approx$ 11kpc and SMC when PSR B1259-63 and the other γ-ray binaries are 2 to 3 kpc away).

What about accretion ? Neutron stars in orbit around massive companions can accrete matter from the stellar wind. The Bondi-Hoyle accretion rate onto a compact object moving with speed v in a medium with density ρ is $\dot{M}_{\rm b} = \pi R_{\rm b}^2 \rho v$, where $R_{\rm b} = 2GM/v^2$ is the capture radius. SPH simulations show good agreement with this estimate (Okazaki, these proceedings). The pressure from accretion grows like $R^{-5/2}$ where R is the distance to the neutron star, whereas the pulsar pressure increases like R^{-2}. Accretion will be held off if the pulsar wind pressure is greater than the accretion pressure at the capture radius. This requires $\dot{E} > 4\dot{M}_{\rm b} vc$ or $P_{\rm ns} \leqslant 0.23\ \dot{M}_{15}^{-1/4} B_{12}^{1/2}$ s. The Bondi accretion rate at 0.1 a.u. in a 1000 km/s, 10^{-6} $M_\odot$ yr^{-1} wind (LS 5039) is about 10^{15} g s^{-1}. It can increase dramatically in a Be disc. Still, a fast-spinning neutron star with a strong field, i.e. a young pulsar, can prevent accretion even at Eddington rates (Dubus 2006). As the pulsar spins down on a timescale of a few 10^5 years the pressure from its wind will decrease until accretion pushes its way to the neutron star and a high-mass X-ray binary turns on. γ-ray binaries are a short lived phase following the SN explosion and preceding the HMXB phase, as anticipated in population synthesis of HMXBs that predict fewer than 100 such systems in our Galaxy.

3. The radiation field of the massive star

The γ-ray orbital modulation can happen either because the VHE particles radiate in a phase-dependent manner or because the injection of those particles is phase-dependent. Radiative processes are well known, unlike particle acceleration processes, so these systems offer a unique chance to separate the two. Understanding the orbital phase-dependent effects of radiative processes carries the hope of constraining where and how the VHE particles get their energy. Leptonic models assume these particles are electrons (and positrons). The main radiative processes are synchrotron emission and inverse Compton radiation. Synchrotron radiation cannot produce radiation above $\approx$ 100 MeV. Beyond this limit, synchrotron losses are so quick that the process through which the electron originally gained energy would have to occur on a fraction of the time it takes to turn around the field line, requiring exotic acceleration mechanisms. Therefore, γ-ray radiation is very likely to be inverse Compton (IC) radiation. Models based on high-energy protons (hadronic models) radiating via *pp* or $p\gamma$ interactions have not been much pursued as they require more energy and do not easily explain the modulations.

Inverse Compton modulation The most abundant seed photons are by far those of the massive star (about 10^{14} photons cm^{-3} at the location of the compact object in LS 5039). IC interactions on this external source readily produces an orbital modulation: there is a maximum when the seed photons are backscattered towards the observer (superior conjunction) and a minimum when they are forward-scattered (inferior conjunction). In addition, there can be spectral changes as the interaction occurs in the Klein-Nishina or Thompson IC regime depending on angle (Dubus *et al.* 2008).

Pair production Once emitted, VHE γ rays can have enough energy to create e^+e^- pairs through their interaction with photons from the star ($\gamma + \gamma \rightarrow e^+ + e^-$). The threshold for this reaction is $\epsilon_\gamma \epsilon_\star \geqslant 2(m_e c^2)^2/(1+\cos\psi)$ where ψ is the angle between the two photons. For a stellar temperature of 40,000 K ($\epsilon_\star \approx 9$ eV), this happens when the energy of the γ ray $\epsilon_\gamma \geqslant 30$ GeV. The cross-section is of order of the Thompson cross-section with maximum around a few 100 GeV, in the range of ground-based Cherenkov arrays. Because of the angle dependence, a constant source of VHE γ-rays orbiting the

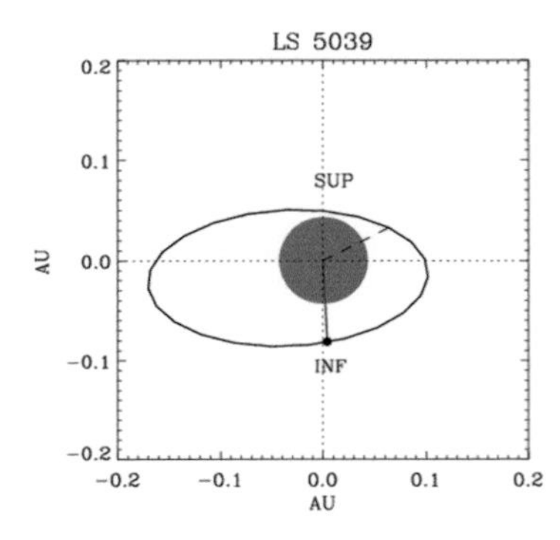

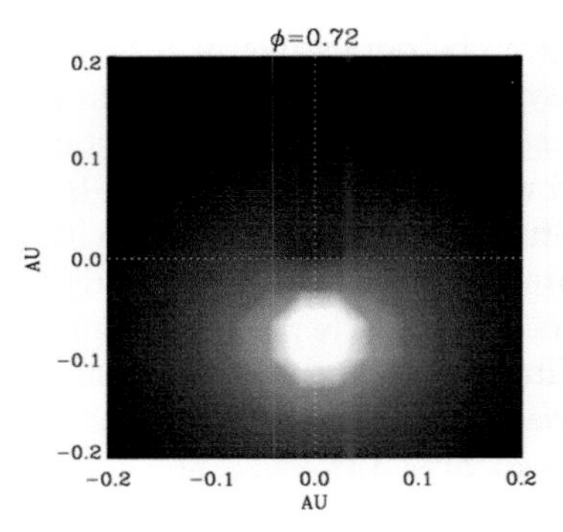

Figure 1. Map of the emission above 100 GeV close to inferior conjunction in LS 5039. This includes inverse Compton emission, pair production and subsequent 3D cascading.

massive star will be modulated by absorption on stellar. Maximum absorption occurs when the two photons interact head-on and minimum when this is tail-on. The absorption maximum will therefore be at superior conjunction and the minimum at inferior conjunction.

Cascade The pairs created when γ rays are 'absorbed' radiate. If they radiate VHE IC emission, the γ rays (of slightly less energy) wills generate new pairs until the energy drops below the threshold for pair production. The geometry and the magnetic field control this cascade emission (Cerutti *et al.* 2010). If the magnetic field is strong the pairs will preferentially radiate synchrotron instead of IC and the cascade process will be stopped. If the magnetic field is very weak then the pairs propagate along the path of the initial γ ray instead of being isotropized (Cerutti *et al.* 2009).

Here, unlike in GRBs or AGNs, the radiation field is set by the massive star. Together with the constrained geometry from the binary orbit, this allows detailed comparisons to be made between models and data. All the processes are at work in LS 5039: in HE there is an IC modulation, modified at VHE by pair production. At superior conjunction, the absorption is less important than expected, suggesting cascading: this requires the magnetic field to be $\leqslant 6$ G in the VHE emitting region. The O star should not have a strong magnetic field. *Fermi* observations of LS 5039 and LS I+61 303 show power law spectra with exponential cutoffs at a few GeV (akin to the *Fermi* pulsars), implying the particle population radiating HE γ rays is different from the one radiating VHE γ rays. In the Be systems the phases of peak emission are difficult to explain with a purely radiative model and other effects are suspected. The interaction with the Be disc can lead to strong orbital variations in particle injection, escape or adiabatic cooling (see also van Soelen, these proceedings). γ-ray binaries are a window into particle acceleration under varying, but controlled, conditions.

4. The stellar wind of the massive star

There is now a reasonable understanding of the radiative processes at work in γ-ray binaries and their dependence with orbital phase (Bosch-Ramon & Khangulyan 2009). There is much less understanding of the dynamics of the interaction between the pulsar wind and the stellar wind. This is accessible through different means.

Absorption in the stellar wind can affect radio and X-ray emission. The free free optical depth at 1 GHz at a distance d from the star is $\tau \approx 3 \cdot 10^4 \dot{M}^2_{\mathrm{w},7} v^{-2}_{w,2000} T_4^{-3/2} \nu^{-2}_{\mathrm{GHz}} d^{-3}_{0.1}$ for a 10^{-7} M$_\odot$ yr^{-1} coasting wind with a speed 2000 km s^{-1} and temperature 10^4 K; d is 0.1 AU, on the order of the orbital separation in LS 5039 and LS I+61 303. This makes it a priori impossible to detect a radio pulsar in these tight systems. PSR B1259-63 has a wider orbit but the radio pulsations are eclipsed for more than a month around periastron. The study of the eclipse show the absorption is best explained as due to the Be disc tilted with respect to the orbital plane Melatos *et al.* (1995). The absorption column N_H in X-rays could also probe the wind interface. There is an orbital modulation of the X-rays from LS 5039 but the observed N_H stays constant. This severely constraints the size of the X-ray source or the stellar wind density: the X-ray source should be $\geqslant 3\ R_\star$ if the stellar wind mass loss rate is 10^{-7} $M_\odot$ yr^{-1} (Szostek & Dubus, submitted).

Variability in the stellar wind can also affect the γ-ray emission. One link is the radio outbursts of PSR B1259-63 (and the other γ-ray binaries with Be companions ?) that are thought to be related to the passage of the pulsar through the Be disc. There is evidence for a long term modulation on a 4 year timescale from the ephemerides of radio outburst in LS I+61 303 as well as from optical lines (Zamanov *et al.* 1999). Be disc precession might be the cause. The orbit-averaged HE γ-ray emission has recently been observed to rise in *Fermi*, at a time in the 4 year cycle when the Hα EW and ΔV is large – suggesting a large Be disc. However, contemporary optical studies shown at this meeting did not report anything unusual. X-ray variability on timescales of minutes has also been reported in LS I +61 303. This has yet to be understood: possibilities include inhomogeneities in the stellar wind/Be disc leading to fluctuations in the wind interface, instabilities at the interface, accretion events or some unrelated source in the field-of-view. Magnetar-like activity has also been reported from LS I+61 303. The IR-optical band gives access to the stellar wind or Be disc and campaigns should include such observations to shed light on these issues.

Morphology of the pulsar wind - stellar wind interaction is accessible via radio observations. VLBI observations of LS I+61 303 (Dhawan *et al.* 2006) and LS 5039 (Ribo *et al.* 2008) show resolved, collimated radio emission on scales of milli-arcseconds (a.u. scales; unfortunately there is no data available for PSR B1259-63). These initially suggested relativistic jets as observed in Cygnus X-3 and other microquasars. However, the position angle of this emission changes with orbital phase. Furthermore, the changes in morphology are reproducible from orbit to orbit. Whereas this is very difficult to explain with a microquasar jet, whose orientation is not expected to change during one orbit, this is rather natural to expect if a pulsar wind is collimated into a comet-like trail by a strong stellar wind. Whether this happens depends on the ratio of wind momentum fluxes, as in colliding wind binaries (Bogovalov *et al.* 2008). The large scale expectation for isotropic winds is an Archimedean spiral with a step $v_w P_{\rm orb}$ where v_w is the terminal speed of the strongest wind (Lamberts, these proc.). What happens when the interaction is with a Be disc is unclear (but see Owocki, Okazaki, these proceedings). Numerical simulations using relativistic MHD are required to obtain the evolution of the magnetic field and speed in the shocked region together with the large scale morphology in order to reproduce the VLBI radio maps.

5. Conclusion: massive stars in extreme environments

More interaction is required between the HE and massive star communities. Many of the techniques used in colliding wind binaries can be applied to γ-ray binaries. The conditions in the wind and constraints on dynamics derived from optical spectra need to be looked at carefully (McSwain, these proceedings). Clumping may have noticeable effects although supergiant fast X-ray transients (a class of HMXBs) are probably better probes for this than γ-ray binaries. The massive star may be substantially modified by this extreme environment. For instance, calculations show the temperature in the wind may be increased to 10^5 K by the γ-rays (Zdziarski *et al.* 2010): what are the consequences on wind acceleration ? If the pulsar wind is strong it may impact directly onto the surface of its companion (as in black widow pulsars, old recycled ms pulsars with very low mass companions): what are the observable consequences, a dearth of UV lines at some phases?

Acknowledgements

We thank the OC for the smooth organization of the conference. This work was supported by the European Community via contract ERC-StG-200911.

References

Abdo, A. A., Ackermann, M., Ajello, M., Atwood, W. B. *et al.* 2010a, *Science*, 329, 817
Abdo, A. A., Ackermann, M., Ajello, M., Atwood, W. B. *et al.* 2010b, *A&A*, 512, A7
The Fermi LAT Collaboration, Abdo, A. A., Ackermann, M., Ajello, M. *et al.* 2009, *Science*, 326, 1512
Bogovalov, S. V., Khangulyan, D. V., Koldoba, A. V., Ustyugova, G. V. *et al.* 2008, *MNRAS*, 387, 63
Bosch-Ramon, V. & Khangulyan, D. 2009, *International Journal of Modern Physics D*, 18, 347
Cerutti, B., Dubus, G., & Henri, G. 2009, *A&A*, 507, 1217
Cerutti, B., Malzac, J., Dubus, G., & Henri, G. 2010, *A&A*, 519A, 81
Dhawan, V., Mioduszewski, A., & Rupen, M. 2006, in: *VI Microquasar Workshop: Microquasars and Beyond*, p. 52
Dubus, G. 2006, *A&A*, 456, 801
Dubus, G., Cerutti, B., & Henri, G. 2008, *A&A*, 477, 691
Holder, J. 2009, ArXiv e-prints 0912.4781
Melatos, A., Johnston, S., & Melrose, D. B. 1995, *MNRAS*, 275, 381
Ribó, M., Paredes, J. M., Moldón, J., Martí, J. *et al.* 2008, *A&A*, 481, 17
Skilton, J. L., Pandey-Pommier, M., Hinton, J. A., Cheung, C. C. *et al.* 2009, *MNRAS*, 399, 317
Zamanov, R. K., Martí, J., Paredes, J. M., Fabregat, J. *et al.* 1999, *A&A*, 351, 543
Zdziarski, A. A., Neronov, A., & Chernyakova, M. 2010, *MNRAS*, 403, 1873

Discussion

McSwain: I just wanted to comment on the properties of the Be star in HESS J0632+057. The Be star has a temperature of 30,000 K and log g about 4.0. We have place constraints on the orbital period and find that it cannot be less than 100 days. These results should appear soon in a paper by Aragona *et al.*

Dubus: This is good to know, a long period could explain the VHE variability. Spectroscopic studies are crucial to understand what is going on.

Okasaki: You mentioned the possibility that a large Be disc eclipses a radio pulsar so that the pulsar is not observed in systems other than PSR B1259-63. However, the Be disc can't be that big if the compact object is a radio pulsar with relativistic wind. Therefore, if there is a pulsar in these systems, it should be detected.

Dubus: The other systems have tighter orbits than PSR B1259-63. The radio pulsar is eclipsed in PSR B1259-63 when it goes round periastron: it is then at a distance from its Be companion that is still greater than the maximum separation in the other systems. Therefore, I don't think the environment will be less dense than for PSR B1259-63 even if the Be disc is not as large. Calculations (Dubus 2006, Zdziarski *et al.* 2010) show the free-free opacity is very large at GHz frequencies. It will be useful if you predict that there are phases with significantly less absorption.

Active OB stars: structure, evolution, mass loss, and critical limits
Proceedings IAU Symposium No. 272, 2010
C. Neiner, G. Wade, G. Meynet & G. Peters, eds.

doi:10.1017/S1743921311011471

Modeling TeV γ-rays from LS 5039: an active OB star at the extreme

Stan P. Owocki[1], **Atsuo T. Okazaki**[2] **and Gustavo Romero**[2]

[1]Bartol Research Institute, Department of Physics & Astronomy, University of Delaware
Newark, DE 19716, USA
email: owocki@udel.edu

[2]Faculty of Engineering, Hokkai-Gakuen University
Toyohira-ku, Sapporo 062-8605, Japan
email: okazaki@elsa.hokkai-s-u.ac.jp

[3]Facultad de Ciencias Astronómicas y Geofísicas, Universidad Nacional de La Plata
Paseo del Bosque, 1900 La Plata, Argentina
email: romero@fcaglp.unlp.edu.ar

Abstract. Perhpas the most extreme examples of "Active OB stars" are the subset of high-mass X-ray binaries – consisting of an OB star plus compact companion – that have recently been observed by *Fermi* and ground-based Cerenkov telescopes like *HESS* to be sources of very high energy (VHE; up to 30 TeV!) γ-rays. This paper focuses on the prominent γ-ray source, LS5039, which consists of a massive O6.5V star in a 3.9-day-period, mildly elliptical ($e \approx 0.24$) orbit with its companion, assumed here to be a black-hole or unmagnetized neutron star. Using 3-D SPH simulations of the Bondi-Hoyle accretion of the O-star wind onto the companion, we find that the orbital phase variation of the accretion follows very closely the simple Bondi-Hoyle-Lyttleton (BHL) rate for the local radius and wind speed. Moreover, a simple model, wherein intrinsic emission of γ-rays is assumed to track this accretion rate, reproduces quite well *Fermi* observations of the phase variation of γ-rays in the energy range 0.1-10 GeV. However for the VHE (0.1-30 TeV) radiation observed by the *HESS* Cerenkov telescope, it is important to account also for photon-photon interactions between the γ-rays and the stellar optical/UV radiation, which effectively attenuates much of the strong emission near periastron. When this is included, we find that this simple BHL accretion model also quite thus making it a strong alternative to the pulsar-wind-shock models commonly invoked to explain such VHE γ-ray emission in massive-star binaries.

1. Introduction

Among the most active OB stars are those in high-mass X-ray binary (HMXB) systems, in which interaction of the massive-star wind with a close compact companion – either a neutron star or black hole – produces hard (> 10 keV) X-ray emission with characteristic, regular modulation over the orbital period. In recent years a small subset of such HMXB's have been found also to be gamma-ray sources, with energies up to ~10 GeV observed by orbiting gamma-ray observatories like *Fermi*, and very-high-energies (VHE) of over a TeV (10^{12} eV) seen by ground-based Cerenkov telescopes like *HESS*, *Veritas*, and *Magic*. For the one case, known as B 1259-63, in which oberved radio pulses show the companion to be a pulsar, the $\gamma-$ray emission seems best explained by a *Pulsar-Wind-Shock* (PWS) model, wherein the interaction of the relativistic pulsar wind with the dense wind of the massive-star produces strong shocks that accelerate electrons to very high energies, with inverse-Compton scattering of the stellar light by these high-energy electrons then producing VHE γ-rays. For the other gamma-ray binaries, pulses have not been detected, and the nature of the companion, and the applicability of the PWS model, are less clear.

Table 1. Model parameters for LS 5039

	Primary	Secondary
Spectral type	O6.5V	Compact
Mass ($M_\odot$)	22.9[a]	3.7[a]
Radius	$9.3R_\odot^{a}$ $(= 0.31a)$	$2.5 \times 10^{-3}a$
Effective temperature $T_{\rm eff}$ (K)	39,000[a]	–
Wind terminal speed V_∞ (km s^{-1})	2440	–
Mass loss rate $\dot{M}_*$ ($M_\odot$ yr^{-1})	5×10^{-7} [a]	–
Orbital period $P_{\rm orb}$ (d)	3.9060[a]	
Orbital eccentricity e	0.24[b]	
Semi-major axis a (cm)	2.17×10^{12}	

[a] Casares *et al.* (2005) [b] Szalai *et al.* (2010)

An alternative*MicroQuasar* (MQ) model instead posits that accretion of circumstellar and/or wind material from the massive star onto the companion – assumed now to be either a black hole or a weakly magnetized neutron star – powers a jet of either relativistic protons, which interact with stellar wind protons to produce pions that quickly decay into γ-rays, or relativistic pair plasma, which inverse Compton scatters stellar photons to γ-rays. A poster paper by Okazaki *et al.* in these proceedings uses Smoothed Particle Hydrodyanmics (SPH) simulations to examine both the PWS and MQ models, applying them respectively for two systems, B 1259-63 and LSI +61 303, in which the massive star is a Be star. In such Be binary systems, the compact companion can interact with the Be star's low-density polar wind, and its dense equatorial decretion disk, and so modeling both types of interaction is key to to understanding their high-energy emission.

In the paper here, we apply the same SPH code to a MQ model for the TeV-binary LS 5039, for which the (non-pulsing) compact companion is in a moderately eccentric ($e \sim 0.24$), 3.9-day orbit around a massive, non-Be primary star of spectral type O6.5V. (See Table 1 for full parameters.) This builds on our previous study (Okazaki *et al.* 2008b) to account now for a fixed wind acceleration, instead of just assuming a constant wind speed. A key result is that the orbital variation of the 3-D SPH accretion rate of the stellar wind flow onto the compact companion follows very closely the analytic Bondi-Hoyle-Lyttleton (BHL) rate that depends on the wind speed and orbital separation at each phase. (See right panel of Figure 1.) Assuming the mass accretion translates promptly into jet power and thus γ-ray emission, we then apply this result to derive predicted light curves at both the GeV energies oberved by *Fermi* and the TeV energies observed by *HESS*. For the former, we find that assuming GeV γ-ray emission tracks closely the BHL accretion rate gives directly a quite good fit to the *Fermi* lightcurve (Abdo *et al.* 2009, see left panel of Figure 3 below). But for the latter case one must also account for the attenuation of emitted TeV γ-rays by photon-photon interaction with the optical and UV radiation of the massive star. When this is included, then the predicted TeV lightcurve also closely matches the *HESS* observations (Aharonian *et al.* 2006, see right panel of Figure 3 below).

2. Bondi-Hoyle-Lyttleton (BHL) Accretion

To provide a physical context for modeling accretion in LS 5039, it is helpful first to review briefly the basic scalings for accretion of an incoming flow onto a gravitating body, as analyzed in pioneering studies by Bondi, Hoyle and Lyttleton (BHL) (see Edgar 2004, for references and a modern review). The left panel of Figure 1 illustrates the basic

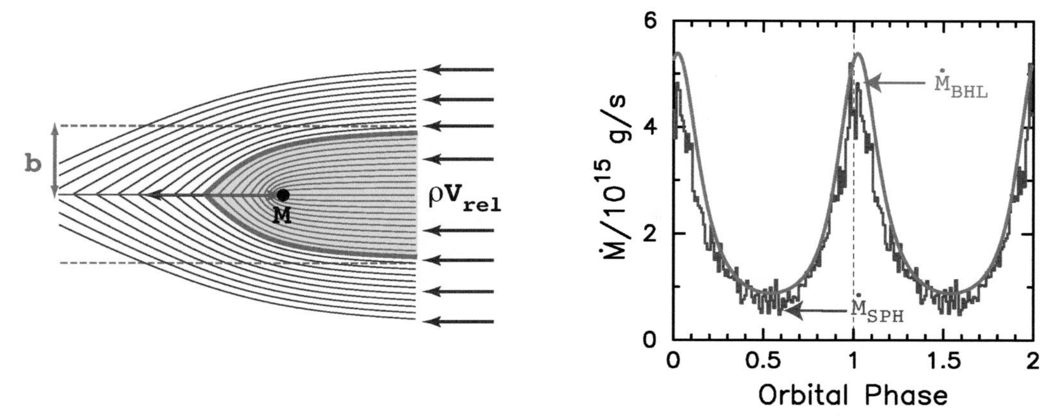

Figure 1. *Left:* Illustration of flow streams for simple planar form of Bondi-Hoyle accretion. The gravitational attraction of mass M focuses an inflow with relative speed V_{rel} onto the horizontal symmetry axis. This causes material impacting within a BHL radius $b = 2GM/V_{rel}^2$ of the axis to be accreted onto M, as outlined here by the bold (red) curve that bounds the lightly shaded region representing accreting material. *Right:* Orbital phase variation of the mass-accretion rate from 3-D SPH simulations (blue), compared to the analytic BHL formula from eqn. (2.2) (red) assuming the parameters given in Table 1.

model. An initially laminar flow with density ρ and relative speed V_{rel} is focussed by the gravity of a mass M onto the downstream side of the flow symmetry axis, whereupon the velocity component normal to the axis is cancelled by collisions among the streams from different azimuths. Material with initial impact parameter equal to a critical value $b \equiv 2GM/V_{rel}^2$ – now typically dubbed the BHL radius, and highlighted by the (red) heavy curve in Figure 1 – arrives on the axis a distance b from the mass, with a parallel flow energy $V_{rel}^2/2$ just equal to the gravitational binding energy GM/b. Since all material with an impact $< b$ (i.e. within the shaded area) should thus have negative total axial energy, it should be eventually accreted onto the central object. This leads to a simple BHL formula for the expected mass accretion rate,

$$\dot{M}_{BHL} = \rho V_{rel} \pi b^2 = \frac{4\pi\rho G^2 M^2}{V_{rel}^3} . \tag{2.1}$$

In a binary system like LS 5039 there arise additional effects from orbital motion and the associated coriolis terms. But if we ignore these and other complexities, we can use eqn. (2.1) to estimate how the accretion rate should change as a result of the changes in binary separation r over the system's elliptical orbit,

$$\dot{M}_{BHL} = \dot{M}_w \frac{G^2 M^2}{r^2 V_w V_{rel}^3} , \tag{2.2}$$

where $\dot{M}_w \equiv 4\pi r^2 \rho(r) V_w(r)$ is the wind mass loss rate, and $V_{rel}(r)$ is the local magnitude of the relative wind and orbital velocity vectors. We assume here that the stellar wind follows a standard 'beta-type' velocity law, $V_w(r) = V_\infty(1 - R_*/r)^\beta$, where R_* is the O-star radius, V_∞ is the wind terminal speed, and we adopt here a standard velocity power index $\beta = 1$.

For the LS 5039 system parameters given in Table 1, the smooth (red) curve in the right panel of Figure 1 plots the orbital phase variation of this analytically predicted BHL-wind-accretion rate. A principal result of this paper is that this is in remarkably close agreement with the jagged (blue) curve, representing the corresponding accretion rate from full, 3-D SPH simulations, the details of which we discuss next.

3. SPH Simulations

The SPH code used here is based on a version originally developed by Benz *et al.* (1990) and Bate *et al.* (1995), with recent extensions to model interacting binaries by Okazaki *et al.* (2008a). It uses a variable smoothing length and integrates the SPH equations with the standard cubic-spline kernel using individual time steps for each particle. The artificial viscosity parameters are set to standard values of $\alpha_{\rm SPH} = 1$ and $\beta_{\rm SPH} = 2$. In the implementation here, the O-star wind is modeled by an ensemble of isothermal gas particles of negligible mass, while the compact object is represented by a sink particle of mass M and radius $R_{acc} = 5 \times 10^9$ cm, i.e. much larger than the $\sim 10^6$ cm radius of a compact object, but still a factor 10 smaller than the minimum (periastron) value of the BHL accretion radius b; if SPH particles fall within this accretion sphere, they are removed from the simulation. The O star's mass exerts a gravitational pull on the binary companion, and the net force of radiative driving vs. gravity of the O-star leads to the outward wind acceleration characterized by the assumed beta=1 velocity law. In addition, the individual SPH particles feel the gravitational attraction from the compact companion. Finally, to optimize the resolution and computational efficiency of our simulations, the wind particles are ejected only in a narrow range of azimuthal and vertical angles toward the companion; figure 1 of Okazaki *et al.* (2008b) shows that this gives quite similar accretion flow structure to what is obtained in a full, spherically symmetric model, with however many fewer required particles.

For the system parameters listed in Table 1, Figure 2 illustrates the nature of the accretion in these SPH simulations, using a time snapshot near periastron, with phase $\phi = 0.06$. The left panel shows the overall, orbitally deflected wind stream that flows radially away from the O-star toward the companion, along with the narrow, dense, gravitationally focussed wind-stream in its wake. The right panel zooms in on this wake on a scale within a BHL accretion radius of the companion, as denoted by the dashed white circle arc. The white arrows show that, within the portions of this dense focal stream wake nearest the companion, the flow becomes directed toward the black sphere with assumed accretion radius R_{acc}.

Averaged over some detailed variations, the overall process of accretion in this fully 3-D SPH simulation is thus remarkably similar to the simple laminar flow picture illustrated in the left panel of Figure 1. The right panel of Figure 1 shows moreover that the SPH accretion rate – averaged over a phase interval 0.01 to smooth over rapid fluctuations – also agrees remarkably well with the simple BHL rate given by eqn. (2.2).

4. Accretion-powered γ-ray emission

In a microquasar model, the accretion onto the compact companion powers a high-energy jet that produces γ-rays. Since the jet acceleration occurs on a scale of a few tens of compact companion radii, it seems reasonble to postulate that the γ-ray emission could promptly track the accretion rate. The left panel of Figure 3 compares the *Fermi* lightcurve for LS 5039 at energies 0.1-10 GeV with a simple emission model that scales directly with the BHL accretion rate. The model fits the data quite well, reproducing

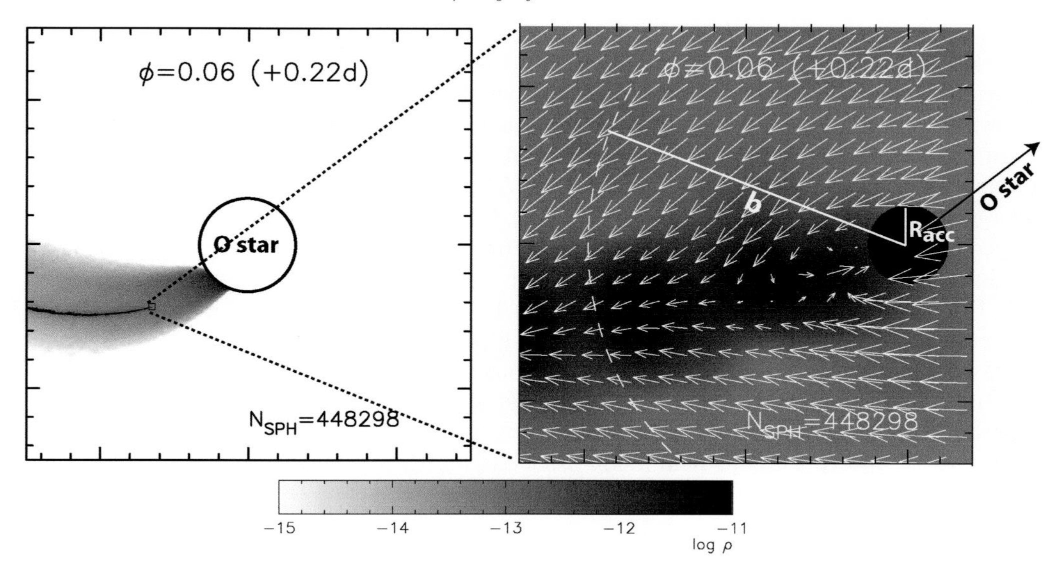

Figure 2. Snapshots near periastron in SPH simulation of LS 5039, with the grayscale showing log of the density (in g/cm^3), and the arrows denoting flow direction and speed. The left panel gives an overall view of the O-star wind stream in the direction of the orbiting companion, which focuses the impingent wind flow into a narrow, dense, downstream wake. The right panel zooms in on this wake at a much smaller scale, within a BHL accretion radius b (marked by the dashed white circle arc) from the companion (represented by the black sphere of assumed accretion radius R_{acc}). The white arrows show how dense material close to the companion and along the gravitational focus axis forms an accretion stream onto the companion, much as in the simple laminar flow picture illustrated in the left panel of Figure 1.

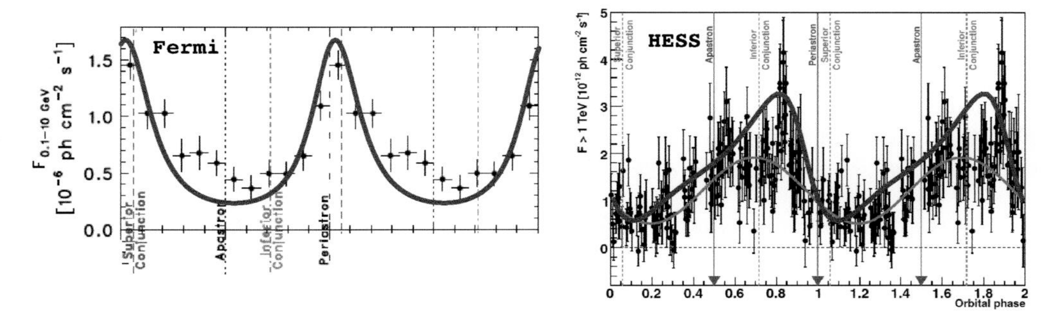

Figure 3. LS 5039 light curves at energies 0.1-10 GeV observed by *Fermi* (Abdo *et al.* 2009, left), and at energies above 1 TeV observed by *HESS* (Aharonian *et al.* 2006, right), plotted vs. orbital phase, and compared against the model results based on BHL accretion, shown as heavy solid (blue) curves. (The lighter (red) curve represents the best-fit sinusoid for the *HESS* data.)

both the roughly factor five variation range of observed γ-rays, as well as the observed flux peak at a phase very near periastron†.

By contrast, the right panel of Figure 3 shows that the TeV γ-rays observed from LS 5039 by *HESS* peak well before periastron, toward the phase associated with inferior conjunction, with the flux minimum occuring just after periastron, near the phase for *superior* conjunction. These phase shifts in the peak and minimum can be explained by

† If instead of an eccentricity $e = 0.24$, we use the larger value $e = 0.35$ suggested by Casares *et al.* (2005), the amplitude of orbital variation in mass accretion rate then becomes stronger than this factor five γ-ray variation observed by *Fermi*.

accounting for the attenuation of the γ-ray rays through photon-photon interaction with the radiation from the O-star. For TeV γ-rays, the geometric mean of the energies of the γ-rays and stellar UV/optical photons exceeds twice the rest-mass-energy of electrons, thus allowing photon-photon production of electron/positron pairs. The thick (blue) curve in Figure 3 shows that a simple model combining a BHL-rate emission with such photon-photon attenuation does indeed match the *HESS* data very well. In contrast, for GeV energy γ-rays the geometric mean with stellar UV/optical light generally falls below the electron mass-energy, and so the *Fermi* lightcurve is unaffected by such attentuation.

Since the cross section for photon-photon is highest near the threshhold of about 0.1 TeV, a pure attenuation model predicts that spectrum should be hardest during the superior conjunction phase of minimum flux and maximum attenuation. However, plots of the photon index for TeV observations by *HESS* show just the opposite, with the minimum flux corresponding to the softest spectum. We are currently investigating whether this can be explained by accounting for the progressive softening of γ-rays during a cascade of absorption and reemission associated with pair production and anniliation.

References

Abdo, A. A., Ackermann, M., Ajello, M., Atwood, W. B. *et al.* 2009, *ApJ* (Letters), 706, L56

Aharonian, F., Akhperjanian, A. G., Bazer-Bachi, A. R., Beilicke, M. *et al.* 2006, *A&A* 460, 743

Bate, M. R., Bonnell, I. A., & Price, N. M. 1995, *MNRAS* 277, 362

Benz, W., Cameron, A. G. W., Press, W. H., & Bowers, R. L. 1990, *ApJ* 348, 647

Casares, J., Ribó, M., Ribas, I., Paredes, J. M. *et al.* 2005, *MNRAS* 364, 899

Edgar, R. 2004, *New Astron. Revs*, 48, 843

Okazaki, A. T., Owocki, S. P., Russell, C. M. P., & Corcoran, M. F. 2008a, *MNRAS* 388, L39

Okazaki, A. T., Romero, G. E., & Owocki, S. P. 2008b, in: *Proceedings of the 7th INTEGRAL Workshop*, p. 74

Szalai, T., Kiss, L. L., & Sarty, G. E. 2010, *Journal of Physics Conference Series*, 218, p. 012028

Discussion

Doug Gies: In Bondi-Hoyle accretion, the gas falls towards the black hole and presumably forms a small accretion disk that powers the relativistic jets. In your model, how long does the gas reside in the disk before being launched into the jets?

Stan Owocki: Our modeling thus far effectively assumes this residence time is short, but this requires further investigation through a more detailed model of the accretion powering of the jet.

Guillaume Dubus: I know of no accretion model showing that gamma ray emission from accretion can promptly follow the orbital variation of the BHL accretion rate. Also, how can the MQ model explain the low fraction of gamma-ray sources among X-ray binaries? The PWS model explains this through the fact they exist for only a short time in a state that can give the required relativistic wind. I thus believe such PWS models are much favored over an MQ model.

Stan Owocki: Our co-author on this paper, Gustavo Romero, who could not attend this meeting, has voiced to us similarly strong arguments in favor of the MQ model over the PWS model for LS5039. I expect there will be a vigorous debate on these issues at the upcoming gamma-ray meeting in Heidelberg this December.

Active OB stars: structure, evolution, mass loss, and critical limits
Proceedings IAU Symposium No. 272, 2010
C. Neiner, G. Wade, G. Meynet & G. Peters, eds.

doi:10.1017/S1743921311011483

Discussion – Normal and active OB stars as extreme condition test beds

summarized by Marc Gagné[1] and Eduardo Janot Pacheco[2]

[1]West Chester University, USA
[2]Universidade de Sao Paulo, Brazil

Discussion

Guillaume Dubus: As far as I can tell your model is exactly the same as standard models for high-mass x-ray binaries, the on difference is that you have a mechanism for generating γ rays. My question is, we know that there are dozens of other systems that are just LS 5039, so why are they not emitting γ rays?

Stan Owocki: I don't know. Atsuo, do you know?

Atsuo Okazaki: With this accretion rate, you have some accreting power so then a jet is emitted. Of course we cannot know why only these four systems show persistent TeV emission.

Dubus: But we *should* know.

Doug Gies: I have a comment and then a question. In Cyg X-1 when the black hole is seen through the wind of the supergiant, you do see these little blips the in the X-ray light curve, so this is a fantastic way of deducing the clumping in the O supergiant.

Owocki: Cyg X-1 did have some high-energy emission, but I don't think it reached TeV energies, and was only a brief detection with Fermi. This is something that Rich Townsend, David Cohen, and I were going to try to model.

Gies: The question I had was about Bondi-Hoyle accretion. In the picture that you showed of Atsuo's model, you could see matter going in, and the scale was too small a scale to follow. But the cartoon showed that there was an accretion disk. So any idea how much time would you need to build up an accretion disk? To make jets, you would probably need an accretion disk.

Owocki: I put the accretion disk there to provide some contrast to the black hole. In the innermost scale, the accretion goes through some innermost scale, but we don't resolve it.

Okazaki: The specific angular momentum of the accreting matter is very, very small. The circularization radius is 10^{-5} to 10^{-6} times the simulation axis, so it's very close to the black, I think.

Owocki: That's why the accretion time scale is very small [compared to the simulation time]. I'm not a modeler of accretion at this point, but maybe I'll become one, I don't know. So I can't really answer the question, but it's a simplifying assumption we're making at this point just to get a result out. But the time scales are very short because the accretion region is on the size scale of a back hole radius.

JOACHIM PULS: In your last cartoon picture of a jet interacting with a clump, wouldn't the jet destroy the clump?

OWOCKI: Yes, but in the meantime, it makes γ rays. You don't care if it destroys the clump. You just want to see if the interaction with wind of the high-energy particles from the jet will make γ rays. That was a big objection the referee had and we had to go back and forth with him a couple of times.

PULS: Maybe I was the referee!

THOMAS RIVINIUS: Joachim mentioned that he has only two stars [HD 64760 and γ Ara, that are] rapidly rotating B supergiants. Just from my observational experience, I would like to add another one that is usually classified as a classical Be star. I took a spectrum of it, and decided I'm not interested, and put it in a drawer. I should have known it was interesting. It is κ Aquilae, which if you look at its optical spectrum and its IUE spectrum, you see that it is identical to γ Ara: it really is a rapidly rotating B0 Ib, with these two little emission bumps left and right. So it would be interesting to look at.

MYRON SMITH: I want to compliment Maria Fernanda-Nieva's work that she displayed for us this morning on getting fundamental parameters for normal B stars and her spectral synthesis work. I also wanted to call her attention to the publication of a B-star far-UV spectral atlas last winter, in which I look at *HST*, *IUE*, *FUSE* and *Copernicus* spectra, and the main point here is not the atlas itself, but the line identification work I did for three spectral types. I identified every known line from the atomic line databases. This would be a very good tool for her to use to extend her work into the UV, to see if God knows she can identify any continuum.

MARIA-FERNANDA NIEVA: Yes, actually, when I started to update the model atoms for the NLTE spectral synthesis, we already started to do the comparisons in the UV, but this is something we need to continue. But the problem in the UV is that you have so many lines that it is difficult to set the continuum. So you really have to model all the lines. In the visual it's easy because you have individual lines, and you can model the abundances independently, but here we have to model everything together.

M. SMITH: Let me add in that case that the intent of the line identification is to only use those lines and no others, because the others make insignificant contributions to the FUV spectra.

GIES: I had a question for Nathan Smith about η Carinae. Nathan in your talk you mentioned this very intriguing possibility that the binary got very close at periastron and dove right into the other star just prior to the great eruption. I was just curious that you said that no merger occurred, but on the other hand it's a fast rotator. Do you think angular momentum might have been transferred?

NATHAN SMITH: Sure.

GAGNÉ: You really want to go home, don't you? Perhaps we can go back to the first talk we heard this morning. Joachim, you told us we should go back and look at γ Ara. Can you remind us why this star is important?

PULS: So it seems that my message didn't come completely along. There are two things actually going on in this rotating star business: the first is this polar versus equatorial enhancement of mass loss. Most stars appear not to have this enhancement. The other question is now when we get close to the $\Omega - \Gamma$ limit where the prediction is that the total mass-loss is increasing, this is important for those people doing stellar evolution and this has not been measured or observed so far, and there is only a very small chance to see this really because you need to see a certain signal above a normal mass-loss rate, and only these two stars, and one other perhaps, have the properties, according to scaling laws, to be observable.

GAGNÉ: And what is the luminosity of γ Ara?

PULS: Just a regular B supergiant, $\log L$ of 5 or 5.5.

SERGIO SIMON-DIAZ: I would like to stress again what Fernanda nicely showed us in her talk about the determination of the parameters in the B stars and abundance, because this can affect our interpretation about these stars. For example, I recently analyzed the stars in the Orion OB1 association. These are four subgroups with different ages. In the past, the abundances were obtained and it was found that the youngest stars near the Orion Nebula were oxygen rich, and this was interpreted as a signature of supernova contamination of the new generation of stars, and so on and so forth. Now, taking into account all these details for the abundances, finally I find that the abundances are homogeneous, so there is no trend with oxygen abundance. So, the details of the analysis are important for our interpretation. I want to stress again the importance of doing things properly.

GAGNÉ: In your opinion, why were the previous abundance determinations incorrect?

SIMON-DIAZ: Because the stellar parameters were derived using photometric indices, and the stellar atmosphere models were not as good as those we have now. These two things together led to the apparent oxygen abundance effect.

NIEVA: I just want to comment on the previous work by Cunha & Lambert that found the abundance anomalies. I did part of my Ph.D. with Katia Cunha, so I know how the work was done in detail. This was my starting point in improving the parameter and abundance determinations of B stars. There are many factors actually. Mainly $\log g$ was derived using Hγ profiles in LTE. You get different $\log g$ values in non-LTE. For $T_{\rm eff}$ you get different values if you use non-ionization equilibria, instead of photometric indices. Many factors contribute: the line selection, the input atomic data of the model atoms. There are many sources of systematic error. It is so difficult to trace back problems in the whole work. They did their best at the time and now there are better techniques, atomic data, computers, so we can do a better job.

GAGNÉ: I think one of the things I took away from your talk though was that there was a rather narrow abundance range that is somewhat consistent with the solar neighbourhood abundance. So, if we are studying the OB stars in open clusters, can we use the the take the effective temperatures and bolometric corrections that you've determined for a given spectral type, and bootstrap them onto our photometric samples, provided we have spectral types?

NIEVA: Actually, I've written a paper on this. The conclusion is that photometry needs some kind of calibration. For example you could use the photometric indices from my multi-ionization equilibria effective temperature calibration, but you have different trends for different evolutionary states of the stars. The ZAMS calibration has one curve but if the star is more evolved, you'll have a different curve. But when you have begin analyzing one star, you don't know $\log g$; you don't know its evolutionary state, so you make an initial guess about $T_{\rm eff}$ and $\log g$. One conclusion of the paper is that you can make a first guess, but you have to take into account the error bar. Whether or not you should interpolate $T_{\rm eff}$ and $\log g$ depends on the application. For example, will errors at the 20 percent level provide useful constraints? In our work, we needed something like 1 percent accuracy. So, it depends on the application, I think.

PULS: I think that future developments in all these respects will provide more and more spectral libraries available, mostly on the net. We are just now finalizing these new developments and though we might not publish all the line lists we use ourselves, we could make available grids for the most commonly used lines, with respect to a $T_{\rm eff}$ and $\log g$. So if an observer can at least get decent spectroscopy of a couple of those stars then you can easily derive these parameters.

NIEVA: The tools are there. There are many published libraries and the tools are there. The thing is that its very tricky to know how they can be used.

PULS: But not all have the same degree of reliability.

NIEVA: Yes, you're right, but tests that can be done. For example, we compared a sample of my work and the work of Sergio. Though we use completely different hydro codes, completely different atomic data, and we sometimes use different lines, we obtain similar results. Perhaps others can use these papers and all the publicly available libraries to perform further tests on the same spectra.

SIMON-DIAZ: To comment about the calibration and determination of stellar parameters, it's true that maybe the grids can be used as a first approach. But this is dangerous because the calibration depends on the models. If use parameters derived with one model and then you go to another stellar atmosphere model, then you introduce differences up to one thousand [degrees] which can significantly affect the derived abundances. For example, in the tests that we have done, the parameters we derived with ATLAS and in detail with FASTWIND were different. But, in the end, the abundances were the same. So if you do your analysis consistently from start to finish, you can obtain parameters which agree within the error bars. If, however you begin to mix things up, this can be dangerous. Also, nowadays we are fortunate to have very high quality spectra for B stars, for example, because we can investigate the behavior of lines of different ions. For example, I discovered that Si II lines often used for the temperature determination in B2 dwarfs, that is the 41, 28, 30 lines, are not correct in the atomic models. So stellar parameters derived from only these lines are bound to be incorrect. So if you have more lines, you can better constrain all of these things.

N. SMITH: I suppose I can expand upon my answer to Doug's question a little bit. When a companion star goes plunging into an extended envelope, inside the envelope of the primary star, you know it's gong to be pretty complicated. So tides could probably spin up (add angular momentum) to the star, but another question that clearly there is also a huge amount of mass lost, right? I mean, that could mean you add angle momentum to the envelope, but then that entire envelope is ejected. I mean, we lose ten solar masses, so the net effect of transferring angular momentum is not clear. Another interesting question that goes along with that though is how does this interaction and mass loss, if it's episodic mass loss, how does that affect the orbit? If you lose mass, imagine you have an eccentric orbit, and you're losing mass steadily in a really strong wind, but at a quasi-steady state, you're going to lose most of the mass when you're moving slowly at apastron, that'll act to circularize and widen the orbit, but if you lose mass preferentially at a burst right at periastron, you're going to make the orbit more eccentric each time. And so, η Car is currently seen with eccentricity of 0.9 or 0.95, so it's very hard to get that way without losing mass right at periastron. This is probably important for forming these Wolf Rayet WC colliding wind binaries, the ones that are very eccentric. It's very hard to get that eccentric without doing something like this.

GAGNÉ: Can I ask you to speculate, or anyone to speculate, on the status of the companion star in η Car?

N. SMITH: A lot of people have been working on this. It's not entirely clear, but the constraints come mainly from the models of the x-ray light curve. From that you get the wind speed, mass-loss rate, and that tells you that it's probably a main-sequence O star or maybe a Wolf Rayet star, but what you also have are upper limits on the ionizing photon luminosity from the amount of Lyman continuum that escapes from the nearby nebulosity. From that it looks like it's not quite such a luminous star: it's less than about 35 to 40 solar masses, but it could be much lower, all the way down to main-sequence stars that are 20 solar masses or so. So there's actually quite a large range of parameter space, but it's consistent with being a fairly normal O star.

GAGNÉ: Like a main sequence O8 or something.

N. SMITH: Like a 30 mass O star. The way, I always describe it is that it's like you take the star that powers the Orion Nebula and you hide it inside the wind of the LBV, so you can't even see it. It's really pretty spectacular.

DIETRICH BAADE: You showed us how the periastron passages line up with certain features in the light curve of η Car. Is this perhaps also a chicken-and-egg situation? On the one hand, it could be, as you said, that the outbursts were triggered by a periastron passage. But it could also be that the outburst was already ongoing, but the companion was diving into the first ejector interacting with it somehow and therefore the star was brightening up. Is that also another possibility?

N. SMITH: Yeah, the picture I was trying to create, is that if you look at the light curve over several hundred years, you see that, over several hundred years, the star is slowly getting brighter, since the 1600s. That picture suggests that you start out with a primary star a lot like what you have now and it's pretty compact and there's no direct interaction between the two. But as something inside the star is driving itself to start an S Doradus-type outburst, the star is getting cooler and bigger on its own, and through that, as it gets bigger, it's more likely each time to interact with the companion as its radius increases. Maybe that's a feedback loop, maybe not, I don't know. But something is driving the star. This eruption is a long-term process – it's driven into that phase where we finally, in the 1820s and 1830s, start seeing these peaks, because the radius has gotten to some critical point, where now the companion is really having some effect, it's either digging a hole in its primary, or it's disrupting the primary so that it allows stored thermo energy envelope to escape, I don't know, but something had to happen in the primary to get it there in the first place.

OWOCKI: Yeah, so, I think about a year and a half ago, when I visited Nathan, we were talking about this just in the context of the current binary, and, because the primary has such a massive wind, I mean, maybe as high as 10^{-3} solar masses, the radius of the apparent photosphere, the wind photosphere, can be quite large and, at periastron at least, there's a good chance that, at least, you're carving a cavity, what we call the borehole effect, and my student, Tom Madura just finished his thesis modeling η Car's spectrum for HST and light curves is able to reproduce a lot of the optical variations with this cavity model, this borehole model. But obviously the giant eruption, that effect will be even much more dramatic. As far as the angular momentum transfer is concerned, if what we are seeing in η Carina, the mass loss is just a larger star because of its wind then it won't be that dynamically coupled onto the actual core hydrostatic star, so i think that one of the opening questions is, in terms of angular momentum transport, is what is the actual size of the hydrostatic core of the star which is thermodynamically coupled? If you distort the wind, you're not going to change the angular momentum of the star, because it's already outflowing at supersonic speeds. So i think it would be very interesting to speculate, or think about, whether during the giant eruption, the hydrostatic core of the star got bigger. That would be probably pretty hard to figure out, because you see the false photosphere so that the optical spectrum, the colors people saw at that time, were form the wind photosphere and not from the core.

GAGNÉ: We will get to you in a second Gloria, but I just want to be clear, I thought in your talk Nathan, though, you were suggesting that when, in the outburst occurred at some point, I'm not sure if it was in the 1840s, this thing was a hyper-giant. Are you suggesting that it was a hyper-giant at the time of the 1840 event?

N. SMITH: So that huge radius I showed was [the one you'd obtain] if you used the present day luminosity and you just made it cooler, so that would be the radius as if it were undergoing an S Doradus outburst, so the radius would have been even bigger. The photosphere would have been even bigger if the luminosity was then increased during the great eruption. What I was trying to show were the conditions as this thing was moving into the great eruption, and so that is the apparent emitting photosphere, the pseudo-photosphere in the wind. What that means, whether that's really the photosphere or whether thats a pseudo-photosphere, we're not sure, but, in any case, the outer envelope of the star is very light and fluffy and a very small amount of mass is in that small outer part. So it's not clear.

FERRER: Let me tell you what happened to HD 5980, which I think is a proxy for η Carina without all the dust around it that obscures it. HD 5980 is a well-known binary with a 19 day period circular eclipsing binary and between 1980 and 1991 it very gradually became brighter by about 0.6 magnitude and its spectral type became slightly cooler, Wolf Rayet spectral type. There is an *Orpheus* spectrum of the star 57 days before the first eruption, and that spectrum is no different than the few spectra that we obtained after and, let's say, maybe a year before the eruption. The eruption occurred extremely violently, extremely abruptly. The eruption had a 3 magnitude brightening increase and lasted about a year, and then it had spin in it and it quickly came down from the large eruption, but it did not come down all the way. It kept on increasing, so we view this in terms of two different phenomena; you have a long-term brightening, which is analogous of the S Doradus activity phase, which will have gone on, probably, for a long time, had it not been for the presence of the binary companion, so we believe that the binary had triggered the rapid and violent eruption, and thereby changed it's future evolution. It is not declining and we find a 40 year timescale for that. So what we think is happening, is that in HD 5980 the star, for some internal reason, which is what will be very interesting for the stellar evolution people to figure out. for some internal reason, it is increasing its radius very slowly and very gradually, and as Nathan mentioned, and the tidal models we have that show this, as you increase the radius, the interaction effects increase enormously until a critical radius is reached, and the critical radius will depend upon the rotation rate, on the orbital separation, and on the radius of the primary itself, and the mass of the companion, and when that critical radius is reached, it will get the violent eruption. So, I think something similar to this, on different scales, has occurred on η Car and I would really urge people to look at HD 5980 as a good proxy for the phenomenon.

OWOCKI: How much mass was lost in the eruption, Gloria?

FERRER: Approximately 10^{-3} solar masses, no more, it lasted, the violent eruption lasted less than a year and, given the velocities and calculations and densities, like 10^{-3}, but the activity long-term activity stage lasted about 20 years. So that part has had an enhanced mass-loss rate compared to the quiescent state.

OWOCKI: I think, 10^{-3} is maybe more a kin to the kind of mass ejections that, cumulative mass ejections, that you might see in S Doradus eruptions, right? So maybe, what you're saying is that it removed it before it got a chance to get to the full S Doradus phase, but it wasn't quite the giant eruptions like you have in η Carina.

GIES: Just one more comment to follow up Nathan's point about how the ejections in η Car are occurring at periastron would increase eccentricity of the orbit. It would seem to me that this would, inevitably, drive the system towards merger at some point, that as you tend towards $e = 1$, the stars will crash into each other at some point.

N. SMITH: Well, if you take the system we have now, and you have another giant eruption where you lose 10 solar masses at periastron, actually, the system will be unbound and it won't be a binary anymore.

Active OB stars: structure, evolution, mass loss, and critical limits
Proceedings IAU Symposium No. 272, 2010
C. Neiner, G. Wade, G. Meynet & G. Peters, eds.

doi:10.1017/S1743921311011495

Ion fractions and the weak wind problem

Matthew J. Austin and Raman K. Prinja

Department of Physics & Astronomy, UCL
Gower Street, London WC1E 6BT, UK
email: mja@star.ucl.ac.uk

Abstract. Some late-type O stars display anomalously weak winds, possibly due to decoupling of the main driving ions from the bulk plasma. This issue and the uncertainty about the nature of wind clumping are a challenge to line-driven wind theory and need resolving in order to fully understand hot stars. We describe the results from the computation of ion fractions for the various elements in O star winds using non-LTE code CMFGEN, including parameterisation of microclumping and X-rays.

Keywords. stars: mass loss, ultraviolet: stars

1. Introduction

There are several spectroscopic diagnostics available for measuring the mass-loss rate of O stars. Each of these has certain advantages and drawbacks, and often requires ill-determined information or parameter knowledge. In the ultraviolet, wind profiles can be matched using the SEI (Sobolev with Exact Integration) method of Lamers *et al.* (1987) to yield the product of the mass-loss rate and the ion fraction of the element. The mass-loss results are therefore very dependent upon the wind ionization balance. We compiled a grid of O star models using the spherical non-LTE model atmosphere code CMFGEN (Hillier & Miller, 1998), producing a variety of model types from O3 to O9.5, with different wind-clumping scenarios and with or without X-rays. These models have a large number of applications, but here we focus on the run of ion fraction with effective stellar temperature, which was derived for each clumping and X-ray scenario, to assess the possible effects of these two phenomena on the ionization balance in O star winds.

2. Model Ion Fractions

For each spectral type a set of mean ion fractions was calculated for each element, normalised to the range 0.2-0.9v_∞. This was done in order to ensure a direct comparison to empirical fits described in a separate paper, in which the very lowest and highest velocity positions are excluded so as to avoid any variable phenomena such as DACs.

Figure 1 shows the results for C^{3+} and C^{4+} for dwarfs, which are predicted to account for most of the total carbon population. N, O, Si, P and S were also processed but for brevity are not shown here; they will be published separately. We focus discussion on carbon, which is pertinent to the project to fit CIV profiles in O dwarfs, published separately. The clumping scenario is either unclumped (smooth wind, volume filling factor 1.0) or moderately clumped (with a volume filling factor of 0.1). The models with X-rays have an X-ray luminosity consistent with the relation $\log L_x/L_{bol} \sim -7$ from e.g. Naze (2009). When changing to a clumped wind in the absence of X-rays, the higher ionization stage reduces in population at higher tempertures as recombination from higher local density forces the balance in favour of the lower stage. In general this balance appears

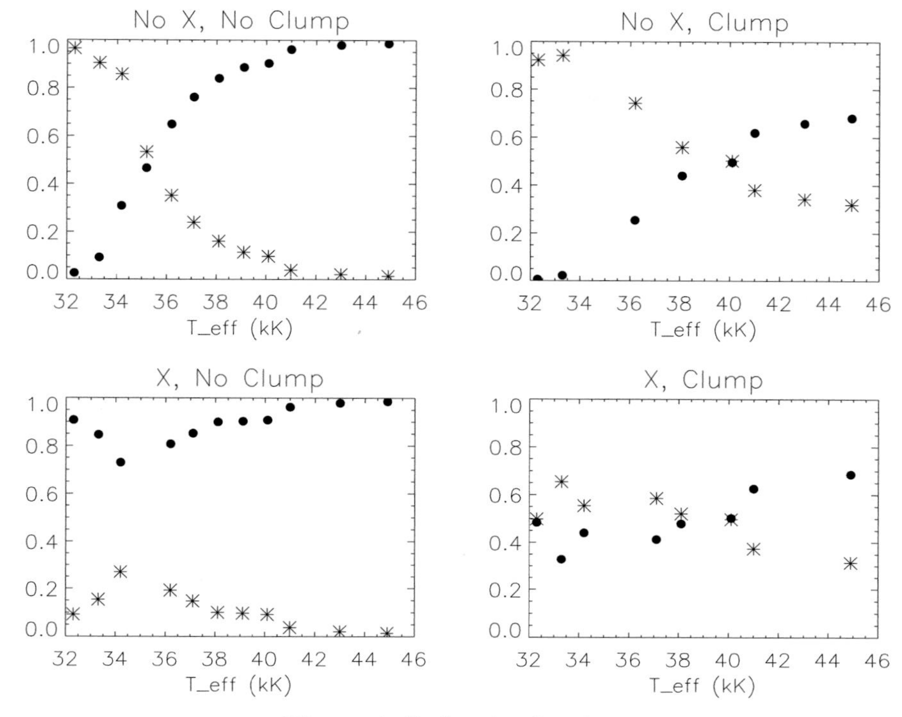

Figure 1. Carbon ion fractions.

more fragile when X-rays are implemented and the effect is greater. Bringing X-rays into the unclumped scenario shifts the balance strongly in favour of the upper stage but with a non-negligible portion still in the lower. The supposedly most detailed description of the wind (i.e. with both clumping and X-rays) yields a somewhat uncertain picture, in which it is unclear for much of the O star range whether either ion becomes dominant. A stark difference in the X-ray models is brought about by introducing moderate clumping. Whilst in the unclumped wind, the X-rays cause C^{4+} to be dominant for the whole O star range, the clumping then changes the ionization balance much more than in the non-X-ray models. In order to distinguish the most likely scenario, the profiles in the corresponding model spectra were examined and compared to observations (*IUE, Copernicus, FUSE*). For later type objects we find that it is crucial to include a treatment of X-rays to get realistic NV and OVI lines. The model profiles also seem to suggest a very low level of clumping to be most likely, implying that the bottom-left panel in Fig. 1 is that which should be used.

References

Hillier, D. J. & Miller, D. L. 1998, *ApJ*, 496, 407
Lamers, H. J. G. L. M., Cerruti-Sola, M., & Perinotto, M. 1987, *ApJ*, 314, 726
Nazé, Y. 2009, *A&A*, 506, 1055

Active OB stars: structure, evolution, mass loss, and critical limits
Proceedings IAU Symposium No. 272, 2010
C. Neiner, G. Wade, G. Meynet & G. Peters, eds.

doi:10.1017/S1743921311011501

Investigation of X-ray transient CI Cam as an unique star with the B[e] phenomenon

Elena A. Barsukova[1], Alexandr N. Burenkov[1], Valentina G. Klochkova[1], Vitalij P. Goranskij[2], Nataly V. Metlova[2], Peter Kroll[3] and Anatoly S. Miroshnichenko[4]

[1]Special Astrophysical Observatory, Russian Academy of Sciences, Nizhnij Arkhyz, Karachai-Cherkesia, 369167, Russia; email: bars@sao.ru, ban@sao.ru, valenta@sao.ru

[2]Sternberg Astronomical Institute, Moscow University, Universitetskii pr. 13, Moscow, 119991, Russia; email: goray@sai.msu.ru, nm@sai.crimea.ua

[3]Sternwarte Sonneberg, Sternwartestrasse, 32, Sonneberg, D-96515, Germany email: pk@4pisysteme.de

[4]University of North Carolina at Greensboro, USA; email: a_mirosh@uncg.edu

Abstract. Double mode pulsations of the B4 component in the system of CI Cam were detected. The photometric 19.4 day orbital period of CI Cam was confirmed with the plates of the Sonneberg collection for a long period of time before the unique 1998 outburst. The amplitude of the periodic component of 0.08 mag before the outburst was larger than that of 0.03 mag after the outburst.

Keywords. binaries: general, novae, cataclysmic variables, stars: oscillations (including pulsations), circumstellar matter

---

We present results of joint optical photometric and spectroscopic investigations of CI Cam (XTE J0421+560) for 11 years. CI Cam was an unclassified star with the B[e] phenomenon. The star was chosen due to its strong outburst in all ranges of electromagnetic spectrum in 1998. The star brightness at the peak of the outburst in optical bands exceeded its brightness in quiescence by at least a factor of 25. This outburst has been treated by researchers as a thermonuclear runaway of hydrogen accumulated on the surface of a white dwarf, so CI Cam may be an sample of a classical nova. Such outbursts of B[e] stars have been observed neither in the Galaxy nor in neighboring galaxies yet.

Our investigations have revealed the following properties of CI Cam. The system consists of a B4III-V star and a white dwarf on an eccentric (e=0.62) orbit with the period of 19.407 day. The white dwarf appears with Doppler shifts of the HeII 4686Å emission line (Barsukova *et al.* 2006). CI Cam demonstrates different types of optical variability including rapid intranight variations and slow quasi-periodic changes on a timescale of several years. We carried out an extensive photometric CCD monitoring (3511 images in the V band) and revealed pulsations of the B-type companion of CI Cam in quiescence with the amplitude up to $0^m.07$ (Goranskij & Barsukova 2009). These pulsations are similar to those of classical Be stars (De Cat 2002). Two waves with periods of $0^d.4152$ ($\pm 0^d.0007$) and $0^d.2667$ ($\pm 0^d.0003$) were dominating with full amplitudes of $0^m.019$ and $0^m.017$, correspondingly. There is an evidence of a resonance between these waves.

We studied also secular and orbital light variations of CI Cam using numerous digitized plates of the Sonneberg Observatory plate collection obtained between 1935 and 1967. We have confirmed the 19.4 day orbital period of CI Cam before the outburst. The amplitude of the periodic component before the outburst (Fig. 1a) is equal to 0.08 mag. After the outburst the amplitude was smaller than before the outburst (0.03 mag) (Fig. 1b).

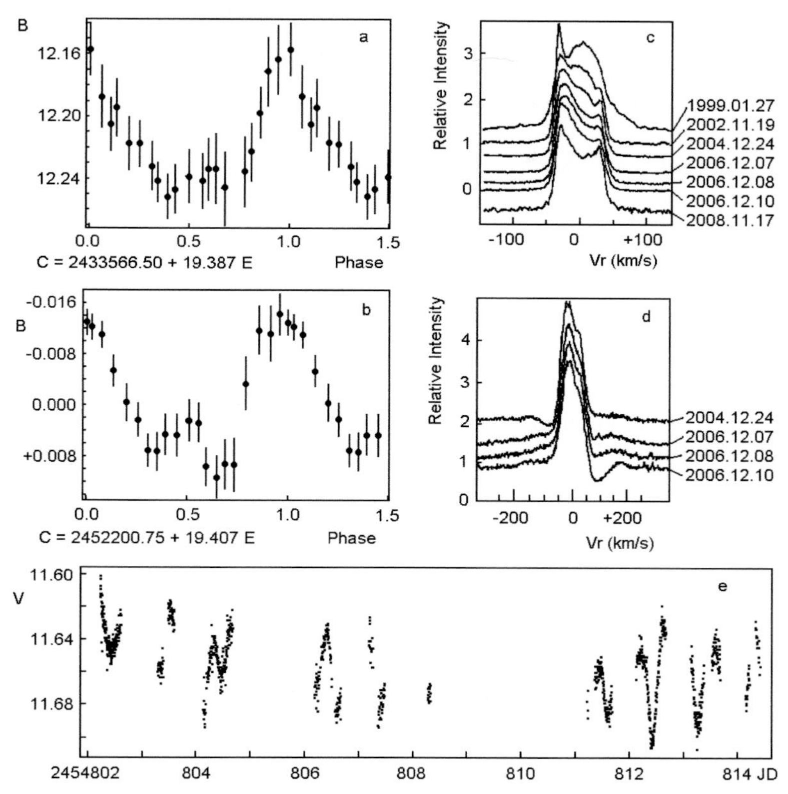

Figure 1. (a) The mean light curve of CI Cam in the B band before the outburst derived from the Sonneberg plates between 1935 and 1967. (b) The mean light curve after the outburst based on photoelectric B-band observations at the SAI Crimean Station. (c) Profile variations in the [NII] 5755Å emission line after the outburst. (d) Examples of direct and reverse P Cyg profiles of the He I 4713Å emission line in the spectrum of CI Cam at a high spectral resolution. (e) Pulsations in the V band.

We analyzed 19 echelle spectra (resolution $FWHM = 0.08 - 0.21\AA$) taken with the Russian 6-m telescope BTA (LYNX and NES spectrographs), and other telescopes obtained from 1999 to 2008. We found that line profiles of Fe II and [N II] emission lines in these spectra shift with time. Radial velocities of these lines do not correlate with the phase of the 19.4 day orbital period, and this shift may be due to gravitational effect from a hypothetical third massive companion (Barsukova *et al.* 2007). We also found temporal variations in emission line profiles. Fig. 1c shows a fading low-velocity component in the [N II] line profiles that is probably connected with dissipation of erupted matter after the outburst. Variations of He I 4713Å line profiles (Fig. 1d) may be caused by pulsations of the B-type star. A light curve showing the B star pulsations is presented in Fig. 1e.

References

De Cat, P. 2002, in: C. Aerts, T. R. Bedding, & J. Christensen-Dalsgaard (eds.), *IAU Colloq. 185: Radial and Nonradial Pulsationsn as Probes of Stellar Physics*, ASP-CS 259, p. 196

Barsukova, E. A., Borisov, N. V., Burenkov, A. N., Goranskii, V. P. *et al.* 2006, *Astron. Rep.*, 50, 664

Barsukova, E.A., Klochkova V.G., Panchuk V.E., Yushkin, M. V. *et al.* 2007, *ATel*, 1036

Goranskij, V. P. & Barsukova, E. A. 2009, *Astrophysical Bulletin*, 64, 50

Active OB stars: structure, evolution, mass loss, and critical limits
Proceedings IAU Symposium No. 272, 2010
C. Neiner, G. Wade, G. Meynet & G. Peters, eds.

doi:10.1017/S1743921311011513

η Carinae long-term variability

Augusto Damineli[1], Mairan Teodoro[1], Michael F. Corcoran[2] and Jose H. Groh[3]

[1]Instituto de Astronomia, Geofísica e Ciências Atmosféricas, Rua do Matão 1226, São Paulo, 05508-900, Brazil
email: damineli@astro.iag.usp.br

[2]NASA/GSFC, Code 662, X-ray Astrophysics, Greenbelt, MD 20771, USA

[3]Max-Planck-Institute für Radioastronomie, Auf dem Hügel 69, D-53121 Bonn, Germany

Abstract. We present preliminary results of our analysis on the long-term variations observed in the optical spectrum of the LBV star η Carinae. Based on the hydrogen line profiles, we conclude that the physical parameters of the primary star did not change in the last 15 years.

Keywords. stars: individual (η Carinae), stars: variables: other, stars: emission-line

---

1. Introduction

Photometric monitoring of η Carinae (Feinstein 1967, Feinstein & Marraco 1974, Sterken *et al.* 1996, Sterken *et al.* 1999, van Genderen *et al.* 2006, Frew 2004, Fernández-Lajús *et al.* 2009) revealed that an increase in brightness at variable rates, since 1950. The mechanism behind such long-term variations are still unclear, however. In this regard, spectroscopic monitoring can put several important constraints to the diagnosis. Unfortunately, frequent spectroscopic observations of this object began just about 2 decades ago, which is not sufficient yet to draw a clear picture of what is happening to the central source.

2. Results and discussion

In order to verify whether or not the central source in η Car is passing through changes, we analized ground-based spectroscopic data taken at the same phase ($\phi \approx 0.3$) of the spectroscopic event, but in different cycles (#9, #10, #11 and #12).

Our analysis revealed that the lines formed in the wind of the primary star – such as the hydrogen lines – do not present any evidence of systematic or significant changes in line profile, as shown in Fig. 1a, for example. In that figure, the Hδ line profile was normalized by the local continuum.

On the other hand, lines with high-ionization potential – such as [Fe III] λ4657 – do show systematic variations, namely, the intensity of the peak of the line's narrow component is decreasing with time, relative to the local continuum and, thus, the equivalent width of the narrow component is *decreasing* with time, as indicated in Fig. 1b (dotted line).

However, since forbidden lines are formed in a more extended region, and we do not know where exactly the increase in brightness is coming from, we converted the equivalent width measurements into line flux by using the B-band magnitudes for each epoch. After that, we normalized the line fluxes by the line flux observed in 1994 Feb 25th ($\phi = 9.31$). The result is shown in Fig. 1b (dashed line).

After correcting the equivalent width of the narrow component of the [Fe III] λ4657 emission line by the flux in the local continuum, the trend changed completely: the line

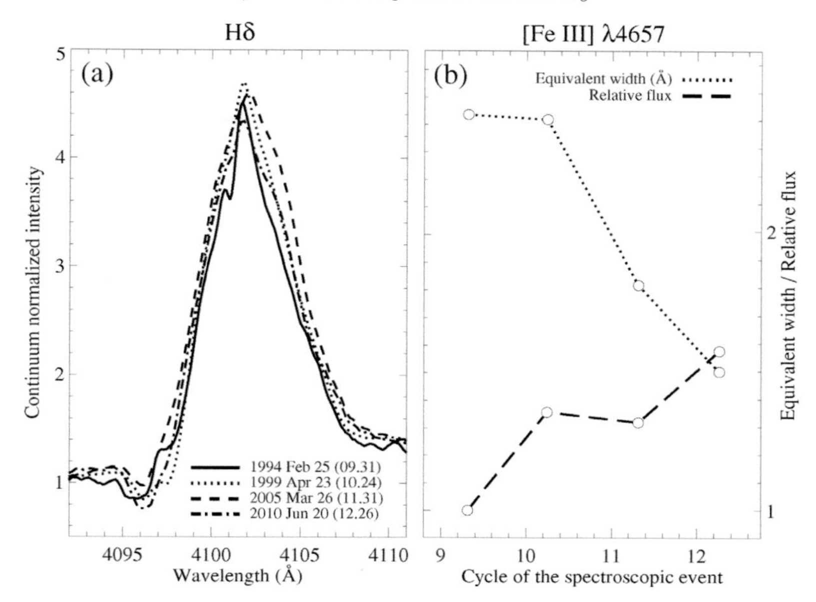

Figure 1. (a) Line profile of Hδ at the epochs indicated in the legend. (b) Equivalent width and relative line flux. The hydrogen line shows no significant variations throughout the last 15 years. On the other hand, the equivalent width of the [Fe III] emision line is systematically decreasing with time, but the relative line flux is increasing.

flux is *increasing* with time. From 1994 to 2010, the narrow component line flux increased by about 60 per cent (in the same period, the continuum flux increased by a factor of 2.5).

We know that the narrow component is formed in the Weigelt's blobs. If the equivalent width of such component is decreasing while the line flux is increasing, then we can conclude that either (1) the total extinction around the central source is decreasing in all directions, not only in our line-of-sight or (2) the effective temperature of the secondary star is increasing (or the wind opacity is decreasing).

Unfortunately, based only on our preliminary results shown in this proceedings, we can only conclude that the wind of the primary star did not change during the last 15 years. At this moment, we cannot point for sure which of the possibilities presented above is the correct one to explain the behavior of the [Fe III] λ4657 line (although we favor the decreasing of extinction in all directions). However, further analysis of other spectral features with high-ionization potential will eventually provide us with more indications on what is happening in the central source of η Car. That will be the subject of a more complete, forthcoming paper.

References

Feinstein, A. 1967, *Observatory*, 87, 287

Feinstein, A. & Marraco, H. G. 1974, *A&A*, 30, 271

Fernández-Lajús, E., Fariña, C., Torres, A. F., Schwartz, M. A. *et al.* 2009, *A&A*, 493, 1093

Frew, D. J. 2004, *Journal of Astronomical Data*, 10, 6

Sterken, C., Freyhammer, L., Arentoft, T., & van Genderen, A. M. 1999, *A&A*, 346, L33

Sterken, C., de Groot, M. J. H., & van Genderen, A. M. 1996, *A&AS*, 116, 9

van Genderen, A. M., Sterken, C., Allen, W. H., & Walker, W. S. G. 2006, *Journal of Astronomical Data*, 12, 3

Active OB stars: structure, evolution, mass loss, and critical limits
Proceedings IAU Symposium No. 272, 2010
C. Neiner, G. Wade, G. Meynet & G. Peters, eds.

doi:10.1017/S1743921311011525

Searching for emission line and OB stars in Cl 1806-20 using a NIR narrow-band technique

Michelle L. Edwards[1,2], Reba M. Bandyopadhyay[2], Stephen S. Eikenberry[2], Valerie J. Mikles[2,3] and Dae-Sik Moon[4]

[1]Gemini Observatory, Southern Operations Center, La Serena, Chile
email: medwards@gemini.edu

[2]Department of Astronomy, University of Florida, Gainesville, FL 32611

[3] Department of Physics and Astronomy, Louisiana State University, Baton Rouge, LA 70803

[4]Department of Astronomy and Astrophysics, University of Toronto, Toronto M5S3H8, Canada

Abstract. We survey the environment of Cl 1806-20 using near-infrared narrow-band imaging to search for Brγ features indicative of evolved massive stars. Using this technique, we successfully detect previously identified massive stars in the cluster. We detect no new emission line stars, establishing a firm upper limit on the number of Wolf Rayets and Luminous Blue Variables; however, we do find several candidate OB supergiants, which likely represent the bulk of the heretofore undiscovered massive star population.

Keywords. stars: emission-line, open clusters and associations: general

---

1. Introduction

Discovered by Fuchs *et al.* (1999), Cl 1806-20 is home to a variety of interesting and rare objects, including a candidate Luminous Blue Variable (LBV 1806-20), multiple Wolf Rayets (WRs), a soft-gamma repeater (SGR 1806-20), and several OB supergiants (Fuchs *et al.* 1999, Eikenberry et al. 2004, Figer *et al.* 2005). Although individual members of Cl 1806-20 have been identified on a case-by-case basis with spectroscopy, no systematic effort to census the cluster's massive stellar population exists in the literature. To better constrain the membership of Cl 1806-20 we performed near-infrared narrow- and broad-band imaging to search for massive candidate cluster members. We focused on Brγ emission lines indicative of stellar winds in massive stars (Hanson et al. 1996, Figer *et al.* 1997, Blum et al. 2001) and Brγ absorption found in OB supergiants (Hanson *et al.* 1996).

2. Observation and Analysis

On 2005 August 26-27, we used the Wide Field Infrared Camera (WIRC) (Wilson *et al.* 2003) on the Palomar 200" telescope to obtain J, K_s, 2.16μm Brγ, and 2.27μm K_{cont} images of an 8.7 arcminute × 8.7 arcminute region around Cl 1806-20. We reduced the data with FATBOY, a PYTHON based data pipeline developed at the University of Florida, and performed astrometry on these images using KOORDS in the KARMA software package. We then completed PSF photometry on our science frames with DAOPHOT II and ALLSTAR (Stetson 1987, Stetson 1992). We calibrated the J and K_s magnitudes for our sources with 2-MASS photometry. Using TOPCAT, the Tool for OPeration on Catalogues and Table we matched data across all four bands.

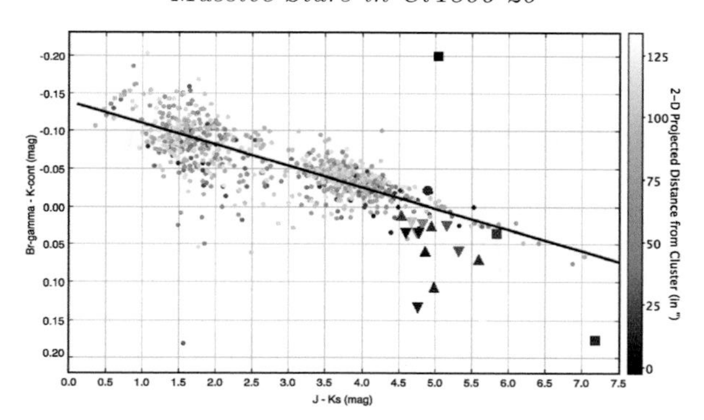

Figure 1. Color-color diagram of stars within an 4pc (2 arcmin) radius of Cl 1806-20. OB supergiants are marked as triangles, WR stars as squares and the LBV as a circle. New OB supergiant candidates are marked as inverted triangles. The solid black line is the narrow-band zeropoint.

Using the resulting 4-band catalogue of sources we created a $J - K_s$ versus $Br\gamma - K_{cont}$ diagram (Fig. 1) of objects within a 2 arcmin radius from the cluster. We calculated the 2-D projected distance between each star and a point close to the center of the cluster.

Using an $A_V = 29 \pm 2$ for Cl 1806-20, (Corbel & Eikenberry 2004) we find $A_K = 3.25 \pm 0.56$ and $A_J = 8.18 \pm 0.22$ yielding a $J - K_s = 4.93 \pm 0.34$ mag for the color of the cluster. We focused our search for cluster members in this region of the diagram.

3. Results

We confirmed the existence of known massive stars in Cl 1806-20. Several of our reported equivalent widths are in good agreement with the literature values. Where discrepancies exist, we explored the reasons. We found that in some cases, insufficient information in the literature prevented quantitative comparison. In other cases, we found literature data or completed follow-up observations that indicated the discrepancy may be a result of intrinsic variations.

We did not detect any previously unknown WR or LBV stars in Cl 1806-20. This finding allows us to place a firm upper limit on the number of very massive stars in the cluster. However, we did find a population of candidate OB supergiants that may represent the bulk of the heretofore undiscovered cluster population. We suggest that these stars should be targeted for future spectroscopic observations.

References

Blum, R. D., Schaerer, D., Pasquali, A., Heydari-Malayeri, M. *et al.* 2001, *AJ*, 122, 1875

Corbel, S. & Eikenberry, S. S. 2004, *A&A*, 419, 191

Eikenberry, S. S., Matthews, K., LaVine, J. L., Garske, M. A. *et al.* 2004, *ApJ*, 616, 506

Figer, D. F., McLean, I. S., & Najarro, F. 1997, *ApJ*, 486, 420

Figer, D. F., Najarro, F., Geballe, T. R., Blum, R. D. *et al.* 2005, *ApJ* (Letters) 622, L49

Fuchs, Y., Mirabel, F., Chaty, S., Claret, A. *et al.* 1999, *A&A*, 350, 891

Hanson, M. M., Conti, P. S., & Rieke, M. J. 1996, *ApJS*, 107, 281

Stetson, P. B. 1987, *PASP*, 99, 191

Stetson, P. B. 1992, in: D. M. Worrall, C. Biemesderfer, & J. Barnes (eds.), *Astronomical Data Analysis Software and Systems I*, ASP-CS 25, p. 297

Wilson, J. C., Eikenberry, S. S., Henderson, C. P., Hayward, T. L. *et al.* 2003, in: M. Iye & A. F. M. Moorwood (eds.), *Instrument Design and Performance for Optical/Infrared Ground-based Telescopes*, SPIE Conference Series 4841, p. 451

Active OB stars: structure, evolution, mass loss, and critical limits
Proceedings IAU Symposium No. 272, 2010
C. Neiner, G. Wade, G. Meynet & G. Peters, eds.

doi:10.1017/S1743921311011537

The *Chandra* survey of Carina OB stars

Marc Gagné[1], Garrett Fehon[1], Michael R. Savoy[1], David H. Cohen[2], Leisa K. Townsley[3], Patrick S. Broos[3], Matthew S. Povich[3], Michael F. Corcoran[4], Nolan R. Walborn[5], Anthony F.J. Moffat[6], Yael Nazé[7], Lidia M. Oskinova[8]

[1]Dept. of Geology and Astronomy, West Chester Univ., West Chester, PA 19383
[2]Dept. of Physics and Astronomy, Swarthmore College, Swarthmore, PA 19081
[3]Dept. of Astronomy & Astrophysics, Penn State Univ., University Park, PA 16802
[4]CRESST and X-ray Astrophysics Laboratory, NASA/GSFC, Greenbelt, MD 20771
[5]Space Telescope Science Institute, Baltimore, MD 21218, USA
[6]Dépt. de Physique, Univ. de Montréal, Succ. Centre-Ville, Montréal, QC, H3C 3J7, Canada
[7]GAPHE, Dépt. AGO, Univ. Liège, Allée du 6 Août 17, Bat. B5C, B4000-Liège, Belgium
[8]Institute for Physics and Astronomy, Univ. of Potsdam, 14476 Potsdam, Germany

Abstract. We have combined 22 deep *Chandra* ACIS-I pointings to map over one square degree of the Carina complex. Our x-ray survey detects 69 of 70 known O-type stars and 61 of 130 known early B stars. The majority of single O stars display soft X-ray spectra and have a mean $\log L_{\rm X}/L_{\rm bol} \approx -7.5$ suggesting shocks embedded in the O-star winds. Over OB stars show unusually high X-ray luminosities, high shock temperatures or time variability, not predicted for embedded wind shocks.

Keywords. X-rays: stars, X-rays: binaries, stars: early-type, stars: pre–main-sequence

---

The *Chandra* Carina Complex Project (CCCP) survey area contains over 200 known massive stars: the LBV η Car, the Wolf Rayet stars WR 22, WR 24 and WR 25, 70 known O stars, and 130 B0-B3 stars with known spectral types. We have constructed a searchable electronic database of x-ray and optical properties for the 200 Carina OB stars. We divide the massive star population in Carina into four main groups based on spectral type and luminosity class: LBV/WR stars (4), O-type binaries (15), O-type single stars (55), and B0-B3 stars (130). We note that none of the B stars are typed as spectroscopic binaries. The goal of this study is to characterize $L_{\rm X}$, $L_{\rm X}/L_{\rm bol}$, the x-ray temperatures kT, and the temporal variability of single and binary OB stars, to look for new candidate colliding wind systems (e.g., Stevens *et al.* 1992) and magnetically confined wind shock sources (e.g., ud-Doula & Owocki 2002).

The *Chandra* CCD spectra of the 78 OB stars with more than 50 ACIS counts were fit in XSPEC using a one- or two-temperature APEC emission model, and a two-component absorption model: a fixed TBABS column density to represent cold, neutral ISM absorption (assuming $A_V/N_{\rm H} = 1.6 \times 10^{21}$ cm^{-2} per mag), and free column density to represent absorption from the overlying stellar wind. $L_{\rm X}$ was calculated correcting only for the ISM column. For single O stars, $\log L_{\rm X}/L_{\rm bol} = -7.59 \pm 0.23$, with no such trend for B stars.

Table 1 highlights three notable single O stars: HD 93250, O4 III(fc), MJ 496, O8.5 V, and MJ 449, O8.5 V((f)), all of which have $kT > 1.7$ keV, $\log L_{\rm X}/L_{\rm bol} > -7.1$, and two of which are time variable. These three stars do not have massive, spectroscopic companions, and their $L_{\rm X}$ is too high to be produced by unseen, lower-mass pre–main-sequence (PMS) companions. For these three we consider two hypotheses: a more distant massive companion (as is the case for HD 93129A), or magnetically confined wind shocks (as in θ^1 Ori C, Gagné *et al.* 2005).

Table 1. Notable stars: XSPEC and time variability parameters

Star name	ACIS name	Spectral type	$\log L_{\rm X}$ (erg s^{-1})	$\log \frac{L_{\rm X}}{L_{\rm bol}}$	$kT_{\rm avg}$ (keV)	$P_{\rm KS}$ (%)
HD 93250	104445.04-593354.6	O4 III(fc)	33.12	-6.41	2.30	20
MJ 496	104508.23-594607.0	O8.5 V	32.09	-6.46	1.70	14
MJ 449	104454.70-595601.8	O8.5 V((f))	31.24	-7.08	3.09	6
HD 93403	104544.13-592428.1	O5.5 I + O7 V	32.93	-6.60	1.00	0
HD 93205	104433.74-594415.4	O3 V + O8 V	32.55	-6.82	0.30	3
HD 93129A	104357.47-593251.3	O2 If*	32.91	-6.85	0.74	46
HD 93343	104512.23-594500.5	O8 V + O7-8.5 V	31.45	-7.19	3.17	30
QZ Car	104422.91-595935.9	O9.7 I + O8 III	32.55	-7.26	1.03	5
FO 15	104536.33-594823.5	O5.5 Vz + O9.5 V	30.62	-8.26	0.50	0
SS73 24	104557.13-595643.1	Be pec	31.70	-5.65	3.13	19
Tr 16 64	104504.75-594053.7	B1.5 Vb	31.31	-6.12	2.73	30
MJ 327	104430.34-593726.8	B0 V	31.24	-7.18	1.68	0
MJ 427	104454.06-594129.4	B1 V	31.10	-6.51	2.94	0
MJ 99	104343.55-593403.4	B2 V	31.09	-6.13	2.74	0
HD 93501	104622.02-600118.8	B1.5 III:	31.09	-6.98	6.50	14
Coll 228 68	104400.17-600607.7	B1 Vn	31.01	-6.52	2.54	2
MJ 224	104405.84-593511.6	B1 V	31.01	-6.72	1.96	12
HD 93190	104419.63-591658.6	B0 IV:ep	30.98	-7.98	2.35	1
MJ 181	104357.96-593353.4	B1.5 V	30.98	-6.40	2.13	40
MJ 126	104345.04-595325.0	B2 V	30.94	-6.44	1.96	0
MJ 289	104422.51-593925.4	B1.5 V	30.90	-6.45	0.98	0
MJ 184	104358.45-593301.5	B1 V	31.07	-6.54	2.38	92
MJ 218	104405.09-593341.4	B1.5 V	30.76	-6.77	2.49	16

Five of the O-type binaries (HD 93403, HD 93205, HD 93129A, HD 93343 and QZ Car) show hard x-rays and high $\log L_{\rm X}/L_{\rm bol} > -7.2$, indicating some colliding wind emission. The remaining O binaries have $-7.2 > \log L_{\rm X}/L_{\rm bol} > -7.7$, much like many of the O single stars, confirming a finding of the XMM survey of the Carina region by Antokhin *et al.* (2008). All of these stars have soft X-ray spectra and show no strong time variability, consistent with the x-ray emission expected from wind shocks embedded in an O-star wind (Owocki & Cohen 1999). One short-period O5.5 Vz + O9.5 V binary (FO 15) has very low $\log L_{\rm X}/L_{\rm bol} = -8.26$. Not only does this star show no evidence of colliding wind shocks, the expected embedded wind shock emission from the O5.5 primary appears to be suppressed by the radiation of the closely orbiting O9.5 secondary.

The X-ray emission for the B stars is more difficult to untangle, partly because their binary status is not known. As a group the early B stars do not follow the $L_{\rm X}/L_{\rm bol}$ trend of most O stars. The most notable of these B stars is a new candidate Herbig Be star, SS73 24, in the Treasure Chest Cluster, with $\log L_{\rm X} = 31.7$ and $\log L_{\rm X}/L_{\rm bol} \approx -5.6$. The early B stars that are detected with *Chandra* often show hard X-rays and higher $L_{\rm X}$ than expected from embedded wind shocks (see Table 1). Some of these may have unseen, lower-mass PMS coronal companions, but as a group, the Carina B stars detected with *Chandra* are too X-ray luminous, even if they all harbored $1 - 3M_{\odot}$ companions. We emphasize that most of the known B stars in Carina are not detected with *Chandra*, or are weak X-ray sources. But the X-rays on the most active B stars must be produced by some intrinsic mechanism, possibly related to magnetic fields.

References

Antokhin, I. I., Rauw, G., Vreux, J.-M., van der Hucht, K. A. *et al.* 2008, *A&A*, 477, 593
Gagné, M., Oksala, M. E., Cohen, D. H., Tonnesen, S. K. *et al.* 2005, *ApJ*, 628, 986
Owocki, S. P. & Cohen, D. H. 1999, *ApJ*, 520, 833
Stevens, I. R., Blondin, J. M. & Pollock, A. M. T. 1992, *ApJ*, 386, 265
ud-Doula, A. & Owocki, S. P. 2002, *ApJ*, 576, 413

Active OB stars: structure, evolution, mass loss, and critical limits
Proceedings IAU Symposium No. 272, 2010
C. Neiner, G. Wade, G. Meynet & G. Peters, eds.

doi:10.1017/S1743921311011549

H_α19 in the galaxy M 33, a high-luminosity massive merging eclipsing binary

Vitaly P. Goranskij[1] **and Elena Bersukova**[2]

[1]Sternberg Astronomical Institute, Moscow University,
Universitetskii pr. 13, Moscow, 119991, Russia
email: goray@sai.msu.ru

[2]Special Astrophysical Observatory, Russian Academy of Sciences,
Nizhnij Arkhyz, Karachai-Cherkesia, 369167, Russia
email: bars@sao.ru

Abstract. CCD photometry reveals a hot spot on the surface of the hot accretion gainer in this supergiant O type binary with a big accretion rate. This spot is as bright as 25000 Suns. The orbital period of this system is 33.108 days. The absorption-line spectrum contains multiple lines of He I, Si III and N II. The star is associated with H II region formed by bipolar gas outflow from the system.

Keywords. binaries: eclipsing, circumstellar matter

We report the results of CCD *BVRc* photometry and medium resolution spectroscopy of OB star No.0712 in the galaxy M 33 (Ivanov et al. 1993) identified with the emission line star H_α19 (Fabrika & Sholukhova 1995). The astrometric position of the star is $1^h33^m39^s.468 + 30°45'40''.25$, Eq. 2000.0. The first photometric and spectroscopic investigations of this star were described in the paper of Sharov, Goranskij & Fabrika (1997). It is an eclipsing binary with the orbital period of 33.108 day. H_α19 is associated with an H II region having two-lobe structure which is seen in two H_α images taken by Courtes *et al.* (1993) using the Russian 6-m telescope BTA. The blue central star and two red lobes of this H II region are seen in the high-resolution color image of M 33 central part taken by King *et al.* (2001) with KPNO 4 m Mayall Telescope. The overall size of H II region is 30 pc. The star is not associated with any X-ray or radio source.

In 2001 – 2009, we performed CCD BVR_C photometry with SAO 1 m reflector and CCD photometer based on the chip EEV 42-40, and SAI Crimean station 60 cm telescope with CCD VersArray manufactured by Princeton Instruments. Light curves, $B-V$ and $V-R_C$ color curves are shown in Fig. 1a and 1b. The range of variability was $16^m.90$ – $17^m.55$ in the V band, with the magnitude of Min II at 17.35. We determined light elements $Min\ I = 2453593.53 + 33^d.108(\pm 0^d.003) \times E$. This formula confirms the results of photographic observations by Sharov, Goranskij & Fabrika (1997). Color indices are the following: $B-V = +0^m.08, V-R_C = +0^m.16$. With these indices, H_α19 is located near the red edge of OB stars' main sequence, and $1^m.5$ lower than the brightest O supergiant in the CM diagram by Massey *et al.* (2006) of M33 galaxy.

The dominating feature of light curves is a hump just before the primary eclipse due to bright extended spot on the surface of a hot luminous companion. This spot is as bright as 25000 Suns. We assume that H_α19 is a merging binary which has a large rate of mass exchange, and therefore the radiative heat transfer has been changed partly by circulation in the volume of the hot gainer. This circulation transports very hot matter from the deep layers of the star envelope to the surface.

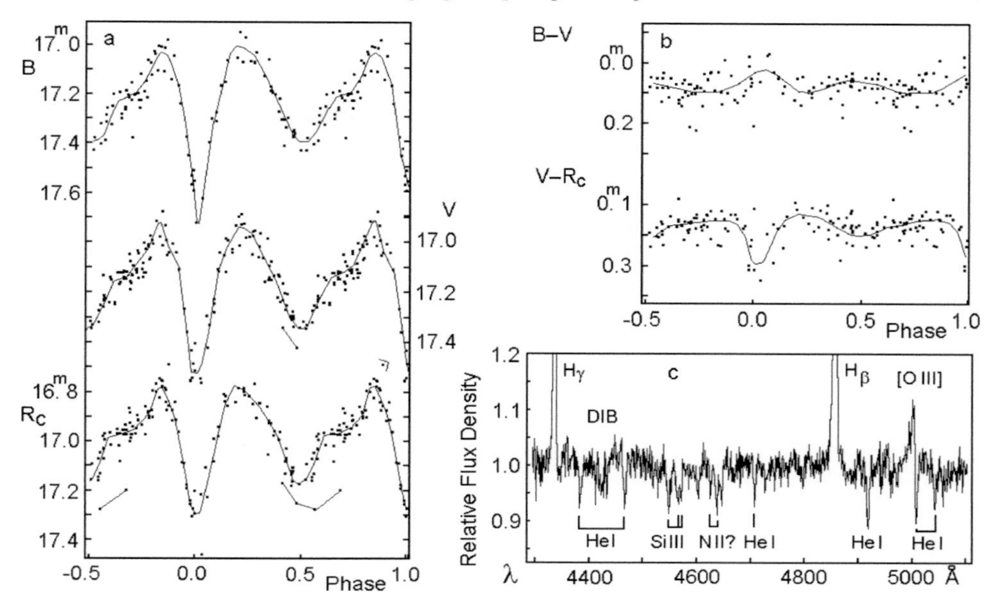

Figure 1. CCD BVR_C light (a) and color (b) curves of H_α 19 plotted with the orbital period 33.108 day. The fragment of stellar spectrum is shown in the box (c).

We analyzed a long-slit spectrum taken with BTA/SCORPIO on 2005 September 8.04 UT with the exposure of 900 s. The slit was located along the lobes. The spectral range λ3960 – 5731 Å, and the spectrum resolution was FWHM = 5.0 Å. A fragment of the spectrum is shown in Fig. 1c. The star has weak absorption lines in the spectrum. He I is represented with lines at λ4009, 4024, 4026, 4143, 4388, 4471, 4713, 4921, 5015 and 5047 Å. We identified also N II absorptions at λ4630, 4643, 5666, 5679 and 5710 Å, and Si III absorptions at λ4552, 4567 and 4574 Å. These features are typical of O type stars. In the spectrum of bipolar nebula, we identified Balmer emissions down to H_ϵ. H_β line width FWHM = 220 km/s corrected for instrumental profile width. This is a low velocity bipolar outflow. EW(H_β) is 128 Å. It is about 10 times larger than previous estimate by S.I. Neizvestnyi with BTA TV scanner on 1991 September 10 (Sharov, Goranskij & Fabrika (1997)). Evidently, the nebula spectrum was not extracted from common light that time. Our nebular spectrum contains also [O III] 5007 Å line with the EW = 11 Å. B. Margon and P. Green didn't found any traces of [O III] λ4959 and 5007 Å lines in their spectra taken on 1991 December 7 and 22.

With M 33 distance module of $24^m.7$, the absolute magnitude M_V of H_α 19 is −8.0, and the bolometric magnitude is −10.5. So, each of merging components of H_α 19 is as massive as 40 - 50 $M_\odot$. H_α 19 represents a rare event of merging massive supergiants.

References

Courtes, G., Petit, H., Petit, M., Sivan, J.-P. *et al.* 1987, *A&A* 174, 28

Fabrika, S. & Sholukhova, O. 1995, *Ap&SS*, 226, 229

Ivanov, G. R., Freedman, W. L., & Madore, B. F. 1993, *ApJS*, 89, 85

King, N. *et al.* 2001, *Astronomy Picture of the Day* (http://antwrp.gsfc.nasa.gov/apod/ap010927.html)

Massey, P., Olsen, K. A. G., Hodge, P. W., Strong, S. B. *et al.* 2006, *AJ*, 131, 2478

Sharov, A. S., Goranskii, V. P., & Fabrika, S. N. 1997, *Astron. Lett.*, 23, 37

Active OB stars: structure, evolution, mass loss, and critical limits
Proceedings IAU Symposium No. 272, 2010
C. Neiner, G. Wade, G. Meynet & G. Peters, eds.

doi:10.1017/S1743921311011550

The mysterious high-latitude O-star HD 93521: new results from XMM-Newton observations

Gregor Rauw and Thierry Morel

Institut d'Astrophysique & Géophysique, University of Liège,
Allée du 6 Août 17, B-4000, Liège, Belgium
email: rauw@astro.ulg.ac.be

Abstract. The O9.5 Vp star HD 93521 is a well known non-radial pulsator located at a high Galactic latitude. The nature (Population I vs. II) of this star has been the subject of controversy for many years. We report on an XMM-Newton observation of the star that sheds new light on its nature.

Keywords. stars: early-type, stars: individual (HD 93521), X-rays: stars

1. Introduction

The high Galactic latitude O9.5 Vp star HD 93521 ($l_{\rm II} = 183.14°$, $b_{\rm II} = 62.15°$), is located 1.4 kpc above the Galactic plane, too far away from any known site of recent star formation for any reasonable value of the runaway velocity. Based on this observation and the fact that the star has a rather low wind velocity, Ebbets & Savage (1982) argued that HD 93521 could instead be a low-mass Population II star. However, Irvine (1989) and Lennon *et al.* (1991) concluded that this is unlikely and that the star must be a Population I O-star that formed in the Galactic halo. HD 93521 is one of the fastest rotators ($v \sin i = 390\,{\rm km\,s^{-1}}$) known among O-stars and its absorption lines display bi-periodic ($\nu_1 = 13.68$, $\nu_2 = 8.31\,{\rm d^{-1}}$) profile variations that are likely due to non-radial pulsations, although an alternative explanation based on the effect of an orbiting (and accreting) compact companion could not be ruled out entirely (see Rauw *et al.* 2008 and references therein).

2. New observations

To help clarify the nature of HD 93521, we have obtained several new observations.

In the optical, we have obtained ELODIE echelle spectra at the 1.93 m telescope of the Haute-Provence Observatory (OHP). The lines are very broad, as expected for a fast rotator. Nevertheless, we can identify a number of metallic lines, in agreement with the finding of Lennon *et al.* (1991) who argued that the metal lines are too strong for a Population II low-mass star.

A 40 ksec XMM-Newton X-ray observation of HD 93521 was obtained in November 2009. HD 93521 is clearly detected, along with a few other sources. We have processed the data with SAS version 8.0 and extracted the EPIC spectra of HD 93521. The spectra can be fitted with a two-temperature thermal plasma model with $kT_1 = 0.29$ and $kT_2 = 3.6$ keV with solar abundances (see Fig. 1) and an interstellar hydrogen column density of $N_{\rm H} = 1.3 \times 10^{20}\,{\rm cm^{-2}}$ (Bohlin *et al.* 1978). A slightly better quality of the fit is obtained with subsolar abundances (0.1× solar). The absorption corrected X-ray flux in

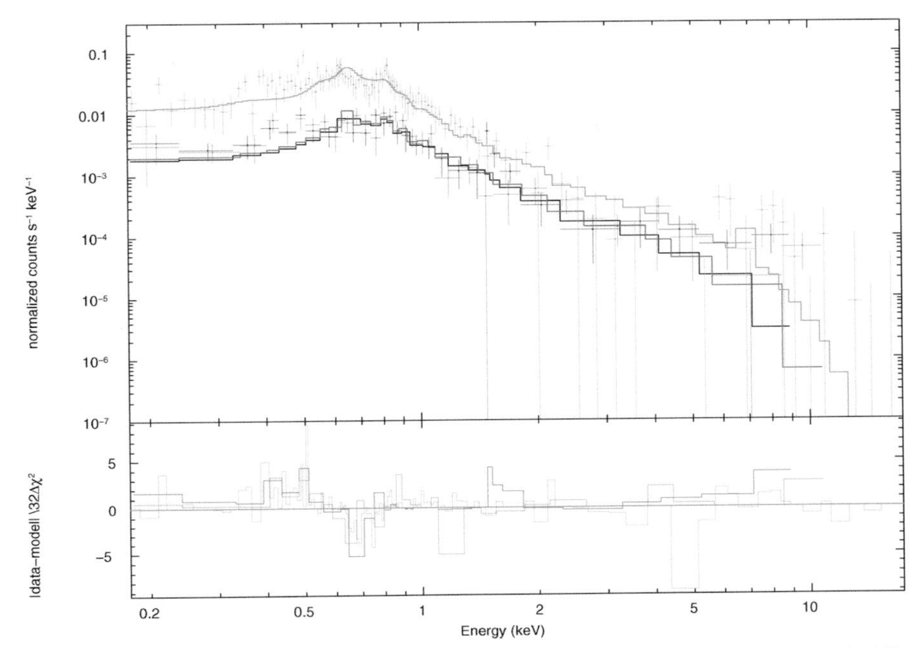

Figure 1. Upper panel: X-ray spectrum of HD 93521 as observed with the three EPIC instruments onboard `XMM-Newton`. The spectrum was fitted with a two-temperature optically thin thermal plasma model with solar metallicity. Bottom panel : residuals of the fit.

the 0.5 - 10 keV range ($5.75 \times 10^{-14}\,\mathrm{erg\,cm^{-2}\,s^{-1}}$) yields an L_X/L_{bol} ratio (essentially independent of the actual abundances and independent of the distance of the star) of 9.5×10^{-8}, towards the lower end of the canonical relation for normal O-type stars (Nazé 2009). The presence of a high-temperature emission at 3.6 keV is somewhat unexpected for a wind terminal velocity of $v_\infty = 400\,\mathrm{km\,s^{-1}}$ (Howarth *et al.* 1997), although the latter value is probably affected by the fast rotation of the star.

3. Conclusions

Our `XMM-Newton` observations of HD 93521 indicate a rather normal X-ray emission for an O-star, with no obvious signature of a compact companion. However, the X-ray spectra do not allow us to conclude about the metallicity of the star. The optical spectra reveal rather strong, but broad metal lines and we are currently analysing the optical spectrum with a model atmosphere code to derive the chemical composition of the star.

References

Bohlin, R. C., Savage, B. D., & Drake, J. F. 1978, *ApJ*, 224, 132
Ebbets, D. C. & Savage, B. D. 1982, *ApJ*, 262, 234
Howarth, I. D., Siebert, K. W., Hussain, G. A. J., & Prinja, R. K. 1997, *MNRAS*, 284, 265
Irvine, N. J. 1989, *ApJ* (Letters), 337, L33
Lennon, D. J., Dufton, P. L., Keenan, F. P., & Holmgren, D. E. 1991, *A&A*, 246, 175
Nazé, Y. 2009, *A&A*, 506, 1055
Rauw, G., De Becker, M., van Winckel, H., Aerts, C. *et al.* 2008, *A&A*, 487, 659

Active OB stars: structure, evolution, mass loss, and critical limits
Proceedings IAU Symposium No. 272, 2010
C. Neiner, G. Wade, G. Meynet & G. Peters, eds.

doi:10.1017/S1743921311011562

X-ray emission from hydrodynamical wind simulations in non-LTE models

Jiri Krtička[1], Achim Feldmeier[2], Lidia M. Oskinova[2], Jiri Kubát[3], Wolf-Rainer Hamann[2]

[1]Masaryk University, Kotlářská 2, CZ-611 37 Brno, Czech Republic
email: krticka@physics.muni.cz

[2]Institut für Physik und Astronomie, Universität Potsdam, Karl-Liebknecht-Straße 24/25, 14476 Potsdam-Golm, Germany
email: afeld, lida, wrh@astro.physik.uni-potsdam.de

[3]Astronomický ústav, Akademie věd České republiky, CZ-251 65 Ondřejov, Czech Republic
email: kubat@sunstel.asu.cas.cz

Abstract. Massive hot stars are strong sources of X-ray emission originating in their winds. Although hydrodynamical wind simulations that are able to predict this X-ray emission are available, the inclusion of X-rays in stationary wind models is usually based on crude approximations. To improve this, we use results from time-dependent hydrodynamical simulations of the line-driven wind instability to derive an analytical approximation of X-ray emission in the stellar wind. We use this approximation in our non-LTE wind models and find that an improved inclusion of X-rays leads to a better agreement between model ionization fractions and those derived from observations. Furthermore, the slope of the Lx-L relation is in better agreement with observations, albeit the X-ray luminosity is underestimated by a factor of three. We propose that a possible solution for this discrepancy is connected with the wind porosity.

Keywords. stars: winds, outflows, stars: early-type, hydrodynamics, X-rays: stars

1. X-rays in non-LTE wind models

Hot stars are known as X-ray sources. These X-rays originate in the stellar wind. As they influence the wind ionization, they should be included in the non-LTE wind models.

There have been earlier attempts to include X-ray emission in non-LTE wind models. They were either based on simplified analytical models, or the X-ray emission was included using free parameters (aka the "filling factor") describing the hot wind part. Here we use the results of hydrodynamical simulations of Feldmeier *et al.* (1997) to describe the X-ray emission in a compact form and include it in our non-LTE wind models.

2. Models: hydrodynamical simulations and non-LTE models

The X-ray emissivity in our models is derived employing hydrodynamical simulation of Feldmeier *et al.* (1997) calculated for ζ Ori A. A turbulent velocity variation at the wind bases was introduced as seed perturbation.

To incorporate the results of hydrodynamical simulations in non-LTE wind code in a manageable way, we approximate the emission from hydrodynamical simulations as a polynomial function. This could be done in two ways. The first way is to approximate the resulting X-ray emission, the second one is to find a polynomial that fits the temperature structure of the simulation (see Krtička *et al.* (2009) for the corresponding fits).

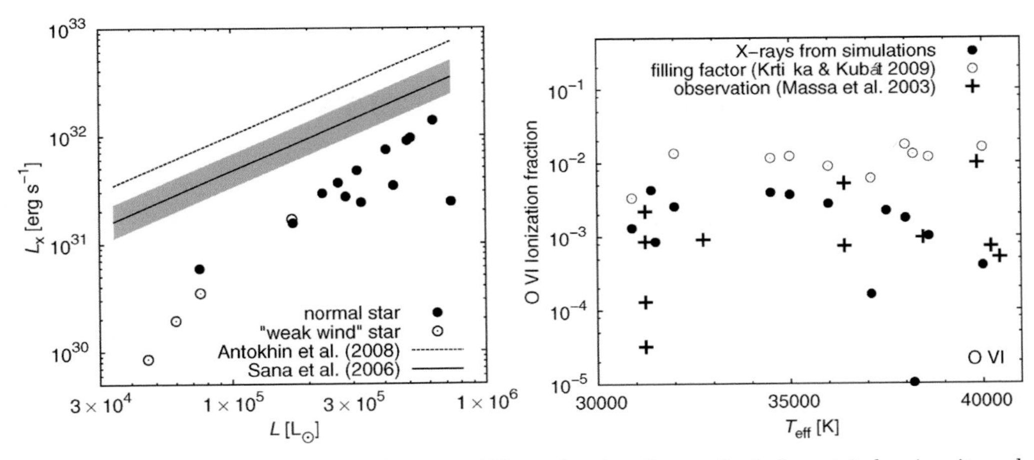

Figure 1. Left: The dependence of the total X-ray luminosity on the bolometric luminosity calculated using non-LTE models with X-ray emissivity from hydrodynamical simulations (filled and empty circles) for individual stars compared with the mean observational relations. Right: Comparison of predicted and observational ionization fraction as a function of the effective temperature. Filled circles refer to the models with X-ray emission from hydrodynamical simulations, and open circles refer to the models with X-ray emission described using filling factor.

Compared to other approximations, the temperature of X-ray emitting gas decreases with radius in the outer wind, and is described by a distribution function, which is more realistic than assuming just one temperature.

We include X-ray emission from hydrodynamical simulations into our stationary, spherically symmetric non-LTE wind model (Krtička & Kubát 2009).

3. Application: $L_{\rm X} - L$ relationship and ionization fractions

The predicted X-ray luminosities for stars with optically thick winds ($L \gtrsim 10^5\,{\rm L}_\odot$) are on average lower roughly by a factor of three than the observed ones (Fig. 1, left panel). This may originate in our neglect of macroclumping, which causes a lower X-ray opacity (Oskinova *et al.* 2004). On the other hand, the derived slope of the $L_{\rm X} - L$ relation for stars with optically thick winds, $L_{\rm X} \sim L^{1.0}$, is in a good agreement with observations.

The X-rays also influence the ionization fractions (Fig. 1, right panel). Although the non-LTE models calculated with X-ray emission from hydrodynamical wind simulations give a too low emergent X-ray luminosity, the ionization structure of these models corresponds in general much better to the trends derived from observations.

Acknowledgements

This research was supported by grant GA ČR 205/08/0003.

References

Feldmeier, A., Puls, J., & Pauldrach, A. W. A. 1997, *A&A*, 322, 878
Krtička, J. & Kubát, J. 2009, *MNRAS*, 394, 2065
Krtička, J., Feldmeier, A., Oskinova, L. M., Kubát, J. *et al.* 2009, *A&A*, 508, 841
Oskinova, L. M., Feldmeier, A., & Hamann, W.-R. 2004, *A&A*, 422, 675

Active OB stars: structure, evolution, mass loss, and critical limits
Proceedings IAU Symposium No. 272, 2010
C. Neiner, G. Wade, G. Meynet & G. Peters, eds.

doi:10.1017/S1743921311011574

High-angular resolution observations of the Pistol star

Christophe Martayan[1,2], Ronny Blomme[3], Jean-Baptiste Le Bouquin[4], Anthony Merand[1], Guillaume Montagnier[1], Fernando Selman[1], Julien Girard[1], Andrew Fox[1], Dietrich Baade[5], Yves Frémat[3], Alex Lobel[3], Fabrice Martins[6], Fabien Patru[1], Thomas Rivinius[1], Hugues Sana[7], Stanislas Štefl[1], Juan Zorec[8] and Thierry Semaan[2]

[1]ESO Chile; [2]GEPI-Observatoire de Meudon, France; [3]Royal Observatory of Belgium, Brussels, Belgium; [4]Laboratoire d'Astrophysique de Grenoble, France; [5]ESO Germany; [6]GRAAL, Montpellier, France; [7]Amsterdam University, The Netherlands; [8]Institut d'Astrophysique de Paris, France

Abstract. First results of near-IR adaptive optics (AO)-assisted imaging, interferometry, and spectroscopy of this Luminous Blue Variable (LBV) are presented. They suggest that the Pistol Star is at least double. If the association is physical, it would reinforce questions concerning the importance of multiplicity for the formation and evolution of extremely massive stars.†

Keywords. binaries: general, circumstellar matter, stars: early-type, stars: mass loss

---

1. Introducing the Pistol Star

At the time of its formation ~2 million years ago, the Pistol Star may have been one of the most massive stars in the Milky Way (Figer *et al.* 1998). Like most other LBV's, it is surrounded by nebulosities, which, therefore, may actually have been ejected by the Pistol Star (cf., e.g., η Car). The shape of the main gas cloud has earned the star its nick name. Owing to its own ejecta but mostly due to its far-away location (~ 8 kpc) in the region of the Galactic Center, observations of the Pistol Star suffer from very high extinction (20-30 magnitudes in the optical).

Among massive stars, the fraction of binaries may approach 100% at the time of formation (cf. Mason *et al.* 2009, Sana *et al.* 2009). In fact, binarity may be one means to enable the formation of massive stars, which with single-star-plus-disk symmetry encounters challenges already at the 10-$M_\odot$ level while observations of LBVs imply original masses that are larger by more than an order of magnitude. One of the most prominent massive stars reported to be a binary is η Car (Damineli, Conti, & Lopez 1997).

2. Observations

In order to more closely examine the nature of the Pistol Star, near-IR observations were undertaken with the VLT AO camera *NACO* and the VLTI spectrograph *AMBER*. The angular resolutions of *NACO*, ranging from natural seeing to near the defraction limit at ~15 mas, and *AMBER*, 20-3 mas, are perfectly complementary to one other. At a distance of 8 kpc, this permits spatial scales to be studied down to ~25 AU. The new VLT spectrograph *X-Shooter* provided spectra with $R \sim 5000 - 11000$ over the range 0.3-2.5 μm. They show a vast number of nebular lines; the stellar continuum begins to

† based on ESO runs 085.D-0182(A), 085.D-0625(A, C)

Figure 1. *NACO* image (2.3x2.2 arcsec) of the Pistol Star. Note the appearance of several nearby point sources, while the Pistol Star itself is surrounded by extended nebulosity, measuring roughly 700 AU across. The structure in the nebulosity is not due to discontinuities in the color coding but believed to be real, possibly evidencing multiple shells.

emerge at ~ 800 nm. At the time of the Symposium, some additional observations were pending, also of other massive stars.

3. First preliminary results

The NACO images resolve most of the stars within 30 arcsec of the Pistol Star into several point sources. Fig. 1 depicts the immediate vicinity of the Pistol Star itself as observed with *NACO* in K band. Apart from several point sources, it reveals an extended nebulosity. At a distance of 8 kpc, its diameter would correspond to more than 6000 AU. Figer *et al.* 1998 find gas velocities with a range of ~50-100 km/s but over a larger area. Combining these numbers leads to a crude estimate of the expansion age of 290 years. If the inference from Fig. 1 of multiple shells is correct, they would differ in age by ~30 years.

A quick preliminary inspection of the *AMBER* interferometry suggests the presence of a further point source at a separation of about 10 mas or ~ 80 AU. For comparison: If the 5.5-year period in η Car is orbital, its companion would be at one-fifth or less of this distance. In the *NACO* images, the inner nebulosity looks less structured than in the case of η Car. An effort will be made to continue the observations in order to establish whether the Pistol Star has a physical companion with effects on formation and evolution of the system.

References

Damineli, A., Conti, P. S., & Lopes, D. F. 1997, *New Astron.*, 2, 107

Figer, D. F., Najarro, F., Morris, M., McLean, I. S. *et al.* 1998, *ApJ*, 506, 384

Mason, B. D., Hartkopf, W. I., Gies, D. R., Henry, T. J. *et al.* 2009, *AJ*, 137, 3358

Sana, H., Gosset, E., & Evans, C. J. 2009, *MNRAS*, 400, 1479

Active OB stars: structure, evolution, mass loss, and critical limits
Proceedings IAU Symposium No. 272, 2010
C. Neiner, G. Wade, G. Meynet & G. Peters, eds.

doi:10.1017/S1743921311011586

Optical spectroscopic observations of the Be/X-Ray binary A0535+262/V725 Tau during the giant outburst in 2009

Yuuki Moritani[1], Daisaku Nogami[2], Atsuo T. Okazaki[3], Akira Imada[4], Eiji Kambe[4], Satoshi Honda[2], Osamu Hashimoto[5] and Kazuhide Ichikawa[1]

[1]Department of Astronomy, Kyoto University, Sakyo-ku, Kyoto 606-8502, Japan
email: moritani@kusastro.kyoto-u.ac.jp

[2]Kwasan Observatory, Kyoto University, Yamashina-ku, Kyoto 607-8471, Japan

[3]Faculty of Engineering, Hokkai-Gakuen University, Toyohira-ku, Sapporo 062-8605, Japan

[4]Okayama Astrophysical Observatory, National Astronomical Observatory of Japan, 3037-5 Honjo, Asakuchi, Okayama 719-0232, Japan

[5]Gunma Astronomical Observatory, Takayama-mura, Gunma 377-0702, Japan

Abstract. A giant outburst occurred in A0535+262/V725 Tau in November 2009, which lasted approximately 30 days. We carried out spectroscopic monitoring at OAO and GAO from November 2009 to March 2010, from before the giant outburst to the rising phase of the normal outburst which occurred after the next periastron. The obtained H-alpha, H-beta and He I emission lines exhibited drastic profile variability during the observations.

Keywords. binaries: spectroscopic, stars: emission-line, Be, stars: individual (A0535+262)

1. Introduction and Observation

Be/X-ray binaries, which dominate the majority of high-mass X-ray binaries, consist of a Be star and a compact object, usually a neutron star (NS). The X-ray activity of Be/X-ray binaries is divided into three states with respect to the luminosity; quiescent ($L_X \lesssim 10^{36}$ $ergs^{-1}$), normal outburst ($L_X \sim 10^{36-37}$ $ergs^{-1}$), and giant outburst ($L_X \gtrsim 10^{37}$ $ergs^{-1}$). These outbursts occur due to mass transfer from the Be disk, the geometrically thin circumstellar envelope formed by the outflow ($\lesssim$ 1 km/s) in the equatorial region and the rapid rotation (at 70–80% of the critical speed), to the neutron star [Okazaki & Negueruela (2001), Negueruela & Okazaki (2001)].

A0535+262/V725 Tau ($P_{orbital}$ = 110.2 days, e$\sim$0.47) is one of the best studied Be/X-ray binaries since its discovery [Rosenberg *et al.* (1975), Coe *et al.* (1975), Finger *et al.* (1994), Moritani *et al.* (2010)]. In November/December 2009, a giant outburst occurred in A0535+262 for the first time since 2005 [Sugizaki *et al.*(2009)]. Swift/BAT team reported that it reached more than 3 times of the Crab in the 15–50 keV band. After the next periastron, in March 2010, a normal outburst occurred.

Optical spectroscopic observations of A0535+262 were carried out from November 2009 to March 2010 at the Okayama Astrophysical Observatory (OAO) with a 188 cm telescope equipped with HIDES (High Dispersion Echelle Spectrograph), and at Gunma Astronomical Observatory (GAO) with a 1.5 m telescope equipped with GAOES (GAO Echelle Spectrograph). The typical wavelength resolution R and the signal to noise ratio S/N of our data around Hα are $R\sim$30000–60000 and $S/N\gtrsim$100. The obtained data were reduced in the standard way using IRAF (see http://iraf.noao.edu) echelle package.

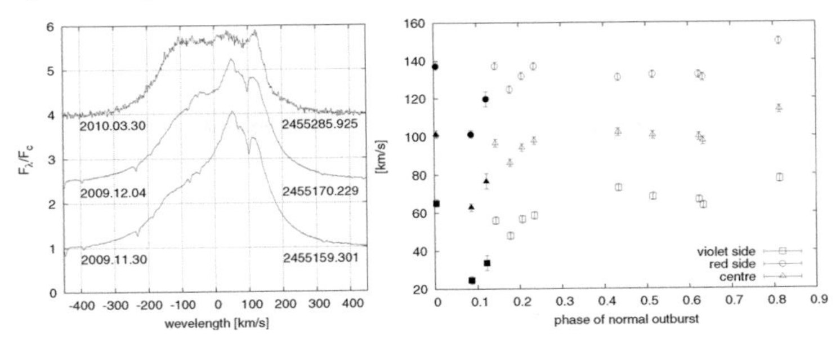

Figure 1. *Left:* representative Hα line profiles. *Right:* radial velocity of redshifted double-peaked component in Hα line profiles.

2. Results and Conclusion

The representative spectra of Hα obtained during the monitoring are shown in Fig. 1. For reasons of clarity, the spectra have linear offsets in the vertical axis from each other. The observation date and HJD of mid exposure time are written below each profile on the left and the right side, respectively.

Balmer and He I emission lines have changed drastically. The absolute value of equivalent width (EW) around the giant outburst were highest in the last five years; EW(Hα) ~ -18 Å, EW(Hβ) ~ -3 Å and EW(He Iλ5875) ~ -2 Å. During the observations, emission line profiles had the redshifted ($\sim$ 100 km/s) enhanced components (double peaked in Hα, and single peaked in other lines). The enhanced component also showed variability along with the X-ray flux.

Optical emission lines come from the Be disk mainly, and therefore observed line profile variability implies that the Be disk changed during the giant outburst and the next normal outburst. However, the contribution of the accretion disk around the NS may not be negligible, since it may grow enough. We then measured the radial velocity of the double-peak component in the red part of Hα to check which is the emitter, the Be disk or the disk around the NS (Fig. 1). As a result, the radial velocity decreased around the periastron. The geometry presented in Coe *et al.* (2006) indicates that the Be star goes back away to an observer at periastron passage. Considering this fact, it is plausible that the enhanced component comes from not the accretion disk but the Be disk.

After the giant outburst, a couple of normal outburst outbursts have occurred [Camero-Arranz *et al.*(2010)] The Be disk of A0535+262 is still active according to its EW, and then the next several periastron passage should be a good chance to chase the interaction of the Be disk and the NS.

References

Camero-Arranz, A., Finger, M. H., Wilson-Hodge, C., & Jenke, P. 2010, *The Astronomer's Telegram* 2705, 1

Coe, M. J., Carpenter, G. F., Engel, A. R., & Quenby, J. J. 1975, *Nature*, 256, 630

Coe, M. J., Reig, P., McBride, V. A., Galache, J. L. *et al.* 2006, *MNRAS*, 368, 447

Finger, M. H., Cominsky, L. R., Wilson, R. B., Harmon, B. A. *et al.* 1994, in: S. Holt & C. S. Day (eds.), *The Evolution of X-ray Binariese*, AIP-CP 308, p. 459

Moritani, Y., Nogami, D., Okazaki, A. T., Imada, A . *et al.* 2010, *MNRAS*, 405, 467

Negueruela, I. & Okazaki, A. T. 2001, *A&A* 369, 108

Okazaki, A. T. & Negueruela, I. 2001, *A&A*, 377, 161

Rosenberg, F. D., Eyles, C. J., Skinner, G. K., & Willmore, A. P. 1975, *Nature*, 256, 628

Sugizaki, M., Mihara, T., Kawai, N., Nakajima, M. *et al.* 2009, *The Astronomer's Telegram* 2277, 1

Active OB stars: structure, evolution, mass loss, and critical limits
Proceedings IAU Symposium No. 272, 2010
C. Neiner, G. Wade, G. Meynet & G. Peters, eds.

doi:10.1017/S1743921311011598

Determination of fundamental parameters and circumstellar properties for a sample of B[e] stars

María F. Muratore[1], **Lydia S. Cidale**[1,2], **María L. Arias**[1,2], **Juan Zorec**[3] **and Andrea F. Torres**[1,2]

[1]Facultad de Ciencias Astronómicas y Geofísicas, Universidad Nacional de La Plata, Paseo del Bosque S/N, 1900 La Plata, Argentina
email: fmuratore@carina.fcaglp.unlp.edu.ar

[2]Instituto de Astrofísica La Plata, CCT La Plata, CONICET, Paseo del Bosque S/N, 1900 La Plata, Argentina

[3]Institut d'Astrophysique de Paris, UPMC, UMR7095 CNRS, Paris, France

Abstract. We develop a simple model to derive theoretical continuum energy distributions for B[e] stars, consisting of a B star surrounded by an envelope made of gas and dust. We select a sample of B[e] objects for which we construct the observed energy distributions, from 0.1 to 100 μm, using available photometric and spectroscopic data. We present some preliminary fittings.

Keywords. stars: emission-line, Be, circumstellar matter, stars: fundamental parameters

1. Introduction

Stars with the B[e] phenomenon are mainly characterized by the presence of forbidden emission lines in the visual wavelength range and strong IR dust emission. This phenomenon is observed in stars in different evolutionary phases (Lamers *et al.* 1998) and many of these objects have "unclear" evolutionary states. In most cases, the determination of the star's properties is rather uncertain, as the circumstellar envelopes frequently blur the photospheric characteristics. Furthermore, the interstellar extinction is difficult to estimate and the unknown or inaccurate distances lead to uncertain luminosities. In this work we present a simple model of an envelope made of gas and dust and analyze its effect on the spectral energy distribution (SED) of a normal B star. Comparing theoretical results with observations we aim at deriving fundamental parameters of the star as well as global physical properties of the gas and dust envelopes.

2. The envelope model

We consider the circumstellar material distributed in a gaseous region located close to the star ($\approx 3\ R_*$) and a dusty region, far from the star ($\gtrsim 100\ R_*$).

Circumstellar gas: The proposed model of the gaseous circumstellar envelope is based on those presented in Cidale & Ringuelet (1989), Moujtahid (1998) and Moujtahid *et al.* (1999). We calculate the emergent flux of a system formed by a star + spherical envelope by considering that the envelope can be reduced to an equivalent shell. We can then apply a plane-parallel solution for the transfer equation. The observed flux at a distance D obtained in this way can be expressed as:

$$f_\lambda^{*+G} = \frac{R_*^2}{D^2}\, F_\lambda^*\, \alpha_\lambda\left(\frac{R_*}{R_G}, \tau_\lambda^G\right) + \frac{R_G^2}{D^2}\, S_\lambda(T_G)\, \beta_\lambda\left(\frac{R_*}{R_G}, \tau_\lambda^G\right) \quad (2.1)$$

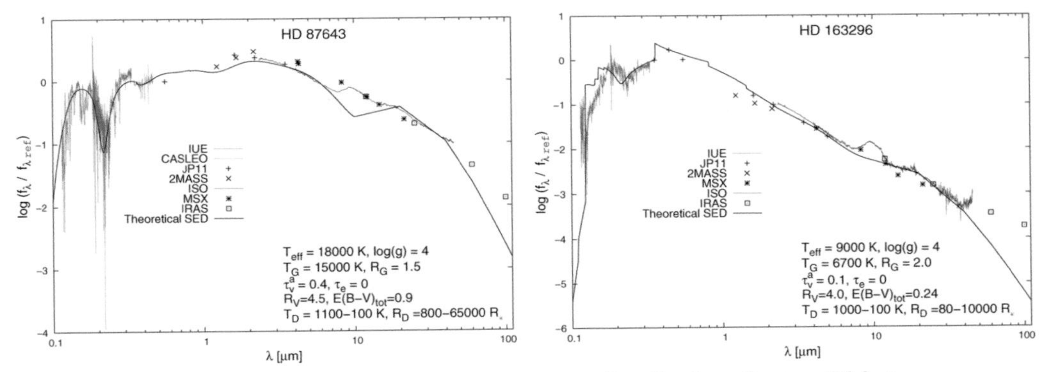

Figure 1. Theoretical and observed energy distributions for two B[e] stars

with τ_λ^G, optical depth of the gaseous shell; R_*, stellar radius; R_G, gaseous shell effective radius; F_λ^*, photospheric stellar flux; $S_\lambda(T_G)$, source function; T_G, electron temperature of the gas, α_λ and β_λ, functions that can be numerically calculated.

<u>*Circumstellar and Interstellar dust:*</u> The circumstellar dust region is treated using the same scheme proposed for the gaseous region. We characterize the dust by a total optical depth τ_λ^D, a temperature T_D and a source function $S_\lambda = B_\lambda(T_D)$. We represent the interstellar extinction by τ_λ^{ISM}. Both, τ_λ^D and τ_λ^{ISM} are related to the absorption $A(\lambda)$ through the expression: $\tau = 0.4ln(10)A(\lambda)$. Using the law given by Cardelli, Clayton & Mathis (1989): $A(\lambda) = [R_v a(1/\lambda) + b(1/\lambda)]E(B-V)$, where R_v is the total to selective extinction and $E(B-V)$ is the color excess. We adopt $R_v^{ISM} = 3.1$ for the interstellar dust, while for the circumstellar dust we try different values of R_v^D greater than 3.1.

<u>*Observed flux:*</u> Finally, we obtain an expression for the observed flux at a distance D, for: star + gaseous shell + dust shell + interstellar extinction given by:

$$f_\lambda^{*+G+D+ISM} = f_\lambda^{*+G} e^{-\tau_\lambda^D} + \frac{R_D^2}{D^2}\, S_\lambda(T_D)\, [1 - 2E_3(2\tau_\lambda^D)]\, e^{-\tau_\lambda^{ISM}} \tag{2.2}$$

In order to eliminate the dependence on the distance D, we normalize the flux to that at a reference wavelength λ_{ref}. Using expression (2.2) we obtain the theoretical SED, from 0.1 to 100 μm, for different sets of the free parameters of our model: R_G, T_G, τ_λ^G, R_D, T_D, R_v^D, $E_D(B-V)$ and $E_{ISM}(B-V)$.

3. Results

Photometric and spectroscopic data for B[e] stars is scarce and/or cover small wavelength ranges. We gathered all the information found in the literature and were able to build observed SEDs that cover a great percentage of the modeled range. In the following figures we present the observed SED together with the best theoretical fit for two objects of our sample: HD87643 and HD163296.

References

Cardelli, J. A., Clayton, G. C., & Mathis, J. S. 1989, *ApJ*, 345, 245

Cidale, L. S. & Ringuelet, A. E. 1989, *PASP*, 101, 417

Lamers, H. J. G. L. M., Zickgraf, F.-J., de Winter, D., Houziaux, L. *et al.* 1998, *A&A*, 340, 117

Moujtahid A. 1998, *Thèse de Doctorat*, Université Paris VI

Moujtahid, A., Zorec, J., & Hubert, A. M. 1999, *A&A*, 349, 151

Active OB stars: structure, evolution, mass loss, and critical limits
Proceedings IAU Symposium No. 272, 2010
C. Neiner, G. Wade, G. Meynet & G. Peters, eds.

doi:10.1017/S1743921311011604

Overall properties of hot, massive stars in the X-ray domain

Yael Nazé

FNRS-ULg, Dept AGO, Allée du 6 Août 17, Bât B5C, B4000-Liège, Belgium,
email: naze@astro.ulg.ac.be

Abstract. Despite the absence of large surveys, the recent X-ray observatories provide X-ray data for hundreds of massive stars (294 OB stars detected in the 2XMM catalog, 129 OB stars detected in the Chandra Carina Complex Project). Analyzing medium-resolution spectra led to new results on the relationship between the X-ray luminosity and the bolometric luminosity, as well as on the typical properties (plasma temperature, variability) of these objects.

Keywords. X-rays: stars, stars: early-type

1. Introduction

The X-ray emission of hot stars was serendipitously discovered in 1978-9 (Harnden *et al.* 1979, Ku *et al.* 1979, Seward *et al.* 1979). A link between the X-ray and the visible/bolometric luminosities was readily suggested (Harnden *et al.* 1979, Pallavicini *et al.* 1981). With time, this relation was investigated theoretically as well as observationally.

Theoretical considerations show that Lx "naturally" scales with $\dot{M}/v_\infty$ and therefore with the bolometric luminosity IF there is "a delicate balance between X-ray emission and absorption" AND "a special form for the radial distribution of wind shocks" (Owocki & Cohen 1999). In fact, the X-ray filling factor should vary with radial distance from the star as r^{-s} (with s=0.25–0.4, but recent fitting of X-ray lines rather yields s=0, Cohen *et al.* 2006).

In-depth observational studies of that $L_X - L_{BOL}$ relation were also performed, with two different approaches. On the one hand, Berghöfer *et al.* (1997) correlated the Rosat All-Sky Survey (RASS) with the Yale bright star catalog (i.e. an heterogeneous population of massive stars, mixing field and cluster objects). Fixing the interstellar absorption, they used the RASS count rates and hardness ratios to derive absorption-corrected X-ray luminosities. They confirmed that $L_X = 10^{-7} \times L_{BOL}$ down to 10^{38} erg s^{-1} (~B1–1.5), though a large scatter (0.4 dex) existed around the relation. On the other hand, XMM-Newton and Chandra took sensitive observations of several clusters (i.e. homogeneous populations). In some cases (e.g. NGC6231 Sana *et al.* 2006, Carina OB1 Antokhin *et al.* 2008), a detailed analysis of the X-ray spectra, taking into account the interstellar absorption (fixed) but also the potentially additional wind absorption, could be done. It provided precise X-ray ! luminosities while, in parallel, precise bolometric luminosities were derived from dedicated monitorings. From these studies, the scatter around the $L_X - L_{BOL}$ relation appears reduced (0.1–0.2 dex). The difference in scatter could have two causes: (1) the improvement in the knowledge of the physical parameters in the recent studies (stellar multiplicity, spectral types and improved photometry from monitorings, detailed X-ray spectral characteristics) or (2) the difference in data analysis (single/several instrument and reduction software, same/different BC and gas-to-dust ratio) and stellar populations (homogeneous vs heterogeneous populations).

2. Hot stars in the 2XMM

The 2XMM survey is not an all-sky survey but a compilation of all archival XMM-Newton datasets (Watson *et al.* 2009). Nevertheless, it permits to combine the characteristics of both approaches, i.e. a detailed fitting and a homogeneous treatment of a large (but heterogeneous) sample of stars. Cross-correlating the Reed catalog (Reed 2003, which takes into account the recent monitorings), 294 OB stars were detected, half of these being bright enough for lightcurve/spectral analyses. The results are (Nazé 2009): X-ray variability appears rare on short timescales but rather common on long timescales; the distribution of the detected O-stars is similar to the overall catalog distribution but there is a lack of detections for late and giant B-stars; the spectral fitting revealed the need for both an additional absorption and a second, hot thermal component (often at ~2keV, which is unexpected!) in O-stars; O+OB binaries display X-ray luminosities quite similar to those of sin! gle objects. Finally, the L_X/L_{BOL} ratio shows a scatter similar to that found in the RASS: cluster-to-cluster differences therefore probably exist.

3. Hot stars in the CCCP

The Chandra Carina Complex Project led to the detection of 129 OB stars, half of them having more than 50 recorded counts (i.e. being fit for at least a rough spectral analysis). Spectral fitting was then performed as for the 2XMM sample, on a large but homogeneous sample of stars. The results (Nazé *et al.* 2010, submitted) confirm the 2XMM findings, i.e. there is a need for additional absorption and 2nd hard thermal component in O-stars and, apart from HD 93403, X-ray bright wind-wind collisions are rare in most O+OB binaries. In addition, there is a smooth transition in the X-ray properties from the late O- to the earliest B-stars; the scatter around the $L_X - L_{BOL}$ relation is small, as for NGC6231; and there is a shallow increasing trend (to be confirmed!) of medium-to-soft flux ratio with larger L_X/L_{BOL} ratio or L_{BOL}. Open questions remain, however, e.g. most sources are soft and not luminous and the few overluminous and hard ones are CWB or MCW candidate! s - but what are the other outliers?

References

Antokhin, I. I., Rauw, G., Vreux, J.-M., van der Hucht, K. A. *et al.* 2008, *A&A*, 477, 593

Berghöfer, T. W., Schmitt, J. H. M. M., Danner, R., & Cassinelli, J. P. 1997, *A&A*, 322, 167

Cohen, D. H., Leutenegger, M. A., Grizzard, K. T., Reed, C. L. *et al.* 2006, *MNRAS*, 368, 1905

Harnden, Jr., F. R., Branduardi, G., Gorenstein, P., Grindlay, J. *et al.* 1979, *ApJ* (Letters), 234, L51

Ku, W. H.-M. & Chanan, G. A. 1979, *ApJ* (Letters), 234, L59

Nazé, Y. 2009, *A&A*, 506, 1055

Owocki, S. P. & Cohen, D. H. 1999, *ApJ*, 520, 833

Pallavicini, R., Golub, L., Rosner, R., Vaiana, G. S. *et al.* 1981, *ApJ*, 248, 279

Reed, B. C. 2003, *AJ*, 125, 2531

Sana, H., Rauw, G., Nazé, Y., Gosset, E. *et al.* 2006, *MNRAS*, 372, 661

Seward, F. D., Forman, W. R., Giacconi, R., Griffiths, R. E. *et al.* 1979, *ApJ* (Letters), 234, L55

Watson, M. G., Schröder, A. C., Fyfe, D., Page, C. G. *et al.* 2009, *A&A*, 493, 339

Active OB stars: structure, evolution, mass loss, and critical limits
Proceedings IAU Symposium No. 272, 2010
C. Neiner, G. Wade, G. Meynet & G. Peters, eds.

doi:10.1017/S1743921311011616

The surprising X-ray emission of Oe stars

Yael Nazé[1], Gregor Rauw[1] and Asif ud-Doula[2]

[1]FNRS-ULg, Dept AGO, Allée du 6 Août 17, Bât B5C, B4000-Liège, Belgium,
email: naze@astro.ulg.ac.be

[2]Penn State Worthington Scranton, 120 Ridge View Drive, Dunmore, PA 18512, USA

Abstract. Oe stars are thought to represent an extension of the Be phenomenon to higher temperatures. Dedicated XMM observations of HD 155806 revealed a surprising X-ray spectrum: soft character, absence of overluminosity, broad X-ray lines. These properties are fully compatible with the wind-shock model, which usually explains the X-rays from "normal", single O-type stars. In contrast, some other Oe/Be stars display a completely different behaviour at high energies.

Keywords. X-rays: stars, stars: early-type, stars: individual (HD 155806), stars: emission-line, Be

---

1. Introduction

The Oe category was defined by Conti & Leep (1974) as O-stars displaying emission in Balmer lines without emission in the He II 4686 and N III 4634/41 lines. Their line profiles being reminiscent of those of Be stars, they are often seen as the continuity of the Be phenomenon at higher temperatures. While direct evidence of disk-like features has been found for Be stars, only indirect evidence can be put forward in the case of Oe stars, such as high projected rotational velocities (Negueruela *et al.* 2004) or long-term spectral variability similar to that of Be objects (e.g. Rauw *et al.* 2007). In this context, X-rays can be useful as they often provide crucial evidence for identifying peculiar wind features. Indeed, if the equatorial regions are optically-thick, some absorption effects should be seen, whereas a strong magnetic confinement would yield X-ray emission very different from that of "normal" O-stars (bright, hard vs faint, soft).

2. HD 155806

HD 155806 is the earliest Oe star (O7.5e), therefore potentially represents an extreme case among the Oe objects. As for Oe/Be stars, variations are detected for HD 155806 but the Hα emission never disappears completely. The star is thought to be single (Garmany *et al.* 1980) and may be magnetic (Hubrig *et al.* 2008, Petit *et al.* 2009).

We investigated its X-ray emission thanks to a sensitive XMM-Newton observation, which yields both medium-resolution spectra (EPIC-MOS and pn) and high-resolution data (RGS). This 35ks exposure was taken in mid-2008, and an archival UVES spectrum obtained close to that date shows strong Hα emission. The data analysis is described in length in Nazé *et al.* (2010). The main results are: the X-ray spectrum is very soft (average plasma temperature ~0.2keV) and well fitted by two "cool" (<0.7keV) thermal components without the need of absorption in addition to ISM; no variations are detected on both short- and long-term ranges; there is no overluminosity ($L_X/L_{BOL} = -6.75$); the X-ray lines are symmetric, broad (with FWHM~2500 km s^{-1} and a strict lower limit of 1000 km s^{-1} for the 90% confidence interval of the worst fit), and with f/i ratios indicating line forming regions $< 10R_*$. All these properties are typical of O stars.

3. Do Oe stars have disks ?

The question of whether Oe stars have disks or not, we believe, is a wide open question. Using spectropolarimetric data, Vink *et al.* (2009) reported the non-detection of any wind asymmetries or disk-like features for a sample of peculiar O-stars (Oe, Of?p,...). However, before drawing such a conclusion, it is worth looking at these results a bit closer. Here, we summarize two main reasons why doubts could be cast (for a detailed rationale, see Nazé & Rauw 2010). First, no asymmetries/disk-like features are detected even for the well-known case of θ^1 Ori C (for which several evidences of a magnetically-confined wind have been found). A putative "inner disk hole" is put forward by the authors, but such a feature is neither detected in the data nor predicted by models (which otherwise reproduce well the observed stellar characteristics). This a posteriori guess of a "hole" should therefore be reconsidered to account for the star's known properties. Second, a similar absence of detection in Oe stars was found and attributed to the lack of disk-like features in the wind. This could at first seem to corroborate our high-energy results, but the authors then propose the presence of "an expanding shell which is spherically symmetric" to explain the double-peaked profiles, characteristics of Oe and Be stars. This scenario would result in strongly variable emission lines on rather short timescales: such changes are not observed, ruling out the shell scenario. The question of absence or presence of disk-like features in Oe stars thus remains open. Any new model should take into account the full properties of Oe stars, including their resemblance with Be objects.

4. X-rays from Oe/Be stars: comparison

For HD 155806, the X-ray emission is comparable to that of "normal" O-stars. There is no evidence for additional absorption or strong magnetic confinement. The physical phenomenon at the origin of the Oe characteristics (a disk-like feature?) thus seems to have no impact whatsoever in the X-ray range. This is very different from the case of θ^1 Ori C, where a magnetically-confined disk-like region clearly rules the X-ray emission.

Only one other Oe has been observed in the high-energy range: HD 119682 (O9.7e), whose high-energy properties were reported by Rakowski *et al.* (2006). Contrary to HD 155806, its X-ray emission is both hard (kT ~10keV) and bright ($L_X/L_{BOL} = -5.4$). It actually resembles that of some Be stars (the so-called "γ-Cas analogs"). At first, such emission was thought to be linked to the Be disk, but doubts have recently been expressed that the harder emission of γ-Cas itself is related to the Be phenomenon (Smith *et al.*, these proceedings). This seems to be supported by our observations, notably: not all Oe/Be stars show these peculiar characteristics.

References

Conti, P. S. & Leep, E. M. 1974, *ApJ*, 193, 113
Garmany, C. D., Conti, P. S., & Massey, P. 1980, *ApJ*, 242, 1063
Hubrig, S., Schöller, M., Schnerr, R. S., González, J. F. *et al.* 2008, *A&A*, 490, 793
Nazé, Y., Rauw, G., & Ud-Doula, A. 2010, *A&A*, 510, A59
Nazé, Y., & Rauw, G. 2010, *arXiv*, 1002.0546
Negueruela, I., Steele, I. A., & Bernabeu, G. 2004, *AN*, 325, 749
Petit, V., Fullerton, A. W., Bagnulo, S., Wade, G. A. *et al.* 2009, in: K. G. Strassmeier, A. G. Kosovichev & J. E. Beckman (eds.), *Cosmic Magnetic Fields: From Planets, to Stars and Galaxies*, IAU Symposium 259, p. 385
Rakowski, C. E., Schulz, N. S., Wolk, S. J., & Testa, P. 2006, *ApJ* (Letters), 649, L111
Rauw, G., Nazé, Y., Marique, P. X., De Becker, M. *et al.* 2007, *IBVS* 5773
Vink, J. S., Davies, B., Harries, T. J., Oudmaijer, R. D. *et al.* 2009, *A&A*, 505, 743

Active OB stars: structure, evolution, mass loss, and critical limits
Proceedings IAU Symposium No. 272, 2010
C. Neiner, G. Wade, G. Meynet & G. Peters, eds.

doi:10.1017/S1743921311011628

The latest developments on Of?p stars

Yael Nazé[1], Asif ud-Doula[2], Maxime Spano[3], Gregor Rauw[1], Michael De Becker[1] and Nolan R. Walborn[4]

[1]FNRS-ULg, Dept AGO, Allée du 6 Août 17, Bât B5C, B4000-Liège, Belgium,
email: naze@astro.ulg.ac.be

[2]Penn State Worthington Scranton, USA; [3]Obs. Genève, Switzerland; [4]STScI, USA

Abstract. In recent years several in-depth investigations of the three prototypical Of?p stars were undertaken, revealing their peculiar properties. To clarify some of the remaining questions, we have continued our monitoring of the prototypical Of?p trio. HD 108 has now reached its quiescent, minimum-emission state, for the first time in 50–60yrs, while new echelle spectra of HD 148937 confirm the presence in several H and He lines of the 7d variations detected previously only in the Hα line. A new XMM observation of HD 191612 clearly shows that its X-ray emission is not modulated by the orbital period of 1542d, but the high-energy variations are rather compatible with the 538d period of the optical changes - it is thus not of colliding-wind origin but linked to the phenomena responsible for the spectral/photometric variations, though our current MHD simulations remain at odds with the observational properties.

Keywords. X-rays: stars, stars: early-type, stars: individual (HD 108, HD 148937, HD 191612), stars: emission-line

---

1. Introduction

The Of?p category was proposed 40 years ago to gather a few O-type stars presenting peculiarities in their optical/UV spectrum. These objects were then known to display strong C III 4650 emission lines but recent studies showed that this is not their sole peculiar characteristics (in fact, at the present time, stars displaying "only" strong C III 4650 emission should be classified as Ofc, see Walborn *et al.* 2010). Of?p stars also possess: narrow P Cygni/emission features in Balmer, HeI,... lines which vary in a recurrent manner; correlated photometric variations; strong magnetic fields; and large X-ray overluminosities (for a review, see Nazé *et al.* 2008a). Five Galactic Of?p stars are known, but only the first three stars which were identified as such in the 1970s (the so-called "prototypical" objects HD 108, HD 148937, HD 191612) have been extensively studied, though some questions remained open about their physical properties. We have obtained additional data with the aim to contribute to the understanding of these peculiar objects (see Nazé *et al.* 2010).

2. HD 108

HD 108 is the first Of?p for which long-term spectral variations were reported (Andrillat *et al.* 1973, Nazé *et al.* 2001). However, the recurrence timescale appeared very long, about 55 yrs (Nazé *et al.* 2006). This value was not well constrained because of the lack of detailed observations over that period (often, only the line shape - emission/absorption/P Cygni - is mentioned). The Liège team has been closely monitoring HD 108 from 1986 on at the Observatoire de Haute-Provence. Since the beginning, the data showed a continuous decline in the emission strengths. This decrease has now stopped, as the spectra taken between 2005 and 2009 appear similar: the star has thus reached its quiescence state for the first time since the 1950s. It remains to be seen when a new cycle will begin. Indeed, the knowledge of the quiescence time will be very useful to better constrain the systems geometry.

3. HD 148937

A previous 3-yrs spectral monitoring revealed small-amplitude (a few percent of the peak's amplitude) variations of the Hα line, with a potential period of 7d (Nazé *et al.* 2008b). While the X-ray data were undistinguishable from those of HD 108 and HD 191612, the optical properties thus appeared quite different from those of the other two "prototypical" Of?p stars (which display large variations of many lines, occurring with a long period and correlated with rather large photometric changes). However, the quality of this first monitoring was quite limited: the data sampling (monthly exposures) and spectral resolution (R$\sim$2300) were not really fit to detect subtle variations occurring on short timescales. A new short-term monitoring with daily exposures taken with higher resolution (R$\sim$55000) was performed at the La Silla Observatory. Variations in some He I and He II 4686 lines were detected (but not in C III 4650) and the 7d period was confirmed for all ! varying lines. HD 148937 is thus similar to the other two "prototypical" Of?p stars, though with a shorter period: the small-amplitude is most probably related to a geometrical effect (e.g. a low inclination of the star's rotation axis in the scenario of an oblique rotator with magnetically-confined winds).

4. HD 191612

A previous X-ray monitoring (Nazé *et al.* 2007) unveiled some variations of the high-energy emission, but their cause was not ascertained: they could either be related to the binary nature of the star ($P_{orb} = 1542d$) through colliding-wind emission or to the line profile variations ($P_{cyc} = 538d$) typical of the Of?p phenomenon. Note that the old data were taken at orbital phases $\phi_{orb} = 0.83$–0.96 and at phases $\phi_{cyc} = 0.09$–0.44 in the line profile cycle. A new XMM dataset was taken at a different orbital phase ($\phi_{orb} = 0.55$) but at a similar phase in the spectral cycle ($\phi_{cyc} = 0.13$). After analysis, the recorded emission appears very similar to the previous data: a colliding-wind origin can thus be rejected.

To explain the variations and the X-ray overluminosity, a scenario involving magnetically-confined winds and oblique rotation was proposed. This can be modelled using MHD simulations, taking into account the known stellar properties. The predicted situation is a bright X-ray emission from a very hot plasma (>10MK), produced close to the star (hence X-ray lines should be narrow). While the overluminosity is indeed observed, the data rather favor a cooler plasma and quite broad X-ray lines. Work still remains to be done until the X-ray emission of the Of?p stars will be fully understood.

References

Andrillat, Y., Fehrenbach, C., Swings, P., & Vreux, J. M. 1973, *A&A*, 29, 171
Nazé, Y., Vreux, J.-M., & Rauw, G. 2001, *A&A*, 372, 195
Nazé, Y., Barbieri, C., Segafredo, A., Rauw, G. *et al.* 2006, *IBVS* 5693
Nazé, Y., Rauw, G., Pollock, A. M. T., Walborn, N. R. *et al.* 2007, *MNRAS*, 375, 145
Nazé, Y., Walborn, N. R., & Martins, F. 2008a, *Rev. Mexicana AyA* 44, 331
Nazé, Y., Walborn, N. R., Rauw, G., Martins, F. *et al.* 2008b, *AJ*, 135, 1946
Nazé, Y., Ud-Doula, A., Spano, M., Rauw, G. *et al.* 2010, *A&A* 520A, 59
Walborn, N. R., Sota, A., Maíz Apellániz, J., Alfaro, E. J. *et al.* 2010, *ApJ* (Letters), 711, L143

Active OB stars: structure, evolution, mass loss, and critical limits
Proceedings IAU Symposium No. 272, 2010
C. Neiner, G. Wade, G. Meynet & G. Peters, eds.

doi:10.1017/S174392131101163X

Interaction between the Be star and the compact companion in TeV γ-ray binaries

Atsuo T. Okazaki[1], Shigehiro Nagataki[2], Tsuguya Naito[3], Akiko Kawachi[4], Kimitake Hayasaki[5], Stanley P. Owocki[6] and Jumpei Takata[7]

[1]Faculty of Engineering, Hokkai-Gakuen University, Toyohira-ku, Sapporo 062-8605, Japan
email: okazaki@elsa.hokkai-s-u.ac.jp

[2]Yukawa Institute for Theoretical Physics, Sakyo-ku, Kyoto 606-8502, Japan
email: nagataki@yukawa.kyoto-u.ac.jp

[3]Faculty of Management Information, Yamanashi Gakuin University, Kofu, Yamanashi 400-8575, Japan
email: tsuguya@ygu.ac.jp

[4]Department of Physics, Tokai University, Hiratsuka, Kanagawa 259-1292, Japan
email: kawachi@icrr.u-tokyo.ac.jp

[5]Department Astronomy, Kyoto University, Sakyo-ku, Kyoto 606-8502, Japan
email: kimi@kusastro.kyoto-u.ac.jp

[6]Bartol Research Institute, University of Delaware, Newark, DE 19716, USA
email: owocki@bartol.udel.edu

[7]Department of Physics, University of Hong Kong, Pokfulam Road, Hong Kong, China
email: takata@hku.hk

Abstract. We report on the results from 3-D SPH simulations of TeV binaries with Be stars. Since there is only one TeV binary (B 1259-63) where the nature of the compact companion has been established, we mainly focus on this Be-pulsar system. From simulations of B 1259-63 around periastron, we find that the pulsar wind dominates the Be-star wind and strips off an outer part of the Be-star disk, causing a strongly asymmetric, phase-dependent structure of the circumstellar material around the Be star. Such a large modulation may be detected by optical, IR, and/or UV observations at phases near periastron. We also discuss the results from simulations of another TeV binary LS I+61 303, for which the nature of the compact object is not yet known.

Keywords. gamma rays: theory, stars: emission-line, Be, stars: winds, outflows, stars: individual (B 1259-63, LS I+61 303)

---

1. Introduction

Recent progress in VHE γ-ray astronomy, driven by ground-based Cherenkov telescopes such as H.E.S.S., MAGIC & VERITAS, has established TeV ($= 10^{12}$ eV) γ-ray binaries as a new class of γ-ray sources. There are only three binaries and one binary candidate that show persistent TeV emission. The nature of the compact object has been established only for one system (B1259−63). Interestingly, three among these four systems have a Be star, which is an early-type star with a polar wind and a dense equatorial disk. Interaction between the Be-star envelope and the compact companion is a key to modeling these systems and understanding physics of high energy emission.

In this paper, we report on the results from 3-D SPH simulations of two TeV binaries, B1259−63 ($P_{\rm orb} = 3.4\,$yr, $e = 0.87$) and LS I+61 303 ($P_{\rm orb} = 26.5\,$d, $e = 0.537$), both of which have a Be star as the primary. In B1259−63 consisting of a B2Ve star and a

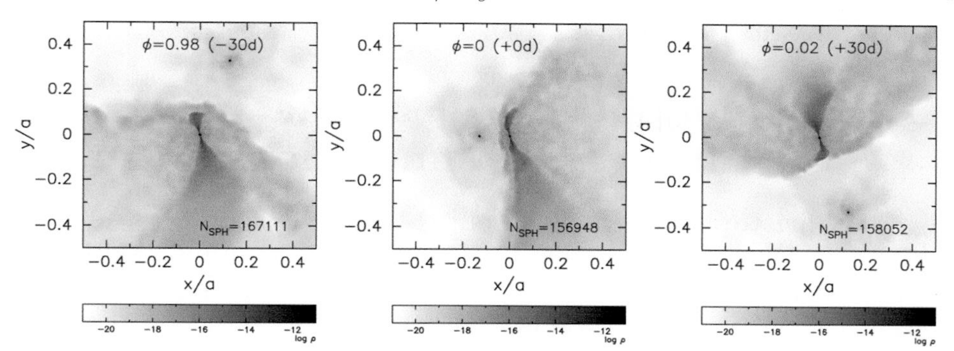

Figure 1. Snapshots of the interaction in B1259−63 around periastron: one month prior to periastron (*left*), periastron (*middle*), and one month after periastron (*right*). The grey-scale plot shows the density on the orbital plane in the logarithmic scale. The Be star has the mass loss rates of $4 \times 10^{-9} M_{\odot}\mathrm{yr}^{-1}$ via the equatorial disk and $10^{-9} M_{\odot}\mathrm{yr}^{-1}$ via the stellar wind, while the pulsar wind has the power of $8.2 \times 10^{35}\,\mathrm{ergs}^{-1}$. Annotated in each panel are the orbital phase and the number of SPH particles.

pulsar, the pulsar wind collides with the Be disk and wind, whereas in LS I+61 303, where the nature of the compact object is not yet known, there is a debate on whether it is a colliding wind binary (Dubus 2006) or a microquasar (Romero et al. 2007).

2. Numerical model

Simulations presented here were performed with a 3-D SPH code, in which optically thin radiative cooling is taken into account. In these simulations, relativistic pulsar wind is emulated by a non-relativistic 10^4 km s^{-1} wind with the adjusted mass-loss rate so as to give the same momentum as a relativistic flow with the assumed energy.

3. B1259−63

Figure 1 shows snapshots of the interaction in this proto-typical TeV binary around periastron. From the figure, we note that the pulsar has a huge influence on the circumstellar environment of the Be star. The wind from the pulsar dominates the Be-star wind and strips off an outer part of the Be-star disk on the side of the pulsar. The strongly asymmetric, phase-dependent structure of the circumstellar material around the Be star may be detected by optical, IR, and/or UV observations.

4. LS I+61 303

In order to see whether two competing models, i.e., a colliding wind model and a microquasar model, are distinguished by studying the Be disk structure, we have carried out a simulation based on each model. Our preliminary results show that the effect of the compact object on the Be disk structure is undistinguishable between these models, as long as the power of the pulsar wind $\lesssim 10^{36}\,\mathrm{ergs}^{-1}$, but this has to be confirmed by further detailed studies.

References

Dubus, G. 2006, *A&A*, 456, 801
Romero, G. E., Okazaki, A. T., Orellana, M., & Owocki, S. P. 2007, *A&A*, 474, 15

Active OB stars: structure, evolution, mass loss, and critical limits
Proceedings IAU Symposium No. 272, 2010
C. Neiner, G. Wade, G. Meynet & G. Peters, eds.

doi:10.1017/S1743921311011641

X-Ray modeling of η Carinae & WR 140 from SPH simulations

Christopher M. P. Russell[1], Michael F. Corcoran[2], Atsuo T. Okazaki[3], Thomas I. Madura[1] and Stanley P. Owocki[1]

[1]Univ. of Delaware, Newark, DE, USA; email: crussell@udel.edu
[2]GSFC/NASA, Greenbelt, MD, USA
[3]Hokkai-Gakuen Univ., Sapporo, Japan

Abstract. The colliding wind binary (CWB) systems η Carinae and WR140 provide unique laboratories for X-ray astrophysics. Their wind-wind collisions produce hard X-rays that have been monitored extensively by several X-ray telescopes, including RXTE. To interpret these RXTE X-ray light curves, we apply 3D hydrodynamic simulations of the wind-wind collision using smoothed particle hydrodynamics (SPH). We find adiabatic simulations that account for the absorption of X-rays from an assumed point source of X-ray emission at the apex of the wind-collision shock cone can closely match the RXTE light curves of both η Car and WR140. This point-source model can also explain the early recovery of η Car's X-ray light curve from the 2009.0 minimum by a factor of 2-4 reduction in the mass loss rate of η Car. Our more recent models account for the extended emission and absorption along the full wind-wind interaction shock front. For WR140, the computed X-ray light curves again match the RXTE observations quite well. But for η Car, a hot, post-periastron bubble leads to an emission level that does not match the extended X-ray minimum observed by RXTE. Initial results from incorporating radiative cooling and radiative forces via an anti-gravity approach into the SPH code are also discussed.

Keywords. X-rays: binaries, stars: individual (η Carinae), hydrodynamics

---

1. Point-Source Emission Model

Our initial attempts to model the 2-10 keV RXTE light curves of both η Car and WR140 have applied a simple model of point-source emission plus line-of-sight wind absorption to 3D, adiabatic, smoothed particle hydrodynamics (SPH) simulations of the binary wind-wind interaction (see Okazaki *et al.* 2008 for details). To match the recent shorter minimum of η Car (Corcoran *et al.* 2010), the primary mass loss rate is reduced by a factor of 2.5 at phase 2.2. Many of the light curve's features are reproduced, including the shorter recent minimum. The WR140 light curve matches remarkably well.

2. Extended Emission Model

Our more recent efforts to model the RXTE light curves of η Car and WR140 relax the point-source approximation. The extended emission comes from the entire wind-wind collision region according to $\rho^2\Lambda(E,T)$, where ρ is the density and $\Lambda(E,T)$ is the emissivity as a function of energy E and temperature T obtained from the MEKAL code (Mewe *et al.* 1995), and the extended wind absorption is now energy dependent. We then use the SPH visualization program SPLASH (Price 2007) to calculate the ray-tracing through the system, which generates images in various X-ray bands that combine to make a 2-10 keV X-ray light curve. Once again, the WR140 light curve matches well (assuming the opacity is 10× the opacity of an O star wind at solar abundances, an assumption that

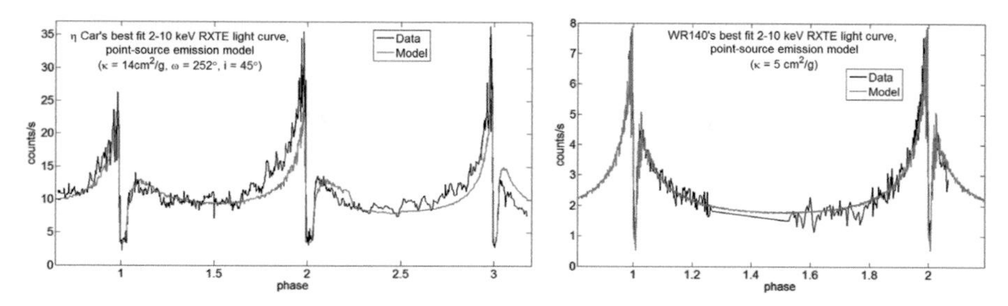

Figure 1. η Car and WR140's best fit RXTE light curves.

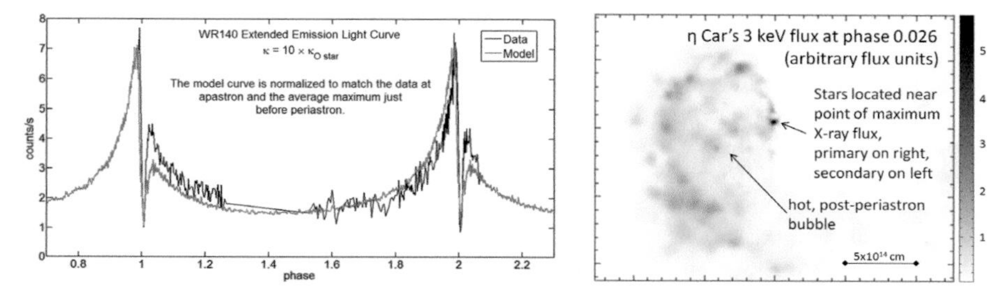

Figure 2. Left: WR140 extended emission light curve. Right: η Car flux at phase 0.026.

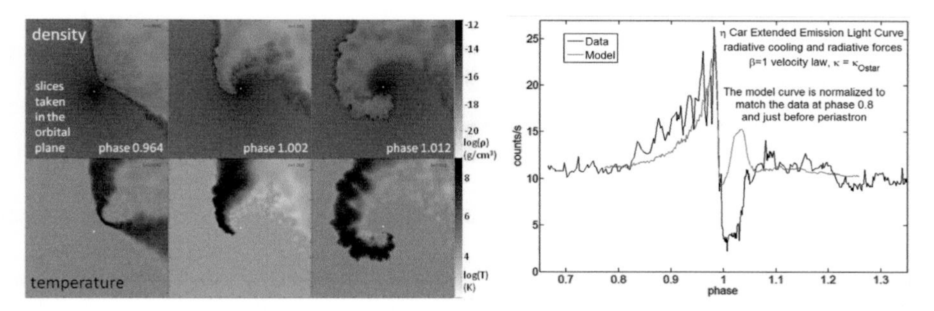

Figure 3. Left: Density and temperature maps. Right: η Car extended emission light curve.

will be relaxed in future work). The same is not true for η Carinae, however, where a hot, post-periastron bubble blown into the slow, dense primary wind by the much faster companion wind prevents the reproduction of the X-ray minimum.

Radiative cooling, via the Exact Integration Scheme (Townsend 2009), and radiative forces, via an anti-gravity approach, have been implemented to improve the SPH code. The acceleration of the secondary wind drastically decreases as the system approaches periastron, so the high temperature shock cone of η Car collapses, but the post-periastron bubble still prevents the reproduction of the minimum of the RXTE light curve with the extended emission model. However, outside the minimum, the model matches fairly well.

References

Corcoran, M. F., Hamaguchi, K., Pittard, J. M., Russell, C. M. P. *et al.* 2010, *ApJ*, submitted
Mewe R., Kaastra J. S., & Liedahl D.A. 1995, *Legacy* (HEASARC), 6, 16
Okazaki, A. T., Owocki, S. P., Russell, C. M. P., & Corcoran, M. F. 2008, *MNRAS*, 388, L39
Price, D. J. 2007, *PASA*, 24, 159
Townsend, R. H. D. 2009, *ApJS*, 181, 391

Active OB stars: structure, evolution, mass loss, and critical limits
Proceedings IAU Symposium No. 272, 2010
C. Neiner, G. Wade, G. Meynet & G. Peters, eds.

doi:10.1017/S1743921311011653

Uniqueness and evolutionary status of MWC 349A

Vladimir Strelnitski[1], **Kamber Schwarz**[1,2], **John Bieging**[2], **Josh T. Fuchs**[1,3], **and Gary Walker**[1]

[1]Maria Mitchell Observatory, USA
[2]University of Arizona, USA
[3] Rhodes College, USA

Abstract. MWC349A, which had remained an ordinary member of the MWC catalog for a few decades, is now known as: (1) the brightest stellar source of radio continuum; (2) the only known high-gain natural maser in hydrogen recombination lines; and (3) the only strictly proven natural high-gain laser (in IR hydrogen recombination lines). These phenomena seem to occur in the circumstellar disk seen almost edge-on. They help us understand the structure and kinematics of the disk. The evolutionary status of MWC 349A is still debated: a young HAeBe star with a pre-planetary disk or an old B[e] star or even a protoplanetary nebula? We discuss new observational data obtained at the Maria Mitchell Observatory and elsewhere which may cast light on this issue.

---

1. New Observational Facts concerning MWC 349A

1. Apparently chaotic variations in BVRI on time scales from days to years, with amplitudes $0.1^{\mathrm{m}} - 0.4^{\mathrm{m}}$ and a power spectrum of $\alpha \approx -0.3 \rightarrow -0.5$ (noise, between "white" and "pink").

2. Correlation of the integrated emission in hydrogen mm recombination lines with the optical emission on the year time scale.

3. Possible periodicity of the peak ratio in the double-peaked maser spectrum, with a period of $238 \pm 8^{\mathrm{d}}$ and possible periodicity of Hα emission with a close period of $223 \pm 7^{\mathrm{d}}$ (Fig. 1) but with unclear phase relations between the two.

4. Possible anti-correlation (at least temporary) of Hα emission with the nearby (645 nm) continuum; some periods of considerable variations of the continuum on the week time scale with almost constant Hα emission.

5. The discovery of a steeper than Keplerian (linear?) velocity gradient in the circumstellar disk (Weintroub *et al.* 2008).

6. The discovery of a few-arcmin hourglass structure at 24 μm around MWC349A (Hora *et al.* 2010), similar in shape and orientation to the sub-arcsec hourglass structures in radio (Tafoya *et al.* 2004). Maps in ^{12}CO and ^{13}CO rotational lines (Strelnitski *et al.* 2010) do not show obvious correlation with the 24 μm structure.

2. Discussion and Conclusions

1. Variability patterns indicate possible complex relationship between the sources of the continuum and emission lines. Accretion from the disk or from a companion of a close binary which had filled its Roche lobe may augment the ordinary hot star/HII region scheme. Periodicity of the maser peak ratio and Hα variations, if confirmed, may indicate a second component, a star or a massive planet, as well as a periodic component in the intrinsic variability of the star.

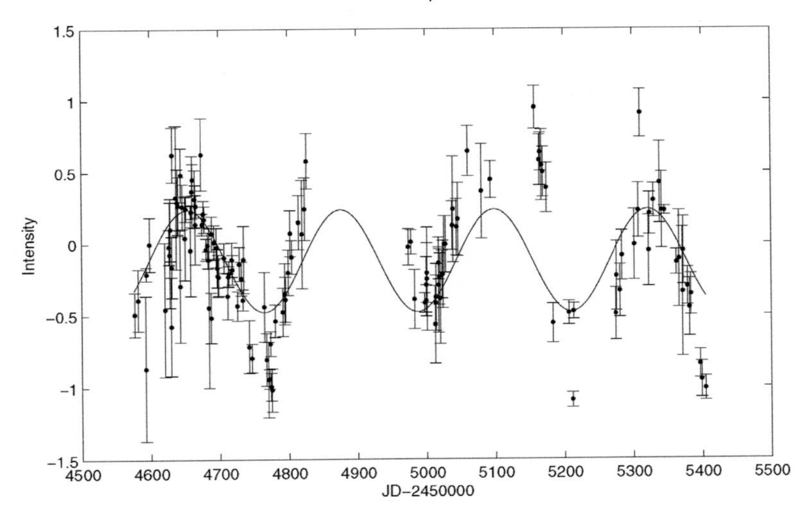

Figure 1. Long-time trend subtracted light curve of Hα emission in MWC349. Observations with the MMO 60 cm telescope and Hα interference filter (Schwarz *et al.*, 2010, BAAS, to be submitted).

2. If the object is young, the observed non-Keplerian (linear) gradient of velocity in the disk can be explained by the spinning up of the inner edge of the disk by magnetic stellar wind from a fast rotating central star. However, if the object is old and its disk is formed from the stellar wind of a fast rotating star, the linear gradient of the disk rotational velocity would be a natural consequence of the angular momentum conservation.

3. The lack of correlation between the 24 μm and CO structures supports the idea that the arc-minute hourglass structure is not shaped by the external clouds but is a result of a recurrent process of mass ejection from the central object, confined into a biconical outflow by a geometrically thick circumstellar disk, as in the prototypical Red Rectangle nebula (Men'shchikov *et al.* 2002).

Acknowledgements

The authors thank the time allocation committee of the 12-m and 10-m ARO radio telescopes and the telescopes operators for help with the observations. JF and KS acknowledge with gratitude their REU internship support by the NSF grant AST-0851892 and by the Maria Mitchell Association.

References

Hora J.L., Gutermuth, R. A., Carey, S., Mizuno, D. *et al.* 2010, in: P.M. Ogle (eds.), *Reionization to Exoplanets: Spitzer's growing legacy*, ASP-CS, in press

Men'shchikov, A. B., Schertl, D., Tuthill, P. G., Weigelt, G. *et al.* 2002, *A&A*, 393, 867

Strelnitski V., Bieging, J., Schwarz, K. *et al.* 2010, in preparation

Tafoya, D., Gómez, Y., & Rodríguez, L. F. 2004, *ApJ*, 610, 827

Weintroub, J., Moran, J. M., Wilner, D. J., Young, K. *et al.* 2008, *ApJ*, 677, 1140

Active OB stars: structure, evolution, mass loss, and critical limits
Proceedings IAU Symposium No. 272, 2010
C. Neiner, G. Wade, G. Meynet & G. Peters, eds.

doi:10.1017/S1743921311011665

Interferometric survey of Be stars with the CHARA array

Yamina N. Touhami[1], Douglas R. Gies[1], Gail H. Schaefer[2], Noel D. Richardson[1], Stephen J. Williams[1], Erika D. Grundstrom[3], and M. Virginia McSwain[4]

[1]Center for High Angular Resolution Astronomy, GSU, Atlanta, GA 30302, USA
[2]The CHARA Array, Mount Wilson Observatory, Mount Wilson, CA 91023, USA
[3]Physics and Astronomy Department, Vanderbilt University, Nashville, TN 37235, USA
[4]Department of Physics, Lehigh University, Bethlehem, PA 18015, USA

Abstract. We present the first spatially resolved observations of circumstellar envelopes of 25 bright northern Be stars. The survey was performed with the CHARA Array interferometer in the K-band at intermediate and long baselines. The interferometric visibilities are well fitted by a viscous disk model where the gas density steeply decreases with the radius. Physical and geometrical parameters such as the density profile, the inclination, and the position angles of the circumstellar disks are determined. We find that the density radial exponent ranges between $n \approx 2.4 - 3.2$, which is consistent with previous IRAS measurements. We have also obtained simultaneous optical and near-IR spectrophotometric measurements, and found that the model reproduces well the observed disk IR-continuum excess emission. By combining the projected rotational velocity of the Be star with the disk inclination derived from interferometry, we give estimates of the equatorial rotational velocities of these Be stars.

Keywords. techniques: photometric, techniques: interferometric, stars: emission-line, Be

---

1. Introduction and Observations

Be stars are rapidly rotating B-type stars that manage to eject gas into a circumstellar disk (observed in H emission lines, an infrared flux excess, and linear polarization; Porter & Rivinius. 2003). The IR flux excess emission results from bound-free and free-free emission, which increases in strength with wavelength (Dougherty *et al.* 1994). We have conducted the first interferometric Be star survey in the K-band using the CHARA Array. These observations cover a wide range of baselines from 30 to 331 meters. Because Be star disks are intrinsically variable on timescales of months to years, it was necessary to obtain contemporaneous spectrophotometry in order to model both the total flux and its angular distribution in the sky. We have obtained simultaneous near-IR spectroscopy of our sample using the NASA Infrared Telescope Facility and SpeX cross-dispersed spectrograph in 2006, and using the Mimir camera/spectrograph and Lowell Observatory Perkins Telescope in 2008 and 2009.

2. Modeling the Visibility Measurments

We calculate the theoretical visibility using a physical thick disk model that predicts given the near-IR emission the disk gas density (Gies *et al.* 2007) and compare it with visibilities for a simple uniform disk and Gaussian ellipsoid models. The disk is assumed to be axisymmetric and isothermal with a disk temperature equal to two thirds the stellar effective temperature. The model generates infrared images of the star surrounded by a

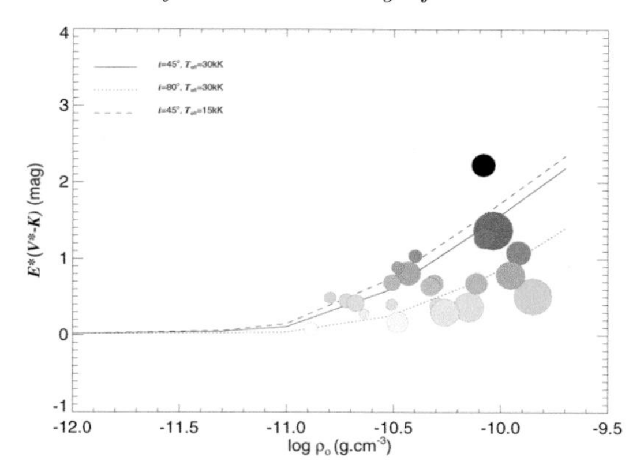

Figure 1. Theoretical IR-excess as a function of the disk base density for several model input parameters. The solid line is the default model with $i = 45^o$ and T_{eff}=30kK. The dotted line represent the case of a higher inclination disk and the dashed line is for a medium inclination and cooler Be star.

circumstellar disk, computes the fast Fourier transform, predicts the visibility curve that is directly comparable to the observational data, and extracts the IR excess emission. Figure 1 shows a plot of the theoretical continuum IR excess as a function of the gas base density. The solid line is the color excess for a disk with an inclination of 45 degrees and an effective temperature of 30kK. The dotted line is for 80 degrees inclination and 30kK effective temperature, the dashed line is for 45 degree inclination and 15kK effective temperature. We also plot in circles the interferometric results with circle sizes proportional to the obtained disk angular sizes.

3. Discussion

Our interferometric survey of nearby Be stars shows that all the stars display strong near-infrared emission that is coming from their circumstellar environments. The observations clearly indicate a resolved circumstellar disk around each star in our sample. We find that the thick disk model agrees well with the interferometric data, and that the best-fit values of the estimated gas densities are consistent with IRAS results. We also determine the angular size for each Be circumstellar disk. We find that the angularly resolved disks appear smaller in the lower opacity K-band continuum compared to that seen in the high opacity Hα line, which is consistent with the K-band excess forming mainly in the inner part of the disk (Touhami *et al.* 2010). The simultaneous spectroscopic survey shows that stars with strong Hα emission display an IR excess relative to the expected photospheric flux distribution. We find a good correlation between the size of the IR excess and the gas base density and the Hα equivalent-width. By combining the derived inclination witht the projected rotational velocity, we determine the equatorial rotational velocity of the central star. We find that Be stars are rapid rotators with equatorial velocities ranging between 0.7-0.9 of their critical velocities.

References

Dougherty, S. M., Waters, L. B. F. M., Burki, G., Cote, J. *et al.* 1994, *A&A*, 290, 609

Gies, D. R., Bagnuolo, Jr., W. G., Baines, E. K., ten Brummelaar, T. A. *et al.* 2007, *ApJ*, 654, 527

Porter, J. M. & Rivinius, T. 2003, *PASP*, 115, 1153

Touhami, Y., Richardson, N. D., Gies, D. R., Schaefer, G. H. *et al.* 2010, *PASP*, 122, 379

Active OB stars: structure, evolution, mass loss, and critical limits
Proceedings IAU Symposium No. 272, 2010
C. Neiner, G. Wade, G. Meynet & G. Peters, eds.

doi:10.1017/S1743921311011677

γ-ray production via IC scattering of the infrared excess from the Be-type star in the binary system PSR B1259-63/SS2883

Brian van Soelen[1] **and Pieter J. Meintjes**[2]

Department of Physics, University of the Free State, Bloemfontein, Republic of South Africa
[1]email: `vansoelenb@ufs.ac.za` [2]email: `MeintjPJ@ufs.ac.za`

Abstract. Un-pulsed γ-ray emission has been detected close to periastron in the pulsar/Be-star binary system PSR B1259-63/SS 2883, believed to originate from the shock front that forms between the stellar and pulsar winds. A likely source of γ-ray production is the inverse Compton up-scattering of target photons from the Be star by relativistic electrons/positrons in the pulsar wind. In this study the influence of the infrared radiation, emanating from the circumstellar disc, on isotropic inverse Compton γ-ray production is investigated. It is shown that the scattering of infrared disc photons can increase the γ-ray flux by a factor ~ 2 in the 1–10 GeV range.

Keywords. stars: emission-line, Be, binaries: general, gamma rays: theory

1. Introduction

The Be X-ray Pulsar binary (Be-XPB) PSR B1259-63/SS 2883 consists of a 48 ms pulsar in orbit around a Be star (Johnston *et al.* 1992). The orbit is eccentric ($e \approx 0.87$) and un-pulsed TeV γ-ray emission has been detected around periastron (Aharonian *et al.* 2005). The system is powered by the spin-down luminosity of the pulsar and the un-pulsed radiation is believed to originate from a stand-off shock-front that forms between the pulsar and stellar wind. A likely production mechanism for the γ-rays is the inverse Compton (IC) scattering of target photons from the Be star (Tavani & Arons 1997). Previous models of IC scattering in PSR B1259-63/SS 2883 have not considered how the infrared (IR) excess from the circumstellar disc will influence the γ-ray production.

2. Modelling

2.1. *Curve of Growth Method*

The Curve of growth method (as outlined by Waters 1986) was used to model the IR excess associated with SS 2883. Under this model it is assumed that the Be star's circumstellar disc has a half-opening angle $\theta = 5°$, extends to a radius $R_{disc} = 50\ R_{star}$, and follows a power-law density profile $\rho \propto (r/R_{star})^{-n}$. A Kurucz atmosphere (Kurucz 1979) was fitted to the optical data (*UBV* from Westerlund & Garnier 1989), giving a temperature of $T_{star} = 25000$ K, while the IR excess was fitted to *2MASS* (Skrutskie *et al.* 2006) and *MSX* data (Price et al 2001) assuming $T_{disc} = 0.5\ T_{star}$.

2.2. *Inverse Compton Scattering*

As a first approximation of the effect of the IR excess on IC scattering the total number of scatters is calculated by integrating over the exact isotropic scattering rate equation given in Blumenthal & Gould (1970). The scattering rate is proportional to the photon distribution, $n(\epsilon)$, which is calculated by using the curve of growth method instead of a

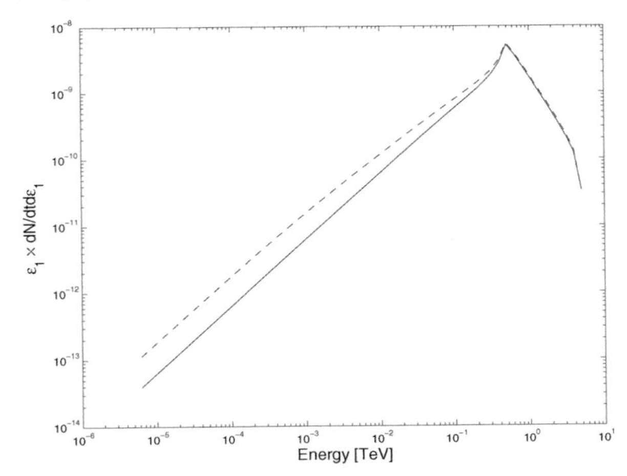

Figure 1. The predicted IC γ-ray flux in arbitrary unites for PSR B 1295-63/SS 2883. Dashed line includes the flux from the star and disc, while the solid line excludes the disc.

blackbody or mono-energetic photon distribution as was used in previous models. Fig. 1 shows the results of the IC scattering when only the flux from the star (solid line) and when the flux from the star plus the disc (dashed line) is considered, assuming a isotropic electron distribution, with a power index $p = 2.2$ between $\gamma = 10^6 - 10^7$. The modelled flux shows an increase of a factor ~ 2 in the 1–10 GeV range.

3. Conclusion

The predicted increase in the γ-ray flux is important for observations of γ-ray Be-XPB systems, especially given the observational range of *Fermi*. Future work will consider the anisotropic effects associated with a finite size star and disc.

Acknowledgements

This publication makes use of data from *2MASS*, a joint project of the University of Massachusetts and the Infrared Processing and Analysis Center/California Institute of Technology, funded by NASA and the NSF. This research made use of data from *MSX*; processing of the data was funded by the Ballistic Missile Defense Organization with additional support from the NASA Ofce of Space Science. This research has made use of the NASA/IPAC Infrared Science Archive, operated by the Jet Propulsion Laboratory, California Institute of Technology, under contract with NASA.

References

Aharonian, F., Akhperjanian, A. G., Aye, K.-M., Bazer-Bachi, A. R. *et al.* 2005, *A&A*, 442, 1
Blumenthal, G. R. & Gould, R. J. 1970, *Reviews of Modern Physics*, 42, 237
Johnston, S., Manchester, R. N., Lyne, A. G., Bailes, M. *et al.* 1992, *ApJ* (Letters), 387, L37
Kurucz, R. L. 1979, *ApJS*, 40, 1
Price, S. D., Egan, M. P., Carey, S. J., Mizuno, D. R. *et al.* 2001, *AJ*, 121, 2819
Skrutskie, M. F., Cutri, R. M., Stiening, R., Weinberg, M. D. *et al.* 2006, *AJ*, 131, 1163
Tavani, M. & Arons, J. 1997, *ApJ*, 477, 439
Waters, L. B. F. M. 1986, *A&A*, 162, 121
Westerlund, B. E. & Garnier, R. 1989, *A&AS*, 78, 203

Active OB stars: structure, evolution, mass loss, and critical limits
Proceedings IAU Symposium No. 272, 2010
C. Neiner, G. Wade, G. Meynet & G. Peters, eds.

doi:10.1017/S1743921311011689

Non-thermal radio emission from colliding-wind binaries: modelling Cyg OB2 No. 8A and No. 9

Delia Volpi[1], Ronny Blomme[1], Michael De Becker[2,3] and Gregor Rauw[2]

[1]Royal Observatory of Belgium, Ringlaan 3, B-1180 Brussels, Belgium,
email: delia.volpi@oma.be

[2]Institut d'Astrophysique, Université de Liège,
Allée du 6 Août, 17, Bât B5c, B-4000 Liège (Sart-Tilman), Belgium

[3]Observatoire de Haute-Provence, F-04870 Saint-Michel l'Observatoire, France

Abstract. Some OB stars show variable non-thermal radio emission. The non-thermal emission is due to synchrotron radiation that is emitted by electrons accelerated to high energies. The electron acceleration occurs at strong shocks created by the collision of radiatively-driven stellar winds in binary systems. Here we present results of our modelling of two colliding wind systems: Cyg OB2 No. 8A and Cyg OB2 No. 9.

Keywords. plasmas, radiation mechanisms: nonthermal, methods: numerical, binaries: spectroscopic, stars: early-type, stars: winds, outflows

---

1. Introduction

During recent years many OB stars have been discovered to be binary systems. Non-thermal radio emission is observed to be produced by some of these binary stars. The non-thermal emissivity is thought to be due to synchrotron emission radiated by relativistic electrons. The electrons are accelerated up to high energies by strong shocks produced by the collision between the two radiatively driven stellar winds (Eichler & Usov 1993). Several parameters of the system can be constrained by the synchrotron emission, among them the mass loss rates from the primary and the secondary. Investigating the synchrotron radiation is thus necessary. We model the non-thermal emission for two colliding wind systems, Cyg OB2 No. 8A and Cyg OB2 No. 9, and compare the obtained results with the observations.

2. Modelling and comparison with the data

The two colliding winds are separated by a contact discontinuity. Its position (which, in our model, we assume to be coincident with the two shocks) is defined as in Antokhin *et al.* (2004). The electrons are accelerated at the shock. We follow them as they advect away and cool down due to adiabatic and inverse Compton losses along the post-shock streamlines. The momenta follow a modified power-law distribution. The synchrotron emissivity from the relativistic electrons is calculated along the post-shock streamlines in the orbital plane. The Razin effect is included. The third dimension is recovered by rotating the orbital plane along the line which connects the two stars. We also include free-free emission and then calculate the fluxes and spectral indices at different orbital phases using Adams method (Adam 1990). For Cyg OB2 No. 8A the parameters are

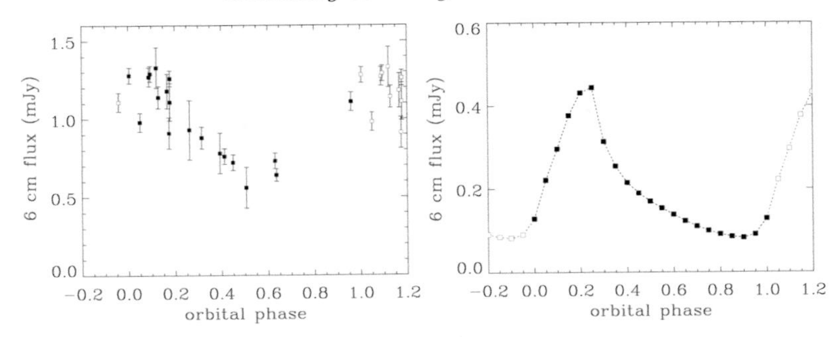

Figure 1. Flux in the radio band at 6 cm for Cyg OB2 No. 8A: on the left the observations from Blomme *et al.* (2010), on the right our simulated results. Periastron is at phase ≈ 0

provided by De Becker *et al.* (2006), for Cyg OB2 No. 9 the orbital parameters are provided by Nazé *et al.* (2010), stellar parameters by Martins *et al.* (2005) and wind parameters by Vink *et al.* (2001).

Cyg OB2 No. 8A results. For Cyg OB2 No. 8A (see Blomme *et al.* 2010) the radio data at 3.6 and 6 cm are obtained with VLA, the X-ray data with XMM and ROSAT. Variability that is locked with the orbital phase is observed in both radio and X-rays. The X-ray and the radio light curves are anti-correlated due to different formation regions. The radio formation region is far out in the wind, along the contact discontinuity, while the X-rays are formed much closer to the apex of the contact discontinuity. The model predicts phase-locked radio variability which is consistent with the observations (see Fig. 1), even if the phases of the flux maximum and minimum do not agree.

Cyg OB2 No. 9 results. Van Loo *et al.* (2008) studied the observed VLA radio fluxes of Cyg OB2 No. 9 at 3.6, 6, and 20 cm. They found a 2.35 yr period from the data. A preliminary 6 cm light curve from our modelling shows variability in the radio flux linked to the orbital period that is the fingerprint of non-thermal radiation. Compared to the observations, the theoretical fluxes are much too high and the maximum occurs too early.

3. Future work

To improve the current results for Cyg OB2 No. 8A and 9, we need to include the orbital motion and solve the hydrodynamics equations. A more detailed study is also necessary to better determine the star and wind parameters of the binary components and to investigate the porosity/clumping problem.

References

Adam, J. 1990, *A&A*, 240, 541
Antokhin, I. I., Owocki, S. P., & Brown, J. C. 2004, *ApJ*, 611, 434
De Becker, M., Rauw, G., Sana, H., Pollock, A. M. T. *et al.* 2006, *MNRAS*, 371, 1280
Blomme, R., De Becker, M., Volpi, D., & Rauw, G. 2010, *A&A* 519A, 111
Eichler, D. & Usov, V. 1993, *ApJ*, 402, 271
Martins, F., Schaerer, D., & Hillier, D. J. 2005, *A&A*, 436, 1049
Nazé, Y., Damerdji, Y., Rauw, G., Kiminki, D. C. *et al.* 2010, *ApJ*, 719, 634
van Loo, S., Blomme, R., Dougherty, S. M., & Runacres, M. C. 2008, *A&A*, 483, 585
Vink, J. S., de Koter, A., & Lamers, H. J. G. L. M. 2001, *A&A*, 369, 574

Active OB stars: structure, evolution, mass loss, and critical limits
Proceedings IAU Symposium No. 272, 2010
C. Neiner, G. Wade, G. Meynet & G. Peters

doi:10.1017/S1743921311011756

Equatorial mass loss from Be stars

Cyril Georgy[1], Sylvia Ekström[1], Anahí Granada[1,2], and Georges Meynet[1]

[1]Geneva University Observatory
Chemin des Maillettes 51, 1290 Versoix, Switzerland

[2]Fac. de Cs. Astr. y Geof. Universidad Nacional de La Plata - IALP CCT La Plata, UNLP-CONICET, Argentina

Abstract. Be stars are thought to be fast rotating stars surrounded by an equatorial disc. The formation, structure and evolution of the disc are still not well understood. In the frame of single star models, it is expected that the surface of an initially fast rotating star can reach its keplerian velocity (critical velocity). The Geneva stellar evolution code has been recently improved, in order to obtain some estimates of the total mass loss and of the mechanical mass loss rates in the equatorial disc during the whole critical rotation phase. We present here the first results of the computation of a grid of fast rotating B stars evolving towards the Be phase, and discuss the first estimates we obtained.

Keywords. stars: evolution, stars: emission-line, Be, stars: mass loss, stars: rotation

1. Introduction

The Be star phenomenon can be explained in terms of a fast rotating star surrounded by an equatorial disc (Porter and Rivinius 2003). In this context, the disc formation occurs when the stellar surface reaches (or, at least, becomes close to) the critical rotation (see *e.g.* Ekström *et al.* 2008). When this occurs, the equatorial mass loss is enhanced by the very low effective gravity at the equator, due to the strong centrifugal acceleration.

Recently, the Geneva stellar evolution code was modified in order to account for the equatorial mass loss when the star reaches the critical velocity. If the surface of the star becomes over-critically rotating because of evolutive processes, the extra angular momentum is evacuated. The mass which is removed is estimated assuming that the mass decouples from the star at the surface in the equatorial plane.

In this work, we present the first results of the computation of a grid of B type stars rotating models, in the mass range from 3 to 15 $M_{\odot}$ with three different initial rotation velocities.

2. Preliminary results

In Fig. 1 (left panel), we show the relative contribution of the equatorial mechanical mass loss during the critical rotation phase compared to the radiative mass loss. For the less massive stars, the mechanical mass loss is the dominant process in these fast rotating models. The radiative mass loss becomes ever larger for higher mass stars, preventing progressively the star to reach the critical velocity. In the right panel, we compare our 5 and 9 $M_{\odot}$ models with observed equatorial mass loss rates from Be stars. We see that our results are marginally compatible with the observations. Note however that the observed values are instantaneous values, whereas the computed ones are averaged values, due to the quite long time step needed for the stellar evolution computations. The main properties of our models of Be stars are summarised in Table 1.

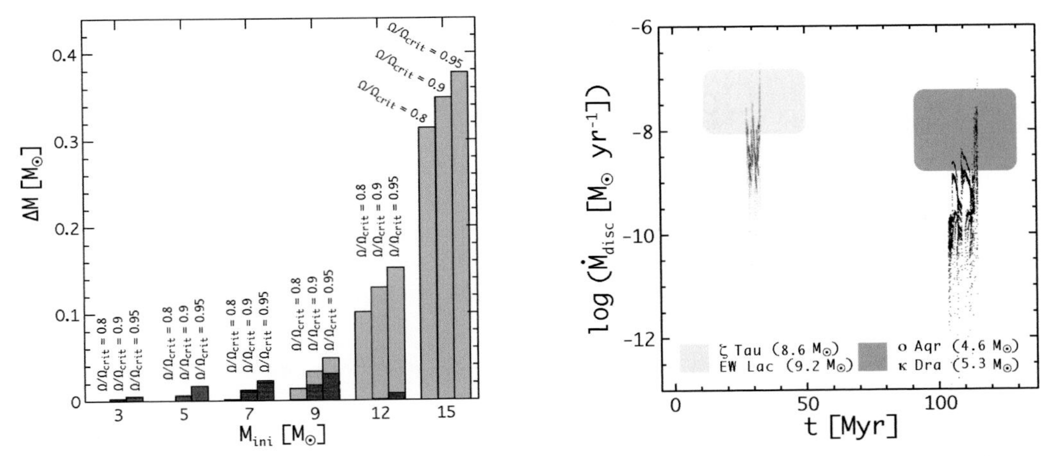

Figure 1. *Left panel*: equatorial mechanical mass loss (dark grey) and radiative mass loss (light grey) as a function of the initial mass and initial rotation velocity. *Right panel:* Comparison of our estimated equatorial mass loss rates with observations for the $5\,M_\odot$ (dark grey points) and $9\,M_\odot$ (light grey points). The shaded areas correspond to the observed mass loss rates of Be stars with similar masses (Rinehart *et al.* 1999; Stee 2003; Zorec *et al.* 2005)

Table 1. Main characteristics of the Be star models. $M_{\rm ini}$ is the initial mass of the model; $\Omega/\Omega_{\rm crit}$ is the initial rotation parameter; $\Delta M_{\rm tot}$ is the total amount of mass lost by the star during the MS; $\Delta M_{\rm mec}$ is the amount of mass lost mechanically in the equatorial disc during the critical-rotation phase; $< \dot{M}_{\rm mec} >$ is the mean equatorial mass loss rate during the phase of critical rotation; $\tau_{\rm cr}/\tau_{\rm MS}$ is the fraction of the MS spent at the critical velocity; $X_{\rm C,cr}$ is the fraction of hydrogen in the burning core when the star reaches the critical velocity for the first time; $< v_{\rm eq} >$ is the mean equatorial velocity during the MS.

$M_{\rm ini}$ [$M_\odot$]	$\Omega/\Omega_{\rm crit}$	$\Delta M_{\rm tot}$ [$M_\odot$]	$\Delta M_{\rm mec}$ [$M_\odot$]	$< \dot{M}_{\rm mec} >$ [$M_\odot\,{\rm yr}^{-1}$]	$\tau_{\rm cr}/\tau_{\rm MS}$	$X_{\rm C,cr}$	$< v_{\rm eq} >$ [km s^{-1}]
3	0.8	0.0	0.0	0.0	0.0	–	219.7
	0.9	$2.36 \cdot 10^{-3}$	$2.36 \cdot 10^{-3}$ (100%)	$5.6 \cdot 10^{-11}$	0.10	0.16	265.8
	0.95	$5.49 \cdot 10^{-3}$	$5.49 \cdot 10^{-3}$ (100%)	$4.7 \cdot 10^{-11}$	0.27	0.33	290.8
5	0.8	0.0	0.0	0.0	0.0	–	248.3
	0.9	$6.34 \cdot 10^{-3}$	$6.34 \cdot 10^{-3}$ (100%)	$5.5 \cdot 10^{-10}$	0.10	0.16	300.5
	0.95	$1.91 \cdot 10^{-2}$	$1.91 \cdot 10^{-2}$ (100%)	$4.0 \cdot 10^{-10}$	0.38	0.40	349.4
7	0.8	$1.55 \cdot 10^{-3}$	$4.93 \cdot 10^{-4}$ (3%)	$1.1 \cdot 10^{-9}$	0.001	0.0	268.3
	0.9	$1.33 \cdot 10^{-2}$	$1.13 \cdot 10^{-2}$ (85%)	$1.5 \cdot 10^{-9}$	0.14	0.20	323.6
	0.95	$2.57 \cdot 10^{-2}$	$2.34 \cdot 10^{-2}$ (91%)	$1.6 \cdot 10^{-9}$	0.27	0.31	356.0
9	0.8	$1.53 \cdot 10^{-3}$	$7.82 \cdot 10^{-4}$ (5%)	$6.2 \cdot 10^{-9}$	0.004	0.01	286.5
	0.9	$3.75 \cdot 10^{-2}$	$1.98 \cdot 10^{-2}$ (53%)	$3.5 \cdot 10^{-9}$	0.17	0.22	347.7
	0.95	$5.48 \cdot 10^{-2}$	$3.45 \cdot 10^{-2}$ (63%)	$3.3 \cdot 10^{-9}$	0.31	0.33	377.0
12	0.8	$1.15 \cdot 10^{-1}$	0.0 (0%)	0.0	0.0	–	300.3
	0.9	$1.47 \cdot 10^{-1}$	$9.08 \cdot 10^{-4}$ (1%)	$3.2 \cdot 10^{-10}$	0.14	0.18	363.2
	0.95	$1.73 \cdot 10^{-1}$	$8.56 \cdot 10^{-3}$ (5%)	$1.5 \cdot 10^{-9}$	0.27	0.29	394.3
15	0.8	$3.57 \cdot 10^{-1}$	0.0 (0%)	0.0	0.0	–	302.5
	0.9	$3.96 \cdot 10^{-1}$	0.0 (0%)	0.0	0.0	–	353.6
	0.95	$4.29 \cdot 10^{-1}$	0.0 (0%)	0.0	0.0	–	384.4

References

Ekström, S., Meynet, G., Maeder, A., & Barblan, F. 2008, *A&A* 478, 467
Porter, J. M. & Rivinius, T. 2003, *PASP* 115, 1153
Rinehart, S. A., Houck, J. R., & Smith, J. D. 1999, *AJ* 118, 2974
Stee, P. 2003, *A&A* 403, 1023
Zorec, J., Frémat, Y., & Cidale, L. 2005, *A&A* 441, 235

Active OB stars: structure, evolution, mass loss, and critical limits
Proceedings IAU Symposium No. 272, 2010
C. Neiner, G. Wade, G. Meynet & G. Peters, eds.

doi:10.1017/S1743921311011690

Concluding Remarks

André Maeder

Geneva Observatory, University of Geneva
email: andre.maeder@unige.ch

Abstract. Highlights of this outstanding meeting are emphasized, as well as important open questions for future research.

Keywords. stars: early-type, stars: rotation, stars: magnetic fields, stars: mass loss, stars: interiors, stars: oscillations (including pulsations)

---

1. Introduction

I remember many years ago at an IAU symposium on QSO in Geneva, Lo Woltjer addressing a speech to the assembly started in this way: my dear colleagues, I am very impressed how little progress has been made over the last 30 years ! Frankly the same cannot be said here. Recent observations in interferometry, asteroseismology, polarimetry and high resolution spectroscopy have brought many new results. In the long tradition of the meetings on massive stars, this one particularly brightens by its new emerging lines on stellar physics and evolution. The global order of the various sessions is respected here, but some topics treated in different sessions have been grouped at one place. These conclusions are not a summary of the various talks, but only the emphasis of some points which appeared, at least to me, as particularly interesting. I often add a few personal comments. I sincerely apologize to those who may consider that I have not well reported on their results, any selection of a limited number of results is evidently disputable. Only the name of the speaker delivering the talk is mentioned; for the full references, please refer to the original contributions in these proceedings.

2. Rapid rotation and mixing in active OB stars

On the theoretical side, many instabilities and transport processes occur in rotating stars as recalled by Jean-Paul Zahn. Among the critical points, there is the horizontal turbulence which is poorly known. It enforces shellular rotation (depending on radius), it also influences shear mixing and meridional circulation, the problem is that the diffusion coefficient associated to horizontal turbulence is uncertain. In this respect and as also stated by Zahn, I want to emphasize that the transport of angular momentum by meridional circulation is not a diffusion process, as often implicitly assumed by many authors. If one does that, even the sign of the effect may be wrong.

The main future developments concern the interaction of rotation with the magnetic field, in particular the two following problems. 1) Is there a dynamo in rotating radiative zones ? Zahn expresses doubts on the closing of the Tayler-Spruit dynamo, however other closing mechanisms have not yet been worked out. The question is a major one, since the field made by such a dynamo may impose rigid body rotation in radiative zones, thus completely modifying the transport of angular momentum and chemical elements. 2) The other question is whether a magnetic field really kills the meridional circulation. Zahn suggests it is the case for fields above 600 G, but the question may still need further investigations. Zahn also emphasizes that convective cores may operate a dynamo, but whether the resulting field may appear to the surface is not known.

Regarding the treatment of convective zones, Adrian Potter discusses the effects of the two different assumptions: 1) Ω = constant and 2) angular momentum constant. The differences are relatively small for the evolution in the HRD, but large for the internal rotation and this may have consequences for the further evolutionary stages. Massive stars have small surface convective zones, as shown by Matteo Cantiello. These zones may play a significant role for microturbulence, wind clumping, magnetic braking and nonradial pulsations. I note the mass in this zone is lost in a few months according to current mass loss rates, thus the processes must be regarded as dynamical.

The observations of rotational mixing are still controversial according to the presentations by Ines Brott and Norbert Langer (see also the posters by Peters and by Dunstall *et al.*). The comparison of theory and observations as usual in Science may either result in the collapse of the theory or to its improvement and reinforcement. For now, it is to early to conclude. However, as shown by Norbert Przybilla the data by Hunter *et al.* (2008) show for the N/C vs. O/N ratio, which is a model independent test, a very scattered diagram instead of a well defined slope of 4 as predicted by the nuclear reactions and demonstrated from high precision spectroscopy. Incidentally, we note that the argument that the scatter is due to uncertainties in C does not hold, since the sum of C+N+O should be constant. Also Maeder *et al.* (2009) have shown that, if stars in a limited domain of masses and ages are considered, the correlation of the N-enrichments with $v \sin i$ is much better than the one found by Hunter *et al.* Awaiting more results from high resolution spectroscopy to solve the debate, we emphasize that 1) the effects of gravity darkening should be accounted in all parameter determinations and 2) the enrichments N/C are a multivariate function

$$\Delta \log(N/H) = f(v \sin i, M, \, age, \, multiplicity, \, Z, \, magnetic \; field) \, . \qquad (2.1)$$

Thus, to properly test the dependence on $v \sin i$, the other variables should be kept as constant as possible.

The interferometric observations globally confirm the Roche model for rotating stars, although some small deviations near the break-up limit may exist as shown by Ming Zhao. The remarkable fact is that the coefficient β of the gravity darkening relation $T_{\rm eff} \sim g^{\beta}$ can now be estimated. Instead of $\beta = 0.25$ in blue stars, lower values are observed in Regulus (B8V, $\beta = 0.178$) and Alderamin (A7V, $\beta = 0.22$), for convective envelopes the lower value $\beta = 0.08$ (Lucy 1967) is supported. In case of fast rotating stars, one may have a variable β over the stellar surface.

Be stars are a crossroad in this meeting. The initial rotation velocities and masses leading to them in the course of MS evolution have been modeled by Sylvia Ekström. She also emphasizes the role of the mechanical mass loss from these stars. This kind of mass loss may play a role in many situations particularly at the lower metallicities and in the first stellar generations. The observation properties have been reviewed by Christophe Martayan, who points out the effects of the line saturation, which leads to an underestimate of the rotation velocities, and the veiling effect from the circumstellar disk. The number frequencies and the pulsation properties change with metallicity Z. Globally the magnetic fields of Be stars are weak, typically lower than 150 G (see also poster by Ruslan Yudin). This is rather expected, otherwise the rotation of these stars would have been slowed down. Interestingly enough, in the higher mass range of Be stars, the Be phenomenon occurs mostly during the early MS phase, while in the lower mass range Be stars tend to concentrate near the end of the MS phase.

The origin of the Be phenomenon has been commented, particularly by Stan Owocki. It is clearly a combination of a long term secular effect and of short term instabilities, for example pulsation. The secular evolution brings the star close enough to the critical break-up velocity, so that the additional velocity field due to an instability may allow some mass ejection.

3. Winds and magnetic fields of active OB stars

The session starts by a useful "survival kit" on magnetic fields by Véronique Petit. Low resolution observations only provide the global longitudinal field, while the high resolution together with the multi-line technique gives the structure and field intensity. For the solar-type stars the magnetic fields are ubiquitous, spatially complex, weak on the average and due to a dynamo. The fields of hot stars are rare, possibly strong, organized and showing no correlation with stellar properties. In the mass range of 1.6 to 5 $M_\odot$, only 2% of the stars show magnetic fields with intensities larger than about 300 G. This limit may be the result of dissipation for weaker fields (poster by Kholtygin). Among OB stars, the frequency of magnetic stars seems to show some differences in clusters. There are 3 over 9 in Orion cluster and 1 over 26 in NGC 2244. The technique of Doppler imaging, as shown by Oleg Kochukhov, also allows one to make a detailed mapping of the field and of the abundance peculiarities over stellar surfaces. I note that one may wonder about the origin of the large differences in the magnetic field of massive stars. What comes from the field of the interstellar medium, what comes from the dynamo in convective regions at the pre-MS stage?

Gregg Wade, pointing out that the magnetic field in astrophysics is like sex in psychiatry, reports on the beautiful MiMeS collaboration, which has received 640 hours of observations on the CFHT (see many posters). The objectives of this collaboration focus on the origin of the fossil fields, the magnetic braking, the interaction with winds and the impact on stellar evolution. Two magnetic stars are particularly remarkable. As reported by Mary Oksala, σ Ori E (B2Vp) has a polar field of 9.6 kG with $v \sin i = 150$ km s^{-1}. HR 7355 (B2Vpn), as reported by Thomas Rivinius and Oksala, has a variable field which reaches 3.2 kG, it has $v \sin i = 310$ km s^{-1}. In these two cases, the wind is magnetically dominated and the ejected particles form structures which corotate with the star. In HR 7355, metals seem to accumulate at the poles. Another similar star, HR 5907, has been detected.

There was unfortunately no remark on the relation between magnetic fields and chemical N–enrichments, except the poster by Thierry Morel. Indeed, it is known from the work by Henrichs *et al.* (2003a) and Henrichs *et al.* (2003b) that some stars with strong magnetic fields and often low rotation velocities show strong nitrogen enrichments by 1 to 2 dex. These interesting observations are recalled in Table 1 below. These observations, although scarce, suggest that there is some mixing effect related to the presence of a magnetic field. What is the mechanism responsible for this mixing? A possibility is that it is due to the meridional circulation, which is not necessarily killed by the magnetic field. Models by Maeder & Meynet (2005) show that the meridional circulation may be enhanced by the fact that a magnetic star rotates near solid body rotation, i.e. out of equilibrium for meridional circulation. This may be the case for moderate fields, while for very large fields circulation may be inhibited. Another possible source of mixing could be a magnetic instability able to make a sufficient transport of the elements. However, such an instability has not yet been identified.

Table 1.

Name	Sp	$v \sin i$	Polar field Bp	$\Delta logN$
β Cep	B1IVe	27 km s^{-1}	360 G $\pm$ 40	1.2
V2052 Oph	B1V	63 km s^{-1}	250 G $\pm$190	1.3
ζ Cas	B2IV	17 km s^{-1}	340 G $\pm$ 90	2.6
ω Ori	B2IVe	172 km s^{-1}	530 G $\pm$200	1.8

On the theoretical side, concerning the fields at stellar surfaces Richard Townsend shows both analytical developments and numerical simulations of magnetospheres and X–ray emissions. He points out two key parameters for magnetospheres:

$$\eta = \frac{B^2 R^2}{\dot{M} v_\infty} \quad \text{and} \quad t_{\rm spindown} = \frac{2}{3} k\, \eta^{-1/2} t_{\rm M-loss} \,. \tag{3.1}$$

The parameter η describes the ratio of the magnetic and wind momenta. A ratio larger than 1 implies the existence of a magnetosphere. For example, in the case of σOri E, the magnetosphere would extend up to 30 stellar radii. The spindown of the star is characterized by the spindown timescale $t_{\rm spindown}$, where k is a constant and $t_{\rm M-loss}$ is the characteristic time of the mass loss of the order of $M/\dot{M}$. It seems to me that the spindown by magnetic fields should soon also be accounted for in evolutionary models. It could account for some slow rotating stars with enhanced mixing.

Stéphane Mathis studies the magnetic effects in stellar interiors. He examines the initial conditions for fossil fields and derives the condition for the stability of such fields. He shows that the ratio of magnetic to gravitational energy must not be too large otherwise instabilities develop. When applied to the complex field structures of magnetic hot stars, these conditions generally indicate stability. Mathis also examines the possibility of a dynamo operating in a radiative zone. He suggests following the paper by Zahn, Brun & Mathis (2007) that the Pitts and Tayler instability exists, however he claims that the dynamo proposed by Spruit does not work. Nevertheless, a loop closing the dynamo may be possible. If it works, as emphasized above, this will be a major effect in stellar evolution, since it will deeply affect the internal rotation and the transport processes.

The winds of OB stars

There are many structures in the winds, as shown by Alex Fullerton. Among the various components, there are the narrow and discrete absorption components (NACS and DACS), which are essentially the same and visible on the left of the big absorption components. They are ubiquitous and persistent for some time. There are also the periodic optical depth modulations (PAMS, "they smile to you"), as well as corotating interacting regions (CIRs) and spiral structures. These result from the interaction of the fast wind with a slower wind component. In addition to the porosity due to the various inhomogeneities, there is also the "vorosity" due to inhomogeneities in the velocity distribution. The important point is that for different mass loss estimates, the dependence in density is different

$$\textit{radio \& } H_\alpha \textit{ determinations} \quad \textit{depend on } \varrho^2 \tag{3.2}$$

$$\textit{UV determinations} \quad \textit{depend on } \varrho \,. \tag{3.3}$$

The presence of clumping leads to large overestimates, up to an order of a magnitude, of the mass loss rates based on radio and H_α observations. Small scale density fluctuations make large effects, while large scales make small effects. On the whole, I think this means that the mass loss rates are still rather uncertain.

As shown by David Cohen in Session 4, the X-ray spectroscopy provides a clumping–independent mean to determine the mass loss rates of O–type stars. The X–rays are formed by shocks in the wind, the plasma reaches temperatures of the order of 10^6 K. With respect to the observer, the more distant part of the shock (contributing to the red side of the lines) is absorbed with respect to the closer part (blue side). The optical depth $\tau = \kappa \dot{M}/(4\pi R v_\infty)$ of this absorption is sensitive to the mass loss rates. The mass loss rates determined in this way are typically a factor of 3 to 5 smaller than the values from radio and H_α determinations.

4. Populations of OB stars in galaxies

Norbert Langer comments on the difficulty to define the end of the MS phase, which seems to extend much outside the formal MS band predicted by evolutionary models. Incidentally, this effect was also found in a study of 23 clusters with ages smaller than $2 \cdot 10^7$ yr in the Galaxy, the LMC and SMC (Meylan & Maeder 1982). They found that 40% of the stars are observed out of the MS band instead of about 8% predicted. Everything is like if the MS band reaches the early types A, thus both studies agree. The relative excesses of A, F, G, K and M stars my increase with decreasing metallicities. Extended mixing and/or binaries may contribute to such an effect, as well as atmospheric effects, however it is clear that no simulations properly reproduce the observations. Langer also discusses the fast decrease of rotation velocities near $\log T_{\rm eff} = 4.3$ as well as the uncertainties concerning the N–enrichments in relation with the $v \sin i$ as nentioned in Sect. 1. He also examine what are the progenitors of GRBs and shows that the number ratio of GRB/SN increases with decreasing Z. Galactic stars are not expected to produce GRBs, because of a too low angular momentum in final stages.

Thibaut Decressin establishes the relation between the chemical properties of stars in globular clusters and the properties of the massive stars in the early stellar generations. Most of the globular clusters show signs of a second stellar generation in the form of a double MS band in the HR diagram. He shows that the abundance anomalies, such as the oxygen–sodium anticorrelation, result from wind enrichment by an earlier generation of fast rotating massive stars, which experienced mechanical ejections as discussed above by Ekström. Dynamical models indicate that following supernova explosions most of the remaining gas is ejected as well as a fraction of the stars of the first generation.

Chris Evans reviews the results on OB stars in nearby galaxies from multi–object spectrographs on large telescopes. Most properties show a strong dependence on metallicity Z, a point which was also nicely emphasized in the introductory talk by Dietrich Baade. An important fact is that the lower Z, the lower the mass loss rates due to the smaller atmospheric and wind opacities. Among other facts, the SMC stars at a given spectral type are hotter. There are more Be stars and the $v \sin i$ are generally larger at lower Z, despite a significant scatter. In this respect, I note that there may be two effects responsible for this property: 1) the effects of the lower mass loss at lower Z and 2) the possibility of faster initial rotations at lower metallicities (Maeder, Grebel & Mermilliod 1999). The binary fraction varies from cluster to cluster (see also Sana, this meeting), but no trend has been identified yet with Z. Evans also comments on recent claims about the existence of very massive stars up to 320 $M_\odot$. As this is is not the first time that such claims are made, some concerns may be expressed whether the announcement will survive further observations. Finally, Evans points out the major interest of I Zw 18, a well known irregular galaxy with a metallicity equal to 1/30 of solar. This is a great step

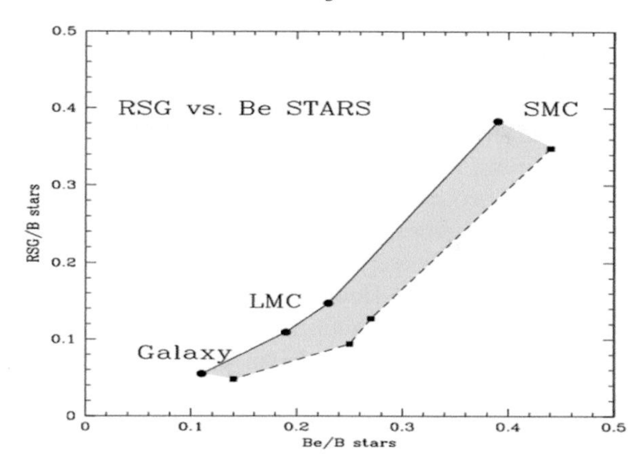

Figure 1. The relation between the ratio of red to blue supergiants and the ratio of Be to all B stars (normal+Be). Two different limits in the definition of RSG are shown. Data collected by the author from the WEBDA cluster data basis and from Maeder, Grebel & Mermilliod (1999).

toward the very low metallicities of the distant Universe. Photometric data have recently been obtained from the HST for a part of this galaxy (Jamet *et al.* 2010).

Indeed, several of the dependences on Z, particularly those concerning number ratios such as the number ratio of WR to O–stars, are the result from the lower mass loss rates at lower Z and/or from the differences in the rotation velocities at the end of the MS phase. Probably not all the possible dependences have been found yet. For example, I note that the ratio of red to blue supergiants is increasing with decreasing metallicities and faster rotations. This in turn implies that there is is a positive relation between the number ratio RSG/BSG of red to blue supergiants and the relative frequency of Be stars among all B stars. I show such a relation in Fig. 1.

Alceste Bonanos reports on an infrared survey of about 5000 stars from Spitzer SAGE. This allows her to study the IR properties of Oe, Be, B[e], LBV, RSG, WR and OH/IR stars. She finds evidences of transition from Be to B stars and vice versa, as well as signatures of dust around the B[e] stars. The survey also provides several interesting statistical informations on the number frequencies. Anatoly Miroshnichenko studies the B[e] stars in the Galaxy and Magellanic Clouds and examines the possible origin of these stars. Is it just some range of masses and rotation velocities which lead to them or is it the effects of binary merging? Dominik Bomans and Kerstin Weis observe the LBV in external galaxies, these are (with the WR stars) the stars which can be identified at the largest distances. From the new LBV discovered, they conclude that the LBV phenomenon is not restricted to high metallicities. Weis emphasizes that at least 50% of the LBV show bipolar nebulae (75% in the Milky way). LBV cover a large range of luminosities. She wonders why not all LBV show bipolar nebulae and whether it is related to their rotation.

5. Circumstellar environment of active OB stars

The observations of circumstellar (CS) discs with various techniques, in particular interferometry, are reviewd by Philippe Stee, by Alex Carciofi and by Christopher Tycner. Properties and processes at the origin of the CS disc of Be stars are reviewed by Stee, who states that each Be star is a special case. Some discs are permanent, some are dissipating. Clearly the ratio Ω/Ω_c of the angular velocity to the critical value is a leading parameter, but there may be other ones, like the beating of nonradial pulsations. Binarity influences the disc properties, for example by making truncations and deformations. Carciofi particularly emphasizes that most disc properties are consistent with the model of viscous decretion disc: the size at various λ, the thin opening angle, the small deviations from a Keplerian disc, the long term variations in the integral light and colors, etc. He also comments on the viscous properties, as well as on the deviations from a steady state viscous disc model. Tycner describes the careful iterations between the long-baseline interferometry and the models necessary to obtain consistent disc models, he provides nice examples with P Cygni and some Be stars.

Rigel (B8Ia) and Deneb (A2Iae) were monitored in the IR by AMBER/VLTI and by VEGA/CHARA as reported by Olivier Chesnau. It is particularly interesting that in the near IR, the stellar diameters are essentially independent of the mass loss rates. His pioneer observations also show the presence of clear signs of activity in these supergiants.

The case of the Herbig Ae/Be stars is presented by Evelyne Alecian. These pre-MS stars show accretion discs. They extend from 0.3 to 1000 AU and may contain PAH as well as various molecules. Generally the discs are less massive around Be stars than around Ae stars, a fact which may be due to higher photoevaporation in the former. Many evidences of planet formation are also present. Noticeably 5% of the Ae/Be stars show signatures of magnetic fields, this fraction is the same as that among the MS stars of the corresponding mass range. The magnetic fields are compatible with flux conservation between these two evolutionary stages.

Be stars generally have IR excesses due to the free–free emission of the ionized gas. In JHK color-color diagram, as shown by Chien–De Lee, a group of Be stars with large excesses behave differently and this is the signature of thermal emission from dust in these stars. Dust seems to form in Be stars near the end of the MS phase.

6. Periodic variations and asteroseismology of OB stars

New data from MOST, CoRoT and Kepler have brought many new facts concerning massive stars, as reported by Peter de Cat. The β Cephei stars (low order p– and g–modes with periods of 2 to 12 h) and Slowly Pulsating B stars (SPB with high–order g–modes and periods of 0.3 to 5 days) are privileged objectives for asteroseismic observations. The observations of V836 Cen, a β Cep star, support an overshooting distance $d_{\rm over}$ equal to 0.10 of the pressure scale height H_P and a core angular velocity 4 times larger than the surface velocity Ω_S. For ν Eri, a SPB star, these values are respectively 0.31 H_P and 3 times Ω_S. Evidences of pulsations are also found in Be and SPBe stars, which both have high rotations. In μ Cen (B2IV–Ve), the outburst coincides with a mode–enhancement, indicating that pulsations may stimulate the ejection (in addition to effects of secular evolution as mentioned above). The variability of the B1Ve star HD 51193 extensively observed by CoRoT, Harps and Sophie is further discussed by Juan Gutierrez–Soto. On the whole I think asteroseismology allows us to make giant steps forward in the study of massive star properties.

In a enlightening discussion of theoretical aspects, Marc–Antoine Dupret shows that low degree g–modes from the core can reach the surface and give information not only on the overshooting, but also on semiconvection and rotational mixing. Indeed current overshooting models predict abrupt transition of the internal distribution of the mean molecular weight μ, while rotation mixing predicts smooth transitions in the distribution of μ. As he reports, Miglio *et al.* (2008) show that in MS models the periods of high-order gravity modes are accurately described by a uniform period spacing superposed to an oscillatory component. The periodicity and amplitude of such a component are related, respectively, to the location and sharpness of the μ–gradient. Observations indicate that there is nos sharp variations of μ above the core, in agreement with models of rotational mixing. Dupret also shows that the great rotational distortion of Be stars demands that 2D models are used to discuss the oscillations. Interestingly enough, new families of instability modes appear in this case. Finally, Dupret calls our attention that due the lower opacities for stars in the LMC and SMC, SPB and β Cep stars should not occur in these galaxies according too the models. The fact they are observed is challenging, particularly regarding the opacities.

The upper part of the HR diagram deserves a particular attention as pointed out by Hideyuki Saio, who shows that above some luminosity limit the strange modes are active. Also, in stars with a convective zone due to the Fe–opacity peak, low degree g–modes may appear at the surface. To what extent do they contribute to turbulence? In the case of LBV, the microvariations are consistent with oscillatory convective and strange modes. The question arises evidently how this is influenced by the boundary conditions, since running waves may form in the wind. The WR stars with their very high L/M ratio are likely to show strange mode pulsations, as reported by André–Nicolas Chené. He reviews the results from MOST, which has observed 6 WR stars. These stars show multimode oscillations mainly in the continuum. A period of 9.8 h has been found by Lefevre *et al.* (2005) for WR123 and two other WR stars show periods of a few days. In addition, some variations may also be due to a corotating interacting region (CIR) in about 20 % of the WR stars, as estimated by Chené.

The interesting case of binaries

Binaries are the sites of many interesting physical processes, in the interior, at the surface as well as in the colliding winds. The review of the observations by Hugues Sana is presented by Chris Evans. According to the domain of the parameter space, different observational techniques are required: radial velocities, interferometry, speckle, adaptative optics and imaging. The fraction of binaries among massive stars varies a lot from cluster to cluster. On the average a fraction of 45 % is found in nearby rich clusters, about 50% have periods smaller than 1000 days and separation smaller than 10 AU. From the statistics given, I note that it would be interesting in relation with models to try to estimate which fraction of all stars experiences binary mass transfer of type A, B or C, as well as synchronisation, tidal mixing, etc.

The evolutionary models of binaries are reviewed by Walter van Rensbergen, who presents a catalog with 561 models of binaries covering a wide range of initial parameters, thus leading to different evolutionary scenarios. Some new physical effects are now accounted for in the models, the spinning up of the gainer as well as the hot spots on its surface due the impacting mass coming from the donor. The rotational energy added to the radiative energy of the hot spots may produce matter ejection from the system (liberal era). A critical parameter in this evolution is the mass fraction called β accreted by the gainer. Van Rensbergen applies his models to the Algol systems issuing from a binary with a B–type primary. The period distribution is well represented, while the models give less Algols (17%) with large mass ratios (0.45-1) than observed (45 %).

7. Normal and active OB stars as extreme conditions test beds

A review of some interesting problems concerning the photospheres and the winds of massive OB stars is given by Joachim Puls. A question concerns the origin of the clumps and the possible role of convection and pulsations. The interest of the B– and A–type supergiants as possible sites of strange modes is underlined. Concerning the $\Omega\Gamma$–limit, he mentions that, while the so-called $g_{\rm eff}$ effect due to gravity darkening is observed in the anisotropic winds, it is not (yet?) the case for the κ–effect, i.e. a jump in the mass loss rates over the stellar surface due to a bistability (a discontinuity in the opacity due to a difference in ionization state). The case of γ Ara would also be interesting to examine, since in principle the $\Omega\Gamma$–limit would predict an enhanced mass loss rate for this star. Finally, Puls emphasizes the problem of the wind momentum of stars with $\log L/L_{\odot} \leqslant 5.5$, which is much lower than predicted and is still a challenging question.

Maria-Fernanda Nieva reports on the results of high precision spectroscopy applied to a well selected sample of 276 B–stars. A non-LTE technique is applied to the treatment of line formation. A very careful checking of all possible systematic effects in the data is performed. It is really quite remarkable how the whole spectrum is well reproduced by a single set of parameters. The method allows her to obtain accurate spectroscopic distances, evolutionary masses, M/L ratios, $T_{\rm eff}$, $\log g$ and chemical compositions. The sample is applied to the discussion of the metal content of the solar neighborhood, as well as to the origin of the chemical mixing.

The extreme case of the LBV

The properties of LBV have been reviewed by Nathan Smith and Jose Groh with different interesting approaches. Both show that the LBV lie between two lines in the HR diagram (Smith *et al.* 2004), which join together at the luminosity level of about a 20 $M_{\odot}$ star. Smith points out that lifetime of the LBV phase is much longer than previously assumed, since the many dormant LBV were not accounted for in previous studies. He proposes a lifetime $t_{\rm LBV} = (2-5)\cdot 10^5$ yr instead of $(2-3)\cdot 10^4$ yr. The very interesting consequence is that this may make the LBV phase the evolutionary stage where occurs most of the mass loss necessary to form WR stars. Thus, it may alleviate or even solve the problem set by the new low mass loss rates which make the formation of WR stars difficult. Interestingly enough, Smith defines a timescale $t_{\rm rad} = t_{\rm erup} \cdot L_{\rm erup}/L_{\rm star}$, this is the timescale for energy supply, i.e. it says how long the star has to store the energy necessary to feed the outburst. For η Carinae, it is of the order of 75 yr.

As mentioned by Groh, the fact that some supernovae (like SN 2008S) might originate from an LBV is an interesting problem, since stellar models do not generally predict SN explosions as LBV. He shows that on the LBV minimum line, i.e. the limiting line on the blue side of the HR diagram, some LBV such as HR Carinae and AG Carinae are about at the critical rotation, while they are much below the critical value (e.g. $\Omega_S/\Omega_C = 0.4$ for AG Car) when they are on the red side of the LBV domain. Thus, rotation certainly is an important effect for at least a part of the LBV. Indeed, Groh suggests that there are two groups of LBV: the fast rotating, highly variable stars with an S Dor cycle and the slow rotating ones with much less variability like P Cygni.

High energies from OB stars

The high energy emissions of OB stars are reviewed by Guillaume Dubus and by Stan Owocki. Dubus examines the interactions in some binaries between the relativistic wind of a pulsar and the wind of an OB star. The wind collision generates a non thermal emission with most of its energy above 1 MeV. Some interesting cases occur when a pulsar moves through the disc of a Be star. Owocki focus on the interesting γ–ray source LS5039, wich consists of an O6.5V star and a compact companion, which may be a black hole. He make 3D SPH simulations of the accretion of the O–star wind on the compact companion and shows that the high–enery observations are correctly described by a Bondi-Hoyle accretion. This new scheme by Owocki offers a valuable alternative to the usual pulsar wind shock model. Interestingly also, some wind properties such the as porosity have an effect of the high energy emissions and might perhaps be tested in this way.

At the end, warm thanks and intense applauses are addressed to Coralie Neiner for this most excellent meeting, both for the outstanding scientific program and for the excellent organization she has conducted throughout with the help of motivated collaborators, to whom we also express our gratitude. This meeting will remain as great step in our understanding of massive stars.

References

Henrichs, H. F., Neiner, C., & Geers, V. C. 2003a, in: K. van der Hucht, A. Herrero, & C. Esteban (eds.), *A Massive Star Odyssey: From Main Sequence to Supernova*, IAU Symposium 212, p. 202

Henrichs, H. F., Neiner, C., & Geers, V. C. 2003b, in: L. A. Balona, H. F. Henrichs, & R. Medupe (eds.), *Magnetic Fields in O, B and A Stars: Origin and Connection to Pulsation, Rotation and Mass Loss*, ASP-CS 305, p. 301

Hunter, I., Brott, I., Lennon, D. J., Langer, N. *et al.* 2008, *ApJ* (Letters), 676, L29

Jamet, L., Cerviño, M., Luridiana, V., Pérez, E., & Yakobchuk, T. 2010, *A&A*, 509A, 10

Lefèvre, L., Marchenko, S. V., Moffat, A. F. J., Chené, A. N. *et al.* 2005, *ApJ* (Letters), 634, L109

Lucy, L. B. 1967, *ZfA*, 65, 89

Maeder, A., Grebel, E. K., & Mermilliod, J.-C. 1999, *A&A*, 346, 459

Maeder, A. & Meynet, G. 2005, *A&A* 440, 1041

Maeder, A., Meynet, G., Ekström, S., & Georgy, C. 2009, *Communications in Asteroseismology* 158, 72

Meylan, G. & Maeder, A. 1982, *A&A*, 108, 148

Miglio, A., Montalbán, J., Eggenberger, P., & Noels, A. 2008, *AN*, 329, 529

Smith, N., Vink, J. S., & de Koter, A. 2004, *ApJ*, 615, 475

Zahn, J.-P., Brun, A. S., & Mathis, S. 2007, *A&A*, 474, 145

Author Index

Object Index

Subject Index

Printed in the United States
by Baker & Taylor Publisher Services